ADVANCED ENGINEERING ELECTROMAGNETICS

ADVANCED ENGINEERING ELECTROMAGNETICS

CONSTANTINE A. BALANIS
Arizona State University

JOHN WILEY & SONS
New York • Chichester • Brisbane • Toronto • Singapore

Copyright © 1989, by John Wiley & Sons, Inc.

All rights reserved. Published simultaneously in Canada.

Reproduction or translation of any part of
this work beyond that permitted by Sections
107 and 108 of the 1976 United States Copyright
Act without the permission of the copyright
owner is unlawful. Requests for permission
or further information should be addressed to
the Permissions Department, John Wiley & Sons.

Library of Congress Cataloging in Publication Data:

Balanis, Constantine A., 1938–
   Advanced engineering electromagnetics.

  Bibliography: p.
  1. Electromagnetism. I. Title.
QC760.B25 1989   537   88-33846
ISBN 0-471-62194-3

Printed in the United States of America

Printed and bound by the Hamilton Printing Company.

10 9 8 7 6

To Helen, Renie, and Stephanie

ABOUT THE AUTHOR

Constantine A. Balanis was born in Trikala, Greece. He received his B.S.E.E. degree from Virginia Polytechnic Institute, Blacksburg, in 1964, his M.E.E. degree from University of Virginia, Charlottesville, in 1966, and his Ph.D. degree in electrical engineering from Ohio State University, Columbus, in 1969.

From 1964 to 1970 he was with the NASA Langley Research Center in Hampton, Virginia, and from 1970 to 1983 he was with the Department of Electrical Engineering, West Virginia University, Morgantown. In 1983 he joined Arizona State University, Tempe, and he is now Regents' Professor of Electrical Engineering and Director of ASU's Telecommunications Research Center. He teaches graduate and undergraduate courses in electromagnetic theory, antennas, and microwave circuits. His research interests are in low- and high-frequency numerical antenna and scattering techniques, electromagnetic wave propagation in microwave-integrated circuit tranmission lines, and reconstruction (inversion) methods. He received the Graduate Teaching Excellence Award, School of Engineering, Arizona State University for 1987–1988, the 1989 JEEE Region 6 Individual Achievement Award, and the 1992 Special Professionalism Award of the JEEE Phoenix Section.

Dr. Balanis is a Fellow of IEEE (Institute of Electrical and Electronics Engineers), and a member of ASEE (American Society for Engineering Education), Sigma Xi, Tau Beta Pi, Eta Kappa Nu, and Phi Kappa Phi. He has served as the Associate Editor of the *IEEE Transactions on Antennas and Propagation* (1974–1977) and *IEEE Transactions on Geoscience and Remote Sensing* (1982–1984), Editor of the Newsletter for the IEEE Geoscience and Remote Sensing Society (1982–1983), and as Second Vice-President of the IEEE Geoscience and Remote Sensing Society (1984). He is also the author of *Antenna Theory: Analysis and Design* (John Wiley & Sons, New York, 1982), and "Horn Antennas," Chapter 8 in *Antenna Handbook: Theory, Applications, and Design* (Y. T. Lo and S. W. Lee, Editors), Van Nostrand Reinhold Co., New York, 1988.

PREFACE

This book is designed for a two-semester sequence in time-harmonic electromagnetics that is beyond an introduction to electromagnetostatics. Although the first part of the book is intended primarily for undergraduates and beginning graduates in electrical engineering and physics, the last part is intended for advanced graduate students and practicing engineers and scientists. The majority of Chapters 1 to 10 can be covered in the first semester, and most of Chapters 11 to 14 can be covered in the second semester. To cover all of the material in the proposed time frame would be, in many instances, a very ambitious task. Sufficient topics have been included, however, to make the text complete and to allow the instructor the flexibility to emphasize, de-emphasize, or omit sections or chapters.

The discussion presumes that the student has a general knowledge of vector analysis, differential and integral calculus, and electromagnetostatics either from an introductory electrical engineering or physics course. Mathematical techniques required for understanding some advanced topics, mostly in the later chapters, are incorporated in the individual chapters or are included as appendixes.

This is a very detailed student-oriented book. The analytical detail, rigor, and thoroughness allows many of the topics to be traced to their origin. In addition to the coverage of traditional classical topics, the book includes state of the art advanced topics on Integral Equations (IE), Moment Method (MM), Geometrical Theory of Diffraction (GTD), and Green's functions. Electromagnetic theorems, as applied to the solution of boundary-value problems, are also included and discussed.

The material is presented in a sequential and unified manner, and each chapter is subdivided into sections or subsections for which the individual heading clearly identifies the topic discussed, examined, or illustrated. The book also includes numerous examples, illustrations, references, and end-of-chapter problems. The examples and end-of-chapter problems have been designed to illustrate basic principles and to challenge the knowledge of the student. An exhaustive list of references is included at the end of each chapter to allow the interested reader to research each topic. A number of appendixes of mathematical identities and special functions, some represented also in tabular and graphical forms, are included to aid the student in the solution of the examples and assigned problems. Moment Method (MM) and Geometrical Theory of Diffraction (GTD) FORTRAN computer programs are included, respectively, at the end of Chapters 12 and 13, and they can be used for the solution of simple and complex problems. Also, information is provided about other national moment method computer programs. A solutions manual for all end-of-chapter problems is available for the instructor.

In Chapter 1 the book covers the classical topics on Maxwell's equations, constitutive parameters and relations, circuit relations, boundary conditions, and power and energy relations. The electrical properties of matter, both direct-current and alternating-current, are covered in Chapter 2, and the wave equation and its solution in rectangular, cylindrical and spherical coordinates are discussed in Chapter 3. Electromagnetic wave propagation and polarization is introduced in Chapter 4. Reflection and transmission at normal and oblique wave incidences are considered in Chapter 5. Chapter 6 covers the auxiliary vector potentials and their use toward the construction of solutions for radiation and scattering problems. The theorems of duality, uniqueness, image, reciprocity, reaction, volume and surface equivalences, induction, and physical and physical optics equivalents are introduced and applied in Chapter 7. Rectangular cross section waveguides and cavities, including dielectric slabs, striplines, and microstrips, are discussed in Chapter 8. Those of circular cross section, including the fiber optics cable, are examined in Chapter 9, and those of spherical geometry are introduced in Chapter 10. Scattering by strips, plates, circular cylinders, wedges, and spheres is analyzed in Chapter 11. Chapter 12 covers the basics and applications of Integral Equations (IE) and Moment Method (MM) and also includes a computer program for wire radiation and scattering. The techniques and applications for the Geometrical Theory of Diffraction (GTD) are introduced and discussed in Chapter 13, including computer programs for diffraction coefficients of conducting wedges. The classic topic of Green's functions is introduced and applied in Chapter 14.

Throughout the book an $e^{j\omega t}$ time convention is assumed, and it is suppressed in almost all of the chapters. The International System of Units, which is an expanded form of the rationalized MKS system, is used throughout the text discussion. In some instances, the units of length are given in meters (or centimeters) and feet (or inches). Numbers in parentheses ( ) refer to equations, whereas those in brackets [ ] refer to references. For emphasis, the most important equations, once they are derived, are boxed.

I acknowledge the invaluable suggestions, corrections, and constructive criticisms of the reviewers of this book: Roger D. Radcliff of Ohio University, Christos G. Christodoulou of University of Central Florida, Prabhakar H. Pathak of Ohio State University, and Thomas E. Tice of Arizona State University. The writing of this book has been a very ambitious task, and its completion would not have been possible without the contributions of many of my graduate students. It is a pleasure to acknowledge those of Kefeng Liu for the development of the moment method computer programs and for the proofreading of parts of the manuscript, and those of Lesley A. Polka, Frank L. Whetten, Mark S. Frank, James P. Gilb, and Craig R. Birtcher for proofreading parts of the manuscript. A special tribute is owed to Thuy Griesser for the expert typing of all phases of the entire manuscript. I am also grateful to Christina Kamra, Wiley editor of electrical engineering and computer science, for her interest in the production and publication of this book. To the companies and authors that provided the copyright permissions, I am most appreciative. In a book of this size, there inevitably will be errors that have been overlooked. I would appreciate having any errors brought to my attention.

Constantine A. Balanis

Tempe, Arizona

CONTENTS

1 TIME-VARYING AND TIME-HARMONIC ELECTROMAGNETIC FIELDS 1

1.1 INTRODUCTION 2
1.2 MAXWELL'S EQUATIONS 2
    1.2.1 Differential Form of Maxwell's Equations 2
    1.2.2 Integral Form of Maxwell's Equations 5
1.3 CONSTITUTIVE PARAMETERS AND RELATIONS 7
1.4 CIRCUIT-FIELD RELATIONS 8
    1.4.1 Kirchhoff's Voltage Law 8
    1.4.2 Kirchhoff's Current Law 9
    1.4.3 Element Laws 11
1.5 BOUNDARY CONDITIONS 13
    1.5.1 Finite Conductivity Media 13
    1.5.2 Infinite Conductivity Media 16
    1.5.3 Sources along Boundaries 19
1.6 POWER AND ENERGY 20
1.7 TIME-HARMONIC ELECTROMAGNETIC FIELDS 23
    1.7.1 Maxwell's Equations in Differential and Integral Forms 24
    1.7.2 Boundary Conditions 24
    1.7.3 Power and Energy 28
REFERENCES 32
PROBLEMS 32

2 ELECTRICAL PROPERTIES OF MATTER 42

2.1 INTRODUCTION 42
2.2 DIELECTRICS, POLARIZATION, AND PERMITTIVITY 44
2.3 MAGNETICS, MAGNETIZATION, AND PERMEABILITY 51
2.4 CURRENT, CONDUCTORS, AND CONDUCTIVITY 59
    2.4.1 Current 59
    2.4.2 Conductors 60
    2.4.3 Conductivity 63
2.5 SEMICONDUCTORS 63
2.6 SUPERCONDUCTORS 68
2.7 LINEAR, HOMOGENEOUS, ISOTROPIC, AND NONDISPERSIVE MEDIA 71
2.8 A.C. VARIATIONS IN MATERIALS 72
    2.8.1 Complex Permittivity 73

2.8.2 Complex Permeability   84
2.8.3 Ferrites   85
REFERENCES   94
PROBLEMS   96

3 WAVE EQUATION AND ITS SOLUTIONS   104

3.1 INTRODUCTION   104
3.2 TIME-VARYING ELECTROMAGNETIC FIELDS   104
3.3 TIME-HARMONIC ELECTROMAGNETIC FIELDS   106
3.4 SOLUTION TO THE WAVE EQUATION   107
   3.4.1 Rectangular Coordinate System   108
      A. Source-Free and Lossless Media   108
      B. Source-Free and Lossy Media   113
   3.4.2 Cylindrical Coordinate System   116
   3.4.3 Spherical Coordinate System   121
REFERENCES   126
PROBLEMS   127

4 WAVE PROPAGATION AND POLARIZATION   129

4.1 INTRODUCTION   129
4.2 TRANSVERSE ELECTROMAGNETIC MODES   129
   4.2.1 Uniform Plane Waves in an Unbounded Lossless Medium—Principal Axis   131
      A. Electric and Magnetic Fields   131
      B. Wave Impedance   133
      C. Phase and Energy (Group) Velocities, Power, and Energy Densities   135
      D. Standing Waves   136
   4.2.2 Uniform Plane Waves in an Unbounded Lossless Medium—Oblique Angle   138
      A. Electric and Magnetic Fields   139
      B. Wave Impedance   142
      C. Phase and Energy (Group) Velocities   143
      D. Power and Energy Densities   144
4.3 TRANSVERSE ELECTROMAGNETIC MODES IN LOSSY MEDIA   145
   4.3.1 Uniform Plane Waves in an Unbounded Lossy Medium—Principal Axis   145
      A. Good Dielectrics $[(\sigma/\omega\varepsilon)^2 \ll 1]$ 149
      B. Good Conductors $[(\sigma/\omega\varepsilon)^2 \gg 1]$ 151
   4.3.2 Uniform Plane Waves in an Unbounded Lossy Medium—Oblique Angle   151

4.4 POLARIZATION   154
    4.4.1 Linear Polarization   156
    4.4.2 Circular Polarization   158
        A. Right-Hand (Clockwise) Circular Polarization   158
        B. Left-Hand (Counterclockwise) Circular Polarization   161
    4.4.3 Elliptical Polarization   163
    4.4.4 Poincaré Sphere   168
REFERENCES   173
PROBLEMS   174

5 REFLECTION AND TRANSMISSION   180

5.1 INTRODUCTION   180
5.2 NORMAL INCIDENCE—LOSSLESS MEDIA   180
5.3 OBLIQUE INCIDENCE—LOSSLESS MEDIA   185
    5.3.1 Perpendicular (Horizontal or E) Polarization   185
    5.3.2 Parallel (Vertical or H) Polarization   189
    5.3.3 Total Transmission–Brewster Angle   193
        A. Perpendicular (Horizontal) Polarization   193
        B. Parallel (Vertical) Polarization   194
    5.3.4 Total Reflection–Critical Angle   196
        A. Perpendicular (Horizontal) Polarization   196
        B. Parallel (Vertical) Polarization   206
5.4 LOSSY MEDIA   206
    5.4.1 Normal Incidence: Conductor–Conductor Interface   207
    5.4.2 Oblique Incidence: Dielectric–Conductor Interface   210
    5.4.3 Oblique Incidence: Conductor–Conductor Interface   214
5.5 REFLECTION AND TRANSMISSION OF MULTIPLE INTERFACES   220
    5.5.1 Reflection Coefficient of a Single Slab Layer   220
    5.5.2 Reflection Coefficient of Multiple Layers   229
        A. Quarter-Wavelength Transformer   230
        B. Binomial (Maximally Flat) Design   231
        C. Tschebyscheff (Equal-Ripple) Design   233
        D. Oblique-Wave Incidence   235
5.6 POLARIZATION CHARACTERISTICS ON REFLECTION   236
REFERENCES   243
PROBLEMS   244

6 AUXILIARY VECTOR POTENTIALS, CONSTRUCTION OF SOLUTIONS, AND RADIATION AND SCATTERING EQUATIONS   254

6.1 INTRODUCTION   254
6.2 THE VECTOR POTENTIAL **A**   256

- 6.3 THE VECTOR POTENTIAL **F** 257
- 6.4 THE VECTOR POTENTIALS **A** AND **F** 259
- 6.5 CONSTRUCTION OF SOLUTIONS 261
    - 6.5.1 Transverse Electromagnetic Modes: Source-Free Region 261
        - A. Rectangular Coordinate System 261
        - B. Cylindrical Coordinate System 266
    - 6.5.2 Transverse Magnetic Modes: Source-Free Region 269
        - A. Rectangular Coordinate System 269
        - B. Cylindrical Coordinate System 272
    - 6.5.3 Transverse Electric Modes: Source-Free Region 273
        - A. Rectangular Coordinate System 274
        - B. Cylindrical Coordinate System 275
- 6.6 SOLUTION OF THE INHOMOGENEOUS VECTOR POTENTIAL WAVE EQUATION 276
- 6.7 FAR-FIELD RADIATION 280
- 6.8 RADIATION AND SCATTERING EQUATIONS 282
    - 6.8.1 Near Field 282
    - 6.8.2 Far Field 285
        - A. Rectangular Coordinate System 288
        - B. Cylindrical Coordinate System 300
- REFERENCES 305
- PROBLEMS 305

7 ELECTROMAGNETIC THEOREMS AND PRINCIPLES 310

- 7.1 INTRODUCTION 310
- 7.2 DUALITY THEOREM 310
- 7.3 UNIQUENESS THEOREM 312
- 7.4 IMAGE THEORY 314
    - 7.4.1 Vertical Electric Dipole 318
    - 7.4.2 Horizontal Electric Dipole 321
- 7.5 RECIPROCITY THEOREM 323
- 7.6 REACTION THEOREM 326
- 7.7 VOLUME EQUIVALENCE THEOREM 327
- 7.8 SURFACE EQUIVALENCE THEOREM: HUYGENS'S PRINCIPLE 329
- 7.9 INDUCTION THEOREM (INDUCTION EQUIVALENT) 334
- 7.10 PHYSICAL EQUIVALENT AND PHYSICAL OPTICS EQUIVALENT 338
- 7.11 INDUCTION AND PHYSICAL EQUIVALENT APPROXIMATIONS 341
- REFERENCES 346
- PROBLEMS 347

8 RECTANGULAR CROSS-SECTION WAVEGUIDES AND CAVITIES 352

- 8.1 INTRODUCTION 352
- 8.2 RECTANGULAR WAVEGUIDE 352
    - 8.2.1 Transverse Electric (TE^z) 353
    - 8.2.2 Transverse Magnetic (TM^z) 362
    - 8.2.3 Dominant TE_{10} Mode 366
    - 8.2.4 Power Density and Power 374
    - 8.2.5 Attenuation 376
        - A. Conduction (Ohmic) Losses 376
        - B. Dielectric Losses 381
        - C. Coupling 384
- 8.3 RECTANGULAR RESONANT CAVITIES 388
    - 8.3.1 Transverse Electric (TE^z) Modes 388
    - 8.3.2 Transverse Magnetic (TM^z) Modes 392
- 8.4 HYBRID (LSE AND LSM) MODES 394
    - 8.4.1 Longitudinal Section Electric (LSE^y) or Transverse Electric (TE^y) or H^y Modes 395
    - 8.4.2 Longitudinal Section Magnetic (LSM^y) or Transverse Magnetic (TM^y) or E^y Modes 397
- 8.5 PARTIALLY FILLED WAVEGUIDE 398
    - 8.5.1 Longitudinal Section Electric (LSE^y) or Transverse Electric (TE^y) 398
    - 8.5.2 Longitudinal Section Magnetic (LSM^y) or Transverse Magnetic (TM^y) 404
- 8.6 TRANSVERSE RESONANCE METHOD 410
    - 8.6.1 Transverse Electric (TE^y) or Longitudinal Section Electric (LSE^y) or H^y 413
    - 8.6.2 Transverse Magnetic (TM^y) or Longitudinal Section Magnetic (LSM^y) or E^y 413
- 8.7 DIELECTRIC WAVEGUIDE 414
    - 8.7.1 Dielectric Slab Waveguide 414
    - 8.7.2 Transverse Magnetic (TM^z) Modes 416
        - A. TM^z (even) 418
        - B. TM^z (odd) 420
        - C. Summary of TM^z (even) and TM^z (odd) Modes 420
        - D. Graphical Solution for TM^z (even) and TM^z (odd) Modes 423
    - 8.7.3 Transverse Electric (TE^z) Modes 427
    - 8.7.4 Ray-Tracing Method 431
        - A. Transverse Magnetic (TM^z) Modes (Parallel Polarization) 436
        - B. Transverse Electric (TE^z) Modes (Perpendicular Polarization) 439
    - 8.7.5 Dielectric Covered Ground Plane 441
- 8.8 STRIPLINE AND MICROSTRIP LINES 444
    - 8.8.1 Stripline 445
    - 8.8.2 Microstrip 449

8.8.3 Microstrip: Boundary-Value Problem 455
8.9 RIDGED WAVEGUIDE 457
REFERENCES 461
PROBLEMS 463

9 CIRCULAR CROSS-SECTION WAVEGUIDES AND CAVITIES 470

9.1 INTRODUCTION 470
9.2 CIRCULAR WAVEGUIDE 470
    9.2.1 Transverse Electric (TE^z) Modes 470
    9.2.2 Transverse Magnetic (TM^z) Modes 477
    9.2.3 Attenuation 485
9.3 CIRCULAR CAVITY 492
    9.3.1 Transverse Electric (TE^z) Modes 492
    9.3.2 Transverse Magnetic (TM^z) Modes 494
    9.3.3 Quality Factor Q 495
9.4 RADIAL WAVEGUIDES 499
    9.4.1 Parallel Plates 499
        A. Transverse Electric (TE^z) Modes 499
        B. Transverse Magnetic (TM^z) Modes 502
    9.4.2 Wedged Plates 504
        A. Transverse Electric (TE^z) Modes 505
        B. Transverse Magnetic (TM^z) Modes 505
9.5 DIELECTRIC WAVEGUIDES AND RESONATORS 506
    9.5.1 Circular Dielectric Waveguide 506
    9.5.2 Circular Dielectric Resonator 517
        A. TE^z Modes 520
        B. TM^z Modes 521
        C. $TE_{01\delta}$ Mode 522
    9.5.3 Optical Fiber Cable 524
    9.5.4 Dielectric Covered Conducting Rod 527
        A. TM^z Modes 528
        B. TE^z Modes 534
REFERENCES 535
PROBLEMS 537

10 SPHERICAL TRANSMISSION LINES AND CAVITIES 543

10.1 INTRODUCTION 543
10.2 CONSTRUCTION OF SOLUTIONS 543
    10.2.1 The Vector Potential $\mathbf{F}$ ($\mathbf{J} = 0$, $\mathbf{M} \neq 0$) 544
    10.2.2 The Vector Potential $\mathbf{A}$ ($\mathbf{J} \neq 0$, $\mathbf{M} = 0$) 546
    10.2.3 The Vector Potentials $\mathbf{F}$ and $\mathbf{A}$ 546
    10.2.4 Transverse Electric (TE) Modes: Source-Free Region 547
    10.2.5 Transverse Magnetic (TM) Modes: Source-Free Region 549
    10.2.6 Solution of the Scalar Helmholtz Wave Equation 550

10.3 BICONICAL TRANSMISSION LINE 552
    10.3.1 Transverse Electric (TE^r) Modes 552
    10.3.2 Transverse Magnetic (TM^r) Modes 554
    10.3.3 Transverse Electromagnetic (TEM^r) Modes 554
10.4 THE SPHERICAL CAVITY 557
    10.4.1 Transverse Electric (TE^r) Modes 557
    10.4.2 Transverse Magnetic (TM^r) Modes 560
    10.4.3 Quality Factor Q 562
REFERENCES 565
PROBLEMS 565

11 SCATTERING 570

11.1 INTRODUCTION 570
11.2 INFINITE LINE-SOURCE CYLINDRICAL WAVE RADIATION 571
    11.2.1 Electric Line Source 571
    11.2.2 Magnetic Line Source 573
    11.2.3 Electric Line Source above Infinite Plane Electric Conductor 574
11.3 PLANE WAVE SCATTERING BY PLANAR SURFACES 577
    11.3.1 TM^z Plane Wave Scattering from a Strip 578
    11.3.2 TE^x Plane Wave Scattering from a Flat Rectangular Plate 586
11.4 CYLINDRICAL WAVE TRANSFORMATIONS AND THEOREMS 595
    11.4.1 Plane Waves in Terms of Cylindrical Wave Functions 595
    11.4.2 Addition Theorem of Hankel Functions 597
    11.4.3 Addition Theorem for Bessel Functions 600
    11.4.4 Summary of Cylindrical Wave Transformations and Theorems 602
11.5 SCATTERING BY CIRCULAR CYLINDERS 602
    11.5.1 Normal Incidence Plane Wave Scattering by Conducting Circular Cylinder: TM^z Polarization 603
        A. Small Radius Approximation 606
        B. Far-Zone Scattered Field 606
    11.5.2 Normal Incidence Plane Wave Scattering by Conducting Circular Cylinder: TE^z Polarization 608
        A. Small Radius Approximation 610
        B. Far-Zone Scattered Field 611
    11.5.3 Oblique Incidence Plane Wave Scattering by Conducting Circular Cylinder: TM^z Polarization 614
        A. Far-Zone Scattered Field 618
    11.5.4 Oblique Incidence Plane Wave Scattering by Conducting Circular Cylinder: TE^z Polarization 620
        A. Far-Zone Scattered Field 624
    11.5.5 Line-Source Scattering by a Conducting Circular Cylinder 626
        A. Electric Line Source (TM^z Polarization) 626
        B. Magnetic Line Source (TE^z Polarization) 630
11.6 SCATTERING BY CONDUCTING WEDGE 634
    11.6.1 Electric Line-Source Scattering by a Conducting Wedge: TM^z Polarization 635

xviii   CONTENTS

                 A. Far-Zone Field   637
                 B. Plane Wave Scattering   638
        11.6.2 Magnetic Line-Source Scattering by a Conducting Wedge: TE^z Polarization   639
        11.6.3 Electric and Magnetic Line-Source Scattering by a Conducting Wedge   642
11.7 SPHERICAL WAVE ORTHOGONALITIES, TRANSFORMATIONS, AND THEOREMS   645
        11.7.1 Vertical Dipole Spherical Wave Radiation   645
        11.7.2 Orthogonality Relationships   647
        11.7.3 Wave Transformations and Theorems   648
11.8 SCATTERING BY A CONDUCTING SPHERE   650
REFERENCES   658
PROBLEMS   660

12 INTEGRAL EQUATIONS AND THE MOMENT METHOD   670

12.1 INTRODUCTION   670
12.2 INTEGRAL EQUATION METHOD   671
        12.2.1 Electrostatic Charge Distribution   671
                 A. Finite Straight Wire   671
                 B. Bent Wire   675
        12.2.2 Integral Equation   677
        12.2.3 Radiation Pattern   680
        12.2.4 Point-Matching (Collocation) Method   681
        12.2.5 Basis Functions   683
                 A. Subdomain Functions   683
                 B. Entire-Domain Functions   685
        12.2.6 Application of Point Matching   687
        12.2.7 Weighting (Testing) Functions   689
        12.2.8 Moment Method   689
12.3 ELECTRIC AND MAGNETIC FIELD INTEGRAL EQUATIONS   695
        12.3.1 Electric Field Integral Equation   696
                 A. Two-Dimensional EFIE: TM^z Polarization   698
                 B. Two-Dimensional EFIE: TE^z Polarization   702
        12.3.2 Magnetic Field Integral Equation   707
                 A. Two-Dimensional MFIE: TM^z Polarization   709
                 B. Two-Dimensional MFIE: TE^z Polarization   710
                 C. Solution of the Two-Dimensional MFIE TE^z Polarization   712
12.4 FINITE DIAMETER WIRES   717
        12.4.1 Pocklington's Integral Equation   718
        12.4.2 Hallén's Integral Equation   720
        12.4.3 Source Modeling   722
                 A. Delta Gap   722
                 B. Magnetic Frill Generator   722
12.5 COMPUTER CODES   726
        12.5.1 Two-Dimensional Radiation and Scattering   727

CONTENTS

        A. Strip   727
        B. Circular, Elliptical, or Rectangular Cylinder   728
  12.5.2 Pocklington's Wire Radiation and Scattering   729
        A. Radiation   729
        B. Scattering   729
  12.5.3 Numerical Electromagnetics Code   729
  12.5.4 Mini-Numerical Electromagnetics Code   730
  12.5.5 Electromagnetic Surface Patch Code   730

REFERENCES   731
PROBLEMS   733
COMPUTER PROGRAM: PWRS   737

13 GEOMETRICAL THEORY OF DIFFRACTION   743

13.1 INTRODUCTION   743
13.2 GEOMETRICAL OPTICS   744
  13.2.1 Amplitude Relation   746
  13.2.2 Phase and Polarization Relations   751
  13.2.3 Reflection From Surfaces   753
13.3 GEOMETRICAL THEORY OF DIFFRACTION: EDGE DIFFRACTION   765
  13.3.1 Amplitude, Phase, and Polarization Relations   765
  13.3.2 Straight Edge Diffraction: Normal Incidence   769
        A. Modal Solution   771
        B. High-Frequency Asymptotic Solution   772
        C. Method of Steepest Descent   777
        D. Geometrical Optics and Diffracted Fields   782
        E. Diffraction Coefficients   785
  13.3.3 Straight Edge Diffraction: Oblique Incidence   807
  13.3.4 Curved Edge Diffraction: Oblique Incidence   814
  13.3.5 Equivalent Currents in Diffraction   824
  13.3.6 Slope Diffraction   828
  13.3.7 Multiple Diffractions   830
13.4 COMPUTER CODES   834
  13.4.1 Wedge Diffraction Coefficients   834
  13.4.2 Fresnel Transition Function   835
  13.4.3 Slope Wedge Diffraction Coefficients   835

REFERENCES   835
PROBLEMS   838
COMPUTER PROGRAM: WDC   848
COMPUTER PROGRAM: SWDC   849
COMPUTER PROGRAM: FTF   850

14 GREEN'S FUNCTIONS   851

14.1 INTRODUCTION   851
14.2 GREEN'S FUNCTIONS IN ENGINEERING   852
  14.2.1 Circuit Theory   852
  14.2.2 Mechanics   855

14.3 STURM–LIOUVILLE PROBLEMS   858
    14.3.1 Green's Function in Closed Form   860
    14.3.2 Green's Function in Series   865
        A. Vibrating String   866
        B. Sturm–Liouville Operator   867
    14.3.3 Green's Function in Integral Form   872

14.4 TWO-DIMENSIONAL GREEN'S FUNCTION IN RECTANGULAR COORDINATES   876
    14.4.1 Static Fields   876
        A. Closed Form   877
        B. Series Form   883
    14.4.2 Time-Harmonic Fields   885

14.5 GREEN'S IDENTITIES AND METHODS   888
    14.5.1 Green's First and Second Identities   888
    14.5.2 Generalized Green's Function Method   890
        A. Nonhomogeneous Partial Differential Equation with Homogeneous Dirichlet Boundary Conditions   891
        B. Nonhomogeneous Partial Differential Equation with Nonhomogeneous Dirichlet Boundary Conditions   892
        C. Nonhomogeneous Partial Differential Equation with Homogeneous Neumann Boundary Conditions   892
        D. Nonhomogeneous Partial Differential Equation with Mixed Boundary Conditions   893

14.6 GREEN'S FUNCTIONS OF THE SCALAR HELMHOLTZ EQUATION   893
    14.6.1 Rectangular Coordinates   893
    14.6.2 Cylindrical Coordinates   897
    14.6.3 Spherical Coordinates   903

14.7 DYADIC GREEN'S FUNCTIONS   907
    14.7.1 Dyadics   907
    14.7.2 Green's Functions   909

REFERENCES   910
PROBLEMS   911

Appendix I   IDENTITIES   917
Appendix II   VECTOR ANALYSIS   920
Appendix III   FRESNEL INTEGRALS   929
Appendix IV   BESSEL FUNCTIONS   934
Appendix V   LEGENDRE POLYNOMIALS AND FUNCTIONS   947
Appendix VI   THE METHOD OF STEEPEST DESCENT (SADDLE-POINT METHOD)   963

INDEX   969

CHAPTER 1

TIME-VARYING AND TIME-HARMONIC ELECTROMAGNETIC FIELDS

1.1 INTRODUCTION

Electromagnetic field theory is a discipline concerned with the study of charges, at rest and in motion, that produce currents and electric–magnetic fields. It is, therefore, fundamental to the study of electrical engineering and physics, and indispensable to the understanding, design, and operation of many practical systems using antennas, scattering, microwave circuits and devices, radio-frequency and optical communications, broadcasting, geosciences and remote sensing, radar, radio astronomy, quantum electronics, solid-state circuits and devices, electromechanical energy conversion, and even computers. Circuit theory, a required area in the study of electrical engineering, is a special case of electromagnetic theory and it is valid when the physical dimensions of the circuit are small compared to the wavelength. Circuit concepts, which deal primarily with lumped elements, must be modified to include distributed elements and coupling phenomena in studies of advanced systems. For example, signal propagation, distortion, and coupling in microstrip lines used in the design of sophisticated systems (such as computers) can be accounted for properly only by understanding the electromagnetic field interactions associated with them.

The study of electromagnetics includes both theoretical and applied concepts. The theoretical concepts are described by a set of basic laws formulated primarily through experiments conducted during the nineteenth century by many scientists—Faraday, Ampere, Gauss, Lenz, Coulomb, Volta, and others. They were then combined into a consistent set of vector equations by Maxwell. These are the widely acclaimed *Maxwell's equations*. The applied concepts of electromagnetics are formulated by applying the theoretical concepts to the design and operation of practical systems.

2 TIME-VARYING AND TIME-HARMONIC ELECTROMAGNETIC FIELDS

In this chapter we will review Maxwell's equations (both in differential and integral forms), describe the relations between electromagnetic field and circuit theories, derive the boundary conditions associated with electric and magnetic field behavior across interfaces, relate power and energy concepts for electromagnetic field and circuit theories, and specialize all these equations, relations, conditions, concepts, and theories to the study of time-harmonic fields.

1.2 MAXWELL'S EQUATIONS

In general, electric and magnetic fields are vector quantities that have both magnitude and direction. The relations and variations of the electric and magnetic fields, charges, and currents associated with electromagnetic waves are governed by physical laws, which are known as Maxwell's equations. These equations, as we have indicated, were arrived at mostly through various experiments carried out by different investigators, but they were put in their final form by James Clerk Maxwell, a Scottish physicist and mathematician. These equations can be written either in differential or in integral form.

1.2.1 Differential Form of Maxwell's Equations

The differential form of Maxwell's equations is the most widely used representation to solve boundary-value electromagnetic problems. It is used to describe and relate the field vectors, current densities, and charge densities *at any point in space at any time*. For these expressions to be valid, it is assumed that the field vectors are *single-valued*, *bounded*, *continuous* functions of position and time and exhibit *continuous derivatives*. Field vectors associated with electromagnetic waves possess these characteristics except where there exist abrupt changes in charge and current densities. Discontinuous distributions of charges and currents usually occur at interfaces between media where there are discrete changes in the electrical parameters across the interface. The variations of the field vectors across such boundaries (interfaces) are related to the discontinuous distributions of charges and currents by what are usually referred to as the *boundary conditions*. Thus a complete description of the field vectors at any point (including discontinuities) at any time requires not only Maxwell's equations in differential form but also the associated *boundary conditions*.

In differential form, Maxwell's equations can be written as

$$\nabla \times \mathscr{E} = -\mathscr{M}_i - \frac{\partial \mathscr{B}}{\partial t} = -\mathscr{M}_i - \mathscr{M}_d = -\mathscr{M}_t \tag{1-1}$$

$$\nabla \times \mathscr{H} = \mathscr{J}_i + \mathscr{J}_c + \frac{\partial \mathscr{D}}{\partial t} = \mathscr{J}_{ic} + \frac{\partial \mathscr{D}}{\partial t} = \mathscr{J}_{ic} + \mathscr{J}_d = \mathscr{J}_t \tag{1-2}$$

$$\nabla \cdot \mathscr{D} = q_{ev} \tag{1-3}$$

$$\nabla \cdot \mathscr{B} = q_{mv} \tag{1-4}$$

where

$$\mathscr{J}_{ic} = \mathscr{J}_i + \mathscr{J}_c \tag{1-5a}$$

$$\mathscr{J}_d = \frac{\partial \mathscr{D}}{\partial t} \tag{1-5b}$$

$$\mathscr{M}_d = \frac{\partial \mathscr{B}}{\partial t} \tag{1-5c}$$

All these field quantities—$\mathscr{E}$, $\mathscr{H}$, $\mathscr{D}$, $\mathscr{B}$, $\mathscr{J}$, $\mathscr{M}$, and q_v—are assumed to be time-varying, and each is a function of the space coordinates and time, that is, $\mathscr{E} = \mathscr{E}(x, y, z; t)$. The definitions and units of the quantities are

- $\mathscr{E}$ = electric field intensity (volts/meter)
- $\mathscr{H}$ = magnetic field intensity (amperes/meter)
- $\mathscr{D}$ = electric flux density (coulombs/square meter)
- $\mathscr{B}$ = magnetic flux density (webers/square meter)
- $\mathscr{J}_i$ = impressed (source) electric current density (amperes/square meter)
- $\mathscr{J}_c$ = conduction electric current density (amperes/square meter)
- $\mathscr{J}_d$ = displacement electric current density (amperes/square meter)
- $\mathscr{M}_i$ = impressed (source) magnetic current density (volts/square meter)
- $\mathscr{M}_d$ = displacement magnetic current density (volts/square meter)
- q_{ev} = electric charge density (coulombs/cubic meter)
- q_{mv} = magnetic charge density (webers/cubic meter)

The electric displacement current density $\mathscr{J}_d = \partial \mathscr{D}/\partial t$ was introduced by Maxwell to complete Ampere's law for statics, $\nabla \times \mathscr{H} = \mathscr{J}$. For free space $\mathscr{J}_d$ was viewed as a motion of bound charges moving in "ether," an ideal weightless fluid pervading all space. Since ether proved to be undetectable and its concept was not totally reasonable with the theory of relativity, it has since been disregarded. Instead, for dielectrics, part of the displacement current density has been viewed as a motion of bound charges creating a true current. Because of this, it is convenient to consider even in free space the entire $\partial \mathscr{D}/\partial t$ term as a displacement current density.

Because of the symmetry of Maxwell's equations, the $\partial \mathscr{B}/\partial t$ term in (1-1) has been designated as a magnetic displacement current density. In addition, impressed (source) magnetic current density $\mathscr{M}_i$ and magnetic charge density q_{mv} have been introduced, respectively, in (1-1) and (1-4) through the "generalized" current concept. Although we have been accustomed to viewing magnetic charges and impressed magnetic current densities as not being physically realizable, they have been introduced to balance Maxwell's equations. Equivalent magnetic charges and currents will be introduced in later chapters to represent physical problems. In addition, impressed magnetic current densities, like impressed electric current densities, can be considered as energy sources that generate the fields and whose field expressions can be written in terms of these current densities. For some electromagnetic problems, their solution can often be aided by the introduction of "equivalent" impressed electric and magnetic current densities. The importance of

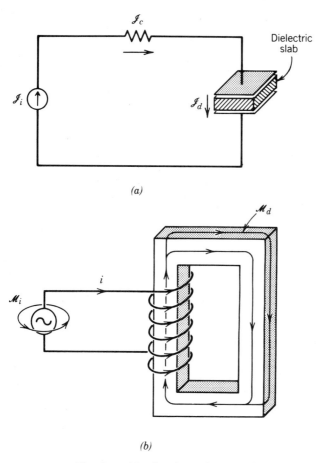

FIGURE 1-1 Circuits with electric and magnetic current densities. (*a*) Electric current density. (*b*) Magnetic current density.

both will become more obvious to the reader as solutions to specific electromagnetic boundary-value problems are considered in later chapters. However, to give the reader an early glimpse of the importance and interpretation of the electric and magnetic current densities, let us consider two familiar circuit examples.

In Figure 1-1a an electric current source is connected in series to a resistor and a parallel-plate capacitor. The electric current density $\mathcal{J}_i$ can be viewed as the current source that generates the conduction current density $\mathcal{J}_c$ through the resistor and the displacement current density $\mathcal{J}_d$ through the dielectric material of the capacitor. In Figure 1-1b a voltage source is connected to a wire that, in turn, is wrapped around a high permeability magnetic core. The voltage source can be viewed as the impressed magnetic current density that generates the displacement magnetic current density through the magnetic material of the core.

In addition to the four Maxwell's equations, there is another equation that relates the variations of the current density $\mathcal{J}_{ic}$ and the charge density q_{ev}. Although not an independent relation, this equation is referred to as the *continuity equation* because it relates the net flow of current out of a small volume (in the limit, a point) to the rate of decrease of charge. It takes the form

$$\nabla \cdot \mathscr{J}_{ic} = -\frac{\partial q_{ev}}{\partial t} \tag{1-6}$$

The continuity equation 1-6 can be derived from Maxwell's equations as given by (1-1) through (1-5c).

1.2.2 Integral Form of Maxwell's Equations

The integral form of Maxwell's equations describes the relations of the field vectors, charge densities, and current densities over an *extended region of space*. They have limited applications and are usually utilized only to solve electromagnetic boundary-value problems that possess complete symmetry (such as rectangular, cylindrical, spherical, etc., symmetries). However, *the fields and their derivatives in question do not need to possess continuous distributions*.

The integral form of Maxwell's equations can be derived from its differential form by utilizing the *Stokes'* and *divergence theorems*. For any arbitrary vector **A** Stokes' theorem states that *the line integral of the vector* **A** *along a closed path C is equal to the integral of the dot product of the curl of the vector* **A** *with the normal to the surface S that has the contour C as its boundary*. In equation form, Stokes' theorem can be written as

$$\oint_C \mathbf{A} \cdot d\mathbf{l} = \iint_S (\nabla \times \mathbf{A}) \cdot d\mathbf{s} \tag{1-7}$$

The divergence theorem states that for any arbitrary vector **A** *the closed surface integral of the normal component of vector* **A** *over a surface S is equal to the volume integral of the divergence of* **A** *over the volume V enclosed by S*. In mathematical form, the divergence theorem is stated as

$$\oiint_S \mathbf{A} \cdot d\mathbf{s} = \iiint_V \nabla \cdot \mathbf{A} \, dv \tag{1-8}$$

Taking the surface integral of both sides of (1-1), we can write

$$\iint_S (\nabla \times \mathscr{E}) \cdot d\mathbf{s} = -\iint_S \mathscr{M}_i \cdot d\mathbf{s} - \iint_S \frac{\partial \mathscr{B}}{\partial t} \cdot d\mathbf{s} = -\iint_S \mathscr{M}_i \cdot d\mathbf{s} - \frac{\partial}{\partial t} \iint_S \mathscr{B} \cdot d\mathbf{s} \tag{1-9}$$

Applying Stokes' theorem, as given by (1-7), on the left side of (1-9) reduces it to

$$\oint_C \mathscr{E} \cdot d\mathbf{l} = -\iint_S \mathscr{M}_i \cdot d\mathbf{s} - \frac{\partial}{\partial t} \iint_S \mathscr{B} \cdot d\mathbf{s} \tag{1-9a}$$

which is referred to as *Maxwell's equation in integral form as derived from Faraday's law*. In the absence of an impressed magnetic current density, Faraday's law states that the electromotive force (emf) appearing at the open-circuited terminals of a loop is equal to the time rate of decrease of magnetic flux linking the loop.

Using a similar procedure, we can show that the corresponding integral form of (1-2) can be written as

$$\oint_C \mathscr{H} \cdot d\mathbf{l} = \iint_S \mathscr{J}_{ic} \cdot d\mathbf{s} + \frac{\partial}{\partial t} \iint_S \mathscr{D} \cdot d\mathbf{s} = \iint_S \mathscr{J}_{ic} \cdot d\mathbf{s} + \iint_S \mathscr{J}_d \cdot d\mathbf{s} \tag{1-10}$$

6 TIME-VARYING AND TIME-HARMONIC ELECTROMAGNETIC FIELDS

which is usually referred to as *Maxwell's equation in integral form as derived from Ampere's law*. Ampere's law states that the line integral of the magnetic field over a closed path is equal to the current enclosed.

The other two Maxwell equations in integral form can be obtained from the corresponding differential forms, using the following procedure. First take the volume integral of both sides of (1-3); that is,

$$\iiint_V \nabla \cdot \mathscr{D}\, dv = \iiint_V q_{ev}\, dv = \mathscr{Q}_e \quad (1\text{-}11)$$

where $\mathscr{Q}_e$ is the total electric charge. Applying the divergence theorem, as given by (1-8), on the left side of (1-11) reduces it to

$$\oiint_S \mathscr{D} \cdot d\mathbf{s} = \iiint_V q_{ev}\, dv = \mathscr{Q}_e \quad (1\text{-}11a)$$

which is usually referred to as *Maxwell's electric field equation in integral form as derived from Gauss's law*. Gauss's law for the electric field states that the total electric flux through a closed surface is equal to the total charge enclosed.

In a similar manner, the integral form of (1-4) is given in terms of the total magnetic charge $\mathscr{Q}_m$ by

$$\oiint_S \mathscr{B} \cdot d\mathbf{s} = \mathscr{Q}_m \quad (1\text{-}12)$$

which is usually referred to as *Maxwell's magnetic field equation in integral form as derived from Gauss's law*. Even though magnetic charge is nonphysical, it is used as an equivalent to represent physical problems. The corresponding integral form of the continuity equation, as given by (1-6) in differential form, can be written as

$$\oiint_S \mathscr{J}_{ic} \cdot d\mathbf{s} = -\frac{\partial}{\partial t} \iiint_V q_{ev}\, dv = -\frac{\partial \mathscr{Q}_e}{\partial t} \quad (1\text{-}13)$$

Maxwell's equations in differential and integral form are summarized and listed in Table 1-1.

TABLE 1-1
Maxwell's equations and the continuity equation in differential and integral forms for time-varying fields

Differential form	Integral form
$\nabla \times \mathscr{E} = -\mathscr{M}_i - \dfrac{\partial \mathscr{B}}{\partial t}$	$\oint_C \mathscr{E} \cdot d\mathbf{l} = -\iint_S \mathscr{M}_i \cdot d\mathbf{s} - \dfrac{\partial}{\partial t}\iint_S \mathscr{B} \cdot d\mathbf{s}$
$\nabla \times \mathscr{H} = \mathscr{J}_i + \mathscr{J}_c + \dfrac{\partial \mathscr{D}}{\partial t}$	$\oint_C \mathscr{H} \cdot d\mathbf{l} = \iint_S \mathscr{J}_i \cdot d\mathbf{s} + \iint_S \mathscr{J}_c \cdot d\mathbf{s} + \dfrac{\partial}{\partial t}\iint_S \mathscr{D} \cdot d\mathbf{s}$
$\nabla \cdot \mathscr{D} = q_{ev}$	$\oiint_S \mathscr{D} \cdot d\mathbf{s} = \mathscr{Q}_e$
$\nabla \cdot \mathscr{B} = q_{mv}$	$\oiint_S \mathscr{B} \cdot d\mathbf{s} = \mathscr{Q}_m$
$\nabla \cdot \mathscr{J}_{ic} = -\dfrac{\partial q_{ev}}{\partial t}$	$\oiint_S \mathscr{J}_{ic} \cdot d\mathbf{s} = -\dfrac{\partial}{\partial t}\iiint_V q_{ev}\, dv = -\dfrac{\partial \mathscr{Q}_e}{\partial t}$

1.3 CONSTITUTIVE PARAMETERS AND RELATIONS

Materials contain charged particles, and when these materials are subjected to electromagnetic fields, their charged particles interact with the electromagnetic field vectors, producing currents and modifying the electromagnetic wave propagation in these media compared to that in free space. A more complete discussion of this is in Chapter 2. To account on a macroscopic scale for the presence and behavior of these charged particles, without introducing them in a microscopic lattice structure, we give a set of three expressions relating the electromagnetic field vectors. These expressions are referred to as the *constitutive relations*, and they will be developed in more detail in Chapter 2.

One of the constitutive relations relates *in the time domain* the electric flux density $\mathscr{D}$ to the electric field intensity $\mathscr{E}$ by

$$\mathscr{D} = \hat{\varepsilon} * \mathscr{E} \tag{1-14}$$

where $\hat{\varepsilon}$ is the time-varying permittivity of the medium (farads/meter) and * indicates convolution. For free space

$$\hat{\varepsilon} = \varepsilon_0 = 8.854 \times 10^{-12} \simeq \frac{10^{-9}}{36\pi} \text{ (farads/meter)} \tag{1-14a}$$

and (1-14) reduces to a product.

Another relation equates *in the time domain* the magnetic flux density $\mathscr{B}$ to the magnetic field intensity $\mathscr{H}$ by

$$\mathscr{B} = \hat{\mu} * \mathscr{H} \tag{1-15}$$

where $\hat{\mu}$ is the time-varying permeability of the medium (henries/meter). For free space

$$\hat{\mu} = \mu_0 = 4\pi \times 10^{-7} \text{ (henries/meter)} \tag{1-15a}$$

and (1-15) reduces to a product.

Finally, the conduction current density $\mathscr{J}_c$ is related *in the time domain* to the electric field intensity $\mathscr{E}$ by

$$\mathscr{J}_c = \hat{\sigma} * \mathscr{E} \tag{1-16}$$

where $\hat{\sigma}$ is the time-varying conductivity of the medium (siemens/meter). For free space

$$\hat{\sigma} = 0 \tag{1-16a}$$

In the frequency domain or for frequency nonvarying constitutive parameters, the relations of (1-14), (1-15) and (1-16) reduce to products. For simplicity of notation, they will be indicated everywhere from now on as products, and the caret (⌃) in the time-varying constitutive parameters will be omitted.

Whereas (1-14), (1-15), and (1-16) are referred to as the *constitutive relations*, $\hat{\varepsilon}$, $\hat{\mu}$, and $\hat{\sigma}$ are referred to as the *constitutive parameters*, which are, in general, functions of the applied field strength, the position within the medium, the direction of the applied field, and the frequency of operation.

The constitutive parameters are used to characterize the electrical properties of a material. In general, materials are characterized as *dielectrics* (*insulators*), *magnetics*, and *conductors* depending on whether *polarization* (electric displacement current density), *magnetization* (magnetic displacement current density), or *conduction* (conduction current density) is the predominant phenomenon. Another class of material is made up of *semiconductors*, which bridge the gap between dielectrics and conductors where neither displacement nor conduction currents are, in general, predominant. In addition, materials are classified as *linear* versus *nonlinear*, *homogeneous* versus *nonhomogeneous* (*inhomogeneous*), *isotropic* versus *nonisotropic* (*anisotropic*), and *dispersive* versus *nondispersive* according to their lattice structure and behavior. All these types of materials will be discussed in detail in Chapter 2.

If all the constitutive parameters of a given medium are not functions of the applied field strength, the material is known as *linear*; otherwise it is *nonlinear*. Media whose constitutive parameters are not functions of position are known as *homogeneous*; otherwise they are referred to as *nonhomogeneous* (*inhomogeneous*). *Isotropic* materials are those whose constitutive parameters are not functions of direction of the applied field; otherwise they are designated as *nonisotropic* (*anisotropic*). Crystals are one form of anisotropic material. Material whose constitutive parameters are functions of frequency are referred to as *dispersive*; otherwise they are known as *nondispersive*. All materials used in our everyday life exhibit some degree of dispersion, although the variations for some may be negligible and for others significant. More details concerning the development of the constitutive parameters can be found in Chapter 2.

1.4 CIRCUIT-FIELD RELATIONS

The differential and integral forms of Maxwell's equations were presented, respectively, in Sections 1.2.1 and 1.2.2. These relations are usually referred to as *field equations* since the quantities appearing in them are all *field quantities*. Maxwell's equations can also be written in terms of what are usually referred to as *circuit quantities*; the corresponding forms are denoted *circuit equations*. The circuit equations are introduced in circuit theory texts, and they are special cases of the more general field equations.

1.4.1 Kirchhoff's Voltage Law

According to Maxwell's equation 1-9a, the left side represents the sum voltage drops (use the convention where positive voltage begins at the start of the path) along a closed path C, which can be written as

$$\sum v = \oint_C \mathcal{E} \cdot d\mathbf{l} \text{ (volts)} \tag{1-17}$$

The right side of (1-9a) must also have the same units (volts) as its left side. Thus in the absence of impressed magnetic current densities ($\mathcal{M}_i = 0$), the right side of (1-9a) can be written as

$$-\frac{\partial}{\partial t}\iint_S \mathcal{B} \cdot d\mathbf{s} = -\frac{\partial \psi_m}{\partial t} = -\frac{\partial}{\partial t}(L_s i) = -L_s \frac{\partial i}{\partial t} \text{ (webers/second = volts)} \tag{1-17a}$$

because by definition $\psi_m = L_s i$ where L_s is an inductance (assumed to be constant) and i is the associated current. Using (1-17) and (1-17a), we can write (1-9a) with $\mathcal{M}_i = 0$ as

$$\sum v = -\frac{\partial \psi_m}{\partial t} = -\frac{\partial}{\partial t}(L_s i) = -L_s \frac{\partial i}{\partial t} \tag{1-17b}$$

Equation 1-17b states that the voltage drops along a closed path of a circuit are equal to the time rate of change of the magnetic flux passing through the surface enclosed by the closed path or equal to the voltage drop across an inductor L_s that

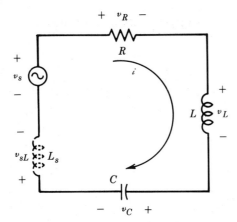

FIGURE 1-2 RLC series network.

is used to represent the *stray inductance* of the circuit. This is the well-known *Kirchhoff loop voltage law* which is used widely in circuit theory, and its form represents a circuit relation. Thus we can write the following field and circuit relations:

Field Relation   *Circuit Relation*

$$\oint_C \mathcal{E} \cdot d\mathbf{l} = -\frac{\partial}{\partial t} \iint_S \mathcal{B} \cdot d\mathbf{s} = -\frac{\partial \psi_m}{\partial t} \Leftrightarrow \sum v = -\frac{\partial \psi_m}{\partial t} = -L_s \frac{\partial i}{\partial t} \quad (1\text{-}17c)$$

In lumped-element circuit analysis, where usually the wavelength is very large (or the dimensions of the total circuit are small compared to the wavelength) and the stray inductance of the circuit is very small, the right side of (1-17b) is very small and it is usually set equal to zero. In these cases (1-17b) states that the voltage drops (or rises) along a closed path are equal to zero, and it represents a very widely used relation to most of us.

To demonstrate Kirchhoff's loop voltage law, let us consider the circuit of Figure 1-2 where a voltage source and three ideal lumped elements (a resistance R, an inductor L, and a capacitor C) are connected in series to form a closed loop. According to (1-17b)

$$-v_s + v_R + v_L + v_C = -L_s \frac{\partial i}{\partial t} = -v_{sL} \quad (1\text{-}18)$$

where L_s, shown dashed in Figure 1-2, represents the total stray inductance associated with the current and the magnetic flux generated by the loop that connects the ideal lumped elements (we assume that the wire resistance is negligible). If the stray inductance L_s of the circuit and the time rate of change of the current is small (the case for low-frequency applications), the right side of (1-18) is small and can be set equal to zero.

1.4.2 Kirchhoff's Current Law

The left side of the integral form of the continuity equation, as given by (1-13), can be written in circuit form as

$$\sum i = \oiint_S \mathcal{J}_{ic} \cdot d\mathbf{s} \quad (1\text{-}19)$$

where Σi represents the sum of the currents passing through closed surface S. Using (1-19) reduces (1-13) to

$$\Sigma i = -\frac{\partial \mathcal{Q}_e}{\partial t} = -\frac{\partial}{\partial t}(C_s v) = -C_s \frac{\partial v}{\partial t} \qquad (1\text{-}19a)$$

since by definition $\mathcal{Q}_e = C_s v$ where C_s is a capacitance (assumed to be constant) and v is the associated voltage.

Equation 1-19a states that the sum of the currents crossing a surface that encloses a circuit is equal to the time rate of change of the total electric charge enclosed by the surface or equal to the current flowing through a capacitor C_s that is used to represent the *stray capacitance* of the circuit. This is the well-known *Kirchhoff node current law* which is widely used in circuit theory, and its form represents a circuit relation. Thus we can write the following field and circuit relations:

Field Relation   *Circuit Relation*

$$\oint_S \mathcal{J}_{ic} \cdot d\mathbf{s} = -\frac{\partial}{\partial t}\iiint_V q_{ev}\, dv = -\frac{\partial \mathcal{Q}_e}{\partial t} \Leftrightarrow \Sigma i = -\frac{\partial \mathcal{Q}_e}{\partial t} = -C_s \frac{\partial v}{\partial t} \qquad (1\text{-}19b)$$

In lumped-element circuit analysis, where the stray capacitance associated with the circuit is very small, the right side of (1-19a) is very small and it is usually set equal to zero. In these cases (1-19a) states that the currents exiting (or entering) a surface enclosing a circuit are equal to zero. This represents a very widely used relation.

To demonstrate Kirchhoff's node current law, let us consider the circuit of Figure 1-3 where a current source and three ideal lumped elements (a resistance R, an inductor L, and a capacitor C) are connected in parallel to form a node. According to (1-19a)

$$-i_s + i_R + i_L + i_C = -C_s \frac{\partial v}{\partial t} = -i_{sC} \qquad (1\text{-}20)$$

where C_s, shown dashed in Figure 1-3, represents the total stray capacitance associated with the circuit of Figure 1-3. If the stray capacitance C_s of the circuit and the time rate of change of the total charge $\mathcal{Q}_e$ are small (the case for low-frequency applications), the right side of (1-20) is small and can be set equal to zero. The current i_{sC} associated with the stray capacitance C_s also includes the displacement (*leakage*) current crossing the closed surface S of Figure 1-3 outside of the wires.

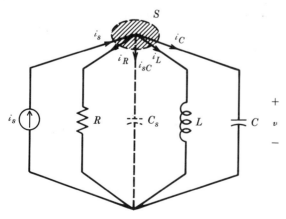

FIGURE 1-3 RLC parallel network.

1.4.3 Element Laws

In addition to Kirchhoff's loop voltage and node current laws as given, respectively, by (1-17b) and (1-19a), there are a number of current element laws that are widely used in circuit theory. One of the most popular is *Ohm's law* for a resistor (or a conductance G), which states that the *voltage drop v_R across a resistor R is equal to the product of the resistor R and the current i_R flowing through it* ($v_R = Ri_R$ or $i_R = v_R/R = Gv_R$). Ohm's law of circuit theory is a special case of the constitutive relation given by (1-16). Thus

Field Relation   *Circuit Relation*

$$\mathscr{J}_c = \sigma \mathscr{E} \quad \Leftrightarrow \quad i_R = \frac{1}{R}v_R = Gv_R \tag{1-21}$$

Another element law is *associated with an inductor L* and states that the *voltage drop across an inductor is equal to the product of L and the time rate of change of the current through the inductor* ($v_L = L\, di_L/dt$). Before proceeding to relate the inductor's voltage drop to the corresponding field relation, let us first define inductance. To do this we state that the magnetic flux ψ_m is equal to the product of the inductance L and the corresponding current i. That is $\psi_m = Li$. The corresponding field equation of this relation is (1-15). Thus

Field Relation   *Circuit Relation*

$$\mathscr{B} = \mu \mathscr{H} \quad \Leftrightarrow \quad \psi_m = Li_L \tag{1-22}$$

Using (1-5c) and (1-15), we can write for a homogeneous and non-time-varying medium that

$$\mathscr{M}_d = \frac{\partial \mathscr{B}}{\partial t} = \frac{\partial}{\partial t}(\mu \mathscr{H}) = \mu \frac{\partial \mathscr{H}}{\partial t} \tag{1-22a}$$

where $\mathscr{M}_d$ is defined as the magnetic displacement current density [analogous to the electric displacement current density $\mathscr{J}_d = \partial \mathscr{D}/\partial t = \partial(\varepsilon\mathscr{E})/\partial t = \varepsilon\, \partial \mathscr{E}/\partial t$]. With the aid of the right side of (1-9a) and the circuit relation of (1-22) we can write

$$\frac{\partial}{\partial t}\iint_S \mathscr{B}\cdot d\mathbf{s} = \frac{\partial \psi_m}{\partial t} = \frac{\partial}{\partial t}(Li_L) = L\frac{\partial i_L}{\partial t} = v_L \tag{1-22b}$$

Using (1-22a) and (1-22b), we can write the following relations:

Field Relation   *Circuit Relation*

$$\mathscr{M}_d = \mu\frac{\partial \mathscr{H}}{\partial t} \quad \Leftrightarrow \quad v_L = L\frac{\partial i_L}{\partial t} \tag{1-22c}$$

Using a similar procedure, for a capacitor C we can write the field and circuit relations analogous to those of (1-22) and (1-22c):

Field Relation   *Circuit Relation*

$$\mathscr{D} = \varepsilon \mathscr{E} \quad \Leftrightarrow \quad \mathscr{Q}_e = Cv_e \tag{1-23}$$

$$\mathscr{J}_d = \varepsilon \frac{\partial \mathscr{E}}{\partial t} \quad \Leftrightarrow \quad i_C = C\frac{\partial v_C}{\partial t} \tag{1-24}$$

A summary of the field theory relations and their corresponding circuit concepts are listed in Table 1-2.

TABLE 1-2
Relations between electromagnetic field and circuit theories

Field theory	Circuit theory
1. $\mathscr{E}$ (electric field intensity)	1. v (voltage)
2. $\mathscr{H}$ (magnetic field intensity)	2. i (current)
3. $\mathscr{D}$ (electric flux density)	3. q_{ev} (electric charge density)
4. $\mathscr{B}$ (magnetic flux density)	4. q_{mv} (magnetic charge density)
5. $\mathscr{J}$ (electric current density)	5. i_e (electric current)
6. $\mathscr{M}$ (magnetic current density)	6. i_m (magnetic current)
7. $\mathscr{J}_d = \varepsilon \dfrac{\partial \mathscr{E}}{\partial t}$ (electric displacement current density)	7. $i = C \dfrac{dv}{dt}$ (current through a capacitor)
8. $\mathscr{M}_d = \mu \dfrac{\partial \mathscr{H}}{\partial t}$ (magnetic displacement current density)	8. $v = L \dfrac{di}{dt}$ (voltage across an inductor)
9. *Constitutive relations*	9. *Element laws*
(a) $\mathscr{J}_c = \sigma \mathscr{E}$ (electric conduction current density)	(a) $i = Gv = \dfrac{1}{R} v$ (Ohm's law)
(b) $\mathscr{D} = \varepsilon \mathscr{E}$ (dielectric material)	(b) $\mathscr{Q}_e = Cv$ (charge in a capacitor)
(c) $\mathscr{B} = \mu \mathscr{H}$ (magnetic material)	(c) $\psi = Li$ (flux of an inductor)
10. $\oint_C \mathscr{E} \cdot d\mathbf{l} = -\dfrac{\partial}{\partial t} \iint_S \mathscr{B} \cdot d\mathbf{s}$ (Maxwell–Faraday equation)	10. $\Sigma v = -L_s \dfrac{\partial i}{\partial t} \simeq 0$ (Kirchhoff's voltage law)
11. $\oiint_S \mathscr{J}_{ic} \cdot d\mathbf{s} = -\dfrac{\partial}{\partial t} \iiint_V q_{ev}\, dv = -\dfrac{\partial \mathscr{Q}_e}{\partial t}$ (continuity equation)	11. $\Sigma i = -C_s \dfrac{\partial v}{\partial t} \simeq 0$ (Kirchhoff's current law)
12. *Power and energy densities*	12. *Power and energy*
(a) $\oiint_S (\mathscr{E} \times \mathscr{H}) \cdot d\mathbf{s}$ (instantaneous power)	(a) $\not{p} = vi$ (power–voltage–current relation)
(b) $\iiint_V \sigma \mathscr{E}^2\, dv$ (dissipated power)	(b) $\not{p}_d = Gv^2 = \dfrac{1}{R} v^2$ (power dissipated in a resistor)
(c) $\tfrac{1}{2} \iiint_V \varepsilon \mathscr{E}^2\, dv$ (electric stored energy)	(c) $\tfrac{1}{2} Cv^2$ (energy stored in a capacitor)
(d) $\tfrac{1}{2} \iiint_V \mu \mathscr{H}^2\, dv$ (magnetic stored energy)	(d) $\tfrac{1}{2} Li^2$ (energy stored in an inductor)

1.5 BOUNDARY CONDITIONS

As previously stated, the differential form of Maxwell's equations are used to solve for the field vectors provided the field quantities are single-valued, bounded, and possess (along with their derivatives) continuous distributions. Along boundaries where the media involved exhibit discontinuities in electrical properties (or there exist sources along these boundaries), the field vectors are also discontinuous and their behavior across the boundaries is governed by the *boundary conditions*.

Maxwell's equations in differential form represent derivatives, with respect to the space coordinates, of the field vectors. At points of discontinuity in the field vectors, the derivatives of the field vectors have no meaning and cannot be properly used to define the behavior of the field vectors across these boundaries. Instead the behavior of the field vectors across discontinuous boundaries must be handled by examining the field vectors themselves and not their derivatives. The dependence of the field vectors on the electrical properties of the media along boundaries of discontinuity is manifested in our everyday life. It has been observed that radio or television reception deteriorates or even ceases as we move from outside to inside an enclosure (such as a tunnel or a well-shielded building). The reduction or loss of the signal is governed not only by its attenuation as it travels through the medium but also by its behavior across the discontinuous interfaces. Maxwell's equations in integral form provide the most convenient formulation for derivation of the boundary conditions.

1.5.1 Finite Conductivity Media

Initially let us consider an interface between two media, as shown in Figure 1-4a, along which there are no charges or sources. These conditions are satisfied provided that neither of the two media is a perfect conductor or that actual sources are not placed there. Media 1 and 2 are characterized, respectively, by the constitutive parameters $\varepsilon_1, \mu_1, \sigma_1$ and $\varepsilon_2, \mu_2, \sigma_2$.

At a given point along the interface, let us choose a rectangular box whose boundary is denoted by C_0 and its areas by S_0. The x, y, z coordinate system is chosen to represent the local geometry of the rectangle. Applying Maxwell's equation 1-9a, with $\mathcal{M}_i = 0$, on the rectangle along C_0 and on S_0, we have

$$\oint_{C_0} \mathcal{E} \cdot d\mathbf{l} = -\frac{\partial}{\partial t} \iint_{S_0} \mathcal{B} \cdot d\mathbf{s} \tag{1-25}$$

As the height Δy of the rectangle becomes progressively shorter, the area S_0 also becomes vanishingly smaller so that the contributions of the surface integral in (1-25) are negligible. In addition, the contributions of the line integral in (1-25) along Δy are also minimal so that in the limit ($\Delta y \to 0$) (1-25) reduces to

$$\mathcal{E}_1 \cdot \hat{a}_x \Delta x - \mathcal{E}_2 \cdot \hat{a}_x \Delta x = 0$$

$$\mathcal{E}_{1t} - \mathcal{E}_{2t} = 0 \Rightarrow \mathcal{E}_{1t} = \mathcal{E}_{2t} \tag{1-26}$$

or

$$\boxed{\hat{n} \times (\mathcal{E}_2 - \mathcal{E}_1) = 0} \qquad \sigma_1, \sigma_2 \text{ are finite} \tag{1-26a}$$

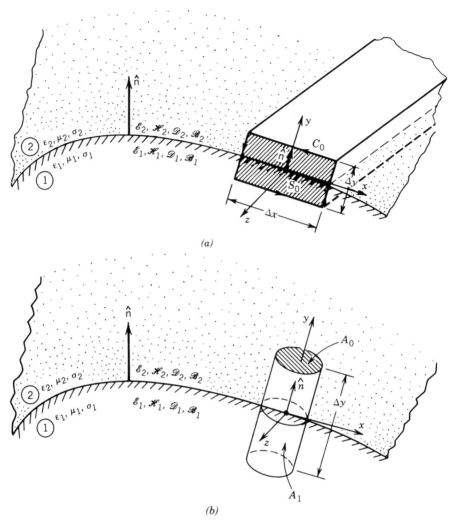

FIGURE 1-4 Geometry for boundary conditions of tangential and normal components. (*a*) Tangential components. (*b*) Normal components.

In (1-26), $\mathscr{E}_{1t}$ and $\mathscr{E}_{2t}$ represent, respectively, the tangential components of the electric field in media 1 and 2 along the interface. Both (1-26) and (1-26a) state that *the tangential components of the electric field across an interface between two media with no impressed magnetic current densities along the boundary of the interface are continuous.*

Using a similar procedure on the same rectangle but for (1-10), assuming $\mathscr{J}_i = 0$, we can write that

$$\mathscr{H}_{1t} - \mathscr{H}_{2t} = 0 \Rightarrow \mathscr{H}_{1t} = \mathscr{H}_{2t} \tag{1-27}$$

or

$$\boxed{\hat{n} \times (\mathscr{H}_2 - \mathscr{H}_1) = 0} \qquad \sigma_1, \sigma_2 \text{ are finite} \tag{1-27a}$$

which state that *the tangential components of the magnetic field across an interface between two media, neither of which is a perfect conductor, are continuous.* This relation also holds if either or both media possess finite conductivity. Equations 1-26a and 1-27a must be modified if either of the two media is a perfect conductor or if there are impressed (source) current densities along the interface. This will be done in the pages that follow.

In addition to the boundary conditions on the tangential components of the electric and magnetic fields across an interface, their normal components are also related. To derive these relations, let us consider the geometry of Figure 1-4b where a cylindrical pillbox is chosen at a given point along the interface. If there are no charges along the interface, which is the case when there are no sources or either of the two media is not a perfect conductor, (1-11a) reduces to

$$\oiint_{A_0, A_1} \mathscr{D} \cdot d\mathbf{s} = 0 \tag{1-28}$$

As the height Δy of the pillbox becomes progressively shorter, the area A_1 also becomes vanishingly smaller so that the contributions to the surface integral of (1-28) by A_1 are negligible. Thus (1-28) can be written in the limit ($\Delta y \to 0$) as

$$\mathscr{D}_2 \cdot \hat{a}_y A_0 - \mathscr{D}_1 \cdot \hat{a}_y A_0 = 0$$
$$\mathscr{D}_{2n} - \mathscr{D}_{1n} = 0 \Rightarrow \mathscr{D}_{2n} = \mathscr{D}_{1n} \tag{1-29}$$

or

$$\boxed{\hat{n} \cdot (\mathscr{D}_2 - \mathscr{D}_1) = 0} \qquad \sigma_1, \sigma_2 \text{ are finite} \tag{1-29a}$$

In (1-29), $\mathscr{D}_{1n}$ and $\mathscr{D}_{2n}$ represent, respectively, the normal components of the electric flux density in media 1 and 2 along the interface. Both (1-29) and (1-29a) state that *the normal components of the electric flux density across an interface between two media, both of which are imperfect electric conductors and where there are no sources, are continuous.* This relation also holds if either or both media possess finite conductivity. Equation 1-29a must be modified if either of the media is a perfect conductor or if there are sources along the interface. This will be done in the pages that follow.

In terms of the electric field intensities, (1-29) and (1-29a) can be written as

$$\varepsilon_2 \mathscr{E}_{2n} = \varepsilon_1 \mathscr{E}_{1n} \Rightarrow \mathscr{E}_{2n} = \frac{\varepsilon_1}{\varepsilon_2} \mathscr{E}_{1n} \Rightarrow \mathscr{E}_{1n} = \frac{\varepsilon_2}{\varepsilon_1} \mathscr{E}_{2n} \tag{1-30}$$

or

$$\boxed{\hat{n} \cdot (\varepsilon_2 \mathscr{E}_2 - \varepsilon_1 \mathscr{E}_1) = 0} \qquad \sigma_1, \sigma_2 \text{ are finite} \tag{1-30a}$$

which state that *the normal components of the electric field intensity across an interface are discontinuous.*

Using a similar procedure on the same pillbox but for (1-12) with no charges along the interface, we can write that

$$\mathscr{B}_{2n} - \mathscr{B}_{1n} = 0 \Rightarrow \mathscr{B}_{2n} = \mathscr{B}_{1n} \tag{1-31}$$

or

$$\hat{n} \cdot (\mathscr{B}_2 - \mathscr{B}_1) = 0 \qquad (1\text{-}31a)$$

which state that *the normal components of the magnetic flux density across an interface between two media where there are no sources are continuous.* In terms of the magnetic field intensities, (1-31) and (1-31a) can be written as

$$\mu_2 \mathscr{H}_{2n} = \mu_1 \mathscr{H}_{1n} \Rightarrow \mathscr{H}_{2n} = \frac{\mu_1}{\mu_2}\mathscr{H}_{1n} \Rightarrow \mathscr{H}_{1n} = \frac{\mu_2}{\mu_1}\mathscr{H}_{2n} \qquad (1\text{-}32)$$

or

$$\hat{n} \cdot (\mu_2 \mathscr{H}_2 - \mu_1 \mathscr{H}_1) = 0 \qquad (1\text{-}32a)$$

which state that *the normal components of the magnetic field intensity across an interface are discontinuous.*

1.5.2 Infinite Conductivity Media

If actual electric sources and charges exist along the interface between the two media, or if either of the two media forming the interface displayed in Figure 1-4 is a perfect electric conductor, the boundary conditions on the tangential components of the magnetic field [stated by (1-27a)] and on the normal components of the electric flux density or normal components of the electric field intensity [stated by (1-29a) or (1-30a)] must be modified to include the sources and charges or the induced linear electric current density ($\mathscr{J}_s$) and surface electric charge density (q_{es}). Similar modifications must be made to (1-26a), (1-31a), and (1-32a) if magnetic sources and charges exist along the interface between the two media, or if either of the two media is a perfect magnetic conductor.

To derive the appropriate boundary conditions for such cases, let us refer first to Figure 1-4a and assume that on a very thin layer along the interface there exists an electric surface charge density q_{es} (C/m^2) and linear electric current density $\mathscr{J}_s$ (A/m). Applying (1-10) along the rectangle of Figure 1-4a, we can write that

$$\oint_{C_0} \mathscr{H} \cdot d\mathbf{l} = \iint_{S_0} \mathscr{J}_{ic} \cdot d\mathbf{s} + \frac{\partial}{\partial t} \iint_{S_0} \mathscr{D} \cdot d\mathbf{s} \qquad (1\text{-}33)$$

In the limit as the height of the rectangle is shrinking, the left side of (1-33) reduces to

$$\lim_{\Delta y \to 0} \oint_{C_0} \mathscr{H} \cdot d\mathbf{l} = (\mathscr{H}_1 - \mathscr{H}_2) \cdot \hat{a}_x \Delta x \qquad (1\text{-}33a)$$

Since the electric current density $\mathscr{J}_{ic}$ is confined on a very thin layer along the interface, the first term on the right side of (1-33) can be written as

$$\lim_{\Delta y \to 0} \iint_{S_0} \mathscr{J}_{ic} \cdot d\mathbf{s}$$
$$= \lim_{\Delta y \to 0} [\mathscr{J}_{ic} \cdot \hat{a}_z \Delta x \Delta y] = \lim_{\Delta y \to 0} [(\mathscr{J}_{ic} \Delta y) \cdot \hat{a}_z \Delta x] = \mathscr{J}_s \cdot \hat{a}_z \Delta x \qquad (1\text{-}33b)$$

Since S_0 becomes vanishingly smaller as $\Delta y \to 0$, the last term on the right side of (1-33) reduces to

$$\lim_{\Delta y \to 0} \frac{\partial}{\partial t} \iint_{S_0} \mathscr{D} \cdot d\mathbf{s} = \lim_{\Delta y \to 0} \frac{\partial}{\partial t} \iint_{S_0} \mathscr{D} \cdot \hat{a}_z \, ds = 0 \qquad (1\text{-}33c)$$

Substituting (1-33a) through (1-33c) into (1-33), we can write it as

$$(\mathscr{H}_1 - \mathscr{H}_2) \cdot \hat{a}_x \Delta x = \mathscr{J}_s \cdot \hat{a}_z \Delta x$$

or

$$(\mathscr{H}_1 - \mathscr{H}_2) \cdot \hat{a}_x - \mathscr{J}_s \cdot \hat{a}_z = 0 \qquad (1\text{-}33d)$$

Since

$$\hat{a}_x = \hat{a}_y \times \hat{a}_z \qquad (1\text{-}34)$$

(1-33d) can be written as

$$(\mathscr{H}_1 - \mathscr{H}_2) \cdot (\hat{a}_y \times \hat{a}_z) - \mathscr{J}_s \cdot \hat{a}_z = 0 \qquad (1\text{-}35)$$

Using the vector identity

$$\mathbf{A} \cdot \mathbf{B} \times \mathbf{C} = \mathbf{C} \cdot \mathbf{A} \times \mathbf{B} \qquad (1\text{-}36)$$

on the first term in (1-35), we can then write it as

$$\hat{a}_z \cdot \left[(\mathscr{H}_1 - \mathscr{H}_2) \times \hat{a}_y \right] - \mathscr{J}_s \cdot \hat{a}_z = 0 \qquad (1\text{-}37)$$

or

$$\left\{ \left[\hat{a}_y \times (\mathscr{H}_2 - \mathscr{H}_1) \right] - \mathscr{J}_s \right\} \cdot \hat{a}_z = 0 \qquad (1\text{-}37a)$$

Equation 1-37a is satisfied provided

$$\hat{a}_y \times (\mathscr{H}_2 - \mathscr{H}_1) - \mathscr{J}_s = 0 \qquad (1\text{-}38)$$

or

$$\hat{a}_y \times (\mathscr{H}_2 - \mathscr{H}_1) = \mathscr{J}_s \qquad (1\text{-}38a)$$

Similar results are obtained if the rectangles chosen are positioned in other planes. Therefore we can write an expression on the boundary conditions of the tangential components of the magnetic field, using the geometry of Figure 1-4a, as

$$\boxed{\hat{n} \times (\mathscr{H}_2 - \mathscr{H}_1) = \mathscr{J}_s} \qquad (1\text{-}39)$$

Equation 1-39 states that *the tangential components of the magnetic field across an interface, along which there exists a surface electric current density $\mathscr{J}_s$ (A/m), are discontinuous by an amount equal to the electric current density.*

If either of the two media is a perfect electric conductor, (1-39) must be reduced to account for the presence of the conductor. Let us assume that medium 1 in Figure 1-4a possesses an infinite conductivity ($\sigma_1 = \infty$). With such conductivity

18 TIME-VARYING AND TIME-HARMONIC ELECTROMAGNETIC FIELDS

$\mathscr{E}_1 = 0$, and (1-26a) reduces to

$$\boxed{\hat{n} \times \mathscr{E}_2 = 0 \Rightarrow \mathscr{E}_{2t} = 0} \tag{1-40}$$

Then (1-1) can be written as

$$\nabla \times \mathscr{E}_1 = 0 = -\frac{\partial \mathscr{B}_1}{\partial t} \Rightarrow \mathscr{B}_1 = 0 \Rightarrow \mathscr{H}_1 = 0 \tag{1-41}$$

provided μ_1 is finite.

In a perfect conductor its free charges are confined to a very thin layer on the surface of the conductor, forming a surface charge density q_{es} (with units of coulombs/square meter). This charge density does not include *bound* (polarization) charges (which contribute to the polarization surface charge density) that are usually found inside and on the surface of dielectric media and that form atomic dipoles having equal and opposite charges separated by an assumed infinitesimal distance. Here instead the surface charge density q_{es} represents actual electric charges separated by finite dimensions from equal quantities of opposite charge.

When the conducting surface is subjected to an applied electromagnetic field, the surface charges are subjected to electric field Lorentz forces. These charges are set in motion and thus create a surface electric current density $\mathscr{J}_s$ with units of amperes per meter. The surface current density $\mathscr{J}_s$ also resides in a vanishingly thin layer on the surface of the conductor so that in the limit as $\Delta y \to 0$ in Figure 1-4a the volume electric current density $\mathscr{J}$ (A/m^2) reduces to

$$\lim_{\Delta y \to 0} (\mathscr{J} \Delta y) = \mathscr{J}_s \tag{1-42}$$

Then the boundary condition of (1-39) reduces, using (1-41) and (1-42), to

$$\boxed{\hat{n} \times \mathscr{H}_2 = \mathscr{J}_s \Rightarrow \mathscr{H}_{2t} = \mathscr{J}_s} \tag{1-43}$$

which states that *the tangential components of the magnetic field intensity are discontinuous next to a perfect electric conductor by an amount equal to the induced linear electric current density.*

The boundary conditions on the normal components of the electric field intensity and the electric flux density on an interface along which a surface current density q_{es} resides on a very thin layer can be derived by applying the integrals of (1-11a) on a cylindrical pillbox shown in Figure 1-4b. Then we can write (1-11a) as

$$\lim_{\Delta y \to 0} \oiint_{A_0, A_1} \mathscr{D} \cdot d\mathbf{s} = \lim_{\Delta y \to 0} \iiint_V q_{ev} \, dv \tag{1-44}$$

Since the cylindrical surface A_1 of the pillbox diminishes as $\Delta y \to 0$, its contributions to the surface integral vanish. Thus we can write (1-44) as

$$(\mathscr{D}_2 - \mathscr{D}_1) \cdot \hat{n} A_0 = \lim_{\Delta y \to 0} \left[(q_{ev} \Delta y) A_0 \right] = q_{es} A_0 \tag{1-45}$$

which reduces to

$$\boxed{\hat{n} \cdot (\mathscr{D}_2 - \mathscr{D}_1) = q_{es} \Rightarrow \mathscr{D}_{2n} - \mathscr{D}_{1n} = q_{es}} \tag{1-45a}$$

Equation (1-45a) states that *the normal components of the electric flux density on an interface along which a surface charge density resides are discontinuous by an amount equal to the surface charge density.*

In terms of the normal components of the electric field intensity, (1-45a) can be written as

$$\boxed{\hat{n} \cdot (\varepsilon_2 \mathscr{E}_2 - \varepsilon_1 \mathscr{E}_1) = q_{es}} \qquad (1\text{-}46)$$

which also indicates that *the normal components of the electric field are discontinuous across a boundary along which a surface charge density resides.*

If either of the media is a perfect electric conductor (assuming that medium 1 possesses infinite conductivity $\sigma_1 = \infty$), (1-45a) and (1-46) reduce, respectively, to

$$\boxed{\hat{n} \cdot \mathscr{D}_2 = q_{es} \Rightarrow \mathscr{D}_{2n} = q_{es}} \qquad (1\text{-}47a)$$

$$\boxed{\hat{n} \cdot \mathscr{E}_2 = q_{es}/\varepsilon_2 \Rightarrow \mathscr{E}_{2n} = q_{es}/\varepsilon_2} \qquad (1\text{-}47b)$$

Both (1-47a) and (1-47b) state that *the normal components of the electric flux density and corresponding electric field intensity are discontinuous next to a perfect electric conductor.*

1.5.3 Sources along Boundaries

If electric and magnetic sources (charges and current densities) are present along the interface between the two media with neither one being a perfect conductor, the boundary conditions on the tangential and normal components of the fields can be written as

$$-\hat{n} \times (\mathscr{E}_2 - \mathscr{E}_1) = \mathscr{M}_s \qquad (1\text{-}48a)$$

$$\hat{n} \times (\mathscr{H}_2 - \mathscr{H}_1) = \mathscr{J}_s \qquad (1\text{-}48b)$$

$$\hat{n} \cdot (\mathscr{D}_2 - \mathscr{D}_1) = q_{es} \qquad (1\text{-}48c)$$

$$\hat{n} \cdot (\mathscr{B}_2 - \mathscr{B}_1) = q_{ms} \qquad (1\text{-}48d)$$

where $(\mathscr{M}_s, \mathscr{J}_s)$ and (q_{ms}, q_{es}) are the magnetic and electric linear (per meter) current and surface (per square meter) charge densities, respectively. The derivation of (1-48a) and (1-48d) proceeds along the same lines, respectively, as the derivation of (1-48b) and (1-48c) in Section 1.5.2, but begins with (1-9a) and (1-12).

A summary of the boundary conditions on all the field components is found in Table 1-3, which also includes the boundary conditions assuming that medium 1 is a perfect magnetic conductor. In general, a magnetic conductor is defined as a material inside of which both time-varying electric and magnetic fields vanish when it is subjected to an electromagnetic field. The tangential components of the magnetic field also vanish next to its surface. In addition, the magnetic charge moves to the surface of the material and creates a magnetic current density which resides on a very thin layer at the surface. Although such materials do not physically exist, they are often used in electromagnetics to develop electrical equivalents that yield the same answers as the actual physical problems.

20 TIME-VARYING AND TIME-HARMONIC ELECTROMAGNETIC FIELDS

TABLE 1-3
Boundary conditions on instantaneous electromagnetic fields

	General	Finite conductivity media, no sources or charges $\sigma_1, \sigma_2 \neq \infty$ $\mathscr{J}_s = 0; q_{es} = 0$ $\mathscr{M}_s = 0; q_{ms} = 0$	Medium 1 of infinite electric conductivity $\sigma_1 = \infty; \sigma_2 \neq \infty$ $\mathscr{M}_s = 0; q_{ms} = 0$	Medium 1 of infinite magnetic conductivity $(\mathscr{H}_{1t} = 0)$ $\mathscr{J}_s = 0; q_{es} = 0$
Tangential electric field intensity	$-\hat{n} \times (\mathscr{E}_2 - \mathscr{E}_1) = \mathscr{M}_s$	$\hat{n} \times (\mathscr{E}_2 - \mathscr{E}_1) = 0$	$\hat{n} \times \mathscr{E}_2 = 0$	$-\hat{n} \times \mathscr{E}_2 = \mathscr{M}_s$
Tangential magnetic field intensity	$\hat{n} \times (\mathscr{H}_2 - \mathscr{H}_1) = \mathscr{J}_s$	$\hat{n} \times (\mathscr{H}_2 - \mathscr{H}_1) = 0$	$\hat{n} \times \mathscr{H}_2 = \mathscr{J}_s$	$\hat{n} \times \mathscr{H}_2 = 0$
Normal electric flux density	$\hat{n} \cdot (\mathscr{D}_2 - \mathscr{D}_1) = q_{es}$	$\hat{n} \cdot (\mathscr{D}_2 - \mathscr{D}_1) = 0$	$\hat{n} \cdot \mathscr{D}_2 = q_{es}$	$\hat{n} \cdot \mathscr{D}_2 = 0$
Normal magnetic flux density	$\hat{n} \cdot (\mathscr{B}_2 - \mathscr{B}_1) = q_{ms}$	$\hat{n} \cdot (\mathscr{B}_2 - \mathscr{B}_1) = 0$	$\hat{n} \cdot \mathscr{B}_2 = 0$	$\hat{n} \cdot \mathscr{B}_2 = q_{ms}$

1.6 POWER AND ENERGY

In a wireless communication system, electromagnetic fields are used to transport information over long distances. To accomplish this, energy must be associated with electromagnetic fields. This transport of energy is accomplished even in the absence of any intervening medium.

To derive the equations that indicate that energy (and forms of it) is associated with electromagnetic waves, let us consider a region V characterized by ε, μ, σ and enclosed by the surface S as shown in Figure 1-5. Within that region there exist electric and magnetic sources represented, respectively, by the electric and magnetic current densities $\mathscr{J}_i$ and $\mathscr{M}_i$. The fields generated by $\mathscr{J}_i$ and $\mathscr{M}_i$ that exist within S are represented by $\mathscr{E}, \mathscr{H}$. These fields obey Maxwell's equations, and we can write using (1-1) and (1-2) that

$$\nabla \times \mathscr{E} = -\mathscr{M}_i - \frac{\partial \mathscr{B}}{\partial t} = -\mathscr{M}_i - \mu\frac{\partial \mathscr{H}}{\partial t} = -\mathscr{M}_i - \mathscr{M}_d \quad (1\text{-}49a)$$

$$\nabla \times \mathscr{H} = \mathscr{J}_i + \mathscr{J}_c + \frac{\partial \mathscr{D}}{\partial t} = \mathscr{J}_i + \sigma\mathscr{E} + \varepsilon\frac{\partial \mathscr{E}}{\partial t} = \mathscr{J}_i + \mathscr{J}_c + \mathscr{J}_d \quad (1\text{-}49b)$$

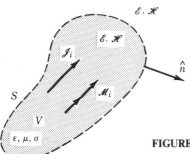

FIGURE 1-5 Electric and magnetic fields within S generated by $\mathscr{J}_i$ and $\mathscr{M}_i$.

Using scalar multiplication on (1-49a) by $\mathcal{H}$ and on (1-49b) by $\mathcal{E}$, we can write that

$$\mathcal{H} \cdot (\nabla \times \mathcal{E}) = -\mathcal{H} \cdot (\mathcal{M}_i + \mathcal{M}_d) \quad (1\text{-}50\text{a})$$

$$\mathcal{E} \cdot (\nabla \times \mathcal{H}) = \mathcal{E} \cdot (\mathcal{J}_i + \mathcal{J}_c + \mathcal{J}_d) \quad (1\text{-}50\text{b})$$

Subtracting (1-50b) from (1-50a) reduces to

$$\mathcal{H} \cdot (\nabla \times \mathcal{E}) - \mathcal{E} \cdot (\nabla \times \mathcal{H}) = -\mathcal{H} \cdot (\mathcal{M}_i + \mathcal{M}_d) - \mathcal{E} \cdot (\mathcal{J}_i + \mathcal{J}_c + \mathcal{J}_d) \quad (1\text{-}51)$$

Using the vector identity

$$\nabla \cdot (\mathbf{A} \times \mathbf{B}) = \mathbf{B} \cdot (\nabla \times \mathbf{A}) - \mathbf{A} \cdot (\nabla \times \mathbf{B}) \quad (1\text{-}52)$$

on the left side of (1-51), we can write that

$$\nabla \cdot (\mathcal{E} \times \mathcal{H}) = -\mathcal{H} \cdot (\mathcal{M}_i + \mathcal{M}_d) - \mathcal{E} \cdot (\mathcal{J}_i + \mathcal{J}_c + \mathcal{J}_d) \quad (1\text{-}53)$$

or

$$\boxed{\nabla \cdot (\mathcal{E} \times \mathcal{H}) + \mathcal{H} \cdot (\mathcal{M}_i + \mathcal{M}_d) + \mathcal{E} \cdot (\mathcal{J}_i + \mathcal{J}_c + \mathcal{J}_d) = 0} \quad (1\text{-}53\text{a})$$

Integrating (1-53) over the volume V leads to

$$\iiint_V \nabla \cdot (\mathcal{E} \times \mathcal{H}) \, dv = -\iiint_V [\mathcal{H} \cdot (\mathcal{M}_i + \mathcal{M}_d) + \mathcal{E} \cdot (\mathcal{J}_i + \mathcal{J}_c + \mathcal{J}_d)] \, dv \quad (1\text{-}54)$$

Applying the divergence theorem of (1-8) on the left side of (1-54) reduces it to

$$\oiint_S (\mathcal{E} \times \mathcal{H}) \cdot d\mathbf{s} = -\iiint_V [\mathcal{H} \cdot (\mathcal{M}_i + \mathcal{M}_d) + \mathcal{E} \cdot (\mathcal{J}_i + \mathcal{J}_c + \mathcal{J}_d)] \, dv \quad (1\text{-}55)$$

or

$$\boxed{\oiint_S (\mathcal{E} \times \mathcal{H}) \cdot d\mathbf{s} + \iiint_V [\mathcal{H} \cdot (\mathcal{M}_i + \mathcal{M}_d) + \mathcal{E} \cdot (\mathcal{J}_i + \mathcal{J}_c + \mathcal{J}_d)] \, dv = 0}$$

$$(1\text{-}55\text{a})$$

Equations 1-53a and 1-55a can be interpreted, respectively, as the differential and integral forms for the *conservation of energy*. To accomplish this, let us consider each of the terms included in (1-55a).

The integrand in the first term of (1-55a) has the form

$$\mathcal{S} = \mathcal{E} \times \mathcal{H} \quad (1\text{-}56)$$

which has the units of power density (watts/square meter), since $\mathcal{E}$ (V/m) and $\mathcal{H}$ (A/m) with $\mathcal{S}$ (V $\cdot$ A/m^2 = W/m^2). Thus the first term of (1-55a), written as

$$\mathcal{P}_e = \oiint_S (\mathcal{E} \times \mathcal{H}) \cdot d\mathbf{s} = \oiint_S \mathcal{S} \cdot d\mathbf{s} \quad (1\text{-}57)$$

represents the total power $\mathcal{P}_e$ exiting the volume V bounded by the surface S.

The other terms in (1-55a), which represent the integrand of the volume integral, can be written as

$$p_s = -(\mathcal{H} \cdot \mathcal{M}_i + \mathcal{E} \cdot \mathcal{J}_i) \tag{1-58a}$$

$$\mathcal{H} \cdot \mathcal{M}_d = \mathcal{H} \cdot \frac{\partial \mathcal{B}}{\partial t} = \mu \mathcal{H} \cdot \frac{\partial \mathcal{H}}{\partial t} = \frac{1}{2}\mu \frac{\partial \mathcal{H}^2}{\partial t} = \frac{\partial}{\partial t}\left(\frac{1}{2}\mu \mathcal{H}^2\right) = \frac{\partial}{\partial t}w_m \tag{1-58b}$$

$$p_d = \mathcal{E} \cdot \mathcal{J}_c = \mathcal{E} \cdot (\sigma \mathcal{E}) = \sigma \mathcal{E}^2 \tag{1-58c}$$

$$\mathcal{E} \cdot \mathcal{J}_d = \mathcal{E} \cdot \frac{\partial \mathcal{D}}{\partial t} = \varepsilon \mathcal{E} \cdot \frac{\partial \mathcal{E}}{\partial t} = \frac{1}{2}\varepsilon \frac{\partial \mathcal{E}^2}{\partial t} = \frac{\partial}{\partial t}\left(\frac{1}{2}\varepsilon \mathcal{E}^2\right) = \frac{\partial}{\partial t}w_e \tag{1-58d}$$

where

$$w_m = \tfrac{1}{2}\mu \mathcal{H}^2 = \text{magnetic energy density } (\text{J/m}^3) \tag{1-58e}$$

$$w_e = \tfrac{1}{2}\varepsilon \mathcal{E}^2 = \text{electric energy density } (\text{J/m}^3) \tag{1-58f}$$

$$p_s = -(\mathcal{H} \cdot \mathcal{M}_i + \mathcal{E} \cdot \mathcal{J}_i) = \text{supplied power density } (\text{W/m}^3) \tag{1-58g}$$

$$p_d = \sigma \mathcal{E}^2 = \text{dissipated power density } (\text{W/m}^3) \tag{1-58h}$$

Integrating each of the terms in (1-58a) through (1-58d), we can write the corresponding forms as

$$\mathcal{P}_s = -\iiint_V (\mathcal{H} \cdot \mathcal{M}_i + \mathcal{E} \cdot \mathcal{J}_i)\, dv = \iiint_V p_s\, dv \tag{1-59a}$$

$$\iiint_V (\mathcal{H} \cdot \mathcal{M}_d)\, dv = \frac{\partial}{\partial t} \iiint_V \frac{1}{2}\mu \mathcal{H}^2\, dv = \frac{\partial}{\partial t} \iiint_V w_m\, dv = \frac{\partial}{\partial t} \mathcal{W}_m \tag{1-59b}$$

$$\mathcal{P}_d = \iiint_V (\mathcal{E} \cdot \mathcal{J}_c)\, dv = \iiint_V \sigma \mathcal{E}^2\, dv = \iiint_V p_d\, dv \tag{1-59c}$$

$$\iiint_V (\mathcal{E} \cdot \mathcal{J}_d)\, dv = \frac{\partial}{\partial t} \iiint_V \frac{1}{2}\varepsilon \mathcal{E}^2\, dv = \frac{\partial}{\partial t} \iiint_V w_e\, dv = \frac{\partial}{\partial t} \mathcal{W}_e \tag{1-59d}$$

where $\mathcal{W}_m$ = magnetic energy (J)
$\mathcal{W}_e$ = electric energy (J)
$\mathcal{P}_s$ = supplied power (W)
$\mathcal{P}_e$ = exiting power (W)
$\mathcal{P}_d$ = dissipated power (W)

Using (1-57) and (1-59a) through (1-59d), we can write (1-55a) as

$$\mathcal{P}_e - \mathcal{P}_s + \mathcal{P}_d + \frac{\partial}{\partial t}(\mathcal{W}_e + \mathcal{W}_m) = 0 \tag{1-60}$$

or

$$\mathcal{P}_s = \mathcal{P}_e + \mathcal{P}_d + \frac{\partial}{\partial t}(\mathcal{W}_e + \mathcal{W}_m) \tag{1-60a}$$

which is the *conservation of power law*. This law states that within a region V

bounded by S the supplied power $\mathscr{P}_s$ is equal to the power $\mathscr{P}_e$ exiting S plus the power $\mathscr{P}_d$ dissipated within that region plus the rate of change (increase if positive) of the electric ($\mathscr{W}_e$) and magnetic ($\mathscr{W}_m$) energies stored within that same region.

A summary of the field theory relations and their corresponding circuit concepts is found listed in Table 1-2.

1.7 TIME-HARMONIC ELECTROMAGNETIC FIELDS

Maxwell's equations in differential and integral forms for general time-varying electromagnetic fields were presented in Sections 1.2.1 and 1.2.2. In addition, various expressions involving and relating the electromagnetic fields (such as the constitutive parameters and relations, circuit relations, boundary conditions, and power and energy) were also introduced in the preceding sections. However in many practical systems involving electromagnetic waves the time variations are of cosinusoidal form and are referred to as *time-harmonic*. In general, such time variations can be represented by[1] $e^{j\omega t}$, and the instantaneous electromagnetic field vectors can be related to their complex forms in a very simple manner. Thus for time-harmonic fields, we can relate the instantaneous fields (represented by script letters) to their complex forms (represented by roman letters) by

$$\mathscr{E}(x, y, z; t) = \mathrm{Re}\left[\mathbf{E}(x, y, z)e^{j\omega t}\right] \tag{1-61a}$$

$$\mathscr{H}(x, y, z; t) = \mathrm{Re}\left[\mathbf{H}(x, y, z)e^{j\omega t}\right] \tag{1-61b}$$

$$\mathscr{D}(x, y, z; t) = \mathrm{Re}\left[\mathbf{D}(x, y, z)e^{j\omega t}\right] \tag{1-61c}$$

$$\mathscr{B}(x, y, z; t) = \mathrm{Re}\left[\mathbf{B}(x, y, z)e^{j\omega t}\right] \tag{1-61d}$$

$$\mathscr{J}(x, y, z; t) = \mathrm{Re}\left[\mathbf{J}(x, y, z)e^{j\omega t}\right] \tag{1-61e}$$

$$\mathscr{q}(x, y, z; t) = \mathrm{Re}\left[q(x, y, z)e^{j\omega t}\right] \tag{1-61f}$$

where $\mathscr{E}, \mathscr{H}, \mathscr{D}, \mathscr{B}, \mathscr{J}$, and $\mathscr{q}$ represent the instantaneous field vectors while $\mathbf{E}, \mathbf{H}, \mathbf{D}, \mathbf{B}, \mathbf{J}$, and q represent the corresponding complex spatial forms which are only a function of position. In this book we have chosen to represent the instantaneous quantities by the real part of the product of the corresponding complex spatial quantities with $e^{j\omega t}$. Another option would be to represent the instantaneous quantities by the imaginary part of the products. It should be stated that throughout this book the magnitudes of the instantaneous fields represent *peak* values that are related to their corresponding root-mean-square (rms) values by the square root of 2 (peak = $\sqrt{2}$ rms). If the complex spatial quantities can be found, it is then a very simple procedure to find their corresponding instantaneous forms by using (1-61a) through (1-61f). In what follows, it will be shown that Maxwell's equations in differential and integral forms for time-harmonic electromagnetic fields can be written in much simpler forms using the complex field vectors.

---

[1]Another representation form of time-harmonic variations is $e^{-j\omega t}$ (most scientists prefer $e^{i\omega t}$ or $e^{-i\omega t}$ where $i = \sqrt{-1}$). Throughout this book we will use the $e^{j\omega t}$ form, which when it is not stated will be assumed. The $e^{-j\omega t}$ fields are related to those of the $e^{j\omega t}$ form by the complex conjugate.

1.7.1 Maxwell's Equations in Differential and Integral Forms

It is a very simple exercise to show that, by substituting (1-61a) through (1-61f) into (1-1) through (1-4) and (1-6), Maxwell's equations and the continuity equation in differential form for time-harmonic fields can be written in terms of the complex field vectors as shown in Table 1-4. Using a similar procedure, we can write the corresponding integral forms of Maxwell's equations and the continuity equation listed in Table 1-1 in terms of the complex spatial field vectors as shown in Table 1-4. Both of these derivations have been assigned as exercises to the reader at the end of the chapter.

By examining the two forms in Table 1-4, we see that one form can be obtained from the other by doing the following.

1. Replace the instantaneous field vectors by the corresponding complex spatial forms, or vice versa.
2. Replace $\partial/\partial t$ by $j\omega$ ($\partial/\partial t \Leftrightarrow j\omega$), or vice versa.

The second step is very similar to that followed in circuit analysis when Laplace transforms are used to analyze RLC a.c. circuits. In these analyses $\partial/\partial t$ is replaced by s ($\partial/\partial t \Leftrightarrow s$). For steady-state conditions $\partial/\partial t$ is replaced by $j\omega$ ($\partial/\partial t \Leftrightarrow s \Leftrightarrow j\omega$). As we all remember, the reason for using Laplace transforms was to transform differential equations to algebraic equations, which are simpler to solve. The same intent is used here to write Maxwell's equations in forms that are easier to solve. Thus if it is desired to solve for the instantaneous field vectors of time-harmonic fields, it is easier to use the following two-step procedure instead of attempting to do it in one step using the general instantaneous forms of Maxwell's equations:

1. Solve for the complex spatial field vectors (**E**, **H**, **D**, **B**, **J**, q), using Maxwell's equations from Table 1-4 that are written in terms of the complex spatial field vectors.
2. Determine the corresponding instantaneous field vectors using the relations 1-61a through 1-61f.

Step 1 is obviously the most difficult, and once it is completed the solution in many instances ceases there. Step 2 is very elementary and straightforward and is often omitted. In practice, the time variations of $e^{j\omega t}$ are stated at the outset, then suppressed.

1.7.2 Boundary Conditions

The boundary conditions for time-harmonic fields are identical to those of general time-varying fields, as derived in Section 1.5, and they can be expressed simply by replacing the instantaneous field vectors in Table 1-3 with their corresponding complex spatial field vectors. A summary of all the boundary conditions for time-harmonic fields, referring to Figure 1-4, is found in Table 1-5.

In addition to the boundary conditions found in Table 1-5, an additional boundary condition on the tangential components of the electric field is often used along an interface when one of the two media is a very good conductor (material that possesses large but finite conductivity). This is illustrated in Figure 1-6 where it is assumed that medium 1 is a very good conductor whose surface, as will be shown

TABLE 1-4
Instantaneous and time-harmonic forms of Maxwell's equations and continuity equation in differential and integral forms

	Instantaneous	Time harmonic
Differential form		
	$\nabla \times \boldsymbol{\mathcal{E}} = -\boldsymbol{\mathcal{M}}_i - \dfrac{\partial \boldsymbol{\mathcal{B}}}{\partial t}$	$\nabla \times \mathbf{E} = -\mathbf{M}_i - j\omega \mathbf{B}$
	$\nabla \times \boldsymbol{\mathcal{H}} = \boldsymbol{\mathcal{J}}_i + \boldsymbol{\mathcal{J}}_c + \dfrac{\partial \boldsymbol{\mathcal{D}}}{\partial t}$	$\nabla \times \mathbf{H} = \mathbf{J}_i + \mathbf{J}_c + j\omega \mathbf{D}$
	$\nabla \cdot \boldsymbol{\mathcal{D}} = q_{ev}$	$\nabla \cdot \mathbf{D} = q_{ev}$
	$\nabla \cdot \boldsymbol{\mathcal{B}} = q_{mv}$	$\nabla \cdot \mathbf{B} = q_{mv}$
	$\nabla \cdot \boldsymbol{\mathcal{J}}_{ic} = -\dfrac{\partial q_{ev}}{\partial t}$	$\nabla \cdot \mathbf{J}_{ic} = -j\omega q_{ev}$
Integral form		
	$\oint_C \boldsymbol{\mathcal{E}} \cdot d\mathbf{l} = -\iint_S \boldsymbol{\mathcal{M}}_i \cdot d\mathbf{s} - \dfrac{\partial}{\partial t}\iint_S \boldsymbol{\mathcal{B}} \cdot d\mathbf{s}$	$\oint_C \mathbf{E} \cdot d\mathbf{l} = -\iint_S \mathbf{M}_i \cdot d\mathbf{s} - j\omega \iint_S \mathbf{B} \cdot d\mathbf{s}$
	$\oint_C \boldsymbol{\mathcal{H}} \cdot d\mathbf{l} = \iint_S \boldsymbol{\mathcal{J}}_i \cdot d\mathbf{s} + \iint_S \boldsymbol{\mathcal{J}}_c \cdot d\mathbf{s} + \dfrac{\partial}{\partial t}\iint_S \boldsymbol{\mathcal{D}} \cdot d\mathbf{s}$	$\oint_C \mathbf{H} \cdot d\mathbf{l} = \iint_S \mathbf{J}_i \cdot d\mathbf{s} + \iint_S \mathbf{J}_c \cdot d\mathbf{s} + j\omega \iint_S \mathbf{D} \cdot d\mathbf{s}$
	$\oiint_S \boldsymbol{\mathcal{D}} \cdot d\mathbf{s} = Q_e$	$\oiint_S \mathbf{D} \cdot d\mathbf{s} = Q_e$
	$\oiint_S \boldsymbol{\mathcal{B}} \cdot d\mathbf{s} = Q_m$	$\oiint_S \mathbf{B} \cdot d\mathbf{s} = Q_m$
	$\oiint_S \boldsymbol{\mathcal{J}}_{ic} \cdot d\mathbf{s} = -\dfrac{\partial Q_e}{\partial t}$	$\oiint_S \mathbf{J}_{ic} \cdot d\mathbf{s} = -j\omega Q_e$

TABLE 1-5
Boundary conditions on time-harmonic electromagnetic fields

	General	Finite-conductivity media, no sources or charges $\sigma_1, \sigma_2 \neq \infty$ $J_s = M_s = 0; q_{es} = q_{ms} = 0$	Medium 1 of infinite electric conductivity $\sigma_1 = \infty; \sigma_2 \neq \infty$ $M_s = 0; q_{ms} = 0$	Medium 1 of infinite magnetic conductivity $(H_{1t} = 0)$ $J_s = 0; q_{es} = 0$
Tangential electric field intensity	$\hat{n} \times (\mathbf{E}_2 - \mathbf{E}_1) = -\mathbf{M}_s$	$\hat{n} \times (\mathbf{E}_2 - \mathbf{E}_1) = 0$	$\hat{n} \times \mathbf{E}_2 = 0$	$\hat{n} \times \mathbf{E}_2 = -\mathbf{M}_s$
Tangential magnetic field intensity	$\hat{n} \times (\mathbf{H}_2 - \mathbf{H}_1) = \mathbf{J}_s$	$\hat{n} \times (\mathbf{H}_2 - \mathbf{H}_1) = 0$	$\hat{n} \times \mathbf{H}_2 = \mathbf{J}_s$	$\hat{n} \times \mathbf{H}_2 = 0$
Normal electric flux density	$\hat{n} \cdot (\mathbf{D}_2 - \mathbf{D}_1) = q_{es}$	$\hat{n} \cdot (\mathbf{D}_2 - \mathbf{D}_1) = 0$	$\hat{n} \cdot \mathbf{D}_2 = q_{es}$	$\hat{n} \cdot \mathbf{D}_2 = 0$
Normal magnetic flux density	$\hat{n} \cdot (\mathbf{B}_2 - \mathbf{B}_1) = q_{ms}$	$\hat{n} \cdot (\mathbf{B}_2 - \mathbf{B}_1) = 0$	$\hat{n} \cdot \mathbf{B}_2 = 0$	$\hat{n} \cdot \mathbf{B}_2 = q_{ms}$

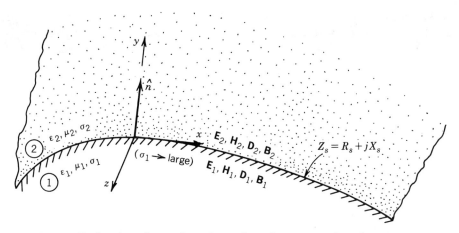

FIGURE 1-6 Surface impedance along the surface of a very good conductor.

in Section 4.3.1, exhibits a surface impedance Z_s (ohms) given approximately by (4-42) or

$$Z_s = R_s + jX_s = (1+j)\sqrt{\frac{\omega\mu}{2\sigma}} \qquad (1\text{-}62)$$

with equal real and imaginary (inductive) parts (σ is the conductivity of the conductor). At the surface there exists a linear current density $\mathbf{J}_s$ (A/m) related to the tangential magnetic field in medium 2 by

$$\mathbf{J}_s \simeq \hat{n} \times \mathbf{H}_2 \qquad (1\text{-}63)$$

Since the conductivity is finite (although large), the most intense current density resides at the surface, and it diminishes (in an exponential form) as the observations are made deeper into the conductor. This is demonstrated in Example 5.7 of Section 5.4.1. In addition, the electric field intensity along the interface cannot be zero (although it may be small). Thus we can write that the tangential component of the electric field in medium 2 along the interface is related to the electric current density $\mathbf{J}_s$ and tangential component of the magnetic field by

$$\mathbf{E}_{t2} = Z_s \mathbf{J}_s = Z_s \hat{n} \times \mathbf{H}_2 = \hat{n} \times \mathbf{H}_2 \sqrt{\frac{\omega\mu}{2\sigma}}(1+j) \qquad (1\text{-}64)$$

For time-harmonic fields all the boundary conditions are not independent from each other since they are related and they are solutions to Maxwell's equations. In fact, if the tangential components of the electric and magnetic field intensities satisfy the boundary conditions, the normal components of the same fields automatically satisfy their appropriate boundary conditions. For example, if the tangential components of the electric field are continuous across a boundary, the derivatives (with respect to the coordinates on the boundary surface) of these tangential components are also continuous. This, in turn, ensures continuity of the normal component of the magnetic field.

To demonstrate that, let us refer to the geometry of Figure 1-6 where the local surface along the interface is described by the x, z coordinates with y being normal

28 TIME-VARYING AND TIME-HARMONIC ELECTROMAGNETIC FIELDS

to the surface. Let us assume that E_x and E_z are continuous, which ensures that their derivatives with respect to x and z ($\partial E_x/\partial x$, $\partial E_x/\partial z$, $\partial E_z/\partial x$, $\partial E_z/\partial z$) are also continuous. Therefore according to Maxwell's curl equation of the electric field

$$\nabla \times \mathbf{E} = \nabla \times (\hat{a}_x E_x + \hat{a}_z E_z) = \begin{vmatrix} \hat{a}_x & \hat{a}_y & \hat{a}_z \\ \dfrac{\partial}{\partial x} & 0 & \dfrac{\partial}{\partial z} \\ E_x & 0 & E_z \end{vmatrix}$$

$$= \hat{a}_x(0) + \hat{a}_y\left(\frac{\partial E_x}{\partial z} - \frac{\partial E_z}{\partial x}\right) + \hat{a}_z(0)$$

$$\nabla \times \mathbf{E} = \hat{a}_y\left(\frac{\partial E_x}{\partial z} - \frac{\partial E_z}{\partial x}\right) = -j\omega\mu\mathbf{H} \qquad (1\text{-}65)$$

or

$$B_y = \mu H_y = -\frac{1}{j\omega}\left(\frac{\partial E_x}{\partial z} - \frac{\partial E_z}{\partial x}\right) \qquad (1\text{-}65a)$$

According to (1-65a), B_y, the normal component of the magnetic flux density along the interface, is continuous across the boundary if $\partial E_x/\partial z$ and $\partial E_z/\partial x$ are also continuous across the boundary.

In a similar manner, it can be shown that continuity of the tangential components of the magnetic field ensures continuity of the normal component of the electric flux density (**D**).

1.7.3 Power and Energy

In Section 1.6 it was shown that power and energy are associated with time-varying electromagnetic fields. The conservation of energy equation, in differential and integral forms, was stated, respectively, by (1-53a) and (1-55a). Similar equations can be derived for time-harmonic electromagnetic fields using the complex spatial forms of the field vectors. Before we attempt this, let us first rewrite the instantaneous Poynting vector $\mathscr{S}$ in terms of the complex field vector.

By definition, the instantaneous Poynting vector was defined by (1-56) and is repeated here as

$$\mathscr{S} = \mathscr{E} \times \mathscr{H} \qquad (1\text{-}66)$$

The electric and magnetic fields of (1-61a) and (1-61b) can also be written as

$$\mathscr{E}(x,y,z;t) = \operatorname{Re}\left[\mathbf{E}(x,y,z)e^{j\omega t}\right] = \tfrac{1}{2}\left[\mathbf{E}e^{j\omega t} + (\mathbf{E}e^{j\omega t})^*\right] \qquad (1\text{-}67a)$$

$$\mathscr{H}(x,y,z;t) = \operatorname{Re}\left[\mathbf{H}(x,y,z)e^{j\omega t}\right] = \tfrac{1}{2}\left[\mathbf{H}e^{j\omega t} + (\mathbf{H}e^{j\omega t})^*\right] \qquad (1\text{-}67b)$$

where the asterisk (*) indicates complex conjugate. Substituting (1-67a) and (1-67b) into (1-66), we have that

$$\mathscr{S} = \mathscr{E} \times \mathscr{H} = \tfrac{1}{2}(\mathbf{E}e^{j\omega t} + \mathbf{E}^*e^{-j\omega t}) \times \tfrac{1}{2}(\mathbf{H}e^{j\omega t} + \mathbf{H}^*e^{-j\omega t})$$

$$= \tfrac{1}{2}\left\{\tfrac{1}{2}[\mathbf{E}\times\mathbf{H}^* + \mathbf{E}^*\times\mathbf{H}] + \tfrac{1}{2}[\mathbf{E}\times\mathbf{H}e^{j2\omega t} + \mathbf{E}^*\times\mathbf{H}^*e^{-j2\omega t}]\right\}$$

$$\mathscr{S} = \tfrac{1}{2}\left\{\tfrac{1}{2}[\mathbf{E}\times\mathbf{H}^* + (\mathbf{E}\times\mathbf{H}^*)^*] + \tfrac{1}{2}[\mathbf{E}\times\mathbf{H}e^{j2\omega t} + (\mathbf{E}\times\mathbf{H}e^{j2\omega t})^*]\right\}$$

$$(1\text{-}68)$$

Using equalities of the (1-67a) or (1-67b) form in reverse order, we can write (1-68) as

$$\mathscr{S} = \tfrac{1}{2}\left[\operatorname{Re}(\mathbf{E} \times \mathbf{H}^*) + \operatorname{Re}(\mathbf{E} \times \mathbf{H} e^{j2\omega t})\right] \quad (1\text{-}69)$$

Since both $\mathbf{E}$ and $\mathbf{H}$ are not functions of time and the time variations of the second term are twice the frequency of the field vectors, the time-average Poynting vector (average power density) over one period is equal to

$$\mathscr{S}_{av} = \mathbf{S} = \tfrac{1}{2}\operatorname{Re}[\mathbf{E} \times \mathbf{H}^*] \quad (1\text{-}70)$$

Since $\mathbf{E} \times \mathbf{H}^*$ is in general complex and the real part of $\mathbf{E} \times \mathbf{H}^*$ represents the real part of the power density, what does the imaginary part represent? As will be seen in what follows, the imaginary part represents the reactive power. With (1-69) and (1-70) in mind, let us now derive the conservation of energy equation in differential and integral forms using the complex forms of the field vector.

From Table 1-4, the first two of Maxwell's equations can be written as

$$\nabla \times \mathbf{E} = -\mathbf{M}_i - j\omega\mu\mathbf{H} \quad (1\text{-}71a)$$

$$\nabla \times \mathbf{H} = \mathbf{J}_i + \mathbf{J}_c + j\omega\varepsilon\mathbf{E} = \mathbf{J}_i + \sigma\mathbf{E} + j\omega\varepsilon\mathbf{E} \quad (1\text{-}71b)$$

Dot multiplying (1-71a) by $\mathbf{H}^*$ and the conjugate of (1-71b) by $\mathbf{E}$, we have that

$$\mathbf{H}^* \cdot (\nabla \times \mathbf{E}) = -\mathbf{H}^* \cdot \mathbf{M}_i - j\omega\mu\mathbf{H} \cdot \mathbf{H}^* \quad (1\text{-}72a)$$

$$\mathbf{E} \cdot (\nabla \times \mathbf{H}^*) = \mathbf{E} \cdot \mathbf{J}_i^* + \sigma\mathbf{E} \cdot \mathbf{E}^* - j\omega\varepsilon\mathbf{E} \cdot \mathbf{E}^* \quad (1\text{-}72b)$$

Subtracting (1-72a) from (1-72b), we can write that

$$\mathbf{E} \cdot (\nabla \times \mathbf{H}^*) - \mathbf{H}^* \cdot (\nabla \times \mathbf{E})$$
$$= \mathbf{H}^* \cdot \mathbf{M}_i + \mathbf{E} \cdot \mathbf{J}_i^* + \sigma\mathbf{E} \cdot \mathbf{E}^* - j\omega\varepsilon\mathbf{E} \cdot \mathbf{E}^* + j\omega\mu\mathbf{H} \cdot \mathbf{H}^* \quad (1\text{-}73)$$

Using the vector identity of (1-52) reduces (1-73) to

$$\nabla \cdot (\mathbf{H}^* \times \mathbf{E}) = \mathbf{H}^* \cdot \mathbf{M}_i + \mathbf{E} \cdot \mathbf{J}_i^* + \sigma|\mathbf{E}|^2 + j\omega\mu|\mathbf{H}|^2 - j\omega\varepsilon|\mathbf{E}|^2 \quad (1\text{-}74)$$

or

$$-\nabla \cdot (\mathbf{E} \times \mathbf{H}^*) = \mathbf{H}^* \cdot \mathbf{M}_i + \mathbf{E} \cdot \mathbf{J}_i^* + \sigma|\mathbf{E}|^2 + j\omega\left(\mu|\mathbf{H}|^2 - \varepsilon|\mathbf{E}|^2\right) \quad (1\text{-}74a)$$

Dividing both sides by 2, we can write that

$$\boxed{-\nabla \cdot \left(\tfrac{1}{2}\mathbf{E} \times \mathbf{H}^*\right) = \tfrac{1}{2}\mathbf{H}^* \cdot \mathbf{M}_i + \tfrac{1}{2}\mathbf{E} \cdot \mathbf{J}_i^* + \tfrac{1}{2}\sigma|\mathbf{E}|^2 + j2\omega\left(\tfrac{1}{4}\mu|\mathbf{H}|^2 - \tfrac{1}{4}\varepsilon|\mathbf{E}|^2\right)} \quad (1\text{-}75)$$

For time-harmonic fields (1-75) represents the *conservation of energy equation* in differential form.

To see that (1-75) represents the conservation of energy equation in differential form, let us first take the volume integral of both sides of (1-75) and then apply the divergence theorem of (1-8) to the left side. Doing both of these steps reduces (1-75) to

$$-\iiint_V \nabla \cdot (\tfrac{1}{2} \mathbf{E} \times \mathbf{H}^*)\, dv = -\oiint_S (\tfrac{1}{2}\mathbf{E} \times \mathbf{H}^*) \cdot d\mathbf{s}$$

$$= \tfrac{1}{2}\iiint_V (\mathbf{H}^* \cdot \mathbf{M}_i + \mathbf{E} \cdot \mathbf{J}_i^*)\, dv$$

$$+ \tfrac{1}{2}\iiint_V \sigma |\mathbf{E}|^2\, dv + j2\omega \iiint_V \left(\tfrac{1}{4}\mu|\mathbf{H}|^2 - \tfrac{1}{4}\varepsilon|\mathbf{E}|^2\right) dv$$

or

$$-\tfrac{1}{2}\iiint_V (\mathbf{H}^* \cdot \mathbf{M}_i + \mathbf{E} \cdot \mathbf{J}_i^*)\, dv = \oiint_S (\tfrac{1}{2}\mathbf{E} \times \mathbf{H}^*) \cdot d\mathbf{s} + \tfrac{1}{2}\iiint_V \sigma|\mathbf{E}|^2\, dv$$

$$+ j2\omega \iiint_V \left(\tfrac{1}{4}\mu|\mathbf{H}|^2 - \tfrac{1}{4}\varepsilon|\mathbf{E}|^2\right) dv \qquad (1\text{-}76)$$

which can be written as

$$P_s = P_e + P_d + j2\omega(\overline{W}_m - \overline{W}_e) \qquad (1\text{-}76a)$$

where

$$P_s = -\tfrac{1}{2}\iiint_V (\mathbf{H}^* \cdot \mathbf{M}_i + \mathbf{E} \cdot \mathbf{J}_i^*)\, dv = \text{supplied complex power (W)} \qquad (1\text{-}76b)$$

$$P_e = \oiint_S (\tfrac{1}{2}\mathbf{E} \times \mathbf{H}^*) \cdot d\mathbf{s} = \text{exiting complex power (W)} \qquad (1\text{-}76c)$$

$$P_d = \tfrac{1}{2}\iiint_V \sigma|\mathbf{E}|^2\, dv = \text{dissipated real power (W)} \qquad (1\text{-}76d)$$

$$\overline{W}_m = \iiint_V \tfrac{1}{4}\mu|\mathbf{H}|^2\, dv = \text{time-average magnetic energy (J)} \qquad (1\text{-}76e)$$

$$\overline{W}_e = \iiint_V \tfrac{1}{4}\varepsilon|\mathbf{E}|^2\, dv = \text{time-average electric energy (J)} \qquad (1\text{-}76f)$$

For an electromagnetic source (represented in Figure 1-5 by electric and magnetic current densities $\mathbf{J}_i$ and $\mathbf{M}_i$) supplying power in a region within S, (1-76) and (1-76a) represent the conservation of energy equation in integral form. Now it is also much easier to accept that (1-75), from which (1-76) was derived, represents the conservation of energy equation in differential form. In (1-76a), P_s and P_e are in general complex and P_d is always real, but the last two terms are always imaginary and represent the reactive power associated, respectively, with magnetic and electric fields. It should be stated that for complex permeabilities and permittivities the contributions from their imaginary parts to the integrals of (1-76e) and (1-76f) should both be transferred and combined with (1-76d), since they both will represent losses associated with the imaginary parts of the permeabilities and permittivities.

The field and circuit theory relations for time-harmonic electromagnetic fields are similar to those found in Table 1-2 for the general time-varying electromagnetic fields, but with the instantaneous field quantities (represented by script letters)

TABLE 1-6
Relations between time-harmonic electromagnetic field and steady-state a.c. circuit theories

Field theory	Circuit theory
1. **E** (electric field intensity)	1. v (voltage)
2. **H** (magnetic field intensity)	2. i (current)
3. **D** (electric flux density)	3. q_{ev} (electric charge density)
4. **B** (magnetic flux density)	4. q_{mv} (magnetic charge density)
5. **J** (electric current density)	5. i_e (electric current)
6. **M** (magnetic current density)	6. i_m (magnetic current)
7. $\mathbf{J}_d = j\omega\varepsilon\mathbf{E}$ (electric displacement current density)	7. $i = j\omega C v$ (current through a capacitor)
8. $\mathbf{M}_d = j\omega\mu\mathbf{H}$ (magnetic displacement current density)	8. $v = j\omega L i$ (voltage across an inductor)
9. *Constitutive relations*	9. *Element laws*
(a) $\mathbf{J}_c = \sigma\mathbf{E}$ (electric conduction current density)	(a) $i = Gv = \dfrac{1}{R}v$ (Ohm's law)
(b) $\mathbf{D} = \varepsilon\mathbf{E}$ (dielectric material)	(b) $Q_e = Cv$ (charge in a capacitor)
(c) $\mathbf{B} = \mu\mathbf{H}$ (magnetic material)	(c) $\psi = Li$ (flux of an inductor)
10. $\oint_C \mathbf{E}\cdot d\mathbf{l} = -j\omega \iint_S \mathbf{B}\cdot d\mathbf{s}$ (Maxwell–Faraday equation)	10. $\Sigma v = -j\omega L_s i \simeq 0$ (Kirchhoff's voltage law)
11. $\oiint_S \mathbf{J}_{ic}\cdot d\mathbf{s} = -j\omega \iiint_V q_{ev}\,dv = -\dfrac{\partial Q_e}{\partial t}$ (continuity equation)	11. $\Sigma i = -j\omega Q_e = -j\omega C_s v \simeq 0$ (Kirchoff's current law)
12. *Power and energy densities*	12. *Power and energy* (v and i represent peak values)
(a) $\tfrac{1}{2}\oiint_S (\mathbf{E}\times\mathbf{H}^*)\cdot d\mathbf{s}$ (**E** and **H** represent peak values)	(a) $P = \tfrac{1}{2}vi$ (power–voltage–current relation)
(b) $\tfrac{1}{2}\iiint_V \sigma|\mathbf{E}|^2\,dv$ (dissipated real power)	(b) $P_d = \dfrac{1}{2}Gv^2 = \dfrac{1}{2}\dfrac{v^2}{R}$ (power dissipated in a resistor)
(c) $\tfrac{1}{4}\iiint_V \varepsilon|\mathbf{E}|^2\,dv$ (time-average electric stored energy)	(c) $\tfrac{1}{4}Cv^2$ (energy stored in a capacitor)
(d) $\tfrac{1}{4}\iiint_V \mu|\mathbf{H}|^2\,dv$ (time-average magnetic stored energy)	(d) $\tfrac{1}{4}Li^2$ (energy stored in an inductor)

replaced by their corresponding complex field quantities (represented by roman and italic letters) and with $\partial/\partial t$ replaced by $j\omega$ ($\partial/\partial t \Leftrightarrow j\omega$). These are shown listed in Table 1-6.

Over the years many excellent introductory books on electromagnetics—[1] through [18]—and advanced books—[19] through [29]—have been published. Some of them can serve both purposes, and a few may not still be in print. Each is contributing to the general knowledge of electromagnetic theory and its applications. The reader is encouraged to consult them for an even better understanding of the subject.

REFERENCES

1. J. D. Kraus, *Electromagnetics*, Third Edition, McGraw-Hill, New York, 1984.
2. S. Ramo, J. R. Whinnery and T. Van Duzer, *Fields and Waves in Communication Electronics*, Second Edition, Wiley, New York, 1984.
3. W. H. Hayt, Jr., *Engineering Electromagnetics*, McGraw-Hill, New York, 1981.
4. D. T. Paris and F. K. Hurd, *Basic Electromagnetic Theory*, McGraw-Hill, New York, 1969.
5. C. T. A. Johnk, *Engineering Electromagnetic Fields and Waves*, Second Edition, Wiley, New York, 1988.
6. H. P. Neff, Jr., *Basic Electromagnetic Fields*, Second Edition, Wiley, New York, 1987.
7. S. V. Marshall and G. G. Skitek, *Electromagnetic Concepts and Applications*, Second Edition, Prentice-Hall, Englewood Cliffs, N.J., 1987.
8. D. K. Cheng, *Field and Wave Electromagnetics*, Addison-Wesley, Reading, Mass., 1983.
9. C. R. Paul and S. A. Nasar, *Introduction to Electromagnetic Fields*, Second Edition, McGraw-Hill, New York, 1987.
10. L. C. Shen and J. A. Kong, *Applied Electromagnetism*, Brooks/Cole Engineering Division, Monterey, Calif., 1983.
11. N. N. Rao, *Elements of Engineering Electromagnetics*, Second Edition, Prentice-Hall, Englewood Cliffs, N.J., 1987.
12. M. A. Plonus, *Applied Electromagnetics*, McGraw-Hill, New York, 1978.
13. A. T. Adams, *Electromagnetics for Engineers*, Ronald Press, New York, 1971.
14. M. Zahn, *Electromagnetic Field Theory*, Wiley, New York, 1979.
15. L. M. Magid, *Electromagnetic Fields, Energy, and Waves*, Wiley, New York, 1972.
16. S. Seely and A. D. Poularikas, *Electromagnetics: Classical and Modern Theory and Applications*, Dekker, New York, 1979.
17. D. M. Cook, *The Theory of the Electromagnetic Field*, Prentice-Hall, Englewood Cliffs, N.J., 1975.
18. R. P. Feynman, R. B. Leighton, and M. Sands, *The Feynman Lectures on Physics: Mainly Electromagnetism and Matter*, Volume II, Addison-Wesley, Reading, Mass., 1964.
19. W. R. Smythe, *Static and Dynamic Electricity*, McGraw-Hill, New York, 1939.
20. J. A. Stratton, *Electromagnetic Theory*, McGraw-Hill, New York, 1941.
21. R. E. Collin, *Field Theory of Guided Waves*, McGraw-Hill, New York, 1960.
22. E. C. Jordan and K. G. Balmain, *Electromagnetic Waves and Radiating Systems*, Second Edition, Prentice-Hall, Englewood Cliffs, N.J., 1968.
23. R. F. Harrington, *Time-Harmonic Electromagnetic Fields*, McGraw-Hill, New York, 1961.
24. J. R. Wait, *Electromagnetic Wave Theory*, Harper & Row, New York, 1985.
25. J. A. Kong, *Theory of Electromagnetic Waves*, Wiley, New York, 1975.
26. C. C. Johnson, *Field and Wave Electrodynamics*, McGraw-Hill, New York, 1965.
27. J. D. Jackson, *Classical Electrodynamics*, Wiley, New York, 1962.
28. D. S. Jones, *Methods in Electromagnetic Wave Propagation*, Oxford Univ. Press (Clarendon), London/New York, 1979.
29. M. Kline (Ed.), *The Theory of Electromagnetic Waves*, Interscience, New York, 1951.

PROBLEMS

1.1. Derive the differential form of the continuity equation, as given by (1-6), from Maxwell's equations 1-1 through 1-4.

1.2. Derive the integral forms of Maxwell's equations and the continuity equation, as listed in Table 1-1, from the corresponding ones in differential form.

1.3. The electric flux density inside a cube is given by (a) $\mathbf{D} = \hat{a}_x(3 + x)$, (b) $\mathbf{D} = \hat{a}_y(4 + y^2)$. Find the total electric charge enclosed inside the cubical volume when the cube is in the first octant with three edges coincident with the x, y, z axes and one corner at the origin. Each side of the cube is 1 m.

1.4. The electric field inside a circular cylinder of radius a and height h is given by

$$\mathbf{E} = \hat{a}_z \left[-\frac{c}{h} + \frac{b}{6\varepsilon_0}(3z^2 - h^2) \right]$$

where c and b are constants. Assuming the medium within the cylinder is free space, find the total charge enclosed within the cylinder.

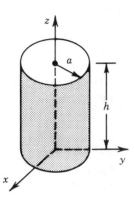

FIGURE P1-4

1.5. The instantaneous electric field inside a source-free, homogeneous, isotropic, and linear medium is given by

$$\mathbf{\mathcal{E}} = \left[\hat{a}_x A(x + y) + \hat{a}_y B(x - y) \right] \cos(\omega t)$$

Determine the relations between A and B.

1.6. The magnetic flux density produced on its plane by a current-carrying circular loop of radius $a = 0.1$ m, placed on the xy plane at $z = 0$, is given by

$$\mathbf{\mathcal{B}} = \hat{a}_z \frac{10^{-12}}{1 + 25\rho} \cos(1500\pi t) \text{ Wb/m}^2$$

where ρ is the radial distance in cylindrical coordinates.
(a) Find the total flux in the z direction passing through the loop.
(b) Find the electric field at any point ρ within the loop. Check your answer by using Maxwell's equation 1-1.

1.7. The instantaneous magnetic flux density in free space is given by

$$\mathbf{\mathcal{B}} = \hat{a}_x B_x \cos(2y) \sin(\omega t - \pi z) + \hat{a}_y B_y \cos(2y) \cos(\omega t - \pi z)$$

where B_x and B_y are constants. Assuming there are no sources, determine the electric displacement current density.

1.8. The displacement current density within a source-free ($\mathbf{\mathcal{J}}_i = 0$) cube centered about the origin is given by

$$\mathbf{\mathcal{J}}_d = \hat{a}_x yz + \hat{a}_y y^2 + \hat{a}_z xyz$$

Each side of the cube is 1 m and the medium within it is free space. Find the displacement current leaving, in the outward direction, through the surface of the cube.

1.9. The electric flux density in free space produced by an oscillating electric charge placed at the origin is given by

$$\mathscr{D} = \hat{a}_r \frac{10^{-9}}{4\pi} \frac{1}{r^2} \cos(\omega t - \beta r)$$

where $\beta = \omega\sqrt{\mu_0 \varepsilon_0}$. Find the time-average charge that produces this electric flux density.

1.10. The electric field radiated at large distances in free space by a current-carrying small circular loop of radius a, placed on the xy plane at $z = 0$, is given by

$$\mathscr{E} = \hat{a}_\phi E_0 \sin\theta \frac{\cos(\omega t - \beta_0 r)}{r}, \qquad r \gg a$$

where E_0 is a constant, $\beta_0 = \omega\sqrt{\mu_0 \varepsilon_0}$, r is the radial distance in spherical coordinates, and θ is the spherical angle measured from the z axis that is perpendicular to the plane of the loop. Determine the corresponding radiated magnetic field at large distances from the loop ($r \gg a$).

1.11. A time-varying voltage source of $v(t) = 10\cos(\omega t)$ is connected across a parallel plate capacitor with polystyrene ($\varepsilon = 2.56\varepsilon_0$, $\sigma = 3.7 \times 10^{-4}$ S/m) between the plates. Assuming a small plate separation of 2 cm and no field fringing, determine at (a) $f = 1$ MHz, (b) $f = 100$ MHz the maximum values of the conduction and displacement current densities within the polystyrene and compare them.

1.12. A dielectric slab of polystyrene ($\varepsilon = 2.56\varepsilon_0$, $\mu = \mu_0$) of height $2h$ is bounded above and below by free space, as shown in Figure P1-12. Assuming the electric field within the slab is given by

$$\mathscr{E} = (\hat{a}_y 5 + \hat{a}_z 10)\cos(\omega t - \beta x)$$

where $\beta = \omega\sqrt{\mu_0 \varepsilon}$, determine (a) the corresponding magnetic field within the slab and (b) the electric and magnetic fields in free space right above and below the slab.

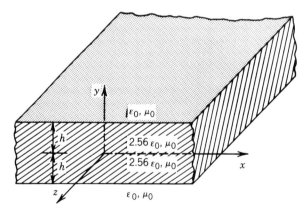

FIGURE P1-12

1.13. A finite conductivity rectangular strip, shown in Figure P1-13, is used to carry electric current. Because of the strip's lossy nature, the current is nonuniformly distributed over the cross section of the strip. The current density on the *upper and lower* sides is given by

$$\mathscr{J} = \hat{a}_z 10^4 \cos(2\pi \times 10^9 t) \text{ A/m}^2$$

and it rapidly decays in an exponential fashion from the lower side toward the

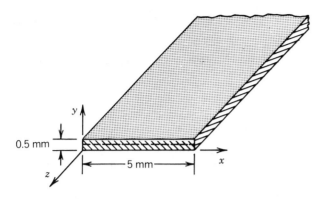

FIGURE P1-13

center by the factor $e^{-10^6 y}$, or

$$\mathscr{J} = \hat{a}_z 10^4 e^{-10^6 y} \cos(2\pi \times 10^9 t) \text{ A/m}^2$$

A similar decay is experienced by the current density from the upper side toward the center. Assuming no variations of the current density with respect to x, determine the total current flowing through the wire.

1.14. The instantaneous electric field inside a conducting rectangular pipe (waveguide) is given by

$$\mathscr{E} = \hat{a}_y E_0 \sin\left(\frac{\pi}{a} x\right) \cos(\omega t - \beta_z z)$$

where β_z is the waveguide's phase constant. Assuming there are no sources within the free-space-filled pipe, determine (a) the corresponding instantaneous magnetic field components inside the conducting pipe and (b) the phase constant β_z.

1.15. The instantaneous electric field intensity inside a source-free coaxial line with inner and outer radii of a and b, respectively, that is filled with a homogeneous dielectric of $\varepsilon = 2.25\varepsilon_0$, $\mu = \mu_0$, and $\sigma = 0$ is given by

$$\mathscr{E} = \hat{a}_\rho \left(\frac{100}{\rho}\right) \cos(10^8 t - \beta z)$$

where β is the phase constant and ρ is the cylindrical radial distance from the center of the coaxial line. Determine (a) the corresponding instantaneous magnetic field $\mathscr{H}$, (b) the phase constant β, and (c) the displacement current density $\mathscr{J}_d$.

1.16. A coaxial line resonator with inner and outer conductors at $a = 5$ mm and $b = 20$ mm, and with conducting plates at $z = 0$ and $z = \ell$, is filled with a dielectric with $\varepsilon_r = 2.56$, $\mu_r = 1$, and $\sigma = 0$. The instantaneous magnetic field intensity inside the source-free dielectric medium is given by

$$\mathscr{H} = \hat{a}_\phi \left(\frac{2}{\rho}\right) \cos\left(\frac{\pi}{\ell} z\right) \cos(4\pi \times 10^8 t)$$

Find the following:
(a) The electric field intensity within the dielectric.
(b) The surface current density $\mathscr{J}_s$ at the conductor surfaces at $\rho = a$ and $\rho = b$.
(c) The displacement current density $\mathscr{J}_d$ at any point within the dielectric.
(d) The total displacement current flowing through the circumferential surface of the resonator.

1.17. Using the instantaneous forms of Maxwell's equation and the continuity equations listed in Tables 1-1 and 1-4, derive the corresponding time-harmonic forms (in

1.18. Show that the electric and magnetic fields of (1-61a) and (1-61b) can be written, respectively, by (1-67a) and (1-67b).

1.19. An electric line source of infinite length and constant current, placed along the z axis, radiates in free space at large distances from the source ($\rho \to$ large) a time-harmonic complex magnetic field given by

$$\mathbf{H} = \hat{a}_\phi H_0 \frac{e^{-j\beta_0 \rho}}{\sqrt{\rho}}, \qquad \rho \to \text{large}$$

where H_0 is a constant, $\beta_0 = \omega\sqrt{\mu_0 \varepsilon_0}$, and ρ is the radial cylindrical distance. Determine the corresponding electric field.

1.20. The time-harmonic complex electric field radiated in free space by a linear radiating element is given by

$$\mathbf{E} = \hat{a}_r E_r + \hat{a}_\theta E_\theta$$

$$E_r = E_0 \frac{\cos\theta}{r^2}\left[1 + \frac{1}{j\beta_0 r}\right] e^{-j\beta_0 r}$$

$$E_\theta = jE_0 \frac{\beta\sin\theta}{2r}\left[1 + \frac{1}{j\beta_0 r} - \frac{1}{(\beta_0 r)^2}\right] e^{-j\beta_0 r}$$

where $\hat{a}_r$ and $\hat{a}_\theta$ are unit vectors in the spherical direction r and θ, E_0 is a constant, and $\beta_0 = \omega\sqrt{\mu_0\varepsilon_0}$. Determine the corresponding spherical magnetic field components.

1.21. The time-harmonic complex electric field radiated by a current-carrying small circular loop in free space is given by

$$\mathbf{E} = \hat{a}_\phi E_0 \frac{\sin\theta}{r}\left[1 + \frac{1}{j\beta_0 r}\right] e^{-j\beta_0 r}$$

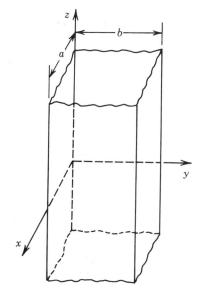

FIGURE P1-22

where $\hat{a}_\phi$ is the spherical unit vector in the ϕ direction, E_0 is a constant, and $\beta_0 = \omega\sqrt{\mu_0\varepsilon_0}$. Determine the corresponding spherical magnetic field components.

1.22. The electric field inside an infinite length rectangular pipe, with all four vertical sides perfectly electric conducting, as shown in Figure P1-22, is given by

$$\mathbf{E} = \hat{a}_z(1+j)\sin\left(\frac{\pi}{a}x\right)\sin\left(\frac{\pi}{b}y\right)$$

Assuming that there are no sources within the box and $a = \lambda_0$, $b = 0.5\lambda_0$, and $\mu = \mu_0$, where λ_0 = free space, infinite medium wavelength, find the (a) conductivity and (b) dielectric constant of the medium within the box.

1.23. A time-harmonic electromagnetic field in free space is perpendicularly incident upon a perfectly conducting semi-infinite planar surface, as shown in Figure P1-23. Assuming the incident $\mathbf{E}^i$ and reflected $\mathbf{E}^r$ complex electric fields on the free-space side of the interface are given by

$$\mathbf{E}^i = \hat{a}_x e^{-j\beta_0 z}$$

$$\mathbf{E}^r = -\hat{a}_x e^{+j\beta_0 z}$$

where

$$\beta_0 = \omega\sqrt{\mu_0\varepsilon_0}$$

determine the current density $\mathbf{J}_s$ induced on the surface of the conducting surface. Evaluate all the constants.

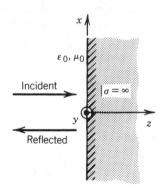

FIGURE P1-23

1.24. The free-space incident $\mathbf{E}^i$ and reflected $\mathbf{E}^r$ fields of a time-harmonic electromagnetic field obliquely incident upon a perfectly conducting semi-infinite planar

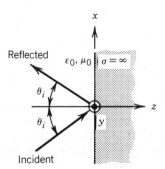

FIGURE P1-24

surface of Figure P1-24 are given by

$$\mathbf{E}^i = \hat{a}_y E_0 e^{-j\beta_0(x \sin \theta_i + z \cos \theta_i)}$$

$$\mathbf{E}^r = \hat{a}_y E_0 \Gamma_h e^{-j\beta_0(x \sin \theta_i - z \cos \theta_i)}$$

where E_0 is a constant and $\beta_0 = \omega\sqrt{\mu_0 \varepsilon_0}$. Determine the coefficient Γ_h.

1.25. For Problem 1.24, determine (a) the corresponding incident and reflected magnetic fields and (b) the electric current density along the interface between the two media.

1.26. Repeat Problem 1.24 when the incident and reflected electric fields are given by

$$\mathbf{E}^i = (\hat{a}_x \cos \theta_i - \hat{a}_z \sin \theta_i) E_0 e^{-j\beta_0(x \sin \theta_i + z \cos \theta_i)}$$

$$\mathbf{E}^r = (\hat{a}_x \cos \theta_i + \hat{a}_z \sin \theta_i) \Gamma_e E_0 e^{-j\beta_0(x \sin \theta_i - z \cos \theta_i)}$$

where E_0 is a constant and $\beta_0 = \omega\sqrt{\mu_0 \varepsilon_0}$. Determine the coefficient Γ_e by applying the boundary conditions on the tangential components.

1.27. Repeat Problem 1.26 except that Γ_e should be determined using the boundary conditions on the normal components. Compare the answer with that obtained in Problem 1.26. Explain.

1.28. For Problem 1.26 determine (a) the corresponding incident and reflected magnetic fields and (b) the electric current density along the interface between the two media.

1.29. A time-harmonic electromagnetic field traveling in free space and perpendicularly incident upon a flat surface of distilled water ($\varepsilon = 81\varepsilon_0$, $\mu = \mu_0$), as shown in Figure P1-29, creates a reflected field on the free-space side of the interface and a transmitted field on the water side of the interface. Assuming the incident ($\mathbf{E}^i$), reflected ($\mathbf{E}^r$), and transmitted ($\mathbf{E}^t$) electric fields are given, respectively, by

$$\mathbf{E}^i = \hat{a}_x E_0 e^{-j\beta_0 z}$$

$$\mathbf{E}^r = \hat{a}_x \Gamma_0 E_0 e^{+j\beta_0 z}$$

$$\mathbf{E}^t = \hat{a}_x T_0 E_0 e^{-j\beta z}$$

determine the coefficients Γ_0 and T_0. E_0 is a constant, $\beta_0 = \omega\sqrt{\mu_0 \varepsilon_0}$, $\beta = \omega\sqrt{\mu_0 \varepsilon}$.

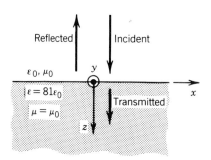

FIGURE P1-29

1.30. When a time-harmonic electromagnetic field is traveling in free space and is obliquely incident upon a flat surface of distilled water ($\varepsilon = 81\varepsilon_0$, $\mu = \mu_0$), it creates a reflected field on the free-space side of the interface and a transmitted field on the water side of the interface. Assume the incident, reflected, and transmitted electric

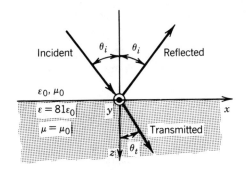

FIGURE P1-30

and magnetic fields are given by

$$\mathbf{E}^i = \hat{a}_y E_0 e^{-j\beta_0(x\sin\theta_i + z\cos\theta_i)}$$

$$\mathbf{H}^i = (-\hat{a}_x \cos\theta_i + \hat{a}_z \sin\theta_i)\sqrt{\frac{\varepsilon_0}{\mu_0}}\, E_0 e^{-j\beta_0(x\sin\theta_i + z\cos\theta_i)}$$

$$\mathbf{E}^r = \hat{a}_y \Gamma_h E_0 e^{-j\beta_0(x\sin\theta_i - z\cos\theta_i)}$$

$$\mathbf{H}^r = (\hat{a}_x \cos\theta_i + \hat{a}_z \sin\theta_i)\sqrt{\frac{\varepsilon_0}{\mu_0}}\, \Gamma_h E_0 e^{-j\beta_0(x\sin\theta_i - z\cos\theta_i)}$$

$$\mathbf{E}^t = \hat{a}_y T_h E_0 e^{-j\beta_0\left(x\sin\theta_i + z\sqrt{\frac{\varepsilon}{\varepsilon_0} - \sin^2\theta_i}\right)}$$

$$\mathbf{H}^t = \left(-\hat{a}_x\sqrt{1 - \frac{\varepsilon_0}{\varepsilon}\sin^2\theta_i} + \hat{a}_z\sqrt{\frac{\varepsilon_0}{\varepsilon}}\sin\theta_i\right)\sqrt{\frac{\varepsilon}{\mu_0}}\, T_h E_0 e^{-j\beta_0\left(x\sin\theta_i + z\sqrt{\frac{\varepsilon}{\varepsilon_0} - \sin^2\theta_i}\right)}$$

where E_0 is a constant and $\beta_0 = \omega\sqrt{\mu_0\varepsilon_0}$. Determine the coefficients Γ_h and T_h by applying the boundary conditions on the tangential components. Evaluate all the constants.

1.31. Repeat Problem 1.30 except that Γ_h and T_h should be determined using the boundary conditions on the normal components. Compare the answers to those obtained in Problem 1.30. Explain.

1.32. Repeat Problem 1.30 when the incident, reflected, and transmitted electric and magnetic fields are given by

$$\mathbf{E}^i = (\hat{a}_x \cos\theta_i - \hat{a}_z \sin\theta_i) E_0 e^{-j\beta_0(x\sin\theta_i + z\cos\theta_i)}$$

$$\mathbf{H}^i = \hat{a}_y \sqrt{\frac{\varepsilon_0}{\mu_0}}\, E_0 e^{-j\beta_0(x\sin\theta_i + z\cos\theta_i)}$$

$$\mathbf{E}^r = (\hat{a}_x \cos\theta_i + \hat{a}_z \sin\theta_i) \Gamma_e E_0 e^{-j\beta_0(x\sin\theta_i - z\cos\theta_i)}$$

$$\mathbf{H}^r = -\hat{a}_y \sqrt{\frac{\varepsilon_0}{\mu_0}}\, \Gamma_e E_0 e^{-j\beta_0(x\sin\theta_i - z\cos\theta_i)}$$

$$\mathbf{E}^t = \left[\hat{a}_x \sqrt{1 - \frac{\varepsilon_0}{\varepsilon}\sin^2\theta_i} - \hat{a}_z\sqrt{\frac{\varepsilon_0}{\varepsilon}}\sin\theta_i\right] T_e E_0 e^{-j\beta_0\left(x\sin\theta_i + z\sqrt{\frac{\varepsilon}{\varepsilon_0} - \sin^2\theta_i}\right)}$$

$$\mathbf{H}^t = \hat{a}_y \sqrt{\frac{\varepsilon}{\mu_0}}\, T_e E_0 e^{-j\beta_o\left(x\sin\theta_i + z\sqrt{\frac{\varepsilon}{\varepsilon_0} - \sin^2\theta_i}\right)}$$

Γ_e and T_e should be determined using the boundary conditions on the tangential components.

1.33. Repeat Problem 1.32 except that Γ_e and T_e should be determined using the boundary conditions on the normal components. Compare the answers to those obtained in Problem 1.32. Explain.

1.34. For Problem 1.10 find (a) the average power density at large distances and (b) the total power exiting through the surface of a large sphere of radius r ($r \gg a$).

1.35. The time-harmonic complex fields inside a source-free conducting pipe of rectangular cross section (waveguide) shown in Figure P1-35 filled with free space are given by

$$\mathbf{E} = \hat{a}_y E_0 \sin\left(\frac{\pi}{a}x\right) e^{-j\beta_z z}, \quad 0 \le x \le a, 0 \le y \le b$$

where

$$\beta_z = \beta_0 \sqrt{1 - \left(\frac{\lambda_0}{2a}\right)^2}$$

E_0 is a constant, and $\beta_0 = 2\pi/\lambda_0 = \omega\sqrt{\mu_0\varepsilon_0}$. For a section of waveguide of length ℓ along the z axis, determine (a) the corresponding complex magnetic field, (b) the supplied complex power, (c) the exiting complex power, (d) the dissipated real power, (e) the time-average magnetic energy, and (f) the time-average electric energy. Ultimately verify that the conservation of energy equation in integral form is satisfied for this set of fields inside this section of the waveguide.

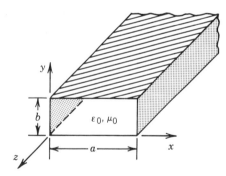

FIGURE P1-35

1.36. For the waveguide and its set of fields of Problem 1.35, verify the conservation of energy equation in differential form for any observation point within the waveguide.

1.37. At microwave frequencies high Q resonant cavities are usually constructed of enclosed conducting pipes (waveguides) of different cross sections. One such cavity is that of rectangular cross section that is enclosed on all six sides, as shown in Figure P1-37. One set of complex fields that can exist inside such a source-free cavity filled with free space is given by

$$\mathbf{E} = \hat{a}_y E_0 \sin\left(\frac{\pi}{a}x\right) \sin\left(\frac{\pi}{c}z\right)$$

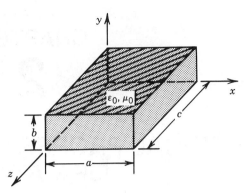

FIGURE P1-37

such that

$$\omega = \omega_r = \frac{1}{\sqrt{\mu_0 \varepsilon_0}} \sqrt{\left(\frac{\pi}{a}\right)^2 + \left(\frac{\pi}{c}\right)^2}$$

where E_0 is a constant and ω_r is referred to as the resonant radial frequency. Within the cavity, determine (a) the corresponding magnetic field, (b) the supplied complex power, (c) the dissipated real power, (d) the time-average magnetic energy, (e) the time-average electric energy. Ultimately verify that the conservation of energy equation in integral form is satisfied for this set of fields inside this resonant cavity.

CHAPTER 2

ELECTRICAL PROPERTIES OF MATTER

2.1 INTRODUCTION

An *atom* of an element consists of a very small but massive nucleus that is surrounded by a number of negatively charged electrons revolving about the nucleus. The nucleus contains *neutrons*, which are neutral particles, and *protons*, which are positively charged particles. All matter is made up of one or more of the 102 different elements that are now known to exist. Of this number, only 92 occur in nature. If the substance in question is a compound, it is composed of two or more different elements. The smallest constituent of a compound is a *molecule*, which is composed of one or more atoms held together by the short-range forces of their electrical charges.

For a given element, each of its atoms contains the same number of protons in its nucleus. Depending on the element, that number ranges from 1 to 102 and represents the *atomic number* of the element. For an atom in its normal state, the number of electrons is also equal to the atomic number. The revolving electrons that surround the nucleus exist in various shells, and they exert forces of repulsion on each other and forces of attraction on the positive charges of the nucleus. The outer shell of an atom is referred to as the *valence shell* (band) and the electrons occupying that shell are known as *valence electrons*. They are of most interest here. The portrayal of an atom by such a model is referred to as the *Bohr* model [1]. Atoms and their charges for some typical elements of interest in electronics (such as hydrogen, aluminum, silicon, and germanium) are shown in Figure 2-1.

For an atom, all the electrons in a given shell (orbit) exist in the same energy level (fixed state). Since there are several shells (orbits) around the nucleus of an atom, there exist several discrete energy levels (fixed states) each representing a given shell (orbit). In general, there are more energy levels than electrons. Therefore some of the energy levels (orbits, shells, bands) are not occupied by electrons. The

INTRODUCTION 43

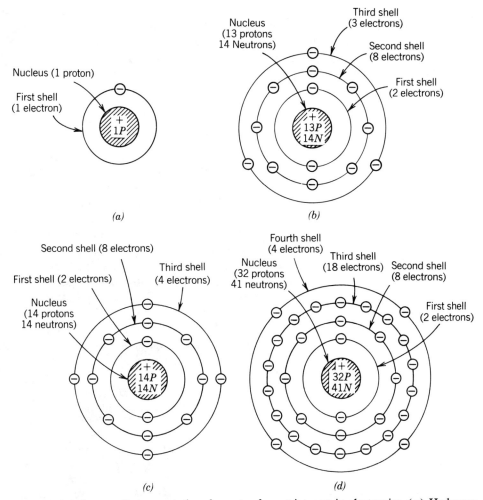

FIGURE 2-1 Atoms of representative elements of most interest in electronics. (*a*) Hydrogen atom. (*b*) Aluminum atom. (*c*) Silicon atom. (*d*) Germanium atom. (*Source:* R. R. Wright and H. R. Skutt, *Electronics: Circuits and Devices*, 1965; reprinted by permission of John Wiley and Sons, Inc.)

Bohr model of an atom states that

1. Electrons of any atom exist only in *discrete* states and possess only *discrete* amounts of energy corresponding to the *discrete* radii of their corresponding orbital shells.
2. If an electron moves from a lower- to a higher-energy level (orbit), it *absorbs* a discrete quantity of energy (referred to as *quanta*).
3. If an electron moves from a higher- to a lower-energy level (orbit), it *radiates* a discrete quantity of energy (referred to as *quanta*).
4. If an electron maintains its energy level (orbit), it neither absorbs nor radiates energy.

When a molecule is formed with two or more atoms, forces between the atoms result in new arrangements of the charges. For an electron to be freed from an atom, it must acquire sufficient energy to allow it to escape its atomic forces and become a free body. This is analogous to the energy required by a projectile to escape the earth's gravity and become a free body.

2.2 DIELECTRICS, POLARIZATION, AND PERMITTIVITY

Dielectrics (insulators) are material whose dominant charges in atoms and molecules are *bound* negative and positive charges that are held in place by atomic and molecular forces, and they are not free to travel. Thus ideal dielectrics do not contain any free charges (such as in conductors), and their atoms and molecules are macroscopically neutral as shown in Figure 2-2a. However, when external fields are applied, these bound negative and positive charges do not move to the surface of the material, as would be the case for conductors, but their respective centroids can shift slightly in positions (assumed to be an infinitesimal distance) relative to each other, thus creating numerous electric dipoles. This is illustrated in Figure 2-2b. In conductors positive and negative charges are separated by macroscopic distances, and they can be separated by a surface of integration. This is not permissible for bound charges and illustrates a fundamental difference between bound charges in dielectrics and true charges in conductors.

For dielectrics, the formation of the electric dipoles is usually referred to as *orientational polarization*. The effect of each electric dipole can be represented by a dipole, as shown in Figure 2-3, with a dipole moment $d\mathbf{p}_i$ given by

$$d\mathbf{p}_i = Q\boldsymbol{\ell}_i \qquad (2\text{-}1)$$

where Q is the magnitude (in coulombs) of each of the negative and positive charges whose centroids are displaced vectorially by distance $\boldsymbol{\ell}_i$.

When a material is subjected to an electric field, the polarization dipoles of the material interact with the applied electromagnetic field. For dielectric (insulating) material, whether they are solids, liquids, or gases, this interaction provides the material the ability to store electric energy, which is accomplished by the shift against restraining forces of their bound charges when they are subjected to external applied forces. This is analogous to stretching a spring or lifting a weight, and it represents potential energy.

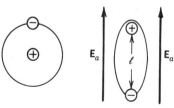

FIGURE 2-2 Typical atom in (a) the absence of and (b) under an applied field.

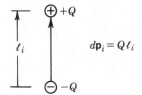

FIGURE 2-3 Formation of a dipole between two opposite charges of equal magnitude Q.

The presence of these dipoles can be accounted for by developing a microscopic model in which each individual charge and dipole as represented by (2-1) is considered. Such a procedure, although accurate if performed properly, is very impractical if applied to a dielectric slab because the spatial position of each atom and molecule of the material must be known. Instead, in practice, the behavior of these dipoles and bound charges is accounted for in a qualitative way by introducing an *electric polarization vector* **P** using a macroscopic scale model involving thousands of atoms and molecules.

The total dipole moment $\mathbf{p}_t$ of a material is obtained by summing the dipole moments of all the orientational polarization dipoles each of which is represented by (2-1). For a volume Δv where there are N_e electric dipoles per unit volume, or a total of $N_e \Delta v$ electric dipoles, we can write that

$$\mathbf{p}_t = \sum_{i=1}^{N_e \Delta v} d\mathbf{p}_i \tag{2-2}$$

The *electric polarization vector* **P** can then be defined as the *dipole moment per unit volume*, or

$$\mathbf{P} = \lim_{\Delta v \to 0} \left[\frac{1}{\Delta v} \mathbf{p}_t \right] = \lim_{\Delta v \to 0} \left[\frac{1}{\Delta v} \sum_{i=1}^{N_e \Delta v} d\mathbf{p}_i \right] \quad (\text{C/m}^2) \tag{2-3}$$

The units of **P** are coulomb-meters per cubic meter or coulombs per square meter, which is representative of a surface charge density. It should be noted that this is a *bound* surface charge density (q_{sp}), and it is not permissible to separate the positive and negative charges by an integration surface. Therefore within a volume an integral (whole) number of positive and negative pairs (dipoles) with an overall zero net charge must exist. Hence the bound surface charge should *not* be included in (1-45a) or (1-46) to determine the boundary conditions on the normal components of the electric flux density (or normal components of the electric field intensity).

Assuming an average dipole moment of

$$d\mathbf{p}_i = d\mathbf{p}_{av} = Q\mathbf{l}_{av} \tag{2-4}$$

per molecule, the electric polarization vector of (2-3) can be written, when all dipoles are aligned in the same direction, as

$$\mathbf{P} = \lim_{\Delta v \to 0} \left[\frac{1}{\Delta v} \sum_{i=1}^{N_e \Delta v} d\mathbf{p}_i \right] = N_e\, d\mathbf{p}_{av} = N_e Q \mathbf{l}_{av} \tag{2-5}$$

Electric polarization for dielectrics can be produced by any of the following three mechanisms, as demonstrated in Figure 2-4 [2]. Few materials involve all three

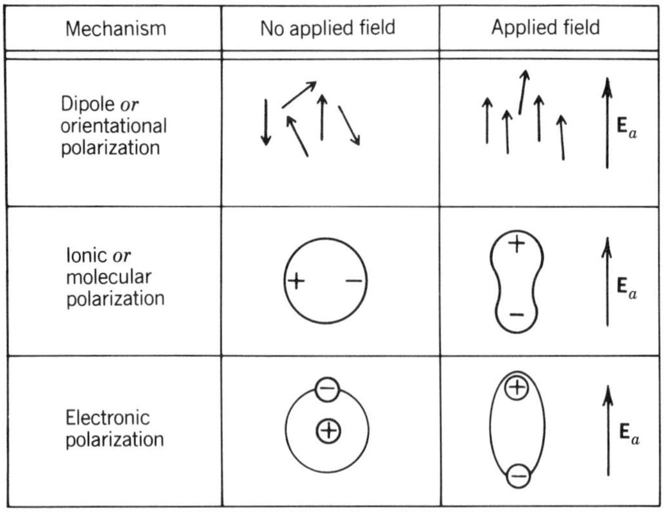

Mechanism	No applied field	Applied field
Dipole or orientational polarization	↓ ↖ ↑ ↘	↑ ↑ ↑ ↑ E_a ↑
Ionic or molecular polarization	(+ −)	(+ / −) E_a ↑
Electronic polarization	(⊖ over ⊕)	(⊕ over ⊖) E_a ↑

FIGURE 2-4 Mechanisms producing electric polarization in dielectrics.

mechanisms:

1. *Dipole or Orientational Polarization*: This polarization is evident in material that, in the absence of an applied field and owing to their structure, possess permanent dipole moments that are randomly oriented. However when an electric field is applied, the dipoles tend to align with the applied field. As will be discussed later, such materials are known as *polar* materials; water is a good example.
2. *Ionic or Molecular Polarization*: This polarization is evident in materials, such as sodium chloride (NaCl), that possess positive and negative ions and that tend to displace themselves when an electric field is applied.
3. *Electronic Polarization*: This polarization is evident in most materials, and it exists when an applied electric field displaces the electric cloud center of an atom relative to the center of the nucleus.

If the charges in a material, in the absence of an applied electric field $\mathbf{E}_a$, are averaged in such a way that positive and negative charges cancel each other throughout the entire material, then there are no individual dipoles formed and the total dipole moment of (2-2) and electric polarization vector **P** of (2-3) are zero. However, when an electric field is applied, it exhibits a net nonzero polarization. Such a material is referred to as *nonpolar*, and it is illustrated in Figure 2-5a. Polar materials are those whose charges in the absence of an applied electric field $\mathbf{E}_a$ are distributed so that there are individual dipoles formed, each with a dipole moment $\mathbf{p}_i$ as given by (2-1) but with a net total dipole moment $\mathbf{p}_t = 0$ and electric polarization vector **P** = 0. This is usually a result of the random orientation of the dipoles as illustrated in Figure 2-5b. Typical dipole moments of polar material are of the order of 10^{-30} C-m. Materials that, in the absence of an applied electric field $\mathbf{E}_a$, possess nonzero net dipole moment and electric polarization vector **P** are referred to as *electrets*.

There is also a class of dielectric materials that are usually referred to as *ferroelectrics* [3]. They exhibit a hysteresis loop of polarization (P) versus electric field (E) that is similar to the hysteresis loop of B versus H for ferromagnetic

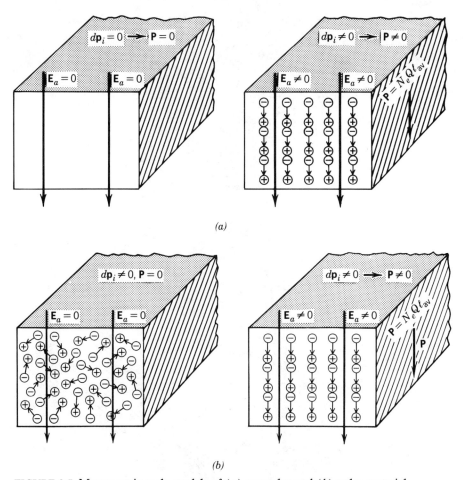

FIGURE 2-5 Macroscopic scale models of (*a*) nonpolar and (*b*) polar material.

material, and it possesses a *remnant polarization* P_r and *coercive electric field* E_c. At some critical temperature, referred to as *ferroelectric Curie temperature*, the spontaneous polarization in ferroelectrics disappears. Above the Curie temperature the relative permittivity varies according to the Curie–Weiss law; below it the electric flux density D and the polarization P are not linear functions of the electric field E [3]. Barium titanate ($BaTiO_3$) is one such material.

When an electric field is applied to a nonpolar or polar dielectric material, as shown in Figures 2-5*a* and 2-5*b*, the charges in each medium are aligned in such a way that individual dipoles with nonzero dipole moments are formed within the material. However, when we examine the material on a microscopic scale, the following items become evident from Figures 2-5*a* and 2-5*b*.

1. On the lower surface there exists a net positive surface charge density q_s^+ (representing bound charges).
2. On the upper surface there exists a net negative surface charge density q_s^- (representing bound charges).
3. The volume charge density q_v inside the material is zero because the positive and negative charges of adjacent dipoles cancel each other.

48 ELECTRICAL PROPERTIES OF MATTER

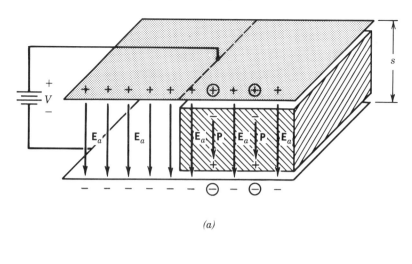

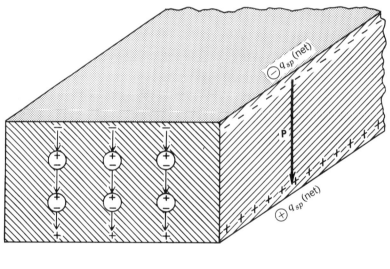

FIGURE 2-6 Dielectric slab subjected to an applied electric field $\mathbf{E}_a$. (*a*) Total charge. (*b*) Net charge.

The preceding items can also be illustrated by macroscopically examining Figure 2-6*a* where a d.c. voltage source is connected and remains across two parallel plates separated by distance *s*. Half of the space between the two plates is occupied by a dielectric material whereas the other half is free space. For a better illustration of this point, let us assume that there are five free charges on each part of the plates separated by free space. The same number appears on the part of the plates separated by the dielectric material. Because of the realignment of the bound charges in the dielectric material and the formation of the electric dipoles and cancellation of adjacent opposite charges shown circled in Figure 2-6*b*, a polarization electric vector **P** is formed within the dielectric material. Thus the polarization vector **P** is a result of the *bound* surface charge density $-q_{sp}$ found on the upper and $+q_{sp}$ found on the lower surface of the dielectric slab. Let us assume that there are two pairs of bound charges that form the bound surface charge density q_{sp} on the surface of the dielectric slab of Figure 2-6a (negative on top and positive on the bottom). Because the surfaces of the slab are assumed to be in contact with the plates of the capacitor, the two negative bound charges on the top surface will tend to cancel two of the positive free charges on the upper capacitor plate; a similar phenomenon occurs at the bottom. If this were to happen, the net number of charges on the top and bottom plates of the capacitor would diminish to three and the electric field intensity in the dielectric material between the plates would be reduced. Since the d.c. voltage supply is maintained across the plates, the net charge on the upper and lower parts of the capacitor and the electric field intensity in the dielectric material between the plates are also maintained by the introduction of two additional free charges on each of the capacitor plates (positive on top and negative on bottom). For identification purposes, these two induced free charges have been circled in each of the two plates in Figure 2-6.

In each of the situations discussed, the net effect is that between the lower and upper surfaces of the dielectric there is a net electric polarization vector **P** directed from the upper toward the lower surfaces, in the same direction as the applied electric field $\mathbf{E}_a$, whose amplitude is given by

$$P = q_{sp} \tag{2-6}$$

Whereas the applied electric field $\mathbf{E}_a$ maintains its value, the electric flux density inside the dielectric material differs from what would exist were the dielectric material replaced by free space. In the free-space part of the parallel plate capacitor of Figure 2-6, the electric flux density $\mathbf{D}_0$ is given by

$$\mathbf{D}_0 = \varepsilon_0 \mathbf{E}_a \tag{2-7}$$

In the dielectric portion, the electric flux density **D** is related to that in free space D_0 by

$$\mathbf{D} = \varepsilon_0 \mathbf{E}_a + \mathbf{P} \tag{2-8}$$

where the magnitude of **P** is given by (2-6). The electric flux density **D** of (2-8) can also be related to the applied electric field intensity $\mathbf{E}_a$ by a parameter that we designate here as ε_s (farads/meter). Thus we can write that

$$\mathbf{D} = \varepsilon_s \mathbf{E}_a \tag{2-9}$$

Comparing (2-8) and (2-9), it is apparent that **P** is also related to $\mathbf{E}_a$ and can be expressed as

$$\mathbf{P} = \varepsilon_0 \chi_e \mathbf{E}_a \tag{2-10}$$

or

$$\chi_e = \frac{1}{\varepsilon_0} \frac{P}{E_a} \tag{2-10a}$$

where χ_e is called the *electric susceptibility* (dimensionless quantity).

Substituting (2-10) into (2-8) and equating the result to (2-9), we can write that

$$\mathbf{D} = \varepsilon_0 \mathbf{E}_a + \varepsilon_0 \chi_e \mathbf{E}_a = \varepsilon_0 (1 + \chi_e) \mathbf{E}_a = \varepsilon_s \mathbf{E}_a \tag{2-11}$$

or that

$$\varepsilon_s = \varepsilon_0(1 + \chi_e) \tag{2-11a}$$

In (2-11a) ε_s is the *static permittivity* of the medium whose relative value ε_{sr} (compared to that of free space ε_0) is given by

$$\varepsilon_{sr} = \frac{\varepsilon_s}{\varepsilon_0} = 1 + \chi_e \tag{2-12}$$

which is usually referred to as the *relative permittivity*, better known in practice as

TABLE 2-1
Approximate static dielectric constants (relative permittivities) of dielectric materials

Material	Static dielectric constant (ε_r)
Air	1.0006
Styrofoam	1.03
Paraffin	2.1
Teflon	2.1
Plywood	2.1
RT/duroid 5880	2.20
Polyethylene	2.26
RT/duroid 5870	2.35
Glass-reinforced teflon (microfiber)	2.32–2.40
Teflon quartz (woven)	2.47
Glass-reinforced teflon (woven)	2.4–2.62
Cross-linked polystyrene (unreinforced)	2.56
Polyphenelene oxide (PPO)	2.55
Glass-reinforced polystyrene	2.62
Amber	3
Soil (dry)	3
Rubber	3
Plexiglas	3.4
Lucite	3.6
Fused silica	3.78
Nylon (solid)	3.8
Quartz	3.8
Sulfur	4
Bakelite	4.8
Formica	5
Lead glass	6
Mica	6
Beryllium oxide (BeO)	6.8–7.0
Marble	8
Sapphire	$\varepsilon_x = \varepsilon_y = 9.4$ $\varepsilon_z = 11.6$
Flint glass	10
Ferrite (Fe_2O_3)	12–16
Silicon (Si)	12
Gallium arsenide (GaAs)	13
Ammonia (liquid)	22
Glycerin	50
Water	81
Rutile (TiO_2)	$\varepsilon_x = \varepsilon_y = 89$ $\varepsilon_z = 173$

the *dielectric constant*. Scientists and engineers usually designate the square root of the relative permittivity as the *index of refraction*. Typical values of dielectric constants at static frequencies of some prominent dielectric materials are listed in Table 2-1.

Thus the dielectric constant of a dielectric material is a parameter that indicates the relative (compared to free-space) charge (energy) storage capabilities of a dielectric material; the larger its value, the greater its ability to store charge (energy). Parallel plate capacitors utilize dielectric material between their plates to increase their charge (energy) storage capacity by forcing extra free charges to be induced on the plates. These free charges neutralize the bound charges on the surface of the dielectric so that the voltage and electric field intensity is maintained constant between the plates.

Example 2-1. The static dielectric constant of water is 81. Assuming the electric field intensity applied to water is 1 V/m, determine the magnitudes of the electric flux density and electric polarization vector within the water.

Solution. Using (2-9), we have

$$D = \varepsilon_s E_a = 81(8.854 \times 10^{-12})(1) = 7.17 \times 10^{-10} \text{ C/m}^2$$

Using (2-12), we have

$$\chi_e = \varepsilon_{sr} - 1 = 81 - 1 = 80$$

Thus the electric polarization vector is given, using (2-10), by

$$P = \varepsilon_0 \chi_e E_a = 8.854 \times 10^{-12}(80)(1) = 7.08 \times 10^{-12} \text{ C/m}^2$$

The permittivity of (2-11a), or its relative form of (2-12), represents values at static or quasistatic frequencies. These values vary as a function of the alternating field frequency. The variations of the permittivity as a function of the frequency of the applied fields are examined in Section 2.8.1.

2.3 MAGNETICS, MAGNETIZATION, AND PERMEABILITY

Magnetic materials are those that exhibit magnetic polarization when they are subjected to an applied magnetic field. The magnetization phenomenon is represented by the alignment of the magnetic dipoles of the material with the applied magnetic field, similar to the alignment of the electric dipoles of the dielectric material with the applied electric field.

Accurate results concerning the behavior of magnetic material when they are subjected to applied magnetic fields can only be predicted by the use of quantum theory. This is usually quite complex and unnecessary for most engineering applications. Quite satisfactory quantitative results can be obtained, however, by using simple atomic models to represent the atomic lattice structure of the material. The atomic models used here represent the electrons as negative charges orbiting around the positive charged nucleus, as shown in Figure 2-7a. Each orbiting electron can be modeled by an equivalent small electric current loop of area *ds* whose current flows in the direction opposite to the electron orbit, as shown in Figure 2-7b. As long as

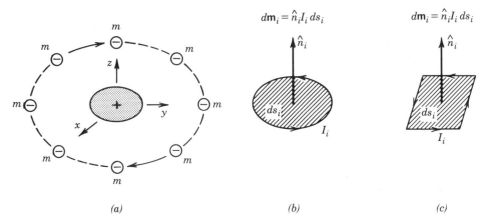

FIGURE 2-7 Atomic models and their equivalents, representing the atomic lattice structure of magnetic material. (*a*) Orbiting electrons. (*b*) equivalent circular electric loop. (*c*) Equivalent square electric loop.

the loop is small, its shape can be circular, square, or any other configuration, as shown in Figure 2-7c. The fields produced by a small loop of electric current at large distances are the same as those produced by a linear bar magnet (magnetic dipole) of length d.

By referring to the equivalent loop models of Figure 2-7, the angular momentum associated with an orbiting electron can be represented by a magnetic dipole moment $d\mathbf{m}_i$ of

$$d\mathbf{m}_i = I_i \, d\mathbf{s}_i = I_i \hat{n}_i \, ds_i = \hat{n}_i I_i \, ds_i \; (\text{A-m}^2) \tag{2-13}$$

For atoms that possess many orbiting electrons, the total magnetic dipole moment $\mathbf{m}_t$ is equal to the vector sum of all the individual magnetic dipole moments each represented by (2-13). Thus we can write that

$$\mathbf{m}_t = \sum_{i=1}^{N_m \Delta v} d\mathbf{m}_i = \sum_{i=1}^{N_m \Delta v} \hat{n}_i I_i \, ds_i \tag{2-14}$$

where N_m is equal to the number of orbiting electrons (equivalent loops) per unit volume. A magnetic polarization (magnetization) vector $\mathbf{M}$ is then defined as

$$\mathbf{M} = \lim_{\Delta v \to 0} \left[\frac{1}{\Delta v} \mathbf{m}_t \right] = \lim_{\Delta v \to 0} \left[\frac{1}{\Delta v} \sum_{i=1}^{N_m \Delta v} d\mathbf{m}_i \right] = \lim_{\Delta v \to 0} \left[\frac{1}{\Delta v} \sum_{i=1}^{N_m \Delta v} \hat{n}_i I_i \, ds_i \right] (\text{A/m}) \tag{2-15}$$

Assuming for each of the loops an average magnetic moment of

$$d\mathbf{m}_i = d\mathbf{m}_{av} = \hat{n}(I \, ds)_{av} \tag{2-16}$$

the magnetic polarization vector $\mathbf{M}$ of (2-15) can be written (assuming all the loops are aligned in the parallel planes) as

$$\mathbf{M} = \lim_{\Delta v \to 0} \left[\frac{1}{\Delta v} \sum_{i=1}^{N_m \Delta v} d\mathbf{m}_i \right] = N_m \, d\mathbf{m}_{av} = \hat{n} N_m (I \, ds)_{av} \tag{2-17}$$

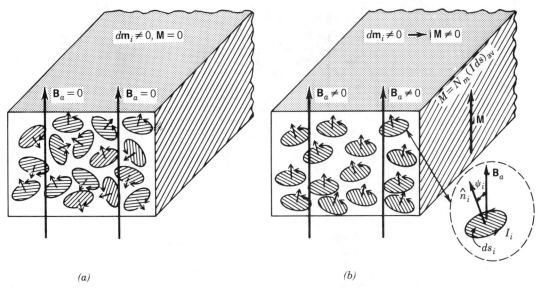

FIGURE 2-8 Random orientation of magnetic dipoles and their alignment (*a*) in the absence of and (*b*) under an applied field.

A magnetic material is represented by a number of magnetic dipoles and thus by many magnetic moments. In the absence of an applied magnetic field the magnetic dipoles and their corresponding electric loops are oriented in a random fashion so that on a macroscopic scale the vector sum of the magnetic moments of (2-14) and the magnetic polarization of (2-15) are equal to zero. The random orientation of the magnetic dipoles and loops is illustrated in Fig. 2-8*a*. When the magnetic material is subjected to an applied magnetic field, represented by the magnetic flux density $\mathbf{B}_a$ in Figure 2-8*b*, the magnetic dipoles of most material will tend to align in the direction of the $\mathbf{B}_a$ since a torque given by

$$|\Delta \mathbf{T}| = |d\mathbf{m}_i \times \mathbf{B}_a| = |d\mathbf{m}_i||\mathbf{B}_a|\sin(\psi_i) = |(\hat{n}_i I_i \, ds_i) \times \mathbf{B}_a| = |I_i \, ds_i \, B_a \sin(\psi_i)| \tag{2-18}$$

will be exerted in each of the magnetic dipole moments. This is shown in the insert to Figure 2-8*b*. Ideally, if there were no other magnetic moments to consider, torque would be exerted. The torque would exist until each of the orbiting electrons shifted in such a way that the magnetic field produced by each of its equivalent electric loops (or magnetic moments) was aligned with the applied field and its value represented by (2-18) vanished. Thus the resultant magnetic field at every point in the material would be greater than its corresponding value at the same point when the material is absent.

The magnetization vector **M** resulting from the realignment of the magnetic dipoles is better illustrated by considering a slab of magnetic material across which a magnetic field $\mathbf{B}_a$ is applied, as shown in Figure 2-9. Ideally, on a microscopic scale, for most magnetic material all the magnetic dipoles will align themselves so that their individual magnetic moments are pointed in the direction of the applied field, as shown in Figure 2-9. In the limit, as the number of magnetic dipoles and their corresponding equivalent electric loops become very large, the currents of the loops found in the interior parts of the slab are canceled by those of the neighboring

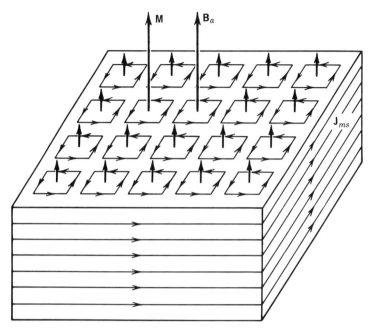

FIGURE 2-9 Magnetic slab subjected to an applied magnetic field and the formation of the magnetization current density $\mathbf{J}_{ms}$.

loops. On a macroscopic scale a net nonzero equivalent magnetic current, resulting in an equivalent magnetic current surface density (A/m), is found on the exterior surface of the slab. This equivalent magnetic current density $\mathbf{J}_{ms}$ is responsible for the introduction of the magnetization vector $\mathbf{M}$ in the direction of $\mathbf{B}_a$.

The magnetic flux density across the slab is increased by the presence of $\mathbf{M}$ so that the net magnetic flux density at any interior point of the slab is given by

$$\mathbf{B} = \mu_0(\mathbf{H}_a + \mathbf{M}) \tag{2-19}$$

It should be pointed out that $\mathbf{M}$, as given by (2-15), has the units of amperes per meter and correspond to those of the magnetic field intensity. In general, we can relate the magnetic flux density to the magnetic field intensity by a parameter that is designated as μ_s (henries/meter). Thus we can write that

$$\mathbf{B} = \mu_s \mathbf{H}_a \tag{2-20}$$

Comparing (2-19) and (2-20) indicates that $\mathbf{M}$ is also related to $\mathbf{H}_a$ by

$$\mathbf{M} = \chi_m \mathbf{H}_a \tag{2-21}$$

where χ_m is called the *magnetic susceptibility* (dimensionless quantity).

Substituting (2-21) into (2-19) and equating the result to (2-20) leads to

$$\mathbf{B} = \mu_0(\mathbf{H}_a + \chi_m \mathbf{H}_a) = \mu_0(1 + \chi_m)\mathbf{H}_a = \mu_s \mathbf{H}_a \tag{2-22}$$

Therefore we can define

$$\mu_s = \mu_0(1 + \chi_m) \tag{2-22a}$$

TABLE 2-2
Approximate static relative permeabilities of magnetic materials

Material	Class	Relative permeability (μ_{sr})
Bismuth	Diamagnetic	0.999834
Silver	Diamagnetic	0.99998
Lead	Diamagnetic	0.999983
Copper	Diamagnetic	0.999991
Water	Diamagnetic	0.999991
Vacuum	Nonmagnetic	1.0
Air	Paramagnetic	1.0000004
Aluminum	Paramagnetic	1.00002
Nickel chloride	Paramagnetic	1.00004
Palladium	Paramagnetic	1.0008
Cobalt	Ferromagnetic	250
Nickel	Ferromagnetic	600
Mild steel	Ferromagnetic	2,000
Iron	Ferromagnetic	5,000
Silicon iron	Ferromagnetic	7,000
Mumetal	Ferromagnetic	100,000
Purified iron	Ferromagnetic	200,000
Supermalloy	Ferromagnetic	1000,000

In (2-22a) μ_s is the *static permeability* of the medium whose relative value μ_{sr} (compared to that of free space μ_0) is given by

$$\mu_{sr} = \frac{\mu_s}{\mu_0} = 1 + \chi_m \qquad (2\text{-}23)$$

Static values of μ_{sr} for some representative material are listed in Table 2-2.

Within the material, a *bound* magnetic current density $\mathbf{J}_m$ is induced that is related to the magnetic polarization vector $\mathbf{M}$ by

$$\mathbf{J}_m = \nabla \times \mathbf{M} \ (\text{A}/\text{m}^2) \qquad (2\text{-}24)$$

To account for this current density, we modify the Maxwell–Ampere equation 1-71b and write it as

$$\nabla \times \mathbf{H} = \mathbf{J}_i + \mathbf{J}_c + \mathbf{J}_m + \mathbf{J}_d = \mathbf{J}_i + \sigma\mathbf{E} + \nabla \times \mathbf{M} + j\omega\varepsilon\mathbf{E} \qquad (2\text{-}24\text{a})$$

On the surface of the material, the *bound* magnetization surface current density $\mathbf{J}_{ms}$ is related to the magnetic polarization vector $\mathbf{M}$ at the surface by

$$\mathbf{J}_{ms} = \mathbf{M} \times \hat{n}|_{\text{surface}} \ (\text{A}/\text{m}) \qquad (2\text{-}25)$$

where $\hat{n}$ is a unit vector normal to the surface of the material. The *bound* magnetization current I_m flowing through a cross section S_0 of the material can be obtained by using

$$I_m = \iint_S \mathbf{J}_m \cdot d\mathbf{s} = \iint_{S_0} (\nabla \times \mathbf{M}) \cdot d\mathbf{s} \ (\text{A}) \qquad (2\text{-}26)$$

56 ELECTRICAL PROPERTIES OF MATTER

In addition to orbiting, the electrons surrounding the nucleus of an atom also spin about their own axis. Therefore magnetic moments of the order of $\pm 9 \times 10^{-24}$ A-m² are also associated with the spinning of the electrons that aid or oppose the applied magnetic field (the + sign is used for addition and the − for subtraction). For atoms that have many electrons in their shells, only the spins associated with the electrons found in shells that are not completely filled will contribute to the magnetic moment of the atoms. A third contributor to the total magnetic moment of an atom is that associated with the spinning of the nucleus, which is referred to as *nuclear spin*. However, this nuclear spin magnetic moment is usually much smaller (typically by a factor of about 10^{-3}) than those attributed to the orbiting and the spinning electrons.

Example 2-2. A bar of magnetic material of finite length, which is placed along the z axis as shown in Figure 2-10, has a cross section of 0.3 m in the x direction ($0 \leq x \leq 0.3$) and 0.2 m in the y direction ($0 \leq y \leq 0.2$). The bar is subjected to a magnetic field so that the magnetization vector inside the bar is given by

$$\mathbf{M} = \hat{a}_z(4y)$$

Determine the volumetric current density $\mathbf{J}_m$ at any point inside the bar, the surface current density $\mathbf{J}_{ms}$ on the surface of each of the four faces, and the total current I_m per unit length flowing through the bar face that is parallel to the y axis at $x = 0.3$ m.

Solution. Using (2-24), we have

$$\mathbf{J}_m = \nabla \times \mathbf{M} = \hat{a}_x \frac{\partial M_z}{\partial y} = \hat{a}_x 4$$

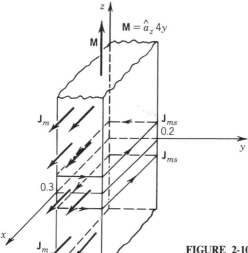

FIGURE 2-10 Magnetic bar of rectangular cross section subjected to a magnetic field.

Using (2-25), we have

$$\mathbf{J}_{ms} = \mathbf{M} \times \hat{n}|_{\text{surface}}$$

Therefore at

$x = 0$:
$$\mathbf{J}_{ms} = (\hat{a}_z 4y) \times (-\hat{a}_x)|_{x=0} = -\hat{a}_y(4y) \qquad \text{for } 0 \leq y \leq 0.2$$

$y = 0$:
$$\mathbf{J}_{ms} = (\hat{a}_z 4y) \times (-\hat{a}_y)|_{y=0} = \hat{a}_x(4y) = 0 \qquad \text{for } 0 \leq x \leq 0.3$$

$x = 0.3$:
$$\mathbf{J}_{ms} = (\hat{a}_z 4y) \times (\hat{a}_x)|_{x=0.3} = \hat{a}_y(4y) \qquad \text{for } 0 \leq y \leq 0.2$$

$y = 0.2$
$$\mathbf{J}_{ms} = (\hat{a}_z 4y) \times (\hat{a}_y)|_{y=0.2} = -\hat{a}_x(4y) = -\hat{a}_x 0.8 \qquad \text{for } 0 \leq x \leq 0.3$$

According to (2-26) the current (per unit length) flowing through the bar face at $x = 0.3$ is given by

$x = 0.3$:
$$I_m = \iint_S \mathbf{J}_m \cdot d\mathbf{s} = \int_0^1 \int_0^{0.2} (\hat{a}_x 4) \cdot (\hat{a}_x \, dy \, dz) = 4(1)(0.2) = 0.8$$

Consistent with the relative permittivity (dielectric constant), the values of μ, and thus μ_r, vary as a function of frequency. These variations will be discussed in Section 2.8.2. The values of μ_r listed in Table 2-2 are representative of frequencies related to static or quasistatic fields. Excluding ferromagnetic material, it is apparent that most relative permeabilities are very near unity, so that for engineering problems a value of unity is almost always used.

According to the direction in which the net magnetization vector **M** is pointing (either aiding or opposing the applied magnetic field), material are classified into two groups, Group A and Group B as shown:

Group A	Group B
Diamagnetic	Paramagnetic
	Ferromagnetic
	Antiferromagnetic
	Ferrimagnetic

In general, for material in Group A the net magnetization vector (although small in magnitude) opposes the applied magnetic field, resulting in a relative permeability slightly smaller than unity. *Diamagnetic* materials fall into that group. For material in Group B the net magnetization vector is aiding the applied magnetic field, resulting in relative permeabilities greater than unity. Some of them (*paramagnetic* and *antiferromagnetic*) have only slightly greater than unity relative permeabilities whereas others (*ferromagnetic* and *ferrimagnetic*) have relative permeabilities much greater than unity.

In the absence of an applied magnetic field, the moments of the electron spins of *diamagnetic* material are opposite to each other as well as to the moments associated with the orbiting electrons so that a zero net magnetic moment $\mathbf{m}_t$ is produced on a macroscopic scale. In the presence of an external applied magnetic field, each atom has a net nonzero magnetic moment, and on a macroscopic scale there is a net total magnetic moment for all the atoms that results in a magnetization vector $\mathbf{M}$. For diamagnetic material, this vector $\mathbf{M}$ is very small, opposes the applied magnetic field, leads to a negative magnetic susceptibility χ_m, and results in values of relative permeability that are slightly less than unity. For example, copper is a diamagnetic material with a magnetic susceptibility $\chi_m = -9 \times 10^{-6}$ and a relative permeability $\mu_r = 0.999991$.

In *paramagnetic* material, the magnetic moments associated with the orbiting and spinning electrons of an atom do not quite cancel each other in the absence of an applied magnetic field. Therefore each atom possesses a small magnetic moment. However, because the orientation of the magnetic moment of each atom is random, the net magnetic moment of a large sample (macroscopic scale) of dipoles, and the magnetization vector $\mathbf{M}$, are zero when there is no applied field. When the paramagnetic material is subjected to an applied magnetic field, the magnetic dipoles align slightly with the applied field to produce a small nonzero $\mathbf{M}$ in its direction and a small increase in the magnetic flux density within the material. Thus the magnetic susceptibilities have small positive values and the relative permeabilities are slightly greater than unity. For example, aluminum possesses a susceptibility of $\chi_m = 2 \times 10^{-5}$ and a relative permeability of $\mu_r = 1.00002$.

The individual atoms of *ferromagnetic* material possess, in the absence of an applied magnetic field, very strong magnetic moments caused primarily by uncompensated electron spin moments. The magnetic moments of many atoms (usually as many as five to six) reinforce one another and form regions called *domains*, which have various sizes and shapes. The dimensions of the domains depend on the material's past magnetic state and history, and range from 1 μm to a few millimeters. On a macroscopic scale, however, the net magnetization vector $\mathbf{M}$ in the absence of an applied field is zero because the domains are randomly oriented and the magnetic moments of the various atoms cancel one another. When a ferromagnetic material is subjected to an applied field, there are not only large magnetic moments associated with the individual atoms, but the vector sum of all the magnetic moments and the associated vector magnetization $\mathbf{M}$ are very large, leading to extreme values of magnetic susceptibility χ_m and relative permeability. Typical values of μ_r for some representative ferromagnetic material are found in Table 2-2. When the applied field is removed, the magnetic moments of the various atoms do not attain a random orientation and a net nonzero residual magnetic moment remains. Since the magnetic moment of a ferromagnetic material on a macroscopic scale is different after the applied field is removed, its magnetic state depends on the material's past history. Therefore a plot of the magnetic flux density $\mathscr{B}$ versus $\mathscr{H}$ leads to a double-valued curve known as the *hysteresis loop*. Material with such properties are very desirable in the design of transformers, induction cores, and coatings for magnetic recording tapes.

Materials that possess strong magnetic moments, but whose adjacent atoms are about equal in magnitude and opposite in direction, with zero net total magnetic moment in the absence of an applied magnetic field, are called *antiferromagnetic*. The presence of an applied magnetic field has a minor effect on the material and leads to relative permeabilities slightly greater than unity.

If the adjacent opposing magnetic moments of a material are very large in magnitude but greatly unequal in the absence of an applied magnetic field, the material is known as *ferrimagnetic*. The presence of an applied magnetic field has a large effect on the material and leads to large permeabilities (but not as large as those of ferromagnetic material). *Ferrites* make up a group of ferrimagnetic materials that have low conductivities (several orders smaller than those of semiconductors). Because of their large resistances, smaller currents are induced in them that result in lower ohmic losses when they are subjected to alternating fields. They find wide applications in the design of nonreciprocal microwave components (isolators, hybrids, gyrators, phase shifters, etc.) and they will be discussed briefly in Section 2.8.3.

CURRENT, CONDUCTORS, AND CONDUCTIVITY

The prominent characteristic of dielectric materials is the electric polarization introduced through the formation of electric dipoles between opposite charges of atoms. Magnetic dipoles, modeled by equivalent small electric loops, were introduced to account for the orbiting of electrons in atoms of magnetic material. This phenomenon was designated as magnetic polarization. Conductors are material whose prominent characteristic is the motion of electric charges and the creation of a current flow.

Current

Let us assume that an electric volume charge density, represented here by q_v, is distributed uniformly in an infinitesimal circular cylinder of cross-sectional area Δs and volume ΔV, as shown in Figure 2-11. The total electric charge ΔQ within the volume ΔV is moving in the z direction with a uniform velocity v_z. Thus we can write that

$$\frac{\Delta Q_e}{\Delta t} = q_v \frac{\Delta V}{\Delta t} = q_v \frac{\Delta s \, \Delta z}{\Delta t} = q_v \, \Delta s \, \frac{\Delta z}{\Delta t} \qquad (2\text{-}27)$$

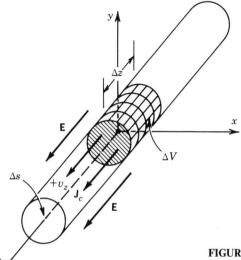

FIGURE 2-11 Charge uniformly distributed in an infinitesimal circular cylinder.

60 ELECTRICAL PROPERTIES OF MATTER

In the limit as $\Delta t \to 0$, (2-27) is used to define the current ΔI (with units of amperes) that flows through Δs. Thus

$$\Delta I = \lim_{\Delta t \to 0} \left[\frac{\Delta Q_e}{\Delta t} \right] = \lim_{\Delta t \to 0} \left[q_v \Delta s \frac{\Delta z}{\Delta t} \right] = q_v v_z \Delta s \qquad (2\text{-}28)$$

Dividing both sides of (2-28) by Δs and taking the limit as $\Delta s \to 0$, we can define the current density J_z (with units of amperes/per square meter) as

$$J_z = \lim_{\Delta s \to 0} \left[\frac{\Delta I}{\Delta s} \right] = q_v v_z \qquad (2\text{-}29)$$

Using a similar procedure for the x- and y-directed currents, we can write in general that

$$\mathbf{J} = q_v \mathbf{v} \; (\text{A}/\text{m}^2) \qquad (2\text{-}30)$$

In (2-30), $\mathbf{J}$ is defined as the *convection current density*. The current density between the cathode and anode of a vacuum tube is a convection current density. It should be noted that for an electric field intensity of $\mathbf{E} = \hat{a}_z E_z$, a positive charge density $+q_v$ will experience a force that will move it in the $+z$ direction. Thus the current density $\mathbf{J}$ will be directed in the $+z$ direction or

$$\mathbf{J} = +q_v(+\hat{a}_z v_z) = \hat{a}_z q_v v_z \qquad (2\text{-}31)$$

If the same electric field $\mathbf{E} = \hat{a}_z E_z$ is subjected to a negative charge density $-q_v$, the field will force the negative charge to move in the negative z direction ($\mathbf{v} = -\hat{a}_z v_z$). However, the electric current density $\mathbf{J}$ is still directed along the $+z$ direction,

$$\mathbf{J} = -q_v(-\hat{a}_z v_z) = \hat{a}_z q_v v_z \qquad (2\text{-}32)$$

since both the charge density and the velocity are negative. If positive (q_v^+) and negative (q_v^-) charges are present, (2-30) can be written as

$$\mathbf{J} = q_v^+ \mathbf{v}^+ + q_v^- \mathbf{v}^- \qquad (2\text{-}33)$$

2.4.2 Conductors

Conductors are material whose atomic outer shell (valence) electrons are not held very tightly and can migrate from one atom to another. These are known as *free electrons*, and for metal conductors they are very large in number. With no applied external field, these free electrons move with different velocities in random directions producing zero net current through the surface of the conductor.

When free charge q_{v0} is placed inside a conductor that is subjected to a static field, the charge density at that point decays exponentially as

$$q_v(t) = q_{v0} e^{-t/t_r} = q_{v0} e^{-(\sigma/\varepsilon)t} \qquad (2\text{-}34)$$

because the charge migrates toward the surface of the conductor. The time it takes for this to occur depends on the conductivity of the material; for metals it is equal to a few time constants. During this time, charges move, currents flow, and nonstatic conditions exist. The time t_r that it takes for the free charge density placed inside a conductor to decay to $e^{-1} = 0.368$, or 36.8 percent of its initial value, is known as the *relaxation time constant*. Mathematically it is represented by

$$t_r = \frac{\varepsilon}{\sigma} \tag{2-35}$$

where ε = permittivity of conductor (F/m)
σ = conductivity of conductor (S/m) (see equation 2-39)

Example 2-3. Find the relaxation time constant for a metal such as copper ($\sigma = 5.76 \times 10^7$ S/m, $\varepsilon = \varepsilon_0$) and a good dielectric such as glass ($\sigma \simeq 10^{-12}$ S/m, $\varepsilon = 6\varepsilon_0$).

Solution. For copper

$$t_r = \frac{\varepsilon}{\sigma} = \frac{8.854 \times 10^{-12}}{5.76 \times 10^7} = 1.54 \times 10^{-19} \text{ s}$$

which is very short. For glass

$$t_r = \frac{\varepsilon}{\sigma} = 6\left(\frac{8.854 \times 10^{-12}}{10^{-12}}\right) = 53.1 \text{ s} \simeq 1 \text{ min}$$

which is comparatively quite long.

The free charges of a very good conductor ($\sigma \to \infty$), which is subjected to an electric field, migrate very rapidly and distribute themselves as surface charge density q_s to the surface of the conductor within an extremely short period of time (several very short relaxation time constants). The surface charge density q_s will induce on the conductor an electric field intensity $\mathbf{E}_i$ so that the total electric field $\mathbf{E}_t$ within the conductor ($\mathbf{E}_i + \mathbf{E}_a = \mathbf{E}_t$ where $\mathbf{E}_a$ is the applied field) is essentially equal to zero. This is illustrated in Figure 2-12. For perfect conductors ($\sigma = \infty$) the electric field within the conductors is exactly equal to zero.

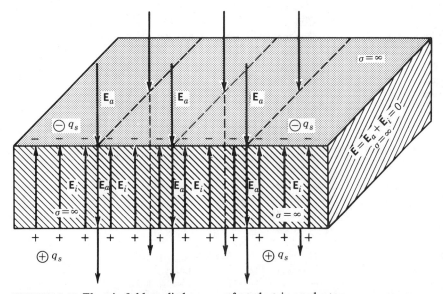

FIGURE 2-12 Electric field applied on a perfect electric conductor.

TABLE 2-3
Typical conductivities of insulators, semiconductors, and conductors

Material	Class	Conductivity σ (S/m)
Fused quartz	Insulator	$\sim 10^{-17}$
Ceresin wax	Insulator	$\sim 10^{-17}$
Sulfur	Insulator	$\sim 10^{-15}$
Mica	Insulator	$\sim 10^{-15}$
Paraffin	Insulator	$\sim 10^{-15}$
Hard rubber	Insulator	$\sim 10^{-15}$
Porcelain	Insulator	$\sim 10^{-14}$
Glass	Insulator	$\sim 10^{-12}$
Bakelite	Insulator	$\sim 10^{-9}$
Distilled water	Insulator	$\sim 10^{-4}$
Fused silica[a]	Semiconductor	$\sim 2.1 \times 10^{-4}$
Cross-linked polystyrene (unreinforced)[a]	Semiconductor	$\sim 3.7 \times 10^{-4}$
Beryllium Oxide (BeO)[a]	Semiconductor	$\sim 3.9 \times 10^{-4}$
Intrinsic silicon	Semiconductor	$\sim 4.39 \times 10^{-4}$
Sapphire[a]	Semiconductor	$\sim 5.5 \times 10^{-4}$
Glass-reinforced Teflon (microfiber)[a]	Semiconductor	$\sim 7.8 \times 10^{-4}$
Teflon quartz (woven)[a]	Semiconductor	$\sim 8.2 \times 10^{-4}$
Dry soil	Semiconductor	$\sim 10^{-4} - 10^{-3}$
Ferrite (Fe_2O_3)[a]	Semiconductor	$\sim 1.3 \times 10^{-3}$
Glass-reinforced Polystyrene[a]	Semiconductor	$\sim 1.45 \times 10^{-3}$
Polyphenelene oxide (PPO)[a]	Semiconductor	$\sim 2.27 \times 10^{-3}$
Glass-reinforced Teflon (woven)[a]	Semiconductor	$\sim 2.43 \times 10^{-3}$
Plexiglas[a]	Semiconductor	$\sim 5.1 \times 10^{-3}$
Gallium arsenide (GaAs)[a]	Semiconductor	$\sim 8 \times 10^{-3}$
Wet soil	Semiconductor	$\sim 10^{-3} - 10^{-2}$
Fresh water	Semiconductor	$\sim 10^{-2}$
Human and animal tissue	Semiconductor	$\sim 0.2 - 0.7$
Intrinsic germanium	Semiconductor	~ 2.227
Seawater	Semiconductor	~ 4
Tellurium	Conductor	$\sim 5 \times 10^2$
Carbon	Conductor	$\sim 3 \times 10^4$
Graphite	Conductor	$\sim 3 \times 10^4$
Cast iron	Conductor	$\sim 10^6$
Mercury	Conductor	10^6
Nichrome	Conductor	10^6
Silicon steel	Conductor	2×10^6
German silver	Conductor	2×10^6
Lead	Conductor	5×10^6
Tin	Conductor	9×10^6
Iron	Conductor	1.03×10^7
Nickel	Conductor	1.45×10^7
Zinc	Conductor	1.7×10^7
Tungsten	Conductor	1.83×10^7
Brass	Conductor	2.56×10^7
Aluminum	Conductor	3.96×10^7
Gold	Conductor	4.1×10^7
Copper	Conductor	5.76×10^7
Silver	Conductor	6.1×10^7

[a] For most semiconductors the conductivities are representative for a frequency of about 10 GHz.

2.4.3 Conductivity

When a conductor is subjected to an electric field, the electrons still move in random directions but drift slowly (with a drift velocity $\mathbf{v}_e$) in the negative direction of the applied electric field, thus creating a conduction current in the conductor. The applied electric field $\mathbf{E}$ and drift velocity $\mathbf{v}_e$ of the electrons are related by

$$\mathbf{v}_e = -\mu_e \mathbf{E} \tag{2-36}$$

where μ_e is defined to be the *electron mobility* [positive quantity with units of m²/(V-s)]. Substituting (2-36) into (2-30), we can write that

$$\mathbf{J} = q_{ve}\mathbf{v}_e = q_{ve}(-\mu_e \mathbf{E}) = -q_{ve}\mu_e \mathbf{E} \tag{2-37}$$

where q_{ve} is the electron charge density. Comparing (2-37) with (1-16), or

$$\mathbf{J} = \sigma_s \mathbf{E} \tag{2-38}$$

we define the static conductivity of a conductor as

$$\sigma_s = -q_{ve}\mu_e \text{ (S/m)} \tag{2-39}$$

Its reciprocal value is called the *resistivity* (ohm-meters).

The conductivity σ_s of a conductor is a parameter that characterizes the free-electron conductive properties of a conductor. As temperature increases, the increased thermal energy of the conductor lattice structure increases lattice vibration. Thus the possibility of the moving free electrons colliding increases, which results in a decrease in the conductivity of the conductor. Materials with a very low value of conductivity are classified as *dielectrics* (*insulators*). The conductivity of ideal dielectrics is zero.

The conductivity of (2-39) is referred to as the static or d.c. conductivity; typical values of several materials are listed in Table 2-3. The conductivity varies as a function of frequency. These variations, along with the mechanisms that result in them, will be discussed in Section 2.8.1.

2.5 SEMICONDUCTORS

Materials whose conductivities bridge the gap between dielectrics (insulators) and conductors (typically the conductivity being 10^{-3} to unity) are referred to as *intrinsic* (pure) semiconductors. A graph illustrating the range of conductivities, from insulators to conductors, is displayed in Figure 2-13. Two such materials of significant importance to electrical engineering are intrinsic *germanium* and intrinsic *silicon*. In intrinsic (pure) semiconductors there are two common carriers: the *free electrons* and the *bound electrons* (referred to as positive *holes*) [4].

As the temperature rises, the mobilities of semiconducting material decrease but their charge densities increase more rapidly. The increases in the charge density more than offset the decreases in mobilities, resulting in a general increase in the conductivity of semiconducting material with rises in temperature. This is one of the

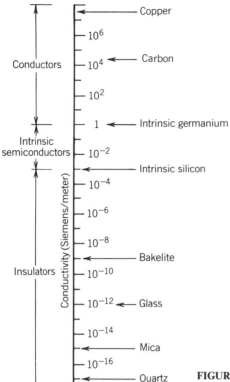

FIGURE 2-13 Range of conductivities of insulators, intrinsic semiconductors, and conductors.

characteristic differences between intrinsic semiconductors and metallic conductors: for semiconductors the conductivity increases with rising temperature whereas for metallic conductors it decreases. Typically the conductivity of germanium will increase by a factor of 10 as the temperature increases from 300 to about 360 K, and it will decrease by the same factor of 10 as the temperature decreases from 300 to 255 K. The conductivity of semiconductors can also be increased by adding impurities to the intrinsic (pure) properties. This process is known as *doping*. Some impurities (such as phosphorus) are called *donors* because they add more electrons and form *n-type* semiconductors, with the electrons being the major carriers. Impurities (such as boron) are called *acceptors* because they add more holes to form *p-type* semiconductors, with the holes being the predominant carriers. When both *n*- and *p*-type regions exist on a single semiconductor, the junction formed between the two regions is used to build diodes and transistors.

At temperatures near absolute zero (0 K or $\simeq -273°C$), the valence electrons of the outer shell of semiconducting material are held very tightly and they are not free to travel. Thus the material behaves as an insulator under those conditions. As the temperature rises, thermal vibration of the lattice structure in a semiconductor material increases, and some of the electrons gain sufficient thermal energy to break away from the tight grip of their atom and become free electrons similar to those in a metallic conductor. As was shown in Figure 2-1, the atoms of silicon and germanium have four valence electrons in their outer shell which are held very tightly at temperatures near absolute zero, but some of them may break away as the temperature rises. The valence electrons of any semiconductor must gain sufficient

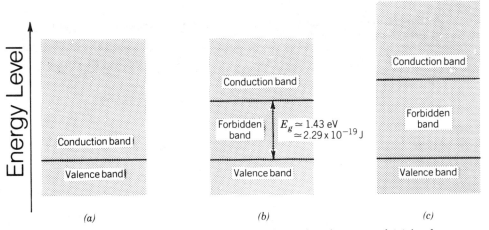

FIGURE 2-14 Energy levels for (*a*) conductors, (*b*) semiconductors, and (*c*) insulators.

energy to allow them to go from the valence band to the conduction band by jumping over the *forbidden* band, as shown in Figure 2-14. For all semiconductors the energy gap of the forbidden band is about $E_g = 1.43$ eV $= 2.29 \times 10^{-19}$ J. The bound electrons must gain at least that much energy, if not more, through increased thermal activity to make the jump.

The electrons that gain sufficient energy to break away from their atoms create vacancies in the other shell, designated as *holes*, which also move in a random fashion. When the semiconducting material is not subjected to an applied electric field, the net current from the bound electrons (which became free electrons) and the bound holes is zero because the net drift velocity of each type of carrier (electrons and holes) is zero (since they move in a random fashion). When an electric field is applied, the electrons move with a nonzero net drift velocity of $\mathbf{v}_{ed}$ (in the direction opposite to the applied field) while the holes move with a nonzero net drift velocity of $\mathbf{v}_{hd}$ (in the same direction as the applied field), thus creating a nonzero current. Therefore we write the conduction current density for the two carriers (electrons and holes) as

$$\mathbf{J}_c = q_{ev}\mathbf{v}_{ed} + q_{hv}\mathbf{v}_{hd} = q_{ev}(-\mu_e\mathbf{E}) + q_{hv}(+\mu_h\mathbf{E})$$

$$\mathbf{J}_c = (-q_{ev}\mu_e + q_{hv}\mu_h)\mathbf{E} = (\sigma_{es} + \sigma_{hs})\mathbf{E} = \sigma_s\mathbf{E} \quad (2\text{-}40)$$

where μ_e = mobility of electrons [m²/(V-s)]
μ_h = mobility of holes [m²/(V-s)]
σ_{es} = static conductivity due to electrons
σ_{hs} = static conductivity due to holes

The static conductivities of the electrons (σ_{es}) and the holes (σ_{hs}) can also be written as

$$\sigma_{es} = -q_{ev}\mu_e = -N_e q_e \mu_e = N_e|q_e|\mu_e \quad (2\text{-}41a)$$

$$\sigma_{hs} = -q_{hv}\mu_h = -N_h q_h \mu_h = N_h|q_h|\mu_h \quad (2\text{-}41b)$$

TABLE 2-4
Charge densities, mobilities, and conductivities for silicon, germanium, aluminum, copper, silver, and gallium arsenide at 300 K

	q_{ev} (C/m^3)	q_{hv} (C/m^3)	μ_e [m^2/(V-s)]	μ_h [m^2/(V-s)]	σ (S/m)
Intrinsic silicon	-2.4×10^{-3}	$+2.4 \times 10^{-3}$	0.135 at 300 K	0.048 at 300 K	0.439×10^{-3}
Intrinsic germanium	-3.84	$+3.84$	0.39 at 300 K	0.19 at 300 K	2.227
Aluminum	-1.8×10^{10}	0	2.2×10^{-3}	0	3.96×10^7
Copper	-1.8×10^{10}	0	3.2×10^{-3}	0	5.76×10^7
Silver	-1.8×10^{10}	0	3.4×10^{-3}	0	6.12×10^7
Gallium arsenide	-2.1	2.1	0.310 at 300 K	0.032 at 300 K	0.79×10^{-2}

where

N_e = free electron density (electrons per cubic meter)

N_h = bound hole density (holes per cubic meter)

$|q_e| = |q_h|$ = charge of an electron (magnitude) = 1.6×10^{-19} (coulombs)

$q_{ev} = N_e q_e = -N_e |q_e|$

$q_{hv} = N_h q_h = +N_h |q_h| = N_h |q_e|$

For comparison, representative values of charge densities, mobilities, and conductivities for intrinsic silicon, intrinsic germanium, aluminum, copper, silver, and gallium arsenide are given in Table 2-4 [5].

Six different materials were chosen to illustrate the formation of conductivity; their conductivity conditions are shown in Figure 2-15 [6]. These, in order, are representative of a dielectric (insulator), plasma (liquid or gas), conductor (metal), pure semiconductor, n-type semiconductor, and p-type semiconductor. It is observed that positively charged particles (holes) travel in the direction of the electric field whereas negatively charged particles (electrons) travel opposite to the electric field. However, both add to the total current.

The temperature variations of the mobilities of germanium, silicon, and gallium arsenide are given approximately by

Silicon [5]:

$$\mu_e \simeq (2.1 \pm 0.2) \times 10^5 T^{-2.5 \pm 0.1} \qquad 160 \leq T \leq 400 \text{ K} \qquad (2\text{-}42a)$$

$$\mu_h \simeq (2.3 \pm 0.1) \times 10^5 T^{-2.7 \pm 0.1} \qquad 150 \leq T \leq 400 \text{ K} \qquad (2\text{-}42b)$$

Germanium [5]:

$$\mu_e \simeq 4.9 \times 10^3 T^{-1.66} \qquad 100 \leq T \leq 300 \text{ K} \qquad (2\text{-}43a)$$

$$\mu_h \simeq 1.05 \times 10^5 T^{-2.33} \qquad 125 \leq T \leq 300 \text{ K} \qquad (2\text{-}43b)$$

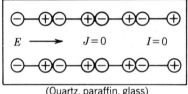
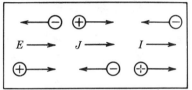
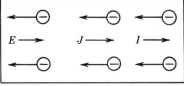
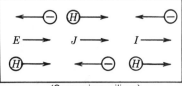
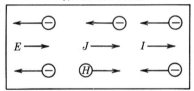
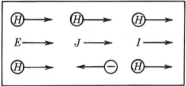

FIGURE 2-15 Conductivity conditions for six different materials representing dielectrics, plasmas, conductors, and semiconductors (*Source:* J. D. Kraus, *Electromagnetics*, 1984, McGraw-Hill Book Co.).

Gallium arsenide:

$$\mu_e \simeq 0.310\left(\frac{300}{T}\right)0.15 \qquad (2\text{-}44a)$$

$$\mu_h \simeq 0.032\left(\frac{300}{T}\right)2.2 \qquad (2\text{-}44b)$$

Example 2-4. For the semiconducting materials silicon and germanium, determine conductivities at a temperature of 10°F. The electron and hole densities for silicon and germanium are, respectively, equal to about 3.03×10^{16} and 1.47×10^{19} electrons or holes per cubic meter.

Solution. At $T = 10°F$, the respective temperatures on the Celsius (°C) and Kelvin (K) scales are

$$°C = \tfrac{5}{9}(°F - 32) = \tfrac{5}{9}(10 - 32) = -12.2$$
$$K = °C + 273.2 = -12.2 + 273.2 = 261$$

The mobilities of silicon and germanium at 10°F (261 K) are approximately equal to

Silicon:

$$\mu_e \simeq 2.1 \times 10^5 T^{-2.5} = 2.1 \times 10^5 (261^{-2.5}) = 0.1908$$
$$\mu_h \simeq 2.3 \times 10^5 T^{-2.7} = 2.3 \times 10^5 (261^{-2.7}) = 0.0687$$

Germanium:

$$\mu_e \simeq 4.9 \times 10^3 T^{-1.66} = 4.9 \times 10^3 (261^{-1.66}) = 0.4771$$
$$\mu_h \simeq 1.05 \times 10^5 T^{-2.33} = 1.05 \times 10^5 (261^{-2.33}) = 0.2457$$

In turn the conductivities are equal to

Silicon:

$$\sigma_e \simeq n_e |q_e| \mu_e = 3.03 \times 10^{16} (1.6 \times 10^{-19})(0.1908) = 0.925 \times 10^{-3} \text{ S/m}$$
$$\sigma_h \simeq n_h |q_h| \mu_h = 3.03 \times 10^{16} (1.6 \times 10^{-19})(0.0687) = 0.333 \times 10^{-3} \text{ S/m}$$
$$\sigma = \sigma_e + \sigma_h \simeq 1.258 \times 10^{-3} \text{ S/m}$$

Germanium:

$$\sigma_e \simeq n_e |q_e| \mu_e = 1.47 \times 10^{19} (1.6 \times 10^{-19})(0.4771) = 1.122 \text{ S/m}$$
$$\sigma_h \simeq n_h |q_h| \mu_h = 1.47 \times 10^{19} (1.6 \times 10^{-19})(0.2457) = 0.578 \text{ S/m}$$
$$\sigma = \sigma_e + \sigma_h \simeq 1.7 \text{ S/m}$$

2.6 SUPERCONDUCTORS

Ideal conductors ($\sigma = \infty$) are usually understood to be material within which an electric field **E** cannot exist at any frequency. Through Maxwell's time-varying equations, this absence of an electric field also assures that there is no time-varying magnetic field. For static fields, however, the magnetic field should not be affected by the conductivity (including infinity) of the material. Therefore for static fields ($f = 0$) a perfect conductor is defined as one that possesses an equipotential on its surface.

In practice no ideal conductors exist. Metallic conductors (such as aluminum, copper, silver, gold, etc.) have very large conductivities (typically 10^7–10^8 S/m), and the rf fields in them decrease very rapidly with depth measured from the surface (being essentially zero at a few skin depths). However, the resistivity of certain metals essentially vanishes (conductivity becomes extremely large, almost infinity) at temperatures near absolute zero ($T = 0$ K or $-273°C$). Such materials are usually called *superconductors*, and the temperature at which this is achieved is referred to as the *critical temperature* (T_c). This phenomenon was discovered in 1911 by Dutch physicist H. Kamerlingh Onnes, who received the Nobel Prize in 1913, and was also observed experimentally in 1933 by Meissner and Ochsenfeld [7, 8]. For example, aluminum becomes superconducting at a critical temperature of 1.2 K, niobium (also called columbium) at 9.2 K, and the intermetallic compound niobium–germanium (Nb_3Ge) at 23 K. For temperatures down to 0.05 K, copper and gold have not yet achieved superconductor status.

Before 1986, it was accepted that if materials could become superconducting at temperatures of 25 K or greater, there would be a major technological breakthrough. The reason for the breakthrough is that materials can be cooled to these temperatures with relatively inexpensive liquid hydrogen, whose boiling temperature is about 20.4 K. Some of the potential applications of superconductivity would be

1. supercomputers becoming smaller, faster, and thus more powerful;
2. more detailed and less expensive magnetic resonance imaging (MRI);
3. economical, efficient, pollution-free, and safe-generating power plants using fusion or magnetohydromagnetic technology;
4. loss-free transmission lines and more efficient power transmission;
5. trains traveling quietly and pollution-free at speeds of over 300 mph as they levitate on a magnetic cushion;
6. improved electronic instrumentation.

From 1911 to 1986, a span of 75 years, research into superconductivity yielded more than one thousand superconductive substances, but the increase in critical temperature was moderate and was accomplished at a very slow pace. Prior to January 1986, the record for the highest critical temperature belonged to niobium–germanium (Nb_3Ge), which in 1973 achieved a T_c of 23 K. In January 1986 a major breakthrough in superconductivity may have provided the spark for which the scientific community had been waiting. Karl Alex Mueller and Johannes Georg Bednorz, IBM Zurich Research Laboratory scientists, observed that a new class of oxide materials exhibited a superconductivity at a critical temperature much higher than anyone had observed before [9, 10]. The material was a ceramic copper oxide containing barium and lanthanum, and it had a critical temperature up to about 35 K, which was substantially higher than the 23 K for niobium–germanium.

Before Mueller and Bednorz's discovery, the best superconducting materials were intermetallic compounds, which included niobium–tin, niobium–germanium, and others. However, Mueller and Bednorz were convinced that the critical temperature could not be raised much higher using such compounds. Therefore they turned their attention to oxides with which they were familiar and which they believed to be better candidates for higher-temperature superconductors. For superconductivity to occur in a material, either the number of electrons that are available to transport current must be high or the electron pairs that are responsible for superconductivity must exhibit strong attractive forces [9]. Usually metals are very good candidates for superconductors because they have many available electrons. Oxides, however, have fewer electrons but it was shown that some metallic oxides of nickel and copper exhibited strong attractive electron-pair forces, and others could be found with even stronger pairing forces. Mueller and Bednorz became aware that some copper oxides behaved like metals in conducting electricity. This led them to the superconducting copper oxide containing barium and lanthanum with a critical temperature of 35 K.

Since then many other groups have reported even higher superconductivities, up to about 90 K in a number of ternary oxides of rare earth elements [10]. One of the main questions still to be answered is why are they superconducting at such high temperatures. Paul C. W. Chu, from the University of Houston, found that by pressurizing a superconducting copper oxide, lanthanum, and barium he could observe critical temperatures of up to 70 K. He reasoned that the pressure brought the layers of the different elements closer together, leading to the higher superconductivity. He also found that by replacing barium with strontium, which is a very similar element but has smaller atoms, brought the layers even closer together and led to even higher temperatures. In February 1987 Dr. Chu also discovered that replacing lanthanum with yttrium resulted in even higher temperatures, up to about 95 K. This was considered another major breakthrough, because it surpassed the barrier of the boiling point of liquid nitrogen (77 K). Liquid nitrogen is relatively inexpensive (by a factor of 50) compared to liquid helium or hydrogen, which are used with superconducting material at lower temperatures.

On January 22, 1988 researchers at the National Research Institute for Metals, Tsukaba, Japan, reported that a compound of bismuth, calcium, strontium-copper, and oxygen had achieved a critical temperature of 105 K. Three days later Dr. Chu announced an identical compound except that it contained one additional element —aluminum. Dr. Chu has indicated that bismuth contains two superconducting phases (chemical structures). This two-phase superconducting condition causes the resistance to drop drastically between 120 and 110 K, but not to reach zero until about 83 K, after a second sudden drop. One of the phases has a transition temperature of about 115 K, and the other phase becomes superconducting at 90 K. Efforts are underway to isolate the two phases, to keep the lower temperature phase from surrounding the higher one. Although the yttrium–copper oxides are very sensitive to oxygen content and a high temperature anneal is consequently needed after the material is made superconducting, bismuth compounds do not lose oxygen when heated. In addition, the bismuth compounds appear less brittle than the yttrium compounds.

Now the march is on to try to understand better the physics of superconductivity and to see whether the critical temperature cannot be raised even further. It is even reasonable to expect that superconductivity may be achieved at room temperature. Several researchers have already observed two to three orders of magnitude decrease in resistivity of mixed-phase samples at temperatures of 240 K. Therefore superconductivity at 240 K may be just around the corner.

Even though practical superconductivity now seems more of a reality, there are many problems that must be overcome. For example, most superconductive materials are difficult to produce consistently. They seem to be stronger in some directions than in others and in general are too brittle to be used for flexible wires. Moreover, they exhibit certain crystal anisotropies as current flow can vary by a factor of 30, depending on the direction. In addition the materials seem to lose their superconductive properties as the current density exceeds certain critical values. These critical current densities are believed to be around 10^5 A/cm^2, although values of 1.8×10^6 A/cm^2 have been reported at Japan's NTT Ibaragi Telecommunication Laboratory [10]. These current densities are about 10 to 100 times greater than reported previously, and they are also about 1000 times the current density of typical household wiring. These values are reassurance that materials would sustain superconductivity at current density levels required for power transmission and generation, electronic circuits, and electromagnets. The critical current density for the yttrium compounds may be too low for many applications; that of the bismuth compounds has not yet been measured.

2.7 LINEAR, HOMOGENEOUS, ISOTROPIC, AND NONDISPERSIVE MEDIA

The electrical behavior of materials when they are subjected to electromagnetic fields is characterized by their constitutive parameters (ε, μ, and σ).

Materials whose constitutive parameters are not functions of the applied field are usually known as *linear*; otherwise they are *nonlinear*. In practice, many material exhibit almost linear characteristics as long as the applied fields are within certain ranges. Beyond those points, the material may exhibit a high degree of nonlinearity. For example, air is nearly linear for applied electric fields up to about 1×10^6 V/m. Beyond that, air breaks down and exhibits a high degree of nonlinearity.

When the constitutive parameters of media are not functions of position, the materials are called *homogeneous*; otherwise they are *inhomogeneous* or *nonhomogeneous*. Almost all materials exhibit some degree of nonhomogeneity; however, for most materials used in practice the nonhomogeneity is so small that the materials are treated as being purely homogeneous.

If the constitutive parameters of a material vary as a function of frequency, they are denoted as being *dispersive*; otherwise they are *nondispersive*. All materials used in practice display some form of dispersion. The permittivities and the conductivities, especially of dielectric material, and the permeabilities of ferromagnetic material and ferrites exhibit rather pronounced dispersive characteristics. These will be discussed in the next two sections.

Anisotropic or *nonisotropic* materials are those whose constitutive parameters are a function of the direction of the applied field; otherwise they are known as *isotropic*. Many materials, especially crystals, exhibit a rather high degree of anisotropy. For example, dielectric materials in which each component of their electric flux density **D** depends on more than one component of the electric field **E**, are called *anisotropic dielectrics*. For such material, the permittivities and susceptibilities cannot be represented by a single value. Instead, for example, $[\bar{\varepsilon}]$ takes the form of a 3×3 tensor, which is known as the *permittivity tensor*. The electric flux density **D** and electric field intensity **E** are not parallel to each other, and they are related by the permittivity tensor $\bar{\varepsilon}$ in a form given by

$$\mathbf{D} = \bar{\varepsilon} \cdot \mathbf{E} \qquad (2\text{-}45)$$

In expanded form (2-45) can be written as

$$\begin{bmatrix} D_x \\ D_y \\ D_z \end{bmatrix} = \begin{bmatrix} \varepsilon_{xx} & \varepsilon_{xy} & \varepsilon_{xz} \\ \varepsilon_{yx} & \varepsilon_{yy} & \varepsilon_{yz} \\ \varepsilon_{zx} & \varepsilon_{zy} & \varepsilon_{zz} \end{bmatrix} \begin{bmatrix} E_x \\ E_y \\ E_z \end{bmatrix} \qquad (2\text{-}46)$$

which reduces to

$$D_x = \varepsilon_{xx} E_x + \varepsilon_{xy} E_y + \varepsilon_{xz} E_z$$

$$D_y = \varepsilon_{yx} E_x + \varepsilon_{yy} E_y + \varepsilon_{yz} E_z$$

$$D_z = \varepsilon_{zx} E_x + \varepsilon_{zy} E_y + \varepsilon_{zz} E_z \qquad (2\text{-}46a)$$

The permittivity tensor $\bar{\varepsilon}$ is written, in general, as a 3 × 3 matrix of the form

$$[\bar{\varepsilon}] = \begin{bmatrix} \varepsilon_{xx} & \varepsilon_{xy} & \varepsilon_{xz} \\ \varepsilon_{yx} & \varepsilon_{yy} & \varepsilon_{yz} \\ \varepsilon_{zx} & \varepsilon_{zy} & \varepsilon_{zz} \end{bmatrix} \quad (2\text{-}47)$$

where each entry may be complex. For anisotropic material, not all the entries of the permittivity tensor are necessarily nonzero. For some only the diagonal terms (ε_{xx}, ε_{yy}, ε_{zz}), referred to as the *principal permittivities*, are nonzero. If that is not the case, for some material a set of new axes (x', y', z') can be selected by rotation of coordinates so that the permittivity tensor referenced to this set of axes possesses only diagonal entries (principal permittivities). This process is known as *diagonalization*, and the new set of axes are referred to as the *principal coordinates*. For physically realizable materials, the entries ε_{ij} of the permittivity tensor satisfy the relation

$$\varepsilon_{ij} = \varepsilon_{ji}^* \quad (2\text{-}48)$$

Matrices whose entries satisfy (2-48) are referred to as *Hermitian*. If the material is lossless (imaginary parts of ε_{ij} are zero) and the entries of the permittivity tensor satisfy (2-48), then the permittivity tensor is also symmetrical.

2.8 a.c. VARIATIONS IN MATERIALS

It has been shown that when a material is subjected to an applied static electric field, the centroids of the positive and negative charges (representing, respectively, the positive charges found in the nucleus of an atom and the negative electrons found in the shells surrounding the nucleus) are displaced relative to each other forming a linear electric dipole. When a material is examined macroscopically, the presence of all the electric dipoles is accounted for by introducing an electric polarization vector **P** [see (2-3) and (2-10)]. Ultimately, the static permittivity ε_s [see (2-11a)] is introduced to account for the presence of **P**. A similar procedure is used to account for the orbiting and spinning of the electrons of atoms (which are represented electrically by small electric current-carrying loops) when magnetic materials are subjected to applied static magnetic fields. When the material is examined macroscopically, the presence of all the loops is accounted for by introducing the magnetic polarization (magnetization) vector **M** [see (2-15) and (2-21)]. In turn the static permeability μ_s [see (2-22(a)] is introduced to account for the presence of **M**.

When the applied fields begin to alternate in polarity, the polarization vectors **P** and **M**, and in turn the permittivities and permeabilities, are affected and they are functions of the frequency of the alternating fields. By this action of the alternating fields, there are simultaneous changes imposed upon the static conductivity σ_s [see (2-39) and (2-40)] of the material. In fact, the incremental changes in the conductivity that are attributable to the reverses in polarity of the applied fields (frequency) are responsible for the heating of materials using microwaves (for example, microwave cooking of food) [11–16].

In the sections that follow, the variations of ε, σ, and μ as a function of frequency of the applied fields will be examined.

2.8.1 Complex Permittivity

Let us assume that each atom of a material in the absence of an applied electric field (unpolarized atom) is represented by positive (representing the nucleus) and negative (representing the electrons) charges whose respective centroids coincide. The electrical and mechanical equivalents of a typical atom are shown in Figure 2-16a [6]. The large positive sphere of a mass M represents the massive nucleus whereas the small negative sphere of mass m and charge $-Q$ represents the electrons. When an electric field is applied, it is assumed that the positive charge remains stationary and the negative charge moves relative to the positive along a platform that exhibits a friction (damping) coefficient d. In addition, the two charges will be connected with a spring whose spring (tension) coefficient is s. The entire mechanical equivalent of a typical atom then consists of the classical mass–spring system moving along a platform with friction.

When an electric field is applied that is directed along the $+x$ direction, the negative charge will be displaced a distance ℓ in the negative x direction, as shown in Figure 2-16b, forming an electric dipole. If the material is not permanently polarized (as are the electrets), the atom will achieve its initial normal position when the applied electric field diminishes to zero, as shown in Figure 2-16c. Now if the applied electric field is polarized in the $-x$ direction, the negative charge will move a distance ℓ in the positive x direction, as shown in Figure 2-16d, forming again an electric dipole in the direction opposite of that in Figure 2-16b.

When a time-harmonic field of angular frequency ω is applied to an atom, the forces of the system that describe the movement of the negative charge of mass m relative to the stationary nucleus and that are opposed by damping (friction) and tension (spring) can be represented by [6, 17]

$$m\frac{d^2\ell}{dt^2} + d\frac{d\ell}{dt} + s\ell = Q\mathscr{E}(t) = QE_0 e^{j\omega t} \tag{2-49}$$

By dividing both sides of (2-49) by m, we can write it as

$$\frac{d^2\ell}{dt^2} + 2\alpha\frac{d\ell}{dt} + \omega_0^2 \ell = \frac{Q}{m}\mathscr{E}(t) = \frac{Q}{m}E_0 e^{j\omega t} \tag{2-50}$$

where
$$\alpha = \frac{d}{2m} \tag{2-50a}$$

$$\omega_0 = \sqrt{\frac{s}{m}} \tag{2-50b}$$

$$Q = \text{dipole charge} \tag{2-50c}$$

The terms on the left side of (2-49) represent, in order, the forces associated with mass times acceleration, damping times velocity, and spring times displacement. The term on the right side represents the driving force of the time-harmonic applied field (of peak value QE_0). Equations 2-49 and 2-50 are second-order differential equations that are also representative of the natural responses of *RLC* circuit systems.

For a source-free series *RLC* network, (2-50) takes the form for the current $i(t)$ of

$$\frac{d^2 i}{dt^2} + 2\alpha \frac{di}{dt} + \omega_0^2 i = 0 \tag{2-51}$$

74 ELECTRICAL PROPERTIES OF MATTER

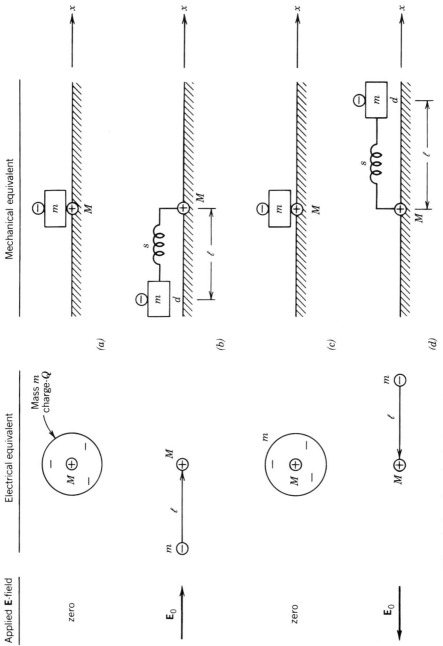

FIGURE 2-16 Electrical and mechanical equivalents of a typical atom in the absence of and under an applied electric field.

where
$$\alpha = \frac{R}{2L} \tag{2-51a}$$

$$\omega_0 = \frac{1}{\sqrt{LC}} \tag{2-51b}$$

In a similar manner, the voltage $v(t)$ for a parallel source-free RLC network can be obtained by writing (2-50) as

$$\frac{d^2v}{dt^2} + 2\alpha \frac{dv}{dt} + \omega_0^2 v = 0 \tag{2-52}$$

where
$$\alpha = \frac{1}{2RC} \tag{2-52a}$$

$$\omega_0 = \frac{1}{\sqrt{LC}} \tag{2-52b}$$

Solutions to (2-51) and (2-52) can be classified as *overdamped*, *critically damped*, or *underdamped* according to the values of the α/ω_0 ratio. That is, the solution to (2-51) for $i(t)$ or (2-52) for $v(t)$ is considered

Classification of Solution		Criterion	
overdamped	if	$\alpha > \omega_0$	(2-53a)
critically damped	if	$\alpha = \omega_0$	(2-53b)
underdamped	if	$\alpha < \omega_0$	(2-53c)

The solution to (2-49) can be obtained by first dividing both of its sides by m. Doing this reduces (2-49) to

$$\frac{d^2\ell}{dt^2} + \frac{d}{m}\frac{d\ell}{dt} + \frac{s}{m}\ell = \frac{Q}{m}E_0 e^{j\omega t} \tag{2-54}$$

The general solution to (2-54) is usually comprised of two parts: a complementary solution ℓ_c and a particular solution ℓ_p. The complementary solution represents the transient response of the system and is obtained by setting the driving force equal to zero. Since (2-54) is a quadratic, the general form of the complementary (transient) solution will be in terms of exponentials whose values vanish as $t \to \infty$. The particular solution represents the steady-state response of the system, and it is of interest here. Thus the particular (steady-state) solution of (2-54) can be written as

$$\ell_p(t) = \ell_0 e^{j\omega t} \tag{2-55}$$

where ℓ_0 is the solution of $\ell_p(t)$ when $t = 0$.
Substituting (2-55) into (2-54) leads to the solution of

$$\ell_0 = \frac{\frac{Q}{m}E_0}{(\omega_0^2 - \omega^2) + j\omega\left(\frac{d}{m}\right)} \tag{2-56}$$

where
$$\omega_0 = \sqrt{\frac{s}{m}} \qquad (2\text{-}56a)$$

Thus (2-55) can be written as

$$\ell_p(t) = \ell_0 e^{j\omega t} = \frac{\dfrac{Q}{m} E_0 e^{j\omega t}}{(\omega_0^2 - \omega^2) + j\omega\left(\dfrac{d}{m}\right)} \qquad (2\text{-}57)$$

and it represents the steady-state displacements of the negative charges (electrons) of an atom relative to those of the positive charges (nucleus).

The resonant (natural) angular frequency ω_d of the system is obtained by setting $E_0 = 0$ in (2-54). Doing this and assuming an underdamped system ($\alpha < \omega_0$ or $d < 2\sqrt{sm}$) leads to

$$\omega_d = \sqrt{\omega_0^2 - \alpha^2} = \sqrt{\frac{s}{m} - \left(\frac{d}{2m}\right)^2} \qquad (2\text{-}58)$$

For a frictionless system ($d = 0$) the resonant angular frequency ω_d reduces to

$$\omega_d|_{d=0} = \omega_0 = \sqrt{\frac{s}{m}} \qquad (2\text{-}58a)$$

Assuming that the oscillating dipoles, which represent the numerous atoms of a material, are all similar and there is no coupling between the dipoles (atoms), the macroscopic steady-state electric polarization $\mathscr{P}$ of (2-5) can be written using (2-57) as

$$\mathscr{P} = \mathscr{P}(t) = N_e Q \ell(t) = \frac{N_e \left(\dfrac{Q^2}{m}\right) E_0 e^{j\omega t}}{(\omega_0^2 - \omega^2) + j\omega\left(\dfrac{d}{m}\right)} = \frac{N_e \left(\dfrac{Q^2}{m}\right) \mathscr{E}(t)}{(\omega_0^2 - \omega^2) + j\omega\left(\dfrac{d}{m}\right)} \qquad (2\text{-}59)$$

where N_e represents the number of dipoles per unit volume. Dividing both sides of (2-59) by $\mathscr{E}(t) = E_0 e^{j\omega t}$ reduces it to

$$\frac{\mathscr{P}}{\mathscr{E}} = \frac{N_e \left(\dfrac{Q^2}{m}\right)}{(\omega_0^2 - \omega^2) + j\omega\left(\dfrac{d}{m}\right)} \qquad (2\text{-}60)$$

In turn the permittivity $\dot{\varepsilon}$ of the medium can be written, using (2-10a) and (2-11a), as

$$\dot{\varepsilon} = \varepsilon_0 + \frac{\mathscr{P}}{\mathscr{E}} = \varepsilon_0 + \frac{N_e \left(\dfrac{Q^2}{m}\right)}{(\omega_0^2 - \omega^2) + j\omega\left(\dfrac{d}{m}\right)} = \varepsilon' - j\varepsilon'' \qquad (2\text{-}61)$$

which is recognized as being complex (with real and imaginary parts, respectively, of

ε' and ε'') as denoted by the dot. Equation 2-61 is also referred to as the *dispersion equation* for the complex permittivity.

The relative complex permittivity $\dot\varepsilon_r$ of the material is obtained by dividing both sides of (2-61) by ε_0 leading to

$$\dot\varepsilon_r = \frac{\dot\varepsilon}{\varepsilon_0} = \varepsilon_r' - j\varepsilon_r'' = 1 + \frac{\dfrac{N_e Q^2}{\varepsilon_0 m}}{(\omega_0^2 - \omega^2) + j\omega\dfrac{d}{m}} \quad (2\text{-}62)$$

The real ε_r' and imaginary ε_r'' parts of (2-62) can be written, respectively, as

$$\varepsilon_r' = 1 + \frac{\dfrac{N_e Q^2}{\varepsilon_0 m}(\omega_0^2 - \omega^2)}{(\omega_0^2 - \omega^2)^2 + \left(\omega\dfrac{d}{m}\right)^2} \quad (2\text{-}63a)$$

$$\varepsilon_r'' = \frac{N_e Q^2}{\varepsilon_0 m}\left[\frac{\omega\dfrac{d}{m}}{(\omega_0^2 - \omega^2)^2 + \left(\omega\dfrac{d}{m}\right)^2}\right] \quad (2\text{-}63b)$$

For nonmagnetic material

$$\dot\varepsilon_r = \dot n^2 \quad (2\text{-}64)$$

where $\dot n$ is the complex index of refraction. For materials with no damping ($d/m = 0$), (2-63a) and (2-63b) reduce to

$$\varepsilon_r' = 1 + \frac{\dfrac{N_e Q^2}{\varepsilon_0 m}}{\omega_0^2 - \omega^2} \quad (2\text{-}65a)$$

$$\varepsilon_r'' = 0 \quad (2\text{-}65b)$$

Since the permittivity of a medium as given by (2-61) [or its relative value as given by (2-62)] is in general complex, the Maxwell–Ampere equation can be written as

$$\nabla \times \mathbf{H} = \mathbf{J}_i + \mathbf{J}_c + j\omega\dot\varepsilon\mathbf{E} = \mathbf{J}_i + \sigma_s\mathbf{E} + j\omega(\varepsilon' - j\varepsilon'')\mathbf{E}$$

$$\nabla \times \mathbf{H} = \mathbf{J}_i + (\sigma_s + \omega\varepsilon'')\mathbf{E} + j\omega\varepsilon'\mathbf{E} = \mathbf{J}_i + \sigma_e\mathbf{E} + j\omega\varepsilon'\mathbf{E} \quad (2\text{-}66)$$

where

σ_e = equivalent conductivity = $\sigma_s + \omega\varepsilon'' = \sigma_s + \sigma_a$ \quad (2-66a)

σ_a = alternating field conductivity = $\omega\varepsilon''$ \quad (2-66b)

σ_s = static field conductivity

$$= \begin{cases} -\mu_e q_{ve} & \text{for conductors} \quad (2\text{-}66c) \\ -\mu_e q_{ve} + \mu_h q_{vh} & \text{for semiconductors} \quad (2\text{-}66d) \end{cases}$$

78 ELECTRICAL PROPERTIES OF MATTER

In (2-66a) σ_e represents the total (referred to here as the equivalent) conductivity composed of the static portion σ_s and the alternating part σ_a caused by the rotation of the dipoles as they attempt to align with the applied field when its polarity is alternating. The phenomenon (rotation of dipoles) that contributes the alternating conductivity σ_a is referred to as *dielectric hysteresis*.

Many dielectric materials (such as glass and plastic) possess very low values of static σ_s conductivities and behave as good insulators. However, when they are subjected to alternating fields, they exhibit very high values of alternating field σ_a conductivities and they consume considerable energy. The heat generated by this radio frequency process is used for industrial heating processes. The best-known process is that of *microwave cooking* [11–16]. Others include selective heating of human tissue for tumor treatment [18–20] and selective heating of certain compounds in materials that possess conductivities higher than the other constituents. For example, pyrite (a form of sulfur considered to be a pollutant), which exhibits higher conductivities than the other minerals of coal, can be heated selectively. This technique has been used as a process to clean coal by extracting, through microwave heating, its sulfur content.

In (2-66), aside from the impressed (source) electric current density $\mathbf{J}_i$, there are two other components: the effective conduction electric current density $\mathbf{J}_{ce}$ and the effective displacement electric current density $\mathbf{J}_{de}$. Thus we can write the total electric current density $\mathbf{J}_t$ as

$$\mathbf{J}_t = \mathbf{J}_i + \mathbf{J}_{ce} + \mathbf{J}_{de} = \mathbf{J}_i + \sigma_e \mathbf{E} + j\omega\varepsilon'\mathbf{E} \tag{2-67}$$

where

$$\mathbf{J}_t = \text{total electric current density} \tag{2-67a}$$

$$\mathbf{J}_i = \text{impressed (source) electric current density} \tag{2-67b}$$

$$\mathbf{J}_{ce} = \text{effective electric conduction current density}$$

$$= \sigma_e \mathbf{E} = (\sigma_s + \omega\varepsilon'')\mathbf{E} \tag{2-67c}$$

$$\mathbf{J}_{de} = \text{effective displacement electric current density}$$

$$= j\omega\varepsilon'\mathbf{E} \tag{2-67d}$$

The total electric current density of (2-67) can also be written as

$$\mathbf{J}_t = \mathbf{J}_i + \sigma_e\mathbf{E} + j\omega\varepsilon'\mathbf{E} = \mathbf{J}_i + j\omega\varepsilon'\left(1 - j\frac{\sigma_e}{\omega\varepsilon'}\right)\mathbf{E} = \mathbf{J}_i + j\omega\varepsilon'(1 - j\tan\delta_e)\mathbf{E} \tag{2-68}$$

where

$$\tan\delta_e = \text{effective electric loss tangent} = \frac{\sigma_e}{\omega\varepsilon'} = \frac{\sigma_s + \sigma_a}{\omega\varepsilon'} = \frac{\sigma_s}{\omega\varepsilon'} + \frac{\sigma_a}{\omega\varepsilon'}$$

$$= \frac{\sigma_s}{\omega\varepsilon'} + \frac{\varepsilon''}{\varepsilon'} = \tan\delta_s + \tan\delta_a = \frac{\varepsilon''_e}{\varepsilon'_e} \tag{2-68a}$$

$$\tan\delta_s = \text{static electric loss tangent} = \frac{\sigma_s}{\omega\varepsilon'} \tag{2-68b}$$

$$\tan\delta_a = \text{alternating electric loss tangent} = \frac{\sigma_a}{\omega\varepsilon'} = \frac{\varepsilon''}{\varepsilon'} \tag{2-68c}$$

The manufacturer of any given material usually specifies either the conductivity (S/m) or the electric loss tangent (tan δ, dimensionless). Although it is usually not stated as such, the specified conductivity σ_e and loss tangent should represent,

TABLE 2-5
Dielectric constants and loss tangents of typical dielectric materials

Material	ε_r'	$\tan \delta$
Air	1.0006	
Alcohol (ethyl)	25	0.1
Aluminum oxide	8.8	6×10^{-4}
Bakelite	4.74	22×10^{-3}
Carbon dioxide	1.001	
Germanium	16	
Glass	4–7	1×10^{-3}
Ice	4.2	0.1
Mica	5.4	6×10^{-4}
Nylon	3.5	2×10^{-2}
Paper	3	8×10^{-3}
Plexiglas	3.45	4×10^{-2}
Polystyrene	2.56	5×10^{-5}
Porcelain	6	14×10^{-3}
Pyrex glass	4	6×10^{-4}
Quartz (fused)	3.8	7.5×10^{-4}
Rubber	2.5–3	2×10^{-3}
Silica (fused)	3.8	7.5×10^{-4}
Silicon	11.8	
Snow	3.3	0.5
Sodium chloride	5.9	1×10^{-4}
Soil (dry)	2.8	7×10^{-2}
Styrofoam	1.03	1×10^{-4}
Teflon	2.1	3×10^{-4}
Titanium dioxide	100	15×10^{-4}
Water (distilled)	80	4×10^{-2}
Water (sea)	81	4.64
Water (dehydrated)	1	0
Wood (dry)	1.5–4	1×10^{-2}

respectively, the effective conductivity and loss tangent $\tan \delta_e$ at a given frequency. Typical values of loss tangent for some materials are listed in Table 2-5.

The effective conduction $\mathbf{J}_{ce}$ and displacement $\mathbf{J}_{de}$ current densities of (2-67) can also be written as

$$\mathbf{J}_{cd} = \mathbf{J}_{ce} + \mathbf{J}_{de} = \sigma_e \mathbf{E} + j\omega\varepsilon'\mathbf{E} = j\omega\varepsilon'\left(1 - j\frac{\sigma_e}{\omega\varepsilon'}\right)\mathbf{E} = j\omega\varepsilon'(1 - j\tan\delta_e)\mathbf{E} \quad (2\text{-}69)$$

In phasor form, these can be represented as shown in Figure 2-17. It is evident that the conduction and displacement current densities are orthogonal to each other. Material can also be classified as good dielectrics or good conductors according to

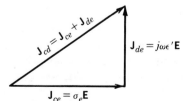

FIGURE 2-17 Phasor representation of effective conduction and displacement current densities.

the values of the $\sigma_e/\omega\varepsilon'$ ratio. That is

1. *Good Dielectrics,* $(\sigma_e/\omega\varepsilon') \ll 1$

$$\mathbf{J}_{cd} = j\omega\varepsilon'\left(1 - j\frac{\sigma_e}{\omega\varepsilon'}\right)\mathbf{E} \overset{\sigma_e/\omega\varepsilon' \ll 1}{\simeq} j\omega\varepsilon'\mathbf{E} \qquad (2\text{-}70a)$$

For these materials, the displacement current density is much greater than the conduction current density, and the total current density is approximately equal to the displacement current density.

2. *Good Conductors,* $(\sigma_e/\omega\varepsilon') \gg 1$

$$\mathbf{J}_{cd} = j\omega\varepsilon'\left(1 - j\frac{\sigma_e}{\omega\varepsilon'}\right)\mathbf{E} \overset{\sigma_e/\omega\varepsilon' \gg 1}{\simeq} \sigma_e\mathbf{E} \qquad (2\text{-}70b)$$

For these materials, the conduction current density is much greater than the displacement current density, and the total current density is approximately equal to the conduction current density.

As discussed in Section 2.2 and demonstrated in Figure 2-4 the electric polarization for dielectrics, as given by (2-3) or (2-5), can be composed of any combination involving the dipole (orientational), ionic (molecular), and electronic polarizations. As a function of frequency, the electric polarization of (2-10) can be written as

$$\mathbf{P}(\omega) = \varepsilon_0 \chi_e(\omega) \mathbf{E}_a(\omega) \qquad (2\text{-}71)$$

where in general

$$\chi_e(\omega) = \chi'_e(\omega) - j\chi''_e(\omega)$$
$$= [\chi'_{ed}(\omega) + \chi'_{ei}(\omega) + \chi'_{ee}(\omega)] - j[\chi''_{ed}(\omega) + \chi''_{ei}(\omega) + \chi''_{ee}(\omega)]$$
$$(2\text{-}71a)$$

$\chi'_{ed}(\omega)$ = dipole real electric susceptibility $\qquad (2\text{-}71b)$

$\chi'_{ei}(\omega)$ = ionic real electric susceptibility $\qquad (2\text{-}71c)$

$\chi'_{ee}(\omega)$ = electronic real electric susceptibility $\qquad (2\text{-}71d)$

$\chi''_{ed}(\omega)$ = dipole loss electric susceptibility $\qquad (2\text{-}71e)$

$\chi''_{ei}(\omega)$ = ionic loss electric susceptibility $\qquad (2\text{-}71f)$

$\chi''_{ee}(\omega)$ = electronic loss electric susceptibility $\qquad (2\text{-}71g)$

It should be noted that, in general,

$$\chi'_e(-\omega) = \chi'_e(\omega) \qquad (2\text{-}72a)$$
$$\chi''_e(-\omega) = -\chi''_e(\omega) \qquad (2\text{-}72b)$$

A general sketch of the variations of the susceptibilities as a function of frequency is given in Figure 2-18 [21, 22]. It should be stated, however, that this does not represent any one particular material, and very few materials exhibit all three mechanisms. Measurements have been made on many materials [22], with some up to 90 GHz, using microwave and millimeter wave techniques [23].

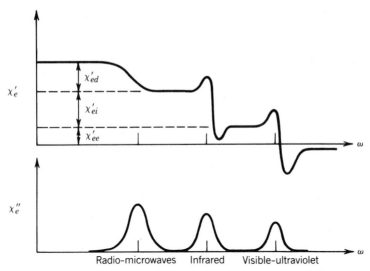

FIGURE 2-18 Electric susceptibility (real and imaginary) variations as a function of frequency for a typical dielectric.

Since the relative permittivity (dielectric constant) is related to the electric susceptibility by (2-12), we should expect similar variations of the dielectric constant as a function of frequency. To demonstrate that, we have plotted in Figure 2-19 as a function of frequency ($0 \leq \omega \leq 10$) the relative complex permittivity (real and imaginary parts and magnitude) of (2-62) or (2-63a) and (2-63b) (assuming $N_e Q^2 / \varepsilon_0 m = 1$ and $d/m = 1$) for

$$\frac{\alpha}{\omega_0} = \frac{d}{2m}\sqrt{\frac{m}{s}} = \frac{1}{5} \text{ (underdamped with } \omega_0 = 2.5 \text{ and } \omega_d = \sqrt{6} = 2.449\text{)}$$

$$\frac{\alpha}{\omega_0} = \frac{d}{2m}\sqrt{\frac{m}{s}} = \frac{1}{10} \text{ (underdamped with } \omega_0 = 5 \text{ and } \omega_d = \sqrt{99}/2 = 4.975\text{)}$$

It is observed that the values of ε_r'' peak at the resonant frequencies, which indicates that the medium attains its most lossy state at the resonant frequency. Multiple variations of this type would also be observed in a given curve at other frequencies if the medium possesses multiple resonant frequencies. For frequencies not near one of the resonant frequencies, the curve representing the variations of $|\dot{\varepsilon}_r|$ exhibits a positive slope and is referred to as *normal* dispersion (because it occurs most commonly). Very near the resonant frequencies there is a small range of frequencies for which the variations of $|\dot{\varepsilon}_r|$ exhibit a negative slope that is referred to as *anomalous* (abnormal) dispersion. Although there is nothing abnormal about this type of dispersion, the name was given because it seemed unusual when it was first observed.

When (2-57) and (2-59) to (2-63b) were derived, it was assumed that the medium possessed only one resonant (natural) frequency presented by one type of harmonic oscillator. In general, however, there are several natural frequencies associated with a particular atom. These can be accounted for in our dispersion equations for ε_r' and ε_r'' by introducing several different kinds of oscillators with no coupling between them. This type of modeling allows the contributions from each

82 ELECTRICAL PROPERTIES OF MATTER

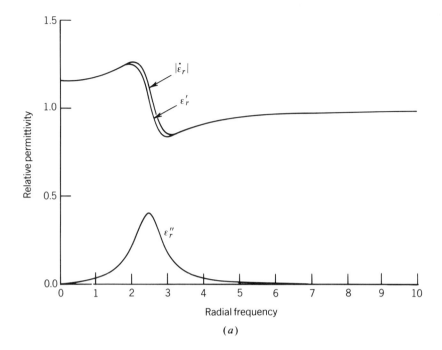

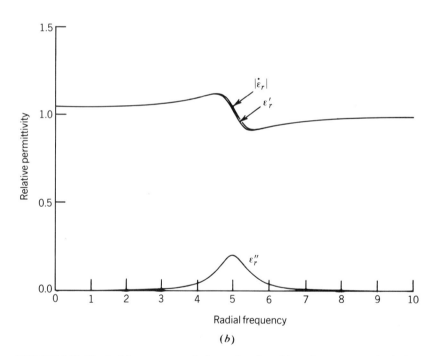

FIGURE 2-19 Typical frequency variations of real and imaginary parts of relative permittivity of dielectrics. (a) $N_e Q^2 / \varepsilon_0 m = 1$, $d/m = 1$, $\alpha/\omega_0 = 1/5$, $\omega_0 = 2.5$. (b) $N_e Q^2 / \varepsilon_0 m = 1$, $d/m = 1$, $\alpha/\omega_0 = 1/10$, $\omega_0 = 5$.

oscillator to be accounted for by a simple addition. Thus for a medium with p natural frequencies (represented by p independent oscillators), we can write (2-60) to (2-63b) as

$$\frac{\mathcal{P}}{\mathcal{E}} = \sum_{s=1}^{p} \frac{N_e \dfrac{Q^2}{m}}{\left(\omega_s^2 - \omega^2\right) + j\dfrac{\omega d}{m}} \tag{2-73a}$$

$$\dot{\varepsilon} = \varepsilon' - j\varepsilon'' = \varepsilon_0 + \sum_{s=1}^{p} \frac{N_e \dfrac{Q^2}{m}}{\left(\omega_s^2 - \omega^2\right) + j\dfrac{\omega d}{m}} \tag{2-73b}$$

$$\dot{\varepsilon}_r = \varepsilon'_r - j\varepsilon''_r = 1 + \sum_{s=1}^{p} \frac{\dfrac{N_e Q^2}{\varepsilon_0 m}}{\left(\omega_s^2 - \omega^2\right) + j\dfrac{\omega d}{m}} \tag{2-73c}$$

$$\varepsilon'_r = 1 + \sum_{s=1}^{p} \frac{\dfrac{N_e Q^2}{\varepsilon_0 m}\left(\omega_s^2 - \omega^2\right)}{\left(\omega_s^2 - \omega^2\right)^2 + \left(\dfrac{\omega d}{m}\right)^2} \tag{2-73d}$$

$$\varepsilon''_r = \sum_{s=1}^{p} \frac{N_e Q^2}{\varepsilon_0 m} \frac{\omega \dfrac{d}{m}}{\left(\omega_s^2 - \omega^2\right)^2 + \left(\dfrac{\omega d}{m}\right)^2} \tag{2-73e}$$

Often the question is asked whether there are any relations between the real and imaginary parts of the complex permittivity. The answer to that is yes. Known as the Kramers–Kronig [24–26] relations, they are given by

$$\varepsilon'_r(\omega) = 1 + \frac{2}{\pi} \int_0^\infty \frac{\omega' \varepsilon''_r(\omega')}{(\omega')^2 - \omega^2} \, d\omega' \tag{2-74a}$$

$$\varepsilon''_r(\omega) = \frac{2\omega}{\pi} \int_0^\infty \frac{1 - \varepsilon'_r(\omega')}{(\omega')^2 - \omega^2} \, d\omega' \tag{2-74b}$$

and they are very similar to the frequency relations between resistance and reactance in circuit theory [26].

In addition to the Kramers–Kronig relations of (2-74a) and (2-74b), there are simple relations that allow the calculation of the real and imaginary parts of the complex relative permittivity for many materials as a function of frequency provided that the real part of the complex permittivity is known at zero frequency (denoted by ε'_{rs}) and at very large (ideally infinity) frequency (denoted by $\varepsilon'_{r\infty}$). These relations are obtained from the well-known *Debye equation* [17, 21, 22] for the complex dielectric constant, which states that

$$\dot{\varepsilon}_r(\omega) = \varepsilon'_r(\omega) - j\varepsilon''_r(\omega) = \varepsilon'_{r\infty} + \frac{\varepsilon'_{rs} - \varepsilon'_{r\infty}}{1 + j\omega\tau_e} \tag{2-75}$$

84 ELECTRICAL PROPERTIES OF MATTER

where τ_e is a *new relaxation time constant* related to *original relaxation time constant* τ by

$$\tau_e = \tau \frac{\varepsilon'_{rs} + 2}{\varepsilon'_{r\infty} + 2} \tag{2-75a}$$

The Debye equation of (2-75) is derived using the *Clausius–Mosotti equation* [21, 22, 27]. The real and imaginary parts of (2-75) can be written as

$$\varepsilon'_r(\omega) = \varepsilon'_{r\infty} + \frac{\varepsilon'_{rs} - \varepsilon'_{r\infty}}{1 + (\omega\tau_e)^2} \tag{2-76a}$$

$$\varepsilon''_r(\omega) = \frac{(\varepsilon'_{rs} - \varepsilon'_{r\infty})\omega\tau_e}{1 + (\omega\tau_e)^2} \tag{2-76b}$$

which can be found at any frequency provided ε'_{rs}, $\varepsilon'_{r\infty}$, and τ are known. The relations of (2-76a) and (2-76b) can be used to estimate the real and imaginary parts of the complex relative permittivity (complex dielectric constant) for many gases, liquids, and solids.

2.8.2 Complex Permeability

As discussed in Section 2.3, the permeability of most dielectric material, including diamagnetic, paramagnetic, and antiferromagnetic material, is nearly the same as that of free space μ_0 ($\mu_0 = 4\pi \times 10^{-7}$ H/m). Ferromagnetic and ferrimagnetic materials exhibit much higher permeability than free space, as is demonstrated by the data of Table 2-2. These classes of material are also magnetically lossy, and their magnetic losses are accounted for by introducing a complex permeability.

In general then, we can write the Maxwell–Faraday equation as

$$\nabla \times \mathbf{E} = -\mathbf{M}_i - j\omega\mu\mathbf{H} = -\mathbf{M}_i - j\omega(\mu' - j\mu'')\mathbf{H}$$
$$= -\mathbf{M}_i - \omega\mu''\mathbf{H} - j\omega\mu'\mathbf{H} = -\mathbf{M}_t \tag{2-77}$$

where

$$\mathbf{M}_t = \mathbf{M}_i + \omega\mu''\mathbf{H} + j\omega\mu'\mathbf{H} \tag{2-77a}$$

$$\mathbf{M}_t = \text{total magnetic current density} \tag{2-77b}$$

$$\mathbf{M}_i = \text{impressed (source) magnetic current density} \tag{2-77c}$$

$$\mathbf{M}_c = \text{conduction magnetic current density} = \omega\mu''\mathbf{H} \tag{2-77d}$$

$$\mathbf{M}_d = \text{displacement magnetic current density} = j\omega\mu'\mathbf{H} \tag{2-77e}$$

Another form of (2-77a) is to write it as

$$\mathbf{M}_t = \mathbf{M}_i + j\omega\mu'\left(1 - j\frac{\mu''}{\mu'}\right)\mathbf{H} = \mathbf{M}_i + j\omega\mu'(1 - j\tan\delta_m)\mathbf{H} \tag{2-78}$$

where

$$\tan\delta_m = \text{alternating magnetic loss tangent} = \frac{\mu''}{\mu'} \tag{2-78a}$$

In addition to being complex, the permeability of ferromagnetic and ferrimagnetic material is often a function of frequency. Thus it should, in general, be written as

$$\dot{\mu} = \mu'(\omega) - j\mu''(\omega) \tag{2-79}$$

or

$$\dot{\mu}_r = \frac{\dot{\mu}}{\mu_0} = \mu'_r(\omega) - j\mu''_r(\omega) \tag{2-79a}$$

Most ferromagnetic materials possess very high relative permeabilities (on the order of several thousand) and good conductivities such that there is a minimum interaction between these materials and the electromagnetic waves propagating through them. As such, they will not be discussed further here. There is, however, a class of ferrimagnetic material, referred to as *ferrites*, that finds wide applications in the design of nonreciprocal microwave components (such as isolators, hybrids, gyrators, phase shifters, etc.). Ferrites become attractive for these applications because at microwave frequencies they exhibit strong magnetic effects which result in anisotropic properties and large resistances (good insulators). These resistances limit the current induced in them and in turn result in lower ohmic losses. Because of the appeal of ferrites to the microwave circuit design, their magnetic properties will be discussed further in the section that follows.

2.8.3 Ferrites

Ferrites are a class of solid ceramic materials that have crystal structures formed by sintering at high temperatures (typically 1000–1500°C) stoichiometric mixtures of certain metal oxides (such as oxygen and iron, and cadmium, lithium, magnesium, nickel, or zinc, or some combination of them). These materials are ferrimagnetic, and they are considered to be good insulators with high permeabilities, dielectric constants between 10 to 15 or greater, and specific resistivities as much as 10^{14} greater than those of metals. In addition, they possess properties that allow strong interaction between the magnetic dipole moment associated with the electron spin, as discussed in Section 2.3, and the microwave electromagnetic fields [28–30]. In contrast to ferromagnetic materials, ferrites have their magnetic ions distributed over at least two interpenetrating sublattices. Within each sublattice all magnetic moments are aligned, but the sublattices are oppositely directed.

As a result of these interactions, ferrites exhibit nonreciprocal properties such as different phase constants and phase velocities for right- and left-hand circularly polarized waves, transmission coefficients that are functions of direction of travel, and permeabilities that are represented by tensors (in the form of a matrix) rather than by a single scalar. These characteristics become important in the design of nonreciprocal microwave devices [31–33]. Although all ferrimagnetic materials possess these properties, it is only in ferrites that they are pronounced and significant. The properties of ferrites will be discussed here by examining the propagation of microwave electromagnetic waves in an unbounded ferrite material.

There are two possible models that can be used to understand the technical properties of magnetic material: the *phenomenological* model and the *atomic* model [30]. For the purposes of this book, the phenomenological model is sufficient to examine the properties of magnetic oxides. As discussed in Section 2.3, the magnetic

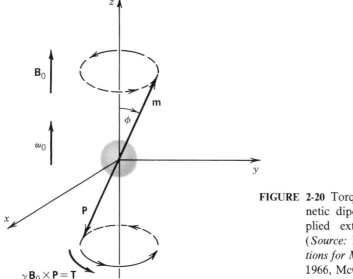

FIGURE 2-20 Torque on a single magnetic dipole caused by an applied external magnetic field (*Source:* R. E. Collin, *Foundations for Microwave Engineering*, 1966, McGraw-Hill Book Co.).

material is replaced by an array of magnetic dipoles that are maintained in a permanent and rigid alignment as shown in Figure 2-8a. When a magnetic field is applied, as shown in Figure 2-8b, the magnetic moments of the dipoles can turn freely in space as long as they turn together. Much of the discussion of this section follows that of [30] and [33].

Under an applied magnetic field each single magnetic dipole rotates with a precession frequency that is referred to as the *Larmor precession frequency*. The precession frequency is altered when one or more dipoles are introduced. The dipoles in the array interact with each other and attempt to achieve an alignment that will minimize the interaction energy. The change in precession frequency is equivalent to introducing an additional demagnetizing field. When many dipole arrays are subjected to d.c., rf, or demagnetizing fields, magnetic resonance is introduced. This is a phenomenon that is of fundamental interest to the design of microwave nonreciprocal components. The discussion here will be that of the phenomenological model.

The magnetic dipole moment **m** of a single magnetic dipole of Figure 2-7a or 2-7b is given by (2-13). When an external magnetic field is applied, as shown in Figure 2-20 for a single dipole, exerted on the dipole is a torque **T** of

$$\mathbf{T} = \mu_0 \mathbf{m} \times \mathbf{H}_0 = \mathbf{m} \times \mathbf{B}_0 \qquad (2\text{-}80)$$

where $\mathbf{m} = \hat{n} I \, ds$ = magnetic dipole moment of single dipole
$\mathbf{H}_0$ = applied magnetic field
$\mathbf{B}_0$ = applied magnetic flux density

The torque will cause the dipole to precess about the z axis, which is parallel to $\mathbf{B}_0$, as shown in Figure 2-20.

The interaction energy W_m between the dipole and the applied field can be expressed as

$$W_m = -\mu_0 m H_0 \cos \phi \qquad (2\text{-}81)$$

$$T = -\frac{\partial W_m}{\partial \phi} \qquad (2\text{-}81a)$$

where ϕ is the angle between the applied magnetic field and the magnetic dipole axis. It is observed that the energy is minimum ($W_m = -\mu_0 m H_0$) when $\phi = 0$ whereas when $\phi = \pi/2$, T is zero and the dipole is in unstable equilibrium.

When electrons of a physically realizable dipole are moving, they create a current whose motion is associated with a circulation of mass (angular momentum) as well as charge. Therefore the magnetic dipole moment of a single electron of charge e, which is moving with a velocity v in a circle of radius a, can be also be expressed as

$$m = I\,ds = \frac{ev}{2\pi a}(\pi a^2) = \frac{1}{2}eva \tag{2-82}$$

and the angular momentum P can be written as

$$P = m_e v a \tag{2-83}$$

where m_e is the mass of the electron. The ratio of the magnetic moment [as given by (2-82)] to the angular momentum [as given by (2-83)] is referred to as the *gyromagnetic ratio* γ, and it is equal to

$$\gamma = \frac{m}{P} = \frac{e}{2m_e} \Rightarrow m = \gamma P \tag{2-84}$$

which is negative because of the negative electron charge e. This makes the angular momentum P of the electron antiparallel to the magnetic dipole moment **m**, as shown in Figure 2-20.

To obtain the equation of motion we set the rate of change (with time) of the angular momentum equal to the torque, that is,

$$\frac{d\mathbf{P}}{dt} = \mathbf{T} = \mu_0 \mathbf{m} \times \mathbf{H}_0 = -\mu_0|\gamma|\mathbf{P} \times \mathbf{H}_0 = -\mathbf{P} \times \boldsymbol{\omega}_0 = \boldsymbol{\omega}_0 \times \mathbf{P} \tag{2-85}$$

or

$$\mu_0|\gamma|PH_0 \sin\phi = \omega_0 P \sin\phi = -\mu_0 m H_0 \sin\phi \tag{2-85a}$$

In (2-85) and (2-85a) $\boldsymbol{\omega}_0$ is the vector precession angular velocity which is directed along $\mathbf{H}_0$, as shown in Figure 2-20. For the free precession of a *single* dipole, the angular velocity ω_0 is referred to as the Larmor precession frequency, which is given by

$$\omega_0 = |\gamma|\mu_0 H_0 = |\gamma| B_0 \tag{2-86}$$

and it is independent of the angle ϕ.

Let us assume that on the static applied field $\mathbf{B}_0$ a small a.c. magnetic field $\mathbf{B}_1$ is superimposed. This additional applied field will impose a forced precession on the magnetic dipole. To examine the effects of the forced precession, let us assume that the a.c. applied magnetic field $\mathbf{B}_1^\pm$ is circularly polarized, either right hand (CW) $\mathbf{B}_1^+$ or left hand (CCW) $\mathbf{B}_1^-$, and it is directed perpendicular to the z axis. As will be shown in Section 4.4.2 these fields can be written as

$$\mathbf{B}_1^+ = (\hat{a}_x - j\hat{a}_y) B_1^+ e^{-j\beta z} \quad \text{right-hand (CW)} \tag{2-86a}$$

$$\mathbf{B}_1^- = (\hat{a}_x + j\hat{a}_y) B_1^- e^{-j\beta z} \quad \text{left-hand (CCW)} \tag{2-86b}$$

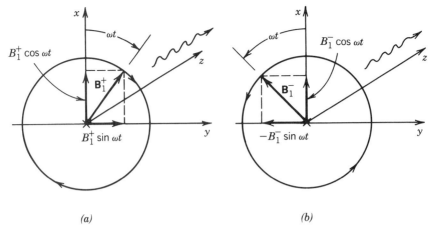

FIGURE 2-21 Rotation of magnetic field, as a function of time, for (a) clockwise and (b) counterclockwise polarizations (*Source:* R. E. Collin, *Foundations for Microwave Engineering*, 1966, McGraw-Hill Book Co.).

The corresponding instantaneous fields obtained using (1-61d) rotate, respectively, in the clockwise and counterclockwise directions when viewed from the rear as they travel in the $+z$ direction. This is demonstrated in Figure 2-21. When each of the a.c. signals are superimposed upon the static field $\mathbf{B}_0$ directed along the z axis, the resultant $\mathbf{B}_t^\pm$ field will be at angle $\theta^\pm$ (measured from the z axis) given by

$$\theta^\pm = \tan^{-1}\left(\frac{B_1^\pm}{B_0}\right) \qquad (2\text{-}87)$$

as shown in Figure 2-22a and 2-22b. The resultant magnetic field $\mathbf{B}_t^\pm$ will rotate about the z axis at a rate of ω^+ in the clockwise direction for $\mathbf{B}_t^+$ and ω^- in the counterclockwise direction for $\mathbf{B}_t^-$, as shown in Figure 2-22. The magnetic dipole will be forced to precess at the same rate about the z axis when steady-state conditions prevail.

For the torque to impose a clockwise precession on $\mathbf{B}_t^+$ and a counterclockwise precession on $\mathbf{B}_t^-$, the precession angle ϕ^+ must be larger than θ^+ (as shown in Figure 2-22a) and ϕ^- must be smaller than θ^- (as shown in Figure 2-22b). Therefore for each case (2-85), the equation of motion, can be written as

$$\frac{d\mathbf{P}^+}{dt} = \mathbf{T}^+ = \mathbf{m}^+ \times \mathbf{B}_t^+ = -|\gamma|\mathbf{P}^+ \times \mathbf{B}_t^+ = \omega^+ \hat{a}_z \times \mathbf{P}^+ \qquad (2\text{-}88a)$$

$$\frac{d\mathbf{P}^-}{dt} = \mathbf{T}^- = \mathbf{m}^- \times \mathbf{B}_t^- = -|\gamma|\mathbf{P}^- \times \mathbf{B}_t^- = -\omega^- \hat{a}_z \times \mathbf{P}^- \qquad (2\text{-}88b)$$

or

$$-|\gamma|P^+ B_t^+ \sin(\phi^+ - \theta^+) = \omega^+ P^+ \sin\phi^+ \qquad (2\text{-}89a)$$

$$-|\gamma|P^- B_t^- \sin(\theta^- - \phi^-) = -\omega^- P^- \sin\phi^- \qquad (2\text{-}89b)$$

Expanding (2-89a) and (2-89b) leads to

$$-|\gamma|\left[(B_t^+ \sin\phi^+)\cos\theta^+ - (B_t^+ \cos\phi^+)\sin\theta^+\right] = \omega^+ \sin\phi^+ \qquad (2\text{-}90a)$$

$$-|\gamma|\left[(B_t^- \sin\theta^-)\cos\phi^- - (B_t^- \cos\theta^-)\sin\phi^-\right] = -\omega^- \sin\phi^- \qquad (2\text{-}90b)$$

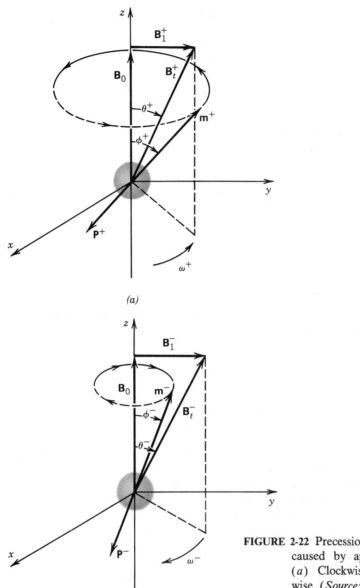

FIGURE 2-22 Precession of spinning electron caused by applied magnetic field. (*a*) Clockwise. (*b*) Counterclockwise. (*Source:* R. E. Collin, *Foundations for Microwave Engineering*, 1966, McGraw-Hill Book Co.)

Since

$$B_t^+ \sin \theta^+ = B_1^+ \tag{2-91a}$$

$$B_t^+ \cos \theta^+ = B_0 \tag{2-91b}$$

$$B_t^- \sin \theta^- = B_1^- \tag{2-91c}$$

$$B_t^- \cos \theta^- = B_0 \tag{2-91d}$$

then (2-90a) and (2-90b) can be reduced, respectively, to

$$\tan \phi^+ = \frac{|\gamma|B_1^+}{|\gamma|B_0 - \omega^+} = \frac{|\gamma|B_1^+}{\omega_0 - \omega^+} \quad (2\text{-}92\text{a})$$

$$\tan \phi^- = \frac{|\gamma|B_1^-}{|\gamma|B_0 + \omega^-} = \frac{|\gamma|B_1^-}{\omega_0 + \omega^-} \quad (2\text{-}92\text{b})$$

According to Figure 2-22 the components $m_t^\pm$ of $\mathbf{m}^\pm$ that rotate in synchronism with their respective $\mathbf{B}_1^\pm$, and $m_z^\pm$ that are directed along the z axis, are given, respectively, by

$$m_t^\pm = m^\pm \sin \phi^\pm = m^\pm \cos \phi^\pm \frac{\sin \phi^\pm}{\cos \phi^\pm} = m^\pm \cos \phi^\pm \tan \phi^\pm = m_0^\pm \tan \phi^\pm \quad (2\text{-}93\text{a})$$

$$m_z^\pm = m^\pm \cos \phi^\pm = m_0^\pm \quad (2\text{-}93\text{b})$$

where

$$m_0^\pm = m^\pm \cos \phi^\pm \quad (2\text{-}93\text{c})$$

Using (2-92a) and (2-92b) we can write the components of $\mathbf{m}^\pm$ that rotate in synchronism with $\mathbf{B}_1^\pm$ as

$$m_t^+ = m_0^+ \tan \phi^+ = \frac{m_0^+ |\gamma| B_1^+}{\omega_0 - \omega^+} \quad (2\text{-}94\text{a})$$

$$m_t^- = m_0^- \tan \phi^- = \frac{m_0^- |\gamma| B_1^-}{\omega_0 + \omega^-} \quad (2\text{-}94\text{b})$$

In the previous discussion we considered the essential properties of single spinning electrons in a magnetic field that is a superposition of a static magnetic field along the z axis and an a.c. circularly polarized field perpendicular to it. Let us now examine macroscopically the properties of N orbiting electrons per unit volume whose density is uniformly and continuously distributed. Doing this we can represent the total magnetization $\mathbf{M}$ of all N electrons as the product of N times that of a single electron ($\mathbf{M} = N\mathbf{m}$), as given by (2-17). In addition the magnetic flux density $\mathbf{M}$ will be related to the magnetic field intensity $\mathbf{H}$ and magnetization vector $\mathbf{M}$ by (2-19). Thus we can write (2-19), using (2-94a) and (2-94b) for the magnetization of the N orbiting electrons superimposed with the circularly polarized a.c. signal of $\mathbf{B}_1^\pm$, as

$$\mathbf{B}^+ = \mu_0(\mathbf{H}_1^+ + \mathbf{M}_1^+) = \mu_0(\mathbf{H}_1^+ + N\mathbf{m}_t^+) = \mu_0\left(\mathbf{H}_1^+ + \frac{Nm_0^+|\gamma|\mathbf{B}_1^+}{\omega_0 - \omega^+}\right)$$

$$= \mu_0\left(1 + \frac{Nm_0^+|\gamma|\mu_0}{\omega_0 - \omega^+}\right)\mathbf{H}_1^+ = \mu_0\left(1 + \frac{\mu_0|\gamma|M_0^+}{\omega_0 - \omega^+}\right)\mathbf{H}_1^+ = \mu_e^+ \mathbf{H}_1^+ \quad (2\text{-}95\text{a})$$

$$\mathbf{B}^- = \mu_0(\mathbf{H}_1^- + \mathbf{M}_1^-) = \mu_0(\mathbf{H}_1^- + N\mathbf{m}_t^-) = \mu_0\left(\mathbf{H}_1^- + \frac{Nm_0^-|\gamma|\mathbf{B}_1^-}{\omega_0 + \omega^-}\right)$$

$$= \mu_0\left(1 + \frac{Nm_0^-|\gamma|\mu_0}{\omega_0 + \omega^-}\right)\mathbf{H}_1^- = \mu_0\left(1 + \frac{\mu_0|\gamma|M_0^-}{\omega_0 + \omega^-}\right)\mathbf{H}_1^- = \mu_e^- \mathbf{H}_1^- \quad (2\text{-}95\text{b})$$

where

$$M_0^+ = Nm_0^+ \tag{2-95c}$$

$$M_0^- = Nm_0^- \tag{2-95d}$$

$$\boxed{\mu_e^+ = \mu_0\left(1 + \frac{\mu_0|\gamma|M_0^+}{\omega_0 - \omega^+}\right)} \tag{2-95e}$$

$$\boxed{\mu_e^- = \mu_0\left(1 + \frac{\mu_0|\gamma|M_0^-}{\omega_0 + \omega^-}\right)} \tag{2-95f}$$

In (2-95e) and (2-95f) μ_e^+ and μ_e^- represent, respectively, the effective permeabilities for clockwise and counterclockwise circularly polarized waves. It is apparent that the two are not equal, which is a fundamental property utilized in the design of nonreciprocal microwave devices.

If the static magnetic field $\mathbf{B}_0$ is much larger than the superimposed a.c. magnetic field $\mathbf{B}_1^\pm$ ($B_0 \gg B_1^\pm$) so that the magnetization of the ferrite material is saturated by the static field, then all the spinning dipoles are tightly coupled and the entire material acts as a large single magnetic dipole. In that case the magnetization vector $\mathbf{M}^\pm$ for the positive (CW) and negative (CCW) circularly polarized fields superimposed on the static field can be approximated by

$$\mathbf{M}^\pm = N\mathbf{m}^\pm \simeq \mathbf{M}_s \simeq \mathbf{M}_0 \tag{2-96}$$

where $\mathbf{M}_s$ is the magnetization vector caused by the static field when no time-varying magnetic field is applied. For those cases the effective permeabilities can be approximated by

$$\boxed{\mu_e^+ \simeq \mu_0\left(1 + \frac{\mu_0|\gamma|M_s}{\omega_0 - \omega^+}\right)} \tag{2-97a}$$

$$\boxed{\mu_e^- = \mu_0\left(1 + \frac{\mu_0|\gamma|M_s}{\omega_0 + \omega^-}\right)} \tag{2-97b}$$

which are not equal. Equations 2-97a and 2-97b are good approximations when the a.c. signals are small compared to the applied static field.

It can be shown (see Chapter 4) that a time-harmonic transverse electromagnetic (TEM) wave can be decomposed into a combination of clockwise and counterclockwise circularly polarized waves. Therefore the implications of (2-97a) and (2-97b) are that when a TEM wave travels through a ferrite material the clockwise circularly polarized portion of the wave will experience the permeability of (2-97a) while the counterclockwise wave will experience that of (2-97b). Since the permeability of a material influences the phase velocity and phase constant (see Chapter 4), the phases associated with (2-97a) and (2-97b) will be different. This is one of the fundamental features of ferrites that is utilized for the design of microwave nonreciprocal devices.

When a unbounded ferrite material is subjected to a static magnetic field $\mathbf{B}_0$ directed along the z axis of

$$\mathbf{B}_0 = \hat{a}_z B_0 = \hat{a}_z \mu_0 H_0 \tag{2-98a}$$

and a time-harmonic magnetic field $\mathscr{B}$ of

$$\mathscr{B} = \mu_0 \mathscr{H} \tag{2-98b}$$

each will induce a magnetization per unit volume vector of $\mathbf{M}_s$ and $\mathscr{M}$, respectively. The script is used to indicate time-varying components. Under these conditions, the equation of motion can be written as

$$\frac{d(\mathbf{M}_s + \mathscr{M})}{dt} = \frac{d\mathscr{M}}{dt} = -|\gamma|[(\mathbf{M}_s + \mathscr{M}) \times (\mathbf{B}_0 + \mathscr{B})] \tag{2-99}$$

or in expanded form as

$$\frac{d\mathscr{M}}{dt} = -|\gamma|\mu_0[(\mathbf{M}_s + \mathscr{M}) \times (\mathbf{H}_0 + \mathscr{H})]$$

$$\frac{d\mathscr{M}}{dt} = -|\gamma|\mu_0(\mathbf{M}_s \times \mathbf{H}_0 + \mathbf{M}_s \times \mathscr{H} + \mathscr{M} \times \mathbf{H}_0 + \mathscr{M} \times \mathscr{H}) \tag{2-99a}$$

If the time-harmonic field $\mathscr{B}$ is small such that

$$|\mathscr{M}| \ll |\mathbf{M}_s| \tag{2-100a}$$

$$|\mathscr{H}| \ll |\mathbf{H}_0| \tag{2-100b}$$

and since the applied magnetic field $\mathbf{B}_0$ is in the same direction as the static saturation magnetization vector $\mathbf{M}_s$, or

$$\mathbf{M}_s \times \mathbf{H}_0 = 0 \tag{2-101}$$

then (2-99a) can be approximated by

$$\frac{d\mathscr{M}}{dt} \simeq -|\gamma|\mu_0(\mathbf{M}_s \times \mathscr{H} + \mathscr{M} \times \mathbf{H}_0) \tag{2-102}$$

If each of the time-harmonic components is written in the form described by (1-61a) through (1-61d), then (2-102) ultimately reduces, using (2-86), to

$$j\omega \mathbf{M} \simeq -|\gamma|\mu_0(\mathbf{M}_s \times \mathbf{H} + \mathbf{M} \times \mathbf{H}_0)$$

$$j\omega \mathbf{M} + |\gamma|\mu_0 \mathbf{M} \times \mathbf{H}_0 \simeq -|\gamma|\mu_0 \mathbf{M}_s \times \mathbf{H}$$

$$j\omega \mathbf{M} + |\gamma|\mathbf{M} \times \mathbf{B}_0 \simeq -|\gamma|\mu_0 \mathbf{M}_s \times \mathbf{H}$$

$$j\omega \mathbf{M} + \mathbf{M} \times (|\gamma|\mathbf{B}_0) \simeq -|\gamma|\mu_0 \mathbf{M}_s \times \mathbf{H}$$

$$j\omega \mathbf{M} + \omega_0 \mathbf{M} \times \hat{a}_z \simeq -|\gamma|\mu_0 \mathbf{M}_s \times \mathbf{H} \tag{2-103}$$

Assuming $\mathbf{M}_s$ has only a z component whereas $\mathbf{H}$ has both x and y components, expanding (2-103) leads to

$$j\omega M_x + \omega_0 M_y \simeq |\gamma|\mu_0 M_s H_y \tag{2-104a}$$

$$-\omega_0 M_x + j\omega M_y \simeq -|\gamma|\mu_0 M_s H_x \tag{2-104b}$$

$$j\omega M_z \simeq 0 \tag{2-104c}$$

Solving (2-104a) through (2-104c) for M_x, M_y, and M_z leads to

$$M_x = \frac{\omega_0|\gamma|\mu_0 M_s H_x + j\omega|\gamma|\mu_0 M_s H_y}{\omega_0^2 - \omega^2} \qquad (2\text{-}105a)$$

$$M_y = \frac{\omega_0|\gamma|\mu_0 M_s H_y - j\omega|\gamma|\mu_0 M_s H_x}{\omega_0^2 - \omega^2} \qquad (2\text{-}105b)$$

$$M_z = 0 \qquad (2\text{-}105c)$$

By introducing the magnetic susceptibility χ, we can write (2-105a) through (2-105c) using the forms of (2-21) and (2-22) as

$$[M] = [\chi][H] \qquad (2\text{-}106)$$

or

$$\begin{bmatrix} M_x \\ M_y \\ M_z \end{bmatrix} = \begin{bmatrix} \chi_{xx} & \chi_{xy} & 0 \\ \chi_{yx} & \chi_{yy} & 0 \\ 0 & 0 & 0 \end{bmatrix} \begin{bmatrix} H_x \\ H_y \\ H_z \end{bmatrix} \qquad (2\text{-}106a)$$

$$[B] = \mu_0[[I] + [\chi]][H] \qquad (2\text{-}107)$$

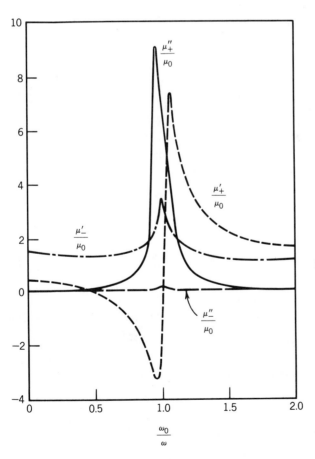

FIGURE 2-23 Frequency variations of real and imaginary parts of complex permeability for circularly polarized waves in a ferrite ($\omega = 20\pi$ GHz, $\omega_m = 11.2\pi$ GHz, $\alpha = 0.05$) (*Source*: R. E. Collin, *Foundations for Microwave Engineering*, 1966, McGraw-Hill Book Co).

or

$$\begin{bmatrix} B_x \\ B_y \\ B_z \end{bmatrix} = \mu_0 \begin{bmatrix} 1 + \chi_{xx} & \chi_{xy} & 0 \\ \chi_{yx} & 1 + \chi_{yy} & 0 \\ 0 & 0 & 1 \end{bmatrix} \begin{bmatrix} H_x \\ H_y \\ H_z \end{bmatrix} \quad (2\text{-}107a)$$

where

$$\chi_{xx} = \chi_{yy} = \frac{\omega_0 |\gamma| \mu_0 M_s}{\omega_0^2 - \omega^2} \quad (2\text{-}107b)$$

$$\chi_{xy} = -\chi_{yx} = j\frac{\omega |\gamma| \mu_0 M_s}{\omega_0^2 - \omega^2} \quad (2\text{-}107c)$$

In (2-106) through (2-107c) χ_{xx}, χ_{yy}, χ_{xy}, and χ_{yx} represent the entries of the susceptibility tensor $\bar{\chi}$ for the ferrite material and $[I]$ is the unit matrix. Equation 2-107a can also be written in a more general form as

$$\mathbf{B} = \bar{\mu} \cdot \mathbf{H} \quad (2\text{-}108)$$

where $\bar{\mu}$ is the permeability tensor written, in general, as a 3 × 3 matrix of the form

$$[\bar{\mu}] = \mu_0 \begin{bmatrix} 1 + \chi_{xx} & \chi_{xy} & 0 \\ \chi_{yx} & 1 + \chi_{yy} & 0 \\ 0 & 0 & 1 \end{bmatrix} \quad (2\text{-}108a)$$

which is a more general form of (2-22a).

Practical ferrite materials also contain magnetic losses. Therefore the permeability of the material will have both real and imaginary parts, as given by (2-79) or (2-79a). A phenomenological model used to derive the variations as a function of frequency of the real $\mu'(\omega)$ and imaginary $\mu''(\omega)$ parts of both (2-95e) and (2-95f) when losses are included is somewhat complex and beyond the treatment pressented here for ferrites. However, the development of this can be found in [33] and [34]. A typical plot as a function of ω_0/ω is shown in Figure 2-23 where $\omega_m = \mu_0|\gamma|M_s$. A resonance phenomenon is indicated when $\omega_0/\omega = 1$.

REFERENCES

1. R. R. Wright and H. R. Skutt, *Electronics: Circuits and Devices*, Ronald Press, New York, 1965.
2. S. V. Marshall and G. G. Skitek, *Electromagnetic Concepts and Applications*, Second Edition, Prentice-Hall, Englewood Cliffs, N.J., 1987.
3. R. M. Rose, L. A. Shepard, and J. Wulff, *Electronic Properties*, Wiley, New York, p. 262, 1966.
4. M. F. Uman, *Introduction to the Physics of Electronics*, Prentice-Hall, Englewood Cliffs, N.J., 1974.
5. E. M. Conwell, "Properties of silicon and germanium: II," *Proc. IRE*, vol. 46, pp. 1281–1300, June 1958.
6. J. D. Kraus, *Electromagnetics*, Third Edition, McGraw-Hill, New York, 1984.
7. F. London, *Superfluids*, Vol. 1, Dover, New York, 1961.
8. A. C. Rose-Innes and E. H. Rhoderick, *Introduction to Superconductivity*, Pergamon, Elmsford, N.Y., 1978.

9. B. C. Fenton, "Superconductivity breakthroughs," *Radio Electronics*, vol. 59, no. 2, pp. 43–45, February 1988.
10. A. Khurana, "Superconductivity seen above the boiling point of nitrogen," *Physics Today*, vol. 40, no. 4, pp. 17–23, April 1987.
11. G. P. de Loor and F. W. Meijboom, "The dielectric constant of foods and other materials with high water content at microwave frequencies," *Journal of Food Technology*, vol. 1, no. 1., pp. 313–322, 1966.
12. W. E. Pace, W. B. Westphal, and S. A. Goldblith, "Dielectric properties of commercial cooking oils," *Journal of Food Science*, vol. 33, p. 30, 1968.
13. D. Van Dyke, D. I. C. Wang, and S. A. Goldblith, "Dielectric loss factor of reconstituted ground beef: The effect of chemical composition," *Journal of Food Technology*, vol. 23, p. 944, 1969.
14. N. E. Bengtsson and P. O. Risman, "Dielectric properties of foods at 3 GHz as determined by a cavity perturbation technique. II. Measurements on food materials," *Journal of Microwave Power*, vol. 6, no. 2, pp. 107–123, 1971.
15. S. S. Stuchly and M. A. K. Hamid, "Physical properties in microwave heating processes," *Journal of Microwave Power*, vol. 7, no. 2, p. 117, 1972.
16. N. E. Bengtsson and T. Ohlsson, "Microwave heating in the food industry," *Proc. IEEE*, vol. 62, no. 1, pp. 44–55, January 1974.
17. P. Debye, *Polar Molecules*, Chem. Catalog Co., New York, 1929.
18. H. F. Cook, "The dielectric behaviour of some types of human tissue at microwave frequencies," *British Journal of Applied Physics*, vol. 2, p. 295, October 1951.
19. A. W. Guy, J. F. Lehmann, and J. B. Stonebridge, "Therapeutic applications of electromagnetic power," *Proc. IEEE*, vol. 62, no. 1, pp. 55–75, January 1974.
20. "Biological Effects of EM Waves," Speical Issue, *Radio Science*, vol. 12, no. 6(S), November–December 1977.
21. A. R. von Hippel, *Dielectrics and Waves*, MIT Press, Cambridge, Mass., 1954.
22. A. R. von Hippel, *Dielectric Materials and Applications*, Wiley, New York, pp. 93–252, 1954.
23. C. A. Balanis, "Dielectric constant and loss tangent measurements at 60 and 90 GHz using the Fabry–Perot interferometer," *Microwave Journal*, vol. 14, pp. 39–44, March 1971.
24. L. D. Landau and E. M. Lifshitz, *Electrodynamics of Continuous Media* (translated by J. B. Sykes and J. S. Bell), Pergamon, Elmsford, N.Y., Chapter IX, pp. 239–268, 1960.
25. J. D. Jackson, *Classical Electrodynamics*, Second Edition, Wiley, New York, p. 311, 1975.
26. S. Ramo, J. R. Whinnery, and T. Van Duzer, *Fields and Waves in Communication*, Second Edition, Wiley, New York, pp. 556–558, 671, 1984.
27. M. C. Lovell, A. J. Avery, and M. W. Vernon, *Physical Properties of Materials*, Van Nostrand–Reinhold, Princeton, N.J., p. 161, 1976.
28. J. L. Snoek, "Non-metallic magnetic material for high frequencies," *Philips Technical Review*, vol. 8, no. 12, pp. 353–384, December 1946.
29. D. Polder, "On the theory of ferromagnetic resonance," *Philosophical Magazine*, vol. 40, pp. 99–115, January 1949.
30. W. H. von Aulock, "Ferrimagnetic materials—phenomenological and atomic models," *Handbook of Microwave Ferrite Materials* (W. H. von Aulock, ed.), Section I, Chapter 1, Academic, New York, 1965.
31. W. von Aulock and J. H. Rowen, "Measurement of dielectric and magnetic properties of ferromagnetic materials at microwave frequencies," *The Bell System Technical Journal*, vol. 36, pp. 427–448, March 1957.
32. W. H. von Aulock, "Selection of ferrite materials for microwave device applications," *IEEE Trans. on Magnetics*, vol. MAG-2, no. 3, pp. 251–255, September 1966.
33. R. E. Collin, *Foundations for Microwave Engineering*, McGraw-Hill, New York, pp. 286–302, 1966.
34. R. F. Soohoo, *Theory and Application of Ferrites*, Chapter 5, Prentice-Hall, Englewood Cliffs, N.J., 1960.

PROBLEMS

2.1. A dielectric slab, shown in Figure P2-1, exhibits an electric polarization vector of
$$\mathbf{P} = \hat{a}_y 2.762 \times 10^{-11} \text{ C/m}^2$$
when it is subjected to an electric field of
$$\mathbf{E} = \hat{a}_y 2 \text{ V/m}$$
Determine

(a) The bound surface charge density q_{sp} in each of its six faces.
(b) The net bound charge Q_p associated with the slab.
(c) The volume bound charge density q_{vp} within the dielectric slab.
(d) The dielectric constant of the material.

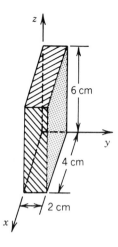

FIGURE P2-1

2.2. A cylindrical dielectric shell of Figure P2-2 with inner and outer radii, respectively, of $a = 2$ cm and $b = 6$ cm, and of length $\ell = 10$ cm exhibits an electric polarization vector of
$$\mathbf{P} = \hat{a}_\rho \frac{2}{\rho} \times 10^{-10} \text{ C/m}^2, \quad a \leq \rho \leq b$$

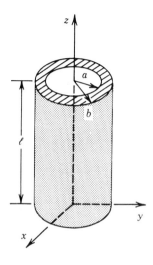

FIGURE P2-2

when it is subjected to an electric field of

$$\mathbf{E} = \hat{a}_\rho \frac{7.53}{\rho} \text{ V/m}, \quad a \leq \rho \leq b$$

Neglecting fringing, find

(a) The bound surface charge density q_{sp} in each of the surfaces.
(b) The net bound charge q_{sp} at the inner, outer, upper, and lower surfaces.
(c) The volume bound charge density q_{vp} within the dielectric.
(d) The dielectric constant of the material.

2.3. A spherical dielectric shell of Figure P2-3 with inner and outer radii $a = 2$ cm and $b = 4$ cm, respectively, exhibits an electric polarization vector of

$$\mathbf{P} = \hat{a}_r \frac{31.87}{r^2} \times 10^{-12} \text{ C/m}^2, \quad a \leq r \leq b$$

when it subjected to an electric field of

$$\mathbf{E} = \hat{a}_r \frac{0.45}{r^2} \text{ V/m}, \quad a \leq r \leq b$$

Determine

(a) The bound surface charge density q_{sp} in each of the surfaces.
(b) The net bound charge Q_p at the inner and outer surfaces.
(c) The volume bound charge density q_{vp} within the dielectric.
(d) The dielectric constant of the material.

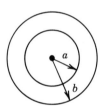

FIGURE P2-3

2.4. Two parallel conducting plates, each having a surface area of 2×10^{-2} m^2 on its sides, form a parallel-plate capacitor. Their separation is 1.25 mm and the medium between them is free space. A 100-V d.c. battery is connected across them, and it is maintained there at all times. Then a dielectric sheet, 1 mm thick and with the same shape and area as the plates, is slipped carefully between the plates so that one of its sides touches one of the conducting plates. After the insertion of the slab and neglecting fringing, if the dielectric constant of the dielectric sheet is $\varepsilon_r = 5$, determine

(a) The electric field intensity between the plates (inside and outside the slab).
(b) The electric flux density between the plates (inside and outside the slab).
(c) The surface charge density in each of the plates.
(d) The total charge in each of the plates.
(e) The capacitance across the slab, the free space, and both of them.
(f) The energy stored in the slab, the free space, and both of them.

2.5. For Problem 2.4, assume that after the 100-V voltage source charges the conducting plates, it is then removed. Then the dielectric sheet is inserted between the plates as indicated in Problem 2.4. After the insertion of the dielectric sheet, find

(a) The total charge Q on the upper and lower plates.
(b) The surface charge density on the upper and lower plates.
(c) The electric flux density in the dielectric slab and free space.
(d) The electric field intensity in the dielectric slab and free space.
(e) The voltage across the slab, the free space, and both of them.
(f) The capacitance across the slab, the free space, and both of them.
(g) The energy stored in the slab, the free space, and both of them.

2.6. A parallel-plate capacitor of Figure P2-6, with plates each of area 64 cm², separation of 4 cm, and free space between them, is charged by a 8-V d.c. source that is kept across the plates at all times. After the charging of the plates a 4-cm dielectric slab of polystyrene ($\varepsilon_r = 2.56$, $\mu_r = 1$) 4 cm in thickness is inserted between the plates and occupies half of the space between them.

Before insertion of the slab, determine

(a) The total charge on the upper and lower plates.
(b) The electric field between the plates.
(c) The electric flux density between the plates.
(d) The capacitance of the capacitor.
(e) The total stored energy in the capacitor.

After insertion of the slab, determine

(f) The total charge on the upper and lower plates in the free space and dielectric parts.
(g) The electric field in the free space and dielectric parts.
(h) The electric flux density in the free space and dielectric parts.
(i) The capacitance of each of the free space and dielectric parts.
(j) The total capacitance (combined free space and dielectric parts).
(k) The stored energy in each of the free space and dielectric parts.
(l) The total stored energy (combined free space and dielectric parts). Compare with that of part (e) and if there is a difference, explain why.

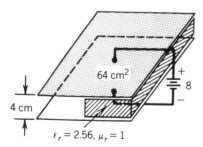

FIGURE P2-6

2.7. Repeat Problem 2.6 except assume that the voltage source is removed after the charging of the plates and before the insertion of the slab.

2.8. A 10-V d.c. voltage source, placed across the inner and outer conductors of a coaxial cylinder as shown in Figure P2-8, is used to charge the conductors and is then removed. The total charge in each conductor is $\pm Q$. The inner conductor has a radius of $a = 2$ cm, the radius of the outer conductor is 4 cm, and the length of the cylinder is $\ell = 6$ cm. Assuming no field fringing and free space between the conductors and neglecting fringing, find

(a) The electric field intensity between the conductors in terms of Q.
(b) The total charge Q on the inner and outer conductors.

(c) The surface charge density on the inner and outer conductors.
(d) The electric flux density between the conductors.
(e) The capacitance between the conductors.
(f) The energy stored between the conductors.

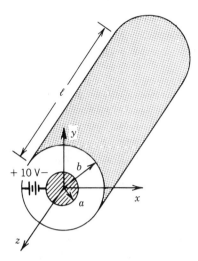

FIGURE P2-8

2.9. For Problem 2.8, assume that after the 10-V source charges the conductors and is removed, a cylindrical dielectric jacket of polystyrene ($\varepsilon_r = 2.56$) of inner radius $a = 2$ cm and outer radius $b = 3$ cm is inserted over the inner conductor of the coaxial cylinder. After the insertion of the jacket and neglecting fringing, find

(a) The total charge Q on the inner and outer conductors.
(b) The surface charge density on the inner and outer conductors.
(c) The electric flux density between the conductors in the dielectric and free space.
(d) The electric field intensity between the conductors in the dielectric and free space.
(e) The voltage between the conductors.
(f) The total capacitance between the conductors.
(g) The total energy stored between the conductors.

2.10. For Problem 2.9 assume that the 10-V source that charges the conductors remains connected at all times. By neglecting fringing, determine

(a) The electric field intensity between the conductors inside and outside the dielectric jacket.
(b) The electric flux density between the conductors inside and outside the dielectric jacket.
(c) The surface charge density in each of the plates.
(d) The total charge in each of the conductors.
(e) The total capacitance between the conductors.
(f) The total energy stored between the conductors.

2.11. A 100-V d.c. voltage source is placed across two parallel-plate sets that are connected in parallel. Each conductor in each parallel-plate set has a surface area of 2×10^{-2} m² on each of its sides which are separated by 4 cm. For one parallel-plate set the medium between them is free space whereas for the other it is lossless polystyrene ($\varepsilon_r = 2.56$, $\mu_r = 1$). For each parallel-plate set, by neglecting fringing, determine

(a) The electric field intensity between the plates.
(b) The electric flux density between the plates.
(c) The total charge on the upper and lower plates.
(d) The total energy stored between the plates.

For the two-set parallel-plate combination determine

(e) The total charge on the two upper and two lower plates.
(f) The total capacitance between the upper and lower plates.
(g) The total energy stored between the plates.

2.12. For the coaxial cylinder of Problem 2.8 assume that half of the space ($0 \leq \phi \leq \pi$) between the conductors is occupied by free space and the other half ($\pi \leq \phi \leq 2\pi$) is taken up by polystyrene ($\varepsilon_r = 2.56$, $\mu_r = 1$). Once the battery charges the conductors and is removed, find, by neglecting fringing,

(a) The electric field intensity between the conductors, in free space and in polystyrene, in terms of Q.
(b) The total charge in each of the inner and outer conductors.
(c) The surface charge density in the inner and outer conductors.
(d) The electric flux density between the inner and outer conductors.
(e) The capacitance between the conductors in free space, in polystyrene, and total.
(f) The energy stored between the conductors in free space, in polystyrene, and total.

2.13. For the coaxial cylinder of Problem 2.8 assume that once the 10-V voltage source charges the conductors and is removed, a curved dielectric slab of polystyrene ($\varepsilon_r = 2.56$, $\mu_r = 1$) of thickness equal to the spacing between the conductors is inserted between the conductors and occupies half of the space ($\pi \leq \phi \leq 2\pi$); the other half is still occupied by free space. By neglecting fringing, determine

(a) The total charge on the inner and outer conductors in free space and in polystyrene.
(b) The surface charge density on inner and outer conductors in free space and in polystyrene.
(c) The electric flux density between the conductors in free space and in polystyrene.
(d) The electric field intensity between the conductors in free space and in polystyrene.
(e) The voltage between the conductors in free space and in polystyrene.
(f) The capacitance between the conductors in free space, in polystyrene, and total.
(g) The energy stored between the conductors in free space, in polystyrene, and total.

2.14. For Problem 2.13 assume that the 10-V charging source is maintained across the conductors at all times. By neglecting fringing, determine:

(a) The electric field intensity between the conductors in free space and in polystyrene.
(b) The electric flux density between the conductors in free space and in polystyrene.
(c) The charge density in each of the conductors in free space and in polystyrene.
(d) The total charge in each of the conductors in free space and in polystyrene.
(e) The capacitance between the conductors in free space, in polystyrene, and total.
(f) The energy stored between the conductors in free space, in polystyrene, and total.

2.15. A rectangular slab of ferrimagnetic material as shown in Figure P2-15 exhibits a magnetization vector of

$$\mathbf{M} = \hat{a}_z 1.245 \times 10^6 \text{ A/m}$$

when it is subjected to a magnetic field intensity of

$$\mathbf{H} = \hat{a}_z 5 \times 10^3 \text{ A/m}$$

Find

(a) The bound magnetization surface current density in all its six faces.
(b) The bound magnetization volume current density within the slab.
(c) The net bound magnetization current associated with the slab.
(d) The relative permeability of the slab.

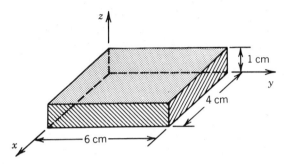

FIGURE P2-15

2.16. A coaxial line of length ℓ with inner and outer conductor radii of 1 and 3 cm, respectively, is filled with a ferromagnetic material, as shown in Figure P2-16. When the material is subjected to a magnetic field intensity of

$$\mathbf{H} = \hat{a}_\phi \frac{0.3183}{\rho} \text{ A/m}$$

it induces a magnetization vector potential of

$$\mathbf{M} = \hat{a}_\phi \frac{190.67}{\rho} \text{ A/m}$$

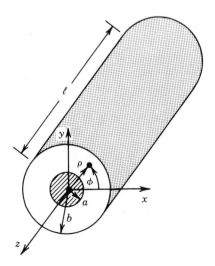

FIGURE P2-16

Determine

(a) The bound magnetization surface current density in all surfaces.
(b) The bound magnetization volume current density within the material.
(c) The net bound magnetization current associated with the coaxial line.
(d) The relative permeability of the material.

2.17. The magnetization vector inside a cylindrical magnetic bar of infinite length and circular cross section of radius $a = 1$ m, as shown in Figure P2-17, is given by
$$\mathbf{M} = \hat{a}_\phi 10 \text{ A/m}$$
Find

(a) The magnetic surface current density at the outside circumferential surface of the bar.
(b) The magnetic volume current density at any point inside the bar.
(c) The total current that flows through the cross section of the bar.

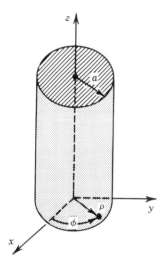

FIGURE P2-17

2.18. The current density through a cylindrical wire of square cross section as shown in Figure P2-18 is given by
$$\mathbf{J} \simeq \hat{a}_z J_0 e^{-10^2[(a-|x|)+(a-|y|)]}$$

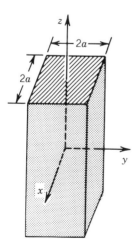

FIGURE P2-18

where J_0 is a constant. Assuming that each side of the wire is 2×10^{-2} m, find the current flow through the cross section of the wire.

2.19. A 10-A current is pushed through a circular cross section of wire of infinite length as shown in Figure P2-19. Assuming that the current density over the cross section of the wire decays from its surface toward its center as

$$\mathbf{J} = \hat{a}_z J_0 e^{-10^4(a-\rho)} \text{ A/m}^2$$

where J_0 is the current density at the surface and the wire radius is $a = 10^{-2}$ m, determine

(a) The current density at the surface of the wire.
(b) The depth from the surface of the wire through which the current density has decayed to 36.8 percent of its value at the surface.

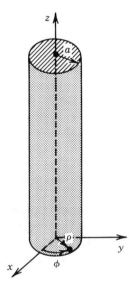

FIGURE P2-19

2.20. Show that the relaxation time constant for copper ($\sigma = 5.76 \times 10^7$ S/m) is much smaller than the period of waves in the microwave (1–10 GHz) region and is comparable to the period of x-rays [$\lambda \simeq 1$–10 Å $= (1 - 10) \times 10^{-8}$ cm]. Consequently, conductors cannot maintain a charge configuration long enough to permit propagation of the wave more than a short distance into the conductor at microwave frequencies. However x-ray propagation is possible because the relaxation time constant is comparable to the period of the wave.

2.21. Aluminum has a static conductivity of about $\sigma = 3.96 \times 10^7$ S/m and an electron mobility of $\mu_e = 2.2 \times 10^{-3}$ m²/(V-s). Assuming that an electric field of $\mathbf{E} = \hat{a}_x 2$ V/m is applied perpendicularly to the square area of an aluminum wafer with cross-sectional area of about 10 cm², find the (a) electron charge density q_{ve}, (b) electron drift velocity $\mathbf{v}_e$, (c) electric current density $\mathbf{J}$, (d) electric current flowing through the square cross section of the wafer, and (e) electron density N_e.

CHAPTER 3

WAVE EQUATION AND ITS SOLUTIONS

3.1 INTRODUCTION

The electromagnetic fields of boundary-value problems are obtained as solutions to Maxwell's equations, which are first-order partial differential equations. However, Maxwell's equations are coupled partial differential equations, which means that each equation has more than one unknown field. These equations can be uncoupled only at the expense of raising their order. For each of the fields, following such a procedure leads to an uncoupled second-order partial differential equation that is usually referred to as the *wave equation*. Therefore electric and magnetic fields for a given boundary-value problem can be obtained either as solutions to Maxwell's or the wave equations. The choice of equations is related to individual problems by convenience and ease of use. In this chapter we will develop the vector wave equations for each of the fields, and then we will demonstrate their solutions in the rectangular, cylindrical, and spherical coordinate systems.

3.2 TIME-VARYING ELECTROMAGNETIC FIELDS

The first two of Maxwell's equations in differential form, as given by (1-1) and (1-2), are first-order, coupled differential equations; that is, both the unknown fields ($\mathscr{E}$ and $\mathscr{H}$) appear in each equation. Usually it is very desirable, for convenience in solving for $\mathscr{E}$ and $\mathscr{H}$, to uncouple these equations. This can be accomplished at the expense of increasing the order of the differential equations to second order. To do this, we repeat (1-1) and (1-2), that is,

$$\nabla \times \mathscr{E} = -\mathscr{M}_i - \mu \frac{\partial \mathscr{H}}{\partial t} \tag{3-1}$$

$$\nabla \times \mathscr{H} = \mathscr{J}_i + \sigma \mathscr{E} + \varepsilon \frac{\partial \mathscr{E}}{\partial t} \tag{3-2}$$

where it is understood in the remaining part of the book that σ represents the effective conductivity σ_e and ε represents ε'. Taking the curl of both sides of each of equations 3-1 and 3-2 and assuming a homogeneous medium, we can write that

$$\nabla \times \nabla \times \mathcal{E} = -\nabla \times \mathcal{M}_i - \mu \nabla \times \left(\frac{\partial \mathcal{H}}{\partial t}\right) = -\nabla \times \mathcal{M}_i - \mu \frac{\partial}{\partial t}(\nabla \times \mathcal{H}) \tag{3-3}$$

$$\nabla \times \nabla \times \mathcal{H} = \nabla \times \mathcal{J}_i + \sigma \nabla \times \mathcal{E} + \varepsilon \nabla \times \left(\frac{\partial \mathcal{E}}{\partial t}\right)$$

$$= \nabla \times \mathcal{J}_i + \sigma \nabla \times \mathcal{E} + \varepsilon \frac{\partial}{\partial t}(\nabla \times \mathcal{E}) \tag{3-4}$$

Substituting (3-2) into the right side of (3-3) and using the vector identity

$$\nabla \times \nabla \times \mathbf{F} = \nabla(\nabla \cdot \mathbf{F}) - \nabla^2 \mathbf{F} \tag{3-5}$$

into the left side, we can rewrite (3-3) as

$$\nabla(\nabla \cdot \mathcal{E}) - \nabla^2 \mathcal{E} = -\nabla \times \mathcal{M}_i - \mu \frac{\partial}{\partial t}\left[\mathcal{J}_i + \sigma \mathcal{E} + \varepsilon \frac{\partial \mathcal{E}}{\partial t}\right]$$

$$\nabla(\nabla \cdot \mathcal{E}) - \nabla^2 \mathcal{E} = -\nabla \times \mathcal{M}_i - \mu \frac{\partial \mathcal{J}_i}{\partial t} - \mu \sigma \frac{\partial \mathcal{E}}{\partial t} - \mu \varepsilon \frac{\partial^2 \mathcal{E}}{\partial t^2} \tag{3-6}$$

Substituting Maxwell's equation 1-3, or

$$\nabla \cdot \mathcal{D} = \varepsilon \nabla \cdot \mathcal{E} = q_{ev} \Rightarrow \nabla \cdot \mathcal{E} = \frac{q_{ev}}{\varepsilon} \tag{3-7}$$

into (3-6) and rearranging its terms, we have that

$$\boxed{\nabla^2 \mathcal{E} = \nabla \times \mathcal{M}_i + \mu \frac{\partial \mathcal{J}_i}{\partial t} + \frac{1}{\varepsilon} \nabla q_{ev} + \mu \sigma \frac{\partial \mathcal{E}}{\partial t} + \mu \varepsilon \frac{\partial^2 \mathcal{E}}{\partial t^2}} \tag{3-8}$$

which is recognized as an uncoupled second-order differential equation for $\mathcal{E}$.

In a similar manner, by substituting (3-1) into the right side of (3-4) and using the vector identity of (3-5) in the left side of (3-4), we can rewrite it as

$$\nabla(\nabla \cdot \mathcal{H}) - \nabla^2 \mathcal{H} = \nabla \times \mathcal{J}_i + \sigma\left(-\mathcal{M}_i - \mu \frac{\partial \mathcal{H}}{\partial t}\right) + \varepsilon \frac{\partial}{\partial t}\left(-\mathcal{M}_i - \mu \frac{\partial \mathcal{H}}{\partial t}\right)$$

$$\nabla(\nabla \cdot \mathcal{H}) - \nabla^2 \mathcal{H} = \nabla \times \mathcal{J}_i - \sigma \mathcal{M}_i - \mu \sigma \frac{\partial \mathcal{H}}{\partial t} - \varepsilon \frac{\partial \mathcal{M}_i}{\partial t} - \mu \varepsilon \frac{\partial^2 \mathcal{H}}{\partial t^2} \tag{3-9}$$

Substituting Maxwell's equation

$$\nabla \cdot \mathcal{B} = \mu \nabla \cdot \mathcal{H} = q_{mv} \Rightarrow \nabla \cdot \mathcal{H} = \left(\frac{q_{mv}}{\mu}\right) \tag{3-10}$$

into (3-9), we have that

$$\nabla^2 \mathcal{H} = -\nabla \times \mathcal{J}_i + \sigma \mathcal{M}_i + \frac{1}{\mu}\nabla(q_{mv}) + \varepsilon\frac{\partial \mathcal{M}_i}{\partial t} + \mu\sigma\frac{\partial \mathcal{H}}{\partial t} + \mu\varepsilon\frac{\partial^2 \mathcal{H}}{\partial t^2}$$

(3-11)

which is recognized as an uncoupled second-order differential equation for $\mathcal{H}$. Thus (3-8) and (3-11) form a pair of uncoupled second-order differential equations that are a by-product of Maxwell's equations as given by (1-1) through (1-4).

Equations 3-8 and 3-11 are referred to as the *vector wave equations* for $\mathcal{E}$ and $\mathcal{H}$. For solving an electromagnetic boundary-value problem, the equations that must be satisfied are Maxwell's equations as given by (1-1) through (1-4) or the wave equations as given by (3-8) and (3-11). Often, the forms of the wave equations are preferred over those of Maxwell's equations.

For source-free regions ($\mathcal{J}_i = q_{ev} = 0$ and $\mathcal{M}_i = q_{mv} = 0$), the wave equations 3-8 and 3-11 reduce, respectively, to

$$\nabla^2 \mathcal{E} = \mu\sigma\frac{\partial \mathcal{E}}{\partial t} + \mu\varepsilon\frac{\partial^2 \mathcal{E}}{\partial t^2}$$

(3-12)

$$\nabla^2 \mathcal{H} = \mu\sigma\frac{\partial \mathcal{H}}{\partial t} + \mu\varepsilon\frac{\partial^2 \mathcal{H}}{\partial t^2}$$

(3-13)

For source-free ($\mathcal{J}_i = q_{ev} = 0$ and $\mathcal{M}_i = q_{mv} = 0$) and lossless media ($\sigma = 0$), the wave equations 3-8 and 3-11 or 3-12 and 3-13 simplify to

$$\nabla^2 \mathcal{E} = \mu\varepsilon\frac{\partial^2 \mathcal{E}}{\partial t^2}$$

(3-14)

$$\nabla^2 \mathcal{H} = \mu\varepsilon\frac{\partial^2 \mathcal{H}}{\partial t^2}$$

(3-15)

Equations 3-14 and 3-15 represent the simplest forms of the vector wave equations.

3.3 TIME-HARMONIC ELECTROMAGNETIC FIELDS

For time-harmonic fields (time variations of the form $e^{j\omega t}$), the wave equations can be derived using a similar procedure as in Section 3.2 for the general time-varying fields, starting with Maxwell's equations as given in Table 1-4. However, instead of going through this process, we find, by comparing Maxwell's equations for the general time-varying fields with those for the time-harmonic fields (both are displayed in Table 1-4), that one set can be obtained from the other by replacing $\partial/\partial t \Leftrightarrow j\omega$, $\partial^2/\partial t^2 \Leftrightarrow (j\omega)^2 = -\omega^2$, and the instantaneous fields ($\mathcal{E}, \mathcal{H}, \mathcal{D}, \mathcal{B}$), respectively, with the complex fields (**E, H, D, B**) and vice versa. Doing this for the wave equations 3-8, 3-11, 3-12, and 3-13, we can write each, respectively, as

$$\nabla^2 \mathbf{E} = \nabla \times \mathbf{M}_i + j\omega\mu\mathbf{J}_i + \frac{1}{\varepsilon}\nabla q_{ev} + j\omega\mu\sigma\mathbf{E} - \omega^2\mu\varepsilon\mathbf{E} \qquad (3\text{-}16a)$$

$$\nabla^2 \mathbf{H} = -\nabla \times \mathbf{J}_i + \sigma\mathbf{M}_i + j\omega\varepsilon\mathbf{M}_i + \frac{1}{\mu}\nabla q_{mv} + j\omega\mu\sigma\mathbf{H} - \omega^2\mu\varepsilon\mathbf{H} \qquad (3\text{-}16b)$$

$$\nabla^2 \mathbf{E} = j\omega\mu\sigma\mathbf{E} - \omega^2\mu\varepsilon\mathbf{E} = \gamma^2\mathbf{E} \qquad (3\text{-}17a)$$

$$\nabla^2 \mathbf{H} = j\omega\mu\sigma\mathbf{H} - \omega^2\mu\varepsilon\mathbf{H} = \gamma^2\mathbf{H} \qquad (3\text{-}17b)$$

where

$$\gamma^2 = j\omega\mu\sigma - \omega^2\mu\varepsilon = j\omega\mu(\sigma + j\omega\varepsilon) \qquad (3\text{-}17c)$$

$$\gamma = \alpha + j\beta = \text{propagation constant} \qquad (3\text{-}17d)$$

$$\alpha = \text{attenuation constant (Np/m)} \qquad (3\text{-}17e)$$

$$\beta = \text{phase constant (rad/m)} \qquad (3\text{-}17f)$$

The constants α, β, and γ will be discussed in more detail in Section 4.3 where α and β are expressed by (4-28c) and (4-28d) in terms of ω, ε, μ, and σ.

Similarly (3-14) and (3-15) can be written, respectively, as

$$\nabla^2 \mathbf{E} = -\omega^2\mu\varepsilon\mathbf{E} = -\beta^2\mathbf{E} \qquad (3\text{-}18a)$$

$$\nabla^2 \mathbf{H} = -\omega^2\mu\varepsilon\mathbf{H} = -\beta^2\mathbf{H} \qquad (3\text{-}18b)$$

where

$$\beta^2 = \omega^2\mu\varepsilon \qquad (3\text{-}18c)$$

In the literature the phase constant β is often represented by k.

3.4 SOLUTION TO THE WAVE EQUATION

The time variations of most practical problems are of the time-harmonic form. Fourier series can be used to express time variations of other forms in terms of a number of time-harmonic terms. Electromagnetic fields associated with a given boundary-value problem must satisfy Maxwell's equations or the vector wave equations. For many cases, the vector wave equations reduce to a number of scalar Helmholtz (wave) equations, and the general solutions can be constructed once solutions to each of the scalar Helmholtz equations are found.

In this section we want to demonstrate at least one method that can be used to solve the scalar Helmholtz equation in rectangular, cylindrical, and spherical coordinates. The method is known as the *separation of variables* [1, 2], and the general solution to the scalar Helmholtz equation using this method can be constructed in 11 three-dimensional orthogonal coordinate systems (including the rectangular, cylindrical, and spherical systems) [3].

The solutions for the instantaneous time-harmonic electric and magnetic field intensities can be obtained by considering the forms of the vector wave equations given either in Section 3.2 or Section 3.3. The approach chosen here will be to use those of Section 3.3 to solve for the complex field intensities **E** and **H** first. The corresponding instantaneous quantities can then be formed using the relations (1-61a) through (1-61f) between the instantaneous time-harmonic fields and their complex counterparts.

3.4.1 Rectangular Coordinate System

In a rectangular coordinate system, the vector wave equations 3-16a through 3-18c can be reduced to three scalar wave (Helmholtz) equations. First we will consider the solutions for source-free and lossless media. This will be followed by solutions for source-free but lossy media.

A. SOURCE-FREE AND LOSSLESS MEDIA

For source-free ($\mathbf{J}_i = \mathbf{M}_i = q_{ve} = q_{vm} = 0$) and lossless ($\sigma = 0$) media, the vector wave equations for the complex electric and magnetic field intensities are those given by (3-18a) through (3-18c). Since (3-18a) and (3-18b) are of the same form, let us examine the solution to one of them. The solution to the other can then be written by an interchange of **E** with **H** or **H** with **E**. We will begin by examining the solution for **E**.

In rectangular coordinates, a general solution for **E** can be written as

$$\mathbf{E}(x, y, z) = \hat{a}_x E_x(x, y, z) + \hat{a}_y E_y(x, y, z) + \hat{a}_z E_z(x, y, z) \quad (3\text{-}19)$$

where x, y, z are the rectangular coordinates, as illustrated in Figure 3-1. Substituting (3-19) into (3-18a) we can write that

$$\nabla^2 \mathbf{E} + \beta^2 \mathbf{E} = \nabla^2(\hat{a}_x E_x + \hat{a}_y E_y + \hat{a}_z E_z) + \beta^2(\hat{a}_x E_x + \hat{a}_y E_y + \hat{a}_z E_z) = 0$$
$$(3\text{-}20)$$

which reduces to three scalar wave equations of

$$\nabla^2 E_x(x, y, z) + \beta^2 E_x(x, y, z) = 0 \quad (3\text{-}20\text{a})$$
$$\nabla^2 E_y(x, y, z) + \beta^2 E_y(x, y, z) = 0 \quad (3\text{-}20\text{b})$$
$$\nabla^2 E_z(x, y, z) + \beta^2 E_z(x, y, z) = 0 \quad (3\text{-}20\text{c})$$

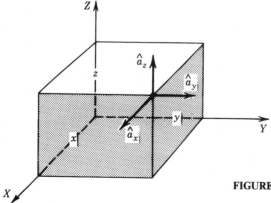

FIGURE 3-1 Rectangular coordinate system and corresponding unit vectors.

because

$$\nabla^2(\hat{a}_x E_x + \hat{a}_y E_y + \hat{a}_z E_z) = \hat{a}_x \nabla^2 E_x + \hat{a}_y \nabla^2 E_y + \hat{a}_z \nabla^2 E_z \quad (3\text{-}21)$$

Equations 3-20a through 3-20c are all of the same form; once a solution of any one of them is obtained, the solutions to the others can be written by inspection. We choose to work first with that for E_x as given by (3-20a).

In expanded form (3-20a) can be written as

$$\nabla^2 E_x + \beta^2 E_x = \frac{\partial^2 E_x}{\partial x^2} + \frac{\partial^2 E_x}{\partial y^2} + \frac{\partial^2 E_x}{\partial z^2} + \beta^2 E_x = 0 \quad (3\text{-}22)$$

Using the *separation of variables method*, we assume that a solution for $E_x(x, y, z)$ can be written in the form of

$$E_x(x, y, z) = f(x)g(y)h(z) \quad (3\text{-}23)$$

where the x, y, z variations of E_x are separable (hence the name). If any inconsistencies are encountered with assuming such a form of solution, another form must be attempted. This is the procedure usually followed in solving differential equations. Substituting (3-23) into (3-22), we can write that

$$gh\frac{\partial^2 f}{\partial x^2} + fh\frac{\partial^2 g}{\partial y^2} + fg\frac{\partial^2 h}{\partial z^2} + \beta^2 fgh = 0 \quad (3\text{-}24)$$

Since $f(x)$, $g(y)$, and $h(z)$ are each a function of only one variable, we can replace the partials in (3-24) by ordinary derivatives. Doing this and dividing each term by fgh, we can write that

$$\frac{1}{f}\frac{d^2 f}{dx^2} + \frac{1}{g}\frac{d^2 g}{dy^2} + \frac{1}{h}\frac{d^2 h}{dz^2} + \beta^2 = 0 \quad (3\text{-}25)$$

or

$$\frac{1}{f}\frac{d^2 f}{dx^2} + \frac{1}{g}\frac{d^2 g}{dy^2} + \frac{1}{h}\frac{d^2 h}{dz^2} = -\beta^2 \quad (3\text{-}25\text{a})$$

Each of the first three terms in (3-25a) is a function of only a single independent variable; hence the sum of these terms can equal $-\beta^2$ only if each term is a constant. Thus (3-25a) separates into three equations of the form

$$\frac{1}{f}\frac{d^2 f}{dx^2} = -\beta_x^2 \Rightarrow \frac{d^2 f}{dx^2} = -\beta_x^2 f \quad (3\text{-}26\text{a})$$

$$\frac{1}{g}\frac{d^2 g}{dy^2} = -\beta_y^2 \Rightarrow \frac{d^2 g}{dy^2} = -\beta_y^2 g \quad (3\text{-}26\text{b})$$

$$\frac{1}{h}\frac{d^2 h}{dz^2} = -\beta_z^2 \Rightarrow \frac{d^2 h}{dz^2} = -\beta_z^2 h \quad (3\text{-}26\text{c})$$

where, in addition,

$$\beta_x^2 + \beta_y^2 + \beta_z^2 = \beta^2 \quad (3\text{-}27)$$

Equation 3-27 is referred to as the *constraint* equation. In addition $\beta_x, \beta_y, \beta_z$ are known as the wave constants (numbers) in the x, y, z directions, respectively, that will be determined using boundary conditions.

The solution to each of equations 3-26a, 3-26b, or 3-26c can take different forms. Some typical valid solutions for $f(x)$ of (3-26a) would be

$$f_1(x) = A_1 e^{-j\beta_x x} + B_1 e^{+j\beta_x x} \tag{3-28a}$$

or

$$f_2(x) = C_1 \cos(\beta_x x) + D_1 \sin(\beta_x x) \tag{3-28b}$$

Similarly the solutions to (3-26b) and (3-26c) for $g(y)$ and $h(z)$ can be written, respectively, as

$$g_1(y) = A_2 e^{-j\beta_y y} + B_2 e^{+j\beta_y y} \tag{3-29a}$$

or

$$g_2(y) = C_2 \cos(\beta_y y) + D_2 \sin(\beta_y y) \tag{3-29b}$$

and

$$h_1(z) = A_3 e^{-j\beta_z z} + B_3 e^{+j\beta_z z} \tag{3-30a}$$

TABLE 3-1
Wave functions, zeroes, and infinities of plane wave functions in rectangular coordinates

Wave type	Wave functions	Zeroes of wave functions	Infinities of wave functions
Traveling waves	$e^{-j\beta x}$ for $+x$ travel $e^{+j\beta x}$ for $-x$ travel	$\beta x \to -j\infty$ $\beta x \to +j\infty$	$\beta x \to +j\infty$ $\beta x \to -j\infty$
Standing waves	$\cos(\beta x)$ for $\pm x$ $\sin(\beta x)$ for $\pm x$	$\beta x = \pm(n + \tfrac{1}{2})\pi$ $\beta x = \pm n\pi$ $n = 0, 1, 2, \ldots$	$\beta x \to \pm j\infty$ $\beta x \to \pm j\infty$
Evanescent waves	$e^{-\alpha x}$ for $+x$ $e^{+\alpha x}$ for $-x$ $\cosh(\alpha x)$ for $\pm x$ $\sinh(\alpha x)$ for $\pm x$	$\alpha x \to +\infty$ $\alpha x \to -\infty$ $\alpha x = \pm j(n + \tfrac{1}{2})\pi$ $\alpha x = \pm jn\pi$ $n = 0, 1, 2, \ldots$	$\alpha x \to -\infty$ $\alpha x \to +\infty$ $\alpha x \to \pm \infty$ $\alpha x \to \pm \infty$
Attenuating traveling waves	$e^{-\gamma x} = e^{-\alpha x} e^{-j\beta x}$ for $+x$ travel $e^{+\gamma x} = e^{+\alpha x} e^{+j\beta x}$ for $-x$ travel	$\gamma x \to +\infty$ $\gamma x \to -\infty$	$\gamma x \to -\infty$ $\gamma x \to +\infty$
Attenuating standing waves	$\cos(\gamma x) = \cos(\alpha x)\cosh(\beta x)$ $\quad -j\sin(\alpha x)\sinh(\beta x)$ for $\pm x$	$\gamma x = \pm j(n + \tfrac{1}{2})\pi$	$\gamma x \to \pm j\infty$
	$\sin(\gamma x) = \sin(\alpha x)\cosh(\beta x)$ $\quad +j\cos(\alpha x)\sinh(\beta x)$ for $\pm x$	$\gamma x = \pm jn\pi$ $n = 0, 1, 2, \ldots$	$\gamma x \to \pm j\infty$

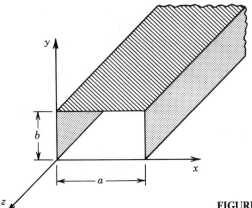

FIGURE 3-2 Rectangular waveguide geometry.

or

$$h_2(z) = C_3 \cos(\beta_z z) + D_3 \sin(\beta_z z) \tag{3-30b}$$

Although all the aforementioned solutions are valid for $f(x)$, $g(y)$, and $h(z)$, the most appropriate form should be chosen to simplify the complexity of the problem at hand. In general, the solutions of (3-28a), (3-29a), and (3-30a) in terms of complex exponentials represent *traveling waves* and the solutions of (3-28b), (3-29b), and (3-30b) represent *standing waves*. Wave functions representing various wave types in rectangular coordinates are found listed in Table 3-1. In Chapter 8 we will consider specific examples and the appropriate solution forms for $f(x)$, $g(y)$, and $h(z)$.

Once the appropriate forms for $f(x)$, $g(y)$, and $h(z)$ have been decided, the solution for the scalar function $E_x(x, y, z)$ of (3-22) can be written as the product of *fgh* as stated by (3-23). To demonstrate that, let us consider a specific example in which it will be assumed that the appropriate solutions for f, g, and h are given, respectively, by (3-28b), (3-29b), and (3-30a). Thus we can write that

$$E_x(x, y, z) = [C_1 \cos(\beta_x x) + D_1 \sin(\beta_x x)][C_2 \cos(\beta_y y) + D_2 \sin(\beta_y y)]$$
$$\times [A_3 e^{-j\beta_z z} + B_3 e^{+j\beta_z z}] \tag{3-31}$$

This is an appropriate solution for any of the electric or magnetic field components inside a rectangular pipe (waveguide), shown in Figure 3-2, that is bounded in the x and y directions and has its length along the z axis. Because the waveguide is bounded in the x and y directions, standing waves, represented by cosine and sine functions, have been chosen as solutions for $f(x)$ and $g(y)$ functions. However, because the waveguide is not bounded in the z direction, traveling waves, represented by complex exponential functions, have been chosen as solutions for $h(z)$. A complete discussion of the fields inside a rectangular waveguide can be found in Chapter 8.

For $e^{j\omega t}$ time variations, which are assumed throughout this book, the first complex exponential term in (3-31) represents a wave that travels in the $+z$ direction; the second exponential represents a wave that travels in the $-z$ direction. To demonstrate this, let us examine the instantaneous form $\mathscr{E}_x(x, y, z; t)$ of the scalar complex function $E_x(x, y, z)$. Since the solution of (3-31) represents the

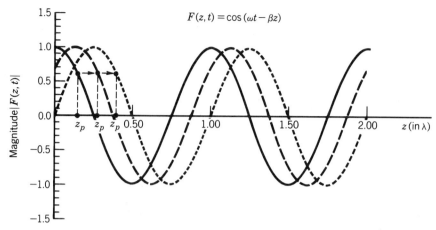

FIGURE 3-3 Variations as a function of distance for different times of positive traveling wave. —— time $t_0 = 0$; – – – – time $t_1 = T/8$; - - - - time $t_2 = T/4$.

complex form of E_x, its instantaneous form can be written as

$$\mathscr{E}_x(x, y, z; t) = \text{Re}\left[E_x(x, y, z)e^{j\omega t}\right] \quad (3\text{-}32)$$

Considering only the first exponential term of (3-31) and assuming all constants are real, we can write the instantaneous form of the $\mathscr{E}_x$ function for that term as

$$\mathscr{E}_x^+(x, y, z; t) = \text{Re}\left[E_x^+(x, y, z)e^{j\omega t}\right]$$
$$= \text{Re}\left\{[C_1 \cos(\beta_x x) + D_1 \sin(\beta_x x)]\right.$$
$$\left.\times [C_2 \cos(\beta_y y) + D_2 \sin(\beta_y y)] A_3 e^{j(\omega t - \beta_z z)}\right\} \quad (3\text{-}33)$$

or, if the constants C_1, D_1, C_2, D_2, and A_3 are real, as

$$\mathscr{E}_x^+(x, y, z; t) = [C_1 \cos(\beta_x x) + D_1 \sin(\beta_x x)]$$
$$\times [C_2 \cos(\beta_y y) + D_2 \sin(\beta_y y)] A_3 \cos(\omega t - \beta_z z) \quad (3\text{-}33a)$$

where the superscript plus is used to denote a positive traveling wave.

A plot of the normalized $\mathscr{E}_x^+(x, y, z; t)$ as a function of z for different times $(t = t_0, t_1, \ldots, t_n, t_{n+1})$ is shown in Figure 3-3. It is evident that as time increases $(t_{n+1} > t_n)$, the waveform of $\mathscr{E}_x^+$ is essentially the same, with the exception of an apparent shift in the $+z$ direction indicating a wave traveling in the $+z$ direction. This shift in the $+z$ direction can also be demonstrated by examining what happens to a given point z_p in the waveform of $\mathscr{E}_x^+$ for $t = t_0, t_1, \ldots, t_n, t_{n+1}$. To follow the point z_p for different values of t, we must maintain constant the amplitude of the last cosine term in (3-33a). This is accomplished by keeping its argument $\omega t - \beta_z z_p$ constant, that is,

$$\omega t - \beta_z z_p = C_0 = \text{constant} \quad (3\text{-}34)$$

which when differentiated with respect to time reduces to

$$\omega(1) - \beta_z \frac{dz_p}{dt} = 0 \Rightarrow \frac{dz_p}{dt} = v_p = +\frac{\omega}{\beta_z} \quad (3\text{-}35)$$

The point z_p is referred to as an *equiphase* point and its velocity is denoted as the *phase velocity*. A similar procedure can be used to demonstrate that the second complex exponential term in (3-31) represents a wave that travels in the $-z$ direction.

B. SOURCE-FREE AND LOSSY MEDIA

When the media in which the waves are traveling are lossy ($\sigma \neq 0$) but source-free ($\mathbf{J}_i = \mathbf{M}_i = q_{ve} = q_{vm} = 0$), the vector wave equations that the complex electric $\mathbf{E}$ and magnetic $\mathbf{H}$ field intensities must satisfy are (3-17a) and (3-17b). As for the lossless case, let us examine the solution to one of them; the solution to the other can then be written by inspection once the solution to the first has been obtained. We choose to consider the solution for the electric field intensity $\mathbf{E}$, which must satisfy (3-17a). An extended presentation of electromagnetic wave propagation in lossy media can be found in [4].

In a rectangular coordinate system, the general solution for $\mathbf{E}(x, y, z)$ can be written as

$$\mathbf{E}(x, y, z) = \hat{a}_x E_x(x, y, z) + \hat{a}_y E_y(x, y, z) + \hat{a}_z E_z(x, y, z) \quad (3\text{-}36)$$

When (3-36) is substituted into (3-17a), we can write that

$$\nabla^2 \mathbf{E} - \gamma^2 \mathbf{E} = \nabla^2(\hat{a}_x E_x + \hat{a}_y E_y + \hat{a}_z E_z) - \gamma^2(\hat{a}_x E_x + \hat{a}_y E_y + \hat{a}_z E_z) = 0 \quad (3\text{-}37)$$

which reduces to three scalar wave equations of

$$\nabla^2 E_x(x, y, z) - \gamma^2 E_x(x, y, z) = 0 \quad (3\text{-}37\text{a})$$

$$\nabla^2 E_y(x, y, z) - \gamma^2 E_y(x, y, z) = 0 \quad (3\text{-}37\text{b})$$

$$\nabla^2 E_z(x, y, z) - \gamma^2 E_z(x, y, z) = 0 \quad (3\text{-}37\text{c})$$

where

$$\gamma^2 = j\omega\mu(\sigma + j\omega\varepsilon) \quad (3\text{-}37\text{d})$$

If we were to allow for positive and negative values of σ

$$\gamma = \pm\sqrt{j\omega\mu(\sigma + j\omega\varepsilon)} = \begin{cases} \pm(\alpha + j\beta) & \text{for } +\sigma \\ \pm(\alpha - j\beta) & \text{for } -\sigma \end{cases} \quad (3\text{-}37\text{e})$$

In (3-37e),

γ = propagation constant

α = attenuation constant (Np/m)

β = phase constant (rad/m)

where α and β are assumed to be real and positive. Although some authors choose to represent the phase constant by k, the symbol β will be used throughout this book.

Examining (3-37e) reveals that there are four possible combinations for the form of γ. That is,

$$\gamma = \begin{cases} +(\alpha + j\beta) & \text{(3-38a)} \\ -(\alpha + j\beta) & \text{(3-38b)} \\ +(\alpha - j\beta) & \text{(3-38c)} \\ -(\alpha - j\beta) & \text{(3-38d)} \end{cases}$$

Of the four combinations, only one will be appropriate for our solution. That form will be selected once the solutions to any of (3-37a) through (3-37c) have been decided.

Since all three equations represented by (3-37a) through (3-37c) are of the same form, let us examine only one of them. We choose to work first with (3-37a) whose solution can be derived using the method of *separation of variables*. Using a similar procedure as for the lossless case, we can write that

$$E_x(x, y, z) = f(x)g(y)h(z) \tag{3-39}$$

where it can be shown that $f(x)$ has solutions of the form

$$f_1(x) = A_1 e^{-\gamma_x x} + B_1 e^{+\gamma_x x} \tag{3-40a}$$

or

$$f_2(x) = C_1 \cosh(\gamma_x x) + D_1 \sinh(\gamma_x x) \tag{3-40b}$$

and $g(y)$ can be expressed as

$$g_1(y) = A_2 e^{-\gamma_y y} + B_2 e^{+\gamma_y y} \tag{3-41a}$$

or

$$g_2(y) = C_2 \cosh(\gamma_y y) + D_2 \sinh(\gamma_y y) \tag{3-41b}$$

and $h(z)$ as

$$h_1(z) = A_3 e^{-\gamma_z z} + B_3 e^{+\gamma_z z} \tag{3-42a}$$

or

$$h_2(z) = C_3 \cosh(\gamma_z z) + D_3 \sinh(\gamma_z z) \tag{3-42b}$$

Whereas (3-40a) through (3-42b) are appropriate solutions for f, g, and h of (3-39), which satisfy (3-37a), the constraint equation takes the form of

$$\boxed{\gamma_x^2 + \gamma_y^2 + \gamma_z^2 = \gamma^2} \tag{3-43}$$

The appropriate forms of f, g, and h chosen to represent the solution of $E_x(x, y, z)$, as given by (3-39), must be made by examining the geometry of the problem in question. As for the lossless case, the exponentials represent attenuating traveling waves and the hyperbolic cosines and sines represent attenuating standing waves. These and other waves types are listed in Table 3-1.

To decide on the appropriate form for any of the γ's (whether it be γ_x, γ_y, γ_z, or γ), let us choose the form of γ_z by examining one of the exponentials in (3-42a). We choose to work with the first one. The four possible combinations for γ_z, according to (3-38a) through (3-38d) will be

$$\gamma_z = \begin{cases} +(\alpha_z + j\beta_z) & \text{(3-44a)} \\ -(\alpha_z + j\beta_z) & \text{(3-44b)} \\ +(\alpha_z - j\beta_z) & \text{(3-44c)} \\ -(\alpha_z - j\beta_z) & \text{(3-44d)} \end{cases}$$

If we want the first exponential in (3-42a) to represent a decaying wave which travels in the $+z$ direction, then by substituting (3-44a) through (3-44d) into it we can write that

$$h_1^+(z) = \begin{cases} A_3 e^{-\gamma_z z} = A_3 e^{-\alpha_z z} e^{-j\beta_z z} & \text{(3-45a)} \\ A_3 e^{-\gamma_z z} = A_3 e^{+\alpha_z z} e^{+j\beta_z z} & \text{(3-45b)} \\ A_3 e^{-\gamma_z z} = A_3 e^{-\alpha_z z} e^{+j\beta_z z} & \text{(3-45c)} \\ A_3 e^{-\gamma_z z} = A_3 e^{+\alpha_z z} e^{-j\beta_z z} & \text{(3-45d)} \end{cases}$$

By examining (3-45a) through (3-45d) and assuming $e^{j\omega t}$ time variations, the following statements can be made:

1. Equation 3-45a represents a wave that travels in the $+z$ direction, as determined by $e^{-j\beta_z z}$, and it decays in that direction, as determined by $e^{-\alpha_z z}$.
2. Equation 3-45b represents a wave that travels in the $-z$ direction, as determined by $e^{+j\beta_z z}$, and it decays in that direction, as determined by $e^{+\alpha_z z}$.
3. Equation 3-45c represents a wave that travels in the $-z$ direction, as determined by $e^{+j\beta_z z}$, and it is increasing in that direction, as determined by $e^{-\alpha_z z}$.
4. Equation 3-45d represents a wave that travels in the $+z$ direction, as determined by $e^{-j\beta_z z}$, and it is increasing in that direction, as determined by $e^{+\alpha_z z}$.

From the preceding statements it is apparent that for $e^{-\gamma_z z}$ to represent a wave that travels in the $+z$ direction and that concurrently also decays (to represent propagation in passive lossy media), and to satisfy the conservation of energy laws, the only correct form of γ_z is that of (3-44a). The same conclusion will result if the second exponential of (3-42a) represents a wave that travels in the $-z$ direction and that concurrently also decays. Thus the general form of any γ_i (whether it be γ_x, γ_y, γ_z, or γ), as given by (3-38a) through (3-38d), is

$$\gamma_i = \alpha_i + j\beta_i \tag{3-46}$$

Whereas the forms of f, g, and h [as given by (3-40a) through (3-42b)] are used to arrive at the solution for the complex form of E_x as given by (3-39), the instantaneous form of $\mathscr{E}_x$ can be obtained by using the relation of (3-32). A similar procedure can be used to derive the solutions of the other components of $\mathbf{E}$ (E_y and E_z), all those of $\mathbf{H}$ (H_x, H_y, and H_z), and of their instantaneous counterparts.

116 WAVE EQUATION AND ITS SOLUTIONS

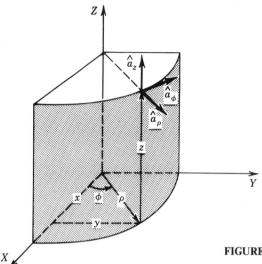

FIGURE 3-4 Cylindrical coordinate system and corresponding unit vectors.

3.4.2 Cylindrical Coordinate System

If the geometry of the system is of a cylindrical configuration, it would be very advisable to solve the boundary-value problem for the **E** and **H** fields using cylindrical coordinates. Maxwell's equations and the vector wave equations, which the **E** and **H** fields must satisfy, should be solved using cylindrical coordinates. Let us first consider the solution for **E** for a source-free and lossless medium. A similar procedure can be used for **H**. To maintain some simplicity in the mathematics, we will examine only lossless media.

In cylindrical coordinates a general solution to the vector wave equation for source-free and lossless media, as given by (3-18a), can be written as

$$\mathbf{E}(\rho, \phi, z) = \hat{a}_\rho E_\rho(\rho, \phi, z) + \hat{a}_\phi E_\phi(\rho, \phi, z) + \hat{a}_z E_z(\rho, \phi, z) \qquad (3\text{-}47)$$

where ρ, ϕ, and z are the cylindrical coordinates as illustrated in Figure 3-4. Substituting (3-47) into (3-18a), we can write that

$$\nabla^2 (\hat{a}_\rho E_\rho + \hat{a}_\phi E_\phi + \hat{a}_z E_z) = -\beta^2 (\hat{a}_\rho E_\rho + \hat{a}_\phi E_\phi + \hat{a}_z E_z) \qquad (3\text{-}48)$$

which does not reduce to three simple scalar wave equations, similar to those of (3-20a) through (3-20c) for (3-20), because

$$\nabla^2 (\hat{a}_\rho E_\rho) \neq \hat{a}_\rho \nabla^2 E_\rho \qquad (3\text{-}49a)$$

$$\nabla^2 (\hat{a}_\phi E_\phi) \neq \hat{a}_\phi \nabla^2 E_\phi \qquad (3\text{-}49b)$$

However, because

$$\nabla^2 (\hat{a}_z E_z) = \hat{a}_z \nabla^2 E_z \qquad (3\text{-}49c)$$

one of the three scalar equations to which (3-48) reduces is

$$\nabla^2 E_z + \beta^2 E_z = 0 \qquad (3\text{-}50)$$

The other two are of more complex form and they will be addressed in what follows.

Before we derive the other two scalar equations [in addition to (3-50)] to which (3-48) reduces, let us attempt to give a physical explanation of (3-49a), (3-49b), and (3-49c). By examining two different points (ρ_1, ϕ_1, z_1) and (ρ_2, ϕ_2, z_2) and their corresponding unit vectors on a cylindrical surface (as shown in Figure 3-4), we see that the directions of $\hat{a}_\rho$ and $\hat{a}_\phi$ have changed from one point to another (they are not parallel) and therefore cannot be treated as constants but rather are functions of ρ, ϕ, and z. In contrast, the unit vector $\hat{a}_z$ at the two points is pointed in the same direction (is parallel). The same is true for the unit vectors $\hat{a}_x$ and $\hat{a}_y$ in Figure 3-1.

Let us now return to the solution of (3-48). Since (3-48) does not reduce to (3-49a) and (3-49b), although it does satisfy (3-49c), how do we solve (3-48)? The procedure that follows can be used to reduce (3-48) to three scalar partial differential equations.

The form of (3-48) written in general as

$$\nabla^2 \mathbf{E} = -\beta^2 \mathbf{E} \tag{3-51}$$

was placed in this form by utilizing the vector identity of (3-5) during its derivation. Generally we are under the impression that we do not know how to perform the Laplacian of a vector ($\nabla^2 \mathbf{E}$) as given by the left side of (3-51). However, by utilizing (3-5) we can rewrite the left side of (3-51) as

$$\nabla^2 \mathbf{E} = \nabla(\nabla \cdot \mathbf{E}) - \nabla \times \nabla \times \mathbf{E} \tag{3-52}$$

whose terms can be expanded in any coordinate system. Using (3-52) we can write (3-51) as

$$\nabla(\nabla \cdot \mathbf{E}) - \nabla \times \nabla \times \mathbf{E} = -\beta^2 \mathbf{E} \tag{3-53}$$

which is an alternate, but not as commonly recognizable, form of the vector wave equation for the electric field in source-free and lossless media.

Assuming a solution for the electric field of the form given by (3-47), we can expand (3-53) and reduce it to three scalar partial differential equations of the form

$$\nabla^2 E_\rho + \left(-\frac{E_\rho}{\rho^2} - \frac{2}{\rho^2}\frac{\partial E_\phi}{\partial \phi}\right) = -\beta^2 E_\rho \tag{3-54a}$$

$$\nabla^2 E_\phi + \left(-\frac{E_\phi}{\rho^2} + \frac{2}{\rho^2}\frac{\partial E_\rho}{\partial \phi}\right) = -\beta^2 E_\phi \tag{3-54b}$$

$$\nabla^2 E_z = -\beta^2 E_z \tag{3-54c}$$

In each of (3-54a) through (3-54c) $\nabla^2 \psi(\rho, \phi, z)$ is the Laplacian of a scalar that in cylindrical coordinates takes the form of

$$\nabla^2 \psi(\rho, \phi, z) = \frac{1}{\rho}\frac{\partial}{\partial \rho}\left(\rho \frac{\partial \psi}{\partial \rho}\right) + \frac{1}{\rho^2}\frac{\partial^2 \psi}{\partial \phi^2} + \frac{\partial^2 \psi}{\partial z^2}$$

$$= \frac{\partial^2 \psi}{\partial \rho^2} + \frac{1}{\rho}\frac{\partial \psi}{\partial \rho} + \frac{1}{\rho^2}\frac{\partial^2 \psi}{\partial \phi^2} + \frac{\partial^2 \psi}{\partial z^2} \tag{3-55}$$

Equations 3-54a and 3-54b are *coupled* (each contains more than one electric field component) second-order partial differential equations, which are the most difficult to solve. However, (3-54c) is an *uncoupled* second-order partial differential equation whose solution will be most useful in the construction of TE^z and TM^z mode solutions of boundary-value problems, as discussed in Chapters 6 and 9.

In expanded form (3-54c) can then be written as

$$\frac{\partial^2 \psi}{\partial \rho^2} + \frac{1}{\rho}\frac{\partial \psi}{\partial \rho} + \frac{1}{\rho^2}\frac{\partial^2 \psi}{\partial \phi^2} + \frac{\partial^2 \psi}{\partial z^2} = -\beta^2 \psi \qquad (3\text{-}56)$$

where $\psi(\rho, \phi, z)$ is a scalar function that can represent a field or a vector potential component. Assuming a separable solution for $\psi(\rho, \phi, z)$ of the form

$$\psi(\rho, \phi, z) = f(\rho)g(\phi)h(z) \qquad (3\text{-}57)$$

and substituting it into (3-56), we can write that

$$gh\frac{\partial^2 f}{\partial \rho^2} + gh\frac{1}{\rho}\frac{\partial f}{\partial \rho} + fh\frac{1}{\rho^2}\frac{\partial^2 g}{\partial \phi^2} + fg\frac{\partial^2 h}{\partial z^2} = -\beta^2 fgh \qquad (3\text{-}58)$$

Dividing both sides of (3-58) by *fgh* and replacing the partials by ordinary derivatives reduces (3-58) to

$$\frac{1}{f}\frac{d^2 f}{d\rho^2} + \frac{1}{f}\frac{1}{\rho}\frac{df}{d\rho} + \frac{1}{g}\frac{1}{\rho^2}\frac{d^2 g}{d\phi^2} + \frac{1}{h}\frac{d^2 h}{dz^2} = -\beta^2 \qquad (3\text{-}59)$$

The last term on the left side of (3-59) is only a function of z. Therefore, using the discussion of Section 3.4.1, we can write that

$$\frac{1}{h}\frac{d^2 h}{dz^2} = -\beta_z^2 \Rightarrow \frac{d^2 h}{dz^2} = -\beta_z^2 h \qquad (3\text{-}60)$$

where β_z is a constant. Substituting (3-60) into (3-59) and multiplying both sides by ρ^2, reduces it to

$$\frac{\rho^2}{f}\frac{d^2 f}{d\rho^2} + \frac{\rho}{f}\frac{df}{d\rho} + \frac{1}{g}\frac{d^2 g}{d\phi^2} + (\beta^2 - \beta_z^2)\rho^2 = 0 \qquad (3\text{-}61)$$

Since the third term on the left side of (3-61) is only a function of ϕ, it can be set equal to a constant $-m^2$. Thus we can write that

$$\frac{1}{g}\frac{d^2 g}{d\phi^2} = -m^2 \Rightarrow \frac{d^2 g}{d\phi^2} = -m^2 g \qquad (3\text{-}62)$$

Letting

$$\beta^2 - \beta_z^2 = \beta_\rho^2 \Rightarrow \beta_\rho^2 + \beta_z^2 = \beta^2 \qquad (3\text{-}63)$$

then using (3-62), and multiplying both sides of (3-61) by f, we can reduce (3-61) to

$$\rho^2 \frac{d^2 f}{d\rho^2} + \rho \frac{df}{d\rho} + \left[(\beta_\rho \rho)^2 - m^2\right]f = 0 \qquad (3\text{-}64)$$

Equation 3-63 is referred to as the constraint equation for the solution to the wave equation in cylindrical coordinates, and equation 3-64 is recognized as the classic *Bessel differential equation* [1–3, 5–10].

In summary then, the partial differential equation (3-56) whose solution was assumed to be separable of the form given by (3-57) reduces to the three differential equations 3-60, 3-62, 3-64 and the constraint equation 3-63. Thus

$$\nabla^2 \psi(\rho, \phi, z) = \frac{\partial^2 \psi}{\partial \rho^2} + \frac{1}{\rho}\frac{\partial \psi}{\partial \rho} + \frac{1}{\rho^2}\frac{\partial^2 \psi}{\partial \phi^2} + \frac{\partial^2 \psi}{\partial z^2} = -\beta^2 \psi \quad (3\text{-}65)$$

where

$$\psi(\rho, \phi, z) = f(\rho) g(\phi) h(z) \quad (3\text{-}65a)$$

reduces to

$$\rho^2 \frac{d^2 f}{d\rho^2} + \rho \frac{df}{d\rho} + \left[(\beta_\rho \rho)^2 - m^2\right] f = 0 \quad (3\text{-}66a)$$

$$\frac{d^2 g}{d\phi^2} = -m^2 g \quad (3\text{-}66b)$$

$$\frac{d^2 h}{dz^2} = -\beta_z^2 h \quad (3\text{-}66c)$$

with

$$\beta_\rho^2 + \beta_z^2 = \beta^2 \quad (3\text{-}66d)$$

Solutions to (3-66a), (3-66b), and (3-66c) take the form, respectively, of

$$f_1(\rho) = A_1 J_m(\beta_\rho \rho) + B_1 Y_m(\beta_\rho \rho) \quad (3\text{-}67a)$$

or

$$f_2(\rho) = C_1 H_m^{(1)}(\beta_\rho \rho) + D_1 H_m^{(2)}(\beta_\rho \rho) \quad (3\text{-}67b)$$

and

$$g_1(\phi) = A_2 e^{-jm\phi} + B_2 e^{+jm\phi} \quad (3\text{-}68a)$$

or

$$g_2(\phi) = C_2 \cos(m\phi) + D_2 \sin(m\phi) \quad (3\text{-}68b)$$

and

$$h_1(z) = A_3 e^{-j\beta_z z} + B_3 e^{+j\beta_z z} \quad (3\text{-}69a)$$

or

$$h_2(z) = C_3 \cos(\beta_z z) + D_3 \sin(\beta_z z) \quad (3\text{-}69b)$$

In (3-67a) $J_m(\beta_\rho \rho)$ and $Y_m(\beta_\rho \rho)$ represent, respectively, the Bessel functions of the first and second kind; $H_m^{(1)}(\beta_\rho \rho)$ and $H_m^{(2)}(\beta_\rho \rho)$ in (3-67b) represent, respectively, the Hankel functions of the first and second kind. A more detailed discussion of Bessel and Hankel functions is found in Appendix IV.

120 WAVE EQUATION AND ITS SOLUTIONS

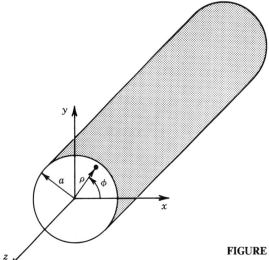

FIGURE 3-5 Cylindrical waveguide of the circular cross section.

Although (3-67a) through (3-69b) are valid solutions for $f(\rho)$, $g(\phi)$, and $h(z)$, the most appropriate form will depend on the problem in question. For example, for the cylindrical waveguide of Figure 3-5 the most convenient solutions for $f(\rho)$, $g(\phi)$, and $h(z)$ are those given, respectively, by (3-67a), (3-68b), and (3-69a). Thus we can write

$$\psi_1(\rho, \phi, z) = f(\rho)g(\phi)h(z)$$
$$= [A_1 J_m(\beta_\rho \rho) + B_1 Y_m(\beta_\rho \rho)]$$
$$\times [C_2 \cos(m\phi) + D_2 \sin(m\phi)][A_3 e^{-j\beta_z z} + B_3 e^{+j\beta_z z}] \quad (3\text{-}70)$$

These forms for $f(\rho)$, $g(\phi)$, and $h(z)$ were chosen in cylindrical coordinates for the following reasons.

1. Bessel functions of (3-67a) are used to represent standing waves whereas Hankel functions of (3-67b) represent traveling waves.
2. Exponentials of (3-68a) represent traveling waves whereas the cosines and sines of (3-68b) represent periodic waves.
3. Exponentials of (3-69a) represent traveling waves whereas the cosines and sines of (3-69b) represent standing waves.

Wave functions representing various radial waves in cylindrical coordinates are found listed in Table 3-2.

Within the circular waveguide of Figure 3-5 standing waves are created in the radial (ρ) direction, periodic waves in the phi (ϕ) direction, and traveling waves in the z direction. For the fields to be finite at $\rho = 0$, where $Y_m(\beta_\rho \rho)$ possesses a singularity, (3-70) reduces to

$$\psi_1'(\rho, \phi, z) = A_1 J_m(\beta_\rho \rho)[C_2 \cos(m\phi) + D_2 \sin(m\phi)][A_3 e^{-j\beta_z z} + B_3 e^{+j\beta_z z}]$$
$$(3\text{-}70a)$$

SOLUTION TO THE WAVE EQUATION   121

TABLE 3-2
Wave functions, zeroes, and infinities for radial wave functions in cylindrical coordinates

Wave type	Wave functions	Zeroes of wave functions	Infinities of wave functions
Traveling waves	$H_m^{(1)}(\beta\rho) = J_m(\beta\rho) + jY_m(\beta\rho)$ for $-\rho$ travel	$\beta\rho \to +j\infty$	$\beta\rho = 0$ $\beta\rho \to -j\infty$
	$H_m^{(2)}(\beta\rho) = J_m(\beta\rho) - jY_m(\beta\rho)$ for $+\rho$ travel	$\beta\rho \to -j\infty$	$\beta\rho = 0$ $\beta\rho \to +j\infty$
Standing waves	$J_m(\beta\rho)$ for $\pm\rho$	Infinite number (see Table 9-2)	$\beta\rho \to \pm j\infty$
	$Y_m(\beta\rho)$ for $\pm\rho$	Infinite number	$\beta\rho = 0$ $\beta\rho \to \pm j\infty$
Evanescent waves	$K_m(\alpha\rho) = \frac{\pi}{2}(-j)^{m+1}H_m^{(2)}(-j\alpha\rho)$ for $+\rho$	$\alpha\rho \to +\infty$	
	$I_m(\alpha\rho) = j^m J_m(-j\alpha\rho)$ for $-\rho$		$\alpha\rho \to +\infty$ for integer orders
Attenuating traveling waves	$H_m^{(1)}(\gamma\rho) = H_m^{(1)}(\alpha\rho + j\beta\rho)$ for $-\rho$ travel	$\gamma\rho \to +j\infty$	$\gamma\rho \to -j\infty$
	$H_m^{(2)}(\gamma\rho) = H_m^{(2)}(\alpha\rho + j\beta\rho)$ for $+\rho$ travel	$\gamma\rho \to -j\infty$	$\gamma\rho \to +j\infty$
Attenuating standing waves	$J_m(\gamma\rho) = J_m(\alpha\rho + j\beta\rho)$ for $\pm\rho$ $Y_m(\gamma\rho) = Y_m(\alpha\rho + j\beta\rho)$ for $\pm\rho$	Infinite number Infinite number	$\gamma\rho \to \pm j\infty$ $\gamma\rho \to \pm j\infty$

To represent the fields in the region outside the cylinder, a typical solution for $\psi(\rho, \phi, z)$ would take the form of

$$\psi_2(\rho, \phi, z) = B_1 H_m^{(2)}(\beta_\rho \rho)[C_2 \cos(m\phi) + D_2 \sin(m\phi)][A_3 e^{-j\beta_z z} + B_3 e^{+j\beta_z z}]$$
(3-70b)

whereby the Hankel function of the second kind $H_m^{(2)}(\beta_\rho \rho)$ has replaced the Bessel function of the first kind $J_m(\beta_\rho \rho)$ because outward traveling waves are formed outside the cylinder, in contrast to the standing waves inside the cylinder.

More details concerning the application and properties of Bessel and Hankel function can be found in Chapter 9.

3.4.3 Spherical Coordinate System

Spherical coordinates should be utilized in solving problems that exhibit spherical geometries. As for the rectangular and cylindrical geometries, the electric and magnetic fields of a spherical geometry boundary-value problem must satisfy the corresponding vector wave equation, which is most conveniently solved in spherical coordinates as illustrated in Figure 3-6.

To simplify the problem, let us assume that the space in which the electric and magnetic fields must be solved is source-free and lossless. A general solution for the

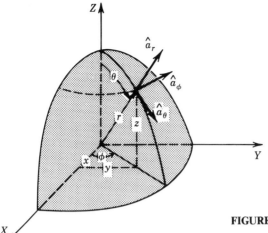

FIGURE 3-6 Spherical coordinate system and corresponding unit vectors.

electric field can then be written as

$$\mathbf{E}(r,\theta,\phi) = \hat{a}_r E_r(r,\theta,\phi) + \hat{a}_\theta E_\theta(r,\theta,\phi) + \hat{a}_\phi E_\phi(r,\theta,\phi) \quad (3\text{-}71)$$

Substituting (3-71) into the vector wave equation of (3-18a), we can write that

$$\nabla^2(\hat{a}_r E_r + \hat{a}_\theta E_\theta + \hat{a}_\phi E_\phi) = -\beta^2(\hat{a}_r E_r + \hat{a}_\theta E_\theta + \hat{a}_\phi E_\phi) \quad (3\text{-}72)$$

Since

$$\nabla^2(\hat{a}_r E_r) \neq \hat{a}_r \nabla^2 E_r \quad (3\text{-}73a)$$

$$\nabla^2(\hat{a}_\theta E_\theta) \neq \hat{a}_\theta \nabla^2 E_\theta \quad (3\text{-}73b)$$

$$\nabla^2(\hat{a}_\phi E_\phi) \neq \hat{a}_\phi \nabla^2 E_\phi \quad (3\text{-}73c)$$

(3-72) does not reduce to three simple scalar wave equations, similar to those of (3-20a) through (3-20c) for (3-20). Therefore the reduction of (3-72) to three scalar partial differential equations must proceed in a different manner. In fact, the method used here will be similar to that utilized in cylindrical coordinates to reduce the vector wave equation to three scalar partial differential equations.

To accomplish this, we first rewrite the vector wave equation of (3-51) in a form given by (3-53) where now all the operators on the left side can be performed in any coordinate system. Substituting (3-71) into (3-53) shows that after some lengthy mathematical manipulations (3-53) reduces to three scalar partial differential equations of the form

$$\nabla^2 E_r - \frac{2}{r^2}\left(E_r + E_\theta \cot\theta + \csc\theta \frac{\partial E_\phi}{\partial \phi} + \frac{\partial E_\theta}{\partial \theta}\right) = -\beta^2 E_r \quad (3\text{-}74a)$$

$$\nabla^2 E_\theta - \frac{1}{r^2}\left(E_\theta \csc^2\theta - 2\frac{\partial E_r}{\partial \theta} + 2\cot\theta \csc\theta \frac{\partial E_\phi}{\partial \phi}\right) = -\beta^2 E_\theta \quad (3\text{-}74b)$$

$$\nabla^2 E_\phi - \frac{1}{r^2}\left(E_\phi \csc^2\theta - 2\csc\theta \frac{\partial E_r}{\partial \phi} - 2\cot\theta \csc\theta \frac{\partial E_\theta}{\partial \phi}\right) = -\beta^2 E_\phi \quad (3\text{-}74c)$$

Unfortunately, all three of the preceding partial differential equations are coupled. This means each contains more than one component of the electric field and would be most difficult to solve in its present form. However, as will be shown in Chapter 10, TE′ and TM′ wave mode solutions can be formed that in spherical coordinates must satisfy the scalar wave equation of

$$\nabla^2 \psi(r, \theta, \phi) = -\beta^2 \psi(r, \theta, \phi) \tag{3-75}$$

where $\psi(r, \theta, \phi)$ is a scalar function that can represent a field or a vector potential component. Therefore it would be advisable here to demonstrate the solution to (3-75) in spherical coordinates.

Assuming a separable solution for $\psi(r, \theta, \phi)$ of the form

$$\psi(r, \theta, \phi) = f(r) g(\theta) h(\phi) \tag{3-76}$$

we can write the expanded form of (3-75)

$$\frac{1}{r^2} \frac{\partial}{\partial r}\left(r^2 \frac{\partial \psi}{\partial r} \right) + \frac{1}{r^2 \sin\theta} \frac{\partial}{\partial \theta}\left(\sin\theta \frac{\partial \psi}{\partial \theta} \right) + \frac{1}{r^2 \sin^2\theta} \frac{\partial^2 \psi}{\partial \phi^2} = -\beta^2 \psi \tag{3-77}$$

as

$$gh \frac{1}{r^2} \frac{\partial}{\partial r}\left(r^2 \frac{\partial f}{\partial r} \right) + fh \frac{1}{r^2 \sin\theta} \frac{\partial}{\partial \theta}\left(\sin\theta \frac{\partial g}{\partial \theta} \right) + fg \frac{1}{r^2 \sin^2\theta} \frac{\partial^2 h}{\partial \phi^2} = -\beta^2 fgh \tag{3-78}$$

Dividing both sides by fgh, multiplying by $r^2 \sin^2\theta$, and replacing the partials by ordinary derivatives reduces (3-78) to

$$\frac{\sin^2\theta}{f} \frac{d}{dr}\left(r^2 \frac{df}{dr} \right) + \frac{\sin\theta}{g} \frac{d}{d\theta}\left(\sin\theta \frac{dg}{d\theta} \right) + \frac{1}{h} \frac{d^2 h}{d\phi^2} = -(\beta r \sin\theta)^2 \tag{3-79}$$

Since the last term on the left side of (3-79) is only a function of ϕ, it can be set equal to

$$\frac{1}{h} \frac{d^2 h}{d\phi^2} = -m^2 \Rightarrow \frac{d^2 h}{d\phi^2} = -m^2 h \tag{3-80}$$

where m is a constant.

Substituting (3-80) into (3-79), dividing both sides by $\sin^2\theta$, and transposing the term from the right to the left side reduces (3-79) to

$$\frac{1}{f} \frac{d}{dr}\left(r^2 \frac{df}{dr} \right) + (\beta r)^2 + \frac{1}{g \sin\theta} \frac{d}{d\theta}\left(\sin\theta \frac{dg}{d\theta} \right) - \left(\frac{m}{\sin\theta} \right)^2 = 0 \tag{3-81}$$

Since the last two terms on the left side of (3-81) are only a function of θ, we can set them equal to

$$\frac{1}{g \sin\theta} \frac{d}{d\theta}\left(\sin\theta \frac{dg}{d\theta} \right) - \left(\frac{m}{\sin\theta} \right)^2 = -n(n+1) \tag{3-82}$$

where n is usually an integer. Equation 3-82 is closely related to the well-known *Legendre differential equation* (see Appendix V) [1–3, 6–10].

Substituting (3-82) into (3-81) reduces it to

$$\frac{1}{f}\frac{d}{dr}\left\{r^2\frac{df}{dr}\right\} + (\beta r)^2 - n(n+1) = 0 \tag{3-83}$$

which is closely related to the Bessel differential equation (see Appendix IV).

In summary then, the scalar wave equation 3-75 whose expanded form in spherical coordinates can be written as

$$\frac{1}{r^2}\frac{\partial}{\partial r}\left\{r^2\frac{\partial \psi}{\partial r}\right\} + \frac{1}{r^2 \sin\theta}\frac{\partial}{\partial \theta}\left\{\sin\theta \frac{\partial \psi}{\partial \theta}\right\} + \frac{1}{r^2 \sin^2\theta}\frac{\partial^2 \psi}{\partial \phi^2} = -\beta^2 \psi \tag{3-84}$$

and whose separable solution takes the form of

$$\boxed{\psi(r,\theta,\phi) = f(r)g(\theta)h(\phi)} \tag{3-85}$$

reduces to the three scalar differential equations

$$\boxed{\frac{d}{dr}\left\{r^2\frac{df}{dr}\right\} + \left[(\beta r)^2 - n(n+1)\right]f = 0} \tag{3-86a}$$

$$\boxed{\frac{1}{\sin\theta}\frac{d}{d\theta}\left\{\sin\theta \frac{dg}{d\theta}\right\} + \left[n(n+1) - \left\{\frac{m}{\sin\theta}\right\}^2\right]g = 0} \tag{3-86b}$$

$$\boxed{\frac{d^2 h}{d\phi^2} = -m^2 h} \tag{3-86c}$$

where m and n are constants (usually integers).

Solutions to (3-86a) through (3-86c) take the forms, respectively, of

$$f_1(r) = A_1 j_n(\beta r) + B_1 y_n(\beta r) \tag{3-87a}$$

or

$$f_2(r) = C_1 h_n^{(1)}(\beta r) + D_1 h_n^{(2)}(\beta r) \tag{3-87b}$$

and

$$g_1(\theta) = A_2 P_n^m(\cos\theta) + B_2 P_n^m(-\cos\theta) \qquad n \neq \text{integer} \tag{3-88a}$$

or

$$g_2(\theta) = C_2 P_n^m(\cos\theta) + D_2 Q_n^m(\cos\theta) \qquad n = \text{integer} \tag{3-88b}$$

and

$$h_1(\phi) = A_3 e^{-jm\phi} + B_3 e^{+jm\phi} \tag{3-89a}$$

or

$$h_2(\phi) = C_3 \cos(m\phi) + D_3 \sin(m\phi) \tag{3-89b}$$

SOLUTION TO THE WAVE EQUATION

TABLE 3-3
Wave functions, zeroes, and infinites for radial waves in spherical coordinates

Wave type	Wave functions	Zeroes of wave functions	Infinities of wave functions
Traveling waves	$h_n^{(1)}(\beta r) = j_n(\beta r) + jy_n(\beta r)$ for $-r$ travel	$\beta r \to +j\infty$	$\beta r = 0$ $\beta r \to -j\infty$
	$h_n^{(2)}(\beta r) = j_n(\beta r) - jy_n(\beta r)$ for $+r$ travel	$\beta r \to -j\infty$	$\beta r = 0$ $\beta r \to +j\infty$
Standing waves	$j_n(\beta r)$ for $\pm r$	Infinite number	$\beta r \to \pm j\infty$
	$y_n(\beta r)$ for $\pm r$	Infinite number	$\beta r = 0$ $\beta r \to \pm j\infty$

In (3-87a) $j_n(\beta r)$ and $y_n(\beta r)$ are referred to, respectively, as the *spherical Bessel functions* of the first and second kind. They are used to represent radial standing waves, and they are related, respectively, to the corresponding regular Bessel functions $J_{n+1/2}(\beta r)$ and $Y_{n+1/2}(\beta r)$ by

$$j_n(\beta r) = \sqrt{\frac{\pi}{2\beta r}} J_{n+1/2}(\beta r) \tag{3-90a}$$

$$y_n(\beta r) = \sqrt{\frac{\pi}{2\beta r}} Y_{n+1/2}(\beta r) \tag{3-90b}$$

In (3-87b) $h_n^{(1)}(\beta r)$ and $h_n^{(2)}(\beta r)$ are referred to, respectively, as the *spherical Hankel functions* of the first and second kind. They are used to represent radial traveling waves, and they are related, respectively, to the regular Hankel functions $H_{n+1/2}^{(1)}(\beta r)$ and $H_{n+1/2}^{(2)}(\beta r)$ by

$$h_n^{(1)}(\beta r) = \sqrt{\frac{\pi}{2\beta r}} H_{n+1/2}^{(1)}(\beta r) \tag{3-91a}$$

$$h_n^{(2)}(\beta r) = \sqrt{\frac{\pi}{2\beta r}} H_{n+1/2}^{(2)}(\beta r) \tag{3-91b}$$

Wave functions used to represent radial traveling and standing waves in spherical coordinates are listed in Table 3-3. More details on the spherical Bessel and Hankel functions can be found in Appendix IV.

In (3-88a) and (3-88b) $P_n^m(\cos\theta)$ and $Q_n^m(\cos\theta)$ are referred to, respectively, as the *associated Legendre functions* of the first and second kind (more details can be found in Appendix V).

The appropriate solution forms of f, g, and h will depend on the problem in question. For example, a typical solution for $\psi(r, \theta, \phi)$ of (3-85) to represent the fields within a sphere as shown in Figure 3-7 may take the form

$$\psi_1(r, \theta, \phi) = [A_1 j_n(\beta r) + B_1 y_n(\beta r)]$$
$$\times [C_2 P_n^m(\cos\theta) + D_2 Q_n^m(\cos\theta)][C_3 \cos(m\phi) + D_3 \sin(m\phi)] \tag{3-92}$$

126 WAVE EQUATION AND ITS SOLUTIONS

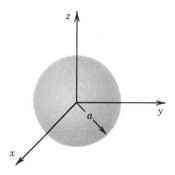

FIGURE 3-7 Geometry of a sphere of radius a.

For the fields to be finite at $r = 0$, where $y_n(\beta r)$ possesses a singularity, and for any value of θ, including $\theta = 0, \pi$ where $Q_n^m(\cos\theta)$ possesses singularities, (3-92) reduces to

$$\psi_1(r,\theta,\phi) = A_{mn} j_n(\beta r) P_n^m(\cos\theta)[C_3 \cos(m\phi) + D_3 \sin(m\phi)] \quad (3\text{-}92a)$$

To represent the fields outside a sphere a typical solution for $\psi(r, \theta, \phi)$ would take the form of

$$\psi_2(r,\theta,\phi) = B_{mn} h_n^{(2)}(\beta r) P_n^m(\cos\theta)[C_3 \cos(m\phi) + D_3 \sin(m\phi)] \quad (3\text{-}92b)$$

whereby the spherical Hankel function of the second kind $h_n^{(2)}(\beta r)$ has replaced the spherical Bessel function of the first kind $j_n(\beta r)$ because outward traveling waves are formed outside the sphere, in contrast to the standing waves inside the sphere.

Other spherical Bessel and Hankel functions that are most often encountered in boundary-value electromagnetic problems are those utilized by Schelkunoff [3, 11]. These spherical Bessel and Hankel functions, denoted in general by $\hat{B}_n(\beta r)$ to represent any of them, must satisfy the differential equation

$$\frac{d^2\hat{B}_n}{dr^2} + \left[\beta^2 - \frac{n(n+1)}{r^2}\right]\hat{B}_n = 0 \quad (3\text{-}93)$$

The spherical Bessel and Hankel functions that are solutions to this equation are related to other spherical Bessel and Hankel functions of (3-90a) through (3-91b), denoted here by $b_n(\beta r)$, and to the regular Bessel and Hankel functions, denoted here by $B_{n+1/2}(\beta r)$, by

$$\hat{B}_n(\beta r) = \beta r\, b_n(\beta r) = \beta r \sqrt{\frac{\pi}{2\beta r}}\, B_{n+1/2}(\beta r) = \sqrt{\frac{\pi \beta r}{2}}\, B_{n+1/2}(\beta r) \quad (3\text{-}94)$$

More details concerning the application and properties of the spherical Bessel and Hankel functions can be found in Chapter 10.

REFERENCES

1. F. B. Hilderbrand, *Advanced Calculus for Applications*, Prentice-Hall, Englewood Cliffs, N.J., 1962.
2. C. R. Wylie, Jr., *Advanced Engineering Mathematics*, McGraw-Hill, New York, 1960.

3. R. F. Harrington, *Time-Harmonic Electromagnetic Fields*, McGraw-Hill, New York, 1961.
4. R. B. Adler, L. J. Chu, and R. M. Fano, *Electromagnetic Energy Transmission and Radiation*, Chapter 8, Wiley, New York, 1960.
5. G. N. Watson, *A Treatise on the Theory of Bessel Functions*, Cambridge Univ. Press, London, 1948.
6. W. R. Smythe, *Static and Dynamic Electricity*, McGraw-Hill, New York, 1941.
7. J. A. Stratton, *Electromagnetic Theory*, McGraw-Hill, New York, 1960.
8. P. M. Morse and H. Feshbach, *Methods of Theoretical Physics*, Parts I and II, McGraw-Hill, New York, 1953.
9. M. Abramowitz and I. A. Stegun (eds.), *Handbook of Mathematical Functions with Formulas, Graphs, and Mathematical Tables*, National Bureau of Standards Applied Mathematics Series-55, U.S. Gov. Printing Office, Washington, D.C., 1966.
10. M. R. Spiegel, *Mathematical Handbook of Formulas and Tables*, Schaum's Outline Series, McGraw-Hill, New York, 1968.
11. S. A. Schelkunoff, *Electromagnetic Waves*, Van Nostrand, Princeton, N.J., 1943.

PROBLEMS

3.1. Derive the vector wave equations 3-16a and 3-16b for time-harmonic fields using the Maxwell equations of Table 1-4 for time-harmonic fields.

3.2. Verify that (3-28a) and (3-28b) are solutions to (3-26a).

3.3. Show that the second complex exponential in (3-31) represents a wave traveling in the $-z$ direction. Determine its phase velocity.

3.4. Using the method of separation of variables show that a solution to (3-37a) of the form (3-39) can be represented by (3-40a) through (3-43).

3.5. Show that the vector wave equation of (3-53) reduces, when **E** has a solution of the form (3-47), to the three scalar wave equations 3-54a through 3-54c.

3.6. Reduce (3-51) to (3-54a) through (3-54c) by expanding $\nabla^2 \mathbf{E}$. Do not use (3-52); rather use the scalar Laplacian in cylindrical coordinates and treat **E** as a vector given by (3-47). Use that

$$\frac{\partial \hat{a}_\rho}{\partial \rho} = \frac{\partial \hat{a}_\phi}{\partial \rho} = \frac{\partial \hat{a}_z}{\partial \rho} = 0 = \frac{\partial \hat{a}_z}{\partial \phi} = \frac{\partial \hat{a}_\rho}{\partial z} = \frac{\partial \hat{a}_\phi}{\partial z} = \frac{\partial \hat{a}_z}{\partial z}$$

$$\frac{\partial \hat{a}_\rho}{\partial \phi} = \hat{a}_\phi \qquad \frac{\partial \hat{a}_\phi}{\partial \phi} = -\hat{a}_\rho$$

3.7. Using large argument asymptotic forms, show that Bessel and Hankel functions represent, respectively, standing and traveling waves in the radial direction.

3.8. Using large argument asymptotic forms and assuming $e^{j\omega t}$ time convention, show that Hankel functions of the first kind represent traveling waves in the $-\rho$ direction whereas Hankel functions of the second kind represent traveling waves in the $+\rho$ direction. The opposite would be true were the time variations of the $e^{-j\omega t}$ form.

3.9. Using large argument asymptotic forms show that Bessel functions of complex argument represent attenuating standing waves.

3.10. Assuming time variations of $e^{j\omega t}$ and using large argument asymptotic forms, show that Hankel functions of the first and second kind with complex arguments represent, respectively, attenuating traveling waves in the $-\rho$ and $+\rho$ directions.

3.11. Show that when **E** has a solution of the form 3-71, the vector wave equation 3-53 reduces to the three scalar wave equations 3-74a through 3-74c.

3.12. Reduce (3-51) to (3-74a) through (3-74c) by expanding $\nabla^2 \mathbf{E}$. Do not use (3-52); rather use the scalar Laplacian in spherical coordinates and treat **E** as a vector given by (3-71). Use that

$$\frac{\partial \hat{a}_r}{\partial r} = \frac{\partial \hat{a}_\theta}{\partial r} = \frac{\partial \hat{a}_\phi}{\partial r} = 0$$

$$\frac{\partial \hat{a}_r}{\partial \theta} = \hat{a}_\theta \qquad \frac{\partial \hat{a}_\theta}{\partial \theta} = -\hat{a}_r \qquad \frac{\partial \hat{a}_\phi}{\partial \theta} = 0$$

$$\frac{\partial \hat{a}_r}{\partial \phi} = \sin\theta \, \hat{a}_\phi \qquad \frac{\partial \hat{a}_\theta}{\partial \phi} = \cos\theta \, \hat{a}_\phi \qquad \frac{\partial \hat{a}_\phi}{\partial \phi} = -\sin\theta \, \hat{a}_r - \cos\theta \, \hat{a}_\theta$$

3.13. Using large argument asymptotic forms show that spherical Bessel functions represent standing waves in the radial direction.

3.14. Show that spherical Hankel functions of the first and second kind represent, respectively, radial traveling waves in the $-\hat{r}$ and $+\hat{r}$ directions. Assume time variations of $e^{j\omega t}$ and large argument asymptotic expansions for the spherical Hankel functions.

3.15. Justify that associated Legendre functions represent standing waves in the θ direction of the spherical coordinate system.

3.16. Verify the relation 3-94 between the various forms of the spherical Bessel and Hankel functions and the regular Bessel and Hankel functions.

CHAPTER 4

WAVE PROPAGATION AND POLARIZATION

4.1 INTRODUCTION

In Chapter 3 we developed the vector wave equations for the electric and magnetic fields in lossless and lossy media. Solutions to the wave equations were also demonstrated in rectangular, cylindrical, and spherical coordinates using the method of *separation of variables*. In this chapter we want to consider solutions for the electric and magnetic fields of time-harmonic waves that travel in infinite lossless and lossy media. In particular, we want to develop expressions for *transverse electromagnetic* (TEM) waves (or modes) traveling along principal axes and oblique angles. The parameters of wave impedance, phase and group velocities, and power and energy densities will be discussed for each.

The concept of wave polarization will be introduced, and the necessary and sufficient conditions to achieve linear, circular, and elliptical polarizations will be discussed and illustrated. The sense of rotation, clockwise (right-hand) or counterclockwise (left-hand), will also be introduced.

4.2 TRANSVERSE ELECTROMAGNETIC MODES

A *mode* is a particular field configuration. For a given electromagnetic boundary-value problem, many field configurations that satisfy the wave equations, Maxwell's equations, and the boundary conditions usually exist. All these different field configurations (solutions) are usually referred to as *modes*.

A TEM mode is one whose field intensities, both **E** (electric) and **H** (magnetic), at every point in space are contained in a local plane, referred to as *equiphase plane*, that is independent of time. In general, the orientations of the local planes

130 WAVE PROPAGATION AND POLARIZATION

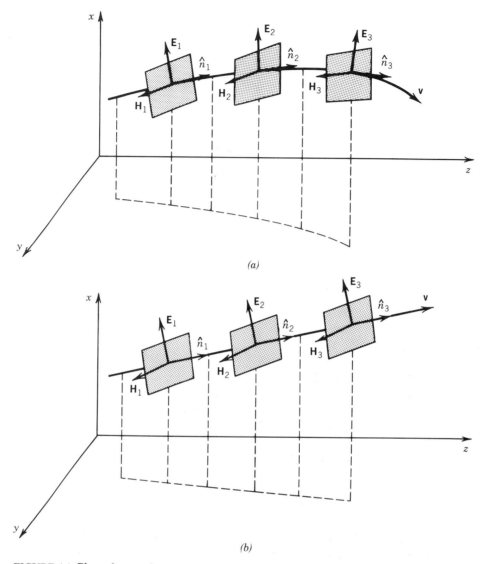

FIGURE 4-1 Phase fronts of (*a*) TEM and (*b*) planes waves.

associated with the TEM wave are different at different points in space. In other words, at point (x_1, y_1, z_1) all the field components are contained in a plane. At another point (x_2, y_2, z_2) all field components are again contained in a plane; however, the two planes need not be parallel. This is illustrated in Figure 4-1*a*.

If the space orientation of the planes for a TEM mode is the same (equiphase planes are parallel), as shown in Figure 4-1*b*, then the fields form *plane waves*. In other words, the equiphase surfaces are parallel planar surfaces. If in addition to having planar equiphases the field has equiamplitude planar surfaces (the amplitude is the same over each plane), then it is called a *uniform plane wave*; that is, the field is not a function of the coordinates that form the equiphase and equiamplitude planes.

4.2.1 Uniform Plane Waves in an Unbounded Lossless Medium—Principal Axis

In this section we will write expressions for the electric and magnetic fields of a uniform plane wave traveling in an unbounded medium. In addition the wave impedance, phase and energy (group) velocities, and power and energy densities of the wave will be discussed.

A. ELECTRIC AND MAGNETIC FIELDS

Let us assume that a time-harmonic uniform plane wave is traveling in an unbounded lossless medium (ε, μ) in the z direction (either positive or negative), as shown in Figure 4-2a. In addition, for simplicity, let us assume the electric field of the wave has only an x component. We want to write expressions for the electric and magnetic fields associated with this wave.

For the electric and magnetic field components to be valid solutions of a time-harmonic electromagnetic wave, they must satisfy Maxwell's equations as given

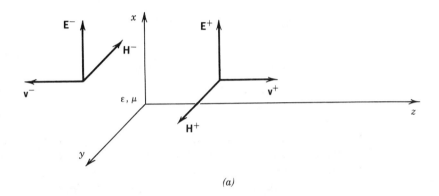

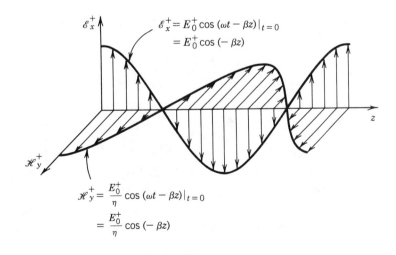

FIGURE 4-2 (a) Complex and (b) instantaneous fields of a uniform plane wave.

132 WAVE PROPAGATION AND POLARIZATION

in Table 1-4 or the corresponding wave equations as given, respectively, by (3-18a) and (3-18b). Here the approach will be to initiate the solution by solving the wave equation for either the electric or magnetic field and then finding the other field using Maxwell's equations. An alternate procedure, which has been assigned as an end of chapter problem, would be to follow the entire solution using only Maxwell's equations.

Since the electric field has only an x component, it must satisfy the scalar wave equation of (3-20a) or (3-22) whose general solution is given by (3-23). Because the wave is a uniform plane wave that travels in the z direction, its solution is not a function of x and y. Therefore (3-23) reduces to

$$E_x(z) = h(z) \tag{4-1}$$

The solutions of $h(z)$ are given by (3-30a) or (3-30b). Since the wave in question is a traveling wave instead of a standing wave, its most appropriate solution is that given by (3-30a). The first term in (3-30a) represents a wave that travels in the $+z$ direction and the second term represents a wave that travels in the $-z$ direction. Therefore the solution of (4-1) using (3-30a) can be written as

$$E_x(z) = A_3 e^{-j\beta z} + B_3 e^{+j\beta z} = E_x^+ + E_x^- \tag{4-2}$$

or

$$E_x(z) = E_0^+ e^{-j\beta z} + E_0^- e^{+j\beta z} = E_x^+ + E_x^- \tag{4-2a}$$

$$E_x^+(z) = E_0^+ e^{-j\beta z} \tag{4-2b}$$

$$E_x^-(z) = E_0^- e^{+j\beta z} \tag{4-2c}$$

since $\beta_z = \beta$ because $\beta_x = \beta_y = 0$. E_0^+ and E_0^- represent, respectively, the amplitudes of the positive and negative (in the z direction) traveling waves.

The corresponding magnetic field must also be a solution of its wave equation 3-18b, and its form will be similar to (4-2). However, since we do not know which components of magnetic field coexist with the x component of the electric field, they are most appropriately determined by using one of Maxwell's equations as given in Table 1-4. Since the electric field is known, as given by (4-2), the magnetic field can best be found using

$$\nabla \times \mathbf{E} = -j\omega\mu\mathbf{H} \tag{4-3}$$

or

$$\mathbf{H} = -\frac{1}{j\omega\mu}\nabla \times \mathbf{E} = -\frac{1}{j\omega\mu} \begin{bmatrix} \hat{a}_x & \hat{a}_y & \hat{a}_z \\ \dfrac{\partial}{\partial x} & \dfrac{\partial}{\partial y} & \dfrac{\partial}{\partial z} \\ E_x & 0 & 0 \end{bmatrix} \tag{4-3a}$$

which using (4-2a) reduces to

$$\mathbf{H} = -\hat{a}_y \frac{1}{j\omega\mu}\left\{\frac{\partial E_x}{\partial z}\right\} = \hat{a}_y \frac{\beta}{\omega\mu}\left\{E_0^+ e^{-j\beta z} - E_0^- e^{+j\beta z}\right\}$$

$$\mathbf{H} = \hat{a}_y \frac{1}{\sqrt{\mu/\varepsilon}}\left\{E_0^+ e^{-j\beta z} - E_0^- e^{+j\beta z}\right\} = \hat{a}_y \frac{1}{\sqrt{\mu/\varepsilon}}\left\{E_x^+ - E_x^-\right\} = \hat{a}_y\left\{H_y^+ + H_y^-\right\}$$

$$\tag{4-3b}$$

where

$$H_y^+ = \frac{1}{\sqrt{\mu/\varepsilon}} E_x^+ \qquad (4\text{-}3c)$$

$$H_y^- = -\frac{1}{\sqrt{\mu/\varepsilon}} E_x^- \qquad (4\text{-}3d)$$

Plots of the instantaneous *positive* traveling electric and magnetic fields at $t = 0$ as a function of z are shown in Figure 4-2b. Similar plots can be drawn for the negative traveling fields.

B. WAVE IMPEDANCE

Since each term for the magnetic field (A/m) in (4-3c) and (4-3d) is individually identical to the corresponding term for the electric field (V/m) in (4-2a), the factor $\sqrt{\mu/\varepsilon}$ in the denominator in (4-3c) and (4-3d) must have units of ohms (V/A). Therefore the factor $\sqrt{\mu/\varepsilon}$ is known as the *wave impedance* Z_w, the ratio of the electric to magnetic field, and it is usually represented by

$$Z_w = \frac{E_x^+}{H_y^+} = -\frac{E_x^-}{H_y^-} = \eta = \sqrt{\frac{\mu}{\varepsilon}} \qquad (4\text{-}4)$$

The wave impedance of (4-4) is identical to a quantity that is referred to as the *intrinsic impedance* $\eta = \sqrt{\mu/\varepsilon}$ of the medium. In general, this is true, not only for uniform plane waves but also for plane and TEM waves; however, it is not true for TE or TM modes.

In (4-3d) it is also observed that a negative sign is found in front of the magnetic field component that travels in the $-z$ direction; a positive sign is noted in front of the positive traveling wave. The general procedure that can be followed to find the magnetic field components, given the electric field components, or to find the electric field components, given the magnetic field components, is the following:

1. Place the fingers of your right hand in the direction of the electric field component.
2. Direct your thumb toward the direction of wave travel (power flow).
3. Rotate your fingers 90° in a direction so that a right-hand screw is formed.
4. The new direction of your fingers is the direction of the magnetic field component.
5. Divide the electric field component by the wave impedance to obtain the corresponding magnetic field component.

The foregoing procedure must be followed for each term of each component of an electric or magnetic field. The results are identical to those that would be obtained by using Maxwell's equations. If the wave impedance is known in advance, as it is for TEM waves, this procedure is simpler and much more rapid than using Maxwell's equations. By following this procedure, the answers (including the signs) in (4-3c) and (4-3d) given (4-2b) and (4-2c) are obvious.

134 WAVE PROPAGATION AND POLARIZATION

To illustrate the procedure, let us consider another example.

Example 4-1. The electric field of a uniform plane wave traveling in free space is given by

$$\mathbf{E} = \hat{a}_y\left(E_0^+ e^{-j\beta z} + E_0^- e^{+j\beta z}\right) = \hat{a}_y\left(E_y^+ + E_y^-\right)$$

where E_0^+ and E_0^- are constants. Find the corresponding magnetic field using the outlined procedure.

Solution. For the electric field component that is traveling in the $+z$ direction, the corresponding magnetic field component is given by

$$\mathbf{H}^+ = -\hat{a}_x \frac{E_0^+}{\eta_0} e^{-j\beta z} \simeq -\hat{a}_x \frac{E_0^+}{377} e^{-j\beta z}$$

where

$$\eta_0 = Z_w = \sqrt{\frac{\mu_0}{\varepsilon_0}} \simeq 377 \text{ ohms}$$

Similarly, for the wave that is traveling in the $-z$ direction we can write that

$$\mathbf{H}^- = \hat{a}_x \frac{E_0^-}{\eta_0} e^{+j\beta z} \simeq \hat{a}_x \frac{E_0^-}{377} e^{+j\beta z}$$

Therefore the total magnetic field is equal to

$$\mathbf{H} = \mathbf{H}^+ + \mathbf{H}^- = \hat{a}_x \frac{1}{\eta_0}\left(-E_0^+ e^{-j\beta z} + E_0^- e^{+j\beta z}\right)$$

The same answer would be obtained if Maxwell's equations were used, and it is assigned as an end of chapter problem.

The term in the expression for the electric field in (4-2a) that identifies the direction of wave travel can also be written in vector notation. This is usually more convenient to use when dealing with waves traveling at oblique angles. Equation 4-2a can therefore take the more general form of

$$E_x(z) = E_0^+ e^{-j\boldsymbol{\beta}^+ \cdot \mathbf{r}} + E_0^- e^{-j\boldsymbol{\beta}^- \cdot \mathbf{r}} \tag{4-5}$$

where

$$\boldsymbol{\beta}^+ = \hat{\boldsymbol{\beta}}^+ \beta = \hat{a}_x \beta_x^+ + \hat{a}_y \beta_y^+ + \hat{a}_z \beta_z^+ \Big|_{\substack{\beta_x^+ = \beta_y^+ = 0 \\ \beta_z^+ = \beta}} = \hat{a}_z \beta \tag{4-5a}$$

$$\boldsymbol{\beta}^- = \hat{\boldsymbol{\beta}}^- \beta = \hat{a}_x \beta_x^- + \hat{a}_y \beta_y^- - \hat{a}_z \beta_z^- \Big|_{\substack{\beta_x^- = \beta_y^- = 0 \\ \beta_z^- = \beta}} = -\hat{a}_z \beta \tag{4-5b}$$

$$\mathbf{r} = \text{position vector} = \hat{a}_x x + \hat{a}_y y + \hat{a}_z z \tag{4-5c}$$

In (4-5a) through (4-5c) β_x, β_y, β_z represent, respectively, the phase constants of the wave in the x, y, z directions, $\mathbf{r}$ represents the position vector in rectangular coordinates, and $\hat{\boldsymbol{\beta}}^+$ and $\hat{\boldsymbol{\beta}}^-$ represent unit vectors in the directions of $\boldsymbol{\beta}^+$ and $\boldsymbol{\beta}^-$.

The notation used in (4-5) through (4-5c) to represent the wave travel will be most convenient to express wave travel at oblique angles, as will be the case in Section 4.2.2.

C. PHASE AND ENERGY (GROUP) VELOCITIES, POWER, AND ENERGY DENSITIES

The expressions for the electric and magnetic fields, as given by (4-2a) and (4-3b), represent the spatial variations of the field intensities. The corresponding instantaneous forms of each can be written, using (1-61a) and (1-61b) and assuming E_0^+ and E_0^- are real constants, as

$$\mathscr{E}_x(z;t) = \mathscr{E}_x^+(z;t) + \mathscr{E}_x^-(z;t) = \text{Re}\left[E_0^+ e^{-j\beta z} e^{j\omega t}\right] + \text{Re}\left[E_0^- e^{+j\beta z} e^{j\omega t}\right]$$
$$= E_0^+ \cos(\omega t - \beta z) + E_0^- \cos(\omega t + \beta z) \quad (4\text{-}6a)$$

$$\mathscr{H}_y(z;t) = \mathscr{H}_y^+(z;t) + \mathscr{H}_y^-(z;t)$$
$$= \frac{1}{\sqrt{\mu/\varepsilon}}\left[E_0^+ \cos(\omega t - \beta z) - E_0^- \cos(\omega t + \beta z)\right] \quad (4\text{-}6b)$$

In each of the fields, as given by (4-6a) and (4-6b), the first term represents, according to (3-34) through (3-35) and Figure 3-3, a wave that travels in the $+z$ direction; the second term represents a wave that travels in the $-z$ direction. To maintain a constant phase in the first term of (4-6a), the velocity must be equal, according to (3-35), to

$$v_p^+ = +\frac{dz}{dt} = \frac{\omega}{\beta} = \frac{\omega}{\omega\sqrt{\mu\varepsilon}} = \frac{1}{\sqrt{\mu\varepsilon}} \quad (4\text{-}7)$$

The corresponding velocity of the second term in (4-6a) is identical in magnitude to (4-7) but with a negative sign to reflect the direction of wave travel. The velocity of (4-7) is referred to as the *phase velocity*, and it represents the velocity that must be maintained in order to keep in step with a constant phase front of the wave. As will be shown for oblique traveling waves, the phase velocity of such waves can exceed the velocity of light. Aside of nonuniform plane waves, also referred to as slow surface waves (see Section 5.3.4A), in general the phase velocity can be equal to or even greater than the speed of light. Variations of the instantaneous positive traveling electric $\mathscr{E}_x^+(z;t)$ and magnetic $\mathscr{H}_y^+(z;t)$ fields as a function of z for $t = 0$ are shown in Figure 4-2b. As time increases, both curves will shift in the positive z direction. A similar set of curves can be drawn for the negative traveling electric $\mathscr{E}_x^-(z;t)$ and magnetic $\mathscr{H}_y^-(z;t)$ fields.

The electric and magnetic energies (W-s/m³) and power densities (W/m²) associated with the positive traveling waves of (4-6a) and (4-6b) can be written, according to (1-58f) and (1-58e), as

$$w_e^+ = \tfrac{1}{2}\varepsilon\mathscr{E}_x^{+2} = \tfrac{1}{2}\varepsilon E_0^{+2}\cos^2(\omega t - \beta z) \quad (4\text{-}8a)$$

$$w_m^+ = \tfrac{1}{2}\mu\mathscr{H}_y^{+2} = \tfrac{1}{2}\mu\left[(\varepsilon/\mu)E_0^{+2}\cos^2(\omega t - \beta z)\right] = \tfrac{1}{2}\varepsilon E_0^{+2}\cos^2(\omega t - \beta z) \quad (4\text{-}8b)$$

$$\mathscr{S}^+ = \mathscr{E}^+ \times \mathscr{H}^+ = \hat{a}_x E_0^+ \cos(\omega t - \beta z) \times \left[\hat{a}_y(1/\sqrt{\mu/\varepsilon})E_0^+ \cos(\omega t - \beta z)\right]$$
$$= \hat{a}_z\mathscr{S}^+ = \hat{a}_z\left(1/\sqrt{\mu/\varepsilon}\right)E_0^{+2}\cos^2(\omega t - \beta z) \quad (4\text{-}8c)$$

The ratio formed by dividing the power density $\mathscr{S}$ (W/m²) by the total energy density $w = w_e + w_m$ (J/m³ = W-s/m³) is referred to as the *energy (group) velocity* v_e, and it is given by

$$v_e^+ = \frac{\mathscr{S}^+}{w^+} = \frac{\mathscr{S}^+}{w_e^+ + w_m^+} = \frac{(1/\sqrt{\mu/\varepsilon})E_0^{+2}\cos^2(\omega t - \beta z)}{\varepsilon E_0^{+2}\cos^2(\omega t - \beta z)} = \frac{1}{\sqrt{\mu\varepsilon}} \quad (4\text{-}9)$$

The energy velocity represents the velocity with which the wave energy is transported. It is apparent that (4-9) is identical to (4-7). In general that is not the case. In fact the energy velocity v_e^+ can be equal to but not exceed the speed of light, and the product of the phase velocity v_p and energy velocity v_e must always be equal to

$$\boxed{v_p^+ v_e^+ = v^{+2} = \frac{1}{\mu\varepsilon}} \quad (4\text{-}10)$$

where $v^+ = 1/\sqrt{\mu\varepsilon}$ is the speed of light. The same holds for the negative traveling waves.

The time-average power density (Poynting vector) associated with the positive traveling wave can be written, using (1-70) and the first terms of (4-2a) and (4-3b), as

$$\mathscr{S}_{av}^+ = \frac{1}{2}\text{Re}(\mathbf{E}^+ \times \mathbf{H}^{+*}) = \hat{a}_z \frac{1}{2\sqrt{\mu/\varepsilon}}|E_x^+|^2 = \hat{a}_z \frac{|E_0^+|^2}{2\sqrt{\mu/\varepsilon}} = \hat{a}_z \frac{|E_0^+|^2}{2\eta} \quad (4\text{-}11)$$

A similar expression is derived for the negative traveling wave.

D. STANDING WAVES

Each of the terms in (4-2a) and (4-3b) represents individually *traveling* waves, the first traveling in the positive z direction and the second in the negative z direction. The two together form a so-called *standing wave*, which is comprised of two oppositely traveling waves.

To examine the characteristics of a standing wave, let us rewrite (4-2a) as

$$E_x(z) = E_0^+ e^{-j\beta z} + E_0^- e^{+j\beta z}$$
$$= E_0^+[\cos(\beta z) - j\sin(\beta z)] + E_0^-[\cos(\beta z) + j\sin(\beta z)]$$
$$= (E_0^+ + E_0^-)\cos(\beta z) - j(E_0^+ - E_0^-)\sin(\beta z)$$
$$= \sqrt{(E_0^+ + E_0^-)^2 \cos^2(\beta z) + (E_0^+ - E_0^-)^2 \sin^2(\beta z)}$$
$$\times \exp\left\{-j\tan^{-1}\left[\frac{(E_0^+ - E_0^-)\sin(\beta z)}{(E_0^+ + E_0^-)\cos(\beta z)}\right]\right\}$$
$$E_x(z) = \sqrt{(E_0^+)^2 + (E_0^-)^2 + 2E_0^+ E_0^- \cos(2\beta z)}$$
$$\times \exp\left\{-j\tan^{-1}\left[\frac{(E_0^+ - E_0^-)}{(E_0^+ + E_0^-)}\tan(\beta z)\right]\right\} \quad (4\text{-}12)$$

The amplitude of the waveform given by (4-12) is equal to

$$|E_x(z)| = \sqrt{(E_0^+)^2 + (E_0^-)^2 + 2E_0^+ E_0^- \cos(2\beta z)} \qquad (4\text{-}12a)$$

By examining (4-12a) it is evident that its maximum and minimum values are given, respectively, by

$$|E_x(z)|_{\max} = |E_0^+| + |E_0^-| \quad \text{when } \beta z = m\pi,\ m = 0, 1, 2, \ldots \qquad (4\text{-}13a)$$

For $|E_0^+| > |E_0^-|$

$$|E_x(z)|_{\min} = |E_0^+| - |E_0^-| \quad \text{when } \beta z = \frac{(2m+1)\pi}{2},\ m = 0, 1, 2, \ldots \qquad (4\text{-}13b)$$

Neighboring maximum and minimum values are separated by a distance of $\lambda/4$ or successive maxima or minima are separated by $\lambda/2$.

The instantaneous field of (4-12) can also be written as

$$\begin{aligned}\mathscr{E}_x(z; t) &= \mathrm{Re}\left[E_x(z) e^{j\omega t}\right] \\ &= \sqrt{(E_0^+)^2 + (E_0^-)^2 + 2E_0^+ E_0^- \cos(2\beta z)} \\ &\quad \times \cos\left[\omega t - \tan^{-1}\left\{\frac{E_0^+ - E_0^-}{E_0^+ + E_0^-} \tan(\beta z)\right\}\right]\end{aligned} \qquad (4\text{-}14)$$

It is apparent that (4-12a) represents the envelope of the maximum values the instantaneous field of (4-14) will ever achieve as a function of time at a given position. Since this envelope of maximum values does not move (change) in position as a function of time, it is referred to as the *standing wave pattern* and the associated wave of (4-12) or (4-14) is referred to as the *standing wave*.

The ratio of the maximum/minimum values of the standing wave pattern of (4-12a), as given by (4-13a) and (4-13b), is referred to as the standing wave ratio (SWR) and it is given by

$$\mathrm{SWR} = \frac{|E_x(z)|_{\max}}{|E_x(z)|_{\min}} = \frac{|E_0^+| + |E_0^-|}{|E_0^+| - |E_0^-|} = \frac{1 + \dfrac{|E_0^-|}{|E_0^+|}}{1 - \dfrac{|E_0^-|}{|E_0^+|}} = \frac{1 + |\Gamma|}{1 - |\Gamma|} \qquad (4\text{-}15)$$

where Γ is the reflection coefficient. Since in transmission lines we usually deal with voltages and currents (instead of electric and magnetic fields), the SWR is usually referred to as the VSWR (voltage standing wave ratio). Plots of the standing wave pattern in terms of E_0^+ as a function of z ($-\lambda \leq z \leq \lambda$) for $|\Gamma| = 0, 0.2, 0.4, 0.6, 0.8$, and 1 are shown in Figure 4-3.

The SWR is a quantity that can be measured with instrumentation [1, 2]. SWR has values in the range of $1 \leq \mathrm{SWR} \leq \infty$. The value of the SWR indicates the amount of interference between the two opposite traveling waves; the smaller the SWR value, the lesser the interference. The minimum SWR value of unity occurs when $|\Gamma| = E_0^-/E_0^+ = 0$, and it indicates that no interference is formed. Thus the standing wave reduces to a pure traveling wave. The maximum SWR of infinity occurs when $|\Gamma| = E_0^-/E_0^+ = 1$, and it indicates that the negative traveling wave is of

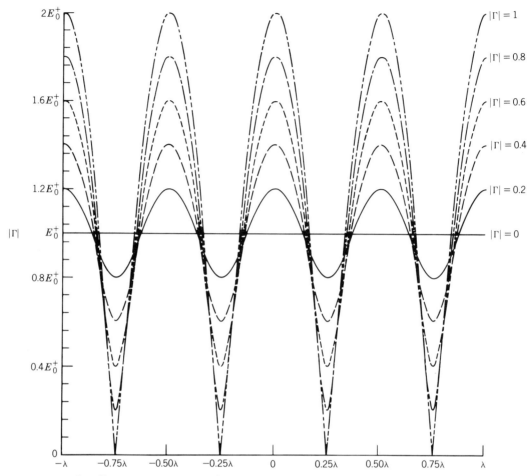

FIGURE 4-3 Standing wave pattern as a function of distance for a uniform plane wave with different reflection coefficients.

the same intensity as the positive traveling wave. This provides the maximum interference, and the wave forms a pure standing wave pattern given by

$$|E_x(z)|_{E_0^+ = E_0^-} = 2E_0^+ |\cos(\beta z)| = 2E_0^- |\cos(\beta z)| \tag{4-16}$$

The pattern of this is a rectified cosine function, and it is represented in Figure 4-3 by the $|\Gamma| = 1$ curve. The pattern exhibits pure nulls and peak values of twice the amplitude of the incident wave.

4.2.2 Uniform Plane Waves in an Unbounded Lossless Medium—Oblique Angle

In this section expressions for the electric and magnetic fields, wave impedance, phase and group velocities, and power and energy densities will be written for uniform plane waves traveling at oblique angles in an unbounded medium. All of these will be done for waves that are uniform plane waves to the direction of travel.

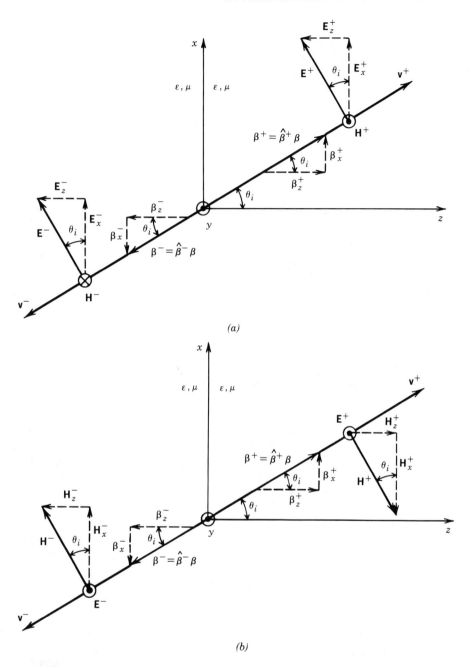

FIGURE 4-4 Transverse electric and magnetic uniform plane waves in an unbounded medium at an oblique angle. (*a*) TE^y mode. (*b*) TM^y mode.

A. ELECTRIC AND MAGNETIC FIELDS

Let us assume that a uniform plane wave is traveling in an unbounded medium in a direction shown in Figure 4-4*a*. The amplitudes of the positive and negative traveling electric fields are E_0^+ and E_0^-, respectively, and the assumed directions of each are also illustrated in Figure 4-4*a*. It is desirable to write expressions for the positive and negative traveling electric and magnetic field components.

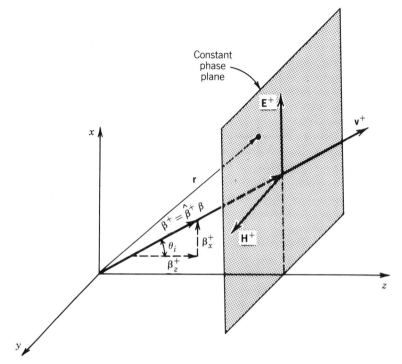

FIGURE 4-5 Phase front of a TEM wave traveling in a general direction.

Since the electric field of the wave of Figure 4-4a does not have a y component, the field configuration is referred to as *transverse electric to y* (TE^y). More detailed discussion on the construction of *transverse electric* (TE) and *transverse magnetic* (TM) field configurations, as well as *transverse electromagnetic* (TEM), can be found in Chapter 6.

Because for the TE^y wave of Figure 4-4a the electric field is pointing along a direction that does not coincide with any of the principal axes, it can be decomposed into components coincident with the principal axes. According to the geometry of Figure 4-4a, it is evident that the electric field can be written as

$$\mathbf{E} = \mathbf{E}^+ + \mathbf{E}^- = E_0^+ (\hat{a}_x \cos\theta_i - \hat{a}_z \sin\theta_i) e^{-j\boldsymbol{\beta}^+ \cdot \mathbf{r}}$$
$$+ E_0^- (\hat{a}_x \cos\theta_i - \hat{a}_z \sin\theta_i) e^{-j\boldsymbol{\beta}^- \cdot \mathbf{r}} \quad (4\text{-}17)$$

where $\mathbf{r}$ is the position vector of (4-5c), and it is displayed graphically in Figure 4-5. Since the phase constants $\boldsymbol{\beta}^+$ and $\boldsymbol{\beta}^-$ can be written, respectively, as

$$\boldsymbol{\beta}^+ = \hat{\boldsymbol{\beta}}^+ \beta = \hat{a}_x \beta_x^+ + \hat{a}_z \beta_z^+ = \beta(\hat{a}_x \sin\theta_i + \hat{a}_z \cos\theta_i) \quad (4\text{-}17a)$$

$$\boldsymbol{\beta}^- = \hat{\boldsymbol{\beta}}^- \beta = \hat{a}_x \beta_x^- + \hat{a}_z \beta_z^- = -\beta(\hat{a}_x \sin\theta_i + \hat{a}_z \cos\theta_i) \quad (4\text{-}17b)$$

(4-17) can be expressed as

$$\mathbf{E} = E_0^+ (\hat{a}_x \cos\theta_i - \hat{a}_z \sin\theta_i) e^{-j\beta(x \sin\theta_i + z \cos\theta_i)}$$
$$+ E_0^- (\hat{a}_x \cos\theta_i - \hat{a}_z \sin\theta_i) e^{+j\beta(x \sin\theta_i + z \cos\theta_i)} \quad (4\text{-}18a)$$

Since the wave is a uniform plane wave, the amplitude of its magnetic field is related to the amplitude of its electric field by the wave impedance (in this case also by the intrinsic impedance) as given by (4-4). Since the magnetic field is traveling in the same direction as the electric field, the exponentials used to indicate its directions of travel are the same as those of the electric field as given in (4-18a). The directions of the magnetic field can be found using the right-hand procedure outlined in Section 4.2.1 and illustrated graphically in Figure 4-2b for the positive traveling wave. Using all of the preceding information, it is evident that the magnetic field corresponding to the electric field of (4-18a) can be written as

$$\mathbf{H} = \mathbf{H}^+ + \mathbf{H}^- = \hat{a}_y \left[\frac{E_0^+}{\eta} e^{-j\beta(x \sin \theta_i + z \cos \theta_i)} - \frac{E_0^-}{\eta} e^{+j\beta(x \sin \theta_i + z \cos \theta_i)} \right] \quad (4\text{-}18b)$$

In vector form (4-18b) can also be written as

$$\mathbf{H} = \frac{1}{\eta} \left[\hat{\beta}^+ \times \mathbf{E}^+ + \hat{\beta}^- \times \mathbf{E}^- \right] \quad (4\text{-}18c)$$

The same form can be used to relate the $\mathbf{E}$ and $\mathbf{H}$ for any TEM wave traveling in any direction. It is apparent that when $\theta_i = 0$ (4-18a) and (4-18b) reduce to the forms of (4-2a) and (4-3b), respectively. The same answer for the magnetic field of (4-18b) can be obtained by applying Maxwell's equation 4-3 to the electric field of (4-18a). This is left for the reader as an end of the chapter exercise.

The planes of constant phase at any time t are obtained by setting the phases of (4-18a) or (4-18b) equal to a constant, that is

$$\boldsymbol{\beta}^+ \cdot \mathbf{r} = \beta_x^+ x + \beta_y^+ y + \beta_z^+ z \big|_{y=0} = \beta(x \sin \theta_i + z \cos \theta_i) = C^+ \quad (4\text{-}19a)$$

$$\boldsymbol{\beta}^- \cdot \mathbf{r} = \beta_x^- x + \beta_y^- y + \beta_z^- z \big|_{y=0} = -\beta(x \sin \theta_i + z \cos \theta_i) = C^- \quad (4\text{-}19b)$$

Each of (4-19a) and (4-19b) are equations of a plane in either the spherical or rectangular coordinates with unit vectors $\hat{\beta}^+$ and $\hat{\beta}^-$ normal to each of the respective surfaces. The respective phase velocities in any direction (r, x, or z) are obtained by letting

$$\boldsymbol{\beta}^+ \cdot \mathbf{r} - \omega t = \beta(x \sin \theta_i + z \cos \theta_i) - \omega t = C_0^+ \quad (4\text{-}19c)$$

$$\boldsymbol{\beta}^- \cdot \mathbf{r} - \omega t = -\beta(x \sin \theta_i + z \cos \theta_i) - \omega t = C_0^- \quad (4\text{-}19d)$$

and taking a derivative with respect to time.

Example 4-2. Another exercise of interest is that in which the electric field is directed along the $+y$ direction and the wave is traveling along an oblique angle θ_i, as shown in Figure 4-4b. This is referred to as a TM^y wave. The objective here is again to write expressions for the positive and negative electric and magnetic field components, assuming the amplitudes of the positive and negative electric field components are E_0^+ and E_0^-, respectively.

Solution. Since this wave only has a y electric field component and it is traveling in the same direction as that of Figure 4-4a, we can write the electric field as

$$\mathbf{E} = \mathbf{E}^+ + \mathbf{E}^- = \hat{a}_y \left[E_0^+ e^{-j\beta(x \sin \theta_i + z \cos \theta_i)} + E_0^- e^{+j\beta(x \sin \theta_i + z \cos \theta_i)} \right]$$

Using the right-hand procedure outlined in Section 4.2.1, the corresponding magnetic field components are pointed along directions indicated in Figure 4-4b. Since the magnetic field is not directed along any of the principal axes, it can be decomposed into components that coincide with the principal axes [as shown in Figure 4-4b]. Doing this and relating the amplitude of the electric and magnetic fields by the intrinsic impedance, we can write the magnetic field as

$$\mathbf{H} = \mathbf{H}^+ + \mathbf{H}^- = \frac{E_0^+}{\eta}(-\hat{a}_x \cos\theta_i + \hat{a}_z \sin\theta_i)e^{-j\beta(x\sin\theta_i + z\cos\theta_i)}$$

$$+ \frac{E_0^-}{\eta}(\hat{a}_x \cos\theta_i - \hat{a}_z \sin\theta_i)e^{+j\beta(x\sin\theta_i + z\cos\theta_i)}$$

The same answers could have been obtained if Maxwell's equation 4-3 were used. Since the magnetic field does not have any y components, this field configuration is referred to as *transverse magnetic to y* (TM^y), which will be discussed more in detail in Chapter 6.

B. WAVE IMPEDANCE

Since the TE^y and TM^y fields of Section 4.2.2A were TEM to the direction of travel, the wave impedance of each in the direction β of wave travel is the same as the intrinsic impedance of the medium. However there are other directional impedances toward the x and z directions. These impedances are obtained by dividing the electric field component by the corresponding orthogonal magnetic field component. These two components are chosen so that the cross product of the electric to the magnetic field, which corresponds to the direction of power flow, is in the direction of the wave travel.

Following the aforementioned procedure, the directional impedances for the TE^y fields of (4-18a) and (4-18b) can be written as

$$\underline{\text{TE}^y}$$

$$Z_x^+ = -\frac{E_z^+}{H_y^+} = \eta \sin\theta_i = Z_x^- = \frac{E_z^-}{H_y^-} \qquad (4\text{-}20a)$$

$$Z_z^+ = \frac{E_x^+}{H_y^+} = \eta \cos\theta_i = Z_z^- = -\frac{E_x^-}{H_y^-} \qquad (4\text{-}20b)$$

In the same manner, the directional impedances of the TM^y fields of Example 4-2 can be written as

$$\underline{\text{TM}^y}$$

$$Z_x^+ = \frac{E_y^+}{H_z^+} = \frac{\eta}{\sin\theta_i} = Z_x^- = -\frac{E_y^-}{H_z^-} \qquad (4\text{-}21a)$$

$$Z_z^+ = -\frac{E_y^+}{H_x^+} = \frac{\eta}{\cos\theta_i} = Z_z^- = \frac{E_y^-}{H_x^-} \qquad (4\text{-}21b)$$

It is apparent from the preceding results that the directional impedances of the TE^y oblique incidence traveling waves are equal to or smaller than the intrinsic impedance and those of the TM^y are equal to or larger than the intrinsic impedance. In addition the positive and negative directional impedances of the same orientation are the same. This is the main principle of the *transverse resonance method* (see Section 8.6), which is used to analyze microwave circuits and antenna systems [3, 4].

C. PHASE AND ENERGY (GROUP) VELOCITIES

The wave velocity v_r of the fields given by (4-18a) and (4-18b) in the direction β of travel is equal to the speed of light v. Since the wave is a plane wave to the direction β of travel, the planes over which the phase is constant (constant phase planes) are perpendicular to the direction β of wave travel. This is illustrated graphically in Figure 4-6. To maintain a constant phase (or to keep in step with a constant phase plane) a velocity equal to the speed of light must be maintained in the direction β of travel. This is referred to as the phase velocity v_{pr} along the direction β of travel. Since the energy also is being transported with the same speed, the energy velocity v_{er} in the direction β of travel is also equal to the speed of light. Thus

$$v_r = v_{pr} = v_{er} = v = \frac{1}{\sqrt{\mu\varepsilon}} \tag{4-22}$$

where v_r = wave velocity in the direction of wave travel
v_{pr} = phase velocity in the direction of wave travel
v_{er} = energy (group) velocity in the direction of wave travel
v = speed of light

To keep in step with a constant phase plane of the wave of Figure 4-6, a velocity in the z direction equal to

$$v_{pz} = \frac{v}{\cos\theta_i} = \frac{1}{\sqrt{\mu\varepsilon}\,\cos\theta_i} \geq v \tag{4-23}$$

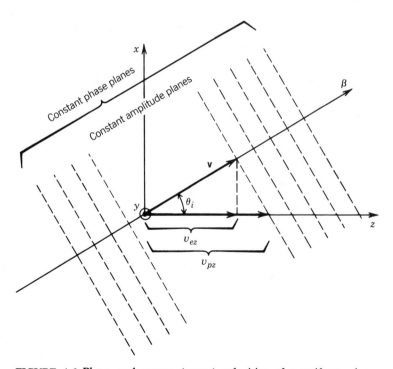

FIGURE 4-6 Phase and energy (group) velocities of a uniform plane wave.

must be maintained. This is referred to as the phase velocity v_{pz} in the z direction, and it is greater than the speed of light. Since nothing travels with speeds greater than the speed of light, it must be remembered that this is a hypothetical velocity that must be maintained in order to keep in step with a constant phase plane of the wave that itself travels with the speed of light in the direction β of travel. The phase velocities of (4-22) and (4-23) can be obtained, respectively, by using (4-19c) and (4-19d). These are left as end of chapter exercises for the reader.

Whereas a velocity greater than the speed of light must be maintained in the z direction to keep in step with a constant phase plane of Figure 4-6, the energy is transported in the z direction with a velocity that is equal to

$$v_{ez} = v \cos \theta_i = \frac{\cos \theta_i}{\sqrt{\mu\varepsilon}} \leq v \tag{4-24}$$

This is referred to as the energy (group) velocity v_{ez} in the z direction, and it is equal to or smaller than the speed of light. Graphically this is illustrated in Figure 4-6.

For any wave, the product of the phase and energy velocities in any direction must be equal to the speed of light squared or

$$v_{pr}v_{er} = v_{pz}v_{ez} = v^2 = \frac{1}{\mu\varepsilon} \tag{4-25}$$

This obviously is satisfied by the previously derived results.

The energy velocity of (4-24) can be derived using (4-18a) and (4-18b) along with the definition of (4-9). This is left for the reader as an end of chapter exercise.

Since the fields of (4-18a) and (4-18b) form a uniform plane wave, the planes over which the amplitude is maintained constant are also constant planes that are perpendicular to the direction β of travel. These are illustrated in Figure 4-6 and coincide with the constant phase planes. For other types of waves, the constant phase and amplitude planes do not in general coincide.

D. POWER AND ENERGY DENSITIES

The average power density associated with the fields of (4-18a) and (4-18b) that travel in the β^+ direction is given by

$$(\mathbf{S}_{av}^+)_r = \frac{1}{2} \operatorname{Re}\left[(\mathbf{E}^+) \times (\mathbf{H}^+)^*\right]$$

$$= \frac{1}{2} \operatorname{Re}\left[E_0^+(\hat{a}_x \cos \theta_i - \hat{a}_z \sin \theta_i) e^{-j\beta(x \sin \theta_i + z \cos \theta_i)}\right.$$

$$\left. \times \hat{a}_y \frac{E_0^{+*}}{\eta} e^{+j\beta(x \sin \theta_i + z \cos \theta_i)}\right]$$

$$(\mathbf{S}_{av}^+)_r = (\hat{a}_x \sin \theta_i + \hat{a}_z \cos \theta_i) \frac{|E_0^+|^2}{2\eta} = \hat{a}_r \frac{|E_0^+|^2}{2\eta} = \hat{a}_x (S_{av}^+)_x + \hat{a}_z (S_{av}^+)_z \tag{4-26}$$

where

$$(S_{av}^+)_x = \sin \theta_i \frac{|E_0^+|^2}{2\eta} = \sin \theta_i (S_{av}^+)_r \tag{4-26a}$$

$$(S_{av}^+)_z = \cos \theta_i \frac{|E_0^+|^2}{2\eta} = \cos \theta_i (S_{av}^+)_r \tag{4-26b}$$

$(S_{av}^+)_r$ represents the average power density along the principal β^+ direction of travel and $(S_{av}^+)_x$ and $(S_{av}^+)_z$ represent the directional power densities of the wave in the $+x$ and $+z$ directions, respectively. Similar expressions can be derived for the wave that travels along the β^- direction.

Example 4-3. For the TM^y fields of Example 4-2, derive expressions for the average power density along the principal β^+ direction of travel and for the directional power densities along the $+x$ and $+z$ directions.

Solution. Using the electric and magnetic fields of the solution of Example 4-2 and following the procedure used to derive (4-26) through (4-26b), it can be shown that

$$(\mathbf{S}_{av}^+)_r = \frac{1}{2}\operatorname{Re}\left[(\mathbf{E}^+) \times (\mathbf{H}^+)^*\right]$$

$$= \frac{1}{2}\operatorname{Re}\left[\hat{a}_y E_0^+ e^{-j\beta(x\cos\theta_i + y\sin\theta_i)}\right.$$

$$\left.\times \frac{(E_0^+)^*}{\eta}(-\hat{a}_x\cos\theta_i + \hat{a}_z\sin\theta_i)e^{+j\beta(x\cos\theta_i + y\sin\theta_i)}\right]$$

$$(\mathbf{S}_{av}^+)_r = (\hat{a}_x\sin\theta_i + \hat{a}_z\cos\theta_i)\frac{|E_0^+|^2}{2\eta} = \hat{a}_r\frac{|E_0^+|^2}{2\eta}$$

$$= \hat{a}_x(S_{av}^+)_x + \hat{a}_z(S_{av}^+)_z$$

where

$$(S_{av}^+)_x = \sin\theta_i\frac{|E_0^+|^2}{2\eta} = \sin\theta_i(S_{av}^+)_r$$

$$(S_{av}^+)_z = \cos\theta_i\frac{|E_0^+|^2}{2\eta} = \cos\theta_i(S_{av}^+)_r$$

(S_{av}^+), $(S_{av}^+)_x$, and $(S_{av}^+)_z$ of this TM^y wave are identical to the corresponding ones of the TE^y wave given by (4-26) through (4-26b).

4.3 TRANSVERSE ELECTROMAGNETIC MODES IN LOSSY MEDIA

In addition to the accumulation of phase, electromagnetic waves that travel in lossy media undergo attenuation. To account for the attenuation, an attenuation constant is introduced as discussed in Chapter 3, Section 3.4.1B. In this section we want to discuss the solution for the electric and magnetic fields of uniform plane waves as they travel in lossy media [5].

4.3.1 Uniform Plane Waves in an Unbounded Lossy Medium—Principal Axis

As for the electromagnetic wave of Section 4.2.1, let us assume that a uniform plane wave is traveling in a lossy medium. Using the coordinate system of Figure 4-1, the electric field is assumed to have an x component and the wave is traveling in the $\pm z$ direction. Since the electric field must satisfy the wave equation for lossy media, its expression takes according to (3-42a) the form of

$$\mathbf{E}(z) = \hat{a}_x E_x(z) = \hat{a}_x \left(E_0^+ e^{-\gamma z} + E_0^- e^{+\gamma z} \right) = \hat{a}_x \left(E_0^+ e^{-\alpha z} e^{-j\beta z} + E_0^- e^{+\alpha z} e^{+j\beta z} \right) \tag{4-27}$$

where $\gamma_x = \gamma_y = 0$ and $\gamma_z = \gamma$. The first term represents the positive traveling wave and the second term represents the negative traveling wave. In (4-27) γ is the propagation constant whose real α and imaginary β parts are defined, respectively, as the attenuation and phase constants. According to (3-37e) and (3-46) γ takes the form of

$$\gamma = \alpha + j\beta = \sqrt{j\omega\mu(\sigma + j\omega\varepsilon)} = \sqrt{-\omega^2\mu\varepsilon + j\omega\mu\sigma} \tag{4-28}$$

Squaring (4-28) and equating real and imaginary from both sides reduces it to

$$\alpha^2 - \beta^2 = -\omega^2\mu\varepsilon \tag{4-28a}$$

$$2\alpha\beta = \omega\mu\sigma \tag{4-28b}$$

Solving (4-28a) and (4-28b) simultaneously, we can write α and β as

$$\boxed{\alpha = \omega\sqrt{\mu\varepsilon}\left\{\frac{1}{2}\left[\sqrt{1 + \left(\frac{\sigma}{\omega\varepsilon}\right)^2} - 1\right]\right\}^{1/2} \quad \text{Np/m}} \tag{4-28c}$$

$$\boxed{\beta = \omega\sqrt{\mu\varepsilon}\left\{\frac{1}{2}\left[\sqrt{1 + \left(\frac{\sigma}{\omega\varepsilon}\right)^2} + 1\right]\right\}^{1/2} \quad \text{rad/m}} \tag{4-28d}$$

In the literature the phase constant β is also represented by k.

The attenuation constant α is often expressed in decibels per meter (dB/m). The conversion between Nepers per meter and decibels per meter is obtained by examining the real exponential in (4-27) that represents the attenuation factor of the wave in a lossy medium. Since that factor represents the relative attenuation of the electric or magnetic field, its conversion to decibels (dB) is obtained by

$$dB = 20 \log_{10}(e^{-\alpha z}) = 20(-\alpha z)\log_{10}(e)$$
$$= 20(-\alpha z)(0.434) = -8.68(\alpha z) \tag{4-28e}$$

or

$$|\alpha \text{ (Np/m)}| = \frac{1}{8.68}|\alpha \text{ (dB/m)}| \tag{4-28f}$$

The magnetic field associated with the electric field of (4-27) can be obtained using Maxwell's equation 4-3 or 4-3a, that is,

$$\mathbf{H} = -\frac{1}{j\omega\mu}\nabla \times \mathbf{E} = -\hat{a}_y \frac{1}{j\omega\mu}\frac{\partial E_x}{\partial z} \tag{4-29}$$

Using (4-27) reduces (4-29) to

$$\mathbf{H} = +\hat{a}_y \frac{\gamma}{j\omega\mu}(E_0^+ e^{-\gamma z} - E_0^- e^{+\gamma z})$$

$$= \hat{a}_y \frac{\sqrt{j\omega\mu(\sigma + j\omega\varepsilon)}}{j\omega\mu}(E_0^+ e^{-\gamma z} - E_0^- e^{+\gamma z})$$

$$= \hat{a}_y \sqrt{\frac{\sigma + j\omega\varepsilon}{j\omega\mu}}(E_0^+ e^{-\gamma z} - E_0^- e^{+\gamma z})$$

$$= \hat{a}_y \frac{1}{Z_w}(E_0^+ e^{-\gamma z} - E_0^- e^{+\gamma z})$$

$$\mathbf{H} = \hat{a}_y \frac{1}{Z_w}(E_0^+ e^{-\gamma z} - E_0^- e^{+\gamma z}) \tag{4-29a}$$

In (4-29a) Z_w is the wave impedance of the wave, and it takes the form of

$$Z_w = \sqrt{\frac{j\omega\mu}{\sigma + j\omega\varepsilon}} = \eta_c \tag{4-30}$$

which is also equal to the intrinsic impedance η_c of the lossy medium. The equality between the wave and intrinsic impedances for TEM waves in lossy media is identical to that for lossless media of Section 4.2.1B.

The average power density associated with the positive traveling fields of (4-27) and (4-29a) can be written as

$$\mathbf{S}^+ = \frac{1}{2}\operatorname{Re}(\mathbf{E}^+ \times \mathbf{H}^{+*}) = \frac{1}{2}\operatorname{Re}\left(\hat{a}_x E_0^+ e^{-\alpha z} e^{-j\beta z} \times \hat{a}_y \frac{E_0^{+*}}{\eta_c^*} e^{-\alpha z} e^{+j\beta z}\right)$$

$$\mathbf{S}^+ = \hat{a}_z \frac{|E_0^+|^2}{2} e^{-2\alpha z}\operatorname{Re}\left[\frac{1}{\eta_c^*}\right] \tag{4-31}$$

Individually each term of (4-27) or (4-29a) represents a traveling wave in its respective direction. The magnitude of each term in (4-27) takes the form of

$$|E_x^+(z)| = |E_0^+|e^{-\alpha z} \tag{4-32a}$$

$$|E_x^-(z)| = |E_0^-|e^{+\alpha z} \tag{4-32b}$$

which, when plotted for $-\lambda \le z \le +\lambda$, take the form shown in Figure 4-7a.

Collectively both terms in each of the fields in (4-27) or (4-29a) represent a standing wave. Using the procedure outlined in Section 4.2.1D, (4-27) can also be written as

$$E_x(z) = \sqrt{(E_0^+)^2 e^{-2\alpha z} + (E_0^-)^2 e^{+2\alpha z} + 2 E_0^+ E_0^- \cos(2\beta z)}$$

$$\times \exp\left\{-j\tan^{-1}\left[\frac{E_0^+ e^{-\alpha z} - E_0^- e^{+\alpha z}}{E_0^+ e^{-\alpha z} + E_0^- e^{+\alpha z}}\tan(\beta z)\right]\right\} \tag{4-33}$$

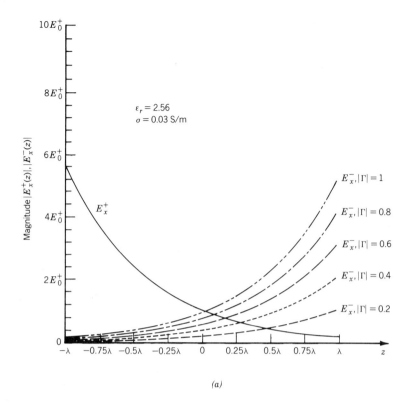

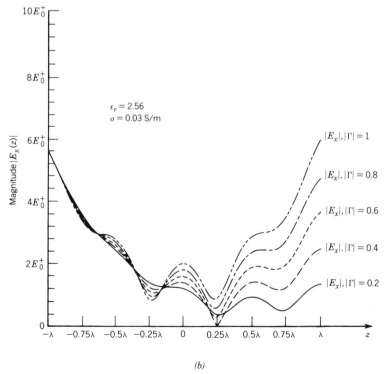

FIGURE 4-7 (*a*) Traveling and (*b*) standing wave patterns of uniform plane waves in a lossy medium.

The standing wave pattern is given by the amplitude term of

$$|E_x(z)| = \sqrt{(E_0^+)^2 e^{-2\alpha z} + (E_0^-)^2 e^{+2\alpha z} + 2 E_0^+ E_0^- \cos(2\beta z)} \quad (4\text{-}33\text{a})$$

which for $|\Gamma| = E_0^-/E_0^+ = 0.2$ through 1 in increments of 0.2 is shown plotted in Figure 4-7b in the range $-\lambda \leq z \leq \lambda$ when $f = 100$ MHz, $\varepsilon_r = 2.56$, $\mu_r = 1$, and $\sigma = 0.03$ S/m.

The distance the wave must travel in a lossy medium to reduce its value to $e^{-1} = 0.368 = 36.8\%$ is defined as the skin depth δ. For each of the terms of (4-27) or (4-29a), this distance is

$$\boxed{\delta = \text{skin depth} = \frac{1}{\alpha} = \frac{1}{\omega\sqrt{\mu\varepsilon}\left\{\frac{1}{2}\left[\sqrt{1 + (\sigma/\omega\varepsilon)^2} - 1\right]\right\}^{1/2}} \text{ m}} \quad (4\text{-}34)$$

In summary, the attenuation constant α, phase constant β, wave Z_w and intrinsic η_c impedances, wavelength λ, velocity v, and skin depth δ for a uniform plane wave traveling in a lossy medium are listed in the second column of Table 4-1. The same expressions are valid for plane and TEM waves. Simpler expressions for each can be derived depending upon the value of the $(\sigma/\omega\varepsilon)^2$ ratio. Media whose $(\sigma/\omega\varepsilon)^2$ is much less than unity $[(\sigma/\omega\varepsilon)^2 \ll 1]$ are referred to as *good dielectrics* and those whose $(\sigma/\omega\varepsilon)^2$ is much greater than unity $[(\sigma/\omega\varepsilon)^2 \gg 1]$ are referred to as *good conductors* [6]; each will now be discussed.

A. GOOD DIELECTRICS $[(\sigma/\omega\varepsilon)^2 \ll 1]$

For source-free lossy media, Maxwell's equation in differential form as derived from Ampere's law takes the form, by referring to Table 1-4, of

$$\nabla \times \mathbf{H} = \mathbf{J}_c + \mathbf{J}_d = \sigma \mathbf{E} + j\omega\varepsilon \mathbf{E} = (\sigma + j\omega\varepsilon)\mathbf{E} \quad (4\text{-}35)$$

where $\mathbf{J}_c$ and $\mathbf{J}_d$ represent, respectively, the conduction and displacement current densities. When $\sigma/\omega\varepsilon \ll 1$, the displacement current density is much greater than the conduction current density; when $\sigma/\omega\varepsilon \gg 1$ the conduction current density is much greater than the displacement current density. For each of these two cases the exact forms of the field parameters of Table 4-1 can be approximated by simpler forms. This will be demonstrated next.

For a good dielectric [when $(\sigma/\omega\varepsilon)^2 \ll 1$] the exact expression for the attenuation constant of (4-28c) can be written using the binomial expansion and it takes the form of

$$\alpha = \omega\sqrt{\mu\varepsilon}\left\{\frac{1}{2}\left[\sqrt{1 + \left(\frac{\sigma}{\omega\varepsilon}\right)^2} - 1\right]\right\}^{1/2}$$

$$\alpha = \omega\sqrt{\mu\varepsilon}\left\{\frac{1}{2}\left[\left(1 + \frac{1}{2}\left(\frac{\sigma}{\omega\varepsilon}\right)^2 - \frac{1}{8}\left(\frac{\sigma}{\omega\varepsilon}\right)^4 \cdots\right) - 1\right]\right\}^{1/2} \quad (4\text{-}36)$$

TABLE 4-1
Propagation constant, wave impedance, wavelength, velocity, and skin depth of TEM wave in lossy media

	Exact	Good dielectric $\left(\dfrac{\sigma}{\omega\varepsilon}\right)^2 \ll 1$	Good conductor $\left(\dfrac{\sigma}{\omega\varepsilon}\right)^2 \gg 1$
Attenuation constant α	$= \omega\sqrt{\mu\varepsilon}\left\{\dfrac{1}{2}\left[\sqrt{1+\left(\dfrac{\sigma}{\omega\varepsilon}\right)^2}-1\right]\right\}^{1/2}$	$\simeq \dfrac{\sigma}{2}\sqrt{\dfrac{\mu}{\varepsilon}}$	$\simeq \sqrt{\dfrac{\omega\mu\sigma}{2}}$
Phase constant β	$= \omega\sqrt{\mu\varepsilon}\left\{\dfrac{1}{2}\left[\sqrt{1+\left(\dfrac{\sigma}{\omega\varepsilon}\right)^2}+1\right]\right\}^{1/2}$	$\simeq \omega\sqrt{\mu\varepsilon}$	$\simeq \sqrt{\dfrac{\omega\mu\sigma}{2}}$
Wave Z_w, intrinsic η_c impedances $Z_w = \eta_c$	$= \sqrt{\dfrac{j\omega\mu}{\sigma+j\omega\varepsilon}}$	$\simeq \sqrt{\dfrac{\mu}{\varepsilon}}$	$\simeq \sqrt{\dfrac{\omega\mu}{2\sigma}}(1+j)$
Wavelength λ	$= \dfrac{2\pi}{\beta}$	$\simeq \dfrac{2\pi}{\omega\sqrt{\mu\varepsilon}}$	$\simeq 2\pi\sqrt{\dfrac{2}{\omega\mu\sigma}}$
Velocity v	$= \dfrac{\omega}{\beta}$	$\simeq \dfrac{1}{\sqrt{\mu\varepsilon}}$	$\simeq \sqrt{\dfrac{2\omega}{\mu\sigma}}$
Skin depth δ	$= \dfrac{1}{\alpha}$	$\simeq \dfrac{2}{\sigma}\sqrt{\dfrac{\varepsilon}{\mu}}$	$\simeq \sqrt{\dfrac{2}{\omega\mu\sigma}}$

Retaining only the first two terms of the infinite series, (4-36) can be approximated by

$$\alpha \simeq \omega\sqrt{\mu\varepsilon}\left[\dfrac{1}{4}\left(\dfrac{\sigma}{\omega\varepsilon}\right)^2\right]^{1/2} = \dfrac{\sigma}{2}\sqrt{\dfrac{\mu}{\varepsilon}} \qquad (4\text{-}36a)$$

In a similar manner it can be shown that by following the same procedure but only retaining the first term of the infinite series, the exact expression for β of (4-28d) can be approximated by

$$\beta \simeq \omega\sqrt{\mu\varepsilon} \qquad (4\text{-}37)$$

For good dielectrics, the wave and intrinsic impedances of (4-30) can be approximated by

$$Z_w = \eta_c = \sqrt{\dfrac{j\omega\mu}{\sigma+j\omega\varepsilon}} = \sqrt{\dfrac{j\omega\mu/j\omega\varepsilon}{\sigma/j\omega\varepsilon+1}} \simeq \sqrt{\dfrac{\mu}{\varepsilon}} \qquad (4\text{-}38)$$

while the skin depth can be represented by

$$\delta = \dfrac{1}{\alpha} \simeq \dfrac{2}{\sigma}\sqrt{\dfrac{\varepsilon}{\mu}} \qquad (4\text{-}39)$$

These and other approximate forms for the parameters of good dielectrics are summarized on the third column of Table 4-1.

B. GOOD CONDUCTORS $[(\sigma/\omega\varepsilon)^2 \gg 1]$

For good conductors, the exact expression for the attenuation constant of (4-28c) can be written using the binomial expansion and takes the form of

$$\alpha = \omega\sqrt{\mu\varepsilon}\left\{\frac{1}{2}\left[\sqrt{\left(\frac{\sigma}{\omega\varepsilon}\right)^2 + 1} - 1\right]\right\}^{1/2}$$

$$\alpha = \omega\sqrt{\mu\varepsilon}\left\{\frac{1}{2}\left[\frac{\sigma}{\omega\varepsilon} + \frac{1}{2}\frac{1}{\sigma/\omega\varepsilon} - \frac{1}{8}\frac{1}{(\sigma/\omega\varepsilon)^3} + \cdots - 1\right]\right\}^{1/2} \quad (4\text{-}40)$$

Retaining only the first term of the infinite series expansion, (4-40) can be approximated by

$$\alpha \simeq \omega\sqrt{\mu\varepsilon}\left[\frac{1}{2}\frac{\sigma}{\omega\varepsilon}\right]^{1/2} = \sqrt{\frac{\omega\mu\sigma}{2}} \quad (4\text{-}40\text{a})$$

Following a similar procedure, the phase constant of (4-28d) can be approximated by

$$\beta \simeq \sqrt{\frac{\omega\mu\sigma}{2}} \quad (4\text{-}41)$$

which is identical to the approximate expression for the attenuation constant of (4-40a).

For good conductors, the wave and intrinsic impedances of (4-30) can be approximated by

$$Z_w = \eta_c = \sqrt{\frac{j\omega\mu}{\sigma + j\omega\varepsilon}} = \sqrt{\frac{j\omega\mu/\omega\varepsilon}{\sigma/\omega\varepsilon + j}} \simeq \sqrt{j\frac{\omega\mu}{\sigma}} = \sqrt{\frac{\omega\mu}{2\sigma}}(1+j) \quad (4\text{-}42)$$

whose real and imaginary parts are identical. For the same conditions, the skin depth can be approximated by

$$\delta = \frac{1}{\alpha} \simeq \sqrt{\frac{2}{\omega\mu\sigma}} \quad (4\text{-}43)$$

This is the most widely recognized form for the skin depth.

4.3.2 Uniform Plane Waves in an Unbounded Lossy Medium—Oblique Angle

For lossy media the difference between principal axes propagation and propagation at oblique angles is that the propagation constant γ_r along the direction β of propagation must be decomposed into its directional components along the principal axes of the coordinate system. In addition, since the propagation constant γ has real (α) and imaginary (β) parts, constant amplitude and constant phase planes are associated with the wave. As discussed in Section 4.2.2C and illustrated in Figure 4-6, the constant phase planes for a uniform plane wave are planes that are parallel to each other, perpendicular to the direction of propagation, and coincide with the constant amplitude planes. The constant amplitude planes are planes over which the amplitude remains constant. For a uniform plane wave traveling in a lossy medium the constant amplitude planes are also parallel to each other, coincide with the constant phase planes, and are perpendicular to the direction of travel. This is illustrated in Figure 4-6 for a uniform plane wave traveling at an oblique angle in a lossless medium.

Let us assume that a uniform plane wave that is also TE^y is traveling in a lossy medium at an angle θ_i, as shown in Figure 4-4a. Following a procedure similar to the lossless case and referring to (4-17a) and (4-17b), the propagation constant of (4-28) can now be written for the positive and negative traveling waves as

$$\gamma^+ = \gamma(\hat{a}_x \sin\theta_i + \hat{a}_z \cos\theta_i) = (\alpha + j\beta)(\hat{a}_x \sin\theta_i + \hat{a}_z \cos\theta_i) \quad (4\text{-}44a)$$
$$\gamma^- = -\gamma(\hat{a}_x \sin\theta_i + \hat{a}_z \cos\theta_i) = -(\alpha + j\beta)(\hat{a}_x \sin\theta_i + \hat{a}_z \cos\theta_i) \quad (4\text{-}44b)$$

where the real (α) and imaginary (β) parts of γ are given by (4-28c) and (4-28d). Using (4-44a) and (4-44b) the electric and magnetic fields can be written by referring to (4-17) through (4-18c) as

$$\mathbf{E} = E_0^+(\hat{a}_x \cos\theta_i - \hat{a}_z \sin\theta_i)e^{-\gamma^+ \cdot \mathbf{r}} + E_0^-(\hat{a}_x \cos\theta_i - \hat{a}_z \sin\theta_i)e^{-\gamma^- \cdot \mathbf{r}}$$
$$\mathbf{E} = E_0^+(\hat{a}_x \cos\theta_i - \hat{a}_z \sin\theta_i)e^{-(\alpha+j\beta)(x\sin\theta_i + z\cos\theta_i)}$$
$$+ E_0^-(\hat{a}_x \cos\theta_i - \hat{a}_z \sin\theta_i)e^{+(\alpha+j\beta)(x\sin\theta_i + z\cos\theta_i)} \quad (4\text{-}45a)$$

$$\mathbf{H} = \hat{a}_y\left[\frac{E_0^+}{\eta_c}e^{-(\alpha+j\beta)(x\sin\theta_i + z\cos\theta_i)} - \frac{E_0^-}{\eta_c}e^{+(\alpha+j\beta)(x\sin\theta_i + z\cos\theta_i)}\right] \quad (4\text{-}45b)$$

Because the wave is a uniform plane wave in the β direction of propagation, the wave impedance Z_{wr} in the direction of propagation is equal to the intrinsic impedance η_c of the lossy medium given by (4-30) or

$$Z_{wr} = \eta_c = \sqrt{\frac{j\omega\mu}{\sigma + j\omega\varepsilon}} \quad (4\text{-}46)$$

However the directional impedances in the x and z directions are given, by referring to (4-20a) and (4-20b), by

$$Z_x^+ = -\frac{E_z^+}{H_y^+} = \eta_c \sin\theta_i = Z_x^- = \frac{E_z^-}{H_y^-} \quad (4\text{-}47a)$$

$$Z_z^+ = \frac{E_x^+}{H_y^+} = \eta_c \cos\theta_i = Z_z^- = -\frac{E_x^-}{H_y^-} \quad (4\text{-}47b)$$

According to (4-22) through (4-24) the phase and energy velocities in the principal β direction of travel and in the z direction are given, respectively, by

$$v_r = v_{pr} = v_{er} = v = \frac{\omega}{\beta} \quad (4\text{-}48a)$$

$$v_{pz} = \frac{v}{\cos\theta_i} = \frac{\omega}{\beta\cos\theta_i} \geq v = \frac{\omega}{\beta} \quad (4\text{-}48b)$$

$$v_{ez} = v\cos\theta_i = \frac{\omega}{\beta}\cos\theta_i \leq v = \frac{\omega}{\beta} \quad (4\text{-}48c)$$

where β for a lossy medium is given by (4-28d) or

$$\beta = \omega\sqrt{\mu\varepsilon}\left\{\frac{1}{2}\left[\sqrt{1+\left(\frac{\sigma}{\omega\varepsilon}\right)^2}+1\right]\right\}^{1/2} \qquad (4\text{-}48\text{d})$$

As for the lossless medium, the product of the phase and energy velocities is equal to the square of the velocity of light v in the lossy medium, or

$$v_{pr}v_{er} = v_{pz}v_{ez} = v^2 \qquad (4\text{-}48\text{e})$$

Using the procedure followed to derive (4-26) through (4-26b) and (4-31), the average power density along the principal direction β of travel and the directional power densities along the x and z directions can be written for the fields of (4-45a) and (4-45b) as

$$(\mathbf{S}_{av}^+)_r = (\hat{a}_x \sin\theta_i + \hat{a}_z \cos\theta_i)\frac{|E_0^+|^2}{2}e^{-2\alpha(x\sin\theta_i + z\cos\theta_i)}\,\text{Re}\left[\frac{1}{\eta_c^*}\right]$$

$$= \hat{a}_r \frac{|E_0^+|^2}{2}e^{-2\alpha r}\,\text{Re}\left[\frac{1}{\eta_c^*}\right] \qquad (4\text{-}49\text{a})$$

$$(\mathbf{S}_{av}^+)_x = \sin\theta_i \frac{|E_0^+|^2}{2}e^{-2\alpha(x\sin\theta_i + z\cos\theta_i)}\,\text{Re}\left[\frac{1}{\eta_c^*}\right]$$

$$= \sin\theta_i \frac{|E_0^+|^2}{2}e^{-2\alpha r}\,\text{Re}\left[\frac{1}{\eta_c^*}\right] \qquad (4\text{-}49\text{b})$$

$$(\mathbf{S}_{av}^+)_z = \cos\theta_i \frac{|E_0^+|^2}{2}e^{-2\alpha(x\sin\theta_i + z\cos\theta_i)}\,\text{Re}\left[\frac{1}{\eta_c^*}\right]$$

$$= \cos\theta_i \frac{|E_0^+|^2}{2}e^{-2\alpha r}\,\text{Re}\left[\frac{1}{\eta_c^*}\right] \qquad (4\text{-}49\text{c})$$

Example 4-4. For a TM^y wave traveling in a lossy medium at an oblique angle θ_i, derive expressions for the fields, wave impedances, phase and energy velocities, and average power densities.

Solution. The solution to this problem can be accomplished by following the procedure used to derive the expressions of the fields and other wave characteristics of a TE^y wave traveling at an oblique angle in a lossy medium, as outlined in this section, and referring to the solution of Examples 4-2 and 4-3. Doing this we can write the fields of a TM^y traveling in a lossy medium at an oblique angle θ_i, the coordinate system of which is illustrated in Figure 4-4b, as

$$\mathbf{E} = \mathbf{E}^+ + \mathbf{E}^- = \hat{a}_y\left[E_0^+ e^{-(\alpha+j\beta)(x\sin\theta_i + z\cos\theta_i)} + E_0^- e^{+(\alpha+j\beta)(x\sin\theta_i + z\cos\theta_i)}\right]$$

$$\mathbf{H} = \mathbf{H}^+ + \mathbf{H}^- = \frac{E_0^+}{\eta_c}(-\hat{a}_x\cos\theta_i + \hat{a}_z\sin\theta_i)e^{-(\alpha+j\beta)(x\sin\theta_i + z\cos\theta_i)}$$

$$+ \frac{E_0^-}{\eta_c}(\hat{a}_x\cos\theta_i - \hat{a}_z\sin\theta_i)e^{+(\alpha+j\beta)(x\sin\theta_i + z\cos\theta_i)}$$

154 WAVE PROPAGATION AND POLARIZATION

In addition the wave impedances are given, by referring to (4-21a) and (4-21b), by

$$Z_x^+ = \frac{E_y^+}{H_z^+} = \frac{\eta_c}{\sin\theta_i} = Z_x^- = -\frac{E_y^-}{H_z^-}$$

$$Z_z^+ = -\frac{E_y^+}{H_x^+} = \frac{\eta_c}{\cos\theta_i} = Z_z^- = \frac{E_y^-}{H_x^-}$$

The phase and energy velocities, and their relationships, are the same as those for the TE^y wave as given by (4-48a) through (4-48e). Similarly the average power densities are those given by (4-49a) through (4-49c).

4.4 POLARIZATION

According to the *IEEE Standard Definitions for Antennas* [7, 8], the *polarization of a radiated wave* is defined as "that property of a radiated electromagnetic wave describing the time-varying direction and relative magnitude of the electric field vector; specifically, the figure traced as a function of time by the extremity of the vector at a fixed location in space, and the sense in which it is traced, as observed along the direction of propagation." In other words, polarization is the curve traced out by the end point of the arrow representing the instantaneous electric field. The

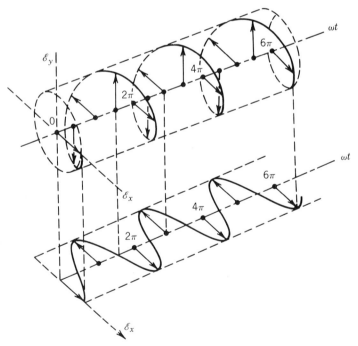

FIGURE 4-8 Rotation of a plane electromagnetic wave at $z = 0$ as a function of time. (*Source:* C. A. Balanis, *Antenna Theory: Analysis and Design*. Copyright © 1982, John Wiley & Sons, Inc. Reprinted by permission of John Wiley & Sons, Inc.)

field must be observed along the direction of propagation. A typical trace as a function of time is shown in Figure 4-8 [8].

Polarization may be classified into three categories: *linear, circular, and elliptical* [8]. If the vector that describes the electric field at a point in space as a function of time is always directed along a line, which is normal to the direction of propagation, the field is said to be *linearly* polarized. In general, however, the figure that the electric field traces is an ellipse, and the field is said to be *elliptically* polarized. Linear and circular polarizations are special cases of elliptical, and they can be obtained when the ellipse becomes a straight line or a circle, respectively. The figure of the electric field is traced in a *clockwise* (CW) or *counterclockwise* (CCW) sense. Clockwise rotation of the electric field vector is also designated as *right-hand polarization* and counterclockwise as *left-hand polarization*. In Figure

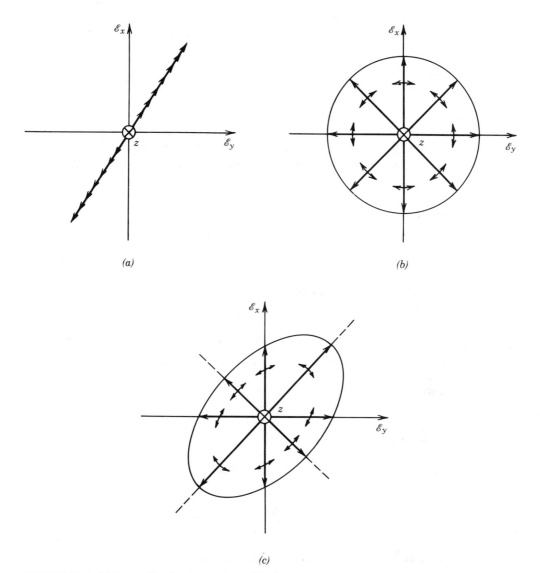

FIGURE 4-9 (*a*) Linear, (*b*) circular, and (*c*) elliptical polarization figure traces of an electric field extremity *as a function of time for a fixed position*.

156 WAVE PROPAGATION AND POLARIZATION

4-9a, b, and c we show the figure traced by the extremity of the time-varying field for linear, circular, and elliptical polarizations.

The mathematical details for defining linear, circular, and elliptical polarizations follow.

4.4.1 Linear Polarization

Let us consider a harmonic plane wave, with x and y electric field components, traveling in the positive z direction (into the page), as shown in Figure 4-10 [8]. The instantaneous electric and magnetic fields are given by

$$\mathscr{E} = \hat{a}_x \mathscr{E}_x + \hat{a}_y \mathscr{E}_y = \text{Re}\left[\hat{a}_x E_x^+ e^{j(\omega t - \beta z)} + \hat{a}_y E_y^+ e^{j(\omega t - \beta z)}\right]$$
$$= \hat{a}_x E_{x_0}^+ \cos(\omega t - \beta z + \phi_x) + \hat{a}_y E_{y_0}^+ \cos(\omega t - \beta z + \phi_y) \quad (4\text{-}50a)$$

$$\mathscr{H} = \hat{a}_y \mathscr{H}_y + \hat{a}_x \mathscr{H}_x = \text{Re}\left[\hat{a}_y \frac{E_x^+}{\eta} e^{j(\omega t - \beta z)} - \hat{a}_x \frac{E_y^+}{\eta} e^{j(\omega t - \beta z)}\right]$$
$$= \hat{a}_y \frac{E_{x_0}^+}{\eta} \cos(\omega t - \beta z + \phi_x) - \hat{a}_x \frac{E_{y_0}^+}{\eta} \cos(\omega t - \beta z + \phi_y) \quad (4\text{-}50b)$$

where E_x^+, E_y^+ are complex and $E_{x_0}^+$, $E_{y_0}^+$ are real.

Let us now examine the variation of the instantaneous electric field vector $\mathscr{E}$ as given by (4-50a) at the $z = 0$ plane. Other planes may be considered, but the $z = 0$ plane is chosen for convenience and simplicity. For the first example, let

$$E_{y_0}^+ = 0 \quad (4\text{-}51)$$

in (4-50a). Then

$$\mathscr{E}_x = E_{x_0}^+ \cos(\omega t + \phi_x)$$
$$\mathscr{E}_y = 0 \quad (4\text{-}51a)$$

The locus of the instantaneous electric field vector is given by

$$\mathscr{E} = \hat{a}_x E_{x_0}^+ \cos(\omega t + \phi_x) \quad (4\text{-}51b)$$

which is a straight line, and it will always be directed along the x axis at all times as shown in Figure 4-10. The field is said to be *linearly polarized in the x direction*.

Example 4-5. Determine the polarization of the wave given by (4-50a) when $E_{x_0}^+ = 0$.

Solution. Since

$$E_{x_0}^+ = 0$$

then

$$\mathscr{E}_x = 0$$
$$\mathscr{E}_y = E_{y_0}^+ \cos(\omega t + \phi_y)$$

The locus of the instantaneous electric field vector is given by

$$\mathscr{E} = \hat{a}_y E_{y_0}^+ \cos(\omega t + \phi_y)$$

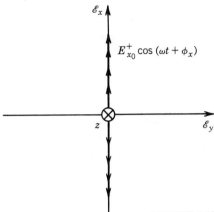

FIGURE 4-10 Linearly polarized field in the x direction.

which again is a straight line but directed along the y axis at all times as shown in Figure 4-11. The field is said to be *linearly polarized in the y direction*.

Example 4-6. Determine the polarization and direction of polarization of the wave given by (4-50a) when $\phi_x = \phi_y = \phi$.

Solution. Since

$$\phi_x = \phi_y = \phi$$

then

$$\mathscr{E}_x = E_{x_0}^+ \cos(\omega t + \phi)$$
$$\mathscr{E}_y = E_{y_0}^+ \cos(\omega t + \phi)$$

The amplitude of the electric field vector is given by

$$\mathscr{E} = \sqrt{\mathscr{E}_x^2 + \mathscr{E}_y^2} = \sqrt{\left(E_{x_0}^+\right)^2 + \left(E_{y_0}^+\right)^2} \cos(\omega t + \phi)$$

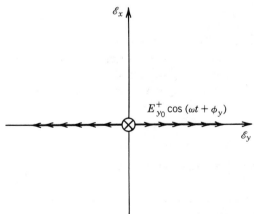

FIGURE 4-11 Linearly polarized field in the y direction.

158   WAVE PROPAGATION AND POLARIZATION

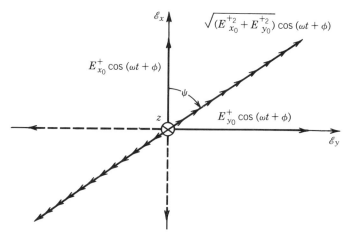

FIGURE 4-12 Linearly polarized field in the ψ direction.

which is a straight line directed at all times along a line that makes an angle ψ with the x axis as shown in Figure 4-12. The angle ψ is given by

$$\psi = \tan^{-1}\left[\frac{\mathscr{E}_y}{\mathscr{E}_x}\right] = \tan^{-1}\left[\frac{E^+_{y_0}}{E^+_{x_0}}\right]$$

The field is said to be *linearly polarized in the ψ direction*.

It is evident from the preceding examples that *a time-harmonic field is linearly polarized at a given point in space if the electric field (or magnetic field) vector at that point is always oriented along the same straight line at every instant of time*. This is accomplished if the field vector (electric or magnetic) possesses (a) only one component or (b) two orthogonal linearly polarized components that are in time phase or 180° out of phase.

4.4.2   Circular Polarization

A wave is said to be *circularly polarized if the tip of the electric field vector traces out a circular locus in space*. At various instants of time, the electric field intensity of such a wave always has the same amplitude and the orientation in space of the electric field vector changes continuously with time in such a manner as to describe a circular locus [8, 9].

A.   RIGHT-HAND (CLOCKWISE) CIRCULAR POLARIZATION

A wave has *right-hand circular polarization* if its electric field vector has a clockwise sense of rotation *when it is viewed along the axis of propagation*. In addition, the electric field vector must trace a circular locus if the wave is to have also a circular polarization.

Let us examine the locus of the instantaneous electric field vector ($\mathscr{E}$) at the $z = 0$ plane at all times. For this particular example, let in (4-50a)

$$\phi_x = 0$$
$$\phi_y = -\pi/2$$
$$E_{x_0}^+ = E_{y_0}^+ = E_R \qquad (4\text{-}52)$$

Then

$$\mathscr{E}_x = E_R \cos(\omega t)$$
$$\mathscr{E}_y = E_R \cos\left(\omega t - \frac{\pi}{2}\right) = E_R \sin(\omega t) \qquad (4\text{-}52\text{a})$$

The locus of the amplitude of the electric field vector is given by

$$\mathscr{E} = \sqrt{\mathscr{E}_x^2 + \mathscr{E}_y^2} = \sqrt{E_R^2(\cos^2 \omega t + \sin^2 \omega t)} = E_R \qquad (4\text{-}52\text{b})$$

and it is directed along a line making an angle ψ with the x axis, which is given by

$$\psi = \tan^{-1}\left[\frac{\mathscr{E}_y}{\mathscr{E}_x}\right] = \tan^{-1}\left[\frac{E_R \sin(\omega t)}{E_R \cos(\omega t)}\right] = \tan^{-1}[\tan(\omega t)] = \omega t \qquad (4\text{-}52\text{c})$$

If we plot the locus of the electric field vector for various times at the $z = 0$ plane, we see that it forms a circle of radius E_R and it rotates clockwise with an angular frequency ω as shown in Figure 4-13. Thus the wave is said to have a *right-hand circular polarization*. Remember that the rotation is viewed from the "rear" of the wave in the direction of propagation. In this example, the wave is traveling in the positive z direction (into the page) so that the rotation is examined from an observation point looking into the page and perpendicular to it.

We can write the instantaneous electric field vector expression as

$$\mathscr{E} = \text{Re}\left[\hat{a}_x E_R e^{j(\omega t - \beta z)} + \hat{a}_y E_R e^{j(\omega t - \beta z - \pi/2)}\right]$$
$$= E_R \text{Re}\left\{[\hat{a}_x - j\hat{a}_y] e^{j(\omega t - \beta z)}\right\} \qquad (4\text{-}52\text{d})$$

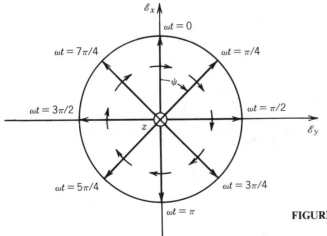

FIGURE 4-13 Right-hand circularly polarized wave.

We note that there is a 90° phase difference between the two orthogonal components of the electric field vector.

Example 4-7. If $\phi_x = +\pi/2$, $\phi_y = 0$, and $E_{x_0}^+ = E_{y_0}^+ = E_R$, determine the polarization and sense of rotation of the wave of (4-50a).

Solution. Since

$$\phi_x = +\frac{\pi}{2}$$

$$\phi_y = 0$$

$$E_{x_0}^+ = E_{y_0}^+ = E_R$$

then

$$\mathscr{E}_x = E_R \cos\left(\omega t + \frac{\pi}{2}\right) = -E_R \sin \omega t$$

$$\mathscr{E}_y = E_R \cos(\omega t)$$

and the locus of the amplitude of the electric field vector is given by

$$\mathscr{E} = \sqrt{\mathscr{E}_x^2 + \mathscr{E}_y^2} = \sqrt{E_R^2(\cos^2 \omega t + \sin^2 \omega t)} = E_R$$

The angle ψ along which the field is directed is given by

$$\psi = \tan^{-1}\left[\frac{\mathscr{E}_y}{\mathscr{E}_x}\right] = \tan^{-1}\left[-\frac{E_R \cos(\omega t)}{E_R \sin(\omega t)}\right] = \tan^{-1}[-\cot(\omega t)] = \omega t + \frac{\pi}{2}$$

The locus of the field vector is a circle of radius E_R, and it rotates clockwise with an angular frequency ω as shown in Figure 4-14, *hence it is a right-hand circular polarization.*

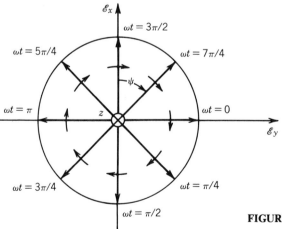

FIGURE 4-14 Right-hand circularly polarized wave.

The expression for the instantaneous electric field vector is

$$\mathscr{E} = \mathrm{Re}\left[\hat{a}_x E_R e^{j(\omega t - \beta z + \pi/2)} + \hat{a}_y E_R e^{j(\omega t - \beta z)}\right]$$
$$= E_R \,\mathrm{Re}\left\{\left[j\hat{a}_x + \hat{a}_y\right] e^{j(\omega t - \beta z)}\right\}$$

Again we note a 90° phase difference between the orthogonal components.

From the previous discussion we see that a right-hand *circular polarization* can be achieved *if and only if its two orthogonal linearly polarized components have equal amplitudes and a 90° phase difference of one relative to the other*. The sense of rotation (*clockwise here*) is determined by rotating the phase-leading component (in this instance $\mathscr{E}_x$) toward the phase-lagging component (in this instance $\mathscr{E}_y$). The field rotation must be viewed as the wave travels away from the observer.

B. LEFT-HAND (COUNTERCLOCKWISE) CIRCULAR POLARIZATION

If the electric field vector has a counterclockwise sense of rotation, the polarization is designated as *left-hand polarization*. To demonstrate this, let in (4-50a)

$$\phi_x = 0$$
$$\phi_y = \frac{\pi}{2}$$
$$E_{x_0}^+ = E_{y_0}^+ = E_L \qquad (4\text{-}53)$$

then

$$\mathscr{E}_x = E_L \cos(\omega t)$$
$$\mathscr{E}_y = E_L \cos\left(\omega t + \frac{\pi}{2}\right) = -E_L \sin(\omega t) \qquad (4\text{-}53\mathrm{a})$$

and the locus of the amplitude is

$$\mathscr{E} = \sqrt{\mathscr{E}_x^2 + \mathscr{E}_y^2} = \sqrt{E_L^2(\cos^2 \omega t + \sin^2 \omega t)} = E_L \qquad (4\text{-}53\mathrm{b})$$

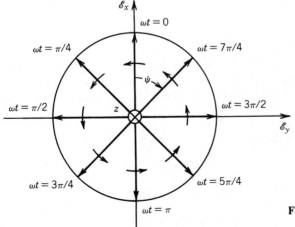

FIGURE 4-15 Left-hand circularly polarized wave.

The angle ψ is given by

$$\psi = \tan^{-1}\left[\frac{\mathscr{E}_y}{\mathscr{E}_x}\right] = \tan^{-1}\left[\frac{-E_L \sin(\omega t)}{E_L \cos(\omega t)}\right] = -\omega t \qquad (4\text{-}53c)$$

The locus of the field vector is a circle of radius E_L, and it rotates counterclockwise with an angular frequency ω as shown in Figure 4-15; *hence it is a left-hand circular polarization.*

The instantaneous electric field vector can be written as

$$\mathscr{E} = \text{Re}\left[\hat{a}_x E_L e^{j(\omega t - \beta z)} + \hat{a}_y E_L e^{j(\omega t - \beta z + \pi/2)}\right]$$
$$= E_L \text{Re}\left\{[\hat{a}_x + j\hat{a}_y] e^{j(\omega t - \beta z)}\right\} \qquad (4\text{-}53d)$$

In (4-53d) we note a 90° phase advance of the $\mathscr{E}_y$ component relative to the $\mathscr{E}_x$ component.

Example 4-8. Determine the polarization and sense of rotation of the wave given by (4-50a) if $\phi_x = -\pi/2$, $\phi_y = 0$, and $E_{x_0}^+ = E_{y_0}^+ = E_L$.

Solution. Since

$$\phi_x = -\frac{\pi}{2}$$
$$\phi_y = 0$$
$$E_{x_0}^+ = E_{y_0}^+ = E_L$$

then

$$\mathscr{E}_x = E_L \cos\left(\omega t - \frac{\pi}{2}\right) = E_L \sin(\omega t)$$
$$\mathscr{E}_y = E_L \cos(\omega t)$$

and the locus of the amplitude is

$$\mathscr{E} = \sqrt{\mathscr{E}_x^2 + \mathscr{E}_y^2} = \sqrt{E_L^2(\sin^2 \omega t + \cos^2 \omega t)} = E_L$$

The angle ψ is given by

$$\psi = \tan^{-1}\left[\frac{\mathscr{E}_y}{\mathscr{E}_x}\right] = \tan^{-1}\left[\frac{E_L \cos(\omega t)}{E_L \sin(\omega t)}\right] = \tan^{-1}[\cot(\omega t)] = \frac{\pi}{2} - \omega t$$

The locus of the electric field vector is a circle of radius E_L, and it rotates counterclockwise with an angular frequency ω as shown in Figure 4-16; *hence, it is a left-hand circular polarization.* For this case we can write the electric field as

$$\mathscr{E} = \text{Re}\left[\hat{a}_x E_L e^{j(\omega t - \beta z - \pi/2)} + \hat{a}_y E_L e^{j(\omega t - \beta z)}\right]$$
$$= E_L \text{Re}\left\{[-j\hat{a}_x + \hat{a}_y] e^{j(\omega t - \beta z)}\right\}$$

and we note a 90° phase delay of the $\mathscr{E}_x$ component relative to $\mathscr{E}_y$.

From the previous discussion we see that *left-hand circular polarization* can be achieved *if and only if its two orthogonal components have equal amplitudes and a 90°*

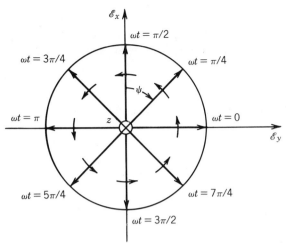

FIGURE 4-16 Left-hand circularly polarized wave.

phase difference of one component relative to the other. The sense of rotation (counter-clockwise here) is determined by rotating the phase-leading component (in this instance $\mathscr{E}_y$) toward the phase-lagging component (in this instance $\mathscr{E}_x$). The field rotation must be viewed as the wave travels away from the observer.

The *necessary and sufficient* conditions for circular polarization are the following:

1. The field must have two orthogonal linearly polarized components.
2. The two components must have the same magnitude.
3. The two components must have a time-phase difference of odd multiples of 90°.

The sense of rotation is always determined by rotating the phase-leading component toward the phase-lagging component and observing the field rotation as the wave is traveling away from the observer. The rotation of the phase-leading component toward the phase-lagging component should be done along the angular separation between the two components that is less than 180°. Phases equal to or greater than 0° and less than 180° should be considered leading whereas those equal to or greater than 180° and less than 360° should be considered lagging.

4.4.3 Elliptical Polarization

A wave is said to be elliptically polarized if the tip of the electric field vector traces an elliptical locus in space. At various instants of time the electric field vector changes continuously with time in such a manner as to describe an elliptical locus. It is right-hand elliptically polarized if the electric field vector of the ellipse rotates clockwise, and it is left-hand elliptically polarized if the electric field vector of the ellipse rotates counterclockwise [8, 10–14].

Let us examine the locus of the instantaneous electric field vector ($\mathscr{E}$) at the $z = 0$ plane at all times. For this particular example, let in (4-50a)

$$\phi_x = \frac{\pi}{2}$$
$$\phi_y = 0$$
$$E_{x_0}^+ = (E_R + E_L)$$
$$E_{y_0}^+ = (E_R - E_L) \quad (4\text{-}54)$$

Then

$$\mathcal{E}_x = (E_R + E_L)\cos\left(\omega t + \frac{\pi}{2}\right) = -(E_R + E_L)\sin\omega t$$
$$\mathcal{E}_y = (E_R - E_L)\cos(\omega t) \tag{4-54a}$$

We can write the locus for the amplitude of the electric field vector as

$$\mathcal{E}^2 = \mathcal{E}_x^2 + \mathcal{E}_y^2 = (E_R + E_L)^2 \sin^2\omega t + (E_R - E_L)^2 \cos^2\omega t$$
$$= E_R^2 \sin^2\omega t + E_L^2 \sin^2\omega t + 2E_R E_L \sin^2\omega t$$
$$+ E_R^2 \cos^2\omega t + E_L^2 \cos^2\omega t - 2E_R E_L \cos^2\omega t$$
$$\mathcal{E}_x^2 + \mathcal{E}_y^2 = E_R^2 + E_L^2 + 2E_R E_L [\sin^2\omega t - \cos^2\omega t] \tag{4-54b}$$

However

$$\sin\omega t = -\mathcal{E}_x/(E_R + E_L)$$
$$\cos\omega t = \mathcal{E}_y/(E_R - E_L) \tag{4-54c}$$

Substituting (4-54c) into (4-54b) reduces to

$$\left\{\frac{\mathcal{E}_x}{E_R + E_L}\right\}^2 + \left\{\frac{\mathcal{E}_y}{E_R - E_L}\right\}^2 = 1 \tag{4-54d}$$

which is the equation for an ellipse with major axis intercept $|\mathcal{E}|_{max} = |E_R + E_L|$ and minor axis intercept $|\mathcal{E}|_{min} = |E_R - E_L|$. As time elapses, the electric vector rotates and its length varies with its tip tracing an ellipse as shown in Figure 4-17. The maximum and minimum lengths of the electric vector are the major and minor axes intercepts given by

$$|\mathcal{E}|_{max} = |E_R + E_L| \quad \text{when } \omega t = (2n+1)\frac{\pi}{2}, \ n = 0, 1, 2, \ldots \tag{4-54e}$$

$$|\mathcal{E}|_{min} = |E_R - E_L| \quad \text{when } \omega t = n\pi, \ n = 0, 1, 2, \ldots \tag{4-54f}$$

The axial ratio (AR) is defined to be the ratio of the major axis (including its sign) of the polarization ellipse to the minor axis, or

$$\text{AR} = -\frac{\mathcal{E}_{max}}{\mathcal{E}_{min}} = -\frac{(E_R + E_L)}{(E_R - E_L)} \tag{4-54g}$$

where E_R and E_L are positive real quantities. As defined in (4-54g), the axial ratio AR can take positive (for left-hand polarization) or negative (for right-hand polarization) values in the range $1 \leq |\text{AR}| \leq \infty$. The instantaneous electric field vector can be written as

$$\mathcal{E} = \text{Re}\left\{\hat{a}_x[E_R + E_L]e^{j(\omega t - \beta z + \pi/2)} + \hat{a}_y[E_R - E_L]e^{j(\omega t - \beta z)}\right\}$$
$$= \text{Re}\left\{[\hat{a}_x j(E_R + E_L) + \hat{a}_y(E_R - E_L)]e^{j(\omega t - \beta z)}\right\}$$
$$\mathcal{E} = \text{Re}\left\{[E_R(j\hat{a}_x + \hat{a}_y) + E_L(j\hat{a}_x - \hat{a}_y)]e^{j(\omega t - \beta z)}\right\} \tag{4-54h}$$

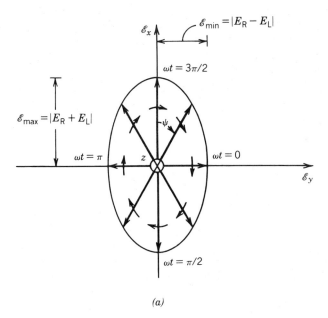

(a)

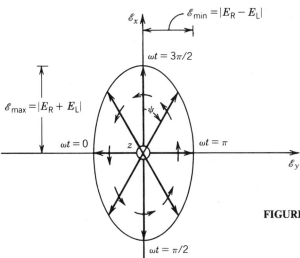

(b)

FIGURE 4-17 Right- and left-hand elliptical polarizations with major axis along the x axis. (a) Right-hand (clockwise) when $E_R > E_L$. (b) Left-hand (counterclockwise) when $E_R < E_L$.

From (4-54h) we see that we can represent an elliptical wave as the sum of a right-hand [first term of (4-54h)] and a left-hand [second term of (4-54h)] circularly polarized waves with amplitudes E_R and E_L, respectively. If $E_R > E_L$, the axial ratio will be negative and the right-hand circular component will be stronger than the left-hand circular component, and the electric vector rotates in the same direction as that of the right circularly polarized wave. Thus a *right-hand elliptically polarized wave* as shown in Figure 4-17a results. If $E_L > E_R$, the axial ratio will be positive and the left-hand circularly polarized component will be stronger than the right-hand circularly polarized component. The electric field vector will rotate in the same direction as that of the left-hand circularly polarized component. Thus a

166 WAVE PROPAGATION AND POLARIZATION

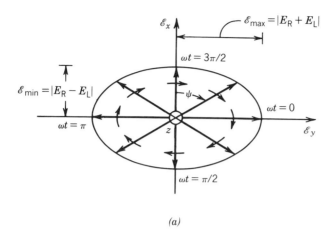

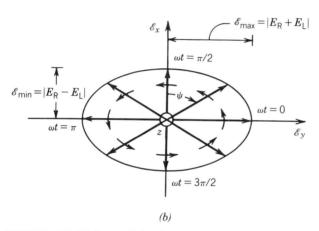

FIGURE 4-18 Right- and left-hand elliptical polarizations with major axis along the y axis. (*a*) Right-hand (clockwise) when $E_R > E_L$. (*b*) Left-hand (counterclockwise) when $E_R < E_L$.

left-hand elliptically polarized wave as shown in Figure 4-17b results. The sign of the axial ratio carries information on the direction of rotation of the electric field vector.

An analogous situation exists when

$$\phi_x = \frac{\pi}{2}$$
$$\phi_y = 0$$
$$E_{x_0}^+ = (E_R - E_L)$$
$$E_{y_0}^+ = (E_R + E_L)$$
(4-55)

The polarization loci are shown in Figure 4-18a and b when $E_R > E_L$ and $E_R < E_L$, respectively.

From (4-54e) and (4-54f) it can be seen that the component of $\mathscr{E}$ measured along the major axis of the polarization ellipse is 90° out of phase with the

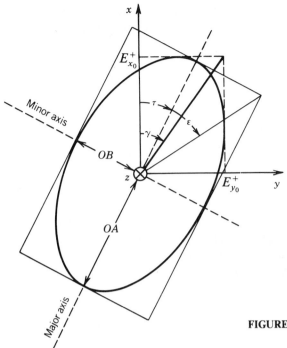

FIGURE 4-19 Rotation of a plane electromagnetic wave and its tilted ellipse at $z = 0$ as a function of time.

component of $\mathscr{E}$ measured along the minor axis. Also with the aid of (4-54b) it can be shown that the electric vector rotates through 90° in space between the instants of time given by (4-54e) and (4-54f) when the vector has maximum and minimum lengths, respectively. Thus the major and minor axes of the polarization ellipse are orthogonal in space, just as we might anticipate.

Since linear polarization is a special kind of elliptical polarization, we can represent a linear polarization as the sum of a right- and a left-hand circularly polarized components of *equal amplitudes*. We see that for this case ($E_R = E_L$) (4-54h) will degenerate into a linear polarization.

A more general orientation of an elliptically polarized locus is the tilted ellipse of Figure 4-19. This is representative of the fields of (4-50a) when

$$\Delta\phi = \phi_x - \phi_y \neq \frac{n\pi}{2} \quad n = 0, 1, 2, 3 \ldots$$

$$\geq 0 \quad \begin{array}{l} \text{for CW if } E_R > E_L \\ \text{for CCW if } E_R < E_L \end{array} \quad (4\text{-}56\text{a})$$

$$\leq 0 \quad \begin{array}{l} \text{for CW if } E_R < E_L \\ \text{for CCW if } E_R > E_L \end{array} \quad (4\text{-}56\text{b})$$

$$E_{x_0}^+ = E_R + E_L$$
$$E_{y_0}^+ = E_R - E_L \quad (4\text{-}56\text{c})$$

Thus the major and minor axes of the ellipse do not coincide with the principal axes of the coordinate system unless the phase difference between the two orthogonal components is equal to odd multiples of $\pm 90°$.

The ratio of the major to the minor axes, which is defined as the axial ratio (AR), is equal to [8]

$$\text{AR} = \pm \frac{\text{major axis}}{\text{minor axis}} = \pm \frac{OA}{OB} \qquad 1 \leq |\text{AR}| \leq \infty \qquad (4\text{-}57)$$

where

$$OA = \left[\tfrac{1}{2} \left\{ (E_{x_0}^+)^2 + (E_{y_0}^+)^2 + \left[(E_{x_0}^+)^4 + (E_{y_0}^+)^4 + 2(E_{x_0}^+)^2 (E_{y_0}^+)^2 \cos(2\Delta\phi) \right]^{1/2} \right\} \right]^{1/2}$$
(4-57a)

$$OB = \left[\tfrac{1}{2} \left\{ (E_{x_0}^+)^2 + (E_{y_0}^+)^2 - \left[(E_{x_0}^+)^4 + (E_{y_0}^+)^4 + 2(E_{x_0}^+)^2 (E_{y_0}^+)^2 \cos(2\Delta\phi) \right]^{1/2} \right\} \right]^{1/2}$$
(4-57b)

$E_{x_0}^+$ and $E_{y_0}^+$ are given by (4-56c). The plus (+) sign in (4-57) is for left-hand and the minus (−) is for right-hand polarization.

The tilt of the ellipse, *relative to the x axis*, is represented by the angle τ given by

$$\tau = \frac{\pi}{2} - \frac{1}{2} \tan^{-1} \left[\frac{2 E_{x_0}^+ E_{y_0}^+}{(E_{x_0}^+)^2 - (E_{y_0}^+)^2} \cos(\Delta\phi) \right] \qquad (4\text{-}57c)$$

4.4.4 Poincaré Sphere

The polarization state, defined here as P, of any wave can be uniquely represented by a point on the surface of a sphere [15–19]. This is accomplished by either of the two sets of angles (γ, δ) and (ε, τ). By referring to (4-50a) and Figure 4-20a, we can define the two sets of angles:

$$(\gamma, \delta) \text{ set}$$

$$\gamma = \tan^{-1}\left[\frac{E_{y_0}^+}{E_{x_0}^+}\right] \text{ or } \tan^{-1}\left[\frac{E_{x_0}^+}{E_{y_0}^+}\right] \qquad 0° \leq \gamma \leq 90° \qquad (4\text{-}58a)$$

$$\delta = \phi_y - \phi_x = \text{phase difference between } \mathscr{E}_y \text{ and } \mathscr{E}_x \qquad -180° \leq \delta \leq 180° \qquad (4\text{-}58b)$$

where 2γ is the great-circle angle drawn from a reference point on the equator and δ is the equator to great-circle angle;

$$(\varepsilon, \tau) \text{ set}$$

$$\varepsilon = \cot^{-1}(\text{AR}) \Rightarrow \text{AR} = \cot(\varepsilon) \qquad -45° \leq \varepsilon \leq +45° \qquad (4\text{-}59a)$$

$$\tau = \text{tilt angle} \qquad 0° \leq \tau \leq 180° \qquad (4\text{-}59b)$$

where

$$2\varepsilon = \text{latitude}$$
$$2\tau = \text{longitude}$$

In (4-58a) the appropriate ratio is the one that satisfies the angular limits of all the Poincaré sphere angles (especially those of ε). The axial ratio AR is positive for

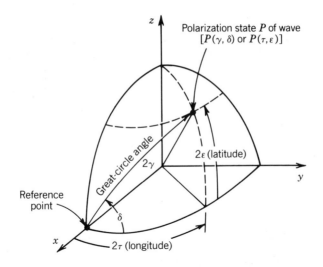

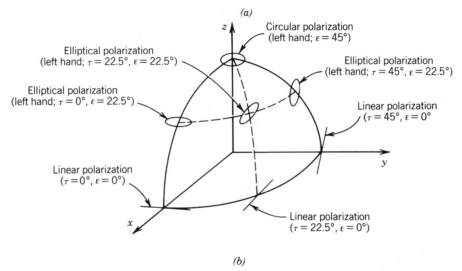

FIGURE 4-20 Poincaré sphere for the polarization state of an electromagnetic wave. (*Source:* J. D. Kraus, *Electromagnetics*, 1984, McGraw-Hill Book Co.) (*a*) Poincaré sphere. (*b*) Polarization state.

left-hand polarization and negative for right-hand polarization. Some polarization states are displayed on the first octant of the Poincaré sphere in Figure 4-20b. The polarization states on a planar surface representation (projection) of the Poincaré sphere ($-45° \leq \varepsilon \leq +45°$, $0° \leq \tau \leq 180°$) are shown in Figure 4-21.

For the polarization ellipse of Figure 4-19, the two sets of angles are related geometrically as shown in Figure 4-20. Analytically it can be shown through spherical trigonometry [20] that the two sets of angles (γ, δ) and (ε, τ) are related by

$$\cos(2\gamma) = \cos(2\varepsilon)\cos(2\tau) \qquad (4\text{-}60a)$$

$$\tan(\delta) = \frac{\tan(2\varepsilon)}{\sin(2\tau)} \qquad (4\text{-}60b)$$

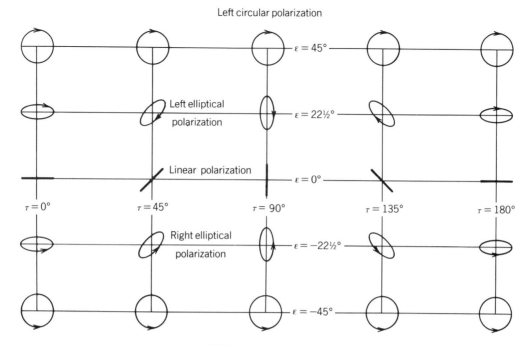

FIGURE 4-21 Polarization states of electromagnetic waves on a planar surface projection of a Poincaré sphere. (*Source:* J. D. Kraus, *Electromagnetics*, 1984, McGraw-Hill Book Co.)

or

$$\sin(2\varepsilon) = \sin(2\gamma)\sin(\delta) \quad (4\text{-}61\text{a})$$
$$\tan(2\tau) = \tan(2\gamma)\cos(\delta) \quad (4\text{-}61\text{b})$$

Thus one set can be obtained by knowing the other.

It is apparent from Figure 4-20 that the linear polarization is always found along the equator; the right-hand circular resides along the south pole and the left-hand circular along the north pole. The remaining surface of the sphere is used to represent elliptical polarization with left-hand elliptical in the upper hemisphere and right-hand elliptical on the lower hemisphere.

Example 4-9. Determine the point on the Poincaré sphere of Figure 4-20 when the wave represented by (4-50a) is such that

$$\mathscr{E}_x = E_{x_0}^+ \cos(\omega t - \beta z + \phi_x)$$
$$\mathscr{E}_y = 0$$

Solution. Using (4-58a) and (4-58b)

$$\gamma = \tan^{-1}\left[\frac{E_{y_0}^+}{E_{x_0}^+}\right] = \tan^{-1}\left[\frac{0}{E_{x_0}^+}\right] = 0°$$

and δ could be of any value $-180° \leq \delta \leq 180°$. The values of ε and τ can now be obtained from (4-61a) and (4-61b), and they are equal to

$$2\varepsilon = \sin^{-1}[\sin(2\gamma)\sin(\delta)] = \sin^{-1}(0) = 0°$$
$$2\tau = \tan^{-1}[\tan(2\gamma)\cos(\delta)] = \tan^{-1}(0) = 0°$$

It is apparent that for this wave, which is obviously linearly polarized, the polarization state (point) is at the reference point of Figure 4-20. The axial ratio is obtained from (4-59a), and it is equal to

$$AR = \cot(\varepsilon) = \cot(0) = \infty$$

An axial ratio of infinity is always used to represent linear polarization.

Example 4-10. Repeat Example 4-9 when the wave of (4-50a) is such that

$$\mathscr{E}_x = 0$$
$$\mathscr{E}_y = E_{y_0}^+ \cos(\omega t - \beta z + \phi_y)$$

Solution. Using (4-58a) and (4-58b),

$$\gamma = \tan^{-1}\left[\frac{E_{y_0}^+}{E_{x_0}^+}\right] = \tan^{-1}(\infty) = 90°$$

and δ could be of any value $-180° \leq \delta \leq 180°$. The values of ε and τ can now be obtained from (4-61a) and (4-61b), and they are equal to

$$2\varepsilon = \sin^{-1}[\sin(2\gamma)\sin(\delta)] = \sin^{-1}(0) = 0°$$
$$2\tau = \tan^{-1}[\tan(2\gamma)\cos(\delta)] = \tan^{-1}(0) = 180°$$

The polarization state (point) of this linearly polarized wave is diametrically opposed to that in Example 4-9. The axial ratio is also infinity.

Example 4-11. Determine the polarization state (point) on the Poincaré sphere of Figure 4-20 when the wave of (4-50a) is such that

$$\mathscr{E}_x = E_{x_0}^+ \cos(\omega t - \beta z + \phi_x) = 2E_0 \cos\left(\omega t - \beta z + \frac{\pi}{2}\right)$$
$$\mathscr{E}_y = E_{y_0}^+ \cos(\omega t - \beta z + \phi_y) = E_0 \cos(\omega t - \beta z)$$

Solution. Using (4-58a) and (4-58b),

$$\gamma = \tan^{-1}\left[\frac{E_{y_0}^+}{E_{x_0}^+}\right] = \tan^{-1}\left[\frac{E_0}{2E_0}\right] = 26.56°$$

$$\delta = \phi_y - \phi_x = -90°$$

The values of ε and τ can now be obtained from (4-61a) and (4-61b), and they are equal to

$$2\varepsilon = \sin^{-1}[\sin(2\gamma)\sin(\delta)] = \sin^{-1}[-\sin(2\gamma)] = -2\gamma = -53.12°$$
$$2\tau = \tan^{-1}[\tan(2\gamma)\cos(\delta)] = \tan^{-1}(0) = 0°$$

Therefore this point is situated on the principal xz plane at an angle of $2\gamma = -2\varepsilon = 53.12°$ from the reference point of the x axis of Figure 4-20. The axial ratio is obtained using (4-59a), and it is equal to

$$AR = \cot(\varepsilon) = \cot(-26.56°) = -2$$

The negative sign indicates that the wave has a right-hand (clockwise) polarization. Therefore the wave is right-hand elliptically polarized with $AR = -2$.

In general, points on the principal xz elevation plane, aside from the two intersecting points on the equator and the north and south poles, are used to represent elliptical polarization when the major and minor axes of the polarization ellipse of Figure 4-19 coincide with the principal axes.

If the polarization state of a wave is defined as P_w and that of an antenna as P_a, then the voltage response of the antenna due to the wave is obtained by [10, 19]

$$V = C\cos\left[\frac{P_w P_a}{2}\right] \quad (4\text{-}62)$$

where C = constant that is a function of the antenna size and field strength of the wave
$P_w P_a$ = angle subtended by a great-circle arc from polarization P_w to P_a
P_w = polarization state of the wave
P_a = polarization state of the antenna

Remember that the polarization of a wave, by IEEE standards [7, 8], is determined as the wave is observed from the rear (is receding). Therefore the polarization of the antenna is determined by its radiated field in the transmitting mode.

Example 4-12. If the polarization states of the wave and antenna are given, respectively, by those of Examples 4-9 and 4-10, determine the voltage response of the antenna due to that wave.

Solution. Since the polarization state P_w of the wave is at the $+x$ axis and that of the antenna P_a is at the $-x$ axis of Figure 4-20, then the angle $P_w P_a$ subtended by a great-circle arc from P_w to P_a is equal to

$$P_w P_a = 180°$$

Therefore the voltage response of the antenna is according to (4-62) equal to

$$V = C\cos\left[\frac{P_w P_a}{2}\right] = C\cos(90°) = 0$$

This is expected since the fields of the wave and those of the antenna are orthogonal (cross-polarized) to each other.

Example 4-13. The polarization of a wave that impinges upon a left-hand counterclockwise) circularly polarized antenna is circularly polarized. Determine the response of the antenna when the sense of rotation of the incident wave is

1. Left-hand (counterclockwise).
2. Right-hand (clockwise).

Solution.

1. Since the antenna is left-hand circularly polarized, its polarization state (point) on the Poincaré sphere is on the north pole ($2\gamma = \delta = 90°$). When the wave is also left-hand circularly polarized, then its polarization state (point) is also on the north pole ($2\gamma = \delta = 90°$). Therefore the subtended angle $P_w P_a$ between the two polarization states is equal to

$$P_w P_a = 0°$$

and the voltage response of the antenna, according to (4-62), is equal to

$$V = C \cos\left[\frac{P_w P_a}{2}\right] = C \cos(0) = C$$

This represents the maximum response of the antenna, and it occurs when the polarization (including sense of rotation) of the wave is the same as that of the antenna.

2. When the sense of rotation of the wave is right-hand circularly polarized, its polarization state (point) is on the south pole ($2\gamma = 90°$, $\delta = -90°$). Therefore the subtended angle $P_w P_a$ between the two polarization states is equal to

$$P_w P_a = 180°$$

and the response of the antenna, according to (4-62), is equal to

$$V = C \cos\left[\frac{P_w P_a}{2}\right] = C \cos\left[\frac{180°}{2}\right] = C \cos(90°) = 0$$

This represents a null response of the antenna, and it occurs when the sense of rotation of the circularly polarized wave is opposite to that of the circularly polarized antenna. This is one technique, in addition to those shown in Example 4-12, that can be used to null the response of an antenna system.

REFERENCES

1. S. F. Adam, *Microwave Theory and Applications*, Prentice-Hall, Englewood Cliffs, N.J., 1969.
2. A. L. Lance, *Introduction to Microwave Theory and Measurements*, McGraw-Hill, New York, 1964.

3. N. Marcuvitz (ed.), *Waveguide Handbook*, McGraw-Hill, New York, 1951, Chapter 8, pp. 387–413.
4. C. H. Walter, *Traveling Wave Antennas*, McGraw-Hill, New York, 1965, pp. 172–187.
5. R. B. Adler, L. J. Chu, and R. M. Fano, *Electromagnetic Energy Transmission and Radiation*, Wiley, New York, 1960, Chapter 8.
6. D. T. Paris and F. K. Hurd, *Basic Electromagnetic Theory*, McGraw-Hill, New York, 1969.
7. "IEEE Standard 145-1983, IEEE Standard Definitions of Terms for Antennas," reprinted in *IEEE Trans. Antennas Propagat.*, vol. AP-31, no. 6, part II, pp. 1–29, November 1983.
8. C. A. Balanis, *Antenna Theory: Analysis and Design*, Wiley, New York, 1982.
9. W. Sichak and S. Milazzo, "Antennas for circular polarization," *Proc. IEEE*, vol. 36, pp. 997–1002, August 1948.
10. G. Sinclair, "The transmission and reception of elliptically polarized waves," *Proc. IRE*, vol. 38, pp. 148–151, February 1950.
11. V. H. Rumsey, G. A. Deschamps, M. L. Kales, and J. I. Bohnert, "Techniques for handling elliptically polarized waves with special reference to antennas," *Proc. IRE*, vol. 39, pp. 533–534, May 1951.
12. V. H. Rumsey, "Part I—Transmission between elliptically polarized antennas," *Proc. IRE*, vol. 39, pp. 535–540, May 1951.
13. M. L. Kales, "Part III—Elliptically polarized waves and antennas," *Proc. IRE*, vol. 39, pp. 544–549, May 1951.
14. J. I. Bohnert, "Part IV—Measurements on elliptically polarized antennas," *Proc. IRE*, vol. 39, pp. 549–552, May 1951.
15. H. Poincaré, *Théorie Mathématique de la Limiére*, Georges Carré, Paris, France, 1892.
16. G. A. Deschamps, "Part II—Geometrical representation of the polarization of a plane electromagnetic wave," *Proc. IRE*, vol. 39, pp. 540–544, May 1951.
17. E. F. Bolinder, "Geometric analysis of partially polarized electromagnetic waves," *IEEE Trans. Antennas Propagat.*, vol. AP-15, no. 1, pp. 37–40, January 1967.
18. G. A. Deschamps and P. E. Mast, "Poincaré sphere representation of partially polarized fields," *IEEE Trans. Antennas Propagat.*, vol. AP-21, no. 4, pp. 474–478, July 1973.
19. J. D. Kraus, *Electromagnetics*, Third Edition, McGraw-Hill, New York, 1984.
20. M. Born and E. Wolf, *Principles of Optics*, Macmillan Co., New York, pp. 24–27, 1964.

PROBLEMS

4.1. A uniform plane wave having only an x component of the electric field is traveling in the $+z$ direction in an unbounded lossless, source-free region. Using Maxwell's equations write expressions for the electric and corresponding magnetic field intensities. Compare your answers to those of (4-2b) and (4-3c).

4.2. Using Maxwell's equations, find the magnetic field components for the wave whose electric field is given in Example 4-1. Compare your answer with that obtained in the solution of Example 4-1.

4.3. The complex **H** field of a uniform plane wave, traveling in an unbounded source-free medium of free space, is given by

$$\mathbf{H} = \frac{1}{120\pi}\left(\hat{a}_x - 2\hat{a}_y\right)e^{-j\beta_0 z}$$

Find (a) the corresponding electric field, (b) the instantaneous power density vector, and (c) the time-average power density.

4.4. The complex **E** field of a uniform plane wave is given by
$$\mathbf{E} = (\hat{a}_x + j\hat{a}_z)e^{-j\beta_0 y} + (2\hat{a}_x - j\hat{a}_z)e^{+j\beta_0 y}$$
Assuming an unbounded source-free, free-space medium, find (a) the corresponding magnetic field, (b) the time-average power density flowing in the $+y$ direction, and (c) the time-average power density flowing in the $-y$ direction.

4.5. The magnetic field of a uniform plane wave in a source-free region is given by
$$\mathbf{H} = 10^{-6}\left[-\hat{a}_x(2+j) + \hat{a}_z(1+j3)\right]e^{+j\beta y}$$
Assuming that the medium is free space, determine (a) the corresponding electric field and (b) the time-average power density.

4.6. The electric field of a uniform plane wave traveling in a source-free region of free space is given by
$$\mathbf{E} = 10^{-3}(\hat{a}_x + j\hat{a}_y)\sin(\beta_0 z)$$
(a) Is this a traveling or a standing wave?
(b) Identify the traveling wave(s) of the electric field and the direction(s) of travel.
(c) Find the corresponding magnetic field.
(d) Determine the time-average power density of the wave.

4.7. The magnetic field of a uniform plane wave traveling in a source-free, free-space region is given by
$$\mathbf{H} = 10^{-6}(\hat{a}_y + j\hat{a}_z)\cos(\beta_0 x)$$
(a) Is this a traveling or a standing wave?
(b) Identify the traveling wave(s) of the magnetic field and the direction(s) of travel.
(c) Find the corresponding electric field.
(d) Determine the time-average power density of the wave.

4.8. A uniform plane wave is traveling in the $-z$ direction inside an unbounded source-free, free-space region. Assuming that the electric field has only an E_x component, its value at $z = 0$ is 4×10^{-3} V/m, and its frequency of operation is 300 MHz, write expressions for (a) the complex electric and magnetic fields, (b) the instantaneous electric and magnetic fields, (c) the time-average and instantaneous power densities, and (d) the time-average and instantaneous electric and magnetic energy densities.

4.9. A uniform plane wave traveling inside an unbounded free-space medium has peak electric and magnetic fields given by
$$\mathbf{E} = \hat{a}_x E_0 e^{-j\beta_0 z}$$
$$\mathbf{H} = \hat{a}_y H_0 e^{-j\beta_0 z}$$
where $E_0 = 1$ mV/m.
(a) Evaluate H_0.
(b) Find the corresponding average power density. Evaluate all the constants.
(c) Determine the volume electric and magnetic energy densities. Evaluate all the constants.

4.10. The complex electric field of a uniform plane wave traveling in an unbounded nonferromagnetic dielectric medium is given by
$$\mathbf{E} = \hat{a}_y 10^{-3} e^{-j2\pi z}$$

where z is measured in meters. Assuming that the frequency of operation is 100 MHz, find the (a) phase velocity of the wave (give units), (b) dielectric constant of the medium, (c) wavelength (in meters), (d) time-average power density, and (e) time-average total energy density.

4.11. The complex electric field of a time-harmonic field in free space is given by

$$\mathbf{E} = \hat{a}_z 10^{-3}(1+j)e^{-j(2/3)\pi x}$$

Assuming the distance x is measured in meters, find (a) the wavelength (in meters), (b) the frequency, and (c) the associated magnetic field.

4.12. A uniform plane wave is traveling inside the earth, which is assumed to be a perfect dielectric, and it is infinite in extent. If the relative permittivity of the earth is 9, find at a frequency of 1 MHz the phase velocity, wave impedance, intrinsic impedance, and wavelength of the wave inside the earth.

4.13. An 11-GHz transmitter radiates its power isotropically in a free-space medium. Assuming its total radiated power is 50 mW, at a distance of 3 km find (a) the time-average power density, (b) rms electric and magnetic fields, and (c) total time-average volume energy densities. In all cases, specify the units.

4.14. The electric field of a time-harmonic wave traveling in free space is given by

$$\mathbf{E} = \hat{a}_x 10^{-4}(1+j)e^{-j\beta_0 z}$$

Find the amount of real power crossing a rectangular aperture whose cross section is perpendicular to the z axis. The area of the aperture is 20 cm^2.

4.15. The following complex electric field of a time-harmonic wave traveling in a source-free, free-space region is given by

$$\mathbf{E} = 5 \times 10^{-3}(4\hat{a}_y + 3\hat{a}_z)e^{j(6y-8z)}$$

Assuming y and z represent their respective distances in meters, determine:
(a) The angle of wave travel (relative to the z axis).
(b) The three phase constants of the wave along its oblique direction of travel, the y axis, and the z axis (in radians per meter).
(c) The three wavelengths of the wave along its oblique direction of travel, the y axis, and the z axis (in meters).
(d) The three phase velocities of the wave along the oblique direction of travel, the y axis, and the z axis (in meters per second).
(e) The three energy velocities of the wave along the oblique direction of travel, the y axis, and the z axis (in meters per second).
(f) The frequency of the wave.
(g) The associated magnetic field.

4.16. Using Maxwell's equations determine the magnetic field of (4-18b) given the electric field of (4-18a).

4.17. Given the electric field of Example 4-2 and using Maxwell's equations, determine the magnetic field. Compare it with that found in the solution of Example 4-2.

4.18. Given (4-19a) and (4-19c), determine the phase velocities of (4-22) and (4-23).

4.19. Derive the energy velocity of (4-24) using the definition of (4-9), (4-18a), and (4-18b).

4.20. A uniform plane wave of 3 GHz is incident upon an unbounded conducting medium of copper that has a conductivity of 5.76×10^7 S/m and $\varepsilon = \varepsilon_0$, $\mu = \mu_0$. Find the approximate (a) intrinsic impedance of copper and (b) skin depth (in meters).

4.21. The magnetic field intensity of a plane wave traveling in a lossy earth is given by

$$\mathbf{H} = (\hat{a}_y + j2\hat{a}_z)H_0 e^{-\alpha x} e^{-j\beta x}$$

where $H_0 = 1$ μA/m. Assuming the lossy earth has a conductivity of 10^{-4} S/m, a dielectric constant of 9, and the frequency of operation is 1 GHz, find inside the earth:

(a) The corresponding electric field vector.
(b) The average power density vector.
(c) The phase constant (radians per meter).
(d) The phase velocity (meters per second).
(e) The wavelength (meters).
(f) The attenuation constant (Nepers per meter).
(g) The skin depth (meters).

4.22. Sea water is an important medium in communication between submerged submarines or between submerged submarines and receiving and transmitting stations located above the surface of the sea. Assuming the constitutive electrical parameters of the sea are $\sigma = 4$ S/m, $\varepsilon_r = 81$, $\mu_r = 1$, and $f = 10^4$ Hz, find:

(a) The complex propagation constant (per meter).
(b) The phase velocity (meters per second).
(c) The wavelength (meters).
(d) The attenuation constant (Nepers per meter).
(e) The skin depth (meters).

4.23. The electrical constitutive parameters of moist earth at a frequency of 1 MHz are $\sigma = 10^{-1}$ S/m, $\varepsilon_r = 4$, and $\mu_r = 1$. Assuming that the electric field of a uniform plane wave at the interface (on the side of the earth) is 3×10^{-2} V/m, find:

(a) The distance through which the wave must travel before the magnitude of the electric field reduces to 1.104×10^{-2} V/m.
(b) The attenuation the electric field undergoes in part (a) (in decibels).
(c) The wavelength inside the earth (in meters).
(d) The phase velocity inside the earth (in meters per second).
(e) The intrinsic impedance of the earth.

4.24. The complex electric field of a uniform plane wave is given by

$$\mathbf{E} = 10^{-2}\left[\hat{a}_x\sqrt{2} + \hat{a}_z(1+j)e^{j\pi/4}\right]e^{-j\beta y}$$

(a) Find the polarization of the wave (linear, circular, or elliptical).
(b) Determine the sense of rotation (clockwise or counterclockwise).
(c) Sketch the figure the electric field traces as a function of ωt.

4.25. The complex magnetic field of a uniform plane wave is given by

$$\mathbf{H} = \frac{10^{-3}}{120\pi}(\hat{a}_x - j\hat{a}_z)e^{+j\beta y}$$

(a) Find the polarization of the wave (linear, circular, or elliptical).
(b) State the direction of rotation (clockwise or counterclockwise) and why?
(c) Sketch the polarization curve denoting the $\mathcal{H}$-field amplitude, and direction of rotation. Indicate on the curve the various times for the rotation of the vector.

4.26. In a source-free, free-space region the complex magnetic field of a time-harmonic field is represented by

$$\mathbf{H} = \left[\hat{a}_x (1+j) + \hat{a}_z \sqrt{2}\, e^{j\pi/4} \right] \frac{E_0}{\eta_0} e^{-j\beta_0 y}$$

where E_0 is a constant and η_0 is the intrinsic impedance of free space. Determine:
(a) The polarization of the wave (linear, circular, or elliptical). Give reasons for your answer.
(b) State the sense of rotation, if any.
(c) The corresponding electric field.

4.27. Show that any linearly polarized wave can be decomposed into two circularly polarized waves (one CW and the other CCW) but both traveling in the same direction as the linearly polarized wave.

4.28. In a source-free, free-space region the complex magnetic field is given by

$$\mathbf{H} = j\left(\hat{a}_y - j\hat{a}_z \right) \frac{E_0}{\eta_0} e^{+j\beta_0 x}$$

where E_0 is a constant and η_0 is the intrinsic impedance of free space. Find:
(a) The polarization of the wave (linear, circular, or elliptical). Justify your answer.
(b) The sense of rotation, if any (CW or CCW). Justify your answer.
(c) The time-average power density.
(d) The polarization of the wave on the Poincaré sphere.

4.29. The electric field of a time-harmonic wave is given by

$$\mathbf{E} = 2 \times 10^{-3} \left(\hat{a}_x + \hat{a}_y \right) e^{-j2z}$$

(a) State the polarization of the wave (linear, circular, or elliptical).
(b) Find the polarization on the Poincaré sphere by identifying the angles δ, γ, τ, and ε (in degrees).
(c) Locate that point on the Poincaré sphere.

4.30. For a uniform plane wave represented by the electric field of

$$\mathbf{E} = E_0 \left(\hat{a}_x - j2\hat{a}_y \right) e^{-j\beta z}$$

where E_0 is constant, do the following.
(a) Determine the longitude angle 2τ, latitude angle 2ε, great-circle angle 2γ, and equator to great-circle angle δ (all in degrees) that are used to identify and locate the polarization of the wave on the Poincaré sphere.
(b) Using the answers from part (a), state the polarization of the wave (linear, circular, or elliptical), its sense of rotation (CW or CCW), and its axial ratio.
(c) Find the signal loss (in decibels) when the wave is received by a right-hand circularly polarized antenna.

4.31. The electric field of (4-50a) has an axial ratio of infinity and a great-circle angle of $2\gamma = 109.47°$.

(a) Find the relative magnitude (ratio) of $E_{y_0}^+$ to $E_{x_0}^+$. Which component is more dominant (E_x or E_y)?
(b) Identify the polarization point on the Poincaré sphere (i.e., find δ, τ, and ε in degrees).
(c) State the polarization of the wave (linear, circular, or elliptical).

4.32. The field radiated by an antenna has electric field components represented by (4-50a) such that $E_{x_0}^+ = E_{y_0}^+$ and its axial ratio is infinity.
(a) Identify the polarization point on the Poincaré sphere (i.e., find γ, δ, τ, and ε in degrees).
(b) If this antenna is used to receive the wave of Problem 4.31, find the polarization loss (in decibels). To do this part, use the Poincaré sphere parameters.

CHAPTER 5

REFLECTION AND TRANSMISSION

5.1 INTRODUCTION

In the previous chapter we discussed solutions to TEM waves in unbounded media. In real world problems, however, the fields encounter boundaries, scatterers, and other objects. Therefore the fields must be found by taking into account these discontinuities.

In this chapter we want to discuss TEM field solutions in two semi-infinite lossless and lossy media bounded by a planar boundary of infinite extent. Reflection and transmission coefficients will be derived to account for the reflection and transmission of the fields by the boundary. These coefficients will be functions of the constitutive parameters of the two media, the direction of wave travel (angle of incidence), and the direction of the electric and magnetic fields (wave polarization).

In general, the reflection and transmission coefficients are complex quantities. It will be demonstrated that their amplitudes and phases can be varied by controlling the direction of wave travel (angle of incidence). In fact for one wave polarization (parallel polarization) the reflection coefficient can be made equal to zero. The angle of incidence when this occurs is known as the *Brewster angle*. This principle is used in the design of many instruments (such as binoculars, etc.).

The magnitude of the reflection coefficient can also be made unity by properly selecting the wave incidence angle. This angle is known as the *critical angle*, and it is independent of wave polarization; however, in order for this angle to occur, the incident wave must exist in the denser medium. The critical angle concept plays a crucial role in the design of transmission lines (such as optical fiber, slab waveguides, and coated conductors; the microstrip is one example).

5.2 NORMAL INCIDENCE—LOSSLESS MEDIA

We begin the discussion of reflection and transmission from planar boundaries of lossless media by assuming the wave travel is perpendicular (*normal incidence*) to the planar interface formed by two semi-infinite lossless media, as shown in Figure 5-1, each characterized by the constitutive parameters of ε_1, μ_1 and ε_2, μ_2. When the incident wave encounters the interface, a fraction of the wave intensity will be reflected into medium 1 and part will be transmitted into medium 2.

Assuming the incident electric field of amplitude E_0 is polarized in the x direction, we can write expressions for its incident, reflected, and transmitted electric field components, respectively, as

$$\mathbf{E}^i = \hat{a}_x E_0 e^{-j\beta_1 z} \qquad (5\text{-}1\text{a})$$

$$\mathbf{E}^r = \hat{a}_x \Gamma^b E_0 e^{+j\beta_1 z} \qquad (5\text{-}1\text{b})$$

$$\mathbf{E}^t = \hat{a}_x T^b E_0 e^{-j\beta_2 z} \qquad (5\text{-}1\text{c})$$

where Γ^b and T^b are used here to represent, respectively, the reflection and transmission coefficients *at the interface*. Presently these coefficients are unknowns and will be determined by applying boundary conditions on the fields along the interface. Since the incident fields are linearly polarized and the reflecting surface is planar, the reflected and transmitted fields will also be linearly polarized. Because we do not know the direction of polarization (positive or negative) of the reflected and transmitted electric fields, they are assumed here to be in the same direction (positive) as the incident electric fields. If that is not the case, it will be corrected by the appropriate signs on the reflection and transmission coefficients.

Using the right-hand procedure outlined in Section 4.2.1 or Maxwell's equations 4-3 or 4-3a, the magnetic field components corresponding to (5-1a) through

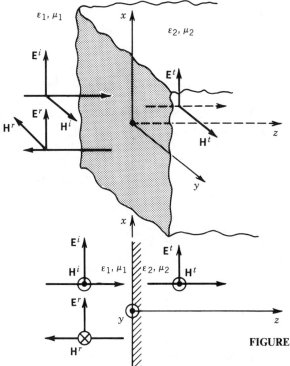

FIGURE 5-1 Wave reflection and transmission at normal incidence by a planar interface.

(5-1c) can be written as

$$\mathbf{H}^i = \hat{a}_y \frac{E_0}{\eta_1} e^{-j\beta_1 z} \tag{5-2a}$$

$$\mathbf{H}^r = -\hat{a}_y \frac{\Gamma^b E_0}{\eta_1} e^{+j\beta_1 z} \tag{5-2b}$$

$$\mathbf{H}^t = \hat{a}_y \frac{T^b E_0}{\eta_2} e^{-j\beta_2 z} \tag{5-2c}$$

The reflection and transmission coefficients will now be determined by enforcing continuity of the tangential components of the electric and magnetic fields across the interface. Using (5-1a) through (5-2c), continuity of the tangential components of the electric and magnetic fields at the interface ($z = 0$) leads, respectively, to

$$1 + \Gamma^b = T^b \tag{5-3a}$$

$$\frac{1}{\eta_1}(1 - \Gamma^b) = \frac{1}{\eta_2} T^b \tag{5-3b}$$

Solving these two equations for Γ^b and T^b, we can write that

$$\boxed{\Gamma^b = \frac{\eta_2 - \eta_1}{\eta_2 + \eta_1} = \frac{E^r}{E^i} = -\frac{H^r}{H^i}} \tag{5-4a}$$

$$\boxed{T^b = \frac{2\eta_2}{\eta_1 + \eta_2} = 1 + \Gamma^b = \frac{E^t}{E^i} = \frac{\eta_2}{\eta_1} \frac{H^t}{H^i}} \tag{5-4b}$$

Therefore the plane wave reflection and transmission coefficients of a planar interface for normal incidence are functions of the constitutive properties, and they are given by (5-4a) and (5-4b). Since the angle of incidence is fixed at normal, the reflection coefficient cannot be equal to zero unless $\eta_2 = \eta_1$. For most dielectric material, aside from ferromagnetics, this implies that $\varepsilon_2 = \varepsilon_1$ since for them $\mu_1 \simeq \mu_2$.

Away from the interface the reflection Γ and transmission T coefficients are related to those at the boundary (Γ^b, T^b) and can be written, respectively, as

$$\boxed{\Gamma(z = -\ell_1) = \left.\frac{E^r(z)}{E^i(z)}\right|_{z=-\ell_1} = \left.\frac{\Gamma^b E_0 e^{+j\beta_1 z}}{E_0 e^{-j\beta_1 z}}\right|_{z=-\ell_1} = \Gamma^b e^{-j2\beta_1 \ell_1}} \tag{5-5a}$$

$$\boxed{T\!\begin{pmatrix} z_2 = \ell_2, \\ z_1 = -\ell_1 \end{pmatrix} = \left.\frac{E^t(z_2)|_{z_2=\ell_2}}{E^i(z_1)|_{z_1=-\ell_1}}\right. = \frac{T^b E_0 e^{-j\beta_2 \ell_2}}{E_0 e^{+j\beta_1 \ell_1}} = T^b e^{-j(\beta_2 \ell_2 + \beta_1 \ell_1)}} \tag{5-5b}$$

where ℓ_1 and ℓ_2 are positive distances measured from the interface to mediums 1 and 2, respectively.

Associated with the electric and magnetic fields of (5-1a) through (5-2c) are corresponding average power densities that can be written as

$$\mathbf{S}_{av}^{i} = \frac{1}{2}\operatorname{Re}(\mathbf{E}^{i} \times \mathbf{H}^{i*}) = \hat{a}_{z}\frac{|E_{0}|^{2}}{2\eta_{1}} \qquad (5\text{-}6a)$$

$$\mathbf{S}_{av}^{r} = \frac{1}{2}\operatorname{Re}(\mathbf{E}^{r} \times \mathbf{H}^{r*}) = -\hat{a}_{z}|\Gamma^{b}|^{2}\frac{|E_{0}|^{2}}{2\eta_{1}} = -\hat{a}_{z}|\Gamma^{b}|^{2}\mathbf{S}_{av}^{i} \qquad (5\text{-}6b)$$

$$\mathbf{S}_{av}^{t} = \frac{1}{2}\operatorname{Re}(\mathbf{E}^{t} \times \mathbf{H}^{t*}) = \hat{a}_{z}|T^{b}|^{2}\frac{|E_{0}|^{2}}{2\eta_{2}} = \hat{a}_{z}|T^{b}|^{2}\frac{\eta_{1}}{\eta_{2}}\frac{|E_{0}|^{2}}{2\eta_{1}}$$

$$= \hat{a}_{z}|T^{b}|^{2}\frac{\eta_{1}}{\eta_{2}}\mathbf{S}_{av}^{i} = \hat{a}_{z}(1 - |\Gamma^{b}|^{2})\mathbf{S}_{av}^{i} \qquad (5\text{-}6c)$$

It is apparent that the ratio of the reflected to the incident power densities is equal to the square of the magnitude of the reflection coefficient. However the ratio of the transmitted to the incident power density is not equal to the square of the magnitude of the transmission coefficient; this is one of the most common errors. Instead the ratio is proportional to the magnitude of the transmission coefficient squared and weighted by the intrinsic impedances of the two media, as given by (5-6c). Remember that the reflection and transmission coefficients relate the reflected and transmitted field intensities to the incident field intensity. Since the total tangential components of these field intensities on either side must be continuous across the boundary, it is expected that the transmitted field can be greater than the incident field and that will require a transmission coefficient greater than unity. However, by the conservation of power, it is well known that the transmitted power density cannot exceed the incident power density.

Example 5-1. A uniform plane wave traveling in free space is incident normally upon a flat semi-infinite lossless medium with a dielectric constant of 2.56 (being representative of polystyrene). Determine the reflection and transmission coefficients and incident, reflected, and transmitted power densities. Assume that the amplitude of the incident electric field at the interface is 1 mV/m.

Solution. Since $\varepsilon_1 = \varepsilon_0$ and $\varepsilon_2 = 2.56\varepsilon_0$,

$$\mu_1 = \mu_2 = \mu_0$$

then

$$\eta_1 = \sqrt{\frac{\mu_1}{\varepsilon_1}} = \sqrt{\frac{\mu_0}{\varepsilon_0}}$$

$$\eta_2 = \sqrt{\frac{\mu_2}{\varepsilon_2}} = \sqrt{\frac{\mu_0}{2.56\varepsilon_0}} = \frac{1}{1.6}\sqrt{\frac{\mu_0}{\varepsilon_0}} = \frac{\eta_1}{1.6}$$

184 REFLECTION AND TRANSMISSION

Thus according to (5-4a) and (5-4b)

$$\Gamma^b = \frac{\eta_2 - \eta_1}{\eta_2 + \eta_1} = \frac{\frac{1}{1.6} - 1}{\frac{1}{1.6} + 1} = \frac{1 - 1.6}{1 + 1.6} = -0.231$$

$$T^b = \frac{2\eta_2}{\eta_1 + \eta_2} = \frac{2\left(\frac{1}{1.6}\right)}{1 + \frac{1}{1.6}} = \frac{2}{2.6} = 0.769$$

In addition the incident, reflected and transmitted power densities are obtained using, respectively, (5-6a), (5-6b), and (5-6c). Thus

$$S_{av}^i = \frac{|E_0|^2}{2\eta_1} = \frac{(10^{-3})^2}{2(376.73)} = 1.327 \times 10^{-9} \text{ W/m}^2 = 1.327 \text{ nW/m}^2$$

$$S_{av}^r = |\Gamma^b|^2 S_{av}^i = |-0.231|^2(1.327) \times 10^{-9} = 0.071 \text{ nW/m}^2$$

$$S_{av}^t = |T^b|^2 \frac{\eta_1}{\eta_2} S_{av}^i = |0.769|^2 \frac{1}{1/1.6}(1.327) \times 10^{-9} = 1.256 \text{ nW/m}^2$$

or

$$S_{av}^t = (1 - |\Gamma^b|^2) S_{av}^i = (1 - |0.231|^2)(1.327) \times 10^{-9} = 1.256 \text{ nW/m}^2$$

In medium 1 the total field is equal to the sum of the incident and reflected fields. Thus for the total electric and magnetic fields in medium 1 we can write that

$$\mathbf{E}^1 = \mathbf{E}^i + \mathbf{E}^r = \hat{a}_x \underbrace{E_0 e^{-j\beta_1 z}}_{\text{traveling wave}} \underbrace{(1 + \Gamma^b e^{+j2\beta_1 z})}_{\text{standing wave}} = \hat{a}_x E_0 e^{-j\beta_1 z}[1 + \Gamma(z)] \quad (5\text{-}7a)$$

$$\mathbf{H}^1 = \mathbf{H}^i + \mathbf{H}^r = \hat{a}_y \underbrace{(E_0/\eta_1) e^{-j\beta_1 z}}_{\text{traveling wave}} \underbrace{(1 - \Gamma^b e^{+j2\beta_1 z})}_{\text{standing wave}} = \hat{a}_y \frac{E_0}{\eta_1} e^{-j\beta_1 z}[1 - \Gamma(z)] \quad (5\text{-}7b)$$

In each expression the factors outside the parentheses represent the *traveling wave part* of the wave and those within the parentheses represent the *standing wave part*. Therefore the total field of two waves is the product of one of the waves times a factor that in this case is the standing wave pattern. This is analogous to the *array multiplication rule* in antennas where the total field of an array of identical elements is equal to the product of the field of a single element times a factor that is referred to as the array factor [1].

As discussed in Section 4.2.1D, the ratio of the maximum value of the electric field magnitude to that of the minimum is defined as the standing wave ratio (SWR), and it is given here by

$$\text{SWR} = \frac{|\mathbf{E}^1|_{\max}}{|\mathbf{E}^1|_{\min}} = \frac{1 + |\Gamma^b|}{1 - |\Gamma^b|} = \frac{1 + \left|\frac{\eta_2 - \eta_1}{\eta_2 + \eta_1}\right|}{1 - \left|\frac{\eta_2 - \eta_1}{\eta_2 + \eta_1}\right|} \quad (5\text{-}8)$$

For two media with identical permeabilities ($\mu_1 = \mu_2$), the SWR can be written as

$$\text{SWR} = \frac{\left|\sqrt{\varepsilon_1} + \sqrt{\varepsilon_2}\right| + \left|\sqrt{\varepsilon_1} - \sqrt{\varepsilon_2}\right|}{\left|\sqrt{\varepsilon_1} + \sqrt{\varepsilon_2}\right| - \left|\sqrt{\varepsilon_1} - \sqrt{\varepsilon_2}\right|} = \begin{cases} \sqrt{\dfrac{\varepsilon_1}{\varepsilon_2}} & \varepsilon_1 > \varepsilon_2 \\ \sqrt{\dfrac{\varepsilon_2}{\varepsilon_1}} & \varepsilon_2 > \varepsilon_1 \end{cases} \quad \begin{array}{l}(5\text{-}9a)\\(5\text{-}9b)\end{array}$$

5.3 OBLIQUE INCIDENCE—LOSSLESS MEDIA

To analyze reflections and transmissions at oblique wave incidence, we need to introduce the *plane of incidence*, which is defined as *the plane formed by a unit vector normal to the reflecting interface and the vector in the direction of incidence*. For a wave whose wave vector is on the xz plane and is incident upon an interface that is parallel to the xy plane as shown in Figure 5-2, the plane of incidence is the xz plane.

To examine reflections and transmissions at oblique angles of incidence for a general wave polarization, it is most convenient to decompose the electric field into its *perpendicular* and *parallel* components (relative to the plane of incidence) and analyze each one of them individually. The total reflected and transmitted field will be the vector sum from each one of these two polarizations.

When the electric field is perpendicular to the plane of incidence, the polarization of the wave is referred to as *perpendicular polarization*. Since the electric field is parallel to the interface, it is also known as *horizontal or E polarization*. When the electric field is parallel to the plane of incidence, the polarization is referred to as *parallel polarization*. Because a component of the electric field is also perpendicular to the interface when the magnetic field is parallel to the interface, it is also known as *vertical or H polarization*. Each type of polarization will be further examined.

5.3.1 Perpendicular (Horizontal or E) Polarization

Let us now assume that the electric field of the uniform plane wave incident on a planar interface at an oblique angle, as shown in Figure 5-2, is oriented perpendicu-

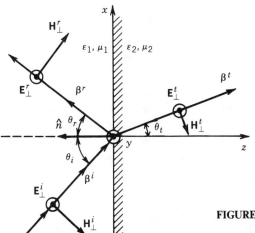

FIGURE 5-2 Perpendicular (horizontal) polarized uniform plane wave incident at an oblique angle on an interface.

186 REFLECTION AND TRANSMISSION

larly to the plane of incidence. As previously stated, this is referred to as the perpendicular polarization.

Using the techniques outlined in Section 4.2.2, the incident electric and magnetic fields can be written as

$$\mathbf{E}_\perp^i = \hat{a}_y E_\perp^i e^{-j\boldsymbol{\beta}^i \cdot \mathbf{r}} = \hat{a}_y E_0 e^{-j\beta_1(x \sin \theta_i + z \cos \theta_i)} \quad (5\text{-}10\text{a})$$

$$\mathbf{H}_\perp^i = (-\hat{a}_x \cos \theta_i + \hat{a}_z \sin \theta_i) H_\perp^i e^{-j\boldsymbol{\beta}^i \cdot \mathbf{r}}$$

$$= (-\hat{a}_x \cos \theta_i + \hat{a}_z \sin \theta_i) \frac{E_0}{\eta_1} e^{-j\beta_1(x \sin \theta_i + z \cos \theta_i)} \quad (5\text{-}10\text{b})$$

where

$$E_\perp^i = E_0 \quad (5\text{-}10\text{c})$$

$$H_\perp^i = \frac{E_\perp^i}{\eta_1} = \frac{E_0}{\eta_1} \quad (5\text{-}10\text{d})$$

Similarly the reflected fields can be expressed as

$$\mathbf{E}_\perp^r = \hat{a}_y E_\perp^r e^{-j\boldsymbol{\beta}^r \cdot \mathbf{r}} = \hat{a}_y \Gamma_\perp^b E_0 e^{-j\beta_1(x \sin \theta_r - z \cos \theta_r)} \quad (5\text{-}11\text{a})$$

$$\mathbf{H}_\perp^r = (\hat{a}_x \cos \theta_r + \hat{a}_z \sin \theta_r) H_\perp^r e^{-j\boldsymbol{\beta}^r \cdot \mathbf{r}}$$

$$= (\hat{a}_x \cos \theta_r + \hat{a}_z \sin \theta_r) \frac{\Gamma_\perp^b E_0}{\eta_1} e^{-j\beta_1(x \sin \theta_r - z \cos \theta_r)} \quad (5\text{-}11\text{b})$$

where

$$E_\perp^r = \Gamma_\perp^b E^i = \Gamma_\perp^b E_0 \quad (5\text{-}11\text{c})$$

$$H_\perp^r = \frac{E_\perp^r}{\eta_1} = \frac{\Gamma_\perp^b E_0}{\eta_1} \quad (5\text{-}11\text{d})$$

Also the transmitted fields can be written as

$$\mathbf{E}_\perp^t = \hat{a}_y E_\perp^t e^{-j\boldsymbol{\beta}^t \cdot \mathbf{r}} = \hat{a}_y T_\perp^b E_0 e^{-j\beta_2(x \sin \theta_t + z \cos \theta_t)} \quad (5\text{-}12\text{a})$$

$$\mathbf{H}_\perp^t = (-\hat{a}_x \cos \theta_t + \hat{a}_z \sin \theta_t) H_\perp^t e^{-j\boldsymbol{\beta}^t \cdot \mathbf{r}}$$

$$= (-\hat{a}_x \cos \theta_t + \hat{a}_z \sin \theta_t) \frac{T_\perp^b E_0}{\eta_2} e^{-j\beta_2(x \sin \theta_t + z \cos \theta_t)} \quad (5\text{-}12\text{b})$$

where

$$E_\perp^t = T_\perp^b E_\perp^i = T_\perp^b E_0 \quad (5\text{-}12\text{c})$$

$$H_\perp^t = \frac{E_\perp^t}{\eta_2} = \frac{T_\perp^b E_0}{\eta_2} \quad (5\text{-}12\text{d})$$

The reflection $\Gamma_\perp^b$ and transmission $T_\perp^b$ coefficients, and the relation between the incident θ_i, reflected θ_r, and transmission (refracted) θ_t angles can be obtained by applying the boundary conditions on the continuity of the tangential components of the electric and magnetic fields. That is

$$\left. (\mathbf{E}_\perp^i + \mathbf{E}_\perp^r) \right|_{\substack{\tan \\ z=0}} = \left. (\mathbf{E}_\perp^t) \right|_{\substack{\tan \\ z=0}} \quad (5\text{-}13\text{a})$$

$$\left. (\mathbf{H}_\perp^i + \mathbf{H}_\perp^r) \right|_{\substack{\tan \\ z=0}} = \left. (\mathbf{H}_\perp^t) \right|_{\substack{\tan \\ z=0}} \quad (5\text{-}13\text{b})$$

Using the appropriate terms of (5-10a) through (5-12d), (5-13a) and (5-13b) can be written, respectively, as

$$e^{-j\beta_1 x \sin\theta_i} + \Gamma_\perp^b e^{-j\beta_1 x \sin\theta_r} = T_\perp^b e^{-j\beta_2 x \sin\theta_t} \qquad (5\text{-}14a)$$

$$\frac{1}{\eta_1}\left(-\cos\theta_i e^{-j\beta_1 x \sin\theta_i} + \Gamma_\perp^b \cos\theta_r e^{-j\beta_1 x \sin\theta_r}\right) = -\frac{T_\perp^b}{\eta_2}\cos\theta_t e^{-j\beta_2 x \sin\theta_t} \qquad (5\text{-}14b)$$

Whereas (5-14a) and (5-14b) represent two equations with four unknowns ($\Gamma_\perp^b$, $T_\perp^b$, θ_r, θ_t), it should be noted that each equation is complex. By equating the corresponding real and imaginary parts of each side, each can be reduced to two equations (a total of four). If this procedure is utilized, it will be concluded that (5-14a) and (5-14b) lead to the following two relations:

$$\theta_r = \theta_i \qquad (\textit{Snell's law of reflection}) \qquad (5\text{-}15a)$$

$$\beta_1 \sin\theta_i = \beta_2 \sin\theta_t \qquad (\textit{Snell's law of refraction}) \qquad (5\text{-}15b)$$

Using (5-15a) and (5-15b) reduces (5-14a) and (5-14b) to

$$1 + \Gamma_\perp^b = T_\perp^b \qquad (5\text{-}16a)$$

$$\frac{\cos\theta_i}{\eta_1}(-1 + \Gamma_\perp^b) = -\frac{\cos\theta_t}{\eta_2} T_\perp^b \qquad (5\text{-}16b)$$

Solving (5-16a) and (5-16b) simultaneously for $\Gamma_\perp^b$ and $T_\perp^b$ leads to

$$\boxed{\Gamma_\perp^b = \frac{E_\perp^r}{E_\perp^i} = \frac{\eta_2 \cos\theta_i - \eta_1 \cos\theta_t}{\eta_2 \cos\theta_i + \eta_1 \cos\theta_t} = \frac{\sqrt{\frac{\mu_2}{\varepsilon_2}}\cos\theta_i - \sqrt{\frac{\mu_1}{\varepsilon_1}}\cos\theta_t}{\sqrt{\frac{\mu_2}{\varepsilon_2}}\cos\theta_i + \sqrt{\frac{\mu_1}{\varepsilon_1}}\cos\theta_t}} \qquad (5\text{-}17a)$$

$$\boxed{T_\perp^b = \frac{E_\perp^t}{E_\perp^i} = \frac{2\eta_2 \cos\theta_i}{\eta_2 \cos\theta_i + \eta_1 \cos\theta_t} = \frac{2\sqrt{\frac{\mu_2}{\varepsilon_2}}\cos\theta_i}{\sqrt{\frac{\mu_2}{\varepsilon_2}}\cos\theta_i + \sqrt{\frac{\mu_1}{\varepsilon_1}}\cos\theta_t}} \qquad (5\text{-}17b)$$

$\Gamma_\perp^b$ and $T_\perp^b$ of (5-17a) and (5-17b) are usually referred to as the plane wave *Fresnel reflection and transmission coefficients* for perpendicular polarization.

Since for most dielectric media (excluding ferromagnetic material) $\mu_1 \simeq \mu_2 \simeq \mu_0$, (5-17a) and (5-17b) reduce by also utilizing (5-15b) to

$$\Gamma_\perp^b\Big|_{\mu_1=\mu_2} = \frac{\cos\theta_i - \sqrt{\frac{\varepsilon_2}{\varepsilon_1}}\sqrt{1 - \left(\frac{\varepsilon_1}{\varepsilon_2}\right)\sin^2\theta_i}}{\cos\theta_i + \sqrt{\frac{\varepsilon_2}{\varepsilon_1}}\sqrt{1 - \left(\frac{\varepsilon_1}{\varepsilon_2}\right)\sin^2\theta_i}} \qquad (5\text{-}18a)$$

$$T_\perp^b\Big|_{\mu_1=\mu_2} = \frac{2\cos\theta_i}{\cos\theta_i + \sqrt{\frac{\varepsilon_2}{\varepsilon_1}}\sqrt{1 - \left(\frac{\varepsilon_1}{\varepsilon_2}\right)\sin^2\theta_i}} \qquad (5\text{-}18b)$$

188 REFLECTION AND TRANSMISSION

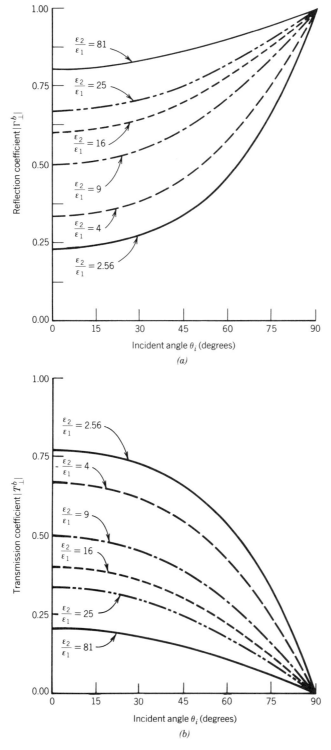

FIGURE 5-3 Magnitude of (*a*) reflection and (*b*) transmission coefficients for perpendicular polarization as a function of incident angle.

Plots of $|\Gamma_\perp^b|$ and $|T_\perp^b|$ of (5-18a) and (5-18b) for $\varepsilon_2/\varepsilon_1 = 2.56, 4, 9, 16, 25$, and 81 as a function of θ_i are shown in Figure 5-3. It is apparent that as the relative ratio of $\varepsilon_2/\varepsilon_1$ increases, the magnitude of the reflection coefficient increases whereas that of the transmission coefficient decreases. This is expected since large ratios of $\varepsilon_2/\varepsilon_1$ project larger discontinuities in the dielectric properties of the media along the interface. Also it is observed that for $\varepsilon_2 > \varepsilon_1$ the magnitude of the reflection coefficient never vanishes regardless of the $\varepsilon_2/\varepsilon_1$ ratio or the angle of incidence.

For $\varepsilon_2/\varepsilon_1 > 1$ both $\Gamma_\perp^b$ and $T_\perp^b$ are real with $\Gamma_\perp^b$ being negative and $T_\perp^b$ being positive for all angles of incidence. Therefore as a function of θ_i the phase of $\Gamma_\perp^b$ is equal to 180° and that of the transmission coefficient $T_\perp^b$ is zero. When $\varepsilon_2/\varepsilon_1 = 1$ the reflection coefficient vanishes and the transmission coefficient reduces to unity. When $\varepsilon_2/\varepsilon_1 < 1$ then both $\Gamma_\perp^b$ and $T_\perp^b$ are real up to an incidence angle $\theta_i = \theta_c$; after that they become complex. The angle θ_i for which $|\Gamma_\perp^b|_{\varepsilon_2/\varepsilon_1 < 1}(\theta_i = \theta_c) = 1$ is referred to as the *critical* angle, and it represents conditions of total internal reflection. More discussion on the critical angle ($\theta_i = \theta_c$) and the wave propagation for $\theta_i > \theta_c$ can be found in Section 5.3.4.

In medium 1 the total electric field can be written as

$$\mathbf{E}_\perp^1 = \mathbf{E}_\perp^i + \mathbf{E}_\perp^r = \hat{a}_y E_0 e^{-j\beta_1(x\sin\theta_i + z\cos\theta_i)}\underbrace{\left[1 + \Gamma_\perp^b e^{+j2\beta_1 z\cos\theta_i}\right]}_{\text{traveling wave} \quad \text{standing wave}}$$

$$= \hat{a}_y E_0 e^{-j\beta_1(x\sin\theta_i + z\cos\theta_i)}\left[1 + \Gamma_\perp(z)\right] \tag{5-19}$$

where

$$\Gamma_\perp(z) = \Gamma_\perp^b e^{+j2\beta_1 z\cos\theta_i} \tag{5-19a}$$

5.3.2 Parallel (Vertical or H) Polarization

For this polarization the electric field is parallel to the plane of incidence and it impinges upon a planar interface as shown in Figure 5-4. The directions of the incident, reflected, and transmitted electric and magnetic fields in Figure 5-4 are chosen so that for the special case of $\theta_i = 0$ they reduce to those of Figure 5-1.

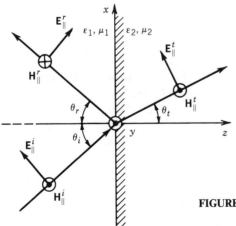

FIGURE 5-4 Parallel (vertical) polarized uniform plane wave incident at an oblique angle on an interface.

Using the techniques outlined in Section 4.2.2, we can write that

$$\mathbf{E}^i_{\parallel} = (\hat{a}_x \cos\theta_i - \hat{a}_z \sin\theta_i) E_0 e^{-j\boldsymbol{\beta}^i \cdot \mathbf{r}}$$
$$= (\hat{a}_x \cos\theta_i - \hat{a}_z \sin\theta_i) E_0 e^{-j\beta_1(x\sin\theta_i + z\cos\theta_i)} \quad (5\text{-}20a)$$

$$\mathbf{H}^i_{\parallel} = \hat{a}_y H^i_{\parallel} e^{-j\boldsymbol{\beta}^i \cdot \mathbf{r}} = \hat{a}_y \frac{E_0}{\eta_1} e^{-j\beta_1(x\sin\theta_i + z\cos\theta_i)} \quad (5\text{-}20b)$$

where

$$E^i_{\parallel} = E_0 \quad (5\text{-}20c)$$

$$H^i_{\parallel} = \frac{E^i_{\parallel}}{\eta_1} = \frac{E_0}{\eta_1} \quad (5\text{-}20d)$$

Similarly,

$$\mathbf{E}^r_{\parallel} = (\hat{a}_x \cos\theta_r + \hat{a}_z \sin\theta_r) E^r e^{-j\boldsymbol{\beta}^r \cdot \mathbf{r}}$$
$$= (\hat{a}_x \cos\theta_r + \hat{a}_z \sin\theta_r) \Gamma^b_{\parallel} E_0 e^{-j\beta_1(x\sin\theta_r - z\cos\theta_r)} \quad (5\text{-}21a)$$

$$\mathbf{H}^r_{\parallel} = -\hat{a}_y H^r_{\parallel} e^{-j\boldsymbol{\beta}^r \cdot \mathbf{r}} = -\hat{a}_y \frac{\Gamma^b_{\parallel} E_0}{\eta_1} e^{-j\beta_1(x\sin\theta_r - z\cos\theta_r)} \quad (5\text{-}21b)$$

where

$$E^r_{\parallel} = \Gamma^b_{\parallel} E^i = \Gamma^b_{\parallel} E_0 \quad (5\text{-}21c)$$

$$H^r_{\parallel} = \frac{E^r_{\parallel}}{\eta_1} = \frac{\Gamma^b_{\parallel} E_0}{\eta_1} \quad (5\text{-}21d)$$

Also,

$$\mathbf{E}^t_{\parallel} = (\hat{a}_x \cos\theta_t - \hat{a}_z \sin\theta_t) E^t_{\parallel} e^{-j\boldsymbol{\beta}^t \cdot \mathbf{r}}$$
$$= (\hat{a}_x \cos\theta_t - \hat{a}_z \sin\theta_t) T^b_{\parallel} E_0 e^{-j\beta_2(x\sin\theta_t + z\cos\theta_t)} \quad (5\text{-}22a)$$

$$\mathbf{H}^t_{\parallel} = \hat{a}_y H^t_{\parallel} e^{-j\boldsymbol{\beta}^t \cdot \mathbf{r}} = \hat{a}_y \frac{T^b_{\parallel} E_0}{\eta_2} e^{-j\beta_2(x\sin\theta_t + z\cos\theta_t)} \quad (5\text{-}22b)$$

where

$$E^t_{\parallel} = T^b_{\parallel} E^i = T^b_{\parallel} E_0 \quad (5\text{-}22c)$$

$$H^t_{\parallel} = \frac{E^t_{\parallel}}{\eta_2} = \frac{T^b_{\parallel} E_0}{\eta_2} \quad (5\text{-}22d)$$

As before, the reflection $\Gamma^b_{\parallel}$ and transmission $T^b_{\parallel}$ coefficients, and the reflection θ_r and transmission (refraction) θ_t angles are the four unknowns. These can be determined and expressed in terms of the incident angle θ_i and the constitutive parameters of the two media by applying the boundary conditions on the continuity across the interface ($z = 0$) of the tangential components of the electric and magnetic fields as given by (5-13a) and (5-13b) and applied to parallel polarization. Using the appropriate terms of (5-20a) through (5-22d), we can write (5-13a) and (5-13b) as applied to parallel polarization, respectively, as

$$\cos\theta_i e^{-j\beta_1 x\sin\theta_i} + \Gamma^b_{\parallel} \cos\theta_r e^{-j\beta_1 x\sin\theta_r} = T^b_{\parallel} \cos\theta_t e^{-j\beta_2 x\sin\theta_t} \quad (5\text{-}23a)$$

$$\frac{1}{\eta_1}\left(e^{-j\beta_1 x\sin\theta_i} - \Gamma^b_{\parallel} e^{-j\beta_1 x\sin\theta_r}\right) = \frac{1}{\eta_2} T^b_{\parallel} e^{-j\beta_2 x\sin\theta_t} \quad (5\text{-}23b)$$

Following the procedure outlined in Section 5.3.1 for the solution of (5-14a) and (5-14b), it can show that (5-23a) and (5-23b) reduce to

$$\theta_r = \theta_i \qquad (\text{Snell's law of reflection}) \qquad (5\text{-}24\text{a})$$

$$\beta_1 \sin \theta_i = \beta_2 \sin \theta_t \qquad (\text{Snell's law of refraction}) \qquad (5\text{-}24\text{b})$$

$$\Gamma_\parallel^b = \frac{-\eta_1 \cos \theta_i + \eta_2 \cos \theta_t}{\eta_1 \cos \theta_i + \eta_2 \cos \theta_t} = \frac{-\sqrt{\dfrac{\mu_1}{\varepsilon_1}} \cos \theta_i + \sqrt{\dfrac{\mu_2}{\varepsilon_2}} \cos \theta_t}{\sqrt{\dfrac{\mu_1}{\varepsilon_1}} \cos \theta_i + \sqrt{\dfrac{\mu_2}{\varepsilon_2}} \cos \theta_t} \qquad (5\text{-}24\text{c})$$

$$T_\parallel^b = \frac{2\eta_2 \cos \theta_i}{\eta_1 \cos \theta_i + \eta_2 \cos \theta_t} = \frac{2\sqrt{\dfrac{\mu_2}{\varepsilon_2}} \cos \theta_i}{\sqrt{\dfrac{\mu_1}{\varepsilon_1}} \cos \theta_i + \sqrt{\dfrac{\mu_2}{\varepsilon_2}} \cos \theta_t} \qquad (5\text{-}24\text{d})$$

$\Gamma_\parallel^b$ and $T_\parallel^b$ of (5-24c) and (5-24d) are usually referred to as the plane wave *Fresnel reflection and transmission coefficients* for parallel polarization.

Excluding ferromagnetic material, (5-24c) and (5-24d) reduce using also (5-24b) to

$$\Gamma_\parallel^b \big|_{\mu_1 = \mu_2} = \frac{-\cos \theta_i + \sqrt{\dfrac{\varepsilon_1}{\varepsilon_2}} \sqrt{1 - \left(\dfrac{\varepsilon_1}{\varepsilon_2}\right) \sin^2 \theta_i}}{\cos \theta_i + \sqrt{\dfrac{\varepsilon_1}{\varepsilon_2}} \sqrt{1 - \left(\dfrac{\varepsilon_1}{\varepsilon_2}\right) \sin^2 \theta_i}} \qquad (5\text{-}25\text{a})$$

$$T_\parallel^b \big|_{\mu_1 = \mu_2} = \frac{2\sqrt{\dfrac{\varepsilon_1}{\varepsilon_2}} \cos \theta_i}{\cos \theta_i + \sqrt{\dfrac{\varepsilon_1}{\varepsilon_2}} \sqrt{1 - \left(\dfrac{\varepsilon_1}{\varepsilon_2}\right) \sin^2 \theta_i}} \qquad (5\text{-}25\text{b})$$

Plots of $|\Gamma_\parallel^b|$ and $|T_\parallel^b|$ of (5-25a) and (5-25b) for $\varepsilon_2/\varepsilon_1 = 2.56, 4, 9, 16, 25$, and 81 as a function of θ_i are shown in Figure 5-5. It is observed in Figure 5-5a that for this polarization there is an angle where the reflection coefficient does vanish. The angle where the reflection coefficient vanishes is referred to as the *Brewster angle* θ_B, and it increases toward 90° as the ratio of $\varepsilon_2/\varepsilon_1$ becomes larger. More discussion on the Brewster angle can be found in the next section (Section 5.3.3).

For $\varepsilon_2/\varepsilon_1 > 1$, $\Gamma_\parallel^b$ and $T_\parallel^b$ are both real. For angles of incidence less than the Brewster angle ($\theta_i < \theta_B$) $\Gamma_\parallel^b$ is negative, indicating a 180° phase as a function of incidence angle; for $\theta_i > \theta_B$ $\Gamma_\parallel^b$ is positive, representing a 0° phase. The transmission coefficient $T_\parallel^b$ is positive for all values of θ_i, indicating a 0° phase. When $\varepsilon_2/\varepsilon_1 = 1$ then the reflection coefficient vanishes and the transmission coefficient reduces to unity. As for the perpendicular polarization, when $\varepsilon_2/\varepsilon_1 < 1$ then both $\Gamma_\perp^b$ and $T_\perp^b$ are real up to an incidence angle $\theta_i = \theta_c$; after that they become complex. The angle for which $|\Gamma_\parallel^b|_{\varepsilon_2/\varepsilon_1 < 1}(\theta_i = \theta_c) = 1$ is again referred to as *critical angle*, and it represents conditions of total internal reflection. Further discussion of

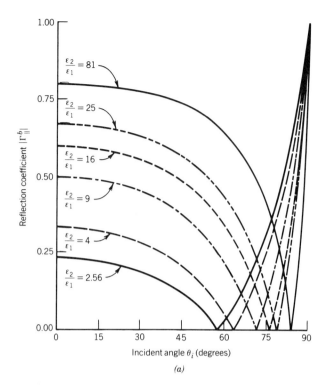

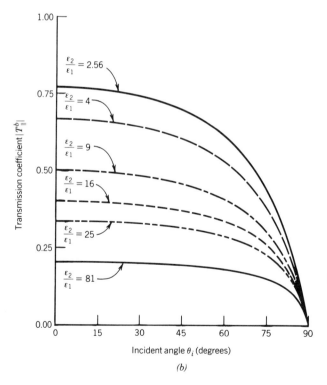

FIGURE 5-5 Magnitude of (a) reflection and (b) transmission coefficients for parallel polarization as a function of incident angle.

the critical angle ($\theta_i = \theta_c$) and the wave propagation for $\theta_i > \theta_c$ can be found in Section 5.3.4. It is evident that the critical angle is not a function of polarization; it occurs only when the wave propagates from the more dense to the less dense medium.

The total electric field in medium 1 can be written as

$$\mathbf{E}_\parallel^1 = \mathbf{E}_\parallel^i + \mathbf{E}_\parallel^r = \hat{a}_x \cos\theta_i \underbrace{E_0 e^{-j\beta_1(x\sin\theta_i + z\cos\theta_i)}}_{\text{traveling wave}} \underbrace{\left[1 + \Gamma_\parallel^b e^{+j2\beta_1 z \cos\theta_i}\right]}_{\text{standing wave}}$$

$$-\hat{a}_z \sin\theta_i \underbrace{E_0 e^{-j\beta_1(x\sin\theta_i + z\cos\theta_i)}}_{\text{traveling wave}} \underbrace{\left[1 - \Gamma_\parallel^b e^{+j2\beta_1 z \cos\theta_i}\right]}_{\text{standing wave}}$$

$$\mathbf{E}_\parallel^1 = \mathbf{E}_x^1 + \mathbf{E}_z^1 = \hat{a}_x \cos\theta_i E_0 e^{-j\beta_1(x\sin\theta_i + z\cos\theta_i)}\left[1 + \Gamma_\parallel(z)\right]$$
$$-\hat{a}_z \sin\theta_i E_0 e^{-j\beta_1(x\sin\theta_i + z\cos\theta_i)}\left[1 - \Gamma_\parallel(z)\right] \quad (5\text{-}26)$$

where

$$\Gamma_\parallel(z) = \Gamma_\parallel^b e^{+j2\beta_1 z \cos\theta_i} \quad (5\text{-}26a)$$

5.3.3 Total Transmission–Brewster Angle

The reflection and transmission coefficients for both perpendicular and parallel polarizations are functions of the constitutive parameters of the two media forming the interface, the angle of incidence, and the angle of refraction that is related to the angle of incidence through Snell's law of refraction. One may ask: "For a given set of constitutive parameters of two media forming an interface, is there an incidence angle that allows no reflection, i.e., $\Gamma = 0$?" To answer this we need to refer back to the expressions for the reflection coefficients as given by (5-17a) and (5-24c).

A. PERPENDICULAR (HORIZONTAL) POLARIZATION

To see the conditions under which the reflection coefficient of (5-17a) will vanish, we set it equal to zero, which leads to

$$\Gamma_\perp^b = \frac{\sqrt{\dfrac{\mu_2}{\varepsilon_2}} \cos\theta_i - \sqrt{\dfrac{\mu_1}{\varepsilon_1}} \cos\theta_t}{\sqrt{\dfrac{\mu_2}{\varepsilon_2}} \cos\theta_i + \sqrt{\dfrac{\mu_1}{\varepsilon_1}} \cos\theta_t} = 0 \quad (5\text{-}27)$$

or

$$\cos\theta_i = \sqrt{\frac{\mu_1}{\mu_2}\left(\frac{\varepsilon_2}{\varepsilon_1}\right)} \cos\theta_t \quad (5\text{-}27a)$$

Using Snell's law of refraction, as given by (5-15b), (5-27a) can be written as

$$\left(1 - \sin^2\theta_i\right) = \frac{\mu_1}{\mu_2}\left(\frac{\varepsilon_2}{\varepsilon_1}\right)\left(1 - \sin^2\theta_t\right)$$

$$\left(1 - \sin^2\theta_i\right) = \frac{\mu_1}{\mu_2}\left(\frac{\varepsilon_2}{\varepsilon_1}\right)\left[1 - \frac{\mu_1}{\mu_2}\left(\frac{\varepsilon_1}{\varepsilon_2}\right)\sin^2\theta_i\right] \quad (5\text{-}28)$$

or

$$\sin\theta_i = \sqrt{\dfrac{\dfrac{\varepsilon_2}{\varepsilon_1} - \dfrac{\mu_2}{\mu_1}}{\dfrac{\mu_1}{\mu_2} - \dfrac{\mu_2}{\mu_1}}} \qquad (5\text{-}28a)$$

Since the sine function cannot exceed unity, (5-28a) exists only if

$$\dfrac{\varepsilon_2}{\varepsilon_1} - \dfrac{\mu_2}{\mu_1} \le \dfrac{\mu_1}{\mu_2} - \dfrac{\mu_2}{\mu_1} \qquad (5\text{-}29)$$

or

$$\dfrac{\varepsilon_2}{\varepsilon_1} \le \dfrac{\mu_1}{\mu_2} \qquad (5\text{-}29a)$$

If however $\mu_1 = \mu_2$, (5-28a) indicates that

$$\sin\theta_i\big|_{\mu_1=\mu_2} = \infty \qquad (5\text{-}29b)$$

Therefore there exists no real angle θ_i under this condition that will reduce the reflection coefficient to zero. Since the permeability for most dielectric material (aside from ferromagnetics) is almost the same and equal to that of free space ($\mu_1 \simeq \mu_2 \simeq \mu_0$), then for *these materials there exists no real incidence angle that will reduce the reflection coefficient for perpendicular polarization to zero.*

B. PARALLEL (VERTICAL) POLARIZATION

To examine the conditions under which the reflection coefficient for parallel polarization will vanish, we set (5-24c) equal to zero; that is

$$\Gamma_\parallel^b = \dfrac{-\sqrt{\dfrac{\mu_1}{\varepsilon_1}}\cos\theta_i + \sqrt{\dfrac{\mu_2}{\varepsilon_2}}\cos\theta_t}{\sqrt{\dfrac{\mu_1}{\varepsilon_1}}\cos\theta_i + \sqrt{\dfrac{\mu_2}{\varepsilon_2}}\cos\theta_t} = 0 \qquad (5\text{-}30)$$

or

$$\cos\theta_i = \sqrt{\dfrac{\mu_2}{\mu_1}\left(\dfrac{\varepsilon_1}{\varepsilon_2}\right)}\cos\theta_t \qquad (5\text{-}30a)$$

Using Snell's law of refraction, as given by (5-24b), (5-30a) can be written as

$$(1 - \sin^2\theta_i) = \dfrac{\mu_2}{\mu_1}\left(\dfrac{\varepsilon_1}{\varepsilon_2}\right)(1 - \sin^2\theta_t)$$

$$(1 - \sin^2\theta_i) = \dfrac{\mu_2}{\mu_1}\left(\dfrac{\varepsilon_1}{\varepsilon_2}\right)\left(1 - \dfrac{\mu_1}{\mu_2}\left(\dfrac{\varepsilon_1}{\varepsilon_2}\right)\sin^2\theta_i\right) \qquad (5\text{-}31)$$

or

$$\sin\theta_i = \sqrt{\dfrac{\dfrac{\varepsilon_2}{\varepsilon_1} - \dfrac{\mu_2}{\mu_1}}{\dfrac{\varepsilon_2}{\varepsilon_1} - \dfrac{\varepsilon_1}{\varepsilon_2}}} \qquad (5\text{-}31a)$$

Since the sine function cannot exceed unity, (5-31a) exists only if

$$\frac{\varepsilon_2}{\varepsilon_1} - \frac{\mu_2}{\mu_1} \leq \frac{\varepsilon_2}{\varepsilon_1} - \frac{\varepsilon_1}{\varepsilon_2} \qquad (5\text{-}32)$$

or

$$\frac{\mu_2}{\mu_1} \geq \frac{\varepsilon_1}{\varepsilon_2} \qquad (5\text{-}32\text{a})$$

If however $\mu_1 = \mu_2$, (5-31a) reduces to

$$\boxed{\theta_i = \theta_B = \sin^{-1}\left(\sqrt{\frac{\varepsilon_2}{\varepsilon_1 + \varepsilon_2}}\right)} \qquad (5\text{-}33)$$

The incidence angle θ_i, as given by (5-31a) or (5-33), which reduces the reflection coefficient for parallel polarization to zero, is referred to as the Brewster angle θ_B. It should be noted that when $\mu_1 = \mu_2$, the incidence Brewster angle $\theta_i = \theta_B$ of (5-33) exists only if the polarization of the wave is parallel (*vertical*).

Other forms of the Brewster angle, besides that given by (5-33), are

$$\boxed{\theta_i = \theta_B = \cos^{-1}\left(\sqrt{\frac{\varepsilon_1}{\varepsilon_1 + \varepsilon_2}}\right)} \qquad (5\text{-}33\text{a})$$

$$\boxed{\theta_i = \theta_B = \tan^{-1}\left(\sqrt{\frac{\varepsilon_2}{\varepsilon_1}}\right)} \qquad (5\text{-}33\text{b})$$

Example 5-2. A parallel polarized electromagnetic wave radiated from a submerged submarine impinges upon a water–air planar interface. Assuming the water is lossless, its dielectric constant is 81, and the wave approximates a plane wave at the interface, determine the angle of incidence to allow complete transmission of the energy.

Solution. The angle of incidence that allows complete transmission of the energy is the Brewster angle. Using (5-33b) the Brewster angle of the water–air interface is

$$\theta_{iwa} = \theta_{Bwa} = \tan^{-1}\left(\sqrt{\frac{\varepsilon_0}{81\varepsilon_0}}\right) = \tan^{-1}\left(\frac{1}{9}\right) = 6.34°$$

This indicates that the Brewster angle is close to the normal to the interface.

Example 5-3. Repeat the problem of Example 5-2 assuming that the same wave is radiated from a spacecraft in air, and it impinges upon the air–water interface.

Solution. The Brewster angle for an air–water interface is

$$\theta_{iaw} = \theta_{Baw} = \tan^{-1}\left(\sqrt{\frac{81\varepsilon_0}{\varepsilon_0}}\right) = \tan^{-1}(9) = 83.66°$$

It is apparent that the sum of the Brewster angle of Example 5-2 (water–air interface) plus that of Example 5-3 (air–water interface) is equal to 90°. That is

$$\theta_{Bwa} + \theta_{Baw} = 6.34° + 83.66° = 90°$$

From trigonometry, it is obvious that the preceding relation is always going to hold no matter what two media form the interface.

5.3.4 Total Reflection–Critical Angle

In Section 5.3.3 we found the angles, satisfying (5-28a) and (5-31a), which allow total transmission, respectively, for perpendicular and parallel polarizations. When the permeabilities of the two media forming the interface are the same ($\mu_1 = \mu_2$), only parallel polarized fields possess an incidence angle that allows total transmission. As before, that angle is known as the Brewster angle and it is given by either (5-33), (5-33a), or (5-33b).

The next question may be: "Is there an incidence angle that allows total reflection of energy at a planar interface?" If this is possible, then $|\Gamma| = 1$. To determine the conditions under which this can be accomplished, we proceed in a similar manner as for the total transmission case of Section 5.3.3.

A. PERPENDICULAR (HORIZONTAL) POLARIZATION

To see the conditions under which the magnitude of the reflection coefficient is equal to unity, we set the magnitude of (5-17a) equal to

$$\frac{\left|\sqrt{\frac{\mu_2}{\varepsilon_2}}\cos\theta_i - \sqrt{\frac{\mu_1}{\varepsilon_1}}\cos\theta_t\right|}{\left|\sqrt{\frac{\mu_2}{\varepsilon_2}}\cos\theta_i + \sqrt{\frac{\mu_1}{\varepsilon_1}}\cos\theta_t\right|} = 1 \qquad (5\text{-}34)$$

This is satisfied provided the second term in the numerator and denominator is imaginary. Using Snell's law of refraction, as given by (5-15b), the second term in the numerator and denominator can be imaginary if

$$\cos\theta_t = \sqrt{1 - \sin^2\theta_t} = \sqrt{1 - \frac{\mu_1\varepsilon_1}{\mu_2\varepsilon_2}\sin^2\theta_i} = -j\sqrt{\frac{\mu_1\varepsilon_1}{\mu_2\varepsilon_2}\sin^2\theta_i - 1} \qquad (5\text{-}35)$$

In order for (5-35) to hold

$$\frac{\mu_1\varepsilon_1}{\mu_2\varepsilon_2}\sin^2\theta_i \geq 1 \qquad (5\text{-}35a)$$

or

$$\boxed{\theta_i \geq \theta_c = \sin^{-1}\left(\sqrt{\frac{\mu_2\varepsilon_2}{\mu_1\varepsilon_1}}\right)} \qquad (5\text{-}35b)$$

The incidence angle θ_i of (5-35b) that allows total reflection is known as the *critical angle*. Since the argument of the sine function cannot exceed unity, then

$$\mu_2 \varepsilon_2 \leq \mu_1 \varepsilon_1 \tag{5-35c}$$

in order for the critical angle (5-35b) to be physically realizable.

If the permeabilities of the two media are the same ($\mu_1 = \mu_2$), then (5-35b) reduces to

$$\boxed{\theta_i \geq \theta_c = \sin^{-1}\left(\sqrt{\frac{\varepsilon_2}{\varepsilon_1}}\right)} \tag{5-36}$$

which leads to a physically realizable angle provided

$$\varepsilon_2 \leq \varepsilon_1 \tag{5-36a}$$

Therefore for two media with identical permeabilities (which is the case for most dielectrics, aside from ferromagnetic material!), *the critical angle exists only if the wave propagates from a more dense to a less dense medium, as stated by* (5-36a).

Example 5-4. A perpendicularly polarized wave radiated from a submerged submarine impinges upon a water–air interface. Assuming the water is lossless, its dielectric constant is 81, and the wave approximates a plane wave at the interface, determine the angle of incidence that will allow complete reflection of the energy at the interface.

Solution. The angle of incidence that allows complete reflection of energy is the critical angle. Since for water $\mu_2 = \mu_0$, the critical angle is obtained using (5-36), which leads to

$$\theta_i \geq \theta_c = \sin^{-1}\left(\sqrt{\frac{\varepsilon_0}{81\varepsilon_0}}\right) = 6.38°$$

Since there is a large difference between the permittivities of the two media forming the interface, the critical angle of this example is very nearly the same as the Brewster angle of Example 5-2.

The next question asked may be: "What happens to the angle of refraction and to the propagation of the wave when the angle of incidence is equal to or greater than the critical angle?"

When the angle of incidence is equal to the critical angle, the angle of refraction reduces, through Snell's law of refraction of (5-15b) and (5-35b), to

$$\theta_t = \sin^{-1}\left(\sqrt{\frac{\mu_1 \varepsilon_1}{\mu_2 \varepsilon_2}} \sin\theta_i\right)\bigg|_{\theta_i=\theta_c} = \sin^{-1}\left(\sqrt{\frac{\mu_1 \varepsilon_1}{\mu_2 \varepsilon_2}} \sqrt{\frac{\mu_2 \varepsilon_2}{\mu_1 \varepsilon_1}}\right) = \sin^{-1}(1) = 90° \tag{5-37}$$

In turn the reflection and transmission coefficients reduce to

$$\Gamma_\perp^b \big|_{\theta_i=\theta_c} = 1 \tag{5-38a}$$

$$T_\perp^b \big|_{\theta_i=\theta_c} = 2 \tag{5-38b}$$

198 REFLECTION AND TRANSMISSION

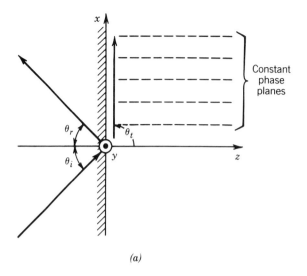

(a)

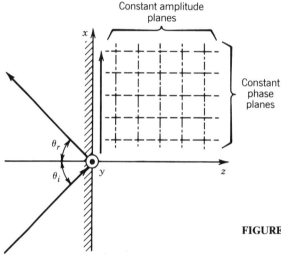

(b)

FIGURE 5-6 Constant phase and amplitude planes for critical ($\theta_i = \theta_c$) and above critical ($\theta_i > \theta_c$) incident angles.

Also the transmitted fields of (5-12a) and (5-12b) can be written as

$$\mathbf{E}^t_\perp = \hat{a}_y 2 E_0 e^{-j\beta_2 x} \tag{5-39a}$$

$$\mathbf{H}^t_\perp = \hat{a}_z \frac{2 E_0}{\eta_2} e^{-j\beta_2 x} \tag{5-39b}$$

which represent a plane wave that travels parallel to the interface in the $+x$ direction as shown in Figure 5-6a. The constant phase planes of the wave are parallel to the z axis. This wave is referred to as a *surface wave* [2].

The average power density associated with the transmitted fields is given by

$$\mathbf{S}^t_{av}\big|_{\theta_i = \theta_c} = \frac{1}{2} \operatorname{Re}\left(\mathbf{E}^t_\perp \times \mathbf{H}^{t*}_\perp\right)\bigg|_{\theta_i = \theta_c} = \hat{a}_x \frac{2|E_0|^2}{\eta_2} \tag{5-40}$$

and it does not contain any component normal to the interface. Therefore there is no transfer of real power across the interface in a direction normal to the boundary; thus all must be reflected. This is also evident by examining the magnitude of the incident and reflected average power densities associated with the fields (5-10a) through (5-11d) under critical angle incidence. These are given by

$$|S^i_{av}|_{\theta_i=\theta_c} = \left|\frac{1}{2}\text{Re}\left(\mathbf{E}^i_\perp \times \mathbf{H}^{i*}_\perp\right)\right|_{\theta_i=\theta_c} = \frac{|E_0|^2}{2\eta_i}|\hat{a}_x \sin\theta_i + \hat{a}_z \cos\theta_i| = \frac{|E_0|^2}{2\eta_1} \quad (5\text{-}41\text{a})$$

$$|S^r_{av}|_{\theta_i=\theta_c} = \left|\frac{1}{2}\text{Re}\left(\mathbf{E}^r_\perp \times \mathbf{H}^{r*}_\perp\right)\right|_{\theta_i=\theta_c} = \frac{|E_0|^2}{2\eta_1}|\hat{a}_x \sin\theta_i - \hat{a}_z \cos\theta_i| = \frac{|E_0|^2}{2\eta_1} \quad (5\text{-}41\text{b})$$

and they are obviously identical.

When the angle of incidence θ_i is greater than the critical angle θ_c ($\theta_i > \theta_c$), then Snell's law of refraction can be written as [3]

$$\sin\theta_t|_{\theta_i>\theta_c} = \frac{\beta_1}{\beta_2}\sin\theta_i\bigg|_{\theta_i>\theta_c} = \sqrt{\frac{\mu_1\epsilon_1}{\mu_2\epsilon_2}}\sin\theta_i\bigg|_{\theta_i>\theta_c} > 1 \quad (5\text{-}42\text{a})$$

which can only be satisfied provided θ_t is complex, that is, $\theta_t = \theta_R + j\theta_X$. Also

$$\cos\theta_t|_{\theta_i>\theta_c} = \sqrt{1-\sin^2\theta_t}\bigg|_{\theta_i>\theta_c} = \sqrt{1-\frac{\mu_1\epsilon_1}{\mu_2\epsilon_2}\sin^2\theta_i}\bigg|_{\theta_i>\theta_c}$$

$$= \pm j\sqrt{\frac{\mu_1\epsilon_1}{\mu_2\epsilon_2}\sin^2\theta_i - 1}\bigg|_{\theta_i>\theta_c} \quad (5\text{-}42\text{b})$$

which again indicates that θ_t is complex.

Therefore when $\theta_i > \theta_c$, there is no physically realizable angle θ_t. If not, what really does happen to the wave propagation? Since under this condition θ_t is complex and not physically realizable, this may be a clue that the wave in medium 2 is again a surface wave. To see this, let us examine the field in medium 2, the reflection and transmission coefficients, and the average power densities.

When the angle of incidence exceeds the critical angle ($\theta_i > \theta_c$), the transmitted E field of (5-12a) can be written, using (5-15b) and (5-35b), as

$$\mathbf{E}^t_\perp|_{\theta_i>\theta_c} = \hat{a}_y T^b_\perp E_0 \exp(-j\beta_2 x \sin\theta_t)\exp(-j\beta_2 z \cos\theta_t)|_{\theta_i>\theta_c}$$

$$= \hat{a}_y T^b_\perp E_0 \exp\left[-j\beta_2 x\left(\sqrt{\frac{\mu_1\epsilon_1}{\mu_2\epsilon_2}}\sin\theta_i\right)\right]\exp\left(-j\beta_2 z\sqrt{1-\sin^2\theta_t}\right)\bigg|_{\theta_i>\theta_c}$$

$$= \hat{a}_y T^b_\perp E_0 \exp\left[-j\beta_2 x\left(\sqrt{\frac{\mu_1\epsilon_1}{\mu_2\epsilon_2}}\sin\theta_i\right)\right]\exp\left(-j\beta_2 z\sqrt{1-\frac{\mu_1\epsilon_1}{\mu_2\epsilon_2}\sin^2\theta_i}\right)\bigg|_{\theta_i>\theta_c}$$

$$= \hat{a}_y T^b_\perp E_0 \exp\left[-j\beta_2 x\left(\sqrt{\frac{\mu_1\epsilon_1}{\mu_2\epsilon_2}}\sin\theta_i\right)\right]\exp\left(-\beta_2 z\sqrt{\frac{\mu_1\epsilon_1}{\mu_2\epsilon_2}\sin^2\theta_i - 1}\right)\bigg|_{\theta_i>\theta_c}$$

$$= \hat{a}_y T^b_\perp E_0 \exp\left[-\beta_2 z\left(\sqrt{\frac{\mu_1\epsilon_1}{\mu_2\epsilon_2}\sin^2\theta_i - 1}\right)\right]\exp\left[-j\beta_2 x\left(\sqrt{\frac{\mu_1\epsilon_1}{\mu_2\epsilon_2}}\sin\theta_i\right)\right]\bigg|_{\theta_i>\theta_c}$$

$$\mathbf{E}^t_\perp|_{\theta_i>\theta_c} = \hat{a}_y T^b_\perp E_0 e^{-\alpha_e z}e^{-j\beta_e x} \quad (5\text{-}43)$$

200 REFLECTION AND TRANSMISSION

where

$$\alpha_e = \beta_2 \sqrt{\frac{\mu_1 \varepsilon_1}{\mu_2 \varepsilon_2} \sin^2 \theta_i - 1} \bigg|_{\theta_i > \theta_c} = \omega \sqrt{\mu_1 \varepsilon_1 \sin^2 \theta_i - \mu_2 \varepsilon_2} \bigg|_{\theta_i > \theta_c} \quad (5\text{-}43a)$$

$$\beta_e = \beta_2 \sqrt{\frac{\mu_1 \varepsilon_1}{\mu_2 \varepsilon_2}} \sin \theta_i \bigg|_{\theta_i > \theta_c} = \omega \sqrt{\mu_1 \varepsilon_1} \sin \theta_i \bigg|_{\theta_i > \theta_c} \quad (5\text{-}43b)$$

$$v_{pe} = \frac{\omega}{\beta_e} = \frac{\omega}{\beta_2 \sqrt{\frac{\mu_1 \varepsilon_1}{\mu_2 \varepsilon_2}} \sin \theta_i} \bigg|_{\theta_i > \theta_c} = \frac{v_{p2}}{\sqrt{\frac{\mu_1 \varepsilon_1}{\mu_2 \varepsilon_2}} \sin \theta_i} \bigg|_{\theta_i > \theta_c} = \frac{1}{\sqrt{\mu_1 \varepsilon_1} \sin \theta_i} < v_{p2}$$
$$(5\text{-}43c)$$

The wave associated with (5-43) also propagates parallel to the interface with constant phase planes that are parallel to the z axis, as shown in Figure 5-6b. The effective phase velocity v_{pe} of the wave is given by (5-43c), and it is less than v_{p2} of an ordinary wave in medium 2. The wave also possesses constant amplitude planes that are parallel to the x axis, as shown in Figure 5-6b. The effective attenuation constant α_e of the wave in the z direction is that given by (5-43a). Its values are such that the wave decays very rapidly, and in a few wavelengths it essentially vanishes. This wave is also a *surface wave*. Since its phase velocity is less than the speed of light, it is a *slow* surface wave. Also since it decays very rapidly in a direction normal to the interface, it is *tightly bound* to the surface or it is a *tightly bound slow surface wave*.

Phase velocities *greater* than the intrinsic phase velocity of an ordinary plane wave in a given medium can be achieved by uniform plane waves at *real oblique angles* of propagation, as illustrated in Section 4.2.2C; phase velocities *smaller* than the intrinsic velocity can only be achieved by uniform plane waves at *complex angles* of propagation. Waves traveling at complex angles are *nonuniform* plane waves oriented so as to provide small phase velocities or large rates of change of phase in a given direction. The price for such large rates of change of phase or small velocities in one direction is associated with large attenuation at right angle directions.

Example 5-5. Since for $\theta_i > \theta_c$ the angle of refraction θ_t is complex ($\theta_t = \theta_R + j\theta_X$), determine the real θ_R and imaginary θ_X parts of θ_t in terms of the constitutive parameters of the two media and the angle of incidence.

Solution. Using (5-42a)

$$\sin \theta_t = \sin(\theta_R + j\theta_X) = \sqrt{\frac{\mu_1 \varepsilon_1}{\mu_2 \varepsilon_2}} \sin \theta_i$$

or

$$\sin(\theta_R) \cosh(\theta_X) + j \cos(\theta_R) \sinh(\theta_X) = \sqrt{\frac{\mu_1 \varepsilon_1}{\mu_2 \varepsilon_2}} \sin \theta_i$$

Since the right side is real, then the only solution that exists is for the imaginary part of the left side to vanish and the real part to be equal to the real part of the right side. Thus

$$\cos(\theta_R)\sinh(\theta_X) = 0 \Rightarrow \theta_R = \frac{\pi}{2}$$

$$\sin(\theta_R)\cosh(\theta_X) = \sqrt{\frac{\mu_1\varepsilon_1}{\mu_2\varepsilon_2}}\sin\theta_i \Rightarrow \theta_X = \cosh^{-1}\left(\sqrt{\frac{\mu_1\varepsilon_1}{\mu_2\varepsilon_2}}\sin\theta_i\right)$$

In turn $\cos\theta_t$ is defined as

$$\cos\theta_t = \cos(\theta_R + j\theta_X) = \cos(\theta_R)\cosh(\theta_X) - j\sin(\theta_R)\sinh(\theta_X)$$

or

$$\cos\theta_t = -j\sinh(\theta_X)$$

which again is shown to be complex as was in (5-42b). When these expressions for $\sin\theta_t$ and $\cos\theta_t$ are used to represent the fields in medium 2, it will be shown that the fields are nonuniform plane waves as illustrated by (5-43).

Under the conditions where the angle of incidence is equal to or greater than the critical angle, the reflection $\Gamma_\perp^b$ and transmission $T_\perp^b$ coefficients of (5-17a) and (5-17b) reduce, respectively, to [3]

$$\Gamma_\perp^b\big|_{\theta_i \geq \theta_c} = \left.\frac{\sqrt{\frac{\mu_2}{\varepsilon_2}}\cos\theta_i - \sqrt{\frac{\mu_1}{\varepsilon_1}}\cos\theta_t}{\sqrt{\frac{\mu_2}{\varepsilon_2}}\cos\theta_i + \sqrt{\frac{\mu_1}{\varepsilon_1}}\cos\theta_t}\right|_{\theta_i \geq \theta_c}$$

$$= \left.\frac{\sqrt{\frac{\mu_2}{\varepsilon_2}}\cos\theta_i - \sqrt{\frac{\mu_1}{\varepsilon_1}}\sqrt{1 - \sin^2\theta_t}}{\sqrt{\frac{\mu_2}{\varepsilon_2}}\cos\theta_i + \sqrt{\frac{\mu_1}{\varepsilon_1}}\sqrt{1 - \sin^2\theta_t}}\right|_{\theta_i \geq \theta_c}$$

$$= \left.\frac{\sqrt{\frac{\mu_2}{\varepsilon_2}}\cos\theta_i - \sqrt{\frac{\mu_1}{\varepsilon_1}}\sqrt{1 - \frac{\mu_1\varepsilon_1}{\mu_2\varepsilon_2}\sin^2\theta_i}}{\sqrt{\frac{\mu_2}{\varepsilon_2}}\cos\theta_i + \sqrt{\frac{\mu_1}{\varepsilon_1}}\sqrt{1 - \frac{\mu_1\varepsilon_1}{\mu_2\varepsilon_2}\sin^2\theta_i}}\right|_{\theta_i \geq \theta_c}$$

$$= \left.\frac{\sqrt{\frac{\mu_2}{\varepsilon_2}}\cos\theta_i + j\sqrt{\frac{\mu_1}{\varepsilon_1}}\sqrt{\frac{\mu_1\varepsilon_1}{\mu_2\varepsilon_2}\sin^2\theta_i - 1}}{\sqrt{\frac{\mu_2}{\varepsilon_2}}\cos\theta_i - j\sqrt{\frac{\mu_1}{\varepsilon_1}}\sqrt{\frac{\mu_1\varepsilon_1}{\mu_2\varepsilon_2}\sin^2\theta_i - 1}}\right|_{\theta_i \geq \theta_c}$$

$$\Gamma_\perp^b\big|_{\theta_i \geq \theta_c} = |\Gamma_\perp^b|e^{j2\psi_\perp} = e^{j2\psi_\perp} \tag{5-44}$$

202 REFLECTION AND TRANSMISSION

where

$$|\Gamma_\perp^b| = 1 \tag{5-44a}$$

$$\psi_\perp = \tan^{-1}\left[\frac{X_\perp}{R_\perp}\right] \tag{5-44b}$$

$$X_\perp = \sqrt{\frac{\mu_1}{\varepsilon_1}}\sqrt{\frac{\mu_1\varepsilon_1}{\mu_2\varepsilon_2}\sin^2\theta_i - 1} \tag{5-44c}$$

$$R_\perp = \sqrt{\frac{\mu_2}{\varepsilon_2}}\cos\theta_i \tag{5-44d}$$

$$T_\perp^b\big|_{\theta_i \geq \theta_c} = \left.\frac{2\sqrt{\dfrac{\mu_2}{\varepsilon_2}}\cos\theta_i}{\sqrt{\dfrac{\mu_2}{\varepsilon_2}}\cos\theta_i + \sqrt{\dfrac{\mu_1}{\varepsilon_1}}\cos\theta_t}\right|_{\theta_i \geq \theta_c}$$

$$= \left.\frac{2\sqrt{\dfrac{\mu_2}{\varepsilon_2}}\cos\theta_i}{\sqrt{\dfrac{\mu_2}{\varepsilon_2}}\cos\theta_i + \sqrt{\dfrac{\mu_1}{\varepsilon_1}}\sqrt{1 - \sin^2\theta_t}}\right|_{\theta_i \geq \theta_c}$$

$$= \left.\frac{2\sqrt{\dfrac{\mu_2}{\varepsilon_2}}\cos\theta_i}{\sqrt{\dfrac{\mu_2}{\varepsilon_2}}\cos\theta_i + \sqrt{\dfrac{\mu_1}{\varepsilon_1}}\sqrt{1 - \dfrac{\mu_1\varepsilon_1}{\mu_2\varepsilon_2}\sin^2\theta_i}}\right|_{\theta_i \geq \theta_c}$$

$$= \left.\frac{2\sqrt{\dfrac{\mu_2}{\varepsilon_2}}\cos\theta_i}{\sqrt{\dfrac{\mu_2}{\varepsilon_2}}\cos\theta_i - j\sqrt{\dfrac{\mu_1}{\varepsilon_1}}\sqrt{\dfrac{\mu_1\varepsilon_1}{\mu_2\varepsilon_2}\sin^2\theta_i - 1}}\right|_{\theta_i \geq \theta_c}$$

$$T_\perp^b\big|_{\theta_i \geq \theta_c} = |T_\perp^b|e^{j\psi_\perp} \tag{5-45}$$

where

$$|T_\perp^b| = \frac{2R_\perp}{\sqrt{R_\perp^2 + X_\perp^2}} \tag{5-45a}$$

In addition the transmitted average power density can now be written, using (5-12a) through (5-12b) and the modified forms (5-43) through (5-43b) for the fields when the incidence angle is equal to or greater than the critical angle, as

$$\mathbf{S}_{av}^t|_{\theta_i \geq \theta_c} = \frac{1}{2} \operatorname{Re}(\mathbf{E}^t \times \mathbf{H}^{t*})_{\theta_i \geq \theta_c}$$

$$= \frac{1}{2} \operatorname{Re}\left[\left(\hat{a}_y T_\perp^b E_0 e^{-\alpha_c z} e^{-j\beta_c x}\right) \times (-\hat{a}_x \cos\theta_t + \hat{a}_z \sin\theta_t) * \frac{(T_\perp^b)^* E_0^*}{\eta_2} e^{-\alpha_c z} e^{+j\beta_c x}\right]_{\theta_i \geq \theta_c}$$

$$= \frac{1}{2} \operatorname{Re}\left\{[\hat{a}_z (\cos\theta_t)^* + \hat{a}_x (\sin\theta_t)^*] \frac{|T_\perp^b|^2 |E_0|^2}{\eta_2} e^{-2\alpha_c z}\right\}_{\theta_i \geq \theta_c}$$

$$= \frac{1}{2} \operatorname{Re}\left\{\left[\hat{a}_z\left(\sqrt{1 - \sin^2\theta_t}\right)^* + \hat{a}_x (\sin\theta_t)^*\right] \frac{|T_\perp^b|^2 |E_0|^2}{\eta_2} e^{-2\alpha_c z}\right\}_{\theta_i \geq \theta_c}$$

$$= \frac{1}{2} \operatorname{Re}\left\{\left[\hat{a}_z\left(\sqrt{1 - \frac{\mu_1 \varepsilon_1}{\mu_2 \varepsilon_2} \sin^2\theta_i}\right)^* \right.\right.$$

$$\left.\left. + \hat{a}_x \left(\sqrt{\frac{\mu_1 \varepsilon_1}{\mu_2 \varepsilon_2}} \sin\theta_i\right)^*\right] \frac{|T_\perp^b|^2 |E_0|^2}{\eta_2} e^{-2\alpha_c z}\right\}_{\theta_i \geq \theta_c}$$

$$= \frac{1}{2} \operatorname{Re}\left\{\left[\hat{a}_z\left(-j\sqrt{\frac{\mu_1 \varepsilon_1}{\mu_2 \varepsilon_2} \sin^2\theta_i - 1}\right)\right.\right.$$

$$\left.\left. + \hat{a}_x \left(\sqrt{\frac{\mu_1 \varepsilon_1}{\mu_2 \varepsilon_2}} \sin\theta_i\right)\right] \frac{|T_\perp^b|^2 |E_0|^2}{\eta_2} e^{-2\alpha_c z}\right\}_{\theta_i \geq \theta_c}$$

$$\mathbf{S}_{av}^t|_{\theta_i \geq \theta_c} = \hat{a}_x \sqrt{\frac{\mu_1 \varepsilon_1}{\mu_2 \varepsilon_2}} \sin\theta_i \frac{|T_\perp^b|^2 |E_0|^2}{2\eta_2} e^{-2\alpha_c z}\bigg|_{\theta_i \geq \theta_c} \quad (5\text{-}46)$$

Again from (5-46) it is apparent that there is no real power transfer across the interface in a direction normal to the boundary. Therefore all the power must be reflected into medium 1. This can also be verified by formulating and examining the incident and reflected average power densities. Doing this, using the fields (5-10a) through (5-11b) where the reflection coefficient is that of (5-44), shows that the magnitudes of the incident and reflected average power densities are those of (5-41a) and (5-41b), which are identical.

The propagation of a wave from a more to a less dense medium ($\varepsilon_2 < \varepsilon_1$ when $\mu_1 = \mu_2$) under oblique incidence can be summarized as follows.

1. When the angle of incidence is smaller than the critical angle [$\theta_i < \theta_c = \sin^{-1}(\sqrt{\varepsilon_2/\varepsilon_1})$] a wave is transmitted into medium 2 at an angle θ_t which is greater than the incident angle θ_i. Real power is transferred into medium 2, and it is directed along angle θ_t as shown in Figure 5-7a.

2. As the angle of incidence increases and reaches the critical angle $\theta_i = \theta_c = \sin^{-1}(\sqrt{\varepsilon_2/\varepsilon_1})$, the refracted angle θ_t, which varies more rapidly than the incident angle θ_i, approaches 90°. Although a wave into medium 2 exists under this condition (which is necessary to satisfy the boundary conditions), the fields form a surface wave that is directed along the x axis (which is parallel to the interface). There is no real power transfer normal to the boundary into medium 2, and all

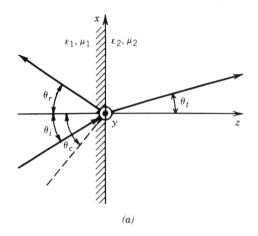

(a)

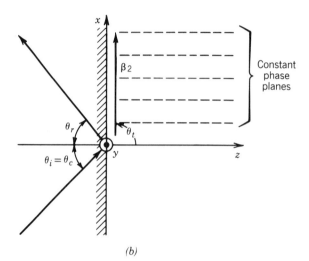

(b)

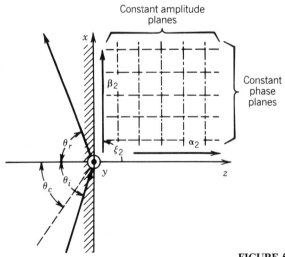

(c)

FIGURE 5-7 Critical angle wave propagation along an interface.

the power is reflected in medium 1 along reflected angle θ_r as shown in Figure 5-7b. The constant phase planes are parallel to the z axis.

3. When the incident angle θ_i exceeds the critical angle θ_c [$\theta_i > \theta_c = \sin^{-1}(\sqrt{\varepsilon_2/\varepsilon_1}\,)$], a wave into medium 2 still exists which travels along the x axis (which is parallel to the interface) and is simultaneously heavily attenuated in the z direction (which is normal to the interface). There is no real power transfer normal to the boundary into medium 2, and all power is reflected into medium 1 along reflection angle θ_r, as shown in Figure 5-7c. Although there is no power transferred into medium 2, a wave exists there that is necessary to satisfy the boundary conditions on the continuity of the tangential components of the electric and magnetic fields. The wave in medium 2 travels parallel to the interface with a phase velocity that is less than that of an ordinary wave in the same medium [as given by (5-43c)], and it is rapidly attenuated in a direction normal to the interface with an attenuation constant given by (5-43a). This wave is *tightly bound* to the surface, and it is referred to as a *tight bound slow surface wave*.

The critical angle is used to design many practical instruments and transmission lines, such as binoculars, dielectric covered ground plane (surface wave) transmission lines, fiber optic cables, etc. To see how critical angle may be utilized, let us consider an example.

Example 5-6. Determine the range of values of the dielectric constant of a dielectric slab of thickness t so that, when a wave is incident on it from one of its ends at an oblique angle $0° \leq \theta_i \leq 90°$, the energy of the wave in the dielectric is contained within the slab. The geometry of the problem is shown in the Figure 5-8.

Solution. We assume that the slab width is infinite. To contain the energy of the wave within the slab, the reflection angle θ_r of the wave bouncing within the slab must be equal to or greater than the critical angle θ_c. By referring to Figure 5-8, the critical angle can be related to the refraction angle θ_t by

$$\sin \theta_r = \sin\left(\frac{\pi}{2} - \theta_t\right) = \cos \theta_t \geq \sin \theta_c = \sqrt{\frac{\varepsilon_0}{\varepsilon_r \varepsilon_0}} = \frac{1}{\sqrt{\varepsilon_r}}$$

or

$$\cos \theta_t \geq \frac{1}{\sqrt{\varepsilon_r}}$$

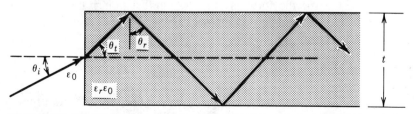

FIGURE 5-8 Dielectric slab of thickness t and wave containment within.

At the interface formed at the leading edge, Snell's law of refraction must be satisfied. That is,

$$\beta_0 \sin\theta_i = \beta_1 \sin\theta_t \Rightarrow \sin\theta_t = \frac{\beta_0}{\beta_1}\sin\theta_i = \frac{1}{\sqrt{\varepsilon_r}}\sin\theta_i$$

Using this, we can write the aforementioned $\cos\theta_t$ as

$$\cos\theta_t = \sqrt{1-\sin^2\theta_t} = \sqrt{1 - \frac{1}{\varepsilon_r}\sin^2\theta_i} \geq \frac{1}{\sqrt{\varepsilon_r}}$$

or

$$\sqrt{1 - \frac{1}{\varepsilon_r}\sin^2\theta_i} \geq \frac{1}{\sqrt{\varepsilon_r}}$$

Solving this leads to

$$\varepsilon_r - \sin^2\theta_i \geq 1$$

or

$$\varepsilon_r \geq 1 + \sin^2\theta_i$$

To accommodate all the angles, the dielectric constant must be

$$\varepsilon_r \geq 2$$

since the smallest and largest values of θ_i are, respectively, 0° and 90°. This is achievable by many practical dielectric material such as Teflon ($\varepsilon_r \simeq 2.1$), polystyrene ($\varepsilon_r \simeq 2.56$), and many others.

B. PARALLEL (VERTICAL) POLARIZATION

The procedure used to derive the critical angle and to examine the properties for perpendicular (horizontal) polarization can be repeated for parallel (vertical) polarization. However, it can be shown that the critical angle is not a function of polarization, and that it exists for both parallel and perpendicular polarizations. The only limitation of the critical angle is that the wave propagation be to a less dense medium ($\mu_2\varepsilon_2 < \mu_1\varepsilon_1$ or $\varepsilon_2 < \varepsilon_1$ when $\mu_1 = \mu_2$).

The expression for the critical angle for parallel polarization is the same as that for perpendicular polarization as given by (5-35b) or (5-36). In addition the wave propagation phenomena that occur for perpendicular polarization when the incidence angle is less than, equal to, or greater than the critical angle are also identical to those for parallel polarization. Although the formulas for the reflection $\Gamma_\parallel^b$ and transmission $T_\parallel^b$ coefficients, and transmitted average power density $\mathbf{S}_\parallel^t$ for parallel polarization are not identical to those of perpendicular polarization as given by (5-44) through (5-46), the principles stated previously are identical here. The derivation of the specific formulas for the parallel polarization for critical angle propagation are left as an end of chapter exercise for the reader.

5.4 LOSSY MEDIA

In the previous sections we examined wave reflection and transmission under normal and oblique wave incidence when both media forming the interface are lossless. Let us now examine the reflection and transmission of waves under normal and oblique incidence when either one or both media are lossy [4]. Although in some cases the formulas will be the same as for the lossless cases, there are differences especially under oblique wave incidence.

5.4.1 Normal Incidence: Conductor–Conductor Interface

When a uniform plane wave is normally incident upon a planar interface formed by two lossy media (as shown in Figure 5-1 but allowing for losses in both media through the conductivity σ), the incident, reflected, and transmitted fields, reflection and transmission coefficients, and average power densities are identical to (5-1a) through (5-6c) except that (a) an attenuation constant must be included in each field and (b) the intrinsic impedances, and attenuation and phases constants must be modified to include the conductivities of the media. Thus we can summarize the results here as

$$\mathbf{E}^i = \hat{a}_x E_0 e^{-\alpha_1 z} e^{-j\beta_1 z} \quad (5\text{-}47\text{a})$$

$$\mathbf{H}^i = \hat{a}_y \frac{E_0}{\eta_1} e^{-\alpha_1 z} e^{-j\beta_1 z} \quad (5\text{-}47\text{b})$$

$$\mathbf{E}^r = \hat{a}_x \Gamma^b E_0 e^{+\alpha_1 z} e^{+j\beta_1 z} \quad (5\text{-}48\text{a})$$

$$\mathbf{H}^r = -\hat{a}_y \frac{\Gamma^b E_0}{\eta_1} e^{+\alpha_1 z} e^{+j\beta_1 z} \quad (5\text{-}48\text{b})$$

$$\mathbf{E}^t = \hat{a}_x T^b E_0 e^{-\alpha_2 z} e^{-j\beta_2 z} \quad (5\text{-}49\text{a})$$

$$\mathbf{H}^t = \hat{a}_y \frac{T^b E_0}{\eta_2} e^{-\alpha_2 z} e^{-j\beta_2 z} \quad (5\text{-}49\text{b})$$

$$\Gamma^b = \frac{\eta_2 - \eta_1}{\eta_2 + \eta_1} \quad (5\text{-}50\text{a})$$

$$T^b = \frac{2\eta_2}{\eta_2 + \eta_1} \quad (5\text{-}50\text{b})$$

$$\mathbf{S}^i_{av} = \hat{a}_z \frac{|E_0|^2}{2} e^{-2\alpha_1 z} \operatorname{Re}\left(\frac{1}{\eta_1^*}\right) \quad (5\text{-}51\text{a})$$

$$\mathbf{S}^r_{av} = -\hat{a}_z |\Gamma^b|^2 \frac{|E_0|^2}{2} e^{+2\alpha_1 z} \operatorname{Re}\left(\frac{1}{\eta_1^*}\right) \quad (5\text{-}51\text{b})$$

$$\mathbf{S}^t_{av} = \hat{a}_z |T^b|^2 \frac{|E_0|^2}{2} e^{-2\alpha_2 z} \operatorname{Re}\left(\frac{1}{\eta_2^*}\right) \quad (5\text{-}51\text{c})$$

For each lossy medium the attenuation constants α_i, phase constants β_i, and intrinsic impedances η_i are related to the corresponding constitutive parameters ε_i, μ_i, and σ_i by the expressions in Table 4-1.

The total electric and magnetic fields in medium 1 can be written as

$$\mathbf{E}^1 = \mathbf{E}^i + \mathbf{E}^r = \hat{a}_x \underbrace{E_0 e^{-\alpha_1 z} e^{-j\beta_1 z}}_{\text{traveling wave}} \underbrace{(1 + \Gamma^b e^{+2\alpha_1 z} e^{+j2\beta_1 z})}_{\text{standing wave}} \quad (5\text{-}52\text{a})$$

$$\mathbf{H}^1 = \mathbf{H}^i + \mathbf{H}^r = \hat{a}_y \underbrace{(E_0/\eta_1) e^{-\alpha_1 z} e^{-j\beta_1 z}}_{\text{traveling wave}} \underbrace{(1 - \Gamma^b e^{+2\alpha_1 z} e^{+j2\beta_1 z})}_{\text{standing wave}} \quad (5\text{-}52\text{b})$$

In each field the factors outside the parentheses form the *traveling wave part* of the total wave; those within the parentheses form the *standing wave part*.

208 REFLECTION AND TRANSMISSION

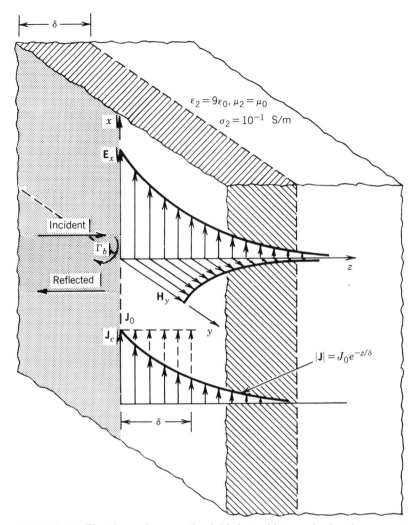

FIGURE 5-9 Electric and magnetic field intensities, and electric current density distributions in a lossy earth.

Example 5-7. A uniform plane wave, whose incident electric field has an x component with an amplitude at the interface of 10^{-3} V/m, is traveling in a free-space medium and is normally incident upon a lossy flat earth as shown in Figure 5-9. Assuming that the constitutive parameters of the earth are $\varepsilon_2 = 9\varepsilon_0$, $\mu_2 = \mu_0$, and $\sigma_2 = 10^{-1}$ S/m, determine the variation of the conduction current density in the earth at a frequency of 1 MHz.

Solution. At $f = 10^6$ Hz

$$\frac{\sigma_2}{\omega\varepsilon_2} = \frac{10^{-1}}{2\pi \times 10^6 (9 \times 10^{-9}/36\pi)} = 2 \times 10^2 \gg 1$$

which classifies the material as a very good conductor.

On either side of the interface, the total electric field is equal to

$$\mathbf{E}^{\text{total}}|_{z=0} = \hat{a}_x \times 10^{-3}|1 + \Gamma^b|$$

where

$$\Gamma^b = \frac{\eta_2 - \eta_1}{\eta_2 + \eta_1} = \frac{\eta_2 - \eta_0}{\eta_2 + \eta_0}$$

$$\eta_2 \simeq \sqrt{\frac{\omega\mu}{2\sigma}}(1+j) = \sqrt{\frac{2\pi \times 10^6(4\pi \times 10^{-7})}{2 \times 10^{-1}}}(1+j) = 2\pi(1+j)$$

Thus

$$\Gamma^b = \frac{2\pi(1+j) - 377}{2\pi(1+j) + 377} = \frac{-370.72 + j2\pi}{383.28 + j2\pi}$$

$$= \frac{370.77 \underline{/179.04°}}{383.33 \underline{/0.94°}} = 0.967 \underline{/178.1°}$$

and

$$\mathbf{E}^{\text{total}}|_{z=0} = \hat{a}_x \times 10^{-3}|1 + 0.967\underline{/178.1°}|$$
$$= \hat{a}_x \times 10^{-3}|0.0335 + j0.0321| = \hat{a}_x(4.64 \times 10^{-5})$$

The conduction current density at the surface of the earth is equal to

$$\mathbf{J}_c|_{z=0} = \hat{a}_x J_0 = \hat{a}_x \sigma E^{\text{total}}|_{z=0} = \hat{a}_x \times 10^{-1}(4.64 \times 10^{-5})$$
$$= \hat{a}_x(4.64 \times 10^{-6})$$

or

$$J_0 = 4.64 \ \mu\text{A/m}^2$$

The magnitude of the current density varies inside the earth as

$$|\mathbf{J}_c| = J_0|e^{-\alpha_2 z}e^{-j\beta_2 z}| = J_0 e^{-\alpha_2 z} = J_0 e^{-z/\delta_2}$$

where

$$\delta_2 = \text{skin depth} = \sqrt{\frac{2}{\omega\mu_2\sigma_2}} = \sqrt{\frac{2}{2\pi \times 10^6(4\pi \times 10^{-7}) \times 10^{-1}}}$$

$$= \frac{10}{2\pi} = 1.5915 \text{ m}$$

The magnitude variations of the current density inside the earth are shown in Figure 5-9 and they exhibit an exponential decay. At one skin depth ($z = \delta_2 = 1.5915$ m), the current density has been reduced to

$$|\mathbf{J}_c|_{z=\delta_2} = J_0 e^{-1} = 0.3679 J_0 = 0.3679(4.64 \times 10^{-6}) = 1.707 \ \mu\text{A/m}^2$$

Therefore at one skin depth the current is reduced to 36.79% of its value at the surface.

If the area under the current density curve is found, it is shown to be equal to

$$J_s = \int_0^\infty |\mathbf{J}_c| \, dz = \int_0^\infty J_0 e^{-z/\delta_2} \, dz = -\delta_2 J_0 e^{-z/\delta_2}\Big|_0^\infty = \delta_2 J_0$$

210 REFLECTION AND TRANSMISSION

The same answer can be obtained by assuming that the current density maintains a constant surface value J_0 to a depth equal to the skin depth and equal to zero thereafter, as shown by the dashed curved in Figure 5-9.

The area under the curve can then be interpreted as the total current density J_s (A/m) per unit width in the y direction. It can be obtained by finding the area formed by maintaining constant surface current density J_0 (A/m^2) through a depth equal to the skin depth.

5.4.2 Oblique Incidence: Dielectric–Conductor Interface

Let us assume that a uniform plane wave is obliquely incident upon a planar interface where medium 1 is a perfect dielectric and medium 2 is lossy, as shown in Figure 5-10 [3]. For either the perpendicular or parallel polarization, the transmitted electric field into medium 2 can be written, using modified forms of either (5-12a) or (5-22a), as

$$\mathbf{E}^t = \mathbf{E}_2 \exp\left[-\gamma_2(x \sin \theta_t + z \cos \theta_t)\right] = \mathbf{E}_2 \exp\left[-(\alpha_2 + j\beta_2)(x \sin \theta_t + z \cos \theta_t)\right] \quad (5\text{-}53)$$

It can be shown that for lossy media Snell's law of refraction can be written as

$$\gamma_1 \sin \theta_i = \gamma_2 \sin \theta_t \quad (5\text{-}54)$$

Therefore for the geometry of Figure 5-10,

$$\sin \theta_t = \frac{\gamma_1}{\gamma_2} \sin \theta_i = \frac{j\beta_1}{\alpha_2 + j\beta_2} \sin \theta_i \quad (5\text{-}55a)$$

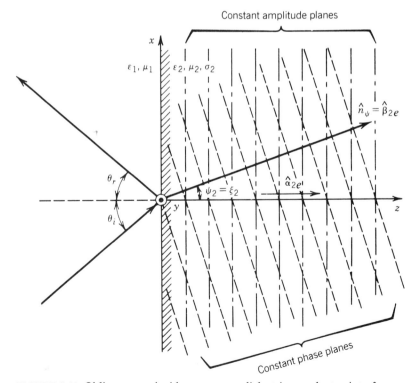

FIGURE 5-10 Oblique wave incidence upon a dielectric–conductor interface.

and

$$\cos\theta_t = \sqrt{1 - \sin^2\theta_t} = \sqrt{1 - \left(\frac{j\beta_1}{\alpha_2 + j\beta_2}\right)^2 \sin^2\theta_i} = se^{j\zeta} = s(\cos\zeta + j\sin\zeta) \quad (5\text{-}55b)$$

Using (5-55a) and (5-55b) we can write (5-53) as

$$\mathbf{E}^t = \mathbf{E}_2 \exp\left\{-(\alpha_2 + j\beta_2)\left[x\frac{j\beta_1}{\alpha_2 + j\beta_2}\sin\theta_i + zs(\cos\zeta + j\sin\zeta)\right]\right\} \quad (5\text{-}56)$$

which reduces to

$$\mathbf{E}^t = \mathbf{E}_2 \exp\left[-zs(\alpha_2 \cos\zeta - \beta_2 \sin\zeta)\right]$$
$$\times \exp\left\{-j\left[\beta_1 x \sin\theta_i + zs(\alpha_2 \sin\zeta + \beta_2 \cos\zeta)\right]\right\}$$
$$\mathbf{E}^t = \mathbf{E}_2 e^{-zp} \exp\left[-j(\beta_1 x \sin\theta_i + zq)\right] \quad (5\text{-}57)$$

where

$$p = s(\alpha_2 \cos\zeta - \beta_2 \sin\zeta) = \alpha_{2e} \quad (5\text{-}57a)$$
$$q = s(\alpha_2 \sin\zeta + \beta_2 \cos\zeta) \quad (5\text{-}57b)$$

It is apparent that (5-57) represents a nonuniform wave.

The instantaneous field of (5-57) can be written, assuming $\mathbf{E}_2$ is real, as

$$\mathcal{E}^t = \operatorname{Re}(\mathbf{E}^t e^{j\omega t}) = \mathbf{E}_2 e^{-zp} \operatorname{Re}\left(\exp\left\{j[\omega t - (\beta_1 x \sin\theta_i + zq)]\right\}\right)$$
$$\mathcal{E}^t = \mathbf{E}_2 e^{-zp} \cos\left[\omega t - (\beta_1 x \sin\theta_i + zq)\right] \quad (5\text{-}58)$$

The constant amplitude planes (z = constant) of (5-58) are parallel to the interface, and they are shown dashed-dotted in Figure 5-10. The constant phase planes $[\omega t - (kx \sin\theta_i + zq) = \text{constant}]$ are inclined at an angle ψ_2 that is no longer θ_t.

To determine the constant phase we write the argument of the exponential or of the cosine function in (5-58) as

$$\omega t - (\beta_1 x \sin\theta_i + zq) = \omega t - \sqrt{(\beta_1 \sin\theta_i)^2 + q^2}$$
$$\times \left[\frac{(\beta_1 \sin\theta_i)x}{\sqrt{(\beta_1 \sin\theta_i)^2 + q^2}} + \frac{qz}{\sqrt{(\beta_1 \sin\theta_i)^2 + q^2}}\right] \quad (5\text{-}59)$$

If we define an angle ψ_2 such that

$$u = \beta_1 \sin\theta_i \quad (5\text{-}60a)$$

$$\sin\psi_2 = \frac{\beta_1 \sin\theta_i}{\sqrt{(\beta_1 \sin\theta_i)^2 + q^2}} = \frac{u}{\sqrt{u^2 + q^2}} \quad (5\text{-}60b)$$

$$\cos\psi_2 = \frac{q}{\sqrt{(\beta_1 \sin\theta_i)^2 + q^2}} = \frac{q}{\sqrt{u^2 + q^2}} \quad (5\text{-}60c)$$

or

$$\psi_2 = \tan^{-1}\left(\frac{\beta_1 \sin \theta_i}{q}\right) = \tan^{-1}\left(\frac{u}{q}\right) \tag{5-60d}$$

we can write (5-59) and in turn (5-58) as

$$\mathscr{E}^t = \mathbf{E}_2 e^{-zp} \operatorname{Re}\left(\exp\left\{j\left[\omega t - \sqrt{u^2+q^2}\left(\frac{ux}{\sqrt{u^2+q^2}} + \frac{qz}{\sqrt{u^2+q^2}}\right)\right]\right\}\right)$$

$$= \mathbf{E}_2 e^{-zp} \operatorname{Re}\left(\exp\left\{j\left[\omega t - \beta_{2e}(x \sin \psi_2 + z \cos \psi_2)\right]\right\}\right)$$

$$\mathscr{E}^t = \mathbf{E}_2 e^{-zp} \operatorname{Re}\left(\exp\left\{j\left[\omega t - \beta_{2e}(\hat{n}_\psi \cdot \mathbf{r})\right]\right\}\right) \tag{5-61}$$

where

$$\hat{n}_\psi = \hat{a}_x \sin \psi_2 + \hat{a}_z \cos \psi_2 \tag{5-61a}$$

$$\beta_{2e} = \sqrt{u^2+q^2} \tag{5-61b}$$

It is apparent from (5-60a) through (5-61a) that

1. The true angle of refraction is ψ_2 and not θ_t (θ_t is complex).
2. The wave travels along a direction defined by unit vector $\hat{n}_\psi$.
3. The constant phase planes are perpendicular to unit vector $\hat{n}_\psi$, and they are shown as dashed lines in Figure 5-10.

The phase velocity of the wave in medium 2 is obtained by setting the exponent of (5-61) to a constant and differentiating it with respect to time. Doing this, we can write the phase velocity v_p of the wave as

$$\omega(1) - \sqrt{u^2+q^2}\left(\hat{n}_\psi \cdot \frac{d\mathbf{r}}{dt}\right) = 0$$

$$\omega(1) - \sqrt{u^2+q^2}\left(\hat{n}_\psi \cdot \frac{d\mathbf{r}}{dt}\right) = \omega - \beta_{2e}(\hat{n}_\psi \cdot \mathbf{v}_p) = 0 \tag{5-62}$$

or

$$v_{pr} = \frac{\omega}{\beta_{2e}} = \frac{\omega}{\sqrt{u^2+q^2}} = \frac{\omega}{\sqrt{(\beta_1 \sin \theta_i)^2 + q^2}} \tag{5-62a}$$

It is evident that the phase velocity is a function of the incidence angle θ_i and the constitutive parameters of the two media.

Example 5-8. A plane wave of either perpendicular or parallel polarization traveling in air is obliquely incident upon a planar interface of copper ($\sigma = 5.76 \times 10^7$ S/m). At a frequency of 10 GHz, determine the angle of refraction and reflection coefficients for each of the two polarizations.

Solution. For copper

$$\frac{\sigma_2}{\omega\varepsilon_2} = \frac{5.8 \times 10^7 (36\pi)}{(2\pi \times 10^{10}) \times 10^{-9}} = 1.037 \times 10^8 \gg 1$$

Therefore according to Table 4-1

$$\alpha_2 \simeq \beta_2 \simeq \sqrt{\frac{\omega\mu_2\sigma_2}{2}}$$

Using (5-55a)

$$\sin\theta_t = \frac{j\beta_1}{\alpha_2 + j\beta_2}\sin\theta_i \simeq \frac{j\beta_1}{\sqrt{\frac{\omega\mu_2\sigma_2}{2}}(1+j)}\sin\theta_i \stackrel{\sigma_2 \gg 1}{\simeq} 0 \Rightarrow \theta_t \simeq 0$$

Therefore (5-55b), (5-57a), and (5-57b) reduce to

$$\cos\theta_t = 1 = se^{j\zeta} \Rightarrow s = 1 \qquad \zeta = 0$$

$$p = s(\alpha_2\cos\zeta - \beta_2\sin\zeta) \simeq \alpha_2 = \sqrt{\frac{\omega\mu_2\sigma_2}{2}}$$

$$q = s(\alpha_2\sin\zeta + \beta_2\cos\zeta) \simeq \beta_2 = \sqrt{\frac{\omega\mu_2\sigma_2}{2}}$$

Using (5-60d) the true angle of refraction is

$$\psi_2 = \tan^{-1}\left(\frac{u}{q}\right) \simeq \tan^{-1}\left(\frac{\beta_1\sin\theta_i}{\beta_2}\right) = \tan^{-1}\left(\frac{\omega\sqrt{\mu_0\varepsilon_0}}{\sqrt{\frac{\omega\mu_0\sigma_2}{2}}}\sin\theta_i\right)$$

$$= \tan^{-1}\left(\sqrt{\frac{2\omega\varepsilon_0}{\sigma_2}}\sin\theta_i\right) \leq \tan^{-1}\left(\sqrt{\frac{2\omega\varepsilon_0}{\sigma_2}}\right) = \tan^{-1}(0.139 \times 10^{-3})$$

$$\psi_2 = \tan^{-1}(0.139 \times 10^{-3}\sin\theta_i) \leq 0.139 \times 10^{-3} \text{ rad} = (7.96 \times 10^{-3})°$$

Using (5-17a) and (5-24c), the reflection coefficients for perpendicular and parallel polarizations reduce to

$$\Gamma_\perp^b = \frac{\eta_2\cos\theta_i - \eta_1\cos\theta_t}{\eta_2\cos\theta_i + \eta_1\cos\theta_t} \simeq \frac{\eta_2\cos\theta_i - \eta_1}{\eta_2\cos\theta_i + \eta_1} = \frac{\cos\theta_i - \eta_1/\eta_2}{\cos\theta_i + \eta_1/\eta_2}$$

$$\Gamma_\parallel^b = \frac{-\eta_1\cos\theta_i + \eta_2\cos\theta_t}{\eta_1\cos\theta_i + \eta_2\cos\theta_t} \simeq \frac{-\eta_1\cos\theta_i + \eta_2}{\eta_1\cos\theta_i + \eta_2} = \frac{-\cos\theta_i + \eta_2/\eta_1}{\cos\theta_i + \eta_2/\eta_1}$$

Since

$$\frac{\eta_1}{\eta_2} = \frac{\sqrt{\frac{\mu_1}{\varepsilon_1}}}{\sqrt{\frac{j\omega\mu_2}{\sigma_2 + j\omega\varepsilon_2}}} \simeq \frac{\sqrt{\frac{\mu_0}{\varepsilon_0}}}{\sqrt{\frac{j\omega\mu_0}{\sigma_2}}} = \sqrt{\frac{\sigma_2}{j\omega\varepsilon_0}}$$

$$\frac{\eta_1}{\eta_2} \simeq 1.02 \times 10^4 e^{-j\pi/4} \gg 1 \geq \cos\theta_i$$

214 REFLECTION AND TRANSMISSION

Then

$$\Gamma^b_\perp \simeq \frac{\cos\theta_i - \eta_1/\eta_2}{\cos\theta_i + \eta_1/\eta_2} \simeq -1$$

$$\Gamma^b_\parallel \simeq \frac{-\cos\theta_i + \eta_2/\eta_1}{\cos\theta_i + \eta_2/\eta_1} \simeq +1$$

Thus for a very good conductor, such as copper, the angle of refraction approaches zero and the magnitude of the reflection coefficients for perpendicular and parallel polarizations approach unity, and they are all essentially independent of the angle of incidence. The same will be true for all other good conductors.

5.4.3 Oblique Incidence: Conductor–Conductor Interface

In Section 5.3.4 it was shown that when a uniform plane wave is incident upon a dielectric–dielectric planar interface at an incidence θ_i equal to or greater than the critical angle θ_c, the transmitted wave produced into medium 2 is a nonuniform plane wave. For this plane wave the constant amplitude planes (which are perpendicular to the α_{2e} vector) of Figure 5-7 are perpendicular to the constant phase planes (which are perpendicular to the β_{2e} vector), or the angle ξ_2 between the α_{2e} and β_{2e} vectors is 90°.

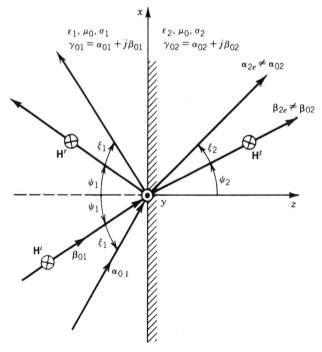

FIGURE 5-11 Geometry for uniform and nonuniform plane wave incidence on a plane boundary between two conducting media.

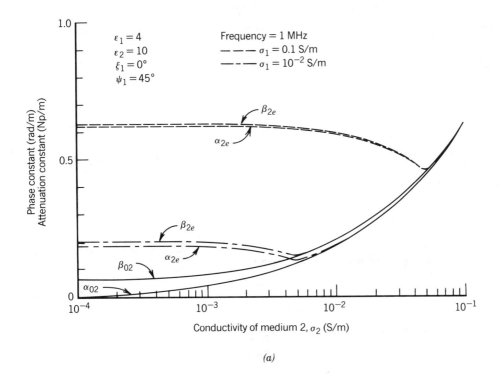

(a)

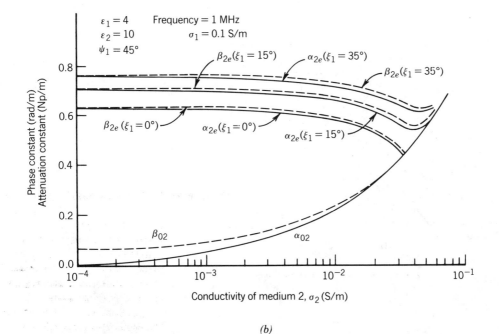

(b)

FIGURE 5-12 Comparison of medium 2 modified (α_{2e}, β_{2e}) with intrinsic (α_{02}, β_{02}) propagation constants for (a) uniform plane wave ($\xi_1 = 0°$). (*Source:* J. J. Holmes and C. A. Balanis, "Refraction of a uniform plane wave incident on a plane boundary between two lossy media," *IEEE Trans. Antennas Propagat.*, © 1978, IEEE.) and (b) nonuniform plane wave ($\xi_1 \neq 0°$) (*Source:* R. D. Radcliff and C. A. Balanis, "Modified propagation constants for nonuniform plane wave transmission through conducting media," *IEEE Trans. Geosci. Remote Sensing*, © 1982, IEEE.)

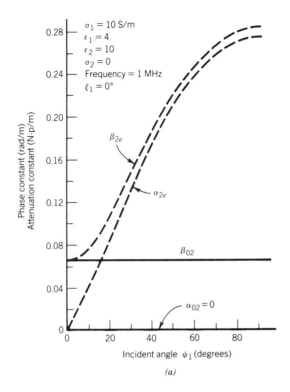

(a)

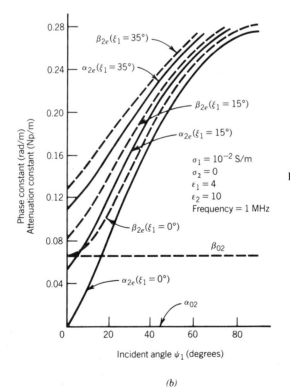

(b)

FIGURE 5-13 Variation of modified (α_{2e}, β_{2e}) and intrinsic (α_{02}, β_{02}) propagation constants as a function of incident angle for (a) uniform plane wave ($\xi_1 = 0°$) and (b) nonuniform plane wave ($\xi_1 \neq 0°$) (*Source:* R. D. Radcliff and C. A. Balanis, "Modified propagation constants for nonuniform plane wave transmission through conducting media," *IEEE Trans. Geosci. Remote Sensing*, © 1982, IEEE.)

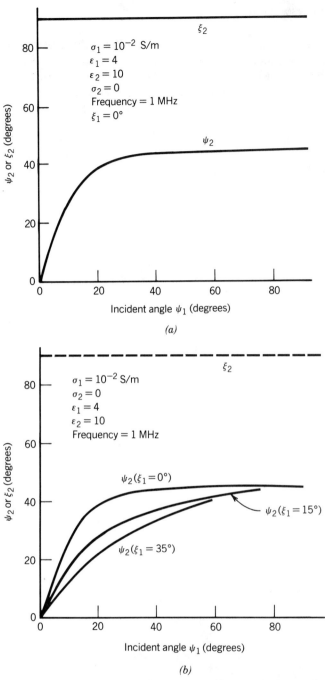

FIGURE 5-14 Variation of the refraction angles ζ_2 and ρ_2 for (a) uniform plane wave ($\xi_1 = 0°$) (*Source:* J. J. Holmes and C. A. Balanis, "Refraction of a uniform plane wave incident on a plane boundary between two lossy media," *IEEE Trans. Antennas Propagat.*, © 1978, IEEE.) and (b) nonuniform plane wave ($\xi_1 \neq 0°$) (*Source:* R. D. Radcliff and C. A. Balanis, "Modified propagation constants for nonuniform plane wave transmission through conducting media," *IEEE Trans. Geosci. Remote Sensing*, © 1982, IEEE.) as a function of the incident angle ζ_1 when medium 2 is a perfect dielectric.

218 REFLECTION AND TRANSMISSION

In Section 5.4.2 it was demonstrated that a uniform plane wave traveling in a lossless medium and obliquely incident upon a lossy medium also produces a nonuniform plane wave where the angle ξ_2 between the α_{2e} and β_{2e} vectors in Figure 5-10 is greater than 0° but less than 90°. In fact for a very good conductor the angle ξ_2 between α_{2e} and β_{2e} is almost zero [for copper with $\sigma = 5.76 \times 10^7$ S/m, $\xi_2 \leq (8 \times 10^{-3})°$]. As the conducting medium becomes less lossy the angle ξ_2 increases and in the limit it approaches 90° for a lossless medium. In fact *for all lossless media, the angle between the effective attenuation constant α_{2e} and phase constant β_{2e} should always be* 90°, *with reactive power flowing along α_{2e} and positive real power along β_{2e}* [4]. This is necessary since there are no real losses associated with the wave propagation along β_{2e}. This was well illustrated in Section 5.3.4 for

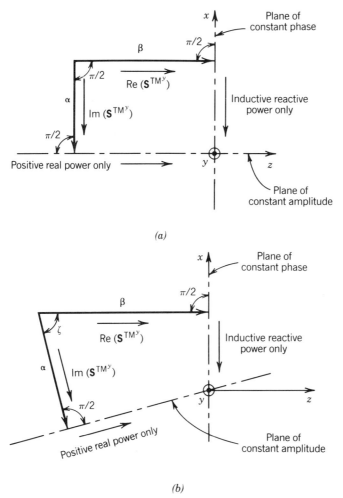

FIGURE 5-15 Analysis of complex Poynting vector $\mathbf{S}^{TM^y}$ for nonuniform TM^y plane wave in (*a*) lossless and (*b*) lossy media . (*Source:* R. B. Adler, L. J. Chu and R. M. Fano, *Electromagnetic Energy Transmission and Radiation*, 1960. Reprinted by permission of John Wiley & Sons, Inc.)

the nonuniform wave produced in a lossless medium when the incidence angle was equal to or greater than the critical angle.

It is very interesting to investigate the field characteristics of uniform or nonuniform plane waves that are obliquely incident upon interfaces comprised of lossy–lossy interfaces. These types of waves have been examined [5, 6], but because of the general complexity of the formulations they will not be repeated here. The

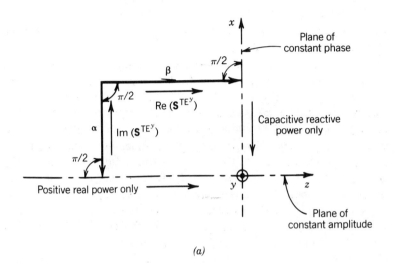

(a)

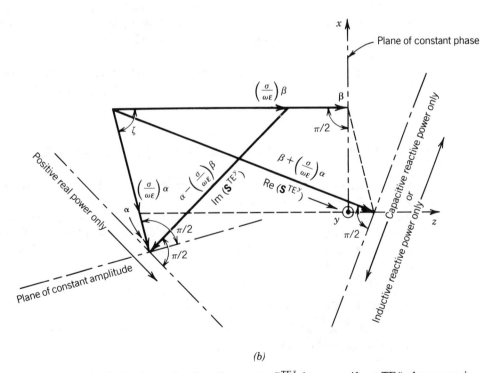

(b)

FIGURE 5-16 Analysis of complex Poynting vector $\mathbf{S}^{TE^y}$ for nonuniform TE^y plane wave in (a) lossless and (b) lossy media. (*Source:* R. B. Adler, L. J. Chu and R. M. Fano, *Electromagnetic Energy Transmission and Radiation*, 1960. Reprinted by permission of John Wiley & Sons, Inc.)

220 REFLECTION AND TRANSMISSION

reader is referred to the literature and to the end of chapter problems. We will summarize the general conclusions derived from these investigations.

1. In medium 2 the wave propagates with an effective attenuation constant α_{2e} and effective phase constant β_{2e}, each of which in general is different, respectively, form the intrinsic attenuation α_{02} and phase β_{02} constants of the medium. For uniform waves the intrinsic attenuation and phase constants were represented by α_2 and β_2, respectively. Referring to the geometry of Figure 5-11, this is illustrated in Figures 5-12 and 5-13 for uniform and nonuniform incident plane waves and typical parameters of the earth.

2. The angle ξ_2 between the effective attenuation constant α_{2e} and effective phase constant β_{2e} is equal to or greater than 0° and equal to or smaller than 90° ($0° \leq \xi_2 \leq 90°$). This is illustrated in Figure 5-14 for uniform and nonuniform plane waves incident on a perfect dielectric. Also in Figure 5-14 the variations of the refraction angle ψ_2 are exhibited as a function of the incidence angle ψ_1.

3. The real and reactive power flow in lossy media depend on the polarization.

   For TM waves ($\mathbf{H} = \hat{a}_y H_y$ for the geometry of Figure 5-11), positive real power flows along directions that are parallel to the constant amplitude planes, or perpendicular to the α_{2e} vector, whereas reactive power flows along directions that are parallel to the constant phase planes, or perpendicular to the β_{2e} vector. A comparison of the complex Poynting vector $\mathbf{S}^{TM^y}$ for TM^y waves in lossless and lossy media is illustrated graphically in Figure 5-15 [4].

   For TE^y waves ($\mathbf{E} = \hat{a}_y E_y$ for the geometry of Figure 5-11), both real and reactive powers flow parallel to the planes of constant amplitude and constant phase. A comparison of the complex Poynting vector $\mathbf{S}^{TE^y}$ for TE^y waves in lossless and lossy media is illustrated graphically in Figure 5-16.

An excellent discussion of uniform and nonuniform plane waves propagating in lossless and lossy media and associated interfaces is found in Chapters 7 and 8 of [4].

5.5 REFLECTION AND TRANSMISSION OF MULTIPLE INTERFACES

Many applications require dielectric interfaces that exhibit specific characteristics as a function of frequency. Accomplishing this often requires multiple interfaces. The objective of this section is to analyze the characteristics of multiple layer interfaces. To reduce the complexity of the problem, we will consider only normal incidence and restrict most of our attention to lossless media. A general formulation for lossy media will also be stated.

5.5.1 Reflection Coefficient of a Single Slab Layer

Section 5.2 showed that for normal incidence the reflection coefficient Γ^b at the boundary of a single planar interface is given by (5-4a) or

$$\boxed{\Gamma^b = \frac{\eta_2 - \eta_1}{\eta_2 + \eta_1}} \quad (5\text{-}63)$$

REFLECTION AND TRANSMISSION OF MULTIPLE INTERFACES

and at a distance $z = -\ell$ from the boundary it is given by (5-5a) or

$$\boxed{\Gamma_{in}(z = -\ell) = \Gamma^b e^{-j2\beta_1 \ell}} \qquad (5\text{-}64)$$

Just to the right of the boundary the input impedance in the $+z$ direction is equal to the intrinsic impedance η_2 of medium 2, that is,

$$Z_{in}(z = 0^+) = \eta_2 = \sqrt{\frac{\mu_2}{\varepsilon_2}} \qquad (5\text{-}65)$$

The input impedance at $z = -\ell$ can be found by using the field expressions (5-1a) through (5-2c). By definition $Z_{in}(z = -\ell)$ is equal to

$$Z_{in}\big|_{z=-\ell} = \frac{E^{total}\big|_{z=-\ell}}{H^{total}\big|_{z=-\ell}} \qquad (5\text{-}66)$$

where

$$E^{total}\big|_{z=-\ell} = (E^i + E^r)\big|_{z=-\ell} = E_0 e^{+j\beta_1 \ell}(1 + \Gamma^b e^{-j2\beta_1 \ell}) = E_0 e^{+j\beta_1 \ell}[1 + \Gamma_{in}(\ell)] \qquad (5\text{-}66a)$$

$$H^{total}\big|_{z=-\ell} = (H^i - H^r)\big|_{z=-\ell} = \frac{E_0}{\eta_1} e^{+j\beta_1 \ell}(1 - \Gamma^b e^{-j2\beta_1 \ell}) = \frac{E_0}{\eta_1} e^{+j\beta_1 \ell}[1 - \Gamma_{in}(\ell)] \qquad (5\text{-}66b)$$

Therefore

$$Z_{in}\big|_{z=-\ell} = \eta_1 \left(\frac{1 + \Gamma^b e^{-j2\beta_1 \ell}}{1 - \Gamma^b e^{-j2\beta_1 \ell}}\right) = \eta_1 \left(\frac{1 + \Gamma_{in}(\ell)}{1 - \Gamma_{in}(\ell)}\right) \qquad (5\text{-}66c)$$

which by using (5-63) can also be written as

$$\boxed{Z_{in}\big|_{z=-\ell} = \eta_1 \left(\frac{1 + \Gamma^b e^{-j2\beta_1 \ell}}{1 - \Gamma^b e^{-j2\beta_1 \ell}}\right) = \eta_1 \left(\frac{1 + \Gamma_{in}(\ell)}{1 - \Gamma_{in}(\ell)}\right) = \eta_1 \left(\frac{\eta_2 + j\eta_1 \tan(\beta_1 \ell)}{\eta_1 + j\eta_2 \tan(\beta_1 \ell)}\right)} \qquad (5\text{-}66d)$$

Equation 5-66d is analogous to the well-known impedance transfer equation that is widely used in transmission line theory [7].

Using the foregoing procedure for normal wave incidence we can derive expressions for multiple layer interfaces [8]. Referring to Figure 5-17a the input impedance at $z = 0^+$ is equal to the intrinsic impedance η_3 of medium 3, that is

$$Z_{in}(z = 0^+) = \eta_3 \qquad (5\text{-}67)$$

In turn the input reflection coefficient at the same interface can be written as

$$\Gamma_{in}(z = 0^-) = \frac{Z_{in}(0^+) - \eta_2}{Z_{in}(0^+) + \eta_2} = \frac{\eta_3 - \eta_2}{\eta_3 + \eta_2} \qquad (5\text{-}67a)$$

222 REFLECTION AND TRANSMISSION

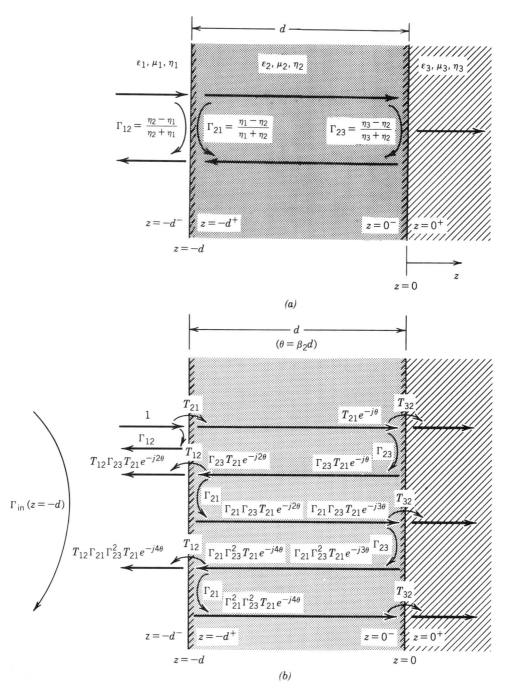

FIGURE 5-17 Impedances and reflection and transmission coefficients for wave propagation in dielectric slab. (*a*) Dielectric slab. (*b*) Reflection and transmission coefficients.

At $z = -d^+$ the input impedance can be written using (5-66d) as

$$Z_{in}(z = -d^+) = \eta_2 \left(\frac{1 + \Gamma_{in}(z=0^-)e^{-j2\beta_2 d}}{1 - \Gamma_{in}(z=0^-)e^{-j2\beta_2 d}} \right) = \eta_2 \left(\frac{(\eta_3 + \eta_2) + (\eta_3 - \eta_2)e^{-j2\beta_2 d}}{(\eta_3 + \eta_2) - (\eta_3 - \eta_2)e^{-j2\beta_2 d}} \right)$$

(5-67b)

and the input reflection coefficient at $z = -d^-$ can be expressed as

$$\Gamma_{in}(z = -d^-) = \frac{Z_{in}(z = -d^+) - \eta_1}{Z_{in}(z = -d^+) + \eta_1}$$

$$= \frac{\eta_2[(\eta_3 + \eta_2) + (\eta_3 - \eta_2)e^{-j2\beta_2 d}] - \eta_1[(\eta_3 + \eta_2) - (\eta_3 - \eta_2)e^{-j2\beta_2 d}]}{\eta_2[(\eta_3 + \eta_2) + (\eta_3 - \eta_2)e^{-j2\beta_2 d}] + \eta_1[(\eta_3 + \eta_2) - (\eta_3 - \eta_2)e^{-j2\beta_2 d}]}$$

(5-67c)

In Figure 5-17a we have defined individual reflection coefficients at each of the boundaries. Here these coefficients are referred to as *intrinsic* reflection coefficients, and they would exist at each boundary if two semi-infinite media form each of the boundaries (neglecting the presence of the other boundaries). Using the intrinsic reflection coefficients defined in Figure 5-17a, the input reflection coefficient of (5-67c) can also be written as

$$\boxed{\Gamma_{in}(z = -d^-) = \frac{\Gamma_{12} + \Gamma_{23} e^{-j2\beta_2 d}}{1 + \Gamma_{12}\Gamma_{23} e^{-j2\beta_2 d}}}$$

(5-67d)

Equation 5-67d can also be derived using the ray-tracing model of Figure 5-17b. Γ_{12} is the intrinsic reflection coefficient of the initial reflection and $T_{12}\Gamma_{23}T_{21}e^{-j2\theta}$, etc., are the contributions to the input reflection that are due to the multiple bounces within the medium 2 slab. The total input reflection coefficient can be written in a geometric series that takes the form of

$$\Gamma_{in}(z = -d^-) = \Gamma_{12} + \frac{T_{12}T_{21}\Gamma_{23}e^{-j2\theta}}{1 - \Gamma_{21}\Gamma_{23}e^{-j2\theta}}$$

(5-68)

where

$$\theta = \beta_2 d$$

(5-68a)

Since according to (5-4a) and (5-4b)

$$\Gamma_{21} = -\Gamma_{12}$$ (5-69a)

$$T_{12} = 1 + \Gamma_{12}$$ (5-69b)

$$T_{21} = 1 + \Gamma_{21} = 1 - \Gamma_{12}$$ (5-69c)

(5-68) can be rewritten and reduced to the form of (5-67d).

If the magnitudes of the intrinsic reflection coefficients $|\Gamma_{12}|$ and $|\Gamma_{23}|$ are low compared to unity, (5-67d) can be approximated by

$$\Gamma_{in}(z = -d^-) = \frac{\Gamma_{12} + \Gamma_{23}e^{-j2\beta_2 d}}{1 + \Gamma_{12}\Gamma_{23}e^{-j2\beta_2 d}} \underset{|\Gamma_{23}| \ll 1}{\overset{|\Gamma_{12}| \ll 1}{\simeq}} \Gamma_{12} + \Gamma_{23}e^{-j2\beta_2 d}$$

(5-70)

The approximate form of (5-70) yields good results if the individual intrinsic reflection coefficients are low. Typically when $|\Gamma_{12}| = |\Gamma_{23}| \leq 0.2$, the error of the approximate form of (5-70) is equal to or less than about 4 percent. The approximate form of (5-70) will be very convenient for representing the input reflection coefficient of multiple interfaces (> 2) when the individual intrinsic reflection coefficients at each interface are low compared to unity.

Example 5-9. A uniform plane wave at a frequency of 10 GHz is incident normally on a dielectric slab of thickness d and bounded on both sides of air. Assume that the dielectric constant of the slab is 2.56.

1. Determine the thickness of the slab so that the input reflection coefficient at 10 GHz is zero.
2. Plot the magnitude of the reflection coefficient as a function of frequency between 5 GHz $\leq f \leq$ 15 GHz when the dielectric slab is 0.9375 cm.

Solution.

1. For the input reflection coefficient to be equal to zero, the reflection coefficient of (5-70) must be set equal to zero. This can be accomplished if

$$|\Gamma_{12} + \Gamma_{23} e^{-j2\beta_2 d}| = 0$$

Since

$$\Gamma_{23} = -\Gamma_{12} = \frac{\eta_1 - \eta_2}{\eta_1 + \eta_2}$$

then

$$|\Gamma_{12}| |1 - e^{-j2\beta_2 d}| = 0 \Rightarrow 2\beta_2 d = 2n\pi \quad n = 0, 1, 2, \ldots$$

For nontrivial solutions, the thickness must be

$$d = \frac{n\pi}{\beta_2} = \frac{n}{2}\lambda_2 \quad n = 1, 2, 3, \ldots$$

where λ_2 is the wavelength inside the dielectric slab. Thus the thickness of the slab must be an integral number of half wavelengths inside the dielectric. At a frequency of 10 GHz and a dielectric constant of 2.56, the wavelength inside the dielectric is

$$\lambda_2 = \frac{30 \times 10^9}{10 \times 10^9 \sqrt{2.56}} = 1.875 \text{ cm}$$

2. At a frequency of 5 GHz, the dielectric slab of thickness 0.9375 cm is equal to

$$d = \frac{0.9375\sqrt{2.56}\,\lambda_2}{30 \times 10^9 / 5 \times 10^9} = 0.25\lambda_2 \Rightarrow 2\beta_2 d = \frac{4\pi}{\lambda_2}\left(\frac{\lambda_2}{4}\right) = \pi$$

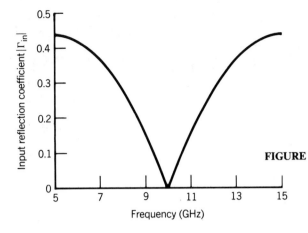

FIGURE 5-18 Input reflection coefficient, as a function of frequency, for wave propagation through a dielectric slab.

and at 15 GHz it is equal to

$$d = \frac{0.9375\sqrt{2.56}\,\lambda_2}{30 \times 10^9/15 \times 10^9} = 0.75\lambda_2 \Rightarrow 2\beta_2 d = \frac{4\pi}{\lambda_2}\left(\frac{3\lambda_2}{4}\right) = 3\pi$$

Since

$$\Gamma_{12} = -\Gamma_{23} = \frac{\eta_2 - \eta_1}{\eta_2 + \eta_1} = \frac{\eta_2/\eta_1 - 1}{\eta_2/\eta_1 + 1} = \frac{1 - \sqrt{\varepsilon_r}}{1 + \sqrt{\varepsilon_r}} = -\frac{0.6}{2.6} = -0.231$$

at $f = 5$ and 15 GHz the input reflection coefficient of (5-70) achieves the maximum magnitude of

$$|\Gamma_{in}(z = d^-)| = \left|\frac{-0.231 - 0.231}{1 - (-0.231)(0.231)}\right| = \frac{2(0.231)}{1 + (0.231)^2} = 0.438$$

A complete plot of $|\Gamma_{in}(z = d^-)|$ for 5 GHz $\leq f \leq$ 15 GHz is shown in the Figure 5-18.

Using the approximate form of (5-70), the magnitude of the input reflection coefficient is equal to

$$|\Gamma_{in}(z = d^-)|_{\substack{f=5,\\15\text{ GHz}}} \simeq |-0.231 - (0.231)| = 0.462$$

The percent error of this is

$$\text{percent error} = \left(\frac{-0.438 + 0.462}{0.438}\right) \times 100 = 5.48$$

Example 5-10. A uniform plane wave is incident normally upon a dielectric slab whose thickness at $f_0 = 10$ GHz is $\lambda_{2_0}/4$ where λ_{2_0} is the wavelength in the dielectric slab. The slab is bounded on the left side by air and on the right side by a semi-infinite medium of dielectric constant $\varepsilon_{r3} = 4$.

1. Determine the intrinsic impedance η_2 and dielectric constant ε_{r2} of the sandwiched slab so that the input reflection coefficient at $f_0 = 10$ GHz is zero.

2. Plot the magnitude response of the input reflection coefficient for $0 \leq f \leq 20$ GHz when the intrinsic impedance and physical thickness of the slab are those found in part 1.
3. Using the ray-tracing model of Figure 5-17b, at $f_0 = 10$ GHz determine the first and next two higher-order terms that contribute to the overall input reflection coefficient. What is the input reflection coefficient using these three terms?

Solution.

1. In order for the input reflection coefficient to vanish, the magnitude of (5-70) must be equal to zero, that is

$$|\Gamma_{12} + \Gamma_{23} e^{-j2\beta_2 d}| = 0$$

Since at $f_0 = 10$ GHz, $d = \lambda_{2_0}/4$, then

$$2\beta_2 d|_{f=10 \text{ GHz}} = 2\left(\frac{2\pi}{\lambda_{2_0}}\right)\left(\frac{\lambda_{2_0}}{4}\right) = \pi$$

Also

$$\Gamma_{12} = \frac{\eta_2 - \eta_1}{\eta_2 + \eta_1}$$

and

$$\Gamma_{23} = \frac{\eta_3 - \eta_2}{\eta_3 + \eta_2}$$

Thus

$$|\Gamma_{12} + \Gamma_{23} e^{-j2\beta_2 d}|_{\substack{d=\lambda_{2_0}/4 \\ f=10 \text{ GHz}}} = \left|\frac{\eta_2 - \eta_1}{\eta_2 + \eta_1} - \frac{\eta_3 - \eta_2}{\eta_3 + \eta_2}\right|$$

$$= \left|\frac{(\eta_2 - \eta_1)(\eta_3 + \eta_2) - (\eta_3 - \eta_2)(\eta_2 + \eta_1)}{(\eta_2 + \eta_1)(\eta_3 + \eta_2)}\right| = 0$$

or

$$2|\eta_2^2 - \eta_1 \eta_3| = 0 \Rightarrow \eta_2 = \sqrt{\eta_1 \eta_3}$$

Since $\eta_1 = \sqrt{\frac{\mu_0}{\varepsilon_0}} = 377$ ohms and $\eta_3 = \sqrt{\frac{\mu_0}{4\varepsilon_0}} = \frac{1}{2}\eta_1 = 188.5$ ohms then

$$\eta_2 = \sqrt{\eta_1 \eta_3} = \frac{\eta_1}{\sqrt{2}} = 0.707 \eta_1 = 0.707(377) = 266.5 \text{ ohms}$$

The dielectric constant of the slab must be equal to

$$\varepsilon_{r2} = 2$$

whereas the physical thickness of the dielectric is

$$d = \frac{\lambda_{2_0}}{4} = \frac{30 \times 10^9}{4(10 \times 10^9)\sqrt{2}} = 0.53 \text{ cm}$$

It is apparent then that whenever the dielectric is bounded by two semi-infinite media and its thickness is a quarter of a wavelength in the dielectric, its intrinsic impedance must always be equal to the square root of the product of the intrinsic impedances of the two media on each of its sides in order for the input reflection coefficient to vanish. This is referred to as the quarter-wavelength transformer which is so popular in transmission line design.

2. Since at $f_0 = 10$ GHz, $d = \lambda_{2_0}/4 = 0.53$ cm, then at $0 \leq f \leq 20$ GHz

$$2\beta_2 d = 2\left(\frac{2\pi}{\lambda_2}\right)\left(\frac{\lambda_{2_0}}{4}\right) = \pi\left(\frac{f}{f_0}\right)$$

also

$$\Gamma_{12} = \frac{\eta_2 - \eta_1}{\eta_2 + \eta_1} = \frac{\eta_2/\eta_1 - 1}{\eta_2/\eta_1 + 1} = \frac{1 - \sqrt{2}}{1 + \sqrt{2}}$$

$$\Gamma_{23} = \frac{\eta_3 - \eta_2}{\eta_3 + \eta_2} = \frac{\eta_3/\eta_2 - 1}{\eta_3/\eta_2 + 1} = \frac{1 - \sqrt{2}}{1 + \sqrt{2}} = \Gamma_{12}$$

Therefore the magnitude of the input reflection coefficient of (5-70) can be written now as

$$|\Gamma_{in}(z = -d^-)| = \left|\frac{\Gamma_{12}(1 + e^{-j\pi f/f_0})}{1 + (\Gamma_{12})^2 e^{-j\pi f/f_0}}\right|$$

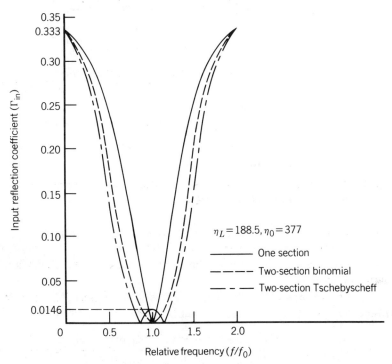

FIGURE 5-19 Responses of single-section, two-section binomial, and two-section Tschebyscheff quarter-wavelength transformers. (*Source:* C. A. Balanis, *Antenna Theory: Analysis and Design.* Copyright © 1982, John Wiley & Sons, Inc. Reprinted by permission of John Wiley & Sons, Inc.)

228 REFLECTION AND TRANSMISSION

whose maximum value, which occurs when $f = 0$ and $2f_0 = 20$ GHz, is approximately equal to

$$|\Gamma_{in}(z = -d^-)|_{max} = \frac{2|\Gamma_{12}|}{(1 + |\Gamma_{12}|^2)} = |\Gamma_{13}| = \left|\frac{\eta_3 - \eta_1}{\eta_3 + \eta_1}\right|$$

$$= 0.333 \simeq 2|\Gamma_{12}| = 0.3431$$

A complete plot of $|\Gamma_{in}(z = -d^-)|_{d=\lambda_{2_0}/4}$ when $0 \leq f \leq 20$ GHz is shown in the Figure 5-19.

It is interesting to note that the magnitude of the input reflection coefficient monotonically decreases from $f = 0$ to f_0, and it monotonically increases from f_0 to $2f_0$. It can also be noted that the bandwidth of the response curve near f_0 is very small, and any deviations of the frequency from f_0 will cause the reflection coefficient to rise sharply.

3. According to Figure 5-17b, the first-order term of the input reflection coefficient is

$$\Gamma_{12} = \frac{\eta_2 - \eta_1}{\eta_2 + \eta_1} = \frac{266.5 - 377}{266.5 + 377} = -0.1717$$

The next two higher terms are equal to

$$T_{12}\Gamma_{23}T_{21}e^{-j2\beta_2 d} = \frac{2\eta_1}{\eta_1 + \eta_2}\left(\frac{\eta_3 - \eta_2}{\eta_3 + \eta_2}\right)\left(\frac{2\eta_2}{\eta_1 + \eta_2}\right)e^{-j\pi}$$

$$= -\frac{2(377)}{377 + 266.5}\left(\frac{188.5 - 266.5}{188.5 + 266.5}\right)\frac{2(266.5)}{377 + 266.5} = +0.1664$$

$$T_{12}\Gamma_{21}\Gamma_{23}^2 T_{21}e^{-j4\beta_2 d} = \frac{2\eta_1}{\eta_1 + \eta_2}\left(\frac{\eta_1 - \eta_2}{\eta_1 + \eta_2}\right)\left(\frac{\eta_3 - \eta_2}{\eta_3 + \eta_2}\right)^2\left(\frac{2\eta_2}{\eta_1 + \eta_2}\right)e^{-j2\pi}$$

$$= \frac{2(377)}{377 + 266.5}\left(\frac{377 - 266.5}{377 + 266.5}\right)\left(\frac{188.5 - 266.5}{188.5 + 266.5}\right)^2\frac{2(266.5)}{377 + 266.5}$$

$$= 0.0049$$

$$\Gamma_{in} \simeq \Gamma_{12} + T_{12}\Gamma_{23}T_{21}e^{-j2\beta_2 d} + T_{12}\Gamma_{21}\Gamma_{23}^2 T_{21}e^{-j4\beta_2 d}$$

$$= -0.1717 + 0.1664 + 0.0049$$

$$\simeq -4 \times 10^{-4} \simeq 0$$

Thus the first three terms, or even the first two terms, provide an excellent approximation to the exact value of zero.

The bandwidth of the response curve can be increased by flattening the curve near f_0. This can be accomplished by increasing the number of layers bounded between the two semi-infinite media. The analysis of such a configuration will be discussed in Section 5.5.2.

If the three media of Figure 5-17 are lossy, then it can be shown that the overall reflection and transmission coefficients can be written as [3]

$$\Gamma_{in} = \frac{E^r}{E^i} = \frac{(1 - Z_{12})(1 + Z_{23}) + (1 + Z_{12})(1 - Z_{23})e^{-2\gamma_2 d}}{(1 + Z_{12})(1 + Z_{23}) + (1 - Z_{12})(1 - Z_{23})e^{-2\gamma_2 d}} \quad (5\text{-}71a)$$

$$T = \frac{E^t}{E^i} = \frac{4}{(1 - Z_{12})(1 - Z_{23})e^{-\gamma_2 d} + (1 + Z_{12})(1 + Z_{23})e^{\gamma_2 d}} \quad (5\text{-}71b)$$

where

$$Z_{ij} = \frac{\mu_i \gamma_j}{\mu_j \gamma_i} \qquad i, j = 1, 2, 3 \tag{5-71c}$$

$$\gamma_k = \pm\sqrt{j\omega\mu_k(\sigma_k + j\omega\varepsilon_k)} \tag{5-71d}$$

The preceding equations are valid for lossless, lossy, or any combination of lossless and lossy media.

5.5.2 Reflection Coefficient of Multiple Layers

The results of Example 5-10 indicate that for normal wave incidence the response of a single dielectric layer sandwiched between two semi-infinite media did not exhibit very broad characteristics around the center frequency f_0, and its overall response was very sensitive to frequency changes. The characteristics of such a response are very similar to the bandstop characteristics of a single section filter or single section quarter-wavelength impedance transformer. To increase the bandwidth of the system under normal wave incidence multiple layers of dielectric slabs, each with different dielectric constant, must be inserted between the two semi-infinite media. Multiple section dielectric layers can be used to design dielectric filters [9]. Coating radar targets with multilayer slabs can also be used to reduce or enhance their scattering characteristics.

When N layers, each with its own thickness and constitutive parameters, are sandwiched between two semi-infinite media as shown in Figure 5-20 the analysis for the overall reflection and transmission coefficients is quite cumbersome, although it is straightforward. However, an approximate form of the input reflection coefficient for the entire system under normal wave incidence can be obtained by utilizing the approximation first introduced to represent (5-70). With this in mind,

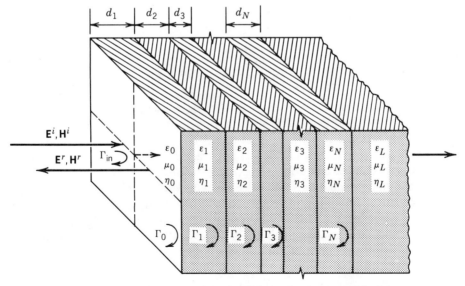

FIGURE 5-20 Normal wave propagation through N layers sandwiched between two media.

the input reflection coefficient under normal wave incidence for the system of Figure 5-20 referenced at the boundary of the leading interface can be written approximately as [1, 8]

$$\Gamma_{in} \simeq \Gamma_0 + \Gamma_1 e^{-j2\beta_1 d_1} + \Gamma_2 e^{-j2(\beta_1 d_1 + \beta_2 d_2)} + \cdots + \Gamma_N e^{-j2(\beta_1 d_1 + \beta_2 d_2 + \cdots + \beta_N d_N)} \quad (5\text{-}72)$$

where

$$\Gamma_0 = \frac{\eta_1 - \eta_0}{\eta_1 + \eta_0} \quad (5\text{-}72a)$$

$$\Gamma_1 = \frac{\eta_2 - \eta_1}{\eta_2 + \eta_1} \quad (5\text{-}72b)$$

$$\Gamma_2 = \frac{\eta_3 - \eta_2}{\eta_3 + \eta_2} \quad (5\text{-}72c)$$

$$\vdots$$

$$\Gamma_N = \frac{\eta_L - \eta_N}{\eta_L + \eta_N} \quad (5\text{-}72d)$$

Expression 5-72 is accurate provided that at each boundary the intrinsic reflection coefficients of (5-72a) through (5-72d) are small in comparison to unity.

A. QUARTER-WAVELENGTH TRANSFORMER

Example 5-10 demonstrated that when a lossless dielectric slab of thickness $\lambda_{2_0}/4$ at a frequency f_0 is sandwiched between two lossless semi-infinite dielectric media, the input reflection coefficient at f_0 is zero provided its intrinsic impedance η_2 is equal to

$$\eta_1 = \sqrt{\eta_0 \eta_L} \quad (5\text{-}73)$$

where η_1 = intrinsic impedance of dielectric slab
η_0 = intrinsic impedance of the input semi-infinite medium
η_L = intrinsic impedance of the load semi-infinite medium.

However as was illustrated in Figure 5-19 the response of the input reflection coefficient as a function of frequency was not very broad near the center frequency f_0.

Matchings that are less sensitive to frequency variations and that provide broader bandwidths require multiple $\lambda/4$ sections. In fact the number of sections and the intrinsic impedance of each section can be designed so that the reflection coefficient follows, within the desired frequency bandwidth, prescribed variations that are symmetrical about the center frequency. This design assumes that the semi-infinite media and the dielectric slabs are all lossless so that their intrinsic impedances are all real. The discussion that follows parallels that of [1] and [8].

Referring to Figure 5-20, the total input reflection coefficient Γ_{in} for an N-section quarter-wavelength transformer with $\eta_L > \eta_0$ can be written using an extension of the approximation used to represent (5-70) as [1, 8]

$$\Gamma_{in}(f) = \Gamma_0 + \Gamma_1 e^{-j2\theta} + \Gamma_2 e^{-j4\theta} + \cdots + \Gamma_N e^{-j2N\theta} = \sum_{n=0}^{N} \Gamma_n e^{-j2n\theta} \quad (5\text{-}74)$$

where Γ_n and θ are represented, respectively, by

$$\Gamma_n = \frac{\eta_{n+1} - \eta_n}{\eta_{n+1} + \eta_n} \tag{5-74a}$$

$$\theta = \beta_n d_n = \frac{2\pi}{\lambda_n}\left(\frac{\lambda_{n_0}}{4}\right) = \frac{\pi}{2}\left(\frac{f}{f_0}\right) \tag{5-74b}$$

In (5-74) Γ_n represents the reflection coefficient at the junction of two infinite lines that have intrinsic impedances η_n and η_{n+1}, f_0 represents the designed center frequency, and f represents the operating frequency. Equation 5-74 is valid provided the Γ_n's at each junction are small (the requirements will be met if $\eta_L \simeq \eta_0$). For lossless dielectrics the η_n's and Γ_n's will all be real.

For a symmetrical transformer ($\Gamma_0 = \Gamma_N$, $\Gamma_1 = \Gamma_{N-1}$, etc.) (5-74) reduces to

$$\Gamma_{in}(f) \simeq 2e^{-jN\theta}[\Gamma_0 \cos N\theta + \Gamma_1 \cos(N-2)\theta + \Gamma_2 \cos(N-4)\theta + \cdots] \tag{5-75}$$

The last term in (5-75) should be

$$\Gamma_{[(N-1)/2]} \cos \theta \qquad \text{for } N = \text{odd integer} \tag{5-75a}$$

$$\tfrac{1}{2}\Gamma_{(N/2)} \qquad \text{for } N = \text{even integer} \tag{5-75b}$$

B. BINOMIAL (MAXIMALLY FLAT) DESIGN

One technique, used to design an N-section $\lambda/4$ transformer, requires that the input reflection coefficient (5-74) have maximally flat passband characteristics. For this method, the junction reflection coefficients (Γ_n's) are derived using the binomial expansion and we can equate (5-74) to [1, 8]

$$\Gamma_{in}(f) = \sum_{n=0}^{N} \Gamma_n e^{-j2n\theta} = e^{-jN\theta}\frac{\eta_L - \eta_0}{\eta_L + \eta_0}\cos^N(\theta)$$

$$= 2^{-N}\frac{\eta_L - \eta_0}{\eta_L + \eta_0}\sum_{n=0}^{N} C_n^N e^{-j2n\theta} \tag{5-76}$$

where

$$C_n^N = \frac{N!}{(N-n)!n!} \qquad n = 0, 1, 2, \ldots, N \tag{5-76a}$$

From (5-76)

$$\Gamma_n = 2^{-N}\frac{\eta_L - \eta_0}{\eta_L + \eta_0}C_n^N \tag{5-77}$$

For this type of design, the fractional bandwidth $\Delta f/f_0$ is given by

$$\frac{\Delta f}{f_0} = 2\frac{f_0 - f_m}{f_0} = 2\left(1 - \frac{f_m}{f_0}\right) = 2\left(1 - \frac{2}{\pi}\theta_m\right) \tag{5-78}$$

Since

$$\theta_m = \frac{2\pi}{\lambda_m}\left(\frac{\lambda_0}{4}\right) = \frac{\pi}{2}\left(\frac{f_m}{f_0}\right) \tag{5-79}$$

232 REFLECTION AND TRANSMISSION

(5-78) reduces using (5-76) to

$$\frac{\Delta f}{f_0} = 2 - \frac{4}{\pi}\cos^{-1}\left|\frac{\Gamma_m}{(\eta_L - \eta_0)/(\eta_L + \eta_0)}\right|^{1/N} \quad (5\text{-}80)$$

where Γ_m is the magnitude of the maximum value of reflection coefficient that can be tolerated within the bandwidth.

The usual design procedure is to specify

1. the load intrinsic impedance η_L
2. the input intrinsic impedance η_0
3. the number of sections N
4. the maximum tolerable reflection coefficient Γ_m (or fractional bandwidth $\Delta f/f_0$)

and to find

1. the intrinsic impedance of each section
2. the fractional bandwidth (or maximum tolerable reflection coefficient Γ_m)

To illustrate the principle let us consider an example.

Example 5-11. Two lossless dielectric slabs each of thickness $\lambda_0/4$ at a center frequency $f_0 = 10$ GHz are sandwiched between air to the left and a lossless semi-infinite medium of dielectric constant $\varepsilon_L = 4$ to the right. Assuming a fractional bandwidth of 0.375 and a binomial design:

1. Determine the intrinsic impedances, dielectric constants, and thicknesses of the sandwiched slabs so that the input reflection coefficient at $f_0 = 10$ GHz is zero.
2. Determine the maximum reflection coefficient and SWR within the fractional bandwidth.
3. Plot the response of the input reflection coefficient for $0 \le f \le 20$ GHz when the intrinsic impedances and physical thicknesses of the slabs are those found in part 1. Compare the response of the two-section binomial design with that of the single section of Example 5-10.

Solution.

1. Using (5-76a) and (5-77)

$$\Gamma_n = 2^{-N}\frac{\eta_L - \eta_0}{\eta_L + \eta_0}C_n^N = 2^{-N}\frac{\eta_L - \eta_0}{\eta_L + \eta_0}\frac{N!}{(N-n)!n!}$$

Since the input dielectric is air and the load dielectric has a dielectric constant $\varepsilon_L = 4$, then

$$\eta_0 = 377$$

$$\eta_L = \sqrt{\frac{\mu_0}{\varepsilon_L \varepsilon_0}} = \frac{377}{2} = 188.5$$

Therefore

$$n=0: \Gamma_0 = \frac{\eta_1 - \eta_0}{\eta_1 + \eta_0} = 2^{-2}\left(\frac{188.5 - 377}{188.5 + 377}\right)\frac{2!}{2!0!} = -\frac{1}{12}$$

$$\Rightarrow \eta_1 = \eta_0\left(\frac{1 - 1/12}{1 + 1/12}\right) = 0.846\eta_0 = 318.94 \text{ ohms}$$

$$\Rightarrow \varepsilon_{r_1} = 1.40 \qquad d_1 = \frac{\lambda_{1_0}}{4} = 0.634 \text{ cm}$$

$$n=1: \Gamma_1 = \frac{\eta_2 - \eta_1}{\eta_2 + \eta_1} = 2^{-2}\left(\frac{188.5 - 377}{188.5 + 377}\right)\frac{2!}{1!1!} = -\frac{1}{6}$$

$$\Rightarrow \eta_2 = \eta_1\left(\frac{1 - 1/6}{1 + 1/6}\right) = 0.714\eta_1 = 227.72 \text{ ohms}$$

$$\Rightarrow \varepsilon_{r_2} = 2.74 \qquad d_2 = \lambda_{2_0}/4 = 0.453 \text{ cm}$$

2. For a fractional bandwidth of 0.375, the magnitude of the maximum reflection coefficient Γ_m is obtained using (5-80) or

$$\frac{\Delta f}{f_0} = 0.375 = 2 - \frac{4}{\pi}\cos^{-1}\left|\frac{\Gamma_m}{(\eta_L - \eta_0)/(\eta_L + \eta_0)}\right|^{1/2}$$

which for $\eta_L = 188.5$ and $\eta_0 = 377$ leads to

$$\Gamma_m = 0.028$$

The maximum standing wave ratio is

$$\text{SWR}_m = \frac{1 + \Gamma_m}{1 - \Gamma_m} = \frac{1 + 0.028}{1 - 0.028} = 1.058$$

3. The magnitude of the input reflection coefficient is given by (5-76) as

$$|\Gamma_{in}| = \left|\frac{\eta_L - \eta_0}{\eta_L + \eta_0}\right|\cos^2\theta = \frac{1}{3}\cos^2\theta = \frac{1}{3}\cos^2\left[\frac{\pi}{2}\left(\frac{f}{f_0}\right)\right]$$

which is shown plotted in Figure 5-19 where it is also compared with that of the one- and two-section Tschebyscheff design to be discussed next.

C. TSCHEBYSCHEFF (EQUAL-RIPPLE) DESIGN

The reflection coefficient can be made to vary within the bandwidth in an oscillatory manner and have equal-ripple characteristics [10–12]. This can be accomplished by making Γ_{in} vary similarly as a Tschebyscheff (Chebyshev) polynomial. For the Tschebyscheff design, the equation that corresponds to (5-76) is [1, 8]

$$\boxed{\Gamma_{in}(f) = e^{-jN\theta}\frac{\eta_L - \eta_0}{\eta_L + \eta_0}\frac{T_N(\sec\theta_m \cos\theta)}{T_N(\sec\theta_m)}} \quad (5\text{-}81)$$

where $T_N(z)$ is the Tschebyscheff polynomial of order N.

The maximum allowable reflection coefficient occurs at the edges of the passband where $\theta = \theta_m$ and $|T_N(\sec\theta_m \cos\theta)|_{\theta=\theta_m} = 1$. Thus

$$\Gamma_m = \left| \frac{\eta_L - \eta_0}{\eta_L + \eta_0} \frac{1}{T_N(\sec\theta_m)} \right| \tag{5-82}$$

The first few Tschebyscheff polynomials can be found in [1, 8]. For $z = \sec\theta_m \cos\theta$, the first three polynomials reduce to

$$T_1(\sec\theta_m \cos\theta) = \sec\theta_m \cos\theta$$

$$T_2(\sec\theta_m \cos\theta) = 2(\sec\theta_m \cos\theta)^2 - 1 = \sec^2\theta_m \cos 2\theta + (\sec^2\theta_m - 1)$$

$$T_3(\sec\theta_m \cos\theta) = 4(\sec\theta_m \cos\theta)^3 - 3(\sec\theta_m \cos\theta)$$

$$= \sec^3\theta_m \cos 3\theta + 3(\sec^3\theta_m - \sec\theta_m)\cos\theta \tag{5-83}$$

The remaining details of the analysis are found in [1, 8].

The design of Example 5-11 using a Tschebyscheff transformer is assigned as an exercise to the reader. However its response is shown plotted in Figure 5-19 for comparison.

In general, multiple sections (either binomial or Tschebyscheff) provide greater bandwidths than a single section. As the number of sections increases, the bandwidth also increases. The advantage of the binomial design is that the reflection coefficient values within the bandwidth monotonically decrease from both ends toward the center. Thus the values are always smaller than an acceptable and designed value that occurs at the "skirts" of the bandwidth. For the Tschebyscheff design, the reflection coefficient values within the designed bandwidth are equal to or smaller than an acceptable and designed value. The number of times the reflection coefficient reaches the maximum value within the bandwidth is determined by the number of sections. In fact, for an even number of sections the

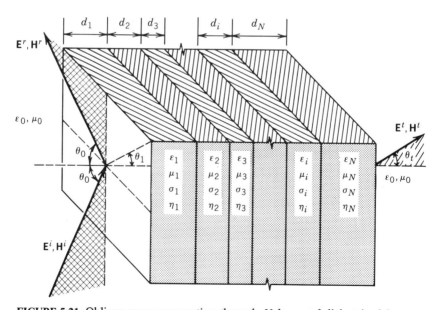

FIGURE 5-21 Oblique wave propagation through N layers of dielectric slabs.

reflection coefficient at the designed center frequency is equal to the maximum allowable value whereas for an odd number of sections it is zero. For a maximum tolerable reflection coefficient, the N-section Tschebyscheff transformer provides a larger bandwidth than a corresponding N-section binomial design, or for a given bandwidth the maximum tolerable reflection coefficient is smaller for a Tschebyscheff design.

D. OBLIQUE-WAVE INCIDENCE

A more general formulation of the reflection and transmission coefficients can be developed by considering the geometry of Figure 5-21 where a uniform plane wave is incident at an oblique angle upon N layers of planar slabs that are bordered on either side by free space. This type of a geometry can be used to approximate the configuration of a radome whose radius of curvature is large in comparison to the wavelength. It can be shown that the overall reflection and transmission coefficients for perpendicular (horizontal) and parallel (vertical) polarizations can be written as [3]

<u>Perpendicular (Horizontal)</u>

$$\Gamma_\perp = \frac{E_\perp^r}{E_\perp^i} = \frac{B_0}{A_0} \tag{5-84a}$$

$$T_\perp = \frac{E_\perp^t}{E_\perp^i} = \frac{1}{A_0} \tag{5-84b}$$

<u>Parallel (Vertical)</u>

$$\Gamma_\| = \frac{E_\|^r}{E_\|^i} = \frac{C_0}{D_0} \tag{5-85a}$$

$$T_\| = \frac{E_\|^t}{E_\|^i} = \frac{1}{D_0} \tag{5-85b}$$

The functions A_0, B_0, C_0, and D_0 are found using the recursive formulas

$$A_j = \frac{e^{\psi_i}}{2}\left[A_{j+1}(1 + Y_{j+1}) + B_{j+1}(1 - Y_{j+1})\right] \tag{5-86a}$$

$$B_j = \frac{e^{-\psi_i}}{2}\left[A_{j+1}(1 - Y_{j+1}) + B_{j+1}(1 + Y_{j+1})\right] \tag{5-86b}$$

$$C_j = \frac{e^{\psi_j}}{2}\left[C_{j+1}(1 + Z_{j+1}) + D_{j+1}(1 - Z_{j+1})\right] \tag{5-86c}$$

$$D_j = \frac{e^{-\psi_j}}{2}\left[C_{j+1}(1 - Z_{j+1}) + D_{j+1}(1 + Z_{j+1})\right] \tag{5-86d}$$

where

$$A_{N+1} = C_{N+1} = 1 \qquad (5\text{-}86\text{e})$$

$$B_{N+1} = G_{N+1} = 0 \qquad (5\text{-}86\text{f})$$

$$Y_{j+1} = \frac{\cos\theta_{j+1}}{\cos\theta_j} \sqrt{\frac{\varepsilon_{j+1}(1 - j\tan\delta_{j+1})}{\varepsilon_j(1 - j\tan\delta_j)}} \qquad (5\text{-}86\text{g})$$

$$Z_{j+1} = \frac{\cos\theta_{j+1}}{\cos\theta_j} \sqrt{\frac{\varepsilon_j(1 - j\tan\delta_j)}{\varepsilon_{j+1}(1 - j\tan\delta_{j+1})}} \qquad (5\text{-}86\text{h})$$

$$\psi_j = d_j \gamma_j \cos\theta_j \qquad (5\text{-}86\text{i})$$

$$\gamma_j = \pm\sqrt{j\omega\mu_j(\sigma_j + j\omega\varepsilon_j)} \qquad (5\text{-}86\text{j})$$

$$\theta_j = \text{complex angle of refraction in the } j\text{th layer} \qquad (5\text{-}86\text{k})$$

5.6 POLARIZATION CHARACTERISTICS ON REFLECTION

When linearly polarized fields are reflected from smooth flat surfaces, the reflected fields maintain their linear polarization characteristics. However, when the reflected surfaces are curved or rough, a linearly polarized component orthogonal to that of the incident field is introduced during reflection. Therefore the total field exhibits two components: one with the same polarization as the incident field (main polarization) and one orthogonal to it (cross polarization). During this process, the field is depolarized owing to reflection.

Circularly polarized fields in free space incident upon flat surfaces

1. Maintain their circular polarization but reverse their sense of rotation when the reflecting surface is perfectly conducting.
2. Are transformed to elliptically polarized fields of opposite sense of rotation when the flat surface is a lossless dielectric and the angle of incidence is smaller than the Brewster angle.

Similarly, elliptically polarized fields in free space upon reflection from flat surfaces

1. Maintain their elliptical polarization and magnitude of axial ratio but reverse their sense of rotation when reflected from a perfectly conducting surface.
2. Maintain their elliptical polarization but change their axial ratio and sense of rotation when the reflecting surface is a dielectric and the angle of incidence is smaller than the Brewster angle.

To analyze the polarization properties of a wave when it is reflected by a surface, let us assume that an elliptically polarized wave is obliquely incident upon a flat surface of infinite extent as shown in Figure 5-22 [7]. Using the localized coordinate system (x', y, z') of Figure 5-22, the incident electric field components

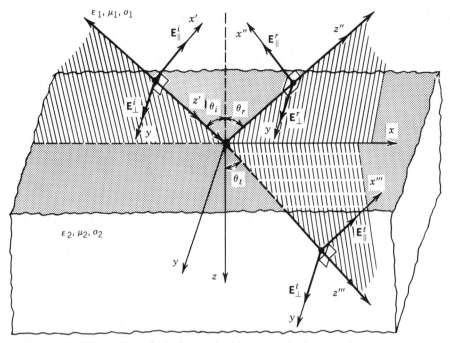

FIGURE 5-22 Elliptically polarized wave incident on a flat lossy surface.

can be written as

$$\mathbf{E}_\parallel^i = \hat{a}_{x'} E_\parallel^i e^{-j\boldsymbol{\beta}^i \cdot \mathbf{r}} = \hat{a}_{x'} E_\parallel^0 e^{-j\boldsymbol{\beta}^i \cdot \mathbf{r}} \quad (5\text{-}87a)$$

$$\mathbf{E}_\perp^i = \hat{a}_y E_\perp^i e^{-j\boldsymbol{\beta}^i \cdot \mathbf{r}} = \hat{a}_y E_\perp^0 e^{-j(\boldsymbol{\beta}^i \cdot \mathbf{r} - \phi_\perp^i)} \quad (5\text{-}87b)$$

where $E_\parallel^0$ and $E_\perp^0$ are assumed to be real.

For this set of field components the Poincaré sphere angles (4-58a) through (4-59b) can be written (assuming that the ratio in (4-58a), selected here to demonstrate the procedure, satisfies the angular limits of all the Poincaré sphere angles) as

$$\gamma^i = \tan^{-1}\left(\frac{E_\perp^0}{E_\parallel^0}\right) \quad (5\text{-}88a)$$

$$\delta^i = \phi_\perp^i - \phi_\parallel^i = \phi_\perp^i \quad (5\text{-}88b)$$

$$\varepsilon^i = \cot^{-1}(AR^i) \quad (5\text{-}88c)$$

$$\tau^i = \text{tilt angle of incident wave} \quad (5\text{-}88d)$$

where δ^i is the phase angle by which the perpendicular component of the incident field leads the parallel component. It is assumed that (AR^i) is positive for left-hand and negative for right-hand polarized fields. These two sets of angles are related to each other by (4-60a) through (4-61b), or

$$\cos(2\gamma^i) = \cos(2\varepsilon^i)\cos(2\tau^i) \quad (5\text{-}89a)$$

$$\tan(\delta^i) = \frac{\tan(2\varepsilon^i)}{\sin(2\tau^i)} \quad (5\text{-}89b)$$

or

$$\sin(2\varepsilon^i) = \sin(2\gamma^i)\sin(\delta^i) \qquad (5\text{-}89\text{c})$$
$$\tan(2\tau^i) = \tan(2\gamma^i)\cos(\delta^i) \qquad (5\text{-}89\text{d})$$

In a similar manner the reflected fields of the elliptically polarized wave can be written according to the localized coordinate system (x'', y, z'') of Figure 5-22 as

$$\mathbf{E}_\|^r = \hat{a}_{x''} E_\|^r e^{-j\boldsymbol{\beta}^r\cdot\mathbf{r}} = -\hat{a}_{x''}\Gamma_\|^b E_\|^0 e^{-j\boldsymbol{\beta}^r\cdot\mathbf{r}} = \hat{a}_{x''}|\Gamma_\|^b|E_\|^0 e^{-j(\boldsymbol{\beta}^r\cdot\mathbf{r}-\pi-\zeta_\|^r)}$$
$$= \hat{a}_{x''}|\Gamma_\|^b|E_\|^0 e^{-j(\boldsymbol{\beta}^r\cdot\mathbf{r}-\phi_\|')} \qquad (5\text{-}90\text{a})$$

$$\mathbf{E}_\perp^r = \hat{a}_y E_\perp^r e^{-j\boldsymbol{\beta}^r\cdot\mathbf{r}} = \hat{a}_y \Gamma_\perp^b E_\perp^0 e^{-j(\boldsymbol{\beta}^r\cdot\mathbf{r}-\phi_\perp')} = \hat{a}_y|\Gamma_\perp^b|E_\perp^0 e^{-j(\boldsymbol{\beta}^r\cdot\mathbf{r}-\delta^i-\zeta_\perp^r)}$$
$$= \hat{a}_y|\Gamma_\perp^b|E_\perp^0 e^{-j(\boldsymbol{\beta}^r\cdot\mathbf{r}-\phi_\perp')} \qquad (5\text{-}90\text{b})$$

where $\zeta_\|^r$ and $\zeta_\perp^r$ are the phases of the reflection coefficients for parallel and perpendicular polarizations, respectively. The Poincaré sphere angles γ^r and δ^r of the reflected field can now be written by referring to (5-90a) and (5-90b) as

$$\boxed{\gamma^r = \tan^{-1}\left(\frac{|\mathbf{E}_\perp^r|}{|\mathbf{E}_\|^r|}\right) = \tan^{-1}\left(\frac{|\Gamma_\perp^b|E_\perp^0}{|\Gamma_\|^b|E_\|^0}\right) = \tan^{-1}\left(\frac{|\Gamma_\perp^b|}{|\Gamma_\|^b|}\tan\gamma^i\right)} \qquad (5\text{-}91\text{a})$$

$$\boxed{\delta^r = \phi_\perp' - \phi_\|' = \left(\delta^i + \zeta_\perp^r\right) - \left(\pi + \zeta_\|^r\right) = \left(\delta^i - \pi\right) + \left(\zeta_\perp^r - \zeta_\|^r\right)} \qquad (5\text{-}91\text{b})$$

where δ^r is the phase angle by which the perpendicular (y) component leads the parallel (x'') component of the reflected field. Using the angles γ^r and δ^r of (5-91a) and (5-91b), the corresponding Poincaré sphere angles ε^r, τ^r (tilt angle of ellipse) and axial ratio $(AR)^r$ of the reflected field can be found using the relations

$$\boxed{\sin(2\varepsilon^r) = \sin(2\gamma^r)\sin(\delta^r)} \qquad (5\text{-}92\text{a})$$

$$\boxed{\tan(2\tau^r) = \tan(2\gamma^r)\cos(\delta^r)} \qquad (5\text{-}92\text{b})$$

$$\boxed{(AR)^r = \cot(\varepsilon^r)} \qquad (5\text{-}92\text{c})$$

Following a similar procedure, the transmitted fields can be expressed as

$$\mathbf{E}_\|^t = \hat{a}_{x'''} E_\|^t e^{-j\boldsymbol{\beta}^t\cdot\mathbf{r}} = \hat{a}_{x'''} T_\|^b E_\|^0 e^{-j\boldsymbol{\beta}^t\cdot\mathbf{r}} = \hat{a}_{x'''}|T_\|^b|E_\|^0 e^{-j(\boldsymbol{\beta}^t\cdot\mathbf{r}-\xi_\|^t)}$$
$$= \hat{a}_{x'''}|T_\|^b|E_\|^0 e^{-j(\boldsymbol{\beta}^t\cdot\mathbf{r}-\phi_\|')} \qquad (5\text{-}93\text{a})$$

$$\mathbf{E}_\perp^t = \hat{a}_y E_\perp^t e^{-j\boldsymbol{\beta}^t\cdot\mathbf{r}} = \hat{a}_y T_\perp^b E_\perp^0 e^{-j(\boldsymbol{\beta}^t\cdot\mathbf{r}-\phi_\perp')} = \hat{a}_y|T_\perp^b|E_\perp^0 e^{-j(\boldsymbol{\beta}^t\cdot\mathbf{r}-\delta^i-\xi_\perp^t)}$$
$$= \hat{a}_y|T_\perp^b|E_\perp^0 e^{-j(\boldsymbol{\beta}^t\cdot\mathbf{r}-\phi_\perp')} \qquad (5\text{-}93\text{b})$$

where $\xi_\|^t$ and $\xi_\perp^t$ are the phases of the transmission coefficients for parallel and perpendicular polarizations, respectively. The Poincaré sphere angles δ^t and γ^t can now be written by referring to (5-93a) and (5-93b) as

$$\boxed{\gamma^t = \tan^{-1}\left(\frac{|\mathbf{E}_\perp^t|}{|\mathbf{E}_\|^t|}\right) = \tan^{-1}\left(\frac{|T_\perp^b|E_\perp^0}{|T_\|^b|E_\|^0}\right) = \tan^{-1}\left(\frac{|T_\perp^b|}{|T_\|^b|}\tan\gamma^i\right)} \qquad (5\text{-}94\text{a})$$

$$\boxed{\delta^t = \phi_\perp' - \phi_\|' = \left(\delta^i + \xi_\perp^t\right) - \xi_\|^t = \delta^i + \left(\xi_\perp^t - \xi_\|^t\right)} \qquad (5\text{-}94\text{b})$$

POLARIZATION CHARACTERISTICS ON REFLECTION 239

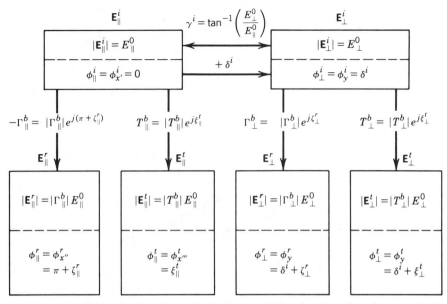

FIGURE 5-23 Block diagram for polarization analysis of reflected and transmitted waves.

where δ^t is the phase angle by which the perpendicular (y) component of the transmitted field leads the parallel (x''') component of the transmitted field. Using the angles γ^t and δ^t of (5-94a) and (5-94b), the corresponding Poincaré sphere angles ε^t, τ^t (tilt angle of ellipse) and axial ratio $(AR)^t$ of the transmitted field can be found using the relations

$$\sin(2\varepsilon^t) = \sin(2\gamma^t)\sin(\delta^t) \quad (5\text{-}95a)$$

$$\tan(2\tau^t) = \tan(2\gamma^t)\cos(\delta^t) \quad (5\text{-}95b)$$

$$(AR)^t = \cot(\varepsilon^t) \quad (5\text{-}95c)$$

The set of (5-90a) through (5-92c) and (5-93a) through (5-95c) can be used to find, respectively, the polarization of the reflected and transmitted fields once the polarization of the incident fields of (5-87a) through (5-88d) has been stated. A block diagram of the relations between the incident, reflected, and transmitted fields is shown in Figure 5-23. The parallel component of the incident field is taken as the reference for the phase of all of the other components.

Example 5-12. A left-hand (CCW) circularly polarized field traveling in free space at an angle of $\theta_i = 30°$ is incident on a flat perfect electric conductor of infinite extent. Find the polarization of the reflected wave.

Solution. A circularly polarized wave is made of two orthogonal linearly polarized components with a 90° phase difference between them. Therefore we can assume that these two orthogonal linearly polarized components represent the perpendicular and parallel polarizations. Since the reflecting surface is

240 REFLECTION AND TRANSMISSION

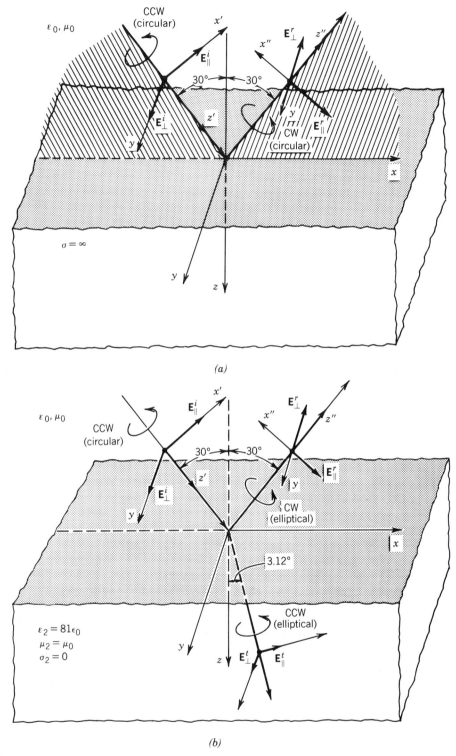

FIGURE 5-24 Circularly polarized wave incident upon flat surfaces with infinite and zero conductivities. (*a*) Infinite conductivity. (*b*) Lossless ocean.

POLARIZATION CHARACTERISTICS ON REFLECTION   241

perfectly conducting ($\eta_2 = 0$), the reflection coefficients of (5-17a) and (5-24c) reduce to

$$\Gamma_\perp^b = -1 = 1 \big/ \pi \Rightarrow |\Gamma_\perp^b| = 1 \qquad \zeta_\perp^r = \pi$$

$$\Gamma_\parallel^b = -1 = 1 \big/ \pi \Rightarrow |\Gamma_\parallel^b| = 1 \qquad \zeta_\parallel^r = \pi$$

Since the incident field is left-hand circularly polarized, then according to (5-87a) through (5-88b)

$$E_\parallel^0 = E_\perp^0$$

$$\delta^i = \phi_\perp^i = \frac{\pi}{2}$$

$$\gamma^i = \tan^{-1}\left(\frac{E_\perp^0}{E_\parallel^0}\right) = \frac{\pi}{4} \Rightarrow \tan \gamma^i = 1$$

Thus according to (5-91a) and (5-91b)

$$\gamma^r = \tan^{-1}\left(\frac{|\Gamma_\perp^b|}{|\Gamma_\parallel^b|} \tan \gamma^i\right) = \frac{\pi}{4}$$

$$\delta^r = \delta^i - \pi + (\zeta_\perp^r - \zeta_\parallel^r) = \frac{\pi}{2} - \pi + (\pi - \pi) = -\frac{\pi}{2}$$

On the Poincaré sphere of Figure 4-20 the angles $\gamma^r = \pi/4$ and $\delta^r = -\pi/2$ define the south pole which represents right-hand (CW) circular polarization. Therefore the reflected field is right-hand (CW) circularly polarized, and it is opposite in rotation to that of the incident field as shown in Figure 5-24a.

Example 5-13. A left-hand (CCW) circularly polarized field traveling in free space at an angle of $\theta_i = 30°$ is incident on a flat lossless ($\sigma_2 = 0$) ocean ($\varepsilon_2 = 81\varepsilon_0$, $\mu_2 = \mu_0$) of infinite extent. Find the polarization of the reflected and transmitted fields.

Solution. Since the incident field is left-hand circularly polarized, then according to (5-87a) through (5-88b)

$$E_\parallel^0 = E_\perp^0$$

$$\delta^i = \phi_\perp^i = \frac{\pi}{2}$$

$$\gamma^i = \tan^{-1}\left(\frac{E_\perp^0}{E_\parallel^0}\right) = \frac{\pi}{4} \Rightarrow \tan \gamma^i = 1$$

To find the polarization of the reflected field, we proceed as follows. Using (5-18a)

$$\Gamma_\perp^b = \frac{\cos(30°) - \sqrt{81}\sqrt{1 - \left(\frac{1}{81}\right)\sin^2(30°)}}{\cos(30°) + \sqrt{81}\sqrt{1 - \left(\frac{1}{81}\right)\sin^2(30°)}} = \frac{0.866 - 9\sqrt{1 - \frac{1}{81}\left(\frac{1}{4}\right)}}{0.866 + 9\sqrt{1 - \frac{1}{81}\left(\frac{1}{4}\right)}}$$

$$= \frac{0.866 - 8.986}{0.866 + 8.986}$$

$$\Gamma_\perp^b = -0.824 \Rightarrow |\Gamma_\perp^b| = 0.824 \qquad \zeta_\perp^r = \pi$$

Using (5-25a)

$$\Gamma_{\|}^{b} = \frac{-\cos(30°) + \sqrt{\frac{1}{81}}\sqrt{1-\left(\frac{1}{81}\right)\sin^{2}(30°)}}{\cos(30°) + \sqrt{\frac{1}{81}}\sqrt{1-\left(\frac{1}{81}\right)\sin^{2}(30°)}} = \frac{-0.866 + \frac{1}{9}\sqrt{1-\frac{1}{81}\left(\frac{1}{4}\right)}}{0.866 + \frac{1}{9}\sqrt{1-\frac{1}{81}\left(\frac{1}{4}\right)}}$$

$$= \frac{-0.866 + 0.111}{0.866 + 0.111}$$

$$\Gamma_{\|}^{b} = -0.773 \Rightarrow |\Gamma_{\|}^{b}| = 0.773 \qquad \zeta_{\|}^{r} = \pi$$

According to (5-91a) and (5-91b)

$$\gamma^{r} = \tan^{-1}\left(\frac{|\Gamma_{\perp}^{b}|}{|\Gamma_{\|}^{b}|}\tan\gamma^{i}\right) = \tan^{-1}\left(\frac{0.824}{0.773}\right) = 46.83° = 0.817 \text{ rad}$$

$$\delta^{r} = \delta^{i} - \pi + (\zeta_{\perp}^{r} - \zeta_{\|}^{r}) = \frac{\pi}{2} - \pi + (\pi - \pi) = -\frac{\pi}{2}$$

Using (5-92a) through (5-92c)

$$2\varepsilon^{r} = \sin^{-1}\left[\sin(2\gamma^{r})\sin(\delta^{r})\right]$$

$$= \sin^{-1}\left[\sin(93.66°)\sin\left(-\frac{\pi}{2}\right)\right] = -86.34°$$

$$\Rightarrow \varepsilon^{r} = -43.17°$$

$$2\tau^{r} = \tan^{-1}\left[\tan(2\gamma^{r})\cos(\delta^{r})\right]$$

$$= \tan^{-1}\left[\tan(93.66°)\cos\left(-\frac{\pi}{2}\right)\right] = 180°$$

$$\Rightarrow \tau^{r} = 90°$$

$$(AR)^{r} = \cot(\varepsilon^{r}) = \cot(-43.17°) = -1.066$$

On the Poincaré sphere of Figure 4-20 the angles $\gamma^{r} = 0.817$ and $\delta^{r} = -\pi/2$ locate a point on the lower hemisphere on the principal xz plane. Therefore the reflected field is right-hand (CW) elliptically polarized, and it has an opposite sense of rotation compared to the left-hand (CCW) circularly polarized incident field as shown in Figure 5-24b. Its axial ratio is -1.066.

To find the polarization of the transmitted field we proceed as follows. Using (5-18b)

$$T_{\perp}^{b} = \frac{2\cos(30°)}{\cos(30°) + \sqrt{81}\sqrt{1-\left(\frac{1}{81}\right)\sin^{2}(30°)}} = \frac{2(0.866)}{0.866 + 8.986}$$

$$= 0.1758 \Rightarrow |T_{\perp}^{b}| = 0.1758 \qquad \xi_{\perp}^{t} = 0$$

Using (5-25b)

$$T_{\|}^{b} = \frac{2\sqrt{\frac{1}{81}}\cos(30°)}{\cos(30°) + \sqrt{\frac{1}{81}}\sqrt{1-\left(\frac{1}{81}\right)\sin^{2}(30°)}} = \frac{2\left(\frac{1}{9}\right)0.866}{0.866 + 0.111}$$

$$= 0.197 \Rightarrow |T_{\|}^{b}| = 0.197 \qquad \xi_{\|}^{t} = 0$$

According to (5-94a) and (5-94b)

$$\gamma^t = \tan^{-1}\left(\frac{|T_\perp^b|}{|T_\parallel^b|}\tan\gamma^i\right) = \tan^{-1}\left(\frac{0.1758}{0.197}\right) = 41.75° = 0.729 \text{ rad}$$

$$\delta^t = \delta^i + (\xi_\perp^t - \xi_\parallel^t) = \frac{\pi}{2} + (0-0) = \frac{\pi}{2}$$

Using (5-95a) through (5-95c)

$$2\varepsilon^t = \sin^{-1}\left[\sin(2\gamma^t)\sin(\delta^t)\right] = \sin^{-1}\left[\sin(83.5°)\sin(90°)\right] = 83.5°$$

$$\Rightarrow \varepsilon^t = 41.75°$$

$$2\tau^t = \tan^{-1}\left[\tan(2\gamma^t)\cos(\delta^t)\right] = \tan^{-1}\left[\tan(83.5°)\cos(90°)\right] = 0$$

$$\Rightarrow \tau^t = 0°$$

$$(AR)^t = \cot(\varepsilon^t) = \cot(41.75°) = 1.12$$

On the Poincaré sphere of Figure 4-20 the angles $\gamma^t = 0.729$ and $\delta = \pi/2$ locate a point on the upper hemisphere on the principal xz plane. Therefore the transmitted field is left-hand (CCW) elliptically polarized, and it is of the same sense of rotation as the left-hand (CCW) circularly polarized incident field as shown in Figure 5-24b. Its axial ratio is 1.12.

REFERENCES

1. C. A. Balanis, *Antenna Theory: Analysis and Design*, Wiley, New York, 1982.
2. M. A. Plonus, *Applied Electromagnetics*, McGraw-Hill, New York, 1978.
3. D. T. Paris and F. K. Hurd, *Basic Electromagnetic Theory*, McGraw-Hill, New York, 1969.
4. R. B. Adler, L. J. Chu, and R. M. Fano, *Electromagnetic Energy Transmission and Radiation*, Chapters 7 and 8, Wiley, New York, 1960.
5. J. J. Holmes and C. A. Balanis, "Refraction of a uniform plane wave incident on a plane boundary between two lossy media," *IEEE Trans. Antennas Propagat.*, vol. AP-26, no. 5, pp. 738–741, September 1978.
6. R. D. Radcliff and C. A. Balanis, "Modified propagation constants for nonuniform plane wave transmission through conducting media," *IEEE Trans. Geoscience Remote Sensing*, vol. GE-20, no. 3, pp. 408–411, July 1982.
7. J. D. Kraus, *Electromagnetics*, Third Edition, McGraw-Hill, New York, 1984.
8. R. E. Collin, *Foundations for Microwave Engineering*, McGraw-Hill, New York, 1966.
9. R. E. Collin and J. Brown, "The design of quarter-wave matching layers for dielectric surfaces," *Proc. IEE*, vol. 103, Part C, pp. 153–158, March 1956.
10. S. B. Cohn, "Optimum design of stepped transmission line transformers," *IRE Trans. Microwave Theory Tech.*, vol. MTT-3, pp. 16–21, April 1955.
11. L. Young, "Optimum quarter-wave transformers," *IRE Trans. Microwave Theory Tech.*, vol. MTT-8, pp. 478–482, September 1960.
12. G. L. Matthaei, L. Young, and E. M. T. Jones, *Microwave Filters, Impedance-Matching Networks and Coupling Structures*, McGraw-Hill, New York, 1964.

PROBLEMS

5.1. A uniform plane wave traveling in a dielectric medium with $\varepsilon_r = 4$ and $\mu_r = 1$ is incident normally upon a free-space medium. If the incident electric field is given by

$$\mathbf{E}^i = \hat{a}_y 2 \times 10^{-3} e^{-j\beta z}$$

write:

(a) The corresponding incident magnetic field.
(b) The reflection and transmission coefficients.
(c) The reflected and transmitted electric and magnetic fields.
(d) The incident, reflected, and transmitted power densities.

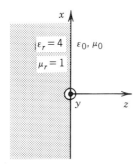

FIGURE P5-1

5.2. The dielectric constant of water is 81. Calculate the percentage of power density reflected and transmitted when a uniform plane wave traveling in air is incident normally upon a calm lake. Assume that the water in the lake is lossless.

5.3. A uniform plane wave propagating in a medium with relative permittivity of 4 is incident normally upon a dielectric medium with dielectric constant of 9. Assuming both media are nonferromagnetic and lossless, determine (a) the reflection and transmission coefficients and (b) the percentage of incident power density that is reflected and transmitted.

5.4. A vertical interface is formed by having free space to its left and a lossless dielectric medium to its right with $\varepsilon = 4\varepsilon_0$ and $\mu = \mu_0$, as shown in Figure P5-4. The incident electric field of a uniform plane wave traveling in the free-space medium and

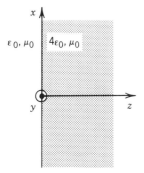

FIGURE P5-4

incident normally upon the interface has a value of 2×10^{-3} V/m right before it strikes the boundary. At a frequency of 3 GHz, find:

(a) The reflection coefficient.
(b) The SWR in the free-space medium.
(c) The positions (in meters) in the free-space medium where the electric field maxima and minima occur.
(d) The maximum and minimum values of the electric field in the free-space medium.

5.5. A time-harmonic electromagnetic wave traveling in free space is incident normally upon a perfect conducting planar surface, as shown in Figure P5-5. Assuming the incident electric field is given by

$$\mathbf{E}^i = \hat{a}_x E_0 e^{-j\beta_0 z}$$

find (a) the reflected electric field, (b) the incident and reflected magnetic fields, and (c) the current density $\mathbf{J}_s$ induced on the conducting surface.

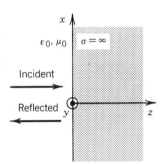

FIGURE P5-5

5.6. A uniform plane wave traveling in air is incident normally on a half space occupied by a lossless dielectric medium of relative permittivity of 4. The reflections can be eliminated by placing another dielectric slab, $\lambda_1/4$ thick, between the air and the original dielectric medium, as shown in Figure P5-6. To accomplish this, the intrinsic impedance η_1 of the slab must be equal to $\sqrt{\eta_0 \eta_2}$ where η_0 and η_2 are, respectively, the intrinsic impedances of air and the original dielectric medium. Assuming that the relative permeabilities of all the media are unity, what should the relative permittivity of the dielectric slab be to accomplish this?

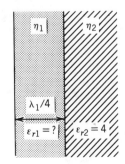

FIGURE P5-6

5.7. A uniform plane wave traveling in free space is incident normally upon a lossless dielectric slab of thickness t, as shown in Figure P5-7. Free space is found on the other side of the slab. Derive expressions for the total reflection and transmission coefficients in terms of the media constitutive electrical parameters and thickness of the slab.

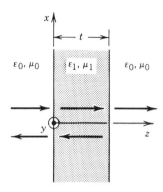

FIGURE P5-7

5.8. The vertical height from the ground to a person's eyes is h, and from his eyes to the top of his head is Δh. A flat mirror of height y is hung vertically at a distance x from the person. The top of the mirror is at a height of $h + (\Delta h/2)$ from the ground, as shown in Figure P5-8. What is the minimum length of the mirror in the vertical direction so that the person *only* sees his entire image in the mirror?

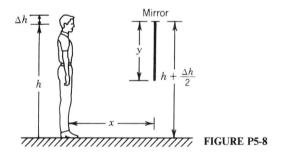

FIGURE P5-8

5.9. A linearly polarized wave is incident on an isosceles right triangle (prism) of glass, and it exits as shown in Figure P5-9. Assuming that the dielectric constant of the prism is 2.25, find the ratio of the exited average power density S_e to that of the incident S_i.

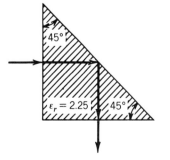

FIGURE P5-9

5.10. A uniform plane wave is obliquely incident at an angle of 30° on a dielectric slab of thickness d with $\varepsilon = 4\varepsilon_0$ and $\mu = \mu_0$ that is embedded in free space, as shown in Figure P5-10. Find the angles θ_2 and θ_3 (in degrees).

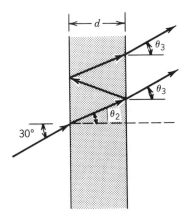

FIGURE P5-10

5.11. A perpendicularly polarized uniform plane wave traveling in free space is obliquely incident on a dielectric with a relative permittivity of 4, as shown in Figure 5-2. What should the incident angle be so that the reflected power density is 25% of the incident power density?

5.12. Repeat Problem 5-11 for a parallel polarized uniform plane wave.

5.13. Find the Brewster angles for the interfaces whose reflection coefficients are shown plotted in Figure 5-5.

5.14. A parallel polarized uniform plane wave is incident obliquely on a lossless dielectric slab that is embedded in a free-space medium, as shown in Figure P5-14. Derive expressions for the total reflection and transmission coefficients in terms of the electrical constitutive parameters, thickness of the slab, and angle of incidence.

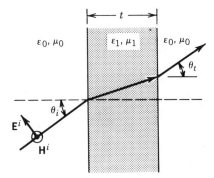

FIGURE P5-14

5.15. Repeat Problem 5-14 for a perpendicularly polarized plane wave, as shown in Figure P5-15.

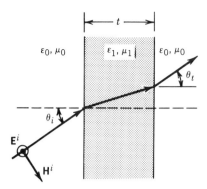

FIGURE P5-15

5.16. A perpendicularly polarized plane wave traveling in a dielectric medium with relative permittivity of 9 is obliquely incident on another dielectric with relative permittivity of 4. Assuming that the permeabilities of both media are the same, find the incident angle (measured from the normal to the interface) that results in total reflection.

5.17. Calculate the Brewster and critical angles for a parallel polarized wave when the plane interface is (a) water to air (ε_r of water is 81), (b) air to water, and (c) high density glass to air (ε_r of glass is 9).

5.18. A uniform plane wave traveling in a lossless dielectric is incident normally on a flat interface formed by the presence of air. For ε_r's of 2.56, 4, 9, 16, 25, and 81:
   (a) Determine the critical angles.
   (b) Find the Brewster angles if the wave is of parallel polarization.
   (c) Compare the critical and Brewster angles found in parts a and b.
   (d) Plot the magnitudes of the reflection coefficients for both perpendicular $|\Gamma_\perp|$ and parallel $|\Gamma_\parallel|$ polarizations versus incidence angle.
   (e) Plot the phase (in degrees) of the reflection coefficients for both perpendicular and parallel polarizations versus incidence angle.

5.19. The transmitting antenna of a ground-to-air communication system is placed at a height of 10 m above the water, as shown in Figure P5-19. For a separation of 10 km between the transmitter and the receiver, which is placed on an airborne platform, find the height h_2 above water of the receiving system so that the wave reflected by the water does not possess a parallel polarized component. Assume that the water surface is flat and lossless.

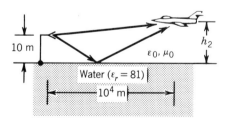

FIGURE P5-19

5.20. For the geometry of Problem 5-19 the transmitter is radiating a right-hand circularly polarized wave. Assuming the aircraft is at a height of 1101.11 m, give the polarization (linear, circular, or elliptical) and sense of rotation (right or left hand) of the following.

(a) A wave reflected by the sea and intercepted by the receiving antenna.
(b) A wave transmitted, at the same reflection point as in part a, into the sea.

5.21. The heights above the earth of a transmitter and receiver are, respectively, 100 and 10 m, as shown in Figure P5-21. Assuming that the transmitter radiates both perpendicular and parallel polarizations, how far apart (in meters) should the transmitter and receiver be placed so that the reflected wave has no parallel polarization? Assume that the reflecting medium is a lossless flat earth with a dielectric constant of 16.

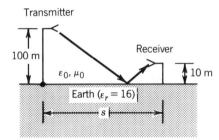

FIGURE P5-21

5.22. A light source that shines isotropically is submerged at a depth d below the surface of water, as shown in Figure P5-22. How far in the x direction (both positive and negative) can an observer (on the water interface) go and still see the light? Assume that the water is flat and lossless with a dielectric constant of 81.

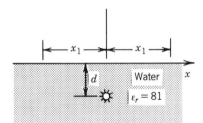

FIGURE P5-22

5.23. The 30° to 60° dielectric prism shown in Figure P5-23 is surrounded by free space.

(a) What is the minimum value of the prism's dielectric constant so that there is no time-average power density transmitted across the hypotenuse when a plane wave is incident on the prism, as shown in the figure?

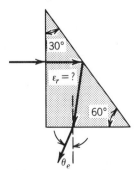

FIGURE P5-23

(b) What is the exiting angle θ_e if the dielectric constant of the prism is that found in part a?

5.24. A uniform plane wave of parallel polarization, traveling in a lossless dielectric medium with relative permittivity of 4, is obliquely incident on a free-space medium. What is the angle of incidence so that the wave results in a complete (a) transmission into the free-space medium and (b) reflection from the free-space medium?

5.25. A fish is swimming in water beneath a circular boat of diameter D, as shown in Figure P5-25.
   (a) Find the largest included angle $2\theta_c$ of an imaginary cone within which the fish can swim and not be seen by an observer at the surface of the water.
   (b) Find the smallest height of the cone.
   Assume that light strikes the boat at grazing incidence $\theta_i = \pi/2$ and refracts into the water.

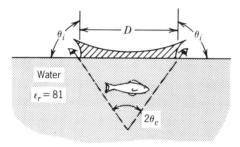

FIGURE P5-25

5.26. Any object above absolute zero temperature (0 K or $-273°C$) emits electromagnetic radiation. According to the reciprocity theorem, the amount of electromagnetic energy emitted by the object toward an angle θ_i is equal to the energy received by the object when an electromagnetic wave is incident at an angle θ_i, as shown in Figure P5-26. The electromagnetic power emitted by the object is sensed by a microwave remote detection system as a brightness temperature T_B given by

$$T_B = eT_m = \left(1 - |\Gamma|^2\right)T_m$$

where e = emissivity of the object (dimensionless)
Γ = reflection coefficient for the interface
T_m = thermal (molecular) temperature of object (water)

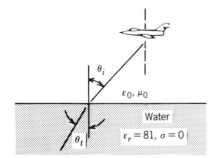

FIGURE P5-26

It is desired to make the brightness temperature T_B equal to the thermal (molecular) temperature T_m.

(a) State the polarization (perpendicular, parallel, or both) that will accomplish this.
(b) At what angle θ_i (in degrees) will this occur when the object is a flat water surface?

5.27. A uniform plane wave at a frequency of 10^4 Hz is traveling in air, and it is incident normally on a large body of salt water with constants of $\sigma = 3$ S/m and $\varepsilon_r = 81$. If the magnitude of the electric field on the salt water side of the interface is 10^{-3} V/m, find the depth (in meters) inside the salt water at which the magnitude of the electric field has been reduced to 0.368×10^{-3} V/m.

5.28. At large observation distances the field radiated by a satellite antenna which is attempting to communicate with a submerged submarine is locally TEM (also assume uniform plane wave), as shown in Figure P5-28. Assuming the incident electric field before it impinges on the water is 1 mV/m and the submarine is directly below the satellite, find at 1 MHz:

(a) The intensity of the reflected E field.
(b) The SWR created in air.
(c) The incident and reflected power densities.
(d) The intensity of the transmitted E field.
(e) The intensity of the transmitted power density.
(f) The depth d (in meters) of the submarine where the intensity of the transmitted electric field is 0.368 of its value immediately after it enters the water.
(g) The depth (in meters) of the submarine so that the distance from the surface of the ocean to the submarine is 20λ (λ in water).
(h) The time (in seconds) it takes the wave to travel from the surface of the ocean to the submarine at a depth of 100 m.
(i) The velocity of the wave in water to that in air (v/v_0).

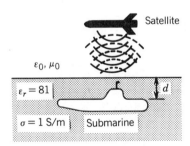

FIGURE P5-28

5.29. A uniform plane wave traveling inside a good conductor with conductivity σ_1 is incident normally on another good conductor with conductivity σ_2 but with $\sigma_1 > \sigma_2$. Determine the ratio of σ_1/σ_2 so that the SWR inside medium 1 near the interface is 1.5.

5.30. A right-hand circularly polarized uniform plane wave traveling in air is incident normally on a flat and smooth water surface with $\varepsilon_r = 81$ and $\sigma = 0.1$ S/m, as shown in Figure P5-30. Assuming a frequency of 1 GHz and an incident electric

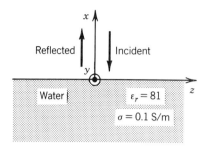

FIGURE P5-30

field of

$$\mathbf{E}^i = \left(\hat{a}_y + \hat{a}_z e^{j\psi}\right) E_0 e^{j\beta_0 x}$$

do the following.

(a) Determine the value of ψ.
(b) Write an expression for the corresponding incident magnetic field.
(c) Write expressions for the reflected electric and magnetic fields.
(d) Determine the polarization (including sense of rotation) of the reflected wave.
(e) Write expressions for the transmitted electric and magnetic fields.
(f) Determine the polarization (including sense of rotation) of the transmitted wave.
(g) Determine the percentage (compared to the incident) of the reflected and transmitted power densities.

5.31. A right-hand circularly polarized wave is incident normally on a perfect conducting flat surface ($\sigma = \infty$).

(a) What is the polarization and sense of rotation of the reflected field?
(b) What is the normalized (maximum unity) output voltage if the reflected wave is received by a right-hand circularly polarized antenna?
(c) Repeat part b if the receiving antenna is left-hand circularly polarized.

5.32. Repeat Problem 5.31 if the reflecting is water ($f = 10$ MHz, $\varepsilon_r = 81$ and $\sigma = 4$ S/m).

5.33. A parallel polarized plane wave traveling in a dielectric medium with ε_1, μ_1 is incident obliquely on a planar interface formed by the dielectric medium with ε_2, μ_2 such that $\varepsilon_2\mu_2 < \varepsilon_1\mu_1$. Assuming that the incident angle θ_i is equal to or greater than the critical angle θ_c of (5-35b), derive expressions for the reflection $\Gamma_{\|}^b$ and transmission $T_{\|}^b$ coefficients, and the incident $\mathbf{S}_{\|}^i$, reflected $\mathbf{S}_{\|}^r$, and transmitted $\mathbf{S}_{\|}^t$ average power densities.

5.34. Using the geometry of Figure 5-11 and assuming a parallel polarized incident *uniform* plane wave traveling at an angle ψ_1, do the following.

(a) Write expressions for the incident, reflected, and refracted magnetic fields.
(b) Derive expressions that can be used to solve for the effective attenuation α_{2e} and the effective phase β_{2e} constants, in terms of the incident angle ψ_1 and the intrinsic propagation constants γ_{01} and γ_{02}.

Hint: Use that $\gamma_{2e} \cdot \gamma_{2e} = (\gamma_{02})^2 = (\alpha_{02} + j\beta_{02})^2$.

5.35. Use the geometry of Figure 5-11 and assume a parallel polarized incident *nonuniform* plane wave traveling at an angle ψ_1.
   (a) Write expressions for the incident, reflected, and refracted magnetic fields.
   (b) Derive expressions that can be used to solve for the effective attenuation α_{2e} and the effective phase β_{2e} constants, in terms of the angles ψ_1 and ξ_1 and the intrinsic propagations constants γ_{01} and γ_{02}.

   Hint: Use that $\gamma_{2e} \cdot \gamma_{2e} = (\gamma_{02})^2 = (\alpha_{02} + j\beta_{02})^2$.

5.36. Two vertical lossless dielectric slabs, each of thickness equal to $\lambda_0/4$ at a center frequency of $f_0 = 2$ GHz, are sandwiched between a lossless semi-infinite medium of dielectric constant $\varepsilon_r = 2.25$ to the left and air to the right. Assume a fractional bandwidth of 0.5 and a binomial design.
   (a) Find the magnitude of the maximum reflection coefficient within the allowable bandwidth.
   (b) Determine the magnitude of the reflection coefficients at each interface (junction).
   (c) Compute the intrinsic impedances, dielectric constants, and thickness (in centimeters) of each dielectric slab.
   (d) Determine the lower and upper frequencies of the bandwidth.
   (e) Plot the magnitude of the reflection coefficient inside the dielectric medium with $\varepsilon_r = 2.25$ as a function of frequency (within the bandwidth).

5.37. It is desired to design a three-layer (each layer of $\lambda_0/4$ thickness) impedance transformer to match a semi-infinite dielectric medium of $\varepsilon_r = 9$ on one of its sides and one with $\varepsilon_r = 2.25$ on the other side. The maximum SWR that can be tolerated inside the dielectric medium with $\varepsilon_r = 9$ is 1.1. Assume a center frequency of $f_0 = 3$ GHz and a binomial design.
   (a) Determine the allowable fractional bandwidth and the lower and upper frequencies of the bandwidth.
   (b) Find the magnitude of reflection coefficients at each junction.
   (c) Compute the magnitude of the maximum reflection coefficient within the bandwidth.
   (d) Determine the intrinsic impedances, dielectric constants, and thicknesses (in centimeters) of each dielectric slab.
   (e) Plot the magnitude of the reflection coefficient inside the dielectric medium with $\varepsilon_r = 9$ as a function of frequency (within the bandwidth).

5.38. Repeat Example 5.11 using a Tschebyscheff design.

5.39. Repeat Problem 5.36 using a Tschebyscheff design.

5.40. Repeat Problem 5.37 using a Tschebyscheff design.

5.41. A right-hand (CW) elliptically polarized wave traveling in free space is obliquely incident at an angle $\theta_i = 30°$, measured from the normal, on a flat perfect electric conductor of infinite extent. If the incident field has an axial ratio of -2, determine the polarization of the reflected field. This is to include the axial ratio as well as its sense of rotation. Assume that the time–phase difference between the components of the incident field is 90°.

5.42. Repeat Problem 5.41 if the reflecting surface is a flat lossless ($\sigma_2 = 0$) ocean ($\varepsilon_2 = 81\varepsilon_0$ and $\mu_2 = \mu_0$) of infinite extent. Also find the polarization of the wave transmitted into the water.

CHAPTER 6

AUXILIARY VECTOR POTENTIALS, CONSTRUCTION OF SOLUTIONS, AND RADIATION AND SCATTERING EQUATIONS

6.1 INTRODUCTION

It is common practice in the analysis of electromagnetic boundary-value problems to use auxiliary vector potentials as aids in obtaining solutions for the electric (**E**) and magnetic (**H**) fields. The most common vector potential functions are the **A**, magnetic vector potential, and **F**, electric vector potential. They are the same pair that were introduced and used extensively in the solution of antenna radiation problems [1]. *Although the electric and magnetic field intensities (**E** and **H**) represent physically measurable quantities, for most engineers the vector potentials are strictly mathematical tools.* The introduction of the potentials often simplifies the solution, even though it may require determination of additional functions. Much of the discussion in this chapter is borrowed from [1].

The Hertz vector potentials Π_e and Π_h make up another pair. The Hertz vector potential Π_e is analogous to **A** and Π_h is analogous to **F**. The functional relation between them is a proportionality constant that is a function of the frequency and the constitutive parameters of the medium. In the solution of a problem, only one set, **A** and **F** or Π_e and Π_h, is required. The author prefers **A** and **F**, and they will be used throughout this book.

The main objective of this book is to obtain electromagnetic field configurations (modes) of boundary-value propagation, radiation, and scattering problems. These field configurations must satisfy Maxwell's equations or the wave equation and the appropriate boundary conditions. The procedure is to specify the electromagnetic boundary-value problem, which may or may not contain sources, and to obtain the field configurations that can exist within the region of the boundary-value problem. This can be accomplished in either of two ways, as shown in Figure 6-1.

INTRODUCTION

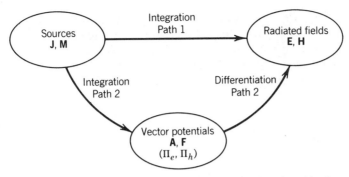

FIGURE 6-1 Block diagram for computing radiated fields from electric and magnetic sources. (*Source:* C. A. Balanis, *Antenna Theory: Analysis and Design.* Copyright © 1982, John Wiley & Sons, Inc. Reprinted by permission of John Wiley & Sons, Inc.)

One procedure for obtaining the electric and magnetic fields of a desired boundary-value problem is to use Maxwell's or the wave equations. This is accomplished essentially in one step, and it is represented in Figure 6-1 by path 1. The formulation using such a procedure is assigned to the reader as an end of chapter problem.

The other procedure requires two steps. In the first step, the vector potentials **A** and **F** (or Π_e and Π_h) are found, once the boundary-value problem is specified. In the second step, the electric and magnetic fields are found, after the vector potentials are determined. The electric and magnetic fields are functions of the vector potentials. This procedure is represented by path 2 of Figure 6-1, and, although it requires two steps, it is often simpler and more straightforward; hence it is often preferred. The mathematical equations of this procedure will be developed next, and they will be utilized in this book to illustrate solutions of boundary-value electromagnetic problems.

In a homogeneous medium, any solution for the time-harmonic electric and magnetic fields must satisfy Maxwell's equations

$$\nabla \times \mathbf{E} = -\mathbf{M} - j\omega\mu\mathbf{H} \tag{6-1a}$$

$$\nabla \times \mathbf{H} = \mathbf{J} + j\omega\varepsilon\mathbf{E} \tag{6-1b}$$

$$\nabla \cdot \mathbf{E} = \frac{q_{ev}}{\varepsilon} \tag{6-1c}$$

$$\nabla \cdot \mathbf{H} = \frac{q_{mv}}{\mu} \tag{6-1d}$$

or the vector wave equations

$$\nabla^2 \mathbf{E} + \beta^2 \mathbf{E} = \nabla \times \mathbf{M} + j\omega\mu\mathbf{J} + \frac{1}{\varepsilon}\nabla q_{ev} \tag{6-2a}$$

$$\nabla^2 \mathbf{H} + \beta^2 \mathbf{H} = -\nabla \times \mathbf{J} + j\omega\varepsilon\mathbf{M} + \frac{1}{\mu}\nabla q_{mv} \tag{6-2b}$$

where

$$\beta^2 = \omega^2\mu\varepsilon \tag{6-2c}$$

256 VECTOR POTENTIALS, SOLUTIONS, AND EQUATIONS

In regions where there are no sources, $\mathbf{J} = \mathbf{M} = q_{ev} = q_{mv} = 0$. In these regions, the preceding equations are of simpler form. Whereas the electric current density $\mathbf{J}$ may represent either actual or equivalent sources, the magnetic current density $\mathbf{M}$ can only represent equivalent sources. Although all of these equations will still be satisfied, an alternate procedure is developed next for the solution of the electric and magnetic fields in terms of the auxiliary vector potentials $\mathbf{A}$ and $\mathbf{F}$.

6.2 THE VECTOR POTENTIAL A

In a source-free region, the magnetic flux density $\mathbf{B}$ is always solenoidal, that is, $\nabla \cdot \mathbf{B} = 0$. Therefore, it can be represented as the curl of another vector because it obeys the vector identity

$$\nabla \cdot \nabla \times \mathbf{A} = 0 \tag{6-3}$$

where $\mathbf{A}$ is an arbitrary vector. Thus we define

$$\mathbf{B}_A = \mu \mathbf{H}_A = \nabla \times \mathbf{A} \tag{6-4}$$

or

$$\boxed{\mathbf{H}_A = \frac{1}{\mu} \nabla \times \mathbf{A}} \tag{6-4a}$$

where subscript A indicates the fields due to the $\mathbf{A}$ potential. Substituting (6-4a) into Maxwell's curl equation

$$\nabla \times \mathbf{E}_A = -j\omega\mu \mathbf{H}_A \tag{6-5}$$

reduces it to

$$\nabla \times \mathbf{E}_A = -j\omega\mu \mathbf{H}_A = -j\omega \nabla \times \mathbf{A} \tag{6-6}$$

which can also be written as

$$\nabla \times [\mathbf{E}_A + j\omega \mathbf{A}] = 0 \tag{6-7}$$

From the vector identity

$$\nabla \times (-\nabla \phi_e) = 0 \tag{6-8}$$

and (6-7), it follows that

$$\mathbf{E}_A + j\omega \mathbf{A} = -\nabla \phi_e \tag{6-9}$$

or

$$\boxed{\mathbf{E}_A = -\nabla \phi_e - j\omega \mathbf{A}} \tag{6-9a}$$

ϕ_e represents an arbitrary electric scalar potential that is a function of position. Taking the curl of both sides of (6-4) and using the vector identity

$$\nabla \times \nabla \times \mathbf{A} = \nabla(\nabla \cdot \mathbf{A}) - \nabla^2 \mathbf{A} \tag{6-10}$$

leads to
$$\nabla \times (\mu \mathbf{H}_A) = \nabla(\nabla \cdot \mathbf{A}) - \nabla^2 \mathbf{A} \qquad (6\text{-}10\text{a})$$

For a homogeneous medium, (6-10a) reduces to
$$\mu \nabla \times \mathbf{H}_A = \nabla(\nabla \cdot \mathbf{A}) - \nabla^2 \mathbf{A} \qquad (6\text{-}11)$$

Equating Maxwell's equation
$$\boxed{\nabla \times \mathbf{H}_A = \mathbf{J} + j\omega\varepsilon \mathbf{E}_A} \qquad (6\text{-}12)$$

to (6-11) leads to
$$\mu \mathbf{J} + j\omega\mu\varepsilon \mathbf{E}_A = \nabla(\nabla \cdot \mathbf{A}) - \nabla^2 \mathbf{A} \qquad (6\text{-}13)$$

Substituting (6-9a) into (6-13) reduces it to
$$\nabla^2 \mathbf{A} + \beta^2 \mathbf{A} = -\mu \mathbf{J} + \nabla(\nabla \cdot \mathbf{A}) + \nabla(j\omega\mu\varepsilon\phi_e) = -\mu \mathbf{J} + \nabla(\nabla \cdot \mathbf{A} + j\omega\mu\varepsilon\phi_e) \qquad (6\text{-}14)$$

where $\beta^2 = \omega^2 \mu\varepsilon$.

In (6-4), the curl of $\mathbf{A}$ was defined. Now we are at liberty to define the divergence of $\mathbf{A}$, which is independent of its curl. Both are required to uniquely define $\mathbf{A}$. In order to simplify (6-14), let

$$\boxed{\nabla \cdot \mathbf{A} = -j\omega\varepsilon\mu\phi_e \Rightarrow \phi_e = -\frac{1}{j\omega\mu\varepsilon}\nabla \cdot \mathbf{A}} \qquad (6\text{-}15)$$

which is known as the *Lorentz condition* (or *gauge*). Other gauges may be defined. Substituting (6-15) into (6-14) leads to

$$\boxed{\nabla^2 \mathbf{A} + \beta^2 \mathbf{A} = -\mu \mathbf{J}} \qquad (6\text{-}16)$$

In addition, (6-9a) reduces to

$$\boxed{\mathbf{E}_A = -\nabla\phi_e - j\omega\mathbf{A} = -j\omega\mathbf{A} - j\frac{1}{\omega\mu\varepsilon}\nabla(\nabla \cdot \mathbf{A})} \qquad (6\text{-}17)$$

Once $\mathbf{A}$ is known, $\mathbf{H}_A$ can be found from (6-4a) and $\mathbf{E}_A$ from (6-17). $\mathbf{E}_A$ can just as easily be found from Maxwell's equation 6-12 by setting $\mathbf{J} = 0$. Since (6-16) is a vector wave equation, solutions for $\mathbf{A}$ in rectangular, cylindrical, and spherical coordinate systems are similar to those for $\mathbf{E}$ in Sections 3.4.1A, 3.4.2, and 3.4.3.

6.3 THE VECTOR POTENTIAL F

In a source-free region, the electric flux density $\mathbf{D}$ is always solenoidal, that is, $\nabla \cdot \mathbf{D} = 0$. Therefore, it can be represented as the curl of another vector because it obeys the vector identity

$$\nabla \cdot (-\nabla \times \mathbf{F}) = 0 \qquad (6\text{-}18)$$

where $\mathbf{F}$ is an arbitrary vector. Thus we can define $\mathbf{D}_F$ by

$$\mathbf{D}_F = -\nabla \times \mathbf{F}$$

or

$$\boxed{\mathbf{E}_F = -\frac{1}{\varepsilon}\nabla \times \mathbf{F}} \qquad (6\text{-}19)$$

where the subscript F indicates the fields due to the $\mathbf{F}$ potential. Substituting (6-19) into Maxwell's curl equation

$$\nabla \times \mathbf{H}_F = j\omega\varepsilon\mathbf{E}_F \qquad (6\text{-}20)$$

reduces it to

$$\nabla \times (\mathbf{H}_F + j\omega\mathbf{F}) = 0 \qquad (6\text{-}21)$$

From the vector identity of (6-8), it follows that

$$\boxed{\mathbf{H}_F = -\nabla\phi_m - j\omega\mathbf{F}} \qquad (6\text{-}22)$$

where ϕ_m represents an arbitrary magnetic scalar potential that is a function of position. Taking the curl of (6-19)

$$\nabla \times \mathbf{E}_F = -\frac{1}{\varepsilon}\nabla \times \nabla \times \mathbf{F} = -\frac{1}{\varepsilon}(\nabla\nabla \cdot \mathbf{F} - \nabla^2\mathbf{F}) \qquad (6\text{-}23)$$

and equating it to Maxwell's equation

$$\boxed{\nabla \times \mathbf{E}_F = -\mathbf{M} - j\omega\mu\mathbf{H}_F} \qquad (6\text{-}24)$$

lead to

$$\nabla^2\mathbf{F} + j\omega\mu\varepsilon\mathbf{H}_F = \nabla\nabla \cdot \mathbf{F} - \varepsilon\mathbf{M} \qquad (6\text{-}25)$$

Substituting (6-22) into (6-25) reduces it to

$$\nabla^2\mathbf{F} + \beta^2\mathbf{F} = -\varepsilon\mathbf{M} + \nabla(\nabla \cdot \mathbf{F} + j\omega\mu\varepsilon\phi_m) \qquad (6\text{-}26)$$

Letting

$$\boxed{\nabla \cdot \mathbf{F} = -j\omega\mu\varepsilon\phi_m \Rightarrow \phi_m = -\frac{1}{j\omega\mu\varepsilon}\nabla \cdot \mathbf{F}} \qquad (6\text{-}27)$$

reduces (6-26) to

$$\boxed{\nabla^2\mathbf{F} + \beta^2\mathbf{F} = -\varepsilon\mathbf{M}} \qquad (6\text{-}28)$$

and (6-22) to

$$\mathbf{H}_F = -j\omega \mathbf{F} - \frac{j}{\omega\mu\varepsilon}\nabla(\nabla \cdot \mathbf{F}) \quad (6\text{-}29)$$

Once $\mathbf{F}$ is known, $\mathbf{E}_F$ can be found from (6-19) and $\mathbf{H}_F$ from (6-29) or (6-24) by setting $\mathbf{M} = 0$. Since (6-28) is a vector wave equation, solutions for $\mathbf{F}$ in rectangular, cylindrical, and spherical coordinate systems are similar to those for $\mathbf{E}$ in Sections 3.4.1A, 3.4.2, and 3.4.3.

6.4 THE VECTOR POTENTIALS A AND F

In the previous two sections, we derived expressions for the $\mathbf{E}$ and $\mathbf{H}$ fields in terms of the vector potentials $\mathbf{A}$ ($\mathbf{E}_A, \mathbf{H}_A$) and $\mathbf{F}$ ($\mathbf{E}_F$ and $\mathbf{H}_F$). In addition, expressions that $\mathbf{A}$ and $\mathbf{F}$ must satisfy were also derived. The total $\mathbf{E}$ and $\mathbf{H}$ fields are obtained by the superposition of the individual fields due to $\mathbf{A}$ and $\mathbf{F}$.

The procedure that can be used to find the fields of path 2 of Figure 6-1 is as follows.

SUMMARY

1. Specify the electromagnetic boundary-value problem, which may or may not contain any sources within its boundaries, and its desired field configurations (modes).
2. a. Solve for $\mathbf{A}$ using (6-16),

$$\nabla^2 \mathbf{A} + \beta^2 \mathbf{A} = -\mu \mathbf{J} \quad \text{where } \beta^2 = \omega^2\mu\varepsilon \quad (6\text{-}30)$$

   Depending on the problem, solutions for $\mathbf{A}$ in rectangular, cylindrical, and spherical coordinate systems take the forms found in Sections 3.4.1A, 3.4.2, and 3.4.3.

   b. Solve for $\mathbf{F}$ using (6-28),

$$\nabla^2 \mathbf{F} + \beta^2 \mathbf{F} = -\varepsilon \mathbf{M} \quad \text{where } \beta^2 = \omega^2\mu\varepsilon \quad (6\text{-}31)$$

   Depending on the problem, solutions for $\mathbf{F}$ in rectangular, cylindrical, and spherical coordinate systems take the forms found in Sections 3.4.1A, 3.4.2, and 3.4.3.

3. a. Find $\mathbf{H}_A$ using (6-4a) and $\mathbf{E}_A$ using (6-17). $\mathbf{E}_A$ can also be found using (6-12) by letting $\mathbf{J} = 0$.

$$\mathbf{H}_A = \frac{1}{\mu}\nabla \times \mathbf{A} \quad (6\text{-}32a)$$

$$\mathbf{E}_A = -j\omega \mathbf{A} - j\frac{1}{\omega\mu\varepsilon}\nabla(\nabla \cdot \mathbf{A}) \quad (6\text{-}32b)$$

or

$$\boxed{\mathbf{E}_A = \frac{1}{j\omega\varepsilon} \nabla \times \mathbf{H}_A} \qquad (6\text{-}32c)$$

b. Find $\mathbf{E}_F$ using (6-19) and $\mathbf{H}_F$ using (6-29). $\mathbf{H}_F$ can also be found using (6-24) by letting $\mathbf{M} = 0$.

$$\boxed{\mathbf{E}_F = -\frac{1}{\varepsilon} \nabla \times \mathbf{F}} \qquad (6\text{-}33a)$$

$$\boxed{\mathbf{H}_F = -j\omega\mathbf{F} - j\frac{1}{\omega\mu\varepsilon} \nabla(\nabla \cdot \mathbf{F})} \qquad (6\text{-}33b)$$

or

$$\boxed{\mathbf{H}_F = -\frac{1}{j\omega\mu} \nabla \times \mathbf{E}_F} \qquad (6\text{-}33c)$$

4. The total fields are then found by the superposition of those given in step 3, that is,

$$\boxed{\mathbf{E} = \mathbf{E}_A + \mathbf{E}_F = -j\omega\mathbf{A} - j\frac{1}{\omega\mu\varepsilon}\nabla(\nabla \cdot \mathbf{A}) - \frac{1}{\varepsilon}\nabla \times \mathbf{F}} \qquad (6\text{-}34)$$

or

$$\boxed{\mathbf{E} = \mathbf{E}_A + \mathbf{E}_F = \frac{1}{j\omega\varepsilon}\nabla \times \mathbf{H}_A - \frac{1}{\varepsilon}\nabla \times \mathbf{F}} \qquad (6\text{-}34a)$$

and

$$\boxed{\mathbf{H} = \mathbf{H}_A + \mathbf{H}_F = \frac{1}{\mu}\nabla \times \mathbf{A} - j\omega\mathbf{F} - j\frac{1}{\omega\mu\varepsilon}\nabla(\nabla \cdot \mathbf{F})} \qquad (6\text{-}35)$$

or

$$\boxed{\mathbf{H} = \mathbf{H}_A + \mathbf{H}_F = \frac{1}{\mu}\nabla \times \mathbf{A} - \frac{1}{j\omega\mu}\nabla \times \mathbf{E}_F} \qquad (6\text{-}35a)$$

Whether (6-32b) or (6-32c) is used to find $\mathbf{E}_A$ and (6-33b) or (6-33c) to find $\mathbf{H}_F$ depends largely on the nature of the problem. In many instances one may be more complex than the other. For computing radiation fields in the far zone, it will be easier to use (6-32b) for $\mathbf{E}_A$ and (6-33b) for $\mathbf{H}_F$ because, as it will be shown, the second term in each expression becomes negligible in this region. The same solution should be obtained using either of the two choices in each case.

6.5 CONSTRUCTION OF SOLUTIONS

For many electromagnetic boundary-value problems, there are usually many field configurations (modes) that are solutions that satisfy Maxwell's equations and the boundary conditions. The most widely known modes are those that are referred to as Transverse ElectroMagnetic (TEM), Transverse Electric (TE), and Transverse Magnetic (TM).

TEM modes are field configurations whose *electric and magnetic* field components are transverse to a given direction. Often, but not necessarily, that direction is the path that the wave is traveling. TE modes are field configurations whose *electric* field components are transverse to a given direction; for TM modes the *magnetic* field components are transverse to a given direction. Here we will illustrate methods that utilize the vector potentials to construct TEM, TE, and TM modes.

6.5.1 Transverse Electromagnetic Modes: Source-Free Region

TEM modes are usually the simplest forms of field configurations, and they are usually referred to as the lowest-order modes. For these field configurations both the electric *and* magnetic field components are transverse to a given direction. To see how these modes can be constructed using the vector potentials, let us illustrate the procedure using the rectangular and cylindrical coordinate systems.

A. RECTANGULAR COORDINATE SYSTEM

According to (6-34) the electric field in terms of the vector potentials **A** and **F** is given by

$$\mathbf{E} = \mathbf{E}_A + \mathbf{E}_F = -j\omega\mathbf{A} - j\frac{1}{\omega\mu\varepsilon}\nabla(\nabla \cdot \mathbf{A}) - \frac{1}{\varepsilon}\nabla \times \mathbf{F} \quad (6\text{-}36)$$

Assuming the vector potentials **A** and **F** have solutions of the form

$$\mathbf{A}(x, y, z) = \hat{a}_x A_x(x, y, z) + \hat{a}_y A_y(x, y, z) + \hat{a}_z A_z(x, y, z) \quad (6\text{-}37)$$

which satisfies (6-30) with **J** = 0

$$\nabla^2 \mathbf{A} + \beta^2 \mathbf{A} = 0 \quad (6\text{-}38)$$

or

$$\nabla^2 A_x + \beta^2 A_x = 0 \quad (6\text{-}38a)$$

$$\nabla^2 A_y + \beta^2 A_y = 0 \quad (6\text{-}38b)$$

$$\nabla^2 A_z + \beta^2 A_z = 0 \quad (6\text{-}38c)$$

and

$$\mathbf{F}(x, y, z) = \hat{a}_x F_x(x, y, z) + \hat{a}_y F_y(x, y, z) + \hat{a}_z F_z(x, y, z) \quad (6\text{-}39)$$

262 VECTOR POTENTIALS, SOLUTIONS, AND EQUATIONS

which satisfies (6-31) with $\mathbf{M} = 0$

$$\nabla^2 \mathbf{F} + \beta^2 \mathbf{F} = 0 \tag{6-40}$$

or

$$\nabla^2 F_x + \beta^2 F_x = 0 \tag{6-40a}$$
$$\nabla^2 F_y + \beta^2 F_y = 0 \tag{6-40b}$$
$$\nabla^2 F_z + \beta^2 F_z = 0 \tag{6-40c}$$

(6-36), when expanded, can be written as

$$\mathbf{E} = \hat{a}_x \left[-j\omega A_x - j\frac{1}{\omega\mu\varepsilon}\left(\frac{\partial^2 A_x}{\partial x^2} + \frac{\partial^2 A_y}{\partial x\,\partial y} + \frac{\partial^2 A_z}{\partial x\,\partial z}\right) - \frac{1}{\varepsilon}\left(\frac{\partial F_z}{\partial y} - \frac{\partial F_y}{\partial z}\right) \right]$$
$$+ \hat{a}_y \left[-j\omega A_y - j\frac{1}{\omega\mu\varepsilon}\left(\frac{\partial^2 A_x}{\partial x\,\partial y} + \frac{\partial^2 A_y}{\partial y^2} + \frac{\partial^2 A_z}{\partial y\,\partial z}\right) - \frac{1}{\varepsilon}\left(\frac{\partial F_x}{\partial z} - \frac{\partial F_z}{\partial x}\right) \right]$$
$$+ \hat{a}_z \left[-j\omega A_z - j\frac{1}{\omega\mu\varepsilon}\left(\frac{\partial^2 A_x}{\partial x\,\partial z} + \frac{\partial^2 A_y}{\partial y\,\partial z} + \frac{\partial^2 A_z}{\partial z^2}\right) - \frac{1}{\varepsilon}\left(\frac{\partial F_y}{\partial x} - \frac{\partial F_x}{\partial y}\right) \right] \tag{6-41}$$

Similarly (6-35)

$$\mathbf{H} = \mathbf{H}_A + \mathbf{H}_F = \frac{1}{\mu}\nabla \times \mathbf{A} - j\omega\mathbf{F} - j\frac{1}{\omega\mu\varepsilon}\nabla(\nabla \cdot \mathbf{F}) \tag{6-42}$$

when expanded using (6-37) and (6-39) can be written as

$$\mathbf{H} = \hat{a}_x \left[-j\omega F_x - j\frac{1}{\omega\mu\varepsilon}\left(\frac{\partial^2 F_x}{\partial x^2} + \frac{\partial^2 F_y}{\partial x\,\partial y} + \frac{\partial^2 F_z}{\partial x\,\partial z}\right) + \frac{1}{\mu}\left(\frac{\partial A_z}{\partial y} - \frac{\partial A_y}{\partial z}\right) \right]$$
$$+ \hat{a}_y \left[-j\omega F_y - j\frac{1}{\omega\mu\varepsilon}\left(\frac{\partial^2 F_x}{\partial x\,\partial y} + \frac{\partial^2 F_y}{\partial y^2} + \frac{\partial^2 F_z}{\partial y\,\partial z}\right) + \frac{1}{\mu}\left(\frac{\partial A_x}{\partial z} - \frac{\partial A_z}{\partial x}\right) \right]$$
$$+ \hat{a}_z \left[-j\omega F_z - j\frac{1}{\omega\mu\varepsilon}\left(\frac{\partial^2 F_x}{\partial x\,\partial z} + \frac{\partial^2 F_y}{\partial y\,\partial z} + \frac{\partial^2 F_z}{\partial z^2}\right) + \frac{1}{\mu}\left(\frac{\partial A_y}{\partial x} - \frac{\partial A_x}{\partial y}\right) \right] \tag{6-43}$$

Example 6-1. Using (6-41) and (6-43) derive expressions for the $\mathbf{E}$ and $\mathbf{H}$ fields, in terms of the components of the $\mathbf{A}$ and $\mathbf{F}$ potentials, that are TEM to the z direction (TEM^z).

Solution. It is apparent by examining (6-41) and (6-43) that TEM^z ($E_z = H_z = 0$) modes can be obtained by any of the following three combinations.

1. Letting

$$A_x = A_y = F_x = F_y = 0 \qquad A_z \neq 0 \qquad F_z \neq 0 \qquad \partial/\partial x \neq 0 \qquad \partial/\partial y \neq 0$$

For this combination, according to (6-41)

$$E_z = -j\omega A_z - j\frac{1}{\omega\mu\varepsilon}\frac{\partial^2 A_z}{\partial z^2} = -j\frac{1}{\omega\mu\varepsilon}\left(\frac{\partial^2}{\partial z^2} + \omega^2\mu\varepsilon\right)A_z = 0$$

provided

$$A_z(x, y, z) = A_z^+(x, y)e^{-j\beta z} + A_z^-(x, y)e^{+j\beta z}$$

Similarly according to (6-43)

$$H_z = -j\omega F_z - j\frac{1}{\omega\mu\varepsilon}\frac{\partial^2 F_z}{\partial z^2} = -j\frac{1}{\omega\mu\varepsilon}\left(\frac{\partial^2}{\partial z^2} + \omega^2\mu\varepsilon\right)F_z = 0$$

provided

$$F_z(x, y, z) = F_z^+(x, y)e^{-j\beta z} + F_z^-(x, y)e^{+j\beta z}$$

Also according to (6-41) and (6-43)

$$E_x = \left(-\frac{1}{\sqrt{\mu\varepsilon}}\frac{\partial A_z^+}{\partial x} - \frac{1}{\varepsilon}\frac{\partial F_z^+}{\partial y}\right)e^{-j\beta z} + \left(\frac{1}{\sqrt{\mu\varepsilon}}\frac{\partial A_z^-}{\partial x} - \frac{1}{\varepsilon}\frac{\partial F_z^-}{\partial y}\right)e^{+j\beta z} = E_x^+ + E_x^-$$

$$E_y = \left(-\frac{1}{\sqrt{\mu\varepsilon}}\frac{\partial A_z^+}{\partial y} + \frac{1}{\varepsilon}\frac{\partial F_z^+}{\partial x}\right)e^{-j\beta z} + \left(\frac{1}{\sqrt{\mu\varepsilon}}\frac{\partial A_z^-}{\partial y} + \frac{1}{\varepsilon}\frac{\partial F_z^-}{\partial x}\right)e^{+j\beta z} = E_y^+ + E_y^-$$

$$H_x = -\sqrt{\frac{\varepsilon}{\mu}}\left(-\frac{1}{\sqrt{\mu\varepsilon}}\frac{\partial A_z^+}{\partial y} + \frac{1}{\varepsilon}\frac{\partial F_z^+}{\partial x}\right)e^{-j\beta z}$$

$$+ \sqrt{\frac{\varepsilon}{\mu}}\left(\frac{1}{\sqrt{\mu\varepsilon}}\frac{\partial A_z^-}{\partial y} + \frac{1}{\varepsilon}\frac{\partial F_z^-}{\partial x}\right)e^{+j\beta z} = H_x^+ + H_x^-$$

$$H_x = -\sqrt{\frac{\varepsilon}{\mu}}(E_y^+) + \sqrt{\frac{\varepsilon}{\mu}}(E_y^-)$$

$$H_y = \sqrt{\frac{\varepsilon}{\mu}}\left(-\frac{1}{\sqrt{\mu\varepsilon}}\frac{\partial A_z^+}{\partial x} - \frac{1}{\varepsilon}\frac{\partial F_z^+}{\partial y}\right)e^{-j\beta z}$$

$$- \sqrt{\frac{\varepsilon}{\mu}}\left(\frac{1}{\sqrt{\mu\varepsilon}}\frac{\partial A_z^-}{\partial x} - \frac{1}{\varepsilon}\frac{\partial F_z^-}{\partial y}\right)e^{+j\beta z} = H_y^+ + H_y^-$$

$$H_y = \sqrt{\frac{\varepsilon}{\mu}}(E_x^+) - \sqrt{\frac{\varepsilon}{\mu}}(E_x^-)$$

Also

$$Z_w^+ = \frac{E_x^+}{H_y^+} = -\frac{E_y^+}{H_x^+} = \sqrt{\frac{\mu}{\varepsilon}}$$

$$Z_w^- = -\frac{E_x^-}{H_y^-} = \frac{E_y^-}{H_x^-} = \sqrt{\frac{\mu}{\varepsilon}}$$

2. Letting

$$A_x = A_y = A_z = F_x = F_y = 0 \qquad F_z \neq 0 \qquad \partial/\partial x \neq 0 \qquad \partial/\partial y \neq 0$$

264 VECTOR POTENTIALS, SOLUTIONS, AND EQUATIONS

For this combination, according to (6-41) and (6-43)

$$E_z = 0$$

$$H_z = -j\omega F_z - j\frac{1}{\omega\mu\varepsilon}\frac{\partial^2 F_z}{\partial z^2} = -j\frac{1}{\omega\mu\varepsilon}\left(\frac{\partial^2}{\partial z^2} + \omega^2\mu\varepsilon\right)F_z = 0$$

provided

$$F_z(x, y, z) = F_z^+(x, y)e^{-j\beta z} + F_z^-(x, y)e^{+j\beta z}$$

Also according to (6-41) and (6-43)

$$E_x = -\frac{1}{\varepsilon}\frac{\partial F_z^+}{\partial y}e^{-j\beta z} - \frac{1}{\varepsilon}\frac{\partial F_z^-}{\partial y}e^{+j\beta z} = E_x^+ + E_x^-$$

$$E_y = +\frac{1}{\varepsilon}\frac{\partial F_z^+}{\partial x}e^{-j\beta z} + \frac{1}{\varepsilon}\frac{\partial F_z^-}{\partial x}e^{+j\beta z} = E_y^+ + E_y^-$$

$$H_x = -\sqrt{\frac{\varepsilon}{\mu}}\left(\frac{1}{\varepsilon}\frac{\partial F_z^+}{\partial x}\right)e^{-j\beta z} + \sqrt{\frac{\varepsilon}{\mu}}\left(\frac{1}{\varepsilon}\frac{\partial F_z^-}{\partial x}\right)e^{+j\beta z} = H_x^+ + H_x^-$$

$$= -\sqrt{\frac{\varepsilon}{\mu}}(E_y^+) + \sqrt{\frac{\varepsilon}{\mu}}(E_y^-)$$

$$H_y = \sqrt{\frac{\varepsilon}{\mu}}\left(-\frac{1}{\varepsilon}\frac{\partial F_z^+}{\partial y}\right)e^{-j\beta z} - \sqrt{\frac{\varepsilon}{\mu}}\left(-\frac{1}{\varepsilon}\frac{\partial F_z^-}{\partial y}\right)e^{+j\beta z} = H_y^+ + H_y^-$$

$$= \sqrt{\frac{\varepsilon}{\mu}}(E_x^+) - \sqrt{\frac{\varepsilon}{\mu}}(E_x^-)$$

Also

$$Z_w^+ = \frac{E_x^+}{H_y^+} = -\frac{E_y^+}{H_x^+} = \sqrt{\frac{\mu}{\varepsilon}}$$

$$Z_w^- = -\frac{E_x^-}{H_y^-} = \frac{E_y^-}{H_x^-} = \sqrt{\frac{\mu}{\varepsilon}}$$

3. Letting

$$A_x = A_y = F_x = F_y = F_z = 0 \qquad A_z \neq 0 \qquad \partial/\partial x \neq 0 \qquad \partial/\partial y \neq 0$$

For this combination, according to (6-41) and (6-43)

$$H_z = 0$$

$$E_z = -j\omega A_z - j\frac{1}{\omega\mu\varepsilon}\frac{\partial^2 A_z}{\partial z^2} = -j\frac{1}{\omega\mu\varepsilon}\left(\frac{\partial^2}{\partial z^2} + \omega^2\mu\varepsilon\right)A_z = 0$$

provided

$$A_z(x, y, z) = A_z^+(x, y)e^{-j\beta z} + A_z^-(x, y)e^{+j\beta z}$$

Also according to (6-41) and (6-43)

$$E_x = -\frac{1}{\sqrt{\mu\varepsilon}} \frac{\partial A_z^+}{\partial x} e^{-j\beta z} + \frac{1}{\sqrt{\mu\varepsilon}} \frac{\partial A_z^-}{\partial x} e^{+j\beta z} = E_x^+ + E_x^-$$

$$E_y = -\frac{1}{\sqrt{\mu\varepsilon}} \frac{\partial A_z^+}{\partial y} e^{-j\beta z} + \frac{1}{\sqrt{\mu\varepsilon}} \frac{\partial A_z^-}{\partial y} e^{+j\beta z} = E_y^+ + E_y^-$$

$$H_x = -\sqrt{\frac{\varepsilon}{\mu}}\left(-\frac{1}{\sqrt{\mu\varepsilon}} \frac{\partial A_z^+}{\partial y}\right)e^{-j\beta z} + \sqrt{\frac{\varepsilon}{\mu}}\left(\frac{1}{\sqrt{\mu\varepsilon}} \frac{\partial A_z^-}{\partial y}\right)e^{+j\beta z} = H_x^+ + H_x^-$$

$$= -\sqrt{\frac{\varepsilon}{\mu}}(E_y^+) + \sqrt{\frac{\varepsilon}{\mu}}(E_y^-)$$

$$H_y = \sqrt{\frac{\varepsilon}{\mu}}\left(-\frac{1}{\sqrt{\mu\varepsilon}} \frac{\partial A_z^+}{\partial x}\right)e^{-j\beta z} - \sqrt{\frac{\varepsilon}{\mu}}\left(\frac{1}{\sqrt{\mu\varepsilon}} \frac{\partial A_z^-}{\partial x}\right)e^{+j\beta z} = H_y^+ + H_y^-$$

$$= \sqrt{\frac{\varepsilon}{\mu}}(E_x^+) - \sqrt{\frac{\varepsilon}{\mu}}(E_x^-)$$

Also

$$Z_w^+ = \frac{E_x^+}{H_y^+} = -\frac{E_y^+}{H_x^+} = \sqrt{\frac{\mu}{\varepsilon}}$$

$$Z_w^- = -\frac{E_x^-}{H_y^-} = \frac{E_y^-}{H_x^-} = \sqrt{\frac{\mu}{\varepsilon}}$$

SUMMARY

From the results of Example 6-1, it is evident that TEM^z modes can be obtained by any of the following three combinations:

$$\underline{\text{TEM}^z}$$

$$\boxed{\begin{aligned} A_x = A_y = F_x = F_y = 0 \quad \partial/\partial x \neq 0 \quad \partial/\partial y \neq 0 \\ A_z = A_z^+(x,y)e^{-j\beta z} + A_z^-(x,y)e^{+j\beta z} \\ F_z = F_z^+(x,y)e^{-j\beta z} + F_z^-(x,y)e^{+j\beta z} \end{aligned}}$$
(6-44)
(6-44a)
(6-44b)

$$\boxed{\begin{aligned} A_x = A_y = A_z = F_x = F_y = 0 \quad \partial/\partial x \neq 0 \quad \partial/\partial y \neq 0 \\ F_z = F_z^+(x,y)e^{-j\beta z} + F_z^-(x,y)e^{+j\beta z} \end{aligned}}$$
(6-45)
(6-45a)

$$\boxed{\begin{aligned} A_x = A_y = F_x = F_y = F_z = 0 \quad \partial/\partial x \neq 0 \quad \partial/\partial y \neq 0 \\ A_z = A_z^+(x,y)e^{-j\beta z} + A_z^-(x,y)e^{+j\beta z} \end{aligned}}$$
(6-46)
(6-46a)

A similar procedure can be used to derive TEM modes in other directions such as TEM^x and TEM^y.

266 VECTOR POTENTIALS, SOLUTIONS, AND EQUATIONS

B. CYLINDRICAL COORDINATE SYSTEM

To derive expressions for TEM modes in a cylindrical coordinate system, a procedure similar to that in the rectangular coordinate system can be used. When (6-34)

$$\mathbf{E} = \mathbf{E}_A + \mathbf{E}_F = -j\omega\mathbf{A} - j\frac{1}{\omega\mu\varepsilon}\nabla(\nabla\cdot\mathbf{A}) - \frac{1}{\varepsilon}\nabla\times\mathbf{F} \qquad (6\text{-}47)$$

and (6-35)

$$\mathbf{H} = \mathbf{H}_A + \mathbf{H}_F = \frac{1}{\mu}\nabla\times\mathbf{A} - j\omega\mathbf{F} - j\frac{1}{\omega\mu\varepsilon}\nabla(\nabla\cdot\mathbf{F}) \qquad (6\text{-}48)$$

are expanded using

$$\mathbf{A}(\rho,\phi,z) = \hat{a}_\rho A_\rho(\rho,\phi,z) + \hat{a}_\phi A_\phi(\rho,\phi,z) + \hat{a}_z A_z(\rho,\phi,z) \qquad (6\text{-}49a)$$
$$\mathbf{F}(\rho,\phi,z) = \hat{a}_\rho F_\rho(\rho,\phi,z) + \hat{a}_\phi F_\phi(\rho,\phi,z) + \hat{a}_z F_z(\rho,\phi,z) \qquad (6\text{-}49b)$$

as solutions, they can be written as

$$\mathbf{E} = \hat{a}_\rho\left\{-j\omega A_\rho - j\frac{1}{\omega\mu\varepsilon}\frac{\partial}{\partial\rho}\left[\frac{1}{\rho}\frac{\partial}{\partial\rho}(\rho A_\rho) + \frac{1}{\rho}\frac{\partial A_\phi}{\partial\phi} + \frac{\partial A_z}{\partial z}\right] - \frac{1}{\varepsilon}\left(\frac{1}{\rho}\frac{\partial F_z}{\partial\phi} - \frac{\partial F_\phi}{\partial z}\right)\right\}$$

$$+\hat{a}_\phi\left\{-j\omega A_\phi - j\frac{1}{\omega\mu\varepsilon}\frac{1}{\rho}\frac{\partial}{\partial\phi}\left[\frac{1}{\rho}\frac{\partial}{\partial\rho}(\rho A_\rho) + \frac{1}{\rho}\frac{\partial A_\phi}{\partial\phi} + \frac{\partial A_z}{\partial z}\right]\right.$$
$$\left.-\frac{1}{\varepsilon}\left(\frac{\partial F_\rho}{\partial z} - \frac{\partial F_z}{\partial\rho}\right)\right\}$$

$$+\hat{a}_z\left\{-j\omega A_z - j\frac{1}{\omega\mu\varepsilon}\frac{\partial}{\partial z}\left[\frac{1}{\rho}\frac{\partial}{\partial\rho}(\rho A_\rho) + \frac{1}{\rho}\frac{\partial A_\phi}{\partial\phi} + \frac{\partial A_z}{\partial z}\right]\right.$$
$$\left.-\frac{1}{\varepsilon}\frac{1}{\rho}\left[\frac{\partial}{\partial\rho}(\rho F_\phi) - \frac{\partial F_\rho}{\partial\phi}\right]\right\} \qquad (6\text{-}50)$$

$$\mathbf{H} = \hat{a}_\rho\left\{-j\omega F_\rho - j\frac{1}{\omega\mu\varepsilon}\frac{\partial}{\partial\rho}\left[\frac{1}{\rho}\frac{\partial}{\partial\rho}(\rho F_\rho) + \frac{1}{\rho}\frac{\partial F_\phi}{\partial\phi} + \frac{\partial F_z}{\partial z}\right] + \frac{1}{\mu}\left(\frac{1}{\rho}\frac{\partial A_z}{\partial\phi} - \frac{\partial A_\phi}{\partial z}\right)\right\}$$

$$+\hat{a}_\phi\left\{-j\omega F_\phi - j\frac{1}{\omega\mu\varepsilon}\frac{1}{\rho}\frac{\partial}{\partial\phi}\left[\frac{1}{\rho}\frac{\partial}{\partial\rho}(\rho F_\rho) + \frac{1}{\rho}\frac{\partial F_\phi}{\partial\phi} + \frac{\partial F_z}{\partial z}\right]\right.$$
$$\left.+\frac{1}{\mu}\left(\frac{\partial A_\rho}{\partial z} - \frac{\partial A_z}{\partial\rho}\right)\right\}$$

$$+\hat{a}_z\left\{-j\omega F_z - j\frac{1}{\omega\mu\varepsilon}\frac{\partial}{\partial z}\left[\frac{1}{\rho}\frac{\partial}{\partial\rho}(\rho F_\rho) + \frac{1}{\rho}\frac{\partial F_\phi}{\partial\phi} + \frac{\partial F_z}{\partial z}\right]\right.$$
$$\left.+\frac{1}{\mu}\frac{1}{\rho}\left[\frac{\partial}{\partial\rho}(\rho A_\phi) - \frac{\partial A_\rho}{\partial\phi}\right]\right\} \qquad (6\text{-}51)$$

Example 6-2. Using (6-50) and (6-51) derive expressions for the **E** and **H** fields, in terms of the components of the **A** and **F** potentials, that are TEM to the ρ direction (TEM$^\rho$).

Solution. It is apparent by examining (6-50) and (6-51) that TEM$^\rho$ ($E_\rho = H_\rho = 0$) modes can be obtained by any of the following three combinations:

1. Letting

$$A_\phi = A_z = F_\phi = F_z = 0 \qquad A_\rho \neq 0 \qquad F_\rho \neq 0 \qquad \partial/\partial\phi \neq 0 \qquad \partial/\partial z \neq 0$$

For this combination, according to (6-50) and (6-51)

$$E_\rho = -j\omega A_\rho - j\frac{1}{\omega\mu\varepsilon}\frac{\partial}{\partial\rho}\left[\frac{1}{\rho}\frac{\partial}{\partial\rho}(\rho A_\rho)\right] = -j\frac{1}{\omega\mu\varepsilon}\left\{\frac{\partial}{\partial\rho}\left[\frac{1}{\rho}\frac{\partial}{\partial\rho}(\rho A_\rho)\right] + \omega^2\mu\varepsilon A_\rho\right\}$$

$$= -j\frac{1}{\omega\mu\varepsilon}\left[\frac{\partial}{\partial\rho}\left(\frac{\partial A_\rho}{\partial\rho} + \frac{A_\rho}{\rho}\right) + \omega^2\mu\varepsilon A_\rho\right]$$

$$= -j\frac{1}{\omega\mu\varepsilon}\left(\frac{\partial^2 A_\rho}{\partial\rho^2} + \frac{1}{\rho}\frac{\partial A}{\partial\rho} - \frac{A_\rho}{\rho^2} + \omega^2\mu\varepsilon A_\rho\right)$$

$$E_\rho = -j\frac{1}{\omega\mu\varepsilon}\left(\frac{\partial^2}{\partial\rho^2} + \frac{1}{\rho}\frac{\partial}{\partial\rho} - \frac{1}{\rho^2} + \beta^2\right)A_\rho = 0$$

provided

$$A_\rho(\rho, \phi, z) = A_\rho^+(\phi, z)H_1^{(2)}(\beta\rho) + A_\rho^-(\phi, z)H_1^{(1)}(\beta\rho)$$

Also

$$H_\rho = -j\omega F_\rho - j\frac{1}{\omega\mu\varepsilon}\frac{\partial}{\partial\rho}\left[\frac{1}{\rho}\frac{\partial}{\partial\rho}(\rho F_\rho)\right]$$

$$= -j\frac{1}{\omega\mu\varepsilon}\left(\frac{\partial^2}{\partial\rho^2} + \frac{1}{\rho}\frac{\partial}{\partial\rho} + \frac{1}{\rho^2} + \beta^2\right)F_\rho = 0$$

provided

$$F_\rho(\rho, \phi, z) = F_\rho^+(\phi, z)H_1^{(2)}(\beta\rho) + F_\rho^-(\phi, z)H_1^{(1)}(\beta\rho)$$

In addition

$$E_\phi = -j\frac{1}{\omega\mu\varepsilon}\frac{1}{\rho}\frac{\partial}{\partial\phi}\left[\frac{1}{\rho}\frac{\partial}{\partial\rho}(\rho A_\rho)\right] - \frac{1}{\varepsilon}\frac{\partial F_\rho}{\partial z}$$

$$E_z = -j\frac{1}{\omega\mu\varepsilon}\frac{\partial}{\partial z}\left[\frac{1}{\rho}\frac{\partial}{\partial\rho}(\rho A_\rho)\right] - \frac{1}{\varepsilon}\left(-\frac{1}{\rho}\frac{\partial F_\rho}{\partial\phi}\right)$$

$$H_\phi = -j\frac{1}{\omega\mu\varepsilon}\frac{1}{\rho}\frac{\partial}{\partial\phi}\left[\frac{1}{\rho}\frac{\partial}{\partial\rho}(\rho F_\rho)\right] + \frac{1}{\mu}\frac{\partial A_\rho}{\partial z}$$

$$H_z = -j\frac{1}{\omega\mu\varepsilon}\frac{\partial}{\partial z}\left[\frac{1}{\rho}\frac{\partial}{\partial\rho}(\rho F_\rho)\right] + \frac{1}{\mu}\left(-\frac{1}{\rho}\frac{\partial A_\rho}{\partial\phi}\right)$$

2. Letting

$$A_\rho = A_\phi = A_z = F_\phi = F_z = 0 \qquad F_\rho \neq 0 \qquad \partial/\partial\phi \neq 0 \qquad \partial/\partial z \neq 0$$

For this combination, according to (6-50) and (6-51)

$$E_\rho = 0$$

$$H_\rho = -j\omega F_\rho - j\frac{1}{\omega\mu\varepsilon}\frac{\partial}{\partial\rho}\left[\frac{1}{\rho}\frac{\partial}{\partial\rho}(\rho F_\rho)\right]$$

$$= -j\frac{1}{\omega\mu\varepsilon}\left(\frac{\partial^2}{\partial\rho^2} + \frac{1}{\rho}\frac{\partial}{\partial\rho} - \frac{1}{\rho^2} + \beta^2\right)F_\rho = 0$$

provided

$$F_\rho(\rho,\phi,z) = F_\rho^+(\phi,z)H_1^{(2)}(\beta\rho) + F_\rho^-(\phi,z)H_1^{(1)}(\beta\rho)$$

In addition

$$E_\phi = -\frac{1}{\varepsilon}\frac{\partial F_\rho}{\partial z}$$

$$E_z = -\frac{1}{\varepsilon}\left(-\frac{1}{\rho}\frac{\partial F_\rho}{\partial\phi}\right)$$

$$H_\phi = -j\frac{1}{\omega\mu\varepsilon}\frac{1}{\rho}\frac{\partial}{\partial\phi}\left[\frac{1}{\rho}\frac{\partial}{\partial\rho}(\rho F_\rho)\right]$$

$$H_z = -j\frac{1}{\omega\mu\varepsilon}\frac{\partial}{\partial z}\left[\frac{1}{\rho}\frac{\partial}{\partial\rho}(\rho F_\rho)\right]$$

3. Letting

$$A_\phi = A_z = F_\rho = F_\phi = F_z = 0 \qquad A_\rho \neq 0 \qquad \partial/\partial\phi \neq 0 \qquad \partial/\partial z \neq 0$$

For this combination, according to (6-50) and (6-51)

$$H_\rho = 0$$

$$E_\rho = -j\omega A_\rho - j\frac{1}{\omega\mu\varepsilon}\frac{\partial}{\partial\rho}\left[\frac{1}{\rho}\frac{\partial}{\partial\rho}(\rho A_\rho)\right]$$

$$= -j\frac{1}{\omega\mu\varepsilon}\left(\frac{\partial^2}{\partial\rho^2} + \frac{1}{\rho}\frac{\partial}{\partial\rho} - \frac{1}{\rho^2} + \beta^2\right)A_\rho = 0$$

provided

$$A_\rho(\rho,\phi,z) = A_\rho^+(\phi,z)H_1^{(2)}(\beta\rho) + A_\rho^-(\phi,z)H_1^{(1)}(\beta\rho)$$

In addition

$$E_\phi = -j\frac{1}{\omega\mu\varepsilon}\frac{1}{\rho}\frac{\partial}{\partial\phi}\left[\frac{1}{\rho}\frac{\partial}{\partial\rho}(\rho A_\rho)\right]$$

$$E_z = -j\frac{1}{\omega\mu\varepsilon}\frac{\partial}{\partial z}\left[\frac{1}{\rho}\frac{\partial}{\partial\rho}(\rho A_\rho)\right]$$

$$H_\phi = \frac{1}{\mu}\left(\frac{\partial A_\rho}{\partial z}\right)$$

$$H_z = \frac{1}{\mu}\left(-\frac{1}{\rho}\frac{\partial A_\rho}{\partial\phi}\right)$$

SUMMARY

From the results of Example 6-2, it is evident that TEM$^\rho$ modes can be obtained by any of the following three combinations:

$$A_\phi = A_z = F_\phi = F_z = 0 \quad \partial/\partial\phi \neq 0 \quad \partial/\partial z \neq 0 \quad (6\text{-}52)$$

$$A_\rho(\rho, \phi, z) = A_\rho^+(\phi, z) H_1^{(2)}(\beta\rho) + A_\rho^-(\phi, z) H_1^{(1)}(\beta\rho) \quad (6\text{-}52\text{a})$$

$$F_\rho(\rho, \phi, z) = F_\rho^+(\phi, z) H_1^{(2)}(\beta\rho) + F_\rho^-(\phi, z) H_1^{(1)}(\beta\rho) \quad (6\text{-}52\text{b})$$

$$A_\rho = A_\phi = A_z = F_\phi = F_z = 0 \quad \partial/\partial\phi \neq 0 \quad \partial/\partial z \neq 0 \quad (6\text{-}53)$$

$$F_\rho(\rho, \phi, z) = F_\rho^+(\phi, z) H_1^{(2)}(\beta\rho) + F_\rho^-(\phi, z) H_1^{(1)}(\beta\rho) \quad (6\text{-}53\text{a})$$

$$A_\phi = A_z = F_\rho = F_\phi = F_z = 0 \quad \partial/\partial\phi \neq 0 \quad \partial/\partial z \neq 0 \quad (6\text{-}54)$$

$$A_\rho(\rho, \phi, z) = A_\rho^+(\phi, z) H_1^{(2)}(\beta\rho) + A_\rho^-(\phi, z) H_1^{(1)}(\beta\rho) \quad (6\text{-}54\text{a})$$

A similar procedure can be used to derive TEM modes in other directions such as TEM$^\phi$ and TEM^z.

6.5.2 Transverse Magnetic Modes: Source-Free Region

Often we seek solutions of higher-order modes, other than transverse electromagnetic (TEM). Some of the higher-order modes, often required to satisfy boundary conditions, are designated as transverse magnetic (TM) and transverse electric (TE). Classical examples of the need for TM and TE modes are modes of propagation in waveguides [2].

Transverse magnetic modes (often also known as transverse magnetic fields) are field configurations whose magnetic field components lie in a plane that is transverse to a given direction. That direction is often chosen to be the path of wave propagation. For example, if the desired fields are TM to z (TM^z), this implies that $H_z = 0$. The other two magnetic field components (H_x and H_y) and three electric field components (E_x, E_y, and E_z) may or may not all exist.

By examining (6-43) and (6-51) it is evident that *to derive the field expressions that are TM to a given direction, independent of the coordinate system, it is sufficient to let the vector potential* **A** *have only a component in the direction in which the fields are desired to be TM. The remaining components of* **A** *as well as all of* **F** *are set equal to zero.*

A. RECTANGULAR COORDINATE SYSTEM

TM^z

To demonstrate the aforementioned procedure, let us assume that we wish to derive field expressions that are TM to z (TM^z). To accomplish this, we let

$$\mathbf{A} = \hat{a}_z A_z(x, y, z) \quad (6\text{-}55\text{a})$$

$$\mathbf{F} = 0 \quad (6\text{-}55\text{b})$$

The vector potential **A** must satisfy (6-30) which reduces from a vector wave equation to a scalar wave equation

$$\nabla^2 A_z(x, y, z) + \beta^2 A_z(x, y, z) = 0 \tag{6-56}$$

Since (6-56) is of the same form as (3-20a), its solution using the *separation of variables method* can be written according to (3-23) as

$$A_z(x, y, z) = f(x)g(y)h(z) \tag{6-57}$$

The solutions of $f(x)$, $g(y)$, and $h(z)$ take the forms given by (3-28a) through (3-30b). The most appropriate forms for $f(x)$, $g(y)$, and $h(z)$ much be chosen judiciously to reduce the complexity of the problem, and they will depend on the configuration of the problem. For the rectangular waveguide of Figure 3-2, for example, the most appropriate forms for $f(x)$, $g(y)$, and $h(z)$ are those given, respectively, by (3-28b), (3-29b), and (3-30a). Thus for the rectangular waveguide, (6-57) can be written as

$$A_z(x, y, z) = [C_1 \cos(\beta_x x) + D_1 \sin(\beta_x x)][C_2 \cos(\beta_y y) + D_2 \sin(\beta_y y)]$$
$$\times (A_3 e^{-j\beta_z z} + B_3 e^{+j\beta_z z}) \tag{6-58}$$

where

$$\beta_x^2 + \beta_y^2 + \beta_z^2 = \beta^2 = \omega^2 \mu \varepsilon \tag{6-58a}$$

Once A_z is found, the next step is to use (6-41) and (6-43) to find the **E** and **H** field components. Doing this, it can be shown that by using (6-55a) and (6-55b) we can reduce (6-41) and (6-43) to

TM^z Rectangular Coordinate System

$$E_x = -j \frac{1}{\omega\mu\varepsilon} \frac{\partial^2 A_z}{\partial x \partial z} \qquad H_x = \frac{1}{\mu} \frac{\partial A_z}{\partial y}$$

$$E_y = -j \frac{1}{\omega\mu\varepsilon} \frac{\partial^2 A_z}{\partial y \partial z} \qquad H_y = -\frac{1}{\mu} \frac{\partial A_z}{\partial x} \tag{6-59}$$

$$E_z = -j \frac{1}{\omega\mu\varepsilon} \left(\frac{\partial^2}{\partial z^2} + \beta^2 \right) A_z \qquad H_z = 0$$

which satisfy the definition of TM^z (i.e., $H_z = 0$).

For the specific example for which the solution of A_z as given by (6-58) is applicable, the unknown constants C_1, D_1, C_2, D_2, A_3, B_3, β_x, β_y, and β_z can be evaluated by substituting A_z of (6-58) into the expressions for **E** and **H** in (6-59) and enforcing the appropriate boundary conditions on the **E** and **H** field components. This will be demonstrated in Chapter 8, and elsewhere, where specific problem configurations are attempted. Following these or similar procedures should lead to the solution of the problem in question.

Expressions for the **E** and **H** field components that are TM^x and TM^y are given, respectively, by

TM^x Rectangular Coordinate System

Let

$$\mathbf{A} = \hat{a}_x A_x(x, y, z) \quad (6\text{-}60a)$$

$$\mathbf{F} = 0 \quad (6\text{-}60b)$$

Then

$$E_x = -j\frac{1}{\omega\mu\varepsilon}\left(\frac{\partial^2}{\partial x^2} + \beta^2\right)A_x \qquad H_x = 0$$

$$E_y = -j\frac{1}{\omega\mu\varepsilon}\frac{\partial^2 A_x}{\partial x \, \partial y} \qquad H_y = \frac{1}{\mu}\frac{\partial A_x}{\partial z} \qquad (6\text{-}61)$$

$$E_z = -j\frac{1}{\omega\mu\varepsilon}\frac{\partial^2 A_x}{\partial x \, \partial z} \qquad H_z = -\frac{1}{\mu}\frac{\partial A_x}{\partial y}$$

where A_x must satisfy the scalar wave equation

$$\nabla^2 A_x(x, y, z) + \beta^2 A_x(x, y, z) = 0 \quad (6\text{-}62)$$

TM^y Rectangular Coordinate System

Let

$$\mathbf{A} = \hat{a}_y A_y(x, y, z) \quad (6\text{-}63a)$$

$$\mathbf{F} = 0 \quad (6\text{-}63b)$$

Then

$$E_x = -j\frac{1}{\omega\mu\varepsilon}\frac{\partial^2 A_y}{\partial x \, \partial y} \qquad H_x = -\frac{1}{\mu}\frac{\partial A_y}{\partial z}$$

$$E_y = -j\frac{1}{\omega\mu\varepsilon}\left(\frac{\partial^2}{\partial y^2} + \beta^2\right)A_y \qquad H_y = 0 \qquad (6\text{-}64)$$

$$E_z = -j\frac{1}{\omega\mu\varepsilon}\frac{\partial^2 A_y}{\partial y \, \partial z} \qquad H_z = \frac{1}{\mu}\frac{\partial A_y}{\partial x}$$

where A_y must satisfy the scalar wave equation of

$$\nabla^2 A_y(x, y, z) + \beta^2 A_y(x, y, z) = 0 \quad (6\text{-}65)$$

The derivations of (6-61) and (6-64) are left to the reader as end of chapter assignments.

The expressions of (6-59), (6-61), and (6-64) are valid forms for the **E** and **H** field components of any problem in a rectangular coordinate system, which are, respectively, TM^z, TM^x, and TM^y. A similar procedure can be used to find expressions for the **E** and **H** field components that are TM to any direction in any coordinate system.

B. CYLINDRICAL COORDINATE SYSTEM

In terms of complexity, the next higher-order coordinate system is that of the cylindrical coordinate system. We will derive expressions that will be valid for TM^z. TM$^\rho$ and TM$^\phi$ are more difficult and are not usually utilized. Therefore they will not be attempted here. The procedure for TM^z in a cylindrical coordinate system is the same as that used for the rectangular coordinate system, as outlined previously in this section.

TM^z

To accomplish this, let

$$\mathbf{A} = \hat{a}_z A_z(\rho, \phi, z) \quad (6\text{-}66a)$$

$$\mathbf{F} = 0 \quad (6\text{-}66b)$$

The vector potential **A** must satisfy (6-30) with **J** = 0 which reduces from its vector form to the scalar wave equation

$$\nabla^2 A_z(\rho, \phi, z) + \beta^2 A_z(\rho, \phi, z) = 0 \quad (6\text{-}67)$$

Since (6-67) is of the same form as (3-54c), its solution using the *separation of variables method* can be written according to (3-57) as

$$A_z(\rho, \phi, z) = f(\rho) g(\phi) h(z) \quad (6\text{-}68)$$

The solutions of $f(\rho)$, $g(\phi)$, and $h(z)$ take the forms given by (3-67a) through (3-69b). The most appropriate forms for $f(\rho)$, $g(\phi)$, and $h(z)$ must be chosen judiciously to reduce the complexity of the problem, and they will depend upon the configuration of the problem. For the cylindrical waveguide of Figure 3-5, for example, the most appropriate forms for $f(\rho)$, $g(\phi)$, and $h(z)$ are those given, respectively, by (3-67a), (3-68b), and (3-69a). Thus for the cylindrical waveguide, (6-68) can be written as

$$A_z(\rho, \phi, z) = \left[A_1 J_m(\beta_\rho \rho) + B_1 Y_m(\beta_\rho \rho) \right] \left[C_2 \cos(m\phi) + D_2 \sin(m\phi) \right]$$
$$\times \left(A_3 e^{-j\beta_z z} + B_3 e^{+j\beta_z z} \right) \quad (6\text{-}69)$$

where

$$\beta_\rho^2 + \beta_z^2 = \beta^2 \quad (6\text{-}69a)$$

Once A_z is found, the next step is to use (6-50) and (6-51) to find the **E** and **H** field components. Then we can show that by using (6-66a) and (6-66b), (6-50) and

(6-51) can be reduced to

$$\boxed{\begin{array}{ll} \textbf{TM}^z \textbf{ Cylindrical Coordinate System} \\[6pt] E_\rho = -j\dfrac{1}{\omega\mu\varepsilon}\dfrac{\partial^2 A_z}{\partial\rho\,\partial z} & H_\rho = \dfrac{1}{\mu}\dfrac{1}{\rho}\dfrac{\partial A_z}{\partial\phi} \\[10pt] E_\phi = -j\dfrac{1}{\omega\mu\varepsilon}\dfrac{1}{\rho}\dfrac{\partial^2 A_z}{\partial\phi\,\partial z} & H_\phi = -\dfrac{1}{\mu}\dfrac{\partial A_z}{\partial\rho} \\[10pt] E_z = -j\dfrac{1}{\omega\mu\varepsilon}\left(\dfrac{\partial^2}{\partial z^2}+\beta^2\right) A_z & H_z = 0 \end{array}}\qquad (6\text{-}70)$$

which also satisfies the TM^z definition (i.e., $H_z = 0$).

For the specific example for which the solution of A_z as given by (6-69) is applicable, the unknown constants A_1, B_1, C_2, D_2, A_3, B_3, β_ρ, and β_z can be evaluated by substituting A_z of (6-69) into the expressions for **E** and **H** in (6-70) and enforcing the appropriate boundary conditions on the **E** and **H** field components. This will be demonstrated in Chapter 9, and elsewhere, where specific problem configurations are attempted. Following these or similar procedures should lead to the solution of the problem in question.

It should be stated that the same TM mode field constructions can be obtained by initiating the procedure with a solution to the scalar wave equation for the electric field component in the direction in which TM mode fields are desired. For example, if TM^z modes are desired, assume a solution for E_z of the same form as the vector potential component A_z. It can then be shown through Maxwell's equations that all the remaining electric and magnetic field components (with $H_z = 0$) can be expressed in terms of E_z. The same can be done for other TM^i modes by beginning with a solution for E_i having the same form as the vector potential component A_i. The only difference between the two formulations, one of which uses the vector potentials adopted in this book and the other that uses the fields themselves, is a normalization constant. For TM^z modes, for example, this normalization constant according to (6-59) is equal to $-j(\partial^2/\partial z^2 + \beta^2)/\omega\mu\varepsilon = -j(\beta^2 - \beta_z^2)/\omega\mu\varepsilon$. The preceding procedure is a very popular method used by many authors, and it is assigned to the reader as end of chapter exercises.

6.5.3 Transverse Electric Modes: Source-Free Region

Transverse electric (TE) modes can be derived in a fashion similar to the TM fields of Section 6.5.2. This time, however, we let the **F** vector potential have a nonvanishing component in the direction in which the TE fields are desired, and all the remaining components of **F** and **A** are set equal to zero. Without going through any of the details, we will list the expressions for the **E** and **H** field components for TE^z, TE^x, and TE^y in rectangular coordinates and TE^z in cylindrical coordinates. The details are left as exercises for the reader.

A. RECTANGULAR COORDINATE SYSTEM

Modes that are TE^z, TE^x, and TE^y are obtained as follows.

TE^z Rectangular Coordinate System

Let

$$\mathbf{A} = 0 \quad (6\text{-}71\text{a})$$

$$\mathbf{F} = \hat{a}_z F_z(x, y, z) \quad (6\text{-}71\text{b})$$

Then

$$\begin{aligned} E_x &= -\frac{1}{\varepsilon}\frac{\partial F_z}{\partial y} & H_x &= -j\frac{1}{\omega\mu\varepsilon}\frac{\partial^2 F_z}{\partial x\,\partial z} \\ E_y &= \frac{1}{\varepsilon}\frac{\partial F_z}{\partial x} & H_y &= -j\frac{1}{\omega\mu\varepsilon}\frac{\partial^2 F_z}{\partial y\,\partial z} \\ E_z &= 0 & H_z &= -j\frac{1}{\omega\mu\varepsilon}\left(\frac{\partial^2}{\partial z^2} + \beta^2\right)F_z \end{aligned} \quad (6\text{-}72)$$

where F_z must satisfy the scalar wave equation of

$$\nabla^2 F_z(x, y, z) + \beta^2 F_z(x, y, z) = 0 \quad (6\text{-}73)$$

TE^x Rectangular Coordinate System

Let

$$\mathbf{A} = 0 \quad (6\text{-}73\text{a})$$

$$\mathbf{F} = \hat{a}_x F_x(x, y, z) \quad (6\text{-}73\text{b})$$

Then

$$\begin{aligned} E_x &= 0 & H_x &= -j\frac{1}{\omega\mu\varepsilon}\left(\frac{\partial^2}{\partial x^2} + \beta^2\right)F_x \\ E_y &= -\frac{1}{\varepsilon}\frac{\partial F_x}{\partial z} & H_y &= -j\frac{1}{\omega\mu\varepsilon}\frac{\partial^2 F_x}{\partial x\,\partial y} \\ E_z &= \frac{1}{\varepsilon}\frac{\partial F_x}{\partial y} & H_z &= -j\frac{1}{\omega\mu\varepsilon}\frac{\partial^2 F_x}{\partial x\,\partial z} \end{aligned} \quad (6\text{-}74)$$

where F_x must satisfy the scalar wave equation of

$$\nabla^2 F_x(x, y, z) + \beta^2 F_x(x, y, z) = 0 \quad (6\text{-}75)$$

TE^y Rectangular Coordinate System

Let

$$\mathbf{A} = 0 \qquad (6\text{-}76a)$$

$$\mathbf{F} = \hat{a}_y F_y(x, y, z) \qquad (6\text{-}76b)$$

Then

$$\begin{aligned} E_x &= \frac{1}{\varepsilon}\frac{\partial F_y}{\partial z} & H_x &= -j\frac{1}{\omega\mu\varepsilon}\frac{\partial^2 F_y}{\partial x\,\partial y} \\ E_y &= 0 & H_y &= -j\frac{1}{\omega\mu\varepsilon}\left(\frac{\partial^2}{\partial y^2}+\beta^2\right)F_y \\ E_z &= -\frac{1}{\varepsilon}\frac{\partial F_y}{\partial x} & H_z &= -j\frac{1}{\omega\mu\varepsilon}\frac{\partial^2 F_y}{\partial y\,\partial z} \end{aligned} \qquad (6\text{-}77)$$

where F_y must satisfy the scalar wave equation of

$$\nabla^2 F_y(x, y, z) + \beta^2 F_y(x, y, z) = 0 \qquad (6\text{-}78)$$

B. CYLINDRICAL COORDINATE SYSTEM

Modes that are TE^z are obtained as follows.

TE^z Cylindrical Coordinate System

Let

$$\mathbf{A} = 0 \qquad (6\text{-}79a)$$

$$\mathbf{F} = \hat{a}_z F_z(\rho, \phi, z) \qquad (6\text{-}79b)$$

Then

$$\begin{aligned} E_\rho &= -\frac{1}{\varepsilon\rho}\frac{\partial F_z}{\partial \phi} & H_\rho &= -j\frac{1}{\omega\mu\varepsilon}\frac{\partial^2 F_z}{\partial \rho\,\partial z} \\ E_\phi &= \frac{1}{\varepsilon}\frac{\partial F_z}{\partial \rho} & H_\phi &= -j\frac{1}{\omega\mu\varepsilon\rho}\frac{\partial^2 F_z}{\partial \phi\,\partial z} \\ E_z &= 0 & H_z &= -j\frac{1}{\omega\mu\varepsilon}\left(\frac{\partial^2}{\partial z^2}+\beta^2\right)F_z \end{aligned} \qquad (6\text{-}80)$$

where F_z must satisfy the scalar wave equation of

$$\nabla^2 F_z(\rho, \phi, z) + \beta^2 F_z(\rho, \phi, z) = 0 \qquad (6\text{-}81)$$

As was suggested earlier for the TM modes, an alternate procedure for construction of TE^i field configurations will be to initiate the procedure with a

276 VECTOR POTENTIALS, SOLUTIONS, AND EQUATIONS

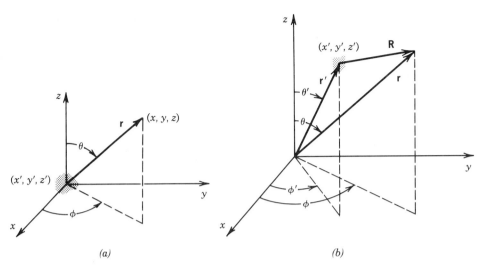

FIGURE 6-2 Coordinate systems for computing radiation fields. (*Source:* C. A. Balanis, *Antenna Theory: Analysis and Design*. Copyright © 1982, John Wiley & Sons, Inc. Reprinted by permission of John Wiley & Sons, Inc.) (*a*) Source at origin. (*b*) Source not at origin.

solution for the H_i component with the same form as the vector potential component F_i. For example, if TE^z modes are desired, assume a solution for H_z of the same form as the vector potential component F_z. It can then be shown through Maxwell's equations that all the remaining electric and magnetic fields (with $E_z = 0$) can be expressed in terms of H_z. The only difference between the two formulations, one that uses the vector potentials adopted in this book and the other that uses the fields themselves, is a normalization. For TE^z modes, for example, this normalization constant according to (6-72) is equal to $-j(\partial^2/\partial z^2 + \beta^2)/\omega\mu\varepsilon = -j(\beta^2 - \beta_z^2)/\omega\mu\varepsilon$. The preceding procedure is also a very popular method used by many authors, and it is assigned to the reader as end of chapter exercises.

6.6 SOLUTION OF THE INHOMOGENEOUS VECTOR POTENTIAL WAVE EQUATION

In Sections 6.2 and 6.3 we derived the inhomogeneous vector wave equations 6-16 and 6-28. In this section we want to derive the solutions to each equation.

Let us assume that a source with current density J_z, which in the limit is an infinitesimal point source, is placed at the origin of a x, y, z coordinate system, as shown in Figure 6-2a. Since the current density J_z is directed along the z axis, only an A_z component will exist. Thus we can write (6-16) as

$$\nabla^2 A_z + \beta^2 A_z = -\mu J_z \qquad (6\text{-}82)$$

At points removed from the source ($J_z = 0$), the wave equation reduces to

$$\nabla^2 A_z + \beta^2 A_z = 0 \qquad (6\text{-}83)$$

Since in the limit the source is a point, it requires that A_z is not a function of direction (θ and ϕ); in a spherical coordinate system, $A_z = A_z(r)$ where r is the radial distance. Thus (6-83) can be written as

$$\nabla^2 A_z(r) + \beta^2 A_z(r) = \frac{1}{r^2} \frac{\partial}{\partial r}\left[r^2 \frac{\partial A_z(r)}{\partial r}\right] + \beta^2 A_z(r) = 0 \qquad (6\text{-}84)$$

which when expanded reduces to

$$\frac{d^2 A_z(r)}{dr^2} + \frac{2}{r}\frac{dA_z(r)}{dr} + \beta^2 A_z(r) = 0 \qquad (6\text{-}84\text{a})$$

The partial derivative has been replaced by the ordinary derivative since A_z is only a function of the radial coordinate.

The differential equation 6-84a has two independent solutions

$$A_{z1} = C_1 \frac{e^{-j\beta r}}{r} \qquad (6\text{-}85\text{a})$$

$$A_{z2} = C_2 \frac{e^{+j\beta r}}{r} \qquad (6\text{-}85\text{b})$$

Equation 6-85a represents an outwardly (in the radial direction) traveling wave and (6-85b) describes an inwardly traveling wave, assuming $e^{j\omega t}$ time variations. For this problem, the source is placed at the origin with the radiated fields traveling in the outward radial direction. Therefore, we choose the solution of (6-85a), or

$$A_z = A_{z1} = C_1 \frac{e^{-j\beta r}}{r} \qquad (6\text{-}86)$$

In the static case, $\omega = 0$, $\beta = 0$, (6-86) simplifies to

$$A_z = \frac{C_1}{r} \qquad (6\text{-}86\text{a})$$

which is a solution to the wave equation 6-83, 6-84, or 6-84a when $\beta = 0$. Thus at points removed from the source, the time-varying and the static solutions of (6-86) and (6-86a) differ only by the $e^{-j\beta r}$ factor, or the time-varying solution of (6-86) can be obtained by multiplying the static solution of (6-86a) by $e^{-j\beta r}$.

In the presence of the source ($J_z \neq 0$) and with $\beta = 0$, the wave equation 6-82 reduces to

$$\nabla^2 A_z = -\mu J_z \qquad (6\text{-}87)$$

This equation is recognized as Poisson's equation whose solution is widely documented. The most familiar equation with Poisson form is that relating the scalar electric potential ϕ to the electric charge density q. This is given by

$$\nabla^2 \phi = -\frac{q}{\varepsilon} \qquad (6\text{-}88)$$

whose solution is

$$\phi = \frac{1}{4\pi\varepsilon} \iiint_V \frac{q}{r} \, dv' \qquad (6\text{-}89)$$

where r is the distance from any point on the charge density to the observation point. Since (6-87) is similar in form to (6-88), its solution is similar to (6-89), or

$$A_z = \frac{\mu}{4\pi} \iiint_V \frac{J_z}{r} \, dv' \tag{6-90}$$

Equation 6-90 represents the solution to (6-82) when $\beta = 0$, the static case. Using the comparative analogy between (6-86) and (6-86a), the time-varying solution of (6-82) can be obtained by multiplying the static solution of (6-90) by $e^{-j\beta r}$. Thus

$$A_z = \frac{\mu}{4\pi} \iiint_V J_z \frac{e^{-j\beta r}}{r} \, dv' \tag{6-91}$$

which is a solution to (6-82).

If the current densities were in the x and y directions (J_x and J_y), the wave equation for each would reduce to

$$\nabla^2 A_x + \beta^2 A_x = -\mu J_x \tag{6-92a}$$
$$\nabla^2 A_y + \beta^2 A_y = -\mu J_y \tag{6-92b}$$

with corresponding solutions similar in form to (6-91), or

$$A_x = \frac{\mu}{4\pi} \iiint_V J_x \frac{e^{-j\beta r}}{r} \, dv' \tag{6-93a}$$

$$A_y = \frac{\mu}{4\pi} \iiint_V J_y \frac{e^{-j\beta r}}{r} \, dv' \tag{6-93b}$$

The solutions of (6-91), (6-93a), and (6-93b) allow us to write the solution to the vector wave equation 6-16 as

$$\mathbf{A} = \frac{\mu}{4\pi} \iiint_V \mathbf{J} \frac{e^{-j\beta r}}{r} \, dv' \tag{6-94}$$

If the source is removed from the origin and placed at a position represented by the primed coordinates (x', y', z'), as shown in Figure 6-2b, (6-94) can be written as

$$\mathbf{A}(x, y, z) = \frac{\mu}{4\pi} \iiint_V \mathbf{J}(x', y', z') \frac{e^{-j\beta R}}{R} \, dv' \tag{6-95a}$$

where the primed coordinates represent the source, the unprimed coordinates represent the observation point, and R represents the distance from any point in the source to the observation point. In a similar fashion we can show that the solution of (6-28) is given by

$$\mathbf{F}(x, y, z) = \frac{\varepsilon}{4\pi} \iiint_V \mathbf{M}(x', y', z') \frac{e^{-j\beta R}}{R} \, dv' \tag{6-95b}$$

SOLUTION OF THE INHOMOGENEOUS VECTOR POTENTIAL WAVE EQUATION

If **J** and **M** represent linear densities (m^{-1}), (6-95a) and (6-95b) reduce to surface integrals, or

$$\mathbf{A} = \frac{\mu}{4\pi}\iint_S \mathbf{J}_s(x', y', z')\frac{e^{-j\beta R}}{R}\,ds' \qquad (6\text{-}96a)$$

$$\mathbf{F} = \frac{\varepsilon}{4\pi}\iint_S \mathbf{M}_s(x', y', z')\frac{e^{-j\beta R}}{R}\,ds' \qquad (6\text{-}96b)$$

For electric and magnetic currents $\mathbf{I}_e$ and $\mathbf{I}_m$, they in turn reduce to line integrals of the form

$$\mathbf{A} = \frac{\mu}{4\pi}\int_C \mathbf{I}_e(x', y', z')\frac{e^{-j\beta R}}{R}\,dl' \qquad (6\text{-}97a)$$

$$\mathbf{F} = \frac{\varepsilon}{4\pi}\int_C \mathbf{I}_m(x', y', z')\frac{e^{-j\beta R}}{R}\,dl' \qquad (6\text{-}97b)$$

Example 6-3. A very thin linear electric current element of very short length ($\ell \ll \lambda$) and with a constant current

$$\mathbf{I}_e(z') = \hat{a}_z I_e$$

such that $I_e \ell$ = constant, is positioned symmetrically at the origin and oriented along the z axis, as shown in Figure 6-2a. Such an element is usually referred to as an *infinitesimal dipole* [1]. Determine the electric and magnetic fields radiated by the dipole.

Solution. The solution will be obtained using the procedure summarized in Section 6.4. Since the element (source) carries only an electric current $\mathbf{I}_e$, the magnetic current $\mathbf{I}_m$ and the vector potential **F** of (6-97b) are both zero. The vector potential **A** of (6-97a) is then written as

$$\mathbf{A}(x, y, z) = \frac{\mu}{4\pi}\int_{-\ell/2}^{+\ell/2} \hat{a}_z I_e \frac{e^{-j\beta R}}{R}\,dz'$$

where R is the distance from any point on the element, $-\ell/2 \le z' \le \ell/2$, to the observation point. Since in the limit as $\ell \to 0$ ($\ell \ll \lambda$), then

$$R = r$$

so that

$$\mathbf{A}(x, y, z) = \hat{a}_z \frac{\mu I_e e^{-j\beta r}}{4\pi r}\int_{-\ell/2}^{+\ell/2} dz' = \hat{a}_z \frac{\mu I_e \ell}{4\pi r}e^{-j\beta r}$$

Transforming the vector potential **A** from rectangular to spherical components using the inverse (in this case also transpose) transformation of (II-9) from

Appendix II, we can write

$$A_r = A_z \cos\theta = \frac{\mu I_e \ell e^{-j\beta r}}{4\pi r} \cos\theta$$

$$A_\theta = -A_z \sin\theta = -\frac{\mu I_e \ell e^{-j\beta r}}{4\pi r} \sin\theta$$

$$A_\phi = 0$$

Using the symmetry of the problem, that is, no variations in ϕ, (6-32a) can be expanded in spherical coordinates and written in simplified form as

$$\mathbf{H} = \hat{a}_\phi \frac{1}{\mu r}\left[\frac{\partial}{\partial r}(rA_\theta) - \frac{\partial A_r}{\partial \theta}\right]$$

which reduces to

$$H_r = H_\theta = 0$$

$$H_\phi = j\frac{\beta I_e \ell \sin\theta}{4\pi r}\left(1 + \frac{1}{j\beta r}\right)e^{-j\beta r}$$

The electric field $\mathbf{E}$ can be found using either (6-32b) or (6-32c), that is,

$$\mathbf{E} = -j\omega\mathbf{A} - j\frac{1}{\omega\mu\varepsilon}\nabla(\nabla\cdot\mathbf{A}) = \frac{1}{j\omega\varepsilon}\nabla\times\mathbf{H}$$

and either leads to

$$E_r = \eta\frac{I_e \ell \cos\theta}{2\pi r^2}\left(1 + \frac{1}{j\beta r}\right)e^{-j\beta r}$$

$$E_\theta = j\eta\frac{\beta I_e \ell \sin\theta}{4\pi r}\left[1 + \frac{1}{j\beta r} - \frac{1}{(\beta r)^2}\right]e^{-j\beta r}$$

$$E_\phi = 0$$

The E- and H-field components are valid everywhere except on the source itself.

6.7 FAR-FIELD RADIATION

The fields radiated by antennas of finite dimensions are spherical waves. For these radiators, a general solution to the vector wave equation 6-16 in spherical components, each as a function of r, θ, and ϕ, takes the general form

$$\mathbf{A} = \hat{a}_r A_r(r,\theta,\phi) + \hat{a}_\theta A_\theta(r,\theta,\phi) + \hat{a}_\phi A_\phi(r,\theta,\phi) \qquad (6\text{-}98)$$

The amplitude variations of r in each component of (6-98) are of the form $1/r^n$, $n = 1, 2, \ldots$ [1]. Neglecting higher-order terms of $1/r^n$ ($1/r^n = 0$, $n = 2, 3, \ldots$) reduces (6-98) to

$$\mathbf{A} \simeq [\hat{a}_r A'_r(\theta,\phi) + \hat{a}_\theta A'_\theta(\theta,\phi) + \hat{a}_\phi A'_\phi(\theta,\phi)]\frac{e^{-j\beta r}}{r} \qquad r\to\infty \qquad (6\text{-}99)$$

The r variations are separable from those of θ and ϕ. This will be demonstrated by many examples in the chapters that follow.

Substituting (6-99) into (6-17) reduces it to

$$\mathbf{E} = \frac{1}{r}\left\{-j\omega e^{-j\beta r}\left[\hat{a}_r(0) + \hat{a}_\theta A'_\theta(\theta,\phi) + \hat{a}_\phi A'_\phi(\theta,\phi)\right]\right\}$$
$$+ \frac{1}{r^2}\{\cdots\} + \cdots \qquad (6\text{-}100\text{a})$$

The radial E-field component has no $1/r$ terms because its contributions from the first and second terms of (6-17) cancel each other.

Similarly, by using (6-99), we can write (6-4a) as

$$\mathbf{H} = \frac{1}{r}\left\{-j\frac{\omega}{\eta}e^{-j\beta r}\left[\hat{a}_r(0) + \hat{a}_\theta A'_\phi(\theta,\phi) - \hat{a}_\phi A'_\theta(\theta,\phi)\right]\right\}$$
$$+ \frac{1}{r^2}\{\cdots\} + \cdots \qquad (6\text{-}100\text{b})$$

where $\eta = \sqrt{\mu/\varepsilon}$ is the intrinsic impedance of the medium.

Neglecting higher-order terms of $1/r^n$, the radiated E and H fields have only θ and ϕ components. They can be expressed as

Far-Field Region

$$\left.\begin{array}{l} E_r \simeq 0 \\ E_\theta \simeq -j\omega A_\theta \\ E_\phi \simeq -j\omega A_\phi \end{array}\right\} \Rightarrow \boxed{\mathbf{E}_A \simeq -j\omega \mathbf{A}} \quad \text{(for the }\theta\text{ and }\phi\text{ components only since } E_r \simeq 0\text{)} \qquad (6\text{-}101\text{a})$$

$$\left.\begin{array}{l} H_r \simeq 0 \\ H_\theta \simeq +j\dfrac{\omega}{\eta}A_\phi = -\dfrac{E_\phi}{\eta} \\ H_\phi \simeq -j\dfrac{\omega}{\eta}A_\theta = +\dfrac{E_\theta}{\eta} \end{array}\right\} \Rightarrow \boxed{\mathbf{H}_A \simeq \dfrac{\hat{a}_r}{\eta}\times\mathbf{E}_A = -j\dfrac{\omega}{\eta}\hat{a}_r\times\mathbf{A}} \quad \text{(for the }\theta\text{ and }\phi\text{ components only since } H_r \simeq 0\text{)} \qquad (6\text{-}101\text{b})$$

Radial field components exist only for higher-order terms of $1/r^n$.

In a similar manner, the far-zone fields that are due to a magnetic source $\mathbf{M}$ (potential $\mathbf{F}$) can be written as

Far-Field Region

$$\left.\begin{array}{l} H_r \simeq 0 \\ H_\theta \simeq -j\omega F_\theta \\ H_\phi \simeq -j\omega F_\phi \end{array}\right\} \Rightarrow \boxed{\mathbf{H}_F \simeq -j\omega \mathbf{F}} \quad \text{(for the }\theta\text{ and }\phi\text{ components only since } H_r \simeq 0\text{)} \qquad (6\text{-}102\text{a})$$

$$\left.\begin{array}{l} E_r \simeq 0 \\ E_\theta \simeq -j\omega\eta F_\phi = +\eta H_\phi \\ E_\phi \simeq +j\omega\eta F_\theta = -\eta H_\theta \end{array}\right\} \Rightarrow \boxed{\mathbf{E}_F = -\eta\hat{a}_r\times\mathbf{H}_F = j\omega\eta\hat{a}_r\times\mathbf{F}} \quad \text{(for the }\theta\text{ and }\phi\text{ components only since } E_r \simeq 0\text{)} \qquad (6\text{-}102\text{b})$$

282 VECTOR POTENTIALS, SOLUTIONS, AND EQUATIONS

Simply stated, the corresponding far-zone **E**- and **H**-field components are orthogonal to each other and form TEM (to r) mode fields. This is a very useful relation, and it will be adopted in the following chapters for the solution of the far-zone radiated fields. The far-zone (far-field) region for a radiator is defined as the region whose smallest radial distance is $2D^2/\lambda$ where D is the largest dimension of the radiator (provided D is large compared to the wavelength) [1].

6.8 RADIATION AND SCATTERING EQUATIONS

In Sections 6.4 and 6.6 it was stated that the fields radiated by sources represented by **J** and **M** in an unbounded medium can be computed using (6-32a) through (6-35a) where **A** and **F** are found using (6-95a) and (6-95b). For (6-95a) and (6-95b) the integration is performed over the entire space occupied by **J** and **M** of Figure 6-2b [or $\mathbf{J}_s$ and $\mathbf{M}_s$ of (6-96a) and (6-96b) or $\mathbf{I}_e$ and $\mathbf{I}_m$ of (6-97a) and (6-97b)]. These equations yield valid solutions for all observation points. For most problems, the main difficulty is the inability to perform the integrations in (6-95a) and (6-95b), (6-96a) and (6-96b), or (6-97a) and (6-97b). However for far-field observations the complexity of the formulation can be reduced.

6.8.1 Near Field

According to Figure 6-2b and equation 6-95a the vector potential **A** that is due to current density **J** is given by

$$\mathbf{A}(x, y, z) = \frac{\mu}{4\pi} \iiint_V \mathbf{J}(x', y', z') \frac{e^{-j\beta R}}{R} dv' \quad (6\text{-}103)$$

where the primed coordinates (x', y', z') represent the source and the unprimed coordinates (x, y, z) represent the observation point. Here we intend to write expressions for the **E** and **H** fields that are due to the potential of (6-103) which would be valid everywhere [3, 4]. The equations will not be in closed form, but will be convenient for computational purposes. The development will be restricted to the rectangular coordinate system.

The magnetic field due to the potential of (6-103) is given by (6-32a) as

$$\mathbf{H}_A = \frac{1}{\mu} \nabla \times \mathbf{A} = \frac{1}{4\pi} \nabla \times \iiint_V \mathbf{J}(x', y', z') \frac{e^{-j\beta R}}{R} dv' \quad (6\text{-}104)$$

Interchanging integration and differentiation we can write (6-104) as

$$\mathbf{H}_A = \frac{1}{4\pi} \iiint_V \nabla \times \left[\mathbf{J}(x', y', z') \frac{e^{-j\beta R}}{R} \right] dv' \quad (6\text{-}104\text{a})$$

Using the vector identity

$$\nabla \times (g\mathbf{F}) = (\nabla g) \times \mathbf{F} + g(\nabla \times \mathbf{F}) \quad (6\text{-}105)$$

we can write

$$\nabla \times \left[\frac{e^{-j\beta R}}{R} \mathbf{J}(x', y', z') \right] = \nabla \left(\frac{e^{-j\beta R}}{R} \right) \times \mathbf{J}(x', y', z') + \frac{e^{-j\beta R}}{R} \nabla \times \mathbf{J}(x', y', z')$$

$$(6\text{-}106)$$

Since **J** is only a function of the primed coordinates and ∇ is a function of the unprimed coordinates,

$$\nabla \times \mathbf{J}(x', y', z') = 0 \quad (6\text{-}106a)$$

Also

$$\nabla \left(\frac{e^{-j\beta R}}{R} \right) = -\hat{R}\left(\frac{1 + j\beta R}{R^2} \right) e^{-j\beta R} \quad (6\text{-}106b)$$

where $\hat{R}$ is a unit vector directed along the line joining any point of the source and the observation point. Using (6-106) through (6-106b) we can write (6-104a) as

$$\mathbf{H}_A(x, y, z) = -\frac{1}{4\pi} \iiint_V (\hat{R} \times \mathbf{J}) \frac{1 + j\beta R}{R^2} e^{-j\beta R} \, dx' \, dy' \, dz' \quad (6\text{-}107)$$

which can be expanded in its three rectangular components [3, 4]

$$H_{Ax} = \frac{1}{4\pi} \iiint_V [(z - z')J_y - (y - y')J_z] \frac{1 + j\beta R}{R^3} e^{-j\beta R} \, dx' \, dy' \, dz' \quad (6\text{-}107a)$$

$$H_{Ay} = \frac{1}{4\pi} \iiint_V [(x - x')J_z - (z - z')J_x] \frac{1 + j\beta R}{R^3} e^{-j\beta R} \, dx' \, dy' \, dz' \quad (6\text{-}107b)$$

$$H_{Az} = \frac{1}{4\pi} \iiint_V [(y - y')J_x - (x - x')J_y] \frac{1 + j\beta R}{R^3} e^{-j\beta R} \, dx' \, dy' \, dz' \quad (6\text{-}107c)$$

Using (6-32b) or Maxwell's equation 6-32c we can write the corresponding electric field components as

$$\mathbf{E}_A = \hat{a}_x E_{Ax} + \hat{a}_y E_{Ay} + \hat{a}_z E_{Az} = -j\omega\mathbf{A} - j\frac{1}{\omega\mu\varepsilon}\nabla(\nabla \cdot \mathbf{A}) = \frac{1}{j\omega\varepsilon}\nabla \times \mathbf{H}_A \quad (6\text{-}108)$$

which with the aid of (6-107a) through (6-107c) reduce to

$$E_{Ax} = -\frac{j\eta}{4\pi\beta} \iiint_V \{G_1 J_x + (x - x')G_2$$
$$\times [(x - x')J_x + (y - y')J_y + (z - z')J_z]\} e^{-j\beta R} \, dx' \, dy' \, dz'$$
$$(6\text{-}108a)$$

$$E_{Ay} = -\frac{j\eta}{4\pi\beta} \iiint_V \{G_1 J_y + (y - y')G_2$$
$$\times [(x - x')J_x + (y - y')J_y + (z - z')J_z]\} e^{-j\beta R} \, dx' \, dy' \, dz'$$
$$(6\text{-}108b)$$

$$E_{Az} = -\frac{j\eta}{4\pi\beta} \iiint_V \{G_1 J_z + (z - z')G_2$$
$$\times [(x - x')J_x + (y - y')J_y + (z - z')J_z]\} e^{-j\beta R} \, dx' \, dy' \, dz'$$
$$(6\text{-}108c)$$

where

$$G_1 = \frac{-1 - j\beta R + \beta^2 R^2}{R^3} \qquad (6\text{-}108\text{d})$$

$$G_2 = \frac{3 + j3\beta R - \beta^2 R^2}{R^5} \qquad (6\text{-}108\text{e})$$

In the same manner, we can write for the vector potential of (6-95b)

$$\mathbf{F}(x, y, z) = \frac{\varepsilon}{4\pi} \iiint_V \mathbf{M}(x', y', z') \frac{e^{-j\beta R}}{R} \, dv' \qquad (6\text{-}109)$$

the electric field components using (6-33a),

$$\mathbf{E}_F = -\frac{1}{\varepsilon} \nabla \times \mathbf{F} \qquad (6\text{-}110)$$

as

$$E_{Fx} = -\frac{1}{4\pi} \iiint_V [(z - z')M_y - (y - y')M_z] \frac{1 + j\beta R}{R^3} e^{-j\beta R} \, dx' \, dy' \, dz' \quad (6\text{-}110\text{a})$$

$$E_{Fy} = -\frac{1}{4\pi} \iiint_V [(x - x')M_z - (z - z')M_x] \frac{1 + j\beta R}{R^3} e^{-j\beta R} \, dx' \, dy' \, dz' \quad (6\text{-}110\text{b})$$

$$E_{Fz} = -\frac{1}{4\pi} \iiint_V [(y - y')M_x - (x - x')M_y] \frac{1 + j\beta R}{R^3} e^{-j\beta R} \, dx' \, dy' \, dz' \quad (6\text{-}110\text{c})$$

Similarly the corresponding magnetic field components can be written using (6-33b) or (6-33c)

$$\mathbf{H}_F = -j\omega \mathbf{F} - j\frac{1}{\omega\mu\varepsilon} \nabla(\nabla \cdot \mathbf{F}) = -\frac{1}{j\omega\mu} \nabla \times \mathbf{E}_F \qquad (6\text{-}111)$$

as

$$H_{Fx} = -\frac{j}{4\pi\beta\eta} \iiint_V \{G_1 M_x + (x - x')G_2$$
$$\times [(x - x')M_x + (y - y')M_y + (z - z')M_z]\} e^{-j\beta R} \, dx' \, dy' \, dz'$$
$$(6\text{-}111\text{a})$$

$$H_{Fy} = -\frac{j}{4\pi\beta\eta} \iiint_V \{G_1 M_y + (y - y')G_2$$
$$\times [(x - x')M_x + (y - y')M_y + (z - z')M_z]\} e^{-j\beta R} \, dx' \, dy' \, dz'$$
$$(6\text{-}111\text{b})$$

$$H_{Fz} = -\frac{j}{4\pi\beta\eta} \iiint_V \{G_1 M_z + (z - z')G_2$$
$$\times [(x - x')M_x + (y - y')M_y + (z - z')M_z]\} e^{-j\beta R} \, dx' \, dy' \, dz'$$
$$(6\text{-}111\text{c})$$

where G_1 and G_2 are given by (6-108d) and (6-108e).

6.8.2 Far Field

It was shown in Section 6.7 that the field equations for far-field ($\beta r \gg 1$) observations simplify considerably. Also in the far zone the **E**- and **H**-field components are orthogonal to each other and form TEM (to r) mode fields. Although the field equations in the far zone simplify, integrations still need to be performed to find the vector potentials of **A** and **F** given, respectively, by (6-95a) and (6-95b), or (6-96a)

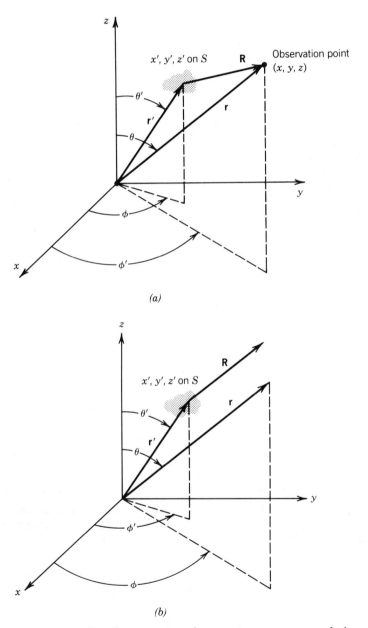

FIGURE 6-3 Coordinate system for aperture antenna analysis. (*Source:* C. A. Balanis, *Antenna Theory: Analysis and Design.* Copyright © 1982, John Wiley & Sons, Inc. Reprinted by permission of John Wiley & Sons, Inc.) (*a*) Near and far fields. (*b*) Far field.

and (6-96b), or (6-97a) and (6-97b). However, the integrations, as will be shown next, can be simplified if the observations are made in the far field.

If the observations are made in the far field ($\beta r \gg 1$), it can be shown [1] that the radial distance **R** of Figure 6-3a from any point on the source or scatterer to the observation point can be assumed to be parallel to the radial distance **r** from the origin to the observation point, as shown in Figure 6-3b. In such cases the relation between the magnitudes of **R** and **r** of Figure 6-3a, given by

$$R = \left[r^2 + (r')^2 - 2rr' \cos \psi \right]^{1/2} \quad (6\text{-}112)$$

can be approximated, according to Figure 6-3b, most commonly by [1]

$$R = \begin{cases} r - r' \cos \psi & \text{for phase variations} \quad (6\text{-}112a) \\ r & \text{for amplitude variations} \quad (6\text{-}112b) \end{cases}$$

where ψ is the angle between **r** and **r**′. These approximations yield a maximum phase error of $\pi/8$ (22.5°) provided the observations are made at distances

$$r \geq \frac{2D^2}{\lambda} \quad (6\text{-}113)$$

where D is the largest dimension of the radiator or scatterer. The distance (6-113) represents the *minimum* distance to the far-field region. The derivation of (6-113), as well as distances for other zones, can be found in [1]. Using (6-112a) and (6-112b) we can write (6-96a) and (6-96b), assuming the current densities reside on the surface of the source, as

$$\mathbf{A} = \frac{\mu}{4\pi} \iint_S \mathbf{J}_s \frac{e^{-j\beta R}}{R} \, ds' \simeq \frac{\mu e^{-j\beta r}}{4\pi r} \mathbf{N} \quad (6\text{-}114a)$$

$$\mathbf{F} = \frac{\varepsilon}{4\pi} \iint_S \mathbf{M}_s \frac{e^{-j\beta R}}{R} \, ds' \simeq \frac{\varepsilon e^{-j\beta r}}{4\pi r} \mathbf{L} \quad (6\text{-}114b)$$

where

$$\mathbf{N} = \iint_S \mathbf{J}_s e^{j\beta r' \cos \psi} \, ds' \quad (6\text{-}114c)$$

$$\mathbf{L} = \iint_S \mathbf{M}_s e^{j\beta r' \cos \psi} \, ds' \quad (6\text{-}114d)$$

It was shown in Section 6.7 that in the far field only the θ and ϕ components of the **E** and **H** fields are dominant. Although the radial components are not necessarily zero, they are negligible compared to the θ and ϕ components. Also it was shown that for (6-32b) and (6-33b), or

$$\mathbf{E}_A = -j\omega \left[\mathbf{A} + \frac{1}{\beta^2} \nabla(\nabla \cdot \mathbf{A}) \right] \quad (6\text{-}115a)$$

$$\mathbf{H}_F = -j\omega \left[\mathbf{F} + \frac{1}{\beta^2} \nabla(\nabla \cdot \mathbf{F}) \right] \quad (6\text{-}115b)$$

where **A** and **F** are given by (6-114a) and (6-114b), the second terms within the brackets only contribute variations of the order $1/r^2$, $1/r^3$, $1/r^4$, etc. Since observations are made in the far field, the dominant variation is of the order $1/r$

and it is contained in the first term of (6-115a) and (6-115b). Thus for far-field observations, (6-115a) and (6-115b) reduce to

$$\mathbf{E}_A \simeq -j\omega\mathbf{A} \quad (\theta \text{ and } \phi \text{ components only}) \tag{6-116a}$$

$$\mathbf{H}_F \simeq -j\omega\mathbf{F} \quad (\theta \text{ and } \phi \text{ components only}) \tag{6-116b}$$

which can be expanded and written as

$$(E_A)_\theta \simeq -j\omega A_\theta \tag{6-117a}$$

$$(E_A)_\phi \simeq -j\omega A_\phi \tag{6-117b}$$

$$(H_F)_\theta \simeq -j\omega F_\theta \tag{6-117c}$$

$$(H_F)_\phi \simeq -j\omega F_\phi \tag{6-117d}$$

The radial components are neglected because they are very small compared to the θ and ϕ components.

To find the remaining **E** and **H** fields contributed by the **F** and **A** potentials, that is $\mathbf{E}_F$ and $\mathbf{H}_A$, we can use (6-33a) and (6-32a), or

$$\mathbf{E}_F = -\frac{1}{\varepsilon}\nabla \times \mathbf{F} \tag{6-118a}$$

$$\mathbf{H}_A = \frac{1}{\mu}\nabla \times \mathbf{A} \tag{6-118b}$$

However, we resort instead to (6-117a) through (6-117d). Since the observations are made in the far field and we know that the E- and H-field components are orthogonal to each other and to the radial direction (plane waves) and are related by the intrinsic impedance of the medium, we can write, using (6-117a) through (6-117d),

$$(E_F)_\theta \simeq +\eta(H_F)_\phi = -j\omega\eta F_\phi \tag{6-119a}$$

$$(E_F)_\phi \simeq -\eta(H_F)_\theta = +j\omega\eta F_\theta \tag{6-119b}$$

$$(H_A)_\theta \simeq -\frac{(E_A)_\phi}{\eta} = +j\omega\frac{A_\phi}{\eta} \tag{6-119c}$$

$$(H_A)_\phi \simeq +\frac{(E_A)_\theta}{\eta} = -j\omega\frac{A_\theta}{\eta} \tag{6-119d}$$

Combining (6-117a) through (6-117d) with (6-119a) through (6-119d) and remembering that the radial components are negligible, we can write the E- and H-field components in the far field as

$$E_r \simeq 0 \tag{6-120a}$$

$$E_\theta \simeq (E_A)_\theta + (E_F)_\theta = -j\omega\left[A_\theta + \eta F_\phi\right] \tag{6-120b}$$

$$E_\phi \simeq (E_A)_\phi + (E_F)_\phi = -j\omega\left[A_\phi - \eta F_\theta\right] \tag{6-120c}$$

$$H_r \simeq 0 \tag{6-120d}$$

$$H_\theta \simeq (H_A)_\theta + (H_F)_\theta = +\frac{j\omega}{\eta}\left[A_\phi - \eta F_\theta\right] \tag{6-120e}$$

$$H_\phi \simeq (H_A)_\phi + (H_F)_\phi = -\frac{j\omega}{\eta}\left[A_\theta + \eta F_\phi\right] \tag{6-120f}$$

Using A_θ, A_ϕ, F_θ, and F_ϕ from (6-114a) through (6-114d), that is,

$$A_\theta = \frac{\mu e^{-j\beta r}}{4\pi r} N_\theta \qquad (6\text{-}121a)$$

$$A_\phi = \frac{\mu e^{-j\beta r}}{4\pi r} N_\phi \qquad (6\text{-}121b)$$

$$F_\theta = \frac{\varepsilon e^{-j\beta r}}{4\pi r} L_\theta \qquad (6\text{-}121c)$$

$$F_\phi = \frac{\varepsilon e^{-j\beta r}}{4\pi r} L_\phi \qquad (6\text{-}121d)$$

we can reduce (6-120a) through (6-120f) to

$$E_r \simeq 0 \qquad (6\text{-}122a)$$

$$E_\theta \simeq -\frac{j\beta e^{-j\beta r}}{4\pi r}(L_\phi + \eta N_\theta) \qquad (6\text{-}122b)$$

$$E_\phi \simeq +\frac{j\beta e^{-j\beta r}}{4\pi r}(L_\theta - \eta N_\phi) \qquad (6\text{-}122c)$$

$$H_r \simeq 0 \qquad (6\text{-}122d)$$

$$H_\theta \simeq +\frac{j\beta e^{-j\beta r}}{4\pi r}\left(N_\phi - \frac{L_\theta}{\eta}\right) \qquad (6\text{-}122e)$$

$$H_\phi \simeq -\frac{j\beta e^{-j\beta r}}{4\pi r}\left(N_\theta + \frac{L_\phi}{\eta}\right) \qquad (6\text{-}122f)$$

A. RECTANGULAR COORDINATE SYSTEM

To find the fields of (6-122a) through (6-122f) the functions N_θ, N_ϕ, L_θ, and L_ϕ must be evaluated from (6-114c) and (6-114d). The evaluation of (6-114c) and (6-114d) can best be accomplished if the most convenient coordinate system is chosen.

For radiators or scatterers whose geometries are most conveniently represented by rectangular coordinates, (6-114c) and (6-114d) can best be expressed as

$$\mathbf{N} = \iint_S \mathbf{J}_s e^{+j\beta r' \cos\psi} ds' = \iint_S (\hat{a}_x J_x + \hat{a}_y J_y + \hat{a}_z J_z) e^{+j\beta r' \cos\psi} ds' \qquad (6\text{-}123a)$$

$$\mathbf{L} = \iint_S \mathbf{M}_s e^{+j\beta r' \cos\psi} ds' = \iint_S (\hat{a}_x M_x + \hat{a}_y M_y + \hat{a}_z M_z) e^{+j\beta r' \cos\psi} ds' \qquad (6\text{-}123b)$$

Using the rectangular-to-spherical component transformation of (II-13a)

$$\begin{bmatrix} \hat{a}_x \\ \hat{a}_y \\ \hat{a}_z \end{bmatrix} = \begin{bmatrix} \sin\theta\cos\phi & \cos\theta\cos\phi & -\sin\phi \\ \sin\theta\sin\phi & \cos\theta\sin\phi & \cos\phi \\ \cos\theta & -\sin\theta & 0 \end{bmatrix} \begin{bmatrix} \hat{a}_r \\ \hat{a}_\theta \\ \hat{a}_\phi \end{bmatrix} \qquad (6\text{-}124)$$

we can reduce (6-123a) and (6-123b) for the θ and ϕ components to

$$N_\theta = \iint_S (J_x \cos\theta \cos\phi + J_y \cos\theta \sin\phi - J_z \sin\theta) e^{+j\beta r' \cos\psi} \, ds' \qquad (6\text{-}125a)$$

$$N_\phi = \iint_S (-J_x \sin\phi + J_y \cos\phi) e^{+j\beta r' \cos\psi} \, ds' \qquad (6\text{-}125b)$$

$$L_\theta = \iint_S (M_x \cos\theta \cos\phi + M_y \cos\theta \sin\phi - M_z \sin\theta) e^{+j\beta r' \cos\psi} \, ds' \qquad (6\text{-}125c)$$

$$L_\phi = \iint_S (-M_x \sin\phi + M_y \cos\phi) e^{+j\beta r' \cos\psi} \, ds' \qquad (6\text{-}125d)$$

Some of the most common and practical radiators and scatterers are represented by rectangular geometries. Because of their configuration, the most convenient coordinate system for expressing the fields or current densities on the structure, and performing the integration over it, would be the rectangular. The three most common and convenient coordinate positions used for the solution of the problem are shown in Figure 6-4. Figure 6-4a, b, and c show, respectively, the structure in the yz plane, in the xz plane, and in the xy plane. For a given field or current density distribution, the analytical forms for the radiated or scattered fields for each of the arrangements would not be the same. However, the computed values will be the same because the problem is physically identical.

For each of the geometries shown in Figure 6-4, the only difference in the analysis will be in the following formulations.

1. The components of the equivalent currents, J_x, J_y, J_z, M_x, M_y, and M_z.
2. The difference in paths from the source to the observation point, $r' \cos\psi$.
3. The differential area ds'.

In general, the nonzero components of $\mathbf{J}_s$ and $\mathbf{M}_s$ will be

$$J_y, J_z, M_y, \text{ and } M_z \quad (\text{Fig. 6-4}a) \qquad (6\text{-}126a)$$

$$J_x, J_z, M_x, \text{ and } M_z \quad (\text{Fig. 6-4}b) \qquad (6\text{-}126b)$$

$$J_x, J_y, M_x, \text{ and } M_y \quad (\text{Fig. 6-4}c) \qquad (6\text{-}126c)$$

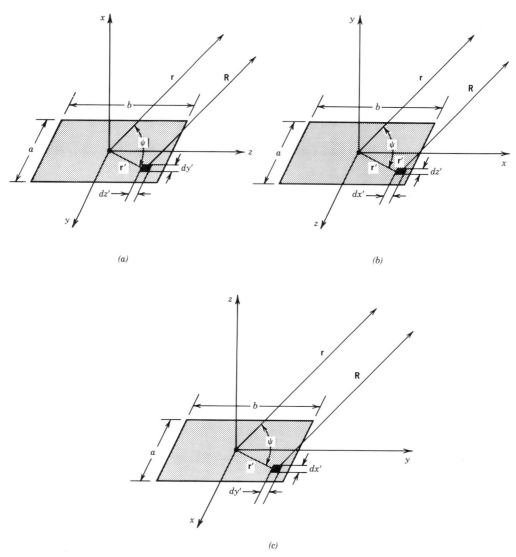

FIGURE 6-4 Rectangular aperture and plate positions for antenna and scattering system analysis. (*Source:* C. A. Balanis, *Antenna Theory: Analysis and Design.* Copyright © 1982, John Wiley & Sons, Inc. Reprinted by permission of John Wiley & Sons, Inc.) (*a*) *yz* plane. (*b*) *xz* plane. (*c*) *xy* plane.

The differential paths will be of the form

$$r'\cos\psi = \mathbf{r}' \cdot \hat{a}_r = (\hat{a}_y y' + \hat{a}_z z') \cdot (\hat{a}_x \sin\theta\cos\phi + \hat{a}_y \sin\theta\sin\phi + \hat{a}_z \cos\theta)$$
$$= y'\sin\theta\sin\phi + z'\cos\theta \quad \text{(Fig. 6-4}a\text{)} \quad (6\text{-}127a)$$

$$r'\cos\psi = \mathbf{r}' \cdot \hat{a}_r = (\hat{a}_x x' + \hat{a}_z z') \cdot (\hat{a}_x \sin\theta\cos\phi + \hat{a}_y \sin\theta\sin\phi + \hat{a}_z \cos\theta)$$
$$= x'\sin\theta\cos\phi + z'\cos\theta \quad \text{(Fig. 6-4}b\text{)} \quad (6\text{-}127b)$$

$$r'\cos\psi = \mathbf{r}' \cdot \hat{a}_r = (\hat{a}_x x' + \hat{a}_y y') \cdot (\hat{a}_x \sin\theta\cos\phi + \hat{a}_y \sin\theta\sin\phi + \hat{a}_z \cos\theta)$$
$$= x'\sin\theta\cos\phi + y'\sin\theta\sin\phi \quad \text{(Fig. 6-4}c\text{)} \quad (6\text{-}127c)$$

and the differential areas of

$$ds' = dy'\,dz' \quad \text{(Fig. 6-4}a) \tag{6-128a}$$
$$ds' = dx'\,dz' \quad \text{(Fig. 6-4}b) \tag{6-128b}$$
$$ds' = dx'\,dy' \quad \text{(Fig. 6-4}c) \tag{6-128c}$$

SUMMARY

To summarize the results, we will outline the procedure that must be followed to solve a problem using the radiation or scattering integrals. Figure 6-3 is used to indicate the geometry.

1. Select a closed surface over which the actual current density $\mathbf{J}_s$ or the equivalent current densities $\mathbf{J}_s$ and $\mathbf{M}_s$ exist.
2. Specify the actual current density $\mathbf{J}_s$ or form the equivalent currents $\mathbf{J}_s$ and $\mathbf{M}_s$ over S using [1, 3, 5]

$$\mathbf{J}_s = \hat{n} \times \mathbf{H}_a \tag{6-129a}$$
$$\mathbf{M}_s = -\hat{n} \times \mathbf{E}_a \tag{6-129b}$$

where $\hat{n}$ = unit vector normal to the surface S
$\mathbf{E}_a$ = total electric field over the surface S
$\mathbf{H}_a$ = total magnetic field over the surface S

3. (*Optional*) Determine the potentials $\mathbf{A}$ and $\mathbf{F}$ using, respectively, (6-103) and (6-109) where the integration is over the surface S of the sources.
4. Determine the corresponding E- and H-field components that are due to $\mathbf{J}_s$ and $\mathbf{M}_s$ using (6-107a) through (6-107c), (6-108a) through (6-108e), (6-110a) through (6-110c), and (6-111a) through (6-111c). Combine the E- and H-field components that are due to both $\mathbf{J}_s$ and $\mathbf{M}_s$ to find the total $\mathbf{E}$ and $\mathbf{H}$ fields.

These steps are valid for all regions (near field and far field) outside the surface S. If, however, the observation point is in the far field, steps 3 and 4 can be replaced by 3' and 4'.

3'. Determine N_θ, N_ϕ, L_θ, and L_ϕ using (6-125a) through (6-125d).
4'. Determine the radiated $\mathbf{E}$ and $\mathbf{H}$ fields using (6-122a) through (6-122f).

This procedure can be used to analyze radiation and scattering problems. The radiation problems most conducive to this procedure are aperture antennas, such as waveguides, horns, reflectors, and others. These aperture antennas are usually best represented by specifying their fields over their apertures.

Example 6-4. The tangential $\mathbf{E}$ and $\mathbf{H}$ fields over a rectangular aperture of dimensions a and b, shown in Figure 6-5, are given by

$$\left. \begin{array}{l} \mathbf{E}_a = \hat{a}_y E_0 \\ \mathbf{H}_a = -\hat{a}_x \dfrac{E_0}{\eta} \end{array} \right\} \quad \begin{array}{l} -\dfrac{a}{2} \leq x' \leq \dfrac{a}{2} \\ -\dfrac{b}{2} \leq y' \leq \dfrac{b}{2} \end{array}$$

$$\mathbf{E}_a \simeq \mathbf{H}_a \simeq 0 \quad \text{elsewhere}$$

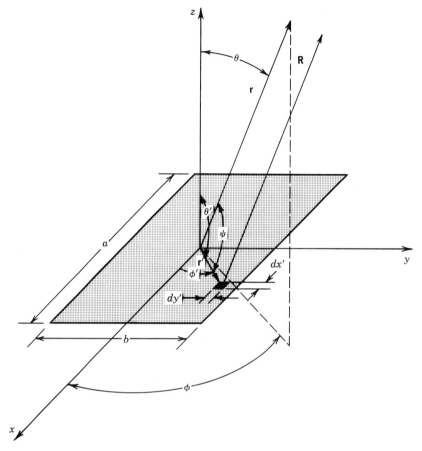

FIGURE 6-5 Rectangular aperture geometry for radiation problem.

Find the far-zone fields radiated by the aperture, and plot the three-dimensional pattern when $a = 3\lambda$ and $b = 2\lambda$. The fields over the aperture and elsewhere have been simplified in order to reduce the complexity of the problem and to avoid having the analytical formulations obscure the analysis procedure.

Solution.

1. The surface of the radiator is defined by $-a/2 \leq x \leq a/2$ and $-b/2 \leq y \leq b/2$.
2. Since the electric and magnetic fields exist only over the bounds of the aperture, the equivalent current densities $\mathbf{J}_s$ and $\mathbf{M}_s$ representing the aperture exist only over the bounds of the aperture. This is a good approximation, and it is derived by the equivalence principle in Chapter 7 [1, 3, 5]. Using (6-129a) and (6-129b) the current densities $\mathbf{J}_s$ and $\mathbf{M}_s$ can be written, by referring to Figure 6-5, as

$$\mathbf{J}_s = \hat{n} \times \mathbf{H}_a = \hat{a}_z \times \left(-\hat{a}_x \frac{E_0}{\eta}\right) = -\hat{a}_y \frac{E_0}{\eta} \Rightarrow J_x = J_z = 0 \qquad J_y = -\frac{E_0}{\eta}$$

$$\mathbf{M}_s = -\hat{n} \times \mathbf{E}_a = -\hat{a}_z \times \hat{a}_y E_0 = \hat{a}_x E_0 \Rightarrow M_x = E_0 \qquad M_y = M_z = 0$$

3. Using (6-125a), (6-127c), and (6-128c) we can reduce N_θ to

$$N_\theta = \iint_S [J_x \cos\theta \cos\phi + J_y \cos\theta \sin\phi - J_z \sin\theta] e^{j\beta r' \cos\psi} \, ds'$$

$$= -\frac{E_0}{\eta} \cos\theta \sin\phi \int_{-b/2}^{b/2} \int_{-a/2}^{a/2} e^{j\beta(x' \sin\theta \cos\phi + y' \sin\theta \sin\phi)} \, dx' \, dy'$$

Using the integral

$$\int_{-c/2}^{c/2} e^{j\alpha z} \, dz = c \left[\frac{\sin\left(\frac{\alpha}{2}c\right)}{\frac{\alpha}{2}c} \right]$$

reduces N_θ to

$$N_\theta = -\frac{abE_0}{\eta} \left\{ \cos\theta \sin\phi \left[\frac{\sin(X)}{X} \right] \left[\frac{\sin(Y)}{Y} \right] \right\}$$

where $X = \dfrac{\beta a}{2} \sin\theta \cos\phi$

$Y = \dfrac{\beta b}{2} \sin\theta \sin\phi$

In a similar manner N_ϕ, L_θ, and L_ϕ of (6-125b), (6-125c), and (6-125d) can be written as

$$N_\phi = \iint_S (-J_x \sin\phi + J_y \cos\phi) e^{j\beta r' \cos\psi} \, ds'$$

$$= -\frac{abE_0}{\eta} \left\{ \cos\phi \left[\frac{\sin(X)}{X} \right] \left[\frac{\sin(Y)}{Y} \right] \right\}$$

$$L_\theta = \iint_S (M_x \cos\theta \cos\phi + M_y \cos\theta \sin\phi - M_z \sin\theta) e^{j\beta r' \cos\psi} \, ds'$$

$$= abE_0 \left\{ \cos\theta \cos\phi \left[\frac{\sin(X)}{X} \right] \left[\frac{\sin(Y)}{Y} \right] \right\}$$

$$L_\phi = \iint_S [-M_x \sin\phi + M_y \cos\phi] e^{j\beta r' \cos\psi} \, ds'$$

$$= -abE_0 \left\{ \sin\phi \left[\frac{\sin(X)}{X} \right] \left[\frac{\sin(Y)}{Y} \right] \right\}$$

The corresponding far-zone **E**- and **H**-field components radiated by the aperture are obtained using (6-122a) through (6-122f), and they can be written as

$$E_r \simeq H_r \simeq 0$$

$$E_\theta \simeq \frac{C}{2} \sin\phi (1 + \cos\theta) \left[\frac{\sin(X)}{X}\right]\left[\frac{\sin(Y)}{Y}\right]$$

$$E_\phi \simeq \frac{C}{2} \cos\phi (1 + \cos\theta) \left[\frac{\sin(X)}{X}\right]\left[\frac{\sin(Y)}{Y}\right]$$

$$H_\theta \simeq -\frac{E_\phi}{\eta}$$

$$H_\phi \simeq +\frac{E_\theta}{\eta}$$

$$C = j\frac{ab\beta E_0 e^{-j\beta r}}{2\pi r}$$

In the principal E and H planes, the electric field components reduce to

E Plane ($\phi = \pi/2$)

$$E_r \simeq E_\phi = 0$$

$$E_\theta = \frac{C}{2}(1 + \cos\theta)\frac{\sin\left(\frac{\beta b}{2}\sin\theta\right)}{\frac{\beta b}{2}\sin\theta}$$

H Plane ($\phi = 0$)

$$E_r \simeq E_\theta = 0$$

$$E_\phi = \frac{C}{2}(1 + \cos\theta)\frac{\sin\left(\frac{\beta a}{2}\sin\theta\right)}{\frac{\beta a}{2}\sin\theta}$$

A three-dimensional plot of the normalized magnitude of the total electric field intensity E ($E \simeq \sqrt{E_\theta^2 + E_\phi^2}$) for an aperture with $a = 3\lambda$, $b = 2\lambda$ as a function of θ, and ϕ ($0° \leq \theta \leq 180°, 0° \leq \phi \leq 360°$) is shown plotted Figure 6-6. Because the aperture is larger in the x direction ($a = 3\lambda$) its pattern in the xz plane exhibits a larger number of lobes compared to the yz plane, as shown also in the two-dimensional E- and H-plane patterns in Figure 6-7.

To demonstrate the application of the techniques to scattering, let us consider a scattering problem.

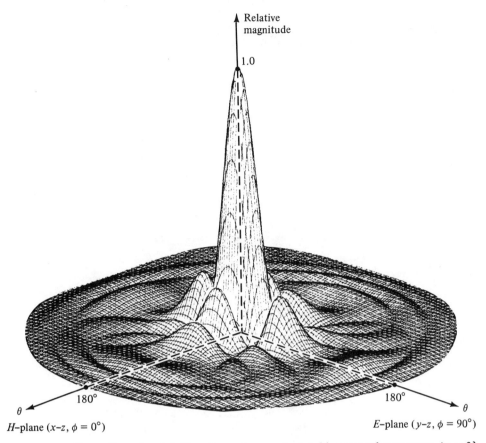

FIGURE 6-6 Three-dimensional field pattern of a constant field rectangular aperture ($a = 3\lambda$ and $b = 2\lambda$). (*Source:* C. A. Balanis, *Antenna Theory: Analysis and Design.* Copyright © 1982, John Wiley & Sons, Inc. Reprinted by permission of John Wiley & Sons, Inc.)

Example 6-5. A parallel polarized uniform plane wave traveling on the yz plane at an angle θ_i from the z axis is incident upon a rectangular electric perfectly conducting flat plate of dimensions a and b, as shown in Figure 6-8. Assuming that the induced current density on the plate is the same as that on an infinite conducting flat plate, find the far-zone spherical scattered electric and magnetic field components in directions specified by θ_s, ϕ_s. Plot the three-dimensional scattering pattern when $a = 3\lambda$ and $b = 2\lambda$.

Solution. Since the incident wave is a parallel polarized uniform plane wave, the incident electric and magnetic fields can be written as

$$\mathbf{E}^i = E_0\left(\hat{a}_y \cos\theta_i + \hat{a}_z \sin\theta_i\right) e^{-j\beta(y' \sin\theta_i - z' \cos\theta_i)}$$

$$\mathbf{H}^i = \frac{E_0}{\eta} \hat{a}_x e^{-j\beta(y' \sin\theta_i - z' \cos\theta_i)}$$

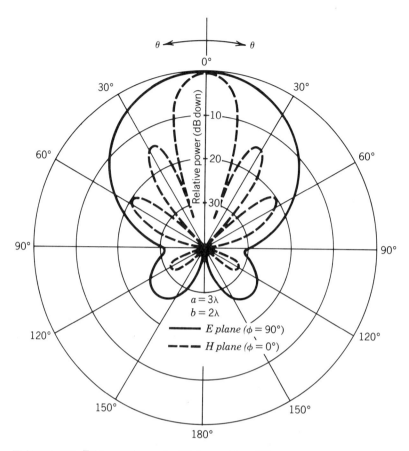

FIGURE 6-7 E-($\phi = 90°$) and H-plane ($\phi = 0°$) power patterns of a rectangular aperture with a uniform field distribution.

The electric current density induced on the surface of the plate is given by

$$\mathbf{J}_s = \hat{n} \times \mathbf{H}^{\text{total}}\big|_{z=0} = \hat{a}_z \times (\mathbf{H}^i + \mathbf{H}^r)\big|_{z=0}$$

According to Figure 5-4 and (5-24c), the reflected magnetic field of (5-21b) can be written as

$$\mathbf{H}^r = -\Gamma_\| \mathbf{H}^i = -(-\mathbf{H}^i) = \mathbf{H}^i$$

Then

$$\mathbf{J}_s = \hat{a}_z \times (\mathbf{H}^i + \mathbf{H}^i)\big|_{z=0} = 2\hat{a}_y H^i\big|_{z=0} = \hat{a}_y 2 \frac{E_0}{\eta} e^{-j\beta y' \sin\theta_i}$$

or

$$J_x = J_z = 0 \qquad \text{everywhere}$$

$$J_y = 2\frac{E_0}{\eta} e^{-jky' \sin\theta_i} \quad \text{for } \begin{cases} -a/2 \leq x' \leq a/2 \\ -b/2 \leq y' \leq b/2 \end{cases} \text{ and zero elsewhere}$$

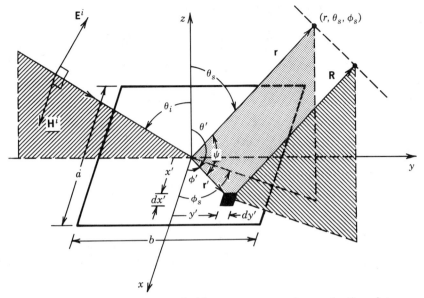

FIGURE 6-8 Uniform plane wave incident on a rectangular conducting plate.

Because the geometry of the plate corresponds to the coordinate system of Figure 6-4c, equation 6-125a can be written using (6-127c) and (6-128c) as

$$N_\theta = \iint_S \left[J_x \cos\theta_s \cos\phi_s + J_y \cos\theta_s \sin\phi_s - J_z \sin\theta_s \right] e^{j\beta r' \cos\psi} \, ds'$$

$$= 2\frac{E_0}{\eta} \cos\theta_s \sin\phi_s \int_{-b/2}^{b/2} \int_{-a/2}^{a/2} e^{j\beta x' \sin\theta_s \cos\phi_s} e^{j\beta y'(\sin\theta_s \sin\phi_s - \sin\theta_i)} \, dx' \, dy'$$

Using the integral

$$\int_{-c/2}^{c/2} e^{j\alpha z} \, dz = c \left[\frac{\sin\left(\frac{\alpha}{2}c\right)}{\frac{\alpha}{2}c} \right]$$

reduces N_θ to

$$N_\theta = 2ab \frac{E_0}{\eta} \left\{ \cos\theta_s \sin\phi_s \left[\frac{\sin(X)}{X} \right] \left[\frac{\sin(Y)}{Y} \right] \right\}$$

where $X = \frac{\beta a}{2} \sin\theta_s \cos\phi_s$

$Y = \frac{\beta b}{2} (\sin\theta_s \sin\phi_s - \sin\theta_i)$

Similarly, according to (6-125b), N_ϕ can be written as

$$N_\phi = \iint_S (-J_x \sin\phi_s + J_y \cos\phi_s) e^{j\beta r' \cos\psi} \, ds'$$

$$= 2ab \frac{E_0}{\eta} \left\{ \cos\phi_s \left[\frac{\sin(X)}{X} \right] \left[\frac{\sin(Y)}{Y} \right] \right\}$$

Because the plate is a perfect electric conductor

$$M_x = M_y = M_z = 0 \quad \text{everywhere}$$

Therefore, according to (6-125c) and (6-125d)

$$L_\theta = L_\phi = 0$$

Thus the scattered electric and magnetic field components can be reduced according to (6-122a) through (6-122f) to

$$E_r^s \simeq H_r^s \simeq 0$$

$$E_\theta^s \simeq -jab\frac{\beta E_0 e^{-jkr}}{2\pi r}\left\{\cos\theta_s \sin\phi_s \left[\frac{\sin(X)}{X}\right]\left[\frac{\sin(Y)}{Y}\right]\right\}$$

$$E_\phi^s \simeq jab\frac{\beta E_0 e^{-jkr}}{2\pi r}\left\{\cos\phi_s \left[\frac{\sin(X)}{X}\right]\left[\frac{\sin(Y)}{Y}\right]\right\}$$

$$H_\theta^s \simeq -\frac{E_\phi^s}{\eta}$$

$$H_\phi^s \simeq +\frac{E_\theta^s}{\eta}$$

In the principal E and H planes, the electric field components reduce, respectively, as follows.

E Plane ($\phi_s = \pi/2, 3\pi/2$)

$$E_r^s \simeq E_\phi^s \simeq 0$$

$$E_\theta^s \simeq -jab\frac{\beta E_0 e^{-jkr}}{2\pi r}\cos\theta_s \frac{\sin\left[\frac{\beta b}{2}(\pm\sin\theta_s - \sin\theta_i)\right]}{\frac{\beta b}{2}(\pm\sin\theta_s - \sin\theta_i)} \quad \begin{array}{l}+\text{for } \phi_s = \pi/2 \\ -\text{for } \phi_s = 3\pi/2\end{array}$$

H Plane ($\phi_s = 0, \pi$)

$$E_r^s \simeq E_\theta^s \simeq 0$$

$$E_\phi^s \simeq +jab\frac{\beta E_0 e^{-jkr}}{2\pi r}\frac{\sin\left(\frac{\beta a}{2}\sin\theta_s\right)}{\frac{\beta a}{2}\sin\theta_s}\frac{\sin\left(\frac{\beta b}{2}\sin\theta_i\right)}{\frac{\beta b}{2}\sin\theta_i}$$

A three-dimensional plot of the normalized magnitude of the total electric field $E^s\left[E^s = \sqrt{(E_\theta^s)^2 + (E_\phi^s)^2}\right]$ for a plate of $a = 3\lambda$ and $b = 2\lambda$ when the incidence angle $\theta_i = 30°$ is shown in Figure 6-9. Its corresponding two-dimensional pattern in the yz plane ($\phi_s = 90°, 270°$) is exhibited in Figure 6-10. It can be observed that the maximum scattered field is directed near $\theta_s = 30°$ which is near

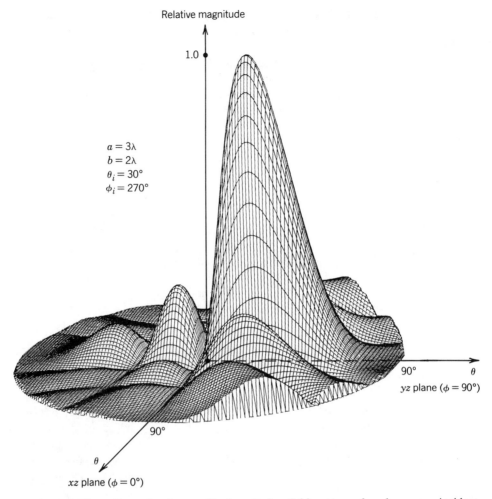

FIGURE 6-9 Three-dimensional normalized scattering field pattern of a plane wave incident on a rectangular ground plane.

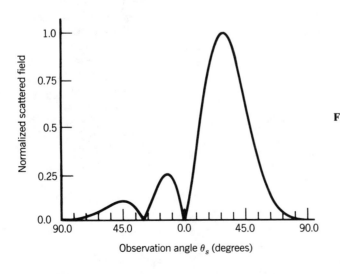

FIGURE 6-10 Two-dimensional normalized electric field scattering pattern for a plane wave incident ($\theta_i = 30°$ and $\phi_s = 90°, 270°$) on a flat conducting plate with $a = 3\lambda$ and $b = 2\lambda$.

the direction of specular reflection (defined as the direction along which the angle of reflection is equal to the angle of incidence). For more details see Section 11.3.

By using such a procedure for plates of finite size, the scattering fields are accurate at and near the specular direction. The angular extent over which the accuracy is acceptable increases as the size of the scatterer increases. Other techniques, such as those discussed in Chapters 12 and 13, can be used to improve the accuracy everywhere.

B. CYLINDRICAL COORDINATE SYSTEM

When the radiating or scattering structure is of circular geometry, the radiation or scattering fields can still be found using (6-122a) through (6-122f). The N_θ, N_ϕ, L_θ, and L_ϕ functions must still be obtained from (6-114c) and (6-114d) but must be expressed in a form that is convenient for cylindrical geometries. Although the general procedure of analysis for circular geometry is identical to that of the rectangular as outlined in the previous section, the primary differences lie in the following.

1. The formulation of the equivalent currents, J_x, J_y, J_z, M_x, M_y, and M_z.
2. The differential paths from the source to the observation point, $r' \cos \psi$.
3. The differential area ds'.

Before we consider an example we will reformulate these differences for the circular aperture.

Because of the circular profile of the aperture, it is often convenient and desirable to adopt cylindrical coordinates for the solution of the fields. In most cases, therefore, the radiated or scattered electric and magnetic field components over the circular geometry will be known in cylindrical form, that is, E_ρ, E_ϕ, E_z, H_ρ, H_ϕ, and H_z. Thus the components of the equivalent currents $\mathbf{M}_s$ and $\mathbf{J}_s$ would also be conveniently expressed in cylindrical form, M_ρ, M_ϕ, M_z, J_ρ, J_ϕ, and J_z. In addition, the required integration over the aperture to find N_θ, N_ϕ, L_θ, and L_ϕ of (6-125a) through (6-125d) should also be done in cylindrical coordinates. It is then desirable to reformulate $r' \cos \psi$ and ds', as given by (6-127a) through (6-128c).

The most convenient position for placing the structure is that shown in Figure 6-11 (structure on xy plane). The transformation between the rectangular and cylindrical components of $\mathbf{J}_s$ is given in Appendix II, equation II-7a, or

$$\begin{bmatrix} J_x \\ J_y \\ J_z \end{bmatrix} = \begin{bmatrix} \cos \phi' & -\sin \phi' & 0 \\ \sin \phi' & \cos \phi' & 0 \\ 0 & 0 & 1 \end{bmatrix} \begin{bmatrix} J_\rho \\ J_\phi \\ J_z \end{bmatrix} \quad (6\text{-}130a)$$

A similar transformation exists for the components of $\mathbf{M}_s$. The rectangular and cylindrical coordinates are related by (see Appendix II)

$$x' = \rho' \cos \phi'$$
$$y' = \rho' \sin \phi' \quad (6\text{-}130b)$$
$$z' = z'$$

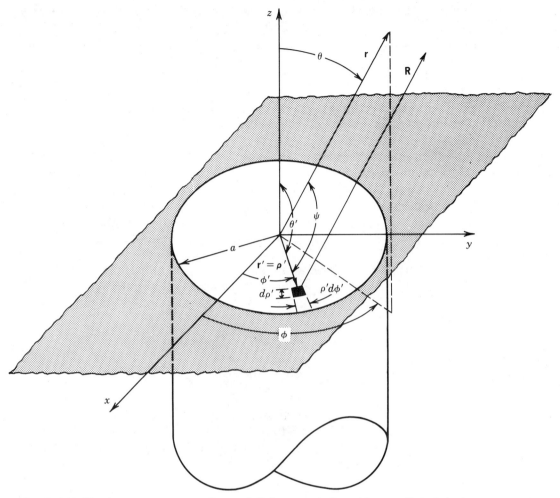

FIGURE 6-11 Circular aperture mounted on an infinite ground plane. (*Source:* C. A. Balanis, *Antenna Theory: Analysis and Design*. Copyright © 1982, John Wiley & Sons, Inc. Reprinted by permission of John Wiley & Sons, Inc.)

Using (6-130a), equations 6-125a through 6-125d can be written as

$$N_\theta = \iint_S [J_\rho \cos\theta \cos(\phi - \phi') + J_\phi \cos\theta \sin(\phi - \phi') - J_z \sin\theta] e^{j\beta r' \cos\psi} \, ds' \quad (6\text{-}131a)$$

$$N_\phi = \iint_S [-J_\rho \sin(\phi - \phi') + J_\phi \cos(\phi - \phi')] e^{j\beta r' \cos\psi} \, ds' \quad (6\text{-}131b)$$

$$L_\theta = \iint_S [M_\rho \cos\theta \cos(\phi - \phi') + M_\phi \cos\theta \sin(\phi - \phi') - M_z \sin\theta] e^{j\beta r' \cos\psi} \, ds' \quad (6\text{-}131c)$$

$$L_\phi = \iint_S [-M_\rho \sin(\phi - \phi') + M_\phi \cos(\phi - \phi')] e^{j\beta r' \cos\psi} \, ds' \quad (6\text{-}131d)$$

where $r'\cos\psi$ and ds' can be written, using (6-127c), (6-128c) and (6-130b), as

$$r'\cos\psi = x'\sin\theta\cos\phi + y'\sin\theta\sin\phi = \rho'\sin\theta\cos(\phi - \phi') \quad (6\text{-}132a)$$

$$ds' = dx'\,dy' = \rho'\,d\rho'\,d\phi' \quad (6\text{-}132b)$$

In summary, for a circular aperture antenna the fields radiated can be obtained by either of the following methods.

1. If the fields over the aperture are known in rectangular components, use the same procedure as for the rectangular aperture except that (6-132a) and (6-132b) should be substituted in (6-125a) through (6-125d).
2. If the fields over the aperture are known in cylindrical components, use the same procedure as for the rectangular aperture with (6-131a) through (6-131d), along (6-132a) and (6-132b), taking the place of (6-125a) through (6-125d).

Example 6-6. To demonstrate the methods, the field radiated by a circular aperture mounted on an infinite ground plane will be formulated. To simplify the mathematical details, the field over the aperture of Figure 6-11 will be assumed to be

$$\left.\begin{array}{l} \mathbf{E}_a = \hat{a}_y E_0 \\ \mathbf{H}_a = -\hat{a}_x \dfrac{E_0}{\eta} \end{array}\right\} \quad \rho' \leq a$$

The objective is to find the far-zone fields radiated by the aperture. The fields over the aperture have been simplified in order to reduce the complexity of the problem and to avoid having the analytical formulations obscure the analysis procedure.

Solution.

1. The surface of the radiating aperture is that defined by $\rho' \leq a$.
2. Since the aperture is mounted on an infinite ground plane, it is shown by the equivalence principle in Chapter 7 and elsewhere [1, 3, 5] that the equivalent current densities that lead to the appropriate radiated fields are given by

$$\mathbf{M}_s = \begin{cases} -2\hat{n} \times \mathbf{E}_a = \hat{a}_x 2E_0 & \rho' \leq a \\ 0 & \text{elsewhere} \end{cases}$$

$$\mathbf{J}_s = 0 \quad \text{elsewhere}$$

This equivalent model for the current densities is valid for any aperture mounted on an infinite perfectly conducting electric ground plane. Thus according to (6-125a) and (6-125b)

$$N_\theta = N_\phi = 0$$

Using (6-125c), (6-132a) and (6-132b)

$$L_\theta = 2E_0 \cos\theta \cos\phi \int_0^a \rho' \left[\int_0^{2\pi} e^{+j\beta\rho'\sin\theta\cos(\phi-\phi')}\,d\phi'\right]d\rho'$$

Because

$$\int_0^{2\pi} e^{+j\beta\rho' \sin\theta \cos(\phi-\phi')} d\phi' = 2\pi J_0(\beta\rho' \sin\theta)$$

we can write L_θ as

$$L_\theta = 4\pi E_0 \cos\theta \cos\phi \int_0^a J_0(\beta\rho' \sin\theta)\rho' \, d\rho'$$

where $J_0(t)$ is the Bessel function of the first kind of order zero. Making the substitution

$$t = \beta\rho' \sin\theta$$
$$dt = \beta \sin\theta \, d\rho'$$

reduces L_θ to

$$L_\theta = \frac{4\pi E_0 \cos\theta \cos\phi}{(\beta \sin\theta)^2} \int_0^{\beta a \sin\theta} t J_0(t) \, dt$$

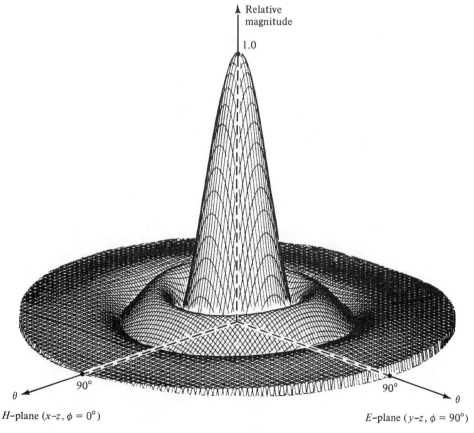

FIGURE 6-12 Three-dimensional field pattern of a constant field circular aperture mounted on an infinite ground plane ($a = 1.5\lambda$). (*Source:* C. A. Balanis, *Antenna Theory: Analysis and Design.* Copyright © 1982, John Wiley & Sons, Inc. Reprinted by permission of John Wiley & Sons, Inc.)

Since

$$\int_0^\delta z J_0(z)\, dz = z J_1(z)\Big|_0^\delta = \delta J_1(\delta)$$

where $J_1(\delta)$ is the Bessel function of order 1, L_θ takes the form of

$$L_\theta = 4\pi a^2 E_0 \left\{ \cos\theta \cos\phi \left[\frac{J_1(\beta a \sin\theta)}{\beta a \sin\theta} \right] \right\}$$

Similarly L_ϕ of (6-125d) reduces to

$$L_\phi = -4\pi a^2 E_0 \left\{ \sin\phi \left[\frac{J_1(\beta a \sin\theta)}{\beta a \sin\theta} \right] \right\}$$

Using N_θ, N_ϕ, L_θ, and L_ϕ previously derived, the electric field components of (6-122a) through (6-122c) can be written as

$$E_r = 0$$

$$E_\theta = j\frac{\beta a^2 E_0 e^{-j\beta r}}{r} \left\{ \sin\phi \left[\frac{J_1(\beta a \sin\theta)}{\beta a \sin\theta} \right] \right\}$$

$$E_\phi = j\frac{\beta a^2 E_0 e^{-j\beta r}}{r} \left\{ \cos\theta \cos\phi \left[\frac{J_1(\beta a \sin\theta)}{\beta a \sin\theta} \right] \right\}$$

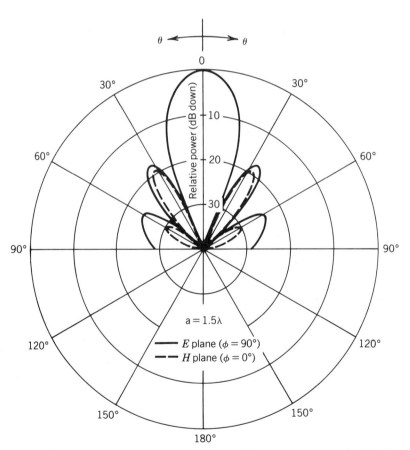

FIGURE 6-13 E-($\phi = 90°$) and H-plane ($\phi = 0°$) power patterns of a circular aperture with a uniform field distribution.

In the principal E and H planes, the electric field components simplify to

E Plane ($\phi = \pi/2$)

$$E_r = E_\phi = 0$$

$$E_\theta = j\frac{\beta a^2 E_0 e^{-j\beta r}}{r}\left[\frac{J_1(\beta a \sin\theta)}{\beta a \sin\theta}\right]$$

H Plane ($\phi = 0$)

$$E_r = E_\theta = 0$$

$$E_\phi = j\frac{\beta a^2 E_0 e^{-j\beta r}}{r}\left\{\cos\theta\left[\frac{J_1(\beta a \sin\theta)}{\beta a \sin\theta}\right]\right\}$$

A three-dimensional plot of the normalized magnitude of the total electric field intensity E ($E \simeq \sqrt{E_\theta^2 + E_\phi^2}$) for an aperture of $a = 1.5\lambda$ as a function of θ and ϕ (0° $\leq \theta \leq$ 90° and 0° $\leq \phi \leq$ 360°) is shown plotted in Figure 6-12, and it seems to be symmetrical. However, closer observation, especially through the two-dimensional E- and H-plane patterns of Figure 6-13, reveals that the pattern is not symmetrical. It does, however, possess characteristics that are almost identical.

REFERENCES

1. C. A. Balanis, *Antenna Theory: Analysis and Design*, Wiley, New York, 1982.
2. R. F. Harrington, *Time-Harmonic Electromagnetic Fields*, McGraw-Hill, New York, 1961.
3. J. H. Richmond, *The Basic Theory of Harmonic Fields, Antennas and Scattering*, Unpublished Notes.
4. R. Mittra (ed.), *Computer Techniques for Electromagnetics*, Chapter 2, Pergamon Press, 1973, pp. 7–95.
5. S. A. Schelkunoff, "Some equivalence theorems of electromagnetics and their application to radiation problems," *Bell System Tech. J.*, vol. 15, pp. 92–112, 1936.

PROBLEMS

6.1. If $\mathbf{H}_e = j\omega\varepsilon\nabla \times \mathbf{\Pi}_e$, where $\mathbf{\Pi}_e$ is the electric Hertzian potential, show that
  (a) $\nabla^2\mathbf{\Pi}_e + \beta^2\mathbf{\Pi}_e = j(1/\omega\varepsilon)\mathbf{J}$.
  (b) $\mathbf{E}_e = \beta^2\mathbf{\Pi}_e + \nabla(\nabla \cdot \mathbf{\Pi}_e)$.
  (c) $\mathbf{\Pi}_e = -j(1/\omega\mu\varepsilon)\mathbf{A}$.

6.2. If $\mathbf{E}_h = -j\omega\mu\nabla \times \mathbf{\Pi}_h$, where $\mathbf{\Pi}_h$ is the magnetic Hertzian potential, show that
  (a) $\nabla^2\mathbf{\Pi}_h + \beta^2\mathbf{\Pi}_h = j(1/\omega\mu)\mathbf{M}$.
  (b) $\mathbf{H}_h = \beta^2\mathbf{\Pi}_h + \nabla(\nabla \cdot \mathbf{\Pi}_h)$.
  (c) $\mathbf{\Pi}_h = -j(1/\omega\mu\varepsilon)\mathbf{F}$.

6.3. Develop expressions for $\mathbf{E}_A$ and $\mathbf{H}_A$ in terms of $\mathbf{J}$ using the path 1 procedure of Figure 6-1. These expressions should be valid everywhere.

6.4. Develop expressions for $\mathbf{E}_F$ and $\mathbf{H}_F$ in terms of $\mathbf{M}$ using the path 1 procedure of Figure 6-1. These expressions should be valid everywhere.

306 VECTOR POTENTIALS, SOLUTIONS, AND EQUATIONS

6.5. In rectangular coordinates derive expressions for **E** and **H**, in terms of the components of the **A** and **F** potentials, that are TEM^x and TEM^y. The procedure should be similar to that of Example 6-1, and it should state all the combinations that lead to the desired modes.

6.6. In cylindrical coordinates derive expressions for **E** and **H**, in terms of the components of the **A** and **F** potentials, that are TEM$^\phi$ and TEM^z. The procedure should be similar to that of Example 6-2, and it should state all the combinations that lead to the desired modes.

6.7. Derive the expressions for the components of **E** and **H** of (6-61) and (6-64), in terms of the components of **A** and **F**, so that the fields are TM^x and TM^y.

6.8. Select one component of **E** and write the other components of **E** and all of **H** in terms of the initial component of **E** so that the fields are TM^x, TM^y, and TM^z. Do this in rectangular coordinates.

6.9. In cylindrical coordinates, select one component of **E** and write the other components of **E** and all of **H** in terms of the initial component of **E** so that the fields are TM^z.

6.10. In rectangular coordinates derive the expressions for the components of **E** and **H** as given by (6-72), (6-74), and (6-77) that are TE^z, TE^x, and TE^y.

6.11. In cylindrical coordinates, derive the expressions for **E** and **H** as given by (6-80) which are TE^z.

6.12. Select one component of **H** and write the other components of **H** and all of **E** in terms of the initial component of **H** so that the fields are TE^x, TE^y, and TE^z. Do this in rectangular coordinates.

6.13. In cylindrical coordinates, select one component of **H** and write the other components of **H** and all of **E** in terms of the initial component of **H** so that the fields are TE^z.

6.14. Verify that (6-85a) and (6-85b) are solutions to (6-84a).

6.15. Show that (6-90) is a solution to (6-87) and that (6-91) is a solution to (6-82).

6.16. For Example 6-3 derive the components of **E**, given the components of **H**.

6.17. Show that for observations made at very large distance ($\beta r \gg 1$) the electric and magnetic fields of Example 6-3 reduce to

$$E_\theta = j\eta \frac{\beta I_e \ell e^{-j\beta r}}{4\pi r} \sin\theta$$

$$H_\phi \simeq \frac{E_\theta}{\eta}$$

$$E_r \simeq 0$$

$$E_\phi = H_r = H_\theta = 0$$

6.18. Verify (6-100a) and (6-100b).

6.19. Show that (6-112) reduces to (6-112a) provided $r \geq 2D^2/\lambda$, where D is the largest dimension of the radiator or scatterer. Such an approximation leads to a phase error that is equal to or smaller than 22.5°.

6.20. The current distribution on a very thin wire dipole antenna of overall length ℓ is given by

$$\mathbf{I}_e = \begin{cases} \hat{a}_z I_0 \sin\left[\beta\left(\dfrac{\ell}{2} - z'\right)\right] & 0 \le z' \le \dfrac{\ell}{2} \\ \hat{a}_z I_0 \sin\left[\beta\left(\dfrac{\ell}{2} + z'\right)\right] & -\dfrac{\ell}{2} \le z' \le 0 \end{cases}$$

where I_0 is a constant. Representing the distance R of (6-112) by the far-field approximations of (6-112a) through (6-112b), derive the far-zone electric and magnetic fields radiated by the dipole using (6-97a) and the far-field formulations of Section 6.7.

6.21. Show that the radiated far-zone electric and magnetic fields derived in Problem 6.20 reduce for a half-wavelength dipole ($\ell = \lambda/2$) to

$$E_\theta \simeq j\eta \frac{I_0 e^{-j\beta r}}{2\pi r} \left[\frac{\cos\left(\dfrac{\pi}{2}\cos\theta\right)}{\sin\theta}\right]$$

$$H_\phi \simeq \frac{E_\theta}{\eta}$$

$$E_r \simeq E_\phi \simeq H_r \simeq H_\theta \simeq 0$$

6.22. Simplify the expressions of Problem 6.3, if the observations are made in the far field.

6.23. Simplify the expressions of Problem 6.4, if the observations are made in the far field.

6.24. The rectangular aperture of Figure 6-4a is mounted on an infinite ground plane that coincides with the xy plane. Assuming that the tangential field over the aperture is given by

$$\mathbf{E}_z = \hat{a}_z E_0 \qquad -a/2 \le y' \le a/2 \qquad -b/2 \le z' \le b/2$$

and the equivalent currents are

$$\mathbf{M}_s = \begin{cases} 2\hat{n} \times \mathbf{E}_a & -a/2 \le y' \le a/2 \qquad -b/2 \le z' \le b/2 \\ 0 & \text{elsewhere} \end{cases}$$

$$\mathbf{J}_s = 0 \quad \text{everywhere}$$

find the far-zone spherical electric and magnetic field components radiated by the aperture.

6.25. Repeat Problem 6.24 when the same aperture is analyzed using the coordinate system of Figure 6-4b. The tangential aperture field distribution is given by

$$\mathbf{E}_a = \hat{a}_x E_0 \qquad -b/2 \le x' \le b/2 \qquad -a/2 \le z' \le a/2$$

and the equivalent currents are

$$\mathbf{M}_s = \begin{cases} 2\hat{n} \times \mathbf{E}_a & -b/2 \le x' \le b/2 \qquad -a/2 \le z' \le a/2 \\ 0 & \text{elsewhere} \end{cases}$$

$$\mathbf{J}_s = 0 \quad \text{everywhere}$$

6.26. Repeat Problem 6.24 when the same aperture is analyzed using the coordinate system of Figure 6-4c. The tangential aperture field distribution is given by
$$\mathbf{E}_z = \hat{a}_y E_0 \quad -a/2 \le x' \le a/2 \quad -b/2 \le y' \le b/2$$
and the equivalent currents are
$$\mathbf{M}_s = \begin{cases} 2\hat{n} \times \mathbf{E}_a & -a/2 \le x' \le a/2 \quad -b/2 \le y' \le b/2 \\ 0 & \text{elsewhere} \end{cases}$$
$$\mathbf{J}_s = 0 \quad \text{everywhere}$$

6.27. Repeat Problem 6.24 when the aperture field distribution is given by
$$\mathbf{E}_a = \hat{a}_z E_0 \cos\left(\frac{\pi}{a} y'\right) \quad -a/2 \le y' \le a/2 \quad -b/2 \le z' \le b/2$$

6.28. Repeat Problem 6.25 when the aperture field distribution is given by
$$\mathbf{E}_a = \hat{a}_x E_0 \cos\left(\frac{\pi}{a} z'\right) \quad -b/2 \le x' \le b/2 \quad -a/2 \le z' \le a/2$$

6.29. Repeat Problem 6.26 when the aperture field distribution is given by
$$\mathbf{E}_a = \hat{a}_y E_0 \cos\left(\frac{\pi}{a} x'\right) \quad -a/2 \le x' \le a/2 \quad -b/2 \le y' \le b/2$$

6.30. For the aperture of Example 6-4 find the angular separation (in degrees) between two points whose radiated electric field value is 0.707 of the maximum (half-power beamwidth). Do this for the radiated fields in the (a) E plane ($\phi = \pi/2$) and (b) H plane ($\phi = 0$). Assume the aperture has dimensions $a = 4\lambda$ and $b = 3\lambda$.

6.31. For the circular aperture of Figure 6-11 derive expressions for the far-zone radiated spherical fields when the aperture field distribution is given by
(a) $\mathbf{E}_a = \hat{a}_y E_0[1 - (\rho'/a)^2]$, $\rho' \le a$.
(b) $\mathbf{E}_a = \hat{a}_y E_0[1 - (\rho'/a)^2]^2$, $\rho' \le a$.

For both cases use equivalent currents $\mathbf{M}_s$ and $\mathbf{J}_s$ such that
$$\mathbf{M}_s = \begin{cases} 2\hat{n} \times \mathbf{E}_a & \rho' \le a \\ 0 & \text{elsewhere} \end{cases}$$
$$\mathbf{J}_s = 0 \quad \text{elsewhere}$$

6.32. A coaxial line of inner and outer radii of a and b, respectively, is mounted on an infinite conducting ground plane. Assuming that the electric field over the aperture

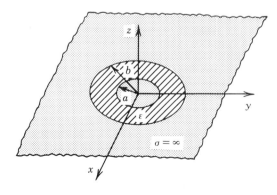

FIGURE P6-32

of the coax is

$$\mathbf{E}_a = -\hat{a}_\rho \frac{V}{\varepsilon \ln(b/a)} \frac{1}{\rho'} \qquad a \le \rho' \le b$$

where V is the applied voltage and ε is the permittivity of medium in the coax, find the far-zone spherical electric and magnetic field components radiated by the aperture. Use equivalent currents $\mathbf{M}_s$ and $\mathbf{J}_s$ such that

$$\mathbf{M}_s = \begin{cases} 2\hat{n} \times \mathbf{E}_a & a \le \rho' \le b \\ 0 & \text{elsewhere} \end{cases}$$

$$\mathbf{J}_s = 0 \quad \text{everywhere}$$

6.33. For the aperture of Example 6-6 find the angular separation (in degrees) between two points whose radiated electric field value is 0.707 of the maximum (half-power beamwidth). Do this for the radiated fields in the (a) E plane ($\phi = \pi/2$) and (b) H plane ($\phi = 0$). Assume that the radius of the aperture is 3λ.

CHAPTER 7

ELECTROMAGNETIC THEOREMS AND PRINCIPLES

7.1 INTRODUCTION

In electromagnetics there are a number of theorems and principles that are fundamental to the understanding of electromagnetic generation, radiation, propagation, scattering, and reception. Many of these are often used to facilitate the solution of interrelated problems. Those that will be discussed here are the theorems of *duality*, *uniqueness*, *image*, *reciprocity*, *reaction*, *volume equivalence*, *surface equivalence*, *induction*, and *physical equivalent* (*physical optics*). When appropriate, examples will be given to illustrate the principles.

7.2 DUALITY THEOREM

When two equations that describe the behavior of two different variables are of the same mathematical form, their solutions will also be identical. The variables in the two equations that occupy identical positions are known as *dual* quantities, and a solution for one can be formed by a systematic interchange of symbols with the other. This concept is known as the *duality theorem*.

Comparison of (6-30), (6-32a), (6-32b), (6-32c), and (6-95a), respectively, to (6-31), (6-33a), (6-33b), (6-33c), and (6-95b), shows that they are dual equations and their variables are dual quantities. Thus if we know the solutions to one set (i.e., $\mathbf{J} \neq 0$, $\mathbf{M} = 0$), the solutions to the other set ($\mathbf{J} = 0$, $\mathbf{M} \neq 0$) can be formed by a proper interchange of quantities. The dual equations and their dual quantities are listed in Tables 7-1 and 7-2 for electric and magnetic sources, respectively. Duality only serves as a guide to forming mathematical solutions. It can be used in an abstract manner to explain the motion of magnetic charges giving rise to magnetic currents, when compared to their dual quantities of moving electric charges creating electric currents [1]. It must, however, be emphasized that this is purely mathemati-

TABLE 7-1
Dual equations for electric (J) and magnetic (M) current sources

Electric sources ($J \neq 0, M = 0$)	Magnetic sources ($J = 0, M \neq 0$)
$\nabla \times \mathbf{E}_A = -j\omega\mu\mathbf{H}_A$	$\nabla \times \mathbf{H}_F = j\omega\varepsilon\mathbf{E}_F$
$\nabla \times \mathbf{H}_A = \mathbf{J} + j\omega\varepsilon\mathbf{E}_A$	$-\nabla \times \mathbf{E}_F = \mathbf{M} + j\omega\mu\mathbf{H}_F$
$\nabla^2 \mathbf{A} + \beta^2 \mathbf{A} = -\mu\mathbf{J}$	$\nabla^2 \mathbf{F} + \beta^2 \mathbf{F} = -\varepsilon\mathbf{M}$
$\mathbf{A} = \dfrac{\mu}{4\pi}\iiint_V \mathbf{J}\dfrac{e^{-j\beta R}}{R}\,dv'$	$\mathbf{F} = \dfrac{\varepsilon}{4\pi}\iiint_V \mathbf{M}\dfrac{e^{-j\beta R}}{R}\,dv'$
$\mathbf{H}_A = \dfrac{1}{\mu}\nabla \times \mathbf{A}$	$\mathbf{E}_F = -\dfrac{1}{\varepsilon}\nabla \times \mathbf{F}$
$\mathbf{E}_A = -j\omega\mathbf{A} - j\dfrac{1}{\omega\mu\varepsilon}\nabla(\nabla \cdot \mathbf{A})$	$\mathbf{H}_F = -j\omega\mathbf{F} - j\dfrac{1}{\omega\mu\varepsilon}\nabla(\nabla \cdot \mathbf{F})$

TABLE 7-2
Dual quantities for electric (J) and magnetic (M) current sources

Electric sources ($J \neq 0, M = 0$)	Magnetic sources ($J = 0, M \neq 0$)
$\mathbf{E}_A$	$\mathbf{H}_F$
$\mathbf{H}_A$	$-\mathbf{E}_F$
$\mathbf{J}$	$\mathbf{M}$
$\mathbf{A}$	$\mathbf{F}$
ε	μ
μ	ε
β	β
η	$1/\eta$
$1/\eta$	η

cal in nature since at present there are no known magnetic charges or currents in nature.

Example 7-1. A very thin linear magnetic current element of very small length ($\ell \ll \lambda$), although nonphysically realizable, is often used to represent the fields of a very small loop radiator. It can be shown that the fields radiated by a small linear magnetic current element are identical to those radiated by a small loop whose area is perpendicular to the length of the dipole [2]. Assume that the magnetic dipole is placed at the origin and is symmetric along the z axis with a constant magnetic current of

$$\mathbf{I}_m = \hat{a}_z I_m$$

Find the fields radiated by the dipole using duality.

Solution. Since the linear magnetic dipole is the dual of the linear electric dipole of Example 6-3, the fields radiated by the magnetic dipole can be written, using the dual quantities of Table 7-2 and the solution of Example

6-3, as

$$E_r = E_\theta = 0$$

$$E_\phi = -j\frac{\beta I_m \ell \sin\theta}{4\pi r}\left(1 + \frac{1}{j\beta r}\right)e^{-j\beta r}$$

$$H_r = \frac{1}{\eta}\frac{I_m \ell \cos\theta}{2\pi r^2}\left(1 + \frac{1}{j\beta r}\right)e^{-j\beta r}$$

$$H_\theta = j\frac{1}{\eta}\frac{\beta I_m \ell \sin\theta}{4\pi r}\left[1 + \frac{1}{j\beta r} - \frac{1}{(\beta r)^2}\right]e^{-j\beta r}$$

$$H_\phi = 0$$

7.3 UNIQUENESS THEOREM

Whenever a problem is solved, it is always gratifying to know that the obtained solution is unique, that is, it is the only solution. If so, we would like to know under what conditions or what information is needed to obtain such solutions.

Given the electric and magnetic sources $\mathbf{J}_i$ and $\mathbf{M}_i$, let us assume that the fields generated in a lossy medium of complex constitutive parameters $\dot{\varepsilon}$ and $\dot{\mu}$ within S are $\mathbf{E}^a, \mathbf{H}^a$ and $\mathbf{E}^b, \mathbf{H}^b$. Each set must satisfy Maxwell's equations

$$-\nabla \times \mathbf{E} = \mathbf{M}_i + j\omega\mu\mathbf{H} \qquad \nabla \times \mathbf{H} = \mathbf{J}_i + \mathbf{J}_c + j\omega\varepsilon\mathbf{E} \qquad (7\text{-}1)$$

or

$$-\nabla \times \mathbf{E}^a = \mathbf{M}_i + j\omega\dot{\mu}\mathbf{H}^a \qquad \nabla \times \mathbf{H}^a = \mathbf{J}_i + \mathbf{J}_c^a + j\omega\dot{\varepsilon}\mathbf{E}^a \qquad (7\text{-}1a)$$

$$-\nabla \times \mathbf{E}^b = \mathbf{M}_i + j\omega\dot{\mu}\mathbf{H}^b \qquad \nabla \times \mathbf{H}^b = \mathbf{J}_i + \mathbf{J}_c^b + j\omega\dot{\varepsilon}\mathbf{E}^b \qquad (7\text{-}1b)$$

Subtracting (7-1b) from (7-1a) we have that

$$-\nabla \times (\mathbf{E}^a - \mathbf{E}^b) = j\omega\dot{\mu}[\mathbf{H}^a - \mathbf{H}^b] \qquad \nabla \times (\mathbf{H}^a - \mathbf{H}^b) = (\sigma + j\omega\dot{\varepsilon})(\mathbf{E}^a - \mathbf{E}^b) \qquad (7\text{-}2)$$

or

$$\left. \begin{array}{l} -\nabla \times \delta\mathbf{E} = j\omega\dot{\mu}\delta\mathbf{H} = \delta\mathbf{M}_t \\ +\nabla \times \delta\mathbf{H} = j\omega\dot{\varepsilon}\delta\mathbf{E} = \delta\mathbf{J}_t \end{array} \right\} \quad \text{within } S \qquad (7\text{-}2a)$$

Thus, the difference fields satisfy the source-free field equations within S. The conditions for uniqueness are those for which $\delta\mathbf{E} = \delta\mathbf{H} = 0$ or $\mathbf{E}^a = \mathbf{E}^b$ and $\mathbf{H}^a = \mathbf{H}^b$.

Let us now apply the conservation of energy equation 1-55a using S as the boundary and $\delta\mathbf{E}$, $\delta\mathbf{H}$, $\delta\mathbf{J}_t$, and $\delta\mathbf{M}_t$ as the sources [1]. For a time-harmonic field (1-55a) can be written as

$$\oiint_S \mathbf{E} \times \mathbf{H}^* \cdot d\mathbf{s} + \iiint_V (\mathbf{E} \cdot \mathbf{J}_t^* + \mathbf{H}^* \cdot \mathbf{M}_t)\, dv' = 0 \qquad (7\text{-}3)$$

which for our case must be

$$\oint_S (\delta\mathbf{E} \times \delta\mathbf{H}^*) \cdot d\mathbf{s} + \iiint_V [\delta\mathbf{E} \cdot (\sigma + j\omega\dot{\varepsilon})^* \delta\mathbf{E}^* + \delta\mathbf{H}^* \cdot (j\omega\dot{\mu})\delta\mathbf{H}] \, dv' = 0 \tag{7-4}$$

or

$$\oint_S (\delta\mathbf{E} \times \delta\mathbf{H}^*) \cdot d\mathbf{s} + \iiint_V [(\sigma + j\omega\dot{\varepsilon})^* |\delta\mathbf{E}|^2 + (j\omega\dot{\mu})|\delta\mathbf{H}|^2] \, dv' = 0 \tag{7-4a}$$

where

$$(\sigma + j\omega\dot{\varepsilon})^* = [\sigma + j\omega(\varepsilon' - j\varepsilon'')]^* = [(\sigma + \omega\varepsilon'') + j\omega\varepsilon']^* = (\sigma + \omega\varepsilon'') - j\omega\varepsilon' \tag{7-4b}$$

$$j\omega\dot{\mu} = j\omega(\mu' - j\mu'') = \omega\mu'' + j\omega\mu' \tag{7-4c}$$

If we can show that

$$\oint_S (\delta\mathbf{E} \times \delta\mathbf{H}^*) \cdot d\mathbf{s} = 0 \tag{7-5}$$

then the volume integral must also be zero, or

$$\iiint_V [(\sigma + j\omega\dot{\varepsilon})^* |\delta\mathbf{E}|^2 + (j\omega\dot{\mu})|\delta\mathbf{H}|^2] \, dv'$$

$$= \mathrm{Re} \iiint_V [(\sigma + j\omega\dot{\varepsilon})^* |\delta\mathbf{E}|^2 + (j\omega\dot{\mu})|\delta\mathbf{H}|^2] \, dv'$$

$$+ \mathrm{Im} \iiint_V [(\sigma + j\omega\dot{\varepsilon})^* |\delta\mathbf{E}|^2 + (j\omega\dot{\mu})|\delta\mathbf{H}|^2] \, dv' = 0 \tag{7-6}$$

Using (7-4b) and (7-4c) reduces (7-6) to

$$\iiint_V [(\sigma + \omega\varepsilon'')|\delta\mathbf{E}|^2 + \omega\mu''|\delta\mathbf{H}|^2] \, dv' = 0 \tag{7-6a}$$

$$\iiint_V [-\omega\varepsilon'|\delta\mathbf{E}|^2 + \omega\mu'|\delta\mathbf{H}|^2] \, dv' = 0 \tag{7-6b}$$

Since $\sigma + \omega\varepsilon''$ and $\omega\mu''$ are positive for dissipative media, the only way for (7-6a) to be zero would be for $|\delta\mathbf{E}|^2 = |\delta\mathbf{H}|^2 = 0$ or $\delta\mathbf{E} = \delta\mathbf{H} = 0$. Therefore we have proved uniqueness. However, all these were based upon the premise that (7-5) applies [1]. Using the vector identity

$$\mathbf{A} \cdot \mathbf{B} \times \mathbf{C} = \mathbf{B} \cdot \mathbf{C} \times \mathbf{A} = \mathbf{C} \cdot \mathbf{A} \times \mathbf{B} \tag{7-7}$$

we can write (7-5) as

$$\oint_S [\delta\mathbf{E} \times \delta\mathbf{H}^*] \cdot \hat{n} \, da = \oint_S (\hat{n} \times \delta\mathbf{E}) \cdot \delta\mathbf{H}^* \, da = \oint_S [\delta\mathbf{H}^* \times \hat{n}] \cdot \delta\mathbf{E} \, da = 0 \tag{7-8}$$

If we can state the conditions under which (7-8) is satisfied, then we have proved uniqueness. This however will only be applicable for dissipative media. *However, we can treat lossless media as special cases of dissipative media as the losses diminish.*

Let us examine some of the important cases where (7-8) is satisfied and uniqueness is obtained in lossy media.

1. A field $\mathbf{E}, \mathbf{H}$ is unique when $\hat{n} \times \mathbf{E}$ is specified on S; then $\hat{n} \times \delta\mathbf{E} = 0$ over S. This results from exact specification of the tangential components of $\mathbf{E}$ and satisfaction of (7-8). No specification on the normal components is necessary.
2. A field $\mathbf{E}, \mathbf{H}$ is unique when $\hat{n} \times \mathbf{H}$ is specified on S; then $\hat{n} \times \delta\mathbf{H} = 0$ over S. This results from exact specification of the tangential component of $\mathbf{H}$ and satisfaction of (7-8).
3. A field $\mathbf{E}, \mathbf{H}$ is unique when $\hat{n} \times \mathbf{E}$ is specified over part of S and $\hat{n} \times \mathbf{H}$ is specified over the rest of S.

SUMMARY

A field in a lossy region, created by sources $\mathbf{J}_i$ and $\mathbf{M}_i$, is unique within the region when one of the following alternatives is specified.

1. *The tangential components of $\mathbf{E}$ over the boundary.*
2. *The tangential components of $\mathbf{H}$ over the boundary.*
3. *The former over part of the boundary and the latter over the rest of the boundary.*

Note: In general, the uniqueness theorem breaks down for lossless media. To justify uniqueness in this case, the fields in a lossless medium, as the dissipation approaches zero, can be considered to be the limit of the corresponding fields in a lossy medium. In some cases, however, unique solutions for lossless problems can be obtained on their own merits without treating them as special cases of lossy solutions.

7.4 IMAGE THEORY

The presence of an obstacle, especially when it is near the radiating element, can significantly alter the overall radiation properties of the radiating system, as illustrated in Chapter 5. In practice the most common obstacle that is always present, even in the absence of anything else, is the ground. Any energy from the radiating element directed toward the ground undergoes reflection. The amount of reflected energy and its direction are controlled by the geometry and constitutive parameters of the ground.

In general the ground is a lossy medium ($\sigma \neq 0$) whose effective conductivity increases with frequency. Therefore it should be expected to act as a very good conductor above a certain frequency, depending primarily upon its moisture content. To simplify the analysis we will assume that the ground is a perfect electric conductor, flat, and infinite in extent. The same procedure can also be used to investigate the characteristics of any radiating element near any other infinite, flat, perfect electric conductor. In practice, it is impossible to have infinite dimensions but we can simulate (electrically) very large obstacles. The effects that finite dimensions have on the radiation properties of a radiating element will be discussed in Chapters 12 and 13.

To analyze the performance of a radiating element near an infinite plane conductor, we will introduce virtual sources (images) that account for the reflections. The discussion here follows that of [2]. As the name implies, these are not real sources but imaginary ones that, in combination with the real sources, form an equivalent system that replaces the actual system for analysis purposes only and gives the same radiated field above the conductor as the actual system itself. Below

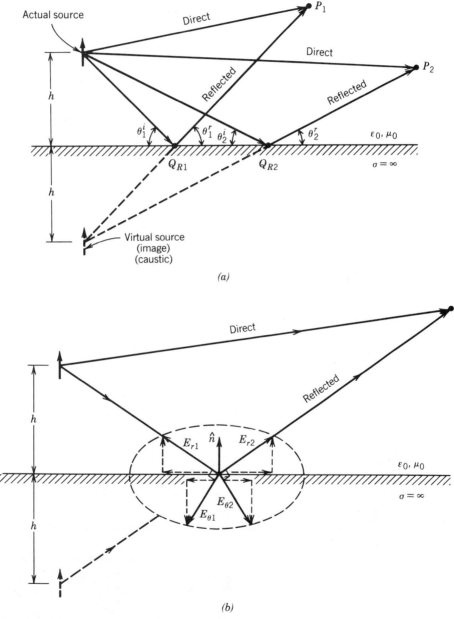

FIGURE 7-1 Vertical dipole, and its image, for reflection from a flat conducting surface of infinite extent. (*a*) Actual source and its image. (*b*) Field components at point of reflection. (*c*) Direct and reflected components.

316   ELECTROMAGNETIC THEOREMS AND PRINCIPLES

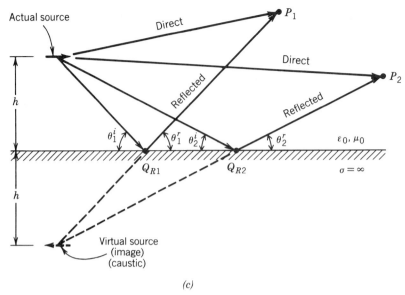

(c)

FIGURE 7-1 (*Continued*)

the conductor the equivalent system does not give the correct field; however, the field there is zero and the equivalent is not necessary.

To begin our discussion let us assume that a vertical electric dipole is placed a distance h above an infinite, flat, perfect electric conductor as shown in Figure 7-1a. Assuming that there is no mutual coupling, energy from the actual source is radiated in all directions in a manner determined by its unbounded medium directional properties. For an observation point P_1, there is a direct wave. In addition, a wave from the actual source radiated toward point R_1 of the interface will undergo reflection with a direction determined by the law of reflection, $\theta_1^r = \theta_1^i$. This follows from the fact that energy in inhomogeneous media travels in straight lines along the shortest paths. The wave will pass through the observation point P_1 and, by extending its actual path below the interface, it will seem to originate from a virtual source positioned a distance h below the boundary. For another observation point P_2 the point of reflection is R_2 but the virtual source is the same as before. The same conclusions can be drawn for all other points above the interface.

The amount of reflection is generally determined by the constitutive parameters of the medium below the interface relative to those above. For a perfect electric conductor below the interface, the incident wave is completely reflected with zero fields below the boundary. According to the boundary conditions, the tangential components of the electric field must vanish at all points along the interface. This condition is used to determine the polarization of the reflected field, compared to the direct wave, as shown in Figure 7-1b. To excite the polarization of the reflected waves, the virtual source must also be vertical and have polarity in the same direction as the actual source. Thus a reflection coefficient of $+1$ is required. Since the boundary conditions on the tangential electric field components are satisfied over a closed surface, in this case along the interface from $-\infty$ to $+\infty$, then the solution is unique according to the uniqueness theorem of Section 7.3.

Another source orientation is to have the radiating element in a horizontal position, as shown in Figure 7-1c. If we follow a procedure similar to that of the

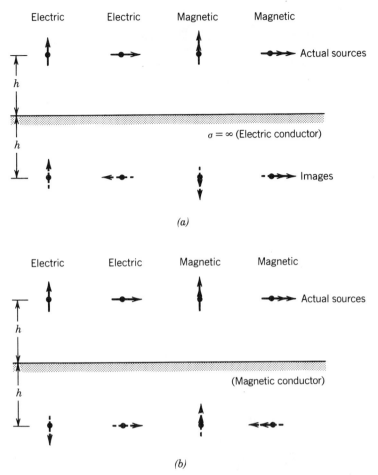

FIGURE 7-2 Electric and magnetic sources and their images near (a) electric and (b) magnetic conductors. (*Source:* C. A. Balanis, *Antenna Theory: Analysis and Design.* Copyright © 1982, John Wiley & Sons, Inc. Reprinted by permission of John Wiley & Sons, Inc.)

vertical dipole, we see that the virtual source (image) is also placed a distance h below the interface but with a 180° polarity difference relative to the actual source, thus requiring a reflection coefficient of -1. Again according to the uniqueness theorem of Section 7.3, the solution is unique because the boundary conditions are satisfied along the closed surface, this time along the interface extending from $-\infty$ to $+\infty$.

In addition to electric sources we have equivalent "magnetic" sources and magnetic conductors, such that tangential components of the magnetic field vanish next to their surface. In Figure 7-2a we have sketched the sources and their images for an electric plane conductor [2]. The single arrow indicates an electric element and the double arrow signifies a magnetic element. The direction of the arrow identifies the polarity. Since many problems can be solved using duality, in Figure 7-2b we have sketched the sources and their images when the obstacle is an infinite, flat, perfect "magnetic" conductor.

7.4.1 Vertical Electric Dipole

In the previous section we graphically illustrated the analysis procedure, using image theory, for vertical and horizontal electric and magnetic elements near infinite electric and magnetic plane conductors. In this section we want to derive the mathematical expressions for the fields of a vertical linear element near a perfect electric conductor, and the derivation will be based on the image solution of Figure 7-1a. For simplicity only far-field observations will be considered.

Let us refer now to the geometry of Figure 7-3a. The far-zone direct component of the electric field, of the infinitesimal dipole of length ℓ, constant current I_0, and observation point P_1, is given according to the dominant terms ($\beta r \gg 1$) of the fields in Example 6-3 by

$$E_\theta^d \stackrel{r \gg \lambda}{\simeq} j\eta \frac{\beta I_0 \ell e^{-j\beta r_1}}{4\pi r_1} \sin\theta_1 \tag{7-9}$$

The reflected component can be accounted for by the introduction of the virtual source (image), as shown in Figure 7-3a, and we can write it as

$$E_\theta^r \stackrel{r \gg \lambda}{\simeq} jR_v \eta \frac{\beta I_0 \ell e^{-j\beta r_2}}{4\pi r_2} \sin\theta_2$$

$$E_\theta^r \stackrel{r \gg \lambda}{\simeq} j\eta \frac{\beta I_0 \ell e^{-j\beta r_2}}{4\pi r_2} \sin\theta_2 \tag{7-10}$$

since the reflection coefficient R_v is equal to unity.

The total field above the interface ($z \geq 0$) is equal to the sum of the incident and reflected components as given by (7-9) and (7-10). Since an electric field cannot exist inside a perfect electric conductor, it is equal to zero below the interface. To simplify the expression for the total electric field, we would like to refer it to the origin of the coordinate system ($z = 0$) and express it in terms of r and θ. In general we can write that

$$r_1 = (r^2 + h^2 - 2rh\cos\theta)^{1/2} \tag{7-11a}$$

$$r_2 = [r^2 + h^2 - 2rh\cos(\pi - \theta)]^{1/2} \tag{7-11b}$$

However, for $r \gg h$ we can simplify and, using the binomial expansion, write [2]

$$r_1 \simeq r - h\cos\theta \tag{7-12a}$$
$$r_2 \simeq r + h\cos\theta \tag{7-12b}$$
for phase variations

$$\theta_1 \simeq \theta_2 \simeq \theta \tag{7-12c}$$

As shown in Figure 7-3b, (7-12a) and (7-12b) geometrically represent parallel lines. Since the amplitude variations are not as critical

$$r_1 \simeq r_2 \simeq r \quad \text{for amplitude variations} \tag{7-12d}$$

IMAGE THEORY 319

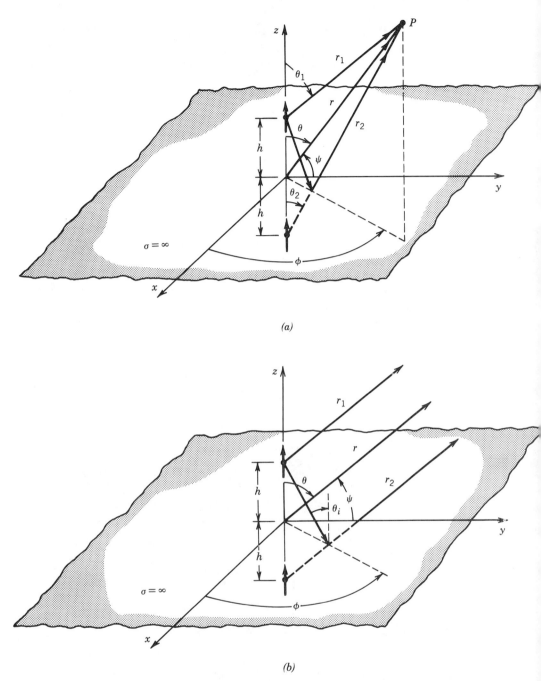

FIGURE 7-3 Vertical electric dipole above an infinite electric conductor. (*Source:* C. A. Balanis, *Antenna Theory: Analysis and Design.* Copyright © 1982, John Wiley & Sons, Inc. Reprinted by permission of John Wiley & Sons, Inc.) (*a*) Vertical electric dipole. (*b*) Far-field observations.

Use of (7-12a) through (7-12d) allows us to write the sum of (7-9) and (7-10) as

$$E_\theta = E_\theta^d + E_\theta^r \simeq j\eta \frac{\beta I_0 \ell e^{-j\beta r}}{4\pi r} \sin\theta \left(e^{+j\beta h \cos\theta} + e^{-j\beta h \cos\theta} \right) \quad z \geq 0$$
$$E_\theta = 0 \quad z < 0 \quad (7\text{-}13)$$

which can be reduced to

$$E_\theta \simeq j\eta \frac{\beta I_0 \ell e^{-j\beta r}}{4\pi r} \sin\theta \left[2\cos(\beta h \cos\theta) \right] \quad z \geq 0$$
$$E_\theta = 0 \quad z < 0 \quad (7\text{-}13a)$$

It is evident that the total electric field is equal to the product of the field of a single source and a factor [within the brackets in (7-13a)] that is a function of the element height h and the observation point θ. This product is referred to as the *pattern multiplication* rule, and the factor is known as the *array factor*. More details can be found in Chapter 6 of [2].

The shape and amplitude of the field is not only controlled by the single element but also by the positioning of the element relative to the ground. To examine the field variations as a function of the height h, we have plotted the power patterns for $h = 2\lambda$ and 5λ in Figure 7-4. Because of symmetry, only half of each

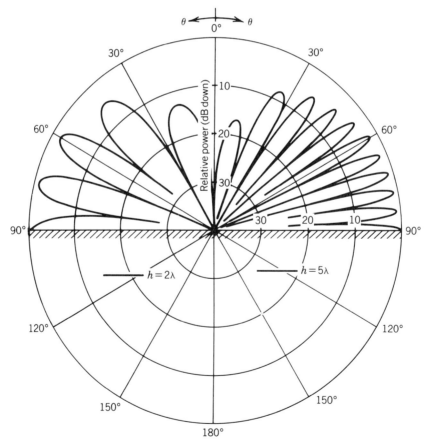

FIGURE 7-4 Elevation plane amplitude patterns of a vertical infinitesimal electric dipole for heights of 2λ and 5λ above an infinite plane electric conductor.

7.4.2 Horizontal Electric Dipole

Another system configuration is to have the linear antenna placed horizontally relative to the infinite electric ground plane, as shown in Figure 7-5a. The analysis procedure is identical to that of the vertical dipole. By introducing an image and assuming far-field observations, as shown in Figure 7-5b, we can write that the dominant terms of the direct component are given by [2]

$$E_\psi^d \stackrel{r \gg \lambda}{\simeq} j\eta \frac{\beta I_0 \ell e^{-j\beta r_1}}{4\pi r_1} \sin \psi \quad (7\text{-}14)$$

and the reflected terms by

$$E_\psi^r \stackrel{r \gg \lambda}{\simeq} jR_h\eta \frac{\beta I_0 \ell e^{-j\beta r_2}}{4\pi r_2} \sin \psi$$

$$E_\psi^r \stackrel{r \gg \lambda}{\simeq} -j\eta \frac{\beta I_0 \ell e^{-j\beta r_2}}{4\pi r_2} \sin \psi \quad (7\text{-}15)$$

since the reflection coefficient is equal to $R_h = -1$.

To find the angle ψ, which is measured from the y axis toward the observation point, we first form

$$\cos \psi = \hat{a}_y \cdot \hat{a}_r = \hat{a}_y \cdot (\hat{a}_x \sin \theta \cos \phi + \hat{a}_y \sin \theta \sin \phi + \hat{a}_z \cos \theta)$$

$$\cos \psi = \sin \theta \sin \phi \quad (7\text{-}16)$$

from which we find

$$\sin \psi = \sqrt{1 - \cos^2 \psi} = \sqrt{1 - \sin^2 \theta \sin^2 \phi} \quad (7\text{-}16\text{a})$$

Since for far-field observations

$$\left.\begin{array}{l} r_1 \simeq r - h \cos \theta \\ r_2 \simeq r + h \cos \theta \end{array}\right\} \text{ for phase variations} \quad (7\text{-}16\text{b})$$

$$\theta_1 \simeq \theta_2 \simeq \theta \quad (7\text{-}16\text{c})$$

$$r_1 \simeq r_2 \simeq r \quad \text{for amplitude variations} \quad (7\text{-}16\text{d})$$

we can write the total field, which is valid only above the ground plane ($z \geq 0$, $0 \leq \theta \leq \pi/2$, $0 \leq \phi \leq 2\pi$), as

$$E_\psi = E_\psi^d + E_\psi^r = j\eta \frac{\beta I_0 \ell e^{-j\beta r}}{4\pi r} \sqrt{1 - \sin^2 \theta \sin^2 \phi} \, [2j \sin(\beta h \cos \theta)] \quad (7\text{-}17)$$

Equation 7-17 again is recognized to consist of the product of the field of a single isolated element placed at the origin and a factor (within the brackets) known as the array factor. This again is the pattern multiplication rule.

322 ELECTROMAGNETIC THEOREMS AND PRINCIPLES

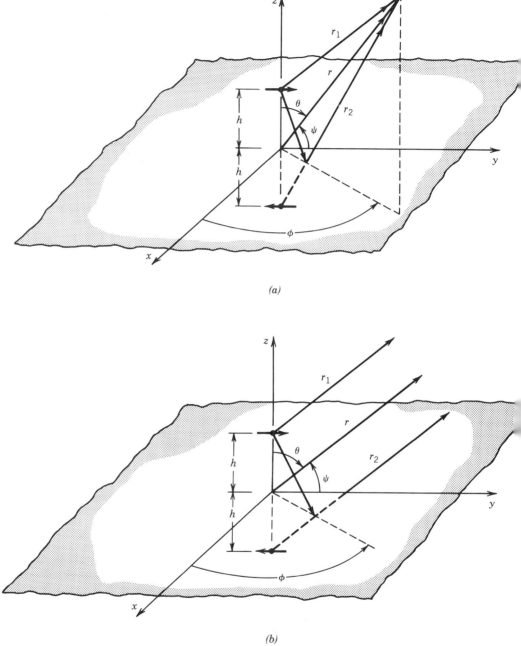

FIGURE 7-5 Horizontal electric dipole above an infinite electric conductor (*Source:* C. A. Balanis, *Antenna Theory: Analysis and Design.* Copyright © 1982, John Wiley & Sons, Inc. Reprinted by permission of John Wiley & Sons, Inc.) (*a*) Horizontal electric dipole. (*b*) Far-field observations.

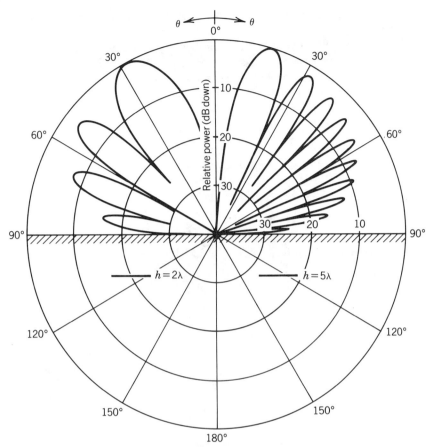

FIGURE 7-6 Elevation plane ($\phi = 90°$) amplitude patterns of a horizontal infinitesimal dipole for heights of 2λ and 5λ above an infinite plane electric conductor.

To examine the variations of the total field as a function of the element height above the ground plane, in Figure 7-6 we have plotted two-dimensional elevation plane patterns for $\phi = 90°$ (yz plane) when $h = 2\lambda$ and 5λ. Again we see that the height of the element above the interface plays a significant role in the radiation pattern of the radiating system.

Problems that require multiple images, such as corner reflectors, are assigned as exercises at the end of the chapter.

7.5 RECIPROCITY THEOREM

We are all well familiar with the reciprocity theorem, as applied to circuits, which states that *in any physical linear network, the positions of an ideal voltage source (zero internal impedance) and an ideal ammeter (infinite internal impedance) can be interchanged without affecting their readings* [5]. Now we want to discuss the reciprocity theorem as it applies to electromagnetic theory [6]. This is done best by using Maxwell's equations. The reciprocity theorem has many applications; one of the most common relating the transmitting and receiving properties of radiating systems [2].

Let us assume that within a linear, isotropic medium, which is not necessarily homogeneous, there exist two sets of sources $\mathbf{J}_1, \mathbf{M}_1$ and $\mathbf{J}_2, \mathbf{M}_2$ that are allowed to radiate simultaneously or individually inside the same medium at the same frequency and produce fields $\mathbf{E}_1, \mathbf{H}_1$ and $\mathbf{E}_2, \mathbf{H}_2$, respectively. For the fields to be valid, they must satisfy Maxwell's equations

$$\nabla \times \mathbf{E}_1 = -\mathbf{M}_1 - j\omega\mu\mathbf{H}_1 \quad \text{(7-18a)}$$
$$\nabla \times \mathbf{H}_1 = \mathbf{J}_1 + j\omega\varepsilon\mathbf{E}_1 \quad \text{(7-18b)}$$
for sources $\mathbf{J}_1, \mathbf{M}_1$

$$\nabla \times \mathbf{E}_2 = -\mathbf{M}_2 - j\omega\mu\mathbf{H}_2 \quad \text{(7-19a)}$$
$$\nabla \times \mathbf{H}_2 = \mathbf{J}_2 + j\omega\varepsilon\mathbf{E}_2 \quad \text{(7-19b)}$$
for sources $\mathbf{J}_2, \mathbf{M}_2$

If we dot multiply of (7-18a) by $\mathbf{H}_2$ and (7-19b) by $\mathbf{E}_1$, we can write that

$$\mathbf{H}_2 \cdot \nabla \times \mathbf{E}_1 = -\mathbf{H}_2 \cdot \mathbf{M}_1 - j\omega\mu\mathbf{H}_2 \cdot \mathbf{H}_1 \quad \text{(7-20a)}$$
$$\mathbf{E}_1 \cdot \nabla \times \mathbf{H}_2 = \mathbf{E}_1 \cdot \mathbf{J}_2 + j\omega\varepsilon\mathbf{E}_1 \cdot \mathbf{E}_2 \quad \text{(7-20b)}$$

Subtracting (7-20a) from (7-20b) reduces to

$$\mathbf{E}_1 \cdot \nabla \times \mathbf{H}_2 - \mathbf{H}_2 \cdot \nabla \times \mathbf{E}_1 = \mathbf{E}_1 \cdot \mathbf{J}_2 + \mathbf{H}_2 \cdot \mathbf{M}_1 + j\omega\varepsilon\mathbf{E}_1 \cdot \mathbf{E}_2 + j\omega\mu\mathbf{H}_2 \cdot \mathbf{H}_1 \quad \text{(7-21)}$$

which by use of the vector identity

$$\nabla \cdot (\mathbf{A} \times \mathbf{B}) = \mathbf{B} \cdot (\nabla \times \mathbf{A}) - \mathbf{A} \cdot (\nabla \times \mathbf{B}) \quad \text{(7-22)}$$

can be written as

$$\nabla \cdot (\mathbf{H}_2 \times \mathbf{E}_1) = -\nabla \cdot (\mathbf{E}_1 \times \mathbf{H}_2)$$
$$= \mathbf{E}_1 \cdot \mathbf{J}_2 + \mathbf{H}_2 \cdot \mathbf{M}_1 + j\omega\varepsilon\mathbf{E}_1 \cdot \mathbf{E}_2 + j\omega\mu\mathbf{H}_2 \cdot \mathbf{H}_1 \quad \text{(7-23)}$$

In a similar manner, if we dot multiply of (7-18b) by $\mathbf{E}_2$ and (7-19a) by $\mathbf{H}_1$, we can write

$$\mathbf{E}_2 \cdot \nabla \times \mathbf{H}_1 = \mathbf{E}_2 \cdot \mathbf{J}_1 + j\omega\varepsilon\mathbf{E}_2 \cdot \mathbf{E}_1 \quad \text{(7-24a)}$$
$$\mathbf{H}_1 \cdot \nabla \times \mathbf{E}_2 = -\mathbf{H}_1 \cdot \mathbf{M}_2 - j\omega\mu\mathbf{H}_1 \cdot \mathbf{H}_2 \quad \text{(7-24b)}$$

Subtraction of (7-24b) from (7-24a) leads to

$$\mathbf{E}_2 \cdot \nabla \times \mathbf{H}_1 - \mathbf{H}_1 \cdot \nabla \times \mathbf{E}_2 = \mathbf{E}_2 \cdot \mathbf{J}_1 + \mathbf{H}_1 \cdot \mathbf{M}_2 + j\omega\varepsilon\mathbf{E}_2 \cdot \mathbf{E}_1 + j\omega\mu\mathbf{H}_1 \cdot \mathbf{H}_2 \quad \text{(7-25)}$$

which by use of (7-22) can be written as

$$\nabla \cdot (\mathbf{H}_1 \times \mathbf{E}_2) = -\nabla \cdot (\mathbf{E}_2 \times \mathbf{H}_1)$$
$$= \mathbf{E}_2 \cdot \mathbf{J}_1 + \mathbf{H}_1 \cdot \mathbf{M}_2 + j\omega\mathbf{E}_2 \cdot \mathbf{E}_1 + j\omega\mu\mathbf{H}_1 \cdot \mathbf{H}_2 \quad \text{(7-26)}$$

Subtraction of (7-26) from (7-23) lead to

$$-\nabla \cdot (\mathbf{E}_1 \times \mathbf{H}_2 - \mathbf{E}_2 \times \mathbf{H}_1) = \mathbf{E}_1 \cdot \mathbf{J}_2 + \mathbf{H}_2 \cdot \mathbf{M}_1 - \mathbf{E}_2 \cdot \mathbf{J}_1 - \mathbf{H}_1 \cdot \mathbf{M}_2 \quad (7\text{-}27)$$

which is called the *Lorentz reciprocity theorem* in differential form [7].

By taking a volume integral of both sides of (7-27) and using the divergence theorem on the left side, we can write (7-27) as

$$-\oiint_S (\mathbf{E}_1 \times \mathbf{H}_2 - \mathbf{E}_2 \times \mathbf{H}_1) \cdot d\mathbf{s}'$$
$$= \iiint_V (\mathbf{E}_1 \cdot \mathbf{J}_2 + \mathbf{H}_2 \cdot \mathbf{M}_1 - \mathbf{E}_2 \cdot \mathbf{J}_1 - \mathbf{H}_1 \cdot \mathbf{M}_2) \, dv' \quad (7\text{-}28)$$

which is known as the Lorentz reciprocity theorem in integral form.

For a source-free ($\mathbf{J}_1 = \mathbf{J}_2 = \mathbf{M}_1 = \mathbf{M}_2 = 0$) region, (7-27) and (7-28) reduce, respectively, to

$$\nabla \cdot (\mathbf{E}_1 \times \mathbf{H}_2 - \mathbf{E}_2 \times \mathbf{H}_1) = 0 \quad (7\text{-}29)$$

$$\oiint_S (\mathbf{E}_1 \times \mathbf{H}_2 - \mathbf{E}_2 \times \mathbf{H}_1) \cdot d\mathbf{s}' = 0 \quad (7\text{-}30)$$

Equations 7-29 and 7-30 are special cases of the Lorentz reciprocity theorem and must be satisfied in source-free regions.

As an example of where (7-29) and (7-30) may be applied and what they would represent, consider a section of a waveguide where two different modes exist with fields $\mathbf{E}_1, \mathbf{H}_1$ and $\mathbf{E}_2, \mathbf{H}_2$. For the expressions of the fields for the two modes to be valid, they must satisfy (7-29) and/or (7-30).

Another useful form of (7-28) is to consider that the fields ($\mathbf{E}_1, \mathbf{H}_1, \mathbf{E}_2, \mathbf{H}_2$) and the sources ($\mathbf{J}_1, \mathbf{M}_1, \mathbf{J}_2, \mathbf{M}_2$) are within a medium that is enclosed by a sphere of infinite radius. Assume that the sources are positioned within a finite region and that the fields are observed in the far field (ideally at infinity). Then the left side of (7-28) is equal to zero, or

$$\oiint_S (\mathbf{E}_1 \times \mathbf{H}_2 - \mathbf{E}_2 \times \mathbf{H}_1) \cdot d\mathbf{s}' = 0 \quad (7\text{-}31)$$

which reduces (7-28) to

$$\iiint_V (\mathbf{E}_1 \cdot \mathbf{J}_2 + \mathbf{H}_2 \cdot \mathbf{M}_1 - \mathbf{E}_2 \cdot \mathbf{J}_1 - \mathbf{H}_1 \cdot \mathbf{M}_2) \, dv' = 0 \quad (7\text{-}32)$$

Equation 7-32 can also be written as

$$\iiint_V (\mathbf{E}_1 \cdot \mathbf{J}_2 - \mathbf{H}_1 \cdot \mathbf{M}_2) \, dv' = \iiint_V (\mathbf{E}_2 \cdot \mathbf{J}_1 - \mathbf{H}_2 \cdot \mathbf{M}_1) \, dv' \quad (7\text{-}32\text{a})$$

The reciprocity theorem, as expressed by (7-32a), is the most useful form.

7.6 REACTION THEOREM

Close observation of (7-28) reveals that it does not, in general, represent relations of power because no conjugates appear. The same is true for (7-30) and (7-32a). Each of the integrals in (7-32a) can be interpreted as a coupling between a set of fields and a set of sources, which produce another set of fields. This coupling has been defined as *reaction* [8, 9] and each of the integrals in (7-32a) has been denoted by

$$\langle 1, 2 \rangle = \iiint_V (\mathbf{E}_1 \cdot \mathbf{J}_2 - \mathbf{H}_1 \cdot \mathbf{M}_2) \, dv' \quad (7\text{-}33a)$$

$$\langle 2, 1 \rangle = \iiint_V (\mathbf{E}_2 \cdot \mathbf{J}_1 - \mathbf{H}_2 \cdot \mathbf{M}_1) \, dv' \quad (7\text{-}33b)$$

The relation $\langle 1, 2 \rangle$ relates the reaction (coupling) of the fields $(\mathbf{E}_1, \mathbf{H}_1)$, which are produced by sources $\mathbf{J}_1, \mathbf{M}_1$, to the sources $(\mathbf{J}_2, \mathbf{M}_2)$, which produce fields $\mathbf{E}_2, \mathbf{H}_2$; $\langle 2, 1 \rangle$ relates the reaction (coupling) of the fields $(\mathbf{E}_2, \mathbf{H}_2)$ to the sources $(\mathbf{J}_1, \mathbf{M}_1)$. A requirement for reciprocity to hold is that the reactions (couplings) of the sources with their corresponding fields must be equal. In equation form

$$\langle 1, 2 \rangle = \langle 2, 1 \rangle \quad (7\text{-}34)$$

The reaction theorem can also be expressed in terms of the voltages and currents induced in one antenna by another [9]. In a general form it can be written as

$$\langle i, j \rangle = V_j I_{ji} = \langle j, i \rangle = V_i I_{ij} \quad (7\text{-}34a)$$

where $V_{i(j)}$ = voltage of source $i(j)$
I_{ij} = current through source j due to source at i

The reactions forms of (7-34) and (7-34a) are most convenient to calculate the mutual impedance and admittance between aperture antennas.

Example 7.2. Derive an expression for the mutual admittance between two aperture antennas. The expression should be in terms of the electric and magnetic fields on the apertures and radiated by the apertures.

Solution. In a multiport network the Y-parameter matrix can be written as

$$[I_i] = [Y_{ij}][V_j]$$

Assuming that the voltages at all ports other than port j are zero, then we can write that the current I_{ij} at port i due to the voltage at port j can be written as

$$I_{ij} = Y_{ij} V_j \Rightarrow Y_{ij} = \frac{I_{ij}}{V_j}$$

Using (7-34a) we can write that

$$I_{ij} = \frac{\langle i, j \rangle}{V_i}$$

This allows us to write the mutual admittance as

$$Y_{ij} = \frac{I_{ij}}{V_j} = \frac{\langle i, j \rangle}{V_i V_j}$$

which by using (7-33a) or (7-33b) can be expressed as

$$Y_{ij} = \frac{\langle i, j \rangle}{V_i V_j} = \frac{1}{V_i V_j} \iiint_V (\mathbf{E}_i \cdot \mathbf{J}_j - \mathbf{H}_i \cdot \mathbf{M}_j) \, dv'$$

Since aperture antennas can be represented by magnetic equivalent currents, then

$$\mathbf{J}_j = 0$$
$$\mathbf{M}_j = -\hat{n} \times \mathbf{E}_j$$

Using these and reducing the volume integral to surface integral over the aperture of the antenna, we can write the mutual admittance as

$$Y_{ij} = -\frac{1}{V_i V_j} \iint_{S_a} (\mathbf{H}_i \cdot \mathbf{M}_j) \, ds' = -\frac{1}{V_i V_j} \iint_{S_a} [\mathbf{H}_i \cdot (-\hat{n} \times \mathbf{E}_j)] \, ds'$$

$$Y_{ij} = \frac{1}{V_i V_j} \iint_{S_a} (\mathbf{E}_j \times \mathbf{H}_i) \cdot \hat{n} \, ds'$$

where $\mathbf{E}_j$ = electric field in aperture j with aperture i shorted
$\mathbf{H}_i$ = magnetic field at shorted aperture i due to excitation of aperture j
$V_{i(j)}$ = voltage amplitudes at each aperture in the absence of the other

7.7 VOLUME EQUIVALENCE THEOREM

Through use of the equivalent electric and magnetic current sources, the volume equivalence theorem can be used to determine the scattered fields when a material obstacle is introduced in a free-space environment where fields $\mathbf{E}_0, \mathbf{H}_0$ were previously generated by sources $\mathbf{J}_i, \mathbf{M}_i$ [7, 10].

To derive the volume equivalence theorem, let us assume that in the free-space environment sources $(\mathbf{J}_i, \mathbf{M}_i)$ generate fields $(\mathbf{E}_0, \mathbf{H}_0)$. These sources and fields must satisfy Maxwell's equations

$$\nabla \times \mathbf{E}_0 = -\mathbf{M}_i - j\omega\mu_0 \mathbf{H}_0 \quad (7\text{-}35a)$$

$$\nabla \times \mathbf{H}_0 = \mathbf{J}_i + j\omega\varepsilon_0 \mathbf{E}_0 \quad (7\text{-}35b)$$

When the same sources $(\mathbf{J}_i, \mathbf{M}_i)$ radiate in a medium represented by (ε, μ), they generate fields $(\mathbf{E}, \mathbf{H})$ that satisfy Maxwell's equations

$$\nabla \times \mathbf{E} = -\mathbf{M}_i - j\omega\mu \mathbf{H} \quad (7\text{-}36a)$$

$$\nabla \times \mathbf{H} = \mathbf{J}_i + j\omega\varepsilon \mathbf{E} \quad (7\text{-}36b)$$

Subtraction of (7-35a) from (7-36a) and (7-35b) from (7-36b), allows us to write that

$$\nabla \times (\mathbf{E} - \mathbf{E}_0) = -j\omega(\mu\mathbf{H} - \mu_0\mathbf{H}_0) \tag{7-37a}$$

$$\nabla \times (\mathbf{H} - \mathbf{H}_0) = j\omega(\varepsilon\mathbf{E} - \varepsilon_0\mathbf{E}_0) \tag{7-37b}$$

Let us define the difference between the fields $\mathbf{E}$ and $\mathbf{E}_0$, and $\mathbf{H}$ and $\mathbf{H}_0$ as the *scattered* (disturbance) fields $\mathbf{E}^s$ and $\mathbf{H}^s$, that is,

$$\mathbf{E}^s = \mathbf{E} - \mathbf{E}_0 \Rightarrow \mathbf{E}_0 = \mathbf{E} - \mathbf{E}^s \tag{7-38a}$$

$$\mathbf{H}^s = \mathbf{H} - \mathbf{H}_0 \Rightarrow \mathbf{H}_0 = \mathbf{H} - \mathbf{H}^s \tag{7-38b}$$

By using the definitions for the scattered fields of (7-38a) and (7-38b), we can write (7-37a) and (7-37b) as

$$\nabla \times \mathbf{E}^s = -j\omega[\mu\mathbf{H} - \mu_0(\mathbf{H} - \mathbf{H}^s)] = -j\omega(\mu - \mu_0)\mathbf{H} - j\omega\mu_0\mathbf{H}^s \tag{7-39a}$$

$$\nabla \times \mathbf{H}^s = j\omega[\varepsilon\mathbf{E} - \varepsilon_0(\mathbf{E} - \mathbf{E}^s)] = j\omega(\varepsilon - \varepsilon_0)\mathbf{E} + j\omega\varepsilon_0\mathbf{E}^s \tag{7-39b}$$

By defining volume equivalent electric $\mathbf{J}_{eq}$ and magnetic $\mathbf{M}_{eq}$ current densities

$$\boxed{\mathbf{J}_{eq} = j\omega(\varepsilon - \varepsilon_0)\mathbf{E}} \tag{7-40a}$$

$$\boxed{\mathbf{M}_{eq} = j\omega(\mu - \mu_0)\mathbf{H}} \tag{7-40b}$$

which exist only in the region where $\varepsilon \neq \varepsilon_0$ and $\mu \neq \mu_0$ (only in the material itself), we can express (7-39a) and (7-39b) as

$$\boxed{\nabla \times \mathbf{E}^s = -\mathbf{M}_{eq} - j\omega\mu_0\mathbf{H}^s} \tag{7-41a}$$

$$\boxed{\nabla \times \mathbf{H}^s = \mathbf{J}_{eq} + j\omega\varepsilon_0\mathbf{E}^s} \tag{7-41b}$$

Equations 7-41a and 7-41b state that the electric $\mathbf{E}^s$ and magnetic $\mathbf{H}^s$ fields scattered by a material obstacle can be generated by using equivalent electric $\mathbf{J}_{eq}$ (A/m^2) and magnetic $\mathbf{M}_{eq}$ (V/m^2) volume current densities that are given by (7-40a) and (7-40b), that exist only within the material, and that radiate in a free-space environment. Although, in principle, the formulation of the problem seems to have been simplified, it is still very difficult to solve because the equivalent current densities are in terms of $\mathbf{E}$ and $\mathbf{H}$ which are unknown. However, the formulation does provide some physical interpretation of scattering and lends itself to development of integral equations for the solution of $\mathbf{E}^s$ and $\mathbf{H}^s$ which are discussed in Chapter 12. The volume equivalent current densities are most useful for finding the fields scattered by dielectric obstacles. The fields scattered by perfectly conducting surfaces can best be determined using surface equivalent densities, especially those discussed in Sections 7.9 and 7.10. The surface equivalence theorem that follows is best utilized for analysis of antenna aperture radiation.

7.8 SURFACE EQUIVALENCE THEOREM: HUYGENS'S PRINCIPLE

The *surface equivalence* theorem is a principle by which actual sources, such as an antenna and transmitter, are replaced by equivalent sources. The fictitious sources are said to be equivalent within a region because they produce within that region the same fields as the actual sources. The formulations of scattering and diffraction problems by the surface equivalence theorem are more suggestive of approximations.

The surface equivalence was introduced in 1936 by Schelkunoff [11], and it is a more rigorous formulation of Huygens's principle [12], which states [13] that "each point on a primary wavefront can be considered to be a new source of a secondary spherical wave and that a secondary wavefront can be constructed as the envelope of these secondary spherical waves." The surface equivalence theorem is based on the uniqueness theorem of Section 7.3, which states [1] that "a field in a lossy region is uniquely specified by the sources within the region plus the tangential components of the electric field over the boundary, or the tangential components of the magnetic field over the boundary, or the former over part of the boundary and the latter over the rest of the boundary." The fields in a lossless medium are considered to be the limit, as the losses go to zero, of the corresponding fields in a lossy medium. Thus if the tangential electric and magnetic fields are completely known over a closed surface, the fields in the source-free region can be determined.

By the surface equivalence theorem, the fields outside an imaginary closed surface are obtained by placing, over the closed surface, suitable electric and magnetic current densities that satisfy the boundary conditions. The current densities are selected so that the fields inside the closed surface are zero and outside are equal to the radiation produced by the actual sources. Thus the technique can be used to obtain the fields radiated outside a closed surface by sources enclosed within it. The formulation is exact but requires integration over the closed surface. The degree of accuracy depends on the knowledge of the tangential components of the fields over the closed surface.

In the majority of applications, the closed surface is selected so that most of it coincides with the conducting parts of the physical structure. This is preferred because the tangential electric field components vanish over the conducting parts of the surface, which results in reduction of the physical limits of integration.

The surface equivalence theorem is developed by considering an actual radiating source, which is represented electrically by current densities $\mathbf{J}_1$ and $\mathbf{M}_1$, as shown in Figure 7-7a. The source radiates fields $\mathbf{E}_1$ and $\mathbf{H}_1$ everywhere. However, we wish to develop a method that will yield the fields outside a closed surface. To accomplish this, a closed surface S is chosen, shown dashed in Figure 7-7a, which encloses the current densities $\mathbf{J}_1$ and $\mathbf{M}_1$. The volume within S is denoted by V_1 and outside S by V_2. The primary task is to replace the original problem, shown in Figure 7-7a, with an equivalent that will yield the same fields $\mathbf{E}_1$ and $\mathbf{H}_1$ outside S (within V_2). The formulation of the problem can be aided immensely if the closed surface is chosen judiciously so that fields over most, if not the entire surface, are known a priori.

An equivalent problem to Figure 7-7a is shown in Figure 7-7b. The original sources $\mathbf{J}_1$ and $\mathbf{M}_1$ are removed, and we assume that there exist fields $\mathbf{E}, \mathbf{H}$ inside S and fields $\mathbf{E}_1, \mathbf{H}_1$ outside S. For these fields to exist within and outside S, they must satisfy the boundary conditions on the tangential electric and magnetic field components of Table 1-5. Thus on the imaginary surface S there must exist the

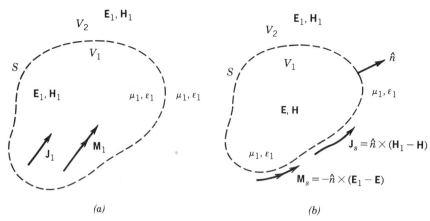

FIGURE 7-7 (*a*) Actual and (*b*) equivalent problem models (*Source:* C. A. Balanis, *Antenna Theory: Analysis and Design*. Copyright © 1982, John Wiley & Sons, Inc. Reprinted with permission of John Wiley & Sons, Inc.)

equivalent sources

$$\mathbf{J}_s = \hat{n} \times (\mathbf{H}_1 - \mathbf{H}) \qquad (7\text{-}42\text{a})$$

$$\mathbf{M}_s = -\hat{n} \times (\mathbf{E}_1 - \mathbf{E}) \qquad (7\text{-}42\text{b})$$

which radiate into an unbounded space (same medium everywhere). The current densities of (7-42a) and (7-42b) are said to be equivalent only within V_2, because they will produce the original field $(\mathbf{E}_1, \mathbf{H}_1)$ only outside S. A field $\mathbf{E}, \mathbf{H}$, different from the original $(\mathbf{E}_1, \mathbf{H}_1)$ will result within V_1. Since the currents of (7-42a) and (7-42b) radiate in an unbounded space, the fields can be determined using (6-30) through (6-35a) and the geometry of Figure 6-3a. In Figure 6-3a, R is the distance from any point on the surface S, where $\mathbf{J}_s$ and $\mathbf{M}_s$ exist, to the observation point.

So far, the tangential components of both $\mathbf{E}$ and $\mathbf{H}$ have been used to set up the equivalent problem. From electromagnetic uniqueness concepts, we know that the tangential components of only $\mathbf{E}$ or $\mathbf{H}$ are needed to determine the field. It will be demonstrated that equivalent problems that require only magnetic currents (tangential $\mathbf{E}$) or only electric currents (tangential $\mathbf{H}$) can be found. This will require modifications to the equivalent problem of Figure 7-7b.

Since the fields $\mathbf{E}, \mathbf{H}$ within S, which is not the region of interest, can be anything, it can be assumed that they are zero. Then the equivalent problem of Figure 7-7b reduces to that of Figure 7-8a with equivalent current densities equal to

$$\mathbf{J}_s = \hat{n} \times (\mathbf{H}_1 - \mathbf{H})|_{\mathbf{H}=0} = \hat{n} \times \mathbf{H}_1 \qquad (7\text{-}43\text{a})$$

$$\mathbf{M}_s = -\hat{n} \times (\mathbf{E}_1 - \mathbf{E})|_{\mathbf{E}=0} = -\hat{n} \times \mathbf{E}_1 \qquad (7\text{-}43\text{b})$$

This form of the field equivalence principle is known as *Love's equivalence principle* [7, 14]. Since the current densities of (7-43a) and (7-43b) radiate in an unbounded medium, that is, have the same μ_1, ε_1 everywhere, they can be used in conjunction with (6-30) through (6-35a) to find the fields everywhere.

Love's equivalence principle in Figure 7-8a produces a null field within the imaginary surface S. Since the value of the $\mathbf{E} = \mathbf{H} = 0$ within S cannot be disturbed if the properties of the medium within it are changed, let us assume that it is

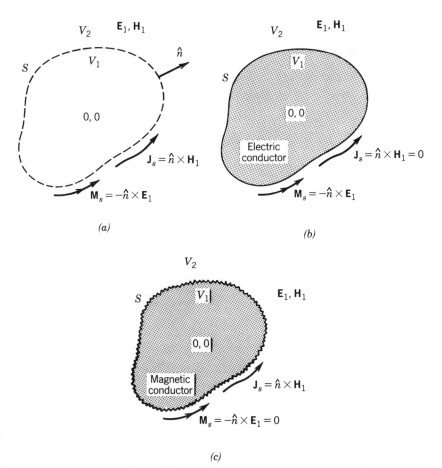

FIGURE 7-8 Equivalence principle models. (*Source:* C. A. Balanis, *Antenna Theory: Analysis and Design.* Copyright © 1982, John Wiley & Sons, Inc. Reprinted with permission of John Wiley & Sons, Inc.) (*a*) Love's equivalent. (*b*) Electric conductor equivalent. (*c*) Magnetic conductor equivalent.

replaced by a perfect electric conductor ($\sigma = \infty$). The introduction of the perfect conductor will have an effect on the equivalent source $\mathbf{J}_s$, and it will prohibit the use of (6-30) through (6-35a) because the current densities no longer radiate into an unbounded medium. Imagine that the geometrical configuration of the electric conductor is identical to the profile of the imaginary surface S, over which $\mathbf{J}_s$ and $\mathbf{M}_s$ exist. As the electric conductor takes its place, as shown in Figure 7-8*b*, the electric current density $\mathbf{J}_s$, which is tangent to the surface S, is short-circuited by the electric conductor. Thus the equivalent problem of Figure 7-8*a* reduces to that of Figure 7-8*b*. Only a magnetic current density $\mathbf{M}_s$ exists over S, and it radiates in the presence of the electric conductor producing the original fields $\mathbf{E}_1, \mathbf{H}_1$ outside S. Within S the fields are zero but, as before, this is not a region of interest. The difficulty in trying to use the equivalent problem of Figure 7-8*b* is that (6-30) through (6-35a) cannot be used, because the current densities do not radiate into an unbounded medium. The problem of a magnetic current radiating in the presence of an electric conducting surface must be solved. Therefore it seems that the equivalent problem is just as difficult as the original problem.

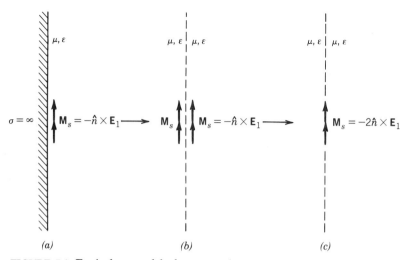

FIGURE 7-9 Equivalent models for magnetic source radiation near a perfect electric conductor. (*Source:* C. A. Balanis, *Antenna Theory: Analysis and Design.* Copyright © 1982, John Wiley & Sons, Inc. Reprinted with permission of John Wiley & Sons, Inc.)

Before some special simple geometries are considered and some suggestions are made for approximating complex geometries, let us introduce another equivalent problem. Refer to Figure 7-8a and assume that instead of placing a perfect electric conductor within S, we introduce a perfect magnetic conductor that will short out the magnetic current and reduce the equivalent problem to that shown in Figure 7-8c. Coincident with the equivalent problem of Figure 7-8b, (6-30) through (6-35a) cannot be used with Figure 7-8c, and the problem is just as difficult as that of Figure 7-8b or the original Figure 7-7a.

To initiate awareness of the utility of the field equivalence principle, especially that of Figure 7-8b, let us assume that the surface of the electric conductor is flat and extends to infinity as shown in Figure 7-9a. For this geometry, the problem is to determine how a magnetic source radiates in the presence of a flat electric conductor. From image theory, this problem reduces to that of Figure 7-9b, where an imaginary source is introduced on the side of the conductor and takes its place (removes the conductor). Since the imaginary source is in the same direction as the equivalent source, the equivalent problem of Figure 7-9b reduces to that of Figure 7-9c. The magnetic current is doubled, it radiates in an unbounded medium, and (6-30) through (6-35a) can be used. The equivalent problem of Figure 7-9c will yield the correct $\mathbf{E}, \mathbf{H}$ fields to the right side of the interface. If the surface of the obstacle is not flat and infinite, but its curvature is large compared to the wavelength, a good approximation will be the equivalent problem of Figure 7-9c.

SUMMARY

In the analysis of electromagnetic problems, many times it is easier to form equivalent problems that will yield the same solution within a region of interest. This is true for scattering, diffraction, and aperture antenna problems. In this section, the main emphasis is on aperture antennas, and concepts will be demonstrated by examples.

The following steps must be used to form an equivalent and solve an aperture problem.

1. Select an imaginary surface that encloses the actual sources (the aperture). The surface must be chosen judiciously so that the tangential components of the electric and/or the magnetic field are known, exactly or approximately, over its entire span. In many cases this surface is a flat plane extending to infinity.
2. Over the imaginary surface form equivalent current densities $\mathbf{J}_s, \mathbf{M}_s$ that take one of the following forms.
   a. $\mathbf{J}_s$ and $\mathbf{M}_s$ over S assuming that the $\mathbf{E}$ and $\mathbf{H}$ fields within S are not zero.
   b. $\mathbf{J}_s$ and $\mathbf{M}_s$ over S assuming that the $\mathbf{E}$ and $\mathbf{H}$ fields within S are zero (Love's theorem).
   c. $\mathbf{M}_s$ over S ($\mathbf{J}_s = 0$) assuming that within S the medium is a perfect electric conductor.
   d. $\mathbf{J}_s$ over S ($\mathbf{M}_s = 0$) assuming that within S the medium is a perfect magnetic conductor
3. Solve the equivalent problem. For forms **a** and **b**, equations 6-30 through 6-35a can be used. For form **c**, the problem of a magnetic current source next to a perfect electric conductor must be solved [(6-30) through (6-35a) cannot be used directly, because the current density does not radiate into an unbounded medium]. If the electric conductor is an infinite flat plane, the problem can be solved exactly by image theory. For form **d**, the problem of an electric current source next to a perfect magnetic conductor must be solved. Again (6-30) through (6-35a) cannot be used directly. If the magnetic conductor is an infinite flat plane, the problem can be solved exactly by image theory.

To demonstrate the usefulness and application of the field equivalence theorem to aperture antenna theory, we consider the following example.

Example 7-3. A waveguide aperture is mounted on an infinite ground plane, as shown in Figure 7-10a. Assume that the tangential components of the electric field over the aperture are known and are given by $\mathbf{E}_a$. Then find an equivalent problem that will yield the same fields $\mathbf{E}, \mathbf{H}$ radiated by the aperture to the right side of the interface.

Solution. First an imaginary closed surface is chosen. For this problem it is appropriate to select a flat plane extending from $-\infty$ to $+\infty$ as shown in Figure 7-10b. Over the infinite plane, the equivalent current densities $\mathbf{J}_s$ and $\mathbf{M}_s$ are formed. Since the tangential components of $\mathbf{E}$ do not exist outside the aperture, because of vanishing boundary conditions, the magnetic current density $\mathbf{M}_s$ is only nonzero over the aperture. The electric current density $\mathbf{J}_s$ is nonzero everywhere and is yet unknown. Now let us assume that an imaginary flat electric conductor approaches the surface S and it shorts out the current density $\mathbf{J}_s$ everywhere. $\mathbf{M}_s$ exists only over the space occupied originally by the aperture, and it radiates in the presence of the conductor (see Figure 7-10c). By image theory, the conductor can be removed and replaced by an imaginary (equivalent) source $\mathbf{M}_s$ as shown in Figure 7-10d, which is analogous to Figure 7-9b. Finally, the equivalent problem of Figure 7-10d reduces to that of Figure 7-10e, which is analogous to that of Figure 7-9c. The original problem

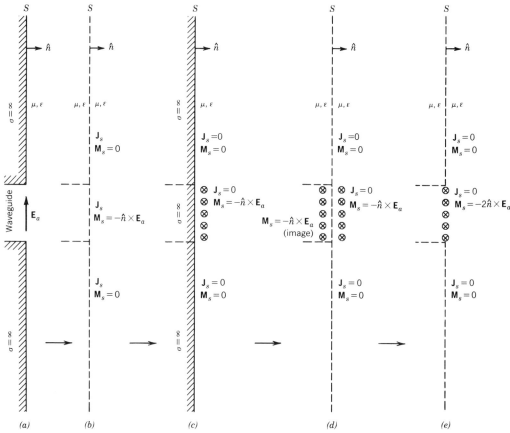

FIGURE 7-10 Equivalent models for a waveguide aperture mounted on an infinite flat electric ground plane. (*Source*: C. A. Balanis, *Antenna Theory: Analysis and Design*. Copyright © 1982, John Wiley & Sons, Inc. Reprinted with permission of John Wiley & Sons, Inc.)

has been reduced to a very simple equivalent, and (6-30) through (6-35a) can be utilized for its solution. For far-field observations, the radiation integrals of Section 6.8.2 can be used instead.

7.9 INDUCTION THEOREM (INDUCTION EQUIVALENT)

Let us now consider a theorem that is closely related to the surface equivalence theorem. It is, however, used more for scattering than for aperture radiation. Equivalent electric and magnetic current densities are introduced to replace physical obstacles. Figure 7-11a shows sources $\mathbf{J}_1$ and $\mathbf{M}_1$ in an unbounded medium with constitutive parameters μ_1 and ε_1 and radiating fields $\mathbf{E}_1$ and $\mathbf{H}_1$ everywhere, including the region V_1 enclosed by the imaginary surface S_1.

Now let us assume that the space within the imaginary surface S_1 is being replaced by another medium with constitutive parameters μ_2, ε_2, which are different from those of the medium outside S_1, as shown in Figure 7-11b. The same sources $\mathbf{J}_1$ and $\mathbf{M}_1$, embedded in the original medium (μ_1, ε_1) outside S_1, are now allowed to radiate in the presence of the obstacle that is occupying region V_1. The total field

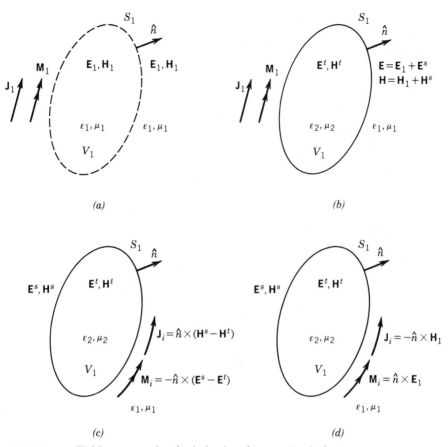

FIGURE 7-11 Field geometry for the induction theorem (equivalent).

outside region V_1, produced by the sources $\mathbf{J}_1$ and $\mathbf{M}_1$, is $\mathbf{E}$ and $\mathbf{H}$ and inside V_1 is $\mathbf{E}^t$ and $\mathbf{H}^t$.

The total field outside V_1 is equal to the original field in the absence of the obstacle ($\mathbf{E}_1$ and $\mathbf{H}_1$) plus a perturbation field ($\mathbf{E}^s, \mathbf{H}^s$), usually referred to as *scattered field*, introduced by the obstacle.

In equation form, we can write

$$\mathbf{E} = \mathbf{E}_1 + \mathbf{E}^s \tag{7-44a}$$

$$\mathbf{H} = \mathbf{H}_1 + \mathbf{H}^s \tag{7-44b}$$

where $\mathbf{E}, \mathbf{H}$ = total electric and magnetic fields in the presence of the obstacle
$\mathbf{E}_1, \mathbf{H}_1$ = total electric and magnetic fields in the absence of the obstacle
$\mathbf{E}^s, \mathbf{H}^s$ = scattered (perturbed) electric and magnetic fields owing to the obstacle

It is assumed here that the original fields $\mathbf{E}_1$ and $\mathbf{H}_1$, in the absence of the obstacle, can be found everywhere. Here we intend to compute $\mathbf{E}, \mathbf{H}$ outside V_1 and $\mathbf{E}^t, \mathbf{H}^t$ inside V_1. It should be pointed out, however, that total $\mathbf{E}, \mathbf{H}$ can be found if we can determine $\mathbf{E}^s, \mathbf{H}^s$ which when added to $\mathbf{E}_1, \mathbf{H}_1$ will give $\mathbf{E}, \mathbf{H}$ [through (7-44a) and (7-44b)].

Let us now formulate an *equivalent* problem that will allow us to determine $\mathbf{E}^s, \mathbf{H}^s$ outside V_1 and $\mathbf{E}^t, \mathbf{H}^t$ inside V_1. Figure 7-11c shows the obstacle occupying region V_1 with fields $\mathbf{E}^s, \mathbf{H}^s$ and $\mathbf{E}^t, \mathbf{H}^t$ outside and inside V_1, respectively. To support such fields and satisfy the boundary conditions of Table 1-5, we must introduce equivalent current densities $\mathbf{J}_i, \mathbf{M}_i$ *on the boundary* such that

$$\mathbf{J}_i = \hat{n} \times (\mathbf{H}^s - \mathbf{H}^t) \qquad (7\text{-}45a)$$

$$\mathbf{M}_i = -\hat{n} \times (\mathbf{E}^s - \mathbf{E}^t) \qquad (7\text{-}45b)$$

Remember that $\mathbf{E}^s, \mathbf{H}^s$ are solutions to Maxwell's equations outside V_1 and $\mathbf{E}^t, \mathbf{H}^t$ are solutions within V_1. Therefore we retain the corresponding media outside and inside V_1.

From Figure 7-11b we also know that the tangential components of $\mathbf{E}$ and $\mathbf{H}$ must be continuous across the boundary, that is,

$$\mathbf{E}_1|_{\tan} + \mathbf{E}^s|_{\tan} = \mathbf{E}^t|_{\tan} \Rightarrow \hat{n} \times (\mathbf{E}_1 + \mathbf{E}^s) = \hat{n} \times \mathbf{E}^t \qquad (7\text{-}46a)$$

$$\mathbf{H}_1|_{\tan} + \mathbf{H}^s|_{\tan} = \mathbf{H}^t|_{\tan} \Rightarrow \hat{n} \times (\mathbf{H}_1 + \mathbf{H}^s) = \hat{n} \times \mathbf{H}^t \qquad (7\text{-}46b)$$

which can also be written as

$$\mathbf{E}^s|_{\tan} - \mathbf{E}^t|_{\tan} = -\mathbf{E}_1|_{\tan} \Rightarrow \hat{n} \times (\mathbf{E}^s - \mathbf{E}^t) = -\hat{n} \times \mathbf{E}_1 \qquad (7\text{-}47a)$$

$$\mathbf{H}^s|_{\tan} - \mathbf{H}^t|_{\tan} = -\mathbf{H}_1|_{\tan} \Rightarrow \hat{n} \times (\mathbf{H}^s - \mathbf{H}^t) = -\hat{n} \times \mathbf{H}_1 \qquad (7\text{-}47b)$$

Substitution of (7-47a) into (7-45b) and (7-47b) into (7-45a), allows us to write the equivalent currents as

$$\mathbf{J}_i = -\hat{n} \times \mathbf{H}_1 \qquad (7\text{-}48a)$$

$$\mathbf{M}_i = \hat{n} \times \mathbf{E}_1 \qquad (7\text{-}48b)$$

Now it is quite clear that the equivalent sources of Figure 7-11c have been written, as shown in (7-48a) and (7-48b), in terms of the tangential components ($-\hat{n} \times \mathbf{H}_1$, $\hat{n} \times \mathbf{E}_1$) of the known fields $\mathbf{E}_1$ and $\mathbf{H}_1$ over the surface occupied by the obstacle.

The equivalent problem of Figure 7-11c is then further reduced to the equivalent problem shown in Figure 7-11d. In words, the equivalent problem of Figure 7-11d states that the scattered fields $\mathbf{E}^s, \mathbf{H}^s$ outside V_1 and the transmitted fields $\mathbf{E}^t, \mathbf{H}^t$ inside V_1 can be computed by placing, along the boundary of the obstacle, equivalent current densities given by (7-48a) and (7-48b) that *radiate in the presence of the obstacle that is occupying region V_1 and that outside V_1 have the original medium* (μ_1, ε_1). The equivalent problem of Figure 7-11d is now no simpler to solve than the original problem because we cannot use (6-30) through (6-35a) which assume that we have the same medium everywhere. However, even though the equivalent problem of Figure 7-11d is just as difficult to solve exactly as that of Figure 7-11b, it does suggest approximate solutions as will be shown later. We call the problem of Figure 7-11d an *induction equivalent* [1].

Let us now assume that the obstacle occupying region V_1 is a perfect electric conductor (PEC) with $\sigma = \infty$. Again we have the medium with parameters μ_1, ε_1 outside V_1 and the sources $\mathbf{J}_1, \mathbf{M}_1$ radiating in the presence of the conductor (obstacle) as shown in Figure 7-12a. We now need to determine the scattered fields

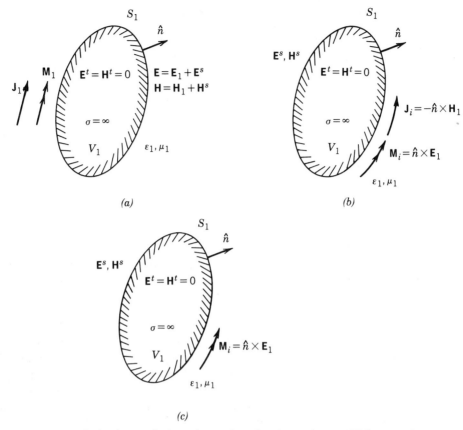

FIGURE 7-12 Induction equivalents for perfect electric conductor (PEC) scattering.

$\mathbf{E}^s$ and $\mathbf{H}^s$ which, when added to $\mathbf{E}_1$ and $\mathbf{H}_1$, will allow us to determine the total fields $\mathbf{E}$ and $\mathbf{H}$.

To compute $\mathbf{E}^s$ and $\mathbf{H}^s$ outside V_1 and $\mathbf{E}^t = \mathbf{H}^t = 0$ inside V_1, we form the equivalent problem of Figure 7-12b, analogous to that of Figure 7-11d with equivalent sources $\mathbf{J}_i = -\hat{n} \times \mathbf{H}_1$ and $\mathbf{M}_i = \hat{n} \times \mathbf{E}_1$ over the boundary. The equivalent problem of Figure 7-12b states that the perturbed fields $\mathbf{E}^s$ and $\mathbf{H}^s$ scattered by the perfect conductor of Figure 7-12a can be computed by placing equivalent current densities $\mathbf{J}_i$ and $\mathbf{M}_i$ given by

$$\mathbf{J}_i = -\hat{n} \times \mathbf{H}_1 \tag{7-49a}$$

$$\mathbf{M}_i = \hat{n} \times \mathbf{E}_1 \tag{7-49b}$$

along the boundary of the conductor and radiating in its presence. However, since an electric current element placed near a perfect electric conductor does not radiate (is shorted by the conductor), then the contributions to $\mathbf{E}^s$ and $\mathbf{H}^s$ that are due to $\mathbf{J}_i$ are zero and the equivalent problem of Figure 7-12b reduces to that of Figure 7-12c. The shorting out of $\mathbf{J}_i$ by the perfect electric conductor can be proved in general by the reciprocity theorem. The problem of Figure 7-12c is an *induction equivalent* for a perfect electric conductor scatterer.

When the surface S_1 is of complex geometry, the exact solution to the equivalent problem of Figure 7-12c is no easier to compute than the original one

338 ELECTROMAGNETIC THEOREMS AND PRINCIPLES

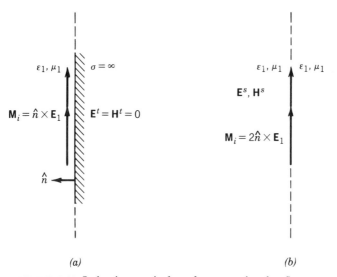

FIGURE 7-13 Induction equivalent for scattering by flat conducting surface of infinite extent.

shown in Figure 7-12a. However, if the obstacle is an infinite, flat, perfect electric conductor (infinite ground plane), then the equivalent problem for computing the scattered fields is that shown in Figure 7-13a. The exact solution to the equivalent problem of Figure 7-13a can be obtained by image theory, which allows us to reduce the equivalent problem of Figure 7-13a to that of Figure 7-13b. The equivalent problem of Figure 7-13b permits the solution for the scattered field $\mathbf{E}^s, \mathbf{H}^s$ reflected by the perfect electric conductor. The fields radiated by the equivalent source of Figure 7-13b can be obtained by using (6-30) through (6-35a) since we have one medium (μ_1, ε_1 everywhere). The fields obtained by using the equivalent problem of Figure 7-13b will give the correct answers for the scattered field only for the region to the left of the boundary S_1.

7.10 PHYSICAL EQUIVALENT AND PHYSICAL OPTICS EQUIVALENT

The problem of Figure 7-12a, scattering of **E** and **H** by a perfect electric conducting obstacle (PEC), is of much practical concern and will also be formulated by an alternate method known as *physical equivalent* [1]. The solutions of the physical equivalent will be compared with those of the induction theorem (induction equivalent) that was discussed in the previous section.

Let us again postulate the problem of Figure 7-12a. In the absence of the obstacle, the fields produced by $\mathbf{J}_1$ and $\mathbf{M}_1$ are $\mathbf{E}_1$ and $\mathbf{H}_1$, which we assume can be calculated. In the presence of the obstacle (perfect conductor in this case), the fields outside the obstacle are **E** and **H** and inside the obstacle are equal to zero. The fields **E** and **H** are related to $\mathbf{E}_1$ and $\mathbf{H}_1$ by

$$\mathbf{E} = \mathbf{E}_1 + \mathbf{E}^s \tag{7-50a}$$

$$\mathbf{H} = \mathbf{H}_1 + \mathbf{H}^s \tag{7-50b}$$

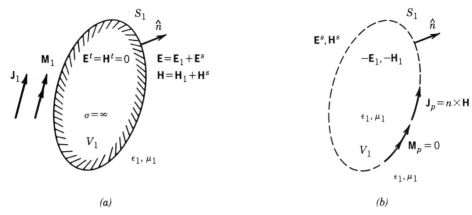

FIGURE 7-14 Physical equivalent for scattering by a perfect electric conductor (PEC). (*a*) Actual problem. (*b*) Physical equivalent.

The original problem is again shown in Figure 7-14*a*. Over the boundary S_1 of the conductor, the total tangential components of the **E** field are equal to zero and the total tangential components of the **H** field are equal to the induced current density $\mathbf{J}_p$. In equation form, we have over S_1,

$$\mathbf{M}_p = -\hat{n} \times (\mathbf{E} - \mathbf{E}^t) = -\hat{n} \times \mathbf{E} = -\hat{n} \times (\mathbf{E}_1 + \mathbf{E}^s) = 0 \quad (7\text{-}51a)$$

or

$$-\hat{n} \times \mathbf{E}_1 = \hat{n} \times \mathbf{E}^s \quad (7\text{-}51b)$$

and

$$\mathbf{J}_p = \hat{n} \times (\mathbf{H} - \mathbf{H}^t) = \hat{n} \times \mathbf{H} = \hat{n} \times (\mathbf{H}_1 + \mathbf{H}^s) \quad (7\text{-}52a)$$

or

$$\mathbf{J}_p = \hat{n} \times \mathbf{H}_1 + \hat{n} \times \mathbf{H}^s \quad (7\text{-}52b)$$

Therefore, the equivalent to the problem of Figure 7-14*a*, computation of $\mathbf{E}^s$ and $\mathbf{H}^s$ outside of S_1, is that of Figure 7-14*b*. Remember that $\mathbf{E}_1, \mathbf{H}_1$ and $\mathbf{E}^s, \mathbf{H}^s$ are *solutions to Maxwell's equations outside V_1, so in the equivalent problems we retain the same medium μ_1, ε_1 inside and outside V_1.* The equivalent of Figure 7-14*b* will give $\mathbf{E}^s, \mathbf{H}^s$ outside of S_1 and $-\mathbf{E}_1, -\mathbf{H}_1$ inside of S_1 because

$$\mathbf{J}_p = \hat{n} \times \mathbf{H} = \hat{n} \times (\mathbf{H}^s + \mathbf{H}_1) = \hat{n} \times [\mathbf{H}^s - (-\mathbf{H}_1)] \quad (7\text{-}53a)$$

$$\mathbf{M}_p = -\hat{n} \times \mathbf{E} = -\hat{n} \times (\mathbf{E}^s + \mathbf{E}_1) = -\hat{n} \times [\mathbf{E}^s - (-\mathbf{E}_1)] = 0 \quad (7\text{-}53b)$$

We call the problem of Figure 7-14*b* the *physical equivalent*. It can be solved by using (6-30) through (6-35a) since we assume that $\mathbf{J}_p$ radiates in one medium (μ_1, ε_1 everywhere). To form $\mathbf{J}_p$ on S_1 we must know the tangential components of **H** on S_1 which are unknown. So the equivalent problem of Figure 7-14*b* has not aided us in solving the problem of Figure 7-14*a*. The exact solution of the problem of Figure 7-14*b* is just as difficult as that of Figure 7-14*a*. However, as will be discussed later, the formulation of Figure 7-14b is more suggestive when it comes time to make approximations.

The physical equivalent of Figure 7-14b is used in Sections 12.3.1 and 12.3.2 to develop electric and magnetic field integral equations designated, respectively, as EFIE and MFIE. These integral equations are then solved for the unknown current density $\mathbf{J}_p$ by representing it with a series of finite terms of known functions (referred to as basis functions) but with unknown amplitude coefficients. This then allows the reduction of the integral equation to a number of algebraic equations that are usually solved by use of either matrix or iterative techniques. To date, the most popular numerical technique in applied electromagnetics for solving these integral equations is the moment method [15] which is discussed in Sections 12.2.4 through 12.2.8. In particular, in Section 12.3.1 the scattered electric field $\mathbf{E}^s$ is written in terms of $\mathbf{J}_p$. When the observations are restricted to the surface of the electric conducting target, the tangential components of $\mathbf{E}^s$ are related to the negative of the tangential components of $\mathbf{E}_1$, as represented by (7-51b). This allows the development of the electric field integral equation (EFIE) for the unknown current density $\mathbf{J}_p$ in terms of the known tangential components of the electric field $\mathbf{E}_1$, as represented by (12-54). In Section 12.3.2 the equivalent of Figure 7-14b, and in particular the relation of (7-52a) or (7-53a), is used to write an expression for the scattered magnetic field $\mathbf{H}^s$ in terms of the tangential components of the magnetic field $\mathbf{H}_1$. This allows the development of the magnetic field integral equation (MFIE) for the unknown current density $\mathbf{J}_p$, as represented by (12-59a).

If the conducting obstacle of Figure 7-14a is an infinite, flat, perfect electric conductor (infinite ground plane), then the physical equivalent problem of Figure 7-14b is that of Figure 7-15 where the electric current $\mathbf{J}_p$ is equal to

$$\mathbf{J}_p = \hat{n} \times \mathbf{H} = \hat{n} \times (\mathbf{H}_1 + \mathbf{H}^s) = 2\hat{n} \times \mathbf{H}_1 \qquad (7\text{-}54)$$

since the tangential components of the scattered $\mathbf{H}^s$ field ($\mathbf{H}^s|_{\tan}$) are in phase and equal in amplitude to the tangential components of the $\mathbf{H}_1$ field ($\mathbf{H}_1|_{\tan}$). The equivalent of Figure 7-15 is also referred to as the *physical optics* [16].

We have until now discussed two different methods, *induction equivalent* and *physical equivalent*, for the solution of the same problem, that is, the determination

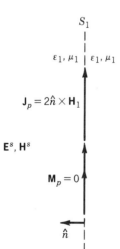

FIGURE 7-15 Physical equivalent of a flat conducting surface of infinite extent.

of the field scattered by a perfect electric conductor. The induction equivalent is shown in Figure 7-12c and the physical equivalent in Figure 7-14b. The question now is whether both give the same result. The answer to this is yes. However, it must be pointed out that when the geometry of the obstacle is complex, neither of the equivalents is easy to use to obtain convenient results. The next question may then be: Why bother introducing the equivalents if they are not easy to apply? There is a two-part answer to this. The first part of the answer is that when the obstacle is an infinite, flat, perfect conductor, the solution to each equivalent is easy to formulate by using "image theory," shown in Figure 7-13a and b for the induction equivalent and in Figure 7-15 for the physical equivalent. The second part of the answer is that the induction and physical equivalent modelings suggest more appropriate approximations or simplifications that can be made when we attempt to solve a problem whose exact solution is not easily obtainable.

The last equation then may be stated as follows: "When making approximations or simplifications to solve an otherwise intractable problem, do both of the equivalents lead to identical approximate results or is one superior to the other? The answer is that the induction equivalent and the physical equivalent do *not*, in general, lead to identical results when simplifications or approximations are made to a given problem. For some special approximations, to be discussed later, they give identical results only when the source and the observer are at the same location (backscattering). However, for any general approximation, *they do not yield identical results even for backscattering. One should then use the method that results in the best approximation for the allowable degree of complexity.*

In an attempt to make use of the equivalents of Figure 7-12c and 7-14b to solve a scattering problem, difficulties are encountered. Here we will summarize these difficulties, and in the next section we will discuss appropriate simplifications that allow us to obtain approximate solutions. The induction equivalent of Figure 7-12c is represented by a *known current* ($\mathbf{M}_i = \hat{n} \times \mathbf{E}_1$) that is placed on the surface of the obstacle and that radiates in its presence. Because the medium within and outside the obstacle is not the same, we *cannot* use (6-30) through (6-35a) to solve for the scattered fields. We must solve a new boundary-value problem, which may be as difficult as the original problem, even though we know the currents on the surface of the obstacle. In other words, we must derive new formulas that will allow us to compute the scattered fields. The physical equivalent of Figure 7-14b is represented by an *unknown current density* ($\mathbf{J}_p = \hat{n} \times \mathbf{H}$) that is placed on the imaginary surface S_1 which represents the geometry of the obstacle. In this case, however, we *can* use (6-30) through (6-35a) to solve for the scattered fields because we have the same medium within and outside S_1. The difficulty here is that we *do not* know the current density on the surface of the obstacle, which in most cases is just as difficult to find as the solution of the original problem, because it requires knowledge of the total $\mathbf{H}$ field which is the answer to the original problem.

7.11 INDUCTION AND PHYSICAL EQUIVALENT APPROXIMATIONS

Let us now concentrate on suggesting appropriate simplifications to be made in the induction and physical equivalent formulations so that we obtain approximate solutions when exact solutions are not feasible. In many cases the approximate solutions will lead to results that are well within measuring accuracies of laboratory experiments.

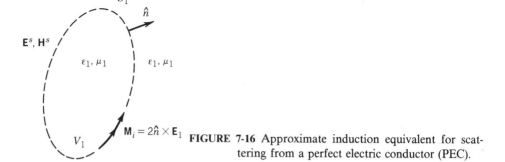

FIGURE 7-16 Approximate induction equivalent for scattering from a perfect electric conductor (PEC).

In the induction equivalent form the difficulty in obtaining a solution arises from the lack of equations that can be used with the known current density. The crudest approximation to the problem is the assumption that the obstacle is large electrically and so we can use image theory to solve the problem. This assumes that locally on the surface of the obstacle each point and its immediate neighbors form a flat surface. The best results with this simplification will be for scatterers whose electrical dimensions are large in comparison to the wavelength. Thus the induction equivalent of Figure 7-12c can be approximated by that in Figure 7-16. Now (6-30) through (6-35a) can be used to compute the scattered fields because we have the same medium inside and outside S_1.

In many cases, even this approximation may not be amenable to a closed form solution because of the inability to integrate over the entire closed surface. To simplify this even further, we may restrict our integration over only part of the surface where the current density is more intense and will provide the major contributions to the scattered field. This surface is usually the part that is "visible" by the transmitter (sources $\mathbf{J}_1$ and $\mathbf{M}_1$).

In the physical equivalent, the difficulty in solving the problem arises because we do not know the current density $\mathbf{J}_p$ ($\mathbf{J}_p = \hat{n} \times \mathbf{H}$) that must be placed along the surface S_1 (see Figure 7-14b). Once we decide on an approximation for the current density, the solution can be carried out because we can use (6-30) through (6-35a). The crudest approximation for this problem is the assumption that the total tangential $\mathbf{H}$ field on the surface of the conductor of Figure 7-14a is equal to twice that of the tangential $\mathbf{H}_1$. Thus the current density to be placed on the surface of the

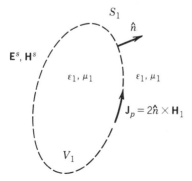

FIGURE 7-17 Approximate physical equivalent for scattering from a perfect electric conductor (PEC).

physical equivalent of Figure 7-14b is

$$\mathbf{J}_p \simeq 2\hat{n} \times \mathbf{H}_1 \tag{7-55}$$

which is a good approximation provided that the scatterer is large electrically (in the limit infinite, flat, perfect conductor). We can then approximate the physical equivalent of Figure 7-14b by that of Figure 7-17. This is usually referred to as the *physical optics approximation* [16], because it is similar to the formulation of the infinite, flat, ground plane. Thus *physical optics approximate only the boundary conditions that concern only the fields on the closed surface* S_1. If a closed form solution still cannot be obtained because of the inability to integrate over the entire surface, then integration over a part of the scattering surface may be sufficient, as was discussed for the induction equivalent.

It should be pointed out that making the aforementioned crude approximations (*image theory for the induction equivalent and physical optics for the physical equivalent*), the two methods lead to identical results only for backscattering. Any further simplifications may lead to solutions that may not be identical even for backscattering. This is discussed in more detail in [17]. The theory can be extended to include imperfect conductors and dielectrics but the formulations become quite complex even when approximations are made.

The best way to illustrate the two different methods, when approximations are made, is to solve the same problem using both methods and compare the results.

Example 7-4. A parallel polarized uniform plane wave on the xy plane, in a free-space medium, is obliquely incident upon a rectangular, flat, perfectly conducting ($\sigma = \infty$) plate, as shown in Figure 7-18a. The dimensions of the plate and a in the y direction and b in the z direction.

Find the electric and magnetic fields scattered by the flat plate, assuming that observations are made in the far zone. Solve the problem by using the *induction equivalent* and *physical equivalent*. Make appropriate simplifications, and compare the results.

Solution.

Induction Equivalent: The simplification to be made in the use of induction equivalent modeling is to assume that the dominant part of the magnetic current $\mathbf{M}_i$ resides only in the front face of the plate and that image theory holds for a finite plate. With these approximations, we reduce the equivalent to that of Figure 7-18b where the magnetic current exists only over the area occupied by the plate. Thus we can write the $\mathbf{E}$ and $\mathbf{H}$ fields as

$$\mathbf{H}^i = \hat{a}_z H_0 e^{+j\beta(x\cos\phi_i + y\sin\phi_i)}$$

$$\mathbf{E}^i = \eta_0 H_0 \left[\hat{a}_x \sin\phi_i - \hat{a}_y \cos\phi_i\right] e^{+j\beta(x\cos\phi_i + y\sin\phi_i)}$$

and the magnetic current density as

$$\mathbf{M}_i = 2\hat{n} \times \mathbf{E}^i\big|_{x=0} = -\hat{a}_z 2\eta_0 H_0 \cos\phi_i e^{+j\beta y \sin\phi_i}$$

or

$$M_x = M_y = 0 \qquad M_z = -2\eta_0 \cos\phi e^{+j\beta y \sin\phi_i}$$

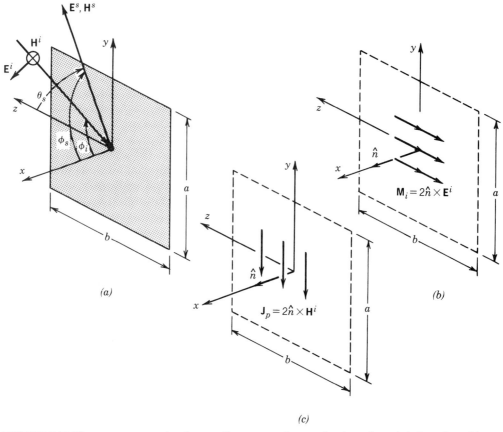

FIGURE 7-18 Plane wave scattering from a flat rectangular conducting plate. (*a*) Actual problem. (*b*) Induction equivalent. (*c*) Physical equivalent.

The scattered electric and magnetic fields, in the far zone, can be written according to (6-122a) through (6-122f), (6-125a) through (6-125d), (6-127a), and (6-128a) as

$$E_\theta^s = 0$$

$$E_\phi^s = +j\frac{\beta e^{-j\beta r}}{4\pi r} L_\theta$$

where

$$L_\theta = \int_{-a/2}^{+a/2}\int_{-b/2}^{+b/2} -M_z \sin\theta_s e^{+j\beta(y'\sin\theta_s \sin\phi_s + z'\cos\theta_s)}\, dz'\, dy'$$

$$L_\theta = 2ab\eta_0 H_0 \cos\phi_i \sin\theta_s \left(\frac{\sin Y}{Y}\right)\left(\frac{\sin Z}{Z}\right)$$

$$Y = \frac{\beta a}{2}(\sin\theta_s \sin\phi_s + \sin\phi_i)$$

$$Z = \frac{\beta b}{2}\cos\theta_s$$

In summary then

$$E_\theta^s = 0$$

$$E_\phi^s = j\frac{ab\beta\eta_0 H_0 e^{-j\beta r}}{2\pi r}\left[\cos\phi_i \sin\theta_s \left(\frac{\sin Y}{Y}\right)\left(\frac{\sin Z}{Z}\right)\right]$$

$$H_\theta^s = -\frac{E_\phi^s}{\eta_0}$$

$$H_\phi^s = \frac{E_\theta^s}{\eta_0} = 0$$

For backscattering observations ($\theta_s = \pi/2$, $\phi_s = \phi_i$), the fields reduce to

$$E_\theta^s = 0$$

$$E_\phi^s = j\frac{ab\beta\eta_0 H_0 e^{-j\beta r}}{2\pi r}\left\{\cos\phi_i\left[\frac{\sin(\beta a \sin\phi_i)}{\beta a \sin\phi_i}\right]\right\}$$

$$H_\theta^s = -\frac{E_\phi^s}{\eta_0}$$

$$H_\phi^s = 0$$

Physical Equivalent: The simplifications for the physical equivalent will be similar to those for the induction equivalent. That is, we will assume that the major contributing current density $\mathbf{J}_p$ resides in the front face of the plate for which the physical equivalent reduces to that of Figure 7-18c. Thus we can write the current density as

$$\mathbf{J}_p = 2\hat{n} \times \mathbf{H}^i|_{x=0} = -\hat{a}_y 2H_0 e^{+j\beta y \sin\phi_i}$$

and the scattered **E** and **H** fields, according to (6-122a) through (6-122f), (6-125a) through (6-125d), (6-127a), and (6-128a), as

$$E_\theta^s = -j\frac{\beta\eta_0 e^{-j\beta r}}{4\pi r}N_\theta$$

$$E_\phi^s = -j\frac{\beta\eta_0 e^{-j\beta r}}{4\pi r}N_\phi$$

where

$$N_\theta = \int_{-a/2}^{a/2}\int_{-b/2}^{b/2} J_y \cos\theta_s \sin\phi_s e^{+j\beta(y'\sin\theta_s \sin\phi_s + z'\cos\theta_s)}\, dz'\, dy'$$

$$N_\theta = -2abH_0\left\{\cos\theta_s \sin\phi_s\left[\frac{\sin(Y)}{Y}\right]\left[\frac{\sin(Z)}{Z}\right]\right\}$$

$$N_\phi = \int_{-a/2}^{a/2}\int_{-b/2}^{b/2} J_y \cos\phi_s e^{+j\beta(y'\sin\theta_s \sin\phi_s + z'\cos\theta_s)}\, dz'\, dy'$$

$$N_\phi = -2abH_0\left\{\cos\phi_s\left[\frac{\sin(Y)}{Y}\right]\left[\frac{\sin(Z)}{Z}\right]\right\}$$

$$Y = \frac{\beta a}{2}(\sin\theta_s \sin\phi_s + \sin\phi_i)$$

$$Z = \frac{\beta b}{2}\cos\theta_s$$

In summary then

$$E_\theta^s = j\frac{ab\beta\eta_0 H_0 e^{-j\beta r}}{2\pi r}\left\{\cos\theta_s \sin\phi_s \left[\frac{\sin(Y)}{Y}\right]\left[\frac{\sin(Z)}{Z}\right]\right\}$$

$$E_\phi^s = j\frac{ab\beta\eta_0 H_0 e^{-j\beta r}}{2\pi r}\left\{\cos\phi_s \left[\frac{\sin(Y)}{Y}\right]\left[\frac{\sin(Z)}{Z}\right]\right\}$$

$$H_\theta^s = -\frac{E_\phi^s}{\eta_0}$$

$$H_\phi^s = \frac{E_\theta^s}{\eta_0}$$

For backscattering observations ($\theta_s = \pi/2$, $\phi_s = \phi_i$), the fields reduce to

$$E_\theta^s = 0$$

$$E_\phi^s = j\frac{ab\beta\eta_0 H_0 e^{-j\beta r}}{2\pi r}\left\{\cos\phi_i \left[\frac{\sin(\beta a \sin\phi_i)}{\beta a \sin\phi_i}\right]\right\}$$

$$H_\theta^s = -\frac{E_\phi^s}{\eta_0}$$

$$H_\phi^s = 0$$

It is quite clear that the solutions of the two different methods do not lead to identical results except for backscatter observations. It seems that the physical equivalent solution gives the best results for general observations because it requires the least simplification in the formulation.

REFERENCES

1. R. F. Harrington, *Time-Harmonic Electromagnetic Fields*, McGraw-Hill, New York, 1961.
2. C. A. Balanis, *Antenna Theory: Analysis and Design*, Wiley, New York, 1982.
3. J. R. Wait, "Characteristics of antennas over lossy earth," in *Antenna Theory Part 2*, Chapter 23, R. E. Collin and F. J. Zucker, eds., McGraw-Hill, New York, 1969.
4. L. E. Vogler and J. L. Noble, "Curves of input impedance change due to ground for dipole antennas," U.S. National Bureau of Standards, Monograph 72, January 31, 1964.
5. W. H. Hayt, Jr., and J. E. Kimmerly, *Engineering Circuit Analysis*, Third Edition, McGraw-Hill, New York, 1978.
6. J. R. Carson, "Reciprocal theorems in radio communication," *Proc. IRE*, vol. 17, pp. 952–956, June 1929.
7. J. H. Richmond, *The Basic Theory of Harmonic Fields, Antennas and Scattering*, Unpublished Notes.
8. V. H. Rumsey, "Reaction concept in electromagnetic theory," *Physical Review*, vol. 94, no. 6, pp. 1483–1491, June 15, 1954.
9. J. H. Richmond, "A reaction theorem and its application to antenna impedance calculations," *IRE Trans. Antennas Propagat.*, vol. AP-9, no. 6, pp. 515–520, November 1961.
10. R. Mittra (ed.), *Computer Techniques for Electromagnetics*, Chapter 2, Pergamon, Elmsford, NY, 1973.
11. S. A. Schelkunoff, "Some equivalence theorems of electromagnetics and their application to radiation problems," *Bell System Tech. J.*, vol. 15, pp. 92–112, 1936.

12. C. Huygens, *Traite de la Lumiere*, Leyeden, 1690. Translated into English by S. P. Thompson, London, 1912 and reprinted by the University of Chicago Press.
13. J. D. Kraus and K. R. Carver, *Electromagnetics*, Second Edition, McGraw-Hill, New York, 1973, pp. 464–467.
14. A. E. H. Love, "The integration of the equations of propagation of electric waves," *Phil. Trans. Roy. Soc. London, Ser. A*, vol. 197, pp. 1–45, 1901.
15. R. F. Harrington, *Field Computation by Moment Methods*, Macmillan, New York, 1968.
16. P. Beckmann, *The Depolarization of Electromagnetic Waves*, The Golem Press, Boulder, CO, 1968, pp. 76–92.
17. R. F. Harrington, "On scattering by large conducting bodies," *IRE Trans. Antennas Propagat.*, vol. AP-7, no. 2, pp. 150–153, April 1959.

PROBLEMS

7.1. For the infinitesimal vertical electric dipole whose far-zone electric field is given by (7-13a):
   (a) Find the corresponding magnetic field.
   (b) Determine the corresponding time-average power density.
   (c) Show that the radiated power, obtained by integrating the power density of part b over a sphere of radius r, can be written as

$$P_{\text{rad}} = \pi \eta \left| \frac{I_0 \ell}{\lambda} \right|^2 \left[\frac{1}{3} - \frac{\cos(2\beta h)}{(2\beta h)^2} + \frac{\sin(2\beta h)}{(2\beta h)^3} \right]$$

7.2. For Problem 7.1 show the following.
   (a) The radiation intensity U defined as $U = r^2 S_{\text{av}}$, where S_{av} is the time-average power density, can be written as

$$U = \frac{\eta}{2} \left| \frac{I_0 \ell}{\lambda} \right|^2 \sin^2 \theta \cos^2(\beta h \cos \theta)$$

   (b) The directivity D_0 of the element, defined as

$$D_0 = \frac{4\pi U_{\text{max}}}{P_{\text{rad}}}$$

   where U_{max} is the maximum radiation intensity, can be written as

$$D_0 = \frac{2}{F(\beta h)}$$

$$F(\beta h) = \left[\frac{1}{3} - \frac{\cos(2\beta h)}{(2\beta h)^2} + \frac{\sin(2\beta h)}{(2\beta h)^3} \right]$$

   (c) The radiation resistance, defined as

$$R_r = \frac{2 P_{\text{rad}}}{|I_0|^2}$$

   can be expressed as

$$R_r = 2\pi \eta \left(\frac{\ell}{\lambda} \right)^2 F(\beta h)$$

   where $F(\beta h)$ is that given in part b.

7.3. Using the electric field of (7-13a), for a fixed r plot the normalized radiation pattern (in dB) versus the angle θ when the height h of the element above the ground is $h = 0, \lambda/8, \lambda/4, 3\lambda/8, \lambda/2,$ and λ.

7.4. A quarter-wavelength ($\ell = \lambda/4$) wire radiator is placed vertically above an infinite electric ground plane and it is fed at its base, as shown in Figure P7-4. This is usually referred to as $\lambda/4$ monopole. Assume that the current on the wire is represented by

$$\mathbf{I} = \hat{a}_z I_0 \sin\left[\beta\left(\frac{\ell}{2} - z'\right)\right] \qquad 0 \le z' \le \ell/2$$

where z' is any point on the monopole and show, using image theory, (6-97a), (6-112a) and (6-112b), and the formulations of Section 6.7, that the far-zone electric and magnetic fields radiated by the element *above the ground plane* are given by

$$E_r \simeq E_\phi \simeq H_r \simeq H_\theta \simeq 0$$

$$E_\theta \simeq j\eta \frac{I_0 e^{-j\beta r}}{2\pi r}\left[\frac{\cos\left(\frac{\pi}{2}\cos\theta\right)}{\sin\theta}\right]$$

$$H_\phi \simeq \frac{E_\phi}{\eta}$$

These expressions are identical to those of Problem 6.21.

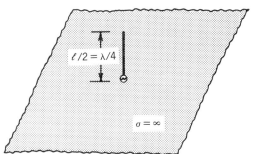

FIGURE P7-4

7.5. A very small ($\ell \ll \lambda$) linear radiating current element is placed between two infinite plates forming a 90° corner reflector. Assume that the length of the element is placed parallel to the plates of the corner reflector.

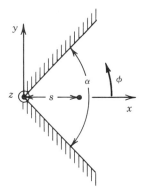

FIGURE P7-5

(a) Determine the number of images, their polarizations, and their positions that are necessary to account for all the reflections from the plates of the reflector and to find the radiated fields within the internal space of the reflector.

(b) Show that the total far-zone radiated fields within the internal region of the reflector can be written as

$$E_\theta^t = E_\theta^0 F(\beta s)$$

$$E_\theta^0 = j\eta \frac{\beta I_0 \ell e^{-j\beta r}}{4\pi r} \sin\theta \quad \begin{cases} 0° \leq \theta \leq 180° \\ 315° \leq \phi \leq 45° \end{cases}$$

$$F(\beta s) = 2[\cos(\beta s \sin\theta \cos\phi) - \cos(\beta s \sin\theta \sin\phi)]$$

where θ is measured from the z axis toward the observation point. E_θ^0 is the far-zone field radiated by a very small ($\ell \ll \lambda$) linear element radiating in an unbounded medium (see Example 6-3) and $F(\beta s)$ is referred to as the array factor representing the array of elements that includes the actual radiating element and its associated images.

7.6. For Problem 7.5 plot the magnitude of $F(\beta s)$ as a function of s ($0 \leq s \leq 10\lambda$) when $\theta = 90°$ and $\phi = 0°$. What is the maximum value of $|F(\beta s)|$? Is the function periodic? If so, what is the period?

7.7. For Problem 7.5 plot the normalized value of the magnitude of $F(\beta s)$ (in dB) as a function of ϕ ($315° \leq \phi \leq 45°$) when $\theta = 90°$. Do this when $s = 0.1\lambda$, $s = 0.7\lambda$, $s = 0.8\lambda$, $s = 0.9\lambda$, and $s = 1.0\lambda$.

7.8. Repeat Problem 7.5 when the included angle α of the corner reflector is $60°$, $45°$, and $30°$, and show that $F(\beta s)$ takes the following forms.
(a) $\alpha = 60°$, $F(\beta s) = 4\sin(X/2)\{\cos(X/2) - \cos[\sqrt{3}(Y/2)]\}$.
(b) $\alpha = 45°$, $F(\beta s) = 2[\cos(X) + \cos(Y) - 2\cos(X/\sqrt{2})\cos(Y/\sqrt{2})]$.
(c) $\alpha = 30°$, $F(\beta s) = 2\{\cos(X) - 2\cos[(\sqrt{3}/2)X]\cos(Y/2) - \cos(Y) + 2\cos(X/2)\cos[(\sqrt{3}/2)Y]\}$.
where $X = \beta s \sin\theta \cos\phi$, $Y = \beta s \sin\theta \sin\phi$

7.9. For Problem 7.8 ($\alpha = 60°$, $45°$, and $30°$) plot the magnitude of $F(\beta s)$ as a function of s ($0 \leq s \leq 10\lambda$) when $\theta = 90°$ and $\phi = 0°$. What is the maximum value $|F(\beta s)|$ will ever achieve if plotted as a function of s? Is the function periodic? If so, what is the period?

7.10. An infinitesimal electric dipole is placed at an angle of $30°$ at a height h above a perfectly conducting electric ground plane. Determine the location and orientation of its image. Do this by sketching the image.

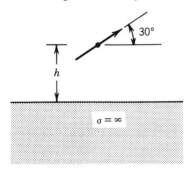

FIGURE P7-10

7.11. A small circular loop of radius a is placed vertically at a height h above a perfectly conducting electric ground plane. Determine the location and direction of current flow of its image. Do this by sketching the image.

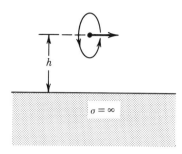

FIGURE P7-11

7.12. A linearly polarized uniform plane wave traveling in free space is incident normally upon a flat dielectric surface. Assume that the incident electric field is given by

$$\mathbf{E} = \hat{a}_x E_0 e^{-j\beta_0 z}$$

Then derive expressions for the equivalent volume electric and magnetic current densities, and the regions over which they exist. These current densities can then be used, in principle, to find the fields scattered by the dielectric surface.

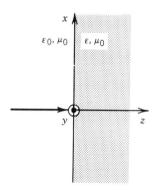

FIGURE P7-12

7.13. The electric and magnetic fields at the aperture of a circular waveguide aperture are given by

$$\left. \begin{array}{l} \mathbf{E}_a = \hat{a}_\rho E_\rho + \hat{a}_\phi E_\phi \\ E_\rho = E_0 J_1\left(\dfrac{\chi'_{11}}{a}\rho'\right)\dfrac{\sin\phi'}{\rho'} \\ E_\phi = E_0 J_1'\left(\dfrac{\chi'_{11}}{a}\rho'\right)\cos\phi' \end{array} \right\} \quad \begin{array}{l} \rho' \le a \\[4pt] \chi'_{11} = 1.841 \\[4pt] ' = \dfrac{\partial}{\partial\rho'} \end{array}$$

$$\mathbf{E}_a = 0 \quad \text{elsewhere}$$

Develop the surface equivalent that can be used to find the fields radiated by the aperture. State the equivalent by giving expressions for the electric $\mathbf{J}_s$ and magnetic $\mathbf{M}_s$ surface current densities and the regions over which they exist.

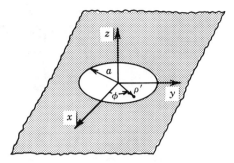

FIGURE P7-13

7.14. A uniform plane wave on the yz plane is incident upon a flat circular conducting plate of radius a. Assume that the incident electric field is given by
$$\mathbf{E}^i = \hat{a}_x E_0 e^{-j\beta_0(y\sin\theta_i - z\cos\theta_i)}$$
Determine the scattered field using (a) the induction equivalent and (b) the physical equivalent. Reduce and compare the expressions for backscatter observations.

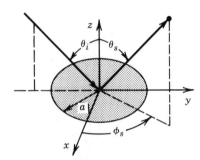

FIGURE P7-14

7.15. Repeat Problem 7.14 when the incident magnetic field is given by
$$\mathbf{H}^i = \hat{a}_x H_0 e^{-j\beta_0(y\sin\theta_i - z\cos\theta_i)}$$

CHAPTER 8

RECTANGULAR CROSS-SECTION WAVEGUIDES AND CAVITIES

8.1 INTRODUCTION

Rectangular transmission lines (such as rectangular waveguides, dielectric slab lines, striplines, and microstrips) and their corresponding cavities represent a significant section of lines used in many practical radio-frequency systems. The objective in this chapter is to introduce and analyze some of them, and to present some data on their propagation characteristics. The parameters of interest include field configurations (modes) that can be supported by such structures and their corresponding cutoff frequencies, guide wavelengths, wave impedances, phase and attenuation constants, and quality factors Q. Because of their general rectilinear geometrical shapes, it is most convenient to use the rectangular coordinate system for the analyses. The field configurations that can be supported by these structures must satisfy Maxwell's equations or the wave equation, and the corresponding boundary conditions.

8.2 RECTANGULAR WAVEGUIDE

Let us consider a rectangular waveguide of lateral dimensions a and b, as shown in Figure 8-1. Initially assume that the waveguide is of infinite length and is empty. It is our purpose to determine the various field configurations (modes) that can exist inside the guide. Although a TEM^z field configuration is of the simplest structure, it cannot satisfy the boundary conditions on the waveguide walls. Therefore, it is not a valid solution. It can be shown that modes TE^x, TM^x, TE^y, TM^y, TE^z, and TM^z satisfy the boundary conditions and are therefore appropriate modes (field configurations) for the rectangular waveguide. We will initially consider TE^z and TM^z; others will follow.

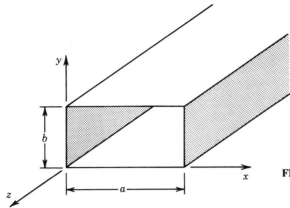

FIGURE 8-1 Rectangular waveguide with its appropriate dimensions.

8.2.1 Transverse Electric (TE^z)

According to (6-71a) through (6-72), TE^z electric and magnetic fields satisfy the following set of equations:

$$E_x = -\frac{1}{\varepsilon}\frac{\partial F_z}{\partial y} \qquad H_x = -j\frac{1}{\omega\mu\varepsilon}\frac{\partial^2 F_z}{\partial x\,\partial z}$$

$$E_y = \frac{1}{\varepsilon}\frac{\partial F_z}{\partial x} \qquad H_y = -j\frac{1}{\omega\mu\varepsilon}\frac{\partial^2 F_z}{\partial y\,\partial z}$$

$$E_z = 0 \qquad H_z = -j\frac{1}{\omega\mu\varepsilon}\left(\frac{\partial^2}{\partial z^2} + \beta^2\right)F_z \qquad (8\text{-}1)$$

where $F_z(x, y, z)$ is a scalar potential function, and it represents the z component of the vector potential function **F**. The potential **F**, and in turn F_z, must satisfy (6-73) or

$$\nabla^2 F_z(x, y, z) + \beta^2 F_z(x, y, z) = 0 \qquad (8\text{-}2)$$

which can be reduced to

$$\frac{\partial^2 F_z}{\partial x^2} + \frac{\partial^2 F_z}{\partial y^2} + \frac{\partial^2 F_z}{\partial z^2} + \beta^2 F_z = 0 \qquad (8\text{-}2a)$$

The solution to (8-2) or (8-2a) is obtained by using the separation of variables method outlined in Section 3.4.1. In general, the solution to $F_z(x, y, z)$ can be written initially as

$$F_z(x, y, z) = f(x)g(y)h(z) \qquad (8\text{-}3)$$

The objective here is to choose judiciously the most appropriate forms for $f(x)$, $g(y)$, and $h(z)$ from (3-28a) through (3-30b).

Since the waveguide is bounded in the x and y directions, the forms of $f(x)$ and $g(y)$ must be chosen to represent standing waves. The most appropriate forms are those of (3-28b) and (3-29b). Thus

$$f(x) = f_2(x) = C_1 \cos(\beta_x x) + D_1 \sin(\beta_x x) \quad (8\text{-}4a)$$

$$g(y) = g_2(y) = C_2 \cos(\beta_y y) + D_2 \sin(\beta_y y) \quad (8\text{-}4b)$$

Because the waveguide is infinite in length, the variations of the fields in the z direction must represent traveling waves as given by (3-30a). Thus

$$h(z) = h_1(z) = A_3 e^{-j\beta_z z} + B_3 e^{+j\beta_z z} \quad (8\text{-}5)$$

Substituting (8-4a) through (8-5) into (8-3), we can write that

$$F_z(x, y, z) = \left[C_1 \cos(\beta_x x) + D_1 \sin(\beta_x x)\right]\left[C_2 \cos(\beta_y y) + D_2 \sin(\beta_y y)\right]$$
$$\times \left[A_3 e^{-j\beta_z z} + B_3 e^{+j\beta_z z}\right] \quad (8\text{-}6)$$

The first exponential in (8-6) represents waves traveling in the $+z$ direction (assuming an $e^{j\omega t}$ time variation) and the second term designates waves traveling in the $-z$ direction. To simplify the notation, assume that the source is located such that the waves are traveling only in the $+z$ direction. Then the second term is not present, or $B_3 = 0$. If the waves are traveling in the $-z$ direction, then the second exponential in (8-6) is appropriate and $A_3 = 0$. If the waves are traveling in both directions, superposition can be used to sum the field expressions for the $+z$ and $-z$ traveling waves.

For $+z$ traveling waves, F_z of (8-6) reduces with $B_3 = 0$ to

$$F_z^+(x, y, z) = \left[C_1 \cos(\beta_x x) + D_1 \sin(\beta_x x)\right]$$
$$\times \left[C_2 \cos(\beta_y y) + D_2 \sin(\beta_y y)\right] A_3 e^{-j\beta_z z} \quad (8\text{-}7)$$

where, according to (3-27)

$$\beta_x^2 + \beta_y^2 + \beta_z^2 = \beta^2 = \omega^2 \mu \varepsilon \quad (8\text{-}7a)$$

C_1, D_1, C_2, D_2, A_3, β_x, β_y, and β_z are constants that will be evaluated by substituting (8-7) into (8-1) and applying the appropriate boundary conditions on the walls of the waveguide.

For the waveguide structure of Figure 8-1, the necessary and sufficient boundary conditions are those that require the tangential components of the electric field to vanish on the walls of the waveguide. Thus, in general, on the bottom and top walls

$$E_x(0 \le x \le a, y = 0, z) = E_x(0 \le x \le a, y = b, z) = 0 \quad (8\text{-}8a)$$

$$E_z(0 \le x \le a, y = 0, z) = E_z(0 \le x \le a, y = b, z) = 0 \quad (8\text{-}8b)$$

and on the left and right walls

$$E_y(x = 0, 0 \le y \le b, z) = E_y(x = a, 0 \le y \le b, z) = 0 \quad (8\text{-}8c)$$

$$E_z(x = 0, 0 \le y \le b, z) = E_z(x = a, 0 \le y \le b, z) = 0 \quad (8\text{-}8d)$$

For the TE^z modes $E_z = 0$, and the boundary conditions of (8-8b) and (8-8d) are automatically satisfied. However, in general, the boundary conditions of (8-8b) and

(8-8d) are not independent, but they represent the same conditions as given, respectively, by (8-8a) and (8-8c). Therefore, the necessary and sufficient independent boundary conditions, in general, will be to enforce either (8-8a) or (8-8b) and either (8-8c) or (8-8d).

Substituting (8-7) into (8-1), we can write the x component of the electric field as

$$E_x^+(x, y, z) = -A_3 \frac{\beta_y}{\varepsilon}[C_1 \cos(\beta_x x) + D_1 \sin(\beta_x x)]$$
$$\times [-C_2 \sin(\beta_y y) + D_2 \cos(\beta_y y)] e^{-j\beta_z z} \qquad (8\text{-}9)$$

Enforcing on (8-9) the boundary condition of (8-8a) on the bottom wall, we have that

$$E_x^+(0 \le x \le a, y = 0, z) = -A_3 \frac{\beta_y}{\varepsilon}[C_1 \cos(\beta_x x) + D_1 \sin(\beta_x x)]$$
$$\times [-C_2(0) + D_2(1)] e^{-j\beta_z z} = 0 \qquad (8\text{-}10)$$

The only way for (8-10) to be satisfied and not lead to a trivial solution will be for $D_2 = 0$. Thus

$$D_2 = 0 \qquad (8\text{-}10\text{a})$$

Now by enforcing on (8-9) the boundary condition of (8-8a) on the top wall, and using (8-10a), we can write that

$$E_x^+(0 \le x \le a, y = b, z) = -A_3 \frac{\beta_y}{\varepsilon}[C_1 \cos(\beta_x x) + D_1 \sin(\beta_x x)]$$
$$\times [-C_2 \sin(\beta_y b)] e^{-j\beta_z z} = 0 \qquad (8\text{-}11)$$

For nontrivial solutions, (8-11) can only be satisfied provided that

$$\sin(\beta_y b) = 0 \qquad (8\text{-}12)$$

which leads to

$$\beta_y b = \sin^{-1}(0) = n\pi \qquad n = 0, 1, 2, \ldots \qquad (8\text{-}12\text{a})$$

or

$$\beta_y = \frac{n\pi}{b} \qquad n = 0, 1, 2, \ldots \qquad (8\text{-}12\text{b})$$

Equation 8-12 is usually referred to as the *eigenfunction* and (8-12b) as the *eigenvalue*.

In a similar manner, we can enforce the boundary conditions on the left and right walls as given by (8-8c). By doing this, it can be shown that

$$D_1 = 0 \qquad (8\text{-}13)$$

and

$$\beta_x = \frac{m\pi}{a} \qquad m = 0, 1, 2, \ldots \qquad (8\text{-}13\text{a})$$

Use of (8-10a), (8-12b), (8-13), and (8-13a) reduces (8-7) to

$$F_z^+(x, y, z) = C_1 C_2 A_3 \cos(\beta_x x) \cos(\beta_y y) e^{-j\beta_z z} \quad (8\text{-}14)$$

or, by combining $C_1 C_2 A_3 = A_{mn}$, to

$$F_z^+(x, y, z) = A_{mn} \cos(\beta_x x) \cos(\beta_y y) e^{-j\beta_z z} \quad (8\text{-}14a)$$

with

$$\left. \begin{array}{l} \beta_x = \dfrac{m\pi}{a} = \dfrac{2\pi}{\lambda_x} \Rightarrow \lambda_x = \dfrac{2a}{m} \quad m = 0, 1, 2, \ldots \\[2mm] \beta_y = \dfrac{n\pi}{b} = \dfrac{2\pi}{\lambda_y} \Rightarrow \lambda_y = \dfrac{2b}{n} \quad n = 0, 1, 2, \ldots \end{array} \right\} \quad m = n \neq 0 \quad (8\text{-}14b)$$

In (8-14b) the $m = n = 0$ is excluded because for that combination, F_z of (8-14a) is a constant and all the components of **E** and **H** as given by (8-1) vanish; thus a trivial solution. Since individually C_1, C_2, and A_3 are constants, their product A_{mn} is also a constant. The subscripts m and n are used to designate the eigenvalues of β_x and β_y and in turn the field configurations (modes). Thus a given combination of m and n in (8-14b) designates a given TE$_{mn}^z$ mode. Since there are infinite combinations of m and n, there are an infinite number of TE$_{mn}^z$ modes.

In (8-14b) β_x and β_y represent the mode wave numbers (eigenvalues) in the x and y directions, respectively. These are related to the wave number in the z direction (β_z) and to that of the unbounded medium (β) by (8-7a). In (8-14b) λ_x and λ_y represent, respectively, the wavelengths of the wave inside the guide in the x and y directions. These are related to the wavelength in the z direction ($\lambda_z = \lambda_g$) and to that in an unbounded medium (λ) according to (8-7a) by

$$\frac{1}{\lambda_x^2} + \frac{1}{\lambda_y^2} + \frac{1}{\lambda_z^2} = \frac{1}{\lambda^2} \quad (8\text{-}14c)$$

In summary then, the appropriate expressions for the TE$_{mn}^z$ modes are, according to (8-1), (8-14a), and (8-14b)

$$\underline{\text{TE}_{mn}^{+z}}$$

$$E_x^+ = A_{mn} \frac{\beta_y}{\varepsilon} \cos(\beta_x x) \sin(\beta_y y) e^{-j\beta_z z} \quad (8\text{-}15a)$$

$$E_y^+ = -A_{mn} \frac{\beta_x}{\varepsilon} \sin(\beta_x x) \cos(\beta_y y) e^{-j\beta_z z} \quad (8\text{-}15b)$$

$$E_z^+ = 0 \quad (8\text{-}15c)$$

$$H_x^+ = A_{mn} \frac{\beta_x \beta_z}{\omega\mu\varepsilon} \sin(\beta_x x) \cos(\beta_y y) e^{-j\beta_z z} \quad (8\text{-}15d)$$

$$H_y^+ = A_{mn} \frac{\beta_y \beta_z}{\omega\mu\varepsilon} \cos(\beta_x x) \sin(\beta_y y) e^{-j\beta_z z} \quad (8\text{-}15e)$$

$$H_z^+ = -j A_{mn} \frac{\beta_c^2}{\omega\mu\varepsilon} \cos(\beta_x x) \cos(\beta_y y) e^{-j\beta_z z} \quad (8\text{-}15f)$$

where

$$\beta_c^2 \equiv \left(\frac{2\pi}{\lambda_c}\right)^2 = \beta^2 - \beta_z^2 = \beta_x^2 + \beta_y^2 = \left(\frac{m\pi}{a}\right)^2 + \left(\frac{n\pi}{b}\right)^2 \quad (8\text{-}15\text{g})$$

The constant β_c is the value of β when $\beta_z = 0$, and it will be referred to as the *cutoff wave number*. Thus

$$\beta_c = \beta|_{\beta_z=0} = \omega\sqrt{\mu\varepsilon}|_{\beta_z=0} = \omega_c\sqrt{\mu\varepsilon} = 2\pi f_c\sqrt{\mu\varepsilon} = \sqrt{\left(\frac{m\pi}{a}\right)^2 + \left(\frac{n\pi}{b}\right)^2}$$

or

$$\boxed{(f_c)_{mn} = \frac{1}{2\pi\sqrt{\mu\varepsilon}}\sqrt{\left(\frac{m\pi}{a}\right)^2 + \left(\frac{n\pi}{b}\right)^2} \quad \begin{matrix} m = 0,1,2,\ldots \\ n = 0,1,2,\ldots \end{matrix} \bigg\} m = n \neq 0} \quad (8\text{-}16)$$

where $(f_c)_{mn}$ represents the cutoff frequency of a given mn mode. Modes that have the same cutoff frequency are called *degenerate*.

To determine the significance of the cutoff frequency, let us examine the values of β_z. Using (8-15g), we can write that

$$\beta_z^2 = \beta^2 - \beta_c^2 = \beta^2 - \left[\left(\frac{m\pi}{a}\right)^2 + \left(\frac{n\pi}{b}\right)^2\right] \quad (8\text{-}17)$$

or

$$(\beta_z)_{mn} = \begin{cases} \pm\sqrt{\beta^2 - \beta_c^2} = \pm\beta\sqrt{1 - \left(\frac{\beta_c}{\beta}\right)^2} \\ \quad = \pm\beta\sqrt{1 - \left(\frac{\lambda}{\lambda_c}\right)^2} = \pm\beta\sqrt{1 - \left(\frac{f_c}{f}\right)^2} \quad \text{for } \beta > \beta_c,\ f > f_c \\ \hfill (8\text{-}17\text{a}) \\ 0 \quad \text{for } \beta = \beta_c,\ f = f_c \\ \hfill (8\text{-}17\text{b}) \\ \pm j\sqrt{\beta_c^2 - \beta^2} = \pm j\beta\sqrt{\left(\frac{\beta_c}{\beta}\right)^2 - 1} \\ \quad = \pm j\beta\sqrt{\left(\frac{\lambda}{\lambda_c}\right)^2 - 1} = \pm j\beta\sqrt{\left(\frac{f_c}{f}\right)^2 - 1} \quad \text{for } \beta < \beta_c,\ f < f_c \\ \hfill (8\text{-}17\text{c}) \end{cases}$$

In order for the waves to be traveling in the $+z$ direction, the expressions for β_z as given by (8-17a) through (8-17c) reduce to

$$(\beta_z)_{mn} = \begin{cases} \beta\sqrt{1 - \left(\dfrac{\lambda}{\lambda_c}\right)^2} = \beta\sqrt{1 - \left(\dfrac{f_c}{f}\right)^2} & \text{for } f > f_c \qquad (8\text{-}18a) \\ 0 & \text{for } f = f_c \qquad (8\text{-}18b) \\ -j\beta\sqrt{\left(\dfrac{\lambda}{\lambda_c}\right)^2 - 1} = -j\beta\sqrt{\left(\dfrac{f_c}{f}\right)^2 - 1} & \text{for } f < f_c \qquad (8\text{-}18c) \end{cases}$$

Substituting the expressions for β_z as given by (8-18a) through (8-18c) in the expressions for **E** and **H** as given by (8-15a) through (8-15f), it is evident that (8-18a) leads to propagating waves, (8-18b) to standing waves, and (8-18c) to *evanescent* (reactive) or nonpropagating waves. Evanescent fields are exponentially decaying fields that do not possess real power. Thus (8-18b) serves as the boundary between propagating and nonpropagating waves, and it is usually referred to as the cutoff which occurs when $\beta_z = 0$. When the frequency of operation is selected to be higher than the value of $(f_c)_{mn}$ for a given mn mode, as given by (8-16), then the fields propagate unattenuated. If, however, f is selected to be smaller then $(f_c)_{mn}$, then the fields are attenuated. Thus the waveguide serves as a high pass filter.

The ratios of E_x/H_y and $-E_y/H_x$ have the units of impedance. Use of (8-15a) through (8-15f) shows that

$$Z_w^{+z}(\text{TE}_{mn}^z) \equiv \frac{E_x}{H_y} = -\frac{E_y}{H_x} = \frac{\omega\mu}{\beta_z} \qquad (8\text{-}19)$$

which can be written by using (8-18a) through (8-18c) as

$$Z_w^{+z}(\text{TE}_{mn}^z) = \begin{cases} \dfrac{\omega\mu}{\beta\sqrt{1 - \left(\dfrac{f_c}{f}\right)^2}} = \dfrac{\sqrt{\dfrac{\mu}{\varepsilon}}}{\sqrt{1 - \left(\dfrac{f_c}{f}\right)^2}} = \dfrac{\eta}{\sqrt{1 - \left(\dfrac{f_c}{f}\right)^2}} & \\ \qquad\qquad\qquad\qquad\qquad\qquad \text{for } f > f_c & (8\text{-}20a) \\ \infty \qquad\qquad\qquad\qquad\qquad\qquad \text{for } f = f_c & (8\text{-}20b) \\ +j\dfrac{\omega\mu}{\beta\sqrt{\left(\dfrac{f_c}{f}\right)^2 - 1}} = +j\dfrac{\sqrt{\dfrac{\mu}{\varepsilon}}}{\sqrt{\left(\dfrac{f_c}{f}\right)^2 - 1}} = +j\dfrac{\eta}{\sqrt{\left(\dfrac{f_c}{f}\right)^2 - 1}} & \\ \qquad\qquad\qquad\qquad\qquad\qquad \text{for } f < f_c & (8\text{-}20c) \end{cases}$$

Z_w^{+z} in (8-20a) through (8-20c) is referred to as the *wave impedance* in the $+z$ direction which is real and greater than the intrinsic impedance η of the medium inside the guide for values of $f > f_c$, infinity at $f = f_c$, and reactively inductive for $f < f_c$. Thus the waveguide for TE_{mn}^z modes behaves as an inductive storage element for $f < f_c$. A plot of Z_w^{+z} for any TE_{mn} mode in the range of $0 \leq f/f_c \leq 3$ is shown in Figure 8-2.

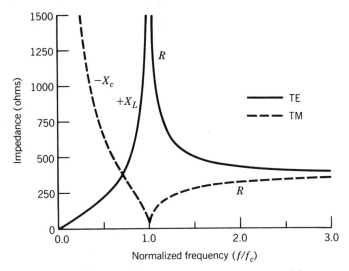

FIGURE 8-2 Wave impedance for a rectangular waveguide.

The expressions of (8-18a) through (8-18c) for β_z can also be used to define a wavelength along the axis of the guide. Thus we can write that

$$\beta_z \equiv \frac{2\pi}{\lambda_z} \Rightarrow \lambda_z = \lambda_g = \frac{2\pi}{\beta_z} \qquad (8\text{-}21)$$

or

$$(\lambda_z)_{mn} = (\lambda_g)_{mn} = \begin{cases} \dfrac{2\pi}{\beta\sqrt{1-\left(\dfrac{f_c}{f}\right)^2}} = \dfrac{\lambda}{\sqrt{1-\left(\dfrac{f_c}{f}\right)^2}} = \dfrac{\lambda}{\sqrt{1-\left(\dfrac{\lambda}{\lambda_c}\right)^2}} & \text{for } f > f_c \quad (8\text{-}21a) \\[2mm] \infty & \text{for } f = f_c \quad (8\text{-}21b) \\[2mm] +j\dfrac{2\pi}{\beta\sqrt{\left(\dfrac{f_c}{f}\right)^2-1}} = +j\dfrac{\lambda}{\sqrt{\left(\dfrac{f_c}{f}\right)^2-1}} = +j\dfrac{\lambda}{\sqrt{\left(\dfrac{\lambda}{\lambda_c}\right)^2-1}} & (8\text{-}21c) \\ & \text{(nonphysical) for } f < f_c \end{cases}$$

In (8-21a) through (8-21c) λ_z represents the wavelength of the wave along the axis of the guide, and it is referred to as the *guide wavelength* λ_g. In the same expressions, λ refers to the wavelength of the wave at the same frequency but traveling in an unbounded medium whose electrical parameters ε and μ are the same as those of the medium inside the waveguide. Cutoff wavelength corresponding to the cutoff frequency f_c is represented by λ_c.

Inspection of (8-21a) through (8-21c) indicates that the guide wavelength λ_g is greater than the unbounded medium wavelength λ for $f > f_c$, it is infinity for $f = f_c$, and has no physical meaning for $f < f_c$ since it is purely imaginary. A plot of λ_g/λ for $1 \leq f/f_c \leq 3$ is shown in Figure 8-3.

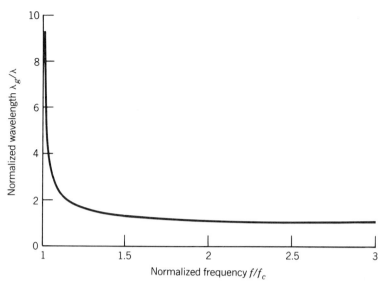

FIGURE 8-3 Normalized wavelength for a rectangular waveguide.

The values of $(f_c)_{mn}$ for different combinations of m and n but fixed values of ε, μ, a, and b determine the cutoff frequencies and order of existence of each mode above its corresponding cutoff frequency. Assuming $a > b$, the mode with the smallest cutoff frequency is that of TE_{10}. Its cutoff frequency is equal to

$$(f_c)_{10} = \frac{1}{2\pi\sqrt{\mu\varepsilon}}\frac{\pi}{a} = \frac{1}{2a\sqrt{\mu\varepsilon}} \qquad (8\text{-}22)$$

In general, the mode with the smallest cutoff frequency is referred to as the *dominant mode*. Thus for a waveguide with $a > b$, the dominant mode is the TE_{10} mode. (If $b > a$, the dominant mode is the TE_{01}.)

The ratio $R_{mn} = (f_c)_{mn}^{TE}/(f_c)_{10}^{TE}$ can be written as

$$R_{mn} = \frac{(f_c)_{mn}^{TE}}{(f_c)_{10}^{TE}} = \sqrt{(m)^2 + \left(\frac{na}{b}\right)^2} \quad \begin{array}{l} m = 0,1,2,3,\ldots \\ n = 0,1,2,3,\ldots \end{array} \bigg\} m = n \neq 0 \quad (8\text{-}23)$$

whose values for $a/b = 10, 5, 2.25, 2$, and 1, for the allowable values of m and n, are listed in Table 8-1. The ratio value R_{mn} of a given m, n combination represents the relative frequency range over which the TE_{10} mode can operate before that m, n mode will begin to appear. For a given a/b ratio, the smallest value of (8-23), above unity, indicates the relative frequency range over which the waveguide can operate in a single TE_{10} mode.

Example 8-1. A rectangular waveguide of dimensions a and b ($a > b$), as shown in Figure 8-1, is to be operated in a single mode. Determine the smallest ratio of the a/b dimensions that will allow the largest bandwidth of the single-mode operation. State the dominant mode and its largest bandwidth of single-mode operation.

TABLE 8-1
Ratio of cutoff frequency of TE_{mn}^z mode to that of TE_{10}^z

$$R_{mn} = \frac{(f_c)_{mn}^{\text{TE}^z}}{(f_c)_{10}^{\text{TE}^z}} = \sqrt{m^2 + \left(\frac{na}{b}\right)^2} \quad \begin{matrix} m = 0, 1, 2, \ldots \\ n = 0, 1, 2, \ldots \end{matrix} \right\} m = n \neq 0$$

$a/b \Rightarrow$	10	5	2.25	2	1
$m, n \Rightarrow$	1, 0	1, 0	1, 0	1, 0	1, 0; 0, 1
$R_{mn} \Rightarrow$	1	1	1	1	1
$m, n \Rightarrow$	2, 0	2, 0	2, 0	2, 0; 0, 1	1, 1
$R_{mn} \Rightarrow$	2	2	2	2	1.414
$m, n \Rightarrow$	3, 0	3, 0	0, 1	1, 1	2, 0
$R_{mn} \Rightarrow$	3	3	2.25	2.236	2
$m, n \Rightarrow$	4, 0	4, 0	1, 1	2, 1	2, 1; 1, 2
$R_{mn} \Rightarrow$	4	4	2.462	2.828	2.236
$m, n \Rightarrow$	5, 0	5, 0; 0, 1	3, 0	3, 0	2, 2
$R_{mn} \Rightarrow$	5	5	3	3	2.828
$m, n \Rightarrow$	6, 0	1, 1	2, 1	3, 1	3, 0; 0, 3
$R_{mn} \Rightarrow$	6	5.099	3.010	3.606	3
$m, n \Rightarrow$	7, 0	2, 1	3, 1	4, 0; 0, 2	3, 1; 1, 3
$R_{mn} \Rightarrow$	7	5.385	3.75	4	3.162
$m, n \Rightarrow$	8, 0	3, 1	4, 0	1, 2	3, 2; 2, 3
$R_{mn} \Rightarrow$	8	5.831	4	4.123	3.606
$m, n \Rightarrow$	9, 0	6, 0	0, 2	4, 1; 2, 2	4, 0; 0, 4
$R_{mn} \Rightarrow$	9	6	4.5	4.472	4
$m, n \Rightarrow$	10, 0; 0, 1	4, 1	4, 1	5, 0; 3, 2	4, 1; 1, 4
$R_{mn} \Rightarrow$	10	6.403	4.589	5	4.123

Solution. According to (8-16), the dominant mode for $a > b$ is the TE_{10} whose cutoff frequency is given by (8-22) or

$$(f_c)_{10} = \frac{1}{2a\sqrt{\mu\varepsilon}}$$

The mode with the next higher cutoff frequency would be either the TE_{20} or TE_{01} mode whose cutoff frequencies are given, respectively, by

$$(f_c)_{20} = \frac{1}{a\sqrt{\mu\varepsilon}} = 2(f_c)_{10}$$

$$(f_c)_{01} = \frac{1}{2b\sqrt{\mu\varepsilon}}$$

It is apparent that the largest bandwidth of single TE_{10} mode operation

would be
$$(f_c)_{10} \le f \le 2(f_c)_{10} = (f_c)_{20} \le (f_c)_{01}$$
and would occur provided
$$2b \le a \Rightarrow b \le a/2 \Rightarrow 2 \le a/b$$

8.2.2 Transverse Magnetic (TM^z)

A procedure similar to that used for the TE^z modes can be used to derive the TM^z fields and the other appropriate parameters for a rectangular waveguide of the geometry shown in Figure 8-1. According to (6-55a) and (6-55b) these can be obtained by letting $\mathbf{A} = \hat{a}_z A_z(x, y, z)$ and $\mathbf{F} = 0$. Without repeating the entire procedure, the most important equations 6-59, 6-56, and 6-58 are summarized:

$$E_x = -j\frac{1}{\omega\mu\varepsilon}\frac{\partial^2 A_z}{\partial x \partial z} \qquad H_x = \frac{1}{\mu}\frac{\partial A_z}{\partial y}$$

$$E_y = -j\frac{1}{\omega\mu\varepsilon}\frac{\partial^2 A_z}{\partial y \partial z} \qquad H_y = -\frac{1}{\mu}\frac{\partial A_z}{\partial x} \qquad (8\text{-}24)$$

$$E_z = -j\frac{1}{\omega\mu\varepsilon}\left(\frac{\partial^2}{\partial z^2} + \beta^2\right)A_z \qquad H_z = 0$$

$$\nabla^2 A_z + \beta^2 A_z = \frac{\partial^2 A_z}{\partial x^2} + \frac{\partial^2 A_z}{\partial y^2} + \frac{\partial^2 A_z}{\partial z^2} + \beta^2 A_z = 0 \qquad (8\text{-}25)$$

$$A_z(x, y, z) = [C_1 \cos(\beta_x x) + D_1 \sin(\beta_x x)][C_2 \cos(\beta_y y) + D_2 \sin(\beta_y y)]$$
$$\times [A_3 e^{-j\beta_z z} + B_3 e^{+j\beta_z z}] \qquad (8\text{-}26)$$

For waves that travel in the $+z$ direction and satisfy the boundary conditions of Figure 8-1 as outlined by (8-8a) through (8-8d), (8-26) reduces to

$$A_z^+(x, y, z) = D_1 D_2 A_3 \sin(\beta_x x) \sin(\beta_y y) e^{-j\beta_z z}$$
$$= B_{mn} \sin(\beta_x x) \sin(\beta_y y) e^{-j\beta_z z} \qquad (8\text{-}26a)$$

where

$$\beta_x \equiv \frac{2\pi}{\lambda_x} = \frac{m\pi}{a} \Rightarrow \lambda_x = \frac{2a}{m} \qquad m = 1, 2, 3, \ldots \qquad (8\text{-}27a)$$

$$\beta_y \equiv \frac{2\pi}{\lambda_y} = \frac{n\pi}{b} \Rightarrow \lambda_y = \frac{2b}{n} \qquad n = 1, 2, 3, \ldots \qquad (8\text{-}27b)$$

Use of (8-26) allows the fields of (8-24) to be written as

$$E_x^+ = -B_{mn}\frac{\beta_x \beta_z}{\omega\mu\varepsilon}\cos(\beta_x x)\sin(\beta_y y)e^{-j\beta_z z} \qquad (8\text{-}28a)$$

$$E_y^+ = -B_{mn}\frac{\beta_y \beta_z}{\omega\mu\varepsilon}\sin(\beta_x x)\cos(\beta_y y)e^{-j\beta_z z} \qquad (8\text{-}28b)$$

$$E_z^+ = -jB_{mn}\frac{\beta_c^2}{\omega\mu\varepsilon}\sin(\beta_x x)\sin(\beta_y y)e^{-j\beta_z z} \qquad (8\text{-}28c)$$

$$H_x^+ = B_{mn}\frac{\beta_y}{\mu}\sin(\beta_x x)\cos(\beta_y y)e^{-j\beta_z z} \qquad (8\text{-}28\text{d})$$

$$H_y^+ = -B_{mn}\frac{\beta_x}{\mu}\cos(\beta_x x)\sin(\beta_y y)e^{-j\beta_z z} \qquad (8\text{-}28\text{e})$$

$$H_z^+ = 0 \qquad (8\text{-}28\text{f})$$

In turn the wave impedance, propagation constant, cutoff frequency, and guide wavelength can be expressed as

$$Z_w^{+z}(\text{TM}_{mn}^z) \equiv \frac{E_x^+}{H_y^+} = -\frac{E_y^+}{H_x^+} = \frac{\beta_z}{\omega\varepsilon} = \begin{cases} +\eta\sqrt{1-\left(\dfrac{f_c}{f}\right)^2} & \text{for } f > f_c \quad (8\text{-}29\text{a}) \\ 0 & \text{for } f = f_c \quad (8\text{-}29\text{b}) \\ -j\eta\sqrt{\left(\dfrac{f_c}{f}\right)^2-1} & \text{for } f < f_c \quad (8\text{-}29\text{c}) \end{cases}$$

$$(\beta_z)_{mn} \equiv \frac{2\pi}{\lambda_z} = \begin{cases} \beta\sqrt{1-\left(\dfrac{f_c}{f}\right)^2} & \text{for } f > f_c \quad (8\text{-}30\text{a}) \\ 0 & \text{for } f = f_c \quad (8\text{-}30\text{b}) \\ -j\beta\sqrt{\left(\dfrac{f_c}{f}\right)^2-1} & \text{for } f < f_c \quad (8\text{-}30\text{c}) \end{cases}$$

$$\beta_c^2 \equiv \left(\frac{2\pi}{\lambda_c}\right)^2 = \beta^2 - \beta_z^2 = \beta_x^2 + \beta_y^2 = \left(\frac{m\pi}{a}\right)^2 + \left(\frac{n\pi}{b}\right)^2 \qquad (8\text{-}31)$$

$$(f_c)_{mn} = \frac{1}{2\pi\sqrt{\mu\varepsilon}}\sqrt{\left(\frac{m\pi}{a}\right)^2 + \left(\frac{n\pi}{b}\right)^2} \qquad \begin{matrix} m = 1,2,3,\ldots \\ n = 1,2,3,\ldots \end{matrix} \qquad (8\text{-}32)$$

$$(\lambda_z)_{mn} = (\lambda_g)_{mn} = \begin{cases} \dfrac{\lambda}{\sqrt{1-\left(\dfrac{f_c}{f}\right)^2}} = \dfrac{\lambda}{\sqrt{1-\left(\dfrac{\lambda}{\lambda_c}\right)^2}} & \text{for } f > f_c \quad (8\text{-}33\text{a}) \\ \infty & \text{for } f = f_c \quad (8\text{-}33\text{b}) \\ j\dfrac{\lambda}{\sqrt{\left(\dfrac{f_c}{f}\right)^2-1}} = j\dfrac{\lambda}{\sqrt{\left(\dfrac{\lambda}{\lambda_c}\right)^2-1}} & \\ & \text{(nonphysical) for } f < f_c \quad (8\text{-}33\text{c}) \end{cases}$$

It is apparent from (8-29c) that below cutoff ($f < f_c$) the waveguide for TM_{mn}^z modes behaves as a capacitive storage element. A plot of Z_w^{+z} for any TM_{mn}^z mode in the range of $0 \le f/f_c \le 3$ is shown in Figure 8-2.

For TM^z, we can classify the modes according to the order of their cutoff frequency. The TM^z mode with the smallest cutoff frequency according to (8-32) is the TM$^z_{11}$ whose cutoff frequency is equal to

$$(f_c)_{11} = \frac{1}{2\sqrt{\mu\varepsilon}}\sqrt{\left(\frac{1}{a}\right)^2 + \left(\frac{1}{b}\right)^2} = \frac{1}{2a\sqrt{\mu\varepsilon}}\sqrt{1 + \left(\frac{a}{b}\right)^2} > \frac{1}{2a\sqrt{\mu\varepsilon}} \quad (8\text{-}34)$$

Since the cutoff frequency of the TM$^z_{11}$ mode, as given by (8-34), is greater than the cutoff frequency of the TE$^z_{10}$, as given by (8-22), then the TE$^z_{10}$ mode is always the dominant mode if $a > b$. If $a = b$, the dominant modes are the TE$^z_{10}$ and TE$^z_{01}$ modes (degenerate), and if $a < b$ the dominant mode is the TE$^z_{01}$ mode.

The order in which the TM$^z_{mn}$ modes occur, relative to the TE$^z_{10}$ mode, can be determined by forming the ratio T_{mn} of the cutoff frequency of any TM$^z_{mn}$ mode to the cutoff frequency of the TE$^z_{10}$ mode. Then we use (8-32) and (8-22) to write that

$$T_{mn} = \frac{(f_c)^{TM}_{mn}}{(f_c)^{TE}_{10}} = \sqrt{m^2 + \left(\frac{na}{b}\right)^2} \qquad \begin{array}{l} m = 1, 2, 3, \ldots \\ n = 1, 2, 3, \ldots \end{array} \quad (8\text{-}35)$$

TABLE 8-2
Ratio of cutoff frequency of TM$^z_{mn}$ mode to that of TE$^z_{10}$

$$T_{mn} = \frac{(f_c)^{TM^z}_{mn}}{(f_c)^{TE^z}_{10}} = \sqrt{m^2 + \left(\frac{na}{b}\right)^2} \qquad \begin{array}{l} m = 1, 2, 3, \ldots \\ n = 1, 2, 3, \ldots \end{array}$$

$a/b \Rightarrow$	10	5	2.25	2	1
$m, n \Rightarrow$	1,1	1,1	1,1	1,1	1,1
$T_{mn} \Rightarrow$	10.05	5.10	2.46	2.23	1.414
$m, n \Rightarrow$	2,1	2,1	2,1	2,1	2,1; 1,2
$T_{mn} \Rightarrow$	10.19	5.38	3.01	2.83	2.236
$m, n \Rightarrow$	3,1	3,1	3,1	3,1	2,2
$T_{mn} \Rightarrow$	10.44	6.00	3.75	3.61	2.828
$m, n \Rightarrow$	4,1	4,1	4,1	1,2	3,1; 1,3
$T_{mn} \Rightarrow$	10.77	6.40	4.59	4.12	3.162
$m, n \Rightarrow$	5,1	5,1	1,2	4,1; 2,2	3,2; 2,3
$T_{mn} \Rightarrow$	11.18	7.07	5.09	4.47	3.606
$m, n \Rightarrow$	6,1	6,1	2,2	3,2	4,1; 1,4
$T_{mn} \Rightarrow$	11.66	7.81	5.38	5.00	4.123
$m, n \Rightarrow$	7,1	7,1	5,1	5,1	3,3
$T_{mn} \Rightarrow$	12.21	8.60	5.48	5.39	4.243
$m, n \Rightarrow$	8,1	8,1	3,2	4,2	4,2; 2,4
$T_{mn} \Rightarrow$	12.81	9.43	5.83	5.66	4.472
$m, n \Rightarrow$	9,1	1,2	4,2	1,3	4,3; 3,4
$T_{mn} \Rightarrow$	13.82	10.04	6.40	6.08	5.00
$m, n \Rightarrow$	10,1	2,2	6,1	2,3	5,1; 1,5
$T_{mn} \Rightarrow$	14.14	10.20	6.41	6.32	5.09

The values of T_{mn} for $a/b = 10, 5, 2.25, 2$, and 1, for the allowable values of m and n, are listed in Table 8-2. Each value of T_{mn} in Table 8-2 represents the relative frequency range over which the TE_{10} mode can operate before that m, n mode will begin to appear.

For a given ratio of a/b, the values of R_{mn} of (8-23) and Table 8-1, and those of T_{mn} of (8-35) and Table 8-2 represent the order, in terms of ascending cutoff

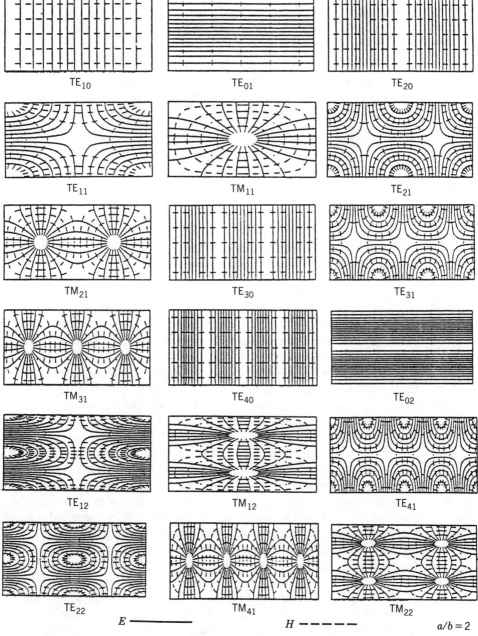

FIGURE 8-4 Field patterns for the first 18 TE^z and/or TM^z modes in a rectangular waveguide with $a/b = 2$ (*Source:* C. S. Lee, S. W. Lee, and S. L. Chuang, "Plot of modal field distribution in rectangular and circular waveguides," *IEEE Trans. Microwave Theory Tech.*, ©, 1985, IEEE.)

frequencies, in which the TE_{mn}^z and TM_{mn}^z modes occur relative to the dominant TE_{10}^z mode.

The xy cross-section field distributions for the first 18 modes [1] of a rectangular waveguide with cross-sectional dimensions $a/b = 2$ are plotted in Figure 8-4. Field configurations of another 18 modes plus the first 30 for a square waveguide ($a/b = 1$) can be found in [1].

Example 8-2. The inner dimensions of an X-band WR90 rectangular waveguide are $a = 0.9$ in. (2.286 cm) and $b = 0.4$ in. (1.016 cm). Assume free space within the guide and determine (in GHz) the cutoff frequencies, in ascending order, of the first 10 TE^z and/or TM^z modes.

Solution. Since $a/b = 0.9/0.4 = 2.25$, then according to Tables 8-1 and 8-2 the cutoff frequencies of the first 10 TE_{mn}^z and/or TM_{mn}^z modes in order of ascending frequency are

1. $\text{TE}_{10} = 6.562$ GHz
2. $\text{TE}_{20} = 13.124$ GHz
3. $\text{TE}_{01} = 14.764$ GHz
4, 5. $\text{TE}_{11} = \text{TM}_{11} = 16.16$ GHz
6. $\text{TE}_{30} = 19.685$ GHz
7, 8. $\text{TE}_{21} = \text{TM}_{21} = 19.754$ GHz
9, 10. $\text{TE}_{31} = \text{TM}_{31} = 24.607$ GHz

8.2.3 Dominant TE_{10} Mode

From the discussion and analysis of the previous two sections, it is evident that there are an infinite number of TE_{mn}^z and TM_{mn}^z modes that satisfy Maxwell's equations and the boundary conditions, and that they can exist inside the rectangular waveguide of Figure 8-1. In addition, other modes, such as TE^x, TM^x, TE^y, and TM^y, can also exist inside that same waveguide. The analysis of the TE^x and TM^x modes has been assigned to the reader as an end of chapter problem while the TE^y and TM^y modes are analyzed in Sections 8.5.1 and 8.5.2.

In a given system, the modes that can exist inside a waveguide depend upon the dimensions of the waveguide, the medium inside it which determines the cutoff frequencies of the different modes, and the excitation and coupling of energy from the source (oscillator) to the waveguide. Since in a multimode waveguide operation the total power is distributed among the existing modes (this will be shown later in this chapter) and the instrumentation (detectors, probes, etc.) required to detect the total power of multimodes is more complex and expensive, it is often most desirable to operate the waveguide in a single mode.

The order in which the different modes enter a waveguide depends upon their cutoff frequency. The modes with cutoff frequencies equal to or smaller than the operational frequency can exist inside the waveguide. Because, in practice, most systems that utilize a rectangular waveguide design require excitation and detection instrumentation for a dominant TE_{10} mode operation, it is prudent at this time to devote some extra effort to examination of the characteristics of this mode.

For the TE_{10}^z mode the pertinent expressions for the field intensities and the various characteristic parameters are obtained from Section 8.2.1 by letting $m = 1$ and $n = 0$. Then we can write the following summary.

$$\underline{\text{TE}_{10}^z \text{ Mode } (m = 1, n = 0)}$$

$$F_z^+(x, z) = A_{10} \cos\left(\frac{\pi}{a}x\right) e^{-j\beta_z z} \qquad (8\text{-}36)$$

$$\beta_x = \frac{\pi}{a} = \frac{2\pi}{\lambda_x} \Rightarrow \lambda_x = 2a \qquad (8\text{-}37\text{a})$$

$$\beta_y = 0 = \frac{2\pi}{\lambda_y} \Rightarrow \lambda_y = \infty \qquad (8\text{-}37\text{b})$$

$$\beta_c = \beta_x = \frac{\pi}{a} = \frac{2\pi}{\lambda_c} \Rightarrow \lambda_c = 2a \qquad (8\text{-}37\text{c})$$

$$\beta_z = \begin{cases} \sqrt{\beta^2 - \left(\frac{\pi}{a}\right)^2} = \beta\sqrt{1 - \left(\frac{\lambda}{2a}\right)^2} \\ = \beta\sqrt{1 - \left(\frac{\lambda}{\lambda_c}\right)^2} = \beta\sqrt{1 - \left(\frac{f_c}{f}\right)^2} & \text{for } f > f_c \quad (8\text{-}38\text{a}) \\ 0 & \text{for } f = f_c \quad (8\text{-}38\text{b}) \\ -j\sqrt{\left(\frac{\pi}{a}\right)^2 - \beta^2} = -j\beta\sqrt{\left(\frac{\lambda}{2a}\right)^2 - 1} \\ = -j\beta\sqrt{\left(\frac{\lambda}{\lambda_c}\right)^2 - 1} = -j\beta\sqrt{\left(\frac{f_c}{f}\right)^2 - 1} & \text{for } f < f_c \quad (8\text{-}38\text{c}) \end{cases}$$

$$E_x^+ = 0 \qquad (8\text{-}39\text{a})$$

$$E_y^+ = -\frac{A_{10}}{\varepsilon}\frac{\pi}{a}\sin\left(\frac{\pi}{a}x\right) e^{-j\beta_z z} \qquad (8\text{-}39\text{b})$$

$$E_z^+ = 0 \qquad (8\text{-}39\text{c})$$

$$H_x^+ = A_{10}\frac{\beta_z}{\omega\mu\varepsilon}\frac{\pi}{a}\sin\left(\frac{\pi}{a}x\right) e^{-j\beta_z z} \qquad (8\text{-}39\text{d})$$

$$H_y^+ = 0 \qquad (8\text{-}39\text{e})$$

$$H_z^+ = -j\frac{A_{10}}{\omega\mu\varepsilon}\left(\frac{\pi}{a}\right)^2 \cos\left(\frac{\pi}{a}x\right) e^{-j\beta_z z} \qquad (8\text{-}39\text{f})$$

$$\mathbf{J}^+ = \hat{n} \times \mathbf{H}^+|_{\text{wall}} = \begin{cases} \hat{a}_y \times (\hat{a}_x H_x^+ + \hat{a}_z H_z^+)|_{y=0} = (+\hat{a}_x H_z^+ - \hat{a}_z H_x^+)|_{y=0} \\ = -\hat{a}_x j \frac{A_{10}}{\omega\mu\varepsilon}\left(\frac{\pi}{a}\right)^2 \cos\left(\frac{\pi}{a}x\right)e^{-j\beta_z z} - \hat{a}_z A_{10}\frac{\beta_z}{\omega\mu\varepsilon}\frac{\pi}{a}\sin\left(\frac{\pi}{a}x\right)e^{-j\beta_z z} \\ \underbrace{\qquad\qquad\qquad\qquad\qquad\qquad}_{\text{for the bottom wall}} \qquad (8\text{-}39\text{g}) \\[1em] = \hat{a}_x \times \hat{a}_z H_z^+|_{x=0} = -\hat{a}_y H_z^+|_{x=0} = \hat{a}_y j \frac{A_{10}}{\omega\mu\varepsilon}\left(\frac{\pi}{a}\right)^2 e^{-j\beta_z z} \\ \underbrace{\qquad\qquad\qquad\qquad\qquad}_{\text{for the left wall}} \qquad (8\text{-}39\text{h}) \end{cases}$$

$$(\lambda_z)_{10} = (\lambda_g)_{10} = \begin{cases} \dfrac{\lambda}{\sqrt{1-\left(\dfrac{f_c}{f}\right)^2}} = \dfrac{\lambda}{\sqrt{1-\left(\dfrac{\lambda}{\lambda_c}\right)^2}} \\ \qquad\qquad = \dfrac{\lambda}{\sqrt{1-\left(\dfrac{\lambda}{2a}\right)^2}} \qquad \text{for } f > f_c \quad (8\text{-}40\text{a}) \\[1em] \infty \qquad\qquad\qquad\qquad\qquad\qquad \text{for } f = f_c \quad (8\text{-}40\text{b}) \\[1em] j\dfrac{\lambda}{\sqrt{\left(\dfrac{f_c}{f}\right)^2 - 1}} = j\dfrac{\lambda}{\sqrt{\left(\dfrac{\lambda}{\lambda_c}\right)^2 - 1}} \\ \qquad\qquad = j\dfrac{\lambda}{\sqrt{\left(\dfrac{\lambda}{2a}\right)^2 - 1}} \quad \begin{array}{l}\text{(nonphysical)} \\ \text{for } f < f_c\end{array} \quad (8\text{-}40\text{c}) \end{cases}$$

$$(f_c)_{10} = \frac{1}{2a\sqrt{\mu\varepsilon}} = \frac{v}{2a} = \frac{v}{(\lambda_c)_{10}} \qquad (8\text{-}41)$$

$$Z_w^{+z}(\text{TE}_{10}^z) = \begin{cases} \dfrac{\eta}{\sqrt{1-\left(\dfrac{f_c}{f}\right)^2}} & \text{for } f > f_c \quad (8\text{-}42\text{a}) \\[1em] \infty & \text{for } f = f_c \quad (8\text{-}42\text{b}) \\[1em] j\dfrac{\eta}{\sqrt{\left(\dfrac{f_c}{f}\right)^2 - 1}} & \text{for } f < f_c \quad (8\text{-}42\text{c}) \end{cases}$$

For the TE_{10} mode at cutoff ($\beta_z = 0 \Rightarrow \lambda_z = \infty$) the wavelength of the wave inside the guide in the x direction (λ_x) is according to (8-14c) equal to the wavelength of the wave in an unbounded medium (λ). That is $(\lambda_x)_{10} = \lambda$ at cutoff.

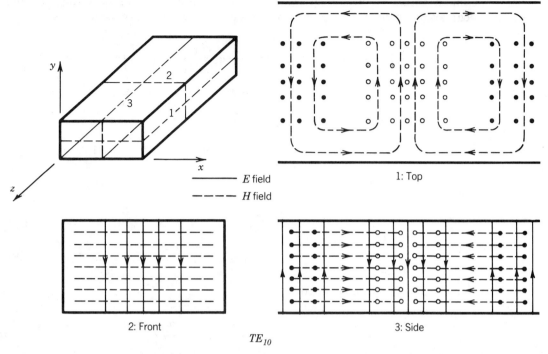

FIGURE 8-5 Electric field patterns for TE_{10} mode in a rectangular waveguide. (*Source:* S. Ramo, J. R. Whinnery, and T. Van Duzer, *Fields and Waves in Communication Electronics*, 1984. Reprinted with permission of John Wiley & Sons, Inc.)

From the preceding information it is evident that the electric field intensity inside the guide has only one component, an E_y. The E- and H-field variations on the top, front, and side views of the guide are shown graphically in Figure 8-5, and the current density and H-field lines on the top and side views are shown in Figure 8-6 [2]. It is instructive at this time to examine the electric field intensity a little closer and attempt to provide some physical interpretation of the propagation

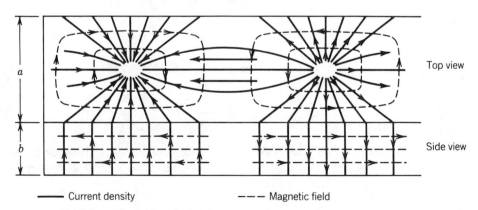

FIGURE 8-6 Magnetic field and electric current density patterns for the TE_{10} mode in a rectangular waveguide. (*Source:* S. Ramo, J. R. Whinnery, and T. Van Duzer, *Fields and Waves in Communication Electronics*, 1984. Reprinted with permission of John Wiley & Sons, Inc.)

characteristics of the waveguide. The total electric field of (8-39a) through (8-39c) can also be written, by representing the sine function with exponentials, as

$$\mathbf{E}^+(x, z) = \hat{a}_y E_y^+(x, z) = -\hat{a}_y \frac{A_{10}}{\varepsilon} \frac{\pi}{a} \sin\left(\frac{\pi}{a}x\right) e^{-j\beta_z z}$$

$$= -\hat{a}_y \frac{A_{10}}{\varepsilon} \frac{\pi}{a} \left[\frac{e^{j[(\pi/a)x - \beta_z z]} - e^{-j[(\pi/a)x + \beta_z z]}}{2j} \right]$$

$$\mathbf{E}^+(x, z) = \hat{a}_y j \frac{A_{10}}{2\varepsilon} \frac{\pi}{a} \left[e^{j[(\pi/a)x - \beta_z z]} - e^{-j[(\pi/a)x + \beta_z z]} \right] \tag{8-43}$$

Letting

$$\frac{\pi}{a} = \beta_x = \beta \sin \psi \tag{8-44a}$$

$$\beta_z = \beta \cos \psi \tag{8-44b}$$

which satisfy the constraint equation 8-7a, or

$$\beta_x^2 + \beta_y^2 + \beta_z^2 = \beta^2 \sin^2 \psi + \beta^2 \cos^2 \psi = \beta^2 \tag{8-45}$$

we can write (8-43) using (8-44a) and (8-44b) as

$$\mathbf{E}^+(x, z) = \hat{a}_y j \frac{A_{10}}{2\varepsilon} \frac{\pi}{a} \left[e^{j\beta(x \sin \psi - z \cos \psi)} - e^{-j\beta(x \sin \psi + z \cos \psi)} \right] \tag{8-46}$$

A close inspection of the two exponential terms inside the brackets indicates, by referring to the contents of Section 4.2.2, that each represents a uniform plane wave traveling in a direction determined by the angle ψ. In Figure 8-7a, which represents a top view of the waveguide of Figure 8-1, the two plane waves representing (8-46) or

$$\mathbf{E}^+(x, z) = \mathbf{E}_1^+(x, z) + \mathbf{E}_2^+(x, z) \tag{8-47}$$

where

$$\mathbf{E}_1^+(x, z) = \hat{a}_y j \frac{A_{10}}{2\varepsilon} \frac{\pi}{a} \left[e^{j\beta(x \sin \psi - z \cos \psi)} \right] \tag{8-47a}$$

$$\mathbf{E}_2^+(x, z) = -\hat{a}_y j \frac{A_{10}}{2\varepsilon} \frac{\pi}{a} \left[e^{-j\beta(x \sin \psi + z \cos \psi)} \right] \tag{8-47b}$$

are indicated as two plane waves that bounce back and forth between the side walls of the waveguide at an angle ψ. There is a 180° phase reversal between the two, which is also indicated in Figure 8-7.

According to (8-44b)

$$\beta_z = \beta \cos \psi \Rightarrow \psi = \cos^{-1}\left(\frac{\beta_z}{\beta}\right) \tag{8-48}$$

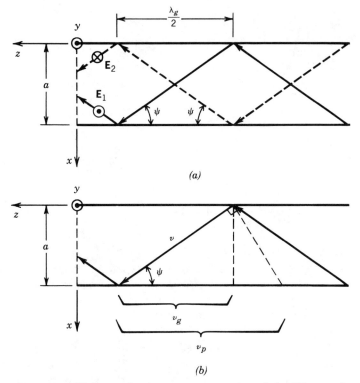

FIGURE 8-7 Uniform plane wave representation of the TE_{10} mode electric field inside a rectangular waveguide. (*a*) Two uniform plane waves. (*b*) Phase and group velocities.

By using (8-38a), we can write (8-48) as

$$\psi = \begin{cases} 0° & \text{for } f = \infty & \text{(8-49a)} \\ \cos^{-1}\left[\sqrt{1 - \left(\dfrac{f_c}{f}\right)^2}\right] & \text{for } f_c \leq f < \infty & \text{(8-49b)} \\ 90° & \text{for } f = f_c & \text{(8-49c)} \end{cases}$$

It is apparent that as $f \to f_c$, the angle ψ approaches 90° and exactly at cutoff ($f = f_c \Rightarrow \psi = 90°$) the plane waves bounce back and forth between the side walls of the waveguide without moving in the z direction. This reduces the fields into standing waves at cutoff.

By using (8-44b) the guide wavelength can be written as

$$\beta_z = \frac{2\pi}{\lambda_z} = \frac{2\pi}{\lambda_g} = \beta \cos\psi \Rightarrow \lambda_g = \frac{2\pi}{\beta \cos\psi} = \frac{\lambda}{\cos\psi} \quad (8\text{-}50)$$

which indicates that as cutoff approaches ($\psi \to 90°$) the guide wavelength approaches infinity. In addition, the phase velocity v_p can also be obtained using (8-44b), that is

$$\beta_z \equiv \frac{\omega}{v_p} = \beta \cos\psi = \frac{\omega}{v}\cos\psi \Rightarrow v_p = \frac{v}{\cos\psi} \quad (8\text{-}51)$$

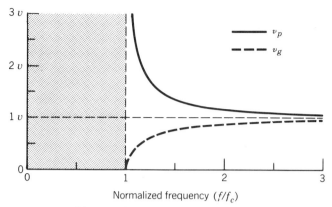

FIGURE 8-8 Phase and group (energy) velocities for the TE_{10} mode in a rectangular waveguide.

where v is the velocity with which the plane wave travels along the direction determined by ψ. Since the phase velocity, as given by (8-51), is greater than the velocity of light, it may be appropriate to illustrate graphically its meaning. By referring to Figure 8-7b, it is evident that whereas v is the velocity of the uniform plane wave along the direction determined by ψ, v_p ($v_p \geq v$) is the phase velocity, that is, the velocity that must be maintained to keep in step with a constant phase front of the wave, and v_g ($v_g \leq v$) is the group velocity, that is, the velocity with which a uniform plane wave travels along the z direction. According to Figure 8-7b

$$v_p = \frac{v}{\cos \psi} \quad (8\text{-}52a)$$

$$v_g = v \cos \psi \quad (8\text{-}52b)$$

and

$$\boxed{v_p v_g = v^2} \quad (8\text{-}52c)$$

These are the same interpretations given to the oblique plane wave propagation in Section 4.2.2C and Figure 4-6. A plot of v_p and v_g as a function of frequency in the range $0 \leq f/f_c \leq 3$ is shown in Figure 8-8.

Above cutoff, the wave impedance of (8-19) can be written in terms of the angle ψ as

$$Z_w^{+z}(TE_{10}) = \frac{\omega \mu}{\beta_z} = \frac{\omega \mu}{\beta \cos \psi} = \frac{\eta}{\cos \psi} \quad (8\text{-}53)$$

whose values are equal to or greater than the intrinsic impedance η of the medium inside the waveguide.

Example 8-3. Design an air-filled rectangular waveguide with dimensions a and b ($a > b$) that will operate in the dominant TE_{10} mode at $f = 10$ GHz. The dimensions a and b of the waveguide should be chosen so that at $f = 10$ GHz the waveguide not only operates on the single TE_{10} mode but also that $f = 10$ GHz is *simultaneously* 25% above the cutoff frequency of the dominant TE_{10} mode and 25% below the next higher-order TE_{01} mode.

Solution. According to (8-22), the cutoff frequency of the TE_{10} mode with a free-space medium in the guide is

$$(f_c)_{10} = \frac{1}{2a\sqrt{\mu_0\varepsilon_0}} = \frac{30 \times 10^9}{2a}$$

Since $f = 10$ GHz must be greater by 25% above the cutoff frequency of the TE_{10} mode, then

$$10 \times 10^9 \geq 1.25\left(\frac{30 \times 10^9}{2a}\right) \Rightarrow a \geq 1.875 \text{ cm} = 0.738 \text{ in.}$$

Since the next higher-order mode is the TE_{01} mode, whose cutoff frequency with a free-space medium in the guide is

$$(f_c)_{01} = \frac{1}{2b\sqrt{\mu_0\varepsilon_0}} = \frac{30 \times 10^9}{2b}$$

then

$$10 \times 10^9 \leq 0.75\left(\frac{30 \times 10^9}{2b}\right) \Rightarrow b \leq 1.125 \text{ cm} = 0.443 \text{ in.}$$

Example 8-4. Design a rectangular waveguide with dimensions a and b ($a > b$) that will operate in a single mode between 9 and 14 GHz. Assuming free space inside the waveguide, determine the waveguide dimensions that will insure single-mode operation over that band.

Solution. Since $a > b$, the dominant mode is the TE_{10} whose cutoff frequency must be

$$(f_c)_{10} = \frac{1}{2a\sqrt{\mu_0\varepsilon_0}} = \frac{30 \times 10^9}{2a} = 9 \times 10^9 \Rightarrow a = 1.667 \text{ cm} = 0.656 \text{ in.}$$

The cutoff frequency of the TE_{20} mode is 18 GHz. Therefore the next higher-order mode is TE_{01}, whose cutoff frequency must be

$$(f_c)_{01} = \frac{1}{2b\sqrt{\mu_0\varepsilon_0}} = \frac{30 \times 10^9}{2b} = 14 \times 10^9 \Rightarrow b = 1.071 \text{ cm} = 0.422 \text{ in.}$$

374 RECTANGULAR CROSS-SECTION WAVEGUIDES AND CAVITIES

8.2.4 Power Density and Power

The fields that are created and propagating inside the waveguide have power associated with them. To find the power flowing down the guide, it is first necessary to find the average power density directed along the axis of the waveguide. The power flowing along the guide can then be found by integrating the axial directed power density over the cross section of the waveguide.

For the waveguide geometry of Figure 8-1, the z-directed power density can be written as

$$(\mathbf{S}_z)_{mn} = \hat{a}_z S_z = \tfrac{1}{2} \operatorname{Re}\left[(\hat{a}_x E_x + \hat{a}_y E_y) \times (\hat{a}_x H_x + \hat{a}_y H_y)^*\right]$$
$$(\mathbf{S}_z)_{mn} = \hat{a}_z S_z = \hat{a}_z \tfrac{1}{2} \operatorname{Re}\left[E_x H_y^* - E_y H_x^*\right] \qquad (8\text{-}54)$$

$$\underline{\text{TE}_{mn}^z \text{ Modes}}$$

Use of the field expressions of (8-15a) through (8-15f), allows the z-directed power density of (8-54) for the TE_{mn}^z modes to be written as

$$(\mathbf{S}_z)_{mn} = \hat{a}_z S_z = \hat{a}_z \frac{|A_{mn}|^2}{2} \operatorname{Re}\left[\frac{\beta_y^2 \beta_z}{\omega\mu\varepsilon^2} \cos^2(\beta_x x)\sin^2(\beta_y y)\right.$$
$$\left.+\frac{\beta_x^2 \beta_z}{\omega\mu\varepsilon^2}\sin^2(\beta_x x)\cos^2(\beta_y y)\right]$$

$$(\mathbf{S}_z)_{mn} = \hat{a}_z S_z = \hat{a}_z |A_{mn}|^2 \frac{\beta_z}{2\omega\mu\varepsilon^2}\left[\beta_y^2 \cos^2(\beta_x x)\sin^2(\beta_y y)\right.$$
$$\left.+\beta_x^2\sin^2(\beta_x x)\cos^2(\beta_y y)\right] \qquad (8\text{-}55)$$

The associated power is obtained by integrating (8-55) over a cross section A_0 of the guide, or

$$P_{mn} = \iint_{A_0} (\mathbf{S}_z)_{mn} \cdot d\mathbf{s} = \int_0^b\!\!\int_0^a (\hat{a}_z S_z)\cdot(\hat{a}_z\,dx\,dy) = \int_0^b\!\!\int_0^a S_z\,dx\,dy \qquad (8\text{-}56)$$

Since

$$\int_0^a \cos^2\!\left(\frac{m\pi}{a}x\right)dx = \begin{cases} a/2 & m\neq 0 \\ a & m=0 \end{cases} \qquad (8\text{-}56a)$$

$$\int_0^a \sin^2\!\left(\frac{m\pi}{a}x\right)dx = \begin{cases} a/2 & m\neq 0 \\ 0 & m=0 \end{cases} \qquad (8\text{-}56b)$$

and similar equalities exist for the y variations, (8-56) reduces by using (8-55), (8-56a), and (8-56b) to

$$P_{mn}^{\text{TE}_z} = |A_{mn}|^2 \frac{\beta_z}{2\omega\mu\varepsilon^2}\left[\beta_y^2\!\left(\frac{a}{\varepsilon_{0m}}\right)\!\left(\frac{b}{\varepsilon_{0n}}\right) + \beta_x^2\!\left(\frac{a}{\varepsilon_{0m}}\right)\!\left(\frac{b}{\varepsilon_{0n}}\right)\right]$$

$$= |A_{mn}|^2 \frac{\beta_z}{2\omega\mu\varepsilon^2}\!\left(\frac{a}{\varepsilon_{0m}}\right)\!\left(\frac{b}{\varepsilon_{0n}}\right)(\beta_x^2+\beta_y^2)$$

$$P_{mn}^{\text{TE}_z} = |A_{mn}|^2 \frac{\beta_z \beta_c^2}{2\omega\mu\varepsilon^2}\!\left(\frac{a}{\varepsilon_{0m}}\right)\!\left(\frac{b}{\varepsilon_{0n}}\right) = |A_{mn}|^2 \frac{\beta_c^2}{2\eta\varepsilon^2}\!\left(\frac{a}{\varepsilon_{0m}}\right)\!\left(\frac{b}{\varepsilon_{0n}}\right)\sqrt{1-\left(\frac{f_{c,mn}}{f}\right)^2}$$

$$(8\text{-}57)$$

where

$$\varepsilon_{0q} = \begin{cases} 1 & q = 0 \\ 2 & q \neq 0 \end{cases} \qquad (8\text{-}57a)$$

TM$^z_{mn}$ Modes

Use of a similar procedure with (8-28a) through (8-28f), allows us to write that for the TM$^z_{mn}$ modes

$$(\mathbf{S}_z)_{mn} = \hat{a}_z S_z = \hat{a}_z |B_{mn}|^2 \frac{\beta_z}{2\omega\varepsilon\mu^2} \left[\beta_x^2 \cos^2(\beta_x x) \sin^2(\beta_y y) \right.$$
$$\left. + \beta_y^2 \sin^2(\beta_x x) \cos^2(\beta_y y) \right] \qquad (8\text{-}58)$$

$$P_{mn}^{TM^z} = |B_{mn}|^2 \frac{\beta_c^2 \eta}{2\mu^2} \left(\frac{a}{2}\right)\left(\frac{b}{2}\right) \sqrt{1 - \left(\frac{f_{c,mn}}{f}\right)^2} \qquad (8\text{-}59)$$

By the use of superposition, the total power associated with a wave is equal to the sum of all the power components associated with each mode that exists inside the waveguide. Thus

$$P_{\text{total}} = \sum_{m,n} P_{mn}^{TE^z} + \sum_{m,n} P_{mn}^{TM^z} \qquad (8\text{-}60)$$

where $P_{mn}^{TE^z}$ and $P_{mn}^{TM^z}$ are given, respectively, by (8-57) and (8-60).

Example 8-5. The inside dimensions of an X-band WR90 waveguide are $a = 0.9$ in. (2.286 cm) and $b = 0.4$ in (1.016 cm). Assume that the waveguide is air-filled and operates in the dominant TE$_{10}$ mode, and that the air will break down when the maximum electric field intensity is 3×10^6 V/m. Find the maximum power that can be transmitted at $f = 9$ GHz in the waveguide before air breakdown occurs.

Solution. Since air will break down when the maximum electric field intensity in the waveguide reaches 3×10^6 V/m, then according to (8-39b)

$$|E_y|_{\max} = \frac{|A_{10}|}{\varepsilon_0} \frac{\pi}{a} \left| \sin\left(\frac{\pi}{a} x\right) \right|_{\max} = \frac{A_{10}}{\varepsilon_0} \frac{\pi}{a} = 3 \times 10^6 \Rightarrow A_{10} = 1.933 \times 10^{-7}$$

By using (8-57),

$$P_{10}^{TE} = |A_{10}|^2 \frac{(\beta_c)_{10}^2}{2\eta_0 \varepsilon_0^2} \left(\frac{a}{\varepsilon_{01}}\right)\left(\frac{b}{\varepsilon_{00}}\right) \sqrt{1 - \left[\frac{(f_c)_{10}}{f}\right]^2}$$

Since the cutoff frequency of the dominant TE$_{10}$ mode is

$$(f_c)_{10} = \frac{1}{2a\sqrt{\mu_0 \varepsilon_0}} = \frac{30 \times 10^9}{2(2.286)} = 6.562 \text{ GHz}$$

and

$$(\beta_c)_{10} = \frac{\pi}{a}$$

then

$$P_{10}^{TE} = (1.933 \times 10^{-7})^2 \frac{(\pi/2.286 \times 10^{-2})^2}{2(377)(8.854 \times 10^{-12})^2} \left(\frac{2.286 \times 10^{-2}}{1}\right)$$

$$\times \left(\frac{1.016 \times 10^{-2}}{2}\right) \sqrt{1 - \left(\frac{6.562}{9}\right)^2}$$

$$P_{10}^{TE} = 948.9 \times 10^3 \text{ W} = 948.9 \text{ kW}$$

8.2.5 Attenuation

Ideally, if the waveguide were made out of a perfect conductor, there would not be any attenuation associated with the guide above cutoff. Below cutoff, the fields reduce to evanescent (nonpropagating) waves that are highly attenuated. In practice, however, no perfect conductors exist, although many (such as metals) are very good conductors, with conductivities on the order of 10^7–10^8 S/m. For waveguides made out of such conductors (metals) there must be some attenuation due to the *conduction (ohmic) losses* in the waveguides themselves. This is accounted for by introducing an attenuation coefficient α_c. Another factor that contributes to the waveguide attenuation is the losses associated with lossy dielectric materials that are inserted inside the guide. These losses are referred to as *dielectric losses* and are accounted for by introducing an attenuation coefficient α_d.

A. CONDUCTION (OHMIC) LOSSES

To find the losses associated with a waveguide whose walls are not perfectly conducting, a new boundary-value problem must be solved. That problem would be the same as Figure 8-1 but with nonperfectly conducting walls. To solve such a problem exactly is an ambitious and complicated task. Instead an alternate procedure is almost always used whereby the solution is obtained using a perturbational method. With that method, it is assumed that the fields inside the waveguide with lossy walls, but of very high conductivity, are slightly perturbed from those of perfectly conducting walls. The differences between the two sets of fields are so small that the fields are usually assumed to be essentially the same. However, the walls themselves are considered as lossy surfaces represented by a surface impedance Z_s given by (4-42), or

$$Z_s = R_s + jX_s = \sqrt{\frac{j\omega\mu}{\sigma + j\omega\varepsilon}} \stackrel{\sigma \gg \omega\varepsilon}{\simeq} \sqrt{\frac{j\omega\mu}{\sigma}} = \sqrt{\frac{\omega\mu}{2\sigma}}(1+j) \qquad (8\text{-}61)$$

The power P_c absorbed and dissipated as heat by each surface (wall) A_m of the waveguide is obtained using an expression analogous to $I^2R/2$ used in lumped

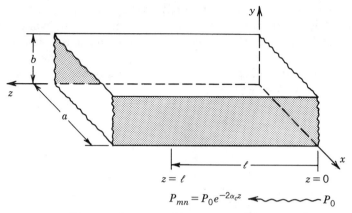

FIGURE 8-9 Rectangular waveguide geometry for attenuation constant derivation.

circuit theory, that is

$$P_c = \frac{R_s}{2} \iint_{A_m} \mathbf{J}_s \cdot \mathbf{J}_s^* \, ds \tag{8-62}$$

where

$$\mathbf{J}_s \simeq \hat{n} \times \mathbf{H}|_{\text{surface}} \tag{8-62a}$$

In (8-62a) $\mathbf{J}_s$ represents the linear current density (A/m) induced on the surface of a lossy conductor as discussed in Example 5-7 and illustrated graphically in Figure 5-9.

Once the total conduction power P_c dissipated as heat on the waveguide has been found by applying (8-62) on all four walls of the guide, the next step is to define and derive an expression for the attenuation coefficient α_c. This can be accomplished by referring to Figure 8-9 which represents the lossy waveguide in its axial direction. If P_0 represents the power at some reference point (i.e., $z = 0$), then the power P_{mn} at some other point z is related to P_0 (P_{mn} at $z = 0$) by

$$P_{mn}(z) = P_{mn}|_{z=0} e^{-2\alpha_c z} = P_0 e^{-2\alpha_c z} \tag{8-63}$$

The negative rate of change (with respect to z) of P_{mn} represents the dissipated power per unit length. Thus for a length ℓ of a waveguide, the total dissipated power P_c is found using

$$P_c = -z \frac{dP_{mn}}{dz}\bigg|_{z=\ell} = -z \frac{d}{dz}\left(P_0 e^{-2\alpha_c z}\right)\bigg|_{z=\ell} = 2\alpha_c z P_0 e^{-2\alpha_c z}\bigg|_{z=\ell}$$

$$P_c = 2\alpha_c \ell P_{mn} \tag{8-64}$$

or

$$\boxed{\alpha_c = \frac{P_c/\ell}{2P_{mn}}} \tag{8-64a}$$

In (8-64a), P_c is obtained using (8-62) and P_{mn} is represented by (8-57) or (8-59).

378 RECTANGULAR CROSS-SECTION WAVEGUIDES AND CAVITIES

The derivation of P_c, as given by (8-62), for any mn TE^z or TM^z mode is a straightforward but tedious process. To reduce the complexity, we will illustrate the procedure for the TE$_{10}$ mode. The derivation for the mn mode is assigned as an end of chapter problem.

Use of the geometry of Figure 8-1 allows the total power P_c dissipated on the walls of the waveguide to be written as

$$(P_c)_{10} = 2\left[\frac{R_s}{2}\iint_{\substack{\text{bottom wall}\\(y=0)}} \mathbf{J}_{sb} \cdot \mathbf{J}_{sb}^* \, ds + \frac{R_s}{2}\iint_{\substack{\text{left wall}\\(x=0)}} \mathbf{J}_{s\ell} \cdot \mathbf{J}_{s\ell}^* \, ds\right]$$

$$(P_c)_{10} = R_s\left[\iint_{\substack{\text{bottom wall}\\(y=0)}} \mathbf{J}_{sb} \cdot \mathbf{J}_{sb}^* \, ds + \iint_{\substack{\text{left wall}\\(x=0)}} \mathbf{J}_{s\ell} \cdot \mathbf{J}_{s\ell}^* \, ds\right] \quad (8\text{-}65)$$

Since the losses on the top wall are the same as those on the bottom and those on the right wall are the same as those on the left, a factor of 2 was used in (8-65) to multiply the losses of the bottom and left walls.

In (8-65) $\mathbf{J}_{sb}$ and $\mathbf{J}_{s\ell}$ represent the linear current densities on the bottom and left walls of the guide which are equal to

$$\mathbf{J}_{sb} = \hat{n} \times \mathbf{H}\big|_{y=0} = \hat{a}_y \times (\hat{a}_x H_x + \hat{a}_z H_z)\big|_{y=0} = (\hat{a}_x H_z - \hat{a}_z H_x)\big|_{y=0}$$

$$= -\hat{a}_x j\frac{A_{10}}{\omega\mu\varepsilon}\left(\frac{\pi}{a}\right)^2 \cos\left(\frac{\pi}{a}x\right)e^{-j\beta_z z} - \hat{a}_z A_{10}\frac{\beta_z}{\omega\mu\varepsilon}\left(\frac{\pi}{a}\right)\sin\left(\frac{\pi}{a}x\right)e^{-j\beta_z z} \quad (8\text{-}66\text{a})$$

$$\mathbf{J}_{s\ell} = \hat{n} \times \mathbf{H}\big|_{x=0} = \hat{a}_x \times (\hat{a}_x H_x + \hat{a}_z H_z)\big|_{x=0} = -\hat{a}_y H_z\big|_{x=0}$$

$$= \hat{a}_y j\frac{A_{10}}{\omega\mu\varepsilon}\left(\frac{\pi}{a}\right)^2 e^{-j\beta_z z} \quad (8\text{-}66\text{b})$$

Use of (8-66a) and (8-66b) allows us to write the losses associated with the bottom–top and left–right walls as given by (8-65) as

$$R_s\iint_{\substack{\text{bottom wall}\\(y=0)}} \mathbf{J}_{sb} \cdot \mathbf{J}_{sb}^* \, ds = R_s|A_{10}|^2\left(\frac{\pi}{a}\right)^2 \frac{1}{(\omega\mu\varepsilon)^2}\left\{\left(\frac{\pi}{a}\right)^2 \int_0^\ell \int_0^a \cos^2\left(\frac{\pi}{a}x\right) dx\, dz\right.$$

$$\left. + \beta_z^2 \int_0^\ell \int_0^a \sin^2\left(\frac{\pi}{a}x\right) dx\, dz\right\}$$

$$= \ell R_s \frac{|A_{10}|^2}{(\omega\mu\varepsilon)^2}\frac{a}{2}\left[\left(\frac{\pi}{a}\right)^2 + \beta_z^2\right]\left(\frac{\pi}{a}\right)^2 \quad (8\text{-}67\text{a})$$

$$R_s\iint_{\substack{\text{left wall}\\(x=0)}} \mathbf{J}_{s\ell} \cdot \mathbf{J}_{s\ell}^* \, ds = R_s|A_{10}|^2 \frac{b\ell}{(\omega\mu\varepsilon)^2}\left(\frac{\pi}{a}\right)^4 \quad (8\text{-}67\text{b})$$

The total dissipated power per unit length can be obtained by combining (8-67a) and (8-67b). Thus it can be shown that

$$\frac{(P_c)_{10}}{\ell} = \frac{aR_s}{2\eta^2}\frac{|A_{10}|^2}{\varepsilon^2}\left(\frac{\pi}{a}\right)^2\left[1 + \frac{2b}{a}\left(\frac{f_c}{f}\right)^2\right] \quad (8\text{-}68)$$

By using (8-57) for $m = 1$ and $n = 0$, and (8-68), the attenuation coefficient of (8-64a) for the TE_{10} mode can be written as

$$(\alpha_c)_{10} = \left[\frac{(P_c)_{10}/\ell}{2P_{mn}}\right]_{\substack{m=1\\n=0}} = \frac{\dfrac{aR_s}{2\eta^2}\dfrac{|A_{10}|^2}{\varepsilon^2}\left(\dfrac{\pi}{a}\right)^2\left[1 + \dfrac{2b}{a}\left(\dfrac{f_c}{f}\right)^2\right]}{|A_{10}|^2\dfrac{\beta_c^2}{\eta\varepsilon^2}\left(\dfrac{a}{2}\right)(b)\sqrt{1-\left(\dfrac{f_c}{f}\right)^2}} \quad (8\text{-}69)$$

which reduces to

$$\boxed{(\alpha_c)_{10} = \frac{R_s}{\eta b}\frac{\left[1 + \dfrac{2b}{a}\left(\dfrac{f_c}{f}\right)^2\right]}{\sqrt{1-\left(\dfrac{f_c}{f}\right)^2}} \quad (\text{Np/m})} \quad (8\text{-}69a)$$

For an X-band WR waveguide with inner dimensions $a = 0.9$ in. (2.286 cm) and $b = 0.4$ in. (1.016 cm), made of copper ($\sigma = 5.7 \times 10^7$ S/m) and filled with a lossless dielectric, we can plot the attenuation coefficient $(\alpha_c)_{10}$ (in Np/m and dB/m) for $\varepsilon = \varepsilon_0$, $2.56\varepsilon_0$, and $4\varepsilon_0$ as shown in Figure 8-10. The attenuation

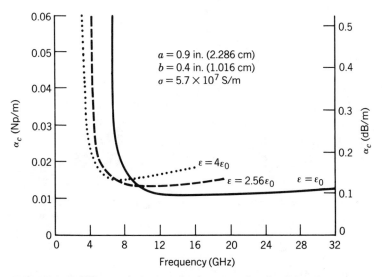

FIGURE 8-10 TE_{10} mode attenuation constant for the X-band rectangular waveguide.

TABLE 8-3
Summary of TE_{mn}^z and TM_{mn}^z mode characteristics of rectangular waveguide

	$\text{TE}_{mn}^z \left(\begin{array}{l} m = 0,1,2,\ldots \\ n = 0,1,2,\ldots \end{array} \; m = n \neq 0 \right)$	$\text{TM}_{mn}^z \left(\begin{array}{l} m = 1,2,3,\ldots \\ n = 1,2,3,\ldots \end{array} \right)$
E_x^+	$A_{mn} \dfrac{n\pi}{b\varepsilon} \cos\left(\dfrac{m\pi}{a}x\right)\sin\left(\dfrac{n\pi}{b}y\right)e^{-j\beta_z z}$	$-B_{mn}\dfrac{m\pi\beta_z}{a\omega\mu\varepsilon}\cos\left(\dfrac{m\pi}{a}x\right)\sin\left(\dfrac{n\pi}{b}y\right)e^{-j\beta_z z}$
E_y^+	$-A_{mn}\dfrac{m\pi}{a\varepsilon}\sin\left(\dfrac{m\pi}{a}x\right)\cos\left(\dfrac{n\pi}{b}y\right)e^{-j\beta_z z}$	$-B_{mn}\dfrac{n\pi\beta_z}{b\omega\mu\varepsilon}\sin\left(\dfrac{m\pi}{a}x\right)\cos\left(\dfrac{n\pi}{b}y\right)e^{-j\beta_z z}$
E_z^+	0	$-jB_{mn}\dfrac{\beta_c^2}{\omega\mu\varepsilon}\sin\left(\dfrac{m\pi}{a}x\right)\sin\left(\dfrac{n\pi}{b}y\right)e^{-j\beta_z z}$
H_x^+	$A_{mn}\dfrac{m\pi\beta_z}{a\omega\mu\varepsilon}\sin\left(\dfrac{m\pi}{a}x\right)\cos\left(\dfrac{n\pi}{b}y\right)e^{-j\beta_z z}$	$B_{mn}\dfrac{n\pi}{b\mu}\sin\left(\dfrac{m\pi}{a}x\right)\cos\left(\dfrac{n\pi}{b}y\right)e^{-j\beta_z z}$
H_y^+	$A_{mn}\dfrac{n\pi\beta_z}{b\omega\mu\varepsilon}\cos\left(\dfrac{m\pi}{a}x\right)\sin\left(\dfrac{n\pi}{b}y\right)e^{-j\beta_z z}$	$-B_{mn}\dfrac{m\pi}{a\mu}\cos\left(\dfrac{m\pi}{a}x\right)\sin\left(\dfrac{n\pi}{b}y\right)e^{-j\beta_z z}$
H_z^+	$-jA_{mn}\dfrac{\beta_c^2}{\omega\mu\varepsilon}\cos\left(\dfrac{m\pi}{a}x\right)\cos\left(\dfrac{n\pi}{b}y\right)e^{-j\beta_z z}$	0
β_c	$\sqrt{\beta_x^2+\beta_y^2} = \sqrt{\left(\dfrac{m\pi}{a}\right)^2+\left(\dfrac{n\pi}{b}\right)^2}$	
f_c	$\dfrac{1}{2\pi\sqrt{\mu\varepsilon}}\sqrt{\left(\dfrac{m\pi}{a}\right)^2+\left(\dfrac{n\pi}{b}\right)^2}$	
λ_c	$\dfrac{2\pi}{\sqrt{\left(\dfrac{m\pi}{a}\right)^2+\left(\dfrac{n\pi}{b}\right)^2}}$	
$\beta_z \; (f \geq f_c)$	$\beta\sqrt{1-\left(\dfrac{f_c}{f}\right)^2}$	
$\lambda_g \; (f \geq f_c)$	$\dfrac{\lambda}{\sqrt{1-\left(\dfrac{f_c}{f}\right)^2}} = \dfrac{\lambda}{\sqrt{1-\left(\dfrac{\lambda}{\lambda_c}\right)^2}}$	
$v_p \; (f \geq f_c)$	$\dfrac{v}{\sqrt{1-\left(\dfrac{f_c}{f}\right)^2}} = \dfrac{v}{\sqrt{1-\left(\dfrac{\lambda}{\lambda_c}\right)^2}}$	
$Z_w \; (f \geq f_c)$	$\dfrac{\eta}{\sqrt{1-\left(\dfrac{f_c}{f}\right)^2}} = \dfrac{\eta}{\sqrt{1-\left(\dfrac{\lambda}{\lambda_c}\right)^2}}$	$\eta\sqrt{1-\left(\dfrac{f_c}{f}\right)^2} = \eta\sqrt{1-\left(\dfrac{\lambda}{\lambda_c}\right)^2}$
$Z_w \; (f \leq f_c)$	$j\dfrac{\eta}{\sqrt{1-\left(\dfrac{f_c}{f}\right)^2}} = j\dfrac{\eta}{\sqrt{1-\left(\dfrac{\lambda}{\lambda_c}\right)^2}}$	$-j\eta\sqrt{1-\left(\dfrac{f_c}{f}\right)^2} = -j\eta\sqrt{1-\left(\dfrac{\lambda}{\lambda_c}\right)^2}$

TABLE 8-3 (*Continued*)

	TE^z_{mn} ($\begin{array}{l} m = 0,1,2,\ldots \\ n = 0,1,2,\ldots \end{array}$ $m = n \neq 0$)	TM^z_{mn} ($\begin{array}{l} m = 1,2,3,\ldots \\ n = 1,2,3,\ldots \end{array}$)
$(\alpha_c)_{mn}$	$\dfrac{2R_s}{b\eta\sqrt{1 - \left(\dfrac{f_{c,mn}}{f}\right)^2}}\left\{\left(\varepsilon_m + \varepsilon_n \dfrac{b}{a}\right)\left(\dfrac{f_{c,mn}}{f}\right)^2 \right.$ $\left. + \dfrac{b}{a}\left[1 - \left(\dfrac{f_{c,mn}}{f}\right)^2\right]\dfrac{m^2ab + (na)^2}{(mb)^2 + (na)^2}\right\}$ where $\varepsilon_p = \begin{cases} 2 & p = 0 \\ 1 & p \neq 0 \end{cases}$	$\dfrac{2R_s}{ab\eta\sqrt{1 - \left(\dfrac{f_{c,mn}}{f}\right)^2}}\dfrac{m^2b^3 + n^2a^3}{(mb)^2 + (na)^2}$

coefficient for any mode (TE_{mn} or TM_{mn}) is given by

$$\underline{\text{TE}^z_{mn}}$$

$$(\alpha_c)_{mn} = \dfrac{2R_s}{\varepsilon_m\varepsilon_n b\eta\sqrt{1 - \left(\dfrac{f_{c,mn}}{f}\right)^2}}\left\{\left(\varepsilon_m + \varepsilon_n \dfrac{b}{a}\right)\left(\dfrac{f_{c,mn}}{f}\right)^2 + \dfrac{b}{a}\left[1 - \left(\dfrac{f_{c,mn}}{f}\right)^2\right]\dfrac{m^2ab + (na)^2}{(mb)^2 + (na)^2}\right\} \quad (8\text{-}70a)$$

where

$$\varepsilon_p = \begin{cases} 2 & p = 0 \\ 1 & p \neq 0 \end{cases} \quad (8\text{-}70b)$$

$$\underline{\text{TM}^z_{mn}}$$

$$(\alpha_c)_{mn} = \dfrac{2R_s}{ab\eta\sqrt{1 - \left(\dfrac{f_{c,mn}}{f}\right)^2}}\dfrac{m^2b^3 + n^2a^3}{(mb)^2 + (na)^2} \quad (8\text{-}70c)$$

The most pertinent equations used to describe the characteristics of TE_{mn} and TM_{mn} modes inside a rectangular waveguide are summarized in Table 8-3.

B. DIELECTRIC LOSSES

When waveguides are filled with lossy dielectric material, an additional attenuation constant must be introduced to account for losses in the dielectric material and are usually designated as *dielectric losses*. Thus the total attenuation constant α_t for the waveguide above cutoff is given by

$$\alpha_t = \alpha_c + \alpha_d \quad (8\text{-}71)$$

382 RECTANGULAR CROSS-SECTION WAVEGUIDES AND CAVITIES

where α_t = total attenuation constant
α_c = ohmic losses attenuation constant [(8-70a) and (8-70c)]
α_d = dielectric losses attenuation constant

To derive α_d, let us refer to the constraint equation of (8-7a), which, for a lossy medium ($\beta = \dot{\beta}_e$), can be written for the complex $\dot{\beta}_z$ as

$$\dot{\beta}_z^2 = \dot{\beta}_e^2 - \left(\beta_x^2 + \beta_y^2\right) = \dot{\beta}_e^2 - \beta_c^2 \tag{8-72}$$

or

$$\dot{\beta}_z = \sqrt{\dot{\beta}_e^2 - \beta_c^2} = \sqrt{\omega^2\mu\dot{\varepsilon}_e - \beta_c^2} = \sqrt{\omega^2\mu(\varepsilon_e' - j\varepsilon_e'') - \beta_c^2}$$

$$= \sqrt{(\omega^2\mu\varepsilon_e' - \beta_c^2) - j\omega^2\mu\varepsilon_e''} = \sqrt{(\beta_e'^2 - \beta_c^2) - j\omega\mu\sigma_e}$$

$$\dot{\beta}_z = \sqrt{\beta_e'^2 - \beta_c^2}\left[1 - j\frac{\omega\mu\sigma_e}{\beta_e'^2 - \beta_c^2}\right]^{1/2} \tag{8-72a}$$

where

$$\beta_e' = \omega\sqrt{\mu\varepsilon_e'} \tag{8-72b}$$

By using the binomial expansion, (8-72a) can be approximated by

$$\dot{\beta}_z \simeq \sqrt{\beta_e'^2 - \beta_c^2}\left[1 - j\frac{\omega\mu\sigma_e}{2(\beta_e'^2 - \beta_c^2)}\right] = \sqrt{\beta_e'^2 - \beta_c^2} - j\frac{\omega\mu\sigma_e}{2\sqrt{\beta_e'^2 - \beta_c^2}}$$

$$\simeq \beta_e'\sqrt{1 - \left(\frac{f_c}{f}\right)^2} - j\frac{\omega\mu\sigma_e}{2\beta_e'\sqrt{1 - \left(\frac{f_c}{f}\right)^2}}$$

$$\simeq \omega\sqrt{\mu\varepsilon_e'}\sqrt{1 - \left(\frac{f_c}{f}\right)^2} - j\frac{\omega\mu\sigma_e}{2\omega\sqrt{\mu\varepsilon_e'}\sqrt{1 - \left(\frac{f_c}{f}\right)^2}}$$

$$\dot{\beta}_z \simeq \omega\sqrt{\mu\varepsilon_e'}\sqrt{1 - \left(\frac{f_c}{f}\right)^2} - j\frac{\eta_e'}{2}\frac{\sigma_e}{\sqrt{1 - \left(\frac{f_c}{f}\right)^2}} \tag{8-73}$$

where

$$\eta_e' = \sqrt{\frac{\mu}{\varepsilon_e'}} \tag{8-73a}$$

Let us define the complex $\dot{\beta}_z$ as

$$\dot{\beta}_z \equiv \beta_d - j\alpha_d \simeq \beta_e'\sqrt{1 - \left(\frac{f_c}{f}\right)^2} - j\frac{\eta_e'}{2}\frac{\sigma_e}{\sqrt{1 - \left(\frac{f_c}{f}\right)^2}} \qquad (8\text{-}74)$$

or

$$\alpha_d \equiv \text{attenuation constant} \simeq \frac{\eta_e'}{2}\frac{\sigma_e}{\sqrt{1 - \left(\frac{f_c}{f}\right)^2}} \qquad (8\text{-}74a)$$

$$\beta_d \equiv \text{phase constant} \simeq \beta_e'\sqrt{1 - \left(\frac{f_c}{f}\right)^2} \qquad (8\text{-}74b)$$

Another form of (8-74a) would be to write it as

$$\alpha_d \simeq \frac{\eta_e'}{2}\frac{\sigma_e}{\sqrt{1 - \left(\frac{f_c}{f}\right)^2}} = \frac{1}{2}\sqrt{\frac{\mu}{\varepsilon_e'}}\frac{\omega\varepsilon_e''}{\sqrt{1 - \left(\frac{f_c}{f}\right)^2}}$$

$$\simeq \frac{1}{2}\frac{\varepsilon_e''}{\varepsilon_e'}\frac{\omega\sqrt{\mu\varepsilon_e'}}{\sqrt{1 - \left(\frac{f_c}{f}\right)^2}} = \frac{1}{2}\frac{\varepsilon_e''}{\varepsilon_e'}\frac{\beta_e'}{\sqrt{1 - \left(\frac{f_c}{f}\right)^2}}$$

$$\simeq \frac{1}{2}\frac{\varepsilon_e''}{\varepsilon_e'}\frac{2\pi}{\lambda\sqrt{1 - \left(\frac{f_c}{f}\right)^2}} = \frac{\pi}{\lambda}\frac{\varepsilon_e''}{\varepsilon_e'}\frac{1}{\sqrt{1 - \left(\frac{f_c}{f}\right)^2}}$$

$$\alpha_d \simeq \frac{\varepsilon_e''}{\varepsilon_e'}\frac{\pi}{\lambda^2}\frac{\lambda}{\sqrt{1 - \left(\frac{f_c}{f}\right)^2}} = \frac{\varepsilon_e''}{\varepsilon_e'}\frac{\pi}{\lambda}\left(\frac{\lambda_g}{\lambda}\right) \text{ Np/m} \quad (\lambda \text{ in meters}) \qquad (8\text{-}75)$$

where λ is the wavelength inside an unbounded infinite lossy dielectric medium and λ_g is the guide wavelength filled with the lossy dielectric material. In decibels, α_d of (8-75) can be written as

$$\alpha_d \simeq 8.68\left(\frac{\varepsilon_e''}{\varepsilon_e'}\right)\frac{\pi}{\lambda}\left(\frac{\lambda_g}{\lambda}\right) = \frac{27.27}{\lambda}\left(\frac{\varepsilon_e''}{\varepsilon_e'}\right)\left(\frac{\lambda_g}{\lambda}\right) \text{ dB/m} \quad (\lambda \text{ in meters}) \qquad (8\text{-}75a)$$

Example 8-6. An X-band (8.2–12.4 GHz) rectangular waveguide is filled with polystyrene whose electrical properties at 6 GHz are $\varepsilon_r = 2.56$ and $\tan \delta_e = 2.55 \times 10^{-4}$. Determine the dielectric attenuation at a frequency of $f = 6$ GHz when the inside dimensions of the waveguide are $a = 0.9$ in. (2.286 cm) and $b = 0.4$ in. (1.016 cm). Assume TE_{10} mode propagation.

Solution. According to (2-68a)

$$\tan \delta_e = \frac{\varepsilon_e''}{\varepsilon_e'} = 2.55 \times 10^{-4}$$

At $f = 6$ GHz,

$$\lambda_0 = \frac{30 \times 10^9}{6 \times 10^9} = 5 \text{ cm} \qquad \lambda = \frac{\lambda_0}{\sqrt{\varepsilon_r}} = \frac{5}{\sqrt{2.56}} = 3.125 \text{ cm}$$

$$(f_c)_{10} = \frac{1}{2a\sqrt{\mu\varepsilon}} = \frac{30 \times 10^9}{2(2.286)(1.6)} = 4.10 \text{ GHz}$$

$$\lambda_g = \frac{\lambda}{\sqrt{1 - \left(\frac{f_c}{f}\right)^2}} = \frac{3.125}{\sqrt{1 - \left(\frac{4.10}{6}\right)^2}} = 4.280 \text{ cm}$$

Thus the dielectric attenuation of (8-75) is equal to

$$\alpha_d = \frac{\varepsilon_e''}{\varepsilon_e'} \frac{\pi}{\lambda} \left(\frac{\lambda_g}{\lambda}\right) = 2.55 \times 10^{-4} \left(\frac{\pi}{3.125}\right)\left(\frac{4.280}{3.125}\right) = 3.511 \times 10^{-4} \text{ Np/cm}$$

$$= 3.511 \times 10^{-2} \text{ Np/m} = 30.476 \times 10^{-2} \text{ dB/m}$$

C. COUPLING

Whenever a given mode is to be excited or detected, the excitation or detection scheme must be such that it maximizes the energy exchange or transfer between the source and the guide or the guide and the receiver. Typically there are a number of techniques that can be used to accomplish this. Some suggested methods that are popular in practice are the following.

1. If the energy exchange is from one waveguide to another, use an iris or hole placed in a location and orientation so that the field distribution of both guides over the extent of the hole or iris are almost identical.
2. If the energy exchange is from a transmission line, such as a coaxial line, to a waveguide, or vice versa, use a linear probe or antenna oriented so that its length is parallel to the electric field lines in the waveguide and placed near the maximum of the electric field mode pattern, as shown in Figure 8-11(a). This is usually referred to as *electric field coupling*. Sometimes the position is varied slightly to achieve better impedance matching.

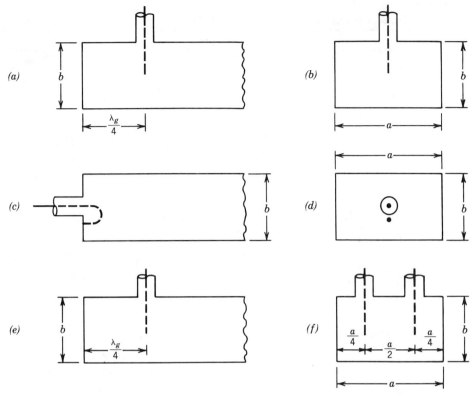

FIGURE 8-11 Coaxial transmission line to rectangular waveguide coupling. (*a*) The side and (*b*) the end views of the coax to waveguide electric field coupling for the TE_{10} mode. (*c*) The side and (*d*) the end view for the coax to waveguide magnetic field coupling for the TE_{10} modes. (*e*) The side and (*f*) the end view for the coax to waveguide electric field coupling for the TE_{20} mode.

3. If the energy transfer is from a transmission line, such as a coaxial line, to a waveguide, or vice versa, use a loop antenna oriented so that the plane of the loop is perpendicular to the magnetic field lines, as shown in Figure 8.11(c). This is usually referred to as *magnetic field coupling*.
4. If the energy transfer is from a transmission line, such as a two conductor line, or other sources to a waveguide, or vice versa, use the transmission line or other sources so that they excite currents on the waveguide that match those of the desired modes in the guide.
5. A number of probes, antennas, or transmission lines properly phased can also be used to excite or detect any mode, especially higher-order modes. Shown in Figure 8-11 are some typical arrangements for coupling energy from a transmission line, such as a coax, to a rectangular waveguide to excite or detect the TE_{10} and TE_{20} modes.

The sizes, flanges, frequency bands, and other parameters pertaining to rectangular waveguides have been standardized so that uniformity is maintained throughout the industry. Standardized reference data on rectangular waveguides are displayed in Table 8-4.

TABLE 8-4
Reference table of rigid rectangular waveguide data and fittings

				Waveguide					Recommended operating range for TE$_{10}$ mode	
				Dimensions (in.)						
EIA designation WR ( )	MDL designation ( ) band	JAN designation RG ( )/U	Material alloy	Inside	Tol.	Outside	Tol.	Wall thickness nominal	Frequency (GHz)	Wavelength (cm)
2300	2300		Alum.	23.000–11.500	±0.020	23.250–11.750	±0.020	0.125	0.32–0.49	93.68–61.18
2100	2100		Alum.	21.000–10.500	±0.020	21.250–10.750	±0.020	0.125	0.35–0.53	85.65–56.56
1800	1800	201	Alum.	18.000–9.000	±0.020	18.250–9.250	±0.020	0.125	0.41–0.625	73.11–47.96
1500	1500	202	Alum.	15.000–7.500	±0.015	15.250–7.750	±0.015	0.125	0.49–0.75	61.18–39.97
1150	1150	203	Alum.	11.500–5.750	±0.015	11.750–6.000	±0.015	0.125	0.64–0.96	46.84–31.23
975	975	204	Alum.	9.750–4.875	±0.010	10.000–5.125	±0.010	0.125	0.75–1.12	39.95–26.76
770	770	205	Alum.	7.700–3.850	±0.005	7.950–4.100	±0.005	0.125	0.96–1.45	31.23–20.67
650	L	69 103	Copper Alum.	6.500–3.250	±0.005	6.660–3.410	±0.005	0.080	1.12–1.70	26.76–17.63
510	510			5.100–2.550	±0.005	5.260–2.710	±0.005	0.080	1.45–2.20	20.67–13.62
430	W	104 105	Copper Alum.	4.300–2.150	±0.005	4.460–2.310	±0.005	0.080	1.70–2.60	17.63–11.53
340	340	112 113	Copper Alum.	3.400–1.700	±0.005	3.560–1.860	±0.005	0.080	2.20–3.30	13.63–9.08
284	S	48 75	Copper Alum.	2.840–1.340	±0.005	3.000–1.500	±0.005	0.080	2.60–3.95	11.53–7.59
229	229			2.290–1.145	±0.005	2.418–1.273	±0.005	0.064	3.30–4.90	9.08–6.12
187	C	49 95	Copper Alum.	1.872–0.872	±0.005	2.000–1.000	±0.005	0.064	3.95–5.85	7.59–5.12
159	159			1.590–0.795	±0.004	1.718–0.923	±0.004	0.064	4.90–7.05	6.12–4.25
137	X_B	50 106	Copper Alum.	1.372–0.622	±0.004	1.500–0.750	±0.004	0.064	5.85–8.20	5.12–3.66
112	X_L	51 68	Copper Alum.	1.122–0.497	±0.004	1.250–0.625	±0.004	0.064	7.05–10.00	4.25–2.99
90	X	52 67	Copper Alum.	0.900–0.400	±0.003	1.000–0.500	±0.003	0.050	8.20–12.40	3.66–2.42
75	75			0.750–0.375	±0.003	0.850–0.475	±0.003	0.050	10.00–15.00	2.99–2.00
62	K_U	91 107	Copper Alum. Silver	0.622–0.311	±0.0025	0.702–0.391	±0.003	0.040	12.4–18.00	2.42–1.66
51	51			0.510–0.255	±0.0025	0.590–0.335	±0.003	0.040	15.00–22.00	2.00–1.36
42	K	53 121 66	Copper Alum. Silver	0.420–0.170	±0.0020	0.500–0.250	±0.003	0.040	18.00–26.50	1.66–1.13
34	34			0.340–0.170	±0.0020	0.420–0.250	±0.003	0.040	22.00–33.00	1.36–0.91
28	K_A	96	Copper Alum. Silver	0.280–0.140	±0.0015	0.360–0.220	±0.002	0.040	26.50–40.00	1.13–0.75
22	Q	97	Copper Silver	0.224–0.112	±0.0010	0.304–0.192	±0.002	0.040	33.00–50.00	0.91–0.60
19	19			0.188–0.094	±0.0010	0.268–0.174	±0.002	0.040	40.00–60.00	0.75–0.50
15	V	98	Copper Silver	0.148–0.074	±0.0010	0.228–0.154	±0.002	0.040	50.00–75.00	0.60–0.40
12	12	99	Copper Silver	0.122–0.061	±0.0005	0.202–0.141	±0.002	0.040	60.00–90.00	0.50–0.33
10	10			0.100–0.050	±0.0005	0.180–0.130	±0.002	0.040	75.00–110.00	0.40–0.27

Source: Microwave Development Laboratories, Inc.
[a] This is an MDL Range Number.

RECTANGULAR WAVEGUIDE

Waveguide						Fittings			
Cutoff for TE$_{10}$ mode		Range in $\frac{2\lambda}{\lambda_c}$	Range in $\frac{\lambda_g}{\lambda}$	Theoretical attenuation lowest to highest frequency (dB / 100 ft)	Theoretical C / W power rating lowest to highest frequency (MW)	Flange		EIA designation WR ( )
Frequency (GHz)	Wavelength (cm)					Choke UG( ) / U	Cover UG( ) / U	
0.256	116.84	1.60–1.05	1.68–1.17	0.051–0.031	153.0–212.0			2300
0.281	106.68	1.62–1.06	1.68–1.18	0.054–0.034	120.0–173.0		FA168A[a]	2100
0.328	91.44	1.60–1.05	1.67–1.18	0.056–0.038	93.4–131.9			1800
0.393	76.20	1.61–1.05	1.62–1.17	0.069–0.050	67.6–93.3			1500
0.513	58.42	1.60–1.07	1.82–1.18	0.128–0.075	35.0–53.8			1150
0.605	49.53	1.61–1.08	1.70–1.19	0.137–0.095	27.0–38.5			975
0.766	39.12	1.60–1.06	1.66–1.18	0.201–0.136	17.2–24.1			770
0.908	33.02	1.62–1.07	1.70–1.18	0.317–0.212 / 0.269–0.178	11.9–17.2		417A / 418A	650
1.157	25.91	1.60–1.05	1.67–1.18					510
1.372	21.84	1.61–1.06	1.70–1.18	0.588–0.385 / 0.501–0.330	5.2–7.5		435A / 437A	430
1.736	17.27	1.58–1.05	1.78–1.22	0.877–0.572 / 0.751–0.492	3.1–4.5		553 / 554	340
2.078	14.43	1.60–1.05	1.67–1.17	1.102–0.752 / 0.940–0.641	2.2–3.2	54A / 585	53 / 584	284
2.577	11.63	1.56–1.05	1.62–1.17					229
3.152	9.510	1.60–1.08	1.67–1.19	2.08–1.44 / 1.77–1.12	1.4–2.0	148B / 406A	149A / 407	187
3.711	8.078	1.51–1.05	1.52–1.19					159
4.301	6.970	1.47–1.05	1.48–1.17	2.87–2.30 / 2.45–1.94	0.56–0.71	343A / 440A	344 / 441	137
5.259	5.700	1.49–1.05	1.51–1.17	4.12–3.21 / 3.50–2.74	0.35–0.46	52A / 137A	51 / 138	112
6.557	4.572	1.60–1.06	1.68–1.18	6.45–4.48 / 5.49–3.83	0.20–0.29	40A / 136A	39 / 135	90
7.868	3.810	1.57–1.05	1.64–1.17					75
9.486	3.160	1.53–1.05	1.55–1.18	9.51–8.31 / 6.14–5.36	0.12–0.16	541 FA190A[a]	419 FA191A[a]	62
11.574	2.590	1.54–1.05	1.58–1.18					51
14.047	2.134	1.56–1.06	1.60–1.18	20.7–14.8 / 17.6–12.6 / 13.3–9.5	0.043–0.058	596 598	595 597	42
17.328	1.730	1.57–1.05	1.62–1.18					34
21.081	1.422	1.59–1.05	1.65–1.17		0.022–0.031	600 FA1241A[a]	599 FA1242A[a]	28
26.342	1.138	1.60–1.05	1.67–1.17	21.9–15.0	0.014–0.020		383	22
31.357	0.956	1.57–1.05	1.63–1.16	31.0–20.9				19
39.863	0.752	1.60–1.06	1.67–1.17		0.0063–0.0090		385	15
48.350	0.620	1.61–1.06	1.68–1.18	52.9–39.1 / 93.3–52.2	0.0042–0.060		387	12
59.010	0.508	1.57–1.06	1.61–1.18					10

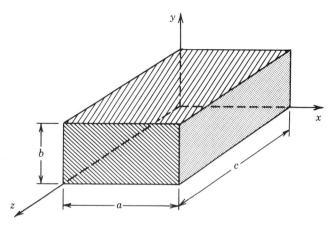

FIGURE 8-12 Geometry for the rectangular cavity.

8.3 RECTANGULAR RESONANT CAVITIES

Waveguide cavities represent a very important class of microwave components. Their applications are numerous and range from use as frequency meters to cavities for measuring the electrical properties of material. The attractive characteristics of waveguide cavities are their very high quality factors Q, typically on the order of 5000–10,000, and their simplicity of construction and use. The most common geometries of cavities are rectangular, cylindrical, and spherical. The rectangular geometry will be discussed in this chapter whereas the cylindrical will be examined in Chapter 9 and the spherical will be analyzed in Chapter 10.

A rectangular waveguide cavity is formed by taking a section of a waveguide and enclosing its front and back faces with conducting plates, as shown in Figure 8-12. Coupling into and out of the cavities is done through coupling probes or holes. The coupling probes may be either electric or magnetic, as shown in Figure 8-11a and c, whereas coupling holes may also be used either on the front, back, top, or bottom walls.

The field configurations inside a rectangular cavity of Figure 8-12 may be either TE^z or TM^z, or any other TE or TM mode, and they are derived in a manner similar to those of the waveguide. The only differences are that we must allow for standing waves, instead of traveling waves, along the length (z axis) of the waveguide, and we must impose additional boundary conditions along the front and back walls. The field forms along the x and y directions and the boundary conditions on the left, right, top, and bottom walls are identical to those of the rectangular waveguide.

8.3.1 Transverse Electric (TE^z) Modes

Since transverse electric (to z) modes for a rectangular cavity must be derived in a manner similar to those of a rectangular waveguide, they must satisfy (8-1) and (8-2). Therefore $F_z(x, y, z)$ must take a form similar to (8-6) except that standing wave functions (sines and cosines) must be used to represent the variations in the z direction. Since the boundary conditions on the bottom, top, left, and right walls are those of (8-8a) to (8-8d), the $F_z(x, y, z)$ function for the rectangular cavity can be

written as

$$F_z(x, y, z) = A_{mn}\cos(\beta_x x)\cos(\beta_y y)[C_3\cos(\beta_z z) + D_3\sin(\beta_z z)] \quad (8\text{-}76)$$

$$\left.\begin{array}{ll} \beta_x = \dfrac{m\pi}{a} & m = 0, 1, 2, \ldots \\[2mm] \beta_y = \dfrac{n\pi}{b} & n = 0, 1, 2, \ldots \end{array}\right\} m = n \neq 0 \quad (8\text{-}76a)$$

which are similar to (8-14a) and (8-14b) except for the standing wave functions representing the z variations. The additional boundary conditions on the front and back walls of the cavity are

$$E_x(0 \leq x \leq a, 0 \leq y \leq b, z = 0) = E_x(0 \leq x \leq a, 0 \leq y \leq b, z = c) = 0$$
$$(8\text{-}77a)$$

$$E_y(0 \leq x \leq a, 0 \leq y \leq b, z = 0) = E_y(0 \leq x \leq a, 0 \leq y \leq b, z = c) = 0$$
$$(8\text{-}77b)$$

The two boundary conditions of (8-77a) and (8-77b) are not independent and either of the two will be sufficient.

By using (8-76) we can write the E_x component according to (8-1) as

$$E_x(x, y, z) = -\frac{1}{\varepsilon}\frac{\partial F_z}{\partial y} = \frac{\beta_y}{\varepsilon} A_{mn}\cos(\beta_x x)\sin(\beta_y y)[C_3\cos(\beta_z z) + D_3\sin(\beta_z z)]$$
$$(8\text{-}78)$$

By applying (8-77a) to (8-78), we can write that

$$E_x(0 \leq x \leq a, 0 \leq y \leq b, z = 0) = \frac{\beta_y}{\varepsilon} A_{mn}\cos(\beta_x x)\sin(\beta_y y)[C_3(1) + D_3(0)] = 0$$
$$\Rightarrow C_3 = 0 \quad (8\text{-}79a)$$

$$E_x(0 \leq x \leq a, 0 \leq y \leq b, z = c) = \frac{\beta_y}{\varepsilon} A_{mn}\cos(\beta_x x)\sin(\beta_y y) D_3 \sin(\beta_z c) = 0$$

$$\Rightarrow \sin(\beta_z c) = 0 \Rightarrow \beta_z c = \sin^{-1}(0) = p\pi$$

$$\beta_z = \frac{p\pi}{c} \quad p = 1, 2, 3 \ldots \quad (8\text{-}79b)$$

Thus (8-76) reduces to

$$F_z(x, y, z) = A_{mn} D_3 \cos(\beta_x x)\cos(\beta_y y)\sin(\beta_z z)$$
$$F_z(x, y, z) = A_{mnp} \cos(\beta_x x)\cos(\beta_y y)\sin(\beta_z z) \quad (8\text{-}80)$$

where

$$\left.\begin{array}{ll} \beta_x = \dfrac{m\pi}{a} & m = 0, 1, 2, \ldots \\[2mm] \beta_y = \dfrac{n\pi}{b} & n = 0, 1, 2, \ldots \\[2mm] \beta_z = \dfrac{p\pi}{c} & p = 1, 2, 3, \ldots \end{array}\right\} m = n \neq 0 \quad (8\text{-}80a)$$

Thus for each mode the dimensions of the cavity in each direction must be an integral number of half wavelengths of the wave in that direction. In addition the electric and magnetic field components of (8-1) can be expressed as

$$E_x = \frac{\beta_y}{\varepsilon} A_{mnp} \cos(\beta_x x) \sin(\beta_y y) \sin(\beta_z z) \tag{8-81a}$$

$$E_y = -\frac{\beta_x}{\varepsilon} A_{mnp} \sin(\beta_x x) \cos(\beta_y y) \sin(\beta_z z) \tag{8-81b}$$

$$E_z = 0 \tag{8-81c}$$

$$H_x = j\frac{\beta_x \beta_z}{\omega\mu\varepsilon} A_{mnp} \sin(\beta_x x) \cos(\beta_y y) \cos(\beta_z z) \tag{8-81d}$$

$$H_y = j\frac{\beta_y \beta_z}{\omega\mu\varepsilon} A_{mnp} \cos(\beta_x x) \sin(\beta_y y) \cos(\beta_z z) \tag{8-81e}$$

$$H_z = -j\frac{A_{mnp}}{\omega\mu\varepsilon}(-\beta_z^2 + \beta^2) \cos(\beta_x x) \cos(\beta_y y) \sin(\beta_z z) \tag{8-81f}$$

By Using (8-80a) we can write (8-7a) as

$$\beta_x^2 + \beta_y^2 + \beta_z^2 = \left(\frac{m\pi}{a}\right)^2 + \left(\frac{n\pi}{b}\right)^2 + \left(\frac{p\pi}{c}\right)^2 = \beta_r^2 = \omega_r^2\mu\varepsilon = (2\pi f_r)^2\mu\varepsilon \tag{8-82}$$

or

$$\boxed{(f_r)_{mnp}^{TE} = \frac{1}{2\pi\sqrt{\mu\varepsilon}}\sqrt{\left(\frac{m\pi}{a}\right)^2 + \left(\frac{n\pi}{b}\right)^2 + \left(\frac{p\pi}{c}\right)^2} \quad \begin{array}{l} m = 0,1,2,\ldots \\ n = 0,1,2,\ldots \\ p = 1,2,3,\ldots \end{array} \bigg\} m = n \neq 0}$$

$$\tag{8-82a}$$

In (8-82a) $(f_r)_{mnp}$ represents the resonant frequency for the TE_{mnp}^z mode. If $c > a > b$, the mode with the lowest-order is the TE_{101}^z mode whose resonant frequency is represented by

$$(f_r)_{101}^{TE} = \frac{1}{2\sqrt{\mu\varepsilon}}\sqrt{\left(\frac{1}{a}\right)^2 + \left(\frac{1}{c}\right)^2} \tag{8-83}$$

In addition to its resonant frequency, one of the most important parameters of a resonant cavity is its quality factor Q defined as

$$Q \equiv \omega \frac{\text{stored energy}}{\text{dissipated power}} = \omega\frac{W_t}{P_d} = \omega\frac{W_e + W_m}{P_d} = \omega\frac{2W_e}{P_d} = \omega\frac{2W_m}{P_d} \tag{8-84}$$

which is proportional to volume and inversely proportional to surface. By using the field expressions of (8-81a) through (8-81f) for the $m = 1$, $n = 0$, and $p = 1$ (101) mode, the total stored energy can be written as

$$W = 2W_e = 2\left[\frac{\varepsilon}{4}\iiint_V |\mathbf{E}|^2 \, dv\right] = \frac{\varepsilon}{2}\left[\frac{|A_{101}|}{\varepsilon}\frac{\pi}{a}\right]^2 \int_0^c \int_0^b \int_0^a \sin^2\left(\frac{\pi}{a}x\right)\sin^2\left(\frac{\pi}{c}z\right) dx\, dy\, dz$$

$$W = \frac{|A_{101}|^2}{\varepsilon}\left(\frac{\pi}{a}\right)^2 \frac{abc}{8} \tag{8-85}$$

The total dissipated power is found by adding the power that is dissipated in each of the six walls of the cylinder. Since the dissipated power on the top wall is the same as that on the bottom, that on the right wall is the same as that on the left, and that on the back is the same as that on the front, we can write the total dissipated power as

$$P_d = \frac{R_s}{2}\left\{2\iint_{\text{bottom}} \mathbf{J}_b \cdot \mathbf{J}_b^* \, ds + 2\iint_{\text{left}} \mathbf{J}_\ell \cdot \mathbf{J}_\ell^* \, ds + 2\iint_{\text{front}} \mathbf{J}_f \cdot \mathbf{J}_f^* \, ds\right\}$$

$$= R_s\left\{\iint_{\text{bottom}} \mathbf{J}_b \cdot \mathbf{J}_b^* \, ds + \iint_{\text{left}} \mathbf{J}_\ell \cdot \mathbf{J}_\ell^* \, ds + \iint_{\text{front}} \mathbf{J}_f \cdot \mathbf{J}_f^* \, ds\right\}$$

$$P_d = P_b + P_\ell + P_f \qquad (8\text{-}86)$$

where

$$P_b = R_s\iint_{\text{bottom}} \mathbf{J}_b \cdot \mathbf{J}_b^* \, ds = R_s\int_0^c\int_0^a |\mathbf{J}_b|^2 \, dx \, dz \qquad (8\text{-}86a)$$

$$P_\ell = R_s\iint_{\text{left}} \mathbf{J}_\ell \cdot \mathbf{J}_\ell^* \, ds = R_s\int_0^c\int_0^b |\mathbf{J}_\ell|^2 \, dy \, dz \qquad (8\text{-}86b)$$

$$P_f = R_s\iint_{\text{front}} \mathbf{J}_f \cdot \mathbf{J}_f^* \, ds = R_s\int_0^b\int_0^a |\mathbf{J}_f|^2 \, dx \, dy \qquad (8\text{-}86c)$$

$$\mathbf{J}_b = \hat{n} \times \mathbf{H}|_{y=0} = -\hat{a}_z j\frac{\pi}{a}\frac{\pi}{c}\frac{A_{101}}{\omega\mu\varepsilon}\sin\left(\frac{\pi}{a}x\right)\cos\left(\frac{\pi}{c}z\right)$$

$$-\hat{a}_x j\left(\frac{\pi}{a}\right)^2\frac{A_{101}}{\omega\mu\varepsilon}\cos\left(\frac{\pi}{a}x\right)\sin\left(\frac{\pi}{c}z\right) \qquad (8\text{-}86d)$$

$$\mathbf{J}_\ell = \hat{n} \times \mathbf{H}|_{x=0} = \hat{a}_y j\frac{A_{101}}{\omega\mu\varepsilon}\left(\frac{\pi}{a}\right)^2\sin\left(\frac{\pi}{c}z\right) \qquad (8\text{-}86e)$$

$$\mathbf{J}_f = \hat{n} \times \mathbf{H}|_{z=c} = \hat{a}_y j\frac{\pi}{a}\frac{\pi}{c}\frac{A_{101}}{\omega\mu\varepsilon}\sin\left(\frac{\pi}{a}x\right) \qquad (8\text{-}86f)$$

Application of the fields of (8-81a) through (8-81f) for $m = 1$, $n = 0$, and $p = 1$ in (8-86) through (8-86f) leads to

$$P_b = R_s\left[\frac{\pi^2}{ac}\frac{A_{101}}{\omega\mu\varepsilon}\right]^2\frac{c}{2}\frac{a}{2} + R_s\left[\left(\frac{\pi}{a}\right)^2\frac{A_{101}}{\omega\mu\varepsilon}\right]^2\frac{c}{2}\frac{a}{2} \qquad (8\text{-}87a)$$

$$P_\ell = R_s\left[\left(\frac{\pi}{a}\right)^2\frac{A_{101}}{\omega\mu\varepsilon}\right]^2 c\frac{b}{2} \qquad (8\text{-}87b)$$

$$P_f = R_s\left[\frac{\pi^2}{ac}\frac{A_{101}}{\omega\mu\varepsilon}\right]^2 b\frac{a}{2} \qquad (8\text{-}87c)$$

$$P_d = \frac{R_s}{4}\frac{|A_{101}|^2}{(\varepsilon\eta)^2}\left(\frac{\pi}{a}\right)^2\frac{1}{a^2+c^2}\left[ac(a^2+c^2) + 2b(a^3+c^3)\right] \qquad (8\text{-}87d)$$

Ultimately then the Q of (8-84) can be expressed, using (8-85) and (8-87d), as

$$(Q)_{101}^{\text{TE}} = \frac{\pi\eta}{2R_s}\left[\frac{b(a^2+c^2)^{3/2}}{ac(a^2+c^2) + 2b(a^3+c^3)}\right] \qquad (8\text{-}88)$$

For a square-based ($a = c$) cavity

$$(Q)_{101}^{TE} = \frac{\pi \eta}{2\sqrt{2} R_s} \left[\frac{1}{1 + \frac{a}{2b}} \right] = 1.1107 \frac{\eta}{R_s} \left[\frac{1}{1 + \frac{a/2}{b}} \right] \quad (8\text{-}88a)$$

Example 8-7. A square-based ($a = c$) cavity of rectangular cross section is constructed of an X-band (8.2–12.4 GHz) copper ($\sigma = 5.7 \times 10^7$ S/m) waveguide that has inner dimensions of $a = 0.9$ in. (2.286 cm) and $b = 0.4$ in. (1.016 cm). For the dominant TE_{101} mode, determine the Q of the cavity. Assume a free-space medium inside the cavity.

Solution. According to (8-82a) the resonant frequency of the TE_{101} mode for the square-based ($a = c$) cavity is

$$(f_r)_{101} = \frac{1}{2\pi\sqrt{\mu\varepsilon}} \sqrt{\left(\frac{\pi}{a}\right)^2 + \left(\frac{\pi}{c}\right)^2} = \frac{\sqrt{2}}{2a\sqrt{\mu\varepsilon}}$$

$$= \frac{1}{\sqrt{2}\, a\sqrt{\mu\varepsilon}} = \frac{30 \times 10^9}{\sqrt{2}\,(2.286)} = 9.28 \text{ GHz}$$

Thus the surface resistance R_s of (8-61) is equal to

$$R_s = \sqrt{\frac{\omega_r \mu}{2\sigma}} = \sqrt{\frac{2\pi(9.28 \times 10^9)(4\pi \times 10^{-7})}{2(5.7 \times 10^7)}}$$

$$= 2\pi\sqrt{\frac{92.8}{5.7}} \times 10^{-3} = 0.0254 \text{ ohms}$$

Therefore the Q of (8-88a) reduces to

$$(Q)_{101} = 1.1107 \frac{377}{0.0254} \left[\frac{1}{1 + \frac{2.286}{2(1.016)}} \right] = 7757.9 \simeq 7758$$

8.3.2 Transverse Magnetic (TM^z) Modes

In addition to TE_{mnp}^z modes inside a rectangular cavity, TM_{mnp}^z modes can also be supported by such a structure. These modes can be derived in a manner similar to the TE_{mnp}^z field configurations.

Using the results of Section 8.2.2 we can write the vector potential component $A_z(x, y, z)$ of (8-26a) for the TM_{mnp}^z modes of Figure 8-12 without applying the boundary conditions on the front and back walls as

$$A_z(x, y, z) = B_{mn} \sin(\beta_x x) \sin(\beta_y y)[C_3 \cos(\beta_z z) + D_3 \sin(\beta_z z)] \quad (8\text{-}89)$$

where

$$\beta_x = \frac{m\pi}{a} \quad m = 1, 2, 3, \ldots \quad (8\text{-}89a)$$

$$\beta_y = \frac{n\pi}{b} \quad n = 1, 2, 3, \ldots \quad (8\text{-}89b)$$

The boundary conditions that have not yet been applied on (8-24) are those of (8-77a) or (8-77b). Using (8-89) we can write the E_x component of (8-24) as

$$E_x(x, y, z) = -j\frac{1}{\omega\mu\varepsilon}\frac{\partial^2 A_z}{\partial x \partial z}$$

$$= -j\frac{\beta_x \beta_z}{\omega\mu\varepsilon} B_{mn} \cos(\beta_x x) \sin(\beta_y y)[-C_3 \sin(\beta_z z) + D_3 \cos(\beta_z z)]$$

(8-90)

Applying the boundary conditions of (8-77a) we can write that

$$E_x(0 \le x \le a, 0 \le y \le b, z = 0) = -j\frac{\beta_x \beta_z}{\omega\mu\varepsilon} B_{mn} \cos(\beta_x x) \sin(\beta_y y)$$

$$\times [-C_3(0) + D_3(1)] = 0 \Rightarrow D_3 = 0 \quad (8\text{-}91a)$$

$$E_x(0 \le x \le a, 0 \le y \le b, z = c) = -j\frac{\beta_x \beta_z}{\omega\mu\varepsilon} B_{mn} \cos(\beta_x x) \sin(\beta_y y)$$

$$\times [-C_3 \sin(\beta_z c)] = 0$$

$$\Rightarrow \sin(\beta_z c) = 0 \Rightarrow \beta_z c = \sin^{-1}(0) = p\pi$$

$$\beta_z = \frac{p\pi}{c} \quad p = 0, 1, 2, 3, \ldots \quad (8\text{-}91b)$$

Thus (8-89) reduces to

$$A_z(x, y, z) = B_{mn}C_3 \sin(\beta_x x) \sin(\beta_y y) \cos(\beta_z z)$$

$$A_z(x, y, z) = B_{mnp} \sin(\beta_x x) \sin(\beta_y y) \cos(\beta_z z) \quad (8\text{-}92)$$

where

$$\beta_x = \frac{m\pi}{a} \quad m = 1, 2, 3, \ldots$$

$$\beta_y = \frac{n\pi}{b} \quad n = 1, 2, 3, \ldots \quad (8\text{-}92a)$$

$$\beta_z = \frac{p\pi}{c} \quad p = 0, 1, 2, \ldots$$

Using (8-7a) and (8-92a) the corresponding resonant frequency can be written as

$$\boxed{(f_r)_{mnp}^{TM} = \frac{1}{2\pi\sqrt{\mu\varepsilon}}\sqrt{\left(\frac{m\pi}{a}\right)^2 + \left(\frac{n\pi}{b}\right)^2 + \left(\frac{p\pi}{c}\right)^2}} \quad \begin{matrix} m = 1, 2, 3, \ldots \\ n = 1, 2, 3, \ldots \\ p = 0, 1, 2, \ldots \end{matrix} \quad (8\text{-}93)$$

394   RECTANGULAR CROSS-SECTION WAVEGUIDES AND CAVITIES

TABLE 8-5
Values of R_{101}^{mnp} for rectangular cavity

$\dfrac{a}{b}$	$\dfrac{c}{b}$	TE_{101}	TE_{011}	TM_{110}	TE_{111} TM_{111}	TE_{102}	TE_{201}	TE_{021}	TE_{012}	TM_{210}	TM_{120}	TE_{112} TM_{112}
1	1	1	1	1	1.22	1.58	1.58	1.58	1.58	1.58	1.58	1.73
1	2	1	1	1.26	1.34	1.26	1.84	1.84	1.26	2.00	2.00	1.55
2	2	1	1.58	1.58	1.73	1.58	1.58	2.91	2.00	2.00	2.91	2.12
2.25	2.25	1	1.74	1.74	1.88	1.58	1.58	3.26	2.13	2.13	3.26	2.24
2	4	1	1.84	2.00	2.05	1.26	1.84	3.60	2.00	2.53	3.68	2.19
2.25	4	1	2.02	2.15	2.20	1.31	1.81	3.95	2.19	2.62	4.02	2.36
4	4	1	2.91	2.91	3.00	1.58	1.58	5.71	3.16	3.16	5.71	3.24
4	8	1	3.62	3.65	3.66	1.26	1.84	7.20	3.65	4.03	7.25	3.82

Since the expression for the resonant frequency of the TM_{mnp} modes is the same as for the TE_{mnp}, the order in which the modes occur can be found by forming the ratio of the resonant frequency of any *mnp* mode (TE or TM) to that of the TE_{101}, that is,

$$R_{101}^{mnp} = \frac{(f_r)_{mnp}}{(f_r)_{101}^{TE^z}} = \sqrt{\frac{\left(\dfrac{m}{a}\right)^2 + \left(\dfrac{n}{b}\right)^2 + \left(\dfrac{p}{c}\right)^2}{\left(\dfrac{1}{a}\right)^2 + \left(\dfrac{1}{c}\right)^2}} \qquad (8\text{-}94)$$

whose values for $c \geq a \geq b$ and different ratios of a/b and c/b are found listed in Table 8-5.

8.4   HYBRID (LSE AND LSM) MODES

For some waveguide configurations, such as partially filled waveguides with the material interface perpendicular to the *x* or *y* axis of Figure 8-1, TE^z or TM^z modes cannot satisfy the boundary conditions of the structure. This will be discussed in the next section. Therefore some other mode configurations may exist within such a structure. It will be shown that field configurations that are combinations of TE^z and TM^z modes can be solutions and satisfy the boundary conditions of such a partially filled waveguide [3]. The modes are referred to as *hybrid modes* or *longitudinal section electric* (LSE) or *longitudinal section magnetic* (LSM), or *H* or *E* modes [4].

In the next section it will be shown that for a partially filled waveguide of the form shown in Figure 8-13*a*, the hybrid modes that are solutions and satisfy the boundary conditions are TE^y (LSE^y or H^y) and/or TM^y (LSM^y or E^y). Here the modes are LSE and/or LSM to a direction that is perpendicular to the interface. Similarly for the configuration of Figure 8-13*b*, the appropriate hybrid modes will be TE^x (LSE^x) and TM^x (LSM^x). Before proceeding with the analysis of these waveguide configurations, let us examine TE^y (LSE^y) and TM^y (LSM^y) for the empty waveguide of Figure 8-1.

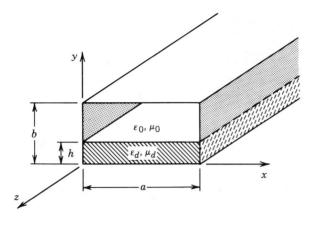

(a)

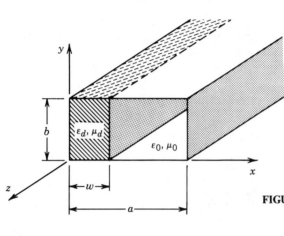

(b)

FIGURE 8-13 Geometry of the dielectric loaded rectangular waveguide. (*a*) Slab along broad wall. (*b*) Slab along narrow wall.

8.4.1 Longitudinal Section Electric (LSE^y) or Transverse Electric (TE^y) or H^y Modes

Just as for other transverse electric modes, TE^y modes are derived using (6-77) or

$$E_x = \frac{1}{\varepsilon}\frac{\partial F_y}{\partial z} \qquad H_x = -j\frac{1}{\omega\mu\varepsilon}\frac{\partial^2 F_y}{\partial x\,\partial y}$$

$$E_y = 0 \qquad H_y = -j\frac{1}{\omega\mu\varepsilon}\left(\frac{\partial^2}{\partial y^2}+\beta^2\right)F_y \qquad (8\text{-}95)$$

$$E_z = -\frac{1}{\varepsilon}\frac{\partial F_y}{\partial x} \qquad H_z = -j\frac{1}{\omega\mu\varepsilon}\frac{\partial^2 F_y}{\partial y\,\partial z}$$

where for the $+z$ traveling wave

$$F_y^+(x,y,z) = [C_1\cos(\beta_x x) + D_1\sin(\beta_x x)]$$
$$\times [C_2\cos(\beta_y y) + D_2\sin(\beta_y y)]A_3 e^{-j\beta_z z} \qquad (8\text{-}95a)$$
$$\beta_x^2 + \beta_y^2 + \beta_z^2 = \beta^2 = \omega^2\mu\varepsilon \qquad (8\text{-}95b)$$

The boundary conditions are those of (8-8a) through (8-8d).

By using (8-95a) we can write the E_z of (8-95) as

$$E_z^+(x, y, z) = -\frac{\beta_x}{\varepsilon}[-C_1 \sin(\beta_x x) + D_1 \cos(\beta_x x)]$$
$$\times [C_2 \cos(\beta_y y) + D_2 \sin(\beta_y y)] A_3 e^{-j\beta_z z} \quad (8\text{-}96)$$

Application of the boundary condition of (8-8b) on (8-96) gives

$$E_z^+(0 \le x \le a, y = 0, z)$$
$$= -\frac{\beta_x}{\varepsilon}[-C_1 \sin(\beta_x x) + D_1 \cos(\beta_x x)][C_2(1) + D_2(0)] A_3 e^{-j\beta_z z} = 0$$
$$\Rightarrow C_2 = 0 \quad (8\text{-}97a)$$

$$E_z^+(0 \le x \le a, y = b, z)$$
$$= -\frac{\beta_x}{\varepsilon}[-C_1 \sin(\beta_x x) + D_1 \cos(\beta_x x)] D_2 \sin(\beta_y b) A_3 e^{-j\beta_z z} = 0$$
$$\Rightarrow \sin(\beta_y b) = 0 \Rightarrow \beta_y b = \sin^{-1}(0) = n\pi$$
$$\beta_y = \frac{n\pi}{b} \quad n = 1, 2, 3, \ldots \quad (8\text{-}97b)$$

By following the same procedure, the boundary condition of (8-8d) leads to

$$E_z^+(x = 0, 0 \le y \le b, z)$$
$$= -\frac{\beta_x}{\varepsilon}[-C_1(0) + D_1(1)] D_2 A_3 \sin(\beta_y y) e^{-j\beta_z z} = 0 \Rightarrow D_1 = 0 \quad (8\text{-}98a)$$

$$E_z^+(x = a, 0 \le y \le b, z) = -\frac{\beta_x}{\varepsilon}[-C_1 \sin(\beta_x a)] D_2 A_3 e^{-j\beta_z z} = 0$$
$$\Rightarrow \sin(\beta_x a) = 0 \Rightarrow \beta_x = \frac{m\pi}{a} \quad m = 0, 1, 2, \ldots \quad (8\text{-}98b)$$

Therefore (8-95a) reduces to

$$F_y^+(x, y, z) = C_1 D_2 A_3 \cos(\beta_x x) \sin(\beta_y y) e^{-j\beta_z z} = A_{mn} \cos(\beta_x x) \sin(\beta_y y) e^{-j\beta_z z} \quad (8\text{-}99)$$

where

$$\beta_x = \frac{m\pi}{a} \quad m = 0, 1, 2, \ldots \quad (8\text{-}99a)$$

$$\beta_y = \frac{n\pi}{b} \quad n = 1, 2, 3, \ldots \quad (8\text{-}99b)$$

By using (8-95b),

$$\beta_z = \pm\sqrt{\beta^2 - (\beta_x^2 + \beta_y^2)} = \pm\sqrt{\beta^2 - \beta_c^2} \quad (8\text{-}100)$$

where

$$\beta_c^2 = \omega_c^2 \mu\varepsilon = (2\pi f_c)^2 \mu\varepsilon = \beta_x^2 + \beta_y^2 = \left(\frac{m\pi}{a}\right)^2 + \left(\frac{n\pi}{b}\right)^2$$

or

$$\boxed{(f_c)_{mn}^{TE^y} = \frac{1}{2\pi\sqrt{\mu\varepsilon}} \sqrt{\left(\frac{m\pi}{a}\right)^2 + \left(\frac{n\pi}{b}\right)^2}} \quad \begin{array}{l} m = 0,1,2,\ldots \\ n = 1,2,3,\ldots \end{array} \quad (8\text{-}100a)$$

The dominant mode is the TE_{01}^y whose cutoff frequency is

$$\boxed{(f_c)_{01}^{TE^y} = \frac{1}{2b\sqrt{\mu\varepsilon}}} \quad (8\text{-}100b)$$

8.4.2 Longitudinal Section Magnetic (LSM^y) or Transverse Magnetic (TM^y) or E^y Modes

By following a procedure similar to that for the TE^y modes of the previous section, it can be shown that for the TM^y modes of Figure 8-1 the field components of (6-64), or

$$\begin{aligned} E_x &= -j\frac{1}{\omega\mu\varepsilon}\frac{\partial^2 A_y}{\partial x\,\partial y} & H_x &= -\frac{1}{\mu}\frac{\partial A_y}{\partial z} \\ E_y &= -j\frac{1}{\omega\mu\varepsilon}\left(\frac{\partial^2}{\partial y^2} + \beta^2\right)A_y & H_y &= 0 \qquad (8\text{-}101) \\ E_z &= -j\frac{1}{\omega\mu\varepsilon}\frac{\partial^2 A_y}{\partial y\,\partial z} & H_z &= \frac{1}{\mu}\frac{\partial A_y}{\partial x} \end{aligned}$$

and the boundary conditions of (8-8a) through (8-8d) lead to

$$A_y(x,y,z) = B_{mn} \sin(\beta_x x)\cos(\beta_y y) e^{-j\beta_z z} \quad (8\text{-}101a)$$

$$\beta_x = \frac{m\pi}{a} \quad m = 1,2,3,\ldots \quad (8\text{-}101b)$$

$$\beta_y = \frac{n\pi}{b} \quad n = 0,1,2,\ldots \quad (8\text{-}101c)$$

$$\boxed{(f_c)_{mn}^{TM^y} = \frac{1}{2\pi\sqrt{\mu\varepsilon}} \sqrt{\left(\frac{m\pi}{a}\right)^2 + \left(\frac{n\pi}{b}\right)^2}} \quad (8\text{-}101d)$$

The dominant mode is the TM_{10}^y whose cutoff frequency is

$$\boxed{(f_c)_{10}^{TM^y} = \frac{1}{2a\sqrt{\mu\varepsilon}}} \quad (8\text{-}101e)$$

398 RECTANGULAR CROSS-SECTION WAVEGUIDES AND CAVITIES

8.5 PARTIALLY FILLED WAVEGUIDE

Let us now consider in detail the analysis of the field configurations in the partially filled waveguide of Figure 8-13a. The analysis of the configuration of Figure 8-13b is left as an end of chapter exercise. It can be shown that for either waveguide configuration neither TE^z nor TM^z modes individually can satisfy the boundary conditions. In fact for the configuration of Figure 8-13a, TE^y (LSE^y) or TM^y (LSM^y) are the appropriate modes whereas TE^x (LSE^x) or TM^x (LSM^x) satisfy the boundary conditions of Figure 8-13b. For either configuration, the appropriate modes are LSE or LSM to a direction that is perpendicular to the material interface.

8.5.1 Longitudinal Section Electric (LSE^y) or Transverse Electric (TE^y)

For the configuration of Figure 8-13a there are two sets of fields: one for the dielectric region ($0 \leq x \leq a, 0 \leq y \leq h, z$), designated by superscript d, and the other for the free-space region ($0 \leq x \leq a, h \leq y \leq b, z$), designated by superscript 0. For each region the TE^y field components are those of (8-95) and the corresponding potential functions are

$$F_y^d(x, 0 \leq y \leq h, z) = \left[C_1^d \cos(\beta_{xd} x) + D_1^d \sin(\beta_{xd} x) \right]$$

$$\times \left[C_2^d \cos(\beta_{yd} y) + D_2^d \sin(\beta_{yd} y) \right] A_3^d e^{-j\beta_z z} \quad (8\text{-}102)$$

$$\beta_{xd}^2 + \beta_{yd}^2 + \beta_z^2 = \beta_d^2 = \omega^2 \mu_d \varepsilon_d \quad (8\text{-}102\text{a})$$

for the dielectric region, and

$$F_y^0(x, h \leq y \leq b, z) = \left[C_1^0 \cos(\beta_{x0} x) + D_1^0 \sin(\beta_{x0} x) \right]$$

$$\times \left\{ C_2^0 \cos[\beta_{y0}(b - y)] + D_2^0 \sin[\beta_{y0}(b - y)] \right\} A_3^0 e^{-j\beta_z z} \quad (8\text{-}103)$$

$$\beta_{x0}^2 + \beta_{y0}^2 + \beta_z^2 = \beta_0^2 = \omega^2 \mu_0 \varepsilon_0 \quad (8\text{-}103\text{a})$$

for the free-space region. In both sets of fields β_z is the same, since for propagation along the interface both sets of fields must be common.

For this waveguide configuration, the appropriate independent boundary conditions are

$$E_z^d(x = 0, 0 \leq y \leq h, z) = E_z^d(x = a, 0 \leq y \leq h, z) = 0 \quad (8\text{-}104\text{a})$$

$$E_z^d(0 \leq x \leq a, y = 0, z) = 0 \quad (8\text{-}104\text{b})$$

$$E_z^d(0 \leq x \leq a, y = h, z) = E_z^0(0 \leq x \leq a, y = h, z) \quad (8\text{-}104\text{c})$$

$$E_z^0(x = 0, h \leq y \leq b, z) = E_z^0(x = a, h \leq y \leq b, z) = 0 \quad (8\text{-}104\text{d})$$

$$E_z^0(0 \leq x \leq a, y = b, z) = 0 \quad (8\text{-}104\text{e})$$

$$H_z^d(0 \leq x \leq a, y = h, z) = H_z^0(0 \leq x \leq a, y = h, z) \quad (8\text{-}104\text{f})$$

Another set of dependent boundary conditions is

$$E_y^d(x = 0, 0 \le y \le h, z) = E_y^d(x = a, 0 \le y \le h, z) = 0 \quad (8\text{-}105a)$$
$$E_x^d(0 \le x \le a, y = 0, z) = 0 \quad (8\text{-}105b)$$
$$E_x^d(0 \le x \le a, y = h, z) = E_x^0(0 \le x \le a, y = h, z) \quad (8\text{-}105c)$$
$$E_y^0(x = 0, h \le y \le b, z) = E_y^0(x = a, h \le y \le b, z) = 0 \quad (8\text{-}105d)$$
$$E_x^0(0 \le x \le a, y = b, z) = 0 \quad (8\text{-}105e)$$
$$H_x^d(0 \le x \le a, y = h, z) = H_x^0(0 \le x \le a, y = h, z) \quad (8\text{-}105f)$$

By using (8-95) and (8-103) we can write that

$$E_z^0 = -\frac{1}{\varepsilon_0}\frac{\partial F_y^0}{\partial x} = -\frac{\beta_{x0}}{\varepsilon_0}\left[-C_1^0 \sin(\beta_{x0}x) + D_1^0 \cos(\beta_{x0}x)\right]$$
$$\times \left\{C_2^0 \cos\left[\beta_{y0}(b-y)\right] + D_2^0 \sin\left[\beta_{y0}(b-y)\right]\right\}A_3^0 e^{-j\beta_z z}$$
$$(8\text{-}106)$$

Application of boundary condition (8-104d) leads to

$$E_z^0(x = 0, h \le y \le b, z)$$
$$= -\frac{\beta_{x0}}{\varepsilon_0}\left[-C_1^0(0) + D_1^0(1)\right]$$
$$\times \left\{C_2^0 \cos\left[\beta_{y0}(b-y)\right] + D_2^0 \sin\left[\beta_{y0}(b-y)\right]\right\}A_3^0 e^{-j\beta_z z} = 0$$
$$\Rightarrow D_1^0 = 0 \quad (8\text{-}106a)$$

$$E_z^0(x = a, h \le y \le b, z)$$
$$= -\frac{\beta_{x0}}{\varepsilon_0}\left[-C_1^0 \sin(\beta_{x0}a)\right]$$
$$\times \left\{C_2^0 \cos\left[\beta_{y0}(b-y)\right] + D_2^0 \sin\left[\beta_{y0}(b-y)\right]\right\}A_3^0 e^{-j\beta_z z} = 0$$
$$\Rightarrow \sin(\beta_{x0}a) = 0 \Rightarrow \beta_{x0} = \frac{m\pi}{a} \quad m = 0, 1, 2, \ldots \quad (8\text{-}106b)$$

Application of (8-104e) leads to

$$E_z^0(0 \le x \le a, y = b, z)$$
$$= -\frac{\beta_{x0}}{\varepsilon_0}\left[-C_1^0 \sin(\beta_{x0}x)\right]\left\{C_2^0(1) + D_2^0(0)\right\}A_3^0 e^{-j\beta_z z} = 0$$
$$\Rightarrow C_2^0 = 0 \quad (8\text{-}106c)$$

Thus (8-103) reduces to

$$F_y^0 = A_{mn}^0 \cos(\beta_{x0}x)\sin\left[\beta_{y0}(b-y)\right]e^{-j\beta_z z} \quad (8\text{-}107)$$

$$\beta_{x0} = \frac{m\pi}{a} \quad m = 0, 1, 2, \ldots \quad (8\text{-}107a)$$

$$\beta_{x0}^2 + \beta_{y0}^2 + \beta_z^2 = \left(\frac{m\pi}{a}\right)^2 + \beta_{y0}^2 + \beta_z^2 = \beta_0^2 = \omega^2 \mu_0 \varepsilon_0 \quad (8\text{-}107b)$$

400 RECTANGULAR CROSS-SECTION WAVEGUIDES AND CAVITIES

with

$$E_z^0 = -\frac{1}{\varepsilon_0}\frac{\partial F_y^0}{\partial x} = \frac{\beta_{x0}}{\varepsilon_0} A_{mn}^0 \sin(\beta_{x0} x) \sin[\beta_{y0}(b-y)] e^{-j\beta_z z} \quad (8\text{-}108)$$

Use of (8-95) and (8-102) gives

$$E_z^d = -\frac{1}{\varepsilon_d}\frac{\partial F_y^d}{\partial x} = -\frac{\beta_{xd}}{\varepsilon_d}\left[-C_1^d \sin(\beta_{xd} x) + D_1^d \cos(\beta_{xd} x)\right]$$

$$\times \left[C_2^d \cos(\beta_{yd} y) + D_2^d \sin(\beta_{yd} y)\right] A_3^d e^{-j\beta_z z} \quad (8\text{-}109)$$

Application of boundary condition (8-104a) leads to

$$E_z^d(x=0, 0 \le y \le h, z)$$
$$= -\frac{\beta_{xd}}{\varepsilon_d}\left[-C_1^d(0) + D_1^d(1)\right]\left[C_2^d \cos(\beta_{yd} y) + D_2^d \sin(\beta_{yd} y)\right] A_3^d e^{-j\beta_z z} = 0$$

$$\Rightarrow D_1^d = 0 \quad (8\text{-}109\text{a})$$

$$E_z^d(x=a, 0 \le y \le h, z)$$
$$= -\frac{\beta_{xd}}{\varepsilon_d}\left[-C_1^d \sin(\beta_{xd} a)\right]\left[C_2^d \cos(\beta_{yd} y) + D_2^d \sin(\beta_{yd} y)\right] A_3^d e^{-j\beta_z z} = 0$$

$$\Rightarrow \sin(\beta_{xd} a) = 0 \Rightarrow \beta_{xd} = \frac{m\pi}{a} \quad m = 0, 1, 2, \ldots \quad (8\text{-}109\text{b})$$

Application of (8-104b) leads to

$$E_z^d(0 \le x \le a, y = 0, z) = -\frac{\beta_{xd}}{\varepsilon_d}\left[-C_1^d \sin(\beta_{xd} x)\right]\left[C_2^d(1) + D_2^d(0)\right] A_3^d e^{-j\beta_z z} = 0$$

$$\Rightarrow C_2^d = 0 \quad (8\text{-}109\text{c})$$

Thus (8-102) reduces to

$$\boxed{F_y^d = A_{mn}^d \cos(\beta_{xd} x) \sin(\beta_{yd} y) e^{-j\beta_z z}} \quad (8\text{-}110)$$

$$\boxed{\beta_{xd} = \frac{m\pi}{a} = \beta_{x0} \quad m = 0, 1, 2, \ldots} \quad (8\text{-}110\text{a})$$

$$\boxed{\beta_{xd}^2 + \beta_{yd}^2 + \beta_z^2 = \left(\frac{m\pi}{a}\right)^2 + \beta_{yd}^2 + \beta_z^2 = \beta_d^2 = \omega^2 \mu_d \varepsilon_d} \quad (8\text{-}110\text{b})$$

with

$$E_z^d = -\frac{1}{\varepsilon_d}\frac{\partial F_y^d}{\partial x} = \frac{\beta_{xd}}{\varepsilon_d} A_{mn}^d \sin(\beta_{xd} x) \sin(\beta_{yd} y) e^{-j\beta_z z}$$

$$= \frac{\beta_{x0}}{\varepsilon_d} A_{mn}^d \sin(\beta_{x0} x) \sin(\beta_{yd} y) e^{-j\beta_z z} \quad (8\text{-}111)$$

Application of boundary condition (8-104c) and use of (8-108) and (8-111) leads to

$$\frac{\beta_{x0}}{\varepsilon_0} A_{mn}^0 \sin(\beta_{x0}x) \sin[\beta_{y0}(b-h)] e^{-j\beta_z z} = \frac{\beta_{x0}}{\varepsilon_d} A_{mn}^d \sin(\beta_{xd}x) \sin(\beta_{yd}h) e^{-j\beta_z z}$$

$$\boxed{\frac{1}{\varepsilon_0} A_{mn}^0 \sin[\beta_{y0}(b-h)] = \frac{1}{\varepsilon_d} A_{mn}^d \sin(\beta_{yd}h)} \quad (8\text{-}112)$$

By using (8-107) and (8-110) the z component of the H field from (8-95) can be written as

$$H_z^0 = -j \frac{1}{\omega\mu_0\varepsilon_0} \frac{\partial^2 F_y^0}{\partial y \partial z} = \frac{\beta_{y0}\beta_z}{\omega\mu_0\varepsilon_0} A_{mn}^0 \cos(\beta_{x0}x) \cos[\beta_{y0}(b-y)] e^{-j\beta_z z} \quad (8\text{-}113a)$$

$$H_z^d = -j \frac{1}{\omega\mu_d\varepsilon_d} \frac{\partial^2 F_y^d}{\partial y \partial z} = -\frac{\beta_{yd}\beta_z}{\omega\mu_d\varepsilon_d} A_{mn}^d \cos(\beta_{xd}x) \cos(\beta_{yd}y) e^{-j\beta_z z} \quad (8\text{-}113b)$$

Application of the boundary condition of (8-104f) reduces with $\beta_{xd} = \beta_{x0}$ to

$$\frac{\beta_{y0}\beta_z}{\omega\mu_0\varepsilon_0} A_{mn}^0 \cos(\beta_{x0}x) \cos[\beta_{y0}(b-h)] e^{-j\beta_z z}$$

$$= -\frac{\beta_{yd}\beta_z}{\omega\mu_d\varepsilon_d} A_{mn}^d \cos(\beta_{xd}x) \cos(\beta_{yd}h) e^{-j\beta_z z}$$

$$\boxed{\frac{\beta_{y0}}{\mu_0\varepsilon_0} A_{mn}^0 \cos[\beta_{y0}(b-h)] = -\frac{\beta_{yd}}{\mu_d\varepsilon_d} A_{mn}^d \cos(\beta_{yd}h)} \quad (8\text{-}114)$$

Division of (8-114) by (8-112) leads to

$$\boxed{\frac{\beta_{y0}}{\mu_0} \cot[\beta_{y0}(b-h)] = -\frac{\beta_{yd}}{\mu_d} \cot(\beta_{yd}h)} \quad (8\text{-}115)$$

$$\boxed{\begin{array}{c} \beta_{x0}^2 + \beta_{y0}^2 + \beta_z^2 = \left(\frac{m\pi}{a}\right)^2 + \beta_{y0}^2 + \beta_z^2 = \beta_0^2 = \omega^2\mu_0\varepsilon_0 \\ m = 0, 1, 2 \ldots \end{array}} \quad (8\text{-}115a)$$

$$\boxed{\begin{array}{c} \beta_{xd}^2 + \beta_{yd}^2 + \beta_z^2 = \left(\frac{m\pi}{a}\right)^2 + \beta_{yd}^2 + \beta_z^2 = \beta_d^2 = \omega^2\mu_d\varepsilon_d \\ m = 0, 1, 2 \ldots \end{array}} \quad (8\text{-}115b)$$

Whereas $\beta_{x0} = \beta_{xd} = m\pi/a$, $m = 0, 1, 2, \ldots$, have been determined, β_{y0}, β_{yd}, and β_z have not yet been found. They can be determined for each mode using (8-115) through (8-115b), and their values vary as a function of frequency. Thus for each frequency a new set of values for β_{y0}, β_{yd}, and β_z must be found that satisfy (8-115) through (8-115b). One procedure that can be used to accomplish this will be to solve (8-115a) for β_{y0} (as a function of β_z and β_0) and (8-115b) for β_{yd} (as a function of β_z and β_d), and then substitute these expressions in (8-115) for β_{y0} and

β_{yd}. The new form of (8-115) will be a function β_z, β_0, and β_d. Thus for a given mode, determined by the value of m, at a given frequency, a particular value of β_z will satisfy the new form of the transcendental equation 8-115; that value of β_z can be found iteratively. The range of β_z will be $\beta_z^0 < \beta_z < \beta_z^d$ where β_z^0 represents the values of the same mode of an air-filled waveguide and β_z^d represents the values of the same mode of a waveguide completely filled with the dielectric. Once β_z has been found at a given frequency for a given mode, the corresponding values of β_{y0} and β_{yd} for the same mode at the same frequency can be determined by using, respectively, (8-115a) and (8-115b). It must be remembered that for each value of m there are infinite values of n ($n = 1, 2, 3, \ldots$). Thus the dominant mode is the one for which $m = 0$, $n = 1$, or TE_{01}^y.

For $m = 0$, the modes will be denoted as TE_{0n}. For these modes (8-115a) and (8-115b) reduce to

$$\beta_{y0}^2 + \beta_z^2 = \omega^2 \mu_0 \varepsilon_0 \Rightarrow \beta_z = \pm\sqrt{\omega^2 \mu_0 \varepsilon_0 - \beta_{y0}^2} \quad (8\text{-}116a)$$

$$\beta_{yd}^2 + \beta_z^2 = \omega^2 \mu_d \varepsilon_d \Rightarrow \beta_z = \pm\sqrt{\omega^2 \mu_d \varepsilon_d - \beta_{yd}^2} \quad (8\text{-}116b)$$

Cutoff occurs when $\beta_z = 0$. Thus at cutoff (8-116a) and (8-116b) reduce to

$$\beta_z = 0 = \pm\sqrt{\omega^2 \mu_0 \varepsilon_0 - \beta_{y0}^2}\bigg|_{\omega = \omega_c} \Rightarrow \omega_c^2 \mu_0 \varepsilon_0 = \beta_{y0}^2 \Rightarrow \beta_{y0} = \omega_c\sqrt{\mu_0 \varepsilon_0} \quad (8\text{-}117a)$$

$$\beta_z = 0 = \pm\sqrt{\omega^2 \mu_d \varepsilon_d - \beta_{yd}^2}\bigg|_{\omega = \omega_c} \Rightarrow \omega_c^2 \mu_d \varepsilon_d = \beta_{yd}^2 \Rightarrow \beta_{yd} = \omega_c\sqrt{\mu_d \varepsilon_d} \quad (8\text{-}117b)$$

which can be used to find β_{y0} and β_{yd} *only at cutoff*, once the cutoff frequency has been determined. By using (8-117a) and (8-117b) we can write (8-115) as

$$\frac{\omega_c\sqrt{\mu_0 \varepsilon_0}}{\mu_0} \cot\left[\omega_c\sqrt{\mu_0 \varepsilon_0}(b - h)\right] = -\frac{\omega_c\sqrt{\mu_d \varepsilon_d}}{\mu_d} \cot\left(\omega_c\sqrt{\mu_d \varepsilon_d}\, h\right)$$

or

$$\boxed{\sqrt{\frac{\varepsilon_0}{\mu_0}} \cot\left[\omega_c\sqrt{\mu_0 \varepsilon_0}(b - h)\right] = -\sqrt{\frac{\varepsilon_d}{\mu_d}} \cot\left(\omega_c\sqrt{\mu_d \varepsilon_d}\, h\right)} \quad (8\text{-}118)$$

which can be used to find the cutoff frequencies of the TE_{0n}^y modes in a partially filled waveguide. A similar expression must be written for the other modes.

For a rectangular waveguide filled completely either with free space or with a dielectric material with ε_d, μ_d the cutoff frequency of the TE_{01}^y mode is given, respectively, according to (8-100b) by

$$(f_c^0)_{01}^{TE^y} = \frac{1}{2b\sqrt{\mu_0 \varepsilon_0}} \quad (8\text{-}119a)$$

$$(f_c^d)_{01}^{TE^y} = \frac{1}{2b\sqrt{\mu_d \varepsilon_d}} \quad (8\text{-}119b)$$

Use of perturbational techniques shows that, in general, the cutoff frequency of the partially filled waveguide (part free space and part dielectric) is greater than the cutoff frequency of the same mode in the same waveguide filled with a dielectric material with ε_d, μ_d and is smaller than the cutoff frequency of the same waveguide filled with free space. Thus the cutoff frequency of the TE_{01}^y mode of a partially filled waveguide (part free space and part dielectric) is greater than (8-119b) and smaller than (8-119a), that is,

$$\frac{1}{2b\sqrt{\mu_d \varepsilon_d}} \leq (f_c)_{01}^{TE^y} \leq \frac{1}{2b\sqrt{\mu_0 \varepsilon_0}} \qquad (8\text{-}120)$$

or

$$\frac{\pi}{b\sqrt{\mu_d \varepsilon_d}} \leq (\omega_c)_{01}^{TE^y} \leq \frac{\pi}{b\sqrt{\mu_0 \varepsilon_0}} \qquad (8\text{-}120a)$$

With this permissible range, the exact values can be found using (8-118). The propagation constant β_z must be solved at each frequency on an individual basis using (8-116a) or (8-116b).

Example 8-8. A WR90 X-band (8.2–12.4 GHz) waveguide of Figure 8-13a with inner dimensions of $a = 0.9$ in. (2.286 cm), $b = 0.4$ in. (1.016 cm), and $a/b = 2.25$, is partially filled with free space and polystyrene ($\varepsilon_d = 2.56\varepsilon_0$, $\mu_d = \mu_0$, and $h = b/3$). For $m = 0$ determine the following.

1. The cutoff frequencies of the hybrid TE_{0n}^y (LSE_{0n}^y) modes for $n = 1, 2, 3$.
2. The corresponding values of β_{y0} and β_{yd} for each mode at the cutoff frequencies.
3. The corresponding values of β_{y0}, β_{yd}, and β_z for the TE_{01}^y mode in the frequency range $(f_c)_{01} \leq f \leq 2(f_c)_{01}$.

TABLE 8-6
Cutoff frequencies and phase constants of partially filled, air-filled, and dielectric-filled rectangular waveguide.[a]

TE_{0n}^y modes		$n = 1$	$n = 2$	$n = 3$
	$(f_c)_{0n}$ (GHz)	12.62	24.03	37.71
Partially filled waveguide	$(\beta_{y0})_{0n}$ at $(f_c)_{0n}$ (rad/m)	264.32	503.38	789.75
	$(\beta_{yd})_{0n}$ at $(f_c)_{0n}$ (rad/m)	422.91	805.41	1263.60
Air-filled waveguide	$(f_c^0)_{0n}$ (GHz)	14.76	29.53	44.29
	$(\beta_y^0)_{0n}$ (rad/m)	309.21	618.42	927.64
Dielectric-filled waveguide	$(f_c^d)_{0n}$ (GHz)	9.23	18.45	27.68
	$(\beta_y^d)_{0n}$ (rad/m)	309.21	618.42	927.64

[a] $a = 0.9$ in. (2.286 cm), $b = 0.4$ in. (1.016 cm), $h = b/3$, $\mu_d = \mu_0$, and $\varepsilon_d = 2.56\varepsilon_0$.

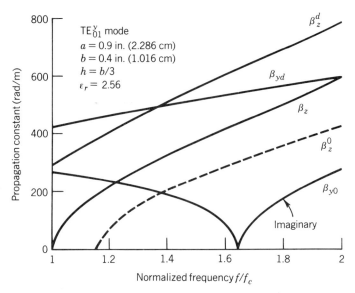

FIGURE 8-14 Propagation constants of TE_{01}^y modes for a partially filled rectangular waveguide.

How do the cutoff frequencies of the first three TE_{0n}^y modes ($n = 1, 2, 3$) of the partially filled waveguide compare with those of the TE_{0n}^y modes of the empty waveguide?

Solution.

1. The cutoff frequencies of the partially filled waveguide are found using (8-118). According to (8-120) and (8-100a), the cutoff frequencies for each of the desired modes must fall in the range of

$$9.23 \text{ GHz} \leq (f_c)_{01} \leq 14.76 \text{ GHz}$$
$$18.45 \text{ GHz} \leq (f_c)_{02} \leq 29.53 \text{ GHz}$$
$$27.68 \text{ GHz} \leq (f_c)_{03} \leq 44.29 \text{ GHz}$$

The actual frequencies are listed in Table 8-6.

2. Once the cutoff frequencies are found, the corresponding wave numbers β_{y0} and β_{yd} *at cutoff* can be found by using (8-117a) and (8-117b). These are also listed in Table 8-6. Also listed in Table 8-6 are the values of the cutoff frequencies and the corresponding wave numbers for the air-filled and dielectric-filled waveguides.

3. Finally the wave numbers β_{y0}, β_{yd}, and β_z for each frequency in the range $(f_c)_{0n} \leq f \leq 2(f_c)_{0n}$ are found by solving (8-115) through (8-115b) as outlined previously. These are shown plotted in Figure 8-14 for the TE_{01}^y mode where they are compared with those of the waveguide filled completely with air (β_z^0) or with the dielectric (β_z^d). The others for the TE_{02}^y and TE_{03}^y modes are assigned as an end of chapter exercise.

8.5.2 Longitudinal Section Magnetic (LSM^y) or Transverse Magnetic (TM^y)

For the waveguide configuration of Figure 8-13a the TM^y field components are those of (8-101) where the corresponding vector potentials in the dielectric and free space regions for the waves traveling in the $+z$ direction are given, respectively, by

$$A_y^d = \left[C_1^d \cos(\beta_{xd} x) + D_1^d \sin(\beta_{xd} x)\right]\left[C_2^d \cos(\beta_{yd} y) + D_2^d \sin(\beta_{yd} y)\right] A_3^d e^{-j\beta_z z} \tag{8-121}$$

$$\beta_{xd}^2 + \beta_{yd}^2 + \beta_z^2 = \beta_d^2 = \omega^2 \mu_d \varepsilon_d \tag{8-121a}$$

$$A_y^0 = \left[C_1^0 \cos(\beta_{x0} x) + D_1^0 \sin(\beta_{x0} x)\right]$$
$$\times \left\{ C_2^0 \cos\left[\beta_{y0}(b-y)\right] + D_2^0 \sin\left[\beta_{y0}(b-y)\right]\right\} A_3^0 e^{-j\beta_z z} \tag{8-122}$$

$$\beta_{x0}^2 + \beta_{y0}^2 + \beta_z^2 = \beta_0^2 = \omega^2 \mu_0 \varepsilon_0 \tag{8-122a}$$

The appropriate boundary conditions are those of (8-104a) through (8-105f).

Application of the boundary conditions of (8-104a) through (8-105f) shows that the following relations follow:

$$A_y^0 = B_{mn}^0 \sin(\beta_{x0} x) \cos\left[\beta_{y0}(b-y)\right] e^{-j\beta_z z} \tag{8-123}$$

$$\beta_{x0} = \frac{m\pi}{a} \qquad m = 1, 2, 3, \ldots \tag{8-123a}$$

$$\beta_{x0}^2 + \beta_{y0}^2 + \beta_z^2 = \left(\frac{m\pi}{a}\right)^2 + \beta_{y0}^2 + \beta_z^2 = \beta_0^2 = \omega^2 \mu_0 \varepsilon_0 \tag{8-123b}$$

$$A_y^d = B_{mn}^d \sin(\beta_{xd} x) \cos(\beta_{yd} y) e^{-j\beta_z z} \tag{8-124}$$

$$\beta_{xd} = \frac{m\pi}{a} \qquad m = 1, 2, 3, \ldots \tag{8-124a}$$

$$\beta_{xd}^2 + \beta_{yd}^2 + \beta_z^2 = \left(\frac{m\pi}{a}\right)^2 + \beta_{yd}^2 + \beta_z^2 = \beta_d^2 = \omega^2 \mu_d \varepsilon_d \tag{8-124b}$$

$$-\frac{\beta_{y0}}{\mu_0 \varepsilon_0} B_{mn}^0 \sin\left[\beta_{y0}(b-h)\right] = \frac{\beta_{yd}}{\mu_d \varepsilon_d} B_{mn}^d \sin(\beta_{yd} h) \tag{8-125}$$

$$\frac{1}{\mu_0} B_{mn}^0 \cos\left[\beta_{y0}(b-h)\right] = \frac{1}{\mu_d} B_{mn}^d \cos(\beta_{yd} h) \tag{8-126}$$

$$\frac{\beta_{y0}}{\varepsilon_0} \tan\left[\beta_{y0}(b-h)\right] = -\frac{\beta_{yd}}{\varepsilon_d} \tan(\beta_{yd} h) \tag{8-127}$$

$$\beta_{x0}^2 + \beta_{y0}^2 + \beta_z^2 = \left(\frac{m\pi}{a}\right)^2 + \beta_{y0}^2 + \beta_z^2 = \beta_0^2 = \omega^2 \mu_0 \varepsilon_0$$
$$m = 1, 2, 3 \ldots \tag{8-127a}$$

$$\beta_{xd}^2 + \beta_{yd}^2 + \beta_z^2 = \left(\frac{m\pi}{a}\right)^2 + \beta_{yd}^2 + \beta_z^2 = \beta_d^2 = \omega^2 \mu_d \varepsilon_d$$
$$m = 1, 2, 3 \ldots \tag{8-127b}$$

Whereas $\beta_{x0} = \beta_{xd} = m\pi/a$, $m = 1, 2, 3, \ldots$, have been determined, β_{y0}, β_{yd}, and β_z have not yet been found. They can be determined by using (8-127) through (8-127b) and following a procedure similar to that outlined in the previous section for the TE^y modes. For each value of m there are infinite values of n ($n = 0, 1, 2, \ldots$). Thus the dominant mode is that for which $m = 1$ and $n = 0$, or the dominant mode is the TM$_{10}^y$.

For $m = 1$, the modes will be denoted as TM$_{1n}^y$. For these modes (8-127a) and (8-127b) reduce to

$$\left(\frac{\pi}{a}\right)^2 + \beta_{y0}^2 + \beta_z^2 = \omega^2 \mu_0 \varepsilon_0 \Rightarrow \beta_z = \pm\sqrt{\omega^2 \mu_0 \varepsilon_0 - \left[\beta_{y0}^2 + \left(\frac{\pi}{a}\right)^2\right]} \quad (8\text{-}128a)$$

$$\left(\frac{\pi}{a}\right)^2 + \beta_{yd}^2 + \beta_z^2 = \omega^2 \mu_d \varepsilon_d \Rightarrow \beta_z = \pm\sqrt{\omega^2 \mu_d \varepsilon_d - \left[\beta_{yd}^2 + \left(\frac{\pi}{a}\right)^2\right]} \quad (8\text{-}128b)$$

Cutoff occurs when $\beta_z = 0$. Thus at cutoff (8-128a) and (8-128b) reduce to

$$\omega_c^2 \mu_0 \varepsilon_0 = \beta_{y0}^2 + \left(\frac{\pi}{a}\right)^2 \Rightarrow \beta_{y0} = \sqrt{\omega_c^2 \mu_0 \varepsilon_0 - \left(\frac{\pi}{a}\right)^2} \quad (8\text{-}129a)$$

$$\omega_c^2 \mu_d \varepsilon_d = \beta_{yd}^2 + \left(\frac{\pi}{a}\right)^2 \Rightarrow \beta_{yd} = \sqrt{\omega_c^2 \mu_d \varepsilon_d - \left(\frac{\pi}{a}\right)^2} \quad (8\text{-}129b)$$

which can be used to find β_{y0} and β_{yd} *only at cutoff* once the cutoff frequency has been determined. By using (8-129a) and (8-129b) we can write (8-127) as

$$\frac{1}{\varepsilon_0}\sqrt{\omega_c^2 \mu_0 \varepsilon_0 - \left(\frac{\pi}{a}\right)^2} \tan\left[\sqrt{\omega_c^2 \mu_0 \varepsilon_0 - \left(\frac{\pi}{a}\right)^2}(b-h)\right]$$

$$= -\frac{1}{\varepsilon_d}\sqrt{\omega_c^2 \mu_d \varepsilon_d - \left(\frac{\pi}{a}\right)^2} \tan\left[h\sqrt{\omega_c^2 \mu_d \varepsilon_d - \left(\frac{\pi}{a}\right)^2}\right]$$

or

$$\boxed{\begin{aligned}\frac{\varepsilon_d}{\varepsilon_0}&\sqrt{\omega_c^2 \mu_0 \varepsilon_0 - \left(\frac{\pi}{a}\right)^2} \tan\left[\sqrt{\omega_c^2 \mu_0 \varepsilon_0 - \left(\frac{\pi}{a}\right)^2}(b-h)\right]\\ &= -\sqrt{\omega_c^2 \mu_d \varepsilon_d - \left(\frac{\pi}{a}\right)^2} \tan\left[h\sqrt{\omega_c^2 \mu_d \varepsilon_d - \left(\frac{\pi}{a}\right)^2}\right]\end{aligned}} \quad (8\text{-}130)$$

which can be used to find the cutoff frequencies of the TM$_{1n}^y$ modes in a partially filled waveguide.

For a rectangular waveguide filled completely either with free space (μ_0, ε_0) or with a dielectric material (ε_d, μ_d) the cutoff frequency of the hybrid TM$_{10}^y$ mode is given, respectively, according to (8-101d) by

$$(f_c^0)_{10}^{\text{TM}^y} = \frac{1}{2a\sqrt{\mu_0 \varepsilon_0}} \quad (8\text{-}131a)$$

$$(f_c^d)_{10}^{\text{TM}^y} = \frac{1}{2a\sqrt{\mu_d \varepsilon_d}} \quad (8\text{-}131b)$$

By using perturbational techniques, it can be shown that, in general, the cutoff frequency of the partially filled waveguide (part free space and part dielectric) is greater than the cutoff frequency of the same mode in the same waveguide filled with a dielectric material with ε_d, μ_d and smaller than the cutoff frequency of the same waveguide filled with free space. Thus the cutoff frequency of the TM_{10}^y mode of a partially filled waveguide (part free space and part dielectric) is greater than (8-131b) and smaller than (8-131a), that is

$$\boxed{\frac{1}{2a\sqrt{\mu_d \varepsilon_d}} \leq (f_c)_{10}^{\text{TM}^y} \leq \frac{1}{2a\sqrt{\mu_0 \varepsilon_0}}} \qquad (8\text{-}132)$$

or

$$\boxed{\frac{\pi}{a\sqrt{\mu_d \varepsilon_d}} \leq (\omega_c)_{10}^{\text{TM}^y} \leq \frac{\pi}{a\sqrt{\mu_0 \varepsilon_0}}} \qquad (8\text{-}132\text{a})$$

With this permissible range, the exact values can be found by using (8-130). The propagation constant β_z must be solved at each frequency on an individual basis using (8-128a) or (8-128b) once the values of β_{y0} and β_{yd} have been determined at that frequency.

Example 8-9. A WR90 X-band (8.2–12.4 GHz) waveguide of Figure 8-13a with inner dimensions of $a = 0.9$ in. (2.286 cm), $b = 0.4$ in. (1.016 cm), and $a/b = 2.25$, is partially filled with free space and polystyrene ($\varepsilon_d = 2.56$, $\mu_d = \mu_0$, and $h = b/3$). For $m = 1$ determine the following.

1. The cutoff frequencies of the hybrid TM_{1n}^y (LSM_{1n}^y) modes for $n = 0, 1, 2$.
2. The corresponding values of β_{y0} and β_{yd} for each of the cutoff frequencies.
3. The corresponding values of β_{y0}, β_{yd}, and β_z for the TM_{10}^y mode in the frequency range $(f_c)_{10} \leq f \leq 2(f_c)_{10}$.

How do the cutoff frequencies of the first three TM_{1n}^y ($n = 0, 1, 2$) of the partially filled waveguide compare with those of the TM_{1n}^y of the empty waveguide?

Solution. Follow a procedure similar to that for the TE_{0n}^y modes, as was done for the solution of Example 8-8. Then the parameters and their associated values are obtained as listed in Table 8-7 and shown Figure 8-15. The parameters versus frequency for the TM_{11}^y and TM_{12}^y modes are assigned as an end of chapter exercise. The cutoff frequencies of the partially filled waveguide are found using (8-130). According to (8-132) and (8-101d), the cutoff frequencies for each of the desired modes must fall in the range of

$$4.101 \text{ GHz} \leq (f_c)_{10} \leq 6.56 \text{ GHz}$$

$$10.098 \text{ GHz} \leq (f_c)_{11} \leq 16.16 \text{ GHz}$$

$$18.905 \text{ GHz} \leq (f_c)_{12} \leq 30.25 \text{ GHz}$$

TABLE 8-7
Cutoff frequencies and phase constants of partially filled, air-filled, and dielectric-filled rectangular waveguide[a]

TM_{1n}^y modes		$n = 0$	$n = 1$	$n = 2$
Partially filled waveguide	$(f_c)_{1n}$ (GHz)	5.79	13.68	24.75
	$(\beta_{y0})_{1n}$ at $(f_c)_{1n}$ (rad/m)	$\pm j64.76$	251.37	499.84
	$(\beta_{yd})_{1n}$ at $(f_c)_{1n}$ (rad/m)	136.84	437.29	817.96
Air-filled waveguide	$(f_c^0)_{1n}$ (GHz)	6.56	16.16	30.25
	$(\beta_y^0)_{1n}$ (rad/m)	0	309.21	618.42
Dielectric-filled waveguide	$(f_c^d)_{1n}$ (GHz)	4.10	10.10	18.90
	$(\beta_y^d)_{1n}$ (rad/m)	0	309.21	618.42

[a] $a = 0.9$ in. (2.286 cm), $b = 0.4$ in. (1.016 cm), $h = b/3$, $\mu_d = \mu_0$, and $\varepsilon_d = 2.56\varepsilon_0$.

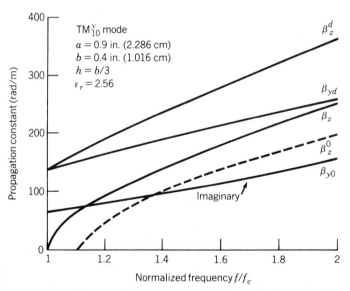

FIGURE 8-15 Propagation constants of TM_{10}^y modes for a partially filled rectangular waveguide.

PARTIALLY FILLED WAVEGUIDE   409

The actual frequencies, as well as the other desired parameters of this problem, are listed in Table 8-7.

When the dielectric properties of the dielectric material inserted into the waveguide are such that $\varepsilon_d \simeq \varepsilon_0$ and $\mu_d \simeq \mu_0$, then the wave constants of the TM_{mn} modes of the partially filled waveguide are approximately equal to the corresponding wave constants of the TM^y_{mn} modes of the totally filled waveguide. Thus according to (8-101c)

$$(\beta_{y0})_{m0} \simeq (\beta_{yd})_{m0} \simeq \text{small} \simeq 0 \qquad (8\text{-}133)$$

Therefore (8-127) can be approximated by

$$\frac{\beta_{y0}}{\varepsilon_0}\tan\left[\beta_{y0}(b-h)\right] \simeq \frac{\beta_{y0}}{\varepsilon_0}\left[\beta_{y0}(b-h)\right] = -\frac{\beta_{yd}}{\varepsilon_d}\tan(\beta_{yd}h) \simeq -\frac{\beta_{yd}}{\varepsilon_d}(\beta_{yd}h)$$

$$\frac{\beta_{y0}^2}{\varepsilon_0}(b-h) \simeq -\frac{\beta_{yd}^2}{\varepsilon_d}h$$

$$\frac{\varepsilon_d}{\varepsilon_0}\beta_{y0}^2(b-h) \simeq -\beta_{yd}^2 h \qquad (8\text{-}134)$$

At cutoff $\beta_z = 0$. Therefore using (8-129a) and (8-129b) we can write (8-134) as

$$\frac{\varepsilon_d}{\varepsilon_0}\left[\omega_c^2\mu_0\varepsilon_0 - \left(\frac{\pi}{a}\right)^2\right](b-h) \simeq -\left[\omega_c^2\mu_d\varepsilon_d - \left(\frac{\pi}{a}\right)^2\right]h \qquad (8\text{-}134a)$$

which reduces to

$$\boxed{(\omega_c)_{10}^{TM^y} \simeq \frac{\pi}{a\sqrt{\varepsilon_r\mu_0\varepsilon_0}}\sqrt{\frac{h+\varepsilon_r(b-h)}{(b-h)+\mu_r h}}} \qquad (8\text{-}135)$$

where

$$\varepsilon_r = \frac{\varepsilon_d}{\varepsilon_0} \qquad (8\text{-}135a)$$

$$\mu_r = \frac{\mu_d}{\mu_0} \qquad (8\text{-}135b)$$

Example 8-10. By using the approximate expression of (8-135), determine the cutoff frequency of the dominant TM^y_{10} hybrid mode for the following cases.

1. $\varepsilon_r = 1$ and $\mu_r = 1$.
2. $h = 0$.
3. $h = b$.
4. $\varepsilon_r = 2.56$, $\mu_r = 1$, $a = 0.9$ in. (2.286 cm), $b = 0.4$ in. (1.016 cm), $a/b = 2.25$, and $h = b/3$.

Solution.

1. When $\varepsilon_r = 1$ and $\mu_r = 1$, (8-135) reduces to

$$(\omega_c)_{10}^{TM^y} \simeq \frac{\pi}{a\sqrt{\mu_0\varepsilon_0}}$$

which is equal to the exact value as predicted by (8-101d).

2. When $h = 0$, (8-135) reduces to

$$(\omega_c)_{10}^{TM^y} \simeq \frac{\pi}{a\sqrt{\mu_0\varepsilon_0}}$$

which again is equal to the exact value predicted by (8-101d).

3. When $h = b$, (8-135) reduces to

$$(\omega_c)_{10}^{TM^y} \simeq \frac{\pi}{a\sqrt{\mu_0\mu_r\varepsilon_0\varepsilon_r}} = \frac{\pi}{a\sqrt{\mu_d\varepsilon_d}}$$

which again is equal to the exact value predicted by (8-101d).

4. When $\varepsilon_r = 2.56$, $\mu_r = 1$, $a = 2.286$ cm, $b = 1.016$ cm, and $h = b/3$, (8-135) reduces to

$$(\omega_c)_{10}^{TM^y} \simeq \frac{\pi}{a\sqrt{\varepsilon_r\mu_0\varepsilon_0}} \sqrt{\frac{b/3 + 2.56(b - b/3)}{(b - b/3) + b/3}}$$

$$= \sqrt{\frac{2.04}{2.56}} \frac{\pi}{a\sqrt{\mu_0\varepsilon_0}} = 0.8927 \frac{\pi}{a\sqrt{\mu_0\varepsilon_0}}$$

$$(f_c)_{10}^{TM^y} \simeq 5.8576 \text{ GHz}$$

whose exact value, according to Example 8-8, is equal to 5.786 GHz. It should be noted that the preceding approximate expression for $(\omega_c)_{10}$ of

$$(\omega_c)_{10} \simeq 0.8927 \frac{\pi}{a\sqrt{\mu_0\varepsilon_0}}$$

falls in the permissible range of

$$\frac{\pi}{a\sqrt{\mu_0\varepsilon_r\varepsilon_0}} = \frac{\pi}{\sqrt{2.56}\,a\sqrt{\mu_0\varepsilon_0}} = 0.6250 \frac{\pi}{a\sqrt{\mu_0\varepsilon_0}} \leq (\omega_c)_{10}^{TM^y} \leq \frac{\pi}{a\sqrt{\mu_0\varepsilon_0}}$$

as given by (8-132a).

8.6 TRANSVERSE RESONANCE METHOD

The transverse resonance method (TRM) is a technique that can be used to find the propagation constant of many practical composite waveguide structures [5, 6], as well as many traveling wave antenna systems [6–8]. By using this method, the cross section of the waveguide or traveling wave antenna structure is represented as a transmission line system. The fields of such a structure must satisfy the transverse wave equation, and the resonances of this transverse network will yield expressions

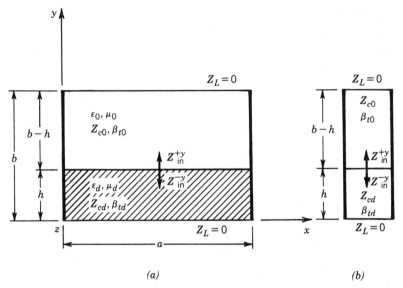

FIGURE 8-16 (*a*) Cross section of rectangular waveguide and (*b*) transmission line equivalent for transverse resonance method (TRM).

for the propagation constants of the waveguide or antenna structure. Whereas the formulations of this method are much simpler when applied to finding the propagation constants, they do not contain the details for finding other parameters of interest (such as field distributions, wave impedances, etc.).

The objective here is to analyze the waveguide geometry of Figure 8-13*a* using the transverse resonance method. Although the method will not yield all the details of the analysis of Sections 8.5.1 and 8.5.2, it will lead to the same characteristic equations 8-115 and 8-127. The problem will be modeled as a two-dimensional structure represented by two transmission lines; one dielectric-filled ($0 \le y \le h$) with characteristic impedance Z_{cd} and wave number β_{td} and the other air-filled ($h \le y \le b$) with characteristic impedance Z_{c0} and wave number β_{t0}, as shown in Figure 8-16. Each line is considered shorted at its load, that is, $Z_L = 0$ at $y = 0$ and $y = b$.

It was shown in Section 3.4.1A that the solution to the scalar wave equation for any of the electric field components, for example, that for E_x of (3-22) as given by (3-23), takes the general form of

$$\nabla^2 \psi + \beta^2 \psi = \frac{\partial^2 \psi}{\partial x^2} + \frac{\partial^2 \psi}{\partial y^2} + \frac{\partial^2 \psi}{\partial z^2} + \beta^2 \psi = 0 \tag{8-136}$$

where

$$\psi(x, y, z) = f(x)g(y)h(z) \tag{8-136a}$$

The scalar function ψ represents any of the electric or magnetic field components. For waves traveling in the z direction the variations of $h(z)$ are represented by exponentials of the form $e^{\pm j\beta_z z}$. Therefore for such waves (8-136) reduces to

$$\nabla^2 \psi + \beta^2 \psi = \left(\nabla_t^2 + \frac{\partial^2}{\partial z^2}\right)\psi + \beta^2 \psi = \left(\nabla_t^2 - \beta_z^2\right)\psi + \beta^2 \psi = 0$$

$$\nabla_t^2 \psi + \left(\beta^2 - \beta_z^2\right)\psi = 0 \tag{8-137}$$

where

$$\nabla_t^2 = \frac{\partial^2}{\partial x^2} + \frac{\partial^2}{\partial y^2} \tag{8-137a}$$

The wave numbers associated with (8-137) are related by

$$(\beta_x^2 + \beta_y^2) + \beta_z^2 = \beta_t^2 + \beta_z^2 = \beta^2 \tag{8-138}$$

where

$$\beta_t^2 = \beta_x^2 + \beta_y^2 \tag{8-138a}$$

The constant β_t is referred to as the *transverse direction wave number* and (8-137) is referred to as the *transverse wave equation*.

Each of the electric and magnetic field components in the dielectric- and air-filled sections of the two-dimensional structure of Figure 8-16 must satisfy the transverse wave equation of (8-137) with corresponding transverse wave numbers of β_{td} and β_{t0} where

$$\beta_{td}^2 + \beta_z^2 = \beta_d^2 = \omega^2 \mu_d \varepsilon_d \tag{8-139a}$$

$$\beta_{t0}^2 + \beta_z^2 = \beta_0^2 = \omega^2 \mu_0 \varepsilon_0 \tag{8-139b}$$

In Section 4.2.2B it was shown by (4-20a) through (4-21b) that the wave impedances of the waves in the positive and negative directions are equal. However, the ratios of the corresponding electric/magnetic field components magnitudes were equal but opposite in direction. Since the input impedance of a line is defined as the ratio of the electric/magnetic field components (or voltage/current), then at any point along the transverse direction of the waveguide structure the input impedance of the transmission line network looking in the positive y direction is equal in magnitude but opposite in phase to that looking in the negative y direction. This follows from the boundary conditions that require continuous tangential components of the electric (**E**) and magnetic (**H**) fields at any point on a plane orthogonal to the transverse structure of the waveguide.

For the transmission line model of Figure 8-16b, the input impedance at the interface looking in the $+y$ direction of the air-filled portion toward the shorted load is given, according to the impedance transfer equation 5-66d, as

$$Z_{in}^{+y} = Z_{c0} \left[\frac{Z_L + jZ_{c0} \tan[\beta_{t0}(b-h)]}{Z_{c0} + jZ_L \tan[\beta_{t0}(b-h)]} \right]_{Z_L=0} = jZ_{c0} \tan[\beta_{t0}(b-h)] \tag{8-140a}$$

In a similar manner, the input impedance at the interface looking in the $-y$ direction of the dielectric-filled portion toward the shorted load is given by

$$Z_{in}^{-y} = jZ_{cd} \tan(\beta_{td}h) \tag{8-140b}$$

Since these two impedances must be equal and opposite, then

$$Z_{in}^{+y} = -Z_{in}^{-y} = jZ_{c0} \tan[\beta_{t0}(b-h)] = -jZ_{cd} \tan(\beta_{td}h)$$

$$\boxed{Z_{c0} \tan[\beta_{t0}(b-h)] = -Z_{cd} \tan(\beta_{td}h)} \tag{8-141}$$

The preceding equation is applicable for both TE and TM modes. It will be applied in the next two sections to examine the TE^y and TM^y modes of the partially filled waveguide of Figure 8-16a.

8.6.1 Transverse Electric (TE^y) or Longitudinal Section Electric (LSE^y) or H^y

The characteristic equation 8-141 will now be applied to examine the TE^y modes of the partially filled waveguide of Figure 8-16a. It was shown in Section 8.2.1 that the wave impedance of the TE^z_{mn} modes is given by (8-19) or

$$Z_w^{TE^z} = \frac{\omega\mu}{\beta_z} \qquad (8\text{-}142)$$

Allow the characteristic impedances for the TE^y modes of the dielectric- and air-filled sections of the waveguide, represented by the two-section transmission line of Figure 8-16b, to be of the same form as (8-142), or

$$Z_{cd} = Z_d^h = \frac{\omega\mu_d}{\beta_{yd}} \qquad (8\text{-}143a)$$

$$Z_{c0} = Z_0^h = \frac{\omega\mu_0}{\beta_{y0}} \qquad (8\text{-}143b)$$

$$\beta_{td} = \beta_{yd} \qquad (8\text{-}143c)$$

$$\beta_{t0} = \beta_{y0} \qquad (8\text{-}143d)$$

Then (8-141) reduces to

$$\frac{\omega\mu_0}{\beta_{y0}} \tan\left[\beta_{y0}(b-h)\right] = -\frac{\omega\mu_d}{\beta_{yd}} \tan\left(\beta_{yd}h\right)$$

or

$$\boxed{\frac{\beta_{y0}}{\mu_0} \cot\left[\beta_{y0}(b-h)\right] = -\frac{\beta_{yd}}{\mu_d} \cot\left(\beta_{yd}h\right)} \qquad (8\text{-}144)$$

Equation 8-144 is identical to (8-115), and it can be solved using the same procedures used in Section 8.5.1 to solve (8-115).

8.6.2 Transverse Magnetic (TM^y) or Longitudinal Section Magnetic (LSM^y) or E^y

The same procedure used in Section 8.6.1 for the TE^y modes can also be used to examine the TM^y modes of the partially filled waveguide of Figure 8-16a. According to (8-29a) the wave impedance of TM^z_{mn} modes is given by

$$Z_w^{TM^z} = \frac{\beta_z}{\omega\varepsilon} \qquad (8\text{-}145)$$

Allow the characteristic impedances for the TM^y modes of the dielectric- and air-filled sections of the waveguide to be of the same form as (8-145), or

$$Z_{cd} = Z_d^e = \frac{\beta_{yd}}{\omega \varepsilon_d} \tag{8-146a}$$

$$Z_{c0} = Z_0^e = \frac{\beta_{y0}}{\omega \varepsilon_0} \tag{8-146b}$$

$$\beta_{td} = \beta_{yd} \tag{8-146c}$$

$$\beta_{t0} = \beta_{y0} \tag{8-146d}$$

Then (8-141) reduces to

$$\frac{\beta_{y0}}{\omega \varepsilon_0} \tan\left[\beta_{y0}(b-h)\right] = -\frac{\beta_{yd}}{\omega \varepsilon_d} \tan\left(\beta_{yd} h\right)$$

or

$$\boxed{\frac{\beta_{y0}}{\varepsilon_0} \tan\left[\beta_{y0}(b-h)\right] = -\frac{\beta_{yd}}{\varepsilon_d} \tan\left(\beta_{yd} h\right)} \tag{8-147}$$

Equation 8-147 is identical to (8-127), and it can be solved using the same procedures used in Section 8.5.2 to solve (8-127).

The transverse resonance method can be used to solve other transmission line discontinuity problems [5] as well as many traveling wave antenna systems [6].

8.7 DIELECTRIC WAVEGUIDE

Transmission lines are used to contain the energy associated with a wave within a given space and guide it in a given direction. Typically many people associate these types of transmission lines with either coaxial and twin lead lines or metal pipes (usually referred to as waveguides) with part or all of their structure being metal. However, dielectric slabs and rods, with or without any associated metal, can also be used to guide waves and serve as transmission lines. Usually these are referred to as *dielectric waveguides*, and the field modes that they can support are known as *surface wave modes* [9].

8.7.1 Dielectric Slab Waveguide

One type of dielectric waveguide is a dielectric slab of height $2h$, as shown in Figure 8-17. To simplify the analysis of the structure, we reduce the problem to a two-dimensional one (its length in the x direction is infinite) so that $\partial/\partial x = 0$. Although in practice the dimensions of the structure are finite, the two-dimensional approximation not only simplifies the analysis but also sheds insight into the characteristics of the structure. Typically the cross section of the slab in Figure 8-17a would be rectangular with height $2h$ and finite width a. We also assume that

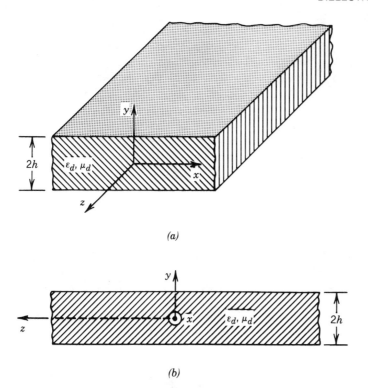

FIGURE 8-17 Geometry for dielectric slab waveguide. (*a*) Perspective. (*b*) Side view.

the waves are traveling in the $\pm z$ direction, and the structure is infinite in that direction as illustrated in Figure 8-17*b*.

Another practical configuration for a dielectric transmission line is a dielectric rod of circular cross section. Because of the cylindrical geometry of the structure, the field solutions will be in terms of Bessel functions. Therefore the discussion of this line will be postponed until Chapter 9.

A very popular dielectric rod waveguide is the fiber optics cable. Typically this cable is made of two different materials, one that occupies the center core and the other that serves as a cladding to the center core. This configuration is usually referred to as the *step index*, and the index of refraction of the center core is slightly greater than that of the cladding. Another configuration has the index of refraction distribution along the cross section of the line graded so that there is a smooth variation in the radial direction from the larger values at the center toward the smaller values at the periphery. This is referred to as the *graded index*. This line is discussed in more detail in Section 9.5.3.

The objective in a dielectric slab waveguide, or any type of a waveguide, is to contain the energy within the structure and direct it toward a given direction. For the dielectric slab waveguide this is accomplished by having the wave bounce back and forth between its upper and lower interfaces at an incidence angle greater than the critical angle. When this is accomplished the refracted fields outside the dielectric form evanescent (decaying) waves and all the real energy is reflected and contained within the structure. The characteristics of this line can be analyzed by treating the structure as a boundary-value problem whose modal solution is obtained by solving the wave equation and enforcing the boundary conditions. The other approach is to examine the characteristics of the line using ray-tracing (geometrical optics) techniques. This approach is simpler and sheds more physical insight onto the propagation characteristics of the line but does not provide the details of the more cumbersome modal solution. Both methods will be examined here. We will begin with the modal solution approach.

It can be shown that the waveguide structure of Figure 8-17 can support TE^z, TM^z, TE^y, and TM^y modes. We will examine here both the TM^z and TE^z modes. We will treat TM^z in detail and then summarize the TE^z.

8.7.2 Transverse Magnetic (TM^z) Modes

The TM^z mode fields that can exist within and outside the dielectric slab of Figure 8-17 must satisfy (8-24) where A_z is the potential function representing the fields either within or outside the dielectric slab. Inside and outside the dielectric region, the fields can be represented by a combination of even and odd modes, as shown in Figure 8-18 [10].

For the fields within the dielectric slab the potential function A_z takes the following form:

$$-h \leq y \leq h$$

$$A_z^d = \left[C_2^d \cos\left(\beta_{yd} y\right) + D_2^d \sin\left(\beta_{yd} y\right) \right] A_3^d e^{-j\beta_z z} = A_{ze}^d + A_{zo}^d \quad (8\text{-}148)$$

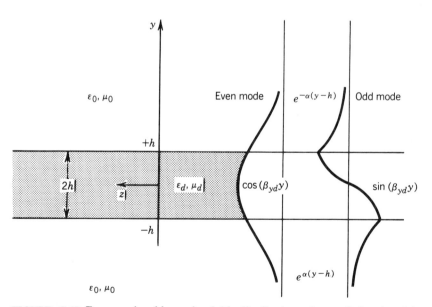

FIGURE 8-18 Even and odd mode field distributions in a dielectric slab waveguide. (*Source:* M. Zahn, *Electromagnetic Field Theory*, 1979. Reprinted with permission of John Wiley & Sons, Inc.)

DIELECTRIC WAVEGUIDE **417**

where

$$A_{ze}^d = C_2^d A_3^d \cos(\beta_{yd} y) e^{-j\beta_z z} = A_{me}^d \cos(\beta_{yd} y) e^{-j\beta_z z} \quad (8\text{-}148\text{a})$$

$$A_{zo}^d = D_2^d A_3^d \sin(\beta_{yd} y) e^{-j\beta_z z} = A_{mo}^d \sin(\beta_{yd} y) e^{-j\beta_z z} \quad (8\text{-}148\text{b})$$

$$\beta_{yd}^2 + \beta_z^2 = \beta_d^2 = \omega^2 \mu_d \varepsilon_d \quad (8\text{-}148\text{c})$$

In (8-148) A_{ze}^d and A_{zo}^d represent, respectively, the even and odd modes. For the slab to function as a waveguide, the fields outside the dielectric slab must be of evanescent form. Therefore the potential function A_z takes the following form:

$$\underline{y \geq h}$$

$$A_z^{0+} = \left(A_{2e}^{0+} e^{-j\beta_{y0} y} + B_{2o}^{0+} e^{-j\beta_{y0} y} \right) A_{3o}^{0+} e^{-j\beta_z z} = A_{ze}^{0+} + A_{zo}^{0+} \quad (8\text{-}149)$$

where

$$A_{ze}^{0+} = A_{2e}^{0+} A_{3o}^{0+} e^{-j\beta_{y0} y} e^{-j\beta_z z} = A_{me}^{0+} e^{-\alpha_{y0} y} e^{-j\beta_z z} \quad (8\text{-}149\text{a})$$

$$A_{zo}^{0+} = B_{2o}^{0+} A_{3o}^{0+} e^{-j\beta_{y0} y} e^{-j\beta_z z} = B_{mo}^{0+} e^{-\alpha_{y0} y} e^{-j\beta_z z} \quad (8\text{-}149\text{b})$$

$$\beta_{y0}^2 + \beta_z^2 = -\alpha_{y0}^2 + \beta_z^2 = \beta_0^2 = \omega^2 \mu_0 \varepsilon_0 \quad (8\text{-}149\text{c})$$

$$\underline{y \leq -h}$$

$$A_z^{0-} = \left(A_{2e}^{0-} e^{+j\beta_{y0} y} + B_{2o}^{0-} e^{+j\beta_{y0} y} \right) A_{3o}^{0-} e^{-j\beta_z z} = A_{ze}^{0-} + A_{zo}^{0-} \quad (8\text{-}150)$$

where

$$A_{ze}^{0-} = A_{2e}^{0-} A_{3o}^{0-} e^{+j\beta_{y0} y} e^{-j\beta_z z} = A_{me}^{0-} e^{+\alpha_{y0} y} e^{-j\beta_z z} \quad (8\text{-}150\text{a})$$

$$A_{zo}^{0-} = B_{2o}^{0-} A_{3o}^{0-} e^{+j\beta_{y0} y} e^{-j\beta_z z} = B_{mo}^{0-} e^{+\alpha_{y0} y} e^{-j\beta_z z} \quad (8\text{-}150\text{b})$$

$$\beta_{y0}^2 + \beta_z^2 = -\alpha_{y0}^2 + \beta_z^2 = \beta_0^2 = \omega^2 \mu_0 \varepsilon_0 \quad (8\text{-}150\text{c})$$

For the fields of (8-149) through (8-150c) to be of evanescent form, α_{y0} must be real.

Since the fields within and outside the slab have been separated into even and odd modes, they can be examined separately and then we will apply superposition. We will examine the even modes first and then the odd. For each mode (even or odd), a number of dependent and independent boundary conditions must be satisfied. A sufficient set of independent boundary conditions chosen here are

$$E_z^d(y = h, z) = E_z^{0+}(y = h, z) \quad (8\text{-}151\text{a})$$

$$E_z^d(y = -h, z) = E_z^{0-}(y = -h, z) \quad (8\text{-}151\text{b})$$

$$H_x^d(y = h, z) = H_x^{0+}(y = h, z) \quad (8\text{-}151\text{c})$$

$$H_x^d(y = -h, z) = H_x^{0-}(y = -h, z) \quad (8\text{-}151\text{d})$$

A. TM^z (EVEN)

By using (8-24) along with the appropriate potential function of (8-148) through (8-150c), we can write the field components as follows.

$$-h \leq y \leq h$$

$$E_{xe}^d = -j\frac{1}{\omega\mu_d\varepsilon_d}\frac{\partial^2 A_{ze}^d}{\partial x\, \partial z} = 0 \tag{8-152a}$$

$$E_{ye}^d = -j\frac{1}{\omega\mu_d\varepsilon_d}\frac{\partial^2 A_{ze}^d}{\partial y\, \partial z} = \frac{\beta_{yd}\beta_z}{\omega\mu_d\varepsilon_d} A_{me}^d \sin(\beta_{yd} y) e^{-j\beta_z z} \tag{8-152b}$$

$$E_{ze}^d = -j\frac{1}{\omega\mu_d\varepsilon_d}\left(\frac{\partial^2}{\partial z^2} + \beta_d^2\right) A_{ze}^d = -j\frac{\beta_d^2 - \beta_z^2}{\omega\mu_d\varepsilon_d} A_{me}^d \cos(\beta_{yd} y) e^{-j\beta_z z} \tag{8-152c}$$

$$H_{xe}^d = \frac{1}{\mu_d}\frac{\partial A_{ze}^d}{\partial y} = -\frac{\beta_{yd}}{\mu_d} A_{me}^d \sin(\beta_{yd} y) e^{-j\beta_z z} \tag{8-152d}$$

$$H_{ye}^d = -\frac{1}{\mu_d}\frac{\partial A_{ze}^d}{\partial x} = 0 \tag{8-152e}$$

$$H_{ze}^d = 0 \tag{8-152f}$$

$$y \geq +h$$

$$E_{xe}^{0+} = -j\frac{1}{\omega\mu_0\varepsilon_0}\frac{\partial^2 A_{ze}^{0+}}{\partial x\, \partial z} = 0 \tag{8-153a}$$

$$E_{ye}^{0+} = -j\frac{1}{\omega\mu_0\varepsilon_0}\frac{\partial^2 A_{ze}^{0+}}{\partial y\, \partial z} = \frac{\alpha_{y0}\beta_z}{\omega\mu_0\varepsilon_0} A_{me}^{0+} e^{-\alpha_{y0} y} e^{-j\beta_z z} \tag{8-153b}$$

$$E_{ze}^{0+} = -j\frac{1}{\omega\mu_0\varepsilon_0}\left(\frac{\partial^2}{\partial z^2} + \beta_0^2\right) A_{ze}^{0+} = -j\frac{\beta_0^2 - \beta_z^2}{\omega\mu_0\varepsilon_0} A_{me}^{0+} e^{-\alpha_{y0} y} e^{-j\beta_z z} \tag{8-153c}$$

$$H_{xe}^{0+} = \frac{1}{\mu_0}\frac{\partial A_{ze}^{0+}}{\partial y} = -\frac{\alpha_{y0}}{\mu_0} A_{me}^{0+} e^{-\alpha_{y0} y} e^{-j\beta_z z} \tag{8-153d}$$

$$H_{ye}^{0+} = -\frac{1}{\mu_0}\frac{\partial A_{ze}^{0+}}{\partial x} = 0 \tag{8-153e}$$

$$H_{ze}^{0+} = 0 \tag{8-153f}$$

$$y \leq -h$$

$$E_{xe}^{0-} = -j\frac{1}{\omega\mu_0\varepsilon_0}\frac{\partial^2 A_{ze}^{0-}}{\partial x\, \partial z} = 0 \tag{8-154a}$$

$$E_{ye}^{0-} = -j\frac{1}{\omega\mu_0\varepsilon_0}\frac{\partial^2 A_{ze}^{0-}}{\partial y\, \partial z} = -\frac{\alpha_{y0}\beta_z}{\omega\mu_0\varepsilon_0} A_{me}^{0-} e^{+\alpha_{y0} y} e^{-j\beta_z z} \tag{8-154b}$$

$$E_{ze}^{0-} = -j\frac{1}{\omega\mu_0\varepsilon_0}\left(\frac{\partial^2}{\partial z^2} + \beta_0^2\right) A_{ze}^{0-} = -j\frac{\beta_0^2 - \beta_z^2}{\omega\mu_0\varepsilon_0} A_{me}^{0-} e^{+\alpha_{y0} y} e^{-j\beta_z z} \tag{8-154c}$$

$$H_{xe}^{0-} = \frac{1}{\mu_0} \frac{\partial A_{ze}^{0-}}{\partial y} = \frac{\alpha_{y0}}{\mu_0} A_{me}^{0-} e^{+\alpha_{y0}y} e^{-j\beta_z z} \tag{8-154d}$$

$$H_{ye}^{0-} = -\frac{1}{\mu_0} \frac{\partial A_{ze}^{0-}}{\partial x} = 0 \tag{8-154e}$$

$$H_{ze}^{0-} = 0 \tag{8-154f}$$

Apply the boundary condition of (8-151a) and use (8-148c) and (8-149c) to obtain

$$-j \frac{\beta_d^2 - \beta_z^2}{\omega \mu_d \varepsilon_d} A_{me}^d \cos(\beta_{yd} h) e^{-j\beta_z z} = -j \frac{\beta_0^2 - \beta_z^2}{\omega \mu_0 \varepsilon_0} A_{me}^{0+} e^{-\alpha_{y0} h} e^{-j\beta_z z}$$

$$\frac{\beta_d^2 - \beta_z^2}{\mu_d \varepsilon_d} A_{me}^d \cos(\beta_{yd} h) = \frac{\beta_0^2 - \beta_z^2}{\mu_0 \varepsilon_0} A_{me}^{0+} e^{-\alpha_{y0} h}$$

$$\frac{\beta_{yd}^2}{\mu_d \varepsilon_d} A_{me}^d \cos(\beta_{yd} h) = -\frac{\alpha_{y0}^2}{\mu_0 \varepsilon_0} A_{me}^{0+} e^{-\alpha_{y0} h} \tag{8-155a}$$

In a similar manner, enforce (8-151b) and use (8-148c) and (8-149c) to obtain

$$\frac{\beta_{yd}^2}{\mu_d \varepsilon_d} A_{me}^d \cos(\beta_{yd} h) = -\frac{\alpha_{y0}^2}{\mu_0 \varepsilon_0} A_{me}^{0-} e^{-\alpha_{y0} h} \tag{8-155b}$$

Comparison of (8-155a) and (8-155b) makes it apparent that

$$A_{me}^{0+} = A_{me}^{0-} = A_{me}^0 \tag{8-155c}$$

Thus (8-155a) and (8-155b) are the same and both can be represented by

$$\boxed{\frac{\beta_{yd}^2}{\mu_d \varepsilon_d} A_{me}^d \cos(\beta_{yd} h) = -\frac{\alpha_{y0}^2}{\mu_0 \varepsilon_0} A_{me}^0 e^{-\alpha_{y0} h}} \tag{8-156}$$

Follow a similar procedure by applying (8-151c) and (8-151d) and using (8-155c). Then we arrive at

$$\boxed{\frac{\beta_{yd}}{\mu_d} A_{me}^d \sin(\beta_{yd} h) = \frac{\alpha_{y0}}{\mu_0} A_{me}^0 e^{-\alpha_{y0} h}} \tag{8-157}$$

Division of (8-157) by (8-156) allows us to write that

$$\frac{\varepsilon_d}{\beta_{yd}} \tan(\beta_{yd} h) = -\frac{\varepsilon_0}{\alpha_{y0}}$$

$$\beta_{yd} \cot(\beta_{yd} h) = -\frac{\varepsilon_d}{\varepsilon_0} \alpha_{y0}$$

$$\boxed{-\frac{\varepsilon_0}{\varepsilon_d} (\beta_{yd} h) \cot(\beta_{yd} h) = \alpha_{y0} h} \tag{8-158}$$

where according to (8-148c) and (8-149c)

$$\beta_{yd}^2 + \beta_z^2 = \beta_d^2 = \omega^2 \mu_d \varepsilon_d \Rightarrow \beta_{yd}^2 = \beta_d^2 - \beta_z^2 = \omega^2 \mu_d \varepsilon_d - \beta_z^2 \quad (8\text{-}158a)$$

$$-\alpha_{y0}^2 + \beta_z^2 = \beta_0^2 = \omega^2 \mu_0 \varepsilon_0 \Rightarrow \alpha_{y0}^2 = \beta_z^2 - \beta_0^2 = \beta_z^2 - \omega^2 \mu_0 \varepsilon_0 \quad (8\text{-}158b)$$

From the free space looking down the slab we can define an impedance that by using (8-153a) through (8-153f) and (8-158b) can be written as

$$Z_w^{-y0} = -\frac{E_{ze}^{0+}}{H_{xe}^{0+}} = \frac{E_{ze}^{0-}}{H_{xe}^{0-}} = -j\frac{\beta_0^2 - \beta_z^2}{\omega \varepsilon_0 \alpha_{y0}} = j\frac{\alpha_{y0}}{\omega \varepsilon_0} \quad (8\text{-}158c)$$

which is inductive, and it indicates that *TM mode surface waves are supported by inductive surfaces*. In fact, surfaces with inductive impedance characteristics, such as dielectric slabs, dielectric covered ground planes, and corrugated surfaces with certain heights and constitutive parameters, are designed to support TM surface waves.

B. TM^z (ODD)

By following a procedure similar to that used for the TM^z (even), utilizing the odd mode TM^z potential functions of (8-148) through (8-150c), it can be shown that the expression corresponding to (8-158) is

$$\frac{\varepsilon_0}{\varepsilon_d}(\beta_{yd}h) \tan(\beta_{yd}h) = \alpha_{y0}h \quad (8\text{-}159)$$

where (8-158a) and (8-158b) also apply for the TM^z odd modes

C. SUMMARY OF TM^z (EVEN) AND TM^z (ODD) MODES

The most important expressions that are applicable for TM^z even and odd modes for a dielectric slab waveguide are (8-158) through (8-159) which are summarized here.

$$-\frac{\varepsilon_0}{\varepsilon_d}(\beta_{yd}h) \cot(\beta_{yd}h) = \alpha_{y0}h \quad \text{TM}^z \text{ (even)} \quad (8\text{-}160a)$$

$$\frac{\varepsilon_0}{\varepsilon_d}(\beta_{yd}h) \tan(\beta_{yd}h) = \alpha_{y0}h \quad \text{TM}^z \text{ (odd)} \quad (8\text{-}160b)$$

$$\beta_{yd}^2 + \beta_z^2 = \beta_d^2 = \omega^2 \mu_d \varepsilon_d \Rightarrow \beta_{yd}^2 = \beta_d^2 - \beta_z^2 = \omega^2 \mu_d \varepsilon_d - \beta_z^2$$

$$\text{TM}^z \text{ (even and odd)} \quad (8\text{-}160c)$$

$$\boxed{-\alpha_{y0}^2 + \beta_z^2 = \beta_0^2 = \omega^2\mu_0\varepsilon_0 \Rightarrow \alpha_{y0}^2 = \beta_z^2 - \beta_0^2 = \beta_z^2 - \omega^2\mu_0\varepsilon_0}$$

TM^z (even and odd) (8-160d)

$$\boxed{Z_w^{-y0} = -\frac{E_z^{0+}}{H_x^{0+}} = \frac{E_z^{0-}}{H_x^{0-}} = j\frac{\alpha_{y0}}{\omega\varepsilon_0}} \quad \mathrm{TM}^z \text{ (even and odd)} \quad (8\text{-}160\mathrm{e})$$

The objective here is to determine which modes can be supported by the dielectric slab when it is used as a waveguide, and to solve for β_{y0}, α_{y0}, β_z, and the cutoff frequencies for each of these modes by using (8-160a) through (8-160d). We will begin by determining the modes and their corresponding frequencies.

It is apparent from (8-160c) and (8-160d) that if β_z is real, then

1. $\beta_z < \beta_0 < \beta_d$:

$$\beta_{yd} = \pm\sqrt{\beta_d^2 - \beta_z^2} = \text{real} \qquad (8\text{-}161\mathrm{a})$$

$$\alpha_{y0} = \pm j\sqrt{\beta_0^2 - \beta_z^2} = \text{imaginary} \qquad (8\text{-}161\mathrm{b})$$

2. $\beta_z > \beta_d > \beta_0$:

$$\beta_{yd} = \pm j\sqrt{\beta_z^2 - \beta_d^2} = \text{imaginary} \qquad (8\text{-}162\mathrm{a})$$

$$\alpha_{y0} = \pm\sqrt{\beta_z^2 - \beta_0^2} = \text{real} \qquad (8\text{-}162\mathrm{b})$$

3. $\beta_0 < \beta_z < \beta_d$:

$$\beta_{yd} = \pm\sqrt{\beta_d^2 - \beta_z^2} = \text{real} \qquad (8\text{-}163\mathrm{a})$$

$$\alpha_{y0} = \pm\sqrt{\beta_z^2 - \beta_0^2} = \text{real} \qquad (8\text{-}163\mathrm{b})$$

For the dielectric slab to perform as a lossless transmission line, β_{yd}, α_{y0}, and β_z must all be real. Therefore for this to occur,

$$\boxed{\omega\sqrt{\mu_0\varepsilon_0} = \beta_0 < \beta_z < \beta_d = \omega\sqrt{\mu_d\varepsilon_d}} \qquad (8\text{-}164)$$

The lowest frequency for which unattenuated propagation occurs is called the cutoff frequency. For the dielectric slab this occurs when $\beta_z = \beta_0$. Thus at cutoff, $\beta_z = \beta_0$, and (8-158a) and (8-158b) reduce to

$$\beta_{yd}\big|_{\beta_z=\beta_0} = \pm\sqrt{\omega^2\mu_d\varepsilon_d - \beta_z^2}\,\Big|_{\beta_z=\beta_0} = \pm\omega_c\sqrt{\mu_d\varepsilon_d - \mu_0\varepsilon_0} = \pm\omega_c\sqrt{\mu_0\varepsilon_0}\sqrt{\mu_r\varepsilon_r - 1}$$
(8-165a)

$$\alpha_{y0}\big|_{\beta_z=\beta_0} = \pm\sqrt{\beta_z^2 - \omega^2\mu_0\varepsilon_0}\,\Big|_{\beta_z=\beta_0} = 0 \qquad (8\text{-}165\mathrm{b})$$

Through the use of (8-165a) and (8-165b) the nonlinear transcendental equations 8-160a and 8-160b are satisfied, respectively, when the following equations hold.

$$\underline{\text{TM}_m^z \text{ (even)}}$$

$$\cot(\beta_{yd} h) = 0 \Rightarrow \beta_{yd} h = \omega_c h \sqrt{\mu_d \varepsilon_d - \mu_0 \varepsilon_0} = \frac{m\pi}{2}$$

$$\boxed{(f_c)_m = \frac{m}{4h\sqrt{\mu_d \varepsilon_d - \mu_0 \varepsilon_0}} \quad m = 1, 3, 5, \ldots} \tag{8-166a}$$

$$\underline{\text{TM}_m^z \text{ (odd)}}$$

$$\tan(\beta_{yd} h) = 0 \Rightarrow \beta_{yd} h = \omega_c h \sqrt{\mu_d \varepsilon_d - \mu_0 \varepsilon_0} = \frac{m\pi}{2}$$

$$\boxed{(f_c)_m = \frac{m}{4h\sqrt{\mu_d \varepsilon_d - \mu_0 \varepsilon_0}} \quad m = 0, 2, 4, \ldots} \tag{8-166b}$$

It is apparent that the cutoff frequency of a given mode is a function of the electrical constitutive parameters of the dielectric slab and its height. The modes are referred to as odd TM_m^z (when $m = 0, 2, 4, \ldots$), and even TM_m^z (when $m = 1, 3, 5, \ldots$). The dominant mode is the TM_0 which is an odd mode and its cutoff frequency is zero. This means that the TM_0 mode will always propagate unattenuated no matter what the frequency of operation. Other higher-order modes can be cut off by selecting a frequency of operation smaller than their cutoff frequencies.

Now that the TM_m^z (even) and TM_m^z (odd) modes and their corresponding cutoff frequencies have been determined, the next step is to find β_{yd}, α_{y0}, and β_z for any TM^z even or odd mode at any frequency above its corresponding cutoff frequency. This is accomplished by solving the transcendental equations 8-160a and 8-160b.

Assume that $\varepsilon_0/\varepsilon_d$, h, and the frequency of operation f are specified. Then (8-160a) and (8-160b) can be solved numerically through the use of iterative techniques by selecting values of β_{yd} and α_{y0} that balance them. Since multiple combinations of β_{yd} and α_{y0} are possible solutions of (8-160a) and (8-160b), each combination corresponds to a given mode. Once the combination of β_{yd} and α_{y0} values that correspond to a given mode is found, the corresponding value of the phase constant β_z is found by using either (8-160c) or (8-160d).

The solution of (8-160a) through (8-160d) for the values of β_{y0}, α_{y0}, and β_z of a given TM^z mode, once $\varepsilon_0/\varepsilon_d$, h, and f are specified, can also be accomplished graphically. Although such a procedure is considered approximate (its accuracy will depend upon the size of the graph), it does shed much more physical insight onto the radiation characteristics of the modes for the dielectric slab waveguide. With such a procedure it becomes more apparent what must be done to limit the number of unattenuated modes that can be supported by the structure and how to control their characteristics. Let us now demonstrate the graphical solution of (8-160a) through (8-160d).

D. GRAPHICAL SOLUTION FOR TM_m^z (EVEN) AND TM_m^z (ODD) MODES

Equations 8-160a through 8-160d can be solved graphically for the characteristics of the TM^z even and odd modes. This is accomplished by referring to Figure 8-19 where the abscissa represents $\beta_{yd}h$ and the ordinate $\alpha_{y0}h$. The procedure can best be illustrated by considering a specific value of $\varepsilon_0/\varepsilon_d$.

Let us assume that $\varepsilon_0/\varepsilon_d = 1/2.56$. With this value of $\varepsilon_0/\varepsilon_d$, (8-160a) and (8-160b) are plotted for $\alpha_{y0}h$ (ordinate) as a function of $\beta_{yd}h$ (abscissa) as shown in Figure 8-19. The next step is to solve graphically (8-160c) and (8-160d). By combining (8-160c) and (8-160d), we can write that

$$\alpha_{y0}^2 + \beta_{yd}^2 = \beta_d^2 - \beta_0^2 = \omega^2(\mu_d\varepsilon_d - \mu_0\varepsilon_0) \qquad (8\text{-}167)$$

By multiplying both sides by h^2 we can write (8-167) as

$$(\alpha_{y0}h)^2 + (\beta_{yd}h)^2 = (\omega h)^2(\mu_d\varepsilon_d - \mu_0\varepsilon_0) = (\omega h)^2 \mu_0\varepsilon_0(\mu_r\varepsilon_r - 1)$$

$$\boxed{(\alpha_{y0}h)^2 + (\beta_{yd}h)^2 = a^2} \quad TM^z \text{ (even and odd)} \qquad (8\text{-}168)$$

where

$$\boxed{a = \omega h\sqrt{\mu_0\varepsilon_0}\sqrt{\mu_r\varepsilon_r - 1} = \beta_0 h\sqrt{\mu_r\varepsilon_r - 1}} \quad TM^z \text{ (even and odd)} \qquad (8\text{-}168a)$$

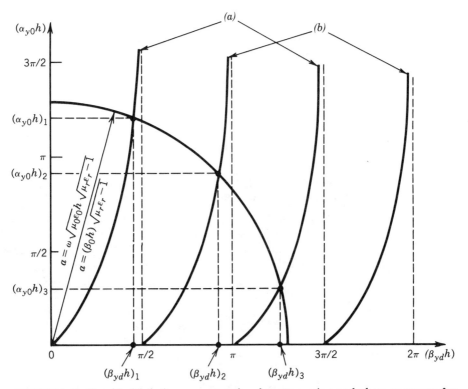

FIGURE 8-19 Graphical solution representation for attenuation and phase constants for a dielectric slab waveguide. (a) TM_m^{zo} odd, $\varepsilon_0/\varepsilon_d(\beta_{yd}h)\tan(\beta_{yd}h)$, $m = 0, 2, 4, \ldots$. (b) TM_m^{ze} even, $-\varepsilon_0/\varepsilon_d(\beta_{yd}h)\cot(\beta_{yd}h)$, $m = 1, 3, 5, \ldots$.

It is recognized that by using the axes $\alpha_{y0}h$ (ordinate) and $\beta_{yd}h$ (abscissa), (8-168) represents a circle with a radius a determined by (8-168a). The extent of the radius is determined by the frequency of operation, the height, and the constitutive electrical parameters of the dielectric slab. The intersections of the circle of (8-168) and (8-168a) with the curves representing (8-160a) and (8-160b), as illustrated in Figure 8-19, determine the modes that propagate unattenuated within the dielectric slab waveguide. For a given intersection representing a given mode, the point of intersection is used to determine the values of $\beta_{yd}h$ and $\alpha_{y0}h$, or β_{yd} and α_{y0} for a specified h, for that mode, as shown in Figure 8-19. Once this is accomplished, the corresponding values of β_z are determined using either (8-160c) or (8-160d). This procedure is followed for each intersection point between the curves representing (8-160a), (8-160b), (8-168), and (8-168a). To illustrate the principles, let us consider a specific example.

Example 8-11. A dielectric slab of polystyrene of half thickness $h = 0.125$ in. (0.3175 cm) and with electrical properties of $\varepsilon_r = 2.56$ and $\mu_r = 1$ is bounded above and below by air. The frequency of operation is 30 GHz.

1. Determine the TM_m^z modes, and their corresponding cutoff frequencies, that propagate unattenuated.
2. Calculate β_{yd} (rad/cm), α_{y0} (Np/cm), β_z (rad/cm), and $(\beta_z/\beta_0)^2$ for the unattenuated TM_m^z modes.

Solution.

1. By using (8-166a) and (8-166b), the cutoff frequencies of the TM_m^z modes that are lower than 30 GHz are

$$(f_c)_0 = 0 \quad TM_0^z \text{ (odd)}$$

$$(f_c)_1 = \frac{30 \times 10^9}{4(0.3175)\sqrt{2.56 - 1}} = 18.913 \text{ GHz} \quad TM_1^z \text{ (even)}$$

The remaining modes have cutoff frequencies that are higher than the desired operational frequency.

2. The corresponding wave numbers for these two modes [TM_0^z (odd) and TM_1^z (even)] will be found by referring to Figure 8-20. Since (8-160a) and (8-160b) are plotted in Figure 8-20 for $\varepsilon_0/\varepsilon_d = 1/2.56$, the only thing that remains is to plot (8-168) where the radius a is given by (8-168a). For $f = 30$ GHz, the radius of (8-168a) that represents the circle of (8-168) is equal to

$$a = \frac{2\pi(30 \times 10^9)(0.3175)}{30 \times 10^9}\sqrt{2.56 - 1} = 2.492$$

This is also plotted in Figure 8-20. The projections from each intersection point to the abscissa ($\beta_{yd}h$ axis) and ordinate ($\alpha_{y0}h$ axis) allow the

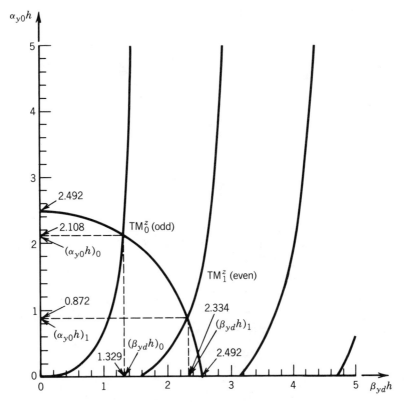

FIGURE 8-20 Graphical solution for attenuation and phase constants of TM_m^z modes in a dielectric slab waveguide ($\varepsilon_r = 2.56$, $\mu_r = 1$, $h = 0.3175$ cm, $f = 30$ GHz).

determination of the corresponding wave numbers. From Figure 8-20

$$\underline{TM_0^z \text{ (odd)}}$$

$$(\beta_{yd}h)_0 = 1.329 \Rightarrow \beta_{yd} = 4.186 \text{ rad/cm}$$

$$(\alpha_{y0}h)_0 = 2.108 \Rightarrow \alpha_{y0} = 6.639 \text{ Np/cm}$$

When these values are substituted in (8-160c) or (8-160d) they lead to

$$\beta_z = 9.140 \text{ rad/cm} \qquad (\beta_z/\beta_0)^2 = 2.116$$

$$\underline{TM_1^z \text{ (even)}}$$

$$(\beta_{yd}h)_1 = 2.334 \Rightarrow \beta_{yd} = 7.351 \text{ rad/cm}$$

$$(\alpha_{y0}h)_1 = 0.872 \Rightarrow \alpha_{y0} = 2.747 \text{ Np/cm}$$

426 RECTANGULAR CROSS-SECTION WAVEGUIDES AND CAVITIES

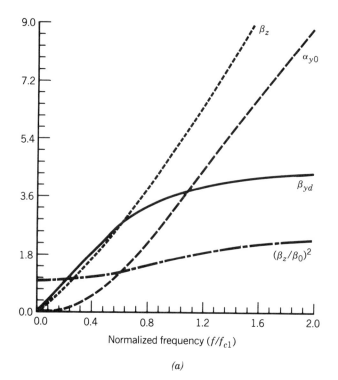

(a)

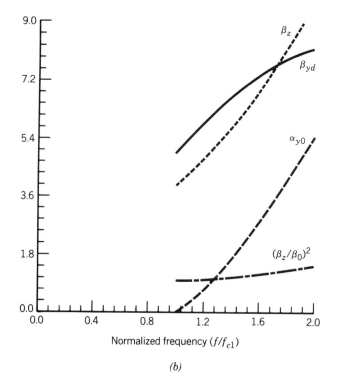

(b)

FIGURE 8-21 Attenuation and phase constants of TM_m^z modes in a dielectric slab waveguide ($\varepsilon_r = 2.56$, $\mu_r = 1$, $h = 0.3175$ cm, $f = 30$ GHz). (a) TM_0^z mode (b) TM_1^z mode.

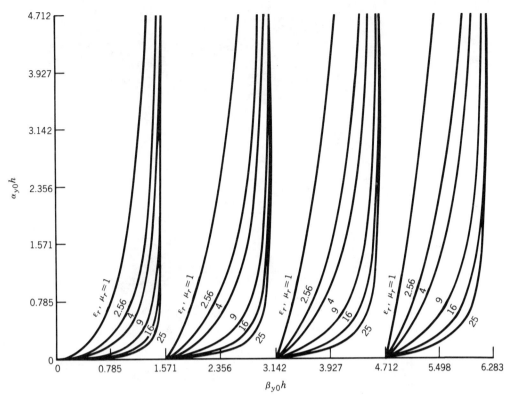

FIGURE 8-22 Curves to be used for graphical solution of attenuation and phase constants for TM_m^z and TE_m^z modes in a dielectric slab waveguide.

When these values are substituted in (8-160c) or (8-160d) they lead to

$$\beta_z = 6.857 \text{ rad/cm} \qquad (\beta_z/\beta_0)^2 = 1.191$$

Equations 8-160a through 8-160d can also be solved simultaneously and analytically for α_{y0}, β_{yd}, and β_z through use of a procedure very similar to that outlined in Section 8.5.1 for the TE^y modes of a partially filled rectangular waveguide. The results are shown, respectively, in Figure 8-21a and b for the TM_0^z and TM_1^z modes of Example 8-11 in the frequency range $0 \leq f \leq 2(f_c)_1$, where $(f_c)_1$ is the cutoff frequency of the TM_1^z mode.

Curves similar to those of Figures 8-19 and 8-20 were generated for $\varepsilon_r = 1$, 2.56, 4, 9, 16, and 25 and are shown in Figure 8-22. These can be used for the solution of TM_m^z problems where the values in the curves will be representing μ_r's instead of ε_r's. This will be seen in the next section.

8.7.3 Transverse Electric (TE^z) Modes

By following a procedure similar to that for the TM^z modes, it can be shown (by leaving out the details) that the critical expressions for the TE^z modes that correspond to those of the TM^z modes of (8-160a) through (8-160d), (8-166a) through (8-166b), (8-168), (8-168a), and (8-160e) are

$$\text{TE}^z \text{ (even) and TE}^z \text{ (odd)}$$

$$\boxed{-\frac{\mu_0}{\mu_d}(\beta_{yd}h)\cot(\beta_{yd}h) = \alpha_{y0}h} \quad \text{TE}^z \text{ (even)} \tag{8-169a}$$

$$\boxed{\frac{\mu_0}{\mu_d}(\beta_{yd}h)\tan(\beta_{yd}h) = \alpha_{y0}h} \quad \text{TE}^z \text{ (odd)} \tag{8-169b}$$

$$\boxed{\beta_{yd}^2 + \beta_z^2 = \beta_d^2 = \omega^2\mu_d\varepsilon_d \Rightarrow \beta_{yd}^2 = \beta_d^2 - \beta_z^2 = \omega^2\mu_d\varepsilon_d - \beta_z^2}$$

$$\text{TE}^z \text{ (even and odd)} \tag{8-169c}$$

$$\boxed{-\alpha_{y0}^2 + \beta_z^2 = \beta_0^2 = \omega^2\mu_0\varepsilon_0 \Rightarrow \alpha_{y0}^2 = \beta_z^2 - \beta_0^2 = \beta_z^2 - \omega^2\mu_0\varepsilon_0}$$

$$\text{TE}^z \text{ (even and odd)} \tag{8-169d}$$

$$\boxed{(f_c)_m = \frac{m}{4h\sqrt{\mu_d\varepsilon_d - \mu_0\varepsilon_0}}} \quad \begin{array}{l} m = 1,3,5,\ldots, \text{ TE}^z \text{ (even)} \\ m = 0,2,4,\ldots, \text{ TE}^z \text{ (odd)} \end{array} \quad \begin{array}{r} (8\text{-}169e) \\ (8\text{-}169f) \end{array}$$

$$\boxed{(\alpha_{y0}h)^2 + (\beta_{yd}h)^2 = a^2} \quad \text{TE}^z \text{ (even and odd)} \tag{8-169g}$$

$$\boxed{a = \omega h\sqrt{\mu_0\varepsilon_0}\sqrt{\mu_r\varepsilon_r - 1} = \beta_0 h\sqrt{\mu_r\varepsilon_r - 1}} \quad \text{TE}^z \text{ (even and odd)} \tag{8-169h}$$

$$\boxed{Z_w^{-y0} = \frac{E_x^{0+}}{H_z^{0+}} = -\frac{E_x^{0-}}{H_z^{0-}} = -j\frac{\omega\mu_0}{\alpha_{y0}}} \quad \text{TE}^z \text{ (even and odd)} \tag{8-169i}$$

Therefore *TE surface waves are capacitive and are supported by capacitive surfaces*, whether they are dielectric slabs, dielectric covered ground planes, or corrugated surfaces.

The solution of these proceeds in the same manner as before. The curves shown in Figure 8-22 must be used where the appropriate value of $\mu_r = \mu_d/\mu_0$ must be selected.

Example 8-12. Repeat the problem of Example 8-11 for the TE_m^z modes.

Solution.

1. By using (8-169e) and (8-169f), the cutoff frequencies of the TE_m^z modes that are smaller than 30 GHz are

$$(f_c)_0 = 0 \quad \text{TE}_0^z \text{ (odd)}$$

$$(f_c)_1 = \frac{30 \times 10^9}{4(0.3175)\sqrt{2.56 - 1}} = 18.913 \text{ GHz} \quad \text{TE}_1^z \text{ (even)}$$

These correspond to the cutoff frequencies of the TM_m^z modes of Example 8-11.

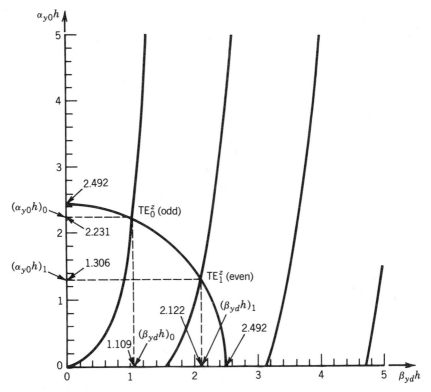

FIGURE 8-23 Graphical solution for attenuation and phase constants of TE_m^z modes in a dielectric slab waveguide ($\varepsilon_r = 2.56$, $\mu_r = 1$, $h = 0.3175$ cm, $f = 30$ GHz).

2. The corresponding wave numbers of the two modes TE_0^z (odd) and TE_1^z (even) are obtained using Figure 8-23. For $f = 30$ GHz the radius of (8-169h), which represents the circle of (8-169g), is the same as that of the TM_m^z modes of Example 8-11, and it is equal to

$$a = \frac{2\pi(30 \times 10^9)(0.3175)}{30 \times 10^9}\sqrt{2.56 - 1} = 2.492$$

This is plotted in Figure 8-23. The projections from each intersection point to the abscissa ($\beta_{yd}h$ axis) and ordinate ($\alpha_{y0}h$ axis) allows the determination of the corresponding wave numbers. From Figure 8-23:

$$\underline{TE_0^z \text{ (odd)}}$$

$$(\beta_{yd}h)_0 = 1.109 \Rightarrow \beta_{yd} = 3.494 \text{ rad/cm}$$
$$(\alpha_{y0}h)_0 = 2.231 \Rightarrow \alpha_{y0} = 7.027 \text{ Np/cm}$$

When these are substituted in (8-169c) or (8-169d),

$$\beta_z = 9.426 \text{ rad/cm} \qquad (\beta_z/\beta_0)^2 = 2.251$$

$$\underline{TE_1^z \text{ (even)}}$$

$$(\beta_{yd}h)_1 = 2.122 \Rightarrow \beta_{yd} = 6.684 \text{ rad/cm}$$
$$(\alpha_{y0}h)_1 = 1.306 \Rightarrow \alpha_{y0} = 4.113 \text{ Np/cm}$$

430 RECTANGULAR CROSS-SECTION WAVEGUIDES AND CAVITIES

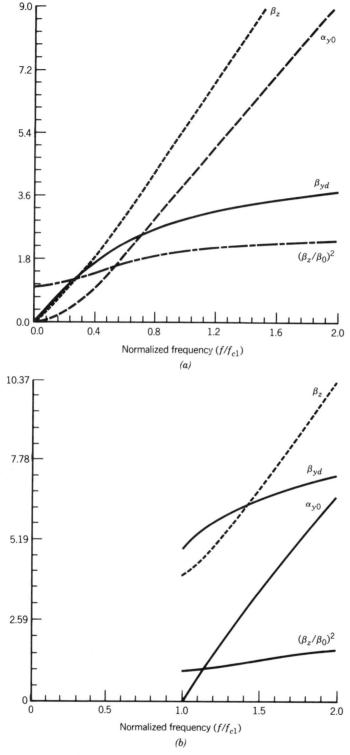

FIGURE 8-24 Attenuation and phase constants of TE_m^z modes in a dielectric slab waveguide ($\varepsilon_r = 2.56$, $\mu_r = 1$, $h = 0.3175$ cm, $f = 30$ GHz). (a) TE_0^z mode (b) TE_1^z mode.

When these are substituted in (8-169c) or (8-169d),

$$\beta_z = 7.510 \text{ rad/cm} \qquad (\beta_z/\beta_0)^2 = 1.428$$

For the TM^z modes, (8-169a) through (8-169d) can be solved simultaneously and analytically for α_{y0}, β_{yd}, and β_z using a procedure very similar to that outlined in Section 8.5.1 for the TE^y modes of a partially filled rectangular waveguide. The results are shown in Figure 8-24a and b for the TE$_0^z$ and TE$_1^z$ modes, respectively, of Example 8-12 in the frequency range $0 \le f \le 2(f_c)_1$ where $(f_c)_1$ is the cutoff frequency of the TE$_1^z$ mode.

8.7.4 Ray-Tracing Method

In Sections 8.7.2 and 8.7.3 we analyzed the dielectric slab waveguide as a boundary-value problem using modal techniques. In this section we want to repeat the analysis of both TE^z and TM^z modes by using a ray-tracing method that sheds more physical insight onto the propagation characteristics of the dielectric slab waveguide but is not as detailed.

A wave beam that is fed into the dielectric slab can propagate into three possible modes [11]. Let us assume that the slab is bounded above by air and below by another dielectric slab as shown in Figure 8-25 such that $\varepsilon_1 > \varepsilon_2 > \varepsilon_0$. Mathematically the problem involves a solution of Maxwell's equations and the appropriate boundary conditions at the two interfaces, as was done in Sections 8.7.2 and 8.7.3. The wave beam has the following properties.

1. It can radiate from the slab into both air and substrate, referred to as the *air-substrate* modes, as shown in Figure 8-25a.
2. It can radiate from the slab only into the substrate, referred to as the *substrate modes*, as shown in Figure 8-25b.
3. It can be bounded and be guided by the slab, referred to as the *waveguide modes*, as shown in Figure 8-25c.

To demonstrate these properties, let us assume that a wave enters the slab which is bounded above by air and below by a substrate such that $\varepsilon_1 > \varepsilon_2 > \varepsilon_0$.

1. *Air-Substrate Modes*: Referring to Figure 8-25a, let us increase θ_1 gradually starting at $\theta_1 = 0$. When θ_1 is small, a wave that enters the slab will be refracted and will exit into the air and the substrate provided that $\theta_1 < (\theta_c)_{10} = \sin^{-1}(\sqrt{\varepsilon_0/\varepsilon_1}) < (\theta_c)_{12} = \sin^{-1}(\sqrt{\varepsilon_2/\varepsilon_1})$. The angles $(\theta_c)_{10}$ and $(\theta_c)_{12}$ represent, respectively, the critical angles at the slab–air and slab–substrate interfaces. In this situation wave energy can propagate freely in all three media (air, slab, and substrate) and can create radiation fields (air–substrate modes).
2. *Substrate Modes*: When θ_1 increases such that it passes the critical angle $(\theta_c)_{10}$ of the slab–air interface but is smaller than the critical angle $(\theta_c)_{12}$ of the slab–substrate interface $[(\theta_c)_{12} = \sin^{-1}(\sqrt{\varepsilon_2/\varepsilon_1}) > \theta_1 > (\theta_c)_{10} = \sin^{-1}(\sqrt{\varepsilon_0/\varepsilon_1})]$, then $\sin\theta_0 > 1$ which indicates that the wave is totally reflected at the slab–air interface. This describes a solution that wave energy in the slab radiates only in the substrate, as shown in Figure 8-25b. These are referred to as *substrate modes*.

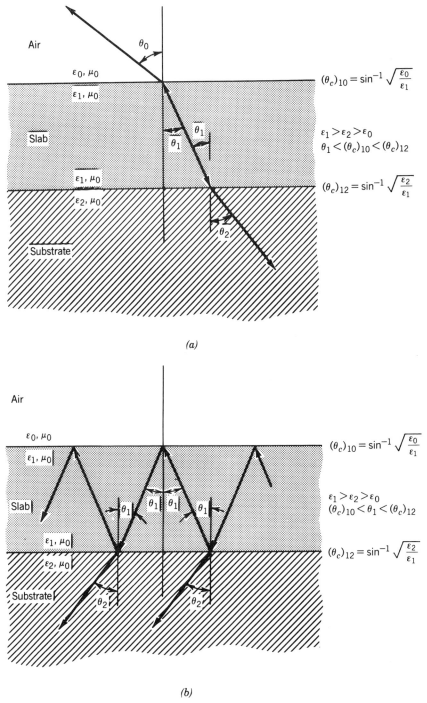

FIGURE 8-25 Propagation modes in a dielectric slab waveguide. (*a*) Air–substrate modes. (*b*) Substrate modes. (*c*) Waveguide modes.

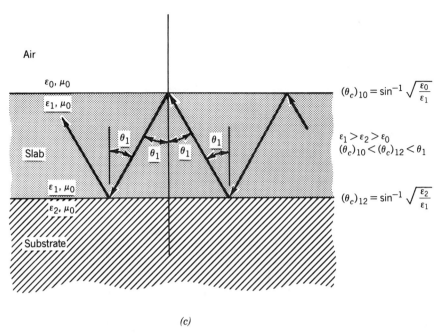

(c)

FIGURE 8-25 (*Continued*).

3. *Waveguide Modes*: Finally when θ_1 is larger than the critical angle $(\theta_c)_{12}$ of the slab–substrate interface $[\theta_1 > (\theta_c)_{12} = \sin^{-1}(\sqrt{\varepsilon_2/\varepsilon_1}) > (\theta_c)_{10} = \sin^{-1}(\sqrt{\varepsilon_0/\varepsilon_1})]$, then $\sin\theta_0 > 1$ and $\sin\theta_2 > 1$ which indicate that the wave is totally reflected in both interfaces. These are referred to as *waveguide modes*, as shown in Figure 8-25c. For these modes the energy is trapped inside the slab, and the waves follow the wave motion pattern represented by two wave vectors $\mathbf{A}_1$ and $\mathbf{B}_1$ as shown in Figure 8-26a. These two vectors are decomposed into their horizontal and vertical components $(\mathbf{A}_{1z}, \mathbf{A}_{1y})$ and $(\mathbf{B}_{1z}, \mathbf{B}_{1y})$.

The horizontal wave vector components are equal which indicates that the waves propagate with a constant velocity in the z direction. However, the vertical components of $\mathbf{A}_1$ and $\mathbf{B}_1$ represent opposite traveling waves which when combined form a standing wave. By changing the angle θ_1, we change the direction of $\mathbf{A}_1$ and $\mathbf{B}_1$. This results in changes in the horizontal and vertical components of $\mathbf{A}_1$ and $\mathbf{B}_1$, in the wave velocity in the z direction, and in the standing wave pattern across the slab.

The wave vectors $\mathbf{A}_1$ and $\mathbf{B}_1$ can be thought to represent a plane wave which bounces back and forth inside the slab. The phase fronts of this plane wave are dashed in Figure 8-26b. An observer who moves in a direction parallel to the z axis does not see the horizontal components of the wave vectors. He does, however, observe a plane wave that bounces upward and downward, which folds one directly on top of the other. In order for the standing wave pattern across the slab to remain the same as the observer travels along the z axis, all multiple reflected waves must add in phase. This is accomplished by having the plane wave that makes one round trip, up and down across the slab, experience a phase shift equal to $2m\pi$, where m is an integer [11]. Otherwise if after the first round trip the wave experiences a small differential phase shift of δ away from $2m\pi$, it will experience differential phase

434 RECTANGULAR CROSS-SECTION WAVEGUIDES AND CAVITIES

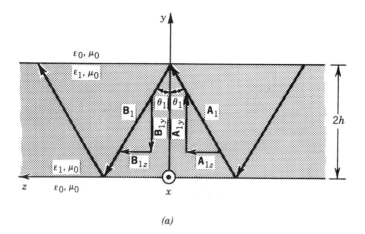

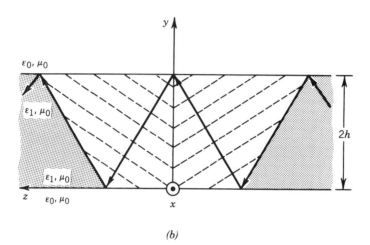

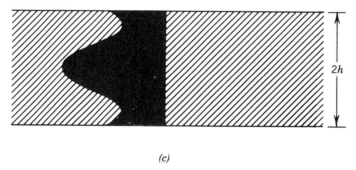

FIGURE 8-26 (a) Reflecting plane wave representation, (b) phase wavefronts, and (c) broadside amplitude pattern of waveguide modes in a dielectric slab waveguide.

shifts $2\delta, 3\delta, \ldots$ after the second, third, ... trips. Therefore these higher-order reflected waves will experience larger differential phase shifts which when added will eventually equal zero and the broadside wave pattern of Figure 8-16c will be a function of the axial position.

A one round trip phase shift must include not only the phase change that is due to the distance traveled by the wave but also the changes in the wave phase that are due to reflections from the upper and lower interface. If the phase constant of the plane wave along wave vectors $\mathbf{A}_1$ and $\mathbf{B}_1$ is β_1, then the wave constant along the vertical direction y is $\beta_1 \cos \theta_1$. Therefore the total phase shift to one round trip (up and down) of wave travel, including the phase changes due to reflection, must be equal to

$$\boxed{4\beta_1 h \cos \theta_1 - \phi_{10} - \phi_{12} = 2m\pi} \qquad m = 0, 1, 2, \ldots \qquad (8\text{-}170)$$

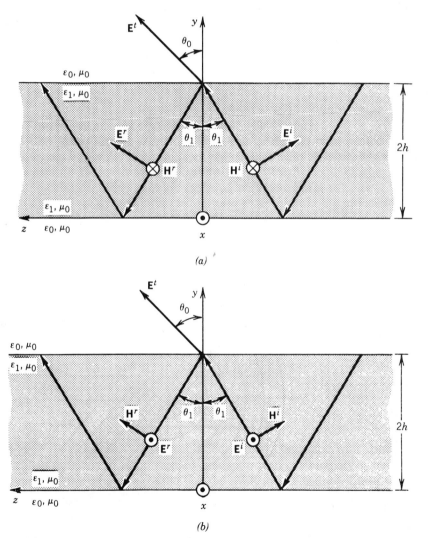

FIGURE 8-27 (*a*) TM^z (parallel polarization) and (*b*) TE^z (perpendicular polarization) modes in a dielectric slab waveguide.

where ϕ_{10} = phase of reflection coefficient at slab–air interface
ϕ_{12} = phase of reflection coefficient at slab–substrate interface

The phases of the reflection coefficients are assumed to be leading. Assume that the media above and below the slab are the same. Then $\phi_{10} = \phi_{12} = \phi$. Thus (8-170) reduces to

$$\boxed{4\beta_1 h \cos\theta_1 - 2\phi = 2m\pi} \qquad m = 0, 1, 2, \ldots \qquad (8\text{-}170a)$$

The preceding equations will now be applied to both TM^z and TE^z modes.

A. TRANSVERSE MAGNETIC (TM^z) MODES (PARALLEL POLARIZATION)

Let us assume that the bouncing plane wave of Figure 8-26a is such that its polarization is TM^z (or parallel polarization) as shown in Figure 8-27a where the slab is bounded on both sides by air. For the orientation of the fields taken as shown in Figure 8-27a, the reflected electric field $\mathbf{E}_\parallel^r$ is related to the incident electric field $\mathbf{E}_\parallel^i$ by

$$\frac{E_\parallel^r}{E_\parallel^i} = -\Gamma_\parallel^b = -\left[\frac{\eta_0 \cos\theta_0 - \eta_1 \cos\theta_1}{\eta_0 \cos\theta_0 + \eta_1 \cos\theta_1}\right] = \frac{\eta_1 \cos\theta_1 - \eta_0 \cos\theta_0}{\eta_1 \cos\theta_1 + \eta_0 \cos\theta_0} \qquad (8\text{-}171)$$

where $\Gamma_\parallel^b$ is the reflection coefficient of (5-24c).

For $\mu_1 = \mu_0$ and for an incidence angle θ_1 greater than the critical angle $(\theta_c)_{10}[\theta_1 > (\theta_c)_{10} = \sin^{-1}(\sqrt{\varepsilon_0/\varepsilon_1})]$, (8-171) reduces using Snell's law of refraction of 5-24b or 5-35 to

$$-\Gamma_\parallel^b = \frac{\cos\theta_1 + j\sqrt{\dfrac{\varepsilon_1}{\varepsilon_0}}\sqrt{\dfrac{\varepsilon_1}{\varepsilon_0}\sin^2\theta_1 - 1}}{\cos\theta_1 - j\sqrt{\dfrac{\varepsilon_1}{\varepsilon_0}}\sqrt{\dfrac{\varepsilon_1}{\varepsilon_0}\sin^2\theta_1 - 1}} = |\Gamma_\parallel^b|\underline{/\phi_\parallel'} = 1\underline{/\phi_\parallel} \qquad (8\text{-}172)$$

where

$$\phi_\parallel = 2\tan^{-1}\left[\frac{\sqrt{\dfrac{\varepsilon_1}{\varepsilon_0}}\sqrt{\dfrac{\varepsilon_1}{\varepsilon_0}\sin^2\theta_1 - 1}}{\cos\theta_1}\right] \qquad (8\text{-}172a)$$

Therefore the transcendental equation that governs these modes is derived by using (8-170a) and (8-172a). Thus

$$4\beta_1 h \cos\theta_1 - 2\phi_\parallel = 4\beta_1 h \cos\theta_1 - 4\tan^{-1}\left[\frac{\sqrt{\dfrac{\varepsilon_1}{\varepsilon_0}}\sqrt{\dfrac{\varepsilon_1}{\varepsilon_0}\sin^2\theta_1 - 1}}{\cos\theta_1}\right] = 2m\pi$$

$$\beta_1 h \cos\theta_1 - \frac{m\pi}{2} = \tan^{-1}\left[\frac{\sqrt{\dfrac{\varepsilon_1}{\varepsilon_0}}\sqrt{\dfrac{\varepsilon_1}{\varepsilon_0}\sin^2\theta_1 - 1}}{\cos\theta_1}\right]$$

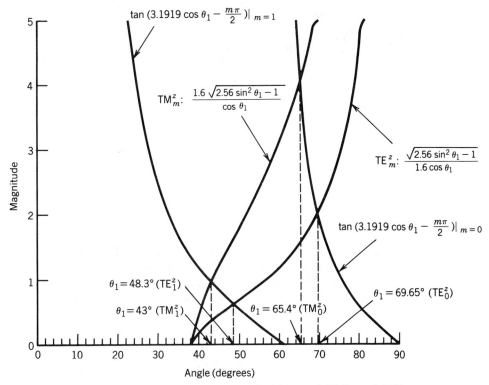

FIGURE 8-28 Graphical solution for angles of incidence of TM_m^z and TE_m^z modes in a dielectric slab waveguide.

$$\tan\left(\beta_1 h \cos\theta_1 - \frac{m\pi}{2}\right) = \left[\frac{\sqrt{\frac{\varepsilon_1}{\varepsilon_0}}\sqrt{\frac{\varepsilon_1}{\varepsilon_0}\sin^2\theta_1 - 1}}{\cos\theta_1}\right]$$

$$\boxed{\tan\left(\beta_1 h \cos\theta_1 - \frac{m\pi}{2}\right) = \frac{\sqrt{\varepsilon_r}\sqrt{\varepsilon_r \sin^2\theta_1 - 1}}{\cos\theta_1}} \qquad m = 0, 1, 2, \ldots \quad (8\text{-}173)$$

Example 8-13. The polystyrene dielectric slab of Example 8-11 of half thickness $h = 0.125$ in. (0.3175 cm) and with electrical properties $\varepsilon_r = 2.56$ and $\mu_r = 1$ is bounded above and below by air. For a frequency of operation of 30 GHz determine the TM^z modes and their corresponding angles of incidence within the slab. Plot the incidence angles as a function of frequency in the range $0 \leq f/(f_c)_1 \leq 2$ where $(f_c)_1$ is the cutoff frequency of the TM_1^z mode.

Solution. The solution for this set of modes is governed by (8-173). For $\varepsilon_r = 2.56$ the right side of (8-173) is plotted in Figure 8-28. For $f = 30$ GHz, the height h of the slab is equal to

$$\lambda_1 = \frac{30 \times 10^9}{\sqrt{2.56}\, 30 \times 10^9} = 0.6250 \text{ cm} \Rightarrow h = \frac{0.3175}{0.6250}\lambda_1 = 0.5080\lambda_1$$

Thus (8-173) reduces to

$$\tan\left(\beta_1 h \cos\theta_1 - \frac{m\pi}{2}\right) = \tan\left[\frac{2\pi}{\lambda_1}(0.5080\lambda_1)\cos\theta_1 - \frac{m\pi}{2}\right]$$

$$= \tan\left(3.1919\cos\theta_1 - \frac{m\pi}{2}\right) = \frac{1.6\sqrt{2.56\sin^2\theta_1 - 1}}{\cos\theta_1}$$

The left side is plotted for $m = 0, 1$ in Figure 8-28, and the intersections of these with that curve representing the right side of (8-173) are solutions. From Figure 8-28, there are two intersections that occur at $\theta_1 = 65.4°$ and $43°$ which represent, respectively, the modes TM_0^z ($\theta_1 = 65.4°$) and TM_1^z ($\theta_1 = 43°$). They agree with the modes of Example 8-11. No other modes are present because curves of the left side of (8-173) for higher orders of m ($m = 2, 3, \ldots$) do not intersect with the curve that represents the right side of (8-173). Remember also that the critical angle for the slab–air interface is equal to

$$(\theta_c)_{10} = \sin^{-1}\left(\sqrt{\varepsilon_0/\varepsilon_1}\right) = \sin^{-1}\left(\sqrt{1/2.56}\right) = 38.68°$$

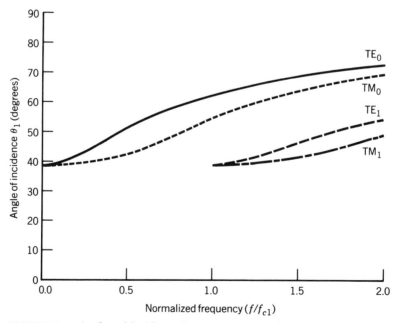

FIGURE 8-29 Angles of incidence for TM_m^z and TE_m^z modes in a dielectric slab waveguide.

DIELECTRIC WAVEGUIDE **439**

and the curve that represents the right side of (8-173) does not exist in Figure 8-28 below 38.68°.

In the slab, the wave number in the direction of incidence is equal to

$$\beta_1 = \frac{2\pi}{\lambda_1} = \frac{2\pi}{0.6250} = 10.053 \text{ rad/cm}$$

Therefore for each mode, the wave numbers in the y direction are equal to

TM_0^z: $\qquad \beta_{yd} = \beta_1 \cos\theta_1 = 10.053 \cos(65.4°) = 4.185 \text{ rad/cm}$

TM_1^z: $\qquad \beta_{yd} = \beta_1 \cos\theta_1 = 10.053 \cos(43°) = 7.352 \text{ rad/cm}$

which closely agree with the corresponding wave numbers obtained graphically in Example 8-11. The angles of incidence as a function of frequency in the range $0 \leq f/(f_c)_1 \leq 2$, where $(f_c)_1$ is the cutoff frequency of the TM_1 mode, are shown plotted in Figure 8-29.

B. TRANSVERSE ELECTRIC (TE^z) MODES (PERPENDICULAR POLARIZATION)

Let us now assume that the bouncing plane wave of Figure 8-26a is such that its polarization is TE^z (or perpendicular polarization) as shown in Figure 8-27b where the slab is bounded on both sides by air. For the orientation of the fields taken as shown in Figure 8-27b, the reflected electric field $\mathbf{E}_\perp^r$ is related to the incident electric field $\mathbf{E}_\perp^i$ by

$$\frac{E_\perp^r}{E_\perp^i} = \Gamma_\perp^b = \frac{\eta_0 \cos\theta_1 - \eta_1 \cos\theta_0}{\eta_0 \cos\theta_1 + \eta_1 \cos\theta_0} \qquad (8\text{-}174)$$

where $\Gamma_\perp^b$ is the reflection coefficient of (5-17a).

For $\mu_1 = \mu_0$ and for an incidence angle θ_1 greater than the critical angle $(\theta_c)_{10}[\theta_1 > (\theta_c)_{10} = \sin^{-1}(\sqrt{\varepsilon_0/\varepsilon_1})]$, (8-174) reduces using Snell's law of refraction of (5-15b) or (5-35) to

$$\Gamma_\perp^b = \frac{\cos\theta_1 + j\sqrt{\sin^2\theta_1 - \frac{\varepsilon_0}{\varepsilon_1}}}{\cos\theta_1 - j\sqrt{\sin^2\theta_1 - \frac{\varepsilon_0}{\varepsilon_1}}} = |\Gamma_\perp^b|\underline{/\phi_\perp} = 1\underline{/\phi_\perp} \qquad (8\text{-}175)$$

where

$$\phi_\perp = 2\tan^{-1}\left[\frac{\sqrt{\sin^2\theta_1 - \frac{\varepsilon_0}{\varepsilon_1}}}{\cos\theta_1}\right] \qquad (8\text{-}175a)$$

Therefore the transcendental equation that governs these modes is derived by using (8-170a) and (8-175a). Thus

$$4\beta_1 h \cos\theta_1 - 2\phi_\perp = 4\beta_1 h \cos\theta_1 - 4\tan^{-1}\left(\frac{\sqrt{\sin^2\theta_1 - \frac{\varepsilon_0}{\varepsilon_1}}}{\cos\theta_1}\right) = 2m\pi$$

$$\beta_1 h \cos\theta_1 - \frac{m\pi}{2} = \tan^{-1}\left(\frac{\sqrt{\sin^2\theta_1 - \frac{\varepsilon_0}{\varepsilon_1}}}{\cos\theta_1}\right)$$

$$\tan\left(\beta_1 h \cos\theta_1 - \frac{m\pi}{2}\right) = \frac{\sqrt{\sin^2\theta_1 - \frac{\varepsilon_0}{\varepsilon_1}}}{\cos\theta_1}$$

$$\tan\left(\beta_1 h \cos\theta_1 - \frac{m\pi}{2}\right) = \frac{\sqrt{\sin^2\theta_1 - \frac{1}{\varepsilon_r}}}{\cos\theta_1}$$

$$\boxed{\tan\left(\beta_1 h \cos\theta_1 - \frac{m\pi}{2}\right) = \frac{\sqrt{\varepsilon_r \sin^2\theta_1 - 1}}{\sqrt{\varepsilon_r}\,\cos\theta_1} \qquad m = 0,1,2,\ldots} \qquad (8\text{-}176)$$

Example 8-14. For the polystyrene slab of Examples 8-12 and 8-13, determine the TE_m^z modes and their corresponding angles of incidence within the slab. Plot the angles of incidence as a function of frequency in the range $0 \le f/(f_c)_1 \le 2$ where $(f_c)_1$ is the cutoff frequency of the TE_1 mode.

Solution. The solution for this set of modes is governed by (8-176). From the solution of Example 8-13

$$h = 0.5080\lambda_1$$

and (8-176) reduces to

$$\tan\left(3.1919\cos\theta_1 - \frac{m\pi}{2}\right) = \frac{\sqrt{2.56\sin^2\theta_1 - 1}}{1.6\cos\theta_1} \qquad m = 0,1,2,\ldots$$

The right side is plotted in Figure 8-28. The left side for $m = 0, 1$ is also plotted in Figure 8-28. The intersections of these curves represent the solutions that from Figure 8-28 correspond, respectively, to the modes TE_0^z ($\theta_1 = 69.65°$) and TE_1^z ($\theta_1 = 48.3°$). These agree with the modes of Example 8-12. In the

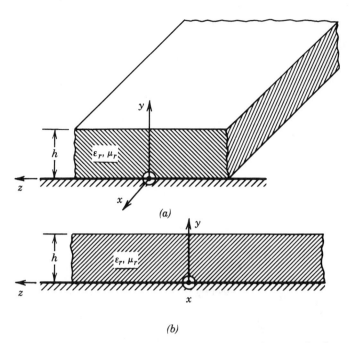

FIGURE 8-30 Geometry for dielectric covered ground plane waveguide. (*a*) Perspective. (*b*) Side view.

slab, the wave number in the direction of incidence is equal to

$$\beta_1 = \frac{2\pi}{\lambda_1} = \frac{2\pi}{0.6250} = 10.053 \text{ rad/cm}$$

Therefore for each mode, the wave numbers in the y direction are equal to

TE$_0^z$: $\beta_{yd} = \beta_1 \cos\theta_1 = 10.053 \cos(69.65°) = 3.496$ rad/cm

TE$_1^z$: $\beta_{yd} = \beta_1 \cos\theta_1 = 10.053 \cos(48.3°) = 6.688$ rad/cm

which closely agree with the corresponding wave numbers obtained graphically in Example 8-12. The angles of incidence as a function of frequency in the range $0 \leq f/(f_c)_1$, where $(f_c)_1$ is the cutoff frequency of the TE$_1$ mode, are shown plotted in Figure 8-29.

8.7.5 Dielectric Covered Ground Plane

Another dielectric type of waveguide is that of a ground plane covered with a dielectric slab of height h, as shown in Figure 8-30. The field analysis of this is similar to the dielectric slab of Sections 8.7.2 and 8.7.3. However, instead of going through all of the details, the solution can be obtained by examining the solutions for the dielectric slab as applied to the dielectric covered ground plane. For $y \geq h$ the main differences between the two geometries are the additional boundary conditions at $y = 0$ for the dielectric covered ground plane.

For the dielectric slab of Figure 8-17 the electric field components *within the dielectric slab* for the TM^z and TE^z modes (even and odd) of Figure 8-18 can be written from Sections 8.7.2 and 8.7.3 as

$$\underline{\text{TM}^z \text{ (even)}}$$

$$\left.\begin{array}{l} E_{xe}^d = 0 \\[6pt] E_{ye}^d = \dfrac{\beta_{yd}\beta_z}{\omega\mu_d\varepsilon_d} A_{me}^d \sin(\beta_{yd}y) e^{-j\beta_z z} \\[10pt] E_{ze}^d = -j\dfrac{\beta_d^2 - \beta_z^2}{\omega\mu_d\varepsilon_d} A_{me}^d \cos(\beta_{yd}y) e^{-j\beta_z z} \end{array}\right\} \quad |y| \le h \qquad (8\text{-}177\text{a})$$

$$\underline{\text{TM}^z \text{ (odd)}}$$

$$\left.\begin{array}{l} E_{xo}^d = 0 \\[6pt] E_{yo}^d = -\dfrac{\beta_{yd}\beta_z}{\omega\mu_d\varepsilon_d} A_{mo}^d \cos(\beta_{yd}y) e^{-j\beta_z z} \\[10pt] E_{zo}^d = -j\dfrac{\beta_d^2 - \beta_z^2}{\omega\mu_d\varepsilon_d} A_{mo}^d \sin(\beta_{yd}y) e^{-j\beta_z z} \end{array}\right\} \quad |y| \le h \qquad (8\text{-}177\text{b})$$

$$\underline{\text{TE}^z \text{ (even)}}$$

$$\left.\begin{array}{l} E_{xe}^d = \dfrac{\beta_{yd}}{\varepsilon_d} B_{me}^d \sin(\beta_{yd}y) e^{-j\beta_z z} \\[10pt] E_{ye}^d = 0 \\[6pt] E_{ze}^d = 0 \end{array}\right\} \quad |y| \le h \qquad (8\text{-}177\text{c})$$

$$\underline{\text{TE}^z \text{ (odd)}}$$

$$\left.\begin{array}{l} E_{xo}^d = -\dfrac{\beta_{yd}}{\varepsilon_d} B_{mo}^d \cos(\beta_{yd}y) e^{-j\beta_z z} \\[10pt] E_{yo}^d = 0 \\[6pt] E_{zo}^d = 0 \end{array}\right\} \quad |y| \le h \qquad (8\text{-}177\text{d})$$

By examining (8-177a) through (8-177d), it is apparent that the tangential electric field components TM^z (odd) of (8-177b) and TE^z (even) of (8-177c) *do* satisfy the boundary conditions of Figure 8-29 at $y = 0$ (vanishing tangential electric components at $y = 0$). However, those TM^z (even) of (8-177a) and TE^z (odd) of (8-177d) *do not* satisfy the boundary conditions of the tangential electric field components at $y = 0$. *Therefore the geometry of Figure 8-30 supports only modes that are TM^z (odd) and TE^z (even).* From Sections 8.7.2 and 8.7.3, the governing equations for the geometry of Figure 8-30 for TM^z (odd) and TE^z (even) modes are

$$\boxed{\frac{\varepsilon_0}{\varepsilon_d}(\beta_{yd}h)\tan(\beta_{yd}h) = (\alpha_{y0}h)} \quad \text{TM}^z \text{ (odd)} \quad (8\text{-}178\text{a})$$

$$\boxed{-\frac{\mu_0}{\mu_d}(\beta_{yd}h)\cot(\beta_{yd}h) = (\alpha_{y0}h)} \quad \text{TE}^z \text{ (even)} \quad (8\text{-}178\text{b})$$

$$\boxed{\beta_{yd}^2 + \beta_z^2 = \beta_d^2 = \omega^2\mu_d\varepsilon_d} \quad \text{TM}^z \text{ (odd), TE}^z \text{ (even)} \quad (8\text{-}178\text{c})$$

$$\boxed{-\alpha_{y0}^2 + \beta_z^2 = \beta_0^2 = \omega^2\mu_0\varepsilon_0} \quad \text{TM}^z \text{ (odd), TE}^z \text{ (even)} \quad (8\text{-}178\text{d})$$

$$\boxed{(f_c)_m = \frac{m}{4h\sqrt{\mu_d\varepsilon_d - \mu_0\varepsilon_0}}} \quad \begin{array}{l} m = 0,2,4,\ldots, \text{TM}^z \text{ (odd)} \quad (8\text{-}178\text{e}) \\ m = 1,3,5,\ldots, \text{TE}^z \text{ (even)} \quad (8\text{-}178\text{f}) \end{array}$$

Thus the dominant mode is the TM_0^z with a zero cutoff frequency. All the modes in a dielectric covered ground plane are usually referred to as *surface wave modes*, and their solutions are obtained in the same manner as outlined in Sections 8.7.2, 8.7.3, and 8.7.4 for the dielectric slab waveguide. The only difference is that for the dielectric covered ground plane we only have TM_m^z (odd) and TE_m^z (even) modes. The structure *cannot* support TM_m^z (even) and TE_m^z (odd) modes.

The attenuation rate of the evanescent fields in air above the dielectric cover is determined by the value of α_{y0} which is found using (8-178d). Above cutoff it is expressed as

$$\alpha_{y0} = \sqrt{\beta_z^2 - \beta_0^2} = \sqrt{\beta_z^2 - \omega^2\mu_0\varepsilon_0} \quad (8\text{-}179)$$

For very thick dielectrics ($h \to$ large) the phase constant β_z approaches β_d ($\beta_z \to \beta_d$). Thus

$$\alpha_{y0}|_{h \to \text{large}} = \sqrt{\beta_z^2 - \omega^2\mu_0\varepsilon_0} \simeq \omega\sqrt{\mu_d\varepsilon_d - \mu_0\varepsilon_0} = \omega\sqrt{\mu_0\varepsilon_0}\sqrt{\frac{\mu_d\varepsilon_d}{\mu_0\varepsilon_0} - 1} \quad (8\text{-}180)$$

which is usually very large.

For very thin dielectrics ($h \to$ small) the phase constant β_z approaches β_0 ($\beta_z \to \beta_0$). Thus from (8-178c) and (8-178d),

$$\beta_{yd}^2|_{h \to \text{small}} = \beta_d^2 - \beta_z^2 \simeq \beta_d^2 - \beta_0^2 = \omega^2(\mu_d\varepsilon_d - \mu_0\varepsilon_0) = \omega^2\mu_0\varepsilon_0\left(\frac{\mu_d\varepsilon_d}{\mu_0\varepsilon_0} - 1\right) \quad (8\text{-}181\text{a})$$

$$\alpha_{y0}^2|_{h \to \text{small}} = \beta_z^2 - \beta_0^2 \simeq \text{small} \quad (8\text{-}181\text{b})$$

For small values of h, (8-178a) reduces for α_{y0} to

$$\alpha_{y0}|_{h \to \text{small}} = \beta_{yd}\frac{\varepsilon_0}{\varepsilon_d}\tan(\beta_{yd}h) \stackrel{h \to 0}{\simeq} h\frac{\varepsilon_0}{\varepsilon_d}(\beta_{yd})^2 \quad (8\text{-}182)$$

Substituting (8-181a) into (8-182) reduces it to

$$\alpha_{y0}|_{h \to \text{small}} \simeq h\frac{\varepsilon_0}{\varepsilon_d}\left[\omega^2\mu_0\varepsilon_0\left(\frac{\mu_d\varepsilon_d}{\mu_0\varepsilon_0} - 1\right)\right] = h\beta_0^2\left(\frac{\mu_d}{\mu_0} - \frac{\varepsilon_0}{\varepsilon_d}\right) = 2\pi\beta_0\left(\frac{\mu_d}{\mu_0} - \frac{\varepsilon_0}{\varepsilon_d}\right)\frac{h}{\lambda_0}$$
(8-182a)

which is usually very small.

Example 8-15. A ground plane is covered with a dielectric sheet of polystyrene of height h. Determine the distance δ (skin depth) above the dielectric–air interface so that the evanescent fields above the sheet will decay to $e^{-1} = 0.368$ of their value at the interface, when h is very large and h is very small ($= 10^{-3}\lambda_0$).

Solution. The distance the wave travels and decays to 36.8% of its value is referred to as the skin depth.

For h very large, according to (8-180),

$$\delta = \frac{1}{\alpha_{y0}} \simeq \frac{1}{\beta_0\sqrt{\frac{\mu_d\varepsilon_d}{\mu_0\varepsilon_0} - 1}} = \frac{\lambda_0}{2\pi\sqrt{2.56 - 1}} = 0.126\lambda_0$$

and the wave is said to be "tightly bound" to the thick dielectric sheet.

For h very small ($h = 10^{-3}\lambda_0$), according to (8-182a),

$$\delta = \frac{1}{\alpha_{y0}} \simeq \frac{\lambda_0}{2\pi\beta_0\left(\frac{\mu_d}{\mu_0} - \frac{\varepsilon_0}{\varepsilon_d}\right)h} = \frac{\lambda_0}{(2\pi)^2\left(1 - \frac{1}{2.56}\right) \times 10^{-3}} = 41.6\lambda_0$$

and the wave is said to be "loosely bound" to the thin dielectric sheet.

8.8 STRIPLINE AND MICROSTRIP LINES

Microwave printed circuit technology has advanced considerably with the introduction of the stripline and microstrip transmission lines [12–27]. These lines are shown, respectively, in Figure 8-31a and b. The stripline consists of a center conductor embedded in a dielectric material that is sandwiched between two conducting plates. The microstrip consists of a thin conducting strip placed above a dielectric material, usually referred to as the substrate, which is supported on its bottom by a conducting plate. Both of these lines have evolved from the coaxial line in stages illustrated in Figure 8-32. In general the stripline and microstrip are lightweight, miniature, easy to fabricate with integrated circuit techniques, and cost effective. Their principal mode of operation is that of the quasi-TEM mode, although higher-order modes, including surface waves, are evident at higher frequencies. In comparison to other popular transmission lines, such as the coax and the waveguide, the stripline and microstrip possess characteristics that are shown listed in Table 8-8. Each of the lines will be discussed by using the most elementary

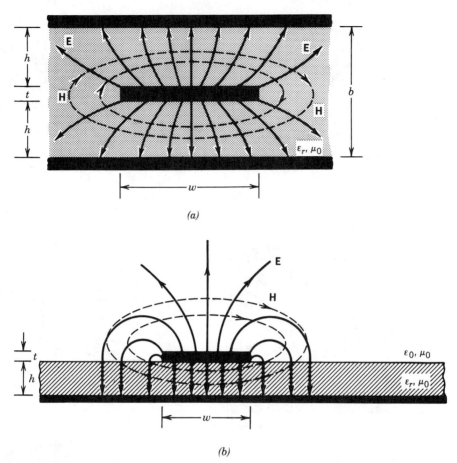

FIGURE 8-31 Geometries for stripline and microstrip transmission lines. (*a*) Stripline. (*b*) Microstrip.

approach to their basic operation. More advanced techniques of analysis can be found in the literature.

8.8.1 Stripline

Two of the most important parameters of any transmission line are its characteristic impedance and phase velocity. Since the basic mode of operation is the TEM, its characteristic impedance Z_c and phase velocity v_p can be written, respectively, as

$$Z_c = \sqrt{\frac{L}{C}} \tag{8-183a}$$

$$v_p = \frac{1}{\sqrt{LC}} = \frac{1}{\sqrt{\mu\varepsilon}} \Rightarrow \sqrt{L} = \frac{1}{v_p\sqrt{C}} = \frac{\sqrt{\mu\varepsilon}}{\sqrt{C}} \tag{8-183b}$$

where L = inductance of line per unit length
C = capacitance of line per unit length

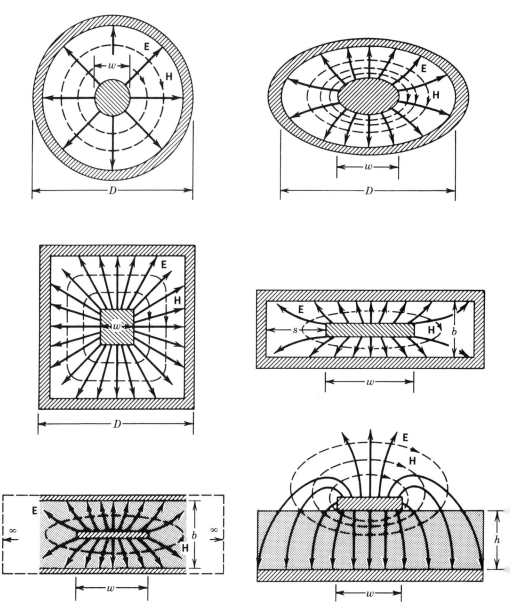

FIGURE 8-32 Evolution of stripline and microstrip transmission lines.

Substituting (8-183b) into (8-183a) reduces to

$$Z_c = \sqrt{\frac{L}{C}} = \frac{1}{v_p C} = \frac{\sqrt{\mu\varepsilon}}{C} = \frac{\sqrt{\mu_0 \varepsilon_0}\sqrt{\mu_r \varepsilon_r}}{C} = \frac{\sqrt{\mu_r \varepsilon_r}}{v_0 C} \qquad (8\text{-}184)$$

where v_0 is the speed of light in free space. Therefore the characteristic impedance can be determined if the capacitance of the line is known.

The total capacitance C_t of a stripline can be modeled as shown in Figure 8-33, and it is given by

$$C_t = 2C_p + 4C_f \qquad (8\text{-}185)$$

TABLE 8-8
Characteristic comparison of popular transmission lines

Characteristic	Coaxial	Waveguide	Stripline	Microstrip
Line losses	Medium	Low	High	High
Unloaded Q	Medium	High	Low	Low
Power Capability	Medium	High	Low	Low
Bandwidth	Large	Small	Large	Large
Miniaturization	Poor	Poor	Very good	Excellent
Volume and weight	Large	Large	Medium	Small
Isolation between neighboring circuits	Very good	Very good	Fair	Poor
Realization of passive circuits	Easy	Easy	Very easy	Very easy
Integration with chip devices	Poor	Poor	Fair	Very good

where

$$C_t = \text{total capacitance per unit length} \qquad (8\text{-}185a)$$

$$C_p = \text{parallel plate capacitance per unit length}$$

(in the absence of fringing)

$$= \varepsilon \frac{2w}{b-t} = 2\varepsilon_r \varepsilon_0 \frac{\dfrac{w}{b}}{1 - \dfrac{t}{b}} \qquad (8\text{-}185b)$$

$$C_f = \text{fringing capacitance per unit length} \qquad (8\text{-}185c)$$

Assume that the dielectric medium between the plates is not ferromagnetic. Then the characteristic impedance of (8-184) can also be written by using (8-185) and (8-185b) as

$$Z_c = \frac{\sqrt{\mu\varepsilon}}{C_t} = \frac{\varepsilon}{C_t}\sqrt{\frac{\mu}{\varepsilon}} = \frac{\varepsilon}{\sqrt{\varepsilon_r}\,C_t}\sqrt{\frac{\mu_0}{\varepsilon_0}} = \frac{120\pi\varepsilon}{\sqrt{\varepsilon_r}\,C_t} \qquad (8\text{-}186)$$

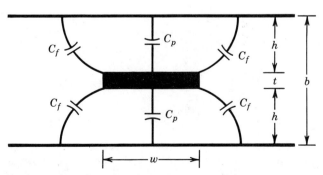

FIGURE 8-33 Capacitance model for stripline transmission line.

or

$$Z_c\sqrt{\varepsilon_r} = \frac{120\pi}{\dfrac{1}{\varepsilon}C_t} = \frac{120\pi}{\dfrac{1}{\varepsilon}(2C_p + 4C_f)} = \frac{30\pi}{\dfrac{w/b}{1 - t/b} + \dfrac{C_f}{\varepsilon}} \qquad (8\text{-}186a)$$

The fringing capacitance of the stripline can be approximated by using

$$\frac{C_f}{\varepsilon} \simeq \frac{1}{\pi}\left\{ \frac{2}{1 - \dfrac{t}{b}} \ln\left(1 + \frac{1}{1 - \dfrac{t}{b}}\right) - \left(\frac{1}{1 - \dfrac{t}{b}} - 1\right)\ln\left[\frac{1}{\left(1 - \dfrac{t}{b}\right)^2} - 1\right] \right\} \qquad (8\text{-}187)$$

which for zero thickness center conductor ($t = 0$) reduces to

$$\frac{C_f}{\varepsilon} \simeq \frac{1}{\pi}[2\ln(2)] = 0.4413 \qquad (8\text{-}187a)$$

For zero thickness center conductor ($t = 0$) an exact solution based on conformal mapping represents the characteristic impedance of (8-186a) by [18]

$$Z_c\sqrt{\varepsilon_r} = \frac{30\pi}{K(k)/K(k')} = 30\pi\left[\frac{K(k')}{K(k)}\right] \qquad (8\text{-}188)$$

where $K(k)$ is an elliptic function of the first kind and it is given by

$$K(k) = \int_0^1 \frac{1}{\sqrt{1 - q^2}}\frac{1}{\sqrt{1 - kq^2}}\,dq = \int_0^{\pi/2} \frac{1}{\sqrt{1 - k\sin^2\psi}}\,d\psi \qquad (8\text{-}188a)$$

$$k = \tanh\left(\frac{\pi}{2}\frac{w}{b}\right) \qquad (8\text{-}188b)$$

$$k' = \sqrt{1 - k^2} = \sqrt{1 - \tanh^2\left(\frac{\pi}{2}\frac{w}{b}\right)} = \text{sech}\left(\frac{\pi}{2}\frac{w}{b}\right) \qquad (8\text{-}188c)$$

It can be shown that the ratio of the elliptic functions in (8-188) can be approximated by

$$\frac{K(k)}{K(k')} \simeq \begin{cases} \dfrac{1}{\pi}\ln\left(2\dfrac{1 + \sqrt{k}}{1 - \sqrt{k}}\right) & \text{when } \dfrac{1}{\sqrt{2}} \leq k = \tanh\left(\dfrac{\pi}{2}\dfrac{w}{b}\right) \leq 1 \quad (8\text{-}189a) \\[2ex] \dfrac{\pi}{\ln\left(2\dfrac{1 + \sqrt{k'}}{1 - \sqrt{k'}}\right)} & \text{when } 0 \leq k = \tanh\left(\dfrac{\pi}{2}\dfrac{w}{b}\right) \leq \dfrac{1}{\sqrt{2}} \quad (8\text{-}189b) \end{cases}$$

Other forms to represent the characteristic impedance of the stripline are available, but the preceding are considered to be sufficiently simple, practical, and accurate.

Example 8-16. Determine the characteristic impedance of a zero thickness center conductor stripline whose dielectric constant is 2.20 and w/b ratio is $w/b = 1$ and $w/b = 0.1$.

Solution. The solution for the characteristic impedance will be based on the more accurate formulation of (8-188) through (8-189b).

Since $w/b = 1$, then according to (8-188b)

$$k = \tanh\left(\frac{\pi}{2}\right) = 0.91715 < 1$$

Thus by using (8-189a),

$$\frac{K(k)}{K(k')} = \frac{1}{\pi}\ln\left(2\frac{1 + \sqrt{0.91715}}{1 - \sqrt{0.91715}}\right) = 1.4411$$

Therefore the characteristic impedance of (8-188) is equal to

$$Z_c = \frac{30\pi}{1.4411\sqrt{2.2}} = 44.09 \text{ ohms}$$

Since $w/b = 0.1$, then according to (8-188b),

$$k = \tanh\left[\frac{\pi}{2}(0.1)\right] = 0.1558$$

and from (8-188c),

$$k' = \sqrt{1 - (0.1558)^2} = 0.98779$$

Thus by using (8-189b),

$$\frac{K(k)}{K(k')} = \frac{\pi}{\ln\left(2\dfrac{1 + \sqrt{0.98779}}{1 - \sqrt{0.98779}}\right)} = 0.4849$$

Therefore the characteristic impedance of (8-188) is equal to

$$Z_c = \frac{30\pi}{0.4849\sqrt{2.2}} = 131.04 \text{ ohms}$$

8.8.2 Microstrip

The early investigations of the microstrip line in the early 1950s did not stimulate its widespread acceptance because of the excitation of radiation and undesired modes caused by lines with discontinuities. However, the rapid rise in miniature microwave circuits, which are usually planar in structure, caused renewed interest in microstrip circuit design. Also the development of high dielectric constant material began to bind the fringing fields more tightly to the center conductor, thus decreasing radiation losses, and simultaneously shrinking the overall circuit dimensions. These developments, plus the advantages of convenient and economical integrated circuit fabrication techniques, tended to lessen the previous concerns and finally allowed microstrip design methods to achieve widespread application.

Because the upper part of the microstrip is usually exposed, some of the fringing field lines will be in air while others will reside within the substrate. Therefore overall the microstrip can be thought of as being a line composed of a homogeneous dielectric whose overall dielectric constant is greater than air but smaller than that of the substrate. The overall dielectric constant is usually referred to as the *effective dielectric constant*. Because most of the field lines reside within the substrate, the effective dielectric constant is usually closer in value to that of the substrate than to that of air; this becomes even more pronounced as the dielectric constant of the substrate increases. Since the microstrip is composed of two different dielectric materials (nonhomogeneous line), it cannot support pure TEM modes. The lowest order modes are quasi-TEM.

There have been numerous investigations of the microstrip ([19–27], and many others). Because of the plethora of information on the microstrip, we will summarize some of the formulations for the characteristic impedance and effective dielectric constant that are simple, accurate, and practical.

At low frequencies the characteristic parameters of the microstrip can be found by using the following expressions:

$$\frac{w_{\text{eff}}(0)}{h} \leq 1$$

$$Z_c(0) = Z_c(f=0) = \frac{60}{\sqrt{\varepsilon_{r,\text{eff}}(0)}} \ln\left[\frac{8h}{w_{\text{eff}}(0)} + \frac{w_{\text{eff}}(0)}{4h}\right] \quad (8\text{-}190\text{a})$$

$$\varepsilon_{r,\text{eff}}(0) = \varepsilon_{r,\text{eff}}(f=0) = \frac{\varepsilon_r + 1}{2} + \frac{\varepsilon_r - 1}{2}$$

$$\times \left\{\left[1 + 12\frac{h}{w_{\text{eff}}(0)}\right]^{-1/2} + 0.04\left[1 - \frac{w_{\text{eff}}(0)}{h}\right]^2\right\} \quad (8\text{-}190\text{b})$$

$$\frac{w_{\text{eff}}(0)}{h} > 1$$

$$Z_c(0) = Z_c(f=0) = \frac{\frac{120\pi}{\sqrt{\varepsilon_{r,\text{eff}}(0)}}}{\frac{w_{\text{eff}}(0)}{h} + 1.393 + 0.667\ln\left[\frac{w_{\text{eff}}(0)}{h} + 1.444\right]} \quad (8\text{-}191\text{a})$$

$$\varepsilon_{r,\text{eff}}(0) = \varepsilon_{r,\text{eff}}(f=0) = \frac{\varepsilon_r + 1}{2} + \frac{\varepsilon_r - 1}{2}\left[1 + 12\frac{h}{w_{\text{eff}}(0)}\right]^{-1/2} \quad (8\text{-}191\text{b})$$

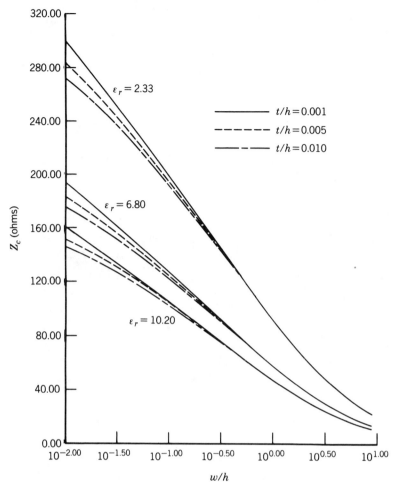

FIGURE 8-34 Characteristic impedance of microstrip line as a function of w/h and t/h.

where

$$\frac{w_{\text{eff}}(0)}{h} = \frac{w_{\text{eff}}(f=0)}{h} = \frac{w}{h} + \frac{1.25}{\pi}\frac{t}{h}\left[1 + \ln\left(\frac{2h}{t}\right)\right] \quad \text{for } \frac{w}{h} \geq \frac{1}{2\pi} \quad (8\text{-}192\text{a})$$

$$\frac{w_{\text{eff}}(0)}{h} = \frac{w_{\text{eff}}(f=0)}{h} = \frac{w}{h} + \frac{1.25}{\pi}\frac{t}{h}\left[1 + \ln\left(\frac{4\pi w}{t}\right)\right] \quad \text{for } \frac{w}{h} < \frac{1}{2\pi} \quad (8\text{-}192\text{b})$$

$\varepsilon_{r,\text{eff}}$ and w_{eff} represent the effective dielectric constant and width of the line, respectively. Plots of the characteristic impedance of (8-190a) or (8-191a) and the effective dielectric constant of (8-190b) or (8-191b) as a function of w/h for three different dielectric constants ($\varepsilon_r = 2.33, 6.80,$ and 10.2) are shown, respectively, in Figures 8-34 and 8-35 [28]. These dielectric constants are representative of common substrates such as RT/Duroid ($\simeq 2.33$), beryllium oxide ($\simeq 6.8$), and alumina ($\simeq 10.2$) used for microstrips. It is evident that the effective dielectric constant is not very sensitive to the thickness of the center strip.

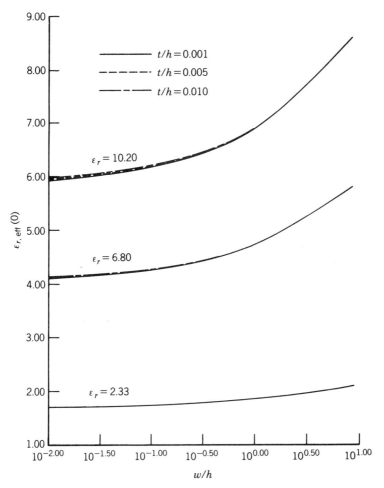

FIGURE 8-35 Effective dielectric constant of microstrip line as a function of w/h and t/h at zero frequency.

Example 8-17. For a microstrip line with $w/h = 1$, $\varepsilon_r = 10$, and $t/h = 0$, calculate at $f \simeq 0$ the effective width, effective dielectric constant and characteristic impedance of the line.

Solution. Since $t/h = 0$, then according to either (8-192a) or (8-192b),

$$\frac{w_{\text{eff}}(0)}{h} = \frac{w}{h} = 1$$

By using (8-190b) the effective dielectric constant is equal to

$$\varepsilon_{r,\text{eff}}(0) = \frac{10 + 1}{2} + \frac{10 - 1}{2}[1 + 12(1)]^{-1/2} = 6.748 < 10$$

The characteristic impedance of (8-190a) is now equal to

$$Z_c(0) = \frac{60}{\sqrt{6.748}} \ln\left[8(1) + \frac{1}{4}(1)\right] = 48.74 \text{ ohms}$$

STRIPLINE AND MICROSTRIP LINES **453**

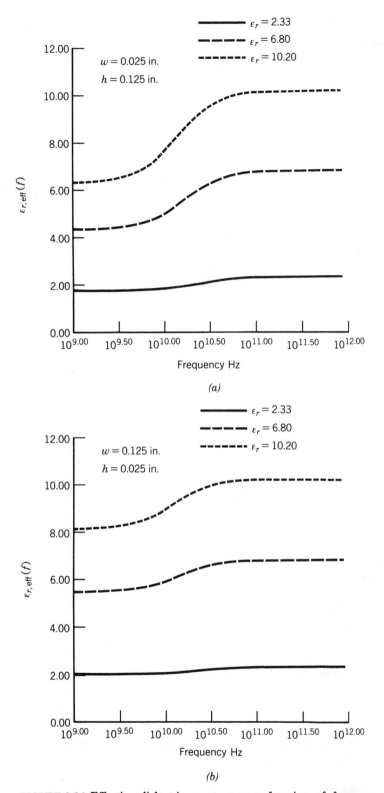

FIGURE 8-36 Effective dielectric constant as a function of frequency for microstrip transmission line. (*a*) $w/h = 0.2$. (*b*) $w/h = 5$.

The microstrip line is considered to be a dispersive transmission line at frequencies about equal to or greater than

$$f_c \geq 0.3\sqrt{\frac{Z_c(0)}{h}\frac{1}{\sqrt{\varepsilon_r - 1}}} \times 10^9 \quad \text{where } h \text{ is in cm} \quad (8\text{-}193)$$

For many typical transmission lines this frequency will be in the 3–10 GHz range. This indicates that the effective dielectric constant, phase velocity, and characteristic impedance will be a function of frequency. In addition pulse wave propagation, whose spectrum spans a wide range of frequencies that depend largely on the width and shape of the pulse, can greatly be affected by the dispersive properties of the line [29–31, 42].

Many models have been developed to predict the dispersive behavior of a microstrip [32–38]. One model which allows simple, accurate and practical values computes the dispersive characteristics using

$$Z_c(f) = Z_c(0)\sqrt{\frac{\varepsilon_{r,\text{eff}}(0)}{\varepsilon_{r,\text{eff}}(f)}} \quad (8\text{-}194\text{a})$$

$$v_p(f) = \frac{1}{\sqrt{\mu \varepsilon_{\text{eff}}(f)}} = \frac{1}{\sqrt{\mu_r \mu_0 \varepsilon_0 \varepsilon_{r,\text{eff}}(f)}} = \frac{v_0}{\sqrt{\mu_r \varepsilon_{r,\text{eff}}(f)}} \quad (8\text{-}194\text{b})$$

$$\lambda_g(f) = \frac{v_p(f)}{f} = \frac{v_0}{f\sqrt{\mu_r \varepsilon_{r,\text{eff}}(f)}} = \frac{\lambda_0}{\sqrt{\mu_r \varepsilon_{r,\text{eff}}(f)}} \quad (8\text{-}194\text{c})$$

$$\varepsilon_{r,\text{eff}}(f) = \varepsilon_r - \left[\frac{\varepsilon_r - \varepsilon_{r,\text{eff}}(0)}{1 + \frac{\varepsilon_{r,\text{eff}}(0)}{\varepsilon_r}\left(\frac{f}{f_t}\right)^2}\right] \quad (8\text{-}194\text{d})$$

$$f_t = \frac{Z_c(0)}{2\mu_0 h} \quad (8\text{-}194\text{e})$$

Typical plots of $\varepsilon_{r,\text{eff}}(f)$ versus frequency for three microstrip lines ($\varepsilon_r = 2.33$, 6.8, and 10.2) are shown in Figure 8-36a and b for $w/h = 0.2$ and 5 [28]. It is evident that for $w/h \gg 1$ the variations are smaller than those for $w/h \ll 1$.

Example 8-18. For a microstrip line with $w/h = 1$, $h = 0.025$ in. (0.0635 cm), $\varepsilon_r = 10$, and $t/h = 0$, calculate the effective dielectric constant, characteristic impedance, phase velocity, and guide wavelength at $f = 3$ and 10 GHz.

Solution. At zero frequency, from Example 8-17,

$$\varepsilon_{r,\text{eff}}(0) = 6.748$$
$$Z_c(0) = 48.74 \text{ ohms}$$

The critical frequency, where dispersion begins to appear, according to (8-193), is equal to or greater than

$$f_c \geq 0.3\sqrt{\frac{48.74}{0.0635\sqrt{10-1}}} \times 10^9 = 4.799 \text{ GHz}$$

By using (8-194e),

$$f_t = \frac{48.74}{2(4\pi \times 10^{-7})(6.35 \times 10^{-4})} = 30.54 \times 10^9$$

$f = 3\ GHz$: By using (8-194d),

$$\varepsilon_{r,\text{eff}}(f = 3\ \text{GHz}) = 10 - \left[\frac{10 - 6.748}{1 + \left(\frac{6.748}{10}\right)\left(\frac{3}{30.54}\right)^2}\right] = 6.7691$$

Thus the characteristic impedance of (8-194a), phase velocity of (8-194b), and guide wavelength of (8-194c) are equal to

$$Z_c(f = 3\ \text{GHz}) = 48.74\sqrt{\frac{6.748}{6.7691}} = 48.664\ \text{ohms}$$

$$v_p(f = 3\ \text{GHz}) = \frac{3 \times 10^8}{\sqrt{6.7691}} = 1.153 \times 10^8\ \text{m/sec}$$

$$\lambda_g(f = 3\ \text{GHz}) = \frac{3 \times 10^8}{3 \times 10^9 \sqrt{6.7691}} = 0.0384\ \text{m} = 3.84\ \text{cm}$$

$f = 10\ GHz$: By repeating the preceding calculations at $f = 10$ GHz we obtain

$$\varepsilon_{r,\text{eff}}(f = 10\ \text{GHz}) = 10 - \left[\frac{10 - 6.748}{1 + \frac{6.748}{10}\left(\frac{10}{30.54}\right)^2}\right] = 6.968$$

$$Z_c(f = 10\ \text{GHz}) = 48.74\sqrt{\frac{6.748}{6.968}} = 47.964\ \text{ohms}$$

$$v_p(f = 10\ \text{GHz}) = \frac{3 \times 10^8}{\sqrt{6.968}} = 1.128 \times 10^8\ \text{m/sec}$$

$$\lambda_g(f = 10\ \text{GHz}) = \frac{3 \times 10^8}{10 \times 10^9 \sqrt{6.968}} = 0.0114\ \text{m} = 1.14\ \text{cm}$$

8.8.3 Microstrip: Boundary-Value Problem

The open microstrip line can be analyzed as a boundary-value problem using modal solutions of the form used for the partially filled waveguide or dielectric covered ground plane. In fact, the open microstrip line can be represented as a partially filled waveguide with the addition of a center conductor placed along the air–dielectric interface, as shown in Figure 8-37. This shielded configuration is considered a good model for the open microstrip provided that the dimensions a and b of the waveguide are equal to or greater than about 10 to 20 times the center conductor width. The fields configurations of this structure that satisfy all the boundary

456 RECTANGULAR CROSS-SECTION WAVEGUIDES AND CAVITIES

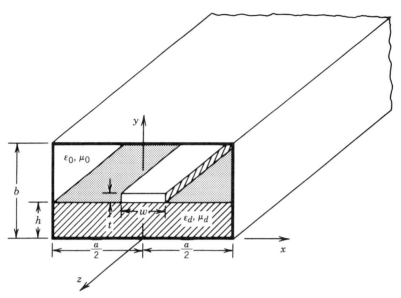

FIGURE 8-37 Shielded configuration of microstrip transmission line.

conditions are hybrid modes that are a superposition of TE^z and TM^z modes [38–42].

Initially the vector potential functions used to represent, respectively, the TE^z and TM^z modes are chosen so that individually they satisfy the field boundary conditions along the metallic periphery of the waveguide. Then the total fields, which are due to the superposition of the TE^z and TM^z fields, must be such that

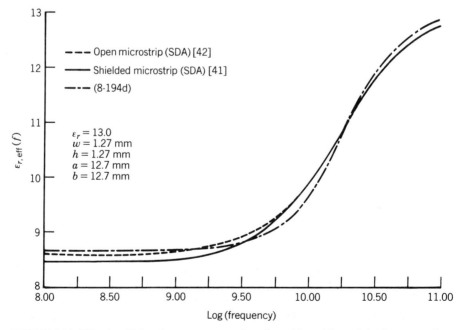

FIGURE 8-38 Effective dielectric constant as a function of logarithm of the frequency for open and shielded microstrip lines.

they satisfy all the additional boundary conditions along the air–dielectric interface ($y = h$), including those at the center metallic strip ($y = h$, $|x| \leq w/2$). The end result of this procedure is an infinite set of coupled homogeneous simultaneous equations that can be solved for the normalized propagation constant along the z direction ($\beta_n = \beta_z/\beta_0$) through the use of various techniques [39–42]. A complete formulation of this problem is very lengthy, and is assigned to the reader as an end of chapter problem.

Another method that can also be used to solve for β_z is to use spectral domain techniques, which transforms the resulting field equations to the spectral domain, and allows for rapid convergence [40–42]. Results obtained with these methods for open and shielded microstrip geometries are shown in Figure 8-38 [42]. The waveguide width and height were chosen to be 10 times greater than the center conductor strip width. As the waveguide width and height are chosen to be even greater, the results of the open and shielded microstrips agree even better [42].

8.9 RIDGED WAVEGUIDE

It was illustrated in Section 8.2.1 that the maximum bandwidth for a dominant single TE_{10} mode operation that can be achieved by a standard rectangular waveguide is 2:1. For some applications, such as coupling, matching, filters, arrays, and so forth, larger bandwidths may be desired. This can be accomplished by using a ridged waveguide.

A ridged waveguide is formed by placing longitudinal metal strip(s) inside a rectangular waveguide, as shown in Figure 8-39. This has the same effect as placing inward ridges on the walls of the waveguide. The most common configurations of a ridged waveguide are those of single, dual, and quadruple ridges, as illustrated in

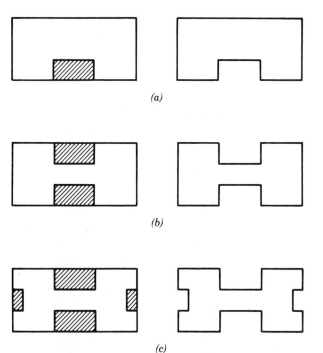

FIGURE 8-39 Various cross sections of ridged waveguide. (*a*) Single. (*b*) Dual. (*c*) Quadruple.

458 RECTANGULAR CROSS-SECTION WAVEGUIDES AND CAVITIES

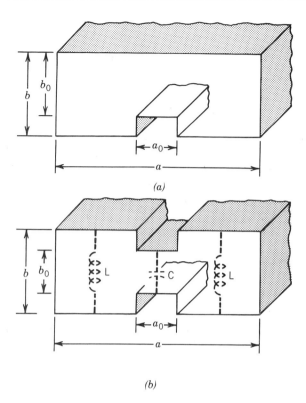

FIGURE 8-40 Geometry for (a) single and (b) dual ridged waveguides.

Figure 8-39. In general the ridges act as uniform distributed loadings which tend to lower the phase velocity and reduce (by a factor of 25 or more) the characteristic impedance. The lowering of the phase velocity is accompanied by a reduction (by a factor as large as 5 to 6) of the cutoff frequency of the TE_{10} mode, an increase of the cutoff frequencies of the higher-order modes, an increase in the attenuation due to losses on the boundary walls, and a decrease in the power-handling capacity. The increases in the bandwidth and attenuation depend upon the dimensions of the ridge compared to those of the waveguide.

The single, dual, and quadruple ridged waveguides of Figures 8-39 and 8-40 have been investigated by many people [43–47]. Since the ridged waveguide possesses an irregular shape, a very appropriate technique that can be used to analyze it is the transverse resonance method of Section 8.6. At cutoff ($\beta_z = 0$) there are no waves traveling along the length (z direction) of the waveguide, and the waves can be thought of as traveling along the transverse directions (x, y directions) of the guide forming standing waves. For the TE_{10} mode, for example, there are field variations only along the x direction and at cutoff the waveguide has a cutoff frequency that is equal to the resonant frequency of a standing plane wave propagating only in the x direction. The transverse dimension (in the x direction) of the waveguide for the TE_{10} at resonance is equal to a half wavelength.

One very approximate equivalent model for representation of the ridged waveguide *at resonance* is that of a parallel LC network [43], shown in Figure 8-40b. The gap between the ridges is represented by the capacitance C whose value for a waveguide of length ℓ can be found by using

$$C = \varepsilon \left(\frac{A_0}{b_0} \right) = \varepsilon \left(\frac{a_0 \ell}{b_0} \right) \tag{8-195}$$

Each side section of the ridged waveguide can be represented by a one-turn solenoidal inductance whose value for a waveguide of length ℓ can be found by using

$$L = \mu\left(\frac{A}{\ell}\right) = \mu\frac{b\dfrac{a-a_0}{2}}{\ell} = \mu\frac{b(a-a_0)}{2\ell} \quad (8\text{-}196)$$

Since the total inductance L_t is the parallel combination of the two L's ($L_t = L/2$), the cutoff frequency is obtained by using

$$\omega_c = 2\pi f_c = \frac{1}{\sqrt{L_t C}} = \sqrt{\frac{2}{LC}} \quad (8\text{-}197)$$

Use of (8-195) and (8-196) reduces the cutoff frequency of (8-197) to

$$f_c = \frac{1}{2a\sqrt{\mu\varepsilon}}\left[\frac{2}{\pi}\sqrt{\frac{a}{a_0}\frac{b_0}{b}\frac{1}{1-\dfrac{a_0}{a}}}\right] \quad (8\text{-}198)$$

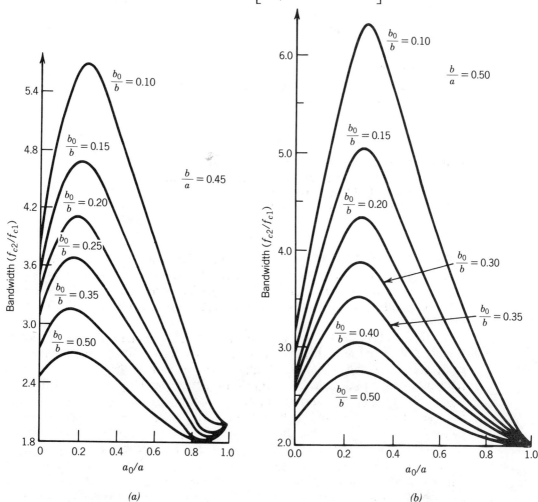

FIGURE 8-41 Bandwidth for (a) single and (b) dual ridged waveguides (*Source:* S. Hopfer, "The design of ridged waveguides," *IRE Trans. Microwave Theory Tech.*, ©, 1955, IEEE.)

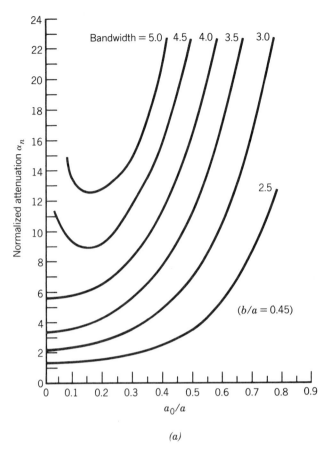

FIGURE 8-42 Normalized attenuation for (*a*) single and (*b*) dual ridged waveguides (*Source:* S. Hopfer, "The design of ridged waveguides," *IRE Trans. Microwave Theory Tech.*, ©, 1955, IEEE.)

which is more valid for the smaller gaps where the b_0/b ratio is very small. More accurate equivalents can be obtained through use of the transverse resonance method where the ridge waveguide can be modelled at resonance as a parallel plate waveguide with a capacitance between them that represents the discontinuity of the ridges.

Curves of available bandwidth of a single TE_{10} mode operation for single (Figure 8-40*a*) and dual (Figure 8-40*b*) ridged waveguides are shown, respectively, in Figures 8-41*a* and *b* [45]. Bandwidth is defined here as the ratio of the cutoff frequency of the next higher-order mode to that of the TE_{10} mode, and it is not necessarily the useful bandwidth. In many applications the lower and upper frequencies of the useful bandwidth are chosen with about a 15 to 25 percent safety factor from the corresponding cutoff frequencies. It is seen from the data in Figure 8-41 that a single-mode bandwidth of about 6:1 is realistic with a ridged rectangular waveguide. However, the penalty in realizing this extended bandwidth is the increase in attenuation. To illustrate this, we have plotted in Figures 8-42*a* and *b* the normalized attenuation α_n for single and dual ridged waveguides, which is defined as the ratio of the ridged waveguide attenuation to that of the rectangular

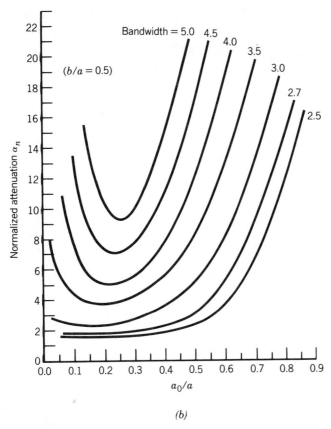

FIGURE 8-42 (*Continued*).

waveguide attenuation, of identical cutoff frequency, evaluated at a frequency of $f = \sqrt{3} f_c$. The curves of Figure 8-42 have been calculated assuming that the ratio b/a of the ridged waveguide, $b/a = 0.45$ for the single ridge and $b/a = 0.5$ for the dual ridge, is the same as that of the rectangular waveguide. The actual attenuation of the ridged waveguide at $f = \sqrt{3} f_c$ can be obtained by multiplying the normalized values of the attenuation coefficient from Figure 8-42 by the attenuation of the rectangular waveguide evaluated at $f = \sqrt{3} f_c$. It should be noted that the power-handling capabilities of ridged waveguides are reduced at the expense of increases in bandwidth.

REFERENCES

1. C. S. Lee, S. W. Lee, and S. L. Chuang, "Plot of modal field distribution in rectangular and circular waveguides," *IEEE Trans. Microwave Theory Tech.*, vol. MTT-33, pp. 271–274, March 1985.
2. S. Ramo, J. R. Whinnery, and T. Van Duzer, *Fields and Waves in Communication Electronics*, Second Edition, Wiley, New York, 1984.
3. R. F. Harrington, *Time-Harmonic Electromagnetic Fields*, McGraw-Hill, New York, 1961.
4. R. E. Collin, *Field Theory of Guided Waves*, McGraw-Hill, New York, 1960.
5. N. Marcuvitz (Ed.), *Waveguide Handbook*, Chapter 8, McGraw-Hill, New York, 1951, pp. 387–413.

6. C. H. Walter, *Traveling Wave Antennas*, McGraw-Hill, New York, 1965, pp. 172–187.
7. L. O. Goldstone and A. A. Oliner, "Leaky wave antennas I: Rectangular waveguides," *IRE Trans. Antennas Propagat.*, vol. AP-7, pp. 307–309, October 1959.
8. L. O. Goldstone and A. A. Oliner, "Leaky wave antennas II: Circular waveguides," *IRE Trans. Antennas Propagat.*, vol. AP-9, pp. 280–290, May 1961.
9. J. H. Richmond, *Reciprocity Theorems and Plane Surface Waves*, Engineering Experiment Station Bulletin, Ohio State University, vol. XXVIII, no. 4, July 1959.
10. M. Zahn, *Electromagnetic Field Theory*, Wiley, New York, 1979.
11. P. K. Tien, "Light waves in thin films and integrated optics," *Applied Optics*, vol. 10, no. 11, pp. 2395–2413, November 1971.
12. R. M. Barrett, "Microwave printed circuits—a historical survey," *IRE Trans. Microwave Theory Tech.*, vol. MTT-3, no. 2, p. 9, March 1955.
13. Special issue on Microwave Strip Circuits, *IRE Trans. Microwave Theory Tech.*, vol. MTT-3, no. 2, March 1955.
14. H. Howe Jr., *Stripline Circuit Design*, Artech House, Dedham, MA, 1974.
15. G. L. Matthaei, L. Young, and E. M. T. Jones, *Microwave Filters Impedance-Matching Networks and Coupling Structures*, McGraw-Hill, New York, 1964.
16. S. B. Cohn, "Characteristic impedance of a shielded-strip transmission line," *IRE Trans. Microwave Theory Tech.*, MTT-2, pp. 52–57, July 1954.
17. S. B. Cohn, "Characteristic impedances of broadside-coupled strip transmission lines," *IRE Trans. Microwave Theory Tech.*, vol. MTT-8, pp. 633–637, November 1960.
18. H. A. Wheeler, "Transmission-line properties of parallel wide strips by a conformal-mapping approximation," *IEEE Trans. Microwave Theory Tech.*, vol. MTT-12, no. 3, pp. 280–289, May 1964.
19. H. A. Wheeler, "Transmission-line properties of parallel strips separated by a dielectric sheet," *IEEE Trans. Microwave Theory Tech.*, vol. MTT-13, no. 2, pp. 172–185, March 1965.
20. T. G. Bryant and J. A. Weiss, "Parameters of microstrip transmission lines and coupled pairs of microstrip lines," *IEEE Trans. Microwave Theory Tech.*, vol. MTT-16, no. 12, pp. 1021–1027, December 1968.
21. M. V. Schneider, "Microstrip lines for microwave integrated circuits," *Bell System Tech. J.*, vol. 48, pp. 1421–1444, May–June 1969.
22. E. O. Hammerstad, "Equations for microstrip circuit design," *Proc. European Microwave Conference*, pp. 268–272, September 1975.
23. H. A. Wheeler, "Transmission-line properties of a strip on a dielectric sheet on a plane," *IEEE Trans. Microwave Theory Tech.*, vol. MTT-25, no. 8, pp. 631–647, August 1977.
24. K. C. Gupta, R. Garg, and I. J. Bahl, *Microstrip Lines and Slotlines*, Artech House, Dedham, MA, 1979.
25. M. V. Schneider, "Dielectric loss in integrated microwave circuits," *Bell System Tech. J.*, vol. 50, pp. 2325–2332, September 1969.
26. R. A. Pucel, D. J. Massé, and C. D. Hartwig, "Losses in microstrip," *IEEE Trans. Microwave Theory Tech.*, vol. MTT-16, pp. 342–350, June 1968; correction vol. MTT-16, p. 1064, December 1968.
27. I. J. Bahl and D. K. Trivedi, "A designer's guide to microstrip line," *Microwaves*, vol. 16, pp. 174–182, May 1977.
28. T. Leung, "Pulse signal distortions in microstrips," MS(EE) Thesis, Department of Electrical and Computer Engineering, Arizona State University, December 1987.
29. R. L. Veghte and C. A. Balanis, "Dispersion of transient signals in microstrip transmission lines," *IEEE Trans. Microwave Theory Tech.*, vol. MTT-34, pp. 1427–1436, December 1986.
30. T. Leung and C. A. Balanis, "Attenuation distortion in microstrips," *IEEE Trans. Microwave Theory Tech.*, vol. MTT-36, no. 4, pp. 765–769, April 1988.
31. E. F. Kuester and D. C. Chang, "An appraisal of methods for computation of the dispersion characteristics of open microstrips," *IEEE Trans. Microwave Theory Tech.*, vol. MTT-27, pp. 691–694, July 1979.

32. M. V. Schneider, "Microstrip dispersion," *IEEE Trans. Microwave Theory Tech.*, vol. MTT-20, pp. 144–146, January 1972.
33. E. J. Denlinger, "A frequency dependent solution for microstrip transmission lines," *IEEE Trans. Microwave Theory Tech.*, vol. MTT-19, no. 1, pp. 30–39, January 1971.
34. W. J. Getsinger, "Microstrip dispersion model," *IEEE Trans. Microwave Theory Tech.*, vol. MTT-22, pp. 34–39, January 1973.
35. H. T. Carlin, "A simplified circuit model for microstrip," *IEEE Trans. Microwave Theory Tech.*, vol. MTT-21, pp. 589–591, September 1973.
36. M. Kobayashi, "Important role of inflection frequency in the dispersive properties of microstrip lines," *IEEE Trans. Microwave Theory Tech.*, vol. MTT-30, pp. 2057–2059, November 1982.
37. P. Pramanick and P. Bhartia, "An accurate description of dispersion in microstrip," *Microwave Journal*, vol. 26, pp. 89–96, December 1983.
38. E. Yamashita, K. Atsuki, and T. Veda, "An approximate dispersion formula of microstrip lines for computer-aided design of microwave integrated circuits," *IEEE Trans. Microwave Theory Tech.*, vol. MTT-27, pp. 1036–1038, December 1979.
39. R. Mittra and T. Itoh, "A new technique for the analysis of the dispersion characteristics of microstrip lines," *IEEE Trans. Microwave Theory Tech.*, vol. MTT-19, no. 1, pp. 47–56, January 1971.
40. T. Itoh and R. Mittra, "Spectral-domain approach for calculating the dispersion characteristics of microstrip lines," *IEEE Trans. Microwave Theory Tech.*, vol. MTT-21, pp. 496–499, July 1973.
41. T. Itoh and R. Mittra, "A technique for computing dispersion characteristics of shielded microstrip lines," *IEEE Trans. Microwave Theory Tech.*, vol. MTT-22, pp. 896–898, October 1974.
42. T. Leung and C. A. Balanis, "Pulse dispersion in open and shielded lines using the spectral-domain method," *IEEE Trans. Microwave Theory Tech.*, vol. MTT-36, no. 7, pp. 1223–1226, July 1988.
43. S. Ramo, J. R. Whinnery, and T. Van Duzer, *Fields and Waves in Communication Electronics*, Second Edition, Wiley, New York, 1984.
44. S. B. Cohn, "Properties of ridge wave guide," *Proc. IRE*, vol. 35, pp. 783–788, August 1947.
45. S. Hopfer, "The design of ridged waveguides," *IRE Trans. Microwave Theory Tech.*, vol. MTT-3, pp. 20–29, October 1955.
46. J. P. Montgomery, "Ridged waveguide phased array elements," *IEEE Trans. Antennas Propagat.*, vol. AP-24, no. 1, pp. 46–53, January 1976.
47. Y. Utsumi, "Variational analysis of ridged waveguide modes," *IEEE Trans. Microwave Theory Tech.*, vol. MTT-33, no. 2, pp. 111–120, February 1985.

PROBLEMS

8.1. An air-filled section of an X-band (8.2–12.4 GHz) rectangular waveguide of length ℓ is used as a delay line. Assume that the inside dimensions of the waveguide are 0.9 in. (2.286 cm), and 0.4 in. (1.016 cm) and that it operates at its dominant mode. Determine its length so that the delay at 10 GHz is 2 μs.

8.2. A standard X-band (8.2–12.4 GHz) rectangular waveguide with inner dimensions of 0.9 in. (2.286 cm) and 0.4 in. (1.016 cm) is filled with lossless polystyrene ($\varepsilon_r = 2.56$). For the lowest-order mode of the waveguide, determine at 10 GHz the following values.

(a) Cutoff frequency (in GHz).
(b) Guide wavelength (in cm).

(c) Wave impedance.
(d) Phase velocity (in m/s).
(e) Group velocity (in m/s).

8.3. An empty X-band (8.2–12.4 GHz) rectangular waveguide, with dimensions of 2.286 cm by 1.016 cm, is to be connected to an X-band waveguide of the same dimensions but filled with lossless polystyrene ($\varepsilon_r = 2.56$). To avoid reflections, an X-band waveguide (of the same dimensions) quarter-wavelength long section is inserted between the two. Assume dominant mode propagation and that matching is to be made at 10 GHz.

(a) Determine the wave impedance of the quarter-wavelength section waveguide.
(b) Determine the dielectric constant of the lossless medium that must be used to fill the quarter-wavelength section waveguide.
(c) Determine the length (in cm) of the quarter-wavelength section waveguide.

8.4. Design a two-section binomial impedance transformer to match an empty ($\varepsilon_r = 1$) X-band waveguide to a dielectric-filled ($\varepsilon_r = 2.56$) X-band waveguide. Use two intermediate X-band waveguide sections, each quarter-wavelength long. Assume dominant mode excitation, $f_0 = 10$ GHz, and waveguide dimensions of 2.286 cm by 1.016 cm.

(a) Determine the wave impedances of each section.
(b) Determine the dielectric constants of the lossless media that must be used to fill the intermediate waveguide sections.
(c) Determine the length (in cm) of each intermediate quarter-wavelength waveguide section.

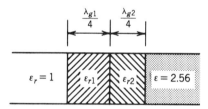

FIGURE P8-4

8.5. Derive expressions for the attenuation coefficient α_c above cutoff for the rectangular waveguide of Figure 8-1, assuming TE$_{mn}^z$ modes and TM$_{mn}^z$ modes. Compare the answers with those found in Table 8-3.

8.6. A parallel-plate waveguide is formed by placing two infinite planar conductors at $y = 0$ and $y = b$.

(a) Show that the electric field

$$E_x = E_0 \sin(\beta_y y) e^{-\gamma z}$$

defines a set of TE$_n$ modes where

$$\gamma = \sqrt{\beta_y^2 - \beta_0^2} \qquad \beta_0 = \omega\sqrt{\mu_0 \varepsilon_0}$$

(b) For the modes of part a, find the allowable eigenvalues, cutoff frequencies, and power transmitted, per unit width in the x direction.

8.7. A rectangular waveguide with dimensions $a = 2.25$ cm and $b = 1.125$ cm, as shown in Figure 8-1, is operating in the dominant mode.

(a) Assume that the medium inside the guide is free space. Then find the cutoff frequency of the dominant mode.
(b) Assume that the physical dimensions of the guide stay the same (as stated) and that we want to reduce the cutoff frequency of the dominant mode of the guide by a factor of 3. Then find the dielectric constant of the medium that must be used to fill the guide to accomplish this.

8.8. If the dielectric constant of the material that is used to construct a dielectric rod waveguide is very large (typically 30 or greater), a good approximation to the boundary conditions is to represent the surface as a perfect magnetic conductor (PMC); see Section 9.5.2. For a PMC surface, the tangential components of the magnetic field vanish. Based on such a model for a rectangular cross-section cylindrical dielectric waveguide and TE^z modes, perform the following tasks.

(a) Write all the boundary conditions on the electric and magnetic fields that must be enforced.
(b) Derive simplified expressions for the vector potential component, the electric and magnetic fields, and the cutoff frequencies.
(c) If $a > b$, identify the lowest-order mode.

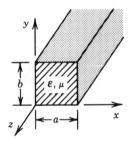

FIGURE P8-8

8.9. Repeat Problem 8.8 for TM^z modes.

8.10. The rectangular waveguide of Figure 8-1 is constructed of two horizontal perfectly electric conducting (PEC) walls at $y = 0$ and $y = b$ and two vertical perfectly magnetic conducting (PMC) walls at $x = 0$ and $x = a$. Derive expressions for the appropriate vector potential, electric and magnetic fields, eigenvalues, cutoff frequencies, phase constant along the z axis, guide wavelength, and wave impedance for TE^z modes and TM^z modes. Identify the lowest-order mode for each set of modes and the dominant mode for both sets.

8.11. Repeat Problem 8.10 for a rectangular waveguide constructed of two horizontal PMC walls at $y = 0$ and $y = b$ and two vertical PEC walls at $x = 0$ and $x = a$.

8.12. An X-band waveguide with dimensions of 0.9 in. (2.286 cm) and 0.4 in. (1.016 cm) is made of copper ($\sigma = 5.76 \times 10^7$ S/m) and it is filled with lossy polystyrene ($\varepsilon'_r = 2.56$, $\tan \delta_e = 4 \times 10^{-4}$). Assume that the frequency of operation is 6.15 GHz. Then determine the attenuation coefficient (in Np/m and dB/m) that accounts for the finite conductivity of the walls and the dielectric losses.

8.13. For the dielectric-filled waveguide of Problem 8.12, assume that the polystyrene is lossless. Determine the following values.
(a) Cutoff frequency of the dominant mode.
(b) Frequency of operation that will allow the plane waves of the dominant mode inside the waveguide to bounce back and forth between its side walls at an angle of 45°.

(c) Guide wavelength (in cm) at the frequency of part b.

(d) Distance (in cm) the wave must travel along the axis of the waveguide to undergo a 360° phase shift at the frequency of part b.

8.14. An air-filled X-band waveguide with dimensions of 0.9 in. (2.286 cm) and 0.4 in. (1.016 cm) is operated at 10 GHz and is radiating into free space.

(a) Find the reflection coefficient (magnitude and phase) at the waveguide aperture junction.

(b) Find the standing wave ratio (SWR) inside the waveguide. Assume that the waveguide is made of a perfect electric conductor.

(c) Find the SWR at distances of $z = 0$, $\lambda_g/4$, and $\lambda_g/2$ from the aperture junction when the waveguide walls are made of copper ($\sigma = 5.76 \times 10^7$ S/m).

8.15. For the rectangular cavity of Figure 8-12, find the length c (in cm) that will resonate the cavity at 10 GHz. Assume dominant mode excitation, $c > a > b$, $a = 2$ cm and $b = 1$ cm, and free space inside the cavity.

8.16. Design a square-base cavity like Figure 8-12, with height one-half the width of the base, to resonate at 1 GHz when the cavity is (a) air-filled and (b) filled with polystyrene ($\varepsilon_r = 2.56$). Assume dominant mode excitation.

8.17. The field between the plates is a linearly polarized (in the y direction) uniform plane wave traveling in the z direction.

(a) Assume that the plates are perfect electric conductors. Then find the **E** and **H** field components between the plates. Neglect the edge effects of the finite plates.

(b) Find the separation d between the plates that creates resonance.

(c) Derive an expression for the Q of the cavity assuming a conductivity of σ for the plates. Neglect any radiation losses through the sides of the cavity.

(d) Compute the Q of the cavity when $f = 60$ GHz and $d = 5\lambda$ and 10λ. Assume a plate conductivity of $\sigma = 5.76 \times 10^7$ S/m.

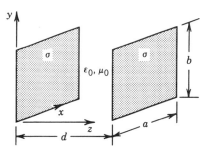

FIGURE P8-17

8.18. An X-band (8.2–12.4-GHz) rectangular waveguide of inner dimensions $a = 0.9$ in. (2.286 cm) and $b = 0.4$ in. (1.016 cm) is partially filled with styrofoam ($\varepsilon_r = 1.1 \simeq 1$), as shown in Figure 8-13a. Assume that the height of the styrofoam is $b/4$. Determine the following for the TM_{10}^y mode.

(a) Phase constants (in rad/cm) in the x direction both in the air and in the styrofoam at any frequency above cutoff.

(b) Approximate phase constants (in rad/cm) in the y direction both in the air and in the styrofoam at any frequency above cutoff.

(c) The approximate value of its cutoff frequency.

(d) The phase constant (in rad/cm) in the z direction at a frequency of $f = 1.25(f_c)_{10}^{TM}$.

PROBLEMS

8.19. For the rectangular waveguide of Figure 8-1 derive expressions for the **E** and **H** fields, eigenvalues, and cutoff frequencies for (a) TE^x (LSE^x) modes and (b) TM^x (LSM^x) modes. Identify the lowest-order mode for each set of modes and the dominant mode for both sets.

8.20. For the partially filled waveguide of Figure 8-13b derive expressions similar to (8-115) through (8-115b) or (8-127) through (8-127b) for (a) LSE^x (TE^x) modes and (b) LSM^x (TM^x) modes.

8.21. For the partially filled waveguide of Figure 8-13a plot on a single figure β_{y0}, β_{yd}, β_z, β_{z0}, and β_{zd}, all in rad/m, versus frequency [$(f_c)_{0n} \leq f \leq 2(f_c)_{0n}$, where $(f_c)_{0n}$ is the cutoff frequency for the TE^y_{0n} mode] for the (a) TE^y_{02} mode and (b) TE^y_{03} mode. Assume $a = 0.9$ in. (2.286 cm), $b = 0.4$ in. (1.016 cm), $h = b/3$, and $\varepsilon_r = 2.56$.

8.22. For the partially filled waveguide of Figure 8-13a plot on a single figure β_{y0}, β_{yd}, β_z, β_{z0}, and β_{zd}, all in rad/m, versus frequency [$(f_c)_{1n} \leq f \leq 2(f_c)_{1n}$, where $(f_c)_{1n}$ is the cutoff frequency for the TM^y_{1n} mode] for the (a) TM^y_{11} mode and (b) TM^y_{12} mode. Assume $a = 0.9$ in. (2.286 cm), $b = 0.4$ in. (1.016 cm), $h = b/3$, and $\varepsilon_r = 2.56$.

8.23. A dielectric slab waveguide, as shown in Figure 8-17, is used to guide electromagnetic energy along its axis. Assume that the slab is 1 cm in height, its dielectric constant is 5, and $\mu = \mu_0$.
   (a) Find the modes that can propagate unattenuated at a frequency of 8 GHz. State their cutoff frequencies.
   (b) Find the respective attenuation (in Np/m) and phase (in rad/m) constants at 8 GHz for the unattenuated modes.
   (c) Find the incidence angles, measured from the normal to the interface, of the bouncing waves within the slab at 8 GHz.

8.24. Design a nonferromagnetic lossless dielectric slab of total height 0.5 in. (1.27 cm) bounded above and below by air so that at $f = 10$ GHz the TE^z_1 mode operates at 10% above its cutoff frequency. Determine the dielectric constant of the slab and the attenuation α_{y0} (in Np/cm) and β_{yd} (in rad/cm) for the TE^z_1 mode at its cutoff frequency.

8.25. A planar perfect electric conductor of infinite dimensions is coated with a dielectric medium of thickness h, as shown in Figure 8-30. Assume that the dielectric constant of the coating is 5, its relative permeability is unity, and its thickness is 5.625 cm.
   (a) Find the cutoff frequencies of the first four TE^z and/or TM^z modes and specify to which group each one belongs.
   (b) For an operating frequency of 1 GHz, find the TE^z modes that can propagate inside the slab unattenuated.
   (c) For each of the TE^z modes found in part b, find the corresponding propagation constant β_z.

   The medium above the coating is free space.

8.26. For the stripline of Example 8-16 find the characteristic impedances based on the approximate formulas of (8-186a), (8-187), and (8-187a). Compare the answers with the more accurate values obtained in Example 8-16, and comment on the comparisons.

8.27. A parallel plate transmission line (waveguide) is formed by two finite width plates placed at $y = 0$ and $y = h$, and it is used to approximate a microstrip. Assume that the electric field between the plates is given by

$$\mathbf{E} \simeq \hat{a}_y E_0 e^{-j\beta z} \quad \text{provided } w/h \gg 1$$

where E_0 is a constant. Derive for the conduction losses an expression for the attenuation constant α_c (in Np/m) in terms of the plate surface resistance R_s, w, h, ε, and μ. The plates are made of metal with conductivity σ, and the medium between the plates is a lossless dielectric.

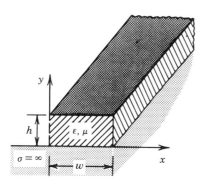

FIGURE P8-27

8.28. A TEM line is composed of a ground plane and a center conductor of width w and thickness t placed at a height h above the ground plane. Assume that the center conductor thickness t is very small. Then the center conductor can be approximated electrically by a wire whose effective radius is

$$a_e \simeq 0.25w$$

Based upon this approximation, derive an approximate expression for the capacitance and for the characteristic impedance of the line.

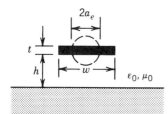

FIGURE P8-28

8.29. Assume that the fields supported by the microstrip line of Figure 8-31b are a combination of TE^z plus TM^z modes. Then derive expressions for the electric and magnetic fields and their associated wave functions and wave numbers by treating the geometry as a boundary-value problem. Do this in the space domain.

8.30. A microstrip transmission line of beryllium oxide ($\varepsilon_r = 6.8$) has a width-to-height ratio of $w/h = 1.5$. Assume that the thickness-to-height ratio is $t/h = 0.01$ and determine the following parameters.

(a) Effective width-to-height ratio at zero frequency.
(b) Effective dielectric constant at zero frequency.
(c) Characteristic impedance at zero frequency.
(d) Approximate frequency where dispersion will begin when $h = 0.05$ cm.
(e) Effective dielectric constant at 15 GHz.
(f) Characteristic impedance at 15 GHz. Compare with the value if dispersion is neglected.
(g) Phase velocity at 15 GHz. Compare with the value if dispersion is neglected.
(h) Guide wavelength at 15 GHz.

CHAPTER 9

CIRCULAR CROSS-SECTION WAVEGUIDES AND CAVITIES

9.1 INTRODUCTION

Cylindrical transmission lines and cavities are very popular geometrical configurations. Cylindrical structures are those that maintain a uniform cross section along their length. Typical cross sections are rectangular, square, triangular, circular, elliptical, and others. Whereas the rectangular and square were analyzed in Chapter 8, the circular cross-section geometries will be discussed in this chapter. This will include transmission lines and cavities (resonators) of conducting walls and dielectric material.

9.2 CIRCULAR WAVEGUIDE

A popular waveguide configuration, in addition to the rectangular one discussed in Chapter 8, is the circular waveguide shown in Figure 9-1. This waveguide is very attractive because of its ease in manufacturing and low attenuation of the TE_{0n} modes. An apparent drawback is its fixed bandwidth between modes. Field configurations (modes) that can be supported inside such a structure are TE^z and TM^z.

9.2.1 Transverse Electric (TE^z) Modes

The transverse electric to z (TE^z) modes can be derived by letting the vector potential **A** and **F** be equal to

$$\mathbf{A} = 0 \tag{9-1a}$$

$$\mathbf{F} = \hat{a}_z F_z(\rho, \phi, z) \tag{9-1b}$$

The vector potential **F** must satisfy the vector wave equation 3-48 which reduces for

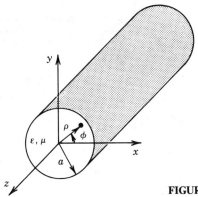

FIGURE 9-1 Cylindrical waveguide of circular cross section.

the **F** of (9-1b) to

$$\nabla^2 F_z(\rho, \phi, z) + \beta^2 F_z(\rho, \phi, z) = 0 \qquad (9\text{-}2)$$

When expanded in cylindrical coordinates, (9-2) reduces to

$$\frac{\partial^2 F_z}{\partial \rho^2} + \frac{1}{\rho}\frac{\partial F_z}{\partial \rho} + \frac{1}{\rho^2}\frac{\partial^2 F_z}{\partial \phi^2} + \frac{\partial^2 F_z}{\partial z^2} + \beta^2 F_z = 0 \qquad (9\text{-}3)$$

whose solution for the geometry of Figure 9-1, according to (3-70), is of the form

$$F_z(\rho, \phi, z) = \left[A_1 J_m(\beta_\rho \rho) + B_1 Y_m(\beta_\rho \rho)\right]$$
$$\times \left[C_2 \cos(m\phi) + D_2 \sin(m\phi)\right]\left[A_3 e^{-j\beta_z z} + B_3 e^{+j\beta_z z}\right] \qquad (9\text{-}4)$$

where, according to (3-66d),

$$\beta_\rho^2 + \beta_z^2 = \beta^2 \qquad (9\text{-}4a)$$

The constants A_1, B_1, C_2, D_2, A_3, B_3, m, β_ρ, and β_z can be found using the boundary conditions of

$$E_\phi(\rho = a, \phi, z) = 0 \qquad (9\text{-}5a)$$

The fields must be finite everywhere (9-5b)

The fields must repeat every 2π radians in ϕ (9-5c)

According to (9-5b), $B_1 = 0$ since $Y_m(\rho = 0) = \infty$. In addition, according to (9-5c),

$$m = 0, 1, 2, 3, \ldots \qquad (9\text{-}6)$$

Consider waves that propagate only in the $+z$ direction. Then (9-4) reduces to

$$F_z^+(\rho, \phi, z) = A_{mn} J_m(\beta_\rho \rho)\left[C_2 \cos(m\phi) + D_2 \sin(m\phi)\right] e^{-j\beta_z z} \qquad (9\text{-}7)$$

TABLE 9-1
Zeroes χ'_{mn} of derivative $J'_m(\chi'_{mn}) = 0$ ($n = 1, 2, 3, \ldots$) of the Bessel function $J_m(x)$

	$m=0$	$m=1$	$m=2$	$m=3$	$m=4$	$m=5$	$m=6$	$m=7$	$m=8$	$m=9$	$m=10$	$m=11$
$n=1$	3.8318	1.8412	3.0542	4.2012	5.3175	6.4155	7.5013	8.5777	9.6474	10.7114	11.7708	12.8264
$n=2$	7.0156	5.3315	6.7062	8.0153	9.2824	10.5199	11.7349	12.9324	14.1155	15.2867	16.4479	17.6003
$n=3$	10.1735	8.5363	9.9695	11.3459	12.6819	13.9872	15.2682	16.5294	17.7740	19.0046	20.2230	21.4309
$n=4$	13.3237	11.7060	13.1704	14.5859	15.9641	17.3129	18.6375	19.9419	21.2291	22.5014	23.7607	25.0085
$n=5$	16.4706	14.8636	16.3475	17.7888	19.1960	20.5755	21.9317	23.2681	24.5872	25.8913	27.1820	28.4609

Use of (6-80) and (9-7) the electric field component of E_ϕ^+ can be written as

$$E_\phi^+ = \frac{1}{\varepsilon}\frac{\partial F_z^+}{\partial \rho} = \beta_\rho \frac{A_{mn}}{\varepsilon} J'_m(\beta_\rho \rho)[C_2 \cos(m\phi) + D_2 \sin(m\phi)]e^{-j\beta_z z} \quad (9\text{-}8)$$

where

$$' = \frac{\partial}{\partial(\beta_\rho \rho)} \quad (9\text{-}8a)$$

Apply the boundary condition of (9-5a) in (9-8). Then we have that

$$E_\phi^+(\rho = a, \phi, z) = \beta_\rho \frac{A_{mn}}{\varepsilon} J'_m(\beta_\rho a)[C_2 \cos(m\phi) + D_2 \sin(m\phi)]e^{-j\beta_z z} = 0 \quad (9\text{-}9)$$

which is only satisfied provided that

$$J'_m(\beta_\rho a) = 0 \Rightarrow \beta_\rho a = \chi'_{mn} \Rightarrow \beta_\rho = \frac{\chi'_{mn}}{a} \quad (9\text{-}10)$$

In (9-10) χ'_{mn} represents the nth zero ($n = 1, 2, 3, \ldots$) of the derivative of the Bessel function J_m of the first kind of order m ($m = 0, 1, 2, 3, \ldots$). An abbreviated list of the zeroes χ'_{mn} of the derivative J'_m of the Bessel function J_m is found in Table 9-1. The smallest value of χ'_{mn} is 1.8412 ($m = 1$, $n = 1$), followed by 3.0542 ($m = 2$, $n = 1$), 3.8318 ($m = 0$, $n = 1$), and so on.

By using (9-4a) and (9-10), β_z of the mn mode can be written as

$$(\beta_z)_{mn} = \begin{cases} \sqrt{\beta^2 - \beta_\rho^2} = \sqrt{\beta^2 - \left(\frac{\chi'_{mn}}{a}\right)^2} & (9\text{-}11a) \\ \quad \text{when } \beta > \beta_\rho = \frac{\chi'_{mn}}{a} & \\ 0 \quad \text{when } \beta = \beta_c = \beta_\rho = \frac{\chi'_{mn}}{a} & (9\text{-}11b) \\ -j\sqrt{\beta_\rho^2 - \beta^2} = -j\sqrt{\left(\frac{\chi'_{mn}}{a}\right)^2 - \beta^2} & (9\text{-}11c) \\ \quad \text{when } \beta < \beta_\rho = \frac{\chi'_{mn}}{a} & \end{cases}$$

Cutoff is defined when $(\beta_z)_{mn} = 0$. Thus, according to (9-11b),

$$\beta_c = \omega_c\sqrt{\mu\varepsilon} = 2\pi f_c\sqrt{\mu\varepsilon} = \beta_\rho = \frac{\chi'_{mn}}{a} \qquad (9\text{-}12)$$

or

$$\boxed{(f_c)_{mn} = \frac{\chi'_{mn}}{2\pi a\sqrt{\mu\varepsilon}}} \qquad (9\text{-}12a)$$

By using (9-12) and (9-12a), we can write (9-11a) through (9-11c) as

$$(\beta_z)_{mn} = \begin{cases} \sqrt{\beta^2 - \beta_\rho^2} = \beta\sqrt{1 - \left(\frac{\beta_\rho}{\beta}\right)^2} = \beta\sqrt{1 - \left(\frac{\beta_c}{\beta}\right)^2} \\ \qquad = \beta\sqrt{1 - \left(\frac{\chi'_{mn}}{\beta a}\right)^2} = \beta\sqrt{1 - \left(\frac{f_c}{f}\right)^2} \qquad (9\text{-}13a) \\ \qquad\qquad\qquad\qquad \text{when } f > f_c = (f_c)_{mn} \\ 0 \qquad\qquad\qquad\qquad \text{when } f = f_c = (f_c)_{mn} \qquad (9\text{-}13b) \\ -j\sqrt{\beta_\rho^2 - \beta^2} = -j\beta\sqrt{\left(\frac{\beta_\rho}{\beta}\right)^2 - 1} = -j\beta\sqrt{\left(\frac{\beta_c}{\beta}\right)^2 - 1} \\ \qquad = -j\beta\sqrt{\left(\frac{\chi'_{mn}}{\beta a}\right)^2 - 1} = -j\beta\sqrt{\left(\frac{f_c}{f}\right)^2 - 1} \qquad (9\text{-}13c) \\ \qquad\qquad\qquad\qquad \text{when } f < f_c = (f_c)_{mn} \end{cases}$$

The guide wavelength λ_g is defined as

$$(\lambda_g)_{mn} = \frac{2\pi}{(\beta_z)_{mn}} \qquad (9\text{-}14)$$

which according to (9-13a) and (9-13b) can be written as

$$(\lambda_g)_{mn} = \begin{cases} \dfrac{2\pi}{\beta\sqrt{1 - \left(\dfrac{f_c}{f}\right)^2}} = \dfrac{\lambda}{\sqrt{1 - \left(\dfrac{f_c}{f}\right)^2}} & \text{when } f > f_c = (f_c)_{mn} \quad (9\text{-}14a) \\ \infty & \text{when } f = (f_c)_{mn} \quad (9\text{-}14b) \end{cases}$$

In (9-14a) λ is the wavelength of the wave in an infinite medium of the kind that exists inside the waveguide. There is no definition of the wavelength below cutoff since the wave is exponentially decaying and there is no repetition of its waveform.

According to (9-12a) and the values of χ'_{mn} in Table 9-1, the order (lower to higher cutoff frequencies) in which the TE^z_{mn} modes occur is TE^z_{11}, TE^z_{21}, TE^z_{01}, etcetera. It should be noted that for a circular waveguide the order in which the TE^z_{mn} modes occur does not change, and the bandwidth between modes is also fixed. For example, the bandwidth of the first single-mode TE_{11}^z operation is $3.042/1.8412 = 1.6588:1$ which is less than $2:1$. This bandwidth is fixed and cannot be varied, as was the case for the rectangular waveguide where the bandwidth between modes was a function of the a/b ratio. In fact, for a rectangular waveguide the maximum bandwidth of a single dominant mode operation was $2:1$ and it occurred when $a/b \geq 2$; otherwise, for $a/b < 2$, the bandwidth of a single dominant mode operation was less than $2:1$. The reason is that in a rectangular waveguide there are two dimensions a and b (2 degrees of freedom) whose relative values can be varied; in the circular waveguide there is only one dimension (the radius a) that can vary. A change in the radius only varies, by the same amount, the absolute values of the cutoff frequencies of all the modes but does not alter their order or relative bandwidth.

The electric and magnetic field components can be written, using (6-80) and (9-7), as

$$E_\rho^+ = -\frac{1}{\varepsilon \rho}\frac{\partial F_z^+}{\partial \phi} = -A_{mn}\frac{m}{\varepsilon \rho}J_m(\beta_\rho \rho)[-C_2 \sin(m\phi) + D_2 \cos(m\phi)]e^{-j\beta_z z}$$
(9-15a)

$$E_\phi^+ = \frac{1}{\varepsilon}\frac{\partial F_z^+}{\partial \rho} = A_{mn}\frac{\beta_\rho}{\varepsilon}J_m'(\beta_\rho \rho)[C_2 \cos(m\phi) + D_2 \sin(m\phi)]e^{-j\beta_z z}$$
(9-15b)

$$E_z^+ = 0$$
(9-15c)

$$H_\rho^+ = -j\frac{1}{\omega\mu\varepsilon}\frac{\partial^2 F_z^+}{\partial \rho \partial z} = -A_{mn}\frac{\beta_\rho \beta_z}{\omega\mu\varepsilon}J_m'(\beta_\rho \rho)[C_2 \cos(m\phi) + D_2 \sin(m\phi)]e^{-j\beta_z z}$$
(9-15d)

$$H_\phi^+ = -j\frac{1}{\omega\mu\varepsilon}\frac{1}{\rho}\frac{\partial^2 F_z^+}{\partial \phi \partial z} = -A_{mn}\frac{m\beta_z}{\omega\mu\varepsilon}\frac{1}{\rho}J_m(\beta_\rho \rho)$$
$$\times[-C_2 \sin(m\phi) + D_2 \cos(m\phi)]e^{-j\beta_z z}$$
(9-15e)

$$H_z^+ = -j\frac{1}{\omega\mu\varepsilon}\left(\frac{\partial^2}{\partial z^2} + \beta^2\right)F_z^+ = -jA_{mn}\frac{\beta_\rho^2}{\omega\mu\varepsilon}J_m(\beta_\rho \rho)$$
$$\times[C_2 \cos(m\phi) + D_2 \sin(m\phi)]e^{-j\beta_z z}$$
(9-15f)

where

$$' = \frac{\partial}{\partial(\beta_\rho \rho)}$$
(9-15g)

By using (9-15a) through (9-15f), the wave impedance $(Z_w^{+z})_{mn}^{\text{TE}}$ of the TE_{mn}^z (H_{mn}^z) modes in the $+z$ direction can be written as

$$Z_{mn}^h = (Z_w^{+z})_{mn}^{\text{TE}} = \frac{E_\rho^+}{H_\phi^+} = -\frac{E_\phi^+}{H_\rho^+} = \frac{\omega\mu}{(\beta_z)_{mn}}$$
(9-16)

With the aid of (9-13a) through (9-13c) the wave impedance of (9-16) reduces to

$$Z_{mn}^h = (Z_w^{+z})_{mn}^{\text{TE}} = \begin{cases} \dfrac{\omega\mu}{\beta\sqrt{1 - \left(\dfrac{f_c}{f}\right)^2}} = \dfrac{\sqrt{\dfrac{\mu}{\varepsilon}}}{\sqrt{1 - \left(\dfrac{f_c}{f}\right)^2}} = \dfrac{\eta}{\sqrt{1 - \left(\dfrac{f_c}{f}\right)^2}} & \text{when } f > f_c = (f_c)_{mn} \quad (9\text{-}16a) \\[2mm] \dfrac{\omega\mu}{0} = \infty & \text{when } f = f_c = (f_c)_{mn} \quad (9\text{-}16b) \\[2mm] \dfrac{\omega\mu}{-j\beta\sqrt{\left(\dfrac{f_c}{f}\right)^2 - 1}} = +j\dfrac{\sqrt{\dfrac{\mu}{\varepsilon}}}{\sqrt{\left(\dfrac{f_c}{f}\right)^2 - 1}} = +j\dfrac{\eta}{\sqrt{\left(\dfrac{f_c}{f}\right)^2 - 1}} & \text{when } f < f_c = (f_c)_{mn} \quad (9\text{-}16c) \end{cases}$$

By examining (9-16a) through (9-16c) we can make the following statements about the impedance.

1. Above cutoff it is real and greater than the intrinsic impedance of the medium inside the waveguide.
2. At cutoff it is infinity.
3. Below cutoff it is imaginary and inductive. This indicates that the waveguide below cutoff behaves as an inductor that is an energy storage element.

The form of Z_w^{+z}, as given by (9-16a) through (9-16c), as a function of f_c/f, is the same as the Z_w^{+z} for the TE^z modes of a rectangular waveguide, as given by (8-20a) through (8-20c). A plot of (9-16a) through (9-16c) for any one TE$_{mn}^z$ mode as a function of f_c/f, where f_c is the cutoff frequency of that mode, is shown in Figure 8-2.

Example 9-1. A circular waveguide of radius $a = 3$ cm that is filled with polystyrene ($\varepsilon_r = 2.56$) is used at a frequency of 2 GHz. For the dominant TE$_{mn}^z$ mode, determine the following:

a. Cutoff frequency.
b. Guide wavelength (in cm). Compare it to the infinite medium wavelength λ.
c. Phase constant β_z (in rad/cm).
d. Wave impedance.
e. Bandwidth over single-mode operation (assuming only TE^z modes).

Solution.

a. The dominant mode is the TE_{11} mode whose cutoff frequency is, according to (9-12a),

$$(f_c)_{11}^{TE^z} = \frac{1.8412}{2\pi a \sqrt{\mu\varepsilon}} = \frac{1.8412(30 \times 10^9)}{2\pi(3)\sqrt{2.56}} = 1.8315 \text{ GHz}$$

b. Since the frequency of operation is 2 GHz, which is greater than the cutoff frequency of 1.8315 GHz, then the guide wavelength of (9-14a) for the TE_{11} mode is

$$\lambda_g = \frac{\lambda}{\sqrt{1 - \left(\frac{f_c}{f}\right)^2}}$$

where

$$\lambda = \frac{\lambda_0}{\sqrt{\varepsilon_r}} = \frac{30 \times 10^9}{2 \times 10^9 \sqrt{2.56}} = 9.375 \text{ cm}$$

$$\sqrt{1 - \left(\frac{f_c}{f}\right)^2} = \sqrt{1 - \left(\frac{1.8315}{2}\right)^2} = 0.4017$$

Thus

$$\lambda_g = \frac{9.375}{0.4017} = 23.34 \text{ cm} \quad \text{where } \lambda = 9.375 \text{ cm}$$

c. The phase constant β_z of the TE_{11} mode is found using (9-13a), or

$$\beta_z = \beta\sqrt{1 - \left(\frac{f_c}{f}\right)^2} = \frac{2\pi}{\lambda}\sqrt{1 - \left(\frac{f_c}{f}\right)^2} = \frac{2\pi}{9.375}(0.4017) = 0.2692 \text{ rad/cm}$$

which can also be obtained using

$$\beta_z = \frac{2\pi}{\lambda_g} = \frac{2\pi}{23.34} = 0.2692 \text{ rad/cm}$$

d. According to (9-16a), the wave impedance of the TE_{11} mode is equal to

$$Z_{11}^h = \frac{\eta}{\sqrt{1 - \left(\frac{f_c}{f}\right)^2}} = \frac{120\pi/\sqrt{2.56}}{0.4017} = 586.56 \text{ ohms}$$

e. Since the next higher-order TE$_{mn}$ mode is the TE$_{21}$, the bandwidth of single TE$_{11}$ mode operation is

$$BW = 3.0542/1.8412 : 1 = 1.6588 : 1$$

9.2.2 Transverse Magnetic (TM^z) Modes

The transverse magnetic to z (TM^z) modes can be derived in a similar manner as the TE^z modes of Section 9.2.1 by letting

$$\mathbf{A} = \hat{a}_z A_z(\rho, \phi, z) \tag{9-17a}$$

$$\mathbf{F} = 0 \tag{9-17b}$$

The vector potential **A** must satisfy the vector wave equation of (3-48) which reduces for the **A** of (9-17a) to

$$\nabla^2 A_z(\rho, \phi, z) + \beta^2 A_z(\rho, \phi, z) = 0 \tag{9-18}$$

The solution of (9-18) is obtained in a manner similar to that of (9-2), as given by (9-4), and it can be written as

$$A_z(\rho, \phi, z) = \left[A_1 J_m(\beta_\rho \rho) + B_1 Y_m(\beta_\rho \rho) \right]$$
$$\times \left[C_2 \cos(m\phi) + D_2 \sin(m\phi) \right] \left[A_3 e^{-j\beta_z z} + B_3 e^{+j\beta_z z} \right] \tag{9-19}$$

with

$$\beta_\rho^2 + \beta_z^2 = \beta^2 \tag{9-19a}$$

The constants A_1, B_1, C_2, D_2, A_3, B_3, m, β_ρ, and β_z can be found using the boundary conditions of

$$E_\phi(\rho = a, \phi, z) = 0 \tag{9-20a}$$

or

$$E_z(\rho = a, \phi, z) = 0 \tag{9-20b}$$

The fields must be finite everywhere (9-20c)

The fields must repeat every 2π radians in ϕ (9-20d)

According to (9-20c), $B_1 = 0$ since $Y_m(\rho = 0) = \infty$. In addition, according to (9-20d),

$$m = 0, 1, 2, 3, \ldots \tag{9-21}$$

Considering waves that propagate only in the $+z$ direction, (9-19) then reduces to

$$A_z^+(\rho, \phi, z) = B_{mn} J_m(\beta_\rho \rho) \left[C_2 \cos(m\phi) + D_2 \sin(m\phi) \right] e^{-j\beta_z z} \tag{9-22}$$

The eigenvalues of β_ρ can be obtained by applying either (9-20a) or (9-20b). Use of

(6-70) and (9-22) allows us to write the electric field component E_z^+ as

$$E_z^+ = -j\frac{1}{\omega\mu\varepsilon}\left(\frac{\partial^2}{\partial z^2} + \beta^2\right)A_z^+$$

$$= -jB_{mn}\frac{\beta_\rho^2}{\omega\mu\varepsilon}J_m(\beta_\rho\rho)[C_2\cos(m\phi) + D_2\sin(m\phi)]e^{-j\beta_z z} \quad (9\text{-}23)$$

Application of the boundary condition of (9-20b) using (9-23) gives

$$E_z^+(\rho = a, \phi, z) = -jB_{mn}\frac{\beta_\rho^2}{\omega\mu\varepsilon}J_m(\beta_\rho a)[C_2\cos(m\phi) + D_2\sin(m\phi)]e^{-j\beta_z z} = 0 \quad (9\text{-}24)$$

which is only satisfied provided that

$$J_m(\beta_\rho a) = 0 \Rightarrow \beta_\rho a = \chi_{mn} \Rightarrow \beta_\rho = \frac{\chi_{mn}}{a} \quad (9\text{-}25)$$

In (9-25) χ_{mn} represents the nth zero ($n = 1, 2, 3, \ldots$) of the Bessel function J_m of the first kind of order m ($m = 0, 1, 2, 3, \ldots$). An abbreviated list of the zeroes χ_{mn} of the Bessel function J_m is found in Table 9-2. The smallest value of χ_{mn} is 2.4049 ($m = 0$, $n = 1$), followed by 3.8318 ($m = 1$, $n = 1$), 5.1357 ($m = 2$, $n = 1$), etcetera.

By using (9-19a) and (9-25), β_z can be written as

$$(\beta_z)_{mn} = \begin{cases} \sqrt{\beta^2 - \beta_\rho^2} = \sqrt{\beta^2 - \left(\frac{\chi_{mn}}{a}\right)^2} & \text{when } \beta > \beta_\rho = \frac{\chi_{mn}}{a} & (9\text{-}26a) \\ 0 & \text{when } \beta = \beta_c = \beta_\rho = \frac{\chi_{mn}}{a} & (9\text{-}26b) \\ -j\sqrt{\beta_\rho^2 - \beta^2} = -j\sqrt{\left(\frac{\chi_{mn}}{a}\right)^2 - \beta^2} & \text{when } \beta < \beta_\rho = \frac{\chi_{mn}}{a} & (9\text{-}26c) \end{cases}$$

By following the same procedure as for the TE^z modes, we can write the expressions for the cutoff frequencies $(f_c)_{mn}$, propagation constant $(\beta_z)_{mn}$, and

TABLE 9-2
Zeroes χ_{mn} of $J_m(\chi_{mn}) = 0$ ($n = 1, 2, 3, \ldots$) of Bessel function $J_m(x)$

	$m=0$	$m=1$	$m=2$	$m=3$	$m=4$	$m=5$	$m=6$	$m=7$	$m=8$	$m=9$	$m=10$	$m=11$
$n=1$	2.4049	3.8318	5.1357	6.3802	7.5884	8.7715	9.9361	11.0864	12.2251	13.3543	14.4755	12.8264
$n=2$	5.5201	7.0156	8.4173	9.7610	11.0647	12.3386	13.5893	14.8213	16.0378	17.2412	18.4335	19.6160
$n=3$	8.6537	10.1735	11.6199	13.0152	14.3726	15.7002	17.0038	18.2876	19.5545	20.8071	22.0470	23.2759
$n=4$	11.7915	13.3237	14.7960	16.2235	17.6160	18.9801	20.3208	21.6415	22.9452	24.2339	25.5095	26.7733
$n=5$	14.9309	16.4706	17.9598	19.4094	20.8269	22.2178	23.5861	24.9349	26.2668	27.5838	28.8874	80.1791

guide wavelength $(\lambda_g)_{mn}$ as

$$(f_c)_{mn} = \frac{\chi_{mn}}{2\pi a\sqrt{\mu\varepsilon}} \qquad (9\text{-}27)$$

$$(\beta_z)_{mn} = \begin{cases} \sqrt{\beta^2 - \beta_\rho^2} = \beta\sqrt{1 - \left(\frac{\beta_\rho}{\beta}\right)^2} = \beta\sqrt{1 - \left(\frac{\beta_c}{\beta}\right)^2} \\ \qquad = \beta\sqrt{1 - \left(\frac{\chi_{mn}}{\beta a}\right)^2} = \beta\sqrt{1 - \left(\frac{f_c}{f}\right)^2} \qquad (9\text{-}28\text{a}) \\ \qquad \text{when } f > f_c = (f_c)_{mn} \\ 0 \qquad \text{when } f = f_c = (f_c)_{mn} \qquad (9\text{-}28\text{b}) \\ -j\sqrt{\beta_\rho^2 - \beta^2} = -j\beta\sqrt{\left(\frac{\beta_\rho}{\beta}\right)^2 - 1} = -j\beta\sqrt{\left(\frac{\beta_c}{\beta}\right)^2 - 1} \\ \qquad = -j\beta\sqrt{\left(\frac{\chi_{mn}}{\beta a}\right)^2 - 1} = -j\beta\sqrt{\left(\frac{f_c}{f}\right)^2 - 1} \qquad (9\text{-}28\text{c}) \\ \qquad \text{when } f < f_c = (f_c)_{mn} \end{cases}$$

$$(\lambda_g)_{mn} = \begin{cases} \dfrac{2\pi}{\beta\sqrt{1 - \left(\dfrac{f_c}{f}\right)^2}} = \dfrac{\lambda}{\sqrt{1 - \left(\dfrac{f_c}{f}\right)^2}} \qquad \text{when } f > f_c = (f_c)_{mn} \qquad (9\text{-}29\text{a}) \\ \infty \qquad \text{when } f = f_c = (f_c)_{mn} \qquad (9\text{-}29\text{b}) \end{cases}$$

According to (9-27) and the values of χ_{mn} of Table 9-2, the order (lower to higher cutoff frequencies) in which the TM^z modes occur is TM_{01}, TM_{11}, TM_{21}, and so forth. The bandwidth of the first single-mode TM_{01}^z operation is $3.8318/2.4049 = 1.5933:1$ which is also less than $2:1$. Comparing the cutoff frequencies of the TE^z and TM^z modes, as given by (9-12a) and (9-27) along with the data of Tables 9-1 and 9-2, the order of the TE_{mn}^z and TM_{mn}^z modes is that of TE_{11} ($\chi'_{11} = 1.8412$), TM_{01} ($\chi_{01} = 2.4049$), TE_{21} ($\chi'_{21} = 3.0542$), TE_{01} ($\chi'_{01} = 3.8318$) = TM_{11} ($\chi_{11} = 3.8318$), TE_{31} ($\chi'_{31} = 4.2012$), and so forth. The dominant mode is TE_{11} and its bandwidth of single-mode operation is $2.4049/1.8412 = 1.3062:1$ which is much smaller than $2:1$. Plots of the field configurations over a cross section of the waveguide, both E and H, for the first 30 TE_{mn}^z and/or TM_{mn}^z modes are shown in Figure 9-2 [1].

It is apparent that the cutoff frequencies of the TE_{0n} and TM_{1n} modes are identical; therefore they are referred to here also as degenerate modes. This is because the zeroes of the derivative of the Bessel function J_0 are identical to the zeroes of the Bessel function J_1. To demonstrate this, let us examine the derivative

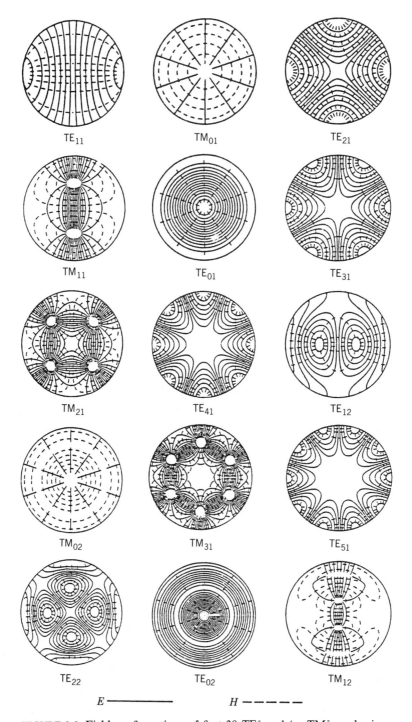

FIGURE 9-2 Field configurations of first 30 TE^z and/or TM^z modes in a circular waveguide. (*Source:* C. S. Lee, S. W. Lee, and S. L. Chuang, "Plot of modal field distribution in rectangular and circular waveguides," *IEEE Trans. Microwave Theory Tech.*, © 1966, IEEE.)

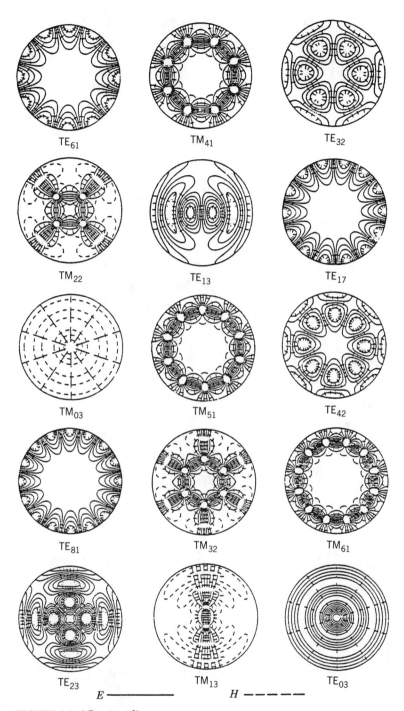

FIGURE 9-2 (*Continued*).

482 CIRCULAR CROSS-SECTION WAVEGUIDES AND CAVITIES

of $J_0(\beta_\rho \rho)$ evaluated at $\rho = a$. Using (IV-19) we can write that

$$\frac{d}{d(\beta_\rho \rho)} J_0(\beta_\rho \rho)\bigg|_{\rho=a} = J_0'(\beta_\rho a) = -J_1(\beta_\rho \rho)\bigg|_{\rho=a} = -J_1(\beta_\rho a) \quad (9\text{-}30)$$

which vanishes when

$$J_0'(\beta_\rho a) = 0 \Rightarrow \beta_\rho a = \chi_{0n}' \quad n = 1, 2, 3, \ldots \quad (9\text{-}30a)$$

or

$$J_1(\beta_\rho a) = 0 \Rightarrow \beta_\rho a = \chi_{1n} \quad n = 1, 2, 3, \ldots \quad (9\text{-}30b)$$

The electric and magnetic field components can be written, using (6-70) and (9-22), as

$$E_\rho^+ = -j\frac{1}{\omega\mu\varepsilon}\frac{\partial^2 A_z^+}{\partial\rho\,\partial z} = -B_{mn}\frac{\beta_\rho \beta_z}{\omega\mu\varepsilon} J_m'(\beta_\rho \rho)[C_2 \cos(m\phi) + D_2 \sin(m\phi)] e^{-j\beta_z z}$$
$$(9\text{-}31a)$$

$$E_\phi^+ = -j\frac{1}{\omega\mu\varepsilon}\frac{1}{\rho}\frac{\partial^2 A_z^+}{\partial\phi\,\partial z} = -B_{mn}\frac{m\beta_z}{\omega\mu\varepsilon\rho} J_m(\beta_\rho \rho)[-C_2 \sin(m\phi) + D_2 \cos(m\phi)] e^{-j\beta_z z}$$
$$(9\text{-}31b)$$

$$E_z^+ = -j\frac{1}{\omega\mu\varepsilon}\left(\frac{\partial^2}{\partial z^2} + \beta^2\right) A_z^+$$

$$= -jB_{mn}\frac{\beta_\rho^2}{\omega\mu\varepsilon} J_m(\beta_\rho \rho)[C_2 \cos(m\phi) + D_2 \sin(m\phi)] e^{-j\beta_z z} \quad (9\text{-}31c)$$

$$H_\rho^+ = \frac{1}{\mu}\frac{1}{\rho}\frac{\partial A_z^+}{\partial\phi} = B_{mn}\frac{m}{\mu}\frac{1}{\rho} J_m(\beta_\rho \rho)[-C_2 \sin(m\phi) + D_2 \cos(m\phi)] e^{-j\beta_z z}$$
$$(9\text{-}31d)$$

$$H_\phi^+ = -\frac{1}{\mu}\frac{\partial A_z^+}{\partial\rho} = -B_{mn}\frac{\beta_\rho}{\mu} J_m'(\beta_\rho \rho)[C_2 \cos(m\phi) + D_2 \sin(m\phi)] e^{-j\beta_z z} \quad (9\text{-}31e)$$

$$H_z^+ = 0 \quad (9\text{-}31f)$$

where

$$' = \frac{\partial}{\partial(\beta_\rho \rho)} \quad (9\text{-}31g)$$

By using (9-31a) through (9-31f), the wave impedance in the $+z$ direction can be written as

$$(Z_w^{+z})_{mn}^{TM} = \frac{E_\rho^+}{H_\phi^+} = -\frac{E_\phi^+}{H_\rho^+} = \frac{(\beta_z)_{mn}}{\omega\varepsilon} \quad (9\text{-}32)$$

With the aid of (9-28a) through (9-28c) the wave impedance of (9-32) reduces to

$$(Z_w^{+z})_{mn}^{TM} = \begin{cases} \dfrac{\beta\sqrt{1-\left(\dfrac{f_c}{f}\right)^2}}{\omega\varepsilon} = \sqrt{\dfrac{\mu}{\varepsilon}}\sqrt{1-\left(\dfrac{f_c}{f}\right)^2} = \eta\sqrt{1-\left(\dfrac{f_c}{f}\right)^2} & \text{when } f > f_c = (f_c)_{mn} \quad (9\text{-}32a) \\[2ex] \dfrac{0}{\omega\varepsilon} = 0 & \text{when } f = f_c = (f_c)_{mn} \quad (9\text{-}32b) \\[2ex] \dfrac{-j\beta\sqrt{\left(\dfrac{f_c}{f}\right)^2-1}}{\omega\varepsilon} = -j\sqrt{\dfrac{\mu}{\varepsilon}}\sqrt{\left(\dfrac{f_c}{f}\right)^2-1} = -j\eta\sqrt{\left(\dfrac{f_c}{f}\right)^2-1} & \text{when } f < f_c = (f_c)_{mn} \quad (9\text{-}32c) \end{cases}$$

Examining (9-32a) through (9-32c) we can make the following statements about the wave impedance for the TM^z modes.

1. Above cutoff it is real and smaller than the intrinsic impedance of the medium inside the waveguide.
2. At cutoff it is zero.
3. Below cutoff it is imaginary and capacitive. This indicates that the waveguide below cutoff behaves as a capacitor that is an energy storage element.

The form of $(Z_w^{+z})_{mn}^{TM}$, as given by (9-32a) through (9-32c), as a function of f_c/f is the same as the $(Z_w^{+z})_{mn}^{TM}$ for the TM^z_{mn} modes of a rectangular waveguide, as given by (8-29a) through (8-29c). A plot of (9-32a) through (9-32c) for any one TM^z_{mn} mode as a function of f_c/f, where f_c is the cutoff frequency of that mode, is shown in Figure 8-2.

Example 9-2. Design a circular waveguide filled with a lossless dielectric medium of dielectric constant 4. The waveguide must operate in a single dominant mode over a bandwidth of 1 GHz.

1. Find its radius (in cm).
2. Determine the lower, center, and upper frequencies of the bandwidth.

Solution.

a. The dominant mode is the TE_{11} mode whose cutoff frequency according to (9-12a) is

$$(f_c)_{11}^{TE^z} = \frac{\chi'_{11}}{2\pi a\sqrt{\mu\varepsilon}} = \frac{1.8412(30 \times 10^9)}{2\pi(a)\sqrt{4}}$$

The next higher-order mode is the TM_{01} mode whose cutoff frequency according to (9-27) is

$$(f_c)_{01}^{TM^z} = \frac{\chi_{01}}{2\pi a\sqrt{\mu\varepsilon}} = \frac{2.4049(30 \times 10^9)}{2\pi(a)\sqrt{4}}$$

The difference between the two must be 1 GHz. To accomplish this, the radius of the waveguide must be equal to

$$\frac{(2.4049 - 1.8412)30 \times 10^9}{2\pi(a)\sqrt{4}} = 1 \times 10^9 \Rightarrow a = 1.3457 \text{ cm}$$

b. The lower, upper, and center frequencies of the bandwidth are now equal to

$$f_\ell = (f_c)_{11}^{TE^z} = \frac{1.8412(30 \times 10^9)}{2\pi(1.3457)2} = 3.2664 \times 10^9 = 3.2664 \text{ GHz}$$

$$f_u = (f_c)_{01}^{TM^z} = \frac{2.4049(30 \times 10^9)}{2\pi(1.3457)2} = 4.2664 \times 10^9 = 4.2664 \text{ GHz}$$

$$f_0 = f_\ell + 0.5 \times 10^9 = f_u - 0.5 \times 10^9 = 3.7664 \times 10^9 = 3.7664 \text{ GHz}$$

Whenever a given mode is desired, it is necessary to design the proper feed to excite the fields within the waveguide and detect the energy associated with such modes. To maximize the energy exchange or transfer, this is accomplished by designing the feed, which is usually a probe or antenna, so that its field pattern matches that of the field configuration of the desired mode. Usually the probe is

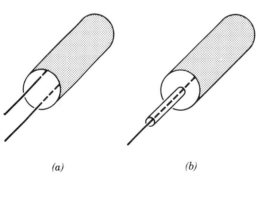

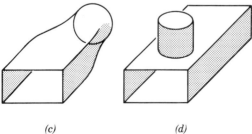

FIGURE 9-3 Excitation of TE_{mn} and TM_{mn} modes in a circular waveguide. (a) TE_{11} mode. (b) TM_{01} mode. (c) TE_{10}-(rectangular)- TE_{11} (circular). (d) TE_{10} (rectangular)-TM_{01}- (circular).

placed near the maximum of the field pattern of the desired mode; however, that position may be varied somewhat in order to achieve some desired matching in the excitation and detection systems. Shown in Figure 9-3 are suggested designs to excite and/or detect the TE_{11} and TM_{01} modes in a circular waveguide, to transition between the TE_{10} of a rectangular waveguide and the TE_{11} mode of a circular waveguide, and to couple between the TE_{10} of a rectangular waveguide and TM_{01} mode of a circular waveguide.

9.2.3 Attenuation

The attenuation in a circular waveguide can be obtained by using techniques similar to those for the rectangular waveguide, as outlined and applied in Section 8.2.5. The basic equation is that of (8-64a), or

$$(\alpha_c)_{mn} = \frac{P_c/\ell}{2P_{mn}} = \frac{P_\ell}{2P_{mn}} \quad (9\text{-}33)$$

which is based on the configuration of Figure 8-9.

It has been shown that the attenuation coefficients of the TE_{0n} ($n = 1, 2, \ldots$) modes in a circular waveguide monotonically decrease as a function of frequency [2, 3]. This is a very desirable characteristic, and because of this the excitation, propagation, and detection of TE_{0n} modes in a circular waveguide have received considerable attention. It can be shown that the attenuation coefficient for the TE^z_{mn} and TM^z_{mn} modes inside a circular waveguide are given, respectively, by

$$\underline{TE^z_{mn}}$$

$$\boxed{(\alpha_c)^{TE^z}_{mn} = \frac{R_s}{a\eta\sqrt{1 - \left(\frac{f_c}{f}\right)^2}} \left[\left(\frac{f_c}{f}\right)^2 + \frac{m^2}{(\chi'_{mn})^2 - m^2}\right] \text{ Np/m}} \quad (9\text{-}34a)$$

$$\underline{TM^z_{mn}}$$

$$\boxed{(\alpha_c)^{TM^z}_{mn} = \frac{R_s}{a\eta} \frac{1}{\sqrt{1 - \left(\frac{f_c}{f}\right)^2}} \text{ Np/m}} \quad (9\text{-}34b)$$

Plots of the attenuation coeficient versus the normalized frequency f/f_c, where f_c is the cutoff frequency of the dominant TE_{11} mode, are shown for six modes in Figure 9-4a and b for waveguide radii of 1.5 and 3 cm, respectively. Within the waveguide is free space and its walls are made of copper ($\sigma = 5.7 \times 10^7$ S/m).

Example 9-3. Derive the attenuation coefficient for the TE_{01} mode inside a circular waveguide of radius a.

486 CIRCULAR CROSS-SECTION WAVEGUIDES AND CAVITIES

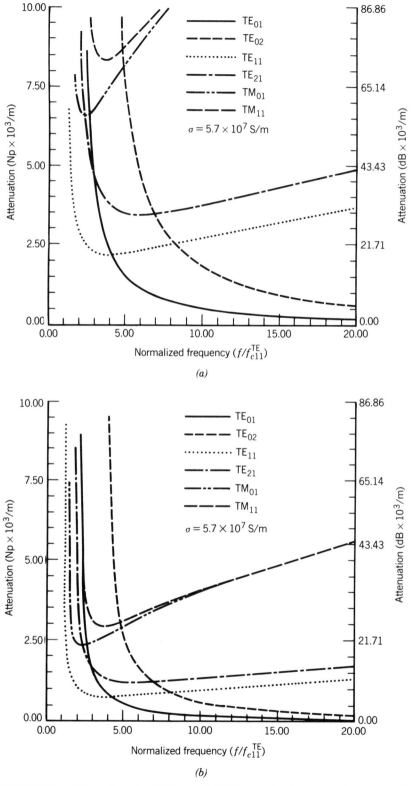

FIGURE 9-4 Attenuation for TE_{mn}^z and TM_{mn}^z modes in a circular waveguide. (*a*) $a = 1.5$ cm. (*b*) $a = 3$ cm.

Solution. According to (9-15a) through (9-15g), the electric and magnetic field components for the TE$_{01}$ ($m = 0$, $n = 1$) mode reduce to

$$E_\rho^+ = E_z^+ = H_\phi^+ = 0$$

$$E_\phi^+ = \beta_\rho \frac{A_{01}}{\varepsilon} J_0'(\beta_\rho \rho) e^{-j\beta_z z}$$

$$H_\rho^+ = -A_{01} \frac{\beta_\rho \beta_z}{\omega \mu \varepsilon} J_0'(\beta_\rho \rho) e^{-j\beta_z z}$$

$$H_z^+ = -jA_{01} \frac{\beta_\rho^2}{\omega \mu \varepsilon} J_0(\beta_\rho \rho) e^{-j\beta_z z}$$

where

$$\beta_\rho = \frac{\chi'_{01}}{a} = \frac{3.8318}{a}$$

Using these equations, the power through a cross section of the waveguide is equal to

$$P_{01} = \frac{1}{2} \iint_{A_0} \mathrm{Re}\left[(\mathbf{E} \times \mathbf{H}^*) \cdot d\mathbf{s}\right]$$

$$= \frac{1}{2} \iint_{A_0} \mathrm{Re}\left[\hat{a}_\phi E_\phi \times (\hat{a}_\rho H_\rho + \hat{a}_z H_z)^*\right] \cdot \hat{a}_z \, ds$$

$$P_{01} = -\frac{1}{2} \mathrm{Re} \int_0^{2\pi} \int_0^a (E_\phi H_\rho^*) \rho \, d\rho \, d\phi = |A_{01}|^2 \frac{\pi \beta_z \beta_\rho^2}{\omega \mu \varepsilon^2} \int_0^a \left[J_0'\left(\frac{\chi'_{01}}{a}\rho\right)\right]^2 \rho \, d\rho$$

Since

$$\frac{dJ_p(cx)}{d(cx)} = -J_{p+1}(cx) + \frac{p}{cx} J_p(cx)$$

then

$$J_0'\left(\frac{\chi'_{01}}{a}\rho\right) = \frac{d}{d(\chi'_{01}\rho/a)} J_0\left(\frac{\chi'_{01}}{a}\rho\right) = -J_1\left(\frac{\chi'_{01}}{a}\rho\right)$$

Thus

$$P_{01} = |A_{01}|^2 \frac{\pi \beta_z}{\omega \mu \varepsilon^2} \left(\frac{\chi'_{01}}{a}\right)^2 \int_0^a J_1^2\left(\frac{\chi'_{01}}{a}\rho\right) \rho \, d\rho$$

Since

$$\int_b^c x J_p^2(cx) \, dx = \frac{x^2}{2}\left[J_p^2(cx) - J_{p-1}(cx) J_{p+1}(cx)\right]_b^c$$

then

$$\int_0^a \rho J_1^2\left(\frac{\chi'_{01}}{a}\rho\right) d\rho = \frac{a^2}{2}\left[J_1^2(\chi'_{01}) - J_0(\chi'_{01}) J_2(\chi'_{01})\right]$$

$$= -\frac{a^2}{2} J_0(\chi'_{01}) J_2(\chi'_{01}) = \frac{a^2}{2} J_0^2(\chi'_{01})$$

because

$$J_1^2(\chi'_{01}) = J_1^2(3.8318) = 0$$
$$J_2(\chi'_{01}) = -J_0(\chi'_{01})$$

Therefore the power of the TE_{01} can be written as

$$P_{01} = |A_{01}|^2 \frac{\pi \beta_z}{2\omega\mu\varepsilon^2}(\chi'_{01})^2 J_0^2(\chi'_{01})$$

The power dissipated on the walls of the waveguide is obtained using

$$P_c = \frac{R_s}{2}\iint_{S_w}(\mathbf{J}_s \cdot \mathbf{J}_s^*)_{\rho=a}\,ds = \frac{R_s}{2}\int_0^\ell\int_0^{2\pi}|\mathbf{J}_s|^2_{\rho=a}\,a\,d\phi\,dz$$

where

$$\mathbf{J}_s|_{\rho=a} = \hat{n}\times\mathbf{H}^+|_{\rho=a} = \hat{a}_\phi H_z^+|_{\rho=a} = -\hat{a}_\phi j\frac{\beta_\rho^2}{\omega\mu\varepsilon}A_{01}J_0(\beta_\rho a)e^{-j\beta_z z}$$

Thus

$$P_c = |A_{01}|^2 \frac{R_s}{2}\left(\frac{\beta_\rho^2}{\omega\mu\varepsilon}\right)^2 aJ_0^2(\chi'_{01})\int_0^\ell\int_0^{2\pi}d\phi\,dz$$

or

$$\frac{P_c}{\ell} = P_\ell = |A_{01}|^2 \frac{\pi R_s}{a^3}\left[\frac{(\chi'_{01})^2}{\omega\mu\varepsilon}\right]^2 J_0^2(\chi'_{01})$$

Therefore the attenuation coefficient of (9-33) for the TE_{01} mode can now be written as

$$\alpha_{01}(TE^z) = \frac{R_s}{a\eta}\frac{\left(\frac{f_c}{f}\right)^2}{\sqrt{1-\left(\frac{f_c}{f}\right)^2}}\quad\text{Np/m}$$

It is evident from the results of the preceding example that as f_c/f becomes smaller the attenuation coefficient decreases monotonically (as shown in Figure 9-4), which is a desirable characteristic. It should be noted that similar monotonically decreasing variations in the attenuation coefficient are evident in all TE_{0n} modes ($n = 1, 2, 3,\ldots$). According to (9-15a) through (9-15f) the only tangential magnetic field components to the conducting surface of the waveguide for all these TE_{0n} ($m = 0$) modes is the H_z component, while the electric field lines are circular. Therefore these modes are usually referred to as circular electric modes. For a constant power in the wave, the H_z component decreases as the frequency increases and approaches zero at infinite frequency. Simultaneously the current density and conductor losses on the waveguide walls also decrease and approach zero. Because of this attractive feature, these modes have received considerable attention for long distance propagation of energy, especially at millimeter wave frequencies. Typically attenuations as low as 1.25 dB/km (2 dB/mi) have been attained [2]. This is to be compared with attenuations of 120 dB/km for WR90 copper rectangular waveguides, and 3 dB/km at 0.85 μm, and less than 0.5 dB/km at 1.3 μm for fiber optics cables.

Although the TE_{0n} modes are very attractive from the attenuation point of view, there are a number of problems associated with their excitation and retention. One of the problems is that the TE_{01} mode, which is the first of the TE_{0n} modes, is not the dominant mode. Therefore in order for this mode to be above its cutoff frequency and propagate in the waveguide, a number of other modes (such as the TE_{11}, TM_{01}, TE_{21}, and TM_{11}) with lower cutoff frequencies can also exist. Additional modes can also be present if the operating frequency is chosen well above the cutoff frequency of the TE_{01} mode in order to provide a margin of safety from being too close to its cutoff frequency.

To support the TE_{01} mode, the waveguide must be oversized and it can support a number of other modes. One of the problems faced with such a guide is how to excite the desired TE_{01} mode with sufficient purity and suppress the others. Another problem is how to prevent coupling between the TE_{01} mode and undesired modes that can exist since the guide is oversized. The presence of the undesired modes causes not only higher losses but dispersion and attenuation distortion to the signal since each exhibits different phase velocities and attenuation. Irregularities in the inner geometry, surface, and direction (such as bends, nonuniform cross sections, etc.) of the waveguide are the main contributors to the coupling to the undesired modes. However, for the guide to be of any practical use, it must be able to sustain and propagate the desired TE_{01} and other TE_{0n} modes efficiently over bends of reasonable curvature. One technique that has been implemented to achieve this is to use mode conversion before entering the corner and another conversion when exiting to convert back to the desired TE_{0n} mode(s).

Another method that has been used to discriminate against undesired modes and avoid coupling to them is to introduce filters inside the guide that cause negligible attenuation to the desired TE_{0n} mode(s). The basic principle of these filters is to introduce cuts that are perpendicular to the current paths of the undesired modes and parallel to the current direction of the desired mode(s). Since the current path of the undesired modes is along the axis (z direction) of the guide and the path of the desired TE_{0n} modes is along the circumference (ϕ direction), a helical wound wire placed on the inside surface of the guide can serve as a filter that discourages any mode that requires an axial component of current flow but propagates the desired TE_{0n} modes [3, 4].

Another filter that can be used to suppress undesired modes is to introduce within the guide very thin baffles of lossy material that will act as attenuating sheets. The surfaces of the baffles are placed in the radial direction of the guide so that they are parallel to the E_ρ and E_z components of the undesired modes (which will be damped) and normal to the E_ϕ component of the TE_{0n} modes that will remain unaffected. Typically two baffles are used and are placed in a crossed pattern over the cross section of the guide.

A summary of the pertinent characteristics of the TE^z_{mn} and TM^z_{mn} modes of a circular waveguide are found listed in Table 9-3.

TABLE 9-3
Summary of TE$_{mn}^z$ and TM$_{mn}^z$ mode characteristics of circular waveguide

	TE$_{mn}^z$ $\begin{pmatrix} m = 0,1,2,\ldots, \\ n = 1,2,3,\ldots \end{pmatrix}$	TM$_{mn}^z$ $\begin{pmatrix} m = 0,1,2,3,\ldots, \\ n = 1,2,3,4,\ldots \end{pmatrix}$
E_ρ^+	$-A_{mn}\dfrac{m}{\varepsilon\rho} J_m(\beta_\rho\rho)[-C_2\sin(m\phi) + D_2\cos(m\phi)]e^{-j\beta_z z}$	$-B_{mn}\dfrac{\beta_\rho\beta_z}{\omega\mu\varepsilon} J_m'(\beta_\rho\rho)[C_2\cos(m\phi) + D_2\sin(m\phi)]e^{-j\beta_z z}$
E_ϕ^+	$A_{mn}\dfrac{\beta_\rho}{\varepsilon} J_m'(\beta_\rho\rho)[C_2\cos(m\phi) + D_2\sin(m\phi)]e^{-j\beta_z z}$	$-B_{mn}\dfrac{m\beta_z}{\omega\mu\varepsilon}\dfrac{1}{\rho} J_m(\beta_\rho\rho)[-C_2\sin(m\phi) + D_2\cos(m\phi)]e^{-j\beta_z z}$
E_z^+	0	$-jB_{mn}\dfrac{\beta_\rho^2}{\omega\mu\varepsilon} J_m(\beta_\rho\rho)[C_2\cos(m\phi) + D_2\sin(m\phi)]e^{-j\beta_z z}$
H_ρ^+	$-A_{mn}\dfrac{\beta_\rho\beta_z}{\omega\mu\varepsilon} J_m'(\beta_\rho\rho)[C_2\cos(m\phi) + D_2\sin(m\phi)]e^{-j\beta_z z}$	$B_{mn}\dfrac{m}{\mu}\dfrac{1}{\rho} J_m(\beta_\rho\rho)[-C_2\sin(m\phi) + D_2\cos(m\phi)]e^{-j\beta_z z}$
H_ϕ^+	$-A_{mn}\dfrac{m\beta_z}{\omega\mu\varepsilon}\dfrac{1}{\rho} J_m(\beta_\rho\rho)[-C_2\sin(m\phi) + D_2\cos(m\phi)]e^{-j\beta_z z}$	$-B_{mn}\dfrac{\beta_\rho}{\mu} J_m'(\beta_\rho\rho)[-C_2\cos(m\phi) + D_2\sin(m\phi)]e^{-j\beta_z z}$
H_z^+	$-jA_{mn}\dfrac{\beta_\rho^2}{\omega\mu\varepsilon} J_m(\beta_\rho\rho)[C_2\cos(m\phi) + D_2\sin(m\phi)]e^{-j\beta_z z}$	0
$'$		$\dfrac{\partial}{\partial(\beta_\rho\rho)}$
$\beta_c = \beta_\rho$	$\dfrac{\chi_{mn}'}{a}$	$\dfrac{\chi_{mn}}{a}$
f_c	$\dfrac{\chi_{mn}'}{2\pi a\sqrt{\mu\varepsilon}}$	$\dfrac{\chi_{mn}}{2\pi a\sqrt{\mu\varepsilon}}$
λ_c	$\dfrac{2\pi a}{\chi_{mn}'}$	$\dfrac{2\pi a}{\chi_{mn}}$

CIRCULAR WAVEGUIDE

TABLE 9-3 (*Continued*).

	$\text{TE}_{mn}^z \left(\begin{array}{l} m = 0, 1, 2, \ldots, \\ n = 1, 2, 3, \ldots \end{array} \right)$	$\text{TM}_{mn}^z \left(\begin{array}{l} m = 0, 1, 2, 3, \ldots, \\ n = 1, 2, 3, 4, \ldots \end{array} \right)$
$\beta_z \ (f \geq f_c)$	$\beta \sqrt{1 - \left(\dfrac{f_c}{f}\right)^2} = \beta \sqrt{1 - \left(\dfrac{\lambda}{\lambda_c}\right)^2}$	
$\lambda_g \ (f \geq f_c)$	$\dfrac{\lambda}{\sqrt{1 - \left(\dfrac{f_c}{f}\right)^2}} = \dfrac{\lambda}{\sqrt{1 - \left(\dfrac{\lambda}{\lambda_c}\right)^2}}$	
$v_p \ (f \geq f_c)$	$\dfrac{v}{\sqrt{1 - \left(\dfrac{f_c}{f}\right)^2}} = \dfrac{v}{\sqrt{1 - \left(\dfrac{\lambda}{\lambda_c}\right)^2}}$	
$Z_w \ (f \geq f_c)$	$\dfrac{\eta}{\sqrt{1 - \left(\dfrac{f_c}{f}\right)^2}} = \dfrac{\eta}{\sqrt{1 - \left(\dfrac{\lambda}{\lambda_c}\right)^2}}$	$\eta \sqrt{1 - \left(\dfrac{f_c}{f}\right)^2} = \eta \sqrt{1 - \left(\dfrac{\lambda}{\lambda_c}\right)^2}$
$Z_w \ (f \leq f_c)$	$j \dfrac{\eta}{\sqrt{\left(\dfrac{f_c}{f}\right)^2 - 1}} = j \dfrac{\eta}{\sqrt{\left(\dfrac{\lambda}{\lambda_c}\right)^2 - 1}}$	$-j\eta \sqrt{\left(\dfrac{f_c}{f}\right)^2 - 1} = -j\eta \sqrt{\left(\dfrac{\lambda}{\lambda_c}\right)^2 - 1}$
α_c	$\dfrac{R_s}{a\eta \sqrt{1 - \left(\dfrac{f_c}{f}\right)^2}} \left[\left(\dfrac{f_c}{f}\right)^2 + \dfrac{m^2}{(\chi'_{mn})^2 - m^2} \right]$	$\dfrac{R_s}{a\eta} \dfrac{1}{\sqrt{1 - \left(\dfrac{f_c}{f}\right)^2}}$

9.3 CIRCULAR CAVITY

As in rectangular waveguides, a circular cavity is formed by closing the two ends of the waveguide with plates, as shown in Figure 9-5. Coupling in and out of the cavity is done using either irises (holes) or probes (antennas), some of which were illustrated in Figure 9-3. Since the boundary conditions along the circumferential surface of the waveguide are the same as those of the cavity, the analysis can begin by modifying only the traveling waves of the z variations of the waveguide in order to obtain the standing waves of the cavity. The radial (ρ) and circumferential (ϕ) variations in both cases will be the same. For the circular cavity both TE^z and TM^z modes can exist and will be examined here.

9.3.1 Transverse Electric (TE^z) Modes

We begin the analysis of the TE^z modes by assuming the vector potential F_z is that of (9-7) modified so that the z variations are standing waves instead of traveling waves. Thus we can write using (9-7) that

$$F_z(\rho, \phi, z) = A_{mn} J_m(\beta_\rho \rho)[C_2 \cos(m\phi) + D_2 \sin(m\phi)]$$
$$\times [C_3 \cos(\beta_z z) + D_3 \sin(\beta_z z)] \qquad (9\text{-}35)$$

where

$$\beta_\rho = \frac{\chi'_{mn}}{a} \qquad (9\text{-}35a)$$

$$m = 0, 1, 2, 3, \ldots \qquad (9\text{-}35b)$$

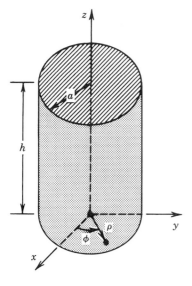

FIGURE 9-5 Geometry for circular cavity.

To determine the permissible values of β_z, we must apply the additional boundary conditions introduced by the presence of the end plates. These additional boundary conditions are

$$E_\rho(0 \le \rho \le a, 0 \le \phi \le 2\pi, z = 0) = E_\rho(0 \le \rho \le a, 0 \le \phi \le 2\pi, z = h) = 0 \tag{9-36a}$$

$$E_\phi(0 \le \rho \le a, 0 \le \phi \le 2\pi, z = 0) = E_\phi(0 \le \rho \le a, 0 \le \phi \le 2\pi, z = h) = 0 \tag{9-36b}$$

Since both are not independent boundary conditions, using either of the two leads to the same results.

Using (6-80) and (9-35), we can write the E_ϕ component as

$$E_\phi = -\frac{1}{\varepsilon}\frac{\partial F_z}{\partial \rho}$$

$$= \beta_\rho \frac{A_{mn}}{\varepsilon} J'_m(\beta_\rho \rho)[C_2 \cos(m\phi) + D_2 \sin(m\phi)][C_3 \cos(\beta_z z) + D_3 \sin(\beta_z z)] \tag{9-37}$$

where

$$' = \frac{\partial}{\partial(\beta_\rho \rho)} \tag{9-37a}$$

Applying (9-36b) leads to

$$E_\phi(0 \le \rho \le a, 0 \le \phi \le 2\pi, z = 0)$$

$$= \beta_\rho \frac{A_{mn}}{\varepsilon} J'_m(\beta_\rho \rho)[C_2 \cos(m\phi) + D_2 \sin(m\phi)][C_3(1) + D_3(0)] = 0$$

$$\Rightarrow C_3 = 0 \tag{9-38a}$$

$$E_\phi(0 \le \rho \le a, 0 \le \phi \le 2\pi, z = h)$$

$$= \beta_\rho \frac{A_{mn}}{\varepsilon} J'_m(\beta_\rho \rho)[C_2 \cos(m\phi) + D_2 \sin(m\phi)] D_3 \sin(\beta_z h) = 0$$

$$\sin(\beta_z h) = 0 \Rightarrow \beta_z h = \sin^{-1}(0) = p\pi$$

$$\beta_z = \frac{p\pi}{h} \quad p = 1, 2, 3, \ldots \tag{9-38b}$$

Thus the resonant frequency is obtained using

$$\beta_\rho^2 + \beta_z^2 = \left(\frac{\chi'_{mn}}{a}\right)^2 + \left(\frac{p\pi}{h}\right)^2 = \beta_r^2 = \omega_r^2 \mu \varepsilon \tag{9-39}$$

or

$$(f_r)_{mnp}^{TE^z} = \frac{1}{2\pi\sqrt{\mu\varepsilon}}\sqrt{\left(\frac{\chi'_{mn}}{a}\right)^2 + \left(\frac{p\pi}{h}\right)^2} \quad \begin{array}{l} m = 0,1,2,3,\ldots \\ n = 1,2,3,\ldots \\ p = 1,2,3,\ldots \end{array} \quad (9\text{-}39a)$$

The values of χ'_{mn} are found listed in Table 9-1. The final form of F_z of (9-35) is

$$F_z(\rho,\phi,z) = A_{mnp}J_m(\beta_\rho\rho)[C_2\cos(m\phi) + D_2\sin(m\phi)]\sin(\beta_z z) \quad (9\text{-}40)$$

9.3.2 Transverse Magnetic (TM^z) Modes

The analysis for the TM^z modes in a circular cavity proceeds in the same manner as for the TE^z modes of the previous section. Using (9-22) we can write that

$$A_z(\rho,\phi,z) = B_{mn}J_m(\beta_\rho\rho)[C_2\cos(m\phi) + D_2\sin(m\phi)]$$
$$\times [C_3\cos(\beta_z z) + D_3\sin(\beta_z z)] \quad (9\text{-}41)$$

where

$$\beta_\rho = \frac{\chi_{mn}}{a} \quad (9\text{-}41a)$$

$$m = 0,1,2,\ldots \quad (9\text{-}41b)$$

Using (6-70) and (9-41), we can write the E_ϕ component as

$$E_\phi(\rho,\phi,z) = -j\frac{1}{\omega\mu\varepsilon}\frac{1}{\rho}\frac{\partial^2 A_z}{\partial\phi\partial z}$$
$$= -jB_{mn}\frac{m\beta_z}{\omega\mu\varepsilon}\frac{1}{\rho}J_m(\beta_\rho\rho)[-C_2\sin(m\phi) + D_2\cos(m\phi)]$$
$$\times [-C_3\sin(\beta_z z) + D_3\cos(\beta_z z)] \quad (9\text{-}42)$$

Applying the boundary conditions of (9-36b) leads to

$$E_\phi(0 \le \rho \le a, 0 \le \phi \le 2\pi, z = 0)$$
$$= -jB_{mn}\frac{m\beta_z}{\omega\mu\varepsilon}\frac{1}{\rho}J_m(\beta_\rho\rho)[-C_2\sin(m\phi) + D_2\cos(m\phi)]$$
$$\times [-C_3(0) + D_3(1)] = 0 \Rightarrow D_3 = 0 \quad (9\text{-}43a)$$

$$E_\phi(0 \le \rho \le a, 0 \le \phi \le 2\pi, z = h)$$
$$= jB_{mn}\frac{m\beta_z}{\omega\mu\varepsilon}\frac{1}{\rho}J_m(\beta_\rho\rho)[-C_2\sin(m\phi) + D_2\cos(m\phi)][C_3\sin(\beta_z h)] = 0$$

$$\sin(\beta_z h) = 0 \Rightarrow \beta_z h = \sin^{-1}(0) = p\pi$$

$$\beta_z = \frac{p\pi}{h} \quad p = 0,1,2,3,\ldots \quad (9\text{-}43b)$$

Thus the resonant frequency is obtained using

$$\beta_\rho^2 + \beta_z^2 = \left(\frac{\chi_{mn}}{a}\right)^2 + \left(\frac{p\pi}{h}\right)^2 = \beta_r^2 = \omega_r^2 \mu\varepsilon \qquad (9\text{-}44)$$

or

$$\boxed{(f_r)^{TM^z}_{mnp} = \frac{1}{2\pi\sqrt{\mu\varepsilon}} \sqrt{\left(\frac{\chi_{mn}}{a}\right)^2 + \left(\frac{p\pi}{h}\right)^2} \quad \begin{array}{l} m = 0,1,2,3,\ldots \\ n = 1,2,3,\ldots \\ p = 0,1,2,3,\ldots \end{array}} \qquad (9\text{-}45)$$

The values of χ_{mn} are found listed in Table 9-2. The final form of A_z of (9-41) is

$$A_z(\rho,\phi,z) = B_{mnp} J_m(\beta_\rho \rho) [C_2 \cos(m\phi) + D_2 \sin(m\phi)] \cos(\beta_z z) \qquad (9\text{-}46)$$

The resonant frequencies of the TE^z_{mnp} and TM^z_{mnp} modes, as given respectively by (9-39a) and (9-45), are functions of the h/a ratio and they are listed in Table 9-4.

The TE^z_{mnp} mode with the smallest resonant frequency is the TE_{111}, and its cutoff frequency is given by

$$(f_r)^{TE^z}_{111} = \frac{1}{2\pi\sqrt{\mu\varepsilon}} \sqrt{\left(\frac{1.8412}{a}\right)^2 + \left(\frac{\pi}{h}\right)^2} \qquad (9\text{-}47a)$$

Similarly the TM^z_{mnp} mode with the smallest resonant frequency is the TM_{010}, and its cutoff frequency is given by

$$(f_r)^{TM^z}_{010} = \frac{1}{2\pi\sqrt{\mu\varepsilon}} \sqrt{\left(\frac{2.4049}{a}\right)^2} \qquad (9\text{-}47b)$$

Equating (9-47a) to (9-47b) indicates that the two are identical (degenerate modes) when

$$\frac{h}{a} = 2.03 \simeq 2 \qquad (9\text{-}48)$$

When $h/a < 2.03$ the dominant mode is the TM_{010} whereas for $h/a > 2.03$ the dominant mode is the TE_{111} mode.

9.3.3 Quality Factor Q

One of the most important parameters of a cavity is its quality factor or better known as the Q, which is defined by (8-84). The Q of the TM_{010} mode, which is the dominant mode when $h/a < 2.03$, is of particular interest and it will be derived here.

For the TM_{010} mode the potential function of (9-46) reduces to

$$A_z = B_{010} J_0(\beta_\rho \rho) \qquad (9\text{-}49)$$

where

$$\beta_\rho = \frac{\chi_{01}}{a} = \frac{2.4049}{a} \qquad (9\text{-}49a)$$

496 CIRCULAR CROSS-SECTION WAVEGUIDES AND CAVITIES

TABLE 9-4
Resonant frequencies for the TE_{mnp} and TM_{mnp} modes of a circular cavity

$$R^{mnp}_{dom} = \frac{(f_r)_{mnp}}{(f_r)_{dom}}$$

$\frac{h}{a}$					R^{mnp}_{dom}					
0	$\frac{TM_{010}}{TM_{010}}$	$\frac{TM_{110}}{TM_{010}}$	$\frac{TM_{210}}{TM_{010}}$	$\frac{TM_{020}}{TM_{010}}$	$\frac{TM_{310}}{TM_{010}}$	$\frac{TM_{120}}{TM_{010}}$	$\frac{TM_{410}}{TM_{010}}$	$\frac{TM_{220}}{TM_{010}}$	$\frac{TM_{030}}{TM_{010}}$	$\frac{TM_{510}}{TM_{010}}$
	1.000	1.593	2.136	2.295	2.653	2.917	3.155	3.500	3.598	3.647
0.5	$\frac{TM_{010}}{TM_{010}}$	$\frac{TM_{110}}{TM_{010}}$	$\frac{TM_{210}}{TM_{010}}$	$\frac{TM_{020}}{TM_{010}}$	$\frac{TM_{310}}{TM_{010}}$	$\frac{TE_{111}}{TM_{010}}$	$\frac{TM_{011}}{TM_{010}}$	$\frac{TE_{211}}{TM_{010}}$	$\frac{TM_{120}}{TM_{010}}$	$\frac{TE_{011}}{TM_{010}}$
	1.000	1.593	2.136	2.295	2.653	2.722	2.797	2.905	2.917	3.060
1.00	$\frac{TM_{010}}{TM_{010}}$	$\frac{TE_{111}}{TM_{010}}$	$\frac{TM_{110}}{TM_{010}}$	$\frac{TM_{011}}{TM_{010}}$	$\frac{TE_{211}}{TM_{010}}$	$\frac{TM_{111}}{TM_{010}}$	$\frac{TE_{011}}{TM_{010}}$	$\frac{TM_{210}}{TM_{010}}$	$\frac{TE_{311}}{TM_{010}}$	$\frac{TM_{020}}{TM_{010}}$
	1.000	1.514	1.593	1.645	1.822	2.060	2.060	2.136	2.181	2.295
2.03	$\frac{TM_{010}}{TM_{010}, TE_{111}}$	$\frac{TE_{111}}{TM_{010}, TE_{111}}$	$\frac{TM_{011}}{TM_{010}, TE_{111}}$	$\frac{TE_{211}}{TM_{010}, TE_{111}}$	$\frac{TE_{212}}{TM_{010}, TE_{111}}$	$\frac{TM_{110}}{TM_{010}, TE_{111}}$	$\frac{TM_{012}}{TM_{010}, TE_{111}}$	$\frac{TE_{011}}{TM_{010}, TE_{111}}$	$\frac{TM_{111}}{TM_{010}, TE_{111}}$	$\frac{TE_{212}}{TM_{010}, TE_{111}}$
	1.000	1.000	1.189	1.424	1.497	1.593	1.630	1.718	1.718	1.808
3.0	$\frac{TE_{111}}{TE_{111}}$	$\frac{TM_{010}}{TE_{111}}$	$\frac{TM_{011}}{TE_{111}}$	$\frac{TE_{112}}{TE_{111}}$	$\frac{TM_{012}}{TE_{111}}$	$\frac{TE_{211}}{TE_{111}}$	$\frac{TE_{113}}{TE_{111}}$	$\frac{TE_{212}}{TE_{111}}$	$\frac{TM_{110}}{TE_{111}}$	$\frac{TM_{013}}{TE_{111}}$
	1.000	1.136	1.238	1.317	1.506	1.524	1.719	1.748	1.809	1.868
4.0	$\frac{TE_{111}}{TE_{111}}$	$\frac{TM_{010}}{TE_{111}}$	$\frac{TE_{112}}{TE_{111}}$	$\frac{TM_{011}}{TE_{111}}$	$\frac{TM_{012}}{TE_{111}}$	$\frac{TE_{113}}{TE_{111}}$	$\frac{TE_{211}}{TE_{111}}$	$\frac{TM_{013}}{TE_{111}}$	$\frac{TE_{212}}{TE_{111}}$	$\frac{TE_{114}}{TE_{111}}$
	1.000	1.202	1.209	1.264	1.435	1.494	1.575	1.682	1.717	1.819
5.0	$\frac{TE_{111}}{TE_{111}}$	$\frac{TE_{112}}{TE_{111}}$	$\frac{TM_{010}}{TE_{111}}$	$\frac{TM_{011}}{TE_{111}}$	$\frac{TE_{113}}{TE_{111}}$	$\frac{TM_{012}}{TE_{111}}$	$\frac{TM_{013}}{TE_{111}}$	$\frac{TE_{114}}{TE_{111}}$	$\frac{TE_{211}}{TE_{111}}$	$\frac{TE_{212}}{TE_{111}}$
	1.000	1.146	1.236	1.278	1.354	1.395	1.571	1.602	1.603	1.698
10.0	$\frac{TE_{111}}{TE_{111}}$	$\frac{TE_{112}}{TE_{111}}$	$\frac{TE_{113}}{TE_{111}}$	$\frac{TE_{114}}{TE_{111}}$	$\frac{TM_{010}}{TE_{111}}$	$\frac{TE_{115}}{TE_{111}}$	$\frac{TM_{011}}{TE_{111}}$	$\frac{TM_{012}}{TE_{111}}$	$\frac{TM_{013}}{TE_{111}}$	$\frac{TE_{116}}{TE_{111}}$
	1.000	1.042	1.107	1.194	1.288	1.296	1.299	1.331	1.383	1.411
∞	$\frac{TE_{11p}}{TE_{111}}$	$\frac{TM_{01p}}{TE_{111}}$	$\frac{TE_{21p}}{TE_{111}}$	$\frac{TE_{01p}}{TE_{111}}$	$\frac{TM_{11p}}{TE_{111}}$	$\frac{TE_{31p}}{TE_{111}}$	$\frac{TM_{21p}}{TE_{111}}$	$\frac{TE_{41p}}{TE_{111}}$	$\frac{TE_{12p}}{TE_{111}}$	$\frac{TM_{02p}}{TE_{111}}$
	1.000	1.306	1.659	2.081	2.081	2.282	2.790	2.888	2.896	2.998

These corresponding electric and magnetic fields are obtained using (6-70), or

$$E_\rho = -j\frac{1}{\omega_r \mu\varepsilon}\frac{\partial^2 A_z}{\partial\rho\,\partial z} = 0$$

$$E_\phi = -j\frac{1}{\omega_r \mu\varepsilon}\frac{1}{\rho}\frac{\partial^2 A_z}{\partial\phi\,\partial z} = 0$$

$$E_z = -j\frac{1}{\omega_r \mu\varepsilon}\left(\frac{\partial^2}{\partial z^2} + \beta_r^2\right)A_z = -j\frac{\beta_r^2}{\omega_r\mu\varepsilon}B_{010}J_0\!\left(\frac{\chi_{01}}{a}\rho\right) \qquad (9\text{-}50)$$

$$H_\rho = \frac{1}{\mu}\frac{1}{\rho}\frac{\partial A_z}{\partial\phi} = 0$$

$$H_\phi = -\frac{1}{\mu}\frac{\partial A_z}{\partial\rho} = -\frac{\chi_{01}}{a}\frac{B_{010}}{\mu}J_0'\!\left(\frac{\chi_{01}}{a}\rho\right)$$

$$H_z = 0$$

The total energy stored in the cavity is given by

$$W = 2W_e = \frac{\varepsilon}{2}\iiint_V |\mathbf{E}|^2\,dv = |B_{010}|^2 \frac{\varepsilon}{2}\left(\frac{\beta_r^2}{\omega_r\mu\varepsilon}\right)^2 \int_0^h\!\int_0^{2\pi}\!\int_0^a J_0^2\!\left(\frac{\chi_{01}}{a}\rho\right)\rho\,d\rho\,d\phi\,dz$$

$$W = |B_{010}|^2 \pi h\varepsilon\left(\frac{\beta_r^2}{\omega_r\mu\varepsilon}\right)^2 \int_0^a J_0^2\!\left(\frac{\chi_{01}}{a}\rho\right)\rho\,d\rho \qquad (9\text{-}51)$$

Since [5]

$$\int_0^a \rho J_0^2\!\left(\frac{\chi_{01}}{a}\rho\right)d\rho = \frac{a^2}{2}J_1^2(\chi_{01}) \qquad (9\text{-}52)$$

then (9-51) reduces to

$$W = |B_{010}|^2 \frac{\pi h\varepsilon}{2}\left(\frac{a\beta_r^2}{\omega_r\mu\varepsilon}\right)^2 J_1^2(\chi_{01}) \qquad (9\text{-}53)$$

Because the medium within the cavity is assumed to be lossless, the total power is dissipated on the conducting walls of the cavity. Thus we can write that

$$P_d = \frac{R_s}{2}\oint_A |\mathbf{H}|^2\,ds = \frac{R_s}{2}\left\{\int_0^{2\pi}\!\int_0^h |\mathbf{H}|^2_{\rho=a}\,a\,d\phi\,dz + 2\int_0^{2\pi}\!\int_0^a |\mathbf{H}|^2_{z=0}\,\rho\,d\rho\,d\phi\right\}$$

$$= |B_{010}|^2 \frac{R_s}{2\mu^2}\left(\frac{\chi_{01}}{a}\right)^2\left\{\int_0^{2\pi}\!\int_0^h [J_0'(\chi_{01})]^2 a\,dz\,d\phi + 2\int_0^{2\pi}\!\int_0^a \left[J_0'\!\left(\frac{\chi_{01}}{a}\rho\right)\right]^2 \rho\,d\rho\,d\phi\right\}$$

$$P_d = |B_{010}|^2 \frac{\pi R_s}{\mu^2}\left(\frac{\chi_{01}}{a}\right)^2\left\{ah[J_0'(\chi_{01})]^2 + 2\int_0^a \left[J_0'\!\left(\frac{\chi_{01}}{a}\rho\right)\right]^2 \rho\,d\rho\right\} \qquad (9\text{-}54)$$

Because

$$J_0'\!\left(\frac{\chi_{01}}{a}\rho\right) = \frac{d}{d(\chi_{01}\rho/a)}\left[J_0\!\left(\frac{\chi_{01}}{a}\rho\right)\right] = -J_1\!\left(\frac{\chi_{01}}{a}\rho\right) \qquad (9\text{-}55a)$$

and at $\rho = a$

$$J_0'(\chi_{01}) = -J_1(\chi_{01}) \tag{9-55b}$$

(9-54) reduces to

$$P_d = |B_{010}|^2 \frac{\pi R_s}{\mu^2} \left(\frac{\chi_{01}}{a}\right)^2 \left\{ ahJ_1^2(\chi_{01}) + 2\int_0^a J_1^2\left(\frac{\chi_{01}}{a}\rho\right)\rho\, d\rho \right\}$$

$$= |B_{010}|^2 \frac{\pi R_s}{\mu^2} \left(\frac{\chi_{01}}{a}\right)^2 \left\{ ahJ_1^2(\chi_{01}) + 2\left[\frac{a^2}{2}J_1^2(\chi_{01})\right]\right\}$$

$$P_d = |B_{010}|^2 \frac{\pi R_s}{\mu^2} \left(\frac{\chi_{01}}{a}\right)^2 a(h+a) J_1^2(\chi_{01}) \tag{9-56}$$

Using the Q definition of (8-84) along with (9-53) and (9-56), we can write that

$$\boxed{Q = \frac{\omega_r W}{P_d} = \frac{\beta_r^4 h a^3}{2\omega_r \varepsilon (h+a) R_s \chi_{01}^2} = \frac{\chi_{01}\sqrt{\frac{\mu}{\varepsilon}}}{2\left(1+\frac{a}{h}\right)R_s} = \frac{1.2025\eta}{R_s\left(1+\frac{a}{h}\right)}} \tag{9-57}$$

since for the TM_{010} mode ($m = 0$, $n = 1$, and $p = 0$)

$$\beta_\rho^2 + \beta_z^2 = \left(\frac{\chi_{01}}{a}\right)^2 = \beta_r^2 \Rightarrow \chi_{01} = \beta_r a \tag{9-57a}$$

Example 9-4. Compare the Q values of a circular cavity operating in the TM_{010} mode to those of a square-based rectangular cavity. The dimensions of each are such that the circular cavity is circumscribed by the square-based rectangular cavity.

Solution. According to (9-57) the Q of a circular cavity of radius a (or diameter d) and height h is given by

$$Q = \frac{1.2025\eta}{R_s\left(1+\frac{a}{h}\right)} = \frac{1.2025\eta}{R_s\left(1+\frac{d/2}{h}\right)}$$

For a square-based rectangular cavity to circumscribe a circular cavity, one of the sides of its base must be equal to the diameter and their heights must be equal. Therefore with a base of $a = c = d$ on each of its sides and a height $b = h$, its Q is equal according to (8-88a) to

$$Q = \frac{1.1107\eta}{R_s}\left[\frac{1}{\left(1+\frac{a/2}{b}\right)}\right] = \frac{1.1107\eta}{R_s}\left[\frac{1}{\left(1+\frac{d/2}{h}\right)}\right]$$

Compare these two expressions and it is evident that the Q of the circular cavity is greater than that of the square-based cavity by

$$\left(\frac{1.2025 - 1.1107}{1.1107}\right) \times 100 = 8.26\%$$

This is expected since the circular cavity does not possess as many sharp corners and edges as the square-based cavity whose volume and surface area are not as well utilized by the interior fields. It should be remembered that the Q of a cavity is proportional to volume and inversely proportional to area.

9.4 RADIAL WAVEGUIDES

For a circular waveguide the waves travel in the $\pm z$ directions and their z variations are represented by the factor $e^{\mp j\beta_z z}$. Their constant phase planes (equiphases) are planes that are parallel to each other and perpendicular to the z direction. If the waves were traveling in the $\pm \phi$ direction, their variations in that direction would be represented by $e^{\mp jm\phi}$. Such waves are usually referred to as *circulating waves*, and their equiphase surfaces are constant ϕ planes. For waves that travel in the $\pm \rho$ (radial) direction, their variations in that direction would be represented by either $H_m^{(2)}(\beta_\rho \rho)$ or $H_m^{(1)}(\beta_\rho \rho)$. Such waves are usually referred to as *radial waves*, and their equiphases are constant ρ (radius) planes. The structures that support radial waves are referred to as radial waveguides, and they will be examined here. Examples are parallel plates, wedged plates (representing horn antennas), and others.

9.4.1 Parallel Plates

When two infinite long parallel plates are excited by a line source placed between them at the center, as shown in Figure 9-6, the excited waves travel in the radial direction and form radial waves. We shall examine here both the TE^z and TM^z modes in the region between the plates.

A. TRANSVERSE ELECTRIC (TE^z) MODES

For the TE^z modes of Figure 9-6, the potential function F_z can be written according to (3-67a) through (3-69b) as

$$F_z(\rho, \phi, z) = \left[C_1 H_m^{(1)}(\beta_\rho \rho) + D_1 H_m^{(2)}(\beta_\rho \rho) \right] \left[C_2 \cos(m\phi) + D_2 \sin(m\phi) \right]$$
$$\times \left[C_3 \cos(\beta_z z) + D_3 \sin(\beta_z z) \right] \quad (9\text{-}58)$$

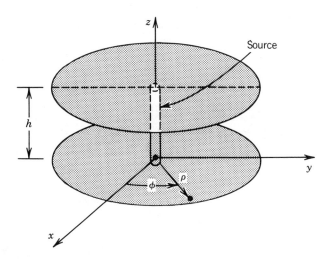

FIGURE 9-6 Geometry for radial waveguide.

500 CIRCULAR CROSS-SECTION WAVEGUIDES AND CAVITIES

where
$$\beta_\rho^2 + \beta_z^2 = \beta^2 \tag{9-58a}$$

The boundary conditions are

$$E_\rho(0 \leq \rho \leq \infty, 0 \leq \phi \leq 2\pi, z = 0) = E_\rho(0 \leq \rho \leq \infty, 0 \leq \phi \leq 2\pi, z = h) = 0 \tag{9-59a}$$

$$E_\phi(0 \leq \rho \leq \infty, 0 \leq \phi \leq 2\pi, z = 0) = E_\phi(0 \leq \rho \leq \infty, 0 \leq \phi \leq 2\pi, z = h) = 0 \tag{9-59b}$$

Since both of the preceding boundary conditions are not independent from each other, either of the two leads to the same results.

Using (6-80) and (9-58), the E_ϕ component can be written as

$$E_\phi(\rho, \phi, z) = \frac{1}{\varepsilon} \frac{\partial F_z}{\partial \rho}$$

$$= \frac{\beta_\rho}{\varepsilon} \left[C_1 H_m^{(1)'}(\beta_\rho \rho) + D_1 H_m^{(2)'}(\beta_\rho \rho) \right] \left[C_2 \cos(m\phi) + D_2 \sin(m\phi) \right]$$

$$\times \left[C_3 \cos(\beta_z z) + D_3 \sin(\beta_z z) \right] \tag{9-60}$$

Applying (9-59b) leads to

$$E_\phi(0 \leq \rho \leq \infty, 0 \leq \phi \leq 2\pi, z = 0)$$

$$= \frac{\beta_\rho}{\varepsilon} \left[C_1 H_m^{(1)'}(\beta_\rho \rho) + D_1 H_m^{(2)'}(\beta_\rho \rho) \right]$$

$$\times \left[C_2 \cos(m\phi) + D_2 \sin(m\phi) \right] \left[C_3(1) + D_3(0) \right] = 0$$

$$\Rightarrow C_3 = 0 \tag{9-61a}$$

$$E_\phi(0 \leq \rho \leq a, 0 \leq \phi \leq 2\pi, z = h)$$

$$= \frac{\beta_\rho}{\varepsilon} \left[C_1 H_m^{(1)'}(\beta_\rho \rho) + D_1 H_m^{(2)'}(\beta_\rho \rho) \right]$$

$$\times \left[C_2 \cos(m\phi) + D_2 \sin(m\phi) \right] D_3 \sin(\beta_z h) = 0$$

$$\Rightarrow \sin(\beta_z h) = 0 \Rightarrow \beta_z h \sin^{-1}(0) = n\pi$$

$$\beta_z = \frac{n\pi}{h} \quad n = 1, 2, 3, \ldots \tag{9-61b}$$

Thus F_z of (9-58) reduces to

$$F_z(\rho, \phi, z) = \left[C_1 H_m^{(1)}(\beta_\rho \rho) + D_1 H_m^{(2)}(\beta_\rho \rho) \right]$$

$$\times \left[C_2 \cos(m\phi) + D_2 \sin(m\phi) \right] D_3 \sin(\beta_z z) \tag{9-62}$$

where

$$\beta_\rho^2 + \beta_z^2 = \beta^2 \Rightarrow \beta_\rho = \pm \sqrt{\beta^2 - \beta_z^2} = \pm \sqrt{\beta^2 - \left(\frac{n\pi}{h}\right)^2} \tag{9-62a}$$

$$\beta_z = \frac{n\pi}{h} \quad n = 1, 2, 3, \ldots \tag{9-62b}$$

$$m = 0, 1, 2, \ldots \quad \text{(because of periodicity of the fields in } \phi \text{)} \tag{9-62c}$$

Cutoff is defined when $\beta_\rho = 0$. Thus using (9-62a)

$$\beta_\rho = \pm \sqrt{\beta^2 - \left(\frac{n\pi}{h}\right)^2}\bigg|_{\substack{f=f_c \\ \beta=\beta_c}} = 0 \Rightarrow \beta_c = \frac{n\pi}{h} \quad (9\text{-}63)$$

or

$$\boxed{(f_c)_n^{TE^z} = \frac{n}{2h\sqrt{\mu\varepsilon}} \quad n = 1, 2, 3, \ldots} \quad (9\text{-}63a)$$

Therefore above, at, and below cutoff β_ρ of (9-62a) takes the forms

$$\beta_\rho = \begin{cases} \sqrt{\beta^2 - \left(\frac{n\pi}{h}\right)^2} = \beta\sqrt{1 - \left(\frac{f_c}{f}\right)^2} & f > f_c \quad (9\text{-}64a) \\ 0 & f = f_c \quad (9\text{-}64b) \\ -j\sqrt{\left(\frac{n\pi}{h}\right)^2 - \beta^2} = -j\beta\sqrt{\left(\frac{f_c}{f}\right)^2 - 1} = -j\alpha & f < f_c \quad (9\text{-}64c) \end{cases}$$

Let us now examine the outward ($+\rho$) traveling waves represented by $H_m^{(2)}(\beta_\rho \rho)$ and those represented simultaneously by the $\cos(m\phi)$ variations of (9-62). In that case (9-62) reduces to

$$F_z^+(\rho, \phi, z) = A_{mn} H_m^{(2)}(\beta_\rho \rho) \cos(m\phi) \sin\left(\frac{n\pi}{h}z\right) \quad (9\text{-}65)$$

The corresponding electric and magnetic fields can be written using (6-80) as

$$E_\rho^+ = -\frac{1}{\varepsilon\rho}\frac{\partial F_z^+}{\partial \phi} = A_{mn}\frac{m}{\varepsilon\rho}H_m^{(2)}(\beta_\rho \rho)\sin(m\phi)\sin\left(\frac{n\pi}{h}z\right) \quad (9\text{-}65a)$$

$$E_\phi^+ = \frac{1}{\varepsilon}\frac{\partial F_z^+}{\partial \rho} = \beta_\rho \frac{A_{mn}}{\varepsilon}H_m^{(2)\prime}(\beta_\rho \rho)\cos(m\phi)\sin\left(\frac{n\pi}{h}z\right) \quad (9\text{-}65b)$$

$$E_z^+ = 0 \quad (9\text{-}65c)$$

$$H_\rho^+ = -j\frac{1}{\omega\mu\varepsilon}\frac{\partial^2 F_z^+}{\partial \rho \partial z} = -jA_{mn}\beta_\rho\frac{n\pi/h}{\omega\mu\varepsilon}H_m^{(2)\prime}(\beta_\rho \rho)\cos(m\phi)\cos\left(\frac{n\pi}{h}z\right) \quad (9\text{-}65d)$$

$$H_\phi^+ = -j\frac{1}{\omega\mu\varepsilon}\frac{1}{\rho}\frac{\partial^2 F_z^+}{\partial \phi \partial z} = jA_{mn}\frac{mn\pi/h}{\omega\mu\varepsilon\rho}H_m^{(2)}(\beta_\rho \rho)\sin(m\phi)\cos\left(\frac{n\pi}{h}z\right) \quad (9\text{-}65e)$$

$$H_z^+ = -j\frac{1}{\omega\mu\varepsilon}\left(\frac{\partial^2}{\partial z^2} + \beta^2\right)F_z^+ = -jA_{mn}\frac{\beta_\rho^2}{\omega\mu\varepsilon}H_m^{(2)}(\beta_\rho \rho)\cos(m\phi)\sin\left(\frac{n\pi}{h}z\right) \quad (9\text{-}65f)$$

$$' = \frac{\partial}{\partial(\beta_\rho \rho)} \quad (9\text{-}65g)$$

The impedance of the wave in the $+\rho$ direction is defined and given by

$$Z_w^{+\rho}(\text{TE}_n^z) = \frac{E_\phi}{H_z} = j\frac{\omega\mu}{\beta_\rho}\frac{H_m^{(2)\prime}(\beta_\rho\rho)}{H_m^{(2)}(\beta_\rho\rho)} \tag{9-66}$$

Since

$$H_m^{(2)\prime}(\beta_\rho\rho) = \frac{\partial}{\partial(\beta_\rho\rho)}\left[H_m^{(2)}(\beta_\rho\rho)\right] \tag{9-67}$$

(9-66) can be written as

$$Z_w^{+\rho}(\text{TE}_n^z) = j\frac{\omega\mu}{\beta_\rho}\frac{\dfrac{\partial}{\partial(\beta_\rho\rho)}\left[H_m^{(2)}(\beta_\rho\rho)\right]}{H_m^{(2)}(\beta_\rho\rho)} \tag{9-68}$$

Below cutoff ($f < f_c$) β_ρ is imaginary, and it is given by (9-64c). Therefore for $f < f_c$

$$H_m^{(2)}(\beta_\rho\rho) = H_m^{(2)}(-j\alpha\rho) \tag{9-69a}$$

$$\frac{d}{d(\beta_\rho\rho)}\left[H_m^{(2)}(\beta_\rho\rho)\right] = \frac{d}{d(-j\alpha\rho)}\left[H_m^{(2)}(-j\alpha\rho)\right] = j\frac{d}{d(\alpha\rho)}\left[H_m^{(2)}(-j\alpha\rho)\right] \tag{9-69b}$$

For complex arguments the Hankel function $H_m^{(2)}$ of the second kind is related to the modified Bessel function K_m of the second kind by

$$H_m^{(2)}(-j\alpha\rho) = \frac{2}{\pi}j^{m+1}K_m(\alpha\rho) \tag{9-70a}$$

$$\frac{d}{d(\alpha\rho)}\left[H_m^{(2)}(-j\alpha\rho)\right] = \frac{2}{\pi}j^{m+1}\frac{d}{d(\alpha\rho)}\left[K_m(\alpha\rho)\right] \tag{9-70b}$$

Thus below cutoff ($f < f_c$) the wave impedance reduces to

$$Z_w^{+\rho}(\text{TE}_n^z)\Big|_{f<f_c} = j\frac{\omega\mu}{-j\alpha}\frac{j\dfrac{d}{d(\alpha\rho)}[K_m(\alpha\rho)]}{K_m(\alpha\rho)} = -j\frac{\omega\mu}{\alpha}\frac{\dfrac{d}{d(\alpha\rho)}[K_m(\alpha\rho)]}{K_m(\alpha\rho)} \tag{9-71}$$

which is always inductive (for $f < f_c$) since $K_m(\alpha\rho) > 0$ and $d/d(\alpha\rho)[K_m(\alpha\rho)] < 0$. Therefore below cutoff the modes are nonpropagating (evanescent) since the waveguide is behaving as an inductive storage element.

B. TRANSVERSE MAGNETIC (TM^z) MODES

The TM^z modes of the radial waveguide structure of Figure 9-6 with the source at the center are derived in a similar manner. Using such a procedure leads to the following results:

$$A_z(\rho, \phi, z) = \left[C_1 H_m^{(1)}(\beta_\rho \rho) + D_1 H_m^{(2)}(\beta_\rho \rho)\right]$$
$$\times \left[C_2' \cos(m\phi) + D_2' \sin(m\phi)\right] C_3' \cos(\beta_z z) \tag{9-72}$$

$$\beta_\rho^2 + \beta_z^2 = \beta^2 \Rightarrow \beta_\rho = \pm\sqrt{\beta^2 - \beta_z^2} = \pm\sqrt{\beta^2 - \left(\frac{n\pi}{h}\right)^2} \tag{9-72a}$$

$$\beta_z = \frac{n\pi}{h} \quad n = 0, 1, 2, \ldots \tag{9-72b}$$

$m = 0, 1, 2, \ldots$ (because of periodicity of the fields in ϕ) (9-72c)

$$\boxed{(f_c)_n^{TM^z} = \frac{n}{2h\sqrt{\mu\varepsilon}} \quad n = 0, 1, 2, \ldots} \tag{9-73}$$

$$\beta_\rho = \begin{cases} \sqrt{\beta^2 - \left(\frac{n\pi}{h}\right)^2} = \beta\sqrt{1 - \left(\frac{f_c}{f}\right)^2} & f > f_c \quad (9\text{-}74a) \\ 0 & f = f_c \quad (9\text{-}74b) \\ -j\sqrt{\left(\frac{n\pi}{h}\right)^2 - \beta^2} = -j\beta\sqrt{\left(\frac{f_c}{f}\right)^2 - 1} = -j\alpha & f < f_c \quad (9\text{-}74c) \end{cases}$$

For outward $(+\rho)$ traveling waves and only $\cos(m\phi)$ variations

$$A_z^+(\rho, \phi, z) = B_{mn} H_m^{(2)}(\beta_\rho \rho) \cos(m\phi) \cos\left(\frac{n\pi}{h} z\right) \tag{9-75}$$

$$E_\rho^+ = -j\frac{1}{\omega\mu\varepsilon}\frac{\partial^2 A_z^+}{\partial \rho \partial z} = jB_{mn}\beta_\rho \frac{n\pi/h}{\omega\mu\varepsilon} H_m^{(2)\prime}(\beta_\rho \rho) \cos(m\phi) \sin\left(\frac{n\pi}{h} z\right) \tag{9-75a}$$

$$E_\phi^+ = -j\frac{1}{\omega\mu\varepsilon}\frac{1}{\rho}\frac{\partial^2 A_z^+}{\partial \phi \partial z} = -jB_{mn}\frac{mn\pi/h}{\omega\mu\varepsilon\rho} H_m^{(2)}(\beta_\rho \rho) \sin(m\phi) \sin\left(\frac{n\pi}{h} z\right) \tag{9-75b}$$

$$E_z^+ = -j\frac{1}{\omega\mu\varepsilon}\left(\frac{\partial^2}{\partial z^2} + \beta^2\right) A_z^+ = -jB_{mn}\frac{\beta_\rho^2}{\omega\mu\varepsilon} H_m^{(2)}(\beta_\rho \rho) \cos(m\phi) \cos\left(\frac{n\pi}{h} z\right) \tag{9-75c}$$

$$H_\rho^+ = \frac{1}{\mu}\frac{1}{\rho}\frac{\partial A_z^+}{\partial \phi} = -B_{mn}\frac{m}{\mu}\frac{1}{\rho} H_m^{(2)}(\beta_\rho \rho) \sin(m\phi) \cos\left(\frac{n\pi}{h} z\right) \tag{9-75d}$$

$$H_\phi^+ = -\frac{1}{\mu}\frac{\partial A_z^+}{\partial \rho} = -\beta_\rho \frac{B_{mn}}{\mu} H_m^{(2)\prime}(\beta_\rho \rho) \cos(m\phi) \cos\left(\frac{n\pi}{h} z\right) \tag{9-75e}$$

$$H_z = 0 \tag{9-75f}$$

$$' = \frac{\partial}{\partial(\beta_\rho \rho)} \tag{9-75g}$$

$$Z_w^{+\rho}(TM_n^z) = \frac{E_z^+}{-H_\phi^+} = -j\frac{\beta_\rho}{\omega\varepsilon}\frac{H_m^{(2)}(\beta_\rho \rho)}{H_m^{(2)\prime}(\beta_\rho \rho)} \tag{9-76}$$

$$Z_w^{+\rho}(TM_n^z)\big|_{f<f_c} = j\frac{\alpha}{\omega\varepsilon}\frac{K_m(\alpha\rho)}{\frac{d}{d(\alpha\rho)}[K_m(\alpha\rho)]} \tag{9-76a}$$

which is always capacitive (for $f < f_c$) since $K_m(\alpha\rho) > 0$ and $d/d(\alpha\rho)[K_m(\alpha\rho)] < 0$. Therefore below cutoff the modes are nonpropagating (evanescent) since the waveguide behaves as a capacitive storage element.

9.4.2 Wedged Plates

Another radial type of waveguide structure is the wedged-plate geometry of Figure 9-7 with plates along $z = 0$, h, and $\phi = 0, \phi_0$. This type of a configuration resembles and can be used to represent the structures of E- and H-plane sectoral horns [6]. In fact the fields within the horns are found using the procedure outlined here. In general, both TE^z and TM^z modes can exist in the space between the plates. The independent sets of boundary conditions of this structure that can be used to solve for the TE_{pn}^z and TM_{pn}^z modes are

$$E_\rho(0 \le \rho \le \infty, 0 \le \phi \le \phi_0, z = 0) = E_\rho(0 \le \rho \le \infty, 0 \le \phi \le \phi_0, z = h) = 0 \tag{9-77a}$$

$$E_\rho(0 \le \rho \le \infty, \phi = 0, 0 \le z \le h) = E_\rho(0 \le \rho \le \infty, \phi = \phi_0, 0 \le z \le h) = 0 \tag{9-77b}$$

or

$$E_\phi(0 \le \rho \le \infty, 0 \le \phi \le \phi_0, z = 0) = E_\phi(0 \le \rho \le \infty, 0 \le \phi \le \phi_0, z = h) = 0 \tag{9-78a}$$

$$E_z(0 \le \rho \le \infty, \phi = 0, 0 \le z \le h) = E_z(0 \le \rho \le \infty, \phi = \phi_0, 0 \le z \le h) = 0 \tag{9-78b}$$

or appropriate combinations of these. Whichever combination of independent boundary conditions is used, the same results are obtained.

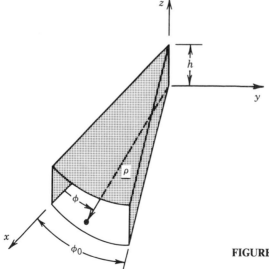

FIGURE 9-7 Geometry of wedged plate radial waveguide.

A. TRANSVERSE ELECTRIC (TE^z) MODES

Since the procedure used to derive this set of TE^z fields is the same as any other TE^z procedure used previously, the results of this set of TE$^z_{pn}$ modes will be summarized here; the details are left as an end of chapter exercise to the reader. Only the outward radial ($+\rho$) parts will be included here.

$$F_z^+(\rho, \phi, z) = A_{pn} H_m^{(2)}(\beta_\rho \rho) \cos(m\phi) \sin(\beta_z z) \qquad (9\text{-}79)$$

$$\boxed{\beta_\rho^2 + \beta_z^2 = \beta^2} \qquad (9\text{-}79a)$$

$$\boxed{\beta_z = \frac{n\pi}{h} \qquad n = 1, 2, 3, \ldots} \qquad (9\text{-}79b)$$

$$\boxed{m = \frac{p\pi}{\phi_0} \qquad p = 0, 1, 2, \ldots} \qquad (9\text{-}79c)$$

$$E_\rho^+ = -\frac{1}{\varepsilon}\frac{1}{\rho}\frac{\partial F_z^+}{\partial \phi} = A_{pn}\frac{p\pi/\phi_0}{\varepsilon\rho} H_m^{(2)}(\beta_\rho \rho) \sin\left(\frac{p\pi}{\phi_0}\phi\right) \sin\left(\frac{n\pi}{h}z\right) \qquad (9\text{-}79d)$$

$$E_\phi^+ = \frac{1}{\varepsilon}\frac{\partial F_z^+}{\partial \rho} = \beta_\rho \frac{A_{pn}}{\varepsilon} H_m^{(2)\prime}(\beta_\rho \rho) \cos\left(\frac{p\pi}{\phi_0}\phi\right) \sin\left(\frac{n\pi}{h}z\right) \qquad (9\text{-}79e)$$

$$E_z^+ = 0 \qquad (9\text{-}79f)$$

$$H_\rho^+ = -j\frac{1}{\omega\mu\varepsilon}\frac{\partial^2 F_z^+}{\partial \rho \partial z} = -jA_{pn}\frac{\beta_\rho \beta_z}{\omega\mu\varepsilon} H_m^{(2)\prime}(\beta_\rho \rho) \cos\left(\frac{p\pi}{\phi_0}\phi\right) \cos\left(\frac{n\pi}{h}z\right) \qquad (9\text{-}79g)$$

$$H_\phi^+ = -j\frac{1}{\omega\mu\varepsilon}\frac{1}{\rho}\frac{\partial^2 F_z^+}{\partial \phi \partial z} = jA_{pn}\frac{\beta_z p\pi/\phi_0}{\omega\mu\varepsilon}\frac{1}{\rho} H_m^{(2)}(\beta_\rho \rho) \sin\left(\frac{p\pi}{\phi_0}\phi\right) \cos\left(\frac{n\pi}{h}z\right) \qquad (9\text{-}79h)$$

$$H_z^+ = -j\frac{1}{\omega\mu\varepsilon}\left(\frac{\partial^2}{\partial z^2} + \beta^2\right) F_z^+ = -jA_{pn}\frac{\beta_\rho^2}{\omega\mu\varepsilon} H_m^{(2)}(\beta_\rho \rho) \cos\left(\frac{p\pi}{\phi_0}\phi\right) \sin\left(\frac{n\pi}{h}z\right) \qquad (9\text{-}79i)$$

$$\boxed{Z_w^{+\rho}(\text{TE}^z_{pn}) = \frac{E_\phi^+}{H_z^+} = j\frac{\omega\mu}{\beta_\rho}\frac{H_m^{(2)\prime}(\beta_\rho \rho)}{H_m^{(2)}(\beta_\rho \rho)} \qquad \prime = \frac{\partial}{\partial(\beta_\rho \rho)}} \qquad (9\text{-}79j)$$

B. TRANSVERSE MAGNETIC (TM^z) MODES

As for the TE$^z_{pn}$ modes, the procedure for deriving the TM$^z_{pn}$ for the wedged plate radial waveguide is the same as that used for TM^z modes of other waveguide configurations. Therefore the results will be summarized here, and the details left as an end of chapter exercise for the reader. Only the outward radial ($+\rho$) parts will be included here.

$$A_z^+(\rho, \phi, z) = B_{pn} H_m^{(2)}(\beta_\rho \rho) \sin(m\phi) \cos(\beta_z z) \qquad (9\text{-}80)$$

$$\boxed{\beta_\rho^2 + \beta_z^2 = \beta^2} \qquad (9\text{-}80a)$$

$$\boxed{\beta_z = \frac{n\pi}{h} \qquad n = 0, 1, 2, \ldots} \qquad (9\text{-}80b)$$

$$\boxed{m = \frac{p\pi}{\phi_0} \qquad p = 1, 2, 3, \ldots} \qquad (9\text{-}80c)$$

$$E_\rho^+ = -j\frac{1}{\omega\mu\varepsilon}\frac{\partial^2 A_z^+}{\partial\rho\,\partial z} = jB_{pn}\frac{\beta_z \beta_\rho}{\omega\mu\varepsilon} H_m^{(2)\prime}(\beta_\rho \rho) \sin\left(\frac{p\pi}{\phi_0}\phi\right) \sin\left(\frac{n\pi}{h}z\right) \qquad (9\text{-}80d)$$

$$E_\phi^+ = -j\frac{1}{\omega\mu\varepsilon}\frac{1}{\rho}\frac{\partial^2 A_z^+}{\partial\phi\,\partial z} = jB_{pn}\frac{\beta_z p\pi/\phi_0}{\omega\mu\varepsilon}\frac{1}{\rho} H_m^{(2)}(\beta_\rho \rho) \cos\left(\frac{p\pi}{\phi_0}\phi\right) \sin\left(\frac{n\pi}{h}z\right) \qquad (9\text{-}80e)$$

$$E_z^+ = -j\frac{1}{\omega\mu\varepsilon}\left(\frac{\partial^2}{\partial z^2} + \beta^2\right) A_z^+ = -jB_{pn}\frac{\beta_\rho^2}{\omega\mu\varepsilon} H_m^{(2)}(\beta_\rho \rho) \sin\left(\frac{p\pi}{\phi_0}\phi\right) \cos\left(\frac{n\pi}{h}z\right) \qquad (9\text{-}80f)$$

$$H_\rho^+ = \frac{1}{\mu}\frac{1}{\rho}\frac{\partial A_z^+}{\partial\phi} = B_{pn}\frac{p\pi/\phi_0}{\mu\rho} H_m^{(2)}(\beta_\rho \rho) \cos\left(\frac{p\pi}{\phi_0}\phi\right) \cos\left(\frac{n\pi}{h}z\right) \qquad (9\text{-}80g)$$

$$H_\phi^+ = -\frac{1}{\mu}\frac{\partial A_z^+}{\partial\rho} = -\beta_\rho\frac{B_{pn}}{\mu} H_m^{(2)\prime}(\beta_\rho \rho) \sin\left(\frac{p\pi}{\phi_0}\phi\right) \cos\left(\frac{n\pi}{h}z\right) \qquad (9\text{-}80h)$$

$$H_z^+ = 0 \qquad (9\text{-}80i)$$

$$\boxed{Z_w^{+\rho}(\text{TM}_{pn}^z) = \frac{E_z^+}{-H_\phi^+} = -j\frac{\beta_\rho}{\omega\varepsilon}\frac{H_m^{(2)}(\beta_\rho \rho)}{H_m^{(2)\prime}(\beta_\rho \rho)} \qquad ' = \frac{\partial}{\partial(\beta_\rho \rho)}} \qquad (9\text{-}80j)$$

9.5 DIELECTRIC WAVEGUIDES AND RESONATORS

Guided electromagnetic propagation by dielectric media has been studied since as early as the 1920s by well known people such as Rayleigh, Sommerfeld, and Debye. Dielectric slabs, strips, and rods have been used as waveguides, resonators, and antennas. Since the 1960s a most well known dielectric waveguide, the fiber optic cable [7–19], has received attention and has played a key role in the general area of communication. Although the subject is very lengthy and involved, we will consider here simplified theories that give the propagation characterisitcs of the cylindrical dielectric rod waveguide, fiber optic cable, and the cylindrical dielectric resonator. Extensive material on each of these topics and others can be found in the literature.

9.5.1 Circular Dielectric Waveguide

The cylindrical dielectric waveguide that will be examined here is that of circular cross section, as shown in Figure 9-8. It usually consists of a high permittivity (ε_d)

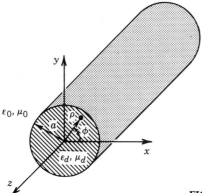

FIGURE 9-8 Geometry of circular dielectric waveguide.

central core dielectric of radius a surrounded by a lower dielectric cladding (which is usually air). For simplicity, we usually assume that both are perfect dielectrics with permeabilities equal to that of free space. Such a structure can support an infinite number of modes. However, for a given set of permittivities and radius a, only a finite number of unattenuated waveguide modes exist with their fields localized in the central dielectric core. Generally the fields within a dielectric waveguide will be TE and/or TM, as was demonstrated in Section 8.7 for the dielectric slab waveguide. However, for the cylindrical dielectric rod of Figure 9-8 pure TE(H) or TM(E) modes exist only when the field configurations are symmetrical and independent of ϕ. However, modes that exhibit angular ϕ variations cannot be pure TE or TM modes. Instead field configurations that are combinations of TE (or H) and TM (or E) modes can be nonsymmetrical and possess angular ϕ variations. Such modes are usually referred to as *hybrid modes*, and are usually designated by IEEE (formerly IRE) Standards [20] as HEM$_{mn}$. In general, mode nomenclature for circular dielectric waveguides and resonators is not well defined, and it is quite confusing. Another designation of the hybrid modes is to denote them as HE [when the TE(H) modes predominate] or EH [when the TM(E) modes predominate]. In general pure TE or TM, or hybrid HEM (HE or EH) modes exhibit cutoff frequencies, below which unattenuating modes cannot propagate. The cutoff frequency is determined by the minimum electrical radius (a/λ) of the dielectric rod; for small values of a/λ the modes cannot propagate unattenuated within the rod. There is, however, one hybrid mode, namely the HEM$_{11}$ (HE$_{11}$), which does not have a cutoff frequency. Because of its zero cutoff frequency, it is referred to as the *dominant mode* and is most widely used in dielectric rod waveguides and end-fire antennas. This HE$_{11}$ mode is also popularly referred to as the *dipole mode*. An excellent reference on dielectric waveguides and resonators is that of [21].

The TE$^z_{mn}$ and TM$^z_{mn}$ electric and magnetic field components in a cylindrical waveguide are given, respectively, by (9-15a) through (9-15g) and (9-31a) through (9-31g). These expressions include both the $\cos(m\phi)$ and $\sin(m\phi)$ angular variations which, in general, both exist. The HEM modes that have only $\cos(m\phi)$ symmetry are combinations of the TE^z and TM^z modes which also exhibit $\cos(m\phi)$ symmetry. Use of the expressions (9-15a) through (9-15g) and (9-31a) through (9-31g) and selection of only the terms that possess simultaneously $\cos(m\phi)$ variations in the E_z component and $\sin(m\phi)$ variations in the H_z component, allows us to write that

<u>HEM Modes ($\rho \leq a$)</u>

$$E_\rho^d = -j\frac{1}{(\beta_\rho^d)^2}\left[m\omega\mu_d A_m \frac{1}{\rho} J_m(\beta_\rho^d \rho) + \beta_z \beta_\rho^d B_m J_m'(\beta_\rho^d \rho)\right]\cos(m\phi)e^{-j\beta_z z} \quad (9\text{-}81\text{a})$$

$$E_\phi^d = j\frac{1}{(\beta_\rho^d)^2}\left[\omega\mu_d \beta_\rho^d A_m J_m'(\beta_\rho^d \rho) + m\beta_z B_m \frac{1}{\rho} J_m(\beta_\rho^d \rho)\right]\sin(m\phi)e^{-j\beta_z z} \quad (9\text{-}81\text{b})$$

$$E_z^d = B_m J_m(\beta_\rho^d \rho)\cos(m\phi)e^{-j\beta_z z} \quad (9\text{-}81\text{c})$$

$$H_\rho^d = -j\frac{1}{(\beta_\rho^d)^2}\left[\beta_z \beta_\rho^d A_m J_m'(\beta_\rho^d \rho) + m\omega\varepsilon_d B_m \frac{1}{\rho} J_m(\beta_\rho^d \rho)\right]\sin(m\phi)e^{-j\beta_z z} \quad (9\text{-}81\text{d})$$

$$H_\phi^d = -j\frac{1}{(\beta_\rho^d)^2}\left[m\beta_z A_m \frac{1}{\rho} J_m(\beta_\rho^d \rho) + \omega\varepsilon_d \beta_\rho^d B_m J_m'(\beta_\rho^d \rho)\right]\cos(m\phi)e^{-j\beta_z z} \quad (9\text{-}81\text{e})$$

$$H_z^d = A_m J_m(\beta_\rho^d \rho)\sin(m\phi)e^{-j\beta_z z} \quad (9\text{-}81\text{f})$$

where

$$(\beta_\rho^d)^2 + \beta_z^2 = \beta_d^2 = \omega^2 \mu_d \varepsilon_d = \omega^2 \mu_0 \varepsilon_0 \varepsilon_r \mu_r \quad (9\text{-}81\text{g})$$

$$A_m = -j\frac{(\beta_\rho^d)^2}{\omega\mu_d \varepsilon_d} A_{mn} D_2 \quad (9\text{-}81\text{h})$$

$$B_m = -j\frac{(\beta_\rho^d)^2}{\omega\mu_d \varepsilon_d} B_{mn} C_2 \quad (9\text{-}81\text{i})$$

$$' = \frac{\partial}{\partial(\beta_\rho^d \rho)} \quad (9\text{-}81\text{j})$$

The coefficients A_m and B_m are not independent of each other, and their relationship can be found by applying the appropriate boundary conditions.

For the dielectric rod to act as a waveguide, the fields outside the rod ($\rho \geq a$) must be of the evanescent type that exhibit a decay in the radial direction. The rate of attenuation is a function of the diameter of the rod. As the diameter of the rod decreases, the following changes occur.

1. The attenuation lessens.
2. The distance to which the fields outside the rod can extend is greater.
3. The propagation constant β_z is only slightly greater than β_0.

As the diameter of the rod increases, the following changes occur.

1. The rate of attenuation also increases.
2. The fields are confined closer to the rod.
3. The propagation constant β_z approaches β_d.

Since in all cases β_z is greater than β_0, the phase velocity is smaller than the velocity of light in free space. For small diameter rods the surface waves are said to be *loosely bound* to the dielectric surface whereas for the larger diameters it is said to be *tightly bound* to the dielectric surface. Therefore the cylindrical functions that are chosen to represent the radial variations of the fields outside the rod must be cylindrical decaying functions. These functions can be either Hankel functions of order m of the first $H_m^{(1)}$ or second $H_m^{(2)}$ kind and of imaginary argument, or modified Bessel functions K_m of the second kind of order m. We choose here to use the modified Bessel functions K_m of the second kind, which are related to the Hankel functions of the first and second kind by

$$K_m(\alpha) = \begin{cases} j^{m+1}\dfrac{\pi}{2}H_m^{(1)}(j\alpha) & \text{(9-82a)} \\ -j^{m+1}\dfrac{\pi}{2}H_m^{(2)}(-j\alpha) & \text{(9-82b)} \end{cases}$$

With (9-81a) through (9-81j), and (9-82a) and (9-82b) as a guide, we can represent the corresponding electric and magnetic field components for the HEM modes outside the dielectric rod ($\rho \geq a$) by

HEM Modes ($\rho \geq a$)

$$E_\rho^0 = j\frac{1}{(\alpha_\rho^0)^2}\left[m\omega\mu_0 C_m\frac{1}{\rho}K_m(\alpha_\rho^0\rho) + \beta_z\alpha_\rho^0 D_m K_m'(\alpha_\rho^0\rho)\right]\cos(m\phi)e^{-j\beta_z z} \quad (9\text{-}83\text{a})$$

$$E_\phi^0 = -j\frac{1}{(\alpha_\rho^0)^2}\left[\omega\mu_0\alpha_\rho^0 C_m K_m'(\alpha_\rho^0\rho) + m\beta_z D_m\frac{1}{\rho}K_m(\alpha_\rho^0\rho)\right]\sin(m\phi)e^{-j\beta_z z} \quad (9\text{-}83\text{b})$$

$$E_z^0 = D_m K_m(\alpha_\rho^0\rho)\cos(m\phi)e^{-j\beta_z z} \quad (9\text{-}83\text{c})$$

$$H_\rho^0 = j\frac{1}{(\alpha_\rho^0)^2}\left[\beta_z\alpha_\rho^0 C_m K_m'(\alpha_\rho^0\rho) + m\omega\varepsilon_0 D_m\frac{1}{\rho}K_m(\alpha_\rho^0\rho)\right]\sin(m\phi)e^{-j\beta_z z} \quad (9\text{-}83\text{d})$$

$$H_\phi^0 = j\frac{1}{(\alpha_\rho^0)^2}\left[m\beta_z C_m\frac{1}{\rho}K_m(\alpha_\rho^0\rho) + \omega\varepsilon_0\alpha_\rho^0 D_m K_m'(\alpha_\rho^0\rho)\right]\cos(m\phi)e^{-j\beta_z z} \quad (9\text{-}83\text{e})$$

$$H_z^0 = C_m K_m(\alpha_\rho^0\rho)\sin(m\phi)e^{-j\beta_z z} \quad (9\text{-}83\text{f})$$

where

$$(j\alpha_\rho^0)^2 + \beta_z^2 = -(\alpha_\rho^0)^2 + \beta_z^2 = \beta_0^2 = \omega^2\mu_0\varepsilon_0 \quad (9\text{-}83\text{g})$$

$$' = \frac{\partial}{\partial(\alpha_\rho^0\rho)} \quad (9\text{-}83\text{h})$$

The coefficients C_m and D_m are not independent of each other or from A_m and B_m, and their relations can be found by applying the appropriate boundary conditions.

The relations between the constants A_m, B_m, C_m, and D_m and the equation, which will be referred to as the eigenvalue equation, that can be used to determine the modes that can be supported by the dielectric rod waveguide are obtained by

510 CIRCULAR CROSS-SECTION WAVEGUIDE AND CAVITIES

applying the boundary conditions of

$$E_\phi^d(\rho = a, 0 \leq \phi \leq 2\pi, z) = E_\phi^0(\rho = a, 0 \leq \phi \leq 2\pi, z) \quad (9\text{-}84\text{a})$$

$$E_z^d(\rho = a, 0 \leq \phi \leq 2\pi, z) = E_z^0(\rho = a, 0 \leq \phi \leq 2\pi, z) \quad (9\text{-}84\text{b})$$

$$H_\phi^d(\rho = a, 0 \leq \phi \leq 2\pi, z) = H_\phi^0(\rho = a, 0 \leq \phi \leq 2\pi, z) \quad (9\text{-}84\text{c})$$

$$H_z^d(\rho = a, 0 \leq \phi \leq 2\pi, z) = H_z^0(\rho = a, 0 \leq \phi \leq 2\pi, z) \quad (9\text{-}84\text{d})$$

Doing this leads to

$$\frac{1}{(\beta_\rho^d)^2} \left[\omega\mu_d \beta_\rho^d A_m J_m'(\beta_\rho^d a) + m\beta_z B_m \frac{1}{a} J_m(\beta_\rho^d a) \right]$$

$$= -\frac{1}{(\alpha_\rho^0)^2} \left[\omega\mu_0 \alpha_\rho^0 C_m K_m'(\alpha_\rho^0 a) + m\beta_z D_m \frac{1}{a} K_m(\alpha_\rho^0 a) \right] \quad (9\text{-}85\text{a})$$

$$B_m J_m(\beta_\rho^d a) = D_m K_m(\alpha_\rho^0 a) \quad (9\text{-}85\text{b})$$

$$-\frac{1}{(\beta_\rho^d)^2} \left[m\beta_z A_m \frac{1}{a} J_m(\beta_\rho^d a) + \omega\varepsilon_d \beta_\rho^d B_m J_m'(\beta_\rho^d a) \right]$$

$$= \frac{1}{(\alpha_\rho^0)^2} \left[m\beta_z C_m \frac{1}{a} K_m(\alpha_\rho^0 a) + \omega\varepsilon_0 \alpha_\rho^0 D_m K_m'(\alpha_\rho^0 a) \right] \quad (9\text{-}85\text{c})$$

$$A_m J_m(\beta_\rho^d a) = C_m K_m(\alpha_\rho^0 a) \quad (9\text{-}85\text{d})$$

where according to (9-81g) and (9-83g)

$$(\beta_\rho^d)^2 + \beta_z^2 = \beta_d^2 \Rightarrow (\beta_\rho^d a)^2 + (\beta_z a)^2 = (\beta_d a)^2 \Rightarrow \beta_z a = \sqrt{(\beta_d a)^2 - (\beta_\rho^d a)^2} \quad (9\text{-}85\text{e})$$

$$-(\alpha_\rho^0)^2 + \beta_z^2 = \beta_0^2 \Rightarrow -(\alpha_\rho^0 a)^2 + (\beta_z a)^2 = (\beta_0 a)^2 \quad (9\text{-}85\text{f})$$

Subtracting (9-85f) from (9-85e) we get that

$$(\beta_\rho^d a)^2 + (\alpha_\rho^0 a)^2 = (\beta_d a)^2 - (\beta_0 a)^2 \Rightarrow \alpha_\rho^0 a = \sqrt{(\beta_d a)^2 - (\beta_0 a)^2 - (\beta_\rho^d a)^2}$$

$$= \sqrt{(\beta_0 a)^2(\varepsilon_r \mu_r - 1) - (\beta_\rho^d a)^2} \quad (9\text{-}86)$$

Using the abbreviated notation of

$$\chi = \beta_\rho^d a \quad (9\text{-}87\text{a})$$

$$\xi = \alpha_\rho^0 a \quad (9\text{-}87\text{b})$$

$$\zeta = \beta_z a \quad (9\text{-}87\text{c})$$

$$\beta_d = \beta_0 \sqrt{\varepsilon_r \mu_r} \quad (9\text{-}87\text{d})$$

we can rewrite (9-85e) and (9-86) as

$$\zeta = \sqrt{(\beta_0 a)^2 \varepsilon_r \mu_r - \chi^2} \tag{9-88a}$$

$$\xi = \sqrt{(\beta_0 a)^2 (\varepsilon_r \mu_r - 1) - \chi^2} \tag{9-88b}$$

With the preceding abbreviated notation and with $\mu_d = \mu_0$, (9-85a) through (9-85d) can be written in matrix form of

$$Fg = 0 \tag{9-89}$$

where F is a 4×4 matrix and g is a column matrix. Each is given by

$$F = \begin{bmatrix} \dfrac{\omega\mu_0 a}{\chi} J_m'(\chi) & \dfrac{m\zeta}{\chi^2} J_m(\chi) & \dfrac{\omega\mu_0 a}{\xi} K_m'(\xi) & \dfrac{m\zeta}{\xi^2} K_m(\xi) \\ 0 & J_m(\chi) & 0 & -K_m(\xi) \\ \dfrac{m\zeta}{\chi^2} J_m(\chi) & \dfrac{\omega\varepsilon_d a}{\chi} J_m'(\chi) & \dfrac{m\zeta}{\xi^2} K_m(\xi) & \dfrac{\omega\varepsilon_0 a}{\xi} K_m'(\xi) \\ J_m(\chi) & 0 & -K_m(\xi) & 0 \end{bmatrix} \tag{9-89a}$$

$$g = \begin{bmatrix} A_m \\ B_m \\ C_m \\ D_m \end{bmatrix} \tag{9-89b}$$

Equation 9-89 has a nontrivial solution provided that the determinant of F of (9-89a) is equal to zero [i.e., $\det(F) = 0$]. Applying this to (9-89a), it can be shown that it leads to [21]

$$\dfrac{\omega\varepsilon_d a}{\chi} J_m'(\chi) K_m(\xi) \left[\dfrac{\omega\mu_0 a}{\xi} J_m(\chi) K_m'(\xi) + \dfrac{\omega\mu_0 a}{\chi} K_m(\xi) J_m'(\chi) \right]$$

$$- \dfrac{m\zeta}{\chi^2} J_m(\chi) K_m(\xi) \left[\dfrac{m\zeta}{\xi^2} J_m(\chi) K_m(\xi) + \dfrac{m\zeta}{\chi^2} J_m(\chi) K_m(\xi) \right]$$

$$+ J_m^2(\chi) \left[-\dfrac{(m\zeta)^2}{\xi^4} K_m^2(\xi) + \dfrac{\omega^2 \mu_0 \varepsilon_0 a^2}{\xi^2} K_m'(\xi) \right]$$

$$+ J_m(\chi) K_m(\xi) \left[\dfrac{\omega\mu_0 \varepsilon_0^2 a}{\chi\xi} [J_m'(\chi)]^2 K_m'(\xi) - \dfrac{(m\zeta)^2}{(\chi\xi)^2} J_m(\chi) K_m(\xi) \right] = 0 \tag{9-90}$$

Dividing all the terms of (9-90) by $K_m^2(\xi)/(\omega^2 \mu_0 \varepsilon a^2)$ and regrouping, it can be shown that (9-90) can be placed in the form of [21]

$$G_1(\chi) G_2(\chi) - G_3^2(\chi) = 0 \tag{9-91}$$

where

$$G_1(\chi) = \frac{J_m'(\chi)}{\chi} + \frac{K_m'(\xi) J_m(\chi)}{\varepsilon_r \xi K_m(\xi)} \qquad (9\text{-}91a)$$

$$G_2(\chi) = \frac{J_m'(\chi)}{\chi} + \frac{K_m'(\xi) J_m(\chi)}{\xi K_m(\xi)} \qquad (9\text{-}91b)$$

$$G_3(\chi) = \frac{m\zeta}{\beta_0 a \sqrt{\varepsilon_r}} J_m(\chi) \left(\frac{1}{\chi^2} + \frac{1}{\xi^2} \right) \qquad (9\text{-}91c)$$

$$\zeta = \sqrt{(\beta_0 a)^2 \varepsilon_r - \chi^2} \qquad (9\text{-}91d)$$

$$\xi = \sqrt{(\beta_0 a)^2 (\varepsilon_r - 1) - \chi^2} \qquad (9\text{-}91e)$$

Equation 9-91 is referred to as the *eigenvalue equation* for the dielectric rod waveguide. The values of χ that are solutions to (9-91) are referred to as the *eigenvalues* for the dielectric rod waveguide.

In order for $\xi = \alpha_\rho^0 a$ to remain real and represent decaying fields outside the dielectric rod waveguide, the values of $\chi = \beta_\rho^d a$ should not exceed a certain maximum value. From (9-88b) this maximum value $\chi_{\max}$ is equal to

$$\chi_{\max} = \left(\beta_\rho^d a \right)_{\max} = \beta_0 a \sqrt{\varepsilon_r - 1} = \omega a \sqrt{\mu_0 \varepsilon_0 (\varepsilon_r - 1)} \qquad (9\text{-}92)$$

For values of $\beta_\rho^d a$ greater than (9-92), the values of $\xi = \alpha_\rho^0 a$ become imaginary and according to (9-82b) the modified Bessel function of the second kind is reduced to a Hankel function of the second kind that represents unattenuated outwardly traveling waves. In this case the dielectric rod is acting as a cylindrical antenna because of energy loss from its side. Therefore for a given value of m, there is a finite number n of χ_{mn}'s (eigenvalues) for which the dielectric rod acts as a waveguide. Each combination of allowable values of m, n that determine a given eigenvalue χ_{mn} represent the hybrid mode HEM$_{mn}$. HEM modes with odd values of the second subscript correspond to HE modes whereas HEM modes with even values of the second subscript correspond to EH modes. Thus HEM$_{m,2n-1}$ ($n = 1, 2, 3, \ldots$) correspond to HE$_{mn}$ modes and HEM$_{m,2n}$ ($n = 1, 2, 3, \ldots$) correspond to EH$_{mn}$ modes. According to (9-88a), if the values of χ exceed $\beta_0 a \sqrt{\varepsilon_r}$ (i.e., $\chi > \beta_0 a \sqrt{\varepsilon_r}$), then $\zeta = \beta_z a$ becomes imaginary and the waves in the dielectric rod become decaying (evanescent) along the axis (z direction) of the rod.

The allowable modes in a dielectric rod waveguide are determined by finding the values of χ, denoted by χ_{mn}, that are solutions to the transcendental equation 9-91. For each value of m there are a finite number of values of n ($n = 1, 2, 3, \ldots$).

Examining (9-91), it is evident that for $m = 0$ (9-91) vanishes when

$$\text{TM}_{0n}$$

$$G_1(\chi_{0n}) = \frac{J_0'(\chi_{0n})}{\chi_{0n}} + \frac{K_0'(\xi_{0n})J_0(\chi_{0n})}{\varepsilon_r \xi_{0n} K_0(\xi_{0n})} = -\frac{J_1(\chi_{0n})}{\chi_{0n}} - \frac{K_1(\xi_{0n})J_0(\chi_{0n})}{\varepsilon_r \xi_{0n} K_0(\xi_{0n})} = 0$$

(9-93a)

$$\text{TE}_{0n}$$

$$G_2(\chi_{0n}) = \frac{J_0'(\chi_{0n})}{\chi_{0n}} + \frac{K_0'(\xi_{0n})J_0(\chi_{0n})}{\xi_{0n} K_0(\xi_{0n})} = -\frac{J_1(\chi_{0n})}{\chi_{0n}} - \frac{K_1(\xi_{0n})J_0(\chi_{0n})}{\xi_{0n} K_0(\xi_{0n})} = 0$$

(9-93b)

since $G_3(\chi_{0n}) = 0$, $J_0'(\chi_{0n}) = -J_1(\chi_{0n})$, and $K_0'(\xi_{0n}) = -K_1(\xi_{0n})$. Equation 9-93a is valid for TM_{0n} modes and (9-93b) is applicable for TE_{0n} modes.

The nonlinear equations 9-91 through 9-91e can be solved for the values of χ_{mn} iteratively by assuming values of χ_{mn} and examining the sign changes of (9-91). It should be noted that in the allowed range of χ_{mn}'s, G_1, G_2, and G_3 are nonsingular. Computed values of χ_{mn} as a function of $\beta_0 a$ for different HEM_{mn} modes are shown in Figure 9-9 for $\varepsilon_r = 20$ and in Figure 9-10 for $\varepsilon_r = 38$ [21]. It should be noted that for a given mn mode the values of χ_{mn} are nonconstant and vary as a function of the radius of the rod. This is in constrast to the circular waveguide with conducting walls whose χ_{mn} or χ'_{mn} values in Tables 9-1 and 9-2 are constant for a given mode.

Once the values of χ_{mn} for a given mode have been found, the corresponding values of $\zeta = \beta_z a$ can be computed using (9-91d). When this is done for a dielectric rod waveguide of polystyrene ($\varepsilon_r = 2.56$), the values of β_z/β_0 for the HE_{11} mode as a function of the radius of the rod are shown in Figure 9-11 [10]. It is apparent that the HE_{11} mode does not possess a cutoff. In the same figure the values of β_z/β_0 for the axially symmetric TE_{01} and TM_{01} surface wave modes, which possess a finite cutoff, are also displayed. Although in principle the HE_{11} mode has zero cutoff, the rate of attenuation the fields outside the slab exhibit decreases as the radius of the rod becomes smaller and the wavenumber β_z approaches β_0. Thus for small radii the fields outside the rod extend to large distances and are said to be loosely bound to the surface. Practically then a minimum radius rod is usually utilized which results in a small but finite cutoff [22]. For the larger dielectric constant material, the fields outside the dielectric waveguide are more tightly bound to the surface since larger values of β_z translate to larger values of the attenuation coefficient α_ρ^0 through (9-83g).

Typical field patterns in the central core for the HE_{11}, TE_{01}, TM_{01}, HE_{21}, EH_{11}, and HE_{31} modes, are shown in Figure 9-12 [23]. Both E- and H-field lines are displayed over the cross section of the central core and over a cutaway a distance $\lambda_g/2$ along its length. For all the plots, the ratio $\tau = \chi^2/(\chi^2 + \xi^2) = (\beta_\rho^d)^2/[(\beta_\rho^d)^2 + (\alpha_\rho^0)^2] = 0.1$.

514 CIRCULAR CROSS-SECTION WAVEGUIDE AND CAVITIES

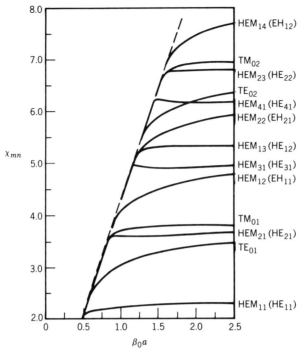

FIGURE 9-9 The first 13 eigenvalues of the dielectric rod waveguide ($\varepsilon_r = 20$). (*Source:* D. Kajfez and P. Guillon (Eds.), *Dielectric Resonators*, 1986, Artech House, Inc.)

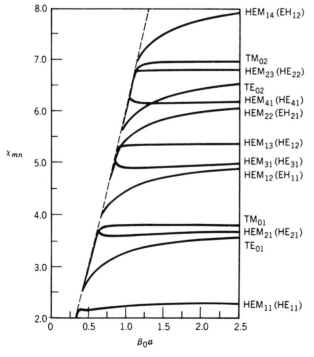

FIGURE 9-10 The first 13 eigenvalues of the dielectric rod waveguide ($\varepsilon_r = 38$). (*Source:* D. Kajfez and P. Guillon (Eds.), *Dielectric Resonators*, 1986, Artech House, Inc.)

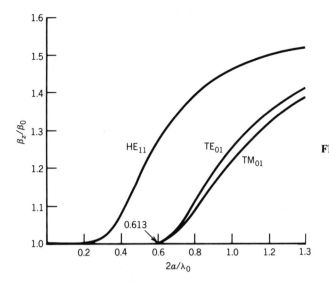

FIGURE 9-11 Ratio of β_z/β_0 for first three surface-wave modes on a polystyrene rod ($\varepsilon_r = 2.56$). (*Source:* R. E. Collin, *Field Theory of Guided Waves*, 1960, McGraw-Hill Book Co.)

For all modes the sum of $\chi^2 + \xi^2$ should be a constant which will be a function of the radius of the dielectric rod and its dielectric constant. Thus according to (9-88b),

$$\chi^2 + \xi^2 = (\beta_0 a)^2(\varepsilon_r - 1) = \text{constant} \quad (9\text{-}94)$$

For all modes, excluding the dominant HE_{11}, this constant should always be equal to or greater than $(2.4049)^2 = 5.7835$; that is, excluding the HE_{11} mode,

$$V^2 = \chi^2 + \xi^2 = (\beta_0 a)^2(\varepsilon_r - 1) \geq (2.4049)^2 \quad (9\text{-}94\text{a})$$

or

$$\frac{2a}{\lambda_0} \geq \frac{1}{\pi} \frac{2.4049}{\sqrt{\varepsilon_r - 1}} \quad (9\text{-}94\text{b})$$

The value of 2.4049 is used because one of the next higher-order modes is the TM_{01} mode which, according to Table 9-2, is $\chi_{mn} = \chi_{01} = 2.4049$. For values of $2a/\lambda_0$ smaller than that of (9-94b) only the dominant HE_{11} dipole mode exists as shown in Figure 9-11.

For a dielectric rod waveguide the first 12 modes (actually first 20 modes if the twofold degeneracy for the HE_{mn} or EH_{mn} is counted), in order of ascending cutoff frequency, along with the vanishing Bessel function and its argument at cutoff, are given:

Mode(s)	$J_m(\chi_{mn}) = 0$	χ_{mn}	Total number of propagating modes
HE_{11} (HEM_{11})	$J_1(\chi_{10}) = 0$	$\chi_{10} = 0$	2
$TE_{01}, TM_{01}, HE_{21}$ (HEM_{21})	$J_0(\chi_{01}) = 0$	$\chi_{01} = 2.4049$	6
HE_{12} (HEM_{13}), EH_{11} (HEM_{12}), HE_{31} (HEM_{31})	$J_1(\chi_{11}) = 0$	$\chi_{11} = 3.8318$	12
EH_{21} (HEM_{22}), HE_{41} (HEM_{41})	$J_2(\chi_{21}) = 0$	$\chi_{21} = 5.1357$	16
$TE_{02}, TM_{02}, HE_{22}$ (HEM_{23})	$J_0(\chi_{02}) = 0$	$\chi_{02} = 5.5201$	20

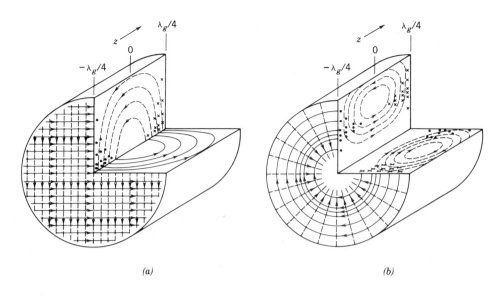
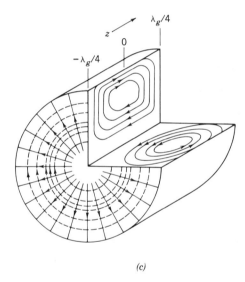

FIGURE 9-12 Field patterns in the central core of a dielectric rod waveguide (in all cases $\tau = 0.1$; **E**: ———, **H**: - - - -). (*Source*: T. Okoshi, *Optical Fibers*, 1982, Academic Press.) (*a*) HEM_{11} (HE_{11}) mode. (*b*) TE_{01} mode. (*c*) TM_{01} mode. (*d*) HEM_{21} (HE_{21}) mode. (*e*) HEM_{12} (EH_{11}) mode. (*f*) HEM_{31} (HE_{31}) mode.

According to (9-94a) a single dominant HE_{11} mode can be maintained within the rod provided the normalized central core radius $V < 2.4049$. This can be accomplished by making the radius a of the central core small and/or choosing, between the central core and the cladding, a small dielectric constant ε_r. However, the smaller the size of the central core, the smaller the rate of attenuation and the less tightly the field outside it is attached to its surface. The normalized diameter over which the $e^{-1} = 0.3679 = 36.79\%$ field point outside the central core extends is shown plotted in Figure 9-13 [16]. Although the fundamental HE_{11} mode does not

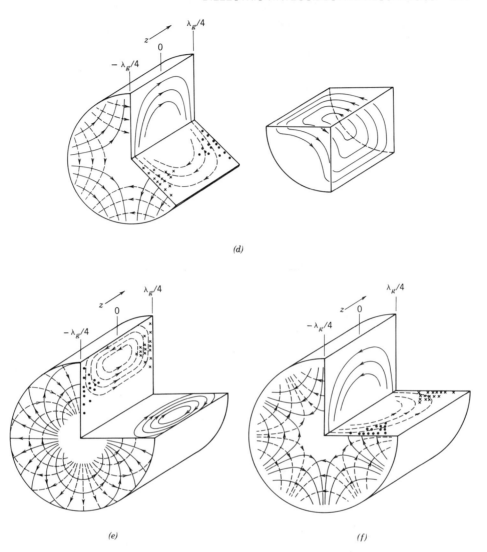

FIGURE 9-12 (*Continued*).

cutoff as the core diameter shrinks, the fields spread out beyond the physical core and become loosely bound.

9.5.2 Circular Dielectric Resonator

Dielectric resonators are unmetalized dielectric objects (spheres, disks, parallelepipeds, etc.) of high dielectric constant (usually ceramic) and high quality factor Q that can function as energy storage devices. Dielectric resonators were first introduced in 1939 by Richtmyer [24], but for almost 25 years his theoretical work failed to generate a continuous and prolonged interest. However, the introduction in the 1960s of material, such as rutile, of high dielectric constant (around 100)

518 CIRCULAR CROSS-SECTION WAVEGUIDE AND CAVITIES

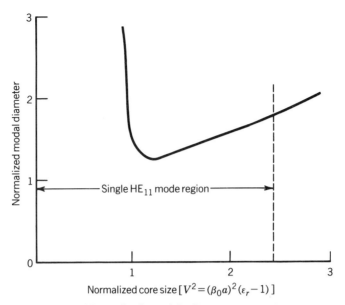

FIGURE 9-13 Normalized modal diameter as a function of normalized core size for dielectric rod waveguide. (*Source:* T. G. Giallorenzi, "Optical communications research and technology: Fiber optics," *Proc. IEEE*, © 1978, IEEE.)

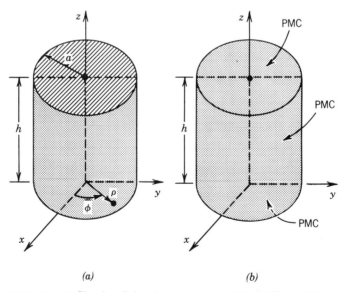

FIGURE 9-14 Circular dielectric resonator and its PMC modeling. (*a*) Dielectric resonator. (*b*) PMC modeling.

renewed the interest in dielectric resonators [25–30]. However, the poor temperature stability of rutile resulted in large resonant frequency changes and prevented the development of practical microwave components. In the 1970s low-loss and temperature-stable ceramics, such as barium tetratitanate and (Zr-Sn)TiO$_4$ were introduced and were used for the design of high performance microwave components such as filters and oscillators. Because dielectric resonators are small, lightweight, temperature stable, high Q, and low cost, they are ideal for design and fabrication of monolithic microwave integrated circuits (MMICs) and general semiconductor devices. Such technology usually requires high Q miniature elements to design and fabricate highly stable frequency oscillators and high performance narrowband filters. Thus dielectric resonators have replaced traditional waveguide resonators, especially in MIC applications, and implementations as high as 94 GHz have been reported. The development of higher dielectric constant material (80 or higher) with stable temperature and low-loss characteristics will have a significant impact on MIC design using dielectric resonators.

In order for the dielectric resonator to function as a resonant cavity, the dielectric constant of the material must be large (usually 30 or greater). Under those conditions the dielectric–air interface acts almost as an open circuit which causes internal reflections and results in the confinement of energy in the dielectric material, thus creating a resonant structure. The plane wave reflection coefficient at the dielectric–air interface is equal to

$$\Gamma = \frac{\eta_0 - \eta}{\eta_0 + \eta} = \frac{\sqrt{\frac{\mu_0}{\varepsilon_0}} - \sqrt{\frac{\mu_0}{\varepsilon}}}{\sqrt{\frac{\mu_0}{\varepsilon_0}} + \sqrt{\frac{\mu_0}{\varepsilon}}} = \frac{\sqrt{\frac{\varepsilon}{\varepsilon_0}} - 1}{\sqrt{\frac{\varepsilon}{\varepsilon_0}} + 1} = \frac{\sqrt{\varepsilon_r} - 1}{\sqrt{\varepsilon_r} + 1} \stackrel{\varepsilon_r \to \text{large}}{\simeq} +1 \quad (9\text{-}95)$$

and it approaches the value of $+1$ as the dielectric constant becomes very large. Under these conditions the dielectric–air interface can be approximated by a hypothetical perfect magnetic conductor (PMC) which requires that the tangential components of the magnetic field (or normal components of the electric field) must vanish (in contrast to the perfect electric conductor, PEC, which requires that the tangential electric field components, or normal components of the magnetic field, must vanish). This of course is a well known and widely used technique in solving boundary-value electomagnetic problems. It is, however, a first order approximation, which usually leads to reasonable results. The magnetic wall model can be used to analyze both the dielectric waveguide and dielectric resonant cavity. Improvements to the magnetic wall approximation have been introduced and resulted in improved data [31, 32].

Although the magnetic wall modeling may not lead to the most accurate data, it will be demonstrated here because it is simple and instructive not only as a first order approximation to this problem but also to other problems including antennas (e.g., microstrip antenna). The geometry of the dielectric resonator is that of Figure 9-14a, whose surface is modeled with the PMC walls of Figure 9-14b which are represented by the independent boundary conditions of

$$H_\phi(\rho = a, 0 \leq \phi \leq 2\pi, 0 \leq z \leq h) = 0 \quad (9\text{-}96a)$$
$$H_\phi(0 \leq \rho \leq a, 0 \leq \phi \leq 2\pi, z = 0) = H_\phi(0 \leq \rho \leq a, 0 \leq \phi \leq 2\pi, z = h) = 0 \quad (9\text{-}96b)$$

or

$$H_z(\rho = a, 0 \leq \phi \leq 2\pi, 0 \leq z \leq h) = 0 \quad (9\text{-}97a)$$

$$H_\rho(0 \leq \rho \leq a, 0 \leq \phi \leq 2\pi, z = 0) = H_\rho(0 \leq \rho \leq a, 0 \leq \phi \leq 2\pi, z = h) = 0 \quad (9\text{-}97b)$$

Either of the preceding sets leads to the same results. Since all the boundary conditions of the first set involve only the H_ϕ component, they will be applied here.

A. TE^z MODES

The TE^z modes can be constructed using the vector potential F_z component of (9-35), or

$$F_z(\rho, \phi, z) = A_{mn} J_m(\beta_\rho^d \rho)[C_2 \cos(m\phi) + D_2 \sin(m\phi)]$$
$$\times [C_3 \cos(\beta_z z) + D_3 \sin(\beta_z z)] \quad (9\text{-}98)$$

where

$$m = 0, 1, 2, \ldots \quad (9\text{-}98a)$$

The H_ϕ component is obtained using (6-80), or

$$H_\phi = -j\frac{1}{\omega_r \mu_d \varepsilon_d}\frac{1}{\rho}\frac{\partial^2 F_z}{\partial \phi \partial z} = -j A_{mn}\frac{m\beta_z}{\omega_r \mu_d \varepsilon_d}\frac{1}{\rho}J_m(\beta_\rho^d \rho)[-C_2 \sin(m\phi) + D_2 \cos(m\phi)]$$
$$\times [-C_3 \sin(\beta_z z) + D_3 \cos(\beta_z z)] \quad (9\text{-}99)$$

Applying (9-96a) leads to

$$H_\phi(\rho = a, 0 \leq \phi \leq 2\pi, 0 \leq z \leq h)$$
$$= -j A_{mn}\frac{m\beta_z}{\omega_r \mu_d \varepsilon_d}\frac{1}{a}J_m(\beta_\rho^d a)[-C_2 \sin(m\phi) + D_2 \cos(m\phi)]$$
$$\times [-C_3 \sin(\beta_z z) + D_3 \cos(\beta_z z)] = 0 \Rightarrow J_m(\beta_\rho^d a) = 0 \Rightarrow \beta_\rho^d a = \chi_{mn}$$

$$\beta_\rho^d = \left(\frac{\chi_{mn}}{a}\right) \quad (9\text{-}100)$$

where χ_{mn} represents the zeroes of the Bessel function of order m, many of which are found in Table 9-2. In a similar manner, the first boundary condition of (9-96b) leads to

$$H_\phi(0 \leq \phi \leq a, 0 \leq \phi \leq 2\pi, z = 0)$$
$$= -j A_{mn}\frac{m\beta_z}{\omega_r \mu_d \varepsilon_d}\frac{1}{\rho}J_m(\beta_\rho^d \rho)[-C_2 \sin(m\phi) + D_2 \cos(m\phi)]$$
$$\times [-C_3(0) + D_3(1)] = 0 \Rightarrow D_3 = 0 \quad (9\text{-}100a)$$

while the second boundary condition of (9-96b) leads to

$$H_\phi(0 \le \rho \le a, 0 \le \phi \le 2\pi, z = h) = -jA_{mn}\frac{m\beta_z}{\omega_r\mu_d\varepsilon_d}\frac{1}{\rho}J_m(\beta_\rho^d\rho)$$

$$\times[-C_2\sin(m\phi) + D_2\cos(m\phi)][-C_3\sin(\beta_z h)] = 0$$

$$\Rightarrow \sin(\beta_z h) = 0 \Rightarrow \beta_z h = p\pi$$

$$\beta_z = \frac{p\pi}{h} \qquad p = 0, 1, 2, \ldots \qquad (9\text{-}100\text{b})$$

Using (9-100) and (9-100b), the resonant frequency is obtained by applying (9-4a) at resonance, that is,

$$\beta_r = \omega_r\sqrt{\mu_d\varepsilon_d} = 2\pi f_r\sqrt{\mu_d\varepsilon_d} = \sqrt{\left(\frac{\chi_{mn}}{a}\right)^2 + \left(\frac{p\pi}{h}\right)^2}$$

or

$$\boxed{(f_r)^{TE^z}_{mnp} = \frac{1}{2\pi\sqrt{\mu_d\varepsilon_d}}\sqrt{\left(\frac{\chi_{mn}}{a}\right)^2 + \left(\frac{p\pi}{h}\right)^2}} \qquad \begin{aligned}m &= 0,1,2,\ldots \\ n &= 1,2,3,\ldots \\ p &= 0,1,2,\ldots\end{aligned} \qquad (9\text{-}101)$$

The dominant TE^z_{mnp} mode is the TE^z_{010} whose resonant frequency is equal to

$$(f_r)^{TE^z}_{010} = \frac{\chi_{01}}{2\pi a\sqrt{\mu_d\varepsilon_d}} = \frac{2.4049}{2\pi a\sqrt{\mu_d\varepsilon_d}} \qquad (9\text{-}101\text{a})$$

B. TM^z MODES

The TM^z modes are obtained using a similar procedure as for the TE^z modes, but starting with the vector potential of (9-41). Doing this leads to the resonant frequency of

$$\boxed{(f_r)^{TM^z}_{mnp} = \frac{1}{2\pi\sqrt{\mu_d\varepsilon_d}}\sqrt{\left(\frac{\chi'_{mn}}{a}\right)^2 + \left(\frac{p\pi}{h}\right)^2}} \qquad \begin{aligned}m &= 0,1,2,\ldots \\ n &= 1,2,3,\ldots \\ p &= 1,2,3,\ldots\end{aligned} \qquad (9\text{-}102)$$

where χ'_{mn} are the zeroes of the derivative of the Bessel function of order m, a partial list of which is found on Table 9-1. The dominant TM^z_{mnp} mode is the TM^z_{111} mode whose resonant frequency is equal to

$$(f_r)^{TM^z}_{111} = \frac{1}{2\pi\sqrt{\mu_d\varepsilon_d}}\sqrt{\left(\frac{\chi'_{11}}{a}\right)^2 + \left(\frac{\pi}{h}\right)^2} = \frac{1}{2\pi\sqrt{\mu_d\varepsilon_d}}\sqrt{\left(\frac{1.8412}{a}\right)^2 + \left(\frac{\pi}{h}\right)^2}$$

(9-102a)

A comparison of the resonant frequencies of (9-101) and (9-102) of the circular dielectric resonator modeled by the PMC surface with those of (9-39a) and (9-45) for the circular waveguide resonator with PEC surface, shows that the TE^z_{mnp} of one are the TM^z_{mnp} of the other, and vice versa.

C. $\text{TE}_{01\delta}$ MODE

Although the dominant mode of the circular dielectric resonator as predicted by the PMC modeling is the TE_{010} mode (provided $h/a < 2.03$), in practice the mode most often used is the $\text{TE}_{01\delta}$ where δ is a noninteger value less than 1. This mode can be modeled using Figure 9-15. For resonators with PEC or PMC walls the third subscript is always an integer (including zero), and it represents the number of half-wavelength variations the field undergoes in the z direction. Since δ is a noninteger and less than unity, the dielectric resonator field of a $\text{TE}_{01\delta}$ mode undergoes less than one half-wavelength variation along its length h. More accurate modelings of the dielectric resonator [31, 33, 34] indicate that $\beta_z h$ of (9-100b) is

$$\beta_z h \simeq \frac{\psi_1}{2} + \frac{\psi_2}{2} + p\pi \qquad p = 0, 1, 2, \ldots \qquad (9\text{-}103)$$

where

$$\frac{\psi_1}{2} = \tan^{-1}\left[\frac{\alpha_1}{\beta_z}\coth(\alpha_1 h_1)\right] \qquad (9\text{-}103\text{a})$$

$$\frac{\psi_2}{2} = \tan^{-1}\left[\frac{\alpha_2}{\beta_z}\coth(\alpha_2 h_2)\right] \qquad (9\text{-}103\text{b})$$

$$\alpha_1 = \sqrt{\left(\frac{\chi_{01}}{a}\right)^2 - \beta_0^2 \varepsilon_{r1}} \qquad (9\text{-}103\text{c})$$

$$\alpha_2 = \sqrt{\left(\frac{\chi_{01}}{a}\right)^2 - \beta_0^2 \varepsilon_{r2}} \qquad (9\text{-}103\text{d})$$

When $p = 0$ in (9-103), the mode is the $\text{TE}_{01\delta}$ where δ is a noninteger less than unity given by

$$\delta = \frac{1}{\pi}\left(\frac{\psi_1}{2} + \frac{\psi_2}{2}\right) \qquad (9\text{-}104)$$

and it signifies the variations of the field between the ends of the dielectric resonator at $z = 0$ and $z = h$. The preceding equations have been derived by modeling the dielectric resonator as shown in Figure 9-15 where PEC plates have been placed at

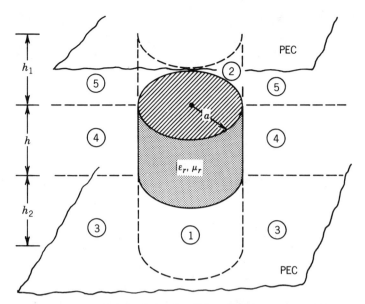

FIGURE 9-15 Modeling of circular dielectric rod resonator. (*Source:* D. Kajfez and P. Guillon (Eds.), *Dielectric Resonators*, 1986, Artech House, Inc.)

$z = -h_2$ and $z = h + h_1$ [21]. The medium in regions 1–5 is usually taken to be air. Thus the dielectric resonator has been sandwiched between two PEC plates each placed at distances h_1 and h_2 from each of its ends. The distances h_1 and h_2 can be chosen to be $0 \leq h_1, h_2 \leq \infty$.

Example 9-5. Find the resonant $TE_{01\delta}$ mode and its resonant frequency when the distances h_1 and h_2 of Figure 9-15 are zero, $h_1 = h_2 = 0$; that is, the dielectric resonator has been sandwiched between two PEC plates, each plate touching each of the ends of the dielectric resonator. Assume the resonator has radius of 5.25 mm, height of 4.6 mm, and dielectric constant of $\varepsilon_r = 38$.

Solution. Since $h_1 = h_2 = 0$, then according to (9-103a) and (9-103b)

$$\frac{\psi_1}{2} = \frac{\psi_2}{2} = \tan^{-1}(\infty) = \frac{\pi}{2}$$

Thus δ of (9-104) is equal to

$$\delta = \frac{1}{\pi}\left(\frac{\pi}{2} + \frac{\pi}{2}\right) = 1$$

and the resonant $TE_{01\delta}$ mode is the TE_{011}. For this mode $\beta_z h$ is, according to

(9-103) with $p = 0$, equal to

$$\beta_z h = \pi \Rightarrow \beta_z = \frac{\pi}{h}$$

Its resonant frequency is identical to (9-101) of the PMC modeling which reduces to

$$(f_r)_{011}^{TE^z} = \frac{1}{2\pi\sqrt{\mu_d \varepsilon_d}} \sqrt{\left(\frac{2.4049}{a}\right)^2 + \left(\frac{\pi}{h}\right)^2}$$

$$= \frac{3 \times 10^{11}}{2\pi\sqrt{38}} \sqrt{\left(\frac{2.4049}{5.25}\right)^2 + \left(\frac{\pi}{4.6}\right)^2} = 6.37 \text{ GHz}$$

When compared to the exact value of 4.82 GHz for a dielectric resonator immersed in free space, the preceding value is $+32\%$ in error. A more accurate result of 4.60 GHz (error of -4.8%) can be obtained using (9-103)–(9-104) by placing the two PECs at infinity ($h_1 = h_2 = \infty$); then the resonator is isolated in free space.

9.5.3 Optical Fiber Cable

Optical communications, which initially started as a speculative research activity, has evolved into a very practical technique which brings new dimensions to miniaturization, data handling capabilities, and signal processing methods. This success has been primarily attributed to the development of fiber optics cables, which in 1970 exhibited attenuations of 20 dB/km and in the early 1980s have been reduced to less than 1 dB/km. The development of suitable solid state diode sources and detectors has certainly eliminated any remaining insurmountable technological barriers and paved the way for widespread implementation of optical communication techniques for commercial and military applications. This technology has spread to the development of integrated optics which provide even greater miniaturization of optical systems, rigid alignment of optical components, and reduced space and weight.

A fiber optics cable is a dielectric waveguide, usually of circular cross section, that guides the electromagnetic wave in discrete modes through internal reflections whose incidence angle at the interface is equal to or greater than the critical angle. The confinement of the energy within the dielectric strucure is described analytically by Maxwell's equations and the boundary conditions at dielectric–dielectric boundaries, in contrast to the traditional metallic–dielectric boundaries in metallic waveguides. Such analyses have already been discussed in Section 8.7 for planar structures and Section 9.5.1 for circular geometries.

The most common geometries of fiber cables are those shown in Figure 9-16. They are classified as *step-index multimode*, *graded-index multimode*, and *single-mode*

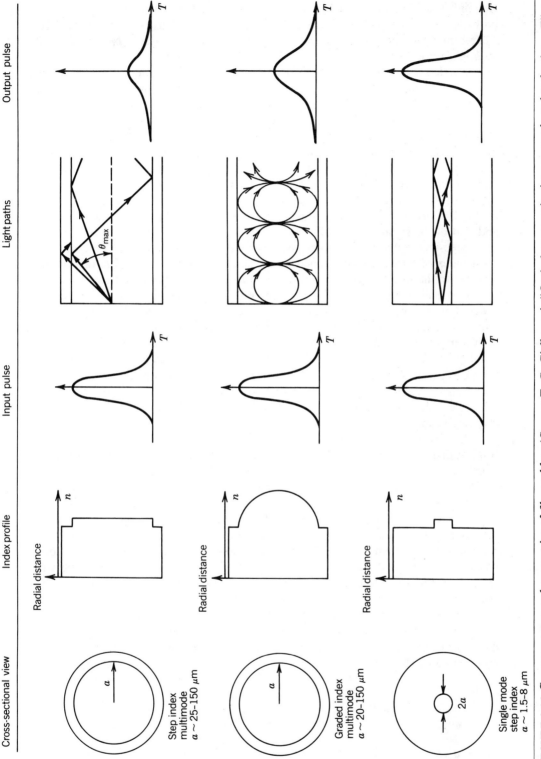

FIGURE 9-16 Common types and geometries of fiber cables. (*Source*: T. G. Giallorenzi, "Optical communications research and technology: Fiber optics," *Proc. IEEE*, © 1978, IEEE.)

step-index [14, 16]. Typical dimensions, index-of-refraction distributions, optical ray paths, and pulse spreading by each type of cable are also illustrated in Figure 9-16. Therefore fiber optics cables can be classified into two cases: *single-mode* and *multimode* fibers.

Single-mode fibers permit wave propagation at a single resolvable angle whereas multimode cables transmit waves that travel at many resolvable angles all within the central core of the fiber. Both single- and multimode fibers are fabricated with a central core of high index of refraction (dielectric constant) surrounded by a cladding of lower index of refraction. The wave is guided in the core by total internal reflection at the core-cladding interface. Single-mode cables usually have a step-index where the diameter of the center core is very small (typically 2–16 μm) and not much larger than the wavelength of the wave it carries. Structurally they are the simplest and exhibit abrupt index-of-refraction discontinuities along the core-cladding interface. Usually the index of refraction of the central core is about 1.471 and that of the cladding is 1.457.

Step-index multimode cables also exhibt a well defined central core of constant index of refraction surrounded by a cladding of lower index of refraction. In contrast to the single-mode step-index the multimode step-index possesses a central core whose diameter is about 25–150 μm. When the wavelength of light is of the order of 10 μm (near infrared region) and the central core diameter is 100 μm, such a cable can support as many as 25,000 modes (usually there are about 200 modes). However, as the central core diameter approaches 10 μm (one wavelength), the number of modes is reduced to one and the cable becomes a single-mode fiber. Because of its multimode field structure, such a cable is very dispersive and provides severe signal distortion to waveforms with broad spectral frequency content. Typically the index of refraction of the central core is about 1.527 and that of the cladding is about 1.517.

Graded-index multimode cables also possess a relatively large central core (typical diameters of 20–150 μm) whose index of refraction continuously decreases from the core center toward the core-cladding interface at which point the core and cladding indexes are identical. For this cable the waves are still contained within the central core and they are continuously refocused toward the central axis of the core by its continuous lensing action. The number of modes supported by such a cable is usually about 2000, and such a multimode structure provides waveform distortion especially to transient signals. All cables possess an outside cladding that adds mechanical strength to the fiber, reduces scattering loss that is due to dielectric discontinuities, and protects the guiding central core from absorbing surface contaminants that the cable may come in contact with.

For the graded-index multimode fiber the index of refraction $n(\rho)$ [$n(\rho) = \sqrt{\varepsilon_r(\rho)}$] of the central core exhibits a nearly parabolic variation in the radial direction from the center of the core toward the cladding. This variation can be represented by [16]

$$n_c(\rho) = n\left[1 + \Delta\left(\frac{a - \rho}{a}\right)^\alpha\right] \qquad 0 \leq \rho \leq a \qquad (9\text{-}105)$$

where $n_c(\rho)$ = index of refraction of the central core (usually ≃ 1.562 on axis)
n = index of refraction of the cladding (usually ≃ 1.540)
a = radius of the central core

Δ = parameter usually much less than unity (usually $0.01 < \Delta < 0.02$)

α = parameter whose value is close to 2 for maximum fiber bandwidth

It has been shown by Example 5-6 that for the dielectric–dielectric interface of the fiber cable to internally reflect the waves of all incidence angles, the ratio of the index of refraction of the central core at the interface to that of the cladding must be equal to or greater than $\sqrt{2}$, that is, $n_c(a)/n \geq \sqrt{2}$. However this is not necessary if the angles of incidence of the waves are not small.

The modes that can be supported by the step-index cable (either single mode or multimode) can be found using techniques outlined in Section 9.5.1 for the dielectric rod waveguide. These modes are, in general, HEM (HE or EH) hybrid modes. The applicable equations are those of (9-81a) through (9-94b) where ε_r should represent the square of the index-of-refraction ratio of the central core to that of the cladding [i.e., $\varepsilon_r = (n_c/n)^2$]. The field configurations of the graded-index cable can be analyzed in terms of Hermite–Gaussian functions [35]. Because of the complexity, the analysis will not be presented here. The interested reader is referred to the literature [11, 16].

9.5.4 Dielectric Covered Conducting Rod

Let us consider the field analysis of a dielectric covered conducting circular rod, as shown in Figure 9-17 [36–38]. The radius of the conducting rod is a while that of

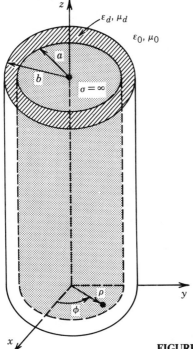

FIGURE 9-17 Geometry of dielectric covered conducting rod.

the dielectric cover is b. The thickness of the dielectric sleeving is denoted by t ($t = b - a$). When the radius of the conducting rod is small, the rod will be representative of a dielectric covered wire. In this section we will consider the TM^z and TE^z modes which, in general, must co-exist to satisfy the boundary conditions. An extended discussion of the modes with azimuthal symmetry, no ϕ variations, will be conducted.

A. TM^z MODES

For the TM^z_{mn} modes, the fields in the radial direction inside the dielectric sleeving must be represented by standing wave functions while those outside the dielectric cover must be decaying in order for the rod to act as a waveguide. Therefore the potential function A_z can be written as

$$a \leq \rho \leq b$$
$$A_z^d = \left[A_1^d J_m(\beta_\rho^d \rho) + B_1^d Y_m(\beta_\rho^d \rho) \right] \left[C_2^d \cos(m\phi) + D_2^d \sin(m\phi) \right]$$
$$\times \left[A_3^d e^{-j\beta_z z} + B_3^d e^{+j\beta_z z} \right] \quad (9\text{-}106a)$$

$$b \leq \rho \leq \infty$$
$$A_z^0 = A_1^0 K_m(\alpha_\rho^0 \rho) \left[C_2^0 \cos(m\phi) + D_2^0 \sin(m\phi) \right] \left[A_3^0 e^{-j\beta_z z} + B_3^0 e^{+j\beta_z z} \right] \quad (9\text{-}106b)$$

For the positive traveling waves of the lowest-order mode that possesses azimuthal symmetry ($m = 0$) and can exist individually, the vector potentials reduce to

$$a \leq \rho \leq b$$
$$A_z^d = \left[A_0^d J_0(\beta_\rho^d \rho) + B_0^d Y_0(\beta_\rho^d \rho) \right] e^{-j\beta_z z} \quad (9\text{-}107)$$

where

$$\left(\beta_\rho^d \right)^2 + \beta_z^2 = \beta_d^2 = \mu_r \varepsilon_r \beta_0^2 \quad (9\text{-}107a)$$

and

$$b \leq \rho \leq \infty$$
$$A_z^0 = A_0^0 K_0(\alpha_\rho^0 \rho) e^{-j\beta_z z} \quad (9\text{-}108)$$

where

$$-\left(\alpha_\rho^0 \right)^2 + \beta_z^2 = \beta_0^2 \quad (9\text{-}108a)$$

The corresponding electric and magnetic fields can be written as

$$a \leq \rho \leq b$$

$$E_\rho^d = -j\frac{1}{\omega\mu_d\varepsilon_d}\frac{\partial^2 A_z^d}{\partial\rho\,\partial z} = -\frac{\beta_z\beta_\rho^d}{\omega\mu_d\varepsilon_d}\left[A_0^d J_0'(\beta_\rho^d\rho) + B_0^d Y_0'(\beta_\rho^d\rho)\right]e^{-j\beta_z z} \quad (9\text{-}109\text{a})$$

$$E_\phi^d = -j\frac{1}{\omega\mu_d\varepsilon_d}\frac{\partial^2 A_z^d}{\partial\phi\,\partial z} = 0 \quad (9\text{-}109\text{b})$$

$$E_z^d = -j\frac{1}{\omega\mu_d\varepsilon_d}\left(\frac{\partial^2}{\partial z^2} + \beta_d^2\right)A_z^d = -j\frac{(\beta_\rho^d)^2}{\omega\mu_d\varepsilon_d}\left[A_0^d J_0(\beta_\rho^d\rho) + B_0^d Y_0(\beta_\rho^d\rho)\right]e^{-j\beta_z z}$$
$$(9\text{-}109\text{c})$$

$$H_\rho^d = \frac{1}{\mu_d}\frac{1}{\rho}\frac{\partial A_z^d}{\partial\phi} = 0 \quad (9\text{-}109\text{d})$$

$$H_\phi^d = -\frac{1}{\mu_d}\frac{\partial A_z^d}{\partial\rho} = -\frac{\beta_\rho^d}{\mu_d}\left[A_0^d J_0'(\beta_\rho^d\rho) + B_0^d Y_0'(\beta_\rho^d\rho)\right]e^{-j\beta_z z} \quad (9\text{-}109\text{e})$$

$$H_z^d = 0 \quad (9\text{-}109\text{f})$$

$$' = \frac{\partial}{\partial(\beta_\rho^d\rho)} \quad (9\text{-}109\text{g})$$

$$b \leq \rho \leq \infty$$

$$E_\rho^0 = -j\frac{1}{\omega\mu_0\varepsilon_0}\frac{\partial^2 A_z^0}{\partial\rho\,\partial z} = \frac{\alpha_\rho^0\beta_z}{\omega\mu_0\varepsilon_0}A_0^0 K_0'(\alpha_\rho^0\rho)e^{-j\beta_z z} \quad (9\text{-}110\text{a})$$

$$E_\phi^0 = -j\frac{1}{\omega\mu_0\varepsilon_0}\frac{1}{\rho}\frac{\partial^2 A_z^0}{\partial\phi\,\partial z} = 0 \quad (9\text{-}110\text{b})$$

$$E_z^0 = -j\frac{1}{\omega\mu_0\varepsilon_0}\left(\frac{\partial^2}{\partial z^2} + \beta_0^2\right)A_z^0 = +j\frac{(\alpha_\rho^0)^2}{\omega\mu_0\varepsilon_0}A_0^0 K_0(\alpha_\rho^0\rho)e^{-j\beta_z z} \quad (9\text{-}110\text{c})$$

$$H_\rho^0 = \frac{1}{\mu_0}\frac{1}{\rho}\frac{\partial A_z^0}{\partial\phi} = 0 \quad (9\text{-}110\text{d})$$

$$H_\phi^0 = -\frac{1}{\mu_0}\frac{\partial A_z^0}{\partial\rho} = -\frac{\alpha_\rho^0}{\mu_0}A_0^0 K_0'(\alpha_\rho^0\rho)e^{-j\beta_z z} \quad (9\text{-}110\text{e})$$

$$H_z^0 = 0 \quad (9\text{-}110\text{f})$$

$$' = \frac{\partial}{\partial(\alpha_\rho^0\rho)} \quad (9\text{-}110\text{g})$$

530   CIRCULAR CROSS-SECTION WAVEGUIDE AND CAVITIES

The vanishing of the tangential electric fields at $\rho = a$ and the continuity of the tangential components of the electric and magnetic fields at $\rho = b$ requires that

$$E_z^d(\rho = a, 0 \leq \phi \leq 2\pi, z) = 0 \qquad (9\text{-}111a)$$

$$E_z^d(\rho = b, 0 \leq \phi \leq 2\pi, z) = E_z^0(\rho = b, 0 \leq \phi \leq 2\pi, z) \qquad (9\text{-}111b)$$

$$H_\phi^d(\rho = b, 0 \leq \phi \leq 2\pi, z) = H_\phi^0(\rho = b, 0 \leq \phi \leq 2\pi, z) \qquad (9\text{-}111c)$$

Applying (9-111a) leads to

$$A_0^d J_0(\beta_\rho^d a) + B_0^d Y_0(\beta_\rho^d a) = 0 \qquad (9\text{-}112)$$

or

$$B_0^d = -A_0^d \frac{J_0(\beta_\rho^d a)}{Y_0(\beta_\rho^d a)} \qquad (9\text{-}112a)$$

whereas (9-111b) leads to

$$-\frac{(\beta_\rho^d)^2}{\omega \mu_d \varepsilon_d}\left[A_0^d J_0(\beta_\rho^d b) + B_0^d Y_0(\beta_\rho^d b)\right] = \frac{(\alpha_\rho^0)^2}{\omega \mu_0 \varepsilon_0} A_0^0 K_0(\alpha_\rho^0 b) \qquad (9\text{-}113)$$

which by using (9-112a) can be written as

$$\boxed{A_0^0 K_0(\alpha_\rho^0 b) = -A_0^d \frac{\mu_0 \varepsilon_0}{\mu_d \varepsilon_d}\left(\frac{\beta_\rho^d}{\alpha_\rho^0}\right)^2 \left[J_0(\beta_\rho^d b) - \frac{J_0(\beta_\rho^d a) Y_0(\beta_\rho^d b)}{Y_0(\beta_\rho^d a)}\right]} \qquad (9\text{-}113a)$$

The continuity of the tangential magnetic field at $\rho = b$, as stated by (9-111c), leads to

$$-\frac{\beta_\rho^d}{\mu_d}\left[A_0^d J_0'(\beta_\rho^d b) + B_0^d Y_0'(\beta_\rho^d b)\right] = -\frac{\alpha_\rho^0}{\mu_0} A_0^0 K_0'(\alpha_\rho^0 b) \qquad (9\text{-}114)$$

which by using (9-112a) can be written as

$$\boxed{A_0^0 K_0'(\alpha_\rho^0 b) = A_0^d \frac{\mu_0}{\mu_d} \frac{\beta_\rho^d}{\alpha_\rho^0}\left[J_0'(\beta_\rho^d b) - \frac{J_0(\beta_\rho^d a) Y_0'(\beta_\rho^d b)}{Y_0(\beta_\rho^d a)}\right]} \qquad (9\text{-}114a)$$

Dividing (9-114a) by (9-113a) leads to

$$\boxed{\frac{K_0'(\alpha_\rho^0 b)}{K_0(\alpha_\rho^0 b)} = -\frac{\varepsilon_d}{\varepsilon_0}\left(\frac{\alpha_\rho^0}{\beta_\rho^d}\right) \frac{\left[J_0'(\beta_\rho^d b) Y_0(\beta_\rho^d a) - J_0(\beta_\rho^d a) Y_0'(\beta_\rho^d b)\right]}{\left[J_0(\beta_\rho^d b) Y_0(\beta_\rho^d a) - J_0(\beta_\rho^d a) Y_0(\beta_\rho^d b)\right]}} \qquad (9\text{-}115a)$$

Subtracting (9-108a) from (9-107a) leads to

$$\boxed{\left(\beta_\rho^d\right)^2 + \left(\alpha_\rho^0\right)^2 = \beta_0^2(\mu_r\varepsilon_r - 1)} \qquad (9\text{-}115\text{b})$$

which along with the transcendental equation 9-115a can be used to solve for β_ρ^d and α_ρ^0. This can be accomplished using graphical or numerical techniques similar to the ones utilized in Section 8.7 for planar structures.

The technique outlined and implemented in Section 8.7 to solve the wave numbers for planar dielectric waveguides is straightforward but complicated when utilized to solve for the wave numbers β_ρ^d and α_ρ^0 of (9-115a) and (9-115b). An approximate solution can be used to solve (9-115a) and (9-115b) simultaneously when the thickness of the dielectric cladding $t = b - a$ is small. Under these conditions the Bessel functions $J_0(\beta_\rho^d b)$ and $Y_0(\beta_\rho^d b)$ can be expanded in a Taylor series about the point $\beta_\rho^d a$, that is

$$J_0(\beta_\rho^d b) \simeq J_0(\beta_\rho^d a) + \left.\frac{dJ_0(\beta_\rho^d b)}{d(\beta_\rho^d b)}\right|_{\beta_\rho^d b = \beta_\rho^d a} \beta_\rho^d(b-a)$$

$$\simeq J_0(\beta_\rho^d a) - \beta_\rho^d(b-a)J_1(\beta_\rho^d a) \qquad (9\text{-}116\text{a})$$

$$Y_0(\beta_\rho^d b) \simeq Y_0(\beta_\rho^d a) + \left.\frac{dY_0(\beta_\rho^d b)}{d(\beta_\rho^d b)}\right|_{\beta_\rho^d b = \beta_\rho^d a} \beta_\rho^d(b-a)$$

$$\simeq Y_0(\beta_\rho^d a) - \beta_\rho^d(b-a)Y_1(\beta_\rho^d a) \qquad (9\text{-}116\text{b})$$

$$J_0'(\beta_\rho^d b) = \left.\frac{dJ_0(\beta_\rho^d \rho)}{d(\beta_\rho^d \rho)}\right|_{\rho=b}$$

$$= -J_1(\beta_\rho^d b) \simeq -\left[J_1(\beta_\rho^d a) + \left.\frac{dJ_1(\beta_\rho^d b)}{d(\beta_\rho^d b)}\right|_{\beta_\rho^d b = \beta_\rho^d a} \beta_\rho^d(b-a)\right]$$

$$\simeq -\left\{J_1(\beta_\rho^d a) + \beta_\rho^d(b-a)\left[J_0(\beta_\rho^d a) - \frac{1}{\beta_\rho^d a}J_1(\beta_\rho^d a)\right]\right\}$$

$$J_0'(\beta_\rho^d b) \simeq -\left[J_1(\beta_\rho^d a)\left(1 - \frac{b-a}{a}\right) + \beta_\rho^d(b-a)J_0(\beta_\rho^d a)\right] \qquad (9\text{-}116\text{c})$$

$$Y_0'(\beta_\rho^d b) \simeq -\left[Y_1(\beta_\rho^d a)\left(1 - \frac{b-a}{a}\right) + \beta_\rho^d(b-a)Y_0(\beta_\rho^d a)\right] \qquad (9\text{-}116\text{d})$$

Therefore the numerator and denominator of (9-115a) can be written, respectively, as

$$J_0'(\beta_\rho^d b) Y_0(\beta_\rho^d a) - J_0(\beta_\rho^d a) Y_0'(\beta_\rho^d b)$$

$$\simeq -Y_0(\beta_\rho^d a)\left[J_1(\beta_\rho^d a)\left(1 - \frac{b-a}{a}\right) + \beta_\rho^d(b-a)J_0(\beta_\rho^d a)\right]$$

$$+ J_0(\beta_\rho^d a)\left[Y_1(\beta_\rho^d a)\left(1 - \frac{b-a}{a}\right) + \beta_\rho^d(b-a)Y_0(\beta_\rho^d a)\right]$$

$$\simeq \left(1 - \frac{b-a}{a}\right)\left[J_0(\beta_\rho^d a)Y_1(\beta_\rho^d a) - Y_0(\beta_\rho^d a)J_1(\beta_\rho^d a)\right] \quad (9\text{-}117a)$$

$$J_0(\beta_\rho^d b) Y_0(\beta_\rho^d a) - J_0(\beta_\rho^d a) Y_0(\beta_\rho^d b)$$

$$\simeq Y_0(\beta_\rho^d a)\left[J_0(\beta_\rho^d a) - \beta_\rho^d(b-a)J_1(\beta_\rho^d a)\right]$$

$$- J_0(\beta_\rho^d a)\left[Y_0(\beta_\rho^d a) - \beta_\rho^d(b-a)Y_1(\beta_\rho^d a)\right]$$

$$\simeq \beta_\rho^d(b-a)\left[J_0(\beta_\rho^d a)Y_1(\beta_\rho^d a) - Y_0(\beta_\rho^d a)J_1(\beta_\rho^d a)\right] \quad (9\text{-}117b)$$

Substituting (9-117a) and (9-117b) into (9-115a) leads to

$$\frac{K_0'(\alpha_\rho^0 b)}{K_0(\alpha_\rho^0 b)} \simeq -\frac{\varepsilon_d}{\varepsilon_0}\left(\frac{\alpha_\rho^0}{\beta_\rho^d}\right)\frac{1 - \frac{b-a}{a}}{\beta_\rho^d(b-a)} = -\frac{\varepsilon_d}{\varepsilon_0}\frac{\alpha_\rho^0}{(\beta_\rho^d)^2}\frac{1 - \frac{b-a}{a}}{b-a} \quad (9\text{-}118)$$

For small values of a (i.e., $a \ll \lambda_0$) the attenuation constant α_ρ^0 is of the same order of magnitude as β_0 and the wave is loosely bound to the surface so that $\alpha_\rho^0 b$ is small for small values of $b - a$. Under these conditions

$$K_0'(\alpha_\rho^0 b) = \left.\frac{dK_0(\alpha_\rho^0 \rho)}{d(\alpha_\rho^0 \rho)}\right|_{\rho=b} = -K_1(\alpha_\rho^0 b) \simeq -\frac{1}{\alpha_\rho^0 b} = -\frac{1}{b\alpha_\rho^0} \quad (9\text{-}119a)$$

$$K_0(\alpha_\rho^0 b) \simeq -\ln(0.89\alpha_\rho^0 b) \quad (9\text{-}119b)$$

so that (9-118) reduces for $\varepsilon_d = \varepsilon_r \varepsilon_0$ to

$$\frac{1}{b\ln(0.89\alpha_\rho^0 b)} \simeq -\varepsilon_r\left(\frac{\alpha_\rho^0}{\beta_\rho^d}\right)^2 \frac{1 - \frac{b-a}{a}}{b-a}$$

$$(\beta_\rho^d)^2(b-a) \simeq -\varepsilon_r b\left(1 - \frac{b-a}{a}\right)(\alpha_\rho^0)^2 \ln(0.89\alpha_\rho^0 b) \quad (9\text{-}120)$$

Substituting (9-115b) into (9-120) for $(\beta_\rho^d)^2$ leads for $t = b - a \ll a$ to

$$\left[\beta_0^2(\varepsilon_r - 1) - (\alpha_\rho^0)^2\right](b - a) \simeq -\varepsilon_r b\left(1 - \frac{b-a}{a}\right)(\alpha_\rho^0)^2 \ln(0.89\alpha_\rho^0 b) \quad (9\text{-}121)$$

or

$$\boxed{\varepsilon_r b (\alpha_\rho^0)^2 \ln(0.89\alpha_\rho^0 b) \simeq \left[-\beta_0^2(\varepsilon_r - 1) + (\alpha_\rho^0)^2\right](b - a)} \quad (9\text{-}121a)$$

Example 9-6. A perfectly conducting wire of radius $a = 0.09$ cm is covered with a dielectric sleeving of polystyrene ($\varepsilon_r = 2.56$) of radius $b = 0.10$ cm. At a frequency of 9.55 GHz, determine the attenuation constants α_ρ^0, β_z, and β_ρ^d and the relative field strength at $\rho = 2\lambda_0$ compared to that at the outside surface of the dielectric sleeving ($\rho = b$).

Solution. At $f = 9.55$ GHz,

$$\lambda_0 = \frac{30 \times 10^9}{9.55 \times 10^9} = 3.1414 \text{ cm} \Rightarrow \beta_0 = \frac{2\pi}{\lambda_0} = 2 \text{ rad/cm}$$

$$\lambda_d = \frac{\lambda_0}{\sqrt{\varepsilon_r}} = \frac{3.1414}{\sqrt{2.56}} = 1.9634 \text{ cm} \Rightarrow \beta_d = \frac{2\pi}{\lambda} = \frac{2\pi}{1.9634} = 3.2 \text{ rad/cm}$$

Since the thickness t of the dielectric sleeving is much smaller than the wavelength,

$$t = b - a = 0.10 - 0.09 = 0.01 \text{ cm} < \lambda_d = 1.9634 \text{ cm} < \lambda_0 = 3.1414 \text{ cm}$$

then the approximate relation of (9-121a) is applicable. Using an iterative procedure, it can be shown that

$$\alpha_\rho^0 = 0.252 \text{ Np/cm}$$

is a solution to (9-121a). Using (9-108a),

$$\beta_z = \sqrt{\beta_0^2 + (\alpha_\rho^0)^2} = \sqrt{(2)^2 + (0.252)^2} = \sqrt{4.0635} = 2.0158 \text{ rad/cm}$$

and β_ρ^d is found using (9-107a) as

$$\beta_\rho^d = \sqrt{\beta_d^2 - \beta_z^2} = \sqrt{(3.2)^2 - (2.0158)^2} = \sqrt{6.1765} = 2.485 \text{ rad/cm}$$

The field outside the sleeving is of decaying form represented by the modified Bessel function $K_0(\alpha_\rho^0 \rho)$ of (9-108). Thus

$$\frac{E(\rho = 2\lambda_0)}{E(\rho = b)} = \frac{E(\rho = 2\pi \text{ cm})}{E(\rho = 0.10 \text{ cm})} = \frac{K_0(\rho = 2\pi \text{ cm})}{K_0(\rho = 0.10 \text{ cm})}$$

$$= \frac{\left.\dfrac{C_0 e^{-\alpha_\rho^0 \rho}}{\sqrt{\alpha_\rho^0 \rho}}\right|_{\rho = 2\pi \text{ cm}}}{\left.\dfrac{C_0 e^{-j\alpha_\rho^0 \rho}}{\sqrt{\alpha_\rho^0 \rho}}\right|_{\rho = 0.10 \text{ cm}}} = \frac{0.1631}{6.1426} = 0.0266 = 2.66\%$$

Therefore at a distance of $\rho = 2\lambda_0$ the relative field has been reduced to a very small value; at points further away, the field intensity is even smaller.

For the dielectric covered wire and the dielectric rod waveguide, we can define an *effective radius* as the radial distance at which point and beyond the relative field intensity is of very low value. If we use the 2.66% value of Example 9-6 as the field value with which we can define the effective radius, then the effective radius for the wire of Example 9-6 is $a_e = 2\lambda_0 = 2\pi$ cm.

The rate of attenuation α_ρ^0 can be increased and the wave can be made more tightly bound to the surface by increasing the dielectric constant and/or thickness of the dielectric sleeving. This, however, results in greater attenuation and larger losses of the wave along the direction of wave travel (axis of the wire) because of the greater field concentration near the conducting boundary.

B. TE^z MODES

Following a procedure similar to that used for the TM$_{mn}^z$ modes, it can be shown that for the dielectric covered conducting rod of Figure 9-17 the TE$_{mn}^z$ positive traveling waves of the lowest-order mode also possess azimuthal symmetry (no ϕ variations, $m = 0$) and can exist individually. The vector potential can be written as

$$a \leq \rho \leq b$$

$$F_z^d = \left[A_0^d J_0(\beta_\rho^d \rho) + B_0^d Y_0(\beta_\rho^d \rho) \right] e^{-j\beta_z z} \tag{9-122}$$

where

$$\left(\beta_\rho^d\right)^2 + \beta_z^2 = \beta_d^2 = \mu_r \varepsilon_r \beta_0^2 \tag{9-122a}$$

$$b \leq \rho \leq \infty$$

$$F_z^0 = A_0^0 K_0(\alpha_\rho^0 \rho) e^{-j\beta_z z} \tag{9-123}$$

where

$$-\left(\alpha_\rho^0\right)^2 + \beta_z^2 = \beta_0^2 \qquad (9\text{-}123a)$$

Leaving the details for the reader as an end of chapter exercise, the equations for the TE_{0n}^z modes corresponding to (9-113a), (9-114a), (9-115a), (9-115b), and (9-121a) for the TM_{0n}^z modes, are given by

$$\boxed{A_0^0 K_0'\left(\alpha_\rho^0 b\right) = A_0^d \frac{\varepsilon_0}{\varepsilon_d} \frac{\beta_\rho^d}{\alpha_\rho^0} \left[J_0'\left(\beta_\rho^d b\right) - \frac{J_0'\left(\beta_\rho^d a\right) Y_0'\left(\beta_\rho^d b\right)}{Y_0'\left(\beta_\rho^d a\right)} \right]} \qquad (9\text{-}124a)$$

$$\boxed{A_0 K_0\left(\alpha_\rho^0 b\right) = -A_0^d \frac{\mu_0}{\mu_d} \frac{\varepsilon_0}{\varepsilon_d} \left(\frac{\beta_\rho^d}{\alpha_\rho^0}\right)^2 \left[J_0\left(\beta_\rho^d b\right) - \frac{J_0'\left(\beta_\rho^d a\right) Y_0\left(\beta_\rho^d b\right)}{Y_0'\left(\beta_\rho^d a\right)} \right]} \qquad (9\text{-}124b)$$

$$\boxed{\frac{K_0'\left(\alpha_\rho^0 b\right)}{K_0\left(\alpha_\rho^0 b\right)} = -\frac{\mu_d}{\mu_0} \left(\frac{\alpha_\rho^0}{\beta_\rho^d}\right) \frac{\left[J_0'\left(\beta_\rho^d b\right) Y_0'\left(\beta_\rho^d a\right) - J_0'\left(\beta_\rho^d a\right) Y_0'\left(\beta_\rho^d b\right)\right]}{\left[J_0\left(\beta_\rho^d b\right) Y_0'\left(\beta_\rho^d a\right) - J_0'\left(\beta_\rho^d a\right) Y_0\left(\beta_\rho^d b\right)\right]}} \qquad (9\text{-}124c)$$

$$\boxed{\left(\beta_\rho^d\right)^2 + \left(\alpha_\rho^0\right)^2 = \beta_0^2 (\mu_r \varepsilon_r - 1)} \qquad (9\text{-}124d)$$

$$\boxed{\mu_r b (b - a) \left(\alpha_\rho^0\right)^2 \ln\left(0.89 \alpha_\rho^0 b\right) \simeq 1} \qquad (9\text{-}124e)$$

REFERENCES

1. C. S. Lee, S. W. Lee, and S. L. Chuang, "Plot of modal field distribution in rectangular and circular waveguides," *IEEE Trans. Microwave Theory Tech.*, vol. MTT-33, no. 3, pp. 271–274, March 1985.
2. S. E. Miller, "Waveguide as a communication medium," *Bell System Tech. J.*, vol. 35, pp. 1347–1384, November 1956.
3. S. P. Morgan and J. A. Young, "Helix waveguide," *Bell System Tech. J.*, vol. 35, pp. 1347–1384, November 1956.
4. S. Ramo, J. R. Whinnery, and T. Van Duzer, *Fields and Waves in Communication Electronics*, Wiley, New York, 1965, pp. 429–439.
5. M. R. Spiegel, *Mathematical Handbook, Schaum's Outline Series*, McGraw-Hill, New York, 1968.
6. C. A. Balanis, *Antenna Theory: Analysis and Design*, Wiley, New York, 1982.

7. C. H. Chandler, "An investigation of dielectric rod as waveguide," *J. Appl. Phys.*, vol. 20, pp. 1188–1192, December 1949.
8. W. M. Elasser, "Attenuation in a dielectric circular rod," *J. Appl. Phys.*, vol. 20, pp. 1193–1196, December 1949.
9. J. W. Duncan and R. H. DuHamel, "A technique for controlling the radiation from dielectric rod waveguides," *IRE Trans. Antennas Propagat.*, vol. AP-5, no. 4, pp. 284–289, July 1957.
10. R. E. Collin, *Field Theory of Guided Waves*, McGraw-Hill, New York, 1960.
11. E. Snitzer, "Cylindrical electric waveguide modes," *J. Optical Soc. America*, vol. 51, no. 5, pp. 491–498, May 1961.
12. A. W. Snyder, "Asymptotic expressions for eigenfunctions and eigenvalues of a dielectric or optical waveguide," *IEEE Trans. Microwave Theory Tech.*, vol. MTT-17, pp. 1130–1138, December 1969.
13. A. W. Snyder, "Excitation and scattering of modes on a dielectric or optical fiber," *IEEE Trans. Microwave Theory Tech.*, vol. MTT-17, pp. 1138–1144, December 1969.
14. S. E. Miller, E. A. J. Marcatili, and T. Li, "Research toward optical-fiber transmission systems. Part I: The transmission medium," *Proc. IEEE*, vol. 61, no. 12, pp. 1703–1726, December 1973.
15. D. Marcuse, *Theory of Dielectric Optical Waveguide*, Academic, New York, 1974.
16. T. G. Giallorenzi, "Optical communications research and technology: Fiber Optics," *Proc. IEEE*, vol. 66, no. 7, pp. 744–780, July 1978.
17. J. Kane, "Fiber optic cables compete with mw relays and coax," *Microwave J.*, vol. 26, pp. 16, 61, January 1979.
18. C. Yeh, "Guided-wave modes in cylindrical optical fibers," *IEEE Trans. Education*, vol. E-30, no. 1, February 1987.
19. R. J. Pieper, "A heuristic approach to fiber optics," *IEEE Trans. Education*, vol. E-30, no. 2, pp. 77–82, May 1987.
20. "IRE standards on antennas and waveguides: definitions of terms, 1953," *Proc. IRE*, vol. 41, pp. 1721–1728, December 1953.
21. D. Kajfez and P. Guillon (Eds.), *Dielectric Resonators*, Artech House, Inc., Dedham, MA, 1986.
22. R. F. Harrington, *Time-Harmonic Electromagnetic Fields*, McGraw-Hill, New York, 1961.
23. T. Okoshi, *Optical Fibers*, Academic, New York, 1982.
24. R. D. Richtmyer, "Dielectric resonator," *J. Appl. Phys.*, vol. 10, pp. 391–398, June 1939.
25. A. Okaya, "The rutile microwave resonator," *Proc. IRE*, vol. 48, p. 1921, November 1960.
26. A. Okaya and L. F. Barash, "The dielectric microwave resonator," *Proc. IRE*, vol. 50, pp. 2081–2092, October 1962.
27. H. Y. Yee, "Natural resonant frequencies of microwave dielectric resonators," *IEEE Trans. Microwave Theory Tech.*, vol. MTT-13, p. 256, March 1965.
28. S. J. Fiedziuszko, "Microwave dielectric resonators," *Microwave J.*, pp. 189–200, September 1980.
29. M. W. Pospieszalski, "Cylindrical dielectric resonators and their applications in TEM line microwave circuits," *IEEE Trans. Microwave Theory Tech.*, vol. MTT-27, no. 3, pp. 233–238, March 1979.
30. K. A. Zaki and C. Chen, "Loss mechanisms in dielectric-loaded resonators," *IEEE Trans. Microwave Theory Tech.*, vol. MTT-33, no. 12, pp. 1448–1452, December 1985.
31. S. B. Cohn, "Microwave bandpass filters containing high-Q dielectric resonators," *IEEE Trans. Microwave Theory Tech.*, vol. MTT-16, pp. 218–227, April 1968.
32. T. Itoh and R. S. Rudokas, "New method for computing the resonant frequencies of dielectric resonators," *IEEE Trans. Microwave Theory Tech.*, vol. MTT-25, pp. 52–54, January 1977.
33. A. W. Glisson, D. Kajfez, and J. James, "Evaluation of modes in dielectric resonators using a surface integral equation formulation," *IEEE Trans. Microwave Theory Tech.*, vol. MTT-31, pp. 1023–1029, December 1983.

34. Y. Kobayashi and S. Tanaka, "Resonant modes of a dielectric rod resonator short-circuited at both ends by parallel conducting plates," *IEEE Trans. Microwave Theory Tech.*, vol. MTT-28, no. 10, pp. 1077–1085, October 1970.
35. R. D. Maurer, "Introduction to optical fiber waveguides," in *Introduction to Integrated Optics*, M. Barnoski (Ed.), Plenum, New York, Chapter 8, 1974.
36. G. Goubau, "Surface waves and their application to transmission lines," *J. Appl. Phys.*, vol. 21, pp. 1119–1128, November 1950.
37. G. Goubau, "Single-conductor surface-wave transmission lines," *Proc. IRE*, vol. 39, pp. 619–624, June 1951.
38. R. E. Collin, *Foundations for Microwave Engineering*, McGraw-Hill, New York, 1966.

PROBLEMS

9.1. Design a circular waveguide filled with a lossless dielectric medium whose relative permeability is unity. The waveguide must operate in a single dominant mode over a bandwidth of 1.5 GHz. Assume that the radius of the guide is 1.12 cm.

 (a) Find the dielectric constant of the medium that must fill the cavity to meet the desired design specifications.

 (b) Find the lower and upper frequencies of operation.

9.2. An air-filled circular waveguide of radius a has a conducting baffle placed along its length at $\phi = 0$ extending from $\rho = 0$ to a, as shown in Figure P9-2. For TE^z modes, derive simplified expressions for the vector potential component, the electric and magnetic fields, and the cutoff frequencies, eigenvalues, phase constant along the axis of the guide, guide wavelength, and wave impedance.

 Also determine the following.

 (a) The cutoff frequencies of the three lowest-order propagating modes in order of ascending cutoff frequency when the radius of the cylinder is 1 cm.

 (b) The wave impedance and guide wavelength (in cm) for the lowest-order mode at $f = 1.5 f_c$ where f_c is the cutoff frequency of the lowest-order mode.

Hint:

$$J'_{1/2}(x) = 0 \quad \text{for } x = 1.1655, 4.6042$$

$$J'_{3/2}(x) = 0 \quad \text{for } x = 2.4605, 6.0293$$

$$J'_{5/2}(x) = 0 \quad \text{for } x = 3.6328$$

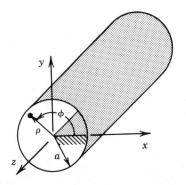

FIGURE P9-2

9.3. Repeat Problem 9.2 for TM^z modes.
Hint:
$$J_{1/2}(x) = 0 \quad \text{for } x = 3.1416, 6.2832$$
$$J_{3/2}(x) = 0 \quad \text{for } x = 4.4934$$
$$J_{5/2}(x) = 0 \quad \text{for } x = 5.7635$$

9.4. The cross section of a cylindrical waveguide is a half circle, as shown in Figure P9-4. Derive simplified expressions for the vector potential component, electric and magnetic fields, eigenvalues, and cutoff frequencies for TE^z modes and TM^z modes.

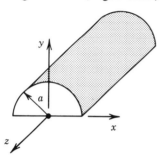

FIGURE P9-4

9.5. Repeat Problem 9.4 for the waveguide cross section of Figure P9-5.

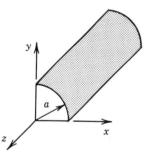

FIGURE P9-5

9.6. Repeat Problem 9.4 when the waveguide cross section is an angular sector as shown in Figure P9-6. Show that the zeroes of $\beta_\rho a$ are obtained using

(a) TE^z: $J'_m(\beta_\rho a) = 0 \quad \beta_\rho = \dfrac{\chi'_{mn}}{a} \quad m = p\left(\dfrac{\pi}{\phi_0}\right) \quad p = 0, 1, 2, \ldots$

(b) TM^z: $J_m(\beta_\rho a) = 0 \quad \beta_\rho = \dfrac{\chi_{mn}}{a} \quad m = p\left(\dfrac{\pi}{\phi_0}\right) \quad p = 1, 2, 3, \ldots$

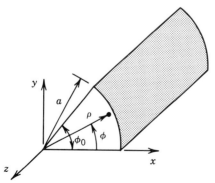

FIGURE P9-6

9.7. The cross section of a cylindrical waveguide is that of a coaxial line with inner radius a and outer radius b, as shown in Figure P9-7. Assume TE^z modes within the waveguide.
  (a) Derive simplified expressions for the vector potential component, and the electric and magnetic fields.
  (b) Show that the eigenvalues are obtained as solutions to
  $$J'_m(\beta_\rho a) Y'_m(\beta_\rho b) - Y'_m(\beta_\rho a) J'_m(\beta_\rho b) = 0$$
  where $m = 0, 1, 2, \ldots$.

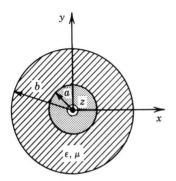

FIGURE P9-7

9.8. Repeat Problem 9.7 for TM^z and show that the eigenvalues are obtained as solutions to
$$J_m(\beta_\rho a) Y_m(\beta_\rho b) - Y_m(\beta_\rho a) J_m(\beta_\rho b) = 0$$
where $m = 0, 1, 2, \ldots$.

9.9. The cross section of a cylindrical waveguide is an annular sector with inner and outer radii of a and b, as shown in Figure P9-9. Assume TE^z modes within the waveguide.
  (a) Derive simplified expressions for the vector potential component, and the electric and magnetic fields.
  (b) Show that the eigenvalues are determined by solving
  $$J'_m(\beta_\rho b) Y'_m(\beta_\rho a) - J'_m(\beta_\rho a) Y'_m(\beta_\rho b) = 0$$
  where $m = p(\pi/\phi_0)$, $p = 0, 1, 2, \ldots$.

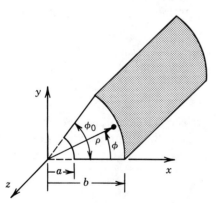

FIGURE P9-9

9.10. Repeat Problem 9.9 for TM^z modes. The eigenvalues are determined by solving
$$J_m(\beta_\rho b) Y_m(\beta_\rho a) - J_m(\beta_\rho a) Y_m(\beta_\rho b) = 0$$
where $m = p(\pi/\phi_0)$, $p = 1, 2, 3, \ldots$.

9.11. A circular waveguide with radius of 3 cm is made of copper ($\sigma = 5.76 \times 10^7$ S/m). For the dominant TE_{11} and low-loss TE_{01} modes, determine their corresponding cutoff frequencies and attenuation constants (in Np/m and dB/m) at a frequency of 7 GHz. Assume that the waveguide is filled with air.

9.12. Derive the attenuation coefficient α_c for the conduction losses of a circular waveguide of radius a for the TM_{01}^z. Show that α_c can be expressed as

$$\alpha_c(TM_{01}) = \frac{R_s}{\eta a \sqrt{1 - \left(\frac{f_c}{f}\right)^2}}$$

where R_s is the surface resistance of the waveguide metal and η is the intrinsic impedance of the medium within the waveguide.

9.13. A circular cavity, as shown in Figure 9-5, has a radius of 6 cm and a height of 10 cm. It is filled with a lossless dielectric with $\varepsilon_r = 4$.
   (a) Find the first four TE^z and/or TM^z modes according to their resonant frequency (in order of ascending values).
   (b) Find the Q of the cavity (assuming dominant mode operation). The walls of the cavity are copper ($\sigma = 5.76 \times 10^7$ S/m).

9.14. Design a circular cavity of radius a and height h such that the resonant frequency of the next higher-order mode is 1.5 times greater than the resonant frequency of the dominant mode. Assume that the radius is 4 cm.
   (a) Find the height of the cavity (in cm).
   (b) Find the resonant frequency of the dominant mode (assume free space within the cavity).
   (c) Find the dielectric constant of the dielectric that must be inserted inside the cavity to reduce the resonant frequency by a factor of 1.5.

9.15. For the radial waveguide of Figure 9-6 determine the maximum spacing h (in m) to insure operation of a single lowest-order mode between the plates up to 300 MHz. Do this individually for TE_n^z modes and TM_n^z modes.

9.16. For the wedged-plate radial waveguide of Figure 9-7, derive for the TE^z modes the expressions for the vector potential component of (9-79), the electric and magnetic fields of (9-79d) through (9-79i), the eigenvalues of (9-79a) through (9-79c), and the wave impedance of (9-79j).

9.17. Given the wedged-plate geometry of Figure 9-7, assume TM^z modes and derive the expressions for the vector potential of (9-80), the electric and magnetic fields of (9-80d) through (9-80i), the eigenvalues of (9-80a) through (9-80c), and the wave impedance of (9-80j).

9.18. If the dielectric constant of the materials that make up the circular dielectric waveguide of Figure 9-8 is very large (usually 30 or greater), the dielectric–air interface along the surface acts almost as an open circuit (see Section 9.5.2). Under these conditions the surface of the dielectric can be approximated by a perfect magnetic conductor (PMC). Assume that the surface of the dielectric rod can be modeled as a PMC and derive, for the TE^z modes, simplified expressions for the vector potential component, electric and magnetic fields, and cutoff frequencies. Assume that the dielectric material is rutile ($\varepsilon_r \simeq 130$). Determine the cutoff frequencies of the lowest two modes when the radius of the rod is 3 cm.

9.19. Repeat Problem 9.18 for the TM^z modes.

9.20. Determine the cutoff frequencies of the first four lowest-order modes (HE, EH, TE, or TM) for a dielectric rod waveguide with radius of 3 cm when the dielectric constant of the material is $\varepsilon_r = 20, 38,$ and 130.

9.21. Design a circular dielectric rod waveguide (find its radius in cm) so that the cutoff frequency of the TE$_{01}$, TM$_{01}$, and HE$_{21}$ modes is 3 GHz when the dielectric constant of the material is $\varepsilon_r = 2.56, 4, 9,$ and 16.

9.22. It is desired to operate a dielectric rod waveguide in the dominant HE$_{11}$ mode over a frequency range of 5 GHz. Design the dielectric rod (find its dielectric constant) to accomplish this when the radius of the rod is $a = 1.315$ and 1.838 cm.

9.23. For the dielectric resonator of Figure 9-14a modeled by PMC walls, as shown in Figure 9-14b, derive simplified expressions for the vector potential component, and the electric and magnetic fields when the modes are TE^z and TM^z.

9.24. Assume that the dielectric resonator of Figure 9-14a is modeled by PMC walls as shown in Figure 9-14b.
   (a) Determine the lowest TE^z or TM^z mode.
   (b) Derive an expression for the Q of the cavity for the lowest-order mode. The only losses associated with the resonator are dielectric losses within the dielectric itself.
   (c) Find the resonant frequency.
   (d) Compute the Q of the cavity.
   The resonator material is rutile ($\varepsilon_r \simeq 130$, $\tan \delta_e \simeq 4 \times 10^{-4}$). The radius of the disk is 0.1148 cm and its height is 0.01148 cm.

9.25. A hybrid dielectric resonator, with a general geometry as shown in Figure 9-14, can be constructed with PEC plates at $z = 0$ and $z = h$ and with open sides. If the dielectric constant of the dielectric is very large, the open dielectric surface can be modeled as a PMC surface. Using such a model for the resonator of Figure 9-14, derive expressions for the resonant frequencies assuming TE^z modes and TM^z modes. This model can also be used as an approximate representation for a circular patch (microstrip) antenna.

9.26. Repeat Problem 9.25 if the top and bottom plates of the resonator are angular sectors each with a subtended angle of ϕ_0, as shown in Figure P9-26. Show that the eigenvalues are obtained as solutions to

(a) TE^z: $J_m(\beta_\rho a) = 0 \quad \beta_\rho = \dfrac{\chi_{mn}}{a} \quad m = p\left(\dfrac{\pi}{\phi_0}\right) \quad p = 1, 2, 3, \ldots$

(b) TM^z: $J'_m(\beta_\rho a) = 0 \quad \beta_\rho = \dfrac{\chi'_{mn}}{a} \quad m = p\left(\dfrac{\pi}{\phi_0}\right) \quad p = 0, 1, 2, \ldots$

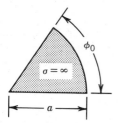

FIGURE P9-26

This model can also be used as an approximate representation for an angular patch (microstrip) antenna.

9.27. Repeat Problem 9.25 if the top and bottom plates of the resonator are annular patches each with inner radius a and outer radius b, as shown in Figure P9-27. Show that the eigenvalues are obtained as solutions to

(a) TE^z: $J_m(\beta_\rho a)Y_m(\beta_\rho b) - Y_m(\beta_\rho a)J_m(\beta_\rho b) = 0 \quad m = 0, 1, 2, \ldots$

(b) TM^z: $J_m'(\beta_\rho a)Y_m'(\beta_\rho b) - Y_m'(\beta_\rho a)J_m'(\beta_\rho b) = 0 \quad m = 0, 1, 2, \ldots$

This model can also be used as an approximate representation for an annular patch (microstrip) antenna.

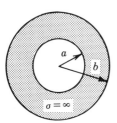

FIGURE P9-27

9.28. Repeat Problem 9.25 if the top and bottom plates of the resonator are annular sectors each with inner radius a and outer radius b, and subtended angle ϕ_0, as shown in Figure P9-28. Show that the eigenvalues are obtained as solutions to

(a) TE^Z: $J_m(\beta_\rho a)Y_m(\beta_\rho b) - Y_m(\beta_\rho a)J_m(\beta_\rho b) = 0 \quad m = p\left(\dfrac{\pi}{\phi_0}\right), \quad p = 1, 2, 3, \ldots$

(b) TM^Z: $J_m'(\beta_\rho a)Y_m'(\beta_\rho b) - Y_m'(\beta_\rho a)J_m'(\beta_\rho b) = 0 \quad m = p\left(\dfrac{\pi}{\phi_0}\right), \quad p = 0, 1, 2, \ldots$

This model can also be used as an approximate representation for an annular sector patch (microstrip) antenna.

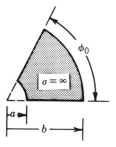

FIGURE P9-28

9.29. For the dielectric covered conducting rod of Figure 9-17, assume TE^z_{mn} modes. For within and outside the dielectric sleeving derive expressions for the vector potential components and the electric and magnetic fields. Also verify (9-124a) through (9-124e).

CHAPTER 10

SPHERICAL TRANSMISSION LINES AND CAVITIES

10.1 INTRODUCTION

Problems involving spherical geometries constitute an important class of electromagnetic boundary-value problems that are used to design transmission lines, cavities, antennas, and scatterers. Some of these may be constructed of metallic walls and others may be of dielectric material. In either case, the field configurations that can be supported by the structure can be obtained by analyzing the structure as a boundary-value problem. We will concern ourselves here with spherical transmission lines and cavities. Scattering by spherical structures will be examined in Chapter 11.

10.2 CONSTRUCTION OF SOLUTIONS

In Chapter 3, Section 3.4.3, we examined the solution of the scalar wave equation in spherical coordinates. It was found that the solution is that of (3-76) where the r, θ, and ϕ variations take the following forms.

1. Radial (r) variations of $f(r)$ can be represented by either
   a. spherical Bessel functions of the first $[j_n(\beta r)]$ and second $[y_n(\beta r)]$ kind, as given by (3-87a) [these functions are used to represent *standing* waves and are related to the regular Bessel functions by (3-90a) and (3-90b)]
   b. or spherical Hankel functions of the first $[h_n^{(1)}(\beta r)]$ and second $[h_n^{(2)}(\beta r)]$ kind, as given by (3-87b) [these are related to the regular Hankel functions by (3-91a) and (3-91b)].
2. θ variations of $g(\theta)$ can be represented by associated Legendre functions of the first $P_n^m(\cos\theta)$ or second $Q_n^m(\cos\theta)$ kind as given by either (3-88a) or (3-88b).
3. ϕ variations of $h(\phi)$ can be represented by either complex exponentials or cosinusoids as given, respectively, by (3-89a) and (3-89b).

544 SPHERICAL TRANSMISSION LINES AND CAVITIES

It must be remembered, however, that the vector wave equation in spherical coordinates, as given by (3-72), does not reduce to three scalar Helmholtz wave equations as stated by (3-73a) through (3-73c). Therefore using the basic approach outlined in Chapter 6, Section 6.5, we cannot construct field solutions that are TE^r and/or TM^r. Hence we must look for other approaches for finding valid field configurations that are supported by spherical structures.

One approach that can be used to find field configurations that are TE^z and/or TM^z will be to represent the potential functions by [1]

$$\underline{TE^z}$$

$$\mathbf{A} = 0 \tag{10-1a}$$

$$\mathbf{F} = \hat{a}_z F_z = (\hat{a}_r \cos\theta - \hat{a}_\theta \sin\theta) F_z \tag{10-1b}$$

$$\underline{TM^z}$$

$$\mathbf{A} = \hat{a}_z A_z = (\hat{a}_r \cos\theta - \hat{a}_\theta \sin\theta) A_z \tag{10-2a}$$

$$\mathbf{F} = 0 \tag{10-2b}$$

where $F_z(r, \theta, \phi)$ and $A_z(r, \theta, \phi)$ are solutions to the scalar Helmholtz equation in the spherical coordinate system. Other field configurations can be formed as superpositions of TE^z and TM^z modes. Although this is a valid approach to the problem, it will not be pursued here; it is assigned to the reader as an end of chapter exercise. Instead an alternate procedure will be outlined for construction of TE^r and TM^r modes.

The procedure outlined in Chapter 6, Sections 6.5.2 and 6.5.3, for the construction, respectively, of TM and TE field configurations was based on the vector potentials **A** and **F** as derived in Sections 6.2 and 6.3, respectively, and summarized in Section 6.4. The final forms, which are summarized in Section 6.4, were based on the selection of the *Lorentz conditions* of (6-15) and (6-27), or

$$\psi_e = -\frac{1}{j\omega\mu\varepsilon} \nabla \cdot \mathbf{A} \tag{10-3a}$$

$$\psi_m = -\frac{1}{j\omega\mu\varepsilon} \nabla \cdot \mathbf{F} \tag{10-3b}$$

to represent, respectively, the scalar potential functions ψ_e and ψ_m. If that choice were not made, then the relations of the **E** and **H** fields to the potentials **A** and **F** would take a slightly different form. These forms will be outlined here.

10.2.1 The Vector Potential F (J = 0, M ≠ 0)

According to (6-19) the electric field is related to the potential **F** by

$$\boxed{\mathbf{E}_F = -\frac{1}{\varepsilon}\nabla \times \mathbf{F}} \tag{10-4}$$

Away from the source **M**, the electric and magnetic fields are related by Maxwell's equation

$$\nabla \times \mathbf{E}_F = -j\omega\mu \mathbf{H}_F \tag{10-5}$$

or

$$\boxed{\mathbf{H}_F = -\frac{1}{j\omega\mu}\nabla \times \mathbf{E}_F = \frac{1}{j\omega\mu\varepsilon}\nabla \times \nabla \times \mathbf{F}} \tag{10-5a}$$

Therefore if the potential **F** can be related to the source (**M**), then $\mathbf{E}_F$ and $\mathbf{H}_F$ can be found using (10-4) and (10-5a). We will attempt to do this next.

Taking the curl of both sides of (10-4) leads to

$$\nabla \times \mathbf{E}_F = -\frac{1}{\varepsilon}\nabla \times \nabla \times \mathbf{F} \tag{10-6}$$

Using Maxwell's equation 6-24, or

$$\nabla \times \mathbf{E}_F = -\mathbf{M} - j\omega\mu \mathbf{H}_F \tag{10-7}$$

and equating it to (10-6) leads to

$$\nabla \times \nabla \times \mathbf{F} = \varepsilon\mathbf{M} + j\omega\mu\varepsilon\mathbf{H}_F \tag{10-8}$$

The electric and magnetic fields are also related by Maxwell's equation 6-20, or

$$\nabla \times \mathbf{H}_F = j\omega\varepsilon\mathbf{E}_F \tag{10-9}$$

Substituting (10-4) into (10-9) and regrouping leads to

$$\nabla \times \mathbf{H}_F = j\omega\varepsilon\left(-\frac{1}{\varepsilon}\nabla \times \mathbf{F}\right) = -j\omega\nabla \times \mathbf{F} \tag{10-10}$$

or

$$\nabla \times (\mathbf{H}_F + j\omega\mathbf{F}) = 0 \tag{10-10a}$$

Using the vector identity of

$$\nabla \times (-\nabla\psi_m) = 0 \tag{10-11}$$

where ψ_m represents an arbitrary scalar potential, and equating it to (10-10a) leads to

$$\mathbf{H}_F + j\omega\mathbf{F} = -\nabla\psi_m \tag{10-12}$$

or

$$\mathbf{H}_F = -j\omega\mathbf{F} - \nabla\psi_m \tag{10-12a}$$

Substituting (10-12a) into (10-8) leads to

$$\nabla \times \nabla \times \mathbf{F} = \varepsilon\mathbf{M} + \omega^2\mu\varepsilon\mathbf{F} - j\omega\mu\varepsilon\nabla\psi_m \tag{10-13}$$

or

$$\boxed{\nabla \times \nabla \times \mathbf{F} - \omega^2\mu\varepsilon\mathbf{F} = \varepsilon\mathbf{M} - j\omega\mu\varepsilon\nabla\psi_m} \tag{10-13a}$$

This is the desired expression which relates the vector potential **F** to the source **M** and the associated scalar potential ψ_m. In a source-free (**M** = 0) region (10-13a) reduces to

$$\nabla \times \nabla \times \mathbf{F} - \omega^2\mu\varepsilon\mathbf{F} = -j\omega\mu\varepsilon\nabla\psi_m \qquad (10\text{-}14)$$

In a source-free region the procedure is to solve (10-14) for **F**, then to use (10-4) and (10-5a) to find, respectively, $\mathbf{E}_F$ and $\mathbf{H}_F$.

10.2.2 The Vector Potential A (J ≠ 0, M = 0)

Following a procedure similar to the one outlined in Section 10.2.1 for the vector potential **F**, it can be shown by referring also to Section 6.2 that the equations for the vector potential **A** analogous to (10-4), (10-5a), and (10-13a) are

$$\mathbf{H}_A = \frac{1}{\mu}\nabla \times \mathbf{A} \qquad (10\text{-}15a)$$

$$\mathbf{E}_A = \frac{1}{j\omega\varepsilon}\nabla \times \mathbf{H}_A = \frac{1}{j\omega\mu\varepsilon}\nabla \times \nabla \times \mathbf{A} \qquad (10\text{-}15b)$$

$$\nabla \times \nabla \times \mathbf{A} - \omega^2\mu\varepsilon\mathbf{A} = \mu\mathbf{J} - j\omega\mu\varepsilon\nabla\psi_e \qquad (10\text{-}15c)$$

In a source-free region (10-15c) reduces to

$$\nabla \times \nabla \times \mathbf{A} - \omega^2\mu\varepsilon\mathbf{A} = -j\omega\mu\varepsilon\nabla\psi_e \qquad (10\text{-}15d)$$

The details are left as end of chapter exercises for the reader.

In a source-free region the procedure is to solve (10-15d) for **A**, then to use (10-15a) and (10-15b) to find, respectively, $\mathbf{H}_A$ and $\mathbf{E}_A$.

10.2.3 The Vector Potentials F and A

The total fields that are due to both potentials **F** and **A** are found as superpositions of the fields of Sections 10.2.1 and 10.2.2. Doing this we have that the total fields are obtained using

$$\mathbf{E} = \mathbf{E}_F + \mathbf{E}_A = -\frac{1}{\varepsilon}\nabla \times \mathbf{F} + \frac{1}{j\omega\varepsilon}\nabla \times \mathbf{H}_A = -\frac{1}{\varepsilon}\nabla \times \mathbf{F} + \frac{1}{j\omega\mu\varepsilon}\nabla \times \nabla \times \mathbf{A} \qquad (10\text{-}16a)$$

$$\mathbf{H} = \mathbf{H}_F + \mathbf{H}_A = -\frac{1}{j\omega\mu}\nabla \times \mathbf{E}_F + \frac{1}{\mu}\nabla \times \mathbf{A} = \frac{1}{j\omega\mu\varepsilon}\nabla \times \nabla \times \mathbf{F} + \frac{1}{\mu}\nabla \times \mathbf{A} \qquad (10\text{-}16b)$$

where **F** and **A** are, respectively, solutions to

$$\nabla \times \nabla \times \mathbf{F} - \omega^2\mu\varepsilon\mathbf{F} = \varepsilon\mathbf{M} - j\omega\mu\varepsilon\nabla\psi_m \quad (10\text{-}16c)$$

$$\nabla \times \nabla \times \mathbf{A} - \omega^2\mu\varepsilon\mathbf{A} = \mu\mathbf{J} - j\omega\mu\varepsilon\nabla\psi_e \quad (10\text{-}16d)$$

which for a source-free region (**M** = **J** = 0) reduce to

$$\nabla \times \nabla \times \mathbf{F} - \omega^2\mu\varepsilon\mathbf{F} = -j\omega\mu\varepsilon\nabla\psi_m \quad (10\text{-}16e)$$

$$\nabla \times \nabla \times \mathbf{A} - \omega^2\mu\varepsilon\mathbf{A} = -j\omega\mu\varepsilon\nabla\psi_e \quad (10\text{-}16f)$$

We will attempt now to form TE^r and TM^r mode field solutions using (10-16a) through (10-16f).

10.2.4 Transverse Electric (TE) Modes: Source-Free Region

It was stated previously in Section 6.5.3 that TE modes to any direction in any coordinate system can be obtained by selecting the vector potential **F** to have only a nonvanishing component in that direction while simultaneously letting **A** = 0. The nonvanishing component of **F** was obtained as a solution to the scalar wave equation 6-31. The same procedure will be used here except that instead of the nonvanishing component of **F** being a solution to (6-31), which does not reduce in spherical coordinates to three scalar noncoupled wave equations, it will be a solution to (10-16e). The nonvanishing component of **F** will be the one that coincides with the direction along which the TE modes are desired. Let us construct solutions that are TE^r in a spherical coordinate system.

TE^r field configurations are constructed by letting the vector potentials **F** and **A** be equal to

TE^r

$$\mathbf{F} = \hat{a}_r F_r(r, \theta, \phi) \quad (10\text{-}17a)$$

$$\mathbf{A} = 0 \quad (10\text{-}17b)$$

Since F_r is not a solution to the scalar Hemlholtz equation

$$\nabla^2 \mathbf{F} = \nabla^2(\hat{a}_r F_r) \neq \hat{a}_r \nabla^2 F_r \quad (10\text{-}18)$$

we will resort, for a source-free region, to (10-16e).

Expanding (10-16e) using (10-17a) leads to

$$\nabla \times \mathbf{F} = \nabla \times (\hat{a}_r F_r) = \hat{a}_\theta \frac{1}{r \sin\theta} \frac{\partial F_r}{\partial \phi} - \hat{a}_\phi \frac{1}{r} \frac{\partial F_r}{\partial \theta} \quad (10\text{-}19a)$$

$$\nabla \times \nabla \times \mathbf{F} = \hat{a}_r \left\{ \frac{1}{r \sin\theta} \left[\frac{\partial}{\partial \theta}\left(-\frac{\sin\theta}{r}\frac{\partial F_r}{\partial \theta}\right) - \frac{\partial}{\partial \phi}\left(\frac{1}{r\sin\theta}\frac{\partial F_r}{\partial \phi}\right)\right]\right\}$$
$$+ \hat{a}_\theta \left[\frac{1}{r}\left(\frac{\partial^2 F_r}{\partial r \partial \theta}\right)\right] + \hat{a}_\phi \left(\frac{1}{r\sin\theta}\frac{\partial^2 F_r}{\partial r \partial \phi}\right) \quad (10\text{-}19b)$$

$$\nabla \psi_m = \hat{a}_r \frac{\partial \psi_m}{\partial r} + \hat{a}_\theta \frac{1}{r}\frac{\partial \psi_m}{\partial \theta} + \hat{a}_\phi \frac{1}{r\sin\theta}\frac{\partial \psi_m}{\partial \phi} \quad (10\text{-}19c)$$

Thus for the r, θ, and ϕ components (10-16e) reduces to

$$\boxed{\frac{1}{r\sin\theta}\left[-\frac{\partial}{\partial\theta}\left(\frac{\sin\theta}{r}\frac{\partial F_r}{\partial\theta}\right) - \frac{\partial}{\partial\phi}\left(\frac{1}{r\sin\theta}\frac{\partial F_r}{\partial\phi}\right)\right] - \beta^2 F_r = -j\omega\mu\varepsilon\frac{\partial\psi_m}{\partial r}}$$
$$(10\text{-}20a)$$

$$\frac{1}{r}\frac{\partial^2 F_r}{\partial r \partial \theta} = -j\frac{\omega\mu\varepsilon}{r}\frac{\partial\psi_m}{\partial\theta} \Rightarrow \boxed{\frac{\partial^2 F_r}{\partial r \partial \theta} = \frac{\partial}{\partial\theta}\left(\frac{\partial F_r}{\partial r}\right) = \frac{\partial}{\partial\theta}(-j\omega\mu\varepsilon\psi_m)} \quad (10\text{-}20b)$$

$$\frac{1}{r\sin\theta}\frac{\partial^2 F_r}{\partial r \partial \phi} = -j\frac{\omega\mu\varepsilon}{r\sin\theta}\frac{\partial\psi_m}{\partial\phi} \Rightarrow \boxed{\frac{\partial^2 F_r}{\partial r \partial \phi} = \frac{\partial}{\partial\phi}\left(\frac{\partial F_r}{\partial r}\right) = \frac{\partial}{\partial\phi}(-j\omega\mu\varepsilon\psi_m)}$$
$$(10\text{-}20c)$$

where $\beta^2 = \omega^2\mu\varepsilon$. The last two equations, (10-20b) and (10-20c), are satisfied simultaneously if

$$\frac{\partial F_r}{\partial r} = -j\omega\mu\varepsilon\psi_m \Rightarrow \boxed{\psi_m = -\frac{1}{j\omega\mu\varepsilon}\frac{\partial F_r}{\partial r}} \quad (10\text{-}21)$$

With the preceding relation for the scalar potential ψ_m, we need to find an *uncoupled* differential equation for F_r. To do this we substitute (10-21) into (10-20a), which leads to

$$-\frac{1}{r^2\sin\theta}\frac{\partial}{\partial\theta}\left(\sin\theta\frac{\partial F_r}{\partial\theta}\right) - \frac{1}{r^2\sin^2\theta}\frac{\partial^2 F_r}{\partial\phi^2} - \beta^2 F_r = \frac{\partial^2 F_r}{\partial r^2} \quad (10\text{-}22)$$

or

$$\frac{\partial^2 F_r}{\partial r^2} + \frac{1}{r^2\sin\theta}\frac{\partial}{\partial\theta}\left(\sin\theta\frac{\partial F_r}{\partial\theta}\right) + \frac{1}{r^2\sin^2\theta}\frac{\partial^2 F_r}{\partial\phi^2} + \beta^2 F_r = 0 \quad (10\text{-}22a)$$

and can also be written in succinct form as

$$\boxed{(\nabla^2 + \beta^2)\frac{F_r}{r} = 0} \quad (10\text{-}22b)$$

Therefore using this procedure the ratio of F_r/r satisfies the scalar Helmholtz wave equation and not F_r itself. A solution of F_r using (10-22b) allows us to find $\mathbf{E}_F$ and $\mathbf{H}_F$ using, respectively, (10-4) and (10-5a).

The solution of (10-22b) will be pursued in Section 10.2.6. In the meantime the electric and magnetic field components can be written in terms of F_r by expanding (10-4) and (10-5a):

$$\text{TE}' \ (\mathbf{F} = \hat{a}_r F_r, \mathbf{A} = 0)$$

$$\boxed{\mathbf{E}_F = -\frac{1}{\varepsilon} \nabla \times \mathbf{F}} \tag{10-23}$$

or

$$E_r = 0 \tag{10-23a}$$

$$E_\theta = -\frac{1}{\varepsilon} \frac{1}{r \sin\theta} \frac{\partial F_r}{\partial \phi} \tag{10-23b}$$

$$E_\phi = \frac{1}{\varepsilon} \frac{1}{r} \frac{\partial F_r}{\partial \theta} \tag{10-23c}$$

$$\boxed{\mathbf{H}_F = \frac{1}{j\omega\mu\varepsilon} \nabla \times \nabla \times \mathbf{F}} \tag{10-24}$$

or

$$H_r = \frac{1}{j\omega\mu\varepsilon} \left(\frac{\partial^2}{\partial r^2} + \beta^2 \right) F_r \tag{10-24a}$$

$$H_\theta = \frac{1}{j\omega\mu\varepsilon} \frac{1}{r} \frac{\partial^2 F_r}{\partial r \partial \theta} \tag{10-24b}$$

$$H_\phi = \frac{1}{j\omega\mu\varepsilon} \frac{1}{r \sin\theta} \frac{\partial^2 F_r}{\partial r \partial \phi} \tag{10-24c}$$

where F_r/r is a solution to (10-22b).

10.2.5 Transverse Magnetic (TM) Modes: Source-Free Region

Following a procedure similar to the one outlined in the previous section for the TE' modes, it can be shown that the TM' fields in spherical coordinates can be constructed by letting the vector potentials $\mathbf{F}$ and $\mathbf{A}$ be equal to

$$\boxed{\mathbf{F} = 0} \tag{10-25a}$$

$$\boxed{\mathbf{A} = \hat{a}_r A_r(r, \theta, \phi)} \tag{10-25b}$$

where the ratio of A_r/r, and not A_r, is a solution to the scalar Helmholtz wave equation

$$\boxed{(\nabla^2 + \beta^2) \frac{A_r}{r} = 0} \tag{10-26}$$

The solution of (10-26) will be pursued in Section 10.2.6. In the meantime the electric and magnetic field components can be written in terms of A_r by expanding (10-15b) and (10-15a) as

$$\text{TM}^r \ (\mathbf{F} = 0, \mathbf{A} = \hat{a}_r A_r)$$

$$\boxed{\mathbf{E}_A = \frac{1}{j\omega\mu\varepsilon} \nabla \times \nabla \times \mathbf{A}} \tag{10-27}$$

or

$$E_r = \frac{1}{j\omega\mu\varepsilon}\left(\frac{\partial^2}{\partial r^2} + \beta^2\right)A_r \tag{10-27a}$$

$$E_\theta = \frac{1}{j\omega\mu\varepsilon}\frac{1}{r}\frac{\partial^2 A_r}{\partial r \partial \theta} \tag{10-27b}$$

$$E_\phi = \frac{1}{j\omega\mu\varepsilon}\frac{1}{r\sin\theta}\frac{\partial^2 A_r}{\partial r \partial \phi} \tag{10-27c}$$

$$\boxed{\mathbf{H}_A = \frac{1}{\mu}\nabla \times \mathbf{A}} \tag{10-28}$$

or

$$H_r = 0 \tag{10-28a}$$

$$H_\theta = \frac{1}{\mu}\frac{1}{r\sin\theta}\frac{\partial A_r}{\partial \phi} \tag{10-28b}$$

$$H_\phi = -\frac{1}{\mu}\frac{1}{r}\frac{\partial A_r}{\partial \theta} \tag{10-28c}$$

where A_r/r is a solution to (10-26).

10.2.6 Solution of the Scalar Helmholtz Wave Equation

To find the TE^r and/or TM^r field of Sections 10.2.4 and 10.2.5 as given, respectively, by (10-23) through (10-24c) and (10-27) through (10-28c), solutions to the scalar Helmholtz wave equations 10-22b and 10-26 must be obtained for F_r/r and A_r/r (and thus F_r and A_r). Both solutions are of the same form

$$(\nabla^2 + \beta^2)\psi = 0 \tag{10-29}$$

where

$$\psi = \begin{cases} \dfrac{F_r}{r} & \text{for TE}^r \text{ modes} \qquad (10\text{-}29\text{a}) \\[6pt] \dfrac{A_r}{r} & \text{for TM}^r \text{ modes} \qquad (10\text{-}29\text{b}) \end{cases}$$

Since the solution of ψ from (10-29) must be multiplied by r to obtain solutions for F_r or A_r, then appropriate solutions for F_r and A_r must be equal to the product of $r\psi$. The solution for F_r or A_r of (10-29) through (10-29b) must take the separable form of

$$\left.\begin{array}{l} F_r(r,\theta,\phi) \\ A_r(r,\theta,\phi) \end{array}\right\} = f(r)g(\theta)h(\phi) \tag{10-30}$$

where $f(r)$, $g(\theta)$, and $h(\phi)$ must be represented by appropriate wave functions that satisfy the wave equation in spherical coordinates. According to (3-88a) or (3-88b) the $g(\theta)$ can be represented by associated Legendre functions of the first $P_n^m(\cos\theta)$ kind or second $Q_n^m(\cos\theta)$ kind whereas $h(\phi)$ can be represented by either complex exponentials or cosinusoids as given, respectively, by (3-89a) and (3-89b).

Since the solution of ψ as given by (10-29) must be multiplied by r in order to obtain solutions to F_r and A_r as given by (10-30), it is most convenient to represent $f(r)$ *not* by spherical Bessel $[j_n(\beta r), y_n(\beta r)]$ or Hankel $[h_n^{(1)}(\beta r), h_n^{(2)}(\beta r)]$ functions, *but by another form* of spherical Bessel and Hankel functions denoted by $\hat{B}_n(\beta r)$ [for either $\hat{J}_n(\beta r)$, $\hat{Y}_n(\beta r)$, $\hat{H}_n^{(1)}(\beta r)$, or $\hat{H}_n^{(2)}(\beta r)$]. These are related to the regular spherical Bessel and Hankel functions denoted by $b_n(\beta r)$ [for either $j_n(\beta r)$, $y_n(\beta r)$, $h_n^{(1)}(\beta r)$, or $h_n^{(2)}(\beta r)$] by

$$\hat{B}_n(\beta r) = \beta r \, b_n(\beta r) = \beta r \sqrt{\frac{\pi}{2\beta r}} B_{n+1/2}(\beta r) = \sqrt{\frac{\pi \beta r}{2}} B_{n+1/2}(\beta r) \tag{10-31}$$

where $B_{n+1/2}(\beta r)$ is used to represent the regular cylindrical Bessel or Hankel functions of $J_{n+1/2}(\beta r)$, $Y_{n+1/2}(\beta r)$, $H_{n+1/2}^{(1)}(\beta r)$, and $H_{n+1/2}^{(2)}(\beta r)$. These new spherical Bessel and Hankel functions were introduced by Schelkunoff [2] and satisfy the differential equation

$$\left[\frac{d^2}{dr^2} + \beta^2 - \frac{n(n+1)}{r^2}\right]\hat{B}_n = 0 \tag{10-32}$$

which is obtained by substituting $b_n(\beta r) = \hat{B}_n(\beta r)/\beta r$ in

$$\frac{d}{dr}\left(r^2 \frac{db_n}{dr}\right) + \left[(\beta r)^2 - n(n+1)\right]b_n = 0 \tag{10-33}$$

Therefore the solutions for $f(r)$ of (10-30) are of the new form of spherical Bessel or Hankel functions denoted by

$$f_1(r) = A_1 \hat{J}_n(\beta r) + B_1 \hat{Y}_n(\beta r) \tag{10-34a}$$

or

$$f_2(r) = C_1 \hat{H}_n^{(1)}(\beta r) + D_1 \hat{H}_n^{(2)}(\beta r) \tag{10-34b}$$

552 SPHERICAL TRANSMISSION LINES AND CAVITIES

which are related to the regular Bessel and Hankel functions by

$$\hat{J}_n(\beta r) = \sqrt{\frac{\pi \beta r}{2}} J_{n+1/2}(\beta r) \tag{10-35a}$$

$$\hat{Y}_n(\beta r) = \sqrt{\frac{\pi \beta r}{2}} Y_{n+1/2}(\beta r) \tag{10-35b}$$

$$\hat{H}_n^{(1)}(\beta r) = \sqrt{\frac{\pi \beta r}{2}} H_{n+1/2}^{(1)}(\beta r) \tag{10-35c}$$

$$\hat{H}_n^{(2)}(\beta r) = \sqrt{\frac{\pi \beta r}{2}} H_{n+1/2}^{(2)}(\beta r) \tag{10-35d}$$

The total solution for F_r or A_r of (10-30) will be the product of the appropriate spherical wave functions representing $f(r)$, $g(\theta)$, and $h(\phi)$.

10.3 BICONICAL TRANSMISSION LINE

One form of a transmission line whose geometry conforms to the spherical orthogonal coordinate system is the biconical structure of Figure 10-1. Typically this configuration is also representative of the biconical antenna [3–9] which exhibits very broad band frequency characteristics. Sets of fields that can be supported by such a structure can be either TE^r, TM^r, or TEM^r. Solutions to these will be examined here.

10.3.1 Transverse Electric (TE^r) Modes

According to the procedure established in Section 10.2.4, transverse electric (to the radial direction) modes (TE^r) can be constructed by choosing the potentials **F** and **A** according to (10-17a) and (10-17b). The scalar component F_r of the vector

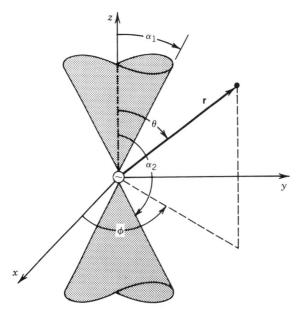

FIGURE 10-1 Geometry of biconical transmission line (*Source:* C. A. Balanis, *Antenna Theory; Analysis and Design.* Copyright © 1982, John Wiley & Sons, Inc. Reprinted by Permission of John Wiley & Sons, Inc.).

potential must satisfy the vector wave equation of (10-22b) whose solution takes the form of (3-85), or

$$F_r(r, \theta, \phi) = f(r)g(\theta)h(\phi) \tag{10-36}$$

where $f(r) = $ a solution to (10-32) as given by either (10-34a) or (10-34b) [the form of (10-34b) is chosen here]

$g(\theta) = $ a solution of (3-86b) as given by either (3-88a) or (3-88b) [the form of (3-88a) is chosen here]

$h(\phi) = $ is a solution to (3-86c) as given by (3-89a) or (3-89b) [the form of (3-89b) is chosen here to represent standing waves]

Therefore F_r of (10-36) can be written, assuming the source is placed at the apex and is generating outwardly traveling waves [$C_1 = 0$ in (10-34b)], as

$$[F_r(r, \theta, \phi)]_{mn} = D_1 \hat{H}_n^{(2)}(\beta r)[A_2 P_n^m(\cos\theta) + B_2 P_n^m(-\cos\theta)] \\ \times [C_3 \cos(m\phi) + D_3 \sin(m\phi)] \tag{10-37}$$

where $m = $ integer ($m = 0, 1, 2, \ldots$)

The corresponding electric and magnetic fields can be found using (10-23) through (10-24c) and the eigenvalues of n can be determined by applying the boundary conditions

$$E_\phi(0 \le r \le \infty, \theta = \alpha_1, 0 \le \phi \le 2\pi) = E_\phi(0 \le r \le \infty, \theta = \alpha_2, 0 \le \phi \le 2\pi) = 0 \tag{10-38}$$

According to (10-23c)

$$E_\phi = \frac{1}{\varepsilon}\frac{1}{r}\frac{\partial F_r}{\partial \theta} = \frac{D_1}{\varepsilon}\frac{1}{r}\hat{H}_n^{(2)}(\beta r)\left[A_2 \frac{dP_n^m(\cos\theta)}{d\theta} + B_2 \frac{dP_n^m(-\cos\theta)}{d\theta}\right] \\ \times [C_3 \cos(m\phi) + D_3 \sin(m\phi)] \tag{10-39}$$

Applying the first boundary condition of (10-38) leads to

$$E_\phi(0 \le r \le \infty, \theta = \alpha_1, 0 \le \phi \le 2\pi) \\ = \frac{D_1}{\varepsilon}\frac{1}{r}\hat{H}_n^{(2)}(\beta r)\left[A_2 \frac{dP_n^m(\cos\theta)}{d\theta} + B_2 \frac{dP_n^m(-\cos\theta)}{d\theta}\right]_{\theta=\alpha_1} \\ \times [C_3 \cos(m\phi) + D_3 \sin(m\phi)] = 0 \\ E_\phi(0 \le r \le \infty, \theta = \alpha_1, 0 \le \phi \le 2\pi) \\ = \frac{D_1}{\varepsilon}\frac{1}{r}\hat{H}_n^{(2)}(\beta r)\left[A_2 \frac{dP_n^m(\cos\alpha_1)}{d\alpha_1} + B_2 \frac{dP_n^m(-\cos\alpha_1)}{d\alpha_1}\right] \\ \times [C_3 \cos(m\phi) + D_3 \sin(m\phi)] = 0 \tag{10-40a}$$

and the second boundary condition of (10-38) leads to

$$E_\phi(0 \le r \le \infty, \theta = \alpha_2, 0 \le \phi \le 2\pi) \\ = \frac{D_1}{\varepsilon}\frac{1}{r}\hat{H}_n^{(2)}(\beta r)\left[A_2 \frac{dP_n^m(\cos\alpha_2)}{d\alpha_2} + B_2 \frac{dP_n^m(-\cos\alpha_2)}{d\alpha_2}\right] \\ \times [C_3 \cos(m\phi) + D_3 \sin(m\phi)] = 0 \tag{10-40b}$$

Equations 10-40a and 10-40b reduce to

$$A_2 \frac{dP_n^m(\cos \alpha_1)}{d\alpha_1} + B_2 \frac{dP_n^m(-\cos \alpha_1)}{d\alpha_1} = 0 \quad (10\text{-}41\text{a})$$

$$A_2 \frac{dP_n^m(\cos \alpha_2)}{d\alpha_2} + B_2 \frac{dP_n^m(-\cos \alpha_2)}{d\alpha_2} = 0 \quad (10\text{-}41\text{b})$$

which are satisfied provided the determinant of (10-41a) and (10-41b) vanishes, that is

$$\boxed{\frac{dP_n^m(\cos \alpha_1)}{d\alpha_1} \frac{dP_n^m(-\cos \alpha_2)}{d\alpha_2} - \frac{dP_n^m(-\cos \alpha_1)}{d\alpha_1} \frac{dP_n^m(\cos \alpha_2)}{d\alpha_2} = 0} \quad (10\text{-}42)$$

Therefore the eigenvalues of n are found as solutions to (10-42), which usually is not a very easy task.

10.3.2 Transverse Magnetic (TM′) Modes

Following a procedure similar to that of the previous section and using the formulations of Section 10.2.5, it can be shown that for TM′ modes the potential component A_r of (10-26) reduces to

$$[A_r(r, \theta, \phi)]_{mn} = D_1 \hat{H}_n^{(2)}(\beta r)[A_2 P_n^m(\cos \theta) + B_2 P_n^m(-\cos \theta)]$$
$$\times [C_3 \cos(m\phi) + D_3 \sin(m\phi)] \quad (10\text{-}43)$$

where m = integer ($m = 0, 1, 2, \ldots$). The values of n are determined by applying the boundary conditions.

The corresponding electric and magnetic fields are obtained using (10-27) through (10-28c). By applying the boundary conditions of

$$E_r(0 \le r \le \infty, \theta = \alpha_1, 0 \le \phi \le 2\pi) = E_r(0 \le r \le \infty, \theta = \alpha_2, 0 \le \phi \le 2\pi) = 0$$
$$(10\text{-}44\text{a})$$

or

$$E_\phi(0 \le r \le \infty, \theta = \alpha_1, 0 \le \phi \le 2\pi) = E_\phi(0 \le r \le \infty, \theta = \alpha_2, 0 \le \phi \le 2\pi) = 0$$
$$(10\text{-}44\text{b})$$

it can be shown that the eigenvalues of n are obtained as solutions to

$$\boxed{P_n^m(\cos \alpha_1) P_n^m(-\cos \alpha_2) - P_n^m(-\cos \alpha_1) P_n^m(\cos \alpha_2) = 0} \quad (10\text{-}45)$$

This usually is not a very easy task.

10.3.3 Transverse Electromagnetic (TEM′) Modes

The lowest-order (dominant) mode of the biconical transmission line is the one for which $m = 0$ and $n = 0$. For this mode both (10-42) and (10-45) are satisfied and the potential components of (10-37) and (10-43) vanish. However, for $m = n = 0$

(10-43) could be redefined as the limit as $n \to 0$. Instead it is usually more convenient to alternately represent the TEM mode by the TM_{00} which is defined, using (3-88b) to represent $g(\theta)$, by [1]

$$(A_r)_{00} = B_{00}\hat{H}_0^{(2)}(\beta r)Q_0(\cos\theta) \tag{10-46}$$

since $P_0^0(\cos\theta) = P_0(\cos\theta) = 1$. The Legendre polynomial $Q_0(\cos\theta)$ can also be represented by

$$Q_0(\cos\theta) = \ln\left[\cot\left(\frac{\theta}{2}\right)\right] \tag{10-47a}$$

and the spherical Hankel function $\hat{H}_0^{(2)}(\beta r)$ can be replaced by its asymptotic form for large arguments of

$$\hat{H}_0^{(2)}(\beta r) \underset{\beta r \to \text{large}}{\simeq} je^{-j\beta r} \tag{10-47b}$$

Using (10-47a) and (10-47b) reduces (10-46) for large observational distances ($\beta r \to$ large) to

$$(A_r)_{00} \simeq jB_{00}\ln\left[\cot\left(\frac{\theta}{2}\right)\right]e^{-j\beta r} \tag{10-48}$$

The corresponding electric and magnetic field components are given, according to (10-27) through (10-28c), by [3]

$$E_r = \frac{1}{j\omega\mu\varepsilon}\left(\frac{\partial^2}{\partial r^2} + \beta^2\right)A_r \simeq 0 \tag{10-49a}$$

$$E_\theta = \frac{1}{j\omega\mu\varepsilon}\frac{1}{r}\frac{\partial^2 A_r}{\partial r \partial\theta} = jB_{00}\frac{\beta}{\omega\mu\varepsilon}\frac{1}{r}\frac{1}{\sin\theta}e^{-j\beta r} \tag{10-49b}$$

$$E_\phi = \frac{1}{j\omega\mu\varepsilon}\frac{1}{r\sin\theta}\frac{\partial^2 A_r}{\partial r \partial\phi} = 0 \tag{10-49c}$$

$$H_r = 0 \tag{10-49d}$$

$$H_\theta = \frac{1}{\mu}\frac{1}{r\sin\theta}\frac{\partial A_r}{\partial\phi} = 0 \tag{10-49e}$$

$$H_\phi = -\frac{1}{\mu}\frac{1}{r}\frac{\partial A_r}{\partial\theta} = jB_{00}\frac{1}{\mu r\sin\theta}e^{-j\beta r} \tag{10-49f}$$

Using these equations we can write the wave impedance in the radial direction as

$$Z_w^{+r} = \frac{E_\theta}{H_\phi} = \frac{\beta}{\omega\varepsilon} = \sqrt{\frac{\mu}{\varepsilon}} = \eta \tag{10-50}$$

which is the same as the intrinsic impedance of the medium.

An impedance of greater interest is the characteristic impedance that is defined in terms of voltages and currents. The voltage between two corresponding points on the cones, a distance r from the origin, is found by

$$V(r) = \int_{\alpha_1}^{\alpha_2} \mathbf{E} \cdot d\ell = \int_{\alpha_1}^{\alpha_2} (\hat{a}_\theta E_\theta) \cdot (\hat{a}_\theta r \, d\theta)$$

$$= \int_{\alpha_1}^{\alpha_2} E_\theta r \, d\theta = jB_{00} \frac{\beta e^{-j\beta r}}{\omega\mu\varepsilon} \int_{\alpha_1}^{\alpha_2} \frac{d\theta}{\sin(\theta)}$$

$$= jB_{00} \frac{\beta e^{-j\beta r}}{\omega\mu\varepsilon} \ln\left[\frac{\cot\left(\frac{\alpha_1}{2}\right)}{\cot\left(\frac{\alpha_2}{2}\right)}\right] \tag{10-51a}$$

The current on the surface of the cones, a distance r from the origin, is found by using (10-49f) as

$$I(r) = \oint_C \mathbf{H} \cdot d\ell = \int_0^{2\pi} (\hat{a}_\phi H_\phi) \cdot (\hat{a}_\phi r \sin\theta \, d\phi) = \int_0^{2\pi} H_\phi r \sin\theta \, d\phi = jB_{00} \frac{2\pi e^{-j\beta r}}{\mu} \tag{10-51b}$$

Taking the ratio of (10-51a) to (10-51b) we can define and write the characteristic impedance as

$$Z_c \equiv \frac{V(r)}{I(r)} = \frac{\beta}{2\pi\omega\varepsilon} \ln\left[\frac{\cot\left(\frac{\alpha_1}{2}\right)}{\cot\left(\frac{\alpha_2}{2}\right)}\right] = \frac{\sqrt{\frac{\mu}{\varepsilon}}}{2\pi} \ln\left[\frac{\cot\left(\frac{\alpha_1}{2}\right)}{\cot\left(\frac{\alpha_2}{2}\right)}\right] \equiv Z_{in} \tag{10-52}$$

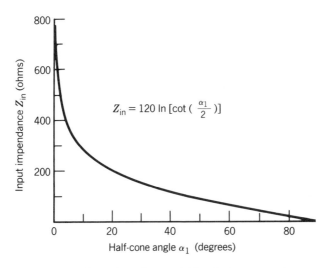

FIGURE 10-2 Input impedance of biconical transmission line (*Source:* C. A. Balanis, *Antenna Theory; Analysis and Design.* Copyright © 1982, John Wiley & Sons, Inc. Reprinted by Permission of John Wiley & Sons, Inc.).

Since the characteristic impedance is not a function of the radial distance r, it also represents the input impedance of the antenna at the feed terminals. For a symmetrical structure ($\alpha_2 = \pi - \alpha_1$) (10-52) reduces to

$$Z_c = \frac{\sqrt{\frac{\mu}{\varepsilon}}}{2\pi} \ln\left[\cot\left(\frac{\alpha_1}{2}\right)\right]^2 = \frac{\sqrt{\frac{\mu}{\varepsilon}}}{\pi} \ln\left[\cot\left(\frac{\alpha_1}{2}\right)\right] = \frac{\eta}{\pi} \ln\left[\cot\left(\frac{\alpha_1}{2}\right)\right] = Z_{in}$$

(10-52a)

It is apparent that the transmission line or alternately the antenna of Figure 10-1 is a very broad band structure since its characteristic or input impedance is only a function of the included angle of the cone. A plot of (10-52a) as a function of α_1 is shown in Figure 10-2.

There are numerous other transmission lines whose geometry can be represented by the spherical orthogonal coordinate systems. They will not be discussed here but some will be assigned to the reader as end of chapter exercises.

10.4 THE SPHERICAL CAVITY

The metallic spherical cavity of Figure 10-3 represents a popular and classic geometry to design resonators. The field configurations that can be supported by such a structure can be TE^r and/or TM^r; both will be examined here. In addition to the field expressions, the resonant frequencies and the quality factors will be the quantities of interest.

10.4.1 Transverse Electric (TE^r) Modes

The TE^r modes in the cavity can be formed by letting the vector potentials **F** and **A** be equal to (10-17a) and (10-17b), respectively. The most appropriate form for the vector potential component F_r is

$$F_r(r, \theta, \phi) = \left[A_1 \hat{J}_n(\beta r) + B_1 \hat{Y}_n(\beta r)\right]\left[C_2 P_n^m(\cos\theta) + D_2 Q_n^m(\cos\theta)\right]$$
$$\times \left[C_3 \cos(m\phi) + D_3 \sin(m\phi)\right] \quad (10\text{-}53)$$

where m and n are integers. The fields must be finite at $r = 0$; thus $B_1 = 0$ since

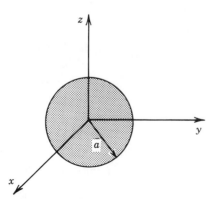

FIGURE 10-3 Geometry of spherical cavity.

$\hat{Y}_n(\beta r)$ possesses a singularity at $r = 0$. Additionally the fields must also be finite at $\theta = 0, \pi$. Therefore $D_2 = 0$ since $Q_n^m(\cos\theta)$ possesses a singularity at $\theta = 0, \pi$. Also because the Legendre polynomial $P_0(w) = 1$, and $P_0^m(w)$ is related to $P_0(w)$ by [10-15]

$$P_0^m(w) = (-1)^m (1-w^2)^{m/2} \frac{d^m P_0(w)}{dw^m} = 0, m = 1, 2, \ldots \quad (10\text{-}54)$$

then for nontrivial (nonzero) solutions $n = 1, 2, 3, \ldots$. Therefore (10-53) reduces to

$$(F_r)_{mnp} = A_{mnp} \hat{J}_n(\beta r) P_n^m(\cos\theta)[C_3 \cos(m\phi) + D_3 \sin(m\phi)] \quad (10\text{-}55)$$

It should also be stated that $P_n^m(\cos\theta) = 0$ if $m > n$.

The corresponding electric and magnetic fields are found using (10-23) through (10-24c). Thus we can write the electric field components of (10-23) through (10-23c), using (10-55), as

$$E_r = 0 \quad (10\text{-}56a)$$

$$E_\theta = -\frac{1}{\varepsilon}\frac{1}{r\sin\theta}\frac{\partial F_r}{\partial \phi}$$

$$= -A_{mnp}\frac{m}{\varepsilon}\frac{1}{r\sin\theta}\hat{J}_n(\beta r)P_n^m(\cos\theta)[-C_3\sin(m\phi) + D_3\cos(m\phi)] \quad (10\text{-}56b)$$

$$E_\phi = \frac{1}{\varepsilon}\frac{1}{r}\frac{\partial F_r}{\partial\theta} = A_{mnp}\frac{1}{\varepsilon}\frac{1}{r}\hat{J}_n(\beta r)P_n^{\prime m}(\cos\theta)[C_3\cos(m\phi) + D_3\sin(m\phi)]$$

$$\quad (10\text{-}56c)$$

where

$$' = \frac{\partial}{\partial\theta} \quad (10\text{-}56d)$$

The boundary conditions that must be satisfied are

$$E_\theta(r = a, 0 \le \theta \le \pi, 0 \le \phi \le 2\pi) = 0 \quad (10\text{-}57a)$$

$$E_\phi(r = a, 0 \le \theta \le \pi, 0 \le \phi \le 2\pi) = 0 \quad (10\text{-}57b)$$

Either condition yields the same eigenfunction and corresponding eigenvalues.

Applying (10-57a) to (10-56b) leads to

$$E_\theta(r = a, 0 \le \theta \le \pi, 0 \le \phi \le 2\pi) = -A_{mnp}\frac{m}{\varepsilon}\frac{1}{a\sin\theta}\hat{J}_n(\beta a)P_n^m(\cos\theta)$$

$$\times [-C_3\sin(m\phi) + D_3\cos(m\phi)] = 0 \Rightarrow \hat{J}_n(\beta a)|_{\beta=\beta_r} = 0 \Rightarrow \beta_r a = \zeta_{np}$$

$$\beta_r = \frac{\zeta_{np}}{a} \quad \begin{array}{l} n = 1, 2, 3, \ldots \\ p = 1, 2, 3, \ldots \end{array} \quad (10\text{-}58)$$

where ζ_{np} represents the p zeroes of the spherical Bessel function of order n. A listing of a limited, but for most applications sufficient, number of ζ_{np}'s is found in Table 10-1.

TABLE 10-1
Zeroes ζ_{np} of spherical Bessel function $\hat{J}_n(\zeta_{np}) = 0$

	n = 1	n = 2	n = 3	n = 4	n = 5	n = 6	n = 7	n = 8
p = 1	4.493	5.763	6.988	8.183	9.356	10.513	11.657	12.791
p = 2	7.725	9.095	10.417	11.705	12.967	14.207	15.431	16.641
p = 3	10.904	12.323	13.698	15.040	16.355	17.648	18.923	20.182
p = 4	14.066	15.515	16.924	18.301	19.653	20.983	22.295	
p = 5	17.221	18.689	20.122	21.525	22.905			
p = 6	20.371	21.854						

The resonant frequencies are found using (10-58) and can be written as

$$\beta_r = \omega_r\sqrt{\mu\varepsilon} = 2\pi f_r\sqrt{\mu\varepsilon} = \frac{\zeta_{np}}{a} \quad (10\text{-}59)$$

or

$$(f_r)_{mnp}^{TE'} = \frac{\zeta_{np}}{2\pi a\sqrt{\mu\varepsilon}} \quad \begin{array}{l} m = 0, 1, 2, \ldots \leq n \\ n = 1, 2, 3, \ldots \\ p = 1, 2, 3, \ldots \end{array} \quad (10\text{-}59a)$$

Since the resonant frequencies of (10-59a) obtained using the ζ_{np}'s of Table 10-1 are independent of the values of m, there are numerous degeneracies (same resonant frequencies) among the modes; for a given n and p there are many m's that have the same resonant frequency. To determine how many m modes exist for each set of n and p, remember that $P_n^m = 0$ if $m > n$. Therefore for P_n^m to be nonzero, $m \leq n$. Thus the order of degeneracy is equal to $m = n$.

According to the values of Table 10-1 the lowest ζ_{np} zeroes in ascending order along with the number of degenerate and total modes are

n, p	ζ_{np}	Degenerate modes	Total number of modes
$n = 1, p = 1$	$\zeta_{11} = 4.493$	$m = 0, 1$ (even, odd)	3
$n = 2, p = 1$	$\zeta_{21} = 5.763$	$m = 0, 1, 2$ (even, odd)	8
$n = 3, p = 1$	$\zeta_{31} = 6.988$	$m = 0, 1, 2, 3$ (even, odd)	15
$n = 1, p = 2$	$\zeta_{12} = 7.725$	$m = 0, 1$ (even, odd)	18
$n = 4, p = 1$	$\zeta_{41} = 8.183$	$m = 0, 1, 2, 3, 4$ (even, odd)	27

The even, odd is used to represent either the $\cos(m\phi)$ or $\sin(m\phi)$ variations of (10-55). For example, for $n = 1, p = 1$ (10-55) has a threefold degeneracy and can be written to represent the following three modes:

$$(F_r)_{011} \text{ (even)} = A_{011}C_3\hat{J}_1(\beta_r r)P_1^0(\cos\theta) = A_{011}C_3\hat{J}_1\left(4.493\frac{r}{a}\right)\cos\theta \quad (10\text{-}60a)$$

$$(F_r)_{111} \text{ (even)} = A_{111}C_3\hat{J}_1(\beta_r r)P_1^1(\cos\theta)\cos\phi = -A_{111}C_3\hat{J}_1\left(4.493\frac{r}{a}\right)\sin\theta\cos\phi \quad (10\text{-}60b)$$

$$(F_r)_{111} \text{ (odd)} = A_{111}D_3\hat{J}_1(\beta_r r)P_1^1(\cos\theta)\sin\phi = -A_{111}D_3\hat{J}_1\left(4.493\frac{r}{a}\right)\sin\theta\sin\phi \quad (10\text{-}60c)$$

since

$$P_1^0(\cos\theta) = P_1(\cos\theta) = \cos\theta \qquad (10\text{-}60\text{d})$$

$$P_1^1(\cos\theta) = -(1-\cos^2\theta)^{1/2} = -\sin\theta \qquad (10\text{-}60\text{e})$$

The modes represented by (10-60b) and (10-60c) are the same except that they are rotated 90° in the ϕ direction from each other. The same is true between (10-60a) and (10-60b) or (10-60c) except that the rotation is in θ and ϕ.

10.4.2 Transverse Magnetic (TM^r) Modes

Following a procedure and justification similar to that for the TE^r modes, it can be shown that the appropriate vector potential component A_r of (10-25b) takes the form of

$$(A_r)_{mnp} = B_{mnp}\hat{J}_n(\beta r)P_n^m(\cos\theta)[C_3\cos(m\phi) + D_3\sin(m\phi)] \qquad (10\text{-}61)$$

The corresponding electric and magnetic fields are found using (10-27) through (10-28c). The boundary conditions are the same as for the TE^r, as given by (10-57a) and (10-57b).

Expanding (10-27b) using (10-61) we can write that

$$E_\theta = \frac{1}{j\omega\mu\varepsilon}\frac{1}{r}\frac{\partial^2 A_r}{\partial r\partial\theta} = B_{mnp}\frac{\beta}{j\omega\mu\varepsilon r}\hat{J}_n'(\beta r)P_n^{'m}(\cos\theta)[C_3\cos(m\phi) + D_3\sin(m\phi)]$$

(10-62)

Applying (10-57a) on (10-62) leads to

$$E_\theta(r=a, 0\le\theta\le\pi, 0\le\phi\le 2\pi) = B_{mnp}\frac{\beta}{j\omega\mu\varepsilon a}\hat{J}_n'(\beta_r a)P_n^{'m}(\cos\theta)$$

$$\times[C_3\cos(m\phi) + D_3\sin(m\phi)] = 0 \Rightarrow \hat{J}_n'(\beta a)|_{\beta=\beta_r} = 0 \Rightarrow \beta_r a = \zeta_{np}'$$

$$\beta_r = \frac{\zeta_{np}'}{a} \qquad \begin{array}{l} n = 1,2,3,\ldots \\ p = 1,2,3,\ldots \end{array} \qquad (10\text{-}63)$$

where ζ_{np}' represents the p zeroes of the derivative of the spherical Bessel function of order n. A listing of a limited, but for most applications sufficient, number of ζ_{np}''s is found in Table 10-2.

The resonant frequencies are found using (10-63) and can be written as

$$\boxed{(f_r)_{mnp}^{TM^r} = \frac{\zeta_{np}'}{2\pi a\sqrt{\mu\varepsilon}}} \qquad \begin{array}{l} m = 0,1,2,\ldots \le n \\ n = 1,2,3,\ldots \\ p = 1,2,3,\ldots \end{array} \qquad (10\text{-}64)$$

As with the TE^r modes there are numerous degeneracies among the modes since the resonant frequencies determined by (10-64) are independent of m. For a given n, the order of degeneracy is $m = n$.

THE SPHERICAL CAVITY

TABLE 10-2
Zeroes ζ'_{np} of derivative of spherical Bessel function $\hat{J}'_n(\zeta'_{np}) = 0$

	$n=1$	$n=2$	$n=3$	$n=4$	$n=5$	$n=6$	$n=7$	$n=8$
$p=1$	2.744	3.870	4.973	6.062	7.140	8.211	9.275	10.335
$p=2$	6.117	7.443	8.722	9.968	11.189	12.391	13.579	14.753
$p=3$	9.317	10.713	12.064	13.380	14.670	15.939	17.190	18.425
$p=4$	12.486	13.921	15.314	16.674	18.009	19.321	20.615	21.894
$p=5$	15.644	17.103	18.524	19.915	21.281	22.626		
$p=6$	18.796	20.272	21.714	23.128				
$p=7$	21.946							

According to the values of Table 10-2 the lowest ζ'_{np} zeroes in ascending order along with the number of degenerate and total modes are

n, p	ζ'_{np}	Degenerate modes	Total number of modes
$n=1, p=1$	$\zeta'_{11} = 2.744$	$m = 0, 1$ (even, odd)	3
$n=2, p=1$	$\zeta'_{21} = 3.870$	$m = 0, 1, 2$ (even, odd)	8
$n=3, p=1$	$\zeta'_{31} = 4.973$	$m = 0, 1, 2, 3$ (even, odd)	15
$n=4, p=1$	$\zeta'_{41} = 6.062$	$m = 0, 1, 2, 3, 4$ (even, odd)	24
$n=1, p=2$	$\zeta'_{12} = 6.117$	$m = 0, 1$ (even, odd)	27

The lowest-order mode is the one found using $n = 1$, $p = 1$, and it has a threefold degeneracy [$m = 0$ (even), $m = 1$ (even), and $m = 1$ (odd)]. For these (10-61) reduces to

$$(A_r)_{011} \text{ (even)} = B_{011} C_3 \hat{J}_1(\beta_r r) P_1^0(\cos\theta) = B_{011} C_3 \hat{J}_1\left(2.744 \frac{r}{a}\right) \cos\theta \quad (10\text{-}65a)$$

$$(A_r)_{111} \text{ (even)} = B_{111} C_3 \hat{J}_1(\beta_r r) P_1^1(\cos\theta) \cos\phi = -B_{111} C_3 \hat{J}_1\left(2.744 \frac{r}{a}\right) \sin\theta \cos\phi \quad (10\text{-}65b)$$

$$(A_r)_{111} \text{ (odd)} = B_{111} D_3 \hat{J}_1(\beta_r r) P_1^1(\cos\theta) \sin\phi = -B_{111} D_3 \hat{J}_1\left(2.744 \frac{r}{a}\right) \sin\theta \sin\phi \quad (10\text{-}65c)$$

Example 10-1. For a spherical cavity of 3 cm radius and filled with air, determine the resonant frequencies (in ascending order) of the first 11 modes (including degenerate modes).

Solution. According to (10-59a) and (10-64), using the values of ζ_{np} and ζ'_{np} from Tables 10-1 and 10-2, and taking into account the degeneracy of the modes in m as well as the even and odd forms in ϕ, we can write the resonant frequencies of the first 11 modes as

1, 2, 3:
$$(f_r)_{011}^{TM} \text{ (even)} = (f_r)_{111}^{TM} \text{ (even)} = (f_r)_{111}^{TM} \text{ (odd)}$$
$$= \frac{2.744(30 \times 10^9)}{2\pi(3)} = 4.367 \times 10^9 \text{ Hz}$$

4, 5, 6, 7, 8:
$$(f_r)_{021}^{TM} \text{ (even)} = (f_r)_{121}^{TM} \text{ (even)} = (f_r)_{121}^{TM} \text{ (odd)} = (f_r)_{221}^{TM} \text{ (even)}$$
$$= (f_r)_{221}^{TM} \text{ (odd)} = \frac{3.870(30 \times 10^9)}{2\pi(3)} = 6.1593 \times 10^9 \text{ Hz}$$

9, 10, 11:
$$(f_r)_{011}^{TE} \text{ (even)} = (f_r)_{111}^{TE} \text{ (even)} = (f_r)_{111}^{TE} \text{ (odd)}$$
$$= \frac{4.493(30 \times 10^9)}{2\pi(3)} = 7.1508 \times 10^9 \text{ Hz}$$

10.4.3 Quality Factor Q

As has already been pointed, the Q of the cavity is probably one of the most important parameters of a cavity, and it is defined by (8-84). To derive the equation for the Q of any mode of a spherical cavity is a most difficult task. However, it is instructive to consider that of the lowest (dominant) mode, which here is any one of the threefold degenerate modes TM_{011} (even), TM_{111} (even), or TM_{111} (odd). Let us consider the TM_{011} (even) mode.

For the TM_{011} (even) mode the potential function of (10-61) reduces to that of (10-65a) which can be written as

$$(A_r)_{011} = B'_{011} \hat{J}_1\left(2.744\frac{r}{a}\right) \cos\theta \tag{10-66}$$

Since the Q of the cavity is defined by (8-84). It is most convenient to find the stored energy and dissipated power by using the magnetic field since it has only one nonzero component (the electric field has two).

The magnetic field components of the TM_{011} mode can be written using (10-28a) through (10-28c) and (10-66) as

$$H_r = 0 \tag{10-67a}$$

$$H_\theta = \frac{1}{\mu} \frac{1}{r \sin\theta} \frac{\partial A_r}{\partial \phi} = 0 \tag{10-67b}$$

$$H_\phi = -\frac{1}{\mu} \frac{1}{r} \frac{\partial A_r}{\partial \theta} = B'_{011} \frac{1}{\mu} \frac{1}{r} \hat{J}_1\left(2.744\frac{r}{a}\right) \sin\theta \tag{10-67c}$$

Therefore at resonance the total stored energy can be found using

$$W = 2W_e = 2W_m = 2\left[\frac{\mu}{4} \iiint_V |\mathbf{H}|^2 \, dv\right] = \frac{\mu}{2} \int_0^{2\pi} \int_0^\pi \int_0^a |H_\phi|^2 r^2 \sin\theta \, dr \, d\theta \, d\phi \tag{10-68}$$

THE SPHERICAL CAVITY

which by substituting (10-67c) reduces to

$$W = \frac{|B'_{011}|^2}{2\mu} \int_0^{2\pi} \int_0^{\pi} \int_0^a \hat{J}_1^2\left(2.744\frac{r}{a}\right) \sin^3\theta \, dr \, d\theta \, d\phi$$

$$= \frac{|B'_{011}|^2}{2\mu}(2\pi)\frac{4}{3}\int_0^a \hat{J}_1^2\left(2.744\frac{r}{a}\right) dr \quad (10\text{-}68\text{a})$$

The integral can be evaluated using the formula

$$\int_0^a \hat{J}_1^2\left(2.744\frac{r}{a}\right) dr = \frac{a}{2}[\hat{J}_1^2(2.744) - \hat{J}_0(2.744)\hat{J}_2(2.744)] \quad (10\text{-}69)$$

where according to (3-94) or (10-31)

$$\hat{J}_1(2.744) = 2.744 j_1(2.744) = 2.744(0.3878) = 1.0640 \quad (10\text{-}69\text{a})$$
$$\hat{J}_0(2.744) = 2.744 j_0(2.744) = 2.744(0.1428) = 0.3919 \quad (10\text{-}69\text{b})$$
$$\hat{J}_2(2.744) = 2.744 j_2(2.744) = 2.744(0.2820) = 0.7738 \quad (10\text{-}69\text{c})$$

Thus (10-69) reduces, using that $\beta_r = 2.744/a$, to

$$\int_0^a \hat{J}_1^2\left(2.744\frac{r}{a}\right) dr = \frac{a}{2}[(1.0640)^2 + 0.3919(0.7738)] = \frac{a}{2}(0.8288)$$

$$= \frac{a}{2.744}\frac{(2.744)(0.8288)}{2} = \frac{1.137}{\beta_r}$$

and (10-68a) to

$$W = \frac{|B'_{011}|^2}{2\mu}(2\pi)\frac{4}{3}\frac{1.137}{\beta_r} \quad (10\text{-}70)$$

The power dissipated on the walls of the cavity can be found using

$$P_d = \frac{R_s}{2}\oiint_S \mathbf{J}_s \cdot \mathbf{J}_s^* \, ds \quad (10\text{-}71)$$

where

$$\mathbf{J}_s = \hat{n} \times \mathbf{H}|_{r=a} = -\hat{a}_r \times \hat{a}_\phi H_\phi|_{r=a} = \hat{a}_\theta H_\phi(r=a) = \hat{a}_\theta B'_{011}\frac{1}{\mu}\frac{1}{a}\hat{J}_1(2.744)\sin\theta$$
$$(10\text{-}71\text{a})$$

Thus (10-71) can be written as

$$P_d = \frac{R_s}{2}\int_0^{2\pi}\int_0^{\pi}|H_\phi(r=a)|^2 a^2 \sin\theta \, d\theta \, d\phi$$

$$= \frac{R_s}{2\mu^2}|B'_{011}|^2 \hat{J}_1^2(2.744)\int_0^{2\pi}\int_0^{\pi}\sin^3\theta \, d\theta \, d\phi$$

$$= |B'_{011}|^2 \frac{R_s}{2\mu^2}(2\pi)\left(\frac{4}{3}\right)\hat{J}_1^2(2.744) = |B'_{011}|^2 \frac{R_s}{2\mu^2}(2\pi)\left(\frac{4}{3}\right)(1.0640)^2$$

$$P_d = \frac{|B'_{011}|^2}{2\mu^2} 1.132 (2\pi)\left(\frac{4}{3}\right)R_s \quad (10\text{-}72)$$

Using (10-70) and (10-72), the Q of the cavity for the TM^r_{011} mode reduces to

$$Q = \omega_r \frac{W}{P_d} = \omega_r \frac{\dfrac{|B'_{011}|^2}{2\mu}(2\pi)\dfrac{4}{3}\dfrac{1.137}{\beta_r}}{\dfrac{|B'_{011}|^2}{2\mu^2}(2\pi)\dfrac{4}{3}1.132 R_s} = \frac{1.137 \omega_r \mu}{1.132 \beta_r R_s} = 1.004 \frac{\omega_r \mu}{\omega_r\sqrt{\mu\varepsilon}\,R_s}$$

$$= 1.004 \frac{\sqrt{\dfrac{\mu}{\varepsilon}}}{R_s} = 1.004 \frac{\eta}{R_s} \tag{10-73}$$

Example 10-2. Compare the Q values of a spherical cavity operating in the dominant TM_{011} (even) mode with those of a circular cylinder and cubical cavities. The dimensions of each are such that the cylindrical and spherical cavities are circumscribed by the cubical cavity.

Solution. According to (10-73) the Q of a spherical cavity of radius a operating in the dominant TM_{011} (even) mode is given by

$$Q = 1.004 \frac{\eta}{R_s}$$

while that of a circular cavity of diameter d and height h operating in the dominant TM_{010} mode (for $h = d$) is given by (9-57) which reduces to

$$Q = 1.2025 \frac{\eta}{R_s} \frac{1}{\left(1 + \dfrac{d/2}{h}\right)} = 0.8017 \frac{\eta}{R_s}$$

For a rectangular cavity operating in the dominant TE_{101} mode the Q is given by (8-88) which for a cubical geometry ($a = b = c$) reduces according to (8-88a) to

$$Q = 1.1107 \frac{\eta}{R_s} \frac{1}{\left(1 + \dfrac{a/2}{b}\right)} = \frac{1.1107}{1.5} \frac{\eta}{R_s} = 0.7405 \frac{\eta}{R_s}$$

Comparing these three expressions, it is evident that the Q of the spherical cavity is greater than that of the circular cavity with $h = d$ by

$$\frac{1.004 - 0.8017}{0.8017} \times 100 = 25.23\%$$

and greater than that of the cubical cavity by

$$\frac{1.004 - 0.7405}{0.7405} \times 100 = 35.58\%$$

This is expected since the spherical cavity does not possess any sharp corners and edges which are evident in the circular cavity and even more in the cubical

REFERENCES

1. R. F. Harrington, *Time-Harmonic Electromagnetic Fields*, McGraw-Hill, New York, 1961.
2. S. A. Schelkunoff, *Electromagnetic Waves*, Van Nostrand, Princeton, N.J. 1943, pp. 51–52.
3. C. A. Balanis, *Antenna Theory: Analysis and Design*, Wiley, New York, 1982.
4. S. A. Schelkunoff, "Theory of antennas of arbitrary size and shape," *Proc. IRE*, vol. 29, no. 9, pp. 493–521, September 1941.
5. S. A. Schelkunoff and C. B. Feldman, "On radiation from antennas," *Proc. IRE*, vol. 30, no. 11, pp. 512–516, November 1942.
6. S. A. Schelkunoff and H. T. Friis, *Antennas: Theory and Practice*, Wiley, New York, 1952.
7. C. T. Tai, "Application of a variational principle to biconical antennas," *J. Appl. Phys.*, vol. 20, no. 11, pp. 1076–1084, November 1949.
8. P. D. Smith, "The conical dipole of wide angle," *J. Appl. Phys.*, vol. 19, no. 1, pp. 11–23, January 1948.
9. L. Bailin and S. Silver, "Exterior electromagnetic boundary value problems for spheres and cones," *IRE Trans. Antennas Propagat.*, vol. AP-4, no. 1, pp. 5–15, January 1956.
10. W. R. Smythe, *Static and Dynamic Electricity*, McGraw-Hill, New York, 1941.
11. J. A. Stratton, *Electromagnetic Theory*, McGraw-Hill, New York, 1960.
12. P. M. Morse and H. Feshbach, *Methods of Theoretical Physics*, Parts I and II, McGraw-Hill, New York, 1953.
13. M. Abramowitz and I. A. Stegun (Eds.), *Handbook of Mathematical Functions with Formulas, Graphs, and Mathematical Tables*, National Bureau of Standards Applied Mathematical Series-55, U.S. Government Printing Office, Washington, D.C., 1966.
14. E. Jahnke and F. Emde, *Tables of Functions*, Dover, New York, 1945.
15. M. R. Spiegel, *Mathematical Handbook of Formulas and Tables*, Schaum's Outline Series, McGraw-Hill, New York, 1968.

PROBLEMS

10.1. In Section 11.7.1 it is shown that the magnetic vector potential for an infinitesimal electric dipole of Figure 11-23a and Example 6-3 is given by (11-209)

$$A_z^{(1)} = -\hat{a}_z j \frac{\mu \beta I_e \Delta \ell}{4\pi} h_0^{(2)}(\beta r)$$

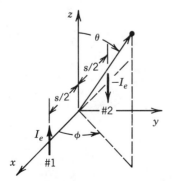

FIGURE P10-1

where $h_0^{(2)}(\beta r)$ is the spherical Hankel function of order zero. Assuming that two such dipoles of equal amplitude but 180° out of phase are displaced along the x axis a distance s apart, as shown in Figure P10-1. The total magnetic potential can be written following a procedure outlined in [1] as

$$A_z^t = A_z^{(1)}\left(x - \frac{s}{2}, y, z\right) - A_z^{(1)}\left(x + \frac{s}{2}, y, z\right) \stackrel{s \to 0}{\simeq} -s\frac{\partial A_z^{(1)}}{\partial x}$$

Show that the total magnetic potential A_z^t can be reduced to

$$A_z^t \stackrel{s \to 0}{\simeq} j\frac{\mu\beta^2 sI_e \Delta\ell}{4\pi} h_0^{(2)\prime}(\beta r)\sin\theta\cos\phi = -j\frac{\mu\beta^2 sI_e \Delta\ell}{4\pi} h_1^{(2)}(\beta r) P_1^1(\cos\theta)\cos\phi$$

where $' = \partial/\partial(\beta r)$.

10.2. Following the procedure of Problem 10.1 show that when the infinitesimal dipoles are displaced along the y axis, as shown in Figure P10-2, the total magnetic potential can be written as

$$A_z^t \stackrel{s \to 0}{\simeq} j\frac{\mu\beta^2 sI_e \Delta\ell}{4\pi} h_1^{(2)}(\beta r) P_1^1(\cos\theta)\sin\phi$$

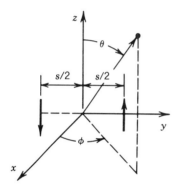

FIGURE P10-2

10.3. Following the procedure of Problem 10.1 show that when the infinitesimal dipoles are displaced along the z axis, as shown in Figure P10-3, the total magnetic potential can be written as

$$A_z^t \stackrel{s \to 0}{\simeq} j\frac{\mu\beta^2 sI_e \Delta\ell}{4\pi} h_1^{(2)}(\beta r) P_1(\cos\theta)$$

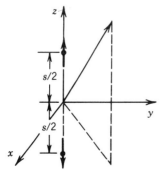

FIGURE P10-3

10.4. Derive expressions for the electric and magnetic field components, in terms of F_z of (10-1b), that are TE^z. F_z should represent a solution to the scalar Helmholtz equation in spherical coordinates.

10.5. Derive expressions for the electric and magnetic field components, in terms of A_z of (10-2a), that are TM^z. A_z should represent a solution to the scalar Helmholtz equation in spherical coordinates.

10.6. For problems with sources of $\mathbf{J} \neq 0$ and $\mathbf{M} = 0$, show that the electric and magnetic fields, and vector potential $\mathbf{A}$ should satisfy (10-15a) through (10-15d).

10.7. Show that using (10-25a) and (10-25b) for TM^r modes in spherical coordinates reduces (10-16f) to (10-26).

10.8. By applying the boundary conditions of (10-44a) or (10-44b) on the electric field components of (10-27a) or (10-27c) where the vector potential A_r is given by (10-43), show that the eigenvalues of n are obtained as solutions to (10-45).

10.9. Use Maxwell's equations

$$\nabla \times \mathbf{E} = -j\omega\mu\mathbf{H}$$
$$\nabla \times \mathbf{H} = j\omega\varepsilon\mathbf{E}$$

and assume TEM modes for the biconical antenna of Figure 10-1 with only E_θ and H_ϕ components, each independent of ϕ. Then show that the H_ϕ component must satisfy the partial differential equation

$$\frac{\partial^2}{\partial r^2}(rH_\phi) = -\beta^2(rH_\phi)$$

whose solution must take the form

$$H_\phi = \frac{H_0}{\sin\theta}\frac{e^{-j\beta r}}{r}$$

whereas that of E_θ must then be written as

$$E_\theta = \eta H_\phi = \eta\frac{H_0}{\sin\theta}\frac{e^{-j\beta r}}{r}$$

10.10. Show that by using the electric and magnetic field components of Problem 10.9 the power radiated by the biconical antenna of Figure 10-1 reduces to

$$P_{\text{rad}} = \oiint \mathbf{S}_{\text{av}} \cdot d\mathbf{s} = 2\pi\eta|H_0|^2 \ln\left[\cot\left(\frac{\alpha}{2}\right)\right]$$

where $\mathbf{S}_{\text{av}}$ represents the average power density and $\alpha = \alpha_1 = \pi - \alpha_2$.

10.11. By using the magnetic field from Problem 10.9, show that the current on the surface of the cone a distance r from the origin is equal to

$$I(r) = 2\pi H_0 e^{-j\beta r}$$

Evaluating the current at the origin $I(r = 0)$, and using the definition for the radiation resistance in terms of the radiated power from Problem 10.10 and the current at the origin, show that the radiation resistance reduces to

$$R_r = \frac{2P_{\text{rad}}}{|I(r=0)|^2} = \frac{\eta}{\pi}\ln\left[\cot\left(\frac{\alpha}{2}\right)\right]$$

where $\alpha = \alpha_1 = \pi - \alpha_2$. This is the same as the characteristic impedance of (10-52a) for a symmetrical biconical transmission line.

10.12. Determine the included angle $\alpha = \alpha_1 = \pi - \alpha_2$ of a symmetrical biconical transmission line so that its characteristic impedance is (a) 300 ohms and (b) 50 ohms.

10.13. For inside ($0 \leq \theta \leq \alpha$) or outside ($0 \leq \theta \leq \alpha$) cones shown, respectively, in Figures P10-13a and b), for TE^r modes:

(a) Show that the electric vector potential reduces to
$$(F_r)_{mnp} = A_2 P_n^m(\cos\theta)\left[C_1 \hat{H}_n^{(1)}(\beta r) + D_1 \hat{H}_n^{(2)}(\beta r)\right]$$
$$\times [C_3 \cos(m\phi) + D_3 \sin(m\phi)] \quad m = 0, 1, 2, \ldots$$

(b) Show that the eigenvalues for n are obtained as solutions to
$$\left.\frac{dP_n(\cos\theta)}{d\theta}\right|_{\theta=\alpha} = 0$$

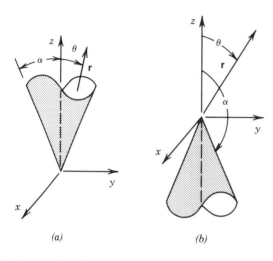

(a)　　　(b)　　　**FIGURE P10-13**

10.14. Repeat Problem 10.13 for TM^r modes.

(a) Show that the magnetic vector potential reduces to
$$(A_r)_{mnp} = A_2 P_n^m(\cos\theta)\left[C_1 \hat{H}_n^{(1)}(\beta r) + D_1 \hat{H}_n^{(2)}(\beta r)\right]$$
$$\times [C_3 \cos(m\phi) + D_3 \sin(m\phi)] \quad m = 0, 1, 2, \ldots$$

(b) Show that the eigenvalues for n are obtained as solutions to
$$P_n^m(\cos\theta)|_{\theta=\alpha} = 0$$

10.15. For the inside ($0 \leq \phi \leq \alpha$) or outside ($\alpha \leq \phi \leq 2\pi$) infinite dimensions wedge of Figure P10-15 derive the reduced vector potential and allowable eigenvalues for TE^r modes.

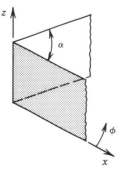

FIGURE P10-15

10.16. Repeat Problem 10.15 for TM^r modes.

10.17. Design a spherical cavity (find its radius in cm) so that the resonant frequency of the dominant mode is 1 GHz and that of the next higher-order mode is approximately 1.41 GHz. The lossless medium within the sphere has electric constitutive parameters of $\varepsilon_r = 2.56$ and $\mu_r = 1$.

10.18. Assume a spherical cavity with 2-cm radius and filled with air.
   (a) Determine the resonant frequency of the dominant degenerate modes.
   (b) Find the bandwidth over which the dominant degenerate modes operate before the next higher-order degenerate modes.
   (c) Determine the dielectric constant that must be used to fill the sphere to reduce the resonant frequency of the dominant degenerate modes by a factor of 2.

10.19. Derive an expression for the Q of a spherical cavity when the fields within it are those of the dominant TM$_{111}$ (even) mode. Compare the expression with (10-73).

10.20. Derive an expression for the Q of a spherical cavity when the fields within it are those of the dominant TM$_{111}$ (odd) mode. Compare the expression with (10-73).

10.21. Determine the Q of the dominant mode of a spherical cavity of 2-cm radius when the medium within the cavity is (a) air and (b) polystyrene with a dielectric constant of 2.56. The cavity is made of copper whose conductivity is 5.76×10^7 S/m.

10.22. It is desired to design a spherical cavity whose Q at the resonant frequency of the dominant mode is 10,000. Assume that the cavity is filled with air and it is made of copper ($\sigma = 5.76 \times 10^7$ S/m). Then determine the resonant frequency of the dominant mode and the radius (in cm) of the cavity. Also find the dielectric constant of the medium that must be used to fill the cavity to reduce its Q by a factor of 3.

10.23. Assume a hemispherical cavity of Figure P10-23 with radius a.
   (a) Determine the dominant mode and the expression for its resonant frequency.
   (b) Show that the Q of the dominant mode is
   $$Q = 0.574 \frac{\eta}{R_s}$$
   (c) Compare the Q of part b with that of the spherical cavity, and those of cylindrical and square-based rectangular cavities with the same height-to-diameter ratios.

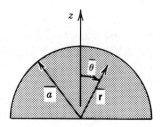

FIGURE P10-23

CHAPTER 11

SCATTERING

11.1 INTRODUCTION

Previously we have considered wave propagation in unbounded media, semi-infinite media forming planar interfaces, and conducting, dielectric, and surface waveguides. Although wave propagation in unbounded media is somewhat idealistic, it serves as a simplified model for examining wave behavior while minimizing mathematical complexities. In general, however, wave propagation must be analyzed when it accounts for the presence of any structures (scatterers), especially when they are in proximity to the wave source and/or receiver.

In this chapter we want to examine wave propagation in the presence of scatterers of various geometries (planar, cylindrical, spherical). This is accomplished by introducing to the total field an additional component, referred to here as the *scattered field*, due to the presence of scatterers. The scattered field ($\mathbf{E}^s, \mathbf{H}^s$) must be such that when it is added, through superposition, to the *incident (direct) field* ($\mathbf{E}^i, \mathbf{H}^i$) the sum represents the total ($\mathbf{E}^t, \mathbf{H}^t$) field, that is,

$$\mathbf{E}^t = \mathbf{E}^i + \mathbf{E}^s \tag{11-1a}$$

$$\mathbf{H}^t = \mathbf{H}^i + \mathbf{H}^s \tag{11-1b}$$

The incident (direct) field $\mathbf{E}^i, \mathbf{H}^i$ will represent the total field produced by the sources *in the absence of any scatterers*.

The direct, scattered and total fields will be obtained using various techniques. In general, geometrical optics (GO), physical optics (PO), modal techniques (MT), integral equations (IE), and diffraction theory [such as the geometrical theory of diffraction (GTD) and physical theory of diffraction (PTD)] can be used to analyze such problems. Typically some of the problems are more conveniently analyzed using particular method(s). The fundamentals of physical optics were introduced in Chapter 7 and modal techniques were utilized in Chapters 8, 9, and 10 to analyze waveguide wave propagation. Integral equations are very popular, and they are introduced in Chapter 12. Geometrical optics and diffraction techniques are introduced and applied in Chapter 13.

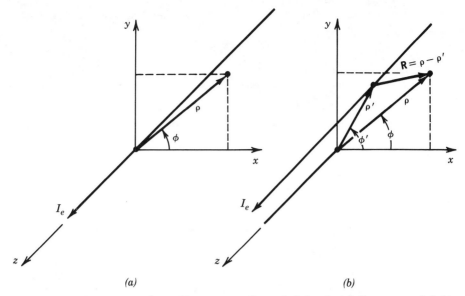

FIGURE 11-1 Geometry and coordinate system for an infinite electric line source. (*a*) At origin. (*b*) Offset.

In this chapter we want to examine scattering primarily by conducting objects. Each scattering problem will be analyzed using image theory, physical optics, or modal techniques. The conveniences and limitations of the applied method to each problem, as well as those of the other methods, will be stated.

11.2 INFINITE LINE-SOURCE CYLINDRICAL WAVE RADIATION

Before we examine the radiation and scattering of sources placed in the presence of scatterers, it is instructive to obtain the fields radiated by an infinite line source (both electric and magnetic) in an unbounded medium. The reason for doing this is that the infinite line source will serve as one type of source for which we will examine radiation properties in the presence of scatterers; its radiation in an unbounded medium will significantly aid in the solution of such problems.

11.2.1 Electric Line Source

The geometry of the line source is that of Figure 11-1*a* and *b* where it is assumed that its length extends to infinity and the electric current is represented by

$$\mathbf{I}_e(z') = \hat{a}_z I_e \quad (11\text{-}2)$$

where I_e is a constant. Since the current is directed along the z axis, the fields radiated by the line source are TM^z and can be obtained by letting

$$\begin{aligned}
\mathbf{F} &= 0 \\
\mathbf{A} &= \hat{a}_z A_z(\rho, \phi, z) \\
&= \hat{a}_z \left[C_1 H_m^{(1)}(\beta_\rho \rho) + D_1 H_m^{(2)}(\beta_\rho \rho) \right] \\
&\quad \times \left[C_2 \cos(m\phi) + D_2 \sin(m\phi) \right] \left(A_3 e^{-j\beta_z z} + B_3 e^{+j\beta_z z} \right)
\end{aligned} \quad (11\text{-}3)$$

Since the line source is infinite in extent, the fields are two-dimensional (no z variations) so that

$$\beta_z = 0 \Rightarrow \beta_\rho^2 + \beta_z^2 = \beta^2 \Rightarrow \beta_\rho = \beta \quad (11\text{-}4)$$

In addition since the waves radiate only in the outward direction and we choose the lowest-order mode, then

$$C_1 = 0 \quad (11\text{-}5a)$$
$$m = 0 \quad (11\text{-}5b)$$

Thus (11-3) reduces to

$$\mathbf{A} = \hat{a}_z A_z(\rho) = \hat{a}_z A_0 H_0^{(2)}(\beta\rho) \quad (11\text{-}6)$$

whose corresponding electric and magnetic fields can be written using (6-70) as

$$E_\rho = -j\frac{1}{\omega\mu\varepsilon}\frac{\partial^2 A_z}{\partial\rho\,\partial z} = 0 \quad (11\text{-}6a)$$

$$E_\phi = -j\frac{1}{\omega\mu\varepsilon}\frac{1}{\rho}\frac{\partial^2 A_z}{\partial\phi\,\partial z} = 0 \quad (11\text{-}6b)$$

$$E_z = -j\frac{1}{\omega\mu\varepsilon}\left(\frac{\partial^2}{\partial z^2} + \beta^2\right)A_z = -j\omega A_0 H_0^{(2)}(\beta\rho) \quad (11\text{-}6c)$$

$$H_\rho = \frac{1}{\mu}\frac{1}{\rho}\frac{\partial A_z}{\partial\phi} = 0 \quad (11\text{-}6d)$$

$$H_\phi = -\frac{1}{\mu}\frac{\partial A_z}{\partial\rho} = -\beta\frac{A_0}{\mu}H_0^{(2)\prime}(\beta\rho) = A_0\frac{\beta}{\mu}H_1^{(2)}(\beta\rho) \quad (11\text{-}6e)$$

$$H_z = 0 \quad (11\text{-}6f)$$

The constant A_0 can be obtained by using

$$I_e = \lim_{\rho\to 0}\oint_C \mathbf{H}\cdot d\ell = \lim_{\rho\to 0}\int_0^{2\pi}(\hat{a}_\phi H_\phi)\cdot(\hat{a}_\phi \rho\, d\phi) = \lim_{\rho\to 0}\int_0^{2\pi} H_\phi \rho\, d\phi \quad (11\text{-}7)$$

Since the integration of (11-7) must be performed in the limit as $\rho \to 0$, it is convenient to represent the Hankel function of (11-6e) by its asymptotic expansion for small arguments. Using (IV-12) we can write that

$$H_1^{(2)}(\beta\rho) = J_1(\beta\rho) - jY_1(\beta\rho) \underset{\beta\rho\to 0}{\simeq} \frac{\beta\rho}{2} + j\frac{2}{\pi}\left(\frac{1}{\beta\rho}\right) \underset{\beta\rho\to 0}{\simeq} j\frac{2}{\pi}\left(\frac{1}{\beta\rho}\right) \quad (11\text{-}8)$$

Therefore (11-7) reduces, using (11-6e) and (11-8), to

$$I_e = \lim_{\rho\to 0}\int_0^{2\pi}\left[A_0\frac{\beta}{\mu}H_1^{(2)}(\beta\rho)\right]\rho\, d\phi \simeq jA_0\frac{2}{\pi\mu}\int_0^{2\pi}\frac{1}{\rho}\rho\, d\phi = jA_0\frac{4}{\mu} \quad (11\text{-}9)$$

or

$$A_0 = -j\frac{\mu}{4}I_e \quad (11\text{-}9a)$$

INFINITE LINE-SOURCE CYLINDRICAL WAVE RADIATION

Thus the nonzero electric and magnetic fields of the electric line source reduce to

$$E_z = -I_e \frac{\omega\mu}{4} H_0^{(2)}(\beta\rho) = -I_e \frac{\beta^2}{4\omega\varepsilon} H_0^{(2)}(\beta\rho) \tag{11-10a}$$

$$H_\phi = -jI_e \frac{\beta}{4} H_1^{(2)}(\beta\rho) \tag{11-10b}$$

Each of the field components is proportional to a Hankel function of the second kind whose argument is proportional to the distance from the source to the observation point. If the source is removed from the origin and it is placed as shown in Figure 11-1b, (11-10a) and (11-10b) can be written as

$$E_z = -I_e \frac{\beta^2}{4\omega\varepsilon} H_0^{(2)}(\beta R) = -I_e \frac{\beta^2}{4\omega\varepsilon} H_0^{(2)}(\beta|\rho - \rho'|) \tag{11-11a}$$

$$H_\psi = -jI_e \frac{\beta}{4} H_1^{(2)}(\beta R) = -jI_e \frac{\beta}{4} H_1^{(2)}(\beta|\rho - \rho'|) \tag{11-11b}$$

where

$$R = |\rho - \rho'| = \sqrt{\rho^2 + (\rho')^2 - 2\rho\rho'\cos(\phi - \phi')} \tag{11-11c}$$

ψ = circumferential angle around the source

For observations at far distances such that $\beta\rho \to$ large, the Hankel functions in (11-10a) and (11-10b) can be approximated by their asymptotic expansions for large argument of

$$H_0^{(2)}(\beta\rho) \overset{\beta\rho \to \text{large}}{\simeq} \sqrt{\frac{2j}{\pi\beta\rho}} e^{-j\beta\rho} \tag{11-12a}$$

$$H_1^{(2)}(\beta\rho) \overset{\beta\rho \to \text{large}}{\simeq} j\sqrt{\frac{2j}{\pi\beta\rho}} e^{-j\beta\rho} \tag{11-12b}$$

Thus (11-10a) and (11-10b) can be simplified for large arguments to

$$E_z = -I_e \frac{\beta^2}{4\omega\varepsilon} H_0^{(2)}(\beta\rho) \overset{\beta\rho \to \text{large}}{\simeq} -\eta I_e \sqrt{\frac{j\beta}{8\pi}} \frac{e^{-j\beta\rho}}{\sqrt{\rho}} \tag{11-13a}$$

$$H_\phi = -jI_e \frac{\beta}{4} H_1^{(2)}(\beta\rho) \overset{\beta\rho \to \text{large}}{\simeq} I_e \sqrt{\frac{j\beta}{8\pi}} \frac{e^{-j\beta\rho}}{\sqrt{\rho}} \tag{11-13b}$$

The ratio (11-13a) to (11-13b) is defined as the wave impedance which reduces to

$$Z_w^{+\rho} = \frac{E_z}{-H_\phi} = \eta \tag{11-14}$$

Since the wave impedance is equal to the intrinsic impedance, the waves radiated by the line source at large distances are TEM$^\rho$.

11.2.2 Magnetic Line Source

Although magnetic sources as presently known are not physically realizable, they are often used to represent virtual sources in equivalent models. This was demonstrated in Chapter 7, Sections 7.7 and 7.8, where the volume and surface fields equivalence theorems were introduced. Magnetic sources can be used to represent radiating apertures.

The fields generated by magnetic sources can be obtained by solutions to Maxwell's equations or the wave equation (subject to the appropriate boundary conditions), or by using the duality theorem of Chapter 7, Section 7.2, once the solution to the same problem but with an electric source excitation is known.

Then using the duality theorem of Section 7.2, the field generated by an infinite magnetic line source of constant current I_m can be obtained using Tables 7-1 and 7-2, (11-10a) through (11-10b), and (11-13a) through (11-13b). These can then be written as

$$E_\phi = +jI_m \frac{\beta}{4} H_1^{(2)}(\beta\rho) \stackrel{\beta\rho \to \text{large}}{\simeq} -I_m \sqrt{\frac{j\beta}{8\pi}} \frac{e^{-j\beta\rho}}{\sqrt{\rho}} \qquad (11\text{-}15a)$$

$$H_z = -I_m \frac{\beta^2}{4\omega\mu} H_0^{(2)}(\beta\rho) \stackrel{\beta\rho \to \text{large}}{\simeq} -\frac{1}{\eta} I_m \sqrt{\frac{j\beta}{8\pi}} \frac{e^{-j\beta\rho}}{\sqrt{\rho}} \qquad (11\text{-}15b)$$

which, when the sources are displaced from the origin, can also be expressed according to (11-11a) through (11-11b) as

$$E_\psi = +jI_m \frac{\beta}{4} H_1^{(2)}(\beta|\rho - \rho'|) \qquad (11\text{-}16a)$$

$$H_z = -I_m \frac{\beta^2}{4\omega\mu} H_0^{(2)}(\beta|\rho - \rho'|) \qquad (11\text{-}16b)$$

11.2.3 Electric Line Source above Infinite Plane Electric Conductor

When an infinite electric line source is placed at a height h above an infinite flat electric conductor, as shown in Figure 11-2a, the solution for the field components must include the presence of the conducting plane. This can be accomplished by using Maxwell's equations or the wave equation subject to the radiation conditions at infinity and boundary condition along the air–conductor interface. Instead of doing this, the same solution is obtained by introducing an equivalent model which leads to the same fields in the region of interest. Since the fields below the interface (in the electric conductor, $y < 0$) are known (they are zero), the equivalent model should be valid and leads to the same fields as the actual physical problem on or above the interface ($y \geq 0$). In this case as long as the equivalent model satisfies the same boundary conditions as the actual physical problem along a closed surface, then according to the uniqueness theorem of Section 7.3 the solution of the equivalent model will be unique and the same as that of the physical problem. For this problem the closed surface that will be chosen is that of the air–conductor interface ($y = 0$) which extends from $-\infty \leq x \leq +\infty$.

The equivalent problem of Figure 11-2a is that of Figure 11-2b where the ground plane has been replaced by an equivalent source (usually referred to as *image* or *virtual source* or *caustic*). According to the theory of Section 7.4 the image source is introduced to account for the reflections from the surface of the ground plane. The magnitude, phase, polarization, and position of the image source must be

INFINITE LINE-SOURCE CYLINDRICAL WAVE RADIATION **575**

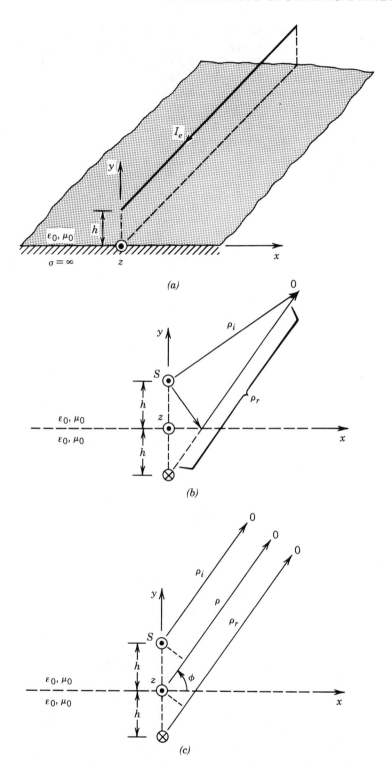

FIGURE 11-2 Electric line source above a flat and infinite electric ground plane. (*a*) Line source. (*b*) Equivalent (near field). (*c*) Equivalent (far field).

such that the boundary conditions of the equivalent problem of Figure 11-2b along $-\infty \leq x \leq +\infty$ are the same as those of the physical problem of Figure 11-2a. In this situation the image must have: the same magnitude as the actual source, its phase must be 180° out of phase from the actual source, it must be placed below the interface at a depth h ($y = -h$) along a line perpendicular to the interface and passing through the actual source, and its length must also be parallel to the z axis. Such a system configuration, as shown in Figure 11-2b, does lead to zero tangential electric field along $-\infty \leq x \leq +\infty$ which is identical to that of Figure 11-2a along the air–conductor interface.

Therefore according to Figure 11-2a or b and equation (11-13a), the total electric field is equal to

$$\mathbf{E}^t = \mathbf{E}^i + \mathbf{E}^r = \begin{cases} -\hat{a}_z \dfrac{\beta^2 I_e}{4\omega\varepsilon} \left[H_0^{(2)}(\beta\rho_i) - H_0^{(2)}(\beta\rho_r) \right] & y \geq 0 \quad (11\text{-}17a) \\ 0 & y < 0 \quad (11\text{-}17b) \end{cases}$$

which for observations at large distances, as shown by Figure 11-2c, reduces using the asymptotic expansion of (11-12a) to

$$\mathbf{E}^t = \mathbf{E}^i + \mathbf{E}^r = \begin{cases} -\hat{a}_z \eta I_e \left(\dfrac{e^{-j\beta\rho_i}}{\sqrt{\rho_i}} - \dfrac{e^{-j\beta\rho_r}}{\sqrt{\rho_r}} \right) \sqrt{\dfrac{j\beta}{8\pi}} & y \geq 0 \quad (11\text{-}18a) \\ 0 & y < 0 \quad (11\text{-}18b) \end{cases}$$

According to Figure 11-2c, for observations made at large distances ($\rho \gg h$)

$$\left. \begin{aligned} \rho_i &\simeq \rho - h \cos\left(\dfrac{\pi}{2} - \phi\right) = \rho - h \sin(\phi) \\ \rho_r &\simeq \rho + h \cos\left(\dfrac{\pi}{2} - \phi\right) = \rho + h \sin(\phi) \end{aligned} \right\} \text{ for phase variations} \quad (11\text{-}19a)$$

$$\rho_i \simeq \rho_r \simeq \rho \quad \text{for amplitude variations} \quad (11\text{-}19b)$$

These approximations are usually referred to in antenna and scattering theory as the *far-field approximations* [1]. Using (11-19a) and (11-19b), we can reduce (11-18a) and (11-18b) to

$$\mathbf{E}^t = \mathbf{E}^i + \mathbf{E}^r = \begin{cases} -\hat{a}_z \eta I_e \sqrt{\dfrac{j\beta}{8\pi}} \left(e^{+j\beta h \sin\phi} - e^{-j\beta h \sin\phi} \right) \dfrac{e^{-j\beta\rho}}{\sqrt{\rho}} \\ \quad = -\hat{a}_z j\eta I_e \sqrt{\dfrac{j\beta}{2\pi}} \sin(\beta h \sin\phi) \dfrac{e^{-j\beta\rho}}{\sqrt{\rho}} & y \geq 0 \quad (11\text{-}20a) \\ 0 & y < 0 \quad (11\text{-}20b) \end{cases}$$

Normalized amplitude patterns (in decibels) for a source placed at a height of $h = 0.25\lambda$ and 0.5λ above the strip are shown in Figure 11-3.

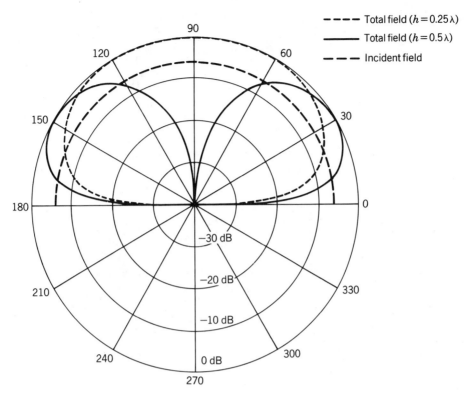

FIGURE 11-3 Radiation patterns of a line source above an infinite electric ground plane for $h = 0.25\lambda_0$ and $0.5\lambda_0$.

11.3 PLANE WAVE SCATTERING BY PLANAR SURFACES

An important parameter in scattering studies is the electromagnetic scattering by a target which is usually represented by its *echo area* or *radar cross section* (RCS) (σ). The echo area or RCS is defined as *the area intercepting the amount of power that, when scattered isotropically, produces at the receiver a density that is equal to the density scattered by the actual target* [1]. For a two-dimensional target the scattering parameter is referred to as the *scattering width* (SW) or alternatively as the *radar cross section per unit length*. In equation form the scattering width and the radar cross section (σ) of a target take the form of

Scattering Width: Two-Dimensional Target

$$\sigma_{2\text{-D}} = \begin{cases} \lim_{\rho \to \infty} \left[2\pi\rho \dfrac{S^s}{S^i} \right] & \text{(11-21a)} \\[1em] \lim_{\rho \to \infty} \left[2\pi\rho \dfrac{|\mathbf{E}^s|^2}{|\mathbf{E}^i|^2} \right] & \text{(11-21b)} \\[1em] \lim_{\rho \to \infty} \left[2\pi\rho \dfrac{|\mathbf{H}^s|^2}{|\mathbf{H}^i|^2} \right] & \text{(11-21c)} \end{cases}$$

578 SCATTERING

$$\sigma_{3\text{-D}} = \begin{cases} \lim_{r \to \infty} \left[4\pi r^2 \dfrac{S^s}{S^i} \right] & (11\text{-}22\text{a}) \\[6pt] \lim_{r \to \infty} \left[4\pi r^2 \dfrac{|\mathbf{E}^s|^2}{|\mathbf{E}^i|^2} \right] & (11\text{-}22\text{b}) \\[6pt] \lim_{r \to \infty} \left[4\pi r^2 \dfrac{|\mathbf{H}^s|^2}{|\mathbf{H}^i|^2} \right] & (11\text{-}22\text{c}) \end{cases}$$

<u>Radar Cross Section: Three-Dimensional Target</u>

where ρ, r = distance from target to observation point
S^s, S^i = scattered, incident power densities
$\mathbf{E}^s, \mathbf{E}^i$ = scattered, incident electric fields
$\mathbf{H}^s, \mathbf{H}^i$ = scattered, incident magnetic fields

For normal incidence the two- and three-dimensional fields, and scattering width and radar cross sections for a target of length ℓ are related by [2–4]

$$E_{3\text{-D}} \simeq \left(E_{2\text{-D}} \dfrac{\ell e^{j\pi/4}}{\sqrt{\lambda \rho}} \right)_{\rho = r} \tag{11-22d}$$

$$\sigma_{3\text{-D}} \simeq \sigma_{2\text{-D}} \dfrac{2\ell^2}{\lambda} \tag{11-22e}$$

The unit of the two-dimensional SW is length (meters in the MKS system) whereas that of the three-dimensional RCS is area (meters squared in the MKS system). A most common reference is *one meter* for the two-dimensional SW and *one meter squared* for the three-dimensional RCS. Therefore a most common designation is dB/m (or dBm) for the two-dimensional SW and dB/(square meter) (or dBsm) for the three-dimensional RCS.

When the transmitter and receiver are at the same location, the RCS is usually referred to as *monostatic (or backscattered)* and it is referred to as *bistatic* when the two are at different locations. Observations made toward directions that satisfy Snell's law of reflection are usually referred to as *specular*. Therefore the RCS of a target is a very important parameter which characterizes its scattering properties. A plot of the RCS as a function of the space coordinates is usually referred to as the *RCS pattern*. The definitions of (11-21a) through (11-22c) all indicate that the SW and RCS of targets are defined under plane wave illumination which in practice can only be approximated when the target is placed in the far field of the source (at least $2D^2/\lambda$) where D is the largest dimension of the target [1].

In this section physical optics (PO) techniques will be used to analyze the scattering from conducting strips and plates of finite width, neglecting edge effects. The edge effects will be considered in Chapters 12 and 13 where, respectively, moment method and geometrical theory of diffraction techniques will be utilized. Physical optics techniques are most accurate at specular directions [5].

11.3.1 TM^z Plane Wave Scattering from a Strip

Let us assume that a TM^z uniform plane wave is incident upon an electric conducting strip of width w and infinite length, as shown in Figure 11-4a and b.

PLANE WAVE SCATTERING BY PLANAR SURFACES **579**

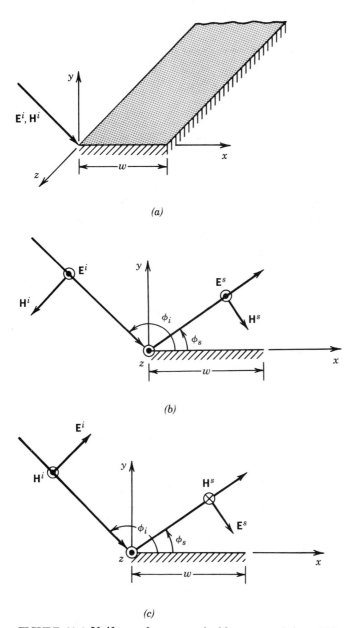

FIGURE 11-4 Uniform plane wave incident on a finite width strip. (*a*) Finite width strip. (*b*) TM^z polarization. (*c*) TE^z polarization.

The incident electric and magnetic fields can be written as

$$\mathbf{E}^i = \hat{a}_z E_0 e^{j\beta(x\cos\phi_i + y\sin\phi_i)} \tag{11-23a}$$

$$\mathbf{H}^i = \frac{E_0}{\eta}\left(-\hat{a}_x \sin\phi_i + \hat{a}_y \cos\phi_i\right) e^{j\beta(x\cos\phi_i + y\sin\phi_i)} \tag{11-23b}$$

where E_0 is a constant and it represents the magnitude of the incident electric field. In Figure 11-4*a* and *b* the angle ϕ_i is shown to be greater than 90°. The reflected

580 SCATTERING

fields can be expressed as

$$\mathbf{E}^r = \hat{a}_z \Gamma_\perp E_0 e^{-j\beta(x\cos\phi_r + y\sin\phi_r)} \tag{11-24a}$$

$$\mathbf{H}^r = \frac{\Gamma_\perp E_0}{\eta}(\hat{a}_x \sin\phi_r - \hat{a}_y \cos\phi_r) e^{-j\beta(x\cos\phi_r + y\sin\phi_r)} \tag{11-24b}$$

where ϕ_r is the reflection angle as determined by enforcing the boundary conditions along the interface, assuming an interface of infinite extent. For the finite width strip the reflection angle ϕ_r is not the same as the scattering angle ϕ_s ($\phi_r \neq \phi_s$). The two coincide for an infinite width strip when geometrical optics and physical optics reduce to each other. Thus we use geometrical optics to determine the reflection coefficient on the surface of the finite width strip, and then apply physical optics to find the scattered fields. According to (5-17a) and (5-15a) the reflection coefficient for a perfectly conducting surface is equal to $\Gamma_\perp = -1$ and $\phi_r = \pi - \phi_i$. Thus (11-24a) and (11-24b) reduce to

$$\mathbf{E}^r = -\hat{a}_z E_0 e^{j\beta(x\cos\phi_i - y\sin\phi_i)} \tag{11-25a}$$

$$\mathbf{H}^r = \frac{E_0}{\eta}(-\hat{a}_x \sin\phi_i - \hat{a}_y \cos\phi_i) e^{j\beta(x\cos\phi_i - y\sin\phi_i)} \tag{11-25b}$$

Using physical optics techniques of (7-54) the current density induced on the surface of the strip can be written as

$$\mathbf{J}_s = \hat{n} \times \mathbf{H}^t \big|_{\substack{y=0 \\ x=x'}} = \hat{n} \times (\mathbf{H}^i + \mathbf{H}^r) \big|_{\substack{y=0 \\ x=x'}}$$

$$= 2\hat{n} \times \mathbf{H}^i \big|_{\substack{y=0 \\ x=x'}} = \hat{a}_y \frac{2E_0}{\eta} \times (-\hat{a}_x \sin\phi_i + \hat{a}_y \cos\phi_i) e^{j\beta x' \cos\phi_i}$$

$$= \hat{a}_z \frac{2E_0}{\eta} \sin\phi_i e^{j\beta x' \cos\phi_i} \tag{11-26}$$

and the far-zone scattered field can be found using (6-96a), (6-101a), and (6-101b) or

$$\mathbf{A} = \frac{\mu}{4\pi} \iint_S \mathbf{J}_s(x', y', z') \frac{e^{-j\beta R}}{R} ds' \tag{11-27a}$$

$$\mathbf{E}_A \simeq -j\omega \mathbf{A} \quad \text{(for θ and ϕ components only)} \tag{11-27b}$$

$$\mathbf{H}_A \simeq \frac{1}{\eta}\hat{a}_r \times \mathbf{E}_A = -j\frac{\omega}{\eta}\hat{a}_r \times \mathbf{A} \quad \text{(for θ and ϕ components only)} \tag{11-27c}$$

where

$$R = \sqrt{(x-x')^2 + (y-y')^2 + (z-z')^2} = \sqrt{(|\boldsymbol{\rho} - \boldsymbol{\rho}'|)^2 + (z-z')^2} \tag{11-27d}$$

Substituting (11-26) into (11-27a) and using (11-27d) we can write that

$$\mathbf{A} = \hat{a}_z \frac{\mu E_0}{2\pi\eta} \sin\phi_i \int_0^w \left\{ \int_{-\infty}^{+\infty} \frac{\exp\left[-j\beta\sqrt{(|\boldsymbol{\rho} - \boldsymbol{\rho}'|)^2 + (z-z')^2}\right]}{\sqrt{(|\boldsymbol{\rho} - \boldsymbol{\rho}'|)^2 + (z-z')^2}} dz' \right\} e^{j\beta x' \cos\phi_i} dx'$$

$$\tag{11-28}$$

Since the integral with the infinite limits can be represented by a Hankel function of the second kind of zero order

$$\int_{-\infty}^{+\infty} \frac{e^{-j\alpha\sqrt{x^2+t^2}}}{\sqrt{x^2+t^2}} \, dt = -j\pi H_0^{(2)}(\alpha x) \tag{11-28a}$$

(11-28) can be reduced to

$$\mathbf{A} = -\hat{a}_z j \frac{\mu E_0}{2\eta} \sin\phi_i \int_0^w H_0^{(2)}(\beta|\boldsymbol{\rho} - \boldsymbol{\rho}'|) e^{j\beta x' \cos\phi_i} \, dx' \tag{11-28b}$$

For far-zone observations

$$|\boldsymbol{\rho} - \boldsymbol{\rho}'| = \sqrt{\rho^2 + (\rho')^2 - 2\rho\rho' \cos(\phi_s - \phi')} \overset{\rho \gg \rho'}{\simeq} \sqrt{\rho^2 - 2\rho\rho' \cos(\phi_s - \phi')}$$

$$\overset{\rho \gg \rho'}{\simeq} \rho\sqrt{1 - 2\left(\frac{\rho'}{\rho}\right)\cos(\phi_s - \phi')} \overset{\rho \gg \rho'}{\simeq} \rho\left[1 - \left(\frac{\rho'}{\rho}\right)\cos(\phi_s - \phi')\right]$$

$$|\boldsymbol{\rho} - \boldsymbol{\rho}'| \overset{\rho \gg \rho'}{\simeq} \rho - \rho' \cos(\phi_s - \phi') \tag{11-29}$$

Since the source (here the current density) exists only over the width of the strip which according to Figure 11-4 lies along the x axis, then $\rho' = x'$ and $\phi' = 0$. Thus for far-field observations (11-29) reduces to

$$|\boldsymbol{\rho} - \boldsymbol{\rho}'| \simeq \begin{cases} \rho - \rho' \cos(\phi_s - \phi') = \rho - x' \cos\phi_s & \text{for phase terms} \tag{11-29a} \\ \rho & \text{for amplitude terms} \tag{11-29b} \end{cases}$$

In turn the Hankel function in the integrand of (11-28b) can be expressed using (11-29a) and (11-29b) as

$$H_0^{(2)}(\beta|\boldsymbol{\rho} - \boldsymbol{\rho}'|) \overset{\rho \gg \rho'}{\simeq} \sqrt{\frac{2j}{\pi\beta\rho}} e^{-j\beta(\rho - x' \cos\phi_s)} = \sqrt{\frac{2j}{\pi\beta}} \frac{e^{-j\beta\rho}}{\sqrt{\rho}} e^{j\beta x' \cos\phi_s} \tag{11-30}$$

Substituting (11-30) into (11-28b) reduces it to

$$\mathbf{A} \simeq -\hat{a}_z j \frac{\mu E_0}{2\eta} \sqrt{\frac{2j}{\pi\beta}} \frac{e^{-j\beta\rho}}{\sqrt{\rho}} \sin\phi_i \int_0^w e^{j\beta x'(\cos\phi_s + \cos\phi_i)} \, dx'$$

$$\mathbf{A} \simeq -\hat{a}_z j \frac{\mu w E_0}{\eta} \sqrt{\frac{j}{2\pi\beta}} e^{j(\beta w/2)(\cos\phi_s + \cos\phi_i)}$$

$$\times \left\{ \sin\phi_i \left[\frac{\sin\left[\frac{\beta w}{2}(\cos\phi_s + \cos\phi_i)\right]}{\frac{\beta w}{2}(\cos\phi_s + \cos\phi_i)} \right] \frac{e^{-j\beta\rho}}{\sqrt{\rho}} \right\} \tag{11-31}$$

Therefore the far-zone scattered spherical components of the electric and magnetic fields of (11-27b) and (11-27c) can be written, using (11-31) and (II-12), as

$$E_\theta^s \simeq j\omega A_z \sin\theta_s = wE_0 \sqrt{\frac{j\beta}{2\pi}}\, e^{j(\beta w/2)(\cos\phi_s + \cos\phi_i)}$$

$$\times \left\{ \sin\theta_s \sin\phi_i \left[\frac{\sin\left[\frac{\beta w}{2}(\cos\phi_s + \cos\phi_i)\right]}{\frac{\beta w}{2}(\cos\phi_s + \cos\phi_i)} \right] \frac{e^{-j\beta\rho}}{\sqrt{\rho}} \right\} \quad (11\text{-}32a)$$

$$H_\phi^s \simeq \frac{E_\theta^s}{\eta} = \frac{wE_0}{\eta}\sqrt{\frac{j\beta}{2\pi}}\, e^{j(\beta w/2)(\cos\phi_s + \cos\phi_i)}$$

$$\times \left\{ \sin\theta_s \sin\phi_i \left[\frac{\sin\left[\frac{\beta w}{2}(\cos\phi_s + \cos\phi_i)\right]}{\frac{\beta w}{2}(\cos\phi_s + \cos\phi_i)} \right] \frac{e^{-j\beta\rho}}{\sqrt{\rho}} \right\} \quad (11\text{-}32b)$$

The bistatic scattering width is obtained using any of (11-21a) through (11-21c), and it is represented at $\theta_s = 90°$ by

$$\sigma_{2\text{-D}}(\text{bistatic}) = \lim_{\rho\to\infty}\left[2\pi\rho\,\frac{|\mathbf{E}^s|^2}{|\mathbf{E}^i|^2}\right] = \frac{2\pi w^2}{\lambda}\left\{ \sin\phi_i \left[\frac{\sin\left[\frac{\beta w}{2}(\cos\phi_s + \cos\phi_i)\right]}{\frac{\beta w}{2}(\cos\phi_s + \cos\phi_i)} \right] \right\}^2 \quad (11\text{-}33)$$

which for the monostatic system configuration ($\phi_s = \phi_i$) reduces to

$$\sigma_{2\text{-D}}(\text{monostatic}) = \frac{2\pi w^2}{\lambda}\left\{ \sin\phi_i \left[\frac{\sin(\beta w \cos\phi_i)}{\beta w \cos\phi_i} \right] \right\}^2 \quad (11\text{-}33a)$$

Computed patterns of the normalized bistatic SW of (11-33) (in dB) for $0° \leq \phi_s \leq 180°$ when $\phi_i = 120°$ and $w = 2\lambda$ and 10λ are shown in Figure 11-5a. It is apparent that the maximum occurs when the $\sin(x)/x$ function reaches its maximum value of unity, that is when $x = \beta w(\cos\phi_s + \cos\phi_i)/2 = 0$. For these examples this occurs when $\phi_s = 180° - \phi_i = 180° - 120° = 60°$ which represents the direction of specular scattering (angle of scattering is equal to the angle of incidence). Away from the direction of maximum radiation the pattern variations are of $\sin(x)/x$ form. The normalized monostatic SW of (11-33a) (in dB) for $w = 2\lambda$ and 10λ are shown plotted in Figure 11-5b for $0° \leq \phi_i \leq 180°$, where the maximum occurs when $\phi_i = 90°$ which is the direction of normal incidence (the strip is viewed perpendicularly to its flat surface). Again away from the maximum radiation the pattern variations are approximately of $\sin(x)/x$ form.

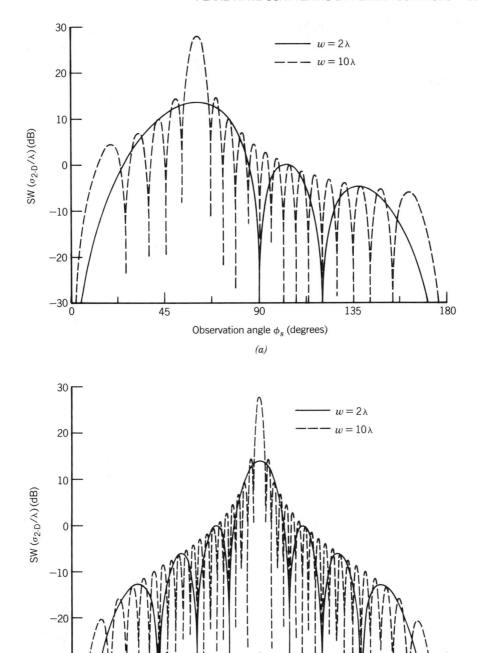

FIGURE 11-5 Bistatic and monostatic scattering width (SW) for a finite width strip. (*a*) Bistatic ($\phi_i = 120°$): TM^z. (*b*) Monostatic: TM^z and TE^z.

Example 11-1. Derive the far-zone scattered fields and the associated scattering width when a TE^z uniform plane wave is incident upon a two-dimensional conducting strip of width w, as shown in Figure 11-4a and c. Use physical optics methods.

Solution. According to Figure 11-4c the incident electric and magnetic field components for a TE^z uniform plane wave can be written as

$$\mathbf{E}^i = \eta H_0 (\hat{a}_x \sin\phi_i - \hat{a}_y \cos\phi_i) e^{j\beta(x\cos\phi_i + y\sin\phi_i)}$$

$$\mathbf{H}^i = \hat{a}_z H_0 e^{j\beta(x\cos\phi_i + y\sin\phi_i)}$$

The current induced on the surface of the finite width strip can be approximated by the physical optics current and is equal to

$$\mathbf{J}_s \simeq 2\hat{n} \times \mathbf{H}^i \Big|_{\substack{y=0 \\ x=x'}} = 2\hat{a}_y \times \hat{a}_z H_z^i \Big|_{\substack{y=0 \\ x=x'}} = 2\hat{a}_x H_z^i \Big|_{\substack{y=0 \\ x=x'}}$$

$$\mathbf{J}_s \simeq \hat{a}_x 2 H_0 e^{j\beta x' \cos\phi_i}$$

$$J_y = J_z = 0, \quad J_x = 2H_0 e^{j\beta x' \cos\phi_i}$$

Using the steps outlined by (11-27a) through (11-31), it can be shown that

$$\mathbf{A} = -\hat{a}_x j \frac{\mu H_0}{2} \int_0^w H_0^{(2)}(\beta|\rho - \rho'|) e^{j\beta x' \cos\phi_i} dx'$$

which for far-zone observations reduces to

$$\mathbf{A} \simeq -\hat{a}_x j\mu w H_0 \sqrt{\frac{j}{2\pi\beta}} e^{j(\beta w/2)(\cos\phi_s + \cos\phi_i)} \left\{ \frac{\sin\left[\frac{\beta w}{2}(\cos\phi_s + \cos\phi_i)\right]}{\frac{\beta w}{2}(\cos\phi_s + \cos\phi_i)} \right\} \frac{e^{-j\beta\rho}}{\sqrt{\rho}}$$

In spherical components this can be written, according to (II-12), as

$$A_r = A_x \sin\theta_s \cos\phi_s$$
$$A_\theta = A_x \cos\theta_s \cos\phi_s$$
$$A_\phi = -A_x \sin\phi_s$$

which for $\theta_s = 90°$ reduce to

$$A_r = A_x \cos\phi_s$$
$$A_\theta = 0$$
$$A_\phi = -A_x \sin\phi_s$$

Thus the far-zone electric and magnetic field components in the $\theta_s = 90°$ plane can be written as

$$E_r^s \simeq E_\theta^s \simeq H_r^s \simeq H_\phi^s \simeq 0$$

$$E_\phi^s \simeq -j\omega A_\phi = j\omega A_x \sin\phi_s$$

$$E_\phi^s \simeq -\eta w H_0 \sqrt{\frac{j\beta}{2\pi}}\, e^{j(\beta w/2)(\cos\phi_s + \cos\phi_i)}$$

$$\times \left\{ \sin\phi_s \left[\frac{\sin\left[\frac{\beta w}{2}(\cos\phi_s + \cos\phi_i)\right]}{\frac{\beta w}{2}(\cos\phi_s + \cos\phi_i)} \right] \frac{e^{-j\beta\rho}}{\sqrt{\rho}} \right\}$$

$$H_\theta^s \simeq -\frac{E_\phi^s}{\eta} = w H_0 \sqrt{\frac{j\beta}{2\pi}}\, e^{j(\beta w/2)(\cos\phi_s + \cos\phi_i)}$$

$$\times \left\{ \sin\phi_s \left[\frac{\sin\left[\frac{\beta w}{2}(\cos\phi_s + \cos\phi_i)\right]}{\frac{\beta w}{2}(\cos\phi_s + \cos\phi_i)} \right] \frac{e^{-j\beta\rho}}{\sqrt{\rho}} \right\}$$

The bistatic and monostatic (backscattering) scattering widths are given by

$$\sigma_{2\text{-D}}(\text{bistatic}) = \lim_{\rho \to \infty}\left[2\pi\rho \frac{|\mathbf{H}^s|^2}{|\mathbf{H}^i|^2} \right]$$

$$= \frac{2\pi w^2}{\lambda}\left\{ \sin\phi_s \left[\frac{\sin\left[\frac{\beta w}{2}(\cos\phi_s + \cos\phi_i)\right]}{\frac{\beta w}{2}(\cos\phi_s + \cos\phi_i)} \right] \right\}^2$$

$$\sigma_{2\text{-D}}(\text{monostatic}) = \frac{2\pi w^2}{\lambda}\left\{ \sin\phi_i \left[\frac{\sin(\beta w \cos\phi_i)}{\beta w \cos\phi_i} \right] \right\}^2$$

The monostatic SW for the TE^z polarization is identical to that of the TM^z as given by (11-33a), and it is shown plotted in Figure 11-5b for $w = 2\lambda$ and 10λ. The bistatic SW, however, differs from that of (11-33) in that the $\sin^2\phi_i$ term is replaced by $\sin^2\phi_s$. Therefore the computed normalized bistatic patterns (in dB) for $0° \leq \phi_s \leq 180°$ when $\phi_1 = 120°$ and $w = 2\lambda$, 10λ are shown, respectively, in Figure 11-6. It is evident that the larger the width of the strip, the larger the maximum value of the SW and the larger the number of minor lobes in its pattern.

586 SCATTERING

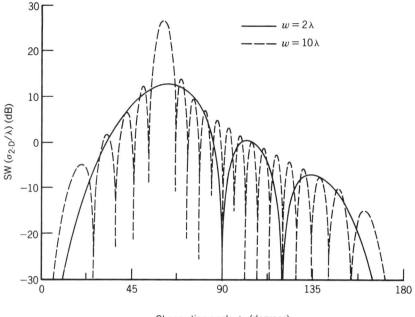

FIGURE 11-6 Bistatic TE^z scattering width (SW) for a finite width strip ($\phi_i = 120°$).

To see that indeed the bistatic SW patterns of the TM^z and TE^z polarizations are different, we have plotted in Figure 11-7a the two for $w = 2\lambda$ when the incidence angle is $\phi_i = 120°$. It is evident that the two are similar but not identical because the $\sin^2 \phi_i$ term is replaced by $\sin^2 \phi_s$, and vice versa. In addition, whereas the maximum for the TM^z occurs at the specular direction ($\phi_s = 60°$ for this example), that of the TE^z occurs at an angle slightly larger than the specular direction. However, as the size of the target becomes very large, the maximum of the TE^z SW moves closer toward the specular direction and matches that of the TM^z, which always occurs at the specular direction [6]. This is illustrated in Figure 11-7b where the bistatic RCS of the two polarizations has been plotted for $w = 10\lambda$ and $\phi_i = 120°$. When the size of the target is very large electrically, the $[\sin(x)/x]^2$ in the bistatic RCS expression of Example 11-1 varies very rapidly as a function of ϕ_s so that the slowly varying $\sin^2 \phi_s$ is essentially a constant near the maximum of the $[\sin(x)/x]^2$ function. This is not true when the size of the target is small electrically, as was demonstrated by the results of Figure 11-7a.

11.3.2 TE^x Plane Wave Scattering from a Flat Rectangular Plate

Let us now consider scattering from a three-dimensional scatterer, specifically uniform plane wave scattering from a rectangular plate as shown in Figure 11-8a. To simplify the details let us assume that the uniform plane wave is TE^x, and that it lies on the yz plane, as shown in Figure 11-8b. The electric and magnetic fields can now be written as

$$\mathbf{E}^i = \eta H_0 \left(\hat{a}_y \cos \theta_i + \hat{a}_z \sin \theta_i \right) e^{-j\beta(y \sin \theta_i - z \cos \theta_i)} \qquad (11\text{-}34a)$$

$$\mathbf{H}^i = \hat{a}_x H_0 e^{-j\beta(y \sin \theta_i - z \cos \theta_i)} \qquad (11\text{-}34b)$$

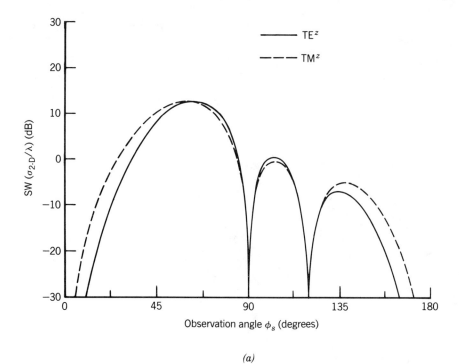

FIGURE 11-7 TE^z and TM^z bistatic scattering widths (SW) for a finite width strip ($\phi_i = 120°$). (a) $w = 2\lambda$. (b) $w = 10\lambda$.

588 SCATTERING

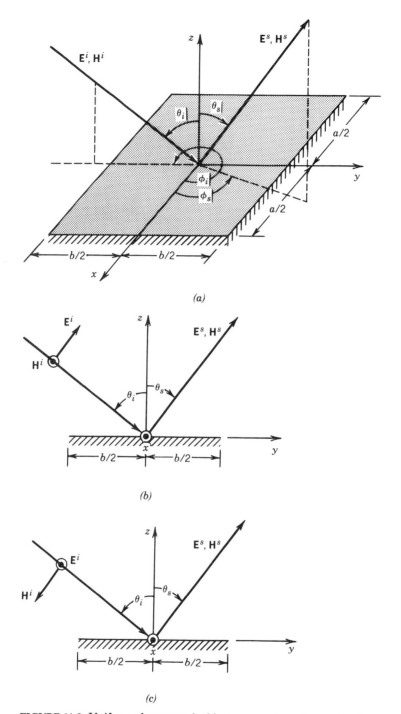

FIGURE 11-8 Uniform plane wave incident on a rectangular conducting plate. (*a*) Rectangular plate. (*b*) *yz* plane: TE^x polarization. (*c*) *yz* plane: TM^x polarization.

PLANE WAVE SCATTERING BY PLANAR SURFACES **589**

where H_0 is a constant and it represents the magnitude of the incident magnetic field.

The scattered field can be found, by neglecting edge effects, using physical optics techniques of Section 7.10 where the current density induced at the surface of the plate is represented on the plate by

$$\mathbf{J}_s \simeq 2\hat{n} \times \mathbf{H}^i\big|_{\substack{z=0 \\ y=y'}} = 2\hat{a}_z \times \hat{a}_x H_x\big|_{\substack{z=0 \\ y=y'}} = \hat{a}_y 2H_0 e^{-j\beta y' \sin\theta_i} \quad (11\text{-}35)$$

Thus

$$J_x = J_z = 0 \quad \text{and} \quad J_y = 2H_0 e^{-j\beta y' \sin\theta_i} \quad (11\text{-}35a)$$

For an infinite plate, the current density of (11-35) or (11-35a) yields exact field solutions. However, for finite size plates it is approximate, and the corresponding scattered fields obtained using it are more accurate toward the specular direction where Snell's law is satisfied. The solutions become less accurate as the observation points are removed further from the specular directions.

The scattered fields are obtained using (6-122a) through (6-122f) and (6-125a) through (6-125b) where the electric current density components are those given by (11-35a). Using (6-125a), (6-127c), (6-128c), and (11-35a), we can write that

$$N_\theta = \iint_S \left[J_x \cos\theta_s \cos\phi_s + J_y \cos\theta_s \sin\phi_s - J_z \sin\theta_s \right]_{J_x = J_z = 0}$$

$$\times e^{j\beta(x' \sin\theta_s \cos\phi_s + y' \sin\theta_s \sin\phi_s)} \, dx' \, dy'$$

$$= 2H_0 \cos\theta_s \sin\phi_s \int_{-b/2}^{+b/2} e^{j\beta y'(\sin\theta_s \sin\phi_s - \sin\theta_i)} \, dy' \int_{-a/2}^{+a/2} e^{j\beta x' \sin\theta_s \cos\phi_s} \, dx' \quad (11\text{-}36)$$

Since

$$\int_{-c/2}^{+c/2} e^{j\alpha z} \, dz = c \left[\frac{\sin\left(\frac{\alpha}{2} c\right)}{\frac{\alpha}{2} c} \right] \quad (11\text{-}37)$$

(11-36) reduces to

$$N_\theta = 2abH_0 \left\{ \cos\theta_s \sin\phi_s \left[\frac{\sin(X)}{X} \right] \left[\frac{\sin(Y)}{Y} \right] \right\} \quad (11\text{-}38)$$

where

$$X = \frac{\beta a}{2} \sin\theta_s \cos\phi_s \quad (11\text{-}38a)$$

$$Y = \frac{\beta b}{2} (\sin\theta_s \sin\phi_s - \sin\theta_i) \quad (11\text{-}38b)$$

In the same manner (6-125b) can be written as

$$N_\phi = \iint_S \left[-J_x \sin\phi_s + J_y \cos\phi_s \right]_{J_x = 0} e^{j\beta(x' \sin\theta_s \cos\phi_s + y' \sin\theta_s \sin\phi_s)} \, dx' \, dy'$$

$$= 2abH_0 \left\{ \cos\phi_s \left[\frac{\sin(X)}{X} \right] \left[\frac{\sin(Y)}{Y} \right] \right\} \quad (11\text{-}39)$$

590 SCATTERING

Therefore the scattered fields are obtained using (6-122a) through (6-122f) and (11-38) through (11-39), and they can be expressed as

$$E_r^s \simeq 0 \tag{11-40a}$$

$$E_\theta^s \simeq -\frac{j\beta e^{-j\beta r}}{4\pi r}(L_\phi + \eta N_\theta)_{L_\phi=0} = C\frac{e^{-j\beta r}}{r}\left\{\cos\theta_s \sin\phi_s \left[\frac{\sin(X)}{X}\right]\left[\frac{\sin(Y)}{Y}\right]\right\} \tag{11-40b}$$

$$E_\phi^s \simeq +\frac{j\beta e^{-j\beta r}}{4\pi r}(L_\theta - \eta N_\phi)_{L_\theta=0} = C\frac{e^{-j\beta r}}{r}\left\{\cos\phi_s \left[\frac{\sin(X)}{X}\right]\left[\frac{\sin(Y)}{Y}\right]\right\} \tag{11-40c}$$

$$H_r^s \simeq 0 \tag{11-40d}$$

$$H_\theta^s \simeq -\frac{E_\phi^s}{\eta} \tag{11-40e}$$

$$H_\phi^s \simeq +\frac{E_\theta^s}{\eta} \tag{11-40f}$$

where

$$C = -j\eta\frac{ab\beta H_0}{2\pi} \tag{11-40g}$$

In the principal E plane ($\phi_s = \pi/2$) and H plane ($\theta_s = \theta_i$, $\phi_s = 0$), the electric field components reduce to

$$\underline{E \text{ Plane } (\phi_s = \pi/2)}$$

$$E_r^s \simeq E_\phi^s \simeq 0 \tag{11-41a}$$

$$E_\theta^s \simeq C\frac{e^{-j\beta r}}{r}\left\{\cos\theta_s \left[\frac{\sin\left[\frac{\beta b}{2}(\sin\theta_s - \sin\theta_i)\right]}{\frac{\beta b}{2}(\sin\theta_s - \sin\theta_i)}\right]\right\} \tag{11-41b}$$

$$\underline{H \text{ Plane } (\phi_s = 0)}$$

$$E_r^s \simeq E_\theta^s \simeq 0 \tag{11-42a}$$

$$E_\phi^s \simeq C\frac{e^{-j\beta r}}{r}\left\{\frac{\sin\left[\frac{\beta b}{2}(\sin\theta_i)\right]}{\frac{\beta b}{2}(\sin\theta_i)}\right\}\left\{\frac{\sin\left[\frac{\beta a}{2}(\sin\theta_s)\right]}{\frac{\beta a}{2}(\sin\theta_s)}\right\} \tag{11-42b}$$

It can be shown that the maximum value of the total scattered field

$$E^s = \sqrt{(E_r^s)^2 + (E_\theta^s)^2 + (E_\phi^s)^2} \simeq \sqrt{(E_\theta^s)^2 + (E_\phi^s)^2} \tag{11-43}$$

for any wave incidence always lies in a scattering plane that is parallel to the incident plane [6]. For the fields of (11-40a) through (11-40f) the scattering plane that contains the maximum scattered field is that defined by $\phi_s = \pi/2, 3\pi/2$ and $0 \leq \theta_s \leq \pi/2$, since the incident plane is that defined by $\phi_i = 3\pi/2$, $0 \leq \theta_i \leq \pi/2$. The electric field components in the plane which contains the maximum reduce to those of (11-41a) and (11-41b) whose maximum value, when $b \gg \lambda$, occurs approximately when $\theta_s = \theta_i$ (specular reflection). For large values of b ($b \gg \lambda$) the $\sin(z)/z$ function in (11-41b) varies very rapidly compared to the $\cos\theta_s$ such that the $\cos\theta_s$ function is essentially constant near the maximum of the $\sin(z)/z$ function. For small values of b the maximum value of (11-41b) can be found iteratively. *Thus for the TE^x polarization the maximum of the scattered field from a flat plate does not occur exactly at the specular direction but it approaches that value as the dimensions of the plate become large compared to the wavelength.* This is analogous to the TE^z polarization of the strip of Example 11-1. It will be shown in the example that follows that *for the TM^x polarization the maximum of the scattered field from a flat plate always occurs at the specular direction no matter what the size of the plate.* This is analogous to the TM^z polarization of the strip of Section 11.3.1.

For the fields of (11-40a) through (11-40g) the radar cross section is obtained using (11-22b) or (11-22c) and can be written as

$$\sigma_{3\text{-D}} = \lim_{r \to \infty} \left[4\pi r^2 \frac{|\mathbf{E}^s|^2}{|\mathbf{E}^i|^2} \right] = \lim_{r \to \infty} \left[4\pi r^2 \frac{|\mathbf{H}^s|^2}{|\mathbf{H}^i|^2} \right]$$

$$= 4\pi \left(\frac{ab}{\lambda} \right)^2 (\cos^2\theta_s \sin^2\phi_s + \cos^2\phi_s) \left[\frac{\sin(X)}{X} \right]^2 \left[\frac{\sin(Y)}{Y} \right]^2 \quad (11\text{-}44)$$

which in the plane that contains the maximum ($\phi_s = \pi/2$) reduces to

Principal Bistatic ($\phi_s = \pi/2$)

$$\sigma_{3\text{-D}} = 4\pi \left(\frac{ab}{\lambda} \right)^2 \cos^2\theta_s \left[\frac{\sin\left[\frac{\beta b}{2}(\sin\theta_s \mp \sin\theta_i)\right]}{\frac{\beta b}{2}(\sin\theta_s \mp \sin\theta_i)} \right]^2 \quad \begin{array}{l} -\text{ for } \phi_s = \dfrac{\pi}{2}, 0 \leq \theta_s \leq \pi/2 \\ \\ +\text{ for } \phi_s = \dfrac{3\pi}{2}, 0 \leq \theta_s \leq \pi/2 \end{array}$$

(11-44a)

while in the backscattering direction ($\phi_s = \phi_i = 3\pi/2$, $\theta_s = \theta_i$) it can be written as

Backscattered

$$\sigma_{3\text{-D}} = 4\pi \left(\frac{ab}{\lambda} \right)^2 \cos^2\theta_i \left[\frac{\sin(\beta b \sin\theta_i)}{\beta b \sin\theta_i} \right]^2 \quad (11\text{-}44\text{b})$$

Plots of (11-44a) for $a = b = 5\lambda$ and $\theta_i = 30°$ ($\phi_s = 90°, 270°$ with $0° \leq \theta_s \leq 90°$) and of (11-44b) for $a = b = 5\lambda$ ($0° \leq \theta_i \leq 90°$) are shown, respectively, in Figures 11-9a and b. It is observed that for Figure 11-9a the maximum occurs when $\phi_s = 90°$ and near $\theta_s \simeq \theta_i = 30°$ while for Figure 11-9b the maximum occurs when $\theta_i = 0°$ (normal incidence).

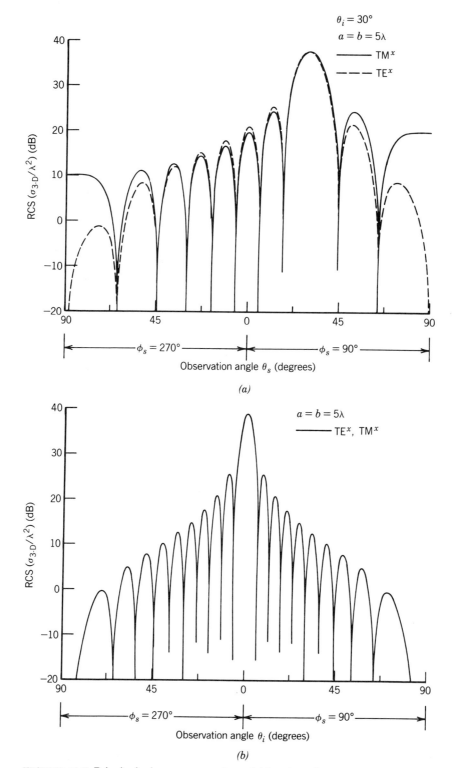

FIGURE 11-9 Principal plane monostatic and bistatic radar cross sections of a rectangular plate. (*a*) Bistatic ($\theta_i = 30°$, $\phi_i = 270°$, $\phi_s = 90°, 270°$). (*b*) Monostatic ($\phi_s = 90°, 270°$).

Example 11-2. Using physical optics techniques, find the scattered fields and radar cross section when a TM^x uniform plane wave is incident upon a flat rectangular plate of dimensions a and b. Assume the incident field lies on the yz plane, as shown in Figure 11-8c.

Solution. According to the geometry of Figure 11-8c, the electric and magnetic field components of the incident uniform plane wave can be written as

$$\mathbf{E}^i = \hat{a}_x E_0 e^{-j\beta(y\sin\theta_i - z\cos\theta_i)}$$

$$\mathbf{H}^i = -\frac{E_0}{\eta}(\hat{a}_y \cos\theta_i + \hat{a}_z \sin\theta_i) e^{-j\beta(y\sin\theta_i - z\cos\theta_i)}$$

Using physical optics techniques the current density induced on the plate can be approximated by

$$\mathbf{J}_s \simeq 2\hat{n} \times \mathbf{H}^i\bigg|_{\substack{z=0\\y=y'}} = 2\hat{a}_z \times (\hat{a}_y H_y + \hat{a}_z H_z)\bigg|_{\substack{z=0\\y=y'}}$$

$$= -2\hat{a}_x H_y\bigg|_{\substack{z=0\\y=y'}} = \hat{a}_x \frac{2E_0}{\eta}\cos\theta_i e^{-j\beta y'\sin\theta_i}$$

or

$$J_y = J_z = 0 \qquad J_x = \frac{2E_0}{\eta}\cos\theta_i e^{-j\beta y'\sin\theta_i}$$

Using (6-125a) and (6-125b) we can write for this problem that

$$N_\theta = \iint_S J_x \cos\theta_s \cos\phi_s e^{j\beta(x'\sin\theta_s \cos\phi_s + y'\sin\theta_s \sin\phi_s)}\,dx'\,dy'$$

$$= \frac{2E_0}{\eta}ab\left\{\cos\theta_i \cos\theta_s \cos\phi_s\left[\frac{\sin(X)}{X}\right]\left[\frac{\sin(Y)}{Y}\right]\right\}$$

$$N_\phi = -\iint_S J_x \sin\phi_s e^{j\beta(x'\sin\theta_s \cos\phi_s + y'\sin\theta_s \sin\phi_s)}\,dx'\,dy'$$

$$= -\frac{2E_0}{\eta}ab\left\{\cos\theta_i \sin\phi_s\left[\frac{\sin(X)}{X}\right]\left[\frac{\sin(Y)}{Y}\right]\right\}$$

where

$$X = \frac{\beta a}{2}\sin\theta_s \cos\phi_s$$

$$Y = \frac{\beta b}{2}(\sin\theta_s \sin\phi_s - \sin\theta_i)$$

The scattered fields are obtained using (6-122a) through (6-122f) and can be

written as

$$E_r^s \simeq H_r^s \simeq 0$$

$$E_\theta^s \simeq -C_1 \frac{e^{-j\beta r}}{r} \left\{ \cos\theta_i \cos\theta_s \cos\phi_s \left[\frac{\sin(X)}{X}\right]\left[\frac{\sin(Y)}{Y}\right]\right\}$$

$$E_\phi^s \simeq C_1 \frac{e^{-j\beta r}}{r} \left\{ \cos\theta_i \sin\phi_s \left[\frac{\sin(X)}{X}\right]\left[\frac{\sin(Y)}{Y}\right]\right\}$$

$$H_\theta^s \simeq -\frac{E_\phi^s}{\eta} \qquad H_\phi^s \simeq +\frac{E_\theta^s}{\eta}$$

$$C_1 = -j\eta \frac{ab\beta E_0}{2\pi}$$

In the principal H plane ($\phi_s = \pi/2$), on which the maximum also lies, the electric field components reduce to

$$E_r^s \simeq E_\theta^s \simeq 0$$

$$E_\phi^s \simeq C_1 \frac{e^{-j\beta r}}{r} \left\{ \cos\theta_i \left[\frac{\sin\left[\frac{\beta b}{2}(\sin\theta_s - \sin\theta_i)\right]}{\frac{\beta b}{2}(\sin\theta_s - \sin\theta_i)}\right]\right\}$$

whose maximum value always occurs when $\theta_s = \theta_i$ (no matter what the size of the ground plane).

The bistatic and monostatic RCSs can be expressed, respectively, as

$$\sigma_{3\text{-D}}(\text{bistatic}) = 4\pi\left(\frac{ab}{\lambda}\right)^2 \left[\cos^2\theta_i\left(\cos^2\theta_s \cos^2\phi_s + \sin^2\phi_s\right)\right]$$
$$\times \left[\frac{\sin(X)}{X}\right]^2\left[\frac{\sin(Y)}{Y}\right]^2$$

$$\sigma_{3\text{-D}}(\text{monostatic}) = 4\pi\left(\frac{ab}{\lambda}\right)^2 \cos^2\theta_i \left[\frac{\sin(\beta b \sin\theta_i)}{\beta b \sin\theta_i}\right]^2$$

They are shown plotted, respectively, in Figure 11-9a and b for a square plate of $a = b = 5\lambda$.

The monostatic RCS of plates using physical optics techniques is insensitive to polarization; it is the same for both polarizations [7–10], as has been demonstrated for the strip and the rectangular plate. Measurements, however, have shown that the monostatic RCS is slightly different for the two polarizations. This is one of the drawbacks of physical optics methods. In addition, the predicted RCSs using physical optics are most accurate at and near the specular directions. However, they begin to become less valid away from the specular directions, especially toward grazing incidences. This will be demonstrated in Example 12-3, and Figures 12-13 and 12-14.

11.4 CYLINDRICAL WAVE TRANSFORMATIONS AND THEOREMS

In scattering it is often most convenient to express wave functions of one coordinate system in terms of wave functions of another coordinate system. An example is a uniform plane wave that can be written in a very simple form in terms of rectilinear wave functions. However, when scattering of plane waves by cylindrical structures is considered, it is most desirable to transform the rectilinear form of the uniform plane wave into terms of cylindrical wave functions. This is desirable because the surface of the cylindrical structure is most conveniently defined using cylindrical coordinates. This and other such transformations are referred to as *wave transformations* [11]. Along with these transformations, certain theorems concerning cylindrical wave functions are very desirable in describing the scattering by cylindrical structures.

11.4.1 Plane Waves in Terms of Cylindrical Wave Functions

The scattering of plane waves by cylindrical structures is considered a fundamental problem in scattering theory. To accomplish this, it is first necessary and convenient to express the plane waves by cylindrical wave functions. To demonstrate that, let us assume that a normalized uniform plane wave traveling in the $+x$ direction, as shown in Figure 11-10, can be written as

$$\mathbf{E} = \hat{a}_z E_z^+ = \hat{a}_z E_0 e^{-j\beta x} = \hat{a}_z e^{-j\beta x} \qquad (11\text{-}45)$$

The plane wave can be represented by an infinite sum of cylindrical wave functions of the form

$$E_z^+ = e^{-j\beta x} = e^{-j\beta \rho \cos \phi} = \sum_{n=-\infty}^{+\infty} a_n J_n(\beta \rho) e^{jn\phi} \qquad (11\text{-}45\text{a})$$

since it must be periodic in ϕ and finite at $\rho = 0$. The next step is to determine the amplitude a_n coefficients.

Multiplying both sides of (11-45a) by $e^{-jm\phi}$, where m is an integer, and integrating from 0 to 2π, we have that

$$\int_0^{2\pi} e^{-j(\beta \rho \cos \phi + m\phi)} d\phi = \int_0^{2\pi} \left[\sum_{n=-\infty}^{+\infty} a_n J_n(\beta \rho) e^{j(n-m)\phi} \right] d\phi \qquad (11\text{-}46)$$

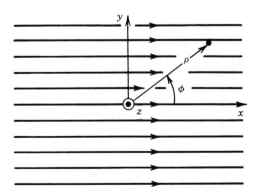

FIGURE 11-10 Uniform plane wave traveling in the $+x$ direction.

Interchanging integration and summation, we have that

$$\int_0^{2\pi} e^{-j(\beta\rho\cos\phi + m\phi)}\,d\phi = \sum_{n=-\infty}^{+\infty} a_n J_n(\beta\rho) \int_0^{2\pi} e^{j(n-m)\phi}\,d\phi \qquad (11\text{-}47)$$

Using the orthogonality condition of

$$\int_0^{2\pi} e^{j(n-m)\phi}\,d\phi = \begin{cases} 2\pi & n = m \\ 0 & n \neq m \end{cases} \qquad (11\text{-}48)$$

the right side of (11-47) reduces to

$$\sum_{n=-\infty}^{+\infty} a_n J_n(\beta\rho) \int_0^{2\pi} e^{j(n-m)\phi}\,d\phi \stackrel{n=m}{=} 2\pi a_m J_m(\beta\rho) \qquad (11\text{-}49)$$

Using the integral of

$$\int_0^{2\pi} e^{j(z\cos\phi + n\phi)}\,d\phi = 2\pi j^n J_n(z) \qquad (11\text{-}50)$$

the left side of (11-47) can be written as

$$\int_0^{2\pi} e^{-j(\beta\rho\cos\phi + m\phi)}\,d\phi = 2\pi j^{-m} J_{-m}(-\beta\rho) \qquad (11\text{-}51)$$

Since

$$J_{-m}(x) = (-1)^m J_m(x) \qquad (11\text{-}52\text{a})$$

and

$$J_m(-x) = (-1)^m J_m(x) \qquad (11\text{-}52\text{b})$$

(11-51) can be written as

$$\int_0^{2\pi} e^{-j(\beta\rho\cos\phi + m\phi)}\,d\phi = 2\pi j^{-m} J_{-m}(-\beta\rho) = 2\pi j^{-m}(-1)^m J_m(-\beta\rho)$$

$$= 2\pi j^{-m}(-1)^m(-1)^m J_m(\beta\rho) = 2\pi j^{-m}(-1)^{2m} J_m(\beta\rho)$$

$$\int_0^{2\pi} e^{-j(\beta\rho\cos\phi + m\phi)}\,d\phi = 2\pi j^{-m} J_{-m}(-\beta\rho) = 2\pi j^{-m} J_m(\beta\rho) \qquad (11\text{-}53)$$

Using (11-49) and (11-53) reduce (11-47) to

$$2\pi j^{-m} J_m(\beta\rho) = 2\pi a_m J_m(\beta\rho) \qquad (11\text{-}54)$$

Thus

$$a_m = j^{-m} \qquad (11\text{-}54\text{a})$$

Therefore (11-45a) can be written as

$$\boxed{E_z^+ = e^{-j\beta x} = e^{-j\beta\rho\cos\phi} = \sum_{n=-\infty}^{+\infty} a_n J_n(\beta\rho) e^{jn\phi} = \sum_{n=-\infty}^{+\infty} j^{-n} J_n(\beta\rho) e^{jn\phi}}$$

$$(11\text{-}55\text{a})$$

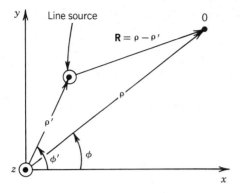

FIGURE 11-11 Geometry for displaced line source.

In a similar manner it can be shown that

$$E_z^- = e^{+j\beta x} = e^{+j\beta\rho\cos\phi} = \sum_{n=-\infty}^{+\infty} j^{+n} J_n(\beta\rho) e^{jn\phi} \qquad (11\text{-}55b)$$

11.4.2 Addition Theorem of Hankel Functions

A transformation that is often convenient and necessary in scattering problems is the *addition theorem of Hankel functions* [11]. Basically it expresses the fields of a cylindrical line source located away from the origin at a radial distance ρ', which are represented by cylindrical wave functions originating at the source, in terms of cylindrical wave functions originating at the origin ($\rho = 0$) of the coordinate system.

To derive this, let us assume that a line source of electric current I_0 is located at $\rho = \rho'$ and $\phi = \phi'$, as shown in Figure 11-11. According to (11-11a), the fields by the line source are given by

$$E_z(\rho, \phi) = -\frac{\beta^2 I_0}{4\omega\varepsilon} H_0^{(2)}(\beta|\rho - \rho'|) \qquad (11\text{-}56)$$

where $|\rho - \rho'|$ is the radial distance from the source to the observation point. Using Figure 11-11 and the law of cosines, (11-56) can also be written as

$$E_z(\rho, \phi) = -\frac{\beta^2 I_0}{4\omega\varepsilon} H_0^{(2)}(\beta|\rho - \rho'|)$$

$$= -\frac{\beta^2 I_0}{4\omega\varepsilon} H_0^{(2)}\left[\beta\sqrt{\rho^2 + (\rho')^2 - 2\rho\rho' \cos(\phi - \phi')}\right] \qquad (11\text{-}56a)$$

Equations 11-56 and 11-56a express the electric field in terms of a Hankel function whose radial distance originates at the source ($|\rho - \rho'| = 0$). For scattering problems, it is very convenient to express the field in terms of cylindrical wave functions, such as Bessel and Hankel functions, whose radial distance originates at the origin ($\rho = 0$). Based upon that and because of reciprocity, the field must be symmetric in terms of the primed and unprimed coordinates. Therefore the permissible wave functions, whose radial distance originates at the origin of the coordinate system, to represent the Hankel function of (11-56a) are of the form

$$\rho \leq \rho'$$

$$f(\beta\rho') J_n(\beta\rho) e^{jn(\phi-\phi')}$$

where n is an integer, because they must be finite at $\rho = 0$ and be periodic in 2π on ϕ.

$$\rho \geq \rho'$$

$$g(\beta\rho') H_n^{(2)}(\beta\rho) e^{jn(\phi-\phi')}$$

where n is an integer, because they must represent outward traveling waves and they must be periodic in 2π on ϕ.

Thus the Hankel function of (11-56a) can be written as

$$H_0^{(2)}(\beta|\rho - \rho'|) = \begin{cases} \sum_{n=-\infty}^{+\infty} c_n f(\beta\rho') J_n(\beta\rho) e^{jn(\phi-\phi')} & \text{for } \rho \leq \rho' \quad (11\text{-}57a) \\ \sum_{n=-\infty}^{+\infty} d_n g(\beta\rho') H_n^{(2)}(\beta\rho) e^{jn(\phi-\phi')} & \text{for } \rho \geq \rho' \quad (11\text{-}57b) \end{cases}$$

Since at $\rho = \rho'$ the fields of the two regions must be continuous, then from (11-57a) and (11-57b)

$$c_n f(\beta\rho') J_n(\beta\rho') = d_n g(\beta\rho') H_n^{(2)}(\beta\rho') \qquad (11\text{-}58)$$

which can be satisfied provided

$$c_n = d_n = b_n \qquad (11\text{-}59a)$$

$$f(\beta\rho') = H_n^{(2)}(\beta\rho') \qquad (11\text{-}59b)$$

$$g(\beta\rho') = J_n(\beta\rho') \qquad (11\text{-}59c)$$

Using (11-59a) through (11-59c), we can write (11-57a) and (11-57b) as

$$H_0^{(2)}(\beta|\rho - \rho'|) = \begin{cases} \sum_{n=-\infty}^{+\infty} b_n J_n(\beta\rho) H_n^{(2)}(\beta\rho') e^{jn(\phi-\phi')} & \text{for } \rho \leq \rho' \quad (11\text{-}60a) \\ \sum_{n=-\infty}^{+\infty} b_n J_n(\beta\rho') H_n^{(2)}(\beta\rho) e^{jn(\phi-\phi')} & \text{for } \rho \geq \rho' \quad (11\text{-}60b) \end{cases}$$

The only remaining part is to evaluate b_n. This can be accomplished by returning to (11-56a), according to which

$$H_0^{(2)}(\beta|\rho - \rho'|) = H_0^{(2)}\left[\beta\sqrt{\rho^2 + (\rho')^2 - 2\rho\rho' \cos(\phi - \phi')}\right] \qquad (11\text{-}61)$$

Moving the source toward infinity ($\rho' \to \infty$) along $\phi' = 0$, the radial distance in (11-61), as represented by the square root, can be approximated using the binomial expansion

$$\sqrt{\rho^2 + (\rho')^2 - 2\rho\rho'\cos(\phi-\phi')} \underset{\rho'\to\infty}{\overset{\phi'=0}{\simeq}} \sqrt{(\rho')^2 - 2\rho\rho'\cos\phi} = \rho'\sqrt{1 - 2\left(\frac{\rho}{\rho'}\right)\cos\phi}$$

$$\simeq \rho'\left(1 - \frac{\rho}{\rho'}\cos\phi\right) = \rho' - \rho\cos\phi \qquad (11\text{-}62)$$

Using (11-62) allows us to write (11-61) as

$$H_0^{(2)}(\beta|\rho - \rho'|) = H_0^{(2)}\left[\beta\sqrt{\rho^2 + (\rho')^2 - 2\rho\rho'\cos(\phi-\phi')}\right]$$

$$\underset{\rho'\to\infty}{\overset{\phi'=0}{\simeq}} H_0^{(2)}[\beta(\rho' - \rho\cos\phi)] \qquad (11\text{-}63)$$

With the aid of the asymptotic form of the Hankel function for large argument

$$H_n^{(2)}(\alpha x) \overset{\alpha x\to\infty}{\simeq} \sqrt{\frac{2j}{\pi\alpha x}}\, j^n e^{-j\alpha x} \qquad (11\text{-}64)$$

(11-63) reduces to

$$H_0^{(2)}(\beta|\rho - \rho'|) \underset{\rho'\to\infty}{\overset{\phi'=0}{\simeq}} H_0^{(2)}[\beta(\rho' - \rho\cos\phi)] \simeq \sqrt{\frac{2j}{\pi\beta(\rho' - \rho\cos\phi)}}\, e^{-j\beta(\rho'-\rho\cos\phi)}$$

$$\simeq \sqrt{\frac{2j}{\pi\beta\rho'}}\, e^{-j\beta\rho'} e^{+j\beta\rho\cos\phi} \qquad (11\text{-}65)$$

Using (11-55b) allows us to write (11-65) as

$$H_0^{(2)}(\beta|\rho - \rho'|) \underset{\rho'\to\infty}{\overset{\phi'=0}{\simeq}} \sqrt{\frac{2j}{\pi\beta\rho'}}\, e^{-j\beta\rho'} \sum_{n=-\infty}^{+\infty} j^n J_n(\beta\rho) e^{jn\phi} \qquad (11\text{-}66)$$

Applying (11-64) to (11-60a) for $\phi' = 0$ and $\rho' \to \infty$, reduces it to

$$H_0^{(2)}(\beta|\rho - \rho'|) \overset{\phi'=0}{=} \sum_{n=-\infty}^{+\infty} b_n J_n(\beta\rho) H_n^{(2)}(\beta\rho') e^{jn\phi}$$

$$\underset{\rho'\to\infty}{\overset{\phi'=0}{\simeq}} \sqrt{\frac{2j}{\pi\beta\rho'}}\, e^{-j\beta\rho'} \sum_{n=-\infty}^{+\infty} b_n j^n J_n(\beta\rho) e^{jn\phi} \qquad (11\text{-}67)$$

Comparing (11-66) and (11-67) leads to

$$b_n = 1 \qquad (11\text{-}68)$$

Thus the final form of (11-60a) and (11-60b) is

$$H_0^{(2)}(\beta|\boldsymbol{\rho} - \boldsymbol{\rho}'|) = \begin{cases} \sum_{n=-\infty}^{+\infty} J_n(\beta\rho) H_n^{(2)}(\beta\rho') e^{jn(\phi-\phi')} & \text{for } \rho \leq \rho' \quad (11\text{-}69\text{a}) \\ \sum_{n=-\infty}^{+\infty} J_n(\beta\rho') H_n^{(2)}(\beta\rho) e^{jn(\phi-\phi')} & \text{for } \rho \geq \rho' \quad (11\text{-}69\text{b}) \end{cases}$$

which can be used to write (11-56) as

$$E_z(\rho, \phi) = -\frac{\beta^2 I_0}{4\omega\varepsilon} H_0^{(2)}(\beta|\boldsymbol{\rho} - \boldsymbol{\rho}'|)$$

$$= -\frac{\beta^2 I_0}{4\omega\varepsilon} \begin{cases} \sum_{n=-\infty}^{+\infty} J_n(\beta\rho) H_n^{(2)}(\beta\rho') e^{jn(\phi-\phi')} & \text{for } \rho \leq \rho' \quad (11\text{-}70\text{a}) \\ \sum_{n=-\infty}^{+\infty} J_n(\beta\rho') H_n^{(2)}(\beta\rho) e^{jn(\phi-\phi')} & \text{for } \rho \geq \rho' \quad (11\text{-}70\text{b}) \end{cases}$$

The procedure can be repeated to expand $H_0^{(1)}(\beta|\boldsymbol{\rho} - \boldsymbol{\rho}'|)$. However, it is obvious from the results of (11-69a) and (11-69b) that

$$H_0^{(1)}(\beta|\boldsymbol{\rho} - \boldsymbol{\rho}'|) = \begin{cases} \sum_{n=-\infty}^{+\infty} J_n(\beta\rho) H_n^{(1)}(\beta\rho') e^{jn(\phi-\phi')} & \text{for } \rho \leq \rho' \quad (11\text{-}71\text{a}) \\ \sum_{n=-\infty}^{+\infty} J_n(\beta\rho') H_n^{(1)}(\beta\rho) e^{jn(\phi-\phi')} & \text{for } \rho \geq \rho' \quad (11\text{-}71\text{b}) \end{cases}$$

11.4.3 Addition Theorem for Bessel Functions

Another theorem that is often useful represents Bessel functions originating at the source, which is located away from the origin, in terms of cylindrical wave functions originating at the origin of the coordinate system. This is usually referred to as the *addition theorem for Bessel functions* [11].

We know that the Hankel functions of the first and second kinds can be written, in terms of the Bessel functions, as

$$H_0^{(1)}(\beta\rho) = J_0(\beta\rho) + jY_0(\beta\rho) \quad (11\text{-}72\text{a})$$

$$H_0^{(2)}(\beta\rho) = J_0(\beta\rho) - jY_0(\beta\rho) \quad (11\text{-}72\text{b})$$

Adding the two, we can write that

$$J_0(\beta\rho) = \tfrac{1}{2}\left[H_0^{(1)}(\beta\rho) + H_0^{(2)}(\beta\rho)\right] \quad (11\text{-}73)$$

Therefore

$$J_0(\beta|\boldsymbol{\rho} - \boldsymbol{\rho}'|) = \tfrac{1}{2}\left[H_0^{(1)}(\beta|\boldsymbol{\rho} - \boldsymbol{\rho}'|) + H_0^{(2)}(\beta|\boldsymbol{\rho} - \boldsymbol{\rho}'|)\right] \quad (11\text{-}74)$$

With the aid of (11-69a), (11-69b) and (11-71a), (11-71b) we can write (11-74) as

$$J_0(\beta|\rho - \rho'|) = \begin{cases} \sum_{n=-\infty}^{+\infty} \tfrac{1}{2}[H_n^{(1)}(\beta\rho') + H_n^{(2)}(\beta\rho')]J_n(\beta\rho)e^{jn(\phi-\phi')} & \text{for } \rho \leq \rho' \quad (11\text{-}75\text{a}) \\ \sum_{n=-\infty}^{+\infty} \tfrac{1}{2}[H_n^{(1)}(\beta\rho) + H_n^{(2)}(\beta\rho)]J_n(\beta\rho')e^{jn(\phi-\phi')} & \text{for } \rho \geq \rho' \quad (11\text{-}75\text{b}) \end{cases}$$

Using the forms of (11-72a) and (11-72b) for nth order Bessel and Hankel functions, we can write (11-75a) and (11-75b) as

$$J_0(\beta|\rho - \rho'|) = \begin{cases} \sum_{n=-\infty}^{+\infty} J_n(\beta\rho')J_n(\beta\rho)e^{jn(\phi-\phi')} & \text{for } \rho \leq \rho' \quad (11\text{-}76\text{a}) \\ \sum_{n=-\infty}^{+\infty} J_n(\beta\rho)J_n(\beta\rho')e^{jn(\phi-\phi')} & \text{for } \rho \geq \rho' \quad (11\text{-}76\text{b}) \end{cases}$$

or that

$$\boxed{J_0(\beta|\rho - \rho'|) = \sum_{n=-\infty}^{+\infty} J_n(\beta\rho)J_n(\beta\rho')e^{jn(\phi-\phi')} \quad \text{for } \rho \lessgtr \rho'} \quad (11\text{-}77)$$

Subtracting (11-72b) from (11-72a), we can write that

$$H_0^{(1)}(\beta\rho) - H_0^{(2)}(\beta\rho) = 2jY_0(\beta\rho)$$

$$Y_0(\beta\rho) = \frac{1}{2j}[H_0^{(1)}(\beta\rho) - H_0^{(2)}(\beta\rho)] \quad (11\text{-}78\text{a})$$

or

$$Y_0(\beta|\rho - \rho'|) = \frac{1}{2j}[H_0^{(1)}(\beta|\rho - \rho'|) - H_0^{(2)}(\beta|\rho - \rho'|)] \quad (11\text{-}78\text{b})$$

Using (11-69a), (11-69b) and (11-71a), (11-71b) we can write (11-78b) as

$$Y_0(\beta|\rho - \rho'|) = \begin{cases} \sum_{n=-\infty}^{\infty} \tfrac{1}{2}[H_n^{(1)}(\beta\rho') - H_n^{(2)}(\beta\rho')]J_n(\beta\rho)e^{jn(\phi-\phi')} & \text{for } \rho \leq \rho' \quad (11\text{-}79\text{a}) \\ \sum_{n=-\infty}^{+\infty} \tfrac{1}{2}[H_n^{(1)}(\beta\rho) - H_n^{(2)}(\beta\rho)]J_n(\beta\rho')e^{jn(\phi-\phi')} & \text{for } \rho \geq \rho' \quad (11\text{-}79\text{b}) \end{cases}$$

Using the forms of (11-72a) and (11-72b) for nth order Hankel functions, we can write (11-79a) and (11-79b) as

$$Y_0(\beta|\boldsymbol{\rho} - \boldsymbol{\rho}'|) = \begin{cases} \sum_{n=-\infty}^{+\infty} \frac{1}{2j}[J_n(\beta\rho') + jY_n(\beta\rho') - J_n(\beta\rho') + jY_n(\beta\rho')]J_n(\beta\rho)e^{jn(\phi-\phi')} \\ \quad = \sum_{n=-\infty}^{+\infty} Y_n(\beta\rho')J_n(\beta\rho)e^{jn(\phi-\phi')} \quad \text{for } \rho \leq \rho' \quad (11\text{-}80a) \\ \sum_{n=-\infty}^{+\infty} \frac{1}{2j}[J_n(\beta\rho) + jY_n(\beta\rho) - J_n(\beta\rho) + jY_n(\beta\rho)]J_n(\beta\rho')e^{jn(\phi-\phi')} \quad (11\text{-}80b) \\ \quad = \sum_{n=-\infty}^{+\infty} Y_n(\beta\rho)J_n(\beta\rho')e^{jn(\phi-\phi')} \quad \text{for } \rho \geq \rho' \end{cases}$$

11.4.4 Summary of Cylindrical Wave Transformations and Theorems

The following are the most prominent cylindrical wave transformations and theorems which are very convenient for scattering from cylindrical scatterers:

$$e^{-j\beta x} = e^{-j\beta\rho\cos\phi} = \sum_{n=-\infty}^{+\infty} j^{-n}J_n(\beta\rho)e^{jn\phi} \quad (11\text{-}81a)$$

$$e^{+j\beta x} = e^{+j\beta\rho\cos\phi} = \sum_{n=-\infty}^{+\infty} j^{+n}J_n(\beta\rho)e^{jn\phi} \quad (11\text{-}81b)$$

$$H_0^{(1,2)}(\beta|\boldsymbol{\rho} - \boldsymbol{\rho}'|) = \begin{cases} \sum_{n=-\infty}^{+\infty} J_n(\beta\rho)H_n^{(1,2)}(\beta\rho')e^{jn(\phi-\phi')} & \rho \leq \rho' \quad (11\text{-}82a) \\ \sum_{n=-\infty}^{+\infty} J_n(\beta\rho')H_n^{(1,2)}(\beta\rho)e^{jn(\phi-\phi')} & \rho \geq \rho' \quad (11\text{-}82b) \end{cases}$$

$$J_0(\beta|\boldsymbol{\rho} - \boldsymbol{\rho}'|) = \sum_{n=-\infty}^{+\infty} J_n(\beta\rho')J_n(\beta\rho)e^{jn(\phi-\phi')} \quad \rho \lessgtr \rho' \quad (11\text{-}83)$$

$$Y_0(\beta|\boldsymbol{\rho} - \boldsymbol{\rho}'|) = \begin{cases} \sum_{n=-\infty}^{+\infty} Y_n(\beta\rho')J_n(\beta\rho)e^{jn(\phi-\phi')} & \rho \leq \rho' \quad (11\text{-}84a) \\ \sum_{n=-\infty}^{+\infty} Y_n(\beta\rho)J_n(\beta\rho')e^{jn(\phi-\phi')} & \rho \geq \rho' \quad (11\text{-}84b) \end{cases}$$

11.5 SCATTERING BY CIRCULAR CYLINDERS

Cylinders represent one of the most important classes of geometrical surfaces. The surface of many practical scatterers, such as the fuselage of airplanes, missiles, and so on, can often be represented by cylindrical structures. The circular cylinder, because of its simplicity and its solution is represented in terms of well known and

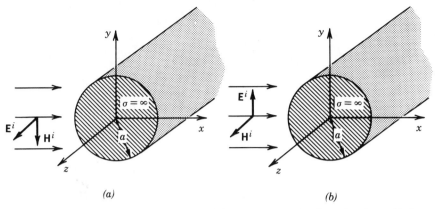

FIGURE 11-12 Uniform plane wave incident on a conducting circular cylinder. (a) TM^z. (b) TE^z.

tabulated functions (such as Bessel and Hankel functions), is probably one of the geometries most widely used to represent practical scatterers [12]. Because of its importance, it will be examined here in some detail. We will consider scattering of both plane and cylindrical waves by circular conducting cylinders of infinite length at normal and oblique incidences. The solutions will be obtained using modal techniques. Scattering from finite length cylinders is obtained by transforming the scattered fields of infinite lengths using approximate relationships. Scattering by dielectric and dielectric covered cylinders will be assigned to the reader as end of chapter exercises.

11.5.1 Normal Incidence Plane Wave Scattering by Conducting Circular Cylinder: TM^z Polarization

Let us assume that a TM^z uniform plane wave is normally incident upon a perfectly conducting circular cylinder of radius a, as shown in Figure 11-12a, and the electric field can be written as

$$\mathbf{E}^i = \hat{a}_z E_z^i = \hat{a}_z E_0 e^{-j\beta x} = \hat{a}_z E_0 e^{-j\beta\rho\cos\phi} \quad (11\text{-}85)$$

which according to the transformation of (11-55a) or (11-81a) can also be expressed as

$$\mathbf{E}^i = \hat{a}_z E_z^i = \hat{a}_z E_0 \sum_{n=-\infty}^{+\infty} j^{-n} J_n(\beta\rho) e^{jn\phi} = \hat{a}_z E_0 \sum_{n=0}^{\infty} (-j)^n \varepsilon_n J_n(\beta\rho) \cos(n\phi) \quad (11\text{-}85\text{a})$$

where

$$\varepsilon_n = \begin{cases} 1 & n = 0 \\ 2 & n \neq 0 \end{cases} \quad (11\text{-}85\text{b})$$

The corresponding magnetic field components can be obtained by using Maxwell's Faraday equation, which in this instance reduces to

$$\mathbf{H}^i = -\frac{1}{j\omega\mu} \nabla \times \mathbf{E}^i = -\frac{1}{j\omega\mu}\left(\hat{a}_\rho \frac{1}{\rho}\frac{\partial E_z^i}{\partial \phi} - \hat{a}_\phi \frac{\partial E_z^i}{\partial \rho}\right) \quad (11\text{-}86)$$

or

$$H_\rho^i = -\frac{1}{j\omega\mu}\frac{1}{\rho}\frac{\partial E_z^i}{\partial \phi} = -\frac{E_0}{j\omega\mu}\frac{1}{\rho}\sum_{n=-\infty}^{+\infty} nj^{-n+1}J_n(\beta\rho)e^{jn\phi} \quad (11\text{-}86\text{a})$$

$$H_\phi^i = \frac{1}{j\omega\mu}\frac{\partial E_z^i}{\partial \rho} = \frac{\beta E_0}{j\omega\mu}\sum_{n=-\infty}^{+\infty} j^{-n}J_n'(\beta\rho)e^{jn\phi} \quad (11\text{-}86\text{b})$$

$$' = \frac{\partial}{\partial(\beta\rho)} \quad (11\text{-}86\text{c})$$

It should be noted here that throughout this chapter the prime indicates partial derivative with respect to the entire argument of the Bessel or Hankel function.

In the presence of the conducting cylinder the total field E_z^t according to (11-1a) can be written as

$$\mathbf{E}^t = \mathbf{E}^i + \mathbf{E}^s \quad (11\text{-}87)$$

where $\mathbf{E}^s$ is the scattered field. Since the scattered fields travel in the outward direction, they must be represented by cylindrical traveling wave functions. Thus we choose to represent $\mathbf{E}^s$ by

$$\mathbf{E}^s = \hat{a}_z E_z^s = \hat{a}_z E_0 \sum_{n=-\infty}^{+\infty} c_n H_n^{(2)}(\beta\rho) \quad (11\text{-}88)$$

where c_n represents the yet unknown amplitude coefficients. Equation 11-88 is chosen to be of similar form to (11-85a) since the two together will be used to represent the total field. This becomes convenient when we attempt to solve for the amplitude coefficients c_n.

The unknown amplitude coefficients c_n can be found by applying the boundary condition of

$$\mathbf{E}^t = \hat{a}_z E_z^t(\rho = a, 0 \le \phi \le 2\pi, z) = 0 \quad (11\text{-}89)$$

Using (11-85a), (11-88), and (11-89) we can write that

$$E_z^t(\rho = a, 0 \le \phi \le 2\pi, z) = E_0 \sum_{n=-\infty}^{\infty} \left[j^{-n}J_n(\beta a)e^{jn\phi} + c_n H_n^{(2)}(\beta a) \right] = 0$$

$$(11\text{-}90)$$

or

$$c_n = -j^{-n}\frac{J_n(\beta a)}{H_n^{(2)}(\beta a)}e^{jn\phi} \quad (11\text{-}90\text{a})$$

Thus the scattered field of (11-88) reduces to

$$E_z^s = -E_0 \sum_{n=-\infty}^{+\infty} j^{-n}\frac{J_n(\beta a)}{H_n^{(2)}(\beta a)}H_n^{(2)}(\beta\rho)e^{jn\phi}$$

$$= -E_0 \sum_{n=0}^{+\infty} (-j)^n \varepsilon_n \frac{J_n(\beta a)}{H_n^{(2)}(\beta a)}H_n^{(2)}(\beta\rho)\cos(n\phi) \quad (11\text{-}91)$$

where ε_n is defined by (11-85b).

SCATTERING BY CIRCULAR CYLINDERS

The corresponding scattered magnetic field components can be obtained by using Maxwell's equation 11-86 which leads to

$$H_\rho^s = -\frac{1}{j\omega\mu}\frac{1}{\rho}\frac{\partial E_z^s}{\partial \phi} = -\frac{E_0}{j\omega\mu}\frac{1}{\rho}\sum_{n=-\infty}^{+\infty} nj^{-n+1}\frac{J_n(\beta a)}{H_n^{(2)}(\beta a)}H_n^{(2)}(\beta\rho)e^{jn\phi} \quad (11\text{-}92a)$$

$$H_\phi^s = \frac{1}{j\omega\mu}\frac{\partial E_z^s}{\partial \rho} = -\frac{\beta E_0}{j\omega\mu}\sum_{n=-\infty}^{+\infty} j^{-n}\frac{J_n(\beta a)}{H_n^{(2)}(\beta a)}H_n^{(2)\prime}(\beta\rho)e^{jn\phi} \quad (11\text{-}92b)$$

Thus the total electric and magnetic field components can be written as

$$E_\rho^t = E_\phi^t = H_z^t = 0 \quad (11\text{-}93a)$$

$$E_z^t = E_0 \sum_{n=-\infty}^{+\infty} j^{-n}\left[J_n(\beta\rho) - \frac{J_n(\beta a)}{H_n^{(2)}(\beta a)}H_n^{(2)}(\beta\rho)\right]e^{jn\phi} \quad (11\text{-}93b)$$

$$H_\rho^t = -\frac{E_0}{j\omega\mu}\frac{1}{\rho}\sum_{n=-\infty}^{+\infty} nj^{-n+1}\left[J_n(\beta\rho) - \frac{J_n(\beta a)}{H_n^{(2)}(\beta a)}H_n^{(2)}(\beta\rho)\right]e^{jn\phi} \quad (11\text{-}93c)$$

$$H_\phi^t = \frac{\beta E_0}{j\omega\mu}\sum_{n=-\infty}^{+\infty} j^{-n}\left[J_n'(\beta\rho) - \frac{J_n(\beta a)}{H_n^{(2)}(\beta a)}H_n^{(2)\prime}(\beta\rho)\right]e^{jn\phi} \quad (11\text{-}93d)$$

On the surface of the cylinder ($\rho = a$) the total tangential magnetic field can be written as

$$H_\phi^t(\rho = a) = \frac{\beta E_0}{j\omega\mu}\sum_{n=-\infty}^{+\infty} j^{-n}\left[J_n'(\beta a) - \frac{J_n(\beta a)}{H_n^{(2)}(\beta a)}H_n^{(2)\prime}(\beta a)\right]e^{jn\phi}$$

$$= \frac{\beta E_0}{\omega\mu}\sum_{n=-\infty}^{+\infty} j^{-n}\left[\frac{J_n(\beta a)Y_n'(\beta a) - J_n'(\beta a)Y_n(\beta a)}{H_n^{(2)}(\beta a)}\right]e^{jn\phi} \quad (11\text{-}94)$$

Using the Wronskian of Bessel functions

$$J_n(\alpha\rho)Y_n'(\alpha\rho) - Y_n(\alpha\rho)J_n'(\alpha\rho) = \frac{2}{\pi\alpha\rho} \quad (11\text{-}95)$$

reduces (11-94) to

$$H_\phi^t(\rho = a) = \frac{2E_0}{\pi a\omega\mu}\sum_{n=-\infty}^{+\infty} j^{-n}\frac{e^{jn\phi}}{H_n^{(2)}(\beta a)} \quad (11\text{-}96)$$

Thus the current induced on the surface of the cylinder can be written as

$$\mathbf{J}_s = \hat{n}\times\mathbf{H}^t|_{\rho=a} = \hat{a}_\rho\times\left(\hat{a}_\rho H_\rho^t + \hat{a}_\phi H_\phi^t\right)|_{\rho=a} = \hat{a}_z H_\phi^t(\rho = a)$$

$$= \hat{a}_z\frac{2E_0}{\pi a\omega\mu}\sum_{n=-\infty}^{+\infty} j^{-n}\frac{e^{jn\phi}}{H_n^{(2)}(\beta a)} \quad (11\text{-}97)$$

A. SMALL RADIUS APPROXIMATION

As the radius of the cylinder increases, more terms in the infinite series of (11-97) are needed to obtain convergence. However, for very small cylinders, like a very thin wire ($a \ll \lambda$), the first term ($n = 0$) in (11-97) is dominant and is often sufficient to represent the induced current. Thus for a very thin wire (11-97) can be approximated by

$$\mathbf{J}_s \stackrel{a \ll \lambda}{\simeq} \hat{a}_z \frac{2E_0}{\pi a \omega \mu} \frac{1}{H_0^{(2)}(\beta a)} \qquad (11\text{-}98)$$

where

$$H_0^{(2)}(\beta a) = J_0(\beta a) - jY_0(\beta a) \stackrel{a \ll \lambda}{\simeq} 1 - j\frac{2}{\pi} \ln\left(\frac{\gamma \beta a}{2}\right) = 1 - j\frac{2}{\pi} \ln\left(\frac{1.781 \beta a}{2}\right)$$

$$\stackrel{a \ll \lambda}{\simeq} -j\frac{2}{\pi} \ln\left(\frac{1.781 \beta a}{2}\right) \qquad (11\text{-}98a)$$

Thus for a very thin wire the current density of (11-98) can be approximated by

$$\mathbf{J}_z \stackrel{a \ll \lambda}{\simeq} \hat{a}_z j \frac{E_0}{a \omega \mu} \frac{1}{\ln\left(\dfrac{1.781 \beta a}{2}\right)} \qquad (11\text{-}98b)$$

B. FAR-ZONE SCATTERED FIELD

One of the most important parameters in scattering is the scattering width which is obtained by knowing the scattered field in the far zone. For this problem this can be accomplished by first reducing the scattered fields for far-zone observation ($\beta \rho \to$ large). Referring to (11-91), the Hankel function can be approximated for observations made in the far field by

$$H_n^{(2)}(\beta \rho) \stackrel{\beta \rho \to \text{large}}{\simeq} \sqrt{\frac{2j}{\pi \beta \rho}}\, j^n e^{-j\beta \rho} \qquad (11\text{-}99)$$

which when substituted in (11-91) reduces it to

$$E_z^s \stackrel{\beta \rho \to \infty}{\simeq} -E_0 \sqrt{\frac{2j}{\pi \beta}} \frac{e^{-j\beta \rho}}{\sqrt{\rho}} \sum_{n=-\infty}^{+\infty} \frac{J_n(\beta a)}{H_n^{(2)}(\beta a)} e^{jn\phi} \qquad (11\text{-}100)$$

The ratio of the far-zone scattered electric field to the incident field can then be written as

$$\frac{|E_z^s|}{|E_z^i|} \stackrel{\beta \rho \to \text{large}}{\simeq} \frac{\left|-E_0 \sqrt{\dfrac{2j}{\pi \beta}} \dfrac{e^{-j\beta \rho}}{\sqrt{\rho}} \sum_{n=-\infty}^{+\infty} \dfrac{J_n(\beta a)}{H_n^{(2)}(\beta a)} e^{jn\phi}\right|}{|E_0 e^{-j\beta x}|}$$

$$= \sqrt{\frac{2}{\pi \beta \rho}} \left|\sum_{n=-\infty}^{+\infty} \frac{J_n(\beta a)}{H_n^{(2)}(\beta a)} e^{jn\phi}\right| \qquad (11\text{-}101)$$

Thus the scattering width of (11-21b) can be expressed as

$$\boxed{\begin{aligned}\sigma_{2\text{-D}} &= \lim_{\rho\to\infty}\left[2\pi\rho\frac{|E_z^s|^2}{|E_z^i|^2}\right] = \frac{4}{\beta}\left|\sum_{n=-\infty}^{+\infty}\frac{J_n(\beta a)}{H_n^{(2)}(\beta a)}e^{jn\phi}\right|^2 \\ &= \frac{2\lambda}{\pi}\left|\sum_{n=0}^{+\infty}\varepsilon_n\frac{J_n(\beta a)}{H_n^{(2)}(\beta a)}\cos(n\phi)\right|^2\end{aligned}}$$

(11-102)

where

$$\varepsilon_n = \begin{cases} 1 & n=0 \\ 2 & n\neq 0 \end{cases}$$

(11-102a)

Plots of the bistatic $\sigma_{2\text{-D}}/\lambda$ computed using (11-102) are shown in Figure 11-13 for cylinder radii of $a = 0.05\lambda$, 0.1λ, 0.2λ, 0.4, and 0.6λ [13]. The backscattered ($\phi = 180°$) patterns of $\sigma_{2\text{-D}}/\lambda$ as a function of the cylinder radius are displayed in Figure 11-14 [13].

For small radii ($a \ll \lambda$) the first term ($n = 0$) in (11-102) is the dominant term, and it is sufficient to represent the scattered field. Thus for small radii the ratio of the Bessel to the Hankel function for $n = 0$ can be approximated using (11-98a) by

$$\frac{J_0(\beta a)}{H_0^{(2)}(\beta a)} \underset{a\ll\lambda}{\simeq} \frac{1}{-j\dfrac{2}{\pi}\ln(0.89\beta a)} = j\frac{\pi}{2}\frac{1}{\ln(0.89\beta a)}$$

(11-103)

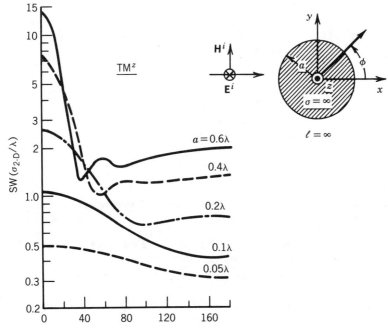

FIGURE 11-13 Two-dimensional TM^z bistatic scattering width (SW) of a circular conducting cylinder. (Courtesy J. H. Richmond, Ohio State University.)

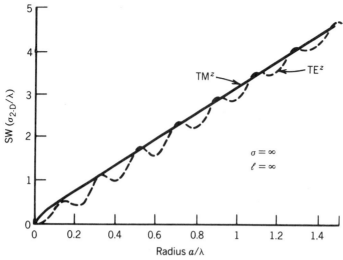

FIGURE 11-14 Two-dimensional monostatic (backscattered) scattering width for a circular conducting cylinder as a function of its radius. (Courtesy J. H. Richmond, Ohio State University.)

and (11-102) can then be reduced to

$$\sigma_{2\text{-D}} \stackrel{a \ll \lambda}{\simeq} \frac{2\lambda}{\pi} \left(\frac{\pi^2}{4} \right) \left| \frac{1}{\ln(0.89\beta a)} \right|^2 = \frac{\pi\lambda}{2} \left| \frac{1}{\ln(0.89\beta a)} \right|^2 \quad (11\text{-}103a)$$

This is independent of ϕ which becomes evident in Figure 11-13 by the curves for the smaller values of a.

For a cylinder of finite length ℓ the three-dimensional radar cross section for normal incidence is related to the two-dimensional scattering width by (11-22e). Thus using (11-102) and (11-103a) we can write the three-dimensional RCS using (11-22e) as

$$\sigma_{3\text{-D}} \simeq \frac{4\ell^2}{\pi} \left| \sum_{n=-\infty}^{+\infty} \frac{J_n(\beta a)}{H_n^{(2)}(\beta a)} e^{jn\phi} \right|^2 \quad (11\text{-}104a)$$

$$\sigma_{3\text{-D}} \stackrel{a \ll \lambda}{\simeq} \pi\ell^2 \left| \frac{1}{\ln(0.89\beta a)} \right|^2 \quad (11\text{-}104b)$$

11.5.2 Normal Incidence Plane Wave Scattering by Conducting Circular Cylinder: TE^z Polarization

Now let us assume that a TE^z uniform plane wave traveling in the $+x$ direction is normally incident upon a perfectly conducting circular cylinder of radius a, as shown in Figure 11-12b. The incident magnetic field can be written as

$$\mathbf{H}^i = \hat{a}_z H_0 e^{-j\beta x} = \hat{a}_z H_0 e^{-j\beta\rho\cos\phi} = \hat{a}_z H_0 \sum_{n=-\infty}^{+\infty} j^{-n} J_n(\beta\rho) e^{jn\phi}$$

$$= \hat{a}_z H_0 \sum_{n=0}^{\infty} (-j)^n \varepsilon_n J_n(\beta\rho) \cos(n\phi) \quad (11\text{-}105)$$

where ε_n is defined by (11-85b). The corresponding incident electric field can be obtained by using Maxwell's Ampere equation, which in this instance reduces to

$$\mathbf{E}^i = \frac{1}{j\omega\varepsilon}\nabla\times\mathbf{H}^i = \frac{1}{j\omega\varepsilon}\left[\hat{a}_\rho\frac{1}{\rho}\frac{\partial H_z^i}{\partial\phi} - \hat{a}_\phi\frac{\partial H_z^i}{\partial\rho}\right] \quad (11\text{-}106)$$

and by using (11-105) leads to

$$E_\rho^i = \frac{1}{j\omega\varepsilon}\frac{1}{\rho}\frac{\partial H_z^i}{\partial\phi} = \frac{H_0}{j\omega\varepsilon}\frac{1}{\rho}\sum_{n=-\infty}^{+\infty} nj^{-n+1}J_n(\beta\rho)e^{jn\phi} \quad (11\text{-}106a)$$

$$E_\phi^i = -\frac{1}{j\omega\varepsilon}\frac{\partial H_z^i}{\partial\rho} = -\frac{\beta H_0}{j\omega\varepsilon}\sum_{n=-\infty}^{+\infty} j^{-n}J_n'(\beta\rho)e^{jn\phi} \quad (11\text{-}106b)$$

The scattered magnetic field takes a form very similar to that of the scattered electric field of (11-88) for the TM^z polarization, and it can be written as

$$\mathbf{H}^s = \hat{a}_z H_z^s = \hat{a}_z H_0\sum_{n=-\infty}^{+\infty} d_n H_n^{(2)}(\beta\rho) \quad (11\text{-}107)$$

where d_n represents the yet unknown amplitude coefficients that will be found by applying the appropriate boundary conditions.

Before the boundary conditions on the vanishing of the total tangential electric field on the surface of the cylinder can be applied, it is necessary to first find the corresponding electric fields. This can be accomplished by using Maxwell's equation 11-106 which for the scattered magnetic field of (11-107) leads to

$$E_\rho^s = \frac{1}{j\omega\varepsilon}\frac{1}{\rho}\frac{\partial H_z^s}{\partial\phi} = \frac{H_0}{j\omega\varepsilon}\frac{1}{\rho}\sum_{n=-\infty}^{+\infty} H_n^{(2)}(\beta\rho)\frac{\partial d_n}{\partial\phi} \quad (11\text{-}108a)$$

$$E_\phi^s = -\frac{1}{j\omega\varepsilon}\frac{\partial H_z^s}{\partial\rho} = -\frac{\beta H_0}{j\omega\varepsilon}\sum_{n=-\infty}^{+\infty} d_n H_n^{(2)'}(\beta\rho) \quad (11\text{-}108b)$$

where $'$ indicates a partial derivate with respect to the entire argument of the Hankel function.

Since the cylinder is perfectly electric conducting, the tangential components of the total electric field must vanish on its surface ($\rho = a$). Thus using (11-106b) and (11-108b) we can write that

$$E_\phi^t(\rho = a, 0 \le \phi \le 2\pi, z) = -\frac{\beta H_0}{j\omega\varepsilon}\sum_{n=-\infty}^{+\infty}\left[j^{-n}J_n'(\beta a)e^{jn\phi} + d_n H_n^{(2)'}(\beta a)\right] = 0 \quad (11\text{-}109)$$

which is satisfied provided

$$d_n = -j^{-n}\frac{J_n'(\beta a)}{H_n^{(2)'}(\beta a)}e^{jn\phi} \quad (11\text{-}109a)$$

Thus the scattered electric and magnetic fields can be written, using (11-107) and

(11-109a), as

$$E_z^s = H_\rho^s = H_\phi^s = 0 \tag{11-110a}$$

$$E_\rho^s = -\frac{H_0}{j\omega\varepsilon}\frac{1}{\rho}\sum_{n=-\infty}^{+\infty} nj^{-n+1}\frac{J_n'(\beta a)}{H_n^{(2)'}(\beta a)}H_n^{(2)}(\beta\rho)e^{jn\phi} \tag{11-110b}$$

$$E_\phi^s = \frac{\beta H_0}{j\omega\varepsilon}\sum_{n=-\infty}^{+\infty} j^{-n}\frac{J_n'(\beta a)}{H_n^{(2)'}(\beta a)}H_n^{(2)'}(\beta\rho)e^{jn\phi} \tag{11-110c}$$

$$H_z^s = -H_0\sum_{n=-\infty}^{+\infty} j^{-n}\frac{J_n'(\beta a)}{H_n^{(2)'}(\beta a)}H_n^{(2)}(\beta\rho)e^{jn\phi} \tag{11-110d}$$

The total electric and magnetic fields can now be expressed [using (11-105), (11-106a), (11-106b), and (11-110a) through (11-110d)] as

$$E_z^t = H_\rho^t = H_\phi^t = 0 \tag{11-111a}$$

$$E_\rho^t = \frac{H_0}{j\omega\varepsilon}\frac{1}{\rho}\sum_{n=-\infty}^{+\infty} nj^{-n+1}\left[J_n(\beta\rho) - \frac{J_n'(\beta a)}{H_n^{(2)'}(\beta a)}H_n^{(2)}(\beta\rho)\right]e^{jn\phi} \tag{11-111b}$$

$$E_\phi^t = -\frac{\beta H_0}{j\omega\varepsilon}\sum_{n=-\infty}^{+\infty} j^{-n}\left[J_n'(\beta\rho) - \frac{J_n'(\beta a)}{H_n^{(2)'}(\beta a)}H_n^{(2)'}(\beta\rho)\right]e^{jn\phi} \tag{11-111c}$$

$$H_z^t = H_0\sum_{n=-\infty}^{+\infty} j^{-n}\left[J_n(\beta\rho) - \frac{J_n'(\beta a)}{H_n^{(2)'}(\beta a)}H_n^{(2)}(\beta\rho)\right]e^{jn\phi} \tag{11-111d}$$

On the surface of the cylinder ($\rho = a$) the total tangential magnetic field can be written as

$$H_z^t(\rho = a) = H_0\sum_{n=-\infty}^{+\infty} j^{-n}\left[J_n(\beta a) - \frac{J_n'(\beta a)}{H_n^{(2)'}(\beta a)}H_n^{(2)}(\beta a)\right]e^{jn\phi}$$

$$= -H_0\sum_{n=-\infty}^{+\infty} j^{-n+1}\left[\frac{J_n(\beta a)Y_n'(\beta a) - J_n'(\beta a)Y_n(\beta a)}{H_n^{(2)'}(\beta a)}\right]e^{jn\phi} \tag{11-112}$$

which reduces, using the Wronskian of (11-95), to

$$H_z^t(\rho = a) = -jH_0\frac{2}{\pi\beta a}\sum_{n=-\infty}^{+\infty} j^{-n}\frac{e^{jn\phi}}{H_n^{(2)'}(\beta a)} \tag{11-112a}$$

Thus the current induced on the surface of the cylinder can be written as

$$\mathbf{J}_s = \hat{n}\times\mathbf{H}^t|_{\rho=a} = \hat{a}_\rho\times\hat{a}_z H_z^t|_{\rho=a} = -\hat{a}_\phi H_z^t(\rho = a)$$

$$= \hat{a}_\phi j\frac{2H_0}{\pi\beta a}\sum_{n=-\infty}^{+\infty} j^{-n}\frac{e^{jn\phi}}{H_n^{(2)'}(\beta a)} \tag{11-113}$$

A. SMALL RADIUS APPROXIMATION

As the radius of the cylinder increases, more terms in the infinite series of (11-113) are needed to obtain convergence. However, for very small cylinders, like very thin wires where $a \ll \lambda$, the first three terms ($n = 0$, $n = \pm 1$) in (11-113) are dominant and are sufficient to represent the induced current. Thus for a very thin wire (11-113) can be approximated by

$$\mathbf{J}_s \stackrel{a \ll \lambda}{\simeq} \hat{a}_\phi j \frac{2H_0}{\pi \beta a} \left[\frac{1}{H_0^{(2)\prime}(\beta a)} + j^{-1} \frac{e^{j\phi}}{H_1^{(2)\prime}(\beta a)} + j^{+1} \frac{e^{-j\phi}}{H_{-1}^{(2)\prime}(\beta a)} \right] \quad (11\text{-}114)$$

where

$$H_0^{(2)\prime}(\beta a) = -H_1^{(2)}(\beta a) = -[J_1(\beta a) - jY_1(\beta a)]$$

$$\stackrel{a \ll \lambda}{\simeq} -\left[\frac{\beta a}{2} + j\frac{1}{\pi}\left(\frac{2}{\beta a}\right) \right] \simeq -j\frac{2}{\pi \beta a} \quad (11\text{-}114a)$$

$$H_1^{(2)\prime}(\beta a) = -H_2^{(2)}(\beta a) + \frac{1}{\beta a} H_1^{(2)}(\beta a)$$

$$= -[J_2(\beta a) - jY_2(\beta a)] + \frac{1}{\beta a}[J_1(\beta a) - jY_1(\beta a)]$$

$$\stackrel{a \ll \lambda}{\simeq} -\left[\frac{1}{2}\left(\frac{\beta a}{2}\right)^2 + j\frac{1}{\pi}\left(\frac{2}{\beta a}\right)^2 \right] + \frac{1}{\beta a}\left[\frac{\beta a}{2} + j\frac{1}{\pi}\left(\frac{2}{\beta a}\right) \right]$$

$$\stackrel{a \ll \lambda}{\simeq} -\left[j\frac{1}{\pi}\left(\frac{2}{\beta a}\right)^2 \right] + \frac{1}{\beta a}\left[j\frac{1}{\pi}\left(\frac{2}{\beta a}\right) \right] = -j\frac{2}{\pi}\frac{1}{(\beta a)^2} \quad (11\text{-}114b)$$

$$H_{-1}^{(2)\prime}(\beta a) = -H_1^{(2)\prime}(\beta a) \stackrel{a \ll \lambda}{\simeq} +j\frac{2}{\pi}\frac{1}{(\beta a)^2} \quad (11\text{-}114c)$$

Therefore (11-114) reduces to

$$\mathbf{J}_s \stackrel{a \ll \lambda}{\simeq} \hat{a}_\phi j \frac{2H_0}{\pi \beta a} \left[-\frac{\pi \beta a}{j2} + j\frac{\pi}{j2}(\beta a)^2 e^{j\phi} + j\frac{\pi}{j2}(\beta a)^2 e^{-j\phi} \right]$$

$$\stackrel{a \ll \lambda}{\simeq} \hat{a}_\phi H_0[-1 + j\beta a(e^{j\phi} + e^{-j\phi})] = \hat{a}_\phi H_0[-1 + j2(\beta a)\cos(\phi)] \quad (11\text{-}114d)$$

B. FAR-ZONE SCATTERED FIELD

Since the scattered field, as given by (11-110a) through (11-110d) has two nonvanishing electric field components and only one magnetic field component, it is most convenient to use the magnetic field to find the far-zone scattered field pattern and the radar cross section. However, the same answer can be obtained using the electric field components.

For far-field observations the scattered magnetic field of (11-110d) can be approximated, using the Hankel function approximation of (11-99), by

$$H_z^s \stackrel{\beta\rho \to \infty}{\simeq} -H_0\sqrt{\frac{2j}{\pi\beta}}\frac{e^{-j\beta\rho}}{\sqrt{\rho}} \sum_{n=-\infty}^{+\infty} \frac{J_n'(\beta a)}{H_n^{(2)\prime}(\beta a)} e^{jn\phi} \quad (11\text{-}115)$$

The ratio of the far-zone scattered magnetic field to the incident field can then be written as

$$\frac{|H_z^s|}{|H_z^i|} \stackrel{\beta\rho \to \infty}{\simeq} \frac{\left|-H_0 \sqrt{\frac{2j}{\pi\beta}} \frac{e^{-j\beta\rho}}{\sqrt{\rho}} \sum_{n=-\infty}^{+\infty} \frac{J_n'(\beta a)}{H_n^{(2)\prime}(\beta a)} e^{jn\phi}\right|}{|H_0 e^{-j\beta x}|}$$

$$= \sqrt{\frac{2}{\pi\beta\rho}} \left|\sum_{n=-\infty}^{+\infty} \frac{J_n'(\beta a)}{H_n^{(2)\prime}(\beta a)} e^{jn\phi}\right| \qquad (11\text{-}116)$$

Thus the scattering width of (11-21c) can be expressed as

$$\boxed{\begin{aligned}\sigma_{2\text{-D}} &= \lim_{\rho \to \infty}\left[2\pi\rho \frac{|H_z^s|^2}{|H_z^i|^2}\right] = \frac{4}{\beta}\left|\sum_{n=-\infty}^{+\infty} \frac{J_n'(\beta a)}{H_n^{(2)\prime}(\beta a)} e^{jn\phi}\right|^2 \\ &= \frac{2\lambda}{\pi}\left|\sum_{n=0}^{+\infty} \varepsilon_n \frac{J_n'(\beta a)}{H_n^{(2)\prime}(\beta a)} \cos(n\phi)\right|^2\end{aligned}} \qquad (11\text{-}117)$$

where

$$\varepsilon_n = \begin{cases} 1 & n = 0 \\ 2 & n \neq 0 \end{cases} \qquad (11\text{-}117a)$$

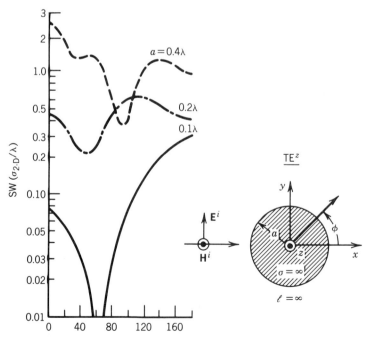

FIGURE 11-15 Two-dimensional TE^z bistatic scattering width (SW) of a circular conducting cylinder (Courtesy J. H. Richmond, Ohio State University.).

Plots of bistatic $\sigma_{2\text{-D}}/\lambda$ computed using (11-117) are shown in Figure 11-15 for cylinder radii of $a = 0.1\lambda$, 0.2λ, and 0.4λ while the backscattered patterns as a function of the cylinder radius are displayed in Figure 11-14 [13].

For small radii ($a \ll \lambda$) the first three terms ($n = 0$, $n = \pm 1$) in (11-117) are the dominant terms, and they are sufficient to represent the scattered field. Thus for small radii (11-117) can be approximated by

$$\sigma_{2\text{-D}} \overset{a \ll \lambda}{\simeq} \frac{2\lambda}{\pi} \left| \frac{J_0'(\beta a)}{H_0^{(2)\prime}(\beta a)} + \frac{J_1'(\beta a)}{H_1^{(2)\prime}(\beta a)} e^{j\phi} + \frac{J_{-1}'(\beta a)}{H_{-1}^{(2)\prime}(\beta a)} e^{-j\phi} \right|^2 \quad (11\text{-}118)$$

where

$$\frac{J_0'(\beta a)}{H_0^{(2)\prime}(\beta a)} = \frac{-J_1(\beta a)}{-H_1^{(2)}(\beta a)} \simeq \frac{\dfrac{\beta a}{2}}{\dfrac{\beta a}{2} + j\dfrac{1}{\pi}\left(\dfrac{2}{\beta a}\right)} \simeq \frac{\dfrac{\beta a}{2}}{j\dfrac{1}{\pi}\left(\dfrac{2}{\beta a}\right)} = -j\frac{\pi}{4}(\beta a)^2$$

(11-118a)

$$\frac{J_1'(\beta a)}{H_1^{(2)\prime}(\beta a)} = \frac{-J_2(\beta a) + \dfrac{1}{\beta a}J_1(\beta a)}{-H_2^{(2)}(\beta a) + \dfrac{1}{\beta a}H_1^{(2)}(\beta a)}$$

$$\simeq \frac{-\dfrac{1}{2}\left(\dfrac{\beta a}{2}\right)^2 + \dfrac{1}{\beta a}\left(\dfrac{\beta a}{2}\right)}{-\left[\dfrac{1}{2}\left(\dfrac{\beta a}{2}\right)^2 + j\dfrac{1}{\pi}\left(\dfrac{2}{\beta a}\right)^2\right] + \dfrac{1}{\beta a}\left[\left(\dfrac{\beta a}{2}\right) + j\dfrac{1}{\pi}\left(\dfrac{2}{\beta a}\right)\right]}$$

$$\simeq \frac{-\dfrac{1}{2}\left(\dfrac{\beta a}{2}\right)^2 + \dfrac{1}{2}}{-j\dfrac{1}{\pi}\left(\dfrac{2}{\beta a}\right)^2 + \dfrac{1}{\beta a}\left[j\dfrac{1}{\pi}\left(\dfrac{2}{\beta a}\right)\right]} \simeq \frac{\dfrac{1}{2}}{-j\dfrac{2}{\pi}\left(\dfrac{1}{\beta a}\right)^2} = j\frac{\pi}{4}(\beta a)^2$$

(11-118b)

$$\frac{J_{-1}'(\beta a)}{H_{-1}^{(2)\prime}(\beta a)} = \frac{J_1'(\beta a)}{H_1^{(2)\prime}(\beta a)} \simeq j\frac{\pi}{4}(\beta a)^2 \quad (11\text{-}118c)$$

Thus (11-118) reduces to

$$\sigma_{2\text{-D}} \overset{a \ll \lambda}{\simeq} \frac{2\lambda}{\pi} \left| -j\frac{\pi}{4}(\beta a)^2 + j\frac{\pi}{4}(\beta a)^2 e^{j\phi} + j\frac{\pi}{4}(\beta a)^2 e^{-j\phi} \right|^2$$

$$= \frac{\pi\lambda}{8}(\beta a)^4 [1 - 2\cos(\phi)]^2 \quad (11\text{-}118\text{d})$$

Even for small radii, σ is a function of ϕ as is evident in Figure 11-15 by the curves for small values of a.

For a cylinder of finite length ℓ the three-dimensional radar cross section for normal incidence is related to the two-dimensional scattering width by (11-21e).

Thus using (11-117) and (11-118d) we can write the three-dimensional RCS using (11-22e) as

$$\sigma_{3\text{-D}} \simeq \frac{4\ell^2}{\pi} \left| \sum_{n=-\infty}^{+\infty} \frac{J_n'(\beta a)}{H_n^{(2)\prime}(\beta a)} e^{jn\phi} \right|^2$$

$$= \frac{4\ell^2}{\pi} \left| \sum_{n=0}^{+\infty} \varepsilon_n \frac{J_n'(\beta a)}{H_n^{(2)\prime}(\beta a)} \cos(n\phi) \right|^2 \quad (11\text{-}119\text{a})$$

$$\sigma_{3\text{-D}} \stackrel{a \ll \lambda}{\simeq} \frac{\pi \ell^2}{4} (\beta a)^4 [1 - 2\cos(\phi)]^2 \quad (11\text{-}119\text{b})$$

11.5.3 Oblique Incidence Plane Wave Scattering by Conducting Circular Cylinder: TM^z Polarization

In the previous two sections we analyzed scattering by a conducting cylinder at normal incidence. Scattering at oblique incidence will be considered here. Let us assume that a TM^z plane wave traveling parallel to the xz plane is incident upon a circular cylinder of radius a, as shown in Figure 11-16. The incident electric field can be written as

$$\mathbf{E}^i = E_0(\hat{a}_x \cos\theta_i + \hat{a}_z \sin\theta_i) e^{-j\beta x \sin\theta_i} e^{+j\beta z \cos\theta_i} \quad (11\text{-}120)$$

Using the transformation of (11-81a), the z component of (11-120) can be expressed as

$$E_z^i = E_0 \sin\theta_i e^{+j\beta z \cos\theta_i} \sum_{n=-\infty}^{+\infty} j^{-n} J_n(\beta\rho \sin\theta_i) e^{jn\phi} \quad (11\text{-}120\text{a})$$

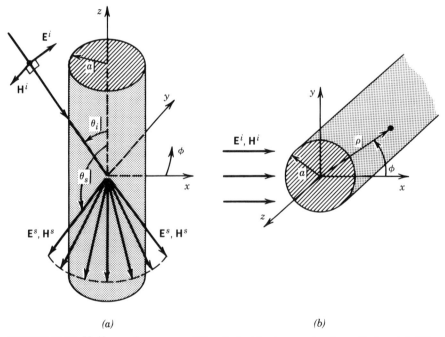

(a) (b)

FIGURE 11-16 Uniform plane wave obliquely incident on a circular cylinder. (*a*) Side view. (*b*) Top view.

The tangential component of the total field is composed of two parts: incident and scattered field components. The z component of the scattered field takes a form similar to (11-120a). By referring to (11-88) and Figure 11-16 where θ_s is shown to be greater than 90°, it can be written as

$$E_z^s = E_0 \sin\theta_s e^{-j\beta z \cos\theta_s} \sum_{n=-\infty}^{+\infty} c_n H_n^{(2)}(\beta\rho \sin\theta_s) \tag{11-121}$$

The Hankel function was chosen to indicate that the scattered field is a wave traveling in the outward radial direction. It should be stated at this time that smooth perfectly conducting infinite cylinders do not depolarize the oblique incident wave (i.e., do not introduce additional components in the scattered field as compared to the incident field). This, however, is not the case for homogeneous dielectric or dielectric coated cylinders which introduce cross polarization under oblique wave incidences.

When the incident electric field is decomposed into its cylindrical components, the E_x^i component of (11-120) will result in E_ρ^i and E_ϕ^i. Similarly scattered E_ρ^s and E_ϕ^s components will also exist. Therefore the boundary conditions on the surface of the cylinder are those of

$$E_z^t(\rho=a, 0\leq\theta_i, \theta_s\leq\pi, 0\leq\phi\leq 2\pi)=0$$
$$= E_z^i(\rho=a, 0\leq\theta_i\leq\pi, 0\leq\phi\leq 2\pi) + E_z^s(\rho=a, 0\leq\theta_s\leq\pi, 0\leq\phi\leq 2\pi) \tag{11-122a}$$

$$E_\phi^t(\rho=a, 0\leq\theta_i, \theta_s\leq\pi, 0\leq\phi\leq 2\pi)=0$$
$$= E_\phi^i(\rho=a, 0\leq\theta_i\leq\pi, 0\leq\phi\leq 2\pi) + E_\phi^s(\rho=a, 0\leq\theta_s\leq\pi, 0\leq\phi\leq 2\pi) \tag{11-122b}$$

Since each is not independent of the other, either one can be used to find the unknown coefficients.

Applying (11-122a) leads to

$$E_0\left[\sin\theta_i e^{+j\beta z\cos\theta_i}\sum_{n=-\infty}^{+\infty} j^{-n}J_n(\beta a\sin\theta_i)e^{jn\phi}\right.$$
$$\left. + \sin\theta_s e^{-j\beta z\cos\theta_s}\sum_{n=-\infty}^{+\infty} c_n H_n^{(2)}(\beta a\sin\theta_s)\right] = 0 \tag{11-123a}$$

which is satisfied provided

$$\theta_s = \pi - \theta_i \tag{11-123b}$$

$$c_n = -j^{-n}\frac{J_n(\beta a \sin\theta_i)}{H_n^{(2)}(\beta a \sin\theta_i)}e^{jn\phi} = j^{-n}a_n e^{jn\phi} \tag{11-123c}$$

$$a_n = -\frac{J_n(\beta a \sin\theta_i)}{H_n^{(2)}(\beta a \sin\theta_i)} \tag{11-123d}$$

Thus the scattered E_z^s component of (11-121) reduces to

$$E_z^s = E_0 \sin\theta_i e^{+j\beta z\cos\theta_i}\sum_{n=-\infty}^{+\infty} j^{-n}a_n H_n^{(2)}(\beta\rho\sin\theta_i)e^{jn\phi} \tag{11-124}$$

616 SCATTERING

It is apparent that the scattered field exists for all ϕ angles (measured from the x axis) along a cone in the forward direction whose half-angle from the z axis is equal to $\theta_s = \pi - \theta_i$.

To find the remaining $\mathbf{E}^s$ and $\mathbf{H}^s$ scattered field components, we expand Maxwell's curl equations as

$$\nabla \times \mathbf{E}^s = -j\omega\mu \mathbf{H}^s \Rightarrow \mathbf{H}^s = -\frac{1}{j\omega\mu}\nabla \times \mathbf{E}^s \qquad (11\text{-}125)$$

or

$$H^s_\rho = -\frac{1}{j\omega\mu}\left(\frac{1}{\rho}\frac{\partial E^s_z}{\partial \phi} - \frac{\partial E^s_\phi}{\partial z}\right) \qquad (11\text{-}125\text{a})$$

$$H^s_\phi = -\frac{1}{j\omega\mu}\left(\frac{\partial E^s_\rho}{\partial z} - \frac{\partial E^s_z}{\partial \rho}\right) \qquad (11\text{-}125\text{b})$$

$$H^s_z = -\frac{1}{j\omega\mu\rho}\left(\frac{\partial}{\partial \rho}(\rho E^s_\phi) - \frac{\partial E^s_\rho}{\partial \phi}\right) \qquad (11\text{-}125\text{c})$$

and

$$\nabla \times \mathbf{H}^s = j\omega\varepsilon \mathbf{E}^s \Rightarrow \mathbf{E}^s = \frac{1}{j\omega\varepsilon}\nabla \times \mathbf{H}^s \qquad (11\text{-}126)$$

or

$$E^s_\rho = \frac{1}{j\omega\varepsilon}\left(\frac{1}{\rho}\frac{\partial H^s_z}{\partial \phi} - \frac{\partial H^s_\phi}{\partial z}\right) \qquad (11\text{-}126\text{a})$$

$$E^s_\phi = \frac{1}{j\omega\varepsilon}\left(\frac{\partial H^s_\rho}{\partial z} - \frac{\partial H^s_z}{\partial \rho}\right) \qquad (11\text{-}126\text{b})$$

$$E^s_z = \frac{1}{j\omega\varepsilon\rho}\left(\frac{\partial}{\partial \rho}(\rho H^s_\phi) - \frac{\partial H^s_\rho}{\partial \phi}\right) \qquad (11\text{-}126\text{c})$$

Since for the TM^z solution $H^s_z = 0$, then (11-126a) reduces to

$$E^s_\rho = -\frac{1}{j\omega\varepsilon}\frac{\partial H^s_\phi}{\partial z} \qquad (11\text{-}127)$$

When substituted into (11-125b) we can write that

$$H^s_\phi = -\frac{1}{j\omega\mu}\left[\frac{\partial}{\partial z}\left(-\frac{1}{j\omega\varepsilon}\frac{\partial H^s_\phi}{\partial z}\right) - \frac{\partial E^s_z}{\partial \rho}\right] = -\frac{1}{\omega^2\mu\varepsilon}\frac{\partial^2 H^s_\phi}{\partial z^2} + \frac{1}{j\omega\mu}\frac{\partial E^s_z}{\partial \rho} \qquad (11\text{-}128)$$

The z variations of all field components are of the same form as in (11-124) (i.e., $e^{+j\beta z \cos \theta_i}$). Thus (11-128) reduces to

$$H_\phi^s = -\frac{(j\beta \cos \theta_i)^2}{\omega^2 \mu \varepsilon} H_\phi^s + \frac{1}{j\omega\mu} \frac{\partial E_z^s}{\partial \rho} = \cos^2 \theta_i H_\phi^s + \frac{1}{j\omega\mu} \frac{\partial E_z^s}{\partial \rho} \quad (11\text{-}129)$$

or

$$H_\phi^s (1 - \cos^2 \theta_i) = \sin^2 \theta_i H_\phi^s = \frac{1}{j\omega\mu} \frac{\partial E_z^s}{\partial \rho} \quad (11\text{-}129\text{a})$$

$$H_\phi^s = \frac{1}{j\omega\mu} \frac{1}{\sin^2 \theta_i} \frac{\partial E_z^s}{\partial \rho} = \frac{\beta}{j\omega\mu} \frac{1}{\sin \theta_i} \frac{\partial E_z^s}{\partial (\beta\rho \sin \theta_i)}$$

$$= -jE_0 \sqrt{\frac{\varepsilon}{\mu}} e^{+j\beta z \cos \theta_i} \sum_{n=-\infty}^{+\infty} j^{-n} a_n H_n^{(2)\prime} (\beta\rho \sin \theta_i) e^{jn\phi} \quad (11\text{-}129\text{b})$$

where

$$' = \frac{\partial}{\partial (\beta\rho \sin \theta_i)} \quad (11\text{-}129\text{c})$$

In a similar manner we can solve for H_ρ^s by first reducing (11-126b) to

$$E_\phi^s = \frac{1}{j\omega\varepsilon} \left(\frac{\partial H_\rho^s}{\partial z} - \frac{\partial H_z^s}{\partial \rho} \right)_{H_z^s = 0} = \frac{1}{j\omega\varepsilon} \frac{\partial H_\rho^s}{\partial z} \quad (11\text{-}130)$$

and then substituting it into (11-125a). Thus

$$H_\rho^s = -\frac{1}{j\omega\mu} \left[\frac{1}{\rho} \frac{\partial E_z^s}{\partial \phi} - \frac{\partial}{\partial z} \left(\frac{1}{j\omega\varepsilon} \frac{\partial H_\rho^s}{\partial z} \right) \right] = -\frac{1}{j\omega\mu} \frac{1}{\rho} \frac{\partial E_z^s}{\partial \phi} - \frac{1}{\omega^2 \mu \varepsilon} \frac{\partial^2 H_\rho^s}{\partial z^2}$$

$$= -\frac{1}{j\omega\mu} \frac{1}{\rho} \frac{\partial E_z^s}{\partial \phi} - \frac{(j\beta \cos \theta_i)^2}{\omega^2 \mu \varepsilon} H_\rho^s = -\frac{1}{j\omega\mu} \frac{1}{\rho} \frac{\partial E_z^s}{\partial \phi} + \cos^2 \theta_i H_\rho^s \quad (11\text{-}131)$$

or

$$(1 - \cos^2 \theta_i) H_\rho^s = \sin^2 \theta_i H_\rho^s = -\frac{1}{j\omega\mu} \frac{1}{\rho} \frac{\partial E_z^s}{\partial \phi} \quad (11\text{-}131\text{a})$$

$$H_\rho^s = -\frac{1}{j\omega\mu\rho \sin^2 \theta_i} \frac{\partial E_z^s}{\partial \phi} = j \frac{E_0 e^{+j\beta z \cos \theta_i}}{\omega\mu\rho \sin \theta_i} \sum_{n=-\infty}^{+\infty} nj^{-n+1} a_n H_n^{(2)} (\beta\rho \sin \theta_i) e^{jn\phi}$$

$$(11\text{-}131\text{b})$$

Expressions for E_ρ^s and E_ϕ^s can be written using (11-126a), (11-126b), (11-129b), and (11-131b). Thus

$$E_\rho^s = -\frac{1}{j\omega\varepsilon}\frac{\partial H_\phi^s}{\partial z} = jE_0\cos\theta_i e^{+j\beta z\cos\theta_i}\sum_{n=-\infty}^{+\infty}j^{-n}a_n H_n^{(2)\prime}(\beta\rho\sin\theta_i)e^{jn\phi} \quad (11\text{-}132)$$

$$E_\phi^s = \frac{1}{j\omega\varepsilon}\frac{\partial H_\rho^s}{\partial z} = jE_0\frac{\cot\theta_i}{\beta\rho}e^{+j\beta z\cos\theta_i}\sum_{n=-\infty}^{+\infty}nj^{-n+1}a_n H_n^{(2)}(\beta\rho\sin\theta_i)e^{jn\phi} \quad (11\text{-}133)$$

In summary then the scattered fields can be written as

$$\underline{\text{TM}^z}$$

$$E_\rho^s = jE_0\cos\theta_i e^{+j\beta z\cos\theta_i}\sum_{n=-\infty}^{+\infty}j^{-n}a_n H_n^{(2)\prime}(\beta\rho\sin\theta_i)e^{jn\phi} \quad (11\text{-}134\text{a})$$

$$E_\phi^s = jE_0\frac{\cot\theta_i}{\beta\rho}e^{+j\beta z\cos\theta_i}\sum_{n=-\infty}^{+\infty}nj^{-n+1}a_n H_n^{(2)}(\beta\rho\sin\theta_i)e^{jn\phi} \quad (11\text{-}134\text{b})$$

$$E_z^s = E_0\sin\theta_i e^{+j\beta z\cos\theta_i}\sum_{n=-\infty}^{+\infty}j^{-n}a_n H_n^{(2)}(\beta\rho\sin\theta_i)e^{jn\phi} \quad (11\text{-}134\text{c})$$

$$H_\rho^s = jE_0\frac{e^{+j\beta z\cos\theta_i}}{\omega\mu\rho\sin\theta_i}\sum_{n=-\infty}^{+\infty}nj^{-n+1}a_n H_n^{(2)}(\beta\rho\sin\theta_i)e^{jn\phi} \quad (11\text{-}134\text{d})$$

$$H_\phi^s = -jE_0\sqrt{\frac{\varepsilon}{\mu}}e^{+j\beta z\cos\theta_i}\sum_{n=-\infty}^{+\infty}j^{-n}a_n H_n^{(2)\prime}(\beta\rho\sin\theta_i)e^{jn\phi} \quad (11\text{-}134\text{e})$$

$$H_z^s = 0 \quad (11\text{-}134\text{f})$$

$$a_n = -\frac{J_n(\beta a\sin\theta_i)}{H_n^{(2)}(\beta a\sin\theta_i)} \quad (11\text{-}134\text{g})$$

$$\prime = \frac{\partial}{\partial(\beta\rho\sin\theta_i)} \quad (11\text{-}134\text{h})$$

A. FAR-ZONE SCATTERED FIELD

Often it is desired to know the scattered fields at large distances. This can be accomplished by approximating in (11-134a) through (11-134f) the Hankel function and its derivative by their corresponding asymptotic expressions for large distances as given by

$$H_n^{(2)}(\alpha x) \stackrel{\alpha x\to\infty}{\simeq} \sqrt{\frac{2j}{\pi\alpha x}}\,j^n e^{-j\alpha x} \quad (11\text{-}135\text{a})$$

$$H_n^{(2)\prime}(\alpha x) = \frac{dH_n^{(2)}(\alpha x)}{d(\alpha x)} \stackrel{\alpha x\to\infty}{\simeq} -\sqrt{\frac{2j}{\pi\alpha x}}\,j^{n+1}e^{-j\alpha x} \quad (11\text{-}135\text{b})$$

Thus we can reduce the scattered magnetic field expressions of (11-134d) and (11-134e) by

$$H_\rho^s \stackrel{\rho \to \infty}{\simeq} -E_0 \frac{1}{\omega\mu} \frac{1}{\rho \sin\theta_i} \sqrt{\frac{2j}{\pi\beta\rho \sin\theta_i}} e^{+j\beta(z\cos\theta_i - \rho\sin\theta_i)} \sum_{n=-\infty}^{+\infty} na_n e^{jn\phi}$$

$$\stackrel{\rho \to \infty}{\simeq} -jE_0 \frac{1}{\omega\mu} \frac{1}{\rho \sin\theta_i} \sqrt{\frac{2j}{\pi\beta\rho \sin\theta_i}} e^{+j\beta(z\cos\theta_i - \rho\sin\theta_i)} \sum_{n=0}^{+\infty} n\varepsilon_n a_n \cos(n\phi)$$

(11-136a)

$$H_\phi^s \stackrel{\rho \to \infty}{\simeq} -E_0 \sqrt{\frac{\varepsilon}{\mu}} \sqrt{\frac{2j}{\pi\beta\rho \sin\theta_i}} e^{+j\beta(z\cos\theta_i - \rho\sin\theta_i)} \sum_{n=-\infty}^{+\infty} a_n e^{jn\phi}$$

$$\stackrel{\rho \to \infty}{\simeq} -E_0 \sqrt{\frac{\varepsilon}{\mu}} \sqrt{\frac{2j}{\pi\beta\rho \sin\theta_i}} e^{+j\beta(z\cos\theta_i - \rho\sin\theta_i)} \sum_{n=0}^{+\infty} \varepsilon_n a_n \cos(n\phi) \quad (11\text{-}136b)$$

where ε_n is defined in (11-102a). A comparison of (11-136a) and (11-136b) indicates that at large distances H_ρ^s is small compared to H_ϕ^s since H_ρ^s varies inversely proportional to $\rho^{3/2}$ while H_ϕ^s is inversely proportional to $\rho^{1/2}$.

The scattering width of (11-21c) can now be expressed as

$$\sigma_{2\text{-D}} = \lim_{\rho \to \infty}\left(2\pi\rho \frac{|H_\phi^s|^2}{|\mathbf{H}^i|^2}\right) = \lim_{\rho \to \infty}\left[2\pi\rho \frac{\frac{|E_0|^2}{\eta^2}\left(\frac{2}{\pi\beta\rho \sin\theta_i}\right)}{\frac{|E_0|^2}{\eta^2}} \left|\sum_{n=-\infty}^{+\infty} a_n e^{jn\phi}\right|^2\right]$$

$$\boxed{\sigma_{2\text{-D}} = \frac{4}{\beta}\frac{1}{\sin\theta_i}\left|\sum_{n=-\infty}^{+\infty} a_n e^{jn\phi}\right|^2 = \frac{2\lambda}{\pi}\frac{1}{\sin\theta_i}\left|\sum_{n=0}^{+\infty} \varepsilon_n a_n \cos(n\phi)\right|^2} \quad (11\text{-}137)$$

where

$$\boxed{a_n = -\frac{J_n(\beta a \sin\theta_i)}{H_n^{(2)}(\beta a \sin\theta_i)}} \quad (11\text{-}137a)$$

$$\boxed{\varepsilon_n = \begin{cases} 1 & n = 0 \\ 2 & n \neq 0 \end{cases}} \quad (11\text{-}137b)$$

which is similar to (11-102) except that β in (11-102) is replaced by $\beta \sin\theta_i$.

From the results of the normal incidence case of Section 11.5.1B we can write by referring to (11-103a) that for small radii the scattering width of (11-137) reduces to

$$\sigma_{2\text{-D}} \stackrel{a \ll \lambda}{\simeq} \frac{\pi\lambda}{2\sin\theta_i}\left|\frac{1}{\ln(0.89\beta a \sin\theta_i)}\right|^2 \quad (11\text{-}138)$$

which is independent of ϕ.

For a cylinder of finite length ℓ the scattered fields of oblique incidence propagate in all directions, in contrast to the infinitely long cylinder where all the energy is along a conical surface formed in the forward direction whose half-angle is equal to θ_i. However, as the length of the cylinder becomes much larger than its radius ($\ell \gg a$), then the scattered fields along $\theta_s = \pi - \theta_i$ will be much greater than those in other directions. When the length of the cylinder is multiples of half wavelength, resonance phenomena are exhibited in the scattered fields [12]. However, as the length increases beyond several wavelengths, the resonance phenomena disappear. For both TM^z and TE^z polarizations the three-dimensional radar cross section for oblique wave incidence is related approximately to the two-dimensional scattering width, by referring to the geometry of Figure 11-16, by [12, 14]

$$\sigma_{3\text{-D}} = \sigma_{2\text{-D}} \left\{ \frac{2\ell^2}{\lambda} \sin^2 \theta_{s,i} \left[\frac{\sin\left[\frac{\beta\ell}{2}(\cos\theta_i + \cos\theta_s)\right]}{\frac{\beta\ell}{2}(\cos\theta_i + \cos\theta_s)} \right]^2 \right\} \qquad \ell \gg a \quad (11\text{-}139)$$

where $\sin^2 \theta_s$ is used for TM^z and $\sin^2 \theta_i$ is used for TE^z. This is analogous to the rectangular plate scattering of Section 11.3.2 and Example 11-2. This indicates that the maximum RCS occurs along the specular direction ($\theta_s = \pi - \theta_i$) and away from it follows the variations exhibited from a flat plate as given by (11-44a). Equation 11-139 yields reasonable good results even for cylinders with lengths near one wavelength ($\ell \simeq \lambda$). Thus using (11-139) converts (11-137) and (11-138) for three-dimensional scatterers to

$$\sigma_{3\text{-D}} \simeq \frac{4\ell^2}{\pi} \frac{\sin^2 \theta_s}{\sin \theta_i} \left| \sum_{n=0}^{\infty} \varepsilon_n a_n \cos(n\phi) \right|^2 \left\{ \frac{\sin\left[\frac{\beta\ell}{2}(\cos\theta_i + \cos\theta_s)\right]}{\frac{\beta\ell}{2}(\cos\theta_i + \cos\theta_s)} \right\}^2 \qquad (11\text{-}140a)$$

$$\sigma_{3\text{-D}} \stackrel{a \ll \lambda}{\simeq} \frac{\pi \ell^2}{\sin \theta_i} \left| \frac{\sin \theta_s}{\ln(0.89 \beta a \sin \theta_i)} \right|^2 \left\{ \frac{\sin\left[\frac{\beta\ell}{2}(\cos\theta_i + \cos\theta_s)\right]}{\frac{\beta\ell}{2}(\cos\theta_i + \cos\theta_s)} \right\}^2 \qquad (11\text{-}140b)$$

11.5.4 Oblique Incidence Plane Wave Scattering by Conducting Circular Cylinder: TE^z Polarization

TE^z scattering by a cylinder at oblique incidence can be analyzed following a procedure similar to that of TM^z scattering as discussed in the previous section. Using the geometry of Figure 11-16 we can write the incident magnetic field, for a plane wave traveling parallel to the xz plane, as

$$\mathbf{H}^i = H_0(\hat{a}_x \cos\theta_i + \hat{a}_z \sin\theta_i) e^{-j\beta x \sin\theta_i} e^{+j\beta z \cos\theta_i} \qquad (11\text{-}141)$$

Using the transformation of (11-81a), it can also be expressed as

$$\mathbf{H}^i = H_0(\hat{a}_x \cos\theta_i + \hat{a}_z \sin\theta_i) e^{+j\beta z \cos\theta_i} \sum_{n=-\infty}^{+\infty} j^{-n} J_n(\beta\rho \sin\theta_i) e^{jn\phi} \quad (11\text{-}141a)$$

Using the transformation from rectangular to cylindrical components of (II-6), or

$$H_\rho = H_x \cos\phi + H_y \sin\phi = H_x \cos\phi \quad (11\text{-}142a)$$
$$H_\phi = -H_x \sin\phi + H_y \cos\phi = -H_x \sin\phi \quad (11\text{-}142b)$$
$$H_z = H_z \quad (11\text{-}142c)$$

reduces (11-141a) to

$$H_\rho^i = H_0 \cos\theta_i \cos\phi\, e^{+j\beta z \cos\theta_i} \sum_{n=-\infty}^{+\infty} j^{-n} J_n(\beta\rho \sin\theta_i) e^{jn\phi} \quad (11\text{-}143a)$$

$$H_\phi^i = -H_0 \cos\theta_i \sin\phi\, e^{+j\beta z \cos\theta_i} \sum_{n=-\infty}^{+\infty} j^{-n} J_n(\beta\rho \sin\theta_i) e^{jn\phi} \quad (11\text{-}143b)$$

$$H_z^i = H_0 \sin\theta_i\, e^{+j\beta z \cos\theta_i} \sum_{n=-\infty}^{+\infty} j^{-n} J_n(\beta\rho \sin\theta_i) e^{jn\phi} \quad (11\text{-}143c)$$

In the source-free region, the corresponding electric field components can be obtained using Maxwell's curl equation

$$\nabla \times \mathbf{H}^i = j\omega\varepsilon \mathbf{E}^i \Rightarrow \mathbf{E}^i = \frac{1}{j\omega\varepsilon} \nabla \times \mathbf{H}^i \quad (11\text{-}144)$$

or

$$E_\rho^i = \frac{1}{j\omega\varepsilon}\left(\frac{1}{\rho}\frac{\partial H_z^i}{\partial \phi} - \frac{\partial H_\phi^i}{\partial z}\right) \quad (11\text{-}144a)$$

$$E_\phi^i = \frac{1}{j\omega\varepsilon}\left(\frac{\partial H_\rho^i}{\partial z} - \frac{\partial H_z^i}{\partial \rho}\right) \quad (11\text{-}144b)$$

$$E_z^i = \frac{1}{j\omega\varepsilon}\frac{1}{\rho}\left[\frac{\partial(\rho H_\phi^i)}{\partial \rho} - \frac{\partial H_\rho^i}{\partial \phi}\right] \quad (11\text{-}144c)$$

To aid in doing this, we also utilize Maxwell's curl equation

$$\nabla \times \mathbf{E}^i = -j\omega\mu \mathbf{H}^i \Rightarrow \mathbf{H}^i = -\frac{1}{j\omega\mu} \nabla \times \mathbf{E}^i \quad (11\text{-}145)$$

which when expanded takes the form, for a TE^z polarization ($E_z^i = 0$), of

$$H_\rho^i = -\frac{1}{j\omega\mu}\left(\frac{1}{\rho}\frac{\partial E_z^i}{\partial \phi} - \frac{\partial E_\phi^i}{\partial z}\right)\bigg|_{E_z^i = 0} = \frac{1}{j\omega\mu}\frac{\partial E_\phi^i}{\partial z} \quad (11\text{-}145a)$$

$$H_\phi^i = -\frac{1}{j\omega\mu}\left(\frac{\partial E_\rho^i}{\partial z} - \frac{\partial E_z^i}{\partial \rho}\right)\bigg|_{E_z^i = 0} = -\frac{1}{j\omega\mu}\frac{\partial E_\rho^i}{\partial z} \quad (11\text{-}145b)$$

$$H_z^i = -\frac{1}{j\omega\mu}\frac{1}{\rho}\left[\frac{\partial(\rho E_\phi^i)}{\partial \rho} - \frac{\partial E_\rho^i}{\partial \phi}\right] \quad (11\text{-}145c)$$

622 SCATTERING

Substituting (11-145a) into (11-144b), we can write that

$$E_\phi^i = \frac{1}{j\omega\varepsilon}\left[\frac{1}{j\omega\mu}\frac{\partial^2 E_\phi^i}{\partial z^2} - \frac{\partial H_z^i}{\partial \rho}\right] = -\frac{1}{\omega^2\mu\varepsilon}\frac{\partial^2 E_\phi^i}{\partial z^2} - \frac{1}{j\omega\varepsilon}\frac{\partial H_z^i}{\partial \rho} \quad (11\text{-}146)$$

Since the z variations of all the field components are of the same form (i.e., $e^{+j\beta z \cos\theta_i}$), as given by (11-141), then (11-146) reduces to

$$E_\phi^i = \frac{\beta^2}{\omega^2\mu\varepsilon}\cos^2\theta_i E_\phi^i - \frac{1}{j\omega\varepsilon}\frac{\partial H_z^i}{\partial \rho} = \cos^2\theta_i E_\phi^i - \frac{1}{j\omega\varepsilon}\frac{\partial H_z^i}{\partial \rho} \quad (11\text{-}147)$$

or

$$\left(1 - \cos^2\theta_i\right)E_\phi^i = \sin^2\theta_i E_\phi^i = -\frac{1}{j\omega\varepsilon}\frac{\partial H_z^i}{\partial \rho}$$

$$E_\phi^i = -\frac{1}{j\omega\varepsilon}\frac{1}{\sin^2\theta_i}\frac{\partial H_z^i}{\partial \rho} = -\frac{\beta}{j\omega\varepsilon}\frac{1}{\sin\theta_i}\frac{\partial H_z^i}{\partial(\beta\rho\sin\theta_i)}$$

$$= j\sqrt{\frac{\mu}{\varepsilon}}\,H_0 e^{+j\beta z\cos\theta_i}\sum_{n=-\infty}^{+\infty}j^{-n}J_n'(\beta\rho\sin\theta_i)e^{jn\phi} \quad (11\text{-}147a)$$

In a similar manner we can solve for E_ρ^i by substituting (11-145b) into (11-144a). Thus

$$E_\rho^i = \frac{1}{j\omega\varepsilon}\left[\frac{1}{\rho}\frac{\partial H_z^i}{\partial \phi} - \left(-\frac{1}{j\omega\mu}\frac{\partial^2 E_\rho^i}{\partial z^2}\right)\right] = \frac{1}{j\omega\varepsilon}\frac{1}{\rho}\frac{\partial H_z^i}{\partial \phi} - \frac{1}{\omega^2\mu\varepsilon}\frac{\partial^2 E_\rho^i}{\partial z^2}$$

$$= \frac{1}{j\omega\varepsilon}\frac{1}{\rho}\frac{\partial H_z^i}{\partial \phi} - \frac{(j\beta\cos\theta_i)^2}{\omega^2\mu\varepsilon}E_\rho^i = \frac{1}{j\omega\varepsilon}\frac{1}{\rho}\frac{\partial H_z^i}{\partial \phi} + \cos^2\theta_i E_\rho^i \quad (11\text{-}148)$$

or

$$\left(1 - \cos^2\theta_i\right)E_\rho^i = \sin^2\theta_i E_\rho^i = \frac{1}{j\omega\varepsilon}\frac{1}{\rho}\frac{\partial H_z^i}{\partial \phi}$$

$$E_\rho^i = \frac{1}{j\omega\varepsilon\rho}\frac{1}{\sin^2\theta_i}\frac{\partial H_z^i}{\partial \phi} = -j\frac{H_0 e^{+j\beta z\cos\theta_i}}{\omega\varepsilon\rho\sin\theta_i}\sum_{n=-\infty}^{+\infty}nj^{-n+1}J_n(\beta\rho\sin\theta_i)e^{jn\phi} \quad (11\text{-}148a)$$

Since the z component of the incident **H** field is given by (11-143c), its scattered field can be written in a form similar to (11-121) or

$$H_z^s = H_0\sin\theta_s e^{-j\beta z\cos\theta_s}\sum_{n=-\infty}^{+\infty}d_n H_n^{(2)}(\beta\rho\sin\theta_s) \quad (11\text{-}149)$$

where d_n represents unknown coefficients to be determined by boundary conditions. According to (11-147a), the ϕ component of the scattered field can be written using

(11-149) as

$$E_\phi^s = -\frac{1}{j\omega\varepsilon}\frac{1}{\sin^2\theta_s}\frac{\partial H_z^s}{\partial\rho} = -\frac{\beta}{j\omega\varepsilon}\frac{1}{\sin\theta_s}\frac{\partial H_z^s}{\partial(\beta\rho\sin\theta_s)}$$

$$= jH_0\sqrt{\frac{\mu}{\varepsilon}}\,e^{-j\beta z\cos\theta_s}\sum_{n=-\infty}^{+\infty}d_n H_n^{(2)\prime}(\beta\rho\sin\theta_s) \quad (11\text{-}150)$$

Applying the boundary condition of

$$E_\phi^t(\rho=a,0\le\theta_i,\theta_s\le\pi,0\le\phi\le 2\pi)=0$$
$$= E_\phi^i(\rho=a,0\le\theta_i\le\pi,0\le\phi\le 2\pi)+E_\phi^s(\rho=a,0\le\theta_s\le\pi,0\le\phi\le 2\pi) \quad (11\text{-}151)$$

leads to

$$jH_0\sqrt{\frac{\mu}{\varepsilon}}\left[e^{+j\beta z\cos\theta_i}\sum_{n=-\infty}^{+\infty}j^{-n}J_n'(\beta a\sin\theta_i)e^{jn\phi}\right.$$
$$\left.+e^{-j\beta z\cos\theta_s}\sum_{n=-\infty}^{+\infty}d_n H_n^{(2)\prime}(\beta a\sin\theta_s)\right]=0 \quad (11\text{-}151\text{a})$$

which is satisfied provided

$$\theta_s = \pi - \theta_i \quad (11\text{-}151\text{b})$$

$$d_n = -j^{-n}\frac{J_n'(\beta a\sin\theta_i)}{H_n^{(2)\prime}(\beta a\sin\theta_i)}e^{jn\phi} = j^{-n}b_n e^{jn\phi} \quad (11\text{-}151\text{c})$$

$$b_n = -\frac{J_n'(\beta a\sin\theta_i)}{H_n^{(2)\prime}(\beta a\sin\theta_i)} \quad (11\text{-}151\text{d})$$

Thus the scattered H_z^s component of (11-149) reduces to

$$H_z^s = H_0\sin\theta_i\,e^{+j\beta z\cos\theta_i}\sum_{n=-\infty}^{+\infty}j^{-n}b_n H_n^{(2)}(\beta\rho\sin\theta_i)e^{jn\phi} \quad (11\text{-}152)$$

Knowing H_z^s the remaining electric and magnetic field components can be found using (11-148a), (11-147a), and (11-145a) through (11-145c).

In summary then the scattered fields can be written as

$$\text{TE}^z$$

$$E_\rho^s = \frac{1}{j\omega\varepsilon\rho}\frac{1}{\sin^2\theta_i}\frac{\partial H_z^s}{\partial\phi} = -j\frac{H_0}{\omega\varepsilon\rho}\frac{e^{+j\beta z\cos\theta_i}}{\sin\theta_i}$$
$$\times\sum_{n=-\infty}^{+\infty}nj^{-n+1}b_n H_n^{(2)}(\beta\rho\sin\theta_i)e^{jn\phi} \quad (11\text{-}153\text{a})$$

$$E_\phi^s = -\frac{1}{j\omega\varepsilon}\frac{1}{\sin^2\theta_i}\frac{\partial H_z^s}{\partial\rho} = jH_0\sqrt{\frac{\mu}{\varepsilon}}\,e^{+j\beta z\cos\theta_i}$$
$$\times\sum_{n=-\infty}^{+\infty}j^{-n}b_n H_n^{(2)\prime}(\beta\rho\sin\theta_i)e^{jn\phi} \quad (11\text{-}153\text{b})$$

$$E_z^s = 0 \quad (11\text{-}153\text{c})$$

$$H_\rho^s = \frac{1}{j\omega\mu}\frac{\partial E_\phi^s}{\partial z} = jH_0\cos\theta_i e^{+j\beta z\cos\theta_i}$$
$$\times \sum_{n=-\infty}^{+\infty} j^{-n} b_n H_n^{(2)\prime}(\beta\rho\sin\theta_i) e^{jn\phi} \quad (11\text{-}153\text{d})$$

$$H_\phi^s = -\frac{1}{j\omega\mu}\frac{\partial E_\rho^s}{\partial z} = jH_0\frac{\cot\theta_i}{\beta\rho}e^{+j\beta z\cos\theta_i}$$
$$\times \sum_{n=-\infty}^{+\infty} nj^{-n+1} b_n H_n^{(2)}(\beta\rho\sin\theta_i) e^{jn\phi} \quad (11\text{-}153\text{e})$$

$$H_z^s = H_0\sin\theta_i e^{+j\beta z\cos\theta_i} \sum_{n=-\infty}^{+\infty} j^{-n} b_n H_n^{(2)}(\beta\rho\sin\theta_i) e^{jn\phi} \quad (11\text{-}153\text{f})$$

$$b_n = -\frac{J_n'(\beta a\sin\theta_i)}{H_n^{(2)\prime}(\beta a\sin\theta_i)} \quad (11\text{-}153\text{g})$$

$$' = \frac{\partial}{\partial(\beta\rho\sin\theta_i)} \quad (11\text{-}153\text{h})$$

A. FAR-ZONE SCATTERED FIELD

The scattered electric fields of (11-153a) through (11-153c) can be approximated in the far zone by replacing the Hankel function and its derivative by their asymptotic forms as given by (11-135a) and (11-135b). Thus

$$E_\rho^s \simeq H_0 \frac{1}{\omega\varepsilon}\frac{1}{\rho\sin\theta_i}\sqrt{\frac{2j}{\pi\beta\rho\sin\theta_i}} e^{+j\beta(z\cos\theta_i-\rho\sin\theta_i)} \sum_{n=-\infty}^{+\infty} nb_n e^{jn\phi}$$

$$\simeq H_0 \frac{1}{\omega\varepsilon}\frac{1}{\rho\sin\theta_i}\sqrt{\frac{2j}{\pi\beta\rho\sin\theta_i}} e^{+j\beta(z\cos\theta_i-\rho\sin\theta_i)} \sum_{n=0}^{+\infty} n\varepsilon_n b_n \cos(n\phi)$$
$$(11\text{-}154\text{a})$$

$$E_\phi^s \simeq H_0\sqrt{\frac{\mu}{\varepsilon}}\sqrt{\frac{2j}{\pi\beta\rho\sin\theta_i}} e^{+j\beta(z\cos\theta_i-\rho\sin\theta_i)} \sum_{n=-\infty}^{+\infty} b_n e^{jn\phi}$$

$$\simeq H_0\sqrt{\frac{\mu}{\varepsilon}}\sqrt{\frac{2j}{\pi\beta\rho\sin\theta_i}} e^{+j\beta(z\cos\theta_i-\rho\sin\theta_i)} \sum_{n=0}^{+\infty} \varepsilon_n b_n \cos(n\phi) \quad (11\text{-}154\text{b})$$

where ε_n is defined in (11-102a). A comparison of (11-154a) and (11-154b) indicates that at large distances E_ρ^s is small compared to E_ϕ^s since E_ρ^s is inversely proportional to $\rho^{3/2}$ whereas E_ϕ^s is inversely proportional to $\rho^{1/2}$.

The scattering width of (11-21b) can now be expressed using (11-154b) and the incident electric field corresponding to (11-141) as

$$\sigma_{2\text{-D}} = \lim_{\rho \to \infty}\left[2\pi\rho \frac{|E_\phi^s|^2}{|E^i|^2}\right] = \lim_{\rho \to \infty}\left[2\pi\rho \frac{|H_0|^2 \frac{\mu}{\varepsilon}\left(\frac{2}{\pi\beta\rho \sin\theta_i}\right)}{|H_0|^2 \frac{\mu}{\varepsilon}}\left|\sum_{n=-\infty}^{+\infty} b_n e^{jn\phi}\right|^2\right]$$

$$\boxed{\sigma_{2\text{-D}} = \frac{4}{\beta} \frac{1}{\sin\theta_i}\left|\sum_{n=-\infty}^{+\infty} b_n e^{jn\phi}\right|^2 = \frac{2\lambda}{\pi} \frac{1}{\sin\theta_i}\left|\sum_{n=0}^{+\infty} \varepsilon_n b_n \cos(n\phi)\right|^2} \quad (11\text{-}155)$$

where

$$\boxed{b_n = -\frac{J_n'(\beta a \sin\theta_i)}{H_n^{(2)\prime}(\beta a \sin\theta_i)}} \quad (11\text{-}155\text{a})$$

$$\boxed{\varepsilon_n = \begin{cases} 1 & n = 0 \\ 2 & n \neq 0 \end{cases}} \quad (11\text{-}155\text{b})$$

which is similar to (11-117) except that β in (11-117) has been replaced by $\beta \sin\theta_i$.

From the results of the normal incidence case of Section 11.5.2B we can write by referring to (11-118d) that for small radii the scattering width of (11-155) reduces to

$$\sigma_{2\text{-D}} \overset{a \ll \lambda}{\simeq} \frac{\pi\lambda}{8} \frac{(\beta a \sin\theta_i)^4}{\sin\theta_i}[1 - 2\cos(\phi)]^2 \quad (11\text{-}156)$$

which is dependent on ϕ even for small radii cylinders.

Using (11-139) the radar cross section at oblique incidence for a finite length ℓ cylinder can be written using (11-155) and (11-156), and referring to the geometry of Figure 11-16, as

$$\sigma_{3\text{-D}} \simeq \frac{4\ell^2}{\pi}\sin\theta_i\left|\sum_{n=0}^{\infty}\varepsilon_n b_n \cos(n\phi)\right|^2\left\{\frac{\sin\left[\frac{\beta\ell}{2}(\cos\theta_i + \cos\theta_s)\right]}{\frac{\beta\ell}{2}(\cos\theta_i + \cos\theta_s)}\right\}^2 \quad (11\text{-}157\text{a})$$

$$\sigma_{3\text{-D}} \overset{a \ll \lambda}{\simeq} \frac{\pi\ell^2}{4}\left[(\beta a \sin\theta_i)^4 \sin\theta_i\right][1 - 2\cos(\phi)]^2\left\{\frac{\sin\left[\frac{\beta\ell}{2}(\cos\theta_i + \cos\theta_s)\right]}{\frac{\beta\ell}{2}(\cos\theta_i + \cos\theta_s)}\right\}^2$$

$$(11\text{-}157\text{b})$$

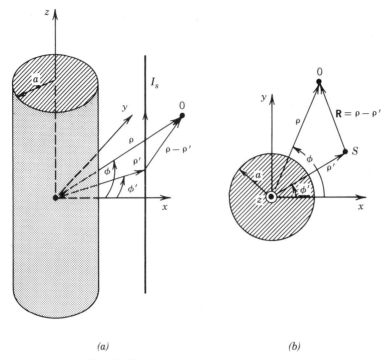

FIGURE 11-17 Electric line source near a circular cylinder. (*a*) Side view. (*b*) Top view.

11.5.5 Line-Source Scattering by a Conducting Circular Cylinder

While in the previous sections we examined plane wave scattering by a conducting circular cylinder, both at normal and oblique wave incidences, a more general problem is that of line-source (both electric and magnetic) scattering. The geometry is that shown in Figure 11-17 where an infinite line of constant current (I_e for electric and I_m for magnetic) is placed in the vicinity of a circular conducting cylinder of infinite length. We will examine here the scattering by the cylinder assuming the source is either electric or magnetic.

A. ELECTRIC LINE SOURCE (TM^z POLARIZATION)

If the line source of Figure 11-17 is of constant electric current I_e, the field generated everywhere by the source in the absence of the cylinder is given, according to (11-11a), by

$$E_z^i = -\frac{\beta^2 I_e}{4\omega\varepsilon} H_0^{(2)}(\beta|\boldsymbol{\rho} - \boldsymbol{\rho}'|) \tag{11-158}$$

which is referred to here as the incident field. By the addition theorem for Hankel functions of (11-69a) through (11-70b) we can write (11-158) as

$$E_z^i = -\frac{\beta^2 I_e}{4\omega\varepsilon} \begin{cases} \displaystyle\sum_{n=-\infty}^{\infty} J_n(\beta\rho) H_n^{(2)}(\beta\rho') e^{jn(\phi-\phi')} & \rho \leq \rho' \quad (11\text{-}158\text{a}) \\ \displaystyle\sum_{n=-\infty}^{+\infty} J_n(\beta\rho') H_n^{(2)}(\beta\rho) e^{jn(\phi-\phi')} & \rho \geq \rho' \quad (11\text{-}158\text{b}) \end{cases}$$

Bessel functions $J_n(\beta\rho)$ were selected to represent the fields for $\rho \leq \rho'$ because the field must be finite everywhere (including $\rho = 0$) and Hankel functions were chosen for $\rho \geq \rho'$ to represent the traveling nature of the wave.

In the presence of the cylinder the total field is composed of two parts: incident and scattered fields. The scattered field is produced by the current induced on the surface of the cylinder that acts as a secondary radiator. The scattered field also has only an E_z component (no cross polarized components are produced), and it can be expressed as

$$E_z^s = -\frac{\beta^2 I_e}{4\omega\varepsilon} \sum_{n=-\infty}^{+\infty} c_n H_n^{(2)}(\beta\rho) \qquad a \leq \rho \leq \rho', \quad \rho \geq \rho' \qquad (11\text{-}159)$$

The same expression is valid for $\rho \leq \rho'$ and $\rho \geq \rho'$ because the scattered field exists only when the cylinder is present, and it is nonzero only when $\rho \geq a$. Since the scattered field emanates from the surface of the cylinder, the Hankel function in (11-159) is chosen to represent the traveling wave nature of the radiation.

The coefficients represented by c_n in (11-159) can be found by applying the boundary condition of

$$E_z^t(\rho = a, 0 \leq \phi, \phi' \leq 2\pi, z)$$
$$= E_z^i(\rho = a, 0 \leq \phi, \phi' \leq 2\pi, z) + E_z^s(\rho = a, 0 \leq \phi, \phi' \leq 2\pi, z) = 0 \quad (11\text{-}160)$$

which by using (11-158a) and (11-159) leads to

$$-\frac{\beta^2 I_e}{4\omega\varepsilon} \sum_{n=-\infty}^{+\infty} \left[H_n^{(2)}(\beta\rho') J_n(\beta a) e^{jn(\phi-\phi')} + c_n H_n^{(2)}(\beta a) \right] = 0 \quad (11\text{-}161)$$

which is satisfied provided

$$c_n = -H_n^{(2)}(\beta\rho') \frac{J_n(\beta a)}{H_n^{(2)}(\beta a)} e^{jn(\phi-\phi')} \qquad (11\text{-}161a)$$

Thus (11-159) can be expressed as

$$E_z^s = +\frac{\beta^2 I_e}{4\omega\varepsilon} \sum_{n=-\infty}^{+\infty} H_n^{(2)}(\beta\rho') \frac{J_n(\beta a)}{H_n^{(2)}(\beta a)} H_n^{(2)}(\beta\rho) e^{jn(\phi-\phi')} \qquad a \leq \rho \leq \rho', \quad \rho \geq \rho'$$
$$(11\text{-}162)$$

The total electric field can then be written as

$$E_\rho^t = E_\phi^t = 0 \qquad (11\text{-}163)$$

$$E_z^t = -\frac{\beta^2 I_e}{4\omega\varepsilon} \begin{cases} \sum_{n=-\infty}^{+\infty} H_n^{(2)}(\beta\rho') \left[J_n(\beta\rho) - \frac{J_n(\beta a)}{H_n^{(2)}(\beta a)} H_n^{(2)}(\beta\rho) \right] \\ \times e^{jn(\phi-\phi')} \qquad a \leq \rho \leq \rho' \qquad (11\text{-}164a) \\ \sum_{n=-\infty}^{+\infty} H_n^{(2)}(\beta\rho) \left[J_n(\beta\rho') - \frac{J_n(\beta a)}{H_n^{(2)}(\beta a)} H_n^{(2)}(\beta\rho') \right] \\ \times e^{jn(\phi-\phi')} \qquad \rho \geq \rho' \qquad (11\text{-}164b) \end{cases}$$

where the first terms within the summations and brackets represent the incident fields and the second terms represent the scattered fields. The corresponding magnetic components can be found using Maxwell's equations 11-86 through 11-86c which can be written as

$$H_\rho^t = -\frac{1}{j\omega\mu}\frac{1}{\rho}\frac{\partial E_z^t}{\partial \phi}$$

$$= -j\frac{I_e}{4\rho}\begin{cases} \sum_{n=-\infty}^{+\infty} jnH_n^{(2)}(\beta\rho')\left[J_n(\beta\rho) - \frac{J_n(\beta a)}{H_n^{(2)}(\beta a)}H_n^{(2)}(\beta\rho)\right] \\ \times e^{jn(\phi-\phi')} \qquad a \le \rho \le \rho' \qquad (11\text{-}165a) \\ \sum_{n=-\infty}^{+\infty} jnH_n^{(2)}(\beta\rho)\left[J_n(\beta\rho') - \frac{J_n(\beta a)}{H_n^{(2)}(\beta a)}H_n^{(2)}(\beta\rho')\right] \\ \times e^{jn(\phi-\phi')} \qquad \rho \ge \rho' \qquad (11\text{-}165b) \end{cases}$$

$$H_\phi^t = \frac{1}{j\omega\mu}\frac{\partial E_z^t}{\partial \rho}$$

$$= j\frac{\beta I_e}{4}\begin{cases} \sum_{n=-\infty}^{+\infty} H_n^{(2)}(\beta\rho')\left[J_n'(\beta\rho) - \frac{J_n(\beta a)}{H_n^{(2)}(\beta a)}H_n^{(2)\prime}(\beta\rho)\right] \\ \times e^{jn(\phi-\phi')} \qquad a \le \rho \le \rho' \qquad (11\text{-}166a) \\ \sum_{n=-\infty}^{+\infty} H_n^{(2)\prime}(\beta\rho)\left[J_n(\beta\rho') - \frac{J_n(\beta a)}{H_n^{(2)}(\beta a)}H_n^{(2)}(\beta\rho')\right] \\ \times e^{jn(\phi-\phi')} \qquad \rho \ge \rho' \qquad (11\text{-}166b) \end{cases}$$

$$\boxed{H_z^t = 0}$$

On the surface of the cylinder the current density can be found to be

$$\mathbf{J}_s = \hat{n} \times \mathbf{H}^t|_{\rho=a} = \hat{a}_\rho \times (\hat{a}_\rho H_\rho^t + \hat{a}_\phi H_\phi^t)|_{\rho=a} = \hat{a}_z H_\phi^t|_{\rho=a}$$

$$= \hat{a}_z j\frac{\beta I_e}{4}\sum_{n=-\infty}^{+\infty} H_n^{(2)}(\beta\rho')\left[J_n'(\beta a) - \frac{J_n(\beta a)}{H_n^{(2)}(\beta a)}H_n^{(2)\prime}(\beta a)\right]e^{jn(\phi-\phi')}$$

$$\mathbf{J}_s = -\hat{a}_z \frac{\beta I_e}{4}\sum_{n=-\infty}^{+\infty} H_n^{(2)}(\beta\rho')\left[\frac{J_n(\beta a)Y_n'(\beta a) - J_n'(\beta a)Y_n(\beta a)}{H_n^{(2)}(\beta a)}\right]e^{jn(\phi-\phi')}$$

(11-168)

which by using the Wronskian of (11-95) reduces to

$$\mathbf{J}_s = -\hat{a}_z \frac{I_e}{2\pi a}\sum_{n=-\infty}^{+\infty} \frac{H_n^{(2)}(\beta\rho')}{H_n^{(2)}(\beta a)}e^{jn(\phi-\phi')} \qquad (11\text{-}168a)$$

For far-field observations ($\beta\rho \gg 1$) the total electric field of (11-164b) can be reduced by replacing the Hankel function $H_n^{(2)}(\beta\rho)$ by its asymptotic expression (11-135a). Doing this reduces (11-164b) to

$$E_z^t \overset{\beta\rho \gg 1}{\simeq} -\frac{\beta^2 I_e}{4\omega\varepsilon} \sqrt{\frac{2j}{\pi\beta}} \frac{e^{-j\beta\rho}}{\sqrt{\rho}} \sum_{n=-\infty}^{+\infty} j^n \left[J_n(\beta\rho') - \frac{J_n(\beta a)}{H_n^{(2)}(\beta a)} H_n^{(2)}(\beta\rho') \right] e^{jn(\phi-\phi')} \quad (11\text{-}169)$$

which can be used to compute more conveniently far-field patterns of an electric line source near a circular conducting cylinder. Plots of the normalized pattern for $a = 5\lambda$, $\phi' = 0$ with $\rho' = 5.25\lambda$ and 5.5λ are shown plotted, respectively, in Figure 11-18a and b where they are compared with that of a planar reflector ($a = \infty$) of Figure 11-3. Because of the finite radius of the cylinder, radiation is allowed to "leak" around the cylinder in terms of "creeping" waves [15–17]; this is not the case for the planar reflector.

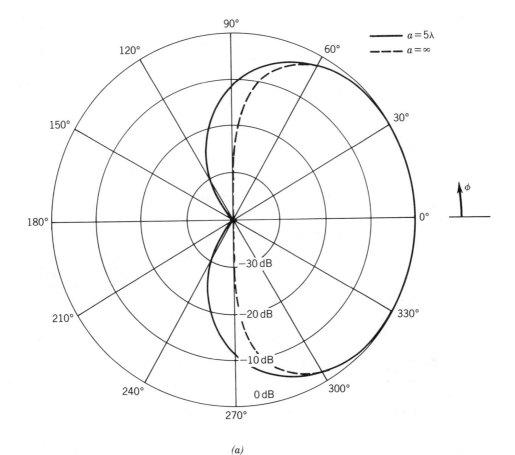

(a)

FIGURE 11-18 Normalized far-field pattern of an electric line source near a circular conducting cylinder. (a) $\rho' = 5.25\lambda$, $\phi' = 0°$. (b) $\rho' = 5.5\lambda$, $\phi' = 0°$.

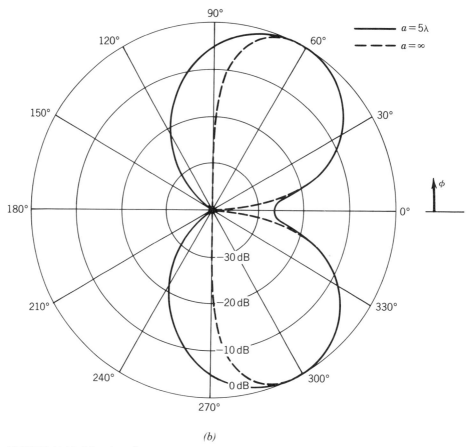

FIGURE 11-18 (*Continued*).

B. MAGNETIC LINE SOURCE (TE^z POLARIZATION)

Magnetic sources, although not physically realizable, are often used as equivalent sources to analyze aperture antennas [1, 18]. If the line source of Figure 11-17 is magnetic and it is allowed to recede to the surface of the cylinder ($\rho' = a$), the total field of the line source in the presence of the cylinder would be representative of a very thin infinite axial slot on the cylinder. Finite width slots can be represented by a number of line sources with some amplitude and phase distribution across the width. Therefore knowing the radiation and scattering by a magnetic line source near a cylinder allows us to solve other physical problems by using it as an equivalent.

If the line of Figure 11-17 is magnetic with a current of I_m, the fields that it radiates in the absence of the cylinder can be obtained from those of an electric line source by the use of duality (Section 7.2). Doing this we can write the incident magnetic field by referring to (11-158) through (11-158b) as

$$H_z^i = -\frac{\beta^2 I_m}{4\omega\mu} H_0^{(2)}(\beta|\rho - \rho'|) \qquad (11\text{-}170)$$

which can also be expressed as

$$H_z^i = -\frac{\beta^2 I_m}{4\omega\mu} \begin{cases} \sum_{n=-\infty}^{\infty} J_n(\beta\rho) H_n^{(2)}(\beta\rho') e^{jn(\phi-\phi')} & \rho \leq \rho' \quad (11\text{-}170a) \\ \sum_{n=-\infty}^{+\infty} J_n(\beta\rho') H_n^{(2)}(\beta\rho) e^{jn(\phi-\phi')} & \rho \geq \rho' \quad (11\text{-}170b) \end{cases}$$

The scattered magnetic field takes a form similar to that of (11-159) and can be written as

$$H_z^s = -\frac{\beta^2 I_m}{4\omega\mu} \sum_{n=-\infty}^{+\infty} d_n H_n^{(2)}(\beta\rho) \qquad a \leq \rho \leq \rho', \ \rho \geq \rho' \quad (11\text{-}171)$$

where d_n is used to represent the coefficients of the scattered field. Thus the total magnetic field can be expressed, by combining (11-170a) through (11-171), as

$$H_z^t = -\frac{\beta^2 I_m}{4\omega\mu} \begin{cases} \sum_{n=-\infty}^{+\infty} \left[H_n^{(2)}(\beta\rho') J_n(\beta\rho) e^{jn(\phi-\phi')} + d_n H_n^{(2)}(\beta\rho) \right] & (11\text{-}172a) \\ \qquad\qquad a \leq \rho \leq \rho' \\ \sum_{n=-\infty}^{+\infty} \left[J_n(\beta\rho') e^{jn(\phi-\phi')} + d_n \right] H_n^{(2)}(\beta\rho) & (11\text{-}172b) \\ \qquad\qquad \rho \geq \rho' \end{cases}$$

The corresponding electric field components can be found using Maxwell's equations 11-106 or 11-106a and 11-106b. Doing this and utilizing (11-172a) through (11-172b) we can write that

$$E_\rho^t = \frac{1}{j\omega\varepsilon} \frac{1}{\rho} \frac{\partial H_z^t}{\partial \phi}$$

$$= j\frac{I_m}{4\rho} \begin{cases} \sum_{n=-\infty}^{+\infty} \left[jn H_n^{(2)}(\beta\rho') J_n(\beta\rho) e^{jn(\phi-\phi')} + H_n^{(2)}(\beta\rho) \frac{\partial d_n}{\partial \phi} \right] & (11\text{-}173a) \\ \qquad\qquad a \leq \rho \leq \rho' \\ \sum_{n=-\infty}^{+\infty} \left[jn J_n(\beta\rho') e^{jn(\phi-\phi')} + \frac{\partial d_n}{\partial \phi} \right] H_n^{(2)}(\beta\rho) & (11\text{-}173b) \\ \qquad\qquad \rho \geq \rho' \end{cases}$$

$$E_\phi^t = -\frac{1}{j\omega\varepsilon} \frac{\partial H_z^t}{\partial \rho}$$

$$= -j\frac{\beta I_m}{4} \begin{cases} \sum_{n=-\infty}^{+\infty} \left[H_n^{(2)}(\beta\rho') J_n'(\beta\rho) e^{jn(\phi-\phi')} + d_n H_n^{(2)\prime}(\beta\rho) \right] & (11\text{-}174a) \\ \qquad\qquad a \leq \rho \leq \rho' \\ \sum_{n=-\infty}^{+\infty} \left[J_n(\beta\rho') e^{jn(\phi-\phi')} + d_n \right] H_n^{(2)\prime}(\beta\rho) & (11\text{-}174b) \\ \qquad\qquad \rho \geq \rho' \end{cases}$$

Applying the boundary condition of

$$E_\phi^t(\rho = a, 0 \le \phi, \phi' \le 2\pi, z)$$
$$= E_\phi^i(\rho = a, 0 \le \phi, \phi' \le 2\pi, z) + E_\phi^s(\rho = a, 0 \le \phi, \phi' \le 2\pi, z) = 0 \quad (11\text{-}175)$$

on (11-174a) leads to

$$d_n = -H_n^{(2)}(\beta\rho') \frac{J_n'(\beta a)}{H_n^{(2)'}(\beta a)} e^{jn(\phi-\phi')} \quad (11\text{-}175a)$$

Thus the total electric and magnetic field components can be written as

$$\underline{\text{TE}^z}$$

$$E_z^t = H_\rho^t = H_\phi^t = 0 \quad (11\text{-}176a)$$

$$E_\rho^t = -\frac{I_m}{4\rho} \begin{cases} \sum_{n=-\infty}^{+\infty} n H_n^{(2)}(\beta\rho') \left[J_n(\beta\rho) - \frac{J_n'(\beta a)}{H_n^{(2)'}(\beta a)} H_n^{(2)}(\beta\rho) \right] \\ \quad \times e^{jn(\phi-\phi')} \qquad a \le \rho \le \rho' \quad (11\text{-}176b) \\ \sum_{n=-\infty}^{+\infty} n H_n^{(2)}(\beta\rho) \left[J_n(\beta\rho') - \frac{J_n'(\beta a)}{H_n^{(2)'}(\beta a)} H_n^{(2)}(\beta\rho') \right] \\ \quad \times e^{jn(\phi-\phi')} \qquad \rho \ge \rho' \quad (11\text{-}176c) \end{cases}$$

$$E_\phi^t = -j\frac{\beta I_m}{4} \begin{cases} \sum_{n=-\infty}^{+\infty} H_n^{(2)}(\beta\rho') \left[J_n'(\beta\rho) - \frac{J_n'(\beta a)}{H_n^{(2)'}(\beta a)} H_n^{(2)'}(\beta\rho) \right] \\ \quad \times e^{jn(\phi-\phi')} \qquad a \le \rho \le \rho' \quad (11\text{-}176d) \\ \sum_{n=-\infty}^{+\infty} H_n^{(2)'}(\beta\rho) \left[J_n(\beta\rho') - \frac{J_n'(\beta a)}{H_n^{(2)'}(\beta a)} H_n^{(2)}(\beta\rho') \right] \\ \quad \times e^{jn(\phi-\phi')} \qquad \rho \ge \rho' \quad (11\text{-}176e) \end{cases}$$

$$H_z^t = -\frac{\beta^2 I_m}{4\omega\mu} \begin{cases} \sum_{n=-\infty}^{+\infty} H_n^{(2)}(\beta\rho') \left[J_n(\beta\rho) - \frac{J_n'(\beta a)}{H_n^{(2)'}(\beta a)} H_n^{(2)}(\beta\rho) \right] \\ \quad \times e^{jn(\phi-\phi')} \qquad a \le \rho \le \rho' \quad (11\text{-}176f) \\ \sum_{n=-\infty}^{+\infty} H_n^{(2)}(\beta\rho) \left[J_n(\beta\rho') - \frac{J_n'(\beta a)}{H_n^{(2)'}(\beta a)} H_n^{(2)}(\beta\rho') \right] \\ \quad \times e^{jn(\phi-\phi')} \qquad \rho \ge \rho' \quad (11\text{-}176g) \end{cases}$$

where the first terms within the summation and brackets represent the incident fields and the second terms represent the scattered fields.

On the surface of the cylinder the current density can be found to be

$$\mathbf{J}_s = \hat{n} \times \mathbf{H}^t|_{\rho=a} = \hat{a}_\rho \times \hat{a}_z H_z^t|_{\rho=a} = -\hat{a}_\phi H_z^t|_{\rho=a}$$

$$= \hat{a}_\phi \frac{\beta^2 I_m}{4\omega\mu} \sum_{n=-\infty}^{+\infty} H_n^{(2)}(\beta\rho') \left[J_n(\beta a) - \frac{J_n'(\beta a)}{H_n^{(2)\prime}(\beta a)} H_n^{(2)}(\beta a) \right] e^{jn(\phi-\phi')}$$

$$\mathbf{J}_s = -j\hat{a}_\phi \frac{\beta^2 I_m}{4\omega\mu} \sum_{n=-\infty}^{+\infty} H_n^{(2)}(\beta\rho') \left[\frac{J_n(\beta a) Y_n'(\beta a) - J_n'(\beta a) Y_n(\beta a)}{H_n^{(2)\prime}(\beta a)} \right] e^{jn(\phi-\phi')} \quad (11\text{-}177)$$

which by using the Wronskian of (11-95) reduces to

$$\mathbf{J}_s = -j\hat{a}_\phi \frac{I_m}{2\eta\pi a} \sum_{n=-\infty}^{+\infty} \frac{H_n^{(2)}(\beta\rho')}{H_n^{(2)\prime}(\beta a)} e^{jn(\phi-\phi')} \quad (11\text{-}177a)$$

For far-field observations ($\beta\rho \gg 1$) the total magnetic field of (11-176g) can be reduced in form by replacing the Hankel function $H_n^{(2)}(\beta\rho)$ by its asymptotic

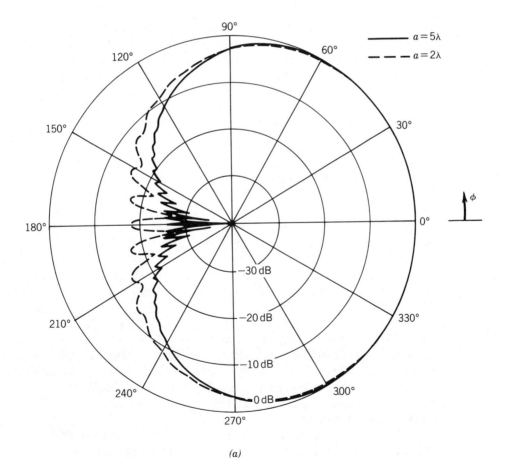

FIGURE 11-19 Normalized far-field amplitude pattern of a very thin axial slot on a circular conducting cylinder ($\phi' = 0°$).

expression (11-135a). Doing this reduces (11-176g) to

$$H_z^t \overset{\beta\rho \gg 1}{\simeq} -\frac{\beta^2 I_m}{4\omega\mu}\sqrt{\frac{2j}{\pi\beta}}\frac{e^{-j\beta\rho}}{\sqrt{\rho}}\sum_{n=-\infty}^{+\infty} j^n \left[J_n(\beta\rho') - \frac{J_n'(\beta a)}{H_n^{(2)'}(\beta a)} H_n^{(2)}(\beta\rho') \right] e^{jn(\phi-\phi')}$$

(11-178)

which can be used to compute far-field patterns of a magnetic line source near a circular conducting cylinder more conveniently. When the line source is moved to the surface of the cylinder ($\rho' = a$), then (11-178) reduces with the aid of the Wronskian of (11-95) to

$$H_z^t \overset{\beta\rho \gg 1}{\simeq} j\frac{\beta^2 I_m}{4\omega\mu}\sqrt{\frac{2j}{\pi\beta}}\frac{e^{-j\beta\rho}}{\sqrt{\rho}}\sum_{n=-\infty}^{+\infty} j^n \left[\frac{J_n(\beta a)Y_n'(\beta a) - J_n'(\beta a)Y_n(\beta a)}{H_n^{(2)'}(\beta a)} \right] e^{jn(\phi-\phi')}$$

$$H_z^t \overset{\beta\rho \gg 1}{\simeq} j\frac{I_m}{\pi} a \sqrt{\frac{\varepsilon}{\mu}}\sqrt{\frac{j}{2\pi\beta}}\frac{e^{-j\beta\rho}}{\sqrt{\rho}}\sum_{n=-\infty}^{+\infty} j^n \frac{e^{jn(\phi-\phi')}}{H_n^{(2)'}(\beta a)}$$

(11-178a)

The pattern of (11-178a) is representative of a very thin (ideally zero width) infinite length axial slot on a circular conducting cylinder, and its normalized form is shown plotted in Figure 11-19 for $a = 2\lambda$ and 5λ. Because of the larger radius of curvature for the $a = 5\lambda$ radius, which results in larger attenuation, less energy is allowed to "creep" around the cylinder compared to that of $a = 2\lambda$.

Scattering by cylinders of other cross sections, and by dielectric and dielectric-covered cylinders can be found in the literature [12, 19–31].

11.6 SCATTERING BY A CONDUCTING WEDGE

Scattering of electromagnetic waves by a two-dimensional conducting wedge has received considerable attention since about the middle 1950s. Because the wedge is a canonical problem that can be used to represent locally (near the edge) the scattering of more complex structures, asymptotic forms of its solution have been utilized to solve numerous practical problems. The asymptotic forms of its solution are obtained by taking the infinite series modal solution and first transforming it into an integral by the so-called *Watson transformation* [19, 32, 33]. The integral is then evaluated by the *method of steepest descent (saddle point method)* (see Appendix VI) [34]. The resulting terms of the integral evaluation can be recognized to represent the geometrical optics fields, both incident and reflected geometrical optics fields, and the diffracted fields, both incident and reflected diffracted fields [35, 36]. These forms of the solution have received considerable attention in the geometrical theory of diffraction (GTD) which has become a generic name in the area of antennas and scattering [37–39].

First we will present the modal solution of the scattering of the wedge. In Chapter 13 we will briefly outline its asymptotic solution, whose form represents the geometrical optics and diffracted fields, and apply it to antenna and scattering problems.

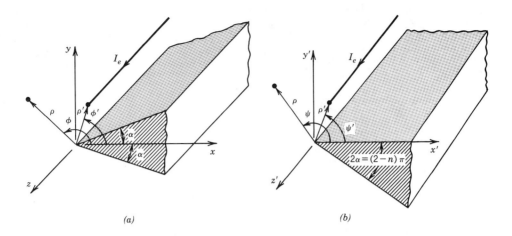

FIGURE 11-20 Electric line source near a two-dimensional conducting wedge. (*a*) Reference at bisector. (*b*) Reference at face.

11.6.1 Electric Line-Source Scattering by a Conducting Wedge: TM^z Polarization

Let us assume that an infinite electric line source of electric current I_e is placed near a conducting wedge whose total inner wedge angle is $WA = 2\alpha$, as shown in Figure 11-20*a*. The incident field produced everywhere by the source in the absence of the wedge can be written according to (11-158a) and (11-158b) as

$$E_z^i = -\frac{\beta^2 I_e}{4\omega\varepsilon} \begin{cases} \displaystyle\sum_{m=-\infty}^{+\infty} J_m(\beta\rho) H_m^{(2)}(\beta\rho') e^{jm(\phi-\phi')} & \rho \le \rho' \quad (11\text{-}179\text{a}) \\ \displaystyle\sum_{n=-\infty}^{+\infty} J_m(\beta\rho') H_m^{(2)}(\beta\rho) e^{jm(\phi-\phi')} & \rho \ge \rho' \quad (11\text{-}179\text{b}) \end{cases}$$

The corresponding *z* component of the scattered field must be chosen so that the sum of the two (incident plus scattered) along the faces of the wedge ($\phi = \alpha$ and $\phi = 2\pi - \alpha$) must vanish and simultaneously satisfy reciprocity (interchanging source and observation points). The ϕ variations must be represented by standing wave functions since in the ϕ direction the waves bounce back and forth between the plates forming the wedge. It can be shown that electric field expressions that satisfy these conditions take the form

$$E_z^t = E_z^i + E_z^s$$
$$= \begin{cases} \displaystyle\sum_v c_v f(\rho') J_v(\beta\rho) \sin[v(\phi'-\alpha)] \sin[v(\phi-\alpha)] & \rho \le \rho' \quad (11\text{-}180\text{a}) \\ \displaystyle\sum_v d_v g(\rho') H_v^{(2)}(\beta\rho) \sin[v(\phi'-\alpha)] \sin[v(\phi-\alpha)] & \rho \ge \rho' \quad (11\text{-}180\text{b}) \end{cases}$$

When $\rho = \rho'$, the two must be identical. Thus

$$\sum_v c_v f(\rho') J_v(\beta\rho') \sin[v(\phi'-\alpha)] \sin[v(\phi-\alpha)]$$
$$= \sum_v d_v g(\rho') H_v^{(2)}(\beta\rho') \sin[v(\phi'-\alpha)] \sin[v(\phi-\alpha)] \quad (11\text{-}181)$$

636 SCATTERING

which is satisfied if

$$c_v f(\rho') J_v(\beta\rho') = d_v g(\rho') H_v^{(2)}(\beta\rho') \tag{11-181a}$$

or

$$a_v = c_v = d_v \tag{11-181b}$$

$$f(\rho') = H_v^{(2)}(\beta\rho') \tag{11-181c}$$

$$g(\rho') = J_v(\beta\rho') \tag{11-181d}$$

Therefore the total electric field of (11-180a) and (11-180b) can be written as

$$E_z^t = E_z^i + E_z^s = \begin{cases} \sum_v a_v J_v(\beta\rho) H_v^{(2)}(\beta\rho') \sin[v(\phi' - \alpha)] \\ \quad \times \sin[v(\phi - \alpha)] \qquad \rho \leq \rho' \quad (11\text{-}182a) \\ \sum_v a_v J_v(\beta\rho') H_v^{(2)}(\beta\rho) \sin[v(\phi' - \alpha)] \\ \quad \times \sin[v(\phi - \alpha)] \qquad \rho \geq \rho' \quad (11\text{-}182b) \end{cases}$$

It is evident from (11-182a) and (11-182b) that when $\phi = \alpha$ the total tangential electric field vanishes. However, when $\phi = 2\pi - \alpha$ the electric field of (11-182a) and (11-182b) vanishes when

$$\sin[v(\phi - \alpha)]_{\phi=2\pi-\alpha} = \sin[v(2\pi - 2\alpha)] = \sin[2v(\pi - \alpha)] = 0 \tag{11-183}$$

or

$$2v(\pi - \alpha) = \sin^{-1}(0) = m\pi$$

$$\boxed{v = \frac{m\pi}{2(\pi - \alpha)} \qquad m = 1, 2, 3\ldots} \tag{11-183a}$$

Thus in (11-182a) and (11-182b) the allowable values of v are those of (11-183a). The values of a_v depend on the type of source.

The magnetic field components can be obtained by using Maxwell's equations 11-86 through 11-86b so that we can write that

$$H_\rho^t = -\frac{1}{j\omega\mu} \frac{1}{\rho} \frac{\partial E_z^t}{\partial \phi}$$

$$= -\frac{1}{j\omega\mu} \frac{1}{\rho} \begin{cases} \sum_v v a_v J_v(\beta\rho) H_v^{(2)}(\beta\rho') \\ \quad \times \sin[v(\phi' - \alpha)] \cos[v(\phi - \alpha)] \qquad \rho \leq \rho' \quad (11\text{-}184a) \\ \sum_v v a_v J_v(\beta\rho') H_v^{(2)}(\beta\rho) \\ \quad \times \sin[v(\phi' - \alpha)] \cos[v(\phi - \alpha)] \qquad \rho \geq \rho' \quad (11\text{-}184b) \end{cases}$$

$$\boxed{\begin{aligned} H_\phi^t &= \frac{1}{j\omega\mu}\frac{\partial E_z^t}{\partial \rho} \\ &= \frac{\beta}{j\omega\mu}\begin{cases} \sum_v a_v J_v'(\beta\rho) H_v^{(2)}(\beta\rho') \\ \quad \times \sin[v(\phi'-\alpha)]\sin[v(\phi-\alpha)] & \rho \le \rho' \quad (11\text{-}185a) \\ \sum_v a_v J_v(\beta\rho') H_v^{(2)'}(\beta\rho) \\ \quad \times \sin[v(\phi'-\alpha)]\sin[v(\phi-\alpha)] & \rho \ge \rho' \quad (11\text{-}185b) \end{cases} \end{aligned}}$$

where
$$' = \frac{\partial}{\partial(\beta\rho)} \qquad (11\text{-}185c)$$

At the source, the current density is obtained using

$$\begin{aligned} \mathbf{J}_s &= \hat{n}\times\mathbf{H}^t = \hat{a}_\rho \times \left(\hat{a}_\rho H_\rho^t + \hat{a}_\phi H_\phi^t\right)_{\rho=\rho_+',\,\rho_-'} \\ &= \hat{a}_z H_\phi^t\big|_{\rho=\rho_+',\rho_-'} = \hat{a}_z\left[H_\phi^t(\rho_+') - H_\phi^t(\rho_-')\right] \\ &= \hat{a}_z\frac{\beta}{j\omega\mu}\sum_v a_v\left[J_v(\beta\rho')H_v^{(2)'}(\beta\rho') - H_v^{(2)}(\beta\rho')J_v'(\beta\rho')\right] \\ &\quad \times \sin[v(\phi'-\alpha)]\sin[v(\phi-\alpha)] \\ \mathbf{J}_s &= \hat{a}_z\frac{\beta}{j\omega\mu}\sum_v a_v(-j)\left[J_v(\beta\rho')Y_v'(\beta\rho') - J_v'(\beta\rho')Y_v(\beta\rho')\right] \\ &\quad \times \sin[v(\phi'-\alpha)]\sin[v(\phi-\alpha)] \qquad (11\text{-}186) \end{aligned}$$

which by using the Wronskian of (11-95) reduces to

$$\mathbf{J}_s = -\hat{a}_z \frac{2}{\pi\omega\mu\rho'}\sum_v a_v \sin[v(\phi'-\alpha)]\sin[v(\phi-\alpha)] \qquad (11\text{-}186a)$$

Since the Fourier series for a current impulse of amplitude I_e located at $\rho=\rho'$ and $\phi=\phi'$ is [11]

$$J_z = \frac{I_e}{(\pi-\alpha)\rho'}\sum_v \sin[v(\phi'-\alpha)]\sin[v(\phi-\alpha)] \qquad (11\text{-}187)$$

then comparing (11-186a) and (11-187) leads to

$$-\frac{2}{\pi\omega\mu}a_v = \frac{I_e}{\pi-\alpha} \Rightarrow \boxed{a_v = -\frac{\pi\omega\mu I_e}{2(\pi-\alpha)}} \qquad (11\text{-}188)$$

A. FAR-ZONE FIELD

When the observations are made in the far zone ($\beta\rho \gg 1$, $\rho > \rho'$) the total electric field of (11-182b) can be written, by replacing the Hankel function $H_v^{(2)}(\beta\rho)$ by its asymptotic expression (11-135a), as

$$E_z^t \overset{\beta\rho \to \infty}{\simeq} \sqrt{\frac{2j}{\pi\beta\rho}} e^{-j\beta\rho} \sum_v a_v j^v J_v(\beta\rho') \sin[v(\phi'-\alpha)] \sin[v(\phi-\alpha)]$$

$$\overset{\beta\rho \to \infty}{\simeq} -I_e \sqrt{\frac{\pi j}{2\beta}} \frac{\omega\mu}{\pi-\alpha} \frac{e^{-j\beta\rho}}{\sqrt{\rho}} \sum_v j^v J_v(\beta\rho') \sin[v(\phi'-\alpha)] \sin[v(\phi-\alpha)]$$

$$E_z^t \overset{\beta\rho \to \infty}{\simeq} f_e(\rho) \sum_v j^v J_v(\beta\rho') \sin[v(\phi'-\alpha)] \sin[v(\phi-\alpha)] \qquad (11\text{-}189)$$

where

$$f_e(\rho) = -I_e \sqrt{\frac{\pi j}{2\beta}} \frac{\omega\mu}{\pi-\alpha} \frac{e^{-j\beta\rho}}{\sqrt{\rho}} \qquad (11\text{-}189a)$$

Therefore, (11-189) represents the total electric field created in the far-zone region by a cylindrical wave source of strength I_e located at ρ', ϕ'.

B. PLANE WAVE SCATTERING

When the source is placed at far distances ($\beta\rho' \gg 1$ and $\rho' > \rho$) and the observations are made at any point, then the total electric field of (11-182a) can be written, by replacing the Hankel function $H_v^{(2)}(\beta\rho')$ by its asymptotic form of (11-135a), as

$$E_z^t \overset{\beta\rho' \to \infty}{\simeq} -I_e \sqrt{\frac{\pi j}{2\beta}} \frac{\omega\mu}{\pi-\alpha} \frac{e^{-j\beta\rho'}}{\sqrt{\rho'}} \sum_v j^v J_v(\beta\rho) \sin[v(\phi'-\alpha)] \sin[v(\phi-\alpha)]$$

$$\overset{\beta\rho' \to \infty}{\simeq} g_e(\rho') \sum_v j^v J_v(\beta\rho) \sin[v(\phi'-\alpha)] \sin[v(\phi-\alpha)]$$

$$E_z^t \overset{\beta\rho' \to \infty}{\simeq} E_0 \sum_v j^v J_v(\beta\rho) \sin[v(\phi'-\alpha)] \sin[v(\phi-\alpha)] \qquad (11\text{-}190)$$

where

$$E_0 = g_e(\rho') = -I_e \sqrt{\frac{\pi j}{2\beta}} \frac{\omega\mu}{\pi-\alpha} \frac{e^{-j\beta\rho'}}{\sqrt{\rho'}} \qquad (11\text{-}190a)$$

It is evident that (11-190) can also be obtained from (11-189) by reciprocity, that is, interchanging source and observation point. This is accomplished by letting $\rho = \rho'$, $\rho' = \rho$, $\phi = \phi'$, and $\phi' = \phi$, or in this situation simply by interchanging ρ and ρ' only.

Equation 11-190 also represents the total electric field of a TM^z uniform plane wave of strength E_0 incident at an angle ϕ' on a conducting wedge of interior angle

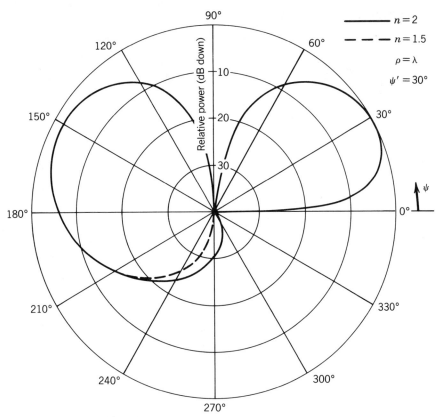

FIGURE 11-21 Normalized amplitude pattern of a TM^z (soft polarization) plane wave incident on a two-dimensional conducting wedge.

2α. When the wedge is a half-plane ($\alpha = 0$), (11-190) reduces to

$$E_z^t \underset{\alpha=0}{\overset{\beta\rho' \to \infty}{\simeq}} E_0 \sum_v j^v J_v(\beta\rho) \sin(v\phi') \sin(v\phi) \qquad (11\text{-}191)$$

which by using (11-183a) can also be expressed as

$$E_z^t \underset{\alpha=0}{\overset{\beta\rho' \to \infty}{\simeq}} E_0 \sum_{m=1}^{\infty} j^{m/2} J_{m/2}(\beta\rho) \sin\left(\frac{m}{2}\phi'\right) \sin\left(\frac{m}{2}\phi\right) \qquad (11\text{-}191a)$$

The normalized scattering patterns at a distance λ ($\rho = \lambda$) from the edge of the wedge formed when a plane wave is incident upon a wedge of $2\alpha = 0°$ ($n = 2$) (half-plane) and $90°$ ($n = 1.5$) are shown in Figure 11-21.

11.6.2 Magnetic Line-Source Scattering by a Conducting Wedge: TE^z Polarization

When the line source of Figure 11-20a is magnetic of current I_m the total magnetic field has only a z component, and it can be written by referring to the forms of

(11-182a) and (11-182b) as

$$
H_z^t = H_z^i + H_z^s
$$
$$
= \begin{cases} \sum_s b_s J_s(\beta\rho) H_s^{(2)}(\beta\rho') \cos[s(\phi' - \alpha)] \cos[s(\phi - \alpha)] & \rho \leq \rho' \quad (11\text{-}192a) \\ \sum_s b_s J_s(\beta\rho') H_s^{(2)}(\beta\rho) \cos[s(\phi' - \alpha)] \cos[s(\phi - \alpha)] & \rho \geq \rho' \quad (11\text{-}192b) \end{cases}
$$

The allowable values of s are obtained by applying the boundary conditions and the coefficients b_s are determined by the type of source.

To apply the boundary conditions, we first need to find the corresponding electric field components (especially the tangential components). This is accomplished by using Maxwell's equations 11-106 through 11-106b. Thus the total radial component of the electric field can be written as

$$
E_\rho^t = \frac{1}{j\omega\varepsilon} \frac{1}{\rho} \frac{\partial H_z^t}{\partial \phi}
$$

$$
= -\frac{1}{j\omega\varepsilon} \frac{1}{\rho} \begin{cases} \sum_s s b_s J_s(\beta\rho) H_s^{(2)}(\beta\rho') \\ \times \cos[s(\phi' - \alpha)] \sin[s(\phi - \alpha)] & \rho \leq \rho' \quad (11\text{-}193a) \\ \sum_s s b_s J_s(\beta\rho') H_s^{(2)}(\beta\rho) \\ \times \cos[s(\phi' - \alpha)] \sin[s(\phi - \alpha)] & \rho \geq \rho' \quad (11\text{-}193b) \end{cases}
$$

The boundary conditions that must be satisfied are

$$
E_\rho^t(0 \leq \rho, \rho' \leq \infty, \phi = \alpha, 0 \leq \phi' \leq 2\pi)
$$
$$
= E_\rho^t(0 \leq \rho, \rho' \leq \infty, \phi = 2\pi - \alpha, 0 \leq \phi' \leq 2\pi) = 0 \quad (11\text{-}194)
$$

The first boundary condition of (11-194) is always satisfied regardless of the values of s. Applying the second boundary condition leads to

$$
E_\rho^t(0 \leq \rho, \rho' \leq \infty, \phi = 2\pi - \alpha, 0 \leq \phi' \leq 2\pi) = 0
$$
$$
= -\frac{1}{j\omega\varepsilon\rho} \sum_s s b_s J_s(\beta\rho) H_s^{(2)}(\beta\rho') \cos[s(\phi' - \alpha)] \sin[2s(\pi - \alpha)]
$$
$$
= -\frac{1}{j\omega\varepsilon\rho} \sum_s s b_s J_s(\beta\rho') H_s^{(2)}(\beta\rho) \cos[s(\phi' - \alpha)] \sin[2s(\pi - \alpha)] \quad (11\text{-}195)
$$

which is satisfied provided

$$
\sin[2s(\pi - \alpha)] = 0 \Rightarrow 2s(\pi - \alpha) = \sin^{-1}(0) = m\pi
$$

$$
\boxed{s = \frac{m\pi}{2(\pi - \alpha)} \qquad m = 0, 1, 2, \ldots} \quad (11\text{-}195a)
$$

SCATTERING BY CONDUCTING WEDGE **641**

Since the source is magnetic, the coefficients b_s take the form of

$$b_s = \varepsilon_s \left[\frac{\pi \omega \varepsilon I_m}{4(\pi - \alpha)} \right] \qquad (11\text{-}196)$$

where

$$\varepsilon_s = \begin{cases} 1 & s = 0 \\ 2 & s \neq 0 \end{cases} \qquad (11\text{-}196a)$$

In the far zone ($\beta\rho \gg 1$) the total field of (11-192b) reduces, by replacing the Hankel function $H_s^{(2)}(\beta\rho)$ with its asymptotic form (11-135a), to

$$H_z^t \underset{\beta\rho \to \infty}{\simeq} I_m \sqrt{\frac{\pi j}{8\beta}} \frac{\omega\varepsilon}{\pi - \alpha} \frac{e^{-j\beta\rho}}{\sqrt{\rho}} \sum_s \varepsilon_s j^s J_s(\beta\rho') \cos[s(\phi' - \alpha)] \cos[s(\phi - \alpha)]$$

$$H_z^t \underset{\beta\rho \to \infty}{\simeq} f_h(\rho) \sum_s \varepsilon_s j^s J_s(\beta\rho') \cos[s(\phi' - \alpha)] \cos[s(\phi - \alpha)] \qquad (11\text{-}197)$$

where

$$f_h(\rho) = I_m \sqrt{\frac{\pi j}{8\beta}} \frac{\omega\varepsilon}{\pi - \alpha} \frac{e^{-j\beta\rho}}{\sqrt{\rho}} \qquad (11\text{-}197a)$$

When the source is removed at far distances ($\beta\rho' \gg 1$ and $\rho' > \rho$), (11-192a) reduces to

$$H_z^t \underset{\beta\rho' \to \infty}{\simeq} g_h(\rho') \sum_s \varepsilon_s j^s J_s(\beta\rho) \cos[s(\phi' - \alpha)] \cos[s(\phi - \alpha)]$$

$$H_z^t \underset{\beta\rho' \to \infty}{\simeq} H_0 \sum_s \varepsilon_s j^s J_s(\beta\rho) \cos[s(\phi' - \alpha)] \cos[s(\phi - \alpha)] \qquad (11\text{-}198)$$

where

$$H_0 = g_h(\rho') = I_m \sqrt{\frac{\pi j}{8\beta}} \frac{\omega\varepsilon}{\pi - \alpha} \frac{e^{-j\beta\rho'}}{\sqrt{\rho'}} \qquad (11\text{-}198a)$$

Equation 11-198 also represents the total magnetic field of a TE^z uniform plane wave of strength H_0 incident at an angle ϕ' on a conducting wedge of interior angle 2α. For a wedge with zero included angle (half-plane $\alpha = 0$), (11-198) reduces to

$$H_z^t \underset{\substack{\beta\rho \to \infty \\ \alpha = 0}}{\simeq} H_0 \sum_s \varepsilon_s j^s J_s(\beta\rho) \cos(s\phi') \cos(s\phi) \qquad (11\text{-}199)$$

which by using (11-195a) can also be expressed as

$$H_z^t \underset{\substack{\beta\rho \to \infty \\ \alpha = 0}}{\simeq} H_0 \sum_{m=0}^{\infty} \varepsilon_{m/2} j^{m/2} J_{m/2}(\beta\rho) \cos\left(\frac{m}{2}\phi'\right) \cos\left(\frac{m}{2}\phi\right) \qquad (11\text{-}199a)$$

642 SCATTERING

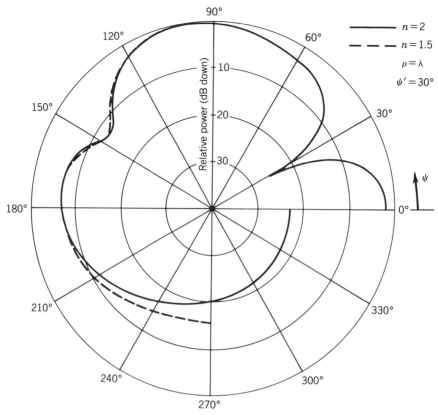

FIGURE 11-22 Normalized amplitude pattern of a TE^z (hard polarization) plane wave incident on a two-dimensional conducting wedge.

The normalized scattering patterns at a distance λ ($\rho = \lambda$) from the edge of the wedge formed when a plane wave is incident upon a wedge of $2\alpha = 0°$ ($n = 2$) (half-plane) and $90°$ ($n = 1.5$) are shown in Figure 11-22.

11.6.3 Electric and Magnetic Line-Source Scattering by a Conducting Wedge

The total electric field of an electric line source near a conducting wedge, as given by (11-182a) and (11-182b), and the total magnetic field of a magnetic line source near a conducting wedge, as given by (11-192a) and (11-192b), can both be represented by the same expression by adopting the coordinate system (x', y', z') of Figure 11-20b, instead of the (x, y, z), which is referenced to the side of the wedge that is illuminated by the source. This is usually more convenient because similar forms of the expression can represent either polarization. Also the interior angle of the wedge will be represented by

$$2\alpha = (2 - n)\pi \Rightarrow n = 2 - \frac{2\alpha}{\pi} \tag{11-200}$$

Thus a given value of n represents a wedge with a specific included angle: values of $n > 1$ represent wedges with included angles of less than $180°$ (referred to as exterior wedges) and values of $n < 1$ represent wedges with included angles of greater than 180° (referred to as interior wedges). Thus the allowable values of v, as given by (11-183a), and those of s, as given by (11-195a), can now be represented by

$$v = \frac{m\pi}{2(\pi - \alpha)}\bigg|_{\alpha = (1-n/2)\pi} = \frac{m}{n} \qquad m = 1, 2, 3, \ldots \qquad (11\text{-}201\text{a})$$

$$s = \frac{m\pi}{2(\pi - \alpha)}\bigg|_{\alpha = (1-n/2)\pi} = \frac{m}{n} \qquad m = 0, 1, 2, \ldots \qquad (11\text{-}201\text{b})$$

In addition the amplitude coefficients a_v, as given by (11-188), and b_s, as given by (11-196), can now be expressed as

$$a_v = -\frac{\pi\omega\mu I_e}{2(\pi - \alpha)}\bigg|_{\alpha = (1-n/2)\pi} = -\frac{\omega\mu I_e}{2}\left(\frac{2}{n}\right) \qquad (11\text{-}202\text{a})$$

$$b_s = \varepsilon_s\left[\frac{\pi\omega\varepsilon I_m}{4(\pi - \alpha)}\right]_{\alpha = (1-n/2)\pi} = \frac{\omega\varepsilon I_m}{2}\left(\frac{\varepsilon_s}{n}\right) \qquad (11\text{-}202\text{b})$$

Using the new coordinate system (x', y', z') of Figure 11-20b we can write that

$$\phi' = \psi' + \alpha \qquad (11\text{-}203\text{a})$$
$$\phi = \psi + \alpha \qquad (11\text{-}203\text{b})$$

Therefore the sine functions of (11-182a) through (11-182b) and the cosine functions of (11-192a) through (11-192b) can be written as

$$\sin[v(\phi' - \alpha)]\sin[v(\phi - \alpha)] = \sin\left[\frac{m}{n}(\psi' + \alpha - \alpha)\right]\sin\left[\frac{m}{n}(\psi + \alpha - \alpha)\right]$$
$$= \sin\left(\frac{m}{n}\psi'\right)\sin\left(\frac{m}{n}\psi\right)$$
$$= \frac{1}{2}\left\{\cos\left[\frac{m}{n}(\psi - \psi')\right] - \cos\left[\frac{m}{n}(\psi + \psi')\right]\right\}$$
$$(11\text{-}204\text{a})$$

$$\cos[s(\phi' - \alpha)]\cos[s(\phi - \alpha)] = \cos\left[\frac{m}{n}(\psi' + \alpha - \alpha)\right]\cos\left[\frac{m}{n}(\psi + \alpha - \alpha)\right]$$
$$= \cos\left(\frac{m}{n}\psi'\right)\cos\left(\frac{m}{n}\psi\right)$$
$$= \frac{1}{2}\left\{\cos\left[\frac{m}{n}(\psi - \psi')\right] + \cos\left[\frac{m}{n}(\psi + \psi')\right]\right\}$$
$$(11\text{-}204\text{b})$$

Using all these new notations we can write for the TM^z polarization the total electric field of (11-182a) through (11-182b) and for the TE^z polarization the total magnetic field of (11-192a) through (11-192b) as

$$\underline{\text{TM}^z}$$

$$E_z^t = -\frac{\omega\mu I_e}{4}\frac{1}{n}\begin{cases}\sum_{m=0,1,\ldots}^{\infty} 2J_{m/n}(\beta\rho)H_{m/n}^{(2)}(\beta\rho')\\ \times\left\{\cos\left[\frac{m}{n}(\psi-\psi')\right]-\cos\left[\frac{m}{n}(\psi+\psi')\right]\right\} \quad \rho\leq\rho' \quad (11\text{-}205a)\\ \sum_{m=0,1,\ldots}^{\infty} 2J_{m/n}(\beta\rho')H_{m/n}^{(2)}(\beta\rho)\\ \times\left\{\cos\left[\frac{m}{n}(\psi-\psi')\right]-\cos\left[\frac{m}{n}(\psi+\psi')\right]\right\} \quad \rho\geq\rho' \quad (11\text{-}205b)\end{cases}$$

$$\underline{\text{TE}^z}$$

$$H_z^t = \frac{\omega\varepsilon I_m}{4}\frac{1}{n}\begin{cases}\sum_{m=0,1,\ldots}^{\infty} \varepsilon_m J_{m/n}(\beta\rho)H_{m/n}^{(2)}(\beta\rho')\\ \times\left\{\cos\left[\frac{m}{n}(\psi-\psi')\right]+\cos\left[\frac{m}{n}(\psi+\psi')\right]\right\} \quad \rho\leq\rho' \quad (11\text{-}206a)\\ \sum_{m=0,1,\ldots}^{\infty} \varepsilon_m J_{m/n}(\beta\rho')H_{m/n}^{(2)}(\beta\rho)\\ \times\left\{\cos\left[\frac{m}{n}(\psi-\psi')\right]+\cos\left[\frac{m}{n}(\psi+\psi')\right]\right\} \quad \rho\geq\rho' \quad (11\text{-}206b)\end{cases}$$

To make the summations in (11-205a) through (11-206b) uniform, the summations of (11-205a) and (11-205b) are noted to begin with $m = 0$, even though the allowable values of m as given by (11-201a) begin with $m = 1$. However, it should be noted that $m = 0$ in (11-205a) and (11-205b) does not contribute anything and the expressions are correct as stated.

It is apparent by comparing (11-205a) and (11-205b) with (11-206a) and (11-206b) that they are of similar forms. Therefore we can write both as

$$E_z^t = -\frac{\omega\mu I_e}{4}G(\rho,\rho',\psi,\psi',n) \quad \text{for TM}^z \quad (11\text{-}207a)$$

$$H_z^t = +\frac{\omega\varepsilon I_m}{4}G(\rho,\rho',\psi,\psi',n) \quad \text{for TE}^z \quad (11\text{-}207b)$$

where

$$G(\rho,\rho',\psi,\psi',n)$$
$$= \frac{1}{n}\begin{cases}\sum_{m=0,1,\ldots}^{\infty} \varepsilon_m J_{m/n}(\beta\rho)H_{m/n}^{(2)}(\beta\rho')\\ \left\{\cos\left[\frac{m}{n}(\psi-\psi')\right]\pm\cos\left[\frac{m}{n}(\psi+\psi')\right]\right\} \quad \rho\leq\rho' \quad (11\text{-}208a)\\ \sum_{m=0,1,\ldots}^{\infty} \varepsilon_m J_{m/n}(\beta\rho')H_{m/n}^{(2)}(\beta\rho)\\ \left\{\cos\left[\frac{m}{n}(\psi-\psi')\right]\pm\cos\left[\frac{m}{n}(\psi+\psi')\right]\right\} \quad \rho\geq\rho' \quad (11\text{-}208b)\end{cases}$$

$$\varepsilon_m = \begin{cases}1 & m=0\\ 2 & m\neq 0\end{cases} \quad (11\text{-}208c)$$

The plus (+) sign between the cosine terms is used for the TE^z polarization and the minus (−) sign is used for the TM^z polarization. Again note that the $m = 0$ terms do not contribute anything for the TM^z polarization.

The forms of (11-207a) through (11-208b) are those usually seen in the geometrical theory of diffraction (GTD) [37–41] where $G(\rho, \rho', \psi, \psi', n)$ is usually referred to as the *Green's function*. Since the summations in (11-208a) and (11-208b) are poorly convergent when the arguments of the Bessel and/or Hankel functions are large, asymptotic forms of them will be derived in Chapter 13 which are much more computationally efficient. The various terms of the asymptotic forms will be associated with incident and reflected geometrical optics and diffracted fields. It is also convenient in diffraction theory to refer to the TM^z polarization as the *soft* polarization; the TE^z is referred to as the *hard* polarization. This is a convenient designation that has been adopted from acoustics.

11.7 SPHERICAL WAVE ORTHOGONALITIES, TRANSFORMATIONS, AND THEOREMS

When dealing with scattering from structures whose geometry best conforms to spherical coordinates, it is often most convenient to transform wave functions (such as plane waves) from one coordinate system to another. This was done in Section 11.4 where uniform plane wave functions in rectilinear form were transformed and represented by cylindrical wave functions. This allowed convenient examination of the scattering of plane waves by cylindrical structures of circular and wedge cross sections. In addition certain theorems concerning cylindrical wave functions were introduced which were helpful in analyzing the scattering by cylindrical structures of circular cross sections of waves emanating from line sources.

In this section we want to introduce some orthogonality relationships, wave transformations, and theorems that are very convenient for examining scattering of plane waves from spherical structures and waves emanating from finite sources placed in the vicinity of spherical scatterers. First of all, however, let us examine radiation from a finite source radiating in an unbounded medium.

11.7.1 Vertical Dipole Spherical Wave Radiation

There are many sources of spherical wave radiation. In fact almost all sources used in practice are considered to excite spherical waves. One of the most prominent is that of a finite length wire whose total radiation can be obtained as a superposition of radiation from a very small linear current element of length $\Delta \ell$ and constant electric current $\mathbf{I}_e = \hat{a}_z I_e$. This is usually referred to as an infinitesimal dipole [1]. The radiation of other sources can be obtained by knowing the radiation from an infinitesimal dipole. Therefore it is important that we briefly examine the radiation from such a source.

It is usually most convenient to place the linear element at the origin of the coordinate system and have its length and current flow along the z axis, as shown in Figure 11-23a. To find the fields radiated by this source we resort to the techniques of Chapter 6, Sections 6.4 and 6.6, where we first specify the currents I_e and I_m of the source. Then we find the potentials $\mathbf{A}$ and $\mathbf{F}$ [using (6-97a) and (6-97b)], and determine the radiated $\mathbf{E}$ and $\mathbf{H}$ [using (6-34) and (6-35)].

Following such a procedure, the electric and magnetic fields radiated by the infinitesimal dipole of Figure 11-23 were derived in Example 6-3. In terms of

646 SCATTERING

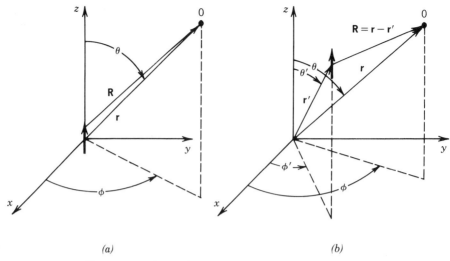

FIGURE 11-23 Geometry and coordinate system for vertical dipole radiation. (*a*) At origin. (*b*) Offset.

spherical wave functions, the vector potential **A** can also be written as

$$\mathbf{A} = \hat{a}_z \frac{\mu I_e \Delta \ell}{4\pi} \frac{e^{-j\beta r}}{r} = -\hat{a}_z j \frac{\mu \beta I_e \Delta \ell}{4\pi} h_0^{(2)}(\beta r) \tag{11-209}$$

where $h_0^{(2)}(\beta r)$ is the spherical Hankel function of order zero that is given by

$$h_0^{(2)}(\beta r) = \frac{e^{-j\beta r}}{-j\beta r} \tag{11-209a}$$

For the dual problem of the linear magnetic current element of current I_m, the potential function **F** takes the form of

$$\mathbf{F} = \hat{a}_z \frac{\varepsilon I_m \Delta \ell}{4\pi} \frac{e^{-j\beta r}}{r} = -\hat{a}_z j \frac{\varepsilon \beta I_m \Delta \ell}{4\pi} h_0^{(2)}(\beta r) \tag{11-210}$$

If the source is removed from the origin, as shown in Figure 11-23b, then the potentials **A** and **F** of (11-209) and (11-210) take the form of

$$\mathbf{A} = -\hat{a}_z j \frac{\mu \beta I_e \Delta \ell}{4\pi} h_0^{(2)}(\beta |\mathbf{r} - \mathbf{r}'|) \tag{11-211a}$$

$$\mathbf{F} = -\hat{a}_z j \frac{\varepsilon \beta I_m \Delta \ell}{4\pi} h_0^{(2)}(\beta |\mathbf{r} - \mathbf{r}'|) \tag{11-211b}$$

where

$$h_0^{(2)}(\beta |\mathbf{r} - \mathbf{r}'|) = \frac{e^{-j\beta |\mathbf{r} - \mathbf{r}'|}}{-j\beta |\mathbf{r} - \mathbf{r}'|} \tag{11-211c}$$

11.7.2 Orthogonality Relationships

When solving electromagnetic wave problems dealing with spherical structures (either waveguides, cavities, or scatterers), the θ variations are represented, as illustrated in Chapters 3 (Section 3.4.3) and 10, by Legendre polynomials $P_n(\cos\theta)$ and associated Legendre functions $P_n^m(\cos\theta)$ (see Appendix V).

The Legendre polynomials $P_n(\cos\theta)$ are often called *zonal harmonics* [42, 43], and they form a complete orthogonal set in the interval of $0 \leq \theta \leq \pi$. Therefore in this interval any arbitrary wave function can be represented by a series of Legendre polynomials. This is similar to the representation of any periodic function by a series of sines and cosines (Fourier series), since Legendre polynomials are very similar in form to cosinusoidal functions. In addition the products of associated Legendre functions $P_n^m(\cos\theta)$ with sines and cosines $[P_n^m(\cos\theta)\cos(m\phi)$ and $P_n^m(\cos\theta)\sin(m\phi)]$ are often referred to as *tesseral harmonics* [42, 43], and they form a complete orthogonal set on the surface of a sphere. Therefore any wave function that is defined over a sphere can be expressed by a series of tesseral harmonics.

Some of the most important and necessary orthogonality relationships that are necessary to solve wave scattering by spheres will be stated here. The interested reader is referred to [11, 42–45] for more details and derivations.

In the interval of $0 \leq \theta \leq \pi$ the integral of the product of Legendre polynomials is equal to

$$\int_0^\pi P_n(\cos\theta) P_m(\cos\theta) \sin\theta \, d\theta = \begin{cases} 0 & n \neq m \quad (11\text{-}212a) \\ \dfrac{2}{2n+1} & n = m \quad (11\text{-}212b) \end{cases}$$

Any function $f(\theta)$ defined in the interval of $0 \leq \theta \leq \pi$ can be represented by a series of Legendre polynomials of

$$f(\theta) = \sum_{n=0}^\infty a_n P_n(\cos\theta) \qquad 0 \leq \theta \leq \pi \qquad (11\text{-}213)$$

where

$$a_n = \frac{2n+1}{2} \int_0^\pi f(\theta) P_n(\cos\theta) \sin\theta \, d\theta \qquad (11\text{-}213a)$$

which is known as the *Fourier–Legendre* series.

Defining the tesseral harmonics by

$$T_{mn}^e(\theta, \phi) = P_n^m(\cos\theta) \cos(m\phi) \qquad (11\text{-}214a)$$

$$T_{mn}^0(\theta, \phi) = P_n^m(\cos\theta) \sin(m\phi) \qquad (11\text{-}214b)$$

and because

$$\int_0^{2\pi} \sin(p\phi) \sin(q\phi) \, d\phi = \int_0^{2\pi} \cos(p\phi) \cos(q\phi) \, d\phi$$

$$= \begin{cases} 0 & p \neq q \quad (11\text{-}215a) \\ \pi & p = q \neq 0 \quad (11\text{-}215b) \end{cases}$$

then it can be shown that

$$\int_0^{2\pi}\left[\int_0^{\pi} T_{mn}^e(\theta,\phi)T_{pq}^0(\theta,\phi)\sin\theta\,d\theta\right]d\phi = 0 \tag{11-216a}$$

$$\int_0^{2\pi}\left[\int_0^{\pi} T_{mn}^i(\theta,\phi)T_{pq}^i(\theta,\phi)\sin\theta\,d\theta\right]d\phi = 0 \quad \begin{array}{l} mn \neq pq \\ i = e \text{ or } 0 \end{array} \tag{11-216b}$$

$$\int_0^{2\pi}\left[\int_0^{\pi}\left[T_{mn}^i(\theta,\phi)\right]^2\sin\theta\,d\theta\right]d\phi$$

$$= \begin{cases} \dfrac{4\pi}{2n+1} & m=0 \quad i=e \tag{11-216c} \\[2mm] \dfrac{2\pi}{2n+1}\dfrac{(n+m)!}{(n-m)!} & m \neq 0 \quad i = e \text{ or } 0 \end{cases} \tag{11-216d}$$

11.7.3 Wave Transformations and Theorems

As was done for cylindrical wave functions in Section 11.4, in scattering it is often most convenient to express wave functions in one coordinate system in terms of wave functions of another coordinate system. The same applies to wave scattering by spherical structures. Therefore it is convenient for plane wave scattering by spherical geometries to express the plane waves, which are most conveniently written in rectilinear form, in terms of spherical wave functions.

To demonstrate that, let us assume that a uniform plane wave is traveling along the $+z$ direction, as shown in Figure 11-24, and it can be written as

$$\mathbf{E}^+ = \hat{a}_x E_x^+ = \hat{a}_x E_0 e^{-j\beta z} = \hat{a}_x e^{-j\beta z} \tag{11-217}$$

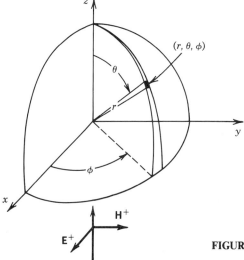

FIGURE 11-24 Uniform plane wave traveling in the $+z$ direction.

The plane wave can be represented by an infinite sum of spherical wave functions of the form

$$E_x^+ = e^{-j\beta z} = e^{-j\beta r \cos\theta} = \sum_{n=0}^{\infty} a_n j_n(\beta r) P_n(\cos\theta) \quad (11\text{-}217a)$$

since it must be independent of ϕ and be finite at the origin ($r = 0$). The next step is to determine the amplitude coefficients a_n. This can be accomplished as follows.

Multiplying both sides of (11-217a) by $P_m(\cos\theta)\sin\theta$ and integrating in θ from 0 to π we have that

$$\int_0^{\pi} e^{-j\beta r \cos\theta} P_m(\cos\theta) \sin\theta \, d\theta = \int_0^{\pi} \left[\sum_{n=0}^{\infty} a_n j_n(\beta r) P_n(\cos\theta) P_m(\cos\theta) \sin\theta \right] d\theta \quad (11\text{-}218)$$

Interchanging integration and summation, we have that

$$\int_0^{\pi} e^{-j\beta r \cos\theta} P_m(\cos\theta) \sin\theta \, d\theta = \sum_{n=0}^{\infty} a_n j_n(\beta r) \int_0^{\pi} P_n(\cos\theta) P_m(\cos\theta) \sin\theta \, d\theta \quad (11\text{-}218a)$$

Using the orthogonality condition of (11-212b) reduces (11-218a) to

$$\int_0^{\pi} e^{-j\beta r \cos\theta} P_m(\cos\theta) \sin\theta \, d\theta = \frac{2a_m}{2m+1} j_m(\beta r) \quad (11\text{-}219)$$

Since the integral of the left side of (11-219) is equal to

$$\int_0^{\pi} e^{-j\beta r \cos\theta} P_m(\cos\theta) \sin\theta \, d\theta = 2 j^{-m} j_m(\beta r) \quad (11\text{-}219a)$$

equating (11-219) and (11-219a) leads to

$$\frac{2a_m}{2m+1} j_m(\beta r) = 2 j^{-m} j_m(\beta r) \Rightarrow a_m = j^{-m}(2m+1) \quad (11\text{-}220)$$

Thus (11-217a) reduces to

$$\boxed{E_x^+ = e^{-j\beta z} = e^{-j\beta r \cos\theta} = \sum_{n=0}^{\infty} a_n j_n(\beta r) P_n(\cos\theta)} \quad (11\text{-}221)$$

where

$$\boxed{a_n = j^{-n}(2n+1)} \quad (11\text{-}221a)$$

In a similar manner it can be shown that

$$\boxed{E_x^- = e^{+j\beta z} = e^{+j\beta r \cos\theta} = \sum_{n=0}^{\infty} b_n j_n(\beta r) P_n(\cos\theta)} \quad (11\text{-}222)$$

where

$$b_n = j^n(2n + 1) \tag{11-222a}$$

When dealing with spherical wave scattering of waves generated by linear dipole radiators, it is convenient to express their radiation, determined using (11-211a) through (11-211c), in terms of spherical wave functions. This can be accomplished using the *addition theorem* [11] of spherical wave functions, which states that (11-211c) can be expressed by referring to the geometry of Figure 11-23b as

$$h_0^{(2)}(\beta|\mathbf{r} - \mathbf{r}'|) = \begin{cases} \sum_{n=0}^{\infty} (2n + 1) h_n^{(2)}(\beta r') j_n(\beta r) P_n(\cos \xi) & r < r' \quad (11\text{-}223a) \\ \sum_{n=0}^{\infty} (2n + 1) h_n^{(2)}(\beta r) j_n(\beta r') P_n(\cos \xi) & r > r' \quad (11\text{-}223b) \end{cases}$$

where

$$\cos \xi = \cos \theta \cos \theta' + \sin \theta \sin \theta' \cos(\phi - \phi') \tag{11-223c}$$

Similarly

$$h_0^{(1)}(\beta|\mathbf{r} - \mathbf{r}'|) = \begin{cases} \sum_{n=0}^{\infty} (2n + 1) h_n^{(1)}(\beta r') j_n(\beta r) P_n(\cos \xi) & r < r' \quad (11\text{-}224a) \\ \sum_{n=0}^{\infty} (2n + 1) h_n^{(1)}(\beta r) j_n(\beta r') P_n(\cos \xi) & r > r' \quad (11\text{-}224b) \end{cases}$$

11.8 SCATTERING BY A CONDUCTING SPHERE

One of the classic problems in scattering is that of plane wave scattering by a conducting sphere. Because of its symmetry, the sphere is often used as a reference scatterer to measure the scattering properties (such as the RCS) of other targets. It is, therefore, very important that we know and understand the scattering characteristics of a conducting sphere. The plane wave scattering by a conducting sphere has been examined by many authors [14, 20, 42, 43, 46–51]. Here we will outline one that parallels that of [11]. Let us assume that the electric field of a uniform plane wave is polarized in the x direction and it is traveling along the z axis as shown in Figure 11-25. The electric field of the incident wave can then be expressed as

$$\mathbf{E}^i = \hat{a}_x E_x^i = \hat{a}_x E_0 e^{-j\beta z} = \hat{a}_x E_0 e^{-j\beta r \cos \theta} \tag{11-225}$$

Using the transformation of (II-12), the x component of (11-225) can be transformed in spherical components to

$$\mathbf{E}^i = \hat{a}_r E_r^i + \hat{a}_\theta E_\theta^i + \hat{a}_\phi E_\phi^i \tag{11-226}$$

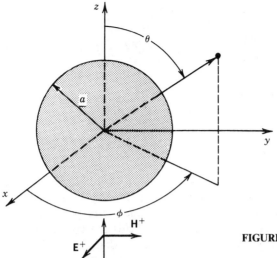

FIGURE 11-25 Uniform plane wave incident on a conducting sphere.

where

$$E_r^i = E_x^i \sin\theta \cos\phi = E_0 \sin\theta \cos\phi e^{-j\beta r\cos\theta} = E_0 \frac{\cos\phi}{j\beta r} \frac{\partial}{\partial\theta}(e^{-j\beta r\cos\theta}) \quad (11\text{-}226a)$$

$$E_\theta^i = E_x^i \cos\theta \cos\phi = E_0 \cos\theta \cos\phi e^{-j\beta r\cos\theta} \quad (11\text{-}226b)$$

$$E_\phi^i = -E_x^i \sin\phi = -E_0 \sin\phi e^{-j\beta r\cos\theta} \quad (11\text{-}226c)$$

Each of the spherical components of the preceding incident electric field can be expressed using the transformation of (11-221) and (11-221a) as

$$E_r^i = E_0 \frac{\cos\phi}{j\beta r} \sum_{n=0}^{\infty} j^{-n}(2n+1) j_n(\beta r) \frac{\partial}{\partial\theta}[P_n(\cos\theta)] \quad (11\text{-}227a)$$

$$E_\theta^i = E_0 \cos\theta \cos\phi \sum_{n=0}^{\infty} j^{-n}(2n+1) j_n(\beta r) P_n(\cos\theta) \quad (11\text{-}227b)$$

$$E_\phi^i = -E_0 \sin\phi \sum_{n=0}^{\infty} j^{-n}(2n+1) j_n(\beta r) P_n(\cos\theta) \quad (11\text{-}227c)$$

Since according to (10-31),

$$j_n(\beta r) = \frac{1}{\beta r} \hat{J}_n(\beta r) \quad (11\text{-}228a)$$

and

$$\frac{\partial P_n}{\partial \theta} = P_n^1(\cos\theta) \quad (11\text{-}228b)$$

$$P_0^1 = 0 \quad (11\text{-}228c)$$

we can rewrite (11-227a) through (11-227c) as

$$E_r^i = -jE_0 \frac{\cos\phi}{(\beta r)^2} \sum_{n=1}^{\infty} j^{-n}(2n+1)\hat{J}_n(\beta r) P_n^1(\cos\theta) \qquad (11\text{-}229a)$$

$$E_\theta^i = E_0 \frac{\cos\theta \cos\phi}{\beta r} \sum_{n=0}^{\infty} j^{-n}(2n+1)\hat{J}_n'(\beta r) P_n^0(\cos\theta) \qquad (11\text{-}229b)$$

$$E_\phi^i = -E_0 \frac{\sin\phi}{\beta r} \sum_{n=0}^{\infty} j^{-n}(2n+1)\hat{J}_n'(\beta r) P_n^0(\cos\theta) \qquad (11\text{-}229c)$$

The incident and scattered fields by the sphere can be expressed as a superposition of TE^r and TM^r as outlined, respectively, in Sections 10.2.4 and 10.2.5. The TE^r fields are constructed by letting the vector potentials **A** and **F** be equal to $\mathbf{A} = 0$ and $\mathbf{F} = \hat{a}_r F_r(r, \theta, \phi)$. The TM^r fields are constructed when $\mathbf{A} = \hat{a}_r A_r(r, \theta, \phi)$ and $\mathbf{F} = 0$. For example, the incident radial electric field component E_r^i can be obtained by expressing it in terms of TM^r modes or A_r^i. Thus using A_r^i we can write according to (10-27a) the incident electric field as

$$E_r^i = \frac{1}{j\omega\mu\varepsilon}\left(\frac{\partial^2}{\partial r^2} + \beta^2\right) A_r^i \qquad (11\text{-}230)$$

Equating (11-230) to (11-229a), it can be shown that A_r^i takes the form of

$$\boxed{A_r^i = E_0 \frac{\cos\phi}{\omega} \sum_{n=1}^{\infty} a_n \hat{J}_n(\beta r) P_n^1(\cos\theta)} \qquad (11\text{-}231)$$

where

$$\boxed{a_n = j^{-n} \frac{(2n+1)}{n(n+1)}} \qquad (11\text{-}231a)$$

This potential component A_r^i will give the correct value of E_r^i, and it will lead to $H_r^i = 0$.

The correct expression for the radial component of the incident magnetic field can be obtained by following a similar procedure but using TE^r modes or F_r^i of Section 10.2.4. This allows us to show that

$$\boxed{F_r^i = E_0 \frac{\sin\phi}{\omega\eta} \sum_{n=1}^{\infty} a_n \hat{J}_n(\beta r) P_n^1(\cos\theta)} \qquad (11\text{-}232)$$

where a_n is given by (11-231a). This expression leads to the correct H_r^i and to $E_r^i = 0$. Therefore the sum of (11-231) and (11-232) will give the correct E_r^i, H_r^i and the remaining electric and magnetic components.

Since the incident electric and magnetic field components of a uniform plane wave can be represented by TM^r and TE^r modes that can be constructed using the potentials A_r^i and F_r^i of (11-231) and (11-232), the scattered fields can also be represented by TM^r and TE^r modes and be constructed using potentials A_r^s and F_r^s.

The forms of A_r^s and F_r^s are similar to those of A_r^i and F_r^i of (11-231) and (11-232), and we can represent them by

$$A_r^s = E_0 \frac{\cos \phi}{\omega} \sum_{n=1}^{\infty} b_n \hat{H}_n^{(2)}(\beta r) P_n^1(\cos \theta) \qquad (11\text{-}233\text{a})$$

$$F_r^s = E_0 \frac{\sin \phi}{\omega \eta} \sum_{n=1}^{\infty} c_n \hat{H}_n^{(2)}(\beta r) P_n^1(\cos \theta) \qquad (11\text{-}233\text{b})$$

where the coefficients b_n and c_n will be found using the appropriate boundary conditions. In (11-233a) and (11-233b) the spherical Hankel function of the second kind $\hat{H}_n^{(2)}(\beta r)$ has replaced the spherical Bessel function $\hat{J}_n(\beta r)$ in (11-231) and (11-232) in order to represent outward traveling waves. Thus all the components of the total field, incident plus scattered, can be found using the sum of (10-23) through (10-24c) and (10-27) through (10-28c), or

$$E_r^t = \frac{1}{j\omega\mu\varepsilon} \left(\frac{\partial^2}{\partial r^2} + \beta^2 \right) A_r^t \qquad (11\text{-}234\text{a})$$

$$E_\theta^t = \frac{1}{j\omega\mu\varepsilon} \frac{1}{r} \frac{\partial^2 A_r^t}{\partial r \partial \theta} - \frac{1}{\varepsilon} \frac{1}{r \sin \theta} \frac{\partial F_r^t}{\partial \phi} \qquad (11\text{-}234\text{b})$$

$$E_\phi^t = \frac{1}{j\omega\mu\varepsilon} \frac{1}{r \sin \theta} \frac{\partial^2 A_r^t}{\partial r \partial \phi} + \frac{1}{\varepsilon} \frac{1}{r} \frac{\partial F_r^t}{\partial \theta} \qquad (11\text{-}234\text{c})$$

$$H_r^t = \frac{1}{j\omega\mu\varepsilon} \left(\frac{\partial^2}{\partial r^2} + \beta^2 \right) F_r^t \qquad (11\text{-}234\text{d})$$

$$H_\theta^t = \frac{1}{\mu} \frac{1}{r \sin \theta} \frac{\partial A_r^t}{\partial \phi} + \frac{1}{j\omega\mu\varepsilon} \frac{1}{r} \frac{\partial^2 F_r^t}{\partial r \partial \theta} \qquad (11\text{-}234\text{e})$$

$$H_\phi^t = -\frac{1}{\mu} \frac{1}{r} \frac{\partial A_r^t}{\partial \theta} + \frac{1}{j\omega\mu\varepsilon} \frac{1}{r \sin \theta} \frac{\partial^2 F_r^t}{\partial r \partial \phi} \qquad (11\text{-}234\text{f})$$

where A_r^t and F_r^t are each equal to the sum of (11-231), (11-232), (11-233a), and (11-233b), or

$$A_r^t = A_r^i + A_r^s = E_0 \frac{\cos \phi}{\omega} \sum_{n=1}^{\infty} \left[a_n \hat{J}_n(\beta r) + b_n \hat{H}_n^{(2)}(\beta r) \right] P_n^1(\cos \theta) \qquad (11\text{-}235\text{a})$$

$$F_r^t = F_r^i + F_r^s = E_0 \frac{\sin \phi}{\omega \eta} \sum_{n=1}^{\infty} \left[a_n \hat{J}_n(\beta r) + c_n \hat{H}_n^{(2)}(\beta r) \right] P_n^1(\cos \theta) \qquad (11\text{-}235\text{b})$$

$$a_n = j^{-n} \frac{2n+1}{n(n+1)} \qquad (11\text{-}235\text{c})$$

654 SCATTERING

To determine the coefficients b_n and c_n, the boundary conditions of

$$E_\theta^t(r = a, 0 \le \theta \le \pi, 0 \le \phi \le 2\pi) = 0 \quad (11\text{-}236a)$$
$$E_\phi^t(r = a, 0 \le \theta \le \pi, 0 \le \phi \le 2\pi) = 0 \quad (11\text{-}236b)$$

must be applied. Using (11-235a) and (11-235b) we can write (11-234b) as

$$E_\theta^t = -j\frac{E_0}{\omega\mu\varepsilon r}\left\{\frac{\beta}{\omega}\cos\phi \sum_{n=1}^{\infty}\left[a_n\hat{J}_n'(\beta r) + b_n\hat{H}_n^{(2)\prime}(\beta r)\right]P_n'^1(\cos\theta)\right\}$$
$$-\frac{E_0}{\varepsilon r \sin\theta}\left\{\frac{1}{\omega\eta}\cos\phi \sum_{n=1}^{\infty}\left[a_n\hat{J}_n(\beta r) + c_n\hat{H}_n^{(2)}(\beta r)\right]P_n^1(\cos\theta)\right\} \quad (11\text{-}237)$$

where

$$\prime = \frac{\partial}{\partial(\beta r)} \quad \text{for the spherical Bessel or Hankel function} \quad (11\text{-}237a)$$

$$\prime = \frac{\partial}{\partial\theta} \quad \text{for the associated Legendre functions} \quad (11\text{-}237b)$$

Using (11-237) the boundary condition of (11-236a) is satisfied provided that

$$a_n\hat{J}_n'(\beta a) + b_n\hat{H}_n^{(2)\prime}(\beta a) = 0 \Rightarrow \boxed{b_n = -a_n\frac{\hat{J}_n'(\beta a)}{\hat{H}_n^{(2)\prime}(\beta a)}} \quad (11\text{-}238a)$$

$$a_n\hat{J}_n(\beta a) + c_n\hat{H}_n^{(2)}(\beta a) = 0 \Rightarrow \boxed{c_n = -a_n\frac{\hat{J}_n(\beta a)}{\hat{H}_n^{(2)}(\beta a)}} \quad (11\text{-}238b)$$

The scattered electric field components can be written using (11-233a) and (11-233b) as

$$E_r^s = -jE_0\cos\phi \sum_{n=1}^{\infty} b_n\left[\hat{H}_n^{(2)\prime\prime}(\beta r) + \hat{H}_n^{(2)}(\beta r)\right]P_n^1(\cos\theta) \quad (11\text{-}239a)$$

$$E_\theta^s = \frac{E_0}{\beta r}\cos\phi \sum_{n=1}^{\infty}\left[jb_n\hat{H}_n^{(2)\prime}(\beta r)\sin\theta P_n'^1(\cos\theta) - c_n\hat{H}_n^{(2)}(\beta r)\frac{P_n^1(\cos\theta)}{\sin\theta}\right] \quad (11\text{-}239b)$$

$$E_\phi^s = \frac{E_0}{\beta r}\sin\phi \sum_{n=1}^{\infty}\left[jb_n\hat{H}_n^{(2)\prime}(\beta r)\frac{P_n^1(\cos\theta)}{\sin\theta} - c_n\hat{H}_n^{(2)}(\beta r)\sin\theta P_n'^1(\cos\theta)\right] \quad (11\text{-}239c)$$

where

$$\prime = \frac{\partial}{\partial(\beta r)} \quad \text{for spherical Hankel functions} \quad (11\text{-}239d)$$

$$\prime\prime = \frac{\partial^2}{\partial(\beta r)^2} \quad \text{for spherical Hankel functions} \quad (11\text{-}239e)$$

$$\prime = \frac{\partial}{\partial(\cos\theta)} = -\frac{1}{\sin\theta}\frac{\partial}{\partial\theta} \quad \text{for associated Legendre functions} \quad (11\text{-}239f)$$

SCATTERING BY A CONDUCTING SPHERE

The spherical Hankel function is related to the regular Hankel function by (10-31) or

$$\hat{H}_n^{(2)}(\beta r) = \sqrt{\frac{\pi \beta r}{2}} \, H_{n+1/2}^{(2)}(\beta r) \tag{11-240}$$

Since for large values of βr the regular Hankel function can be represented by

$$H_{n+1/2}^{(2)}(\beta r) \overset{\beta r \to \infty}{\simeq} \sqrt{\frac{2j}{\pi \beta r}} \, j^{n+1/2} e^{-j\beta r} = j \sqrt{\frac{2}{\pi \beta r}} \, j^n e^{-j\beta r} \tag{11-241}$$

then the spherical Hankel function of (11-240) and its partial derivatives can be approximated by

$$\hat{H}_n^{(2)}(\beta r) \overset{\beta r \to \infty}{\simeq} j^{n+1} e^{-j\beta r} \tag{11-241a}$$

$$\hat{H}_n^{(2)\prime}(\beta r) = \frac{\partial \hat{H}_n^{(2)}(\beta r)}{\partial (\beta r)} \overset{\beta r \to \infty}{\simeq} -j^2 j^n e^{-j\beta r} = j^n e^{-j\beta r} \tag{11-241b}$$

$$\hat{H}_n^{(2)\prime\prime}(\beta r) = \frac{\partial^2 \hat{H}_n^{(2)}(\beta r)}{\partial (\beta r)^2} \overset{\beta r \to \infty}{\simeq} -j^{n+1} e^{-j\beta r} \tag{11-241c}$$

For far-field observations ($\beta r \to$ large), the electric field components of (11-239a) through (11-239c) can be simplified using the approximations of (11-241a) through (11-241c). Since the radial component E_r^s of (11-239a) reduces with the approximations of (11-241a) through (11-241c) to zero, then in the far zone (11-239a) through (11-239c) can be approximated by

<u>Far-Field Observations ($\beta r \to$ large)</u>

$$E_r^s \simeq 0 \tag{11-242a}$$

$$E_\theta^s \simeq jE_0 \frac{e^{-j\beta r}}{\beta r} \cos\phi \sum_{n=1}^\infty j^n \left[b_n \sin\theta P_n^{\prime 1}(\cos\theta) - c_n \frac{P_n^1(\cos\theta)}{\sin\theta} \right] \tag{11-242b}$$

$$E_\phi^s \simeq jE_0 \frac{e^{-j\beta r}}{\beta r} \sin\phi \sum_{n=1}^\infty j^n \left[b_n \frac{P_n^1(\cos\theta)}{\sin\theta} - c_n \sin\theta P_n^{\prime 1}(\cos\theta) \right] \tag{11-242c}$$

where b_n and c_n are given by (11-238a) and (11-238b).

The bistatic radar cross section is obtained using (11-22b), and it can be written using (11-225) and (11-242a) through (11-242c) as

$$\boxed{ \sigma(\text{bistatic}) = \lim_{r \to \infty} \left[4\pi r^2 \frac{|E^s|^2}{|E^i|^2} \right] = \frac{\lambda^2}{\pi} \left[\cos^2\phi |A_\theta|^2 + \sin^2\phi |A_\phi|^2 \right] } \tag{11-243}$$

where

$$|A_\theta|^2 = \left| \sum_{n=1}^\infty j^n \left[b_n \sin\theta P_n'^1(\cos\theta) - c_n \frac{P_n^1(\cos\theta)}{\sin\theta} \right] \right|^2 \quad (11\text{-}243a)$$

$$|A_\phi|^2 = \left| \sum_{n=1}^\infty j^n \left[b_n \frac{P_n^1(\cos\theta)}{\sin\theta} - c_n \sin\theta P_n'^1(\cos\theta) \right] \right|^2 \quad (11\text{-}243b)$$

The monostatic radar cross section can be found by first reducing the field expressions for observations toward $\theta = \pi$. In that direction the scattered electric field of interest is the copolar component E_x^s, and it can be found using (11-242a) through (11-242c) and the transformation of (II-13b) by evaluating either

$$E_x^s = E_\theta^s \cos\theta \cos\phi \big|_{\substack{\theta=\pi \\ \phi=\pi}} = E_\theta^s \big|_{\substack{\theta=\pi \\ \phi=\pi}} \quad (11\text{-}244a)$$

or

$$E_x^s = -E_\phi^s \sin\phi \big|_{\substack{\theta=\pi \\ \phi=3\pi/2}} = E_\phi^s \big|_{\substack{\theta=\pi \\ \phi=3\pi/2}} \quad (11\text{-}244b)$$

To accomplish either (11-244a) or (11-244b) we need to first evaluate the associated Legendre function and its derivative when $\theta = \pi$. It can be shown that [11, 42]

$$\frac{P_n^1(\cos\theta)}{\sin\theta}\bigg|_{\theta=\pi} = (-1)^n \frac{n(n+1)}{2} \quad (11\text{-}245a)$$

$$\sin\theta P_n'^1(\cos\theta)\big|_{\theta=\pi} = \sin\theta \frac{dP_n^1}{d(\cos\theta)} = -\frac{dP_n^1(\cos\theta)}{d\theta} = (-1)^n \frac{n(n+1)}{2}$$
$$(11\text{-}245b)$$

Thus (11-242b) can be expressed using (11-235c), (11-238a) through (11-238b) and (11-245a) through (11-245b) as

$$E_\theta^s \big|_{\substack{\theta=\pi \\ \phi=\pi}} = jE_0 \frac{e^{-j\beta r}}{\beta r} \sum_{n=1}^\infty j^n (-1)^n \frac{n(n+1)}{2} [b_n - c_n]$$

$$= -jE_0 \frac{e^{-j\beta r}}{\beta r} \sum_{n=1}^\infty j^n (-1)^n \frac{n(n+1)}{2} a_n \left[\frac{\hat{J}_n'(\beta a)}{\hat{H}_n^{(2)'}(\beta a)} - \frac{\hat{J}_n(\beta a)}{\hat{H}_n^{(2)}(\beta a)} \right]$$

$$E_\theta^s \big|_{\substack{\theta=\pi \\ \phi=\pi}} = -jE_0 \frac{e^{-j\beta r}}{\beta r} \sum_{n=1}^\infty (-1)^n \frac{(2n+1)}{2} \left[\frac{\hat{J}_n'(\beta a)\hat{H}_n^{(2)}(\beta a) - \hat{J}_n(\beta a)\hat{H}_n^{(2)'}(\beta a)}{\hat{H}_n^{(2)'}(\beta a)\hat{H}_n^{(2)}(\beta a)} \right]$$
$$(11\text{-}246)$$

which reduces, using the Wronskian for spherical Bessel functions of

$$\hat{J}_n'(\beta a)\hat{H}_n^{(2)}(\beta a) - \hat{J}_n(\beta a)\hat{H}_n^{(2)'}(\beta a) = j[\hat{J}_n(\beta a)\hat{Y}_n'(\beta a) - \hat{J}_n'(\beta a)\hat{Y}_n(\beta a)] = j$$
$$(11\text{-}246a)$$

to

$$E_\theta^s \big|_{\substack{\theta=\pi \\ \phi=\pi}} = E_0 \frac{e^{-j\beta r}}{2\beta r} \sum_{n=1}^\infty \frac{(-1)^n (2n+1)}{\hat{H}_n^{(2)'}(\beta a)\hat{H}_n^{(2)}(\beta a)} \quad (11\text{-}246b)$$

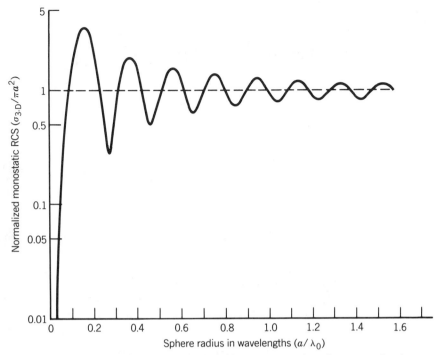

FIGURE 11-26 Normalized monostatic radar cross section for a conducting sphere as a function of its radius. (*Source:* G. T. Ruck, D. E. Barrick, W. D. Stuart, and C. K. Krichbaum, *Radar Cross Section Handbook*, Vol. 1, 1970, Plenum Publishing Co.).

Thus the monostatic radar cross section of (11-22b) can be expressed using (11-246b) by

$$\sigma_{3\text{-D}}(\text{monostatic}) = \lim_{r \to \infty}\left[4\pi r^2 \frac{|\mathbf{E}^s|^2}{|\mathbf{E}^i|^2}\right] = \frac{\lambda^2}{4\pi}\left|\sum_{n=1}^{\infty} \frac{(-1)^n(2n+1)}{\hat{H}_n^{(2)\prime}(\beta a)\hat{H}_n^{(2)}(\beta a)}\right|^2$$

(11-247)

A plot of (11-247) as a function of the sphere radius is shown in Figure 11-26 [50]. This is a classic signature that can be found in any literature dealing with electromagnetic scattering. The total curve can be subdivided into three regions; the *Rayleigh*, the *Mie* (or *resonance*), and the *optical* regions. The Rayleigh region represents the part of the curve for small values of the radius ($a < 0.1\lambda$) and the optical region represents the RCS of the sphere for large values of the radius (typically $a > 2\lambda$). The region between those two extremes is the Mie or resonance region. It is apparent that for small values of the radius the RCS is linear, for intermediate values it is oscillatory about the value of πa^2, and for large values it approaches the value of πa^2 that is the physical area of the cross section of the sphere.

For very small values of the radius a, the first term of (11-247) is sufficient to accurately represent the RCS. Doing this we can approximate (11-247) by

$$\sigma_{3\text{-D}}(\text{monostatic}) \stackrel{a \to 0}{\simeq} \frac{\lambda^2}{4\pi}\left|\frac{3}{\hat{H}_1^{(2)\prime}(\beta a)\hat{H}_1^{(2)}(\beta a)}\right|^2 \qquad (11\text{-}248)$$

Since

$$\hat{H}_1^{(2)}(\beta a) \stackrel{a \to 0}{\simeq} -j\hat{Y}_1(\beta a) = -j\sqrt{\frac{\pi \beta a}{2}} Y_{3/2}(\beta a) \simeq -j\sqrt{\frac{\pi \beta a}{2}} \left[-\frac{\frac{1}{2}!}{\pi} \left(\frac{2}{\beta a}\right)^{3/2} \right] = j\frac{1}{\beta a}$$
(11-248a)

$$\hat{H}_1^{(2)\prime}(\beta a) \stackrel{a \to 0}{\simeq} -j\hat{Y}_1'(\beta a) \simeq -j\frac{1}{(\beta a)^2}$$
(11-248b)

$$\frac{1}{2}! = \sqrt{\frac{\pi}{2}}$$
(11-248c)

(11-248) reduces to

$$\boxed{\sigma_{3\text{-D}}(\text{monostatic}) \stackrel{a \to 0}{\simeq} \frac{9\lambda^2}{4\pi}(\beta a)^6}$$
(11-248d)

which is representative of the Rayleigh region scattering.

For very large values of the radius a, we can approximate the spherical Hankel function and its derivative in (11-247) by their asymptotic forms (11-241a) and (11-241b), or

$$\hat{H}_n^{(2)}(\beta a) \stackrel{\beta a \to \infty}{\simeq} \frac{e^{-j[\beta a(\sin\alpha - \alpha\cos\alpha) - \pi/4]}}{\sqrt{\sin\alpha}}$$
(11-249a)

$$\hat{H}_n^{(2)\prime}(\beta a) \stackrel{\beta a \to \infty}{\simeq} \sqrt{\sin\alpha}\, e^{-j[\beta a(\sin\alpha - \alpha\cos\alpha) + \pi/4]}$$
(11-249b)

$$\cos\alpha = (n + 1/2)/\beta a$$
(11-249c)

Thus (11-247) reduces for very large values of the radius a to

$$\boxed{\sigma_{3\text{-D}}(\text{monostatic}) = \frac{\lambda^2}{4\pi} \left| \sum_{n=1}^{\infty} \frac{(-1)^n (2n+1)}{\hat{H}_n^{(2)\prime}(\beta a)\hat{H}_n^{(2)}(\beta a)} \right|^2 \stackrel{a \to \infty}{\simeq} \pi a^2}$$

which is representative of the optical region scattering, and is also equal to the physical area of the cross section of the sphere.

REFERENCES

1. C. A. Balanis, *Antenna Theory: Analysis and Design*, Wiley, New York, 1982.
2. K. M. Siegel, "Far field scattering from bodies of revolution," *Appl. Sci. Res., Sec. B*, vol. 7, pp. 293–328, 1958.
3. E. F. Knott, V. V. Liepa, and T. B. A. Senior, "Non-specular radar cross section study," Technical Report AFAL-TR-73-70, University of Michigan, April 1973.
4. T. Griesser and C. A. Balanis, "Dihedral corner reflector backscatter using higher-order reflections and diffractions," *IEEE Trans. Antennas Propagat.*, vol. AP-35, no. 11, pp. 1235–1247, November 1987.
5. J. S. Asvestas, "The physical optics method in electromagnetic scattering," *J. Math. Phys.*, vol. 21, pp. 290–299, 1980.
6. J. S. Asvestas, "Physical optics and the direction of maximization of the far-field average power," *IEEE Trans. Antennas Propagat.*, vol. AP-34, no. 12, pp. 1459–1460, December 1986.

7. E. F. Knott, "RCS reduction of dihedral corners," *IEEE Trans. Antennas Propagat.*, vol. AP-25, no. 3, pp. 406–409, May 1977.
8. R. A. Ross, "Radar cross section of rectangular plates as a function of aspect angle," *IEEE Trans. Antennas Propagat.*, vol. AP-14, no. 3, pp. 329–335, May 1966.
9. T. Griesser and C. A. Balanis, "Backscatter analysis of dihedral corner reflectors using physical optics and the physical theory of diffraction," *IEEE Trans. Antennas Propagat.*, vol. AP-35, no. 10, pp. 1137–1147, October 1987.
10. D. P. Marsland, C. A. Balanis, and S. Brumley, "Higher order diffractions from a circular disk," *IEEE Trans. Antennas Propagat.*, vol. AP-35, no. 12, pp. 1436–1444, December 1987.
11. R. F. Harrington, *Time-Harmonic Electromagnetic Fields*, McGraw-Hill, New York, 1961.
12. D. E. Barrick, "Cylinders," in *Radar Cross Section Handbook*, vol. 1, G. T. Ruck, D. E. Barrick, W. D. Stuart, and C. K. Krichbaum (Eds.), Plenum, New York, 1970, Chapter 4, pp. 205–339.
13. J. H. Richmond, *The Basic Theory of Harmonic Fields, Antennas and Scattering*, Ohio State University, unpublished notes.
14. H. C. Van de Hulst, *Light Scattering by Small Particles*, Wiley, New York, 1957, pp. 304–307.
15. C. A. Balanis and L. Peters, Jr., "Analysis of aperture radiation from an axially slotted circular conducting cylinder using geometrical theory of diffraction," *IEEE Trans. Antennas Propagat.*, vol. AP-17, no. 1, pp. 93–97, January 1969.
16. C. A. Balanis and L. Peters, Jr., "Aperture radiation from an axially slotted elliptical conducting cylinder using geometrical theory of diffraction," *IEEE Trans. Antennas Propagat.*, vol. AP-17, no. 4, pp. 507–513, July 1969.
17. P. H. Pathak and R. G. Kouyoumjian, "An analysis of the radiation from apertures in curved surfaces by the geometrical theory of diffraction," *Proc. IEEE*, vol. 62, no. 11, pp. 1438–1447, November 1974.
18. S. A. Schelkunoff, "Some equivalence theorems of electromagnetics and their application to radiation problems," *Bell Syst. Tech. J.*, vol. 15, pp. 92–112, 1936.
19. J. R. Wait, *Electromagnetic Radiation from Cylindrical Structures*, Pergamon, New York, 1959.
20. J. J. Bowman, T. B. A. Senior, and P. L. E. Uslenghi (Eds.), *Electromagnetic and Acoustic Scattering by Simple Shapes*, North-Holland, Amsterdam, 1969.
21. A. W. Adey, "Scattering of electromagnetic waves by coaxial cylinders," *Can. J. Phys.*, vol. 34, pp. 510–520, May 1956.
22. C. C. H. Tang, "Backscattering from dielectric-coated infinite cylindrical obstacles," *J. Appl. Phys.*, vol. 28, pp. 628–633, May 1957.
23. B. R. Levy, "Diffraction by an elliptic cylinder," *J. Math. Mech.*, vol. 9, pp. 147–165, 1960.
24. R. D. Kodis, "The scattering cross section of a composite cylinder, geometrical optics," *IEEE Trans. Antennas Propagat.*, vol. AP-11, no. 1, pp. 86–93, January 1963.
25. K. Mei and J. Van Bladel, "Low-frequency scattering by a rectangular cylinder," *IEEE Trans. Antennas Propagat.*, vol. AP-11, no. 1, pp. 52–56, January 1963.
26. R. D. Kodis and T. T. Wu, "The optical model of scattering by a composite cylinder," *IEEE Trans. Antennas Propagat.*, vol. AP-11, no. 6, pp. 703–705, November 1963.
27. J. H. Richmond, "Scattering by a dielectric cylinder of arbitrary cross section shape," *IEEE Trans. Antennas Propagat.*, vol. AP-13, no. 3, pp. 334–341, May 1965.
28. W. V. T. Rusch and C. Yeh, "Scattering by an infinite cylinder coated with an inhomogeneous and anisotropic plasma sheath," *IEEE Trans. Antennas Propagat.*, vol. AP-15, no. 3, pp. 452–457, May 1967.
29. J.-C. Sureau, "Reduction of scattering cross section of dielectric cylinder by metallic core loading," *IEEE Trans. Antennas Propagat.*, vol. AP-15, no. 5, pp. 657–662, September 1967.
30. H. E. Bussey and J. H. Richmond, "Scattering by a lossy dielectric circular cylindrical multilayer, numerical values," *IEEE Trans. Antennas Propagat.*, vol. AP-23, no. 5, pp. 723–725, September 1975.

31. R. E. Eaves, "Electromagnetic scattering from a conducting circular cylinder covered with a circumferentially magnetized ferrite," *IEEE Trans. Antennas Propagat.*, vol. AP-24, no. 2, pp. 190–197, March 1976.
32. G. N. Watson, "The diffraction of electrical waves by the earth," *Proc. Roy. Soc. (London)*, vol. A95, pp. 83–99, 1918.
33. R. G. Kouyoumjian, "Asymptotic high-frequency methods," *Proc. IEEE*, vol. 53, pp. 864–876, August 1965.
34. L. B. Felsen and N. Marcuvitz, *Radiation and Scattering of Waves*, Prentice-Hall, Englewood Cliffs, N.J., 1973.
35. W. Pauli, "On asymptotic series for functions in the theory of diffraction of light," *Phys. Rev.*, vol. 34, pp. 924–931, December 1938.
36. F. Oberhettinger, "On asymptotic series occurring in the theory of diffraction of waves by a wedge," *Math. Phys.*, vol. 34, pp. 245–255, 1956.
37. J. B. Keller, "Geometrical theory of diffraction," *J. Opt. Soc. Amer.*, vol. 52, no. 2, pp. 116–130, February 1962.
38. R. G. Kouyoumjian and P. H. Pathak, "A uniform geometrical theory of diffraction for an edge in a perfectly conducting surface," *Proc. IEEE*, vol. 62, no. 11, pp. 1448–1461, November 1974.
39. G. L. James, *Geometrical Theory of Diffraction for Electromagnetic Waves*, Third Edition Revised, Peregrinus, London, 1986.
40. D. L. Hutchins, "Asymptotic series describing the diffraction of a plane wave by a two-dimensional wedge of arbitrary angle," Ph.D. dissertation, Dept. of Electrical Engineering, Ohio State University, 1967.
41. P. H. Pathak and R. G. Kouyoumjian, "The dyadic diffraction coefficient for a perfectly conducting wedge," Technical Report 2183-4 (AFCRL-69-0546), ElectroScience Laboratory, Ohio State University, June 5, 1970.
42. J. A. Stratton, *Electromagnetic Theory*, McGraw-Hill, New York, 1941.
43. P. M. Morse and H. Feshbach, *Methods of Theoretical Physics*, Part II, McGraw-Hill, New York, 1953.
44. W. R. Smythe, *Static and Dynamic Electricity*, McGraw-Hill, New York, 1939.
45. M. R. Spiegel, *Mathematical Handbook of Formulas and Tables*, Schaum's Outline Series, McGraw-Hill, New York, 1968.
46. A. L. Aden, "Scattering from spheres with sizes comparable to the wavelength," *J. Appl. Phys.*, vol. 12, 1951.
47. S. I. Rubinow and T. T. Wu, "First correction to the geometrical-optics scattering cross section from cylinders and spheres," *J. Appl. Phys.*, vol. 27, pp. 1032–1039, 1956.
48. J. Rheinstein, "Scattering of electromagnetic waves from dielectric coated conducting spheres," *IEEE Trans. Antennas Propagat.*, vol. AP-12, no. 3, pp. 334–340, May 1964.
49. R. G. Kouyoumjian, L. Peters, Jr., and D. T. Thomas, "A modified geometrical optics method for scattering by dielectric bodies," *IEEE Trans. Antennas Propagat.*, vol. AP-11, no. 6, pp. 690–703, November 1963.
50. D. E. Barrick, "Spheres," in *Radar Cross Section Handbook*, vol. 1, G. T. Ruck, D. E. Barrick, W. D. Stuart, and C. K. Krichbaum (Eds.), Plenum, New York, 1970, Chapter 3, pp. 141–204.
51. D. E. Kerr and H. Goldstein, "Radar targets and echoes," in *Propagation of Short Radio Waves*, D. E. Kerr (Ed.), McGraw-Hill, 1951, Chapter 6, pp. 445–469.

PROBLEMS

11.1. Two constant current, infinite length electric line sources, are displaced along the x axis a distance s apart as shown in Figure P11-1. Use superposition and neglect mutual coupling between the lines.

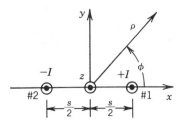

FIGURE P11-1

(a) Show that the magnetic vector potential for the two lines can be written as

$$A_z^t = A_z^{(1)}\left(x - \frac{s}{2}, y\right) - A_z^{(2)}\left(x + \frac{s}{2}, y\right)$$

(b) Show that for small spacings (in the limit as $s \to 0$) the vector potential of part a reduces to

$$A_z^t \overset{s \to 0}{\simeq} \frac{\mu \beta I s}{4j} H_1^{(2)}(\beta \rho) \cos \phi$$

(c) Determine the electric and magnetic field components associated with the two line sources when $s \to 0$.

11.2. Four constant current, infinite length electric line sources, of phase as indicated, are displaced along the x axis, as shown in Figure P11-2. Assume that the spacings s_1 and s_2 are very small.

(a) Find an approximate closed form expression for the magnetic vector potential for the entire array by using the procedure of Problem 11.1. First consider the pairs on each side as individual arrays and then combine the results to form a new array.

(b) Determine in terms of ρ and ϕ the electric and magnetic field components associated with the four line sources when $\rho \gg s_1$ and s_2.

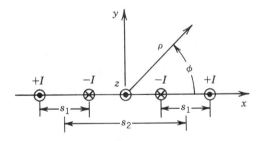

FIGURE P11-2

11.3. Four constant current, infinite length electric line sources, of phase as indicated, are displaced along the x and y axes as shown in Figure P11-3. Assume that the spacings s and h are very small and neglect any mutual coupling between the lines.

(a) Show, by using the procedure of Problem 11.1, that the magnetic vector potential for the entire array can be written as

$$A_z \overset{s \to 0}{\underset{h \to 0}{\simeq}} -\frac{\mu \beta s h I}{4j} \frac{\partial}{\partial y}\left[H_1^{(2)}(\beta \rho) \cos \phi\right] = \frac{\mu \beta^2 s h I}{8j} H_2^{(2)}(\beta \rho) \sin(2\phi)$$

(b) Determine in terms of ρ and ϕ the electric and magnetic field components associated with the four line sources when $\rho \gg s$ and h.

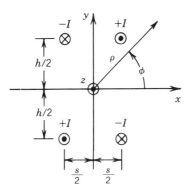

FIGURE P11-3

11.4. Two constant current, infinite length electric line sources are placed above an infinite electric ground plane as shown in Figure P11-4. Assume that the spacings s_1 and s_2 are very small and neglect any mutual coupling between the lines.

(a) Show, by using the procedure of Problem 11.1, that the magnetic vector potential for the entire array can be written as

$$A_z \underset{h \to 0}{\overset{s \to 0}{\simeq}} -\frac{\mu \beta s h I}{4j} \frac{\partial}{\partial y}\left[H_1^{(2)}(\beta \rho) \cos \phi\right] = \frac{\mu \beta^2 s h I}{8j} H_2^{(2)}(\beta \rho) \sin(2\phi)$$

(b) Determine in terms of ρ and ϕ the electric and magnetic field components associated with the two line sources and ground plane when $\rho \gg s$ and h.

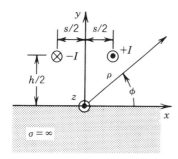

FIGURE P11-4

11.5. An infinite length and constant current electric line source is placed parallel to the plates of a 90° conducting corner reflector, as shown in Figure P11-5. Assume that

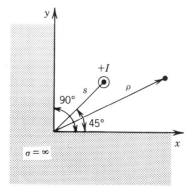

FIGURE P11-5

the plates of the wedge are infinite in extent and the distance s from the apex to the source is very small ($s \ll \lambda$).

(a) Show that the magnetic vector potential for the line source and corner reflector can be written, using the procedure of Problem 11.1, as

$$A_z = \frac{\mu \beta^2 s^2 I}{4j} H_2^{(2)}(\beta \rho) \sin(2\phi)$$

(b) Determine the electric and magnetic field components.
(c) Find the angles ϕ of observation (for a constant value of ρ) where the electric field vanishes.

11.6. Show that for normal incidence the two-dimensional scattering width and three-dimensional RCS are related by (11-22e).

11.7. A uniform plane wave on the yz plane is obliquely incident at an angle θ_i from the vertical z axis upon a perfectly electric conducting circular ground plane of radius a as shown in Figure P11-7. Assume TE^x polarization for the incident field.

(a) Determine the physical optics current density induced on the plate.
(b) Determine the far-zone bistatic scattered electric and magnetic fields based on the physical optics current density of part a.
(c) Determine the bistatic and monostatic RCSs of the plate. Plot the normalized monostatic RCS ($\sigma_{3\text{-D}}/\lambda^2$) in decibels for plates with radii of $a = \lambda$ and 5λ.

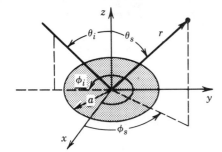

FIGURE P11-7

11.8. Repeat Problem 11.7 for TM^x plane wave incidence.

11.9. Show that for normal incidence the monostatic RCS of a flat plate of area A and any cross section, based on physical optics, is equal to $\sigma_{3\text{-D}} = 4\pi(A/\lambda)^2$.

11.10. A uniform plane wave traveling in the $-z$ direction, as shown in Figure P11-10, is incident upon a perfectly electric conducting curved surface with radii of curvature

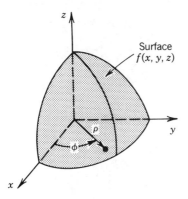

FIGURE P11-10

sufficiently large, usually greater than about one wavelength, so that at each point the surface can be considered locally flat. For such a surface the induced currents and the fields radiated from each infinitesimal area can be represented if the same area were part of an infinite plane that was tangent to the surface at the same location.

(a) Show that the monostatic RCS can be written as

$$\sigma_{3\text{-D}} = \frac{4\pi}{\lambda^2} \left| \iint_A e^{+j2\beta z} \, dA \right|^2$$

where z is any point on the surface of the scatterer and the integration on dA is performed in the xy plane.

(b) If in part a the differential area in the xy plane is expressed in terms of the polar coordinates ρ, ϕ, then show that the monostatic RCS can be expressed as

$$\sigma_{3\text{-D}} = \frac{\pi}{\lambda^2} \left| \int_0^{2\pi} \int_0^{z(\phi)} \frac{d\rho^2}{dz} e^{+j2\beta z} \, dz \, d\phi \right|^2$$

11.11. If the scattering conducting curved surface of Problem 11.10, Figure P11-10, is a quadric surface, the integration to find the RCS can be performed in closed form. Assume that the scattering surface is an elliptic paraboloid opening downward along the positive z axis and that it is represented by

$$\left(\frac{x}{a}\right)^2 + \left(\frac{y}{b}\right)^2 = -\frac{z}{c}$$

where a, b and c are constants. Show the following.

(a) The radii of curvature a_1 and a_2 in the xz and yz planes are given by

$$a_1 = \frac{a^2}{2c} \qquad a_2 = \frac{b^2}{2c}$$

(b) The equation of the elliptic paraboloid surface transformed in polar coordinates can be expressed as

$$z = -\frac{\rho^2}{2a_1}\left[1 - \left(1 - \frac{a_1}{a_2}\right)\sin^2\phi\right]$$

(c) The RCS of Problem 11.10 reduces to

$$\sigma_{3\text{-D}} = \pi a_1 a_2 |e^{-j2\beta h} - 1| = 4\pi a_1 a_2 \sin^2(\beta h)$$

(d) The RCS of part c reduces to $\sigma_{3\text{-D}} = \pi a_1 a_2$ if the height of the paraboloid is cut very irregular so that the contributions from the last zone(s) would tend to cancel. To account for this the exponential term in part c disappears.

11.12. If the scattering conducting curved surface of Problem 11.10, Figure P11-10, is a closed surface, such as an ellipsoid represented by

$$\left(\frac{x}{a}\right)^2 + \left(\frac{y}{b}\right)^2 + \left(\frac{z}{c}\right)^2 = 1$$

demonstrate the following.

(a) The distance z at any point on the surface can be represented by

$$z^2 = c^2 - \rho^2 \frac{c}{a_1}\left[1 - \left(1 - \frac{a_1}{a_2}\right)\sin^2\phi\right]$$

where $a_1 = a^2/c$ and $a_2 = b^2/c$.

(b) The RCS of Problem 11.10 from the upper part of the ellipsoid from $z = h$ to $z = c$ reduces, neglecting $(2\beta c)^{-1}$ terms, to

$$\sigma_{3\text{-D}} = \pi a_1 a_2 \left\{ 1 + \left(\frac{h}{c}\right)^2 - 2\frac{h}{c} \cos\left[2\beta(c - h)\right] \right\}$$

where h ($h < c$) is the distance along the z axis from the origin to the point of integration. If $h = 0$ or if h is very irregular around the periphery of the ellipsoid, the preceding equation reduces to $\sigma_{3\text{-D}} = \pi a_1 a_2$. If $a_1 = a_2 = a$, like for a sphere, then $\sigma_{3\text{-D}} = \pi a^2$.

11.13. Verify (11-55a).

11.14. Verify (11-71a) and (11-71b).

11.15. Verify that the infinite summation from minus to plus infinity for the incident electric field of (11-85a) can also be written as an infinite summation from $n = 0$ to $n = \infty$.

11.16. Write the current density expression of (11-97) as an infinite summation from $n = 0$ to $n = \infty$.

11.17. Refer to Figure 11-12a for the TM^z uniform plane wave scattering by a circular conducting cylinder.
  (a) Determine the normalized induced current density based on the physical optics approximation of Section 7.10.
  (b) Plot and compare for $0° \leq \phi \leq 180°$ the normalized induced current density based on the physical optics approximation and on the modal solution of (11-97). Do this for cylinders with radii of $a = \lambda$ and 5λ.

11.18. For the TM^z uniform plane wave scattering by a circular conducting cylinder of Figure 11-12a, plot and compare the normalized induced current density based on the exact modal solution of (11-97) and its small argument approximation of (11-98b) for radii of $a = 0.01\lambda, 0.1\lambda$, and λ.

11.19. Verify that the infinite summation from minus to plus infinity for the scattering width of (11-102) can also be written as an infinite summation from zero to infinity.

11.20. Using the definition of (11-21c), instead of (11-21b), show that the TM^z polarization radar cross section reduces to (11-102).

11.21. Write the current density expression of (11-113) as an infinite summation from $n = 0$ to $n = \infty$.

11.22. Refer to Figure 11-12b for the TE^z uniform plane wave scattering by a circular conducting cylinder.
  (a) Determine the normalized induced current density based on the physical optics approximation of Section 7.10.
  (b) Plot and compare for $0° \leq \phi \leq 180°$ the normalized induced current density based on the physical optics approximation and on the modal solution of (11-113). Do this for cylinders with radii of $a = \lambda$ and 5λ.

11.23. For the TE^z uniform plane wave scattering by a circular conducting cylinder of Figure 11-12b, plot and compare the normalized induced current density based on the exact modal solution of (11-113) and its small argument approximation of (11-114d) for radii of $a = 0.01\lambda, 0.1\lambda$, and λ.

11.24. Verify that the infinite summation from minus to plus infinity for the scattering width of (11-117) can also be written as an infinite summation from zero to infinity.

11.25. Using the definition of (11-21b), instead of (11-21c), show that the TE^z polarization radar cross section reduces to (11-117).

11.26. A TM^z uniform plane wave traveling in the $+x$ direction in free space is incident normally on a lossless dielectric circular cylinder of radius a as shown in Figure P11-26. Assume that the incident, scattered, and transmitted (into the cylinder) electric fields can be written as

$$\mathbf{E}^i = \hat{a}_z E_0 \sum_{n=-\infty}^{+\infty} j^{-n} J_n(\beta_0 \rho) e^{jn\phi}$$

$$\mathbf{E}^s = \hat{a}_z E_0 \sum_{n=-\infty}^{+\infty} a_n H_n^{(2)}(\beta_0 \rho) e^{jn\phi}$$

$$\mathbf{E}^d = \hat{a}_z E_0 \sum_{n=-\infty}^{+\infty} [b_n J_n(\beta_1 \rho) + c_n Y_n(\beta_1 \rho)] e^{jn\phi}$$

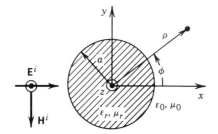

FIGURE P11-26

(a) Derive expressions for the incident, scattered, and transmitted magnetic field components.

(b) Show that the wave amplitude coefficients are equal to

$$c_n = 0$$

$$a_n = j^{-n} \frac{J_n'(\beta_0 a) J_n(\beta_1 a) - \sqrt{\varepsilon_r/\mu_r} J_n(\beta_0 a) J_n'(\beta_1 a)}{\sqrt{\varepsilon_r/\mu_r} J_n'(\beta_1 a) H_n^{(2)}(\beta_0 a) - J_n(\beta_1 a) H_n^{(2)'}(\beta_0 a)}$$

$$b_n = j^{-n} \frac{J_n(\beta_0 a) H_n^{(2)'}(\beta_0 a) - J_n'(\beta_0 a) H_n^{(2)}(\beta_0 a)}{J_n(\beta_1 a) H_n^{(2)'}(\beta_0 a) - \sqrt{\varepsilon_r/\mu_r} J_n'(\beta_1 a) H_n^{(2)}(\beta_0 a)}$$

11.27. A TE^z uniform plane wave traveling in the $+x$ direction in free space is incident normally upon a lossless dielectric circular cylinder of radius a as shown in Figure P11-27. Assume that the incident, scattered, and transmitted (into the cylinder) magnetic fields can be written as

$$\mathbf{H}^i = \hat{a}_z H_0 \sum_{n=-\infty}^{+\infty} j^{-n} J_n(\beta_0 \rho) e^{jn\phi}$$

$$\mathbf{H}^s = \hat{a}_z H_0 \sum_{n=-\infty}^{+\infty} a_n H_n^{(2)}(\beta_0 \rho) e^{jn\phi}$$

$$\mathbf{H}^d = \hat{a}_z H_0 \sum_{n=-\infty}^{+\infty} [b_n J_n(\beta_1 \rho) + c_n Y_n(\beta_1 \rho)] e^{jn\phi}$$

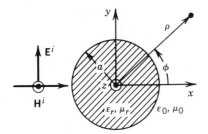

FIGURE P11-27

(a) Derive expressions for the incident, scattered, and transmitted electric field components.
(b) Show that the wave amplitude coefficients are equal to

$$c_n = 0$$

$$a_n = j^{-n} \frac{J_n'(\beta_0 a) J_n(\beta_1 a) - \sqrt{\mu_r/\varepsilon_r} J_n(\beta_0 a) J_n'(\beta_1 a)}{\sqrt{\mu_r/\varepsilon_r} J_n'(\beta_1 a) H_n^{(2)}(\beta_0 a) - J_n(\beta_1 a) H_n^{(2)'}(\beta_0 a)}$$

$$b_n = j^{-n} \frac{J_n(\beta_0 a) H_n^{(2)'}(\beta_0 a) - J_n'(\beta_0 a) H_n^{(2)}(\beta_0 a)}{J_n(\beta_1 a) H_n^{(2)'}(\beta_0 a) - \sqrt{\mu_r/\varepsilon_r} J_n'(\beta_1 a) H_n^{(2)}(\beta_0 a)}$$

11.28. A TM^z uniform plane wave traveling in the $+x$ direction in free space is incident normally upon a dielectric-coated conducting circular cylinder of radius a as shown in Figure P11-28. The thickness of the lossless dielectric coating is $b - a$. Assume that the incident, reflected, and transmitted (into the coating) electric fields can be written as shown in Problem 11.26.

(a) Write expressions for the incident, scattered, and transmitted magnetic field components.
(b) Determine the wave amplitude coefficients a_n, b_n, and c_n. Write them in their simplest forms.

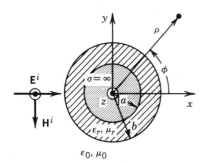

FIGURE P11-28

11.29. Repeat Problem 11.28 for a TE^z uniform plane wave incidence. Assume that the incident, scattered, and transmitted (into the coating) magnetic fields can be written as shown in Problem 11.27.

11.30. Using the definition of (11-21b), instead of (11-21c), show that the TM^z polarization scattering width reduces to (11-137) through (11-137b).

11.31. Using the definition of (11-21c), instead of (11-21b), show that the TE^z polarization scattering width reduces to (11-155) through (11-155b).

11.32. Two infinite length line sources of constant current I and of the same phase are placed near a conducting cylinder along the x axis (one on each side) a distance s from the center of the cylinder, as shown in Figure P11-32.

(a) Neglecting coupling between the sources, write an expression for the total electric field for both sources (assume $\rho > s$).

(b) Assuming the observations are made at large distances from the cylinder ($\rho \gg s$) and the radius of the cylinder as well as the distance s are very small ($a \ll \lambda$ and $s \ll \lambda$), find the distance s that the sources must be placed so that the electric field at any observation point will vanish. Explain.

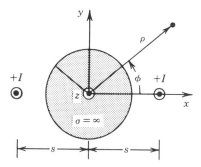

FIGURE P11-32

11.33. Three infinite length line sources carrying constant magnetic currents of I_m, $2I_m$, and I_m, respectively, are positioned a distance b near a perfect electric conducting cylinder, as shown in Figure P11-33. Neglecting mutual coupling between the sources, find the following.

(a) The total scattered magnetic field when $\rho > b$.

(b) The magnitude of the ratio of the scattered to the incident magnetic field for $\rho > b$.

(c) The normalized total magnetic field pattern when $\beta\rho \to$ large.

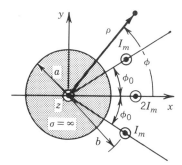

FIGURE P11-33

11.34. A TM^z uniform plane wave is incident at an angle ϕ' upon a half plane, as shown in Figure P11-34. Show that the current density on the upper side of the half plane is

$$J_z = \frac{E_0}{j2\omega\mu\rho} \sum_{m=1}^{\infty} mj^{m/2} J_{m/2}(\beta\rho) \sin\left(\frac{m\phi'}{2}\right) \quad \text{for any } \rho$$

$$J_z \simeq \frac{E_0}{2\eta} \sqrt{\frac{2}{j\pi\beta\rho}} \sin\left(\frac{\phi'}{2}\right) \quad \text{for } \beta\rho \to 0$$

where E_0 is the amplitude of the incident electric field.

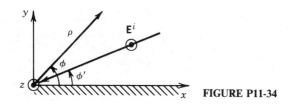

FIGURE P11-34

11.35. Derive (11-198) from (11-192a).

11.36. Repeat Problem 11.34 for TE^z uniform plane wave incidence. Show that the current density on the upper side of the half plane is

$$J_\rho = H_0 \sum_{m=0}^{\infty} \varepsilon_m j^{m/2} J_{m/2}(\beta\rho) \cos\left(\frac{m\phi'}{2}\right) \quad \text{for any } \rho$$

$$J_\rho = H_0 \quad \text{for } \beta\rho \to 0$$

where H_0 is the amplitude of the incident field and ε_m is defined by (11-196a).

11.37. Verify (11-222) through (11-222a).

11.38. A uniform plane wave is incident upon a conducting sphere of radius a. Assume that the diameter of the sphere is 1.128 m and the frequency is 8.5 GHz.
   (a) Determine the monostatic radar cross section of the sphere in decibels per square meter.
   (b) Find the area (in square meters) of a flat plate whose normal incidence monostatic RCS is the same as that of the sphere in part a.

CHAPTER 12

INTEGRAL EQUATIONS AND THE MOMENT METHOD

12.1 INTRODUCTION

In Chapter 11 we discussed scattering from conducting objects, such as plates, circular cylinders, and spheres, using geometrical optics, physical optics, and modal solutions. For the plates and cylinders we assumed that their dimensions were of infinite extent. In practice, however, the dimensions of the objects are always finite, although some of them may be very large. Expressions for the radar cross section of finite size scatterers were introduced in the previous chapter. These, however, represent approximate forms, and more accurate expressions are sometimes desired.

The physical optics method of Chapter 7, Section 7.10, was used in the previous chapter to approximate the current induced on the surface of a finite size target, such as the strip and rectangular plate. Radiation integrals were then used to find the field scattered by the target. To derive a more accurate representation of the current induced on the surface of the finite size target, and thus of the scattered fields, two methods will be examined here.

One method, referred to here as the *integral equation* (IE) technique, casts the solution for the induced current in the form of an integral equation (hence its name) where the unknown induced current density is part of the integrand. Numerical techniques such as the *moment method* (MM) [1–6] can then be used to solve for the current density. Once this is accomplished, the fields scattered by the target can be found using the traditional radiation integrals. The total induced current density will be the sum of the physical optics current density and a *fringe wave* current density [7–13] which can be thought of as a perturbation current density introduced by the edge diffractions of the finite size structure. This method will be introduced and applied in this chapter.

The other method, referred to here as the *geometrical theory of diffraction* (GTD) [14–17], is an extension of geometrical optics and accounts for the contributions from the edges of the finite structure using diffraction theory. This method will be introduced and applied in Chapter 13. More extensive discussions of each can be found in the open literature.

12.2 INTEGRAL EQUATION METHOD

The objective of the integral equation (IE) method for scattering is to cast the solution for the unknown current density, which is induced on the surface of the scatterer, in the form of an integral equation where the unknown induced current density is part of the integrand. The integral equation is then solved for the unknown induced current density using numerical techniques such as the *moment method* (MM). To demonstrate the technique, we will initially consider some specific problems. We will start with an electrostatics problem and follow it with time-harmonic problems.

12.2.1 Electrostatic Charge Distribution

In electrostatics, the problem of finding the potential that is due to a given charge distribution is often considered. In physical situations, however, it is seldom possible to specify a charge distribution. Whereas we may connect a conducting body to a voltage source, and thus specify the potential throughout the body, the distribution of charge is obvious only for a few rotationally symmetric geometries. In this section we will consider an integral equation approach to solve for the electric charge distribution once the electric potential is specified. Some of the material here and in other sections is drawn from [18, 19].

From statics we know that a linear electric charge distribution $\rho(\mathbf{r}')$ will create an electric potential, $V(\mathbf{r})$, according to [20]

$$V(\mathbf{r}) = \frac{1}{4\pi\varepsilon_0} \int_{\substack{\text{source} \\ \text{(charge)}}} \frac{\rho(\mathbf{r}')}{R} \, d\ell' \tag{12-1}$$

where $\mathbf{r}'(x', y', z')$ denotes the source coordinates, $\mathbf{r}(x, y, z)$ denotes the observation coordinates, $d\ell'$ is the path of integration, and R is the distance from any point on the source to the observation point, which is generally represented by

$$R(\mathbf{r}, \mathbf{r}') = |\mathbf{r} - \mathbf{r}'| = \sqrt{(x - x')^2 + (y - y')^2 + (z - z')^2} \tag{12-1a}$$

We see that (12-1) may be used to calculate the potentials that are due to any known line charge density. However, the charge distribution on most configurations of practical interest, i.e., complex geometries, is not usually known, even when the potential on the source is given. It is the nontrivial problem of determining the charge distribution, for a specified potential, that is to be solved here using an integral equation approach.

A. FINITE STRAIGHT WIRE

Consider a straight wire of length ℓ and radius a, placed along the y axis, as shown in Figure 12-1a. The wire is given a normalized constant electric potential of 1 V.

Note that (12-1) is valid everywhere, including on the wire itself ($V_{\text{wire}} = 1$ V). Thus, choosing the observation along the wire axis ($x = z = 0$) and representing the charge density on the surface of the wire, (12-1) can be expressed as

$$1 = \frac{1}{4\pi\varepsilon_0} \int_0^\ell \frac{\rho(y')}{R(y, y')} \, dy' \qquad 0 \leq y \leq \ell \tag{12-2}$$

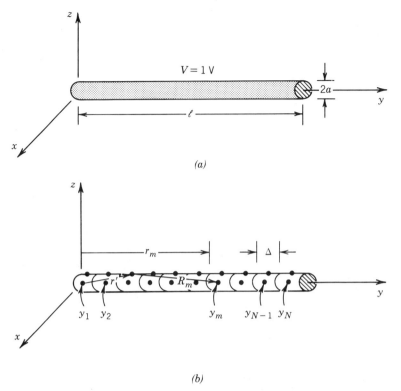

FIGURE 12-1 (*a*) Straight wire of constant potential and (*b*) its segmentation.

where

$$R(y, y') = R(\mathbf{r},\mathbf{r}')|_{x=z=0} = \sqrt{(y-y')^2 + \left[(x')^2 + (z')^2\right]} = \sqrt{(y-y')^2 + a^2} \quad (12\text{-}2a)$$

The observation point is chosen along the wire axis and the charge density is represented along the surface of the wire to avoid $R(y, y') = 0$, which would introduce a singularity in the integrand of (12-2).

It is necessary to solve (12-2) for the unknown $\rho(y')$ (an inversion problem). Equation 12-2 is an integral equation that can be used to find the charge density $\rho(y')$ based on the 1-V potential. The solution may be reached numerically by reducing (12-2) to a series of linear algebraic equations that may be solved by conventional matrix equation techniques. To facilitate this, let us approximate the unknown charge distribution $\rho(y')$ by an expansion of N known terms with constant, but unknown, coefficients, that is

$$\rho(y') = \sum_{n=1}^{N} a_n g_n(y') \quad (12\text{-}3)$$

Thus (12-2) may be written, using (12-3), as

$$4\pi\varepsilon_0 = \int_0^\ell \frac{1}{R(y, y')} \left[\sum_{n=1}^{N} a_n g_n(y') \right] dy' \quad (12\text{-}4)$$

Because (12-4) is a nonsingular integral, its integration and summation can be interchanged and it can be written as

$$4\pi\varepsilon_0 = \sum_{n=1}^{N} a_n \int_0^{\ell} \frac{g_n(y')}{\sqrt{(y-y')^2 + a^2}} \, dy' \qquad (12\text{-}4a)$$

The wire is now divided into N uniform segments, each of length $\Delta = \ell/N$, as illustrated in Figure 12-1b. The $g_n(y')$ functions in the expansion (12-3) are chosen for their ability to accurately model the unknown quantity, while minimizing computation. They are often referred to as *basis* (or expansion) functions, and they will be discussed further in Section 12.2.5. To avoid complexity in this solution, subdomain piecewise constant (or "pulse") functions will be used. These functions, shown in Figure 12-6, are defined to be of a constant value over one segment and zero elsewhere, or

$$g_n(y') = \begin{cases} 0 & y' < (n-1)\Delta \\ 1 & (n-1)\Delta \leq y' \leq n\Delta \\ 0 & n\Delta < y' \end{cases} \qquad (12\text{-}5)$$

Many other basis functions are possible, some of which will be introduced later in Section 12.2.5.

Replacing y in (12-4) by a fixed point on the surface of the wire, such as y_m, results in an integrand that is solely a function of y', so the integral may be evaluated. Obviously (12-4) leads to one equation with N unknowns a_n written as

$$4\pi\varepsilon_0 = a_1 \int_0^{\Delta} \frac{g_1(y')}{R(y_m, y')} \, dy' + a_2 \int_{\Delta}^{2\Delta} \frac{g_2(y')}{R(y_m, y')} \, dy' + \cdots$$

$$+ a_n \int_{(n-1)\Delta}^{n\Delta} \frac{g_n(y')}{R(y_m, y')} \, dy' + \cdots + a_N \int_{(N-1)\Delta}^{\ell} \frac{g_N(y')}{R(y_m, y')} \, dy' \qquad (12\text{-}6)$$

In order to obtain a solution for these N amplitude constants, N linearly independent equations are necessary. These equations may be produced by choosing an observation point y_m on the surface of the wire and at the center of each Δ length element as shown in Figure 12-1b. This will result in one equation of the form of (12-6) corresponding to each observation point. For N such points we can reduce (12-6) to

$$4\pi\varepsilon_0 = a_1 \int_0^{\Delta} \frac{g_1(y')}{R(y_1, y')} \, dy' + \cdots + a_N \int_{(N-1)\Delta}^{\ell} \frac{g_N(y')}{R(y_1, y')} \, dy'$$

$$\vdots$$

$$4\pi\varepsilon_0 = a_1 \int_0^{\Delta} \frac{g_1(y')}{R(y_N, y')} \, dy' + \cdots + a_N \int_{(N-1)\Delta}^{\ell} \frac{g_N(y')}{R(y_N, y')} \, dy' \qquad (12\text{-}6a)$$

We may write (12-6a) more concisely using matrix notation as

$$[V_m] = [Z_{mn}][I_n] \qquad (12\text{-}7)$$

where each Z_{mn} term is equal to

$$Z_{mn} = \int_0^\ell \frac{g_n(y')}{\sqrt{(y_m - y')^2 + a^2}} \, dy' = \int_{(n-1)\Delta}^{n\Delta} \frac{1}{\sqrt{(y_m - y')^2 + a^2}} \, dy' \quad (12\text{-}7a)$$

and

$$[I_n] = [a_n] \quad (12\text{-}7b)$$

$$[V_m] = [4\pi\varepsilon_0] \quad (12\text{-}7c)$$

The V_m column matrix has all terms equal to $4\pi\varepsilon_0$, and the $I_n = a_n$ values are the unknown charge distribution coefficients. Solving (12-7) for $[I_n]$ gives

$$[I_n] = [a_n] = [Z_{mn}]^{-1}[V_m] \quad (12\text{-}8)$$

Either (12-7) or (12-8) may readily be solved on a digital computer by using any of a number of matrix inversion or equation solving routines. Whereas the integrals involved here may be evaluated in closed form by making appropriate approximations, this is not usually possible with more complicated problems. Efficient numerical integration computer subroutines are commonly available in easy-to-use forms.

One closed form evaluation of (12-7a) is to reduce the integral and represent it by

$$Z_{mn} = \begin{cases} 2\ln\left(\dfrac{\dfrac{\Delta}{2} + \sqrt{a^2 + \left(\dfrac{\Delta}{2}\right)^2}}{a}\right) & m = n \quad (12\text{-}9a) \\[2ex] \ln\left(\dfrac{d_{mn}^+ + \left[(d_{mn}^+)^2 + a^2\right]^{1/2}}{d_{mn}^- + \left[(d_{mn}^-)^2 + a^2\right]^{1/2}}\right) & m \neq n \text{ but } |m-n| \leq 2 \quad (12\text{-}9b) \\[2ex] \ln\left(\dfrac{d_{mn}^+}{d_{mn}^-}\right) & |m-n| > 2 \quad (12\text{-}9c) \end{cases}$$

where

$$d_{mn}^+ = \ell_m + \frac{\Delta}{2} \quad (12\text{-}9d)$$

$$d_{mn}^- = \ell_m - \frac{\Delta}{2} \quad (12\text{-}9e)$$

ℓ_m is the distance between the mth matching point and the center of the nth source point.

In summary, the solution of (12-2) for the charge distribution on a wire has been accomplished by approximating the unknown with some basis functions, dividing the wire into segments, and then sequentially enforcing (12-2) at the center of each segment to form a set of linear equations.

Even for the relatively simple straight wire geometry we have discussed, the exact form of the charge distribution is not intuitively apparent. To illustrate the principles of the numerical solution, an example is now presented.

Example 12-1. A 1-m long straight wire of radius $a = 0.001$ m is maintained at a constant potential of 1 V. Determine the linear charge distribution on the wire by dividing the length into 5 and 20 uniform segments. Assume subdomain pulse basis functions.

Solution.

1. $N = 5$. When the 1-m long wire is divided into five uniform segments each of length $\Delta = 0.2$ m, (12-7) reduces to

$$\begin{bmatrix} 10.60 & 1.10 & 0.51 & 0.34 & 0.25 \\ 1.10 & 10.60 & 1.10 & 0.51 & 0.34 \\ 0.51 & 1.10 & 10.60 & 1.10 & 0.51 \\ 0.34 & 0.51 & 1.10 & 10.60 & 1.10 \\ 0.25 & 0.34 & 0.51 & 1.10 & 10.60 \end{bmatrix} \begin{bmatrix} a_1 \\ a_2 \\ a_3 \\ a_4 \\ a_5 \end{bmatrix} = \begin{bmatrix} 1.11 \times 10^{-10} \\ 1.11 \times 10^{-10} \\ \vdots \\ 1.11 \times 10^{-10} \end{bmatrix}$$

Inverting this matrix leads to the amplitude coefficients and subsequent charge distribution of

$$a_1 = 8.81 \text{ pC/m}$$
$$a_2 = 8.09 \text{ pC/m}$$
$$a_3 = 7.97 \text{ pC/m}$$
$$a_4 = 8.09 \text{ pC/m}$$
$$a_5 = 8.81 \text{ pC/m}$$

The charge distribution is shown in Figure 12-2*a*.

2. $N = 20$. Increasing the number of segments to 20 results in a much smoother distribution, as shown plotted in Figure 12-2*b*. As more segments are used, a better approximation of the actual charge distribution is attained, which has smaller discontinuities over the length of the wire.

B. BENT WIRE

In order to illustrate the solution for a more complex structure, let us analyze a body composed of two noncollinear straight wires, that is, a bent wire. If a straight wire is bent, the charge distribution will be altered, although the solution to find it will differ only slightly from the straight wire case. We will assume a bend of angle α, which remains in the yz plane, as shown in Figure 12-3.

For the first segment ℓ_1 of the wire the distance R can be represented by (12-2a). However, for the second segment ℓ_2 we can express the distance as

$$R = \sqrt{(y - y')^2 + (z - z')^2} \qquad (12\text{-}10)$$

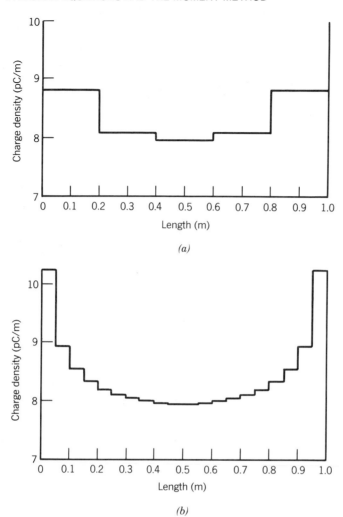

FIGURE 12-2 Charge distribution on a 1-m straight wire at 1 V. (*a*) $N = 5$. (*b*) $N = 20$.

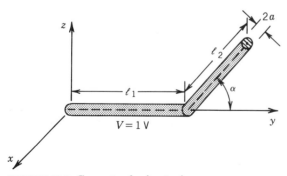

FIGURE 12-3 Geometry for bent wire.

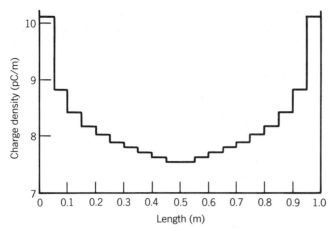

FIGURE 12-4 Charge distribution on a 1-m bent wire ($\alpha = 90°$, $N = 20$).

Also because of the bend, the integral in (12-7a) must be separated into two parts of

$$Z_{mn} = \int_0^{\ell_1} \frac{\rho_n(\ell_1')}{R} d\ell_1' + \int_0^{\ell_2} \frac{\rho_n(\ell_2')}{R} d\ell_2' \qquad (12\text{-}11)$$

where ℓ_1 and ℓ_2 are measured along the corresponding straight sections from their left ends.

Example 12-2. Repeat Example 12-1 assuming that the wire has been bent 90° at its midpoint. Subdivide the wire into 20 uniform segments.

Solution. The charge distribution for this case, calculated using (12-10) and (12-11), is plotted in Figure 12-4 for $N = 20$ segments. Note that the charge is relatively more concentrated near the ends of this structure than was the case for a straight wire of Figure 12-2b. Further, the overall charge density, and thus capacitance, on the structure has decreased.

Arbitrary wire configurations, including numerous bends and even curved sections, may be analyzed by the methods already outlined here. As with the simple bent wire, the only alterations generally necessary are those required to describe the geometry analytically.

12.2.2 Integral Equation

Now that we have demonstrated the numerical solution of a well known electrostatics integral equation, we will derive and solve a time-harmonic integral equation for an infinite line source above a two-dimensional conducting strip, as shown in Figure 12-5a. Once this is accomplished, we will generalize the integral equation formulation for three-dimensional problems in Section 12.3.

Referring to Figure 12-5a, the field radiated by a line source of constant current I_z in the absence of the strip (referred to as E_z^d) is given by (11-10a) or

$$E_z^d(\rho) = -\frac{\beta^2 I_z}{4\omega\varepsilon} H_0^{(2)}(\beta\rho) \qquad (12\text{-}12)$$

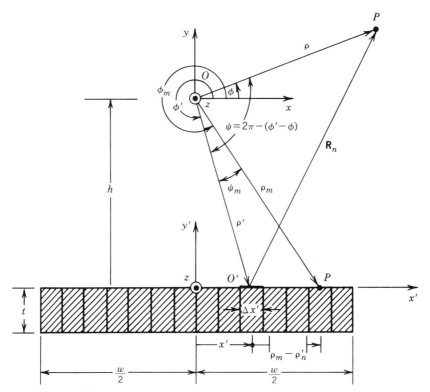

FIGURE 12-5 Geometry of a line source above a two-dimensional finite width strip. (*a*) Boundary conditions and integration on the same surface. (*b*) Boundary conditions and integration not on the same surface.

where $H_0^{(2)}(\beta\rho)$ is the Hankel function of the second kind of order zero. Part of the field given by (12-12) is directed toward the strip, and it induces on it a linear current density J_z (amperes per meter) such that

$$J_z(x')\,\Delta x' = \Delta I_z(x') \tag{12-13}$$

which as $\Delta x' \to 0$ can be written as

$$J_z(x')\,dx' = dI_z(x') \tag{12-13a}$$

The induced current of (12-13a) reradiates and produces an electric field component that will be referred to as *reflected* (or *scattered*) and designated as $E_z^r(\rho)$ [or $E_z^s(\rho)$]. If the strip is subdivided into N segments, each of width $\Delta x_n'$ as shown in Figure 12-5b, the scattered field can be written according to (12-12) as

$$E_z^s(\rho) = -\frac{\beta^2}{4\omega\varepsilon}\sum_{n=1}^{N} H_0^{(2)}(\beta R_n)\,\Delta I_z(x_n') = -\frac{\beta^2}{4\omega\varepsilon}\sum_{n=1}^{N} H_0^{(2)}(\beta R_n)\,J_z(x_n')\,\Delta x_n' \tag{12-14}$$

where x_n' is the position of the nth segment. In the limit, as each segment becomes

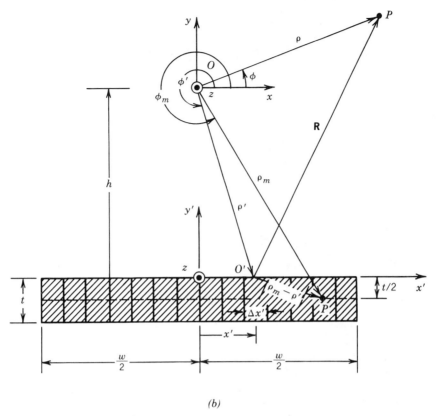

(b)

FIGURE 12-5 (*Continued*).

very small ($\Delta x_n \to$ small), (12-14) can be written as

$$E_z^s(\rho) = -\frac{\beta^2}{4\omega\varepsilon} \int_{\text{strip}} H_0^{(2)}(\beta R)\, dI_z = -\frac{\beta^2}{4\omega\varepsilon} \int_{-w/2}^{w/2} J_z(x') H_0^{(2)}(\beta|\rho - \rho'|)\, dx'$$

(12-15)

since

$$R = |\rho - \rho'| = \sqrt{\rho^2 + (\rho')^2 - 2\rho\rho' \cos(\phi - \phi')}$$
(12-15a)

The total field at any observation point, including the strip itself, will be the sum of the direct E_z^d of (12-12) and the scattered E_z^s of (12-15) components. However, to determine the scattered component we need to know the induced current density $J_z(x')$. The objective here then will be to find an equation, which in this case will be in terms of an integral and will be referred as an *integral equation*, that can be used to determine $J_z(x')$. This can be accomplished by choosing the observation point on the strip itself. Doing this, we have that for any observation point $\rho = \rho_m$ on the strip, the total tangential electric field vanishes and it is given by

$$E_z^t(\rho = \rho_m)\big|_{\text{strip}} = \left[E_z^d(\rho = \rho_m) + E_z^s(\rho = \rho_m)\right]_{\text{strip}} = 0 \quad (12\text{-}16)$$

or

$$E_z^s(\rho = \rho_m)\big|_{\text{strip}} = -E_z^d(\rho = \rho_m)\big|_{\text{strip}} \quad (12\text{-}16a)$$

Using (12-12) and (12-15), we can write (12-16a) as

$$-\frac{\beta^2 I_z}{4\omega\varepsilon} H_0^{(2)}(\beta\rho_m) = +\frac{\beta^2}{4\omega\varepsilon} \int_{-w/2}^{w/2} J_z(x') H_0^{(2)}(\beta|\rho_m - \rho'|)\, dx' \quad (12\text{-}17)$$

which for a unit current I_z (i.e., $I_z = 1$) reduces to

$$\boxed{H_0^{(2)}(\beta\rho_m) = -\int_{-w/2}^{w/2} J_z(x') H_0^{(2)}(\beta|\rho_m - \rho'|)\, dx'} \quad (12\text{-}17a)$$

Equation 12-17a is the *electric field integral equation* (EFIE) for the line source above the strip, and it can be used to find the current density $J_z(x')$ based upon a unit current I_z. If I_z is of any other constant value, then all the values of $J_z(x')$ must be multiplied by that same constant value. Electric field integral equations (EFIE) and magnetic field integral equations (MFIE) are discussed in more general forms in Section 12.3.

12.2.3 Radiation Pattern

Once J_z is found, we can then determine the total radiated field of the entire system for any observation point. The total field is composed of two parts: the field radiated from the line source itself (E_z^d) and that which is scattered (reradiated) from the strip (E_z^s). Thus, using (12-12) and (12-15) we can write the total field as

$$E_z^t(\rho) = E_z^d(\rho) + E_z^s(\rho)$$
$$= -\frac{\beta^2 I_z}{4\omega\varepsilon} H_0^{(2)}(\beta\rho) - \frac{\beta^2}{4\omega\varepsilon} \int_{-w/2}^{w/2} J_z(x') H_0^{(2)}(\beta|\rho - \rho'|)\, dx' \quad (12\text{-}18)$$

which for a unit amplitude current I_z ($I_z = 1$) reduces to

$$\boxed{E_z^t(\rho) = -\frac{\beta^2}{4\omega\varepsilon}\left[H_0^{(2)}(\beta\rho) + \int_{-w/2}^{w/2} J_z(x') H_0^{(2)}(\beta|\rho - \rho'|)\, dx'\right]} \quad (12\text{-}18a)$$

Equation 12-18a can be used to find the total field at any observation point, near or far field. The current density $J_z(x')$ can be found using (12-17a). However, for far-field observations, (12-18a) can be approximated and written in a more simplified form. In general, the distance R is given by (12-15a). However, for far-field observations ($\rho \gg \rho'$), (12-15a) reduces using the binomial expansion to

$$R \simeq \begin{cases} \rho - \rho' \cos(\phi - \phi') & \text{for phase terms} \\ \rho & \text{for amplitude terms} \end{cases} \quad \begin{array}{c} (12\text{-}19a) \\ (12\text{-}19b) \end{array}$$

For large arguments, the Hankel functions in (12-18a) can be replaced by their asymptotic form of

$$H_n^{(2)}(\beta z) \stackrel{\beta z \to \infty}{\simeq} \sqrt{\frac{2j}{\pi \beta z}} \, j^n e^{-j\beta z} \tag{12-20}$$

For $n = 0$, (12-20) reduces to

$$H_0^{(2)}(\beta z) \simeq \sqrt{\frac{2j}{\pi \beta z}} \, e^{-j\beta z} \tag{12-20a}$$

Using (12-19a) through (12-20a), we can write the Hankel functions found in (12-18a) as

$$H_0^{(2)}(\beta \rho) \simeq \sqrt{\frac{2j}{\pi \beta \rho}} \, e^{-j\beta \rho} \tag{12-21a}$$

$$H_0^{(2)}(\beta|\rho - \rho'|) \simeq \sqrt{\frac{2j}{\pi \beta \rho}} \, e^{-j\beta[\rho - \rho' \cos(\phi - \phi')]}$$

$$\simeq \sqrt{\frac{2j}{\pi \beta \rho}} \, e^{-j\beta \rho + j\beta \rho' \cos(\phi - \phi')} \tag{12-21b}$$

When (12-21a) and (12-21b) are substituted into (12-18a), they reduce it to

$$E_z^t(\rho) \simeq -\frac{\beta^2}{4\omega \varepsilon} \sqrt{\frac{2j}{\pi \beta \rho}} \, e^{-j\beta \rho} \left[1 + \int_{-w/2}^{+w/2} J_z(x') e^{j\beta \rho' \cos(\phi - \phi')} \, dx' \right] \tag{12-22}$$

which in normalized form can be written as

$$\boxed{E_z^t \text{ (normalized)} \simeq 1 + \int_{-w/2}^{w/2} J_z(x') e^{j\beta \rho' \cos(\phi - \phi')} \, dx'} \tag{12-22a}$$

Equation 12-22a represents the normalized pattern of the line above the strip. It is based on the linear current density $J_z(x')$ that is induced by the source on the strip. The current density can be found using approximate methods or more accurately it can be determined using the integral equation (12-17a).

12.2.4 Point-Matching (Collocation) Method

The next step will be to use a numerical technique to solve the integral equation 12-17a for the unknown current density $J_z(x')$. We first expand $J_z(x')$ into a finite series of the form

$$J_z(x') \simeq \sum_{n=1}^{N} a_n g_n(x') \tag{12-23}$$

where $g_n(x')$ represents *basis (expansion)* functions [1, 2]. When (12-23) is substituted into (12-17a), we can write it as

$$H_0^{(2)}(\beta\rho_m) = -\int_{-w/2}^{w/2} \sum_{n=1}^{N} a_n g_n(x') H_0^{(2)}(\beta|\rho_m - \rho_n'|)\, dx'$$

$$H_0^{(2)}(\beta\rho_m) = -\sum_{n=1}^{N} a_n \int_{-w/2}^{w/2} g_n(x') H_0^{(2)}(\beta|\rho_m - \rho_n'|)\, dx' \quad (12\text{-}24)$$

which takes the general form

$$\boxed{h = \sum_{n=1}^{N} a_n F(g_n)} \quad (12\text{-}25)$$

where

$$h = H_0^{(2)}(\beta\rho_m) \quad (12\text{-}25a)$$

$$F(g_n) = -\int_{-w/2}^{w/2} g_n(x') H_0^{(2)}(\beta|\rho_m - \rho_n'|)\, dx' \quad (12\text{-}25b)$$

In (12-25) F is referred to as a *linear integral operator*, g_n represents the response function, and h is the known excitation function.

Equation 12-17a is an integral equation derived by enforcing the boundary conditions of vanishing total tangential electric field on the surface of the conducting strip. A numerical solution of (12-17a) is (12-24) or (12-25) through (12-25b) which for a given observation point $\rho = \rho_m$ leads to one equation with N unknowns. This can be repeated N times by choosing N observation points. Such a procedure leads to a system of N linear equations each with N unknowns of the form

$$[H_0^{(2)}(\beta\rho_m)] = \left\{ \sum_{n=1}^{N} a_n \left[-\int_{-w/2}^{w/2} g_n(x') H_0^{(2)}(\beta|\rho_m - \rho_n'|)\, dx' \right] \right\}$$

$$m = 1, 2, \ldots, N \quad (12\text{-}26)$$

which can also be written as

$$\boxed{V_m = \sum_{n=1}^{N} a_n Z_{mn}} \quad (12\text{-}27)$$

where

$$\boxed{V_m = H_0^{(2)}(\beta\rho_m)} \quad (12\text{-}27a)$$

$$\boxed{I_n = a_n} \quad (12\text{-}27b)$$

$$\boxed{Z_{mn} = -\int_{-w/2}^{w/2} g_n(x') H_0^{(2)}(\beta|\rho_m - \rho_n'|)\, dx'} \quad (12\text{-}27c)$$

In matrix form, (12-27) can be expressed as

$$[V_m] = [Z_{mn}][I_n] \qquad (12\text{-}28)$$

where the unknown is $[I]_n$ and can be found by solving (12-28), or

$$[I_n] = [Z_{mn}]^{-1}[V_n] \qquad (12\text{-}28a)$$

Since the system of N linear equations each with N unknowns, as given by (12-26), (12-27), or (12-28), was derived by applying the boundary conditions at N discrete points, the technique is referred to as the *point-matching* (or *collocation*) *method* [1, 2].

Thus by finding the elements of the $[V]$ and $[Z]$, and then the inverse $[Z]^{-1}$, matrices, we can then determine the elements a_n of the $[I]$ matrix. This in turn allows us to approximate $J_z(x')$ using (12-23) which can then be used in (12-18a) to find the total field everywhere. However, for far-field observations, the total field can be found more easily using (12-22) or in normalized form using (12-22a).

12.2.5 Basis Functions

One very important step in any numerical solution is the choice of basis functions. In general, one chooses as basis functions the set that has the ability to accurately represent and resemble the anticipated unknown function, while minimizing the computational effort required to employ it [21–23]. Do not choose basis functions with smoother properties than the unknown being represented.

Theoretically, there are many possible basis sets. However, only a limited number are used in practice. These sets may be divided into two general classes. The first class consists of subdomain functions, which are nonzero only over a part of the domain of the function $g(x')$; its domain is the surface of the structure. The second class contains entire domain functions that exist over the entire domain of the unknown function. The entire domain basis function expansion is analogous to the well known Fourier series expansion method.

A. SUBDOMAIN FUNCTIONS

Of the two types of basis functions, subdomain functions are the most common. Unlike entire domain bases, they may be used without prior knowledge of the nature of the function that they must represent.

The subdomain approach involves subdivision of the structure into N nonoverlapping segments, as illustrated on the axis in Figure 12-6a. For clarity, the segments are shown here to be collinear and of equal length, although neither condition is necessary. The basis functions are defined in conjunction with the limits of one or more of the segments.

Perhaps the most common of these basis functions is the conceptually simple piecewise constant, or "pulse" function, shown in Figure 12-6a. It is defined by

Piecewise Constant

$$g_n(x') = \begin{cases} 1 & x'_{n-1} \leq x' \leq x'_n \\ 0 & \text{elsewhere} \end{cases} \qquad (12\text{-}29)$$

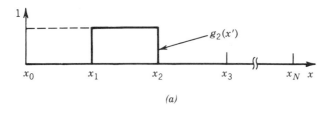

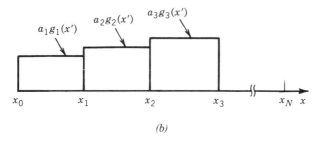

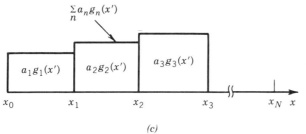

FIGURE 12-6 Piecewise constant subdomain functions. (a) Single. (b) Multiple. (c) Function representation.

Once the associated coefficients are determined, this function will produce a staircase representation of the unknown function, similar to that in Figure 12-6b and c.

Another common basis set is the piecewise linear, or "triangle," functions seen in Figure 12-7a. These are defined by

$$\text{Piecewise Linear}$$

$$g_n(x') = \begin{cases} \dfrac{x' - x'_{n-1}}{x'_n - x'_{n-1}} & x'_{n-1} \leq x' \leq x'_n \\ \dfrac{x'_{n+1} - x'}{x'_{n+1} - x'_n} & x'_n \leq x' \leq x'_{n+1} \\ 0 & \text{elsewhere} \end{cases} \qquad (12\text{-}30)$$

and are seen to cover two segments, and overlap adjacent functions (Figure 12-7b). The resulting representation (Figure 12-7c) is smoother than that for "pulses," but at the cost of increased computational complexity.

Increasing the sophistication of subdomain basis functions beyond the level of the "triangle" may not be warranted by the possible improvement in representation accuracy. However, there are cases where more specialized functions are useful for other reasons. For example, some integral operators may be evaluated without numerical integration when their integrands are multiplied by a $\sin(kx')$ or $\cos(kx')$ function, where x' is the variable of integration. In such examples, considerable

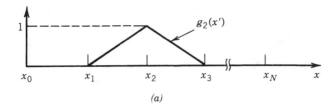

(a)

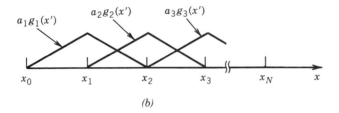

(b)

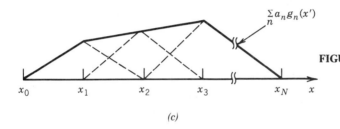

(c)

FIGURE 12-7 Piecewise linear subdomain functions. (*a*) Single. (*b*) Multiple. (*c*) Function representation.

advantages in computation time and resistance to errors can be gained by using basis functions like the piecewise sinusoid of Figure 12-8 or truncated cosine of Figure 12-9. These functions are defined by

Piecewise Sinusoid

$$g_n(x') = \begin{cases} \dfrac{\sin[\beta(x' - x'_{n-1})]}{\sin[\beta(x'_n - x'_{n-1})]} & x'_{n-1} \leq x' \leq x'_n \\ \dfrac{\sin[\beta(x'_{n+1} - x')]}{\sin[\beta(x'_{n+1} - x'_n)]} & x'_n \leq x' \leq x'_{n+1} \\ 0 & \text{elsewhere} \end{cases} \quad (12\text{-}31)$$

Truncated Cosine

$$g_n(x') = \begin{cases} \cos\left[\beta\left(x' - \dfrac{x'_n - x'_{n-1}}{2}\right)\right] & x'_{n-1} \leq x' \leq x'_n \\ 0 & \text{elsewhere} \end{cases} \quad (12\text{-}32)$$

B. ENTIRE-DOMAIN FUNCTIONS

Entire domain basis functions, as their name implies, are defined and are nonzero over the entire length of the structure being considered. Thus no segmentation is involved in their use.

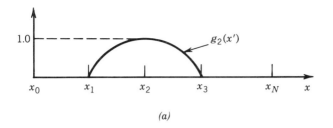

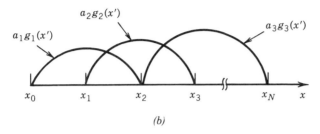

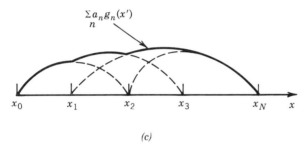

FIGURE 12-8 Piecewise sinusoids subdomain functions. (*a*) Single. (*b*) Multiple. (*c*) Function representation.

A common entire domain basis set is that of sinusoidal functions, where

$$\underline{\text{Entire Domain}}$$

$$g_n(x') = \cos\left[\frac{(2n-a)\pi x'}{\ell}\right] \qquad -\frac{\ell}{2} \leq x' \leq \frac{\ell}{2} \qquad (12\text{-}33)$$

Note that this basis set would be particularly useful for modeling the current distribution on a wire dipole, which is known to have primarily sinusoidal distribution. The main advantage of entire domain basis functions lies in problems where the unknown function is assumed a priori to follow a known pattern. Such entire-domain functions may render an acceptable representation of the unknown while using far fewer terms in the expansion of (12-23) than would be necessary for subdomain bases. Representation of a function by entire domain cosine and/or sine functions is similar to the Fourier series expansion of arbitrary functions.

Because we are constrained to use a finite number of functions (or modes, as they are sometimes called), entire domain basis functions usually have difficulty in modeling arbitrary or complicated unknown functions.

Entire domain basis functions, sets like (12-33), can be generated using Tschebyscheff, Maclaurin, Legendre, and Hermite polynomials, or other convenient functions.

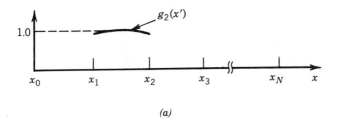

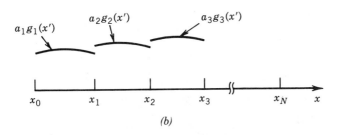

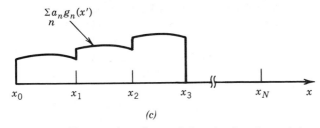

FIGURE 12-9 Truncated cosines subdomain functions. (*a*) Single. (*b*) Multiple. (*c*) Function representation.

12.2.6 Application of Point Matching

If each of the expansion functions $g_n(x')$ in (12-23) is of the subdomain type where each exists only over one segment of the structure, then Z_{mn} of (12-27c) reduces to

$$Z_{mn} = -\int_{x_n}^{x_{n+1}} g_n(x') H_0^{(2)}(\beta|\rho_m - \rho_n'|)\, dx' \qquad (12\text{-}34)$$

where x_n and x_{n+1} represent, respectively, the lower and upper limits of the segment over which each of the subdomain expansion functions $g_n(x')$ exists. If, in addition, the g_n are subdomain pulse expansion functions of the form

$$g_n(x') = \begin{cases} 1 & x_n \leq x' \leq x_{n+1} \\ 0 & \text{elsewhere} \end{cases} \qquad (12\text{-}35)$$

then (12-34) reduces to

$$Z_{mn} = -\int_{x_n}^{x_{n+1}} H_0^{(2)}(\beta|\rho_m - \rho_n'|)\, dx' \qquad (12\text{-}36)$$

The preceding integral cannot be evaluated exactly in closed form. However, there exist various approximations for its evaluation.

In solving (12-17a) using (12-24) or (12-26), there are few problems that must be addressed. Before we do that, let us first state in words what (12-24) and (12-26) represent. Each equation is a solution to (12-17a), which was derived by enforcing the boundary conditions. These conditions required the total tangential electric field to vanish on the surface of the conductor. For each observation point, the total field consists of the sum of the direct (E_z^d) and scattered (E_z^s) components. Thus, to find the total scattered field at *each observation point* we must add the contributions of the scattered field components *from all* the segments of the strip, which also includes those coming from the segment where the observations are made (referred to as *self-terms*). When the contributions from the segment over which the observation point lies are considered, then the distance $R_{mn} = R_{mm} = |\rho_m - \rho'_m|$ used for evaluating the self-term Z_{mm} in (12-27c) will become zero. This introduces a singularity in the integrand of (12-36) because the Hankel function defined as

$$H_0^{(2)}(\beta\rho) = J_0(\beta\rho) - jY_0(\beta\rho) \tag{12-37}$$

is infinite because $Y_0(0) = \infty$.

For finite thickness strips, the easiest way to get around the problem of evaluating the Hankel function for the self-terms will be to choose observation points away from the surface of the strip over which the integration in (12-36) is performed. For example, the observation points can be selected at the center of each segment along a line that divides the thickness of the strip, while the integration is performed along the upper surface of the strip. These points are designated in Figure 12-5b by the distance ρ_m.

Even if the aforementioned procedure is implemented for the evaluation of all the terms of Z_{mn}, including the self-terms, the distance $R_{mn} = |\rho_m - \rho'_n|$ for the self-terms (and some from the neighboring elements) will still sometime be sufficiently small that standard algorithms for computing Bessel functions, and thus Hankel functions, may not be very accurate. For these instances the Hankel functions can be evaluated using asymptotic expressions for small arguments. That is, for cases where the argument of the Hankel functions in (12-36) is small, which may include the self-terms and some of the neighboring elements, the Hankel function can be computed using [24]

$$H_0^{(2)}(\beta\rho) = J_0(\beta\rho) - jY_0(\beta\rho) \stackrel{\beta\rho \to 0}{\simeq} 1 - j\frac{2}{\pi}\ln\left(\frac{1.781\beta\rho}{2}\right) \tag{12-38}$$

The integral of (12-36) can be evaluated approximately in closed form, even if the observation and source points are chosen to be along the same line. This can be done not only for diagonal (self, $m = n$) but also for the nondiagonal ($m \neq n$) terms. For the diagonal terms ($m = n$) the Hankel function of (12-36) has an integrable singularity, and the integral can be evaluated analytically in closed form using the small argument approximation of (12-38) for the Hankel function. When (12-38) is used, it can be shown that (12-36) reduces to [2]

<u>Diagonal Terms Approximation</u>

$$Z_{nn} \simeq -\Delta x_n\left[1 - j\frac{2}{\pi}\ln\left(\frac{1.781\beta\Delta x_n}{4e}\right)\right] \quad m = n \tag{12-39}$$

where

$$\Delta x_n = x_{n+1} - x_n \tag{12-39a}$$
$$e = 2.718 \tag{12-39b}$$

For evaluation of the nondiagonal terms of (12-36) the crudest approximation would be to consider the Hankel function over each segment to be essentially constant [2]. To minimize the error using such an approximation, it is recommended that the argument of the Hankel function in (12-36) be represented by its average value over each segment. For straight line segments that average value will be representative of the distance from the center of the segment to the observation point. Thus for the nondiagonal terms, (12-36) can be approximated by

<u>Nondiagonal Terms Approximation</u>

$$Z_{mn} \simeq -\Delta x_n H_0^{(2)}(\beta |R_{mn}|_{\text{av}}) = -\Delta x_n H_0^{(2)}(\beta |\rho_m - \rho_n'|_{\text{av}}) \qquad m \neq n \quad (12\text{-}40)$$

The average value approximation for the distance R_{mn} in the Hankel function evaluation of (12-36) can also be used for curved surface scattering whereby each segment has been basically approximated by a straight line segment. Crude as it may seem, the average value approximation for the distance yields good results.

12.2.7 Weighting (Testing) Functions

Expansion of (12-24) leads to one equation with N unknowns. It alone is not sufficient to determine the N unknown a_n ($n = 1, 2, \ldots, N$) constants. To resolve the N constants, it is necessary to have N linearly independent equations. This can be accomplished by evaluating (12-24) (e.g., applying boundary conditions) at N different points, as represented by (12-26). To improve the point-matching solution, however, an inner product $\langle w, g \rangle$ can be defined which is a scalar operation satisfying the laws of

$$\langle w, g \rangle = \langle g, w \rangle \qquad (12\text{-}41a)$$
$$\langle bf + cg, w \rangle = b\langle f, w \rangle + c\langle g, w \rangle \qquad (12\text{-}41b)$$
$$\langle g^*, g \rangle > 0 \quad \text{if } g \neq 0 \qquad (12\text{-}41c)$$
$$\langle g^*, g \rangle = 0 \quad \text{if } g = 0 \qquad (12\text{-}41d)$$

where b and c are scalars and the asterisk (*) indicates complex conjugation. A typical, but not unique, inner product is

$$\langle \mathbf{w}, \mathbf{g} \rangle = \iint_S \mathbf{w}^* \cdot \mathbf{g}\, ds \qquad (12\text{-}42)$$

where the w's are the *weighting* (*testing*) functions and S is the surface of the structure being analyzed. Note that the functions w and g can be vectors. This technique is better known as the *moment method* or *method of moments* (MM) [1, 2].

12.2.8 Moment Method

The collocation (point-matching) method is a numerical technique whose solutions satisfy the electromagnetic boundary conditions (e.g., vanishing tangential electric fields on the surface of an electric conductor) only at discrete points. Between these points the boundary conditions may not be satisfied, and we define the deviation as

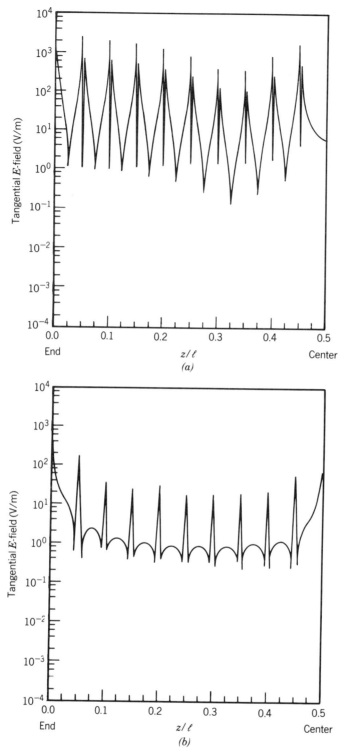

FIGURE 12-10 Tangential electric field on the conducting surface of the $\lambda/2$ dipole. (*Source:* E. K. Miller and F. J. Deadrick, "Some computational aspects of thin-wire modeling" in *Numerical and Asymptotic Techniques in Electromagnetics*, 1975, Springer-Verlag.) (*a*) Pulse basis—point matching. (*b*) Piecewise sinusoids—Galerkin's method.

a *residual* [e.g., residual = $\Delta E|_{\tan} = E(\text{scattered})|_{\tan} + E(\text{incident})|_{\tan} \neq 0$ on the surface of an electric conductor]. For a half-wavelength dipole, a typical residual is shown in Figure 12-10a for pulse basis functions and point matching and Figure 12-10b exhibits the residual for piecewise sinusoids–Galerkin method [25]. As expected, the pulse basis point matching exhibits the most ill-behaved residual and the piecewise sinusoids–Galerkin method indicates an improved residual. To minimize the residual in such a way that its overall average over the entire structure approaches zero, the method of *weighted residuals* is utilized in conjunction with the inner product of (12-42). This technique, referred as *moment method* (MM), does not lead to a vanishing residual at every point on the surface of a conductor, but it forces the boundary conditions to be satisfied in an average sense over the entire surface.

To accomplish this, we define a set of N *weighting* (or testing) functions $\{w_m\} = w_1, w_2, \ldots, w_N$ in the domain of the operator F. Forming the inner product between each of these functions, (12-25) results in

$$\boxed{\langle w_m, h \rangle = \sum_{n=1}^{N} a_n \langle w_m, F(g_n) \rangle \qquad m = 1, 2, \ldots, N} \qquad (12\text{-}43)$$

This set of N equations may be written in matrix form as

$$\boxed{[h_m] = [F_{mn}][a_n]} \qquad (12\text{-}44)$$

where

$$[F_{mn}] = \begin{bmatrix} \langle w_1, F(g_1) \rangle & \langle w_1, F(g_2) \rangle & \cdots \\ \langle w_2, F(g_1) \rangle & \langle w_2, F(g_2) \rangle & \\ \vdots & & \vdots \end{bmatrix} \qquad (12\text{-}44a)$$

$$[a_n] = \begin{bmatrix} a_1 \\ a_2 \\ \vdots \\ a_N \end{bmatrix} \qquad h_m = \begin{bmatrix} \langle w_1, h \rangle \\ \langle w_2, h \rangle \\ \vdots \\ \langle w_N, h \rangle \end{bmatrix} \qquad (12\text{-}44b)$$

The matrix of (12-44) may be solved for the a_n by inversion, and it can be written as

$$\boxed{[a_n] = [F_{mn}]^{-1}[h_m]} \qquad (12\text{-}45)$$

The choice of weighting functions is important in that the elements of $\{w_n\}$ must be linearly independent, so that the N equations in (12-43) will be linearly independent [1–3, 22, 23]. Further, it will generally be advantageous to choose weighting functions that minimize the computations required to evaluate the inner product.

The condition of linear independence between elements and the advantage of computational simplicity are also important characteristics of basis functions. Because of this, similar types of functions are often used for both weighting and expansion. A particular choice of functions may be to let the weighting and basis function be the same, that is, $w_n = g_n$. This technique is known as *Galerkin's method* [26].

It should be noted that there are N^2 terms to be evaluated in (12-44a). Each term usually requires two or more integrations; at least one to evaluate each $F(g_n)$, and one to perform the inner product of (12-42). When these integrations are to be done numerically, as is often the case, vast amounts of computation time may be necessary.

There is, however, a unique set of weighting functions that reduce the number of required integrations. This is the set of Dirac delta weighting functions

$$\boxed{[w_m] = [\delta(p - p_m)] = [\delta(p - p_1), \delta(p - p_2), \ldots]} \quad (12\text{-}46)$$

where p specifies a position with respect to some reference (origin), and p_m represents a point at which the boundary condition is enforced. Using (12-42) and (12-46) reduces (12-43) to

$$\langle \delta(p - p_m), h \rangle = \sum_n a_n \langle \delta(p - p_m), F(g_n) \rangle \quad m = 1, 2, \ldots, N$$

$$\iint_S \delta(p - p_m) h \, ds = \sum_n a_n \iint_S \delta(p - p_m) F(g_n) \, ds \quad m = 1, 2, \ldots, N$$

$$\boxed{h|_{p = p_m} = \sum_n a_n F(g_n)|_{p = p_m} \quad m = 1, 2, \ldots, N} \quad (12\text{-}47)$$

Hence, the only remaining integrations are those specified by $F(g_n)$. This simplification may make possible some solutions that would be impractical if other weighting functions were used. Physically, the use of Dirac delta weighting functions is seen as the relaxation of boundary conditions so that they are enforced only at discrete points on the surface of the structure, hence the name point matching.

An important consideration when using point matching is the positioning of the N points (p_m). While equally spaced points often yield good results, much depends on the basis functions used. When using subsectional basis functions in conjunction with point matching, one match point should be placed on each segment (to maintain linear independence). Placing the points at the center of the segments usually produces the best results. It is important that a match point does not coincide with the "peak" of a triangle or a similar discontinuous function, where the basis function is not differentiably continuous. This may cause errors in some situations.

Because it provides acceptable accuracy along with obvious computational advantages, point matching is easily the most popular testing technique for moment method solutions to electromagnetics problems. The analysis presented here, along with most problems considered in the literature, proceed via point matching.

For the strip problem, a convenient inner product of (12-42) would be

$$\langle w_m, g_n \rangle = \int_{-w/2}^{w/2} w_m^*(x) g_n(x)\, dx \tag{12-48}$$

Applying the inner product of (12-48) on both sides of (12-26), we can write it as

$$V_m' = \sum_{n=1}^{N} I_n Z_{mn}' \qquad m = 1, 2, \ldots, N \tag{12-49}$$

where

$$V_m' = \int_{-w/2}^{w/2} w_m^*(x) H_0^{(2)}(\beta \rho_m)\, dx \tag{12-49a}$$

$$Z_{mn}' = -\int_{-w/2}^{w/2} w_m^*(x) \left[\int_{-w/2}^{w/2} g_n(x') H_0^{(2)}(\beta |\rho_m - \rho_n'|)\, dx' \right] dx \tag{12-49b}$$

or in matrix form as

$$[V']_m = [Z']_{mn}[I]_n \tag{12-50}$$

If the w_m weighting functions are Dirac delta functions [i.e., $w_m(y) = \delta(y - y_m)$], then (12-49) reduces to (12-27) or

$$V_m' = V_m \tag{12-51a}$$

and

$$Z_{mn}' = Z_{mn} \tag{12-51b}$$

The *method of weighted residuals* (moment method) was introduced to minimize the average deviation, from the actual values, of the boundary conditions over the entire structure. However, it is evident that it has complicated the formulation by requiring an integration in the evaluation of the elements of the V' matrix [as given by (12-49a)] and an additional integration in the evaluation of the elements of the Z_{mn}' matrix [as given by (12-49b)]. Therein lies the penalty that is paid to improve the solution.

If both the expansion g_n and the weighting w_m functions are of the subdomain type, each of which exists only over one of the strip segments, then (12-49b) can be written as

$$Z_{mn}' = -\int_{x_m}^{x_{m+1}} w_m^*(x) \left[\int_{x_n'}^{x_{n+1}'} g_n(x') H_0^{(2)}(\beta |\rho_m - \rho'|)\, dx' \right] dx \tag{12-52}$$

where (x_m, x_{m+1}) and (x_n', x_{n+1}') represent, respectively, the lower and upper limits of the strip segments over which the weighting w_m and expansion g_n functions exist. To evaluate the mnth element of Z'_{mn} from (12-49b) or (12-52), we first choose the weighting function w_m, and the region of the segment over which it exists, and weigh the contributions from the g_n expansion function over the region in which it exists. To find the next element $Z'_{m(n+1)}$, we maintain the same weighting function w_m, and the region over which it exists, and weigh the contributions from the g_{n+1} expansion function. We repeat this until the individual contributions from all the N expansion functions (g_n) are weighted by the w_m weighting function. Then we choose the w_{m+1} weighting function, and the region over which it exists, and we weigh individually the contributions from each of the N expansion functions (g_n). We repeat this until all the N weighting functions (w_m), and the regions of the strip over which they exist, are individually weighted by the N expansion functions (g_n). This procedure allows us to form N linear equations, each with N unknowns, that can be solved using matrix inversion methods.

Example 12-3. For the electric line source of Figure 12-5 with $w = 2\lambda$, $t = 0.01\lambda$, and $h = 0.5\lambda$ perform the following.

1. Compute the equivalent current density induced on the open surface of the strip. This equivalent current density is representative of the vector sum of the current densities that flow on the opposite sides of the strip. Use subdomain pulse expansion functions and point matching. Subdivide the strip into 150 segments.
2. Compare the current density of part 1 with the physical optics current density.
3. Compute the normalized far-field amplitude pattern of (12-22a) using the current densities of parts 1 and 2. Compare these patterns with those obtained using a combination of geometrical optics (GO) and geometrical theory of diffraction (GTD) techniques of Chapter 13 and physical optics (PO) and physical theory of diffraction (PTD) techniques of [13].

Solution.

1. Utilizing (12-27) through (12-27b) and (12-36) the current density of (12-23) is computed using (12-28a). It is plotted in Figure 12-11. It is observed that the current density exhibits singularities toward the edges of the strip.
2. The physical optics current density is found using

$$\mathbf{J}_s^{PO} \simeq 2\hat{n} \times \mathbf{H}^i$$

which reduces using (11-10b) to

$$\mathbf{J}_s^{PO} \simeq 2\hat{a}_y \times \hat{a}_\phi H_\phi^i \big|_{\text{strip}} = 2\hat{a}_y \times (-\hat{a}_x \sin\phi + \hat{a}_y \cos\phi) H_\phi^i \big|_{\text{strip}}$$

$$= \hat{a}_z 2 \sin\phi H_\phi^i \big|_{\text{strip}} \simeq -j\hat{a}_z I_z \frac{\beta}{2} \sin\phi_m H_1^{(2)}(\beta\rho_m)$$

$$\mathbf{J}_s^{PO} = -j\hat{a}_z I_z \frac{\beta}{2} \left(\frac{y_m}{\rho_m}\right) H_1^{(2)}(\beta\rho_m)$$

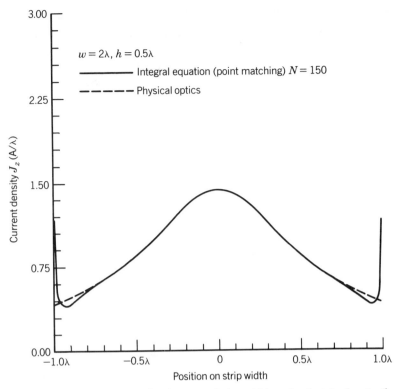

FIGURE 12-11 Current density on a finite width strip that is due to the electric line source above the strip.

The normalized value of this has also been plotted in Figure 12-11 so that it can be compared with the more accurate one obtained in part 1 using the integral equation.

3. The far-field amplitude patterns, based on the current densities of parts 1 and 2, are plotted in Figure 12-12. In addition to the normalized radiation patterns obtained using the current densities of parts 1 and 2, the pattern obtained using geometrical optics (GO) plus first-order diffractions by the geometrical theory of diffraction (GTD), to be discussed in Chapter 13, is also displayed in Figure 12-12. An excellent agreement is indicated between the IE and the GO plus GTD patterns. The pattern obtained using physical optics supplemented by first-order diffractions of the physical theory of diffraction (PTD) [13] is also displayed in Figure 12-12 for comparison purposes. It also compares extremely well with the others. As expected, the only one that does not compare well with the others is that of PO. Its largest differences are in the back lobes.

12.3 ELECTRIC AND MAGNETIC FIELD INTEGRAL EQUATIONS

The key to the solution of any antenna or scattering problem is a knowledge of the physical or equivalent current density distributions on the volume or surface of the antenna or scatterer. Once these are known then the radiated or scattered fields

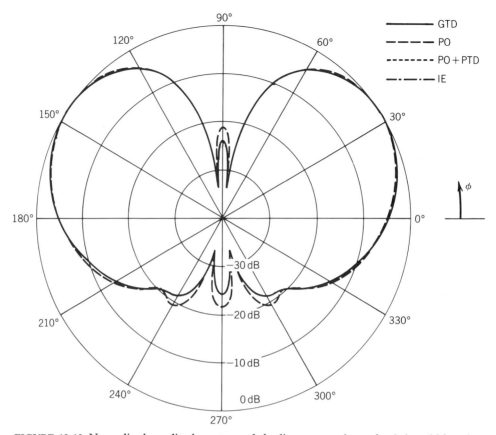

FIGURE 12-12 Normalized amplitude pattern of the line source above the finite width strip ($w = 2\lambda$, $h = 0.5\lambda$).

can be found using the standard radiation integrals. A main objective then of any solution method is to be able to predict accurately the current densities over the antenna or scatterer. This can be accomplished by the integral equation (IE) method. One form of IE, for a two-dimensional structure, was discussed in Section 12.2.2 and represented by the integral equation 12-17a.

In general there are many forms of integral equations. Two of the most popular for time-harmonic electromagnetics are the *electric field integral equation* (EFIE) and the *magnetic field integral equation* (MFIE). The EFIE enforces the boundary condition on the tangential electric field and the MFIE enforces the boundary condition on the tangential components of the magnetic field. Both of these will be discussed here as they applied to perfectly conducting structures.

12.3.1 Electric Field Integral Equation

The electric field integral equation (EFIE) is based on the boundary condition that the total tangential electric field on a perfectly electric conducting (PEC) surface of an antenna or scatterer is zero. This can be expressed as

$$\mathbf{E}_t^t(r = r_s) = \mathbf{E}_t^i(r = r_s) + \mathbf{E}_t^s(r = r_s) = 0 \quad \text{on } S \qquad (12\text{-}53)$$

or

$$\mathbf{E}_t^s(r = r_s) = -\mathbf{E}_t^i(r = r_s) \quad \text{on } S \qquad (12\text{-}53a)$$

where S is the conducting surface of the antenna or scatterer and $r = r_s$ is the distance from the origin to any point on the surface of the antenna or scatterer. The subscript t indicates tangential components.

The incident field that impinges on the surface S of the antenna or scatterer induces on it an electric current density $\mathbf{J}_s$ which in turn radiates the scattered field. If $\mathbf{J}_s$ is known, the scattered field everywhere that is due to $\mathbf{J}_s$ can be found using (6-32b), or

$$\mathbf{E}^s(r) = -j\omega \mathbf{A} - j\frac{1}{\omega\mu\varepsilon}\nabla(\nabla \cdot \mathbf{A}) = -j\frac{1}{\omega\mu\varepsilon}\left[\omega^2\mu\varepsilon \mathbf{A} + \nabla(\nabla \cdot \mathbf{A})\right] \quad (12\text{-}54)$$

where according to (6-96a)

$$\mathbf{A}(r) = \frac{\mu}{4\pi}\iint_S \mathbf{J}_s(r')\frac{e^{-j\beta R}}{R}\,ds' = \mu\iint_S \mathbf{J}_s(r')\frac{e^{-j\beta R}}{4\pi R}\,ds' \quad (12\text{-}54a)$$

Equations 12-54 and 12-54a can also be expressed by referring to Figure 6-2b as

$$\mathbf{E}^s(r) = -j\frac{\eta}{\beta}\left[\beta^2\iint_S \mathbf{J}_s(r')G(\mathbf{r},\mathbf{r}')\,ds' + \nabla\iint_S \nabla' \cdot \mathbf{J}_s(r')G(\mathbf{r},\mathbf{r}')\,ds'\right] \quad (12\text{-}55)$$

where

$$G(\mathbf{r},\mathbf{r}') = \frac{e^{-j\beta R}}{4\pi R} = \frac{e^{-j\beta|\mathbf{r}-\mathbf{r}'|}}{4\pi|\mathbf{r}-\mathbf{r}'|} \quad (12\text{-}55a)$$

$$R = |\mathbf{r} - \mathbf{r}'| \quad (12\text{-}55b)$$

In (12-55) ∇ and ∇' are, respectively, the gradients with respect to the observation (unprimed) and source (primed) coordinates and $G(\mathbf{r},\mathbf{r}')$ is referred to as the Green's function for a three-dimensional radiator or scatterer.

If the observations are restricted on the surface of the antenna or scatterer ($r = r_s$), then (12-55) through (12-55b) can be expressed using (12-53a) as

$$\boxed{j\frac{\eta}{\beta}\left[\beta^2\iint_S \mathbf{J}_s(r')G(\mathbf{r}_s,\mathbf{r}')\,ds' + \nabla\iint_S \nabla' \cdot \mathbf{J}_s(r')G(\mathbf{r}_s,\mathbf{r}')\,ds'\right] = \mathbf{E}_t^i(r = r_s)}$$

(12-56)

Because the right side of (12-56) is expressed in terms of the known incident electric field, it is referred to as the *electric field integral equation* (EFIE). It can be used to find the current density $\mathbf{J}_s(r')$ at any point $r = r'$ on the antenna or scatterer. It should be noted that (12-56) is actually an integrodifferential equation, but usually it is referred to as an integral equation.

Equation 12-56 can be used for closed or open surfaces. The scattered field is found, once $\mathbf{J}_s$ is determined, by using (6-32b) and (6-96a) or (12-54) and (12-54a) which assume that $\mathbf{J}_s$ radiates in one medium. Because of this $\mathbf{J}_s$ in (12-56) represents the physical equivalent electric current density of (7-53a) in Section 7.10. For open surfaces $\mathbf{J}_s$ is also the physical equivalent current density that represents the vector sum of the equivalent current densities on the opposite sides of the surface. Whenever this equivalent current density represents open surfaces, then a

698 INTEGRAL EQUATIONS AND THE MOMENT METHOD

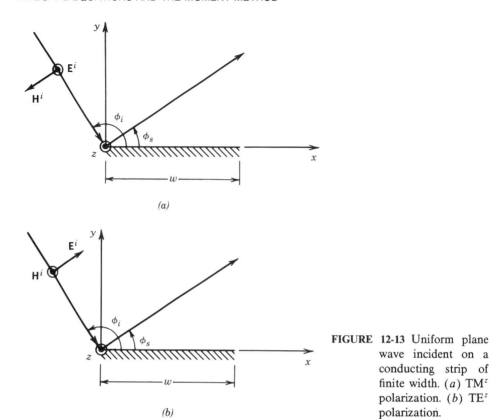

FIGURE 12-13 Uniform plane wave incident on a conducting strip of finite width. (*a*) TM^z polarization. (*b*) TE^z polarization.

boundary condition supplemental to (12-56) must be enforced to yield a unique solution for the normal component of the current density to vanish on S.

Equation 12-56 is a general surface EFIE for three-dimensional problems and its form can be simplified for two-dimensional geometries. To demonstrate this, let us derive the two-dimensional EFIEs for both TM^z and TE^z polarizations.

A. TWO-DIMENSIONAL EFIE: TM^Z POLARIZATION

The best way to demonstrate the derivation of the two-dimensional EFIE for TM^z polarization is to consider a specific example. Its form can then be generalized to more complex geometries. The example to be examined here is that of a TM^z uniform plane wave incidence on a finite width strip, as shown in Figure 12-13*a*.

By referring to Figure 12-13*a* the incident electric field can be expressed as

$$\mathbf{E}^i = \hat{a}_z E_0 e^{-j\boldsymbol{\beta}^i \cdot \mathbf{r}} = \hat{a}_z E_0 e^{j\beta(x \cos \phi_i + y \sin \phi_i)} \qquad (12\text{-}57)$$

which at the surface of the strip ($y = 0, 0 \leq x \leq w$) reduces to

$$\mathbf{E}^i(y = 0, 0 \leq x \leq w) = \hat{a}_z E_0 e^{j\beta x \cos \phi_i} \qquad (12\text{-}57a)$$

Since the incident electric field has only a z component, the scattered and total fields each also has only a z component which is independent of z variations (two dimensional). Therefore the scattered field can be found by expanding (12-54) assuming $\mathbf{A}$ has only a z component which is independent of z variations. Doing

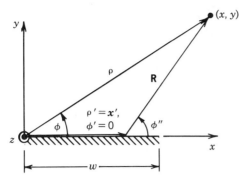

FIGURE 12-14 Geometry of the finite width strip for scattering.

this reduces (12-54) to

$$\mathbf{E}^s = -\hat{a}_z j\omega A_z \tag{12-58}$$

The vector potential component A_z is obtained using (12-54a) which in conjunction with (11-27d) and (11-28a) reduces to

$$A_z = \frac{\mu}{4\pi} \iint_S J_z(x') \frac{e^{-j\beta R}}{R} ds' = \frac{\mu}{4\pi} \int_0^w J_z(x') \left[\int_{-\infty}^{+\infty} \frac{e^{-j\beta\sqrt{|\rho-\rho'|^2+(z-z')^2}}}{\sqrt{|\rho-\rho'|^2+(z-z')^2}} dz' \right] dx'$$

$$= -j\frac{\mu}{4} \int_0^w J_z(x') H_0^{(2)}(\beta|\rho-\rho'|) \, dx' \tag{12-59}$$

where J_z is a linear current density (amperes per meter). Thus we can write the scattered electric field at any observation point, using the geometry of Figure 12-14, as

$$\mathbf{E}^s = -\hat{a}_z j\omega A_z = -\hat{a}_z \frac{\omega\mu}{4} \int_0^w J_z(x') H_0^{(2)}(\beta|\rho-\rho'|) \, dx'$$

$$= -\hat{a}_z \frac{\beta\eta}{4} \int_0^w J_z(x') H_0^{(2)}(\beta|\rho-\mathbf{x}'|) \, dx' \tag{12-60}$$

For far-field observations we can reduce (12-60) using the Hankel function approximation of (12-21b) for $\phi' = 0$ to

$$\boxed{\mathbf{E}^s \simeq -\hat{a}_z \eta \sqrt{\frac{j\beta}{8\pi}} \frac{e^{-j\beta\rho}}{\sqrt{\rho}} \int_0^w J_z(x') e^{j\beta x' \cos\phi} \, dx'} \tag{12-60a}$$

To evaluate the integral in (12-60a) in order to find the scattered field, we must know the induced current density $J_z(x')$ over the extent of the strip $(0 \le x' \le w)$. This can be accomplished by observing the field on the surface of the strip $(\rho = x_m)$. Under those conditions the total field over the strip must vanish. Thus

$$E_z^t(0 \le x_m \le w, y = 0) = E_z^i(0 \le x_m \le w, y = 0) + E_z^s(0 \le x_m \le w, y = 0) = 0 \tag{12-61}$$

or

$$E_z^s(0 \le x_m \le w, y = 0) = -E_z^i(0 \le x_m \le w, y = 0)$$

Since over the strip $\rho = x_m$, we can write the scattered field over the strip as

$$E_z^s(0 \le x_m \le w, y = 0) = -E_z^i(0 \le x_m \le w, y = 0)$$
$$= -\frac{\beta \eta}{4} \int_0^w J_z(x') H_0^{(2)}(\beta|x_m - x'|) \, dx' \quad (12\text{-}62)$$

or

$$\frac{\beta \eta}{4} \int_0^w J_z(x') H_0^{(2)}(\beta|x_m - x'|) \, dx'$$
$$= E_z^i(0 \le x_m \le w, y = 0) = E_0 e^{j\beta x_m \cos \phi_i} \quad (12\text{-}62a)$$

For a normalized field of unity amplitude ($E_0 = 1$) this reduces to

$$\boxed{\frac{\beta \eta}{4} \int_0^w J_z(x') H_0^{(2)}(\beta|x_m - x'|) \, dx' = e^{j\beta x_m \cos \phi_i}} \quad (12\text{-}63)$$

This is the desired two-dimensional electric field integral equation (EFIE) for the TM^z polarization of the conducting strip and it is equivalent to (12-56) for the general three-dimensional case. This EFIE can be solved for $J_z(x')$ using techniques similar to those used to solve the EFIE of (12-17a). It must be used to solve for the induced current density $J_z(x')$ over the surface of the strip. Since the surface of the strip is open, the aforementioned $J_z(x')$ represents the equivalent vector current density that flows on the opposite sides of the surface. For a more general geometry the EFIE of (12-63) can be written as

$$\boxed{\frac{\beta \eta}{4} \int_C J_z(\rho') H_0^{(2)}(\beta|\rho_m - \rho'|) \, dc' = E_z^i(\rho_m)} \quad (12\text{-}64)$$

where ρ_m = any observation point on the scatterer
ρ' = any source point on the scatterer
C = perimeter of scatterer

The solution of the preceding integral equations for the equivalent linear current density can be accomplished by using either the point-matching (collocation) method of Section 12.2.4 or the moment method of Section 12.2.8. However, using either method for the solution of the integral equation 12-63 for the strip of Figure 12-13a, we encounter the same problems as for the evaluation of the integral equation 12-17a for the finite strip of Figure 12-5, which are outlined in Section 12.2.6. However, these problems are overcome here using the same techniques as outlined in Section 12.2.6, namely, choosing the observation points along the bisector of the width of the strip, or using the approximations of (12-39) and (12-40).

Example 12-4. For the TM^z plane wave incidence on the conducting strip of Figure 12-13a perform the following.

1. Plot the induced equivalent current density for normal ($\phi_i = 90°$) incidence obtained using the EFIE of (12-63). Assume the strip has a width of

$w = 2\lambda$ and zero thickness. Use subdomain pulse expansion functions and point matching. Subdivide the strip into 250 segments.

2. Compare the current density of part 1 with the physical optics current density.

3. Compute the monostatic scattering width pattern for $0 \leq \phi_i \leq 180°$ using current density obtained using the EFIE of (12-63). Compare this pattern with those obtained with physical optics (PO), geometrical theory of diffraction (GTD) of Chapter 13, and physical optics (PO) plus physical theory of diffraction (PTD) techniques [13].

Solution.

1. Using the EFIE of (12-63) and applying point-matching methods with subdomain pulse expansion functions, the current density of Figure 12-15 for $\phi_i = 90°$ is obtained for a strip of $w = 2\lambda$. It is observed that the current density exhibits singularities toward the edges of the strip.

2. The physical optics current density is represented by

$$\mathbf{J}_s^{PO} \simeq 2\hat{n} \times \mathbf{H}^i\big|_{\text{strip}} = 2\hat{a}_y \times \left(\hat{a}_x H_x^i + \hat{a}_y H_y^i\right)\big|_{\text{strip}}$$

$$= -\hat{a}_z 2 H_x^i\big|_{\text{strip}} = \hat{a}_z 2 \frac{E_0}{\eta} \sin\phi_i e^{j\beta x \cos\phi_i}$$

which is shown plotted in Figure 12-15 for $\phi_i = 90°$. It is apparent that the

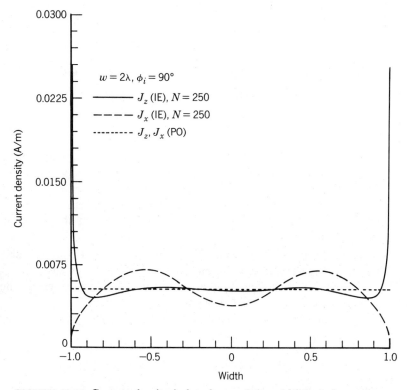

FIGURE 12-15 Current density induced on a finite width strip by a plane wave at normal incidence.

702 INTEGRAL EQUATIONS AND THE MOMENT METHOD

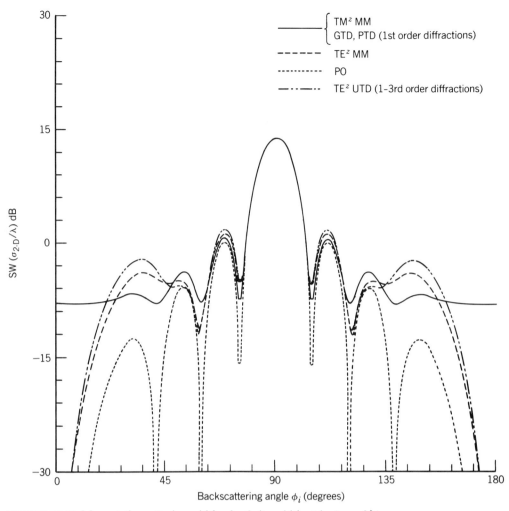

FIGURE 12-16 Monostatic scattering width of a finite width strip ($w = 2\lambda$).

PO current density does not compare well with that obtained using the IE, especially toward the edges of the strip. Therefore it does not provide a good representation of the equivalent current density induced on the strip. In Figure 12-15 we also display the equivalent current density for TE^z polarization which will be discussed in the next section and Example 12-5.

3. The monostatic scattering width patterns for $0° \leq \phi_i \leq 180°$ obtained using the methods of IE, PO, GTD, and PO plus PTD are all shown in Figure 12-16. As expected, the only one that differs from the others is that due to the PO; the other three are indistinguishable from each other and are represented by the solid curve. The pattern for the TE^z polarization for the IE method is also displayed in Figure 12-16. This will be discussed in the next section and Example 12-5.

B. TWO-DIMENSIONAL EFIE: TE^Z POLARIZATION

As in the previous section, the derivation of the EFIE for TE^z polarization is best demonstrated by considering a uniform plane wave incidence on the strip, as shown in Figure 12-13b. Its form can then be generalized to more complex geometries.

By referring to Figure 12-13b the incident electric field can be expressed as

$$\mathbf{E}^i = E_0(\hat{a}_x \sin\phi_i - \hat{a}_y \cos\phi_i)e^{-j\boldsymbol{\beta}_i \cdot \mathbf{r}} = E_0(\hat{a}_x \sin\phi_i - \hat{a}_y \cos\phi_i)e^{j\beta(x\cos\phi_i + y\sin\phi_i)} \quad (12\text{-}65)$$

which at the surface of the strip ($y = 0$, $0 \leq x \leq w$) reduces to

$$\mathbf{E}^i = E_0(\hat{a}_x \sin\phi_i - \hat{a}_y \cos\phi_i)e^{j\beta x \cos\phi_i} \quad (12\text{-}65a)$$

At the surface of the strip ($0 \leq x \leq w$, $y = 0$) the tangential components of the total field, incident plus scattered, must vanish. This can be written as

$$\hat{n} \times \mathbf{E}^t \big|_{\text{strip}} = \hat{n} \times (\mathbf{E}^i + \mathbf{E}^s)\big|_{\text{strip}}$$
$$= \hat{a}_y \times \left[(\hat{a}_x E_x^i + \hat{a}_y E_y^i) + (\hat{a}_x E_x^s + \hat{a}_y E_y^s)\right]_{\text{strip}} = 0 \quad (12\text{-}66)$$

which leads to

$$-\hat{a}_z(E_x^i + E_x^s)\big|_{\text{strip}} = 0 \Rightarrow E_x^s(0 \leq x \leq w, y = 0) = -E_x^i(0 \leq x \leq w, y = 0) \quad (12\text{-}66a)$$

or

$$E_x^s(0 \leq x \leq w, y = 0) = -E_x^i(0 \leq x \leq w, y = 0) = -E_0 \sin\phi_i e^{j\beta x \cos\phi_i} \quad (12\text{-}66b)$$

The x and y components of the scattered electric field, which are independent of z variations, are obtained using (12-54) which when expanded reduce to

$$E_x^s = -j\omega A_x - j\frac{1}{\omega\mu\varepsilon}\frac{\partial^2 A_x}{\partial x^2} = -j\frac{1}{\omega\mu\varepsilon}\left[\beta^2 A_x + \frac{\partial^2 A_x}{\partial x^2}\right]$$
$$= -j\frac{1}{\omega\mu\varepsilon}\left[\left(\beta^2 + \frac{\partial^2}{\partial x^2}\right)A_x\right] \quad (12\text{-}67a)$$

$$E_y^s = -j\frac{1}{\omega\mu\varepsilon}\frac{\partial^2 A_x}{\partial x \partial y} \quad (12\text{-}67b)$$

The vector potential A_x is obtained using (12-54a) which in conjunction with (11-27d) and (11-28a) reduces to

$$A_x = -j\frac{\mu}{4}\int_0^w J_x(x')H_0^{(2)}(\beta|\boldsymbol{\rho}-\boldsymbol{\rho}'|)\,dx' \quad (12\text{-}68)$$

Thus we can write that the x and y components of the scattered field can be expressed as

$$E_x^s = -j\frac{1}{\omega\mu\varepsilon}\left(-j\frac{\mu}{4}\right)\left[\left(\beta^2 + \frac{\partial^2}{\partial x^2}\right)\int_0^w J_x(x')H_0^{(2)}(\beta|\boldsymbol{\rho}-\boldsymbol{\rho}'|)\,dx'\right]$$
$$= -\frac{\eta}{4\beta}\left[\left(\beta^2 + \frac{\partial^2}{\partial x^2}\right)\int_0^w J_x(x')H_0^{(2)}(\beta|\boldsymbol{\rho}-\boldsymbol{\rho}'|)\,dx'\right] \quad (12\text{-}69a)$$

$$E_y^s = -\frac{\eta}{4\beta}\frac{\partial^2}{\partial x \partial y}\int_0^w J_x(x')H_0^{(2)}(\beta|\boldsymbol{\rho}-\boldsymbol{\rho}'|)\,dx' \quad (12\text{-}69b)$$

Interchanging integration and differentiation and letting $\rho' = x'$ we can rewrite the x and y components as

$$E_x^s = -\frac{\eta}{4\beta} \int_0^w J_x(x') \left\{ \left[\frac{\partial^2}{\partial x^2} + \beta^2 \right] H_0^{(2)}(\beta R) \right\} dx' \qquad (12\text{-}70\text{a})$$

$$E_y^s = -\frac{\eta}{4\beta} \int_0^w J_x(x') \left\{ \frac{\partial^2}{\partial x \, \partial y} H_0^{(2)}(\beta R) \right\} dx' \qquad (12\text{-}70\text{b})$$

where

$$R = |\rho - x'| \qquad (12\text{-}70\text{c})$$

It can be shown using the geometry of Figure 12-14 that

$$\left[\frac{\partial^2}{\partial x^2} + \beta^2 \right] H_0^{(2)}(\beta R) = \frac{\beta^2}{2} \left[H_0^{(2)}(\beta R) + H_2^{(2)}(\beta R) \cos(2\phi'') \right] \qquad (12\text{-}71\text{a})$$

$$\frac{\partial^2}{\partial x \, \partial y} H_0^{(2)}(\beta R) = \frac{\beta^2}{2} H_2^{(2)}(\beta R) \sin(2\phi'') \qquad (12\text{-}71\text{b})$$

Thus the x and y components of the electric field can be reduced to

$$E_x^s = -\frac{\beta \eta}{8} \int_0^w J_x(x') \left[H_0^{(2)}(\beta R) + H_2^{(2)}(\beta R) \cos(2\phi'') \right] dx' \qquad (12\text{-}72\text{a})$$

$$E_y^s = -\frac{\beta \eta}{8} \int_0^w J_x(x') H_2^{(2)}(\beta R) \sin(2\phi'') \, dx' \qquad (12\text{-}72\text{b})$$

The next objective is to solve for the induced current density that can then be used to find the scattered field. This can be accomplished by applying the boundary conditions on the x component of the electric field. When the observations are restricted to the surface of the strip ($\rho = x_m$), the x component of the scattered field over the strip can be written as

$$E_x^s(0 \leq x_m \leq w, \, y = 0)$$
$$= -E_x^i(0 \leq x_m \leq w, \, y = 0) = -E_0 \sin \phi_i e^{j\beta x_m \cos \phi_i}$$
$$= -\frac{\beta \eta}{8} \int_0^w J_x(x') \left[H_0^{(2)}(\beta R_m) + H_2^{(2)}(\beta R_m) \cos(2\phi_m'') \right] dx' \qquad (12\text{-}73)$$

or

$$\frac{\beta \eta}{8} \int_0^w J_x(x') \left[H_0^{(2)}(\beta R_m) + H_2^{(2)}(\beta R_m) \cos(2\phi_m'') \right] dx' = E_0 \sin \phi_i e^{j\beta x_m \cos \phi_i}$$
$$(12\text{-}73\text{a})$$

For a normalized field of unity amplitude ($E_0 = 1$) this reduces to

$$\frac{\beta\eta}{8}\int_0^w J_x(x')\left[H_0^{(2)}(\beta R_m) + H_2^{(2)}(\beta R_m)\cos(2\phi_m'')\right]dx' = \sin\phi_i e^{j\beta x_m \cos\phi_i}$$

(12-74)

where

$$R_m = |\boldsymbol{\rho}_m - \mathbf{x}'| \qquad (12\text{-}74a)$$

This is the desired two-dimensional electric field integral equation (EFIE) for the TE^z polarization of the conducting strip, and it is equivalent to (12-56) for the general three-dimensional case. This EFIE must be used to solve for the induced current density $J_x(x')$ over the surface of the strip using techniques similar to those used to solve the EFIE of (12-17a). For a more general geometry the EFIE can be written as

$$\frac{\eta}{4\beta}\left\{\beta^2\int_C J_c(\boldsymbol{\rho}')\left[\hat{c}_m \cdot \hat{c}' H_0^{(2)}(\beta|\boldsymbol{\rho}_m - \boldsymbol{\rho}'|)\right]dc' + \frac{d}{dc}\left[\nabla \cdot \int_C J_c(\boldsymbol{\rho}')\left[\hat{c}' H_0^{(2)}(\beta|\boldsymbol{\rho}_m - \boldsymbol{\rho}'|)\right]dc'\right]\right\} = -E_c^i(\boldsymbol{\rho}_m)$$

(12-75)

where $\boldsymbol{\rho}_m$ = any observation point on the scatterer
$\boldsymbol{\rho}'$ = any source point on the scatterer
C = perimeter of the scatterer
$\hat{c}_m, \hat{c}'$ = unit vector tangent to scatterer perimeter at observation, source points

The linear current density J_x is obtained by solving the integral equation 12-74 using either the point-matching (collocation) method of Section 12.2.4 or the moment method of Section 12.2.8. Using either method the solution of the preceding integral equation for J_x is more difficult than that of the TM^z polarization of the previous example. There exist various approaches (either exact or approximate) that can be used to accomplish this.

To demonstrate this, we will discuss one method that can be used to solve the integral equation 12-74. Let us assume that the current density $J_x(x')$ is expanded into a finite series similar to (12-23). Then the integral equation can be written as

$$\sin\phi_i e^{j\beta x_m \cos\phi_i} = \frac{\beta\eta}{8}\sum_{n=1}^N a_n \int_0^w g_n(x')\left[H_0^{(2)}(\beta R_m) + H_2^{(2)}(\beta R_m)\cos(2\phi_m'')\right]dx'$$

(12-76)

If in addition the basis functions are subdomain pulse functions, as defined by (12-35), then (12-76) using point matching reduces for each observation point to

$$\sin\phi_i e^{j\beta x_m \cos\phi_i} = \frac{\beta\eta}{8}\sum_{n=1}^N a_n \int_{x_n}^{x_{n+1}}\left[H_0^{(2)}(\beta R_{mn}) + H_2^{(2)}(\beta R_{mn})\cos(2\phi_{mn}'')\right]dx'$$

(12-77)

If N observations are selected, then we can write (12-77) as

$$\left[\sin\phi_i e^{j\beta x_m \cos\phi_i}\right] = \sum_{n=1}^{N} a_n \left\{ \frac{\beta\eta}{8} \int_{x_n}^{x_{n+1}} \left[H_0^{(2)}(\beta R_{mn}) + H_2^{(2)}(\beta R_{mn})\cos(2\phi''_{mn}) \right] dx' \right\}$$

$$m = 1, 2, \ldots, N \qquad (12\text{-}78)$$

or

$$[V_m] = [Z_{mn}][I_n] \qquad (12\text{-}78a)$$

where

$$V_m = \sin\phi_i e^{j\beta x_m \cos\phi_i} \qquad (12\text{-}78b)$$

$$I_n = a_n \qquad (12\text{-}78c)$$

$$Z_{mn} = \frac{\beta\eta}{8} \int_{x_n}^{x_{n+1}} \left[H_0^{(2)}(\beta R_{mn}) + H_2^{(2)}(\beta R_{mn})\cos(2\phi''_{mn}) \right] dx' \qquad (12\text{-}78d)$$

One of the tasks here will be the evaluation of the integral for Z_{mn}. We will examine one technique that requires Z_{mn} to be evaluated using three different expressions depending upon the position of the segment relative to the observation point. We propose here that Z_{mn} is evaluated using

$$Z_{mn} = \begin{cases} \dfrac{\beta\eta\,\Delta x_n}{8}\left\{1 - j\dfrac{1}{\pi}\left[-1 + 2\ln\left(\dfrac{1.781\beta\,\Delta x_n}{4e}\right) + \dfrac{16}{(\beta\,\Delta x_n)^2}\right]\right\} & (12\text{-}79a) \\ \qquad e = 2.718 \qquad m = n \\[1em] \dfrac{\beta\eta\,\Delta x_n}{8}\left\{1 + j\dfrac{4}{\pi\beta^2}\dfrac{1}{|x_m - x_n|^2 - \dfrac{(\Delta x_n)^2}{4}}\right\} & (12\text{-}79b) \\ \qquad |m-n| \leq 2 \quad m \neq n \\[1em] \dfrac{\beta\eta}{4}\int_{-\Delta x_n/2}^{\Delta x_n/2} \dfrac{H_1^{(2)}\left[\beta(|x_m - x_n| + x')\right]}{\beta(|x_m - x_n| + x')} dx' & (12\text{-}79c) \\ \qquad |m - n| > 2 \end{cases}$$

where x_m and x_n are measured from the center of their respective segments.

The current density J_x obtained from the preceding integral equation also represents the total current density J_s induced on the strip. This is evident from the induced current density equation

$$\mathbf{J}_s = \hat{n} \times \mathbf{H}^t = \hat{a}_y \times \hat{a}_z H_z^t = \hat{a}_x H_z^t = \hat{a}_x\left(H_z^i + H_z^s\right) \qquad (12\text{-}80)$$

Example 12-5. For the TE^z plane wave incidence on the conducting strip of Figure 12-13b, perform the following tasks.

1. Plot the induced equivalent current density for normal ($\phi_i = 90°$) incidence obtained using the EFIE of (12-74) or (12-78) through (12-78d). Assume a

width of $w = 2\lambda$ and zero thickness. Use subdomain pulse expansion functions and point matching. Subdivide the strip into 250 segments.

2. Compare the current density of part 1 with the physical optics current density.
3. Compute the monostatic scattering width pattern for $0° \leq \phi_i \leq 180°$ using the current density obtained using the EFIE of (12-78) through (12-78d). Compare this pattern with those obtained with physical optics (PO), geometrical theory of diffraction (GTD) of Chapter 13, and physical optics (PO) plus physical theory of diffraction (PTD) techniques [13].

Solution.

1. Using the EFIE of (12-78) through (12-78d), the current density of Figure 12-15 for $\phi_i = 90°$ is obtained for a strip of $w = 2\lambda$. It is observed that the current density vanishes toward the edges of the strip.
2. The physical optics current density is represented by

$$\mathbf{J}_s^{PO} \simeq 2\hat{n} \times \mathbf{H}^i\big|_{\text{strip}} = 2\hat{a}_y \times \hat{a}_z H_z^i\big|_{\text{strip}} = \hat{a}_x 2 H_z^i\big|_{\text{strip}} = \hat{a}_x 2 \frac{E_0}{\eta} e^{j\beta x \cos\phi_i}$$

which for normal incidence ($\phi_i = 90°$) is identical to that for the TM^z polarization, and it is shown plotted in Figure 12-15. As for the TM^z polarization, the PO TE^z polarization current density does not compare well with that obtained using the IE method. Therefore it does not provide a good representation of the equivalent current density induced on the strip. In Figure 12-15 the TE^z polarization current density is compared with that of the TM^z polarization using the different methods.

3. The monostatic scattering width pattern for $0° \leq \phi_i \leq 180°$ obtained using the IE method is shown plotted in Figure 12-16 where it is compared to those obtained by PO, PO plus PTD (first-order diffractions), and GTD (first-order diffractions) techniques. It is observed that the patterns of PO, PO plus PTD, and GTD (using first-order diffractions only) are insensitive to polarization whereby those of the integral equation with moment method solution vary with polarization. The SW patterns should vary with polarization. Therefore those obtained using the integral equation method are more accurate. It can be shown that if higher-order diffractions are included, the patterns of the PO plus PTD, and GTD will also vary with polarization. Higher-order diffractions are greater contributors to the overall scattering pattern for the TE^z polarization than for the TM^z. This is demonstrated by including in Figure 12-16 the monostatic SW for TE^z polarization obtained using higher-order GTD (UTD) diffractions [27]. It is apparent that this pattern agrees quite well with that of the IE method.

12.3.2 Magnetic Field Integral Equation

The magnetic field integral equation (MFIE) is expressed in terms of the known incident magnetic field. It is based on the boundary condition that expresses the total electric current density induced at any point $r = r'$ on the surface of a conducting surface S

$$\mathbf{J}_s(r') = \mathbf{J}_s(r = r') = \hat{n} \times \mathbf{H}^t(r = r') = \hat{n} \times \left[\mathbf{H}^i(r = r') + \mathbf{H}^s(r = r')\right] \quad (12\text{-}81)$$

Once the current density is known or determined, the scattered magnetic field can be obtained using (6-32a) and (6-96a), or

$$\mathbf{H}^s(r) = \frac{1}{\mu} \nabla \times \mathbf{A} = \nabla \times \iint_S \mathbf{J}_s(r') \frac{e^{-j\beta R}}{4\pi R} ds' = \nabla \times \iint_S \mathbf{J}_s(r') G(\mathbf{r},\mathbf{r}') ds' \quad (12\text{-}82)$$

where $G(\mathbf{r},\mathbf{r}')$ is the Green's function of (12-55a). Interchanging differentiation with integration and using the vector identity

$$\nabla \times (\mathbf{J}_s G) = G \nabla \times \mathbf{J}_s - \mathbf{J}_s \times \nabla G \quad (12\text{-}83)$$

where

$$\nabla \times \mathbf{J}_s(r') = 0 \quad (12\text{-}83\text{a})$$

$$\nabla G = -\nabla' G \quad (12\text{-}83\text{b})$$

(12-82) reduces to

$$\mathbf{H}^s(r) = \iint_S \mathbf{J}_s(r') \times [\nabla' G(\mathbf{r},\mathbf{r}')] ds' \quad (12\text{-}84)$$

On the surface S of the conductor the tangential magnetic field is discontinuous by the amount of the current density induced on the surface of the conductor. Therefore the current density is determined by (12-81a) but with $\mathbf{H}^s$ found using (12-84). Thus we can write that

$$\mathbf{J}_s(r') = \hat{n} \times \mathbf{H}^i(r = r') + \lim_{r \to S} [\hat{n} \times \mathbf{H}^s(r = r')]$$

$$= \hat{n} \times \mathbf{H}^i(r = r') + \lim_{r \to S} \left\{ \hat{n} \times \iint_S \mathbf{J}_s(r') \times [\nabla' G(\mathbf{r},\mathbf{r}')] ds' \right\} \quad (12\text{-}85)$$

or

$$\boxed{\mathbf{J}_s(r') - \lim_{r \to S} \left\{ \hat{n} \times \iint_S \mathbf{J}_s(r') \times [\nabla' G(\mathbf{r},\mathbf{r}')] ds' \right\} = \hat{n} \times \mathbf{H}^i(r = r')} \quad (12\text{-}85\text{a})$$

where $r \to S$ indicates that S is approached by r from the outside.

Equation 12-85a is referred to as the *magnetic field integral equation* (MFIE) because its right side is in terms of the incident magnetic field, and it is valid only for *closed* surfaces. Once the current density distribution can be found using (12-85a), then the scattered fields can be found using standard radiation integrals. It should be noted that the integral of (12-85a) must be carefully evaluated. The MFIE is the most popular for TE^z polarizations, although it can be used for both TE^z and TM^z cases. Since (12-85a) is only valid for closed surfaces, the current density obtained using (12-85a) is the actual current density induced on the surface of the conducting obstacle.

Whereas (12-85a) is a general MFIE for three-dimensional problems, its form can be simplified for two-dimensional MFIEs for both TM^z and TE^z polarizations.

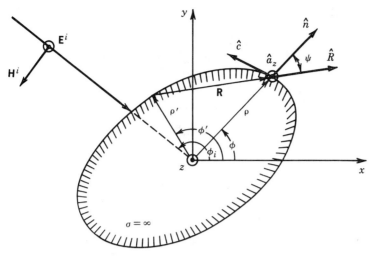

FIGURE 12-17 Geometry for two-dimensional MFIE TM^z polarization scattering.

A. TWO-DIMENSIONAL MFIE: TM^Z POLARIZATION

The best way to demonstrate the derivation of the two-dimensional MFIE for TM^z polarization is to consider a TM^z uniform plane wave incident upon a two-dimensional smooth curved surface, as shown in Figure 12-17.

Since the incident field has only a z component of the electric field, and x and y components of the magnetic field, the electric current density induced on the surface of the scatterer will only have a z component. That is

$$\mathbf{J}_s(\rho) = \hat{a}_z J_z(\rho)|_C \qquad (12\text{-}86)$$

On the surface of the scatterer the current density is related to the incident and scattered magnetic fields by (12-81), which for the geometry of Figure 12-17 can be written as

$$\mathbf{J}_s(\rho)|_C = \hat{a}_z J_z(\rho)|_C = \hat{n} \times (\mathbf{H}^i + \mathbf{H}^s)|_C = \hat{n} \times \mathbf{H}^i + \lim_{\rho \to C}(\hat{n} \times \mathbf{H}^s) \quad (12\text{-}87)$$

where $\rho \to C$ indicates that the boundary C is approached by ρ from the outside. Since the left side of (12-87) has only a z component, then the right side of (12-87) must also have only z components. Therefore the only component of $\mathbf{H}^i$ that contributes to (12-87) is that which is tangent to C and coincides with the surface of the scatterer. Thus we can rewrite (12-87) as

$$J_z(\rho)|_C = H_c^i(\rho)|_C + \lim_{\rho \to C}[\hat{a}_z \cdot (\hat{n} \times \mathbf{H}^s)] \qquad (12\text{-}88)$$

The scattered magnetic field $\mathbf{H}^s$ can be expressed according to (12-82) as

$$\mathbf{H}^s = \frac{1}{\mu}\nabla \times \mathbf{A} = \frac{1}{\mu}\nabla \times \left[\frac{\mu}{4\pi}\int_C\int_{-\infty}^{+\infty}\mathbf{J}_s(\rho')\frac{e^{-j\beta R}}{R}dz'\,dc'\right]$$

$$= \frac{1}{4\pi}\nabla \times \left\{\int_C \mathbf{J}_s(\rho')\left[\int_{-\infty}^{+\infty}\frac{e^{-j\beta R}}{R}dz'\right]dc'\right\} = -j\frac{1}{4}\nabla \times \int_C \mathbf{J}_s(\rho')H_0^{(2)}(\beta R)\,dc'$$

$$\mathbf{H}^s = -j\frac{1}{4}\int_C \nabla \times [\mathbf{J}_s(\rho')H_0^{(2)}(\beta R)]\,dc' \qquad (12\text{-}89)$$

Using (12-83) and (12-83a) reduces (12-89) to

$$\mathbf{H}^s = j\frac{1}{4}\int_C \mathbf{J}_s(\rho') \times \nabla H_0^{(2)}(\beta R)\, dc' \tag{12-90}$$

Since $\mathbf{J}_s(\rho')$ has only a z component, then the second term within the brackets on the right side of (12-88) can be written using (12-90) as

$$\hat{a}_z \cdot (\hat{n} \times \mathbf{H}^s) = \hat{a}_z \cdot \hat{n} \times \left\{ j\frac{1}{4}\int_C [\hat{a}'_z J_z(\rho')] \times [\nabla H_0^{(2)}(\beta R)]\, dc' \right\}$$

$$= j\frac{1}{4}\int_C J_z(\rho')\left\{ \hat{a}_z \cdot [\hat{n} \times \hat{a}_z \times \nabla H_0^{(2)}(\beta R)] \right\} dc' \tag{12-91}$$

since $\hat{a}'_z = \hat{a}_z$. Using the vector identity of

$$\mathbf{A} \times (\mathbf{B} \times \mathbf{C}) = (\mathbf{A} \cdot \mathbf{C})\mathbf{B} - (\mathbf{A} \cdot \mathbf{B})\mathbf{C} \tag{12-92}$$

we can write that

$$\hat{n} \times [\hat{a}_z \times \nabla H_0^{(2)}(\beta R)] = \hat{a}_z[\hat{n} \cdot \nabla H_0^{(2)}(\beta R)] - (\hat{n} \cdot \hat{a}_z)\nabla H_0^{(2)}(\beta R)$$

$$= \hat{a}_z[\hat{n} \cdot \nabla H_0^{(2)}(\beta R)] \tag{12-93}$$

since $\hat{n} \cdot \hat{a}_z = 0$. Substituting (12-93) into (12-91) reduces it to

$$\hat{a}_z \cdot (\hat{n} \times \mathbf{H}^s) = j\frac{1}{4}\int_C J_z(\rho')[\hat{n} \cdot \nabla H_0^{(2)}(\beta R)]\, dc'$$

$$= j\frac{1}{4}\int_C J_z(\rho')[-\beta \cos\psi H_1^{(2)}(\beta R)]\, dc'$$

$$\hat{a}_z \cdot (\hat{n} \times \mathbf{H}^s) = -j\frac{\beta}{4}\int_C J_z(\rho') \cos\psi H_1^{(2)}(\beta R)\, dc' \tag{12-94}$$

where the angle ψ is defined in Figure 12-17. Thus we can write (12-88) using (12-94) as

$$J_z(\rho)|_C = H_z^i(\rho)|_C + \lim_{\rho \to C}\left[-j\frac{\beta}{4}\int_C J_z(\rho') \cos\psi H_1^{(2)}(\beta R)\, dc' \right] \tag{12-95}$$

or

$$\boxed{\; J_z(\rho)\Big|_C + j\frac{\beta}{4}\lim_{\rho \to C}\left[\int_C J_z(\rho') \cos\psi H_1^{(2)}(\beta R)\, dc' \right] = H_z^i(\rho)\Big|_C \;} \tag{12-95a}$$

B. TWO-DIMENSIONAL MFIE: TE^z POLARIZATION

To derive the MFIE for the TE^z polarization, let us consider a TE^z uniform plane wave incidence upon a two-dimensional curved surface, as shown in Figure 12-18. Since the incident field has only a z component of the magnetic field, the current induced on the surface of the scatterer will have only a component that is tangent to C and it will coincide with the surface of the scatterer. That is

$$\mathbf{J}_s = \hat{c} J_c(\rho) \tag{12-96}$$

ELECTRIC AND MAGNETIC FIELD INTEGRAL EQUATIONS   711

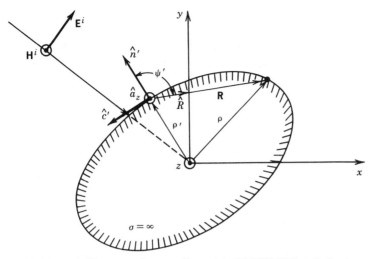

FIGURE 12-18 Geometry for two-dimensional MFIE TE^z polarization scattering.

On the surface of the scatterer the current density is related to the incident and scattered magnetic fields by (12-81), which for the geometry of Figure 12-18 can be written as

$$\mathbf{J}_s\big|_C = \hat{c} J_c(\rho)\big|_C = \hat{n} \times (\mathbf{H}^i + \mathbf{H}^s)\big|_C = \hat{n} \times \mathbf{H}^i + \lim_{\rho \to C}(\hat{n} \times \mathbf{H}^s)$$

$$= \hat{n} \times \hat{a}_z H^i\big|_C + \lim_{\rho \to C}(\hat{n} \times \mathbf{H}^s)$$

$$\mathbf{J}_s\big|_C = \hat{c} J_c(\rho)\big|_C = -\hat{c} H_z^i\big|_C + \lim_{\rho \to C}(\hat{n} \times \mathbf{H}^s) \qquad (12\text{-}97)$$

where $\rho \to C$ indicates that the boundary C is approached by ρ from the outside. Since the left side and the first term of the right side of (12-97) have only c components, then the second term of the right side of (12-97) must also have only a c component. Thus we can write (12-97) as

$$J_c(\rho)\big|_C = -H_z^i(\rho)\big|_C + \lim_{\rho \to C}\left[\hat{c} \cdot (\hat{n} \times \mathbf{H}^s)\right] \qquad (12\text{-}98)$$

Using the scattered magnetic field of (12-90) we can write the second term within the brackets of (12-98) as

$$\hat{c} \cdot (\hat{n} \times \mathbf{H}^s) = \hat{c} \cdot \hat{n} \times \left\{ j\frac{1}{4} \int_C \left[\hat{c}' J_c(\rho') \times \nabla H_0^{(2)}(\beta R)\right]\right\}$$

$$= j\frac{1}{4} \int_C J_c(\rho')\left\{\hat{c} \cdot \hat{n} \times \left[\hat{c}' \times \nabla H_0^{(2)}(\beta R)\right]\right\} dc' \qquad (12\text{-}99)$$

Since from Figure 12-18

$$\hat{c}' = -\hat{n}' \times \hat{a}'_z = -\hat{n}' \times \hat{a}_z \qquad (12\text{-}100)$$

then with the aid of (12-92)

$$\hat{c}' \times \nabla H_0^{(2)}(\beta R) = (-\hat{n}' \times \hat{a}_z) \times \nabla H_0^{(2)}(\beta R) = \nabla H_0^{(2)}(\beta R) \times (\hat{n}' \times \hat{a}_z)$$
$$= -\hat{a}_z[\hat{n}' \cdot \nabla H_0^{(2)}(\beta R)] + \hat{n}'[\hat{a}_z \cdot \nabla H_0^{(2)}(\beta R)]$$
$$= -\hat{a}_z[\hat{n} \cdot \nabla H_0^{(2)}(\beta R)] \qquad (12\text{-}100a)$$

since $\hat{a}_z \cdot \nabla H_0^{(2)}(\beta R) = 0$. Thus the terms within the brackets in (12-99) can be written as

$$\hat{c} \cdot \hat{n} \times [\hat{c}' \times \nabla H_0^{(2)}(\beta R)]$$
$$= -\hat{c} \cdot (\hat{n} \times \hat{a}_z)[\hat{n}' \cdot \nabla H_0^{(2)}(\beta R)] = (\hat{c} \cdot \hat{c})[\hat{n}' \cdot \nabla H_0^{(2)}(\beta R)]$$
$$= \hat{n}' \cdot \nabla H_0^{(2)}(\beta R) \qquad (12\text{-}101)$$

since $-\hat{c} = \hat{n} \times \hat{a}_z$. Substituting (12-101) into (12-99) reduces it to

$$\hat{c} \cdot \hat{n} \times \mathbf{H}^s = j\frac{1}{4}\int_C J_c(\rho')[\hat{n}' \cdot \nabla H_0^{(2)}(\beta R)] \, dc' = j\frac{1}{4}\int_C J_c(\rho')[-\beta \cos \psi' H_1^{(2)}(\beta R)] \, dc'$$

$$\hat{c} \cdot \hat{n} \times \mathbf{H}^s = -j\frac{\beta}{4}\int_C J_c(\rho') \cos \psi' H_1^{(2)}(\beta R) \, dc' \qquad (12\text{-}102)$$

where the angle ψ' is defined in Figure 12-18. Thus we can write (12-98) using (12-102) as

$$J_c(\rho)\big|_C = -H_z^i(\rho)\big|_C + \lim_{\rho \to C}\left[-j\frac{\beta}{4}\int_C J_c(\rho') \cos \psi' H_1^{(2)}(\beta R) \, dc'\right] \qquad (12\text{-}103)$$

or

$$\boxed{J_c(\rho)\big|_C + j\frac{\beta}{4}\lim_{\rho \to C}\left[\int_C J_c(\rho') \cos \psi' H_1^{(2)}(\beta R) \, dc'\right] = -H_z^i\big|_C} \qquad (12\text{-}103a)$$

C. SOLUTION OF THE TWO-DIMENSIONAL MFIE TE^z POLARIZATION

The two-dimensional MFIEs of (12-95a) for TM^z polarization and (12-103a) for TE^z polarization are of identical form and their solutions are then similar. Since TM^z polarizations are very conveniently solved using the EFIE, usually the MFIEs are mostly applied to TE^z polarization problems where the magnetic field has only a z component. Therefore we will demonstrate here the solution to the TE^z MFIE of (12-103a).

In the evaluation of the scattered magnetic field at $\rho = \rho_m$ from all points on C (including the point $\rho = \rho_m$ where the observation is made), the integral of (12-103a) can be split into two parts; one part coming from ΔC and the other part outside $\Delta C(C - \Delta C)$, as shown in Figure 12-19. Thus we can write the integral of (12-103a) as

$$j\frac{\beta}{4}\lim_{\rho \to C}\int_C J_c(\rho') \cos \psi' H_1^{(2)}(\beta R) \, dc'$$
$$= j\frac{\beta}{4}\lim_{\rho \to C}\left\{\int_{\Delta C} J_c(\rho') \cos \psi' H_1^{(2)}(\beta R) \, dc' + \int_{C-\Delta C} J_c(\rho') \cos \psi' H_1^{(2)}(\beta R) \, dc'\right\}$$
$$(12\text{-}104)$$

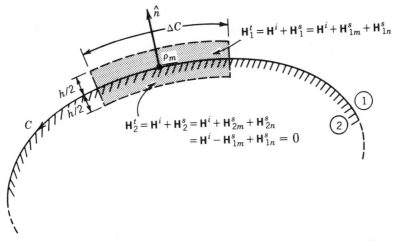

FIGURE 12-19 Geometry and fields along the scattering surface for a two-dimensional MFIE.

In a solution of (12-103a) where C is subdivided into segments, ΔC would typically represent one segment (the self-term) and $C - \Delta C$ would represent the other segments (the non-self-terms). Let us now examine the evaluation of each of the integrals in (12-104).

At any point the total magnetic field is equal to the sum of the incident and scattered parts. Within the scattered conducting obstacle the total magnetic field is zero whereby above the conducting surface the field is nonzero. The discontinuity of the two along C is used to represent the current density along C. Within the thin rectangular box with dimensions of h and ΔC (in the limit $h \to 0$), the total magnetic field above ($\mathbf{H}_1^t$) and below ($\mathbf{H}_2^t$) the interface can be written as

$$\mathbf{H}_1^t = \mathbf{H}^i + \mathbf{H}_1^s = \mathbf{H}^i + (\mathbf{H}_{1m}^s + \mathbf{H}_{1n}^s) \tag{12-105a}$$

$$\mathbf{H}_2^t = \mathbf{H}^i + \mathbf{H}_2^s = \mathbf{H}^i + (\mathbf{H}_{2m}^s + \mathbf{H}_{2n}^s) = \mathbf{H}^i + (-\mathbf{H}_{1m}^s + \mathbf{H}_{1n}^s) = 0 \tag{12-105b}$$

or $\quad \mathbf{H}_{1n}^s = \mathbf{H}_{1m}^s - \mathbf{H}^i \tag{12-105c}$

where $\mathbf{H}_{1m}^s$ ($\mathbf{H}_{2m}^s$) = scattered field in region 1 (2) within the box that is due to ΔC (self-term) which is discontinuous across the boundary along ΔC

$\mathbf{H}_{1n}^s$ ($\mathbf{H}_{2n}^s$) = scattered field in region 1 (2) within the box that is due to $C - \Delta C$ (non-self-terms) which is continuous across the boundary along ΔC

It is assumed here that ΔC along C becomes a straight line as the segment becomes small. The current density along ΔC can then be represented using (12-105a) and (12-105b) by

$$\mathbf{J}_c(\rho)\big|_{\Delta C} = \hat{n} \times (\mathbf{H}_1^t - \mathbf{H}_2^t)\big|_{\Delta C} = \hat{n} \times (\mathbf{H}_{1m}^s + \mathbf{H}_{1m}^s) = 2\hat{n} \times \mathbf{H}_{1m}^s = -2\hat{c}\mathbf{H}_{1m}^s \tag{12-106}$$

or

$$\mathbf{J}_c(\rho_m) = -2H_{1m}^s(\rho_m) \Rightarrow H_{1m}^s(\rho_m) = -\frac{J_c(\rho_m)}{2} \tag{12-106a}$$

Therefore the integral along ΔC in (12-104), which can be used to represent the scattered magnetic field at $\rho = \rho_m$ that is due to the ΔC, can be replaced by (12-106a). The non-self terms can be found using the integral along $C - \Delta C$ in (12-104). Thus using (12-104) and (12-106a) we can reduce (12-103a) for $\rho = \rho_m$ to

$$J_c(\rho_m) - \frac{J_c(\rho_m)}{2} + j\frac{\beta}{4}\int_{C-\Delta C} J_c(\rho')\cos\psi'_m H_1^{(2)}(\beta R_m)\,dc' = -H_z^i(\rho_m) \quad (12\text{-}107)$$

or

$$\boxed{\frac{J_c(\rho_m)}{2} + j\frac{\beta}{4}\int_{C-\Delta C} J_c(\rho')\cos\psi'_m H_1^{(2)}(\beta R_m)\,dc' = -H_z^i(\rho_m)} \quad (12\text{-}107\text{a})$$

An analogous procedure can be used to reduce (12-95a) to a form similar to that of (12-107a).

Let us now represent the current density $J_c(\rho)$ of (12-107a) by the finite series of (12-23)

$$J_c(\rho) \simeq \sum_{n=1}^{N} a_n g_n(\rho) \quad (12\text{-}108)$$

where $g_n(\rho)$ represents the basis (expansion) functions. Substituting (12-108) into (12-107a) and interchanging integration and summation, we can write that at any point $\rho = \rho_m$ on C (12-107a) can be written as

$$-H_z^i(\rho_m) = \frac{1}{2}\sum_{n=1}^{N} a_n g_n(\rho_m) + j\frac{\beta}{4}\sum_{n=1}^{N} a_n \int_{C-\Delta C} g_n(\rho')\cos\psi'_m H_1^{(2)}(\beta R_m)\,dc' \quad (12\text{-}109)$$

If the g_n's are subdomain piecewise constant pulse functions with each basis function existing only over its own segment, then (12-109) reduces to

$$-H_z^i(\rho_m) = \frac{\delta_{mn}}{2}a_n + j\frac{\beta}{4}\sum_{\substack{n=1\\n\neq m}}^{N} a_n \int_{\rho_n}^{\rho_{n+1}} \cos\psi'_{mn} H_1^{(2)}(\beta R_{mn})\,dc' \quad (12\text{-}110)$$

or

$$-H_z^i(\rho_m) = \sum_{n=1}^{N} a_n\left[\frac{\delta_{mn}}{2} + j\frac{\beta}{4}\int_{\substack{\rho_n\\n\neq m}}^{\rho_{n+1}} \cos\psi'_{mn} H_1^{(2)}(\beta R_{mn})\,dc'\right] \quad (12\text{-}110\text{a})$$

where δ_{mn} is the Kronecker delta function defined by

$$\delta_{mn} = \begin{cases} 1 & m = n \\ 0 & m \neq n \end{cases} \quad (12\text{-}110\text{b})$$

The Kronecker delta function is used to indicate that for a given observation point m only the segment itself ($n = m$) contributes to the first term on the right side of (12-110a).

If (12-110a) is applied to m points on C, then it can be written as

$$\left[-H_z^i(\rho_m)\right] = \sum_{n=1}^{N} a_n \left[\frac{\delta_{mn}}{2} + j\frac{\beta}{4} \int_{\rho_n}^{\rho_{n+1}} \cos\psi'_{mn} H_1^{(2)}(\beta R_{mn}) \, dc'\right] \quad (12\text{-}111)$$
$$n \neq m$$
$$m = 1, 2, \ldots, N$$

In general matrix notation (12-111) can be expressed as

$$[V_m] = [Z_{mn}][I_n] \quad (12\text{-}112)$$

where

$$V_m = -H_z^i(\rho_m) \quad (12\text{-}112\text{a})$$

$$Z_{mn} = \left[\frac{\delta_{mn}}{2} - j\frac{\beta}{4}\int_{\rho_n}^{\rho_{n+1}} \cos\psi'_{mn} H_1^{(2)}(\beta R_{mn}) \, dc'\right] \quad (12\text{-}112\text{b})$$

$$I_n = a_n \quad (12\text{-}112\text{c})$$

To demonstrate the applicability of (12-111), let us consider an example.

Example 12-6. A TE^z uniform plane wave is normally incident upon a circular conducting cylinder of radius a, as shown in Figure 12-20.

1. Using the MFIE of (12-107a) determine and plot the current density induced on the surface of the cylinder when $a = 2\lambda$. Assume the incident magnetic field is of unity amplitude. Use subdomain piecewise constant pulse functions. Subdivide the circumference into 540 segments. Compare the current density obtained using the IE with the exact modal solution of (11-113).
2. Based on the electric current density, derive and then plot the normalized ($\sigma_{2\text{-D}}/\lambda$) bistatic scattering width (in decibels) for $0° \leq \phi \leq 360°$ when

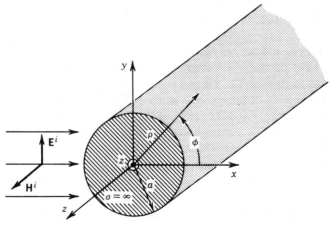

FIGURE 12-20 TE^z uniform plane wave incident on a circular conducting cylinder.

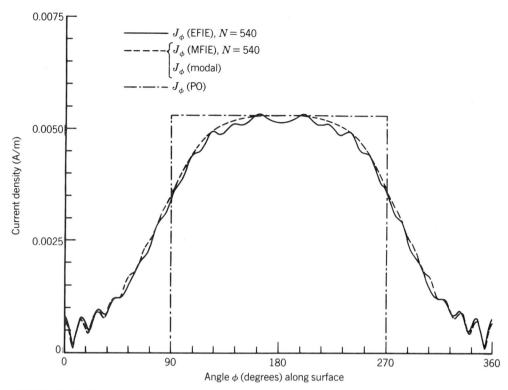

FIGURE 12-21 Current density induced on the surface of a circular conducting cylinder by TE^z plane wave incidence ($a = 2\lambda$).

$a = 2\lambda$. Compare these values with those obtained using the exact modal solution of (11-117).

Solution.

1. Since for subdomain piecewise constant pulse functions (12-103a) or (12-107a) reduces to (12-111) or (12-112) through (12-112c), then a solution of (12-112) for I_n leads to the current density shown in Figure 12-21. In the same figure we have plotted the current density of (11-113) based on the modal solution, and an excellent agreement between the two is indicated. We also have plotted the current densities based on the EFIE for TE^z polarization of Section 12.3.1B and on the physical optics of (7-54) over the illuminated portion of the cylinder surface. The results of the EFIE do not agree with the modal as accurately as those of the MFIE. However, they still are very good. As expected, the physical optics current density is not representative of the true solution.

2. Based on the current densities obtained in part 1, the far-zone scattered field was derived and the corresponding bistatic scattering width was formulated. The computed SW results are shown in Figure 12-22. Besides the results based on the physical optics approximation, the other three (MFIE, EFIE, and modal solution) give almost indistinguishable data and are shown in Figure 12-22 almost as one curve.

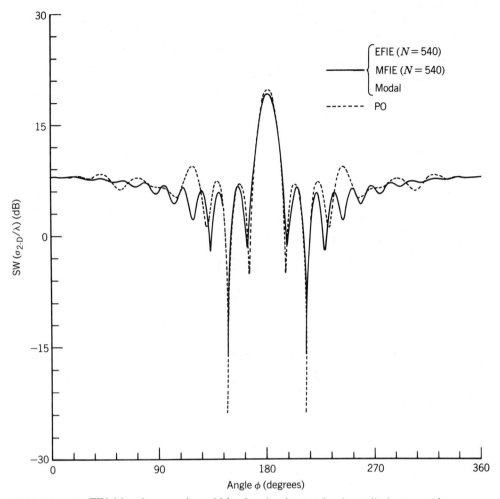

FIGURE 12-22 TE^z bistatic scattering width of a circular conducting cylinder ($a = 2\lambda$).

12.4 FINITE DIAMETER WIRES

In this section we want to derive and apply two classic three-dimensional integral equations, referred to as *Pocklington's integrodifferential equation* and *Hallén's integral equation* [28–36], that can be used most conveniently to find the current distribution on conducting wires. Hallén's equation is usually restricted to the use of a *delta-gap* voltage source model at the feed of a wire antenna. Pocklington's equation, however, is more general and it is adaptable to many types of feed sources (through alteration of its excitation function or excitation matrix), including a magnetic frill [37]. In addition, Hallén's equation requires the inversion of an $N + 1$ order matrix (where N is the number of divisions of the wire) while Pocklington's equation requires the inversion of an N order matrix.

For very thin wires, the current distribution is usually assumed to be of sinusoidal form [24]. For finite diameter wires (usually diameters d of $d > 0.05\lambda$), the sinusoidal current distribution is representative but not accurate. To find a more accurate current distribution on a cylindrical wire, an integral equation is usually derived and solved. Previously, solutions to the integral equation were obtained using iterative methods [30]; presently, it is most convenient to use moment method techniques [1–3].

If we know the voltage at the feed terminals of a wire antenna and find the current distribution, the input impedance and radiation pattern can then be obtained. Similarly if a wave impinges upon the surface of a wire scatterer, it induces a current density that in turn is used to find the scattered field. Whereas the linear wire is simple, most of the information presented here can be readily extended to more complicated structures.

12.4.1 Pocklington's Integral Equation

In deriving Pocklington's integral equation, the integral equation approach of Section 12.3.1 will be used. However, each step, as applied to the wire scatterer, will be repeated here to indicate the simplicity of the method.

Refer to Figure 12-23a. Let us assume that an incident wave impinges on the surface of a conducting wire. The total tangential electric field (E_z) at the surface of the wire is given by (12-53) or (12-53a), that is

$$E_z^t(r = r_s) = E_z^i(r = r_s) + E_z^s(r = r_s) = 0 \tag{12-113}$$

or

$$E_z^s(r = r_s) = -E_z^i(r = r_s) \tag{12-113a}$$

At any observation point, the field scattered by the induced current density on the surface of the wire is given by (12-54). However, for observations at the wire surface only the z component of (12-54) is needed and we can write it as

$$E_z^s(r) = -j\frac{1}{\omega\mu\varepsilon}\left(\beta^2 A_z + \frac{\partial^2 A_z}{\partial z^2}\right) \tag{12-114}$$

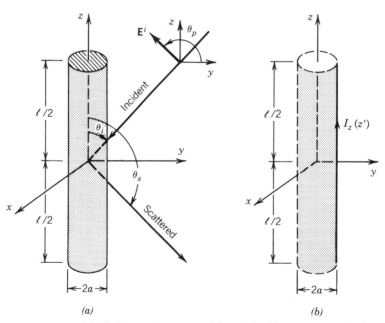

FIGURE 12-23 (a) Uniform plane wave obliquely incident on a conducting wire and (b) equivalent current.

According to (12-54a) and neglecting edge effects

$$A_z = \frac{\mu}{4\pi}\iint_S J_z \frac{e^{-j\beta R}}{R} ds' = \frac{\mu}{4\pi}\int_{-\ell/2}^{+\ell/2}\int_0^{2\pi} J_z \frac{e^{-j\beta R}}{R} a\, d\phi'\, dz' \quad (12\text{-}115)$$

If the wire is very thin, the current density J_z is not a function of the azimuthal angle ϕ, and we can write it as

$$2\pi a J_z = I_z(z') \Rightarrow J_z = \frac{1}{2\pi a} I_z(z') \quad (12\text{-}116)$$

where $I_z(z')$ is assumed to be an equivalent filament line-source current located a radial distance $\rho = a$ from the z axis, as shown in Figure 12-23b. Thus (12-115) reduces to

$$A_z = \frac{\mu}{4\pi}\int_{-\ell/2}^{+\ell/2}\left[\frac{1}{2\pi a}\int_0^{2\pi} I_z(z') \frac{e^{-j\beta R}}{R} a\, d\phi'\right] dz' \quad (12\text{-}117)$$

$$R = \sqrt{(x-x')^2 + (y-y')^2 + (z-z')^2}$$
$$= \sqrt{\rho^2 + a^2 - 2\rho a \cos(\phi - \phi') + (z-z')^2} \quad (12\text{-}117a)$$

where ρ is the radial distance to the observation point and a is the radius.

Because of the symmetry of the scatterer, the observations are not a function of ϕ. For simplicity let us then choose $\phi = 0$. For observations at the surface $\rho = a$ of the scatterer (12-117) and (12-117a) reduce to

$$A_z(\rho = a) = \mu\int_{-\ell/2}^{+\ell/2} I_z(z')\left(\frac{1}{2\pi}\int_0^{2\pi} \frac{e^{-j\beta R}}{4\pi R} d\phi'\right) dz' = \mu\int_{-\ell/2}^{+\ell/2} I_z(z') G(z, z')\, dz' \quad (12\text{-}118)$$

$$\boxed{G(z, z') = \frac{1}{2\pi}\int_0^{2\pi} \frac{e^{-j\beta R}}{4\pi R} d\phi'} \quad (12\text{-}118a)$$

$$\boxed{R(\rho = a) = \sqrt{4a^2 \sin^2\left(\frac{\phi'}{2}\right) + (z-z')^2}} \quad (12\text{-}118b)$$

Thus for observations at the surface $\rho = a$ of the scatterer the z component of the scattered electric field can be expressed as

$$E_z^s(\rho = a) = -j\frac{1}{\omega\varepsilon}\left(\beta^2 + \frac{d^2}{dz^2}\right)\int_{-\ell/2}^{+\ell/2} I_z(z') G(z, z')\, dz' \quad (12\text{-}119)$$

which by using (12-113a) reduces to

$$-j\frac{1}{\omega\varepsilon}\left(\frac{d^2}{dz^2} + \beta^2\right)\int_{-\ell/2}^{+\ell/2} I_z(z') G(z, z')\, dz' = -E_z^i(\rho = a) \quad (12\text{-}120)$$

or

$$\left(\frac{d^2}{dz^2} + \beta^2\right)\int_{-\ell/2}^{+\ell/2} I_z(z') G(z, z')\, dz' = -j\omega\varepsilon E_z^i(\rho = a) \quad (12\text{-}120a)$$

Interchanging integration with differentiation, we can rewrite (12-120a) as

$$\int_{-\ell/2}^{+\ell/2} I_z(z') \left[\left(\frac{\partial^2}{\partial z^2} + \beta^2 \right) G(z, z') \right] dz' = -j\omega\varepsilon E_z^i(\rho = a) \quad (12\text{-}121)$$

where $G(z, z')$ is given by (12-118a).

Equation 12-121 is referred to as *Pocklington's integrodifferential equation* [28], and it can be used to determine the equivalent filamentary line-source current of the wire, and thus current density on the wire, by knowing the incident field on the surface of the wire. It is a simplified form of (12-56) as applied to a wire scatterer, and it could have been derived directly from (12-56).

If we assume that the wire is very thin ($a \ll \lambda$) such that (12-118a) reduces to

$$G(z, z') = G(R) = \frac{e^{-j\beta R}}{4\pi R} \quad (12\text{-}122)$$

(12-121) can also be expressed in a more convenient form as [32]

$$\int_{-\ell/2}^{+\ell/2} I_z(z') \frac{e^{-j\beta R}}{4\pi R^5} \left[(1 + j\beta R)(2R^2 - 3a^2) + (\beta a R)^2 \right] dz' = -j\omega\varepsilon E_z^i(\rho = a)$$

$$(12\text{-}123)$$

where for observations along the center of the wire ($\rho = 0$)

$$R = \sqrt{a^2 + (z - z')^2} \quad (12\text{-}123a)$$

In (12-121) or (12-123) $I(z')$ represents the equivalent filamentary line-source current located on the surface of the wire, as shown in Figure 12-23b, and it is obtained by knowing the incident electric field at the surface of the wire. By point-matching techniques this is solved by matching the boundary conditions at discrete points on the surface of the wire. Often it is easier to choose the matching points to be at the interior of the wire, especially along the axis as shown in Figure 12-24a, where $I_z(z')$ is located on the surface of the wire. By reciprocity the configuration of Figure 12-24a is analogous to that of Figure 12-24b where the equivalent filamentary line-source current is assumed to be located along the center axis of the wire and the matching points are selected on the surface of the wire. Either of the two configurations can be used to determine the equivalent filamentary line-source current $I_z(z')$; the choice is left to the individual.

12.4.2 Hallén's Integral Equation

Referring again to Figure 12-23a let us assume that the length of the cylinder is much larger than its radius ($\ell \gg a$) and its radius is much smaller than the wavelength ($a \ll \lambda$) so that the effects of the end faces of the cylinder can be neglected. Therefore the boundary conditions for a wire with infinite conductivity are those of vanishing total tangential E fields on the surface of the cylinder and vanishing current at the ends of the cylinder [$I_z(z' = \pm\ell/2) = 0$].

Since only an electric current density flows on the cylinder and it is directed

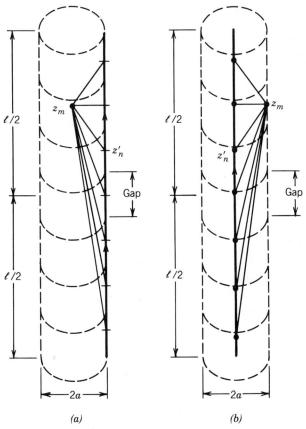

FIGURE 12-24 Dipole segmentation and its equivalent current (*a*) on the surface and (*b*) along its center.

along the z axis ($\mathbf{J} = \hat{a}_z J_z$), then according to (6-30) and (6-96a) $\mathbf{A} = \hat{a}_z A_z(z')$, which for small radii is assumed to be only a function of z'. Thus (6-34) reduces for $\mathbf{F} = 0$ to

$$E_z^t = -j\omega A_z - j\frac{1}{\omega\mu\varepsilon}\frac{\partial^2 A_z}{\partial z^2} = -j\frac{1}{\omega\mu\varepsilon}\left[\frac{d^2 A_z}{dz^2} + \omega^2\mu\varepsilon A_z\right] \quad (12\text{-}124)$$

Since the total tangential electric field E_z^t vanishes on the surface of the cylinder, (12-124) reduces to

$$\frac{d^2 A_z}{dz^2} + \beta^2 A_z = 0 \quad (12\text{-}124a)$$

Because the current density on the cylinder is symmetrical [$J_z(z') = J_z(-z')$], the potential A_z is also symmetrical [i.e., $A_z(z') = A_z(-z')$]. Thus the solution of (12-124a) is given by

$$A_z(z) = -j\sqrt{\mu\varepsilon}\,[B_1\cos(\beta z) + C_1\sin(\beta|z|)] \quad (12\text{-}125)$$

where B_1 and C_1 are constants. For a current-carrying wire, its potential is also given by (6-97a). Equating (12-125) to (6-97a) leads to

$$\int_{-\ell/2}^{+\ell/2} I_z(z') \frac{e^{-j\beta R}}{4\pi R} dz' = -j\sqrt{\frac{\varepsilon}{\mu}} \left[B_1 \cos(\beta z) + C_1 \sin(\beta|z|) \right] \quad (12\text{-}126)$$

If a voltage V_i is applied at the input terminals of the wire, it can be shown that the constant $C_1 = V_i/2$. The constant B_1 is determined from the boundary condition that requires the current to vanish at the end points of the wire.

Equation 12-126 is referred to as *Hallén's integral equation* for a perfectly conducting wire. It was derived by solving the differential equation 6-34 or 12-124a with the enforcement of the appropriate boundary conditions.

12.4.3 Source Modeling

Let us assume that the wire of Figure 12-23 is symmetrically fed by a voltage source, as shown in Figure 12-25a, and the element acting as a dipole antenna. To use, for example, Pocklington's integrodifferential equation 12-121 or 12-123 we need to know how to express $E_z^i(\rho = a)$. Traditionally there have been two methods used to model the excitation to represent $E_z^i(\rho = a, 0 \le \phi \le 2\pi, -\ell/2 \le z \le +\ell/2)$ at all points on the surface of the dipole: One is referred to as the *delta-gap* excitation and the other as the *equivalent magnetic ring current* (better known as *magnetic frill generator*) [37].

A. DELTA GAP

The delta-gap source modeling is the simplest and most widely used of the two, but it is also the least accurate, especially for impedances. Usually it is most accurate for smaller width gaps. Using the delta gap, it is assumed that the excitation voltage at the feed terminals is of a constant V_i value, and zero elsewhere. Therefore the incident electric field $E_z^i(\rho = a, 0 \le \phi \le 2\pi, -\ell/2 \le z \le +\ell/2)$ is also a constant (V_i/Δ where Δ is the gap width) over the feed gap and zero elsewhere, hence the name delta gap. For the delta-gap model, the feed gap Δ of Figure 12-25a is replaced by a narrow band of strips of equivalent magnetic current density of

$$\mathbf{M}_g = -\hat{n} \times \mathbf{E}^i = -\hat{a}_\rho \times \hat{a}_z \frac{V_i}{\Delta} = \hat{a}_\phi \frac{V_i}{\Delta} \qquad -\frac{\Delta}{2} \le z' \le \frac{\Delta}{2} \quad (12\text{-}127)$$

The magnetic current density $\mathbf{M}_g$ is sketched in Figure 12-25a.

B. MAGNETIC FRILL GENERATOR

The magnetic frill generator was introduced to calculate the near- as well as the far-zone fields from coaxial apertures [37]. To use this model the feed gap is replaced with a circumferentially directed magnetic current density that exists over an annular aperture with inner radius a, which is usually chosen to be the radius of the wire, and an outer radius b, as shown in Figure 12-25b. Since the dipole is usually fed by transmission lines, the outer radius b of the equivalent annular aperture of the magnetic frill generator is found using the expression for the characteristic impedance of the transmission line.

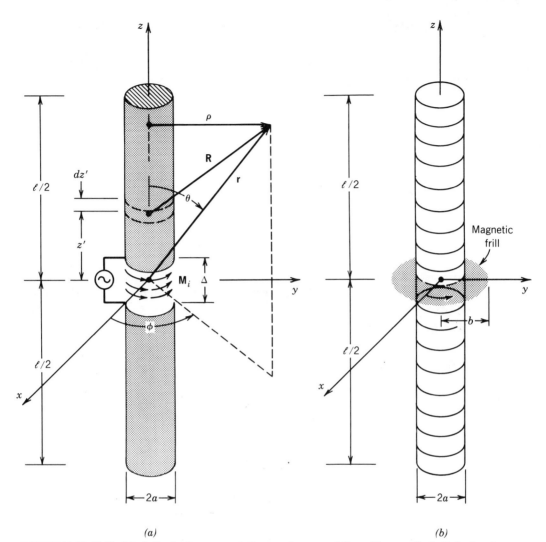

FIGURE 12-25 Cylindrical dipole, its segmentation, and gap modeling. (*Source:* C. A. Balanis, *Antenna Theory: Analysis and Design*, copyright © 1982, John Wiley & Sons, Inc. Reprinted by permission of John Wiley & Sons, Inc.) (*a*) Cylindrical dipole. (*b*) Segmented dipole.

Over the annular aperture of the magnetic frill generator the electric field is represented by the TEM mode field distribution of a coaxial transmission line given by

$$\mathbf{E}_f = \hat{a}_\rho \frac{V_i}{2\rho' \ln(b/a)} \quad (12\text{-}128)$$

Therefore the corresponding equivalent magnetic current density $\mathbf{M}_f$ for the magnetic frill generator used to represent the aperture is equal to

$$\mathbf{M}_f = -2\hat{n} \times \mathbf{E}_f = -2\hat{a}_z \times \hat{a}_\rho E_\rho = -\hat{a}_\phi \frac{V_i}{\rho' \ln(b/a)} \quad (12\text{-}129)$$

The fields generated by the magnetic frill generator of (12-129) on the surface of the wire are found using [37]

$$E_z^i\left(\rho = a, 0 \le \phi \le 2\pi, -\frac{\ell}{2} \le z \le \frac{\ell}{2}\right)$$
$$\simeq -V_i \left(\frac{\beta(b^2-a^2)e^{-j\beta R_0}}{8 \ln(b/a) R_0^2} \left\{ 2\left[\frac{1}{\beta R_0} + j\left(1 - \frac{b^2-a^2}{2R_0^2}\right)\right] \right. \right.$$
$$\left. \left. + \frac{a^2}{R_0} \left[\left(\frac{1}{\beta R_0} + j\left(1 - \frac{(b^2+a^2)}{2R_0^2}\right)\right)\left(-j\beta - \frac{2}{R_0}\right) + \left(-\frac{1}{\beta R_0^2} + j\frac{b^2+a^2}{R_0^3}\right)\right] \right\} \right) \quad (12\text{-}130)$$

where

$$R_0 = \sqrt{z^2 + a^2} \quad (12\text{-}130\text{a})$$

The fields generated on the surface of the wire computed using (12-130) can be approximated by those found along the axis ($\rho = 0$). Doing this leads to a simpler expression of the form [37]

$$E_z^i\left(\rho = 0, -\frac{\ell}{2} \le z \le \frac{\ell}{2}\right) = -\frac{V_i}{2\ln(b/a)}\left[\frac{e^{-j\beta R_1}}{R_1} - \frac{e^{-j\beta R_2}}{R_2}\right] \quad (12\text{-}131)$$

where

$$R_1 = \sqrt{z^2 + a^2} \quad (12\text{-}131\text{a})$$
$$R_2 = \sqrt{z^2 + b^2} \quad (12\text{-}131\text{b})$$

To compare the results using the two source modelings (delta gap and magnetic frill generator), an example will be performed.

Example 12-7. Assume a center-fed linear dipole of $\ell = 0.47\lambda$ and $a = 0.005\lambda$.

1. Determine the voltage and normalized current distribution over the length of the dipole using $N = 21$ segments to subdivide the length. Plot the current distribution.
2. Determine the input impedance using segments of $N = 7, 11, 21, 29, 41, 51, 61, 71$, and 79.

Use Pocklington's integrodifferential equation 12-123 with piecewise constant subdomain basis functions and point matching to solve the problems, model the gap with one segment, and use both the delta gap and magnetic frill generator to model the excitation. Use (12-131) for the magnetic frill generator. Because the current at the ends of the wire vanishes, the piecewise constant subdomain basis functions are not the most judicious choice. However, because of their simplicity, they are chosen here to illustrate the principles even though the results are not the most accurate. Assume that the characteristic impedance of the annular aperture is 50 ohms and the excitation voltage V_i is 1 V.

Solution.

1. Since the characteristic impedance of the annular aperture is 50 ohms, then

$$Z_c = \sqrt{\frac{\mu_0}{\epsilon_0}} \frac{\ln(b/a)}{2\pi} = 50 \Rightarrow \frac{b}{a} = 2.3$$

Subdividing the total length ($\ell = 0.47\lambda$) of the dipole to 21 segments makes the gap and each segment equal to

$$\Delta = \frac{0.47\lambda}{21} = 0.0224\lambda$$

Using (12-131) to compute E_z^i, the corresponding induced voltages obtained by multiplying the value of $-E_z^i$ at each segment by the length of the segment are found listed in Table 12-1, where they are compared with those of the delta gap. $N = 1$ represents the outermost segment and $N = 11$ represents the center segment. Because of the symmetry, only values for the center segment and half of the other segments are shown. Although the two distributions are not identical, the magnetic frill distribution voltages decay quite rapidly away from the center segment and they very quickly reach almost vanishing values.

The corresponding unnormalized and normalized currents are obtained using (12-123) with piecewise constant pulse functions and the point-matching technique for both the delta gap and magnetic frill generator.

TABLE 12-1
Unnormalized and normalized dipole induced voltage[a] differences for delta gap and magnetic frill generator ($\ell = 0.47\lambda$, $a = 0.005\lambda$, $N = 21$)

Segment number n	Delta gap voltage		Magnetic frill generator voltage	
	Unnormalized	Normalized	Unnormalized	Normalized
1	0	0	$1.11 \times 10^{-4} \; / -26.03°$	$7.30 \times 10^{-5} \; / -26.03°$
2	0	0	$1.42 \times 10^{-4} \; / -20.87°$	$9.34 \times 10^{-5} \; / -20.87°$
3	0	0	$1.89 \times 10^{-4} \; / -16.13°$	$1.24 \times 10^{-4} \; / -16.13°$
4	0	0	$2.62 \times 10^{-4} \; / -11.90°$	$1.72 \times 10^{-4} \; / -11.90°$
5	0	0	$3.88 \times 10^{-4} \; / -8.23°$	$2.55 \times 10^{-4} \; / -8.23°$
6	0	0	$6.23 \times 10^{-4} \; / -5.22°$	$4.10 \times 10^{-4} \; / -5.22°$
7	0	0	$1.14 \times 10^{-3} \; / -2.91°$	$7.5 \times 10^{-4} \; / -2.91°$
8	0	0	$2.52 \times 10^{-3} \; / -1.33°$	$1.66 \times 10^{-3} \; / -1.33°$
9	0	0	$7.89 \times 10^{-3} \; / -0.43°$	$5.19 \times 10^{-3} \; / -0.43°$
10	0	0	$5.25 \times 10^{-2} \; / -0.06°$	$3.46 \times 10^{-2} \; / -0.06°$
11	1	1	$1.52 \; / 0°$	$1.0 \; / 0°$

[a] Voltage differences as defined here represent the product of the incident electric field at the center of each segment and the corresponding segment length.

TABLE 12-2
Dipole input impedance for delta gap and magnetic frill generator using Pocklington's integral equation ($\ell = 0.47\lambda$, $a = 0.005\lambda$)

N	Delta gap	Magnetic frill
7	$122.8 + j113.9$	$26.8 + j24.9$
11	$94.2 + j49.0$	$32.0 + j16.7$
21	$77.7 - j0.8$	$47.1 - j0.2$
29	$75.4 - j6.6$	$57.4 - j4.5$
41	$75.9 - j2.4$	$68.0 - j1.0$
51	$77.2 + j2.4$	$73.1 + j4.0$
61	$78.6 + j6.1$	$76.2 + j8.5$
71	$79.9 + j7.9$	$77.9 + j11.2$
79	$80.4 + j8.8$	$78.8 + j12.9$

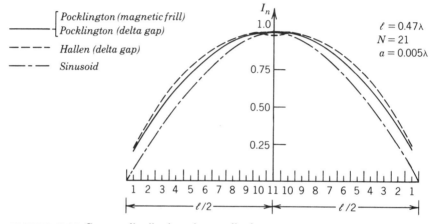

FIGURE 12-26 Current distribution along a dipole antenna.

The normalized magnitudes of the se currents are shown plotted in Figure 12-26. It is apparent that the two distributions are almost identical in shape, and they resemble that of the ideal sinusoidal current distribution which is more valid for very thin wires and very small gaps. The distributions obtained using Pocklington's integral equation do not vanish at the ends because of the use of piecewise constant subdomain basis functions.

2. The input impedances computed using both the delta gap and the magnetic frill generator are shown listed in Table 12-2. It is evident that the values begin to stabilize and compare favorably to each other once 61 or more segments are used.

12.5 COMPUTER CODES

With the advent of the computer there has been a proliferation of computer program development. Many of these programs are based on algorithms that are suitable for efficient computer programming for the analysis and synthesis of electromagnetic boundary-value problems. Some of these computer programs are very sophisticated and can be used to solve complex radiation and scattering problems. Others are much simpler and have limited applications. Many programs are public domain; others are restricted.

Five computer programs based on integral equation formulations and moment method solutions will be described here. The first computes the radiation or scattering by a two-dimensional perfectly electric conducting (PEC) body. It is referred to here as TDRS (two-dimensional radiation and scattering), and it is based on the two-dimensional formulations of the electric field integral equation (EFIE) of Section 12.3.1. It can be used for both electric and magnetic line-source excitation or TM^z and TE^z plane wave incidence. The listing of this program is included in the solutions manual available to the instructors. It can also be obtained from the author. The second program, referred to here as PWRS (Pocklington's wire radiation and scattering) is based on Pocklington's integral equation of Section 12.4.1, and it is used for both radiation and scattering by a perfect electric conducting (PEC) wire. The listing of this program is found at the end of this chapter.

The remaining three programs are more general, public domain moment method programs. A very brief description of these programs is given here. Information as to where these programs can be obtained is also included. It should be stated, however, that there are numerous other codes, public domain and restricted, that utilize moment method and other techniques, such as geometrical optics, geometrical theory of diffraction, physical optics, and physical theory of diffraction, which are too numerous to mention here.

12.5.1 Two-Dimensional Radiation and Scattering

The two-dimensional radiation and scattering (TDRS) program is used to analyze four different two-dimensional perfectly electric conducting problems: the strip, and the circular, elliptical, and rectangular cylinders. The algorithm is based on the electric field integral equation of Section 12.3.1, and it is used for both electric and magnetic line-source excitation, or plane wave incidence of arbitrary polarization. For simplicity, piecewise constant pulse expansion functions and point-matching techniques have been adopted. The listing of this program is included in the solutions manual available to the instructors. It can also be obtained from the author.

A. STRIP

For the strip problem, the program can analyze either of the following:

1. A line source (electric or magnetic). It computes the electric current density over the width of the strip and the normalized radiation amplitude pattern (in decibels) for $0° \leq \phi \leq 360°$. The user must specify the width of the strip (in wavelengths), the type of line source (either electric or magnetic), and the location x_s, y_s of the source (in wavelengths).
2. Plane wave incidence of arbitrary polarization. The program can analyze either monostatic or bistatic scattering.

For monostatic scattering the program computes the two-dimensional normalized (with respect to λ) monostatic SW $\sigma_{2\text{-D}}/\lambda$ (in decibels) for all angles of incidence ($0° \leq \phi \leq 360°$). The program starts at $\phi = 0°$ and then completes the entire 360° monostatic scattering pattern. The user must specify the width w of the strip (in wavelengths) and the polarization angle θ_p (in degrees) of the incident plane wave. The polarization of the incident wave is specified by the direction θ_p of the incident electric field relative to the z axis ($\theta_p = 0°$ implies TM^z; $\theta_p = 90°$ implies TE^z; otherwise arbitrary polarization). The polarization angle θ_p needs to be specified only when the polarization is neither TM^z nor TE^z.

For bistatic scattering, the program computes for the specified incidence angle the current density over the width of the strip and the two-dimensional normalized (with respect to λ) bistatic SW $\sigma_{2\text{-D}}/\lambda$ (in decibels) for all angles of observation ($0° \leq \phi_s \leq 360°$). The user must specify the width w of the strip (in wavelengths), the angle of incidence ϕ_i (in degrees), and the polarization angle θ_p (in degrees) of the incident plane wave. The polarization angle of the incident wave is specified in the same manner as for the monostatic case.

B. CIRCULAR, ELLIPTICAL, OR RECTANGULAR CYLINDER

For the cylinder program, the program can analyze either a line source (electric or magnetic) or plane wave scattering of arbitrary polarization by a two-dimensional circular, or elliptical or rectangular cylinder.

1. For the line source excitation, the program computes the current distribution over the entire surface of the cylinder and the normalized radiation amplitude pattern (in decibels). The user must specify for each cylinder the type of line source (electric or magnetic), the location x_s, y_s of the line source, and the size of the cylinder. For the circular cylinder the size is specified by its radius a (in wavelengths) and for the elliptical and rectangular cylinders by the principal semiaxes lengths a and b (in wavelengths), with a measured along the x axis and b along the y axis.
2. For the plane wave incidence the program computes monostatic or bistatic scattering of arbitrary polarization by a circular, elliptical, or rectangular cylinder.

For monostatic scattering the program computes the two-dimensional normalized (with respect to λ) monostatic SW $\sigma_{2\text{-D}}/\lambda$ (in decibels) for all angles of incidence ($0° \leq \phi \leq 360°$). The program starts at $\phi = 0°$ and then computes the entire 360° monostatic scattering pattern. The user must specify the size of the cylinder, as was done for the line-source excitation, and the polarization angle θ_p (in degrees) of the incident plane wave. The polarization of the incident wave is specified by the direction θ_p of the incident electric field relative to the z axis ($\theta_p = 0°$ implies TM^z; $\theta_p = 90°$ implies TE^z; otherwise arbitrary polarization). The polarization angle θ_p needs to be specified only when the polarization is neither TM^z nor TE^z.

For bistatic scattering, the program computes for the specified incidence angle the current density over the entire surface of the cylinder and the two-dimensional normalized (with respect to λ) bistatic SW $\sigma_{2\text{-D}}/\lambda$ (in decibels) for all angles of observation ($0° \leq \phi_s \leq 360°$). The user must specify the size of the cylinder, as was done for the line-source excitation, the incidence angle ϕ_i (in degrees), and the polarization angle θ_p (in degrees) of the incident plane wave. The polarization angle of the incident wave is specified in the same manner as for the monostatic case.

12.5.2 Pocklington's Wire Radiation and Scattering

Pocklington's wire radiation and scattering (PWRS) program computes the radiation characteristics of a center-fed wire antenna and the scattering characteristics of a perfectly electric conducting (PEC) wire, each of radius a and length ℓ. Both are based on Pocklington's integral equation 12-123.

A. RADIATION

For the wire antenna of Figure 12-25 the excitation is modeled by either a delta gap or a magnetic frill feed modeling, and it computes the current distribution, normalized amplitude radiation pattern, and the input impedance. The user must specify the length of the wire, its radius (both in wavelengths), and the type of feed modeling (delta gap or magnetic frill). A computer program based on Hallén's integral equation can be found at the end of Chapter 7 of [24].

B. SCATTERING

The geometry for the plane wave scattering by the wire is shown in Figure 12-23(a). The program computes the monostatic or bistatic scattering of arbitrary polarization.

For monostatic scattering the program computes the normalized (with respect to m²) RCS $\sigma_{3\text{-D}}/m^2$ (in dBsm) for all angles of incidence ($0° \leq \theta_i \leq 180°$). The program starts at $\theta_i = 0°$ and then computes the entire 180° monostatic scattering pattern. The user must specify the length and radius of the wire (both in wavelengths) and the polarization angle θ_p (in degrees) of the incident plane wave. The polarization of the incident wave is specified by the direction θ_p of the incident electric field relative to the plane of incidence, where the plane of incidence is defined as the plane that contains the vector of the incident wave and the wire scatterer ($\theta = 0°$ implies that the electric field is on the plane of incidence; $\theta = 90°$ implies that the electric field is perpendicular to the plane of incidence and to the wire; thus no scattering occurs for this case).

For bistatic scattering, the program computes for the specified incidence angle the current distribution over the length of the wire and the normalized (with respect to m²) bistatic RCS $\sigma_{3\text{-D}}/m^2$ (in dBsm) for all angles of observation ($0° \leq \theta_s \leq 180°$). The user must specify the length and radius of the wire (both in wavelengths), the angle of incidence θ_i (in degrees), and the polarization angle θ_p of the incident plane wave. The polarization angle is specified in the same manner as for the monostatic case.

12.5.3 Numerical Electromagnetics Code

The numerical electromagnetics code (NEC) [38] is a user-oriented program developed at Lawrence Livermore Laboratory. It is a moment method code for analyzing the interaction of electromagnetic waves with arbitrary structures consisting of conducting wires and surfaces. It combines an integral equation for smooth surfaces with one for wires to provide convenient and accurate modeling for a wide range of applications. The code can model nonradiating networks and transmission lines, perfect and imperfect conductors, lumped element loading, and perfect and imperfect conducting ground planes. It uses the electric field integral equation (EFIE) for thin wires and the magnetic field integral equation (MFIE) for surfaces. The excitation can be either an applied voltage source or an incident plane wave. The program computes induced currents and charges, near- or far-zone electric and magnetic fields, radar cross section, impedances or admittances, gain and directivity, power budget, and antenna to antenna coupling.

Information concerning the code and its availability can be directed to:

>Professor Richard W. Adler
>Naval Postgraduate School
>Code 62 AB
>Monterey, California 93943

12.5.4 Mini-Numerical Electromagnetics Code

The mini-numerical electromagnetics code (MININEC) [39, 40] is a user-oriented compact version of NEC developed at the Naval Ocean Systems Center. It is also a moments method code, but coded in BASIC, and has retained the most frequently used options of NEC. It is intended to be used in mini, micro, and personal computers, and it is most convenient to analyze wire antennas. It computes currents, and near- and far-field patterns. It also optimizes the feed excitation voltages that yield a desired radiation patterns.

Information concerning the MININEC and its availability can be directed to:

>Professor Richard W. Adler
>Naval Postgraduate School
>Code 62 AB
>Monterey, California 93943

or

>Artech House, Inc.
>685 Canton Street
>Norwood, Massachusetts 02062

12.5.5 Electromagnetic Surface Patch Code

The electromagnetic surface patch (ESP) [41] code is a user-oriented program developed at the ElectroScience Laboratory at Ohio State University. It is a moment method, surface patch code based on the piecewise sinusoidal reaction formulation, which is basically equivalent to the electric field integral equation (EFIE). It can treat (perfectly conducting or thin dielectric) geometries that consist of thin wires, rectangular or polygonal plates, wire-plate or plate-plate attachments, and open or closed surfaces. The excitation can be either by a delta gap voltage generator or an incident plane wave. The program computes current distribution, input impedance, radiation efficiency, mutual coupling, near- or far-field gain patterns, and near- or far-field radar cross section patterns. ESP also incorporates an efficient impedance matrix interpolation scheme for obtaining data over a wide frequency bandwidth [42].

Information concerning the code and its availability can be directed to:

> Dr. Edward H. Newman
> The Ohio State University
> ElectroScience Laboratory
> 1320 Kinnear Road
> Columbus, Ohio 43212

REFERENCES

1. R. F. Harrington, "Matrix methods for field problems," *Proc. IEEE*, vol. 55, no. 2, pp. 136–149, February 1967.
2. R. F. Harrington, *Field Computation By Moment Methods*, Macmillan, New York, 1968.
3. J. H. Richmond, "Digital computer solutions of the rigorous equations for scattering problems," *Proc. IEEE*, vol. 53, pp. 796–804, August 1965.
4. L. L. Tsai, "Moment methods in electromagnetics for undergraduates," *IEEE Trans. on Education*, vol. E-21, no. 1, pp. 14–22, February 1978.
5. R. Mittra (Ed.), *Computer Techniques for Electromagnetics*, Pergamon, New York, 1973.
6. J. Moore and R. Pizer, *Moment Methods in Electromagnetics*, Wiley, New York, 1984.
7. P. Y. Ufimtsev, "Method of edge waves in the physical theory of diffraction," translated by U.S. Air Force Foreign Technology Division, Wright-Patterson AFB, Ohio, September 1971.
8. P. Y. Ufimtsev, "Approximate computation of the diffraction of plane electromagnetic waves at certain metal bodies," *Sov. Phys.–Tech. Phys.*, vol. 27, pp. 1708–1718, 1957.
9. P. Y. Ufimtsev, "Secondary diffraction of electromagnetic waves by a strip," *Sov. Phys.–Tech. Phys.*, vol. 3, pp. 535–548, 1958.
10. K. M. Mitzner, "Incremental length diffraction coefficients," Tech. Rep. AFAL-TR-73-296, Northrop Corp., Aircraft Division, April 1974.
11. E. F. Knott and T. B. A. Senior, "Comparison of three high-frequency diffraction techniques," *Proc. IEEE*, vol. 62, no. 11, pp. 1468–1474, November 1974.
12. E. F. Knott, "A progression of high-frequency RCS prediction techniques," *Proc. IEEE*, vol. 73, no. 2, pp. 252–264, February 1985.
13. T. Griesser and C. A. Balanis, "Backscatter analysis of dihedral corner reflectors using physical optics and the physical theory of diffraction," *IEEE Trans. Antennas Propagat.*, vol. AP-35, no. 10, pp. 1137–1147, October 1987.
14. J. B. Keller, "Diffraction by an aperture," *J. Appl. Phys.*, vol. 28, no. 4, pp. 426–444, April 1957.
15. J. B. Keller, "Geometrical theory of diffraction," *J. Opt. Soc. Amer.*, vol. 52, no. 2, pp. 116–130, February 1962.
16. R. G. Kouyoumjian and P. H. Pathak, "A uniform geometrical theory of diffraction for an edge in a prefectly conducting surface," *Proc. IEEE*, vol. 62, no. 11, pp. 1448–1461, November 1974.
17. G. L. James, *Geometrical Theory of Diffraction for Electromagnetic Waves*, Third Edition Revised, Peregrinus, London, 1986.
18. J. D. Lilly, "Application of the moment method to antenna analysis," MSEE Thesis, Department of Electrical Engineering, West Virginia University, 1980.
19. J. D. Lilly and C. A. Balanis, "Current distributions, input impedances, and radiation patterns of wire antennas," North American Radio Science Meeting of URSI, Université Laval, Quebec, Canada, June 2–6, 1980.
20. D. K. Cheng, *Field and Wave Electromagnetics*, Addison-Wesley, Reading, Mass., 1983, p. 88.

21. R. Mittra and C. A. Klein, "Stability and convergence of moment method solutions," in *Numerical and Asymptotic Techniques in Electromagnetics*, R. Mittra (Ed.), Springer-Verlag, New York, 1975, Chapter 5, pp. 129–163.
22. T. K. Sarkar, "A note on the choice weighting functions in the method of moments," *IEEE Trans. Antennas Propagat.*, vol. AP-33, no. 4, pp. 436–441, April 1985.
23. T. K. Sarkar, A. R. Djordjević, and E. Arvas, "On the choice of expansion and weighting functions in the numerical solution of operator equations," *IEEE Trans. Antennas Propagat.*, vol. AP-33, no. 9, pp. 988–996, September 1985.
24. C. A. Balanis, *Antenna Theory: Analysis and Design*, Wiley, New York, 1982.
25. E. K. Miller and F. J. Deadrick, "Some computational aspects of thin-wire modeling," in *Numerical and Asymptotic Techniques in Electromagnetics*, R. Mittra (Ed.), Springer-Verlag, New York, 1975, Chapter 4, pp. 89–127.
26. L. Kantorovich and G. Akilov, *Functional Analysis in Normed Spaces*, Pergamon, Oxford, pp. 586–587, 1964.
27. D. P. Marsland, C. A. Balanis, and S. Brumley, "Higher order diffractions from a circular disk," *IEEE Trans. Antennas Propagat.*, vol. AP-35, no. 12, pp. 1436–1444, December 1987.
28. H. C. Pocklington, "Electrical oscillations in wire," *Cambridge Philos. Soc. Proc.*, vol. 9, pp. 324–332, 1897.
29. E. Hallén, "Theoretical investigations into the transmitting and receiving qualities of antennae," *Nova Acta Regiae Soc. Sci. Upsaliensis*, Ser. IV, no. 4, pp. 1–44, 1938.
30. R. King and C. W. Harrison, Jr., "The distribution of current along a symmetrical center-driven antenna," *Proc. IRE*, vol. 31, pp. 548–567, October 1943.
31. J. H. Richmond, "A wire-grid model for scattering by conducting bodies," *IEEE Trans. Antennas Propagat.*, vol. AP-14, no. 6, pp. 782–786, November 1966.
32. G. A. Thiele, "Wire antennas," in *Computer Techniques for Electromagnetics*, R. Mittra (Ed.), Pergamon, New York, Chapter 2, pp. 7–70, 1973.
33. C. M. Butler and D. R. Wilton, "Evaluation of potential integral at singularity of exact kernel in thin-wire calculations," *IEEE Trans. Antennas Propagat.*, vol. AP-23, no. 2, pp. 293–295, March 1975.
34. L. W. Pearson and C. M. Butler, "Inadequacies of collocation solutions to Pocklington-type models of thin-wire structures," *IEEE Trans. Antennas Propagat.*, vol. AP-23, no. 2, pp. 293–298, March 1975.
35. C. M. Butler and D. R. Wilton, "Analysis of various numerical techniques applied to thin-wire scatterers," *IEEE Trans. Antennas Propagat.*, vol. AP-23, no. 4, pp. 534–540, July 1975.
36. D. R. Wilton and C. M. Butler, "Efficient numerical techniques for solving Pocklington's equation and their relationships to other methods," *IEEE Trans. Antennas Propagat.*, vol. AP-24, no. 1, pp. 83–86, January 1976.
37. L. L. Tsai, "A numerical solution for the near and far fields of an annular ring of magnetic current," *IEEE Trans. Antennas Propagat.*, vol. AP-20, no. 5, pp. 569–576, September 1972.
38. G. J. Burke and A. J. Poggio, "Numerical electromagnetics code (NEC)–method of moments," Technical Document 116, Naval Ocean Systems Center, San Diego, Calif., January 1981.
39. A. J. Julian, J. M. Logan, and J. W. Rockway, "MININEC: A mini-numerical electromagnetics code," Technical Document 516, Naval Ocean Systems Center, San Diego, Calif., September 6, 1982.
40. J. Rockway, J. Logan, D. Tam, and S. Li, *The MININEC SYSTEM: Microcomputer Analysis of Wire Antennas*, Artech House, Inc., 1988.
41. E. H. Newman and D. L. Dilsavor, "A user's manual for the electromagnetic surface patch code: ESP version III," Technical Report No. 716148-19, ElectroScience Laboratory, The Ohio State University, May 1987.
42. E. H. Newman, "Generalization of wideband data from the method of moments by interpolating the impedance matrix," *IEEE Trans. Antennas Propagat.*, to be published.

PROBLEMS

12.1. A circular loop of radius $a = 0.2$ m is constructed out of a wire of radius $b = 10^{-3}$ m, as shown in Figure P12-1. The entire loop is maintained at a constant potential of 1 V. Using integral equation techniques, determine and plot for $0° \leq \phi \leq 360°$ the surface charge density on the wire. Assume that at any given angle the charge is uniformly distributed along the circumference of the wire.

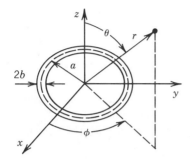

FIGURE P12-1

12.2. Repeat Problem 12.1 when the loop is split into two parts; one part (from 0 to 180°) is maintained at a constant potential of 1 V and the other part (from 180 to 360°) is maintained at a constant potential of 2 V.

12.3. Repeat Example 12.3 for a strip with $w = 2\lambda$, $h = 0.25\lambda$, and $t = 0.01\lambda$.

12.4. An infinite electric line source of constant current I_e is placed next to a circular conducting cylinder of radius a, as shown in Figure P12-4. The line source is positioned a distance b ($b > a$) from the center of the cylinder. Use the EFIE, piecewise constant subdomain basis functions, and point-matching techniques.

(a) Formulate the problem current density induced on the surface of the cylinder.
(b) Compute the induced current density when $a = 5\lambda$ and $b = 5.25\lambda$. Assume a unity line-source current. Compare with the modal solution current density of (11-168a).
(c) Compute for part b the normalized far-zone amplitude pattern (in decibels). Normalize so that the maximum is 0 dB.

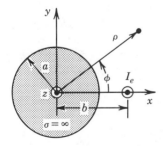

FIGURE P12-4

12.5. Repeat Problem 12.4 for $a = 5\lambda$ and $b = 5.5\lambda$.

12.6. An infinite electric line source of constant current I_e is placed next to a rectangular cylinder of dimensions a and b, as shown in the Figure P12-6. The line source is positioned a distance c ($c > a$) from the center of the cylinder along the x axis. Use

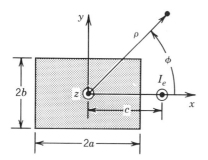

FIGURE P12-6

the EFIE and piecewise subdomain basis functions and point-matching techniques, and do the following.

(a) Compute the induced current on the surface of the cylinder when $a = 5\lambda$, $b = 2.5$, and $c = 5.25\lambda$. Assume a unity line-source current.
(b) Compute for part a the normalized far-zone amplitude pattern (in decibels). Normalize so that the maximum is 0 dB.

12.7. Repeat Problem 12.6 for an electric line source near an elliptic cylinder with $a = 5\lambda$, $b = 2.5\lambda$, and $c = 5.25\lambda$.

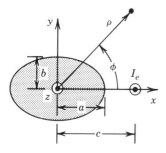

FIGURE P12-7

12.8. A TM^z uniform plane wave traveling in the $+x$ direction is normally incident upon a conducting circular cylinder of radius a, as shown in Figure P12-8. Use the EFIE, piecewise constant subdomain basis functions, and point-matching techniques, write your own program, and do the following.

(a) Plot the current density induced on the surface of the cylinder when $a = 2\lambda$. Assume the incident field is of unity amplitude. Compare with the modal solution current density of (11-97).
(b) Plot the normalized $\sigma_{2\text{-D}}/\lambda$ bistatic scattering width (in decibels) for $0° \leq \phi \leq 360°$ when $a = 2\lambda$. Compare with the modal solution of (11-102).

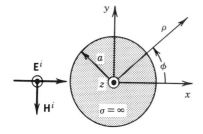

FIGURE P12-8

12.9. A TM^z uniform plane wave traveling in the $+x$ direction is normally incident upon a conducting rectangular cylinder of dimensions a and b, as shown in Figure P12-9. Use the EFIE, piecewise constant subdomain basis functions and point-matching techniques, write your own program, and do the following.

(a) Compute the induced current density on the surface of the cylinder when $a = 5\lambda$ and $b = 2.5\lambda$. Assume a unity line-source current.

(b) Compute and plot for part a the two-dimensional normalized $\sigma_{\text{2-D}}/\lambda$ bistatic scattering width (in decibels) for $0° \leq \phi \leq 360°$.

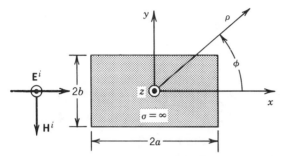

FIGURE P12-9

12.10. Repeat Problem 12.9 for a TM^z uniform plane wave impinging upon an elliptic conducting cylinder with $a = 5\lambda$ and $b = 2.5\lambda$.

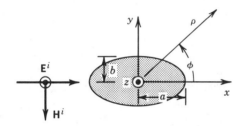

FIGURE P12-10

12.11. Using the geometry of Figure 12-14 verify (12-71a) and (12-71b), and that (12-70a) reduces to (12-72a) and (12-70b) to (12-72b).

12.12. Show that the integral of (12-78d) can be evaluated using (12-79a) through (12-79c).

12.13. Repeat Problem 12.4 for a magnetic line source of constant current $I_m = 1$ when $a = b = 5\lambda$. This problem is representative of a very thin axial slot on the surface of the cylinder. Compare the current density on the surface of the cylinder from part b with that of the modal solution of (11-177a).

12.14. Repeat Problem 12.6 for a magnetic line source of constant current $I_m = 1$ when $a = 5\lambda$, $b = 2.5\lambda$, and $c = 5\lambda$. This problem is representative of a very thin axial slot on the surface of the cylinder.

12.15. Repeat Problem 12.7 for a magnetic line source of constant current $I_m = 1$ when $a = 5\lambda$, $b = 2.5\lambda$, and $c = 5\lambda$. This problem is representative of a very thin axial slot on the surface of the cylinder.

12.16. Repeat Problem 12.8 for a TE^z uniform plane wave of unity amplitude. Compare the current density with the modal solution of (11-113) and the normalized $\sigma_{\text{2-D}}/\lambda$ bistatic scattering width with the modal solution of (11-117).

12.17. Repeat Problem 12.9 for a TE^z uniform plane wave at unity amplitude.

12.18. Repeat Problem 12.10 for a TE^z uniform plane wave of unity amplitude.

12.19. Using the geometry of Figure 12-17 show that
$$\hat{n} \cdot \nabla H_0^{(2)}(\beta R) = -\beta \cos \psi H_1^{(2)}(\beta R)$$

12.20. Repeat Problem 12.4 using the MFIE.

12.21. Repeat Problem 12.8 using the MFIE. You must write your own computer program to solve this problem.

12.22. Using the geometry of Figure 12-18 show that
$$\hat{n}' \cdot \nabla H_0^{(2)}(\beta R) = -\beta \cos \psi' H_1^{(2)}(\beta R)$$

12.23. Repeat Problem 12.13 using the MFIE.

12.24. Derive Pocklington's integral equation 12-123 using (12-121) and (12-122).

12.25. Derive the solution of (12-125) to the differential equation of (12-124a). Show that Hallén's integral equation can be written as (12-126).

12.26. Show that the incident tangential electric field (E_z^i) generated on the surface of a wire of radius a by a magnetic field generator of (12-129) is given by (12-130).

12.27. Reduce (12-130) to (12-131) valid only along the z axis ($\rho = 0$).

12.28. For the center-fed dipole of Example 12-7, write the $[Z]$ matrix for $N = 21$ using for the gap the delta-gap generator and the magnetic frill generator.

12.29. For an infinitesimal center fed dipole of $\ell = \lambda/50$ of radius $a = 0.005\lambda$ derive the input impedance using Pocklington's integral equation with piecewise constant subdomain basis functions and point matching. Use $N = 21$ and model the gap as a delta-gap generator and as a magnetic-frill generator. Use the PWRS computer program at the end of the chapter.

12.30. A conducting wire of length $\ell = 0.47\lambda$ and radius $a = 0.005\lambda$ is placed symmetrically along the z axis. Assuming a TM^z uniform plane wave is incident on the wire at an angle $\theta_i = 30°$ from the z axis, do the following.
   (a) Compute and plot the current induced on the surface of the wire.
   (b) Compute and plot the bistatic RCS for $0° \leq \theta_s \leq 180°$.
   (c) Compute and plot the monostatic RCS for $0° \leq \theta_i = \theta_s \leq 180°$.
   The amplitude of the incident electric field is 10^{-3} V/m. Use Pocklington's integral equation and the PWRS computer program. Determine the number of segments that lead to a stable solution.

12.31. Repeat Problem 12.30 for a TE^z uniform plane wave incidence.

COMPUTER PROGRAM PWRS

```
C***********************************************************************
C    THIS PROGRAM USES POCKLINGTON'S INTEGRAL EQUATIONS  OF ( 12 - 123 )*
C    ON A SYMMETRICAL DIPOLE, AND IT COMPUTES THE CURRENT DISTRIBTUION, *
C    INPUT IMPEDANCE, NORMALIZED AMPLITUDE RADIATION PATTERN, AND       *
C    SCATTERING PATTERNS                                                *
C                                                                       *
C    GEOMETRY                                                           *
C    HL  —  HALF OF THE DIPOLE LENGTH (IN WAVELENGTHS)                  *
C    RA  —  RADIUS OF THE WIRE (IN WAVELENGTHS)                         *
C    NM  —  TOTAL NUMBER OF SUBSECTIONS (MUST BE AN ODD INTEGER)        *
C                                                                       *
C    IOPT— OPTIONS TO USE POCKLINGTON'S FORMULATION OF ( 12 - 123 )     *
C    TO SOLVE THE WIRE ANTENNA PROBLEM OR THE WIRE SCATTERER PROBLEM.   *
C                                                                       *
C        IOPT=1 :  ANTENNA PROBLEM                                      *
C        IOPT=2 :  WIRE SCATTERING PROBLEM                              *
C           *** IGNORE OPTION ISCAT WHEN IOPT=1 ***                     *
C                  ISCAT=1, MONOSTATIC RADAR CROSS SECTON               *
C                  ISCAT=2, BISTATIC RADAR CROSS SECTION                *
C                          NEEDS TO SPECIFY INCIDENT ANGLE              *
C    IEX  —  OPTION TO USE EITHER MAGNETIC-FRILL GENERATOR OR DELTA GAP *
C           *** IGNORE OPTION IEX WHEN IOPT=2 ***                       *
C        IEX =1 :  MAGNETIC-FRILL GENERATOR                             *
C        IEX =2 :  DELTA GAP                                            *
C                                                                       *
C           *** IGNORE POLRD, AND THETD WHEN IOPT=1 ***                 *
C    POLRD— ELECTRIC FIELD POLARIZATION RELATIVE TO THE PLANE OF        *
C           INCIDENCE, WHICH IS DEFINED AS THE PLANE CONTAINING THE     *
C           INCIDENT WAVE VECTOR AND THE WIRE SCATTERER.  REFER TO      *
C           FIGURE 12 - 23(A) FOR THE SCATTERER'S GEOMETRY(IN DEGREES). *
C    THETD— THE INCIDENT ANGLE RELATIVE TO THE Z-AXIS (IN DEGREES).     *
C           *** THETD IS NEEDED FOR BISTATIC CASE ONLY ***              *
C                                                                       *
C    THIS PROGRAM USES PULSE EXPANSION FOR THE ELECTRIC CURRENT MODE    *
C    AND POINT-MATCHING THE ELECTRIC FIELD AT THE CENTER OF EACH        *
C    WIRE SEGMENT.                                                      *
C                                                                       *
C***********************************************************************
C   EXAMPLE A: HOW TO SPECIFY THE NUMBER OF SUBSECTIONS OF THE ANTENNA
C              OR SCATTERER.  NM=21 FOR THIS EXAMPLE.
      PARAMETER  ( NM=21, NMT=2*NM-1 )
      COMMON/SIZE/HL,RA,DZ,ZM,ZN,NMH
      COMMON/CONST/BETA,ETA,RAD,J
      COMPLEX ZMN(NMT),WA(NMT),CGA(NM),ZIN,J,CRT
      DIMENSION INDEX(NM),ETMM(181)
      EXTERNAL CGP
      DATA POLRD,THETD,ISCAT/0.0,0.0,0/
C   EXAMPLE B: HOW TO SPECIFY THE ANTENNA PROBLEM USING MAGNETIC FRILL
C      IOPT=1
C      IEX=1
C   EXAMPLE C: HOW TO SPECIFY THE ANTENNA PROBLEM USING DELTA GAP
C      IOPT=1
C      IEX=2
C   EXAMPLE D: HOW TO SPECIFY THE MONOSTATIC SCATTERING PROBLEM
C              AT A POLARIZATION ANGLE OF 45 DEGREES
C      IOPT=2
C      ISCAT=1
C      POLRD=45.0
C   EXAMPLE E: HOW TO SPECIFY THE BISTATIC SCATTERING PROBLEM
C              AT A POLARIZATION ANGLE OF 45 DEGREES,
C              AND AN INCIDENT ANGLE OF 60 DEGREES
C      IOPT=2
C      ISCAT=2
```

```
C       POLRD=45.0
C       THETD=60.0
C    THE PRESET EXAMPLE HERE IS THE WIRE ANTENNA PROBLEM OF MAGNETIC
C    FRILL GENERATOR MODEL
        IOPT=1
C       IEX=1
C.. GEOMETRY DATA
C.. EXAMPLE: DIPOLE HALF LENGTH  OF 0.235 WAVELENGTHS
C..            AND WIRE RADIUS OF 0.005 WAVELENGTHS.
        HL=.235
        RA=.005
C.. SOME CONSTANTS
        PI=3.14159265
        RAD =PI/180.
        BETA=2.0*PI
        ETA =120.*PI
        NMH=0.5*(NM+1)
        J=CMPLX(0.,1.)
        DZ=2.*HL/NM
        IF(IOPT.EQ.1) THEN
          WRITE(6,50) HL,RA
          IF(IEX.EQ.1) WRITE(6,100)
          IF(IEX.EQ.2) WRITE(6,102)
        ELSE
          WRITE(6,52) HL,RA
        ENDIF
        WRITE(6,54) NM
C.. THE IMPEDANCE MATRIX HAS A TOEPLITZ PROPERTY, THEREFORE ONLY
C.. NM ELEMENTS NEED TO BE COMPUTED, AND THE MATRIX IS FILLED IN
C.. A FORM THAT CAN BE SOLVED BY A TOEPLITZ MATRIX SOLVING SUBROUTINE
        ZM=HL-0.5*DZ
        B=0.5*DZ
        A=-0.5*DZ
        DO 4 I=1,NM
          ZN=HL-(I-0.5)*DZ
          CALL CSINT(CGP,A,B,79,CRT)
          ZMN(I)=CRT
          IF(I.EQ.1) GOTO 4
          ZMN(NM+I-1)=CRT
    4   CONTINUE
        IF(IOPT.EQ.2.AND.ISCAT.EQ.1) GOTO 60
        IF(IOPT.EQ.1) THEN
          RB=2.3*RA
          TLAB=2.*ALOG(2.3)
          DO 10 I=1,NM
          ZI=HL-(I-0.5)*DZ
          R1=BETA*SQRT(ZI*ZI+RA*RA)
          R2=BETA*SQRT(ZI*ZI+RB*RB)
          IF(IEX.EQ.1) THEN
            CGA(I)=-J*BETA**2/(ETA*TLAB)*(CEXP(-J*R1)/R1-CEXP(-J*R2)/R2)
          ELSE
          IF(IEX.NE.2) WRITE(6,999)
            IF(I.NE.NMH) THEN
              CGA(I)=0.
            ELSE
              CGA(I)=-J*BETA/(ETA*DZ)
            ENDIF
          ENDIF
   10     CONTINUE
        CALL TSLZ(ZMN,CGA,WA,NM)
C.. OUTPUT THE CURRENT DISTRIBUTION ALONG OF THE DIPOLE
```

```
      WRITE(6,104)
      DO 12 I=1,NMH
        XI=HL-(I-.5)*DZ
        YI=CABS(CGA(I))
        WRITE(6,106)I, XI,CGA(I),YI
 12   CONTINUE
C..   COMPUTATION OF THE INPUT IMPEDANCE
      ZIN=1./CGA(NMH)
      WRITE(6,108) ZIN
C..   COMPUTATION OF AMPLITUDE RADIATION PATTERN OF THE ANTENNA
      CALL PATN(CGA,NM,ETMM,IOPT)
      WRITE(6,110)
      DO 14 I=1,181
        XI=I-1.
 14     WRITE(6,112)  XI,ETMM(I)
      ELSE
C..   WIRE SCATTERER PROBLEM, BISTATIC CASE
        IF(IOPT.NE.2) WRITE(6,999)
        IF(ISCAT.NE.2) WRITE(6,999)
        CTH=COS(THETD*RAD)
        STH=SIN(THETD*RAD)
        CSPL=COS(POLRD*RAD)
        DO 15 I=1,NM
        ZI=HL- (I-0.5)*DZ
 15     CGA(I)=-J*BETA/ETA*CSPL*STH*CEXP(J*BETA*ZI*CTH)
C..   NOW SOLVE FOR THE CURRENT DISTRIBUTION
        CALL TSLZ(ZMN,CGA,WA,NM)
C..   COMPUTE THE PATTERN
        WRITE(6,120)
        WRITE(6,122) THETD,POLRD
        DO 20 I=1,NM
        XI=HL-(I-0.5)*DZ
        YI=CABS(CGA(I))
        WRITE(6,124) XI,CGA(I),YI
 20     CONTINUE
C..   COMPUTATION OF BISTATIC RCS  PATTERNS
        CALL PATN(CGA,NM,ETMM,IOPT)
        WRITE(6,126)
        DO 40 I=1,181
        XI=I-1.
        WRITE(6,128)   XI,ETMM(I)
 40     CONTINUE
      ENDIF
      GOTO 200
C..   THE MONOSTATIC CASE
 60   WRITE(6,140)
      DO 70 M=1,91
      THETA=(M-1.)*RAD
        CTH=COS(THETA)
        STH=SIN(THETA)
        CSPL=COS(POLRD*RAD)
        DO 62 I=1,NM
        ZI=HL-(I-0.5)*DZ
 62     CGA(I)=-J*BETA/ETA*CSPL*STH*CEXP(J*BETA*ZI*CTH)
      CALL TSLZ(ZMN,CGA,WA,NM)
      IF(ABS(CTH).LE.1.E-3) THEN
        FT=1.
      ELSE
        FT=SIN(BETA*DZ*CTH*.5)/(BETA*DZ*CTH*.5)
      ENDIF
      CRT=0.
      DO 64 I=1,NM
        ZI=HL-(I-.5)*DZ
        CRT=CRT+CEXP(J*BETA*ZI*CTH)*FT*CGA(I)*DZ
```

```
   64   CONTINUE
        PTT=CABS(CRT)*STH*STH*ETA*0.5
        PTT=PTT*PTT*BETA*2.
        IF(PTT.LE.1.E-10) PTT=1.E-10
        PTT=10.*ALOG10(PTT)
        ETMM(M)=PTT
   70   ETMM(182-M)=PTT
        WRITE(6,142) POLRD
        WRITE(6,144)
        DO 72 I=1,181
        XI=I-1
   72   WRITE(6,146) XI,ETMM(I)
   50   FORMAT(15X,'WIRE ANTENNA PROBLEM'//5X,'LENGTH = 2 X ',F6.4,
       +' (WLS)', 4X,'RADIUS OF THE WIRE=',F6.4,' (WLS) '/)
   52   FORMAT(15X,'WIRE SCATTERER PROBLEM'//5X,'LENGTH = 2 X ',F6.4,
       +' (WLS)', 4X,'RADIUS OF THE WIRE=',F6.4,' (WLS) '/)
   54   FORMAT(15X,'NUMBER OF SUBSECTIONS = ', I3/)
  100   FORMAT(5X,'POCKLINGTON''S EQUATION AND MAGNETIC FRILL MODEL'/)
  102   FORMAT(5X,'POCKLINGTON''S EQUATION AND DELTA GAP MODEL'/)
  104   FORMAT(10X,'CURRENT DISTRIBUTION ALONG ONE HALF OF THE DIPOLE'/
       +8X,'POSITION Z',3X,'REAL PART',3X,'IMAGINARY ',3X,'MAGNITUDE'/)
  106   FORMAT(3X,I3,4X,F6.4,5X,F9.6,3X,F9.6,3X,F9.6)
  108   FORMAT(/3X,'INPUT IMPEDANCE = ',F7.1,'+ J',F7.1,' (OHMS)')
  110   FORMAT(/3X,'RADIATION POWER PATTERN VS OBSERVATION ANGLE THETA'//
       +3X,'THETA (IN DEGREES)',2X,'POWER (IN DB)')
  112   FORMAT(8X,F6.1,8X,F8.2)
  120   FORMAT(4X,'BISTATIC WIRE SCATTERER PROBLEM WITH POCKLINGTON''S',
       +' EQUATION'//)
  122   FORMAT(8X,'INCIDENT ANGLE=',F5.1,' DEGREES, POLARIZATION=',F5.1,
       +' DEGREEES'//10X,'CURRENT DISTRIBUTION ALONG THE DIPOLE'//
       +8X,'POSITION Z',3X,'REAL PART',3X,'IMAGINARY ',3X,'MAGNITUDE'/)
  124   FORMAT(10X,F6.4,5X,F9.6,3X,F9.6,3X,F9.6)
  126   FORMAT(/4X,'BISTATIC RADAR CROSS SECTION PATTERN VS OBSERVATION',
       +' ANGLE THETA '//8X,'THETA(IN DEGREES) ',2X,'RCS (IN DBSM)')
  128   FORMAT(12X,F6.1,8X,F10.2)
  140   FORMAT(4X,'MONOSTATIC WIRE SCATTERER PROBLEM WITH POCKLINGTON''S',
       +' EQUATION'/)
  142   FORMAT(4X,'THE POLARIZATION TO THE PLANE OF INCIDENCE = ',F5.1,
       +' DEGREES '/)
  144   FORMAT(14X,'THE MONOSTATIC RADAR CROSS SECTION PATTERN'
       +//10X,'    INCIDENT ANGLE (THETA)',5X,' RCS IN (DBSM)')
  146   FORMAT(20X,F7.2,15X,F10.2)
  999   FORMAT(5X,'******WARNING: NO SUCH OPTION. CHOOSE A VALID OPTION'/
       +20X,'AND TRY AGAIN.')
  200   STOP
        END
C..
        COMPLEX FUNCTION CGP(Z)
C.. POCKLINGTON'S KERNEL
        COMMON/SIZE/HL,RA,DZ,ZM,ZN,NMH
        COMMON/CONST/BETA,ETA,RAD,J
        COMPLEX J
        Z1=ZN-ZM +Z
        R=SQRT(RA*RA+Z1*Z1)
        CGP=CEXP(-J*BETA*R)*((1.+J*BETA*R)*(2.*R*R-3.*RA*RA)+
       +(BETA*RA*R)**2)/(2.*BETA*R**5)
        RETURN
        END
C.
        SUBROUTINE PATN(CGA,NM,ETMM,IOPT)
C.. THE SUBROUTINE TO COMPUTE THE RADIATION PATTERN
        COMMON/SIZE/HL,RA,DZ,ZM,ZN,NMH
        COMMON/CONST/BETA,ETA,RAD,J
        COMPLEX CGA(NM),J,CRT
```

```fortran
      DIMENSION ETMM(181)
      DO 4 I=1,181
      THETA=(I-1.)*RAD
      CTH=COS(THETA)
      STH=SIN(THETA)
      IF(ABS(CTH).LE.1.E-3) THEN
        FT=1.
      ELSE
        FT=SIN(BETA*DZ*CTH*.5)/(BETA*DZ*CTH*.5)
      ENDIF
      CRT=0.
      DO 2 M=1,NM
        ZM=HL-(M-.5)*DZ
        CRT=CRT+CEXP(J*BETA*ZM*CTH)*FT*CGA(M)*DZ
    2 CONTINUE
      PTT=CABS(CRT)*STH*STH*ETA*0.5
    4 ETMM(I)=PTT
      IF(IOPT.EQ.1) THEN
       AMAX=ETMM(1)
       DO 6 I=2,181
       IF(ETMM(I).GT.AMAX) AMAX=ETMM(I)
    6 CONTINUE
      DO 8 I=1,181
      PTT=ETMM(I)/AMAX
      IF(PTT.LE.1.E-5) PTT=1.E-5
    8 ETMM(I)=20.*ALOG10(PTT)
      ELSE
      DO 10 I=1,181
      PTT=ETMM(I)**2*BETA*2.
      IF(PTT.LT.1.E-10) PTT=1.E-10
   10 ETMM(I)=10.*ALOG10(PTT)
      ENDIF
      RETURN
      END
C************************************************************************
C     SUBROUTINE TSLZ          NETLIB
C     INPUT:
C        (C)A(2*M - 1)         THE FIRST ROW OF THE T-MATRIX FOLLOWED BY
C                              ITS FIRST COLUMN BEGINNING WITH THE SECOND
C                              ELEMENT.  ON RETURN A IS UNALTERED.
C        (C)B(M)               THE RIGHT HAND SIDE VECTOR B.
C        (C)WA(2*M-2)          A WORK AREA VECTOR
C        (I)M                  ORDER OF MATRIX A.
C     OUTPUT:
C        (C)B(M)               THE SOLUTION VECTOR.
C     PURPOSE:
C        SOLVE A SYSTEM OF EQUATIONS DESCRIBED BY A TOEPLITZ MATRIX.
C        A * X = B
C     SUBROUTINES AND FUNCTIONS:
C        TOEPLITZ PACKAGE ... TSLZ1
C************************************************************************
      SUBROUTINE TSLZ(A,B,WA,M)
      INTEGER M
      COMPLEX A(2*M-1),B(M),WA(2*M-2)
      CALL TSLZ1(A,A(M+1),B,B,WA,WA(M-1),M)
      RETURN
      END
C..
      SUBROUTINE TSLZ1(A1,A2,B,X,C1,C2,M)
      INTEGER M
      COMPLEX A1(M),A2(M-1),B(M),X(M),C1(M-1),C2(M-1)
      INTEGER I1,I2,N,N1,N2
      COMPLEX R1,R2,R3,R5,R6
      R1 = A1(1)
```

```
          X(1) = B(1)/R1
          IF (M .EQ. 1) GO TO 80
          DO 70 N = 2, M
             N1 = N - 1
             N2 = N - 2
             R5 = A2(N1)
             R6 = A1(N)
             IF (N .EQ. 2) GO TO 20
                C1(N1) = R2
                DO 10 I1 = 1, N2
                   I2 = N - I1
                   R5 = R5 + A2(I1)*C1(I2)
                   R6 = R6 + A1(I1+1)*C2(I1)
 10             CONTINUE
 20          CONTINUE
             R2 = -R5/R1
             R3 = -R6/R1
             R1 = R1 + R5*R3
             IF (N .EQ. 2) GO TO 40
                R6 = C2(1)
                C2(N1) = (0.0D0,0.0D0)
                DO 30 I1 = 2, N1
                   R5 = C2(I1)
                   C2(I1) = C1(I1)*R3 + R6
                   C1(I1) = C1(I1) + R6*R2
                   R6 = R5
 30             CONTINUE
 40          CONTINUE
             C2(1) = R3
             R5 = (0.0D0,0.0D0)
             DO 50 I1 = 1, N1
                I2 = N - I1
                R5 = R5 + A2(I1)*X(I2)
 50          CONTINUE
             R6 = (B(N) - R5)/R1
             DO 60 I1 = 1, N1
                X(I1) = X(I1) + C2(I1)*R6
 60          CONTINUE
             X(N) = R6
 70       CONTINUE
 80       CONTINUE
          RETURN
          END
C..
          SUBROUTINE CSINT(CF,XL,XU,N,CRT)
C..   FAST ALGORITHM FORM OF THE SIMPSON'S INTEGRAL ROUTINE
          IMPLICIT COMPLEX (C)
          CRT=CF(XL)+CF(XU)
          HD=(XU-XL)/(N+1)
          DO 20 I=1,N
            XI=XL+I*HD
            IF(MOD(I,2).NE.0) THEN
              CRT=CRT+4.*CF(XI)
            ELSE
              CRT=CRT+2.*CF(XI)
            ENDIF
 20       CONTINUE
          CRT=CRT*HD*0.33333333
          RETURN
          END
```

CHAPTER 13

GEOMETRICAL THEORY OF DIFFRACTION

13.1 INTRODUCTION

The treatment of the radiation and scattering characteristics from radiating and scattering systems using modal solutions is limited to objects whose surfaces can be described by orthogonal curvilinear coordinates. Moreover, most of the solutions are in the form of infinite series which are poorly convergent when the dimensions of the object are greater than about a wavelength. These limitations, therefore, exclude rigorous analyses of many practical radiating and scattering systems.

A method that describes the solution in the form of an integral equation has received considerable attention. Whereas arbitrary shapes can be handled by this method, it mostly requires the use of a digital computer for numerical computations and therefore is most convenient for objects that are not too many wavelengths in size because of the capacity limitations of present-day computers. This method is usually referred to as the *integral equation* (IE) method, and its solution is generally accomplished by the *moment method* (MM) [1–4]. These were discussed in Chapter 12.

When the dimensions of the radiating or scattering object are many wavelengths, high-frequency asymptotic techniques can be used to analyze many problems that are otherwise mathematically intractable. Two such techniques, which have received considerable attention in the past few years, are the *geometrical theory of diffraction* (GTD) and the *physical theory of diffraction* (PTD). The GTD, originated by Keller [5, 6] and extended by Kouyoumjian and Pathak [7–10], is an extension of the classical *geometrical optics* (GO) (direct, reflected and refracted rays), and it overcomes some of the limitations of geometrical optics by introducing a diffraction mechanism [11]. The PTD, introduced by Ufimtsev [12–14], supplements *physical optics* (PO) to provide corrections that are due to diffractions at edges of conducting surfaces. Ufimtsev suggested the existence of nonuniform ("fringe") edge currents in addition to the uniform physical optics surface currents [15–18]. The PTD bears some resemblance to GTD in method of application.

At high frequencies diffraction, like reflection and refraction, is a local phenomenon and it depends on two things:

1. The geometry of the object at the point of diffraction (edge, vertex, curved surface).
2. The amplitude, phase, and polarization of the incident field at the point of diffraction.

A field is associated with each diffracted ray, and the total field at a point is the sum of all the rays at that point. Some of the diffracted rays enter the shadow regions and account for the field intensity there. The diffracted field, which is determined by a generalization of *Fermat's principle* [6, 7], is initiated at points on the surface of the object which create a discontinuity in the incident GO field (incident and reflected shadow boundaries).

The phase of the field on a ray is assumed to be equal to the product of the optical length of the ray from some reference point and the wave number of the medium. Appropriate phase jumps must be added as rays pass through caustics (defined in Section 13.2.1). The amplitude is assumed to vary in accordance with the principle of conservation of energy in a narrow tube of rays.

The initial value of the field on a diffracted ray is determined from the incident field with the aid of an appropriate diffraction coefficient that is a dyadic for electromagnetic fields. This is analogous to the manner reflected fields are determined using the reflection coefficient. The rays also follow paths that make the optical distance from the source to the observation point an extremum (usually a minimum). This leads to straight line propagation within homogeneous media and along geodesics (surface extrema) on smooth surfaces. The field intensity also attenuates exponentially as it travels along surface geodesics.

The diffraction and attenuation coefficients are usually determined from the asymptotic solutions of the simplest boundary-value problems, which have the same local geometry at the points of diffraction as the object at the points of interest. Geometries of this type are referred to as *canonical* problems. One of the simplest geometries that will be discussed in this chapter is a conducting wedge. The primary objective in using the GTD is to resolve each problem to smaller components [19–25], each representing a canonical geometry of a known solution. The ultimate solution is a superposition of the contributions from each canonical problem.

Some of the advantages of GTD are given in the following list.

1. It is simple to apply.
2. It can be used to solve complicated problems that do not have exact solutions.
3. It provides physical insight into the radiation and scattering mechanisms from the various parts of the structure.
4. It yields accurate results that compare quite well with experiments and other methods.
5. It can be combined with other techniques such as the moment method [26–28].

13.2 GEOMETRICAL OPTICS

Geometrical optics (GO) is an approximate high-frequency method for determining wave propagation for incident, reflected, and refracted fields. Because it uses ray concepts, it is often referred to as *ray optics*. Originally geometrical optics was developed to analyze the propagation of light at sufficiently high frequencies where it was not necessary to consider the wave nature of light. Instead the transport of energy from one point to another in an isotropic lossless medium is accomplished using the conservation of energy flux in a tube of rays. For reflection problems geometrical optics approximates the scattered fields only toward specular directions as determined by Snell's law of reflection: the angle of reflection is equal to the angle of incidence. For sufficiently high frequencies geometrical optics fields may dominate the scattering phenomena and may not require any corrections. This is more evident for backscattering from smooth curved surfaces whose curvature is large compared to the wavelength.

According to classical geometrical optics, the rays between any two points P_1 and P_2 follow a path that makes the optical distance between them an extremum (usually a minimum). In equation form this is expressed as

$$\delta \int_{P_1}^{P_2} n(s)\, ds = 0 \qquad (13\text{-}1)$$

where δ represents what is referred to in the calculus of variations as the *variational differential* and $n(s)$ is the index of refraction of the medium, $\beta(s)/\beta_0 = n(s)$. If the medium is homogeneous, $n(s) = n =$ constant, the paths are straight lines. Equation 13-1 is a mathematical representation of *Fermat's principle*. In addition, the light intensity, power per unit solid angle, between any two points is also governed by the conservation of energy flux in a tube of rays.

To demonstrate the principles of geometrical optics, let us consider a *primary wave front* surface ψ_0, as shown in Figure 13-1, formed at $t = t_0$ by the motion of

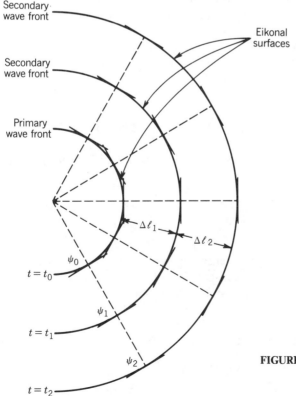

FIGURE 13-1 Primary and secondary wave front (eikonal surfaces) of a radiated wave.

light propagating in an isotropic lossless medium. The objectives here are:

1. To determine the *secondary wave front* surfaces ψ_n formed at $t = t_{n+1} > t_n$, $n = 0, 1, 2, 3 \ldots$.
2. To relate the power density and field intensity on the secondary wave fronts to those of the primary or previous wave fronts.

The secondary wave fronts can be determined by first selecting a number of discrete points on the primary wave front. If the medium of wave propagation is also assumed to be homogeneous, ray paths from the primary to the secondary wave front are drawn as straight lines that at each point are normal to the surface of the primary wave front.

Since the wave travels in the medium with the speed of light given by $v = c/n$, where c is the speed of light in free space and n is the index of refraction, then at $\Delta t = t_1 - t_0$ ($t_1 > t_0$) the wave would have traveled a distance $\Delta \ell = v \Delta t$. Along each of the normal rays a distance $\Delta \ell$ is marked, and a surface perpendicular to each ray is drawn. The surfaces normal to each of the rays are then connected to form the secondary wave front ψ_1, as shown in Figure 13-1. The same procedure can be repeated to determine the subsegment secondary wave front surfaces $\psi_2, \psi_3, \ldots$.

The family of wave front surfaces $\psi_n(x, y, z)$, $n = 0, 1, 2, 3, \ldots$, that are normal to each of the radial rays is referred to as the *eikonal* surfaces, and they can be determined using the *eikonal equation* of [7]

$$|\nabla \psi_n(x, y, z)|^2 = \left\{ \frac{\partial \psi_n}{\partial x} \right\}^2 + \left\{ \frac{\partial \psi_n}{\partial y} \right\}^2 + \left\{ \frac{\partial \psi_n}{\partial z} \right\}^2 = n^2(s) \quad (13\text{-}2)$$

Since the rays normal to the wave fronts and the eikonal surfaces are uniquely related, it is only necessary to know one or the other when dealing with geometrical optics.

Extending this procedure to approximate the wave motion of electromagnetic waves of lower frequencies, it is evident that:

1. The eikonal surfaces for plane waves are planar surfaces perpendicular to the direction of wave travel.
2. The eikonal surfaces for cylindrical waves are cylindrical surfaces perpendicular to the cylindrical radial vectors.
3. The eikonal surfaces for spherical waves are spherical surfaces perpendicular to the spherical radial vectors.

Each of these is demonstrated, respectively, in Figure 13-2a, b, and c.

13.2.1 Amplitude Relation

In geometrical optics, the light intensity (power per unit solid angle) between two points is also governed by the conservation of energy flux in a tube of rays. To demonstrate that let us assume that a point source, as shown in Figure 13-3, emanates isotropically spherical waves. Within a tube of rays, the cross-sectional areas at some reference point $s = 0$ and at s are given, respectively, by dA_0 and dA. The radiation density S_0 at $s = 0$ is related to the radiation density S at s by

$$S_0 \, dA_0 = S \, dA \quad (13\text{-}3)$$

or

$$\frac{S(s)}{S_0(0)} = \frac{dA_0}{dA} \tag{13-3a}$$

It has been assumed that S_0 and S are constant, respectively, throughout the cross-sectional areas dA_0 and dA, and that no power flows across the sides of the conical tube.

For electromagnetic waves the far-zone electric field $\mathbf{E}(r, \theta, \phi)$ is related to the radiation density $S(r, \theta, \phi)$ by [7]

$$S(r, \theta, \phi) = \frac{1}{2\eta}|\mathbf{E}(r, \theta, \phi)|^2 = \frac{1}{2}\sqrt{\frac{\varepsilon}{\mu}}|\mathbf{E}(r, \theta, \phi)|^2 \tag{13-4}$$

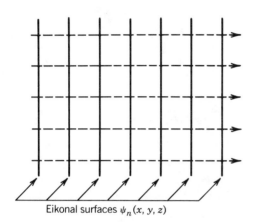

(a)

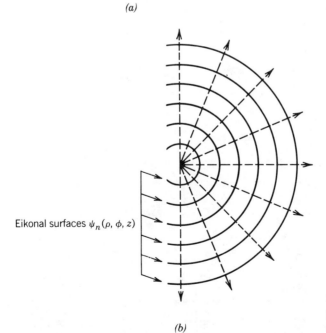

(b)

13-2 Eikonal surfaces for (a) plane, (b) cylindrical, and (c) spherical waves.

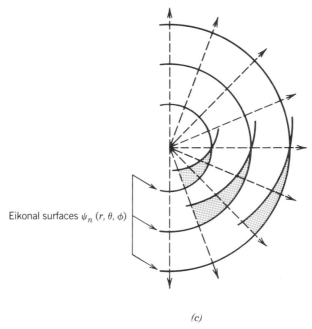

Eikonal surfaces $\psi_n(r, \theta, \phi)$

(c)

FIGURE 13-2 (*Continued*).

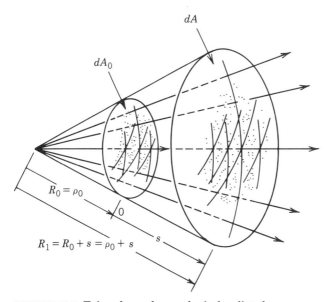

FIGURE 13-3 Tube of rays for a spherical radiated wave.

Therefore (13-3a) can also be written, using (13-4), as

$$\frac{|\mathbf{E}|^2}{|\mathbf{E}_0|^2} = \frac{dA_0}{dA} \tag{13-5}$$

or

$$\frac{|\mathbf{E}|}{|\mathbf{E}_0|} = \sqrt{\frac{dA_0}{dA}} \tag{13-5a}$$

Since in the tube of rays in Figure 13-3 the differential surface areas dA_0 and dA are patches of spherical surfaces with radii of $R_0 = \rho_0$ and $R_1 = R_0 + s = \rho_0 + s$, respectively, then (13-5a) can be written in terms of the radii of curvature of the wave fronts at $s = 0$ and s. Thus (13-5a) reduces to

$$\boxed{\frac{|E|}{|E_0|} = \sqrt{\frac{dA_0}{dA}} = \sqrt{\frac{4\pi R_0^2/C_0}{4\pi R_1^2/C_0}} = \frac{R_0}{R_1} = \frac{\rho_0}{\rho_0 + s}} \qquad (13\text{-}6)$$

and it indicates that the electric field varies, as expected, inversely proportional to the distance of travel.

If the eikonal surfaces of the radiated fields are cylindrical surfaces, representing the wave fronts of cylindrical waves, then the field relation of (13-5a) takes the form

$$\boxed{\frac{|E|}{|E_0|} = \sqrt{\frac{dA_0}{dA}} = \sqrt{\frac{2\pi R_0/C_1}{2\pi R_1/C_1}} = \sqrt{\frac{R_0}{R_1}} = \sqrt{\frac{\rho_0}{\rho_0 + s}}} \qquad (13\text{-}7)$$

The relation of (13-7) indicates that the electric field for cylindrical waves varies, as expected, inversely to the square root of the distance of travel. For planar eikonal surfaces, representing plane waves, (13-5a) simplifies to

$$\boxed{\frac{|E|}{|E_0|} = 1} \qquad (13\text{-}8)$$

For the previous three cases the eikonal surfaces were, respectively, spherical, cylindrical, and planar. Let us now consider a more general configuration in which the eikonal surfaces (wave fronts) are not necessarily spherical. This is illustrated in Figure 13-4a where the wave front is represented by a radius of curvature R_1 in the xz and R_2 in the yz planes which are not equal ($R_1 \neq R_2$). To determine the focusing characteristics of such a surface, let us trace the focusing diagram of rays 1, 2, 3, and 4 from the four corners of the wave front. It is apparent that the rays focus (cross) at different points. For this example, rays 1 and 2 focus at P, rays 3 and 4 focus at P', rays 2 and 3 focus at Q', and rays 1 and 4 focus at Q. This system of a tube of rays is referred to as *astigmatic* (not meeting at a single point) and the lines PP' and QQ' are called *caustics*.[1]

Referring to the geometry of Figure 13-4b, it can be shown that for a wave whose eikonal surface (wave front) forms an astigmatic tube of rays the electric field intensity from one surface relative to that of another, as related by (13-5a), takes the form of

$$\boxed{\frac{|E|}{|E_0|} = \sqrt{\frac{dA_0}{dA}} = \sqrt{\frac{\rho_1 \rho_2}{(\rho_1 + s)(\rho_2 + s)}}} \qquad (13\text{-}9)$$

---

[1] A *caustic* is a point, a line, or a surface through which all the rays of a wave pass. Examples of it are the focal point of a paraboloid (parabola of revolution) and the focal line of a parabolic cylinder. The field at a caustic is infinite, in principle, because an infinite number of rays pass through it.

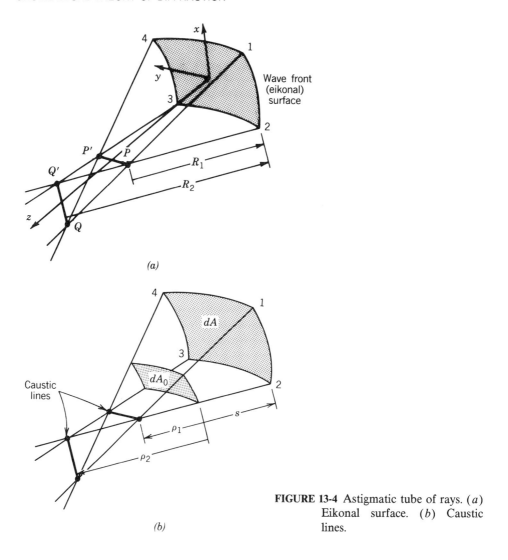

FIGURE 13-4 Astigmatic tube of rays. (*a*) Eikonal surface. (*b*) Caustic lines.

It is apparent that (13-9) reduces to the following equations,

1. (13-6) if the wave front is spherical ($\rho_1 = \rho_2 = \rho_0$).
2. (13-7) if the wave front is cylindrical ($\rho_1 = \infty$, $\rho_2 = \rho_0$ or $\rho_2 = \infty$, $\rho_1 = \rho_0$).
3. (13-8) if the wave front is planar ($\rho_1 = \rho_2 = \infty$).

Expressions (13-6) through (13-9) correctly relate the *magnitudes* of the high-frequency electric field at one wave front surface to that of another. These were derived using geometrical optics based on the principle of conservation of energy flux through a tube of rays. Although these may be valid high-frequency approximations for light waves, they are not accurate relations for electromagnetic waves of lower frequencies. Two apparent missing properties in these relations are those of *phase* and *polarization*.

13.2.2 Phase and Polarization Relations

Phase and polarization information can be introduced to the relations of (13-6) through (13-9) by examining the approach introduced by Luneberg [29] and Kline [30, 31] to develop high-frequency solutions of electromagnetic problems. The works of Luneberg and Kline, referred to as the *Luneberg–Kline high-frequency expansion*, best bridge the gap between geometrical (ray) optics and wave propagation phenomena.

The Luneberg–Kline series expansion solution begins by assuming that the electric field for large ω can be written in a series of

$$\mathbf{E}(\mathbf{R}, \omega) = e^{-j\beta_0 \psi(\mathbf{R})} \sum_{m=0}^{\infty} \frac{\mathbf{E}_m(\mathbf{R})}{(j\omega)^m} \tag{13-10}$$

where $\mathbf{R}$ = position vector
β_0 = phase constant for free-space

Substituting (13-10) into the wave equation

$$\nabla^2 \mathbf{E} + \beta^2 \mathbf{E} = 0 \tag{13-11}$$

subject to Maxwell's equation

$$\nabla \cdot \mathbf{E} = 0 \tag{13-12}$$

it can be shown, by equating like powers of ω, that one obtains the following.

1. The eikonal equation 13-2 or

$$|\nabla \psi|^2 = n^2 \tag{13-13a}$$

where ψ = eikonal (wave front) surface
n = index of refraction

2. The transport equations

$$\frac{\partial \mathbf{E}_0}{\partial s} + \frac{1}{2}\left\{\frac{\nabla^2 \psi}{n}\right\} \mathbf{E}_0 = 0 \quad \text{for first-order terms} \tag{13-13b}$$

$$\frac{\partial \mathbf{E}_m}{\partial s} + \frac{1}{2}\left\{\frac{\nabla^2 \psi}{n}\right\} \mathbf{E}_m = \frac{v_p}{2} \nabla^2 \mathbf{E}_{m-1} \quad \text{for higher-order terms} \tag{13-13c}$$

where $m = 1, 2, 3\ldots$
v_p = velocity of light in medium

3. The conditional equations

$$\hat{s} \cdot \mathbf{E}_0 = 0 \quad \text{for first-order terms} \tag{13-13d}$$

$$\hat{s} \cdot \mathbf{E}_m = v_p \nabla \cdot \mathbf{E}_{m-1} \quad \text{for higher-order terms} \tag{13-13e}$$

$$m = 1, 2, 3\ldots$$

where

$$\hat{s} = \frac{\nabla \psi}{n} = \text{unit vector in the direction path (normal to the wave front } \psi) \quad (13\text{-}13f)$$

s = distance along the ray path

At the present time we are interested mainly in first-order solutions for the electric field of (13-10) which can be approximated and take the form of

$$\mathbf{E}(s) \simeq e^{-j\beta_0 \psi(s)} \mathbf{E}_0(s=0) \quad (13\text{-}14)$$

Integrating the first-order transport equation 13-13b along s and referring to the geometry of Figure 13-4a, it can be shown that (13-14) can be written as [30, 31]

$$\mathbf{E}(s) \simeq \mathbf{E}_0(0) e^{-j\beta_0 \psi(0)} \sqrt{\frac{\rho_1 \rho_2}{(\rho_1 + s)(\rho_2 + s)}} e^{-j\beta s} \quad (13\text{-}15)$$

where $s = 0$ is taken as a reference point. Since $\mathbf{E}_0(0)$ is complex, the phase term $e^{-j\beta_0 \psi(0)}$ can be combined with $\mathbf{E}_0(0)$ and (13-15) rewritten as

$$\mathbf{E}(s) = \underbrace{\mathbf{E}_0'(0) e^{j\phi_0(0)}}_{\substack{\text{Field at reference} \\ \text{point } (s=0)}} \underbrace{\sqrt{\frac{\rho_1 \rho_2}{(\rho_1 + s)(\rho_2 + s)}}}_{\substack{\text{Spatial attenuation} \\ \text{(divergence, spreading)} \\ \text{factor}}} \underbrace{e^{-j\beta s}}_{\substack{\text{Phase} \\ \text{factor}}} \quad (13\text{-}15a)$$

where $\mathbf{E}_0'(0)$ = field amplitude at reference point ($s=0$)
$\phi_0(0)$ = field phase at reference point ($s=0$)

Comparing (13-15a) to (13-9), it is evident that the leading term of the Luneberg–Kline series expansion solution for large ω predicts the spatial attenuation relation between the electric fields of two points as obtained by classical geometrical optics, as given by (13-9) which ignores both the polarization and the wave motion (phase) of electromagnetic fields. It also predicts their phase and polarization relations, as given by (13-15a). Obviously (13-15a) could have been obtained from (13-9) by artificially converting the magnitudes of the fields to vectors (to account for polarization) and by introducing a complex exponential to account for the phase delay of the field from $s = 0$ to s. This was not necessary since (13-15a) was derived here rigorously using the leading term of the Luneberg–Kline expansion series for large ω subject to the wave and Maxwell's equations. It should be pointed out, however, that (13-15a) is only a high-frequency approximation and it becomes more accurate as the frequency approaches infinity. However, for many practical engineering problems it does predict quite accurate results which compare well with measurements.

In principle more accurate expressions to the geometrical optics approximation can be obtained by retaining higher-order terms $\mathbf{E}_1(\mathbf{R}_1), \mathbf{E}_2(\mathbf{R}_2), \ldots$ in the Luneberg–Kline series expansion (13-10), and in the transport (13-13c) and conditional (13-13e) equations. However, such a procedure is very difficult. In addition, the resulting terms do not remove the discontinuities introduced by geometrical optics fields along the incident and reflection boundaries, and the method does not lend itself to other improvements in the geometrical optics, such as those of diffraction. Therefore no such procedure will be pursued here.

It should be noted that when the observation point is chosen so that $s = -\rho_1$ or $s = -\rho_2$, (13-15a) possesses singularities representing the congruence of the rays at the caustic lines PP' and QQ'. Therefore (13-15a) is not valid along the caustics and not very accurate near them, and it should not be used in those regions. Other methods should be utilized to find the fields at and near caustics [32–36]. In addition, it is observed that when $-\rho_2 < s < -\rho_1$ the sign in the $(\rho_1 + s)$ term of the denominator of (13-15a) changes. Similar changes of sign occur in the $(\rho_1 + s)$ and $(\rho_2 + s)$ terms when $s < -\rho_2 < -\rho_1$. Therefore (13-15a) correctly predicts $+90°$ phase jumps each time a caustic is crossed in the direction of propagation.

13.2.3 Reflection from Surfaces

Geometrical optics can be used to compute high-frequency approximations to the fields reflected from surfaces, the directions of which are determined by Snell's law of reflection. To demonstrate the procedure, let us assume that a field impinges on a smooth conducting surface S where it undergoes a reflection at point Q_R. This is illustrated in Figure 13-5a where $\hat{s}^i$ is the unit vector in the direction of incidence, $\hat{s}^r$ is the unit vector in the direction of reflection, $\hat{e}^i_\parallel, \hat{e}^r_\parallel$ are unit vectors, for incident and reflected electric fields, parallel to the planes of incidence and reflection, and $\hat{e}^i_\perp, \hat{e}^r_\perp$ are unit vectors, for incident and reflected electric fields, perpendicular to the planes of incidence and reflection. The plane of incidence is formed by the unit vector $\hat{n}$ normal to the surface at the point of reflection Q_R and the unit vector $\hat{s}^i$, and the plane of reflection is formed by the unit vectors $\hat{n}$ and $\hat{s}^r$. The angle of incidence θ_i is measured between $\hat{n}$ and $\hat{s}^i$ whereas θ_r is measured between $\hat{n}$ and $\hat{s}^r$, and they are equal ($\theta_i = \theta_r$).

The polarization unit vectors are chosen so that

$$\hat{e}^i_\perp \times \hat{s}^i = \hat{e}^i_\parallel \tag{13-16a}$$

$$\hat{e}^r_\perp \times \hat{s}^r = \hat{e}^r_\parallel \tag{13-16b}$$

and the incident and reflected electric fields can be expressed as

$$\mathbf{E}^i_0 = \hat{e}^i_\parallel E^i_{0\parallel} + \hat{e}^i_\perp E^i_{0\perp} \tag{13-17a}$$

$$\mathbf{E}^r_0 = \hat{e}^r_\parallel E^r_{0\parallel} + \hat{e}^r_\perp E^r_{0\perp} \tag{13-17b}$$

The incident and reflected fields at the point of reflection can be related by applying the boundary conditions of vanishing tangential components of the electric field at the point of reflection (Q_R). Doing this we can write that

$$\mathbf{E}^r_0(s=0) = \mathbf{E}^i_0(Q_R) \cdot \overline{\mathbf{R}} = \mathbf{E}^i_0(Q_R) \cdot \left[\hat{e}^i_\parallel \hat{e}^r_\parallel - \hat{e}^i_\perp \hat{e}^r_\perp\right] \tag{13-18}$$

where $\mathbf{E}^r_0(s=0)$ = reflected field at the point of reflection (the reference point for the reflected ray is taken on the reflecting surface so that $s = 0$)
$\mathbf{E}^i_0(Q_R)$ = incident field at the point of reflection Q_R
$\overline{\mathbf{R}}$ = dyadic reflection coefficient

754 GEOMETRICAL THEORY OF DIFFRACTION

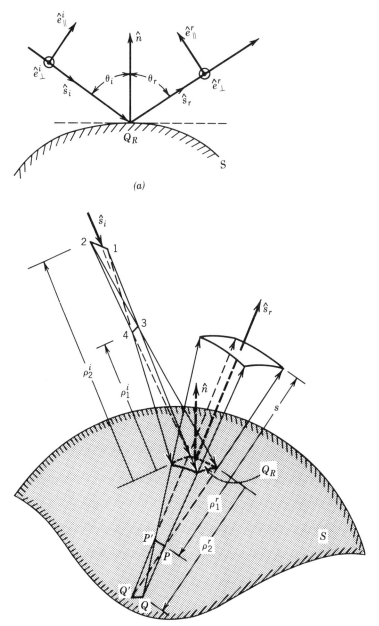

FIGURE 13-5 Reflection from a curved surface. (*a*) Reflection point. (*b*) Astigmatic tube of rays.

In matrix notation the reflection coefficient can be written as

$$R = \begin{bmatrix} 1 & 0 \\ 0 & -1 \end{bmatrix} \qquad (13\text{-}19)$$

which is identical to the Fresnel reflection coefficients of electromagnetic plane waves reflected from plane, perfectly conducting surfaces. This is quite acceptable in practice since at high frequencies reflection, as well as diffraction, is a local phenomenon, and it depends largely on the geometry of the surface in the immediate neighborhood of the reflection point. Therefore near the reflection point Q_R, the following approximations can be made:

1. The reflecting surface can be approximated by a plane tangent at Q_R.
2. The wave front of the incident field can be assumed to be planar.

With the aid of (13-15a) and (13-18), it follows that the reflected field $\mathbf{E}^r(s)$ at a distance s from the point of reflection Q_R can be written as

$$\mathbf{E}^r(s) = \underbrace{\mathbf{E}^i(Q_R)}_{\substack{\text{Field at} \\ \text{reference} \\ \text{point } (Q_r)}} \cdot \underbrace{\overline{\mathbf{R}}}_{\substack{\text{Reflection} \\ \text{coefficient}}} \underbrace{\sqrt{\frac{\rho_1^r \rho_2^r}{(\rho_1^r + s)(\rho_2^r + s)}}}_{\substack{\text{Spatial attenuation} \\ \text{(divergence, spreading)} \\ \text{factor}}} \underbrace{e^{-j\beta s}}_{\substack{\text{Phase} \\ \text{factor}}}$$

(13-20)

where ρ_1^r, ρ_2^r = principal radii of curvature of the reflected wave front at the point of reflection

The astigmatic tube of rays for the reflected fields are shown in Figure 13-5b where the reference surface is taken at the reflecting surface.

The principal radii of curvature of the reflected wave front ρ_1^r and ρ_2^r are related to the principal radii of curvature of the incident wave front ρ_1^i and ρ_2^i, the aspect of wave incidence, and the curvature of the reflecting surface at Q_R. It can be shown that ρ_1^r and ρ_2^r can be expressed as [10]

$$\frac{1}{\rho_1^r} = \frac{1}{2}\left(\frac{1}{\rho_1^i} + \frac{1}{\rho_2^i}\right) + \frac{1}{f_1} \quad (13\text{-}21a)$$

$$\frac{1}{\rho_2^r} = \frac{1}{2}\left(\frac{1}{\rho_1^i} + \frac{1}{\rho_2^i}\right) + \frac{1}{f_2} \quad (13\text{-}21b)$$

where ρ_1^i, ρ_2^i = principal radii of curvature of incident wave front ($\rho_1^i = \rho_2^i = s'$ for spherical incident wave front; $\rho_1^i = \rho', \rho_2^i = \infty$ or $\rho_1^i = \infty, \rho_2^i = \rho'$ for cylindrical incident wave front, and $\rho_1^i = \rho_2^i = \infty$ for planar incident wave front).

Equations 13-21a and 13-21b are similar in form to the simple lens and mirror formulas of elementary physics. In fact when the incident ray is spherical ($\rho_1^i = \rho_2^i = s'$), f_1 and f_2 represent focal distances that are independent of the source range that is creating the spherical wave.

When the incident field has a spherical wave front, $\rho_1^i = \rho_2^i = s'$, then f_1 and f_2 simplify to

$$\frac{1}{f_1} = \frac{1}{\cos\theta_i}\left(\frac{\sin^2\theta_2}{R_1} + \frac{\sin^2\theta_1}{R_2}\right) + \sqrt{\frac{1}{\cos^2\theta_i}\left(\frac{\sin^2\theta_2}{R_1} + \frac{\sin^2\theta_1}{R_2}\right)^2 - \frac{4}{R_1 R_2}}$$

(13-22a)

$$\frac{1}{f_2} = \frac{1}{\cos\theta_i}\left(\frac{\sin^2\theta_2}{R_1} + \frac{\sin^2\theta_1}{R_2}\right) - \sqrt{\frac{1}{\cos^2\theta_i}\left(\frac{\sin^2\theta_2}{R_1} + \frac{\sin^2\theta_1}{R_2}\right)^2 - \frac{4}{R_1 R_2}}$$

(13-22b)

where R_1, R_2 = radii of curvature of the reflecting surface
  θ_1 = angle between the direction of the incident ray $\hat{s}^i$ and $\hat{u}_1$
  θ_2 = angle between the direction of the incident ray $\hat{s}^i$ and $\hat{u}_2$
  $\hat{u}_1$ = unit vector in the principal direction of S at Q_R with principal radius of curvature R_1
  $\hat{u}_2$ = unit vector in the principal direction of S at Q_R with principal radius of curvature R_2

The geometrical arrangement of these is exhibited in Figure 13-6.

If the incident wave form is a plane wave, then $\rho_1^i = \rho_2^i = \infty$ and according to (13-21a) and (13-21b)

$$\frac{1}{\rho_1^r \rho_2^r} = \frac{1}{f_1 f_2} = \frac{4}{R_1 R_2} \qquad (13\text{-}23)$$

or

$$\rho_1^r \rho_2^r = \frac{R_1 R_2}{4} \qquad (13\text{-}23a)$$

The relation of (13-23a) is very useful to calculate the far-zone reflected fields. Now if either R_1 and/or R_2 becomes infinite, as is the case for flat plates or cylindrical scatterers, the geometrical optics field fails to predict the scattered field when the incident field is a plane wave.

To give some physical insight into the principal radii of curvature R_1 and R_2 of a reflecting surface, let us assume that the reflecting surface is well behaved (smooth and continuous). At each point on the surface there exists a unit normal vector. If through that point a plane intersects the reflecting surface, it generates on the reflecting surface a curve as shown in Figure 13-6b. If in addition the intersecting plane contains the unit normal to the surface at that point, the curve generated on the reflecting surface by the intersecting plane is known as the *normal section curve*, as shown in Figure 13-6b. If the intersecting plane is rotated about the surface normal at that point, a number of unique normal sections are generated, one on each orientation of the intersecting plane. Associated with each normal section curve, there is a radius of curvature. It can be shown that for each point on an arbitrary well behaved curved reflecting surface, there is one intersecting plane that maximizes the radius of curvature of its corresponding normal section curve while there is another intersecting plane at the same point that minimizes the radius of curvature of its corresponding normal section curve. For each point on the reflecting

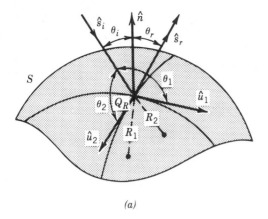

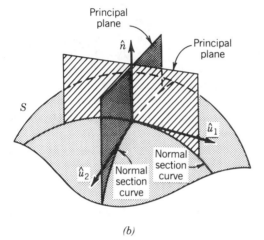

FIGURE 13-6 Geometry for reflection from a three-dimensional curved surface. (*a*) Principal radii of curvature. (*b*) Normal section curves and principal planes.

surface there are two normal section radii of curvature, denoted here as R_1 and R_2 and referred to as *principal radii of curvature*, and the two corresponding planes are known as the *principal planes*. For an arbitrary surface, the two principal planes are perpendicular to each other. Also for each principal plane we can define unit vectors $\hat{u}_1$ and $\hat{u}_2$ that are tangent to each normal section curve generated by each intersecting principal plane. Also the unit vectors $\hat{u}_1$ and $\hat{u}_2$, which lie in the principal plane, point along the principal directions whose normal section curves have radii of curvature R_1 and R_2, respectively. Expressions for determining the principal unit vectors $\hat{u}_1, \hat{u}_2$ and the principal radii of curvature R_1, R_2 of arbitrary surfaces of revolution are given in Problem 13.2 at the end of the chapter.

In general, however, f_1 and f_2 can be obtained using [10]

$$\frac{1}{f_{1(2)}} = \frac{\cos\theta_i}{|\theta|^2}\left(\frac{\theta_{22}^2 + \theta_{12}^2}{R_1} + \frac{\theta_{21}^2 + \theta_{11}^2}{R_2}\right)$$

$$\pm \frac{1}{2}\left\{\left(\frac{1}{\rho_1^i} - \frac{1}{\rho_2^i}\right)^2 + \left(\frac{1}{\rho_1^i} - \frac{1}{\rho_2^i}\right)\frac{4\cos\theta_i}{|\theta|^2}\left(\frac{\theta_{22}^2 - \theta_{12}^2}{R_1} + \frac{\theta_{21}^2 - \theta_{11}^2}{R_2}\right)\right.$$

$$\left. + \frac{4\cos^2\theta_i}{|\theta|^4}\left[\left(\frac{\theta_{22}^2 + \theta_{12}^2}{R_1} + \frac{\theta_{21}^2 + \theta_{11}^2}{R_2}\right)^2 - \frac{4|\theta|^2}{R_1 R_2}\right]\right\}^{1/2} \quad (13\text{-}24)$$

where the plus sign is used for f_1 and the minus for f_2. In (13-24) $|\theta|$ is the determinant of

$$[\theta] = \begin{bmatrix} \hat{X}_1^i \cdot \hat{u}_1 & \hat{X}_1^i \cdot \hat{u}_2 \\ \hat{X}_2^i \cdot \hat{u}_1 & \hat{X}_2^i \cdot \hat{u}_2 \end{bmatrix} \quad (13\text{-}24\text{a})$$

or

$$|\theta| = (\hat{X}_1^i \cdot \hat{u}_1)(\hat{X}_2^i \cdot \hat{u}_2) - (\hat{X}_2^i \cdot \hat{u}_1)(\hat{X}_1^i \cdot \hat{u}_2) \quad (13\text{-}24\text{b})$$

and

$$\theta_{jk} = \hat{X}_j^i \cdot \hat{u}_k \quad (13\text{-}24\text{c})$$

The vectors $\hat{X}_1^i$ and $\hat{X}_2^i$ represent the principal directions of the incident wave front at the reflection point Q_R with principal radii of curvature ρ_1^i and ρ_2^i.

Equations 13-24 through 13-24c can be used to find single, first-order reflections by a reflecting surface. The process, using basically the same set of equations, must be repeated if second- and higher-order reflections are required. However, to accomplish this the principal plane directions $\hat{X}_1^r$ and $\hat{X}_2^r$ of the reflected fields from the previous reflection must be known. For example, second-order reflections can be found provided the principal plane directions $\hat{X}_1^r$ and $\hat{X}_2^r$ of the first-order reflected field are found. To do this, we first introduce

$$Q^r = \begin{bmatrix} Q_{11}^r & Q_{12}^r \\ Q_{12}^r & Q_{22}^r \end{bmatrix} \quad (13\text{-}25)$$

where Q^r is defined as the curvature matrix for the reflected wave front whose entries are

$$Q_{11}^r = \frac{1}{\rho_1^i} + \frac{2\cos\theta_i}{|\theta|^2} \left(\frac{\theta_{22}^2}{R_1} + \frac{\theta_{21}^2}{R_2} \right) \quad (13\text{-}26\text{a})$$

$$Q_{12}^r = -\frac{2\cos\theta_i}{|\theta|^2} \left(\frac{\theta_{22}\theta_{12}}{R_1} + \frac{\theta_{11}\theta_{21}}{R_2} \right) \quad (13\text{-}26\text{b})$$

$$Q_{22}^r = \frac{1}{\rho_2^i} + \frac{2\cos\theta_i}{|\theta|^2} \left(\frac{\theta_{12}^2}{R_1} + \frac{\theta_{11}^2}{R_2} \right) \quad (13\text{-}26\text{c})$$

Then the principal directions $\hat{X}_1^r$ and $\hat{X}_2^r$ of the reflected wave front, *with respect to the x_1^r and x_2^r coordinates*, can be written as

$$\hat{X}_1^r = \frac{\left(Q_{22}^r - \frac{1}{\rho_1^r}\right)\hat{x}_1^r - Q_{12}^r \hat{x}_2^r}{\sqrt{\left(Q_{22}^r - \frac{1}{\rho_1^r}\right)^2 + (Q_{12}^r)^2}} \quad (13\text{-}27\text{a})$$

$$\hat{X}_2^r = -\hat{s}^r \times \hat{X}_1^r \quad (13\text{-}27\text{b})$$

where $\hat{x}_1^r$ and $\hat{x}_2^r$ are unit vectors perpendicular to the reflected ray, and they are determined using

$$\hat{x}_1^r = \hat{X}_1^i - 2(\hat{n} \cdot \hat{X}_1^i)\hat{n} \qquad (13\text{-}28\text{a})$$

$$\hat{x}_2^r = \hat{X}_2^i - 2(\hat{n} \cdot \hat{X}_2^i)\hat{n} \qquad (13\text{-}28\text{b})$$

with $\hat{n}$ being a unit vector normal to the surface at the reflection point.

To demonstrate the application of these formulations, let us consider a problem that is classified as a classic example in scattering.

Example 13-1. A linearly polarized uniform plane wave of amplitude E_0 is incident on a conducting sphere of radius a. Use geometrical optics methods.

1. Determine the far-zone ($s \gg \rho_1^r$ and ρ_2^r) fields that are reflected from the surface of the sphere.
2. Determine the backscatter radar cross section.

Solution. For a linearly polarized uniform plane wave incident upon a conducting sphere (13-20) reduces in the far zone to

$$E^r(s) = E_0(-1)\sqrt{\frac{\rho_1^r \rho_2^r}{(\rho_1^r + s)(\rho_2^r + s)}} e^{-j\beta s} \overset{s \gg \rho_1^r, \rho_2^r}{\simeq} -E_0 \frac{\sqrt{\rho_1^r \rho_2^r}}{s} e^{-j\beta s}$$

According to (13-23a)

$$\rho_1^r \rho_2^r = \frac{a^2}{4}$$

Thus

$$E^r(s) = -E_0 \frac{a}{2s} e^{-j\beta s} = -\frac{E_0}{2}\left(\frac{a}{s}\right) e^{-j\beta s}$$

In turn the backscatter radar cross section, according to (11-22b), can be written as

$$\sigma = \lim_{s \to \infty}\left[4\pi s^2 \frac{|E^r(s)|^2}{|E^i(Q_R)|^2}\right] \simeq 4\pi s^2 \frac{\left|-\frac{E_0}{2}\left(\frac{a}{s}\right)e^{-j\beta s}\right|^2}{|E_0|^2} = \pi a^2$$

It is recognized that the geometrical optics radar cross section of a sphere is equal to its physical cross-sectional area. This is a well known relation, and it is valid when the radius of the sphere is large compared to the wavelength.

The variation of the normalized radar cross section of a sphere as a function of its radius is displayed in Figure 11-26, and it is obtained by solving the wave equation in exact form. The Rayleigh, Mie (resonance), and geometrical optics regions represent three regimes in the figure.

For a cylindrical reflected field with radii of curvature $\rho_1^r = \rho^r$ and $\rho_2^r = \infty$, the reflected field of (13-20) reduces to

$$\mathbf{E}^r(s) = \mathbf{E}^i(Q_R) \cdot \overline{\mathbf{R}} \sqrt{\frac{\rho^r}{\rho^r + s}}\, e^{-j\beta s} \qquad (13\text{-}29)$$

where ρ^r is the radius of curvature of the reflected field wave front.

An expression for ρ^r can be derived [37] by assuming a cylindrical wave, radiated by a line source at ρ_0, incident upon a two-dimensional curved surface S with positive radius of curvature ρ_a, as shown in Figure 13-7a. The rays that are reflected from the surface S diverge for positive values of ρ_a, as shown in Figure 13-7b, and appear to be emanating from a caustic a distance ρ^r from the reflecting surface S. It can be shown that the wave front curvature of the reflected field can be determined using

$$\frac{1}{\rho^r} = \frac{1}{\rho_0} + \frac{2}{\rho_a \cos\theta_i} \qquad (13\text{-}30)$$

This is left as an end of chapter exercise for the reader.

The expression for ρ^r has been developed for a cylindrical wave incident on a two-dimensional convex scattering surface corresponding to positive values of ρ_a. For this arrangement the caustic resides within the reflecting surface, and the rays seem to emanate from a virtual source (image) located at the caustic. This is equivalent to the Cassegrain reflector arrangement where the virtual feed focal point for the main reflector is behind the convex subreflector as shown in Figures 13-28 and 13-29 of [38]. If the curved scattering surface is concave, corresponding to negative values of ρ_a, the value of ρ^r is obtained by making ρ_a negative. In this case the caustic resides outside the reflecting surface, and the rays seem to emanate from

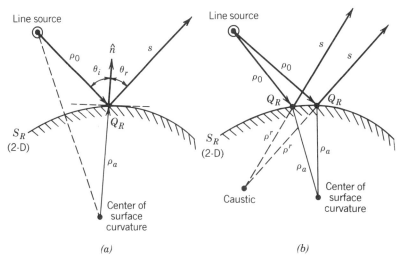

FIGURE 13-7 Line source near a two-dimensional curved surface. (*a*) Reflection point. (*b*) Caustic.

GEOMETRICAL OPTICS 761

that location. This is analogous to the Gregorian reflector arrangement where the effective focal point of the main reflector is between the concave subreflector and main reflector as shown in Figure 13-29 of [38]. Although ρ^r of (13-30) was derived for cylindrical wave incidence on a two-dimensional convex curved surface, it can also be used when the plane of incidence coincides with any of the principal planes of curvature of the reflecting surface.

The application of (13-29) and (13-30) can best be demonstrated by an example.

Example 13-2. An electric line source of infinite length and constant current I_0 is placed symmetrically a distance h above an electric conducting strip of width w and infinite length, as shown in Figure 13-8a. The length of the line is placed parallel to the z axis. Assuming a free-space medium and far-field observations ($\rho \gg w$, $\rho \gg h$), derive expressions for the incident and reflected electric field components. Then compute and plot the normalized

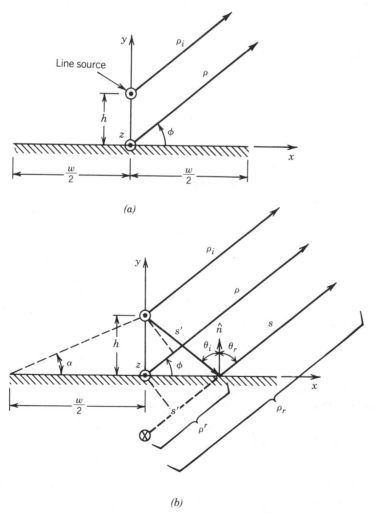

FIGURE 13-8 Line source above a finite width strip. (*a*) Coordinate system. (*b*) Reflection geometry.

amplitude distribution (in decibels) of the incident, reflected, and incident plus reflected geometrical optics fields for $h = 0.50\lambda$ when $w =$ infinite and $w = 2\lambda$. Normalize the fields with respect to the maximum of the total geometrical optics field.

Solution. The analysis begins by first determining the incident (direct) field radiated by the source in the absence of the strip that is given by (11-10a), or

$$E_z^i = -\frac{\beta^2 I_0}{4\omega\varepsilon} H_0^{(2)}(\beta\rho_i)$$

where $H_0^{(2)}(\beta\rho_i)$ is the Hankel function of the second kind and of order zero, and ρ_i is the distance from the source to the observation point.

For far-zone observations ($\beta\rho_i \to$ large) the Hankel function can be replaced by its asymptotic expansion

$$H_0^{(2)}(\beta\rho_i) \stackrel{\beta\rho_i \to \text{large}}{\simeq} \sqrt{\frac{2j}{\pi\beta\rho_i}} e^{-j\beta\rho_i}$$

This allows us to write the incident field as

$$E_z^i = E_0 \frac{e^{-j\beta\rho_i}}{\sqrt{\rho_i}}$$

where

$$E_0 = \left\{ -\frac{\beta^2 I_0}{4\omega\varepsilon} \sqrt{\frac{2j}{\pi\beta}} \right\}$$

Using (13-29) and referring to Figure 13-8b, the reflected field can be written as

$$E_z^r = E_z^i(\rho_i = s')(-1)\sqrt{\frac{\rho^r}{\rho^r + s}} e^{-j\beta s} = -E_0 \frac{e^{-j\beta s'}}{\sqrt{s'}} \sqrt{\frac{\rho^r}{\rho^r + s}} e^{-j\beta s}$$

The wave front radius of curvature of the reflected field can be found using (13-30), that is,

$$\frac{1}{\rho^r} = \frac{1}{s'} + \frac{2}{\infty \cos\theta_i} = \frac{1}{s'} \Rightarrow \rho^r = s'$$

Therefore the reflected field can now be written as

$$E_z^r = -E_0 \frac{e^{-j\beta s'}}{\sqrt{s'}} \sqrt{\frac{s'}{s' + s}} e^{-j\beta s} = -E_0 \frac{e^{-j\beta(s+s')}}{\sqrt{s+s'}} = -E_0 \frac{e^{-j\beta\rho_r}}{\sqrt{\rho_r}}$$

It is apparent from Figure 13-8b that the reflected rays seem to emanate from a virtual (image) source which is also a caustic for the reflected fields. The wave front radius of curvature $\rho^r = s'$ of the reflected field also represents the distance of the caustic (image) from the point of reflection. The reflected field could also have been obtained very simply by using image theory. However, we chose to use the equations of geometrical optics to demonstrate the principles and applications of geometrical optics.

For far-field observations ($\rho_i \gg w$, $\rho_i \gg h$)

$$\left. \begin{array}{l} \rho_i = \rho - h \cos\left(\dfrac{\pi}{2} - \phi\right) = \rho - h \sin\phi \\[2mm] \rho_r = \rho + h \cos\left(\dfrac{\pi}{2} - \phi\right) = \rho + h \sin\phi \end{array} \right\} \quad \text{for phase variations}$$

$$\rho_i \simeq \rho_r \simeq \rho \qquad \text{for amplitude variations}$$

Therefore the incident (direct) and reflected fields can be reduced to

$$E_z^i = E_0 e^{+j\beta h \sin\phi} \frac{e^{-j\beta\rho}}{\sqrt{\rho}} \qquad 0 \le \phi \le \pi + \alpha, \quad 2\pi - \alpha \le \phi \le 2\pi$$

$$E_z^r = -E_0 e^{-j\beta h \sin\phi} \frac{e^{-j\beta\rho}}{\sqrt{\rho}} \qquad \alpha \le \phi \le \pi - \alpha$$

and the total field can be written as

$$E_z^t = \begin{cases} E_z^i = E_0 e^{+j\beta h \sin\phi} \dfrac{e^{-j\beta\rho}}{\sqrt{\rho}} & \begin{array}{l} 0 \le \phi \le \alpha, \\ \pi - \alpha \le \phi \le \pi + \alpha, \\ 2\pi - \alpha \le \phi \le 2\pi \end{array} \\[4mm] E_z^i + E_z^r = 2jE_0 \sin(\beta h \sin\phi)\dfrac{e^{-j\beta\rho}}{\sqrt{\rho}} & \alpha \le \phi \le \pi - \alpha \\[4mm] 0 & \pi + \alpha \le \phi \le 2\pi - \alpha \end{cases}$$

Normalized amplitude patterns for $w = 2\lambda$ and $h = 0.50\lambda$ computed using the preceding geometrical optics fields in their respective regions are plotted (in decibels) in Figure 13-9 where discontinuities created along the incident and reflected shadow boundaries by the geometrical optics fields are apparent. The total amplitude pattern assuming an infinite ground plane is also displayed in Figure 13-9.

In summary, geometrical optics methods approximate the fields by the leading term of the Luneberg–Kline expansion for large ω, but they fail along caustics. Improvements can be incorporated into the solutions by finding higher-order terms $\mathbf{E}_1(\mathbf{R}_1), \mathbf{E}(\mathbf{R}_2), \ldots$ in the Luneberg–Kline series expansion. Higher-order Luneberg–Kline expansions have been derived for fields scattered from spheres, cylinders, and other curved surfaces with simple geometries [30, 31]. These solutions exhibit the following tendencies.

1. They improve the high-frequency field approximations if the observation specular point is not near edges, shadow boundaries, or other surface discontinuities.
2. They become singular as the observation specular point approaches a shadow boundary on the surface.

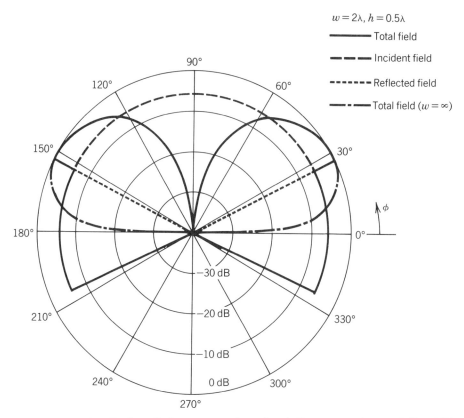

FIGURE 13-9 Amplitude radiation pattern of an electric line source above a finite width strip.

3. They do not correct for geometrical optics discontinuities along incident and reflection shadow boundaries.
4. They do not describe the diffracted fields in the shadow region.

Because of some of these deficiencies, in addition to being quite complex, higher-order Luneberg–Kline expansion methods cannot be used to treat diffraction. Therefore other approaches, usually somewhat heuristic in nature, must be used to introduce diffraction in order to improve geometrical optics approximations. It should be noted, however, that for sufficiently large ω geometrical optics fields may dominate the scattering phenomena and may alone provide results that often agree quite well with measurements. This is more evident for backscattering from smooth curved surfaces with large radii of curvature. In those instances corrections to the fields predicted by geometrical optics methods may not be necessary. However, for other situations where such solutions are inaccurate, corrections are usually provided by including diffraction. Therefore a combination of geometrical optics and diffraction techniques often leads to solutions of many practical engineering problems whose results agree extremely well with measurements. This has been demonstrated in many applications [19–25, 39]. Because of their extreme importance and ease of application, diffraction techniques will be next introduced, discussed, and applied.

13.3 GEOMETRICAL THEORY OF DIFFRACTION: EDGE DIFFRACTION

Examining high-frequency diffraction problems has revealed that their solutions contain terms of fractional power that are not always included in the geometrical optics expression of (13-15a) or in the Luneberg–Kline series solution. For example, geometrical optics fail to account for the energy diffracted into the shadow region when the incident rays are tangent to the surface of a curved object and for the diffracted energy when the surface contains an edge, vertex, or corner. In addition, caustics of the diffracted fields are located at the boundary surface. Therefore some semi-heuristic approaches must be used to provide correction factors that improve the geometrical optics approximation.

13.3.1 Amplitude, Phase, and Polarization Relations

To introduce diffraction, let us assume that the smooth surface S of Figure 13-5a has a curved edge as shown in Figure 13-10a. When an electromagnetic wave impinges on this curved edge, diffracted rays emanate from the edge whose leading term of the high-frequency solution for the electric field takes the form of

$$\mathbf{E}^d(\mathbf{R}) \simeq \frac{e^{-j\beta\psi_d(\mathbf{R})}}{\sqrt{\beta}} \mathbf{A}(\mathbf{R}) \qquad (13\text{-}31)$$

where ψ_d = eikonal surface for the diffracted rays
$\mathbf{A}(\mathbf{R})$ = field factor for the diffracted rays

Substituting (13-31) into (13-11) and (13-12), and referring to the geometry of Figure 13-10a, it can be shown that the diffracted field can be written as

$$\mathbf{E}^d(s) = \left[\frac{\mathbf{A}(0')}{\sqrt{\beta}} e^{-j\beta\psi_d(0')}\right] \sqrt{\frac{\rho_c' \rho_c}{(\rho_c' + s)(\rho_c + s)}} e^{-j\beta s}$$

$$\mathbf{E}^d(s) = \left[\mathbf{E}^d(0')\right] \sqrt{\frac{\rho_c' \rho_c}{(\rho_c' + s)(\rho_c + s)}} e^{-j\beta s} \qquad (13\text{-}32)$$

where $\mathbf{E}^d(0')$ = diffracted field at the reference point $0'$
s = distance along the diffracted ray from the reference point $0'$
ρ_c' = distance from diffraction point Q_D (first caustic of diffracted field) to reference point $0'$
ρ_c = distance between the second caustic of diffracted field and reference point $0'$

It would have been more convenient to choose the reference point $0'$ to coincide with the point of diffraction point Q_D located on the diffracting edge. However, like the geometrical optics rays, the diffracted rays form an astigmatic tube of the form shown in Figure 13-10b where the caustic line PP' coincides with the diffracting edge. Because the diffraction point is a caustic of the diffracted field, it is initially more straightforward to choose the reference point away from the edge diffraction caustic Q_D. However, the diffracted field of (13-32) should be independent of the location of the reference point $0'$, including $\rho_c' = 0$. Therefore the

766 GEOMETRICAL THEORY OF DIFFRACTION

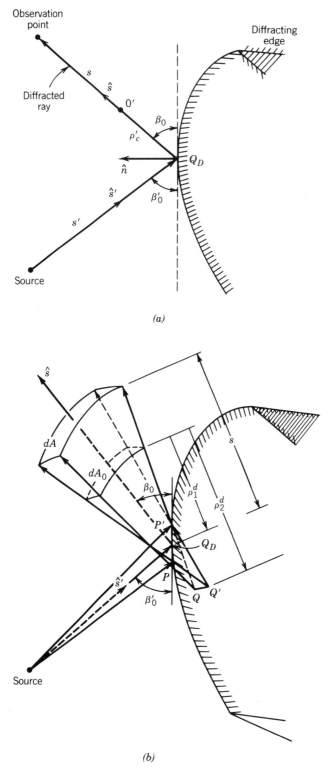

FIGURE 13-10 Geometry for diffraction by a curved edge. (*a*) Diffraction point. (*b*) Astigmatic tube of rays.

diffracted field of (13-32) must be such that

$$\lim_{\rho_c' \to 0} \mathbf{E}^d(0')\sqrt{\rho_c'} = \text{finite} \qquad (13\text{-}33)$$

and must be equal to

$$\lim_{\rho_c' \to 0} \mathbf{E}^d(0')\sqrt{\rho_c'} = \mathbf{E}^i(Q_D) \cdot \overline{\mathbf{D}} \qquad (13\text{-}33a)$$

where $\mathbf{E}^i(Q_D)$ = incident field at the point of diffraction
$\overline{\mathbf{D}}$ = dyadic diffraction coefficient (analogous to dyadic reflection coefficient)

Using (13-33a), the diffracted field of (13-32) reduces to

$$\mathbf{E}^d(s) = \lim_{\rho_c' \to 0} \left\{ \left[\mathbf{E}^d(0')\sqrt{\rho_c'}\right] \sqrt{\frac{\rho_c}{(\rho_c' + s)(\rho_c + s)}}\, e^{-j\beta s} \right\}$$

$$\mathbf{E}^d(s) = \mathbf{E}^i(Q_D) \cdot \overline{\mathbf{D}} \sqrt{\frac{\rho_c}{s(\rho_c + s)}}\, e^{-j\beta s} \qquad (13\text{-}34)$$

which has the form of

$$\underbrace{\mathbf{E}^d(s)}_{} = \underbrace{\mathbf{E}^i(Q_D)}_{\substack{\text{Field at} \\ \text{reference} \\ \text{point}}} \cdot \underbrace{\overline{\mathbf{D}}}_{\substack{\text{Diffraction} \\ \text{coefficient} \\ \text{(usually a} \\ \text{dyadic)}}} \underbrace{A(\rho_c, s)}_{\substack{\text{Spatial} \\ \text{attenuation} \\ \text{(spreading,} \\ \text{divergence)} \\ \text{factor}}} \underbrace{e^{-j\beta s}}_{\substack{\text{Phase} \\ \text{factor}}} \qquad (13\text{-}34a)$$

and compares with that of (13-20) for reflection. In (13-34) and (13-34a)

$$A(\rho_c, s) = \sqrt{\frac{\rho_c}{s(\rho_c + s)}} \qquad (13\text{-}34b)$$

= spatial attenuation (spreading, divergence) factor for a curved surface

ρ_c = distance between the reference point $Q_D(s = 0)$ at the edge (also first caustic of the diffracted rays) and the second caustic of the diffracted rays.

In general ρ_c is a function of the following.

1. Wave front curvature of the incident field.
2. Angles of incidence and diffraction, relative to unit vector normal to edge at the point Q_D of diffraction.
3. Radius of curvature of diffracting edge at point Q_D of diffraction.

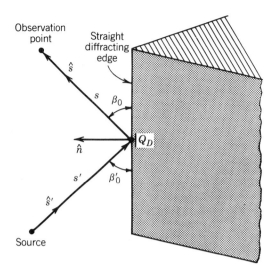

FIGURE 13-11 Diffraction by a wedge with a straight edge.

An expression for ρ_c is given by (13-100a) in Section 13.3.4.

The diffracted field, which is determined by a generalization of Fermat's principle [7, 9], is initiated at a point on the surface of the object where discontinuities are formed along the incident and reflected shadow boundaries. As represented in (13-34) and (13-34a), the initial value of the field of a diffracted ray is determined from the incident field with the aid of an appropriate diffraction coefficient $\overline{\mathbf{D}}$, which in general is a dyadic for electromagnetic fields. The amplitude is assumed to vary in accordance with the principle of conservation of energy flux along a tube of rays. Appropriate phase jumps of $+90°$ are added each time a ray passes through a caustic at $s = 0$ and $s = -\rho_c$, as properly accounted for in (13-34) and (13-34a). The phase of the field on a diffracted ray is assumed to be equal to the product of the optical lengths of the ray, from some reference point Q_D, and the phase constant β of the medium.

When the edge is straight, the source is located a distance s' from the point of diffraction, and the observations are made at a distance s from it, as shown in Figure 13-11. The diffracted field can then be written as

$$\boxed{\mathbf{E}^d(s) = \mathbf{E}^i(Q_D) \cdot \overline{\mathbf{D}} A(s', s) e^{-j\beta s}} \qquad (13\text{-}35)$$

where

$$\boxed{A(s', s) = \begin{cases} \dfrac{1}{\sqrt{s}} & \text{for plane and conical wave incidences} \\[6pt] \dfrac{1}{\sqrt{\rho}}, \ \rho = s \sin \beta_0 & \text{for cylindrical wave incidence} \\[6pt] \sqrt{\dfrac{s'}{s(s+s')}} \underset{s \gg s'}{\simeq} \dfrac{\sqrt{s'}}{s} & \text{for spherical wave incidence} \end{cases}}$$

$(13\text{-}35a)$
$(13\text{-}35b)$
$(13\text{-}35c)$

The $A(s', s)$ formulas of (13-35a) through (13-35c) are obtained by letting ρ_c in

(13-34a) be equal to $\rho_c = \infty$ for plane, cylindrical, and conical wave incidence and $\rho_c = s'$ for spherical wave incidence.

The diffraction coefficients are usually determined from the asymptotic solutions of canonical problems that have the same local geometry at the points of diffraction as the object(s) of investigation. One of the simplest geometries, which will be discussed in this chapter, is a conducting wedge [40–42]. Another is that of a conducting, smooth, and convex surface [43–46]. The main objectives in the remaining part of this chapter are to introduce and apply the diffraction coefficients for the canonical problem of the conducting wedge. Curved surface diffraction is derived in [43–46]. In some of the sections the discussion on diffraction follows that in [38].

13.3.2 Straight Edge Diffraction: Normal Incidence

In order to examine the manner in which fields are diffracted by edges, it is necessary to have a diffraction coefficient available. To derive a diffraction coefficient for an edge, we need to consider a canonical problem, one that has the same local geometry near the edge, like that shown in Figure 13-12a.

Let us begin by assuming that a source is placed near the two-dimensional electric conducting wedge of included angle $(2 - n)\pi$ radians. If observations are made on a circle of constant radius ρ from the edge of the wedge, it is quite clear that in addition to the direct ray (OP) there are rays that are reflected from the side of the wedge (OQ_RP) which contribute to the intensity at point P. These rays obey Fermat's principle, that is, they minimize the path between points O and P by including points on the side of the wedge, and deduce Snell's law of reflection. It would then seem appropriate to extend the class of such points to include in the trajectory rays that pass through the edge of the wedge (OQ_DP) leading to the generalized Fermat principle [7]. This class of rays is designated as diffracted rays and they lead to the *law of diffraction*.

By considering rays that obey only geometrical optics radiation mechanisms (direct and reflected), we can separate the space surrounding the wedge into three different field regions. Using the geometrical coordinates of Figure 13-12b, the following geometrical optics fields will contribute to the corresponding regions:

Region I	Region II	Region III
$0 < \phi < \pi - \phi'$	$\pi - \phi' < \phi < \pi + \phi'$	$\pi + \phi' < \phi < n\pi$
Direct	Direct	...
Reflected		

With these fields, it is evident that the following will occur:

1. Discontinuities in the field will be formed along the RSB boundary separating regions I and II (reflected shadow boundary, $\phi = \pi - \phi'$), and along the ISB boundary separating regions II and III (incident shadow boundary, $\phi = \pi + \phi'$).
2. No field will be present in region III (shadow region).

Since neither of the preceding results should be present in a physically realizable field, modifications and/or additions need to be made.

To *remove* the discontinuities along the boundaries and to *modify* the fields in *all* three regions, diffracted fields must be included. To obtain expressions for the

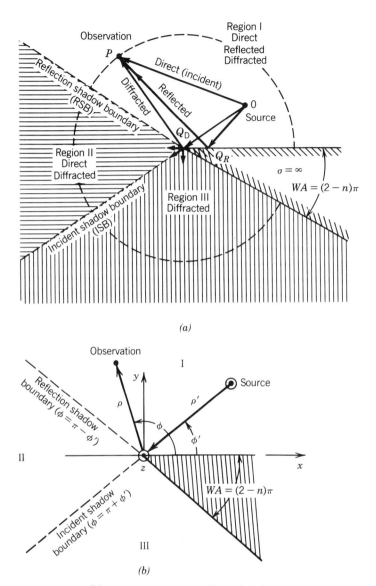

FIGURE 13-12 Line source near a two-dimensional conducting wedge. (*a*) Region separation. (*b*) Coordinate system. Region I: $0 \leq \phi < \pi - \phi'$. Region II: $\pi - \phi' < \phi < \pi + \phi'$. Region III: $\pi + \phi' < \phi < n\pi$.

diffracted field distribution, we assume that the source in Figure 13-12 is an infinite line source (either electric or magnetic).

The fields of an electric line source satisfy the homogeneous Dirichlet boundary conditions $E_z = 0$ on both faces of the wedge and the fields of the magnetic line source satisfy the homogeneous Neumann boundary condition $\partial E_z/\partial \phi = 0$ or $H_z = 0$ on both faces of the wedge. The faces of the wedge are formed by two semi-infinite intersecting planes. The infinitely long line source is parallel to the edge of the wedge, and its position is described by the coordinate (ρ', ϕ'). The typical field point is denoted by (ρ, ϕ), as shown in Figure 13-12*b*. The line source is assumed to have constant current.

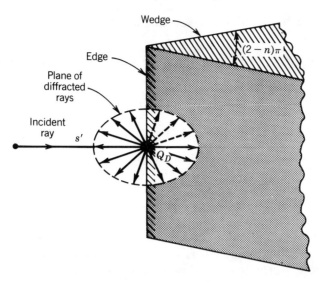

FIGURE 13-13 Wedge plane of diffraction for normal incidence.

Initially only normal incidence diffraction by a straight edge, as shown in Figure 13-13, will be considered. For this situation, the plane of diffraction is perpendicular to the edge of the wedge. Oblique incidence diffraction (Figure 13-30) and curved-edge diffraction (Figure 13-34) will be discussed, respectively, in Sections 13.3.3 and 13.3.4.

The diffraction coefficient for the geometries of Figures 13-12 and 13-13 is obtained by [42]:

1. Finding the Green's function solution in the form of an infinite series using modal techniques, and then approximating it for large values of $\beta\rho$ (far-field observations).
2. Converting the infinite series Green's function solution into an integral.
3. Performing, on the integral Green's function, a high-frequency asymptotic expansion (in inverse powers of $\beta\rho$) using standard techniques, such as the *method of steepest descent*.

An abbreviated derivation of this procedure will now be presented.

A. MODAL SOLUTION

Using modal techniques, the total radiation electric field for an electric line source of current I_e was found in Chapter 11, Section 11.6.3, to be that of (11-207a), or

$$E_z^e = -\frac{\omega\mu I_e}{4}G \Rightarrow \mathbf{H}^e = -\frac{1}{j\omega\mu}\nabla \times \mathbf{E} \qquad (13\text{-}36)$$

and the total magnetic field for a magnetic line source of current I_m to be that of (11-207b), or

$$H_z^m = \frac{\omega\varepsilon I_m}{4}G \Rightarrow \mathbf{E}^m = +\frac{1}{j\omega\varepsilon}\nabla \times \mathbf{H} \qquad (13\text{-}37)$$

where G is referred to as the Green's function, and it is given by (11-208a) as

$$G = \frac{1}{n} \sum_{m=0}^{\infty} \varepsilon_m J_{m/n}(\beta\rho) H_{m/n}^{(2)}(\beta\rho') \left[\cos\frac{m}{n}(\phi - \phi') \pm \cos\frac{m}{n}(\phi + \phi') \right] \quad \text{for } \rho \leq \rho' \tag{13-38}$$

$$\varepsilon_m = \begin{cases} 1 & m = 0 \\ 2 & m \neq 0 \end{cases} \tag{13-38a}$$

For the case $\rho \geq \rho'$, ρ and ρ' are interchanged. The plus sign between the two cosine terms is used if the boundary condition is of the homogeneous Neumann type, $\partial G/\partial \phi = 0$ on both faces of the wedge. For the homogeneous Dirichlet boundary condition, $G = 0$ on both faces of the wedge, the minus sign is used. In acoustic terminology the Neumann boundary condition is referred to as *hard* polarization and the Dirichlet is referred to as the *soft* polarization. This series is an exact solution to the time-harmonic, inhomogeneous wave equation of a radiating line source and wedge embedded in a linear, isotropic, homogeneous, lossless medium.

B. HIGH-FREQUENCY ASYMPTOTIC SOLUTION

Many times it is necessary to determine the total radiation field when the line source is far removed from the vertex of the wedge. In such cases, (13-38) can be simplified by replacing the Hankel function by the first term of its asymptotic expansion, that is, by the relation

$$H_{m/n}^{(2)}(\beta\rho') \underset{\beta\rho' \to \infty}{\simeq} \sqrt{\frac{2}{\pi\beta\rho'}} e^{-j[\beta\rho' - \pi/4 - (m/n)(\pi/2)]} \tag{13-39}$$

This substitution reduces G to

$$G = \sqrt{\frac{2}{\pi\beta\rho'}} e^{-j(\beta\rho' - \pi/4)} \frac{1}{n} \sum_{m=0}^{\infty} \varepsilon_m J_{m/n}(\beta\rho) e^{+j(m/n)(\pi/2)}$$

$$\times \left[\cos\frac{m}{n}(\phi - \phi') \pm \cos\frac{m}{n}(\phi + \phi') \right]$$

$$G = \sqrt{\frac{2}{\pi\beta\rho'}} e^{-j(\beta\rho' - \pi/4)} F(\beta\rho) \tag{13-40}$$

where

$$F(\beta\rho) = \frac{1}{n} \sum_{m=0}^{\infty} \varepsilon_m J_{m/n}(\beta\rho) e^{+j(m/n)(\pi/2)} \left[\cos\frac{m}{n}(\phi - \phi') \pm \cos\frac{m}{n}(\phi + \phi') \right] \tag{13-40a}$$

$F(\beta\rho)$ is used to represent either the total E_z^e, when the source is an electric line (soft polarization), or H_z^m, when the source is a magnetic line (hard polarization).

The infinite series of (13-40a) converges rapidly for small values of $\beta\rho$. For example if $\beta\rho$ is less than 1, less than 15 terms are required to achieve a 5 significant figure accuracy. However, at least 40 terms should be included when $\beta\rho$ is 10 to achieve the 5 significant figure accuracy. To demonstrate the variations of (13-40a),

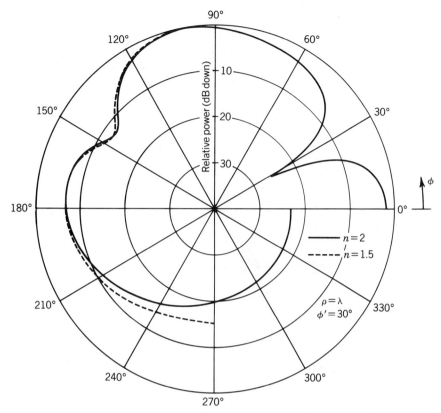

FIGURE 13-14 Normalized amplitude pattern of a hard polarization plane wave incident normally on a two-dimensional conducting wedge.

the patterns of a unit amplitude plane wave of hard polarization incident upon a half plane ($n = 2$) and 90° wedge ($n = 3/2$), computed using (13-40a), are shown in Figure 13-14. Computed patterns for the soft polarization are shown in Figure 13-15.

Whereas (13-40a) represents the total field, in the space around a wedge of included angle $WA = (2 - n)\pi$, of a unity amplitude plane wave incident upon the wedge, the field of a unit amplitude cylindrical wave incident upon the same wedge and observations made at very large distances can be obtained by the reciprocity principle. Graphically this is illustrated in Figures 13-16a and b. Analytically the cylindrical wave incidence fields of Figure 13-16b can be obtained from (13-40a) and Figure 13-16a by substituting in (13-40a) $\rho = \rho'$, $\phi = \phi'$, and $\phi' = \phi$.

A high-frequency asymptotic expansion for $F(\beta\rho)$ in inverse powers of $\beta\rho$ is very useful for computational purposes, because of the slow convergence of (13-40a) for large values of $\beta\rho$. In order to derive an asymptotic expression for $F(\beta\rho)$ by the *conventional method of steepest descent* for isolated poles and saddle points (see Appendix VI), it must first be transformed into an integral or integrals of the form

$$P(\beta\rho) = \int_C H(z) e^{\beta\rho h(z)} \, dz \qquad (13\text{-}41)$$

and then evaluated for large $\beta\rho$ by means of the method of steepest descent [8, 47].

774 GEOMETRICAL THEORY OF DIFFRACTION

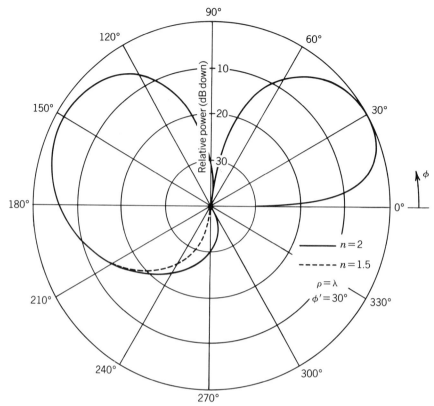

FIGURE 13-15 Normalized amplitude pattern of a soft polarization plane wave incident normally on a two-dimensional conducting wedge.

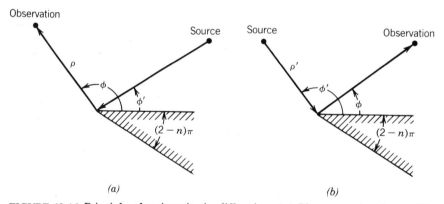

FIGURE 13-16 Principle of reciprocity in diffraction. (*a*) Plane wave incidence. (*b*) Cylindrical wave incidence.

To accomplish this, first the cosine terms are expressed in complex form by

$$\cos\left(\frac{m}{n}\xi^{\mp}\right) = \frac{1}{2}[e^{j(m/n)\xi^{\mp}} + e^{-j(m/n)\xi^{\mp}}] \tag{13-42}$$

where

$$\xi^{\mp} = \phi \mp \phi' \tag{13-42a}$$

and the Bessel functions are replaced by contour integrals in the complex z plane of the form

$$J_{m/n}(\beta\rho) = \frac{1}{2\pi}\int_C e^{j[\beta\rho\cos z + m/n(z-\pi/2)]}\,dz \tag{13-43a}$$

or

$$J_{m/n}(\beta\rho) = \frac{1}{2\pi}\int_{C'} e^{j[\beta\rho\cos z - m/n(z+\pi/2)]}\,dz \tag{13-43b}$$

where the paths C and C' are shown in Figure 13-17a. Doing this and interchanging the order of integration and summation, it can be shown that (13-40a) can be written as

$$F(\beta\rho) = I(\beta\rho, \phi - \phi', n) \pm I(\beta\rho, \phi + \phi', n) \tag{13-44}$$

where

$$I(\beta\rho, \xi^{\mp}, n) = \frac{1}{2\pi n}\int_C e^{j\beta\rho\cos z}\sum_{m=1}^{\infty} e^{+j(m/n)(\xi^{\mp}+z)}\,dz$$

$$+ \frac{1}{2\pi n}\int_{C'} e^{j\beta\rho\cos z}\sum_{m=0}^{\infty} e^{-j(m/n)(\xi^{\mp}+z)}\,dz \tag{13-44a}$$

with $\xi^{\mp} = \phi \mp \phi'$.

Using the series expansions of

$$-\frac{1}{1-x^{-1}} = x(1 + x + x^2 + x^3 + \cdots) = \sum_{m=1}^{\infty} x^m \tag{13-45a}$$

$$\frac{1}{1-x^{-1}} = 1 + x^{-1} + x^{-2} + x^{-3} + \cdots = \sum_{m=0}^{\infty} x^{-m} \tag{13-45b}$$

and that

$$\frac{1}{1 - e^{-j(\xi^{\mp}+z)/n}} = \frac{e^{j(\xi^{\mp}+z)/2n}}{e^{j(\xi^{\mp}+z)/2n} - e^{-j(\xi^{\mp}+z)/2n}} = \frac{1}{2} + \frac{1}{2j}\cot\left(\frac{\xi^{\mp}+z}{2n}\right) \tag{13-46}$$

we can ultimately write (13-44) or (13-40a) as

$$F(\beta\rho) = \frac{1}{4\pi jn}\int_{(C'-C)} \cot\left(\frac{\phi - \phi' + z}{2n}\right) e^{j\beta\rho\cos z}\,dz$$

$$\pm \frac{1}{4\pi jn}\int_{(C'-C)} \cot\left(\frac{\phi + \phi' + z}{2n}\right) e^{j\beta\rho\cos z}\,dz \tag{13-47}$$

where the negative sign before C indicates that this integration path is to be traversed in the direction opposite to that shown in Figure 13-17a. It is now clear that $F(\beta\rho)$ has been written in an integral of the form of (13-41) which can be evaluated asymptotically by contour integration and by the method of steepest descent.

776 GEOMETRICAL THEORY OF DIFFRACTION

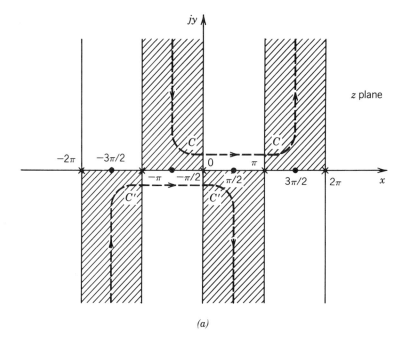

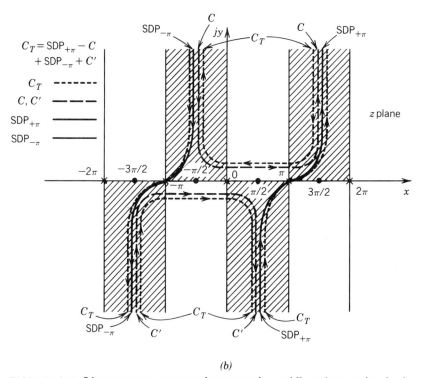

FIGURE 13-17 Line contours, steepest descent paths, saddle points, and poles in complex z plane for asymptotic evaluation of wedge diffraction formulas. (*a*) Contours for Bessel function. (*b*) Steepest descent paths, saddle points, and poles.

C. METHOD OF STEEPEST DESCENT

Equation 13-47 can also be written as

$$F(\beta\rho) = F_1(\beta\rho) \pm F_2(\beta\rho) \tag{13-48}$$

$$F_1(\beta\rho) = \frac{1}{4\pi jn} \int_{(C'-C)} H_1(z) e^{\beta\rho h_1(z)} \, dz \tag{13-48a}$$

$$F_2(\beta\rho) = \frac{1}{4\pi jn} \int_{(C'-C)} H_2(z) e^{\beta\rho h_2(z)} \, dz \tag{13-48b}$$

$$H_1(z) = \cot\left[\frac{(\phi - \phi') + z}{2n}\right] = \cot\left(\frac{\xi^- + z}{2n}\right) \tag{13-48c}$$

$$H_2(z) = \cot\left[\frac{(\phi + \phi') + z}{2n}\right] = \cot\left(\frac{\xi^+ + z}{2n}\right) \tag{13-48d}$$

$$h_1(z) = h_2(z) = j\cos(z) \tag{13-48e}$$

The evaluation of (13-48) can be accomplished by evaluating separately (13-48a) and (13-48b) and summing the results. Let us examine first the evaluation of (13-48a) in detail. A similar procedure can be used for (13-48b).

Using the complex z-plane closed contour C_T, we can write by referring to Figure 13-17b that

$$\frac{1}{4\pi jn} \oint_{C_T} H_1(z) e^{\beta\rho h_1(z)} \, dz = \frac{1}{4\pi jn} \int_{(C'-C)} H_1(z) e^{\beta\rho h_1(z)} \, dz$$

$$+ \frac{1}{4\pi jn} \int_{\text{SDP}_{+\pi}} H_1(z) e^{\beta\rho h_1(z)} \, dz$$

$$+ \frac{1}{4\pi jn} \int_{\text{SDP}_{-\pi}} H_1(z) e^{\beta\rho h_1(z)} \, dz \tag{13-49}$$

The closed contour C_T is equal to the sum of

$$C_T = C' + \text{SDP}_{+\pi} - C + \text{SDP}_{-\pi} \tag{13-49a}$$

where $\text{SDP}_{\pm\pi}$ is used to represent the steepest descent paths passing through the saddle points $\pm\pi$. We can rewrite (13-49) as

$$F_1(\beta\rho) = \frac{1}{4\pi jn} \int_{(C'-C)} H_1(z) e^{\beta\rho h_1(z)} \, dz = \frac{1}{4\pi jn} \oint_{C_T} H_1(z) e^{\beta\rho h_1(z)} \, dz$$

$$- \frac{1}{4\pi jn} \int_{\text{SDP}_{+\pi}} H_1(z) e^{\beta\rho h_1(z)} \, dz - \frac{1}{4\pi jn} \int_{\text{SDP}_{-\pi}} H_1(z) e^{\beta\rho h_1(z)} \, dz \tag{13-50}$$

which is the same as (13-48a). Therefore (13-48a) can be integrated by evaluating the three terms on the right side of (13-50). The closed contour of the first term on the right side of (13-50) is evaluated using residue complex calculus and the other two terms are evaluated using the method of steepest descent. When evaluated, it will be shown that the first term on the right side of (13-50) will represent the incident geometrical optics and the other two terms will represent what will be referred to as the incident diffracted field. A similar interpretation will be given when (13-48b) is evaluated; it represents the reflected geometrical optics and reflected diffracted fields.

GEOMETRICAL THEORY OF DIFFRACTION

Using residue complex calculus, we can write the first term on the right side of (13-50) as [48]

$$\frac{1}{4\pi jn}\oint_{C_T} H_1(z)e^{\beta\rho h_1(z)}\,dz = 2\pi j\sum_p \text{Res}(z=z_p)$$

$$= 2\pi j\sum_p (\text{residues of the poles enclosed by } C_T) \quad (13\text{-}51)$$

To evaluate (13-51) we first rewrite it as

$$\frac{1}{4\pi jn}\oint_{C_T} H_1(z)e^{\beta\rho h_1(z)}\,dz$$

$$= \frac{1}{4\pi jn}\oint_{C_T}\cot\left[\frac{(\phi-\phi')+z}{2n}\right]e^{j\beta\rho\cos(z)}\,dz$$

$$= \oint_{C_T}\frac{1}{4\pi jn}\cot\left[\frac{(\phi-\phi')+z}{2n}\right]e^{j\beta\rho\cos(z)}\,dz = \oint_{C_T}\frac{N(z)}{D(z)}\,dz \quad (13\text{-}52)$$

where

$$\frac{N(z)}{D(z)} = \frac{\cos\left[\dfrac{(\phi-\phi')+z}{2n}\right]e^{j\beta\rho\cos(z)}}{4\pi jn\sin\left[\dfrac{(\phi-\phi')+z}{2n}\right]} \quad (13\text{-}52\text{a})$$

$$N(z) = \cos\left[\frac{(\phi-\phi')+z}{2n}\right]e^{j\beta\rho\cos(z)} \quad (13\text{-}52\text{b})$$

$$D(z) = 4\pi jn\sin\left[\frac{(\phi-\phi')+z}{2n}\right] \quad (13\text{-}52\text{c})$$

Equation 13-52 has simple poles that occur when

$$\left[\frac{(\phi-\phi')+z}{2n}\right]_{z=z_p} = \pi N \qquad N=0,\pm 1,\pm 2,\ldots \quad (13\text{-}53)$$

or

$$z_p = -(\phi-\phi') + 2\pi nN \quad (13\text{-}53\text{a})$$

provided that

$$-\pi \le z_p = -(\phi-\phi') + 2\pi nN \le +\pi \quad (13\text{-}53\text{b})$$

Using residue complex calculus, the residues of (13-51) or (13-52) for simple poles (no branch points, etc.) can be found using [48]

$$\text{Res}(z=z_p) = \left.\frac{N(z)}{\dfrac{dD(z)}{dz}}\right|_{z=z_p} = \left.\frac{N(z)}{D'(z)}\right|_{z=z_p} \quad (13\text{-}54)$$

where

$$N(z)|_{z=z_p} = \cos\left[\frac{(\phi - \phi') + z}{2n}\right] e^{j\beta\rho\cos(z)}\bigg|_{z=-(\phi-\phi')+2\pi nN}$$

$$= \cos(\pi N) e^{j\beta\rho\cos[-(\phi-\phi')+2\pi nN]} \qquad (13\text{-}54a)$$

$$D'(z)|_{z=z_p} = 2\pi j \cos\left[\frac{(\phi - \phi') + z}{2n}\right]\bigg|_{z=-(\phi-\phi')+2\pi nN} = 2\pi j \cos(\pi N) \quad (13\text{-}54b)$$

Thus (13-54) and (13-51) can be written, respectively, as

$$\text{Res}(z = z_p) = \frac{1}{2\pi j} e^{j\beta\rho\cos[-(\phi-\phi')+2\pi nN]} \qquad (13\text{-}55a)$$

$$\frac{1}{4\pi jn}\oint_{C_T} H_1(z) e^{\beta\rho h_1(z)}\,dz = e^{j\beta\rho\cos[-(\phi-\phi')+2\pi nN]} U\left[\pi - |-(\phi - \phi') + 2\pi nN|\right]$$

$$(13\text{-}55b)$$

The $U(t - t_0)$ function in (13-55b) is a unit step function defined as

$$U(t - t_0) = \begin{cases} 1 & t > t_0 \\ \frac{1}{2} & t = t_0 \\ 0 & t < t_0 \end{cases} \qquad (13\text{-}56)$$

The unit step function is introduced in (13-55b) so that (13-53b) is satisfied. When $z = \pm\pi$, (13-53b) is expressed as

$$\boxed{2\pi nN^+ - (\phi - \phi') = +\pi} \qquad (13\text{-}57a)$$

for $z_p = +\pi$ and

$$\boxed{2\pi nN^- - (\phi - \phi') = -\pi} \qquad (13\text{-}57b)$$

for $z_p = -\pi$. For the principal value of N ($N^\pm = 0$) (13-55b) reduces to

$$\boxed{F_1(\beta\rho)|_{C_T} = \frac{1}{4\pi jn}\oint_{C_T} H_1(z) e^{\beta\rho h_1(z)}\,dz = e^{j\beta\rho\cos(\phi-\phi')} U\left[\pi - |-(\phi - \phi')|\right]}$$

$$(13\text{-}58)$$

This is referred to as the *incident geometrical optics* field, and it exists provided $|\phi - \phi'| < \pi$. This completes the evaluation of (13-51). Let us now evaluate the other two terms on the right side of (13-50).

In evaluating the last two terms on the right side of (13-50) the contributions from all the saddle points along the steepest descent paths must be accounted for. In this situation, however, only saddle points at $z = z_s = \pm\pi$ occur. These are found by taking the derivative of (13-48e) and setting it equal to zero. Doing this leads to

$$h_1'(z)|_{z=z_s} = -j\sin(z)|_{z=z_s} = 0 \Rightarrow z_s = \pm\pi \qquad (13\text{-}59)$$

The form used to evaluate the last two terms on the right side of (13-50) depends on whether the poles of (13-53a), which contribute to the geometrical optics field, are near or far removed from the saddle points at $z = z_s = \pm\pi$. Let us first evaluate each of the last two terms on the right side of (13-50) when the poles of (13-53a) are far removed from the saddle point of (13-59).

When the poles of (13-53a) are far removed from the saddle points of (13-59), then the last two terms on the right side of (13-50) are evaluated using the *conventional steepest descent method* for isolated poles and saddle points. Doing this we can write the last two terms on the right side of (13-50), using the saddle points of (13-59), as [8, 47]

$$\frac{1}{4\pi jn}\int_{\mathrm{SDP}_{+\pi}} H_1(z)e^{\beta\rho h_1(z)}\,dz \stackrel{\beta\rho \to \mathrm{large}}{\simeq} \frac{1}{4\pi jn}\left|\sqrt{\frac{2\pi}{-\beta\rho h_1''(z_s=+\pi)}}\right|e^{j\pi/4}$$

$$\times H_1(z_s = +\pi)e^{\beta\rho h_1(z_s=\pi)} = \frac{e^{-j\pi/4}}{2n\sqrt{2\pi\beta}}\cot\left[\frac{\pi+(\phi-\phi')}{2n}\right]\frac{e^{-j\beta\rho}}{\sqrt{\rho}} \quad (13\text{-}60\mathrm{a})$$

$$\frac{1}{4\pi jn}\int_{\mathrm{SDP}_{-\pi}} H_1(z)e^{\beta\rho h_1(z)}\,dz \stackrel{\beta\rho \to \mathrm{large}}{\simeq} \frac{1}{4\pi jn}\left|\sqrt{\frac{2\pi}{-\beta\rho h_1''(z_s=-\pi)}}\right|e^{-j3\pi/4}$$

$$\times H_1(z_s = -\pi)e^{\beta\rho h_1(z_s=-\pi)} = \frac{e^{-j\pi/4}}{2n\sqrt{2\pi\beta}}\cot\left[\frac{\pi-(\phi-\phi')}{2n}\right]\frac{e^{-j\beta\rho}}{\sqrt{\rho}} \quad (13\text{-}60\mathrm{b})$$

Combining (13-60a) and (13-60b), it can be shown that the sum of the two can be written as

$$\boxed{\begin{aligned}F_1(\beta\rho)|_{\mathrm{SDP}_{\pm\pi}} &= -\frac{1}{4\pi jn}\int_{\mathrm{SDP}_{+\pi}} H_1(z)e^{\beta\rho h_1(z)}\,dz - \frac{1}{4\pi jn}\int_{\mathrm{SDP}_{-\pi}} H_1(z)e^{\beta\rho h_1(z)}\,dz \\ &\simeq \frac{e^{-j\pi/4}}{\sqrt{2\pi\beta}}\frac{\frac{1}{n}\sin\left(\frac{\pi}{n}\right)}{\cos\left(\frac{\pi}{n}\right) - \cos\left(\frac{\phi-\phi'}{n}\right)}\frac{e^{-j\beta\rho}}{\sqrt{\rho}}\end{aligned}}$$

(13-61)

This is referred to as the *incident diffracted field*, and its form as given by (13-61) is valid provided that the poles of (13-53a) are not near the saddle points of (13-59). Another way to say this is that (13-61) is valid provided that the observations are not made at or near the incident shadow boundary of Figure 13-12. When the observations are made at the incident shadow boundary $\phi - \phi' = \pi$, (13-61) becomes infinite. Therefore another form must be used for such situations.

Following a similar procedure for the evaluation of (13-48b), it can be shown that its contributions along C_T and the saddle points can be written in forms corresponding to (13-58) and (13-61) for the evaluation of (13-48a). Thus we can write that

$$\boxed{\begin{aligned}F_2(\beta\rho)|_{C_T} &= \frac{1}{4\pi jn}\oint_{C_T} H_2(z)e^{\beta\rho h_2(z)}\,dz \\ &= e^{j\beta\rho\cos(\phi+\phi')}U[\pi-(\phi+\phi')]\end{aligned}} \quad (13\text{-}62\mathrm{a})$$

$$F_2(\beta\rho)|_{\text{SDP}_{\pm\pi}} = -\frac{1}{4\pi jn}\int_{\text{SDP}_{+\pi}} H_2(z)e^{\beta\rho h_2(z)}\,dz - \frac{1}{4\pi jn}\int_{\text{SDP}_{-\pi}} H_2(z)e^{\beta\rho h_2(z)}\,dz$$

$$\simeq \frac{e^{-j\pi/4}}{\sqrt{2\pi\beta}}\,\frac{\dfrac{1}{n}\sin\left(\dfrac{\pi}{n}\right)}{\cos\left(\dfrac{\pi}{n}\right) - \cos\left(\dfrac{\phi+\phi'}{n}\right)}\,\frac{e^{-j\beta\rho}}{\sqrt{\rho}}$$

(13-62b)

Equation 13-62b is valid provided the observations are not made at or near the reflection shadow boundary of Figure 13-12. When the observations are made at the reflection shadow boundary $\phi + \phi' = \pi$, (13-62b) becomes infinite. Another form must be used for such cases.

If the poles of (13-53a) are near the saddle points of (13-59), then the conventional steepest descent method of (13-60a) and (13-60b) cannot be used for the evaluation of the last two terms of (13-50) for $F_1(\beta\rho)$, and similarly for $F_2(\beta\rho)$ of (13-48b). One method that can be used for such cases is the so-called *Pauli–Clemmow modified method of steepest descent* [8, 47]. The main difference between the two methods is that the solution provided by the Pauli–Clemmow modified method of steepest descent has an additional discontinuous function that compensates for the singularity along the corresponding shadow boundaries introduced by the conventional steepest descent method for isolated poles and saddle points. This factor is usually referred to as the *transition function*, and it is proportional to a Fresnel integral. Away from the corresponding shadow boundaries these transition functions are nearly unity, and the Pauli–Clemmow modified method of steepest descent reduces to the conventional method of steepest descent.

It can be shown that by using the Pauli–Clemmow modified method of steepest descents (13-60a) and (13-60b) are evaluated using

$$\frac{1}{4\pi jn}\int_{\text{SDP}_{+\pi}} H_1(z)e^{\beta\rho h_1(z)}\,dz \underset{\beta\rho\to\text{large}}{\simeq} \frac{1}{4\pi jn}\left|\sqrt{\frac{2\pi}{-\beta\rho h_1''(z_s = +\pi)}}\right|$$

$$\times e^{j\pi/4}H_1(z_s = +\pi)e^{\beta\rho h_1(z_s=\pi)}F[\beta\rho g^+(\xi^-)]$$

$$= \frac{e^{-j\pi/4}}{2n\sqrt{2\pi\beta}}\cot\left[\frac{\pi + (\phi - \phi')}{2n}\right]$$

$$\times F[\beta\rho g^+(\phi - \phi')]\frac{e^{-j\beta\rho}}{\sqrt{\rho}} \qquad (13\text{-}63a)$$

$$\frac{1}{4\pi jn}\int_{\text{SDP}_{-\pi}} H_1(z)e^{\beta\rho h_1(z)}\,dz \underset{\beta\rho\to\text{large}}{\simeq} \frac{1}{4\pi jn}\left|\sqrt{\frac{2\pi}{-\beta\rho h_1''(z_s = -\pi)}}\right|$$

$$\times e^{-j3\pi/4}H_1(z_s = -\pi)e^{\beta\rho h_1(z_s=-\pi)}F[\beta\rho g^-(\xi^-)]$$

$$= \frac{e^{-j\pi/4}}{2n\sqrt{2\pi\beta}}\cot\left[\frac{\pi - (\phi - \phi')}{2n}\right]$$

$$\times F[\beta\rho g^-(\phi - \phi')]\frac{e^{-j\beta\rho}}{\sqrt{\rho}} \qquad (13\text{-}63b)$$

where

$$F\left[\beta\rho g^{\pm}(\phi - \phi')\right] \equiv j\left[h_1(z_s) - h_1(z_p)\right]$$

$\equiv$ measure of separation between saddle points and poles

$$= 2j\left|\sqrt{\beta\rho g^{\pm}(\phi - \phi')}\right| \int_{\sqrt{\beta\rho g^{\pm}(\phi-\phi')}}^{\infty} e^{-j\tau^2} d\tau \quad (13\text{-}63c)$$

$$g^{\pm}(\phi - \phi') = 1 + \cos\left[(\phi - \phi') - 2\pi n N^{\pm}\right] \quad (13\text{-}63d)$$

$$2\pi n N^{+} - (\phi - \phi') = +\pi \quad (13\text{-}63e)$$

$$2\pi n N^{-} - (\phi - \phi') = -\pi \quad (13\text{-}63f)$$

Similar forms are used for the evaluation of $F_2(\beta\rho)$ along the steepest descent path using the Pauli–Clemmow modified method of steepest descent. In (13-63e) and (13-63f) $N^{\pm}$ represents integer values that most closely satisfy the equalities. Such a procedure accounts for the poles that are nearest to the saddle point at $x = \pm\pi$, either from outside or within $-\pi \leq x \leq +\pi$, of Figure 13-17b. In general there are two such poles associated with $F_1(\beta\rho)$ and two with $F_2(\beta\rho)$. More details about the transition function will follow.

D. GEOMETRICAL OPTICS AND DIFFRACTED FIELDS

After the contour integration and the method of steepest descent have been applied in the evaluation of (13-47), as discussed in the previous section, it can be shown that for large values of $\beta\rho$ (13-47) is separated into

$$F(\beta\rho) = F_G(\beta\rho) + F_D(\beta\rho) \quad (13\text{-}64)$$

where $F_G(\beta\rho)$ and $F_D(\beta\rho)$ represent, respectively, the total geometrical optics and total diffracted fields created by the incidence of a unit amplitude plane wave upon a two-dimensional wedge, as shown in Figure 13-12.

In summary then, the geometrical optics fields (F_G) and the diffracted fields (F_D) are represented, respectively, by

Geometrical Optics Fields			
	Incident GO	Reflected GO	Region
$F_G(\beta\rho) = \begin{cases} \\ \\ \\ \end{cases}$	$e^{j\beta\rho \cos(\phi-\phi')}$	$\pm e^{j\beta\rho \cos(\phi+\phi')}$	$0 < \phi < \pi - \phi'$
	$e^{j\beta\rho \cos(\phi-\phi')}$		$\pi - \phi' < \phi < \pi + \phi'$
	0		$\pi + \phi' < \phi < n\pi$

(13-65)

Diffracted Fields			
Total diffracted field	=	Incident diffracted	$\pm$ Reflected diffracted
$F_D(\beta\rho) = F_D(\rho,\phi,\phi',n) = V_B(\rho,\phi,\phi',n)$		$= V_B^i(\rho,\phi-\phi',n)$	$\pm V_B^r(\rho,\phi+\phi',n)$

(13-66)

where

$$V_B^{i,r}(\rho, \phi \mp \phi', n) = I_{-\pi}(\rho, \phi \mp \phi', n) + I_{+\pi}(\rho, \phi \mp \phi', n) \qquad (13\text{-}66a)$$

$$I_{\pm\pi}(\rho, \phi \mp \phi', n) \simeq \frac{e^{-j(\beta\rho + \pi/4)}}{jn\sqrt{2\pi}} \sqrt{g^{\pm}} \cot\left[\frac{\pi \pm (\phi \mp \phi')}{2n}\right]$$

$$\times e^{+j\beta\rho g^{\pm}} \int_{\sqrt{\beta\rho g^{\pm}}}^{\infty} e^{-j\tau^2} d\tau + \text{(higher-order terms)} \qquad (13\text{-}66b)$$

$$g^{+} = 1 + \cos\left[(\phi \mp \phi') - 2n\pi N^{+}\right] \qquad (13\text{-}66c)$$

$$g^{-} = 1 + \cos\left[(\phi \mp \phi') - 2n\pi N^{-}\right] \qquad (13\text{-}66d)$$

with N^+ or N^- being a positive or negative integer or zero which most closely satisfies the equation

$$2n\pi N^{+} - (\phi \mp \phi') = +\pi \quad \text{for } g^{+} \qquad (13\text{-}66e)$$

$$2n\pi N^{-} - (\phi \mp \phi') = -\pi \quad \text{for } g^{-} \qquad (13\text{-}66f)$$

Each of the diffracted fields (incident and reflected) exists in all space surrounding the wedge. Equation 13-66b contains the leading term of the diffracted field plus higher-order terms that are negligible for large values of $\beta\rho$. The integral in (13-66b) is a Fresnel integral (see Appendix III). In (13-65) and (13-66) the plus (+) sign is used for the hard polarization and the minus sign is used for the soft polarization.

If observations are made away from each of the shadow boundaries so that $\beta\rho g^{\pm} \gg 1$, (13-66a) and (13-66b) reduce, according to (13-61) and (13-62b), to

$$V_B^{h,s}(\rho, \phi \mp \phi', n) = V_B^i(\rho, \phi - \phi', n) \pm V_B^r(\rho, \phi + \phi', n)$$

$$= \frac{e^{-j\pi/4}}{\sqrt{2\pi\beta}} \frac{1}{n} \sin\left(\frac{\pi}{n}\right) \left[\frac{1}{\cos\left(\frac{\pi}{n}\right) - \cos\left(\frac{\phi - \phi'}{n}\right)} \pm \frac{1}{\cos\left(\frac{\pi}{n}\right) - \cos\left(\frac{\phi + \phi'}{n}\right)}\right] \frac{e^{-j\beta\rho}}{\sqrt{\rho}}$$

(13-67)

which is an expression of much simpler form, even for computational purposes. It is quite evident that when the observations are made at the incident shadow boundary (ISB, where $\phi = \pi + \phi'$), $V_B^i(\rho, \phi - \phi', n)$ of (13-67) becomes infinite because $\phi - \phi'$ is equal to π and the two cosine terms in the denominator of (13-67) are identical. Similarly $V_B^r(\rho, \phi + \phi', n)$ of (13-67) becomes infinite when observations are made at the reflected shadow boundary (RSB, where $\phi = \pi - \phi'$). The incident and reflected diffraction functions of the (13-67) form are referred to as *Keller's diffraction functions* and possess singularities along the incident and reflection shadow boundaries. The diffraction functions of (13-66a) through (13-66f) are representatives of the *uniform theory of diffraction* (UTD). The regions in the neighborhood of the incident and reflection shadow boundaries are referred to as the *transition regions*, and in these regions the fields undergo their most rapid changes.

The functions g^+ and g^- of (13-66c) and (13-66d) are representative of the angular separation between the observation point and the incident or reflection shadow boundary. In fact when observations are made along the shadow boundaries, the $g^{\pm}$ functions are equal to zero. For *exterior wedges* ($1 \leq n \leq 2$), the values of N^+ and N^- in (13-66e) and (13-66f) are equal to $N^+ = 0$ or 1 and

784 GEOMETRICAL THEORY OF DIFFRACTION

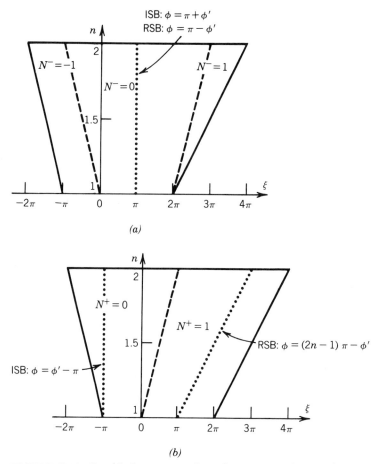

FIGURE 13-18 Graphical representation of (a) N^- and (b) N^+ as a function of ξ and n (*Source:* R. G. Kouyoumjian and P. H. Pathak, "A uniform geometrical theory of diffraction for an edge in a perfectly conducting surface," *Proc. IEEE,* © 1974, IEEE.)

$N^- = -1, 0,$ or 1. The values of n are plotted, as a function of $-2\pi \leq \xi^{\pm} = \phi \pm \phi' \leq 4\pi$; in Figure 13-18a for $N^- = -1, 0, 1$ and in Figure 13-18b for $N^+ = 0, 1$. These integral values of $N^{\pm}$ are particularly important along the shadow boundaries which are represented by the dotted lines. The variations of $N^{\pm}$ as a function of ϕ near the shadow boundaries are not abrupt, and this is a desirable property. The permissible values of $\xi^{\pm} = \phi \pm \phi'$ for $0 \leq \phi, \phi' \leq n\pi$ when $1 \leq n \leq 2$ are those bounded by the trapezoids formed by the solid straight lines in Figure 13-18a and b.

In order for (13-67) to be valid, $\beta \rho g^{\pm} \gg 1$. This can be achieved by having one of the following conditions:

1. $\beta \rho$ and $g^{\pm}$ large. This is satisfied if the distance ρ to the observation point is large and the observation angle ϕ is far away from either of the two shadow boundaries.

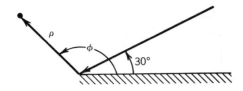

FIGURE 13-19 Plane wave incident on half-plane.

2. $\beta\rho$ large and $g^{\pm}$ small. This is satisfied if the distance ρ to the observation point is large and the observation angle ϕ is near either one or both of the shadow boundaries.

3. $\beta\rho$ small and $g^{\pm}$ large. This is satisfied if the distance ρ to the observation point is small and the observation angle ϕ is far away from either of the two shadow boundaries.

To demonstrate, we consider the problem of a plane wave of incidence angle $\phi' = 30°$ impinging on a half-plane ($n = 2$) as shown in Figure 13-19. In Figure 13-20a and b we have plotted as a function of ϕ, respectively, the magnitude of the incident V_B^i and the reflected V_B^r diffracted fields, as given by (13-66a) through (13-66f) for $\rho = 1\lambda$, and 100λ when $\phi' = 30°$ and $n = 2$. In the same figures, these results are compared with those obtained using (13-67). It is apparent that as the observation distance ρ increases the angular sector near the incident and reflected shadow boundaries over which (13-67) becomes invalid decreases; in the limit as $\rho \to \infty$ both give the same results.

The expression of (13-66) represents the diffraction of the unity strength incident plane wave with observations made at $P(\rho, \phi)$, as shown in Figure 13-12. Diffraction solutions of cylindrical waves, with observations at large distances, can be obtained by the use of the reciprocity principle along with the solution of the diffraction of an incident plane wave by a wedge as given by (13-66) through (13-66f). Using the geometry of Figure 13-16b, it is evident that cylindrical wave incidence diffraction can be obtained by substituting in (13-66) through (13-66f) $\rho = \rho'$, $\phi = \phi'$, and $\phi' = \phi$.

E. DIFFRACTION COEFFICIENTS

The incident V_B^i diffraction function of (13-66) can also be written using (13-66a) through (13-66f) as

$$\boxed{V_B^i(\rho, \phi - \phi', n) = V_B^i(\rho, \xi^-, n) = \frac{e^{-j\beta\rho}}{\sqrt{\rho}} D^i(\rho, \xi^-, n)} \quad (13\text{-}68)$$

where

$$\xi^- = \phi - \phi' \quad (13\text{-}68a)$$

786 GEOMETRICAL THEORY OF DIFFRACTION

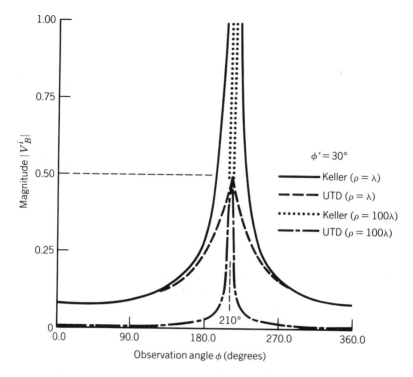

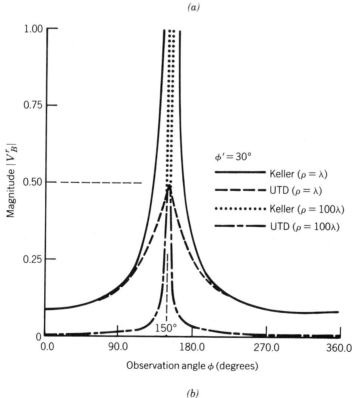

FIGURE 13-20 (*a*) Incident diffracted field and (*b*) reflected diffracted field for plane wave diffraction by a half-plane.

$$\boxed{D^i(\rho, \xi^-, n) = -\frac{e^{-j\pi/4}}{2n\sqrt{2\pi\beta}} \{C^+(\xi^-, n) F[\beta\rho g^+(\xi^-)] + C^-(\xi^-, n) F[\beta\rho g^-(\xi^-)]\}}$$

(13-68b)

$$C^+(\xi^-, n) = \cot\left(\frac{\pi + \xi^-}{2n}\right) \quad \text{(13-68c)}$$

$$C^-(\xi^-, n) = \cot\left(\frac{\pi - \xi^-}{2n}\right) \quad \text{(13-68d)}$$

$$F[\beta\rho g^+(\xi^-)] = 2j\sqrt{\beta\rho g^+(\xi^-)}\, e^{+j\beta\rho g^+(\xi^-)} \int_{\sqrt{\beta\rho g^+(\xi^-)}}^{\infty} e^{-j\tau^2}\, d\tau \quad \text{(13-68e)}$$

$$F[\beta\rho g^-(\xi^-)] = 2j\sqrt{\beta\rho g^-(\xi^-)}\, e^{+j\beta\rho g^-(\xi^-)} \int_{\sqrt{\beta\rho g^-(\xi^-)}}^{\infty} e^{-j\tau^2}\, d\tau \quad \text{(13-68f)}$$

The functions $g^+(\xi^-)$ and $g^-(\xi^-)$ are given by (13-66c) through (13-66f). In (13-68), $D^i(\rho, \xi^-, n)$ is defined as the diffraction coefficient for the incident diffracted field, and it will be referred to as the *incident diffraction coefficient* for a unit amplitude incident plane wave.

In a similar manner, the reflected V_B^r diffraction function of (13-66) can also be written using (13-66a) through (13-66f) as

$$\boxed{V_B^r(\rho, \phi + \phi', n) = V_B^r(\rho, \xi^+, n) = \frac{e^{-j\beta\rho}}{\sqrt{\rho}} D^r(\rho, \xi^+, n)} \quad (13\text{-}69)$$

where

$$\xi^+ = \phi + \phi' \quad \text{(13-69a)}$$

$$\boxed{D^r(\rho, \xi^+, n) = -\frac{e^{-j\pi/4}}{2n\sqrt{2\pi\beta}} \{C^+(\xi^+, n) F[\beta\rho g^+(\xi^+)] + C^-(\xi^+, n) F[\beta\rho g^-(\xi^+)]\}}$$

(13-69b)

$$C^+(\xi^+, n) = \cot\left(\frac{\pi + \xi^+}{2n}\right) \quad \text{(13-69c)}$$

$$C^-(\xi^+, n) = \cot\left(\frac{\pi - \xi^+}{2n}\right) \quad \text{(13-69d)}$$

$$F[\beta\rho g^+(\xi^+)] = 2j\sqrt{\beta\rho g^+(\xi^+)}\, e^{+j\beta\rho g^+(\xi^+)} \int_{\sqrt{\beta\rho g^+(\xi^+)}}^{\infty} e^{-j\tau^2}\, d\tau \quad \text{(13-69e)}$$

$$F[\beta\rho g^-(\xi^+)] = 2j\sqrt{\beta\rho g^-(\xi^+)}\, e^{+j\beta\rho g^-(\xi^+)} \int_{\sqrt{\beta\rho g^-(\xi^+)}}^{\infty} e^{-j\tau^2}\, d\tau \quad \text{(13-69f)}$$

$D^r(\rho, \xi^+, n)$ will be referred to as the *reflection diffraction coefficient* for a unit amplitude incident plane wave.

With the aid of (13-68) and (13-69), the total diffraction function $V_B(\rho, \phi \mp \phi', n) = V_B^i(\rho, \phi - \phi', n) \mp V_B^r(\rho, \phi + \phi', n)$ of (13-66) can now be written as

$$V_B(\rho, \phi, \phi', n) = V_B^i(\rho, \phi - \phi', n) \mp V_B^r(\rho, \phi + \phi', n)$$

$$= \frac{e^{-j\beta\rho}}{\sqrt{\rho}} \left[D^i(\rho, \phi - \phi', n) \mp D^r(\rho, \phi + \phi', n) \right] \quad (13\text{-}70)$$

or

$$V_{Bs}(\rho, \phi, \phi', n) = V_B^i(\rho, \phi - \phi', n) - V_B^r(\rho, \phi + \phi', n)$$

$$= \frac{e^{-j\beta\rho}}{\sqrt{\rho}} \left[D^i(\rho, \phi - \phi', n) - D^r(\rho, \phi + \phi', n) \right]$$

$$= \frac{e^{-j\beta\rho}}{\sqrt{\rho}} D_s(\rho, \phi, \phi', n) \quad (13\text{-}70a)$$

$$V_{Bh}(\rho, \phi, \phi', n) = V_B^i(\rho, \phi - \phi', n) + V_B^r(\rho, \phi + \phi', n)$$

$$= \frac{e^{-j\beta\rho}}{\sqrt{\rho}} \left[D^i(\rho, \phi - \phi', n) + D^r(\rho, \phi + \phi', n) \right]$$

$$= \frac{e^{-j\beta\rho}}{\sqrt{\rho}} D_h(\rho, \phi, \phi', n) \quad (13\text{-}70b)$$

$$D_s(\rho, \phi, \phi', n) = D^i(\rho, \phi - \phi', n) - D^r(\rho, \phi + \phi', n) \quad (13\text{-}70c)$$

$$D_h(\rho, \phi, \phi', n) = D^i(\rho, \phi - \phi', n) + D^r(\rho, \phi + \phi', n) \quad (13\text{-}70d)$$

where $V_{Bs} = V_B^i - V_B^r$ = soft (polarization) diffraction function
$V_{Bh} = V_B^i + V_B^r$ = hard (polarization) diffraction function
V_B^i = incident diffraction function
V_B^r = reflection diffraction function
$D_s = D^i - D^r$ = soft (polarization) diffraction coefficient
$D_h = D^i + D^r$ = hard (polarization) diffraction coefficient
D^i = incident diffraction coefficient
D^r = reflection diffraction coefficient

Using (13-68b) and (13-69b) in expanded form, the soft D_s and hard D_h diffraction coefficients can ultimately be written, respectively, as

$$D_s(\rho, \phi, \phi', n) = -\frac{e^{-j\pi/4}}{2n\sqrt{2\pi\beta}}$$
$$\times \left(\left\{ \cot\left[\frac{\pi + (\phi - \phi')}{2n}\right] F[\beta\rho g^+(\phi - \phi')] + \cot\left[\frac{\pi - (\phi - \phi')}{2n}\right] F[\beta\rho g^-(\phi - \phi')] \right\} \right.$$
$$\left. - \left\{ \cot\left[\frac{\pi + (\phi + \phi')}{2n}\right] F[\beta\rho g^+(\phi + \phi')] + \cot\left[\frac{\pi - (\phi + \phi')}{2n}\right] F[\beta\rho g^-(\phi + \phi')] \right\} \right)$$

(13-71a)

$$D_h(\rho, \phi, \phi', n) = -\frac{e^{-j\pi/4}}{2n\sqrt{2\pi\beta}}$$
$$\times \left(\left\{ \cot\left[\frac{\pi + (\phi - \phi')}{2n}\right] F[\beta\rho g^+(\phi - \phi')] + \cot\left[\frac{\pi - (\phi - \phi')}{2n}\right] F[\beta\rho g^-(\phi - \phi')] \right\} \right.$$
$$\left. + \left\{ \cot\left[\frac{\pi + (\phi + \phi')}{2n}\right] F[\beta\rho g^+(\phi + \phi')] + \cot\left[\frac{\pi - (\phi + \phi')}{2n}\right] F[\beta\rho g^-(\phi + \phi')] \right\} \right)$$

(13-71b)

where

$$\xi^- = \phi - \phi' \qquad (13\text{-}71c)$$

$$\xi^+ = \phi + \phi' \qquad (13\text{-}71d)$$

The formulations of (13-68) through (13-71d) are part of the often referred to *uniform theory of diffraction* (UTD) [10] which are extended to include oblique incidence and curved edge diffraction, which will be discussed in Sections 13.3.3 and 13.3.4.

A Fortran computer program designated WDC, wedge diffraction coefficients, is included at the end of the chapter to compute the soft and hard polarization diffraction coefficients of (13-71a) and (13-71b) (*actually the diffraction coefficients normalized by* $\sqrt{\lambda}$). It was developed and reported in [53]. The program also accounts for oblique wave incidence, and it is based on the more general formulation of (13-89a) through (13-90b). The main difference in the two sets of equations is the $\sin \beta_0'$ function found in the denominator of (13-90a) and (13-90b); it has been introduced to account for the oblique wave incidence. In (13-90a) and (13-90b) L is used as the distance parameter and ρ is used in (13-71a) and (13-71b).

Therefore to use the subroutine WDC to compute (13-71a) and (13-71b) let the oblique incidence angle β_0', referred to as BTD, be 90°. The distance parameter R should represent ρ (in wavelengths). This program uses the complex function FTF (Fresnel transition function), whose listing is also included at the end of the chapter, to complete its computations. The FTF program computes (13-68e), (13-68f) and (13-69e), (13-69f) based on the asymptotic expressions of (13-74a), (13-74b), and a linear interpolation for intermediate arguments. Computations of the Fresnel integral can also be made on an algorithm reported in [49] as well as on approximate expressions of [50].

To use the WDC subroutine the user must specify $R = \rho$ (in wavelengths), PHID = ϕ (in degrees), PHIPD = ϕ' (in degrees), BTD = β_0' (in degrees), and FN = n (dimensionless). The program subroutine computes the normalized (with respect to $\sqrt{\lambda}$) diffraction coefficients CDCS = D_s and CDCH = D_h. The angles represented by ϕ and ϕ' should be referenced from the face of the wedge, as shown in Figure 13-12(b). For normal incidence, $\beta_0' = 90°$. This computer subroutine has been used successfully in a multitude of problems; it is very efficient, and the user is encouraged to utilize it effectively.

Following a similar procedure, the incident and reflected diffraction functions and the incident, reflected, soft and hard diffraction coefficients using Keller's diffraction functions of (13-67) can be written, respectively, as

$$V_B^i(\rho, \phi - \phi', n) = V_B^i(\rho, \xi^-, n) = \frac{e^{-j\beta\rho}}{\sqrt{\rho}} D^i(\rho, \phi - \phi', n) = \frac{e^{-j\beta\rho}}{\sqrt{\rho}} D^i(\rho, \xi^-, n)$$

(13-72a)

$$V_B^r(\rho, \phi + \phi', n) = V_B^r(\rho, \xi^+, n) = \frac{e^{-j\beta\rho}}{\sqrt{\rho}} D^r(\rho, \phi + \phi', n) = \frac{e^{-j\beta\rho}}{\sqrt{\rho}} D^r(\rho, \xi^+, n)$$

(13-72b)

$$\boxed{D^i(\rho, \phi - \phi', n) = \frac{e^{-j\pi/4}}{\sqrt{2\pi\beta}} \frac{\frac{1}{n} \sin\left(\frac{\pi}{n}\right)}{\cos\left(\frac{\pi}{n}\right) - \cos\left(\frac{\phi - \phi'}{n}\right)}}$$

(13-72c)

$$\boxed{D^r(\rho, \phi + \phi', n) = \frac{e^{-j\pi/4}}{\sqrt{2\pi\beta}} \frac{\frac{1}{n} \sin\left(\frac{\pi}{n}\right)}{\cos\left(\frac{\pi}{n}\right) - \cos\left(\frac{\phi + \phi'}{n}\right)}}$$

(13-72d)

$$\boxed{\begin{aligned} D_s(\rho, \phi, \phi', n) &= D^i(\rho, \phi - \phi', n) - D^r(\rho, \phi + \phi', n) \\ &= \frac{e^{-j\pi/4} \frac{1}{n} \sin\left(\frac{\pi}{n}\right)}{\sqrt{2\pi\beta}} \left[\frac{1}{\cos\left(\frac{\pi}{n}\right) - \cos\left(\frac{\phi - \phi'}{n}\right)} - \frac{1}{\cos\left(\frac{\pi}{n}\right) - \cos\left(\frac{\phi + \phi'}{n}\right)} \right] \end{aligned}}$$

(13-72e)

$$\boxed{\begin{aligned} D_h(\rho, \phi, \phi', n) &= D^i(\rho, \phi - \phi', n) + D^r(\rho, \phi + \phi', n) \\ &= \frac{e^{-j\pi/4} \frac{1}{n} \sin\left(\frac{\pi}{n}\right)}{\sqrt{2\pi\beta}} \left[\frac{1}{\cos\left(\frac{\pi}{n}\right) - \cos\left(\frac{\phi - \phi'}{n}\right)} + \frac{1}{\cos\left(\frac{\pi}{n}\right) - \cos\left(\frac{\phi + \phi'}{n}\right)} \right] \end{aligned}}$$

(13-72f)

The diffraction coefficients of (13-72c) through (13-72f) are referred to as *Keller's diffraction coefficients* which possess singularities along the incident and reflection shadow boundaries.

The wedge diffraction coefficients described previously assume that the orientational direction for the incident and diffracted electric and magnetic field components of the soft and hard polarized fields are those shown, respectively, in Figures 13-21a and b. A negative value in the diffraction coefficients will reverse the directions of the appropriate fields.

The function $F(X)$ of (13-63c), (13-68e), (13-68f), (13-69e), and (13-69f) is known as a *Fresnel transition function* and it involves a Fresnel integral. Its magnitude and phase for $0.001 \leq X \leq 10$ are shown plotted in Figure 13-22. It is evident that

$$\left. \begin{array}{l} |F(X)| \leq 1 \\ 0 \leq \text{Phase of } F(X) \leq \pi/4 \end{array} \right\} \quad 0.001 \leq X \leq 10 \quad (13\text{-}73\text{a})$$

and

$$F(X) \simeq 1 \quad X > 10 \quad (13\text{-}73\text{b})$$

Thus if the argument X of the transition function exceeds 10, it can be replaced by

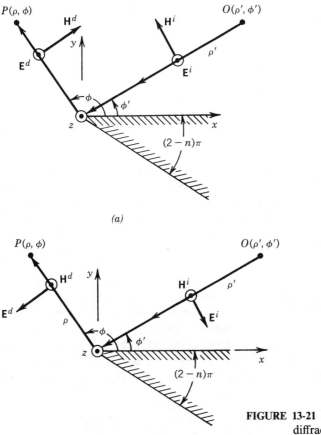

FIGURE 13-21 Polarization of incident and diffracted fields. (*a*) Soft polarization. (*b*) Hard polarization.

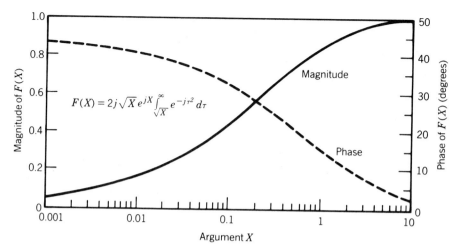

FIGURE 13-22 Variations of magnitude and phase of transition function $F(X)$ as a function of X. (*Source:* R. G. Kouyoumjian and P. H. Pathak, "A uniform geometrical theory of diffraction for an edge in a perfectly conducting surface," *Proc. IEEE,* © 1974, IEEE.)

unity. Then the expressions for the diffraction coefficients of (13-68b), (13-69b), (13-71a), and (13-71b) reduce, respectively, to those of (13-72c) through (13-72f). Asymptotic expressions for the transition function $F(X)$ are [51]:

<u>For Small X ($X < 0.3$)</u>

$$F(X) \simeq \left[\sqrt{\pi X} - 2Xe^{j\pi/4} - \frac{2}{3}X^2 e^{-j\pi/4}\right] e^{j(\pi/4 + X)} \qquad (13\text{-}74a)$$

<u>For Large X ($X > 5.5$)</u>

$$F(X) \simeq \left[1 + j\frac{1}{2X} - \frac{3}{4}\frac{1}{X^2} - j\frac{15}{8}\frac{1}{X^3} + \frac{75}{16}\frac{1}{X^4}\right] \qquad (13\text{-}74b)$$

To facilitate the reader in the computations, a Fortran computer function program designated as FTF, <u>F</u>resnel <u>t</u>ransition <u>f</u>unction, is included at the end of the chapter to compute the wedge transition function $F(X)$ of (13-68e), (13-68f), (13-69e), or (13-69f). The algorithm is based on the approximations of (13-74a) for small arguments ($X < 0.3$) and on (13-74b) for large arguments ($X > 5.5$). For intermediate values ($0.3 \leq X \leq 5.5$) a linear interpolation scheme is used. The program was developed and reported in [53].

To have a better understanding of the UTD diffraction coefficients $D^i(\rho, \phi - \phi', n)$ and $D^r(\rho, \phi + \phi', n)$ of (13-68b) and (13-69b), let us examine their behavior around the incident and reflection shadow boundaries. This will be accomplished by considering only exterior wedges ($n \geq 1$) and examining separately the situation where the incident wave illuminates the $\phi = 0$ side of the wedge $[\phi' \leq (n-1)\pi]$ as shown in Figure 13-23a and the situation where the incident wave illuminates the $\phi = n\pi$ side of the wedge $[\phi' \geq (n-1)\pi]$ as shown in Figure 13-23b.

<u>Case A $[\phi' \leq (n-1)\pi]$, Figure 13-23a</u>

For this case the incident shadow boundary (ISB) occurs when $\phi = \pi + \phi'$

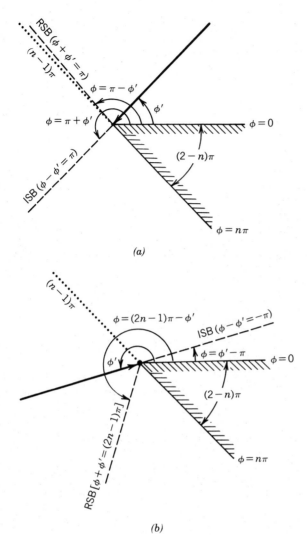

FIGURE 13-23 Incident and reflection shadow boundaries for wedge diffraction. (*a*) Case A: $\phi' \leq (n-1)\pi$. (*b*) Case B: $\phi' \geq (n-1)\pi$.

(or $\phi - \phi' = \pi$) and the reflection shadow boundary (RSB) occurs when $\phi = \pi - \phi'$ (or $\phi + \phi' = \pi$).

1. Along the ISB ($\phi = \pi + \phi'$ or $\phi - \phi' = \pi$) the second cotangent function of (13-68b) becomes singular and the first cotangent function in (13-68b) and both of them in (13-69b) remain bounded. That is, from (13-68b),

$$C^-(\xi^-, n)\big|_{\phi-\phi'=\pi} = \cot\left(\frac{\pi - \xi^-}{2n}\right)\bigg|_{\phi-\phi'=\pi} = \cot\left[\frac{\pi - (\phi - \phi')}{2n}\right]\bigg|_{\phi-\phi'=\pi} = \infty \quad (13\text{-}75a)$$

In addition, the value of N^- from (13-66f) is equal to $N^- = 0$. That is,

$$2n\pi N^- - (\phi - \phi')\big|_{\phi-\phi'=\pi} = 2n\pi N^- - \pi = -\pi \Rightarrow N^- = 0 \quad (13\text{-}75b)$$

2. Along the RSB ($\phi = \pi - \phi'$ or $\phi + \phi' = \pi$) the second cotangent function of (13-69b) becomes singular and the first cotangent function in (13-69b) and both of them in (13-68b) remain bounded. That is, from (13-69b),

$$C^-(\xi^+, n)\big|_{\phi+\phi'=\pi} = \cot\left(\frac{\pi - \xi^+}{2n}\right)\bigg|_{\phi+\phi'=\pi} = \cot\left[\frac{\pi - (\phi + \phi')}{2n}\right]\bigg|_{\phi+\phi'=\pi} = \infty \quad (13\text{-}76\text{a})$$

In addition the value of N^- from (13-66f) is equal to $N^- = 0$. That is,

$$2n\pi N^- - (\phi + \phi')\big|_{\phi+\phi'=\pi} = 2n\pi N^- - \pi = -\pi \Rightarrow N^- = 0 \quad (13\text{-}76\text{b})$$

<u>Case B $[\phi' \geq (n-1)\pi]$</u>, Figure 13-23b

For this case the incident shadow boundary (ISB) occurs when $\phi = \phi' - \pi$ (or $\phi - \phi' = -\pi$) and the reflection shadow boundary (RSB) occurs when $\phi = (2n-1)\pi - \phi'$ [or $\phi + \phi' = (2n-1)\pi$].

1. Along the ISB ($\phi = \phi' - \pi$ or $\phi - \phi' = -\pi$) the first cotangent function in (13-68b) becomes singular and the second cotangent function of (13-68b) and both of them in (13-69b) remain bounded. That is, from (13-68b),

$$C^+(\xi^-, n)\big|_{\phi-\phi'=-\pi} = \cot\left(\frac{\pi + \xi^-}{2n}\right)\bigg|_{\phi-\phi'=-\pi}$$

$$= \cot\left[\frac{\pi + (\phi - \phi')}{2n}\right]\bigg|_{\phi-\phi'=-\pi} = \infty \quad (13\text{-}77\text{a})$$

In addition the value of N^+ from (13-66e) is equal to $N^+ = 0$. That is,

$$2n\pi N^+ - (\phi - \phi')\big|_{\phi-\phi'=-\pi} = 2n\pi N^+ + \pi = +\pi \Rightarrow N^+ = 0 \quad (13\text{-}77\text{b})$$

2. Along the RSB $[\phi = (2n-1)\pi - \phi'$ or $\phi + \phi' = (2n-1)\pi]$ the first cotangent function of (13-69b) becomes singular and the second cotangent function of (13-69b) and both of them in (13-68b) remain bounded. That is, from (13-69b),

$$C^+(\xi^+, n)\big|_{\phi+\phi'=(2n-1)\pi} = \cot\left(\frac{\pi + \xi^+}{2n}\right)\bigg|_{\phi+\phi'=(2n-1)\pi}$$

$$= \cot\left[\frac{\pi + (\phi + \phi')}{2n}\right]\bigg|_{\phi+\phi'=(2n-1)\pi} = \infty \quad (13\text{-}78\text{a})$$

In addition the value of N^+ from (13-66e) is equal to $N^+ = 1$. That is,

$$2n\pi N^+ - (\phi + \phi')\big|_{\phi+\phi'=(2n-1)\pi} = 2n\pi N^+ - (2n-1)\pi = +\pi \Rightarrow N^+ = 1 \quad (13\text{-}78\text{b})$$

The results for Cases A and B are summarized in Table 13-1.

Whereas in the diffraction coefficients of UTD one of the cotangent functions becomes singular along the incident or reflection shadow boundary, while the other three cotangent functions are bounded, the product of the cotangent function along with its corresponding Fresnel transition function along that shadow boundary is

TABLE 13-1
Cotangent function behavior and values of $N^\pm$ along the shadow boundaries

	The cotangent function becomes singular when	Value of $N^\pm$ at the shadow boundary
$\cot\left[\dfrac{\pi - (\phi - \phi')}{2n}\right]$	$\phi = \pi + \phi'$ or $\phi - \phi' = \pi$ ISB of Case A Figure 13-23a	$N^- = 0$
$\cot\left[\dfrac{\pi - (\phi + \phi')}{2n}\right]$	$\phi = \pi - \phi'$ or $\phi + \phi' = \pi$ RSB of Case A Figure 13-23a	$N^- = 0$
$\cot\left[\dfrac{\pi + (\phi - \phi')}{2n}\right]$	$\phi = \phi' - \pi$ or $\phi - \phi' = -\pi$ ISB of Case B Figure 13-23b	$N^+ = 0$
$\cot\left[\dfrac{\pi + (\phi + \phi')}{2n}\right]$	$\phi = (2n-1)\pi - \phi'$ or $\phi + \phi' = (2n-1)\pi$ RSB of Case B Figure 13-23b	$N^+ = 1$

discontinuous but bounded. It is this finite discontinuity created by the singular cotangent term and its corresponding Fresnel transition function that removes the bounded geometrical optics discontinuity along that boundary.

To demonstrate, let us consider one of the four shadow boundaries created in Figure 13-23. We choose the ISB ($\phi = \pi + \phi'$ or $\phi - \phi' = \pi$) of Figure 13-23a where $\phi' \leq (n-1)\pi$. Similar results are found for the other three choices. At the ISB of Figure 13-23a

$$\xi^- = \phi - \phi' = \pi \tag{13-79}$$

and in the neighborhood of it

$$\xi^- = \phi - \phi' = \pi - \varepsilon \tag{13-79a}$$

where ε is positive on the illuminated side of the incident shadow boundary. Using (13-79a), we can write (13-66f) as

$$2n\pi N^- - (\phi - \phi') = 2n\pi N^- - (\pi - \varepsilon) = (2nN^- - 1)\pi + \varepsilon = -\pi \tag{13-80}$$

For this situation the cotangent function that becomes singular is that shown in the first row of Table 13-1 whose N^- value is $N^- = 0$. Therefore that cotangent function near the ISB of Figure 13-23a can be written using (13-80) with $N^- = 0$ (or $\phi - \phi' = \pi - \varepsilon$) as

$$C^-(\phi - \phi', n) = \cot\left[\dfrac{\pi - (\phi - \phi')}{2n}\right]$$

$$= \cot\left[\dfrac{\pi - \pi + \varepsilon}{2n}\right] = \cot\left(\dfrac{\varepsilon}{2n}\right) \simeq \dfrac{2n}{\varepsilon} = \dfrac{2n}{|\varepsilon|\,\text{sgn}(\varepsilon)} \tag{13-80a}$$

where sgn is the sign function. According to (13-66d)

$$g^-(\xi^-) = g^-(\phi - \phi') = 1 + \cos[(\phi - \phi') - 2\pi n N^-] = 1 + \cos(\phi - \phi')$$

$$g^-(\phi - \phi') = 1 + \cos(\pi - \varepsilon) = 1 - \cos(\varepsilon) \overset{\varepsilon \to 0}{\simeq} 1 - \left(1 - \frac{\varepsilon^2}{2}\right) = \frac{\varepsilon^2}{2} \quad (13\text{-}80\text{b})$$

The transition function of (13-68f) can also be written using (13-80b) as

$$F[\beta\rho g^-(\xi^-)] = F[\beta\rho g^-(\phi - \phi')] = F\left[\beta\rho\left(\frac{\varepsilon^2}{2}\right)\right] \quad (13\text{-}80\text{c})$$

which for small values of its argument can be approximated by the first term of its small argument asymptotic form (13-74a). That is,

$$F\left(\frac{\beta\rho\varepsilon^2}{2}\right) \simeq \sqrt{\frac{\pi\beta\rho\varepsilon^2}{2}} e^{j\pi/4} = |\varepsilon|\sqrt{\frac{\pi\beta\rho}{2}} e^{j\pi/4} \quad (13\text{-}81)$$

Thus the product of $C^-(\phi - \phi', n)F[\beta\rho g^-(\phi - \phi')]$, as each is given by (13-80a) and (13-81), can be approximated by

$$\cot\left[\frac{\pi - (\phi - \phi')}{2n}\right] F[\beta\rho g^-(\phi - \phi')] = n\sqrt{2\pi\beta\rho} \operatorname{sgn}(\varepsilon) e^{j\pi/4} \quad (13\text{-}82)$$

It is apparent that (13-82) exhibits a finite discontinuity which is positive along the illuminated side of the incident shadow boundary and negative on the other side.

The corresponding incident diffracted field (13-68) along the incident shadow boundary can be approximated using (13-82) and only the second term within the brackets in (13-68b) as

$$\boxed{V_B^i(\rho, \phi - \phi' = \pi - \varepsilon, n) \simeq \frac{e^{-j\beta\rho}}{\sqrt{\rho}}\left[-\frac{e^{-j\pi/4}}{2n\sqrt{2\pi\beta}} n\sqrt{2\pi\beta\rho} \operatorname{sgn}(\varepsilon) e^{j\pi/4}\right] = -\frac{e^{-j\beta\rho}}{2} \operatorname{sgn}(\varepsilon)}$$

(13-83)

Aside from the phase factor, this function is equal to -0.5, on the illuminated side of the incident shadow boundary and $+0.5$ on the other side. Clearly such a bounded discontinuity possesses the proper magnitude and polarity to compensate for the discontinuity created by the geometrical optics field. A similar procedure can be used to demonstrate the discontinuous nature of the diffracted field along the other shadow boundaries of Figure 13-23.

To illustrate the principles of geometrical optics (GO) and geometrical theory of diffraction (GTD), an example will be considered next.

Example 13-3. A plane wave of unity amplitude is incident upon a half-plane ($n = 2$) at an incidence angle of $\phi' = 30°$, as shown in the Figure 13-19. At a distance of one wavelength ($\rho = \lambda$) from the edge of the wedge, compute and plot the amplitude and phase of the following:

1. The total (incident plus reflected) geometrical optics field.
2. The incident diffracted field.

3. The reflected diffracted field.
4. The total field (geometrical optics plus diffracted).

Do these for both soft and hard polarizations.

Solution. The geometrical optics field components are computed using (13-65); incident and reflected diffracted fields are computed using (13-68) through (13-69f). These are plotted in Figure 13-24a and b for soft and hard polarizations, respectively.

In Figure 13-24a the amplitude patterns of the geometrical optics and diffracted fields for the soft polarization are displayed as follows.

1. The dashed curve (– – – –) represents the total geometrical optics field (incident and reflected) computed using (13-65).
2. The dash-dot curve (– - – - –) represents the amplitude of the incident diffracted (ID) field V_B^i computed using (13-68) through (13-68f).
3. The dotted curve ($\cdots$) represents the amplitude of the reflected diffracted (RD) field V_B^r computed using (13-69) through (13-69f).
4. The solid curve (—) represents the total amplitude pattern for soft polarization computed using results from parts 1 through 3.

Observing the data of Figure 13-24a it is evident that

1. The GO field is discontinuous at the reflection shadow boundary (RSB) ($\phi = 180° - \phi' = 180° - 30° = 150°$) and at the incident shadow boundary (ISB) ($\phi = 180° + \phi' = 180° + 30° = 210°$).
2. The GO field in the shadow region ($210° < \phi < 360°$) is zero.
3. The reflected diffracted (RD) field, although it exists everywhere, predominates around the reflection shadow boundary ($\phi = 150°$) with values of -0.5 for $\phi = (150°)^-$ and $+0.5$ for $\phi = (150)^+$. The total discontinuity at $\phi = 150°$ occurs because the phase undergoes a phase jump of $180°$.
4. The incident diffracted (ID) field also exists everywhere but it predominates around the incident shadow boundary ($\phi = 210°$) with values of -0.5 for $\phi = (210°)^-$ and $+0.5$ for $\phi = (210°)^+$. The total discontinuity at $\phi = 210°$ occurs because the phase undergoes a phase jump of $180°$.
5. The total amplitude field pattern is continuous everywhere with the discontinuities of the GO field compensated with the inclusion of the diffracted fields. It should be emphasized that the GO discontinuity at the RSB was removed by the inclusion of the reflected diffracted (RD) field and that at the ISB was compensated by the incident diffracted (ID) field. The GO field was also modified in all space with the addition of the diffracted fields, and radiation intensity is present in the shadow region ($210° < \phi < 360°$).

Computations for the same geometry were also carried out for the hard polarization and the amplitude is shown in Figure 13-24b. The same phenomena observed for soft polarization are also evident for the hard polarization.

The geometrical optics fields of Example 13-2 and Figure 13-8, displayed in Figure 13-9, exhibit discontinuities. To remove the discontinuities, diffracted fields must be included. This can be accomplished using the formulations for diffracted fields that have been developed up to this point.

798 GEOMETRICAL THEORY OF DIFFRACTION

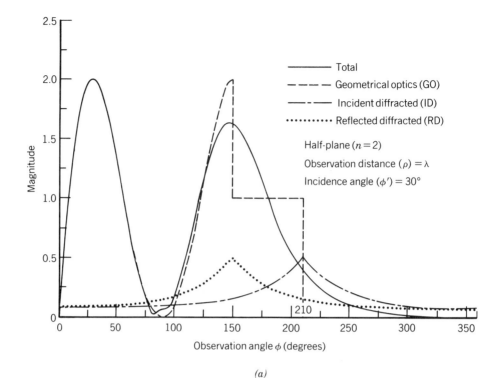

(a)

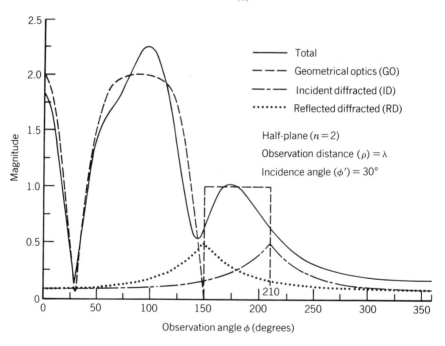

(b)

FIGURE 13-24 Field distribution of various components of a plane wave incident normally on a conducting half-plane. (*a*) Soft polarization. (*b*) Hard polarization. (*Source:* C. A. Balanis, *Antenna Theory: Analysis and Design*, copyright © 1982, John Wiley & Sons, Inc. Reprinted by permission of John Wiley & Sons, Inc.).

Example 13-4. For the geometry of Figure 13-8 repeat the formulations of Example 13-2 including the fields diffracted from the edges of the strip.

Solution. According to the solution of Example 13-2, the normalized incident (direct) and reflected fields of the line source above an infinite width strip are given, respectively, by

$$E_z^i = E_0 \frac{e^{-j\beta\rho_i}}{\sqrt{\rho_i}}$$

and

$$E_z^r = -E_0 \frac{e^{-j\beta\rho_r}}{\sqrt{\rho_r}}$$

where ρ_i and ρ_r are, respectively, the distances from the source and image (caustic) to the observation point, as shown in Figure 13-25a.

To take into account the finite width of the strip, we assume that the far-zone geometrical optics field components (direct and reflected) are the same as for the infinite width strip and the field intensity at the edges of the strip is the same as for the infinite strip.

These assumptions, which become more valid for larger width strips, allow us to determine the diffraction contributions from each of the edges. Because of the geometrical symmetry, we can separate the space surrounding the strip only into four regions, as shown in Figure 13-25a. The angular bounds and the components that contribute to each are as follows:

Region	Angular space	Components
I	$\alpha \leq \phi \leq \pi - \alpha$	Direct, reflected, diffracted (1 & 2)
II	$2\pi - \alpha \leq \phi \leq 2\pi, 0 \leq \phi \leq \alpha$	Direct, diffracted (1 & 2)
III	$\pi + \alpha \leq \phi \leq 2\pi - \alpha$	Diffracted (1 & 2)
IV	$\pi - \alpha \leq \phi \leq \pi + \alpha$	Direct, diffracted (1 & 2)

Because of symmetry, we need only consider half of the total space for computations.

To determine the first-order diffractions from each of the edges, we also assume that each forms a wedge (in this case a half space) that initially is isolated from the other. This allows us to use for each the diffraction properties of the canonical problem (wedge) discussed in the previous section. Thus the field diffracted from wedge 1 is equal to the product of:

1. The direct (incident) field E_z evaluated at the point of diffraction.
2. The diffraction coefficient as given by (13-70c).
3. The spatial attenuation factor as given by (13-35b).
4. The phase factor as given by (13-34a).

In equation form it is similar to (13-34a), and it is written as

$$E_{z1}^d(\rho_1, \phi) = E_z^i(\rho_d = s') D_s(s', \psi_1, \alpha, 2) A_1(\rho_1) e^{-j\beta\rho_1}$$

800 GEOMETRICAL THEORY OF DIFFRACTION

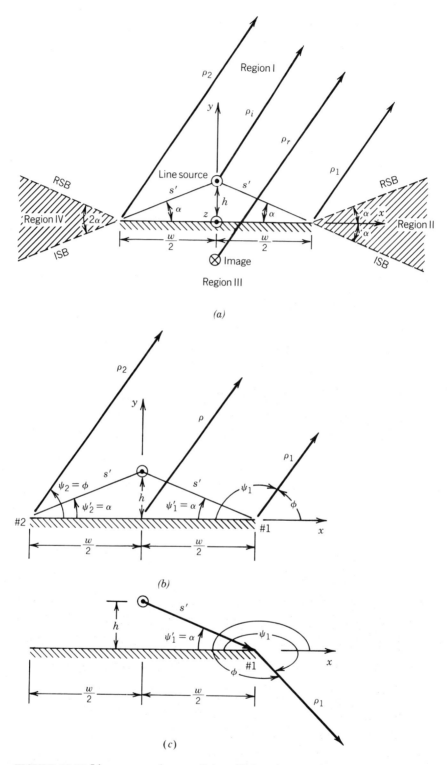

FIGURE 13-25 Line source above a finite width strip. (*a*) Region separation. (*b*) Diffraction by edges 1 and 2. (*c*) Diffraction by edge 1 in region III.

where

$$E_z^i(\rho_d, s') = E_0 \frac{e^{-j\beta s'}}{\sqrt{s'}}$$

$$D_s(s', \psi_1, \alpha, 2) = D^i(s', \psi_1 - \alpha, 2) - D^r(s', \psi_1 + \alpha, 2)$$

$$A_1(\rho_1) = \frac{1}{\sqrt{\rho_1}}$$

Using the preceding equations, we can write that

$$E_{z1}^d(\rho_1, \phi) = E_0 \left\{ \frac{e^{-j\beta s'}}{\sqrt{s'}} [D^i(s', \psi_1 - \alpha, 2) - D^r(s', \psi_1 + \alpha, 2)] \right\} \frac{e^{-j\beta \rho_1}}{\sqrt{\rho_1}}$$

$$E_{z1}^d(\rho_1, \phi) = E_0 [V_B^i(s', \psi_1 - \alpha, 2) - V_B^r(s', \psi_1 + \alpha, 2)] \frac{e^{-j\beta \rho_1}}{\sqrt{\rho_1}}$$

where V_B^i and V_B^r are the diffraction functions of (13-68) and (13-69). According to the geometry of Figure 13-25b and c

$$\psi_1 = \begin{cases} \pi - \phi & 0 \leq \phi \leq \pi \text{ (Figure 13-25b)} \\ 3\pi - \phi & \pi \leq \phi \leq 2\pi \text{ (Figure 13-25c)} \end{cases}$$

In a similar manner it can be shown that the field diffracted by wedge 2 is given by

$$E_{z2}^d = E_0 [V_B^i(s', \psi_2 - \alpha, 2) - V_B^r(s', \psi_2 + \alpha, 2)] \frac{e^{-j\beta \rho_2}}{\sqrt{\rho_2}}$$

$$\psi_2 = \phi \quad \text{for } 0 < \phi < 2\pi \text{ (Figure 13-25b)}$$

For far-field observations

$$\left. \begin{array}{l} \rho_i \simeq \rho - h \sin \phi \\ \rho_r \simeq \rho + h \sin \phi \\ \rho_1 \simeq \rho - \dfrac{w}{2} \cos \phi \\ \rho_2 \simeq \rho + \dfrac{w}{2} \cos \phi \end{array} \right\} \text{ for phase variations}$$

$$\rho_i \simeq \rho_r \simeq \rho_1 \simeq \rho_2 \simeq \rho \quad \text{for amplitude variations}$$

which allow the fields to be written as

$$E_z^i = E_0 e^{+j\beta h \sin \phi} \quad 0 < \phi < \pi + \alpha, \ 2\pi - \alpha < \phi < 2\pi \text{ (direct)}$$

$$E_z^r = -E_0 e^{-j\beta h \sin \phi} \quad \alpha < \phi < \pi - \alpha \text{ (reflected)}$$

$$E_{z1}^d = E_0 [V_B^i(s', \psi_1 - \alpha, 2) - V_B^r(s', \psi_1 + \alpha, 2)] e^{+j(\beta w/2) \cos \phi}$$

(diffracted from wedge #1)

$$\psi_1 = \begin{cases} \pi - \phi & 0 < \phi < \pi \\ 3\pi - \phi & \pi < \phi < 2\pi \end{cases}$$

$$E_{z2}^d = E_0 [V_B^i(s', \psi_2 - \alpha, 2) - V_B^r(s', \psi_2 + \alpha, 2)] e^{-j(\beta w/2) \cos \phi}$$

(diffracted from wedge #2)

$$\psi_2 = \phi \quad 0 < \phi < 2\pi$$

where the $e^{-j\beta \rho}/\sqrt{\rho}$ factor has been suppressed.

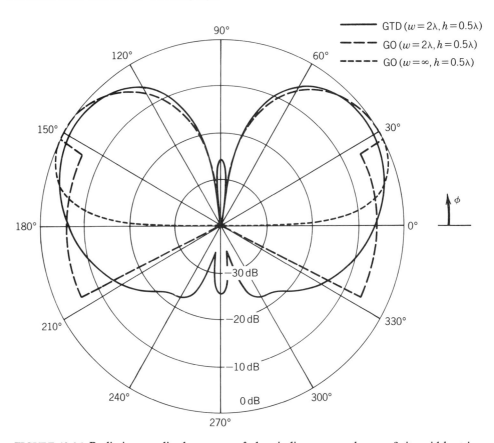

FIGURE 13-26 Radiation amplitude pattern of electric line source above a finite width strip.

It should be stated that the preceding equations represent only first-order diffractions which usually provide sufficient accuracy for many high-frequency applications. Multiple diffractions between the edges occur and should be included when the strip is electrically small and when more accurate results are required.

For a strip of $w = 2\lambda$ and with the source at a height of $h = 0.5\lambda$, the normalized pattern computed using these equations is shown in Figure 13-26 where it is compared with GO patterns for infinite and finite width strips. It is evident that there is a significant difference between the three, especially in the lower hemisphere, where for the most part the GO pattern exhibits no radiation, and in the regions where there are discontinuities in the GO pattern.

In addition to antenna pattern prediction, diffraction techniques are extremely well suited for scattering problems. To demonstrate the applicability and versatility of diffraction techniques to scattering, let us consider such an example.

Example 13-5. A soft polarized uniform plane wave, whose electric field amplitude is E_0, is incident upon a two-dimensional electrically conducting strip of width w, as shown in Figure 13-27*a*.

1. Determine the backscattered ($\phi = \phi'$) electric field and its backscattered scattering width (SW).

2. Compute and plot the normalized SW ($\sigma_{2\text{-D}}/\lambda$) in dB when $w = 2\lambda$ and the SW ($\sigma_{2\text{-D}}$) in dB/m (dBm) when $w = 2\lambda$ and $f = 10$ GHz.

Solution. For a soft polarized field the incident electric field can be written, according to the geometry of Figure 13-27a, as

$$\mathbf{E}^i = \hat{a}_z E_0 e^{-j\boldsymbol{\beta}^i \cdot \mathbf{r}} = \hat{a}_z E_0 e^{j\beta(x \cos \phi' + y \sin \phi')}$$

The backscattered field diffracted from wedge 1 can be written, by referring to the geometry of Figure 13-27b, as

$$\mathbf{E}_1^d = \mathbf{E}^i(Q_1) \cdot \overline{\mathbf{D}}_1^s A_1(\rho_1) e^{-j\beta \rho_1}$$

where

$$\mathbf{E}^i(Q_1) = \mathbf{E}^i \Big|_{\substack{x=w/2 \\ y=0 \\ \phi=\phi'}} = \hat{a}_z E_0 e^{j(\beta w/2) \cos \phi}$$

$$\overline{\mathbf{D}}_1^s = \hat{a}_z \hat{a}_z \frac{e^{-j\pi/4} \sin\left(\dfrac{\pi}{n}\right)}{n\sqrt{2\pi\beta}} \left[\frac{1}{\cos\left(\dfrac{\pi}{n}\right) - \cos\left(\dfrac{\psi_1 - \psi_1'}{n}\right)} \right.$$

$$\left. - \frac{1}{\cos\left(\dfrac{\pi}{n}\right) - \cos\left(\dfrac{\psi_1 + \psi_1'}{n}\right)} \right]_{\substack{n=2 \\ \psi_1 = \psi_1' = \pi - \phi}}$$

$$= -\hat{a}_z \hat{a}_z \frac{e^{-j\pi/4}}{2\sqrt{2\pi\beta}} \left(1 + \frac{1}{\cos \phi}\right)$$

$$A_1(\rho_1) = \frac{1}{\sqrt{\rho_1}}$$

Keller's diffraction form has been used because at very large distances (ideally infinity) the UTD formulations reduce to those of Keller. Thus the backscattered field diffracted from wedge 1 reduces to

$$\mathbf{E}_1^d = -\hat{a}_z E_0 \frac{e^{-j\pi/4} e^{j(\beta w/2) \cos \phi}}{2\sqrt{2\pi\beta}} \left(1 + \frac{1}{\cos \phi}\right) \frac{e^{-j\beta \rho_1}}{\sqrt{\rho_1}}$$

In a similar manner, the fields diffracted from wedge 2 can be written by referring to the geometry of Figure 13-27b as

$$\mathbf{E}_2^d = -\hat{a}_z E_0 \frac{e^{-j\pi/4} e^{-j(\beta w/2) \cos \phi}}{2\sqrt{2\pi\beta}} \left(1 - \frac{1}{\cos \phi}\right) \frac{e^{-j\beta \rho_2}}{\sqrt{\rho_2}}$$

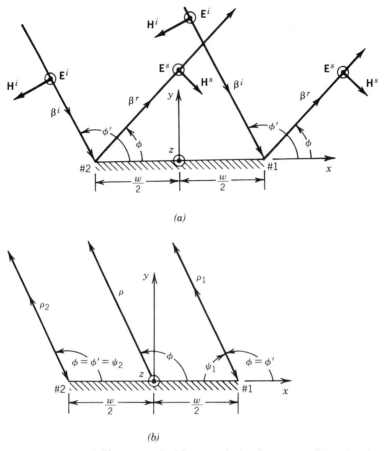

FIGURE 13-27 (*a*) Plane wave incidence and (*b*) plane wave diffraction by a finite width strip.

When both of the diffracted fields are referred to the center of the coordinate system, they can be written using

$$\left.\begin{array}{l}\rho_1 \simeq \rho + \dfrac{w}{2}\cos(\pi - \phi) = \rho - \dfrac{w}{2}\cos(\phi) \\ \rho_2 \simeq \rho - \dfrac{w}{2}\cos(\pi - \phi) = \rho + \dfrac{w}{2}\cos(\phi)\end{array}\right\} \text{for phase terms}$$

$$\rho_1 \simeq \rho_2 \simeq \rho \qquad \text{for amplitude terms}$$

as

$$\mathbf{E}_1^d = -\hat{a}_z E_0 \frac{e^{-j\pi/4}}{2\sqrt{2\pi\beta}}\left(1 + \frac{1}{\cos\phi}\right) e^{j\beta w \cos\phi} \frac{e^{-j\beta\rho}}{\sqrt{\rho}}$$

$$\mathbf{E}_2^d = -\hat{a}_z E_0 \frac{e^{-j\pi/4}}{2\sqrt{2\pi\beta}}\left(1 - \frac{1}{\cos\phi}\right) e^{-j\beta w \cos\phi} \frac{e^{-j\beta\rho}}{\sqrt{\rho}}$$

When the two diffracted fields are combined, the sum can be expressed as

$$\mathbf{E}^d = \mathbf{E}_1^d + \mathbf{E}_2^d = -\hat{a}_z E_0 \frac{e^{-j\pi/4}}{2\sqrt{2\pi\beta}} \left[\left(e^{j\beta w \cos\phi} + e^{-j\beta w \cos\phi} \right) \right.$$
$$\left. + \frac{1}{\cos\phi} \left(e^{j\beta w \cos\phi} - e^{-j\beta w \cos\phi} \right) \right] \frac{e^{-j\beta\rho}}{\sqrt{\rho}}$$

$$\mathbf{E}^d = -\hat{a}_z E_0 \frac{e^{-j\pi/4}}{\sqrt{2\pi\beta}} \left[\cos(\beta w \cos\phi) + j\beta w \frac{\sin(\beta w \cos\phi)}{(\beta w \cos\phi)} \right] \frac{e^{-j\beta\rho}}{\sqrt{\rho}}$$

Since there are no geometrical optics fields in the backscattered direction (Snell's law is not satisfied) when $\phi = \phi' \neq \pi/2$, the total diffracted field also represents the total field. In the limit as $\phi = \phi' = \pi/2$ each diffracted field exhibits a singularity; however, the total diffracted field is finite because the singularity of one diffracted field compensates for the singularity of the other. This is always evident at normal incidence as long as the edges of the two diffracted wedges are parallel to each other, even though the included angles of the two wedges are not necessarily the same [39]. In addition, the limiting value of the total diffracted field at normal incidence reduces and represents also the geometrical optics scattered (reflected) field.

The two-dimensional backscattered scattering width $\sigma_{2\text{-D}}$ of (11-21b) can now be written as

$$\sigma_{2\text{-D}} = \lim_{\rho \to \infty} \left[2\pi\rho \frac{|\mathbf{E}^s|^2}{|\mathbf{E}^i|^2} \right] = \frac{\lambda}{2\pi} \left| \cos(\beta w \cos\phi) + j\beta w \frac{\sin(\beta w \cos\phi)}{\beta w \cos\phi} \right|^2$$

The limiting value, as $\phi \to \pi/2$, reduces to

$$\left. \sigma_{2\text{-D}} \right|_{\phi=\pi/2} = \frac{\lambda}{2\pi} |1 + j\beta w|^2 = \frac{\lambda}{2\pi} \left[1 + (\beta w)^2 \right] \stackrel{\beta w \gg 1}{\simeq} \beta w^2$$

which agrees with the physical optics expression. Computed results for $\sigma_{2\text{-D}}/\lambda$ (in decibels) and $\sigma_{2\text{-D}}$ (in decibels per meter or dBm) at $f = 10$ GHz when $w = 2\lambda$ are shown in Figure 13-28.

Before proceeding to discuss other topics in diffraction, such as oblique incidence, curved edge diffraction, equivalent currents, slope diffraction, and multiple diffraction, let us complete our two-dimensional diffraction by addressing some modifications and extensions to the concepts covered in this section.

For a plane wave incidence, ρ in (13-68) through (13-68f) and in (13-69) through (13-69f) represents the distance from the edge of the wedge to the observation point. According to the principle of reciprocity illustrated in Figure 13-16, ρ in (13-68) through (13-69f) must be replaced by ρ' to represent the diffraction of a cylindrical wave whose source is located a distance ρ' from the edge of the wedge and the observations made in the far zone (ideally at infinity). If both the source and observation point are located at finite distances from the edge of the wedge, represented, respectively, by ρ' and ρ, then a better estimate of the distance would be to introduce a so-called *distance parameter L* which in this case takes the

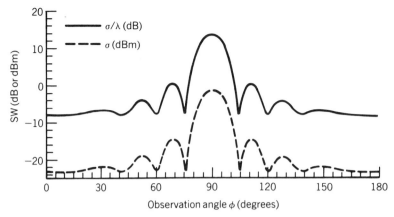

FIGURE 13-28 Two-dimensional monostatic scattering width for soft polarization of a finite width strip of $w = 2\lambda$ at $f = 10$ GHz.

form of

$$L = \frac{\rho\rho'}{\rho + \rho'} \begin{cases} \rho' \to \infty \\ \simeq \rho \\ \rho \to \infty \\ \simeq \rho' \end{cases} \qquad (13\text{-}84)$$

Thus the incident and reflected diffracted fields and coefficients of (13-68) and (13-69) can be written as

$$V_B^i(L, \phi - \phi', n) = V_B^i(L, \xi^-, n) = \frac{e^{-j\beta\rho}}{\sqrt{\rho}} D^i(L, \xi^-, n) \qquad (13\text{-}84\text{a})$$

$$V_B^r(L, \phi + \phi', n) = V_B^i(L, \xi^+, n) = \frac{e^{-j\beta\rho}}{\sqrt{\rho}} D^r(L, \xi^+, n) \qquad (13\text{-}84\text{b})$$

It is observed in (13-68) and (13-69) that for grazing angle incidence, $\phi' = 0$ or $\phi' = n\pi$, then $\xi^- = \xi^+ = \phi - \phi' = \phi + \phi'$ and $V_B^i = V_B^r$, $D^i = D^r$. Therefore, here D_s of (13-70c) or (13-72e) is equal to zero ($D_s = 0$) and D_h of (13-70d) or (13-72f) is equal to twice D^i or twice $D^r(D_h = 2D^i = 2D^r)$. Since grazing is a limiting situation, the incident and reflected fields combine to give the total geometrical optics field effectively incident at the observation point. Therefore one-half of the total field propagating along the face of the wedge toward the edge is the incident field and the other one-half represents the reflected field. The diffracted fields for this case can properly be accounted for by doing *either* of the following:

1. Let the total GO field represent the incident GO field but multiply the diffraction coefficients by a factor of $\frac{1}{2}$.
2. Multiply the total GO field by a factor of $\frac{1}{2}$ and let the product represent the incident field. The diffraction coefficients should not be modified.

Either procedure produces the same results, and the choice is left to the reader.

For grazing incidence, the diffraction coefficient of (13-70c) or (13-72e) are equal to zero. These diffraction coefficients, as well as those of (13-70d) and

(13-72f), account for the diffracted fields based on the value of the field at the point of diffraction. This is formulated using (13-34a). There are other higher-order diffraction coefficients that account for the diffracted fields based on the *rate of change (slope)* of the field at the point of diffraction. These diffraction coefficients are referred to as the *slope diffraction coefficients* [52], and they yield nonzero (even though small) fields for soft polarization at grazing incidence. The slope diffraction coefficients exist also for hard polarization, but they are not as dominant as they are for the soft polarization.

Up to now we have restricted our attention to exterior wedge ($1 < n \leq 2$) diffraction. However, the theory of diffraction can be applied also to interior wedge diffraction ($0 \leq n \leq 1$). When $n = 1$, the wedge reduces to an infinite flat plate and the diffraction coefficients reduce to zero [as seen better by examining (13-72e) and (13-72f)] since $\sin(\pi/n) = 0$. The incident and reflected fields for $n = 1$ (ground plane), $n = \frac{1}{2}$ (90° interior wedge), and $n = 1/M$, $M = 3, 4, \ldots$ (acute interior wedges) can be found exactly by image theory. In each of these, the number of finite images is determined by the included angle of the interior wedge [38]. As $n \to 0$ the geometrical optics (incident and reflected) fields become more dominant as compared to the nonvanishing diffracted fields.

13.3.3 Straight Edge Diffraction: Oblique Incidence

The normal incidence and diffraction formulations of the previous section are convenient to analyze radiation characteristics of antennas and structures primarily in principal planes. However, a complete analysis of an antenna or scatterer requires examination not only in principal planes but also in nonprincipal planes, as shown in Figure 13-29 for an aperture and a horn antenna each mounted on a finite size ground plane.

Whereas the diffraction of a normally incident wave discussed in the previous section led to scalar diffraction coefficients, the diffraction of an obliquely incident wave by a two-dimensional wedge can be derived using the geometry of Figure 13-30. To accomplish this it is most convenient to define ray-fixed coordinate systems (s', β_0', ϕ') for the source and (s, β_0, ϕ) for the observation point [8, 10], in contrast to the edge-fixed coordinate system (ρ', ϕ', z'; ρ, ϕ, z). By doing this it can be shown that the diffracted field in a general form can be written as

$$\boxed{E^d(s) = E^i(Q_D) \cdot \overline{\mathbf{D}}(L; \phi, \phi'; n; \beta_0') \sqrt{\frac{s'}{s(s'+s)}} e^{-j\beta s}} \qquad (13\text{-}85)$$

where $\overline{\mathbf{D}}(L; \phi, \phi'; n; \beta_0')$ is the dyadic edge diffraction coefficient for illumination of the wedge by plane, cylindrical, conical, or spherical waves.

Introducing an edge-fixed plane of incidence with the unit vectors $\hat{\beta}_0'$ and $\hat{\phi}'$ parallel and perpendicular to it, and a plane of diffraction with the unit vectors $\hat{\beta}_0$ and $\hat{\phi}$ parallel and perpendicular to it, we can write the radial unit vectors of incidence and diffraction, respectively, as

$$\hat{s}' = \hat{\phi}' \times \hat{\beta}_0' \qquad (13\text{-}86a)$$

$$\hat{s} = \hat{\phi} \times \hat{\beta}_0 \qquad (13\text{-}86b)$$

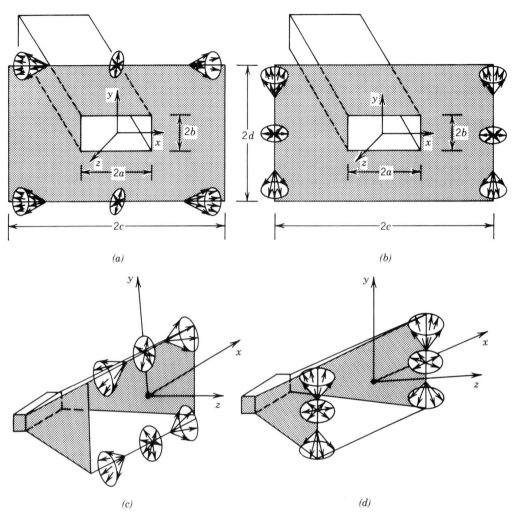

FIGURE 13-29 *E*- and *H*-plane diffraction by rectangular waveguide and pyramidal horn. Waveguide: (*a*) *E*-plane diffraction. (*b*) *H*-plane diffraction. Horn: (*c*) *E*-plane diffraction. (*d*) *H*-plane diffraction.

where $\hat{s}'$ points toward the point of diffraction. With the adoption of the ray-fixed coordinate systems, the dyadic diffraction coefficient can be represented by

$$\overline{\mathbf{D}}(L;\phi,\phi';n;\beta_0') = -\hat{\beta}_0'\hat{\beta}_0 D_s(L;\phi,\phi';n;\beta_0') - \hat{\phi}'\hat{\phi}D_h(L;\phi,\phi';n;\beta_0')$$

(13-87)

where D_s and D_h are, respectively, the scalar diffraction coefficients for soft and hard polarizations. If an edge-fixed coordinate system were adopted, the dyadic coefficient would be the sum of seven dyads which in matrix notation would be represented by a 3×3 matrix with seven nonvanishing elements instead of the 2×2 matrix with two nonvanishing elements.

For the diffraction shown in Figure 13-30, we can write in matrix form the diffracted *E*-field components that are parallel ($E_{\beta_0}^d$) and perpendicular (E_ϕ^d) to the

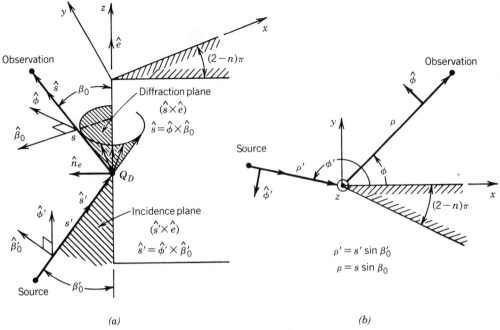

FIGURE 13-30 Oblique incidence wedge diffraction. (*a*) Oblique incidence. (*b*) Top view.

plane of diffraction as

$$\begin{bmatrix} E_{\beta_0}^d(s) \\ E_\phi^d(s) \end{bmatrix} = -\begin{bmatrix} D_s & 0 \\ 0 & D_h \end{bmatrix} \begin{bmatrix} E_{\beta_0'}^i(Q_D) \\ E_{\phi'}^i(Q_D) \end{bmatrix} A(s', s) e^{-j\beta s} \qquad (13\text{-}88)$$

where

$$E_{\beta_0'}^i(Q_D) = \hat{\beta}_0' \cdot \mathbf{E}_i$$
= component of the incident E field parallel to the plane of incidence at the point of diffraction Q_D (13-88a)

$$E_{\phi'}^i(Q_D) = \hat{\phi}' \cdot \mathbf{E}_i$$
= component of the incident E field perpendicular to the plane of incidence at the point of diffraction Q_D (13-88b)

D_s and D_h are the scalar diffraction coefficients which take the form

$$\boxed{D_s(L; \phi, \phi'; n; \beta_0') = D^i(L, \phi - \phi', n, \beta_0') - D^r(L, \phi + \phi', n, \beta_0')}$$
(13-89a)

$$\boxed{D_h(L; \phi, \phi'; n; \beta_0') = D^i(L, \phi - \phi', n, \beta_0') + D^r(L, \phi + \phi', n, \beta_0')}$$
(13-89b)

where

$$D^i(L, \phi - \phi', n, \beta_0')$$
$$= -\frac{e^{-j\pi/4}}{2n\sqrt{2\pi\beta}\sin\beta_0'}\left\{\cot\left[\frac{\pi + (\phi - \phi')}{2n}\right]F[\beta L g^+(\phi - \phi')]\right.$$
$$\left. + \cot\left[\frac{\pi - (\phi - \phi')}{2n}\right]F[\beta L g^-(\phi - \phi')]\right\}$$
(13-90a)

$$D^r(L, \phi + \phi', n, \beta_0')$$
$$= -\frac{e^{-j\pi/4}}{2n\sqrt{2\pi\beta}\sin\beta_0'}\left\{\cot\left[\frac{\pi + (\phi + \phi')}{2n}\right]F[\beta L g^+(\phi + \phi')]\right.$$
$$\left. + \cot\left[\frac{\pi - (\phi + \phi')}{2n}\right]F[\beta L g^-(\phi + \phi')]\right\}$$
(13-90b)

To facilitate the reader in the computations a Fortran computer subroutine designated WDC, <u>w</u>edge <u>d</u>iffraction <u>c</u>oefficient, is included at the end of the chapter to compute the normalized (with respect to $\sqrt{\lambda}$) wedge diffraction coefficients based on (13-89a) through (13-90b). The subroutine utilizes the Fresnel transition function program FTF found also at the end of the chapter. Both programs were developed and reported in [53].

In general, L is a distance parameter that can be found by satisfying the condition that the total field (the sum of the geometrical optics and the diffracted fields) must be continuous along the incident and reflection shadow boundaries. Doing this, it can be shown that a general form of L is

$$L = \frac{s(\rho_e^i + s)\rho_1^i\rho_2^i \sin^2\beta_0'}{\rho_e^i(\rho_1^i + s)(\rho_2^i + s)} \qquad (13\text{-}91)$$

where ρ_1^i, ρ_2^i = radii of curvature of the incident wave front at Q_D
ρ_e^i = radius of curvature of the incident wave front in the edge-fixed plane of incidence

For oblique incidence upon a wedge as shown in Figure 13-30, the distance parameter can be expressed in the ray-fixed coordinate system as

$$L = \begin{cases} s\sin^2\beta_0' & \text{plane wave incidence} \\ \dfrac{\rho\rho'}{\rho + \rho'} & \text{cylindrical wave incidence} \\ & (\rho = s\sin\beta_0, \rho' = s'\sin\beta_0') \\ \dfrac{ss'\sin^2\beta_0'}{s + s'} & \text{conical and spherical wave incidences} \end{cases} \qquad (13\text{-}92)$$

The spatial attenuation factor $A(s', s)$, which describes how the field intensity varies along the diffracted ray, is given by

$$A(s', s) = \begin{cases} \dfrac{1}{\sqrt{s}} & \text{plane and conical wave incidences} \\ \dfrac{1}{\sqrt{\rho}} & \rho = s \sin \beta_0; \text{ cylindrical wave incidence} \\ \sqrt{\dfrac{s'}{s(s' + s)}} & \text{spherical wave incidence} \end{cases} \quad (13\text{-}93)$$

If the observations are made in the far field ($s \gg s'$ or $\rho \gg \rho'$), the distance parameter L and spatial attenuation factor $A(s', s)$ reduce, respectively, to

$$L = \begin{cases} s \sin^2 \beta_0' & \text{plane wave incidence} \\ \rho' & \text{cylindrical wave incidence} \\ s' \sin^2 \beta_0' & \text{conical and spherical wave incidences} \end{cases} \quad (13\text{-}94)$$

$$A(s', s) = \begin{cases} \dfrac{1}{\sqrt{s}} & \text{plane and conical wave incidences} \\ \dfrac{1}{\sqrt{\rho}} & \rho = s \sin \beta_0; \text{ cylindrical wave incidence} \\ \dfrac{\sqrt{s'}}{s} & \text{spherical wave incidence} \end{cases} \quad (13\text{-}95)$$

For normal incidence, $\beta_0 = \beta_0' = \pi/2$.

To demonstrate the principles of this section, an example will be considered.

Example 13-6. To determine the far-zone elevation plane pattern, in the principal planes, of a $\lambda/4$ monopole mounted on a finite size square ground plane of width w on each of its sides, refer to Figure 13-31a. Examine the contributions from all four edges.

Solution. In addition to the direct and reflected field contributions (referred to as geometrical optics, GO), there are diffracted fields from the edges of the ground plane. The radiation mechanisms from the two edges which are perpendicular to the principal plane of observation are illustrated graphically in Figure 13-31b. It is apparent that from these two edges only two points contribute to the radiation in the principal plane. These two points occur at the intersection of the principal plane with the edges.

The incident and reflected fields are obtained by assuming the ground plane is infinite in extent. Using the coordinate system of Figure 13-31a and the image theory of Section 7.4, the total geometrical optics field of the $\lambda/4$ monopole above the ground plane can be written as [38]

$$E_{\theta G}(r, \theta) = E_0 \left[\frac{\cos\left(\dfrac{\pi}{2} \cos \theta\right)}{\sin \theta} \right] \frac{e^{-j\beta r}}{r} \quad 0 \leq \theta \leq \pi/2$$

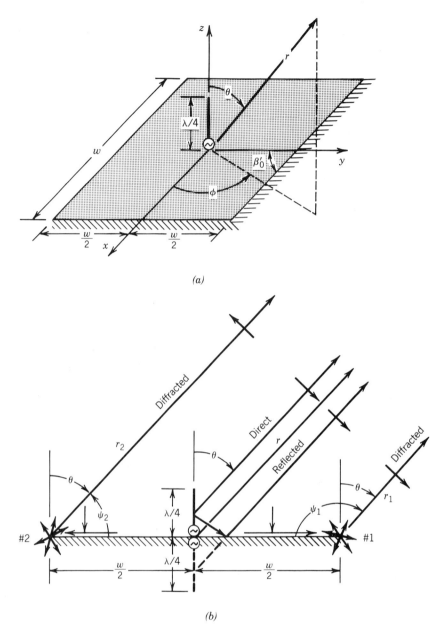

FIGURE 13-31 Vertical monopole on a square ground plane, and reflection and diffraction mechanisms. (*a*) Monopole on ground plane. (*b*) Reflection and diffraction mechanisms.

The field diffracted from wedge 1 can be obtained using the formulation of (13-88) through (13-95). Referring to the geometry of Figure 13-31*b*, the direct field is incident normally ($\beta_0' = \pi/2$) on the edge of the ground plane along the principal planes and that the diffracted field from wedge 1 can be written as

$$E_{\theta 1}^d(\theta) = +E^i(Q_1) D_h(L, \xi_1^\pm, \beta_0' = \pi/2, n = 2) A_1(w, r_1) e^{-j\beta r_1}$$

The total field can be assumed to all emanate from the base of the monopole. This is a good approximation whose modeling has agreed well with measurements. Thus

$$E^i(Q_1) = \frac{1}{2}E_{\theta G}\left(r = \frac{w}{2}, \theta = \frac{\pi}{2}\right) = \frac{E_0}{2}\frac{e^{-j\beta w/2}}{w/2}$$

$$D_h\left(L, \xi_1^{\pm}, \beta_0' = \frac{\pi}{2}, n = 2\right) = D^i(L, \xi_1^-, n = 2) + D^r(L, \xi_1^+, n = 2)$$

Since the incident wave is of spherical waveform and observations are made in the far field, the distance parameter L and spatial attenuation factor $A_1(w, r_1)$ can be expressed, according to (13-94) and (13-95) for $\beta_0' = \pi/2$, as

$$L = s'\sin^2\beta_0'\Big|_{\substack{s'=w/2 \\ \beta_0'=\pi/2}} = \frac{w}{2}$$

$$A_1(w, r_1) = \frac{\sqrt{s'}}{s}\Big|_{\substack{s'=w/2 \\ s=r_1}} = \frac{\sqrt{w/2}}{r_1}$$

Since the angle of incident ψ_0 from the main source toward the point of diffraction Q_1 is zero degrees ($\psi_0 = 0$), then

$$\xi_1^- = \psi_1 - \psi_0 = \psi_1 = \theta + \frac{\pi}{2} = \xi_1$$

$$\xi_1^+ = \psi_1 + \psi_0 = \psi_1 = \theta + \frac{\pi}{2} = \xi_1$$

Therefore

$$D_h\left(L, \xi_1^{\pm}, \beta_0' = \frac{\pi}{2}, n = 2\right) = 2D^i\left(\frac{w}{2}, \theta + \frac{\pi}{2}, n = 2\right)$$
$$= 2D^r\left(\frac{w}{2}, \theta + \frac{\pi}{2}, n = 2\right)$$

The total diffracted field can now be written as

$$E_{\theta 1}^d(\theta) = \frac{E_0}{2}\frac{e^{-j\beta w/2}}{w/2}2D^{i,r}\left(\frac{w}{2}, \theta + \frac{\pi}{2}, n = 2\right)\frac{\sqrt{w/2}}{r_1}e^{-j\beta r_1}$$

$$= E_0\left[\frac{e^{-j\beta w/2}}{\sqrt{w/2}}D^{i,r}\left(\frac{w}{2}, \theta + \frac{\pi}{2}, n = 2\right)\right]\frac{e^{-j\beta r_1}}{r_1}$$

$$E_{\theta 1}^d(\theta) = E_0 V_B^{i,r}\left(\frac{w}{2}, \theta + \frac{\pi}{2}, n = 2\right)\frac{e^{-j\beta r_1}}{r_1}$$

Using a similar procedure, the field diffracted from wedge 2 can be written, by referring to the geometry of Figure 13-31b, as

$$E_{\theta 2}^d(\theta) = -E_0\left[\frac{e^{-j\beta w/2}}{\sqrt{w/2}}D^{i,r}\left(\frac{w}{2}, \xi_2, n = 2\right)\right]\frac{e^{-j\beta r_2}}{r_2}$$

$$E_{\theta 2}^d(\theta) = -E_0 V_B^{i,r}\left(\frac{w}{2}, \xi_2, n = 2\right)\frac{e^{-j\beta r_2}}{r_2}$$

where

$$\xi_2 = \psi_2 = \begin{cases} \dfrac{\pi}{2} - \theta & 0 \leq \theta \leq \dfrac{\pi}{2} \\ \dfrac{5\pi}{2} - \theta & \dfrac{\pi}{2} < \theta < \pi \end{cases}$$

For far-field observations

$$\left.\begin{array}{l} r_1 \simeq r - \dfrac{w}{2}\cos\left(\dfrac{\pi}{2} - \theta\right) = r - \dfrac{w}{2}\sin\theta \\ r_2 \simeq r + \dfrac{w}{2}\cos\left(\dfrac{\pi}{2} - \theta\right) = r + \dfrac{w}{2}\sin\theta \end{array}\right\} \text{for phase terms}$$

$$r_1 \simeq r_2 \simeq r \qquad \text{for amplitude terms}$$

Therefore the diffracted fields from wedges 1 and 2 reduce to

$$E_{\theta 1}^d(\theta) = +E_0 V_B^{i,r}\left(\dfrac{w}{2}, \theta + \dfrac{\pi}{2}, n = 2\right) e^{j(\beta w/2)\sin\theta} \dfrac{e^{-j\beta r}}{r}$$

$$E_{\theta 2}^d(\theta) = -E_0 V_B^{i,r}\left(\dfrac{w}{2}, \xi_2, n = 2\right) e^{-j(\beta w/2)\sin\theta} \dfrac{e^{-j\beta r}}{r}$$

It should be noted that there are oblique incidence diffractions from the other two sides of the ground plane that are parallel to the principal plane of observation. However, the diffracted field from these edges is primarily cross-polarized (E_ϕ component) to the incident E_θ field and to the E_θ field produced in the principal plane. The cross-polarized E_ϕ components produced by diffractions from these two sides cancel each other out so that in the principal plane there is primarily an E_θ component.

Using the total geometrical optics field and the field diffracted from wedges 1 and 2, a normalized amplitude pattern was computed for a $\lambda/4$ monopole mounted on a square ground plane of $w = 4$ ft $= 1.22$ m at a frequency of $f = 1$ GHz. This pattern is shown in Figure 13-32 where it is compared with the computed GO (assuming an infinite ground plane) and measured patterns. A very good agreement is seen between the GO + GTD and measured patterns, which are quite different from that of the GO pattern.

13.3.4 Curved Edge Diffraction: Oblique Incidence

The edges of many practical antenna or scattering structures are not straight, as demonstrated in Figure 13-33 by the edges of a circular ground plane, a paraboloidal reflector, and a conical horn. In order to account for the diffraction phenomenon from the edges of these structures, even in their principal planes, curved edge diffraction must be utilized.

Curved edge diffraction can be derived by assuming an oblique wave incidence (at an angle β_0') on a curved edge, as shown in Figure 13-34, where the surfaces (sides) forming the curved edge in general may be convex, concave, or plane. Since diffraction is a local phenomenon, the curved edge geometry can be approximated at the point of diffraction Q_D by a wedge whose straight edge is tangent to the curved

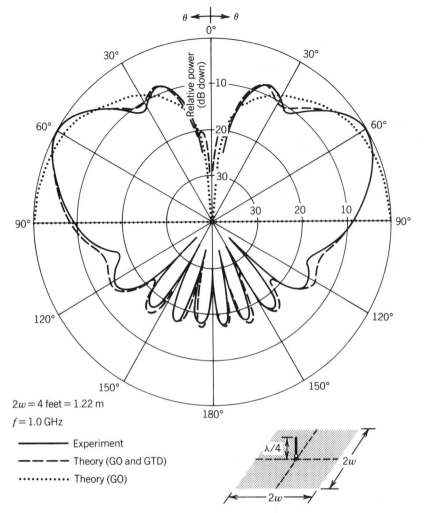

FIGURE 13-32 Measured and computed principal elevation plane amplitude patterns of a $\lambda/4$ monopole above infinite and finite square ground planes. (*Source:* C. A. Balanis, *Antenna Theory: Analysis and Design*, copyright © 1982, John Wiley & Sons, Inc. Reprinted by permission of John Wiley & Sons, Inc.)

edge at that point and whose plane surfaces are tangent to the curved surfaces forming the curved edge. This allows wedge diffraction theory to be applied directly to curved edge diffraction by simply representing the curved edge by an equivalent wedge. Analytically this is accomplished simply by generalizing the expressions for the distance parameter L that appear in the arguments of the transition functions.

The general form of oblique incidence curved edge diffraction can be expressed in matrix form as in (13-88). However, the diffraction coefficients, distance parameters, and spatial spreading factor must be modified to account for the curvature of the edge and its curved surfaces (sides).

The diffraction coefficients D_s and D_h are those of (13-89a) and (13-89b) where D^i and D^r can be found by imposing the continuity conditions on the total field across the incident and reflection shadow boundaries. Doing this we can show

816 GEOMETRICAL THEORY OF DIFFRACTION

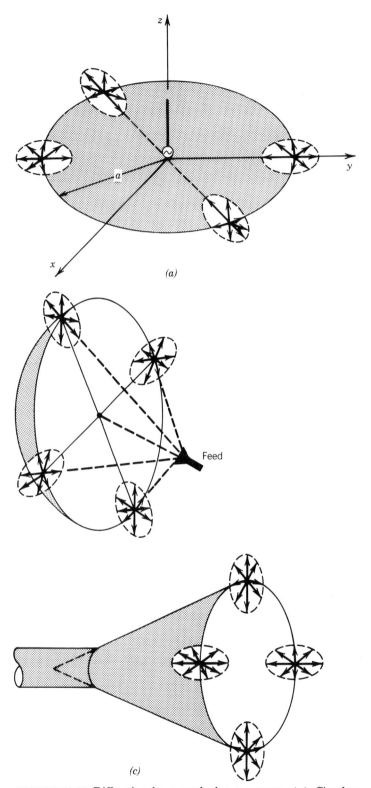

FIGURE 13-33 Diffraction by curved-edge structures. (*a*) Circular ground plane. (*b*) Paraboloidal reflector. (*c*) Conical horn.

GEOMETRICAL THEORY OF DIFFRACTION: EDGE DIFFRACTION

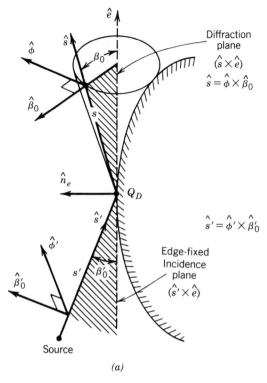

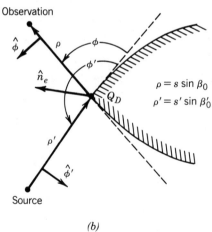

FIGURE 13-34 Oblique incidence diffraction by a curved edge. (*a*) Oblique incidence. (*b*) Top view.

that D^i and D^r of (13-89a) and (13-89b) take the form of [10]

$$\begin{aligned}
D^i(L^i, &\phi - \phi', n) \\
= -&\frac{e^{-j\pi/4}}{2n\sqrt{2\pi\beta}\sin\beta_0'}\left\{\cot\left[\frac{\pi + (\phi - \phi')}{2n}\right]F\left[\beta L^i g^+(\phi - \phi')\right]\right. \\
&\left.+\cot\left[\frac{\pi - (\phi - \phi')}{2n}\right]F\left[\beta L^i g^-(\phi - \phi')\right]\right\}
\end{aligned} \quad (13\text{-}96a)$$

$$D'(L^i, \phi + \phi', n)$$
$$= -\frac{e^{-j\pi/4}}{2n\sqrt{2\pi\beta}\sin\beta_0'}\left\{\cot\left[\frac{\pi + (\phi + \phi')}{2n}\right]F[\beta L^{rn}g^+(\phi + \phi')]\right.$$
$$\left.+ \cot\left[\frac{\pi - (\phi + \phi')}{2n}\right]F[\beta L^{ro}g^+(\phi + \phi')]\right\} \quad (13\text{-}96\text{b})$$

where

$$L^i = \frac{s(\rho_e^i + s)\rho_1^i\rho_2^i \sin^2 \beta_0'}{\rho_e^i(\rho_1^i + s)(\rho_2^i + s)} \quad (13\text{-}97\text{a})$$

$$L^{ro,\,rn} = \frac{s(\rho_e^r + s)\rho_1^r\rho_2^r \sin^2 \beta_0'}{\rho_e^r(\rho_1^r + s)(\rho_2^r + s)} \quad (13\text{-}97\text{b})$$

ρ_1^i, ρ_2^i = radii of curvature of the incident wave front at Q_D

ρ_e^i = radius of curvature of the incident wave front in the edge fixed plane of incidence

ρ_1^r, ρ_2^r = principal radii of curvature of the reflected wave front at Q_D [found using (13-21a) and (13-21b)]

ρ_e^r = radius of curvature of the reflected wave front in the plane containing the diffracted ray and edge

The superscripts ro, rn in (13-96a) and (13-96b) denote that the radii of curvature ρ_1^r, ρ_2^r and ρ_e^r must be calculated for ro at the reflection boundary $\pi - \phi'$ of Figure 13-23a and for rn at the reflection boundary $(2n - 1)\pi - \phi'$ of Figure 13-23b. For far-field observation where $s \gg \rho_e^i, \rho_1^i, \rho_2^i, \rho_e^r, \rho_1^r, \rho_2^r$ (13-97a) and (13-97b) simplify to

$$L^i = \frac{\rho_1^i\rho_2^i}{\rho_e^i}\sin^2\beta_0' \quad (13\text{-}98\text{a})$$

$$L^{ro,\,rn} = \frac{\rho_1^r\rho_2^r}{\rho^r}\sin^2\beta_0' \quad (13\text{-}98\text{b})$$

If the intersecting curved surfaces forming the curved edge in Figure 13-34 are plane surfaces that form an ordinary wedge, then the distance parameters in (13-97a) and (13-97b) or (13-98a) and (13-98b) are equal, that is,

$$L^{ro} = L^{rn} = L^i \quad (13\text{-}99)$$

Using the geometries of Figure 13-35, it can be shown that the spatial spreading factor $A(\rho_c, s)$ of (13-35) for the curved edge diffraction takes the form

$$A(\rho_c, s) = \sqrt{\frac{\rho_c}{s(\rho_c + s)}} \overset{s \gg \rho_c}{\simeq} \frac{1}{s}\sqrt{\rho_c} \qquad (13\text{-}100)$$

$$\frac{1}{\rho_c} = \frac{1}{\rho_e} - \frac{\hat{n}_e \cdot (\hat{s}' - \hat{s})}{\rho_g \sin^2 \beta_0'} \qquad (13\text{-}100a)$$

where ρ_c = distance between caustic at edge and second caustic of diffracted ray
ρ_e = radius of curvature of incidence wave front in the edge-fixed plane of incidence which contains unit vectors $\hat{s}'$ and $\hat{e}$ (infinity for plane, cylindrical, and conical waves; $\rho_e = s'$ for spherical waves)
ρ_g = radius of curvature of the edge at the diffraction point
$\hat{n}_e$ = unit vector normal to the edge at Q_D and directed away from the center of curvature
$\hat{s}'$ = unit vector in the direction of incidence
$\hat{s}$ = unit vector in the direction of diffraction
β_0' = angle between $\hat{s}'$ and tangent to the edge at the point of diffraction
$\hat{e}$ = unit vector tangent to the edge at the point of diffraction

For normal incidence $\beta_0' = \pi/2$.

The spatial attenuation factor of (13-100) creates additional caustics, other than the ones that occur at the points of diffraction. Each caustic occurs at a distance ρ_c from the one at the diffraction point. Diffracted fields in the regions of the caustics must be corrected to remove the discontinuities and inaccuracies from them.

To demonstrate the principles of curved edge diffraction, let us consider an example.

Example 13-7. Determine the far-zone elevation plane pattern of a $\lambda/4$ monopole mounted on a circular electrically conducting ground plane of radius a, as shown in Figure 13-36a.

Solution. Because of the symmetry of the structure, the diffraction mechanism in any of the elevation planes is the same. Therefore the principal yz plane is chosen here. For observations made away from the symmetry axis of the ground plane ($\theta \neq 0°$ and $180°$), it can be shown [33] that most of the diffraction radiation from the rim of the ground plane comes from the two diametrically opposite points of the rim that coincide with the observation plane. Therefore for points removed from the symmetry axis ($\theta \neq 0°$ and $180°$) the overall formulation of this problem and that of Example 13-6 is identical other than the amplitude spreading factor which now must be computed using (13-100) and (13-100a) instead of (13-95).

820 GEOMETRICAL THEORY OF DIFFRACTION

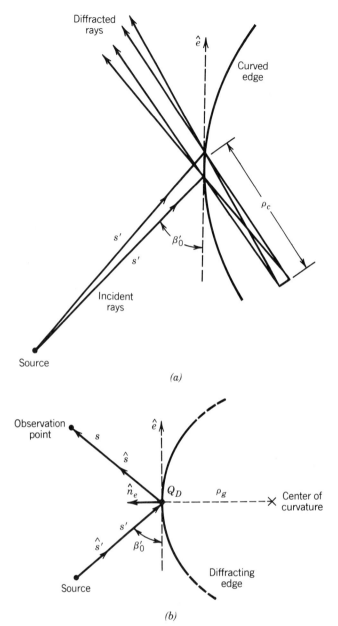

FIGURE 13-35 Caustic distance and center of curvature for curved-edge diffraction. (*Source:* C. A. Balanis, *Antenna Theory: Analysis and Design*, copyright © 1982, John Wiley & Sons, Inc. Reprinted by permission of John Wiley & Sons, Inc.) (*a*) Caustic distance. (*b*) Center of curvature.

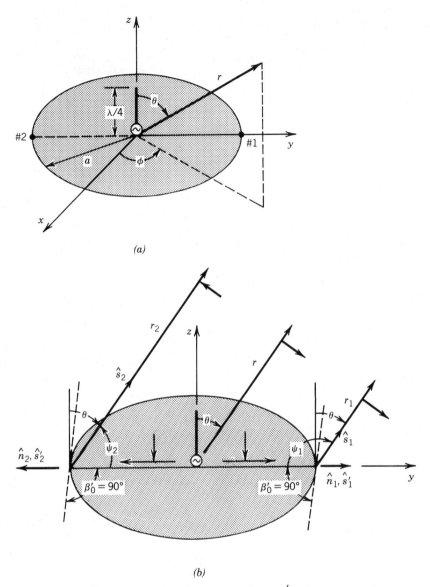

FIGURE 13-36 Quarter-wavelength monopole on a circular ground plane and diffraction mechanism. (*a*) $\lambda/4$ monopole. (*b*) Diffraction mechanism.

Referring to the geometry of Figure 13-36*b* and using (13-100) and (13-100a), the amplitude spreading factor for wedge 1 can be written as

$$A_1(r_1, a) = \frac{1}{r_1}\sqrt{\rho_{c1}}$$

where

$$\frac{1}{\rho_{c1}} = \frac{1}{a} - \frac{\hat{n}_1 \cdot (\hat{s}_1' - \hat{s}_1)}{a} = \frac{1 - \left[1 - \cos\left(\frac{\pi}{2} - \theta\right)\right]}{a} = \frac{\sin\theta}{a} \Rightarrow \rho_{c1} = \frac{a}{\sin\theta}$$

Therefore

$$A_1(r_1, a) = \frac{1}{r_1}\sqrt{\frac{a}{\sin\theta}} \simeq \frac{1}{r}\sqrt{\frac{a}{\sin\theta}}$$

In a similar manner, the amplitude spreading factor for wedge 2 can be expressed as

$$A_2(r_2, a) = \frac{1}{r_2}\sqrt{\rho_{c2}}$$

where

$$\frac{1}{\rho_{c2}} = \frac{1}{a} - \frac{\hat{n}_2 \cdot (\hat{s}_2' - \hat{s}_2)}{a} = \frac{1 - \left[1 - \cos\left(\frac{\pi}{2} + \theta\right)\right]}{a} = -\frac{\sin\theta}{a} \Rightarrow \rho_{c2} = -\frac{a}{\sin\theta}$$

This reduces the amplitude spreading factor to

$$A_2(r_2, a) = \frac{1}{r_2}\sqrt{-\frac{a}{\sin\theta}} \simeq \frac{1}{r}\sqrt{-\frac{a}{\sin\theta}}$$

Using the results from Example 13-6, the fields for this problem can be written as

$$E_{\theta G}(r,\theta) = E_0 \left[\frac{\cos\left(\frac{\pi}{2}\cos\theta\right)}{\sin\theta}\right] \frac{e^{-j\beta r}}{r} \qquad 0 \le \theta \le \pi/2$$

$$E_{\theta 1}^d(r,\theta) = E_0 V_B^{i,r}\left(a, \theta + \frac{\pi}{2}, n=2\right)\frac{e^{j\beta a \sin\theta}}{\sqrt{\sin\theta}}\frac{e^{-j\beta r}}{r} \qquad \theta_0 \le \theta \le \pi - \theta_0$$

$$E_{\theta 2}^d(r,\theta) = -E_0 V_B^{i,r}(a, \xi_2, n=2)\frac{e^{-j\beta a \sin\theta}}{\sqrt{-\sin\theta}}\frac{e^{-j\beta r}}{r} \qquad \theta_0 \le \theta \le \pi - \theta_0$$

where

$$\xi_2 = \psi_2 = \begin{cases} \frac{\pi}{2} - \theta & \theta_0 \le \theta \le \frac{\pi}{2} \\ \frac{5\pi}{2} - \theta & \frac{\pi}{2} < \theta \le \pi - \theta_0 \end{cases}$$

It is noted that at $\theta = 0°$ or $180°$ the diffracted fields become singular because along these directions there are caustics for the diffracted fields. The rim of the ground plane acts as a ring radiator. Toward $\theta = 0°$ and $180°$ the infinite number of diffracted rays from the rim are identical in amplitude and phase and lead to the caustics. Therefore the diffracted fields from the aforementioned two points of the rim are invalid within a cone of half included angle θ_0 which is primarily a function of the radius of curvature of the rim. For most moderate size ground planes, θ_0 is in the range of $15° < \theta_0 < 30°$.

GEOMETRICAL THEORY OF DIFFRACTION: EDGE DIFFRACTION **823**

To make corrections for the diffracted field singularity and inaccuracy at and near the symmetry axis ($\theta = 0°$ and $180°$), due to axial caustics, the rim of the ground plane must be modeled as a ring radiator [32, 33]. This can be accomplished by using "equivalent" current concepts in diffraction which will be discussed in the next section.

A pattern based on the formulations of the preceding two-point diffraction was computed for a ground plane of 4.064λ diameter. This pattern is shown in Figure 13-37 where it is compared with measurements. It should be noted that this pattern was computed using the two-point diffraction for $30° \leq \theta \leq 150°$ ($\theta_0 \simeq 30°$); the remaining parts were computed using equivalent current concepts which will be discussed next.

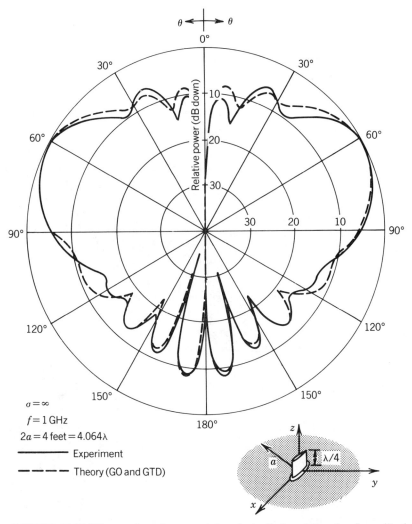

FIGURE 13-37 Measured and computed principal elevation plane amplitude patterns of a $\lambda/4$ monopole (blade) above a circular ground plane. (*Source:* C. A. Balanis, *Antenna Theory: Analysis and Design*, copyright © 1982, John Wiley & Sons, Inc. Reprinted by permission of John Wiley & Sons, Inc.)

824   GEOMETRICAL THEORY OF DIFFRACTION

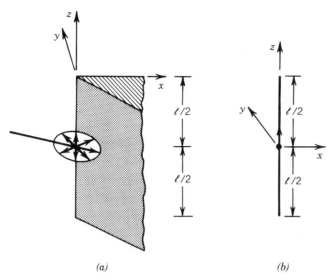

FIGURE 13-38 Wedge diffraction at normal incidence and its equivalent. (*a*) Actual wedge. (*b*) Equivalent.

13.3.5 Equivalent Currents in Diffraction

In contrast to diffraction by straight edges, diffraction by curved edges creates caustics. If observations are not made at or near caustics, ordinary diffraction techniques can be applied; otherwise, corrections must be made.

One technique that can be used to correct for caustic discontinuities and inaccuracies is the concept of the *equivalent currents* [32–36, 54–61]. To apply this principle, the two-dimensional wedge of Figure 13-13 is replaced by one of the following two forms:

1. An equivalent two-dimensional electric line source of equivalent electric current I^e, for soft polarization diffraction.
2. An equivalent two-dimensional magnetic line source of equivalent magnetic current I^m, for hard polarization diffraction.

This is illustrated in Figure 13-38. The equivalent currents I^e and I^m are adjusted so that the field radiated by each of the line sources is equal to the diffracted field of the corresponding polarization.

The electric field radiated by a two-dimensional electric line source placed along the z axis with a constant current I_z^e is given by (11-10a), or

$$E_z = -\frac{\beta^2 I_z^e}{4\omega\varepsilon} H_0^{(2)}(\beta\rho) \overset{\beta\rho \to \text{large}}{\simeq} -I_z^e \frac{\eta\beta}{2} \sqrt{\frac{j}{2\pi\beta}} \frac{e^{-j\beta\rho}}{\sqrt{\rho}} \quad (13\text{-}101a)$$

where $H_0^{(2)}(\beta\rho)$ is the Hankel function of the second kind of zero order. The approximate form of (13-101a) is valid for large distances of observation (far field), and it is obtained by replacing the Hankel function by its asymptotic formula for large argument (see Appendix IV, Equation IV-17).

The magnetic field radiated by a two-dimensional magnetic line source placed along the z axis with a constant current I_z^m can be obtained using the duality theorem (Section 7.2, Table 7-2) and (13-101a). Thus

$$H_z = -\frac{\beta^2 I_z^m}{4\omega\mu} H_0^{(2)}(\beta\rho) \overset{\beta\rho \to \text{large}}{\simeq} -I_z^m \frac{\beta}{2\eta} \sqrt{\frac{j}{2\pi\beta}} \frac{e^{-j\beta\rho}}{\sqrt{\rho}} \qquad (13\text{-}101\text{b})$$

To determine the equivalent electric current I_z^e, (13-101a) is equated to the field diffracted by a wedge when the incident field is of soft polarization. A similar procedure is used for the equivalent magnetic I_z^m of (13-101b). Using (13-34), (13-34a), (13-95), (13-101a), and (13-101b), and assuming normal incidence, we can write that

$$E_z^i(Q_d) D_s(\beta^-, \beta^+, n) \frac{e^{-j\beta\rho}}{\sqrt{\rho}} = -I_z^e \frac{\eta\beta}{2} \sqrt{\frac{j}{2\pi\beta}} \frac{e^{-j\beta\rho}}{\sqrt{\rho}} \qquad (13\text{-}102\text{a})$$

$$H_z^i(Q_d) D_h(\beta^-, \beta^+, n) \frac{e^{-j\beta\rho}}{\sqrt{\rho}} = -I_z^m \frac{\beta}{2\eta} \sqrt{\frac{j}{2\pi\beta}} \frac{e^{-j\beta\rho}}{\sqrt{\rho}} \qquad (13\text{-}102\text{b})$$

where $E_z^i(Q)$ = incident electric field at the diffraction point Q_d
$H_z^i(Q)$ = incident magnetic field at the diffraction point Q_d
D_s = diffraction coefficient for soft polarization [(13-71a) or (13-72e)]
D_h = diffraction coefficient for hard polarization [(13-71b) or (13-72f)]

Solving (13-102a) and (13-102b) for I_z^e and I_z^m respectively, leads to

$$\boxed{I_z^e = -\frac{\sqrt{8\pi\beta}}{\eta\beta} e^{-j\pi/4} E_z^i(Q) D_s(\beta^-, \beta^+, n)} \qquad (13\text{-}103\text{a})$$

$$\boxed{I_z^m = -\frac{\eta\sqrt{8\pi\beta}}{\beta} e^{-j\pi/4} H_z^i(Q) D_h(\beta^-, \beta^+, n)} \qquad (13\text{-}103\text{b})$$

If the wedge of Figure 13-38 is of finite length ℓ, its equivalent current will also be of finite length. The far-zone field radiated by each can be obtained by using techniques similar to those of Chapter 4 of [38]. Assuming the edge is along the z axis, the far-zone electric field radiated by an electric line source of length ℓ can be written using (4-58a) of [38] as

$$\boxed{E_\theta^e = j\eta \frac{\beta e^{-j\beta r}}{4\pi r} \sin\theta \int_{-\ell/2}^{\ell/2} I_z^e(z') e^{j\beta z' \cos\theta} dz'} \qquad (13\text{-}104\text{a})$$

Using duality, the magnetic field of a magnetic line source can be written as

$$\boxed{H_\theta^m = j \frac{\beta e^{-j\beta r}}{4\pi\eta r} \sin\theta \int_{-\ell/2}^{\ell/2} I_z^m(z') e^{j\beta z' \cos\theta} dz'} \qquad (13\text{-}104\text{b})$$

826   GEOMETRICAL THEORY OF DIFFRACTION

For a constant equivalent current, the integrals in (13-104a) and (13-104b) reduce to a $\sin(\zeta)/\zeta$ form.

If the equivalent current is distributed along a circular loop of radius a and it is parallel to the xy plane, the field radiated by each of the equivalent currents can be obtained using the techniques of Chapter 5, Section 5.3, of [38]. Thus

$$E_\phi^e = \frac{-j\omega\mu a e^{-j\beta r}}{4\pi r} \int_0^{2\pi} I_\phi^e(\phi') \cos(\phi - \phi') e^{j\beta a \sin\theta \cos(\phi - \phi')} d\phi' \qquad (13\text{-}105a)$$

$$H_\phi^m = \frac{-j\omega\varepsilon a e^{-j\beta r}}{4\pi r} \int_0^{2\pi} I_\phi^m(\phi') \cos(\phi - \phi') e^{j\beta a \sin\theta \cos(\phi - \phi')} d\phi' \qquad (13\text{-}105b)$$

If the equivalent currents are constant, the field is not a function of the azimuthal observation angle ϕ and (13-105a) and (13-105b) reduce to

$$E_\phi^e = \frac{a\omega\mu e^{-j\beta r}}{2r} I_\phi^e J_1(\beta a \sin\theta) \qquad (13\text{-}106a)$$

$$H_\phi^m = \frac{a\omega\varepsilon e^{-j\beta r}}{2r} I_\phi^m J_1(\beta a \sin\theta) \qquad (13\text{-}106b)$$

where $J_1(x)$ is the Bessel function of the first kind of order 1.

For diffraction by an edge of finite length, the equivalent current concept for diffraction assumes that each incremental segment of the edge radiates as would a corresponding segment of an infinite length two-dimensional edge. Similar assumptions are used for diffraction from finite length curved edges. The concepts, although approximate, have been shown to yield very good results.

For oblique plane wave incidence diffraction by a finite length ℓ wedge, as shown in Figure 13-39, the equivalent currents of (13-103a) and (13-103b) take the form

$$I_z^e = -\frac{\sqrt{8\pi\beta}}{\eta\beta} e^{-j\pi/4} E_z^i(Q_D) D_s(\xi^-, \xi^+, n; \beta_0')$$

$$\stackrel{s' \gg z'}{\simeq} -\frac{\sqrt{8\pi\beta}}{\eta\beta} e^{-j\pi/4} E_z^i(0) D_s(\xi^-, \xi^+, n; \beta_0') e^{-j\beta z' \cos\beta_0'} \qquad (13\text{-}107a)$$

$$I_z^m = -\frac{\eta\sqrt{8\pi\beta}}{\beta} e^{-j\pi/4} H_z^i(Q_D) D_h(\xi^-, \xi^+, n; \beta_0')$$

$$\stackrel{s' \gg z'}{\simeq} -\frac{\eta\sqrt{8\pi\beta}}{\beta} e^{-j\pi/4} H_z^i(0) D_h(\xi^-, \xi^+, n; \beta_0') e^{-j\beta z' \cos\beta_0'} \qquad (13\text{-}107b)$$

where $-\ell/2 \leq z' \leq \ell/2$ (ℓ = length of wedge) and D_s and D_h are formed by

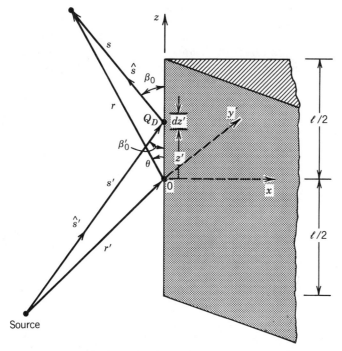

FIGURE 13-39 Oblique incidence diffraction by a finite length wedge.

(13-96a) and (13-96b). The far-zone fields associated with the equivalent currents of (13-107a) and (13-107b) can be found using, respectively, (13-104a) and (13-104b).

To demonstrate the technique of curved edge diffraction and the equivalent current concept, the radiation of a $\lambda/4$ monopole (blade) mounted on a circular ground plane was modeled. The analytical formulation is assigned as a problem at the end of the chapter. The computed pattern is shown in Figure 13-37 where it is compared with measurements.

To make corrections for the diffracted field discontinuity and inaccuracy at and near the symmetry axis ($\theta = 0°$ and $180°$), due to axial caustics, the rim of the ground plane was modeled as a ring radiator [32, 33]. Equivalent currents were used to compute the pattern in the region of $0° \leq \theta \leq \theta_0$ and $180° - \theta_0 \leq \theta \leq 180°$. In the other space, a two-point diffraction was used. The two points were taken diametrically opposite to each other, and they were contained in the plane of observation. The value of θ_0 depends upon the curvature of the ground plane. For most moderate size ground planes, θ_0 is in the range $15° < \theta_0 < 30°$.

A very good agreement between theory and experiment is exhibited in Figure 13-37. For this size ground plane, the blending of the two-point diffraction pattern and the pattern from the ring source radiator was performed at $\theta_0 \simeq 30°$. It should be noted that the minor lobes near the symmetry axis ($\theta \simeq 0°$ and $\theta \simeq 180°$) for the circular ground plane are more intense than the corresponding ones for the square plane of Figure 13-32. In addition, the back lobe nearest $\theta = 180°$ is of greater magnitude than the one next to it. These effects are due to the ring source radiation by the rim [33] of the circular ground plane toward the symmetry axis.

13.3.6 Slope Diffraction

Until now the field diffracted by an edge has been found based on (13-34a), (13-85), or (13-88) where $\mathbf{E}^i(Q_D)$ represents the incident field at the point of diffraction. This type of formulation indicates that if the incident field $\mathbf{E}^i(Q_D)$ at the point of diffraction Q_D is zero, then the diffracted field will be zero. In addition to this type of diffraction there is an additional diffraction term that is based not on the magnitude of the incident field at the point of diffraction but rather on the slope (rate of change, or normal derivative) of the incident field at the point of diffraction. This is a higher-order diffraction, and it becomes more significant when the incident field at the point of diffraction vanishes. It is referred to as *slope diffraction*, and it creates currents on the wedge surface that result in a diffracted field [52].

By referring to the geometry of Figure 13-40, the slope diffracted field can be computed using

Soft Polarization

$$E^d = \frac{1}{j\beta}\left[\frac{\partial E^i(Q_D)}{\partial n}\right]\left(\frac{\partial D_s}{\partial \phi'}\right)\sqrt{\frac{\rho_c}{s(\rho_c + s)}} e^{-j\beta s} \quad (13\text{-}108)$$

$$\frac{\partial E^i(Q_D)}{\partial n} = \frac{1}{s'}\left.\frac{\partial E^i}{\partial \phi'}\right|_{Q_D} = \text{slope of the incident field} \quad (13\text{-}108a)$$

$$\frac{\partial D_s}{\partial \phi'} = \text{slope diffraction coefficient} \quad (13\text{-}108b)$$

Hard Polarization

$$H^d = \frac{1}{j\beta}\left[\frac{\partial H^i(Q_D)}{\partial n}\right]\left(\frac{\partial D_h}{\partial \phi'}\right)\sqrt{\frac{\rho_c}{s(\rho_c + s)}} e^{-j\beta s} \quad (13\text{-}109)$$

$$\frac{\partial H^i(Q_D)}{\partial n} = \frac{1}{s'}\left.\frac{\partial H^i}{\partial \phi'}\right|_{Q_D} = \text{slope of the incident field} \quad (13\text{-}109a)$$

$$\frac{\partial D_h}{\partial \phi'} = \text{slope diffraction coefficient} \quad (13\text{-}109b)$$

Therefore, in general, the total diffracted field can be found using

$$U^d = \left[U^i(Q_D)D_{s,h} + \frac{1}{j\beta}\frac{\partial U^i(Q_D)}{\partial n}\frac{\partial D_{s,h}}{\partial \phi'}\right]\sqrt{\frac{\rho_c}{s(\rho_c + s)}} e^{-j\beta s} \quad (13\text{-}110)$$

where the first term represents the contribution to the total diffracted field due to the magnitude of the incident field and the second accounts for the contribution due to the slope (rate of change) of the incident field. In (13-110) U represents the electric field for soft polarization and the magnetic field for hard polarization.

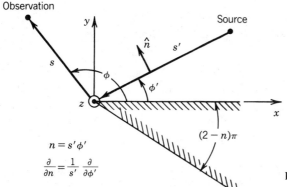

FIGURE 13-40 Wedge geometry for slope diffraction.

Similarly $D_{s,h}$ is used to represent D_s for soft polarization and D_h for hard polarization.

The slope diffraction coefficients for soft and hard polarizations can be written, respectively, as [52]

$$\frac{\partial D_s(\phi, \phi', n; \beta_0')}{\partial \phi'}$$

$$= -\frac{e^{-j\pi/4}}{4n^2\sqrt{2\pi\beta} \sin \beta_0'} \left(\left\{ \csc^2\left[\frac{\pi + (\phi - \phi')}{2n}\right] F_s[\beta L g^+(\phi - \phi')] \right.\right.$$

$$\left. - \csc^2\left[\frac{\pi - (\phi - \phi')}{2n}\right] F_s[\beta L g^-(\phi - \phi')] \right\}$$

$$+ \left\{ \csc^2\left[\frac{\pi + (\phi + \phi')}{2n}\right] F_s[\beta L g^+(\phi + \phi')] \right.$$

$$\left.\left. - \csc^2\left[\frac{\pi - (\phi + \phi')}{2n}\right] F_s[\beta L g^-(\phi + \phi')] \right\} \right) \quad \text{(13-111a)}$$

$$\frac{\partial D_h(\phi, \phi', n; \beta_0')}{\partial \phi'}$$

$$= -\frac{e^{-j\pi/4}}{4n^2\sqrt{2\pi\beta} \sin \beta_0'} \left(\left\{ \csc^2\left[\frac{\pi + (\phi - \phi')}{2n}\right] F_s[\beta L g^+(\phi - \phi')] \right.\right.$$

$$\left. - \csc^2\left[\frac{\pi - (\phi - \phi')}{2n}\right] F_s[\beta L g^-(\phi - \phi')] \right\}$$

$$- \left\{ \csc^2\left[\frac{\pi + (\phi + \phi')}{2n}\right] F_s[\beta L g^+(\phi + \phi')] \right.$$

$$\left.\left. - \csc^2\left[\frac{\pi - (\phi + \phi')}{2n}\right] F_s[\beta L g^-(\phi + \phi')] \right\} \right) \quad \text{(13-111b)}$$

where

$$F_s(X) = 2jX\left[1 - j2\sqrt{X}e^{jX}\int_{\sqrt{X}}^{\infty}e^{-j\tau^2}d\tau\right] = 2jX[1 - F(X)] \quad (13\text{-}111c)$$

A computer subroutine designated as SWDC, slope wedge diffraction coefficients, which computes the normalized (with respect to $\sqrt{\lambda}$) slope diffraction coefficients based on (13-111a) through (13-111c), is available at the end of the chapter. It was developed and reported in [53]. This program uses the complex function FTF (Fresnel transition function), whose listing is also included here, to complete its computations.

To use the subroutine the user must specify $R = L$ (in wavelengths), PHID = ϕ (in degrees), PHIPD = ϕ' (in degrees), BTD = β_0' (in degrees), and FN = n (dimensionless) and the program subroutine computes the normalized with respect to $\sqrt{\lambda}$ slope diffraction coefficients CSDCS = $\partial D_s/\partial \phi'$ and CSDCH = $\partial D_h/\partial \phi'$.

13.3.7 Multiple Diffractions

Until now we have considered single-order diffractions from each of the edges of a structure. If the structure is comprised of multiple edges (as is the case for infinitely thin strips, rectangular and circular ground planes, etc.), then coupling between the edges will take place. For finite thickness ground planes coupling is evident not only between diametrically opposite edges but also between edges on the same side of the ground plane. Coupling plays a bigger role when the separation between the edges is small, and it should then be taken into account.

For structures with multiple edges, coupling is introduced in the form of higher-order diffractions. To illustrate this point, let us refer to Figure 13-41a where a source is placed in the vicinity of a structure that is comprised of two wedges. These two wedges can represent, for example, the edges of a finite thickness ground plane. The radiation mechanism of this system can be outlined as follows: Radiation from the main source is diffracted by wedge 1, as shown in Figure 13-41b. Energy diffracted from wedge 1 in the direction of wedge 2 will be diffracted again, this time by wedge 2 as shown in Figure 13-41c. This is referred to as second-order diffraction, because it is the result of diffraction from diffraction. In turn diffractions from wedge 2 in the direction of wedge 1 undergo another diffraction, as shown in Figure 13-41b, and it is referred to as third-order diffraction. Second- and higher-order diffractions are all referred to as *higher-order diffractions*, and they result in coupling between the edges.

The procedure for formulating higher-order diffractions is the same as that outlined for first-order diffraction, and it should be followed when up to second-order diffractions are of interest. It becomes apparent, however, that the procedure for accounting for higher-order diffractions, especially for third and higher orders, can be very tedious, although straightforward. It is recommended that when third- and even higher-order diffractions are of interest, a procedure be adopted that accounts for all (infinite) orders of diffraction. This procedure is known as the *self-consistent method* [62] which is used in scattering theory [63]. It can be shown that the interactions between the edges can also be expressed in terms of a geometrical progression which in scattering theory is known as the *successive scattering procedure* [63].

Let us now illustrate the self-consistent method as applied to the diffractions of Figure 13-41a. According to Figure 13-41a and b diffractions by wedge 1 that are

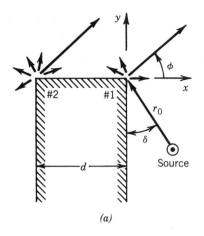
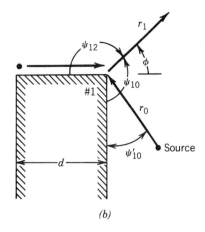
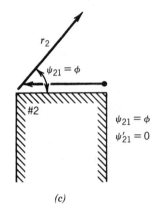

FIGURE 13-41 Finite thickness edge for multiple diffractions. (a) Source incidence. (b) Diffraction by edge 1. (c) Diffraction by edge 2.

due to radiation from the source and that are due to all orders of diffraction from wedge 2 can be written as

$$U_1^{s,h}(r_1, \phi) = U_0^{s,h}(Q_1) D_{10}^{s,h}\left(L_{10}, \psi_{10} = \frac{\pi}{2} + \phi, \psi'_{10} = \delta, n_1\right) A_{10}(r_1) e^{-j\beta r_1}$$

$$+ \frac{1}{2}\left[U_2^{s,h}(r_2 = d, \phi = 0)\right] D_{12}^{s,h}(L_{12}, \psi_{12} = \pi - \phi, \psi'_{12} = 0, n_1)$$

$$\times A_{12}(r_1) e^{-j\beta r_1} \qquad (13\text{-}112a)$$

where $U^{s,h}$ is used to represent here the electric field for soft polarization and the magnetic field for hard polarization. In (13-112a),

$U_1^{s,h}(r_1, \phi)$ = total diffracted field by wedge 1

$U_0^{s,h}(Q_1)$ = field from source at wedge 1

$U_2^{s,h}(r_2 = d, \phi = 0)$ = total diffracted field (including all orders of diffraction) by wedge 2 toward wedge 1

$D_{10}^{s,h}$ = diffraction coefficient (for soft or hard polarization) of wedge 1 that is due to radiation from the source

$D_{12}^{s,h}$ = diffraction coefficient (for soft or hard polarization) of wedge 1 that is due to radiation from wedge 2

A_{10} = amplitude spreading factor of wedge 1 that is due to radiation from the source

A_{12} = amplitude spreading factor of wedge 1 that is due to radiation from wedge 2

The unknown part in (13-112a) is $U_2^{s,h}(r_2 = d, \phi = 0)$, and the self-consistent method will be used to determine it.

Using a similar procedure and referring to Figure 13-41c, the total diffracted field by wedge 2 that is due to all orders of diffraction from wedge 1 can be written as

$$U_2^{s,h}(r_2, \phi) = \tfrac{1}{2}\left[U_1^{s,h}(r_1 = d, \phi = \pi)\right]$$
$$\times D_{21}^{s,h}(L_{21}, \psi_{21} = \phi, \psi'_{21} = 0, n_2) A_{21}(r_2) e^{-j\beta r_2} \quad (13\text{-}112b)$$

where $U_2^{s,h}(r_2, \phi)$ = total diffracted field by wedge 2

$U_1^{s,h}(r_1 = d, \phi = \pi)$ = total diffracted field (including all orders of diffraction) by wedge 1 toward wedge 2

$D_{21}^{s,h}$ = diffraction coefficient (for soft or hard polarization) of wedge 2 due to radiation from wedge 1

A_{21} = amplitude spreading factor of wedge 2 due to radiation from wedge 1

In (13-112b) the unknown part is $U_1^{s,h}(r_1 = d, \phi = \pi)$, and it will be determined using the self-consistent method.

Equations 13-112a and 13-112b form a consistent pair where there are two unknowns, that is $U_2^{s,h}(r_2 = d, \phi = 0)$ in (13-112a) and $U_1^{s,h}(r_1 = d, \phi = \pi)$ in (13-112b). If these two unknowns can be found, then (13-112a) and (13-112b) can be used to predict the diffracted fields from each of the wedges taking into account all (infinite) orders of diffraction. These two unknowns can be found by doing the following. At the position of wedge 2 ($r_1 = d, \phi = \pi$) the total diffracted field by wedge 1, as given by (13-112a), can be reduced to

$$\left[U_1^{s,h}(r_1 = d, \phi = \pi)\right]$$
$$= U_0^{s,h}(Q_1)\left\{D_{10}^{s,h}\left(L_{10}, \psi_{10} = \frac{3\pi}{2}, \psi'_{10} = \delta, n_1\right) A_{10}(r_1 = d) e^{-j\beta d}\right\}$$
$$+ \left[U_2^{s,h}(r_2 = d, \phi = 0)\right]\left\{\frac{1}{2}D_{12}^{s,h}(L_{12}, \psi_{12} = 0, \psi'_{12} = 0, n_1) A_{12}(r_1 = d) e^{-j\beta d}\right\}$$
$$(13\text{-}113a)$$

In a similar manner at the position of wedge 1 ($r_2 = d, \phi = 0$) the total diffracted field by wedge 2, as given by (13-112b), can be reduced to

$$[U_2^{s,h}(r_2 = d, \phi = 0)] = U_1^{s,h}(r_1 = d, \phi = \pi)$$
$$\times \{\tfrac{1}{2} D_{21}^{s,h}(L_{21}, \psi_{21} = 0, \psi'_{21} = 0, n_2) A_{21}(r_2 = d) e^{-j\beta d}\} \quad (13\text{-}113b)$$

Equations 13-113a and 13-113b can be rewritten, respectively, in simplified form as

$$\boxed{[U_1^{s,h}(r_1 = d, \phi = \pi)] = U_0^{s,h}(Q_1) T_{10}^{s,h} + [U_2^{s,h}(r_2 = d, \phi = 0)] R_{12}^{s,h}} \quad (13\text{-}114a)$$

$$\boxed{[U_2^{s,h}(r_2 = d, \phi = 0)] = [U_1^{s,h}(r_1 = d, \phi = \pi)] R_{21}^{s,h}} \quad (13\text{-}114b)$$

where

$$T_{10}^{s,h} = D_{10}^{s,h}\left(L_{10}, \psi_{10} = \frac{3\pi}{2}, \psi'_{10} = \delta, n_1\right) A_{10}(r_1 = d) e^{-j\beta d}$$
= transmission coefficient from wedge 1 toward wedge 2 due to radiation from main source (13-114c)

$$R_{12}^{s,h} = \tfrac{1}{2} D_{12}^{s,h}(L_{12}, \psi_{12} = 0, \psi'_{12} = 0, n_1) A_{12}(r_1 = d) e^{-j\beta d}$$
= reflection coefficient from wedge 1 toward wedge 2 due to diffractions from wedge 2 (13-114d)

$$R_{21}^{s,h} = \tfrac{1}{2} D_{21}^{s,h}(L_{21}, \psi_{21} = 0, \psi'_{21} = 0, n_2) A_{21}(r_2 = d) e^{-j\beta d}$$
= reflection coefficient from wedge 2 toward wedge 1 due to diffractions from wedge 1 (13-114e)

The self-consistent pair of (13-114a) and (13-114b) contains the two unknowns that are needed to predict the total diffracted field as given by (13-112a) and (13-112b). Solving (13-114a) and (13-114b) for $U_1^{s,h}(r_1 = d, \phi = \pi)$ and $U_2^{s,h}(r_2 = d, \phi = 0)$, we can show that

$$\boxed{U_1^{s,h}(r_1 = d, \phi = \pi) = U_0^{s,h}(Q_1) \frac{T_{10}^{s,h}}{1 - R_{21}^{s,h} R_{12}^{s,h}}} \quad (13\text{-}115a)$$

$$\boxed{U_2^{s,h}(r_2 = d, \phi = 0) = U_0^{s,h}(Q_1) \frac{T_{10}^{s,h} R_{21}^{s,h}}{1 - R_{21}^{s,h} R_{12}^{s,h}}} \quad (13\text{-}115b)$$

When expanded, it can be shown that (13-115a) and (13-115b) can be written as a geometrical progression series of the form

$$U_1^{s,h}(r_1 = d, \phi = \pi) = U_0^{s,h}(Q_1) T_{10}^{s,h}[1 + x_0 + x_0^2 + \cdots] \quad (13\text{-}116a)$$
$$U_2^{s,h}(r_2 = d, \phi = 0) = U_0^{s,h}(Q_1) T_{10}^{s,h} R_{21}^{s,h}[1 + x_0 + x_0^2 + \cdots] \quad (13\text{-}116b)$$

where

$$x_0 = R_{21}^{s,h} R_{12}^{s,h} \tag{13-116c}$$

Each term of the geometrical progression series can be related to an order of diffraction by the corresponding wedge.

In matrix form, the self-consistent set of equations as given by (13-114a) and (13-114b) can be written as

$$\begin{Bmatrix} 1 & -R_{12}^{s,h} \\ -R_{21}^{s,h} & 1 \end{Bmatrix} \begin{Bmatrix} U_1^{s,h} \\ U_2^{s,h} \end{Bmatrix} = \begin{Bmatrix} U_0^{s,h} T_{10}^{s,h} \\ 0 \end{Bmatrix} \tag{13-117}$$

which can be solved using standard matrix inversion methods.

The outlined self-consistent method can be extended and applied to the interactions between a larger number of edges. However, the system of equations to be solved will also increase, and its order will be equal to the number of interactions between the various edge combinations.

13.4 COMPUTER CODES

Using geometrical optics and wedge diffraction techniques, a number of computer codes have been developed over the years to compute the radiation and scattering characteristics of simple and complex antenna and scattering systems. Some are in the form of subroutines which are used primarily to compute wedge diffraction coefficients and associated functions. Others are very sophisticated codes that can be used to analyze very complex radiation and scattering problems. We will describe here two wedge diffraction subroutines and a subfunction which the reader can utilize to either develop his own codes or to solve problems of interest.

13.4.1 Wedge Diffraction Coefficients

The wedge diffraction coefficients (WDC) subroutines, whose listing is included at the end of this chapter, computes the soft and hard polarization wedge diffraction coefficients based on (13-89a) through (13-90b). To complete the computations, the program uses the complex function FTF (Fresnel transition function) whose listing is also included at the end of this chapter. This program was developed at the ElectroScience Laboratory at the Ohio State University, and it was reported in [53].

To use this subroutine, the user must specify

$R = L$ (in wavelengths) $=$ distance parameter

PHID $= \phi$ (in degrees) $=$ observation angle

PHIPD $= \phi'$ (in degrees) $=$ incident angle

BTD $= \beta_0'$ (in degrees) $=$ oblique angle

FN $= n$ (dimensionless) $=$ wedge angle factor

and the program computes the complex diffraction coefficients

CDCS $= D_s =$ complex diffraction coefficient (soft polarization)

CDCH $= D_h =$ complex diffraction coefficient (hard polarization)

13.4.2 Fresnel Transition Function

The Fresnel transition function (FTF), whose listing is included at the end of this chapter, computes the Fresnel transition functions $F(X)$ of (13-68e), (13-68f) and (13-69e), (13-69f). The program is based on the asymptotic expression of (13-74a) for small arguments ($X < 0.3$) and on (13-74b) for large arguments ($X > 5.5$). For intermediate values ($0.3 \leq X \leq 5.5$) a linear interpolation scheme is used. The program was developed at the ElectroScience Laboratory of the Ohio State University, and it was reported in [53].

13.4.3 Slope Wedge Diffraction Coefficients

The slope wedge diffraction coefficients (SWDC) subroutine, whose listing is included at the end of this chapter, computes the soft and hard polarization wedge diffraction coefficients based on (13-111a) through (13-111c). To complete the computations, the program uses the complex function FTF (Fresnel transition function) whose listing is also included at the end of this chapter. This program was developed at the ElectroScience Laboratory at the Ohio State University, and it was reported in [53].

To use this subroutine, the user must specify

$$R = L \text{ (in wavelengths)} = \text{distance parameter}$$
$$\text{PHID} = \phi \text{ (in degrees)} = \text{observation angle}$$
$$\text{PHIPD} = \phi' \text{ (in degrees)} = \text{incident angle}$$
$$\text{BTD} = \beta_0' \text{ (in degrees)} = \text{oblique angle}$$
$$\text{FN} = n \text{ (dimensionless)} = \text{wedge angle factor}$$

and the program computes the complex slope diffraction coefficients

$$\text{CSDCS} = \frac{\partial D_s}{\partial \phi'} = \text{complex slope diffraction coefficient (soft polarization)}$$

$$\text{CSDCH} = \frac{\partial D_h}{\partial \phi'} = \text{complex slope diffraction coefficient (hard polarization)}$$

REFERENCES

1. R. F. Harrington, "Matrix methods for field problems," *Proc. IEEE*, vol. 55, no. 2, pp. 136–149, February 1967.
2. R. F. Harrington, *Field Computation by Moment Methods*, Macmillan, New York, 1968.
3. J. H. Richmond, "Digital computer solutions of the rigorous equations for scattering problems," *Proc. IEEE*, vol. 53, pp. 796–804, August 1965.
4. J. Moore and R. Pizer, *Moment Methods in Electromagnetics*, Wiley, New York, 1984.
5. J. B. Keller, "Diffraction by an aperture," *J. Appl. Phys.*, vol. 28, no. 4, pp. 426–444, April 1957.
6. J. B. Keller, "Geometrical theory of diffraction," *J. Opt. Soc. Amer.*, vol. 52, no. 2, pp. 116–130, February 1962.

7. R. G. Kouyoumjian, "Asymptotic high-frequency methods," *Proc. IEEE*, vol. 53, pp. 864–876, August 1965.
8. P. H. Pathak and R. G. Kouyoumjian, "The dyadic diffraction coefficient for a perfectly conducting wedge," Technical Report 2183-4 (AFCRL-69-0546), Ohio State University ElectroScience Lab., June 5, 1970.
9. P. H. Pathak and R. G. Kouyoumjian, "An analysis of the radiation from apertures on curved surfaces by the geometrical theory of diffraction," *Proc. IEEE*, vol. 62, no. 11, pp. 1438–1447, November 1974.
10. R. G. Kouyoumjian and P. H. Pathak, "A uniform geometrical theory of diffraction for an edge in a perfectly conducting surface," *Proc. IEEE*, vol. 62, no. 11, pp. 1448–1461, November 1974.
11. G. L. James, *Geometrical Theory of Diffraction for Electromagnetic Waves*, Third Edition Revised, Peregrinus, London, 1986.
12. P. Y. Ufimtsev, "Method of edge waves in the physical theory of diffraction," translated by U.S. Air Force Foreign Technology Division, Wright-Patterson AFB, OH, September 1971.
13. P. Y. Ufimtsev, "Approximate computation of the diffraction of plane electromagnetic waves at certain metal bodies," *Sov. Phys.—Tech. Phys.*, pp. 1708–1718, 1957.
14. P. Y. Ufimtsev, "Secondary diffraction of electromagnetic waves by a strip," *Sov. Phys.—Tech. Phys.*, vol. 3, pp. 535–548, 1958.
15. K. M. Mitzner, "Incremental length diffraction coefficients," Technical Report AFAL-TR-73-296, Northrop Corp., Aircraft Division, April 1974.
16. E. F. Knott and T. B. A. Senior, "Comparison of three high-frequency, diffraction techniques," *Proc. IEEE*, vol. 62, no. 11, pp. 1468–1474, November 1974.
17. E. F. Knott, "A progression of high-frequency RCS prediction techniques," *Proc. IEEE*, vol. 73, no. 2, pp. 252–264, February 1985.
18. T. Griesser and C. A. Balanis, "Backscatter analysis of dihedral corner reflectors using physical optics and physical theory of diffraction," *IEEE Trans. Antennas Propagat.*, vol. AP-35, no. 10, pp. 1137–1147, October 1987.
19. P. M. Russo, R. C. Rudduck, and L. Peters, Jr., "A method for computing E-plane patterns of horn antennas," *IEEE Trans. Antennas Propagat.*, vol. AP-13, no. 2, pp. 219–224, 1965.
20. R. C. Rudduck and L. L. Tsai, "Aperture reflection coefficient of TEM and TE_{01} mode parallel-plate waveguide," *IEEE Trans. Antennas Propagat.*, vol. AP-16, no. 1, pp. 83–89, January 1968.
21. C. A. Balanis and L. Peters, Jr., "Analysis of aperture radiation from an axially slotted circular conducting cylinder using geometrical theory of diffraction," *IEEE Trans. Antennas Propagat.*, vol. AP-17, no. 1, pp. 93–97, January 1969.
22. C. A. Balanis and L. Peters, Jr., "Equatorial plane pattern of an axial-TEM slot on a finite size ground plane," *IEEE Trans. Antennas Propagat.*, vol. AP-17, no. 3, pp. 351–353, May 1969.
23. C. A. Balanis, "Radiation characteristics of current elements near a finite length cylinder," *IEEE Trans. Antennas Propagat.*, vol. AP-18, no. 3, pp. 352–359, May 1970.
24. C. A. Balanis, "Analysis of an array of line sources above a finite ground plane," *IEEE Trans. Antennas Propagat.*, vol. AP-19, no. 2, pp. 181–185, March 1971.
25. C. L. Yu, W. D. Burnside, and M. C. Gilreath, "Volumetric pattern analysis of airborne antennas," *IEEE Trans. Antennas Propagat.*, vol. AP-26, no. 5, pp. 636–641, September 1978.
26. G. A. Thiele and T. H. Newhouse, "A hybrid technique for combining moment methods with the geometrical theory of diffraction," *IEEE Trans. Antennas Propagat.*, vol. AP-23, no. 1, pp. 62–69, 1975.
27. W. D. Burnside, C. L. Yu, and R. J. Marhefka, "A technique to combine the geometrical theory of diffraction and the moment method," *IEEE Trans. Antennas Propagat.*, vol. AP-23, no. 4, pp. 551–558, July 1975.

28. J. N. Sahalos and G. A. Thiele, "On the application of the GTD-MM technique and its limitations," *IEEE Trans. Antennas Propagat.*, vol. AP-29, no. 5, pp. 780–786, September 1981.
29. R. K. Luneberg, "Mathematical theory of optics," Brown University Notes, Providence, RI, 1944.
30. M. Kline, "An asymptotic solution of Maxwell's equations," in *The Theory of Electromagnetic Waves*, Interscience, New York, 1951.
31. M. Kline and I. Kay, *Electromagnetic Theory and Geometrical Optics*, Interscience, New York, 1965.
32. C. E. Ryan, Jr., and L. Peters, Jr., "Evaluation of edge-diffracted fields including equivalent currents for the caustic regions," *IEEE Trans. Antennas Propagat.*, vol. AP-17, pp. 292–299, May 1969; erratum, vol. AP-18, p. 275, March 1970.
33. C. A. Balanis, "Radiation from conical surfaces used for high-speed spacecraft," *Radio Science*, vol. 7, pp. 339–343, February 1972.
34. E. F. Knott, T. B. A. Senior, and P. L. E. Uslenghi, "High-frequency backscattering from a metallic disc," *Proc. IEEE*, vol. 118, no. 12, pp. 1736–1742, December 1971.
35. W. D. Burnside and L. Peters, Jr., "Edge diffracted caustic fields," *IEEE Trans. Antennas Propagat.*, vol. AP-22, no. 4, pp. 620–623, July 1974.
36. D. P. Marsland, C. A. Balanis, and S. Brumley, "Higher order diffractions from a circular disk," *IEEE Trans. Antennas Propagat.*, vol. AP-35, no. 12, pp. 1436–1444, December 1987.
37. R. G. Kouyoumjian, L. Peters, Jr., and D. T. Thomas, "A modified geometrical optics method for scattering by dielectric bodies," *IRE Trans. Antennas Propagat.*, vol. AP-11, no. 6, pp. 690–703, November 1963.
38. C. A. Balanis, *Antenna Theory: Analysis and Design*, Wiley, New York, 1982.
39. T. Griesser and C. A. Balanis, "Dihedral corner reflector backscatter using higher-order reflections and diffractions," *IEEE Trans. Antennas Propagat.*, vol. AP-35, no. 11, pp. 1235–1247, November 1987.
40. W. Pauli, "On asymptotic series for functions in the theory of diffraction of light," *Physical Review*, vol. 34, pp. 924–931, December 1938.
41. F. Oberhettinger, "On asymptotic series occurring in the theory of diffraction of waves by a wedge," *J. Math. Phys.*, vol. 34, pp. 245–255, 1956.
42. D. L. Hutchins, "Asymptotic series describing the diffraction of a plane wave by a two-dimensional wedge of arbitrary angle," Ph.D. dissertation, Dept. of EE, Ohio State University, 1967.
43. W. Franz and K. Deppermann, "Theorie der beugung am zylinder unter berücksichtigung der kriechwelle," *Ann. Phys.*, 6 Folge, Bd. 10, Heft 6-7, pp. 361–373, 1952.
44. B. R. Levy and J. B. Keller, "Diffraction by a smooth object," *Commun. Pure Appl. Math.*, vol. XII, no.1, pp. 159–209, February 1959.
45. J. B. Keller and B. R. Levy, "Decay exponents and diffraction coefficients for surface waves of nonconstant curvature," *IRE Trans. Antennas Propagat.*, vol. AP-7 (special suppl.), pp. S52–S61, December 1959.
46. D. R. Voltmer, "Diffraction by doubly curved convex surfaces," Ph.D. dissertation, Dept. of EE, Ohio State University, 1970.
47. L. B. Felsen and N. Marcuvitz, *Radiation and Scattering of Waves*, Prentice-Hall, Englewood Cliffs, NJ, 1973.
48. F. B. Hildebrand, *Advanced Calculus for Applications*, Prentice-Hall, Englewood Cliffs, NJ, 1962.
49. J. Boersma, "Computation of Fresnel integrals," *J. Math. Comp.*, vol. 14, p. 380, 1960.
50. G. L. James, "An approximation to the Fresnel integral," *Proc. IEEE*, vol. 67, no. 4, pp. 677–678, April 1979.
51. P. C. Clemmow, *The Plane Wave Spectrum Representation of Electromagnetic Fields*, Pergamon, Elmsford, NY, 1966.

52. R. G. Kouyoumjian, "The geometrical theory of diffraction and its application," in *Numerical and Asymptotic Techniques in Electromagnetics*, R. Mittra (Ed.), Springer, New York, 1975, Chapter 6.
53. "The modern geometrical theory of diffraction," vol. 1, *Short Course Notes*, ElectroScience Lab., Ohio State University.
54. R. F. Millar, "An approximate theory of the diffraction of an electromagnetic wave by an aperture in a plane screen," *Proc. IEE*, Monograph No. 152R, vol. 103 (pt. C), pp. 117–185, October 1955.
55. R. F. Millar, "The diffraction of an electromagnetic wave by a circular aperture," *Proc. IEE*, Monograph No. 196R, vol. 104 (pt. C), pp. 87–95, September 1956.
56. R. F. Millar, "The diffraction of an electromagnetic wave by a large aperture," *Proc. IEE*, Monograph No. 213R (pt. C), pp. 240–250, December 1956.
57. W. D. Burnside and L. Peters, Jr., "Axial-radar cross section of finite cones by the equivalent-current concept with higher-order diffraction," *Radio Science*, vol. 7, no. 10, pp. 943–948, October 1982.
58. E. F. Knott, "The relationship between Mitzner's ILDC and Michaeli's equivalent currents," *IEEE Trans. Antennas Propagat.*, vol. AP-33, no. 1, pp. 112–114, January 1985.
59. A. Michaeli, "Equivalent edge currents for arbitrary aspects of observation," *IEEE Trans. Antennas Propagat.*, vol. AP-32, no. 3, pp. 252–258, March 1984; erratum, vol. AP-33, no. 2, p. 227, February 1985.
60. A. Michaeli, "Elimination of infinities in equivalent edge currents, Part I: Fringe current components," *IEEE Trans. Antennas Propagat.*, vol. AP-34, no. 7, pp. 912–918, July 1986.
61. A. Michaeli, "Elimination of infinities in equivalent edge currents, Part II: Physical optics components," *IEEE Trans. Antennas Propagat.*, vol. AP-34, no. 8, pp. 1034–1037, August 1986.
62. R. C. Rudduck and J. S. Yu, "Higher-order diffraction concept applied to parallel-plate waveguide patterns," Report No. 1691-16, The Antenna Lab. (now ElectroScience Lab.), Ohio State University, October 15, 1965.
63. V. Twersky, "Multiple scattering of waves and optical phenomena," *J. Opt. Soc. Amer.*, vol. 52, no. 2, pp. 145–171, February 1962.

PROBLEMS

13.1. Using the geometry of Figure 13-4, derive (13-9).

13.2. An arbitrary surface of revolution can be represented by
$$z = g(u) \quad \text{where} \quad u = \frac{x^2 + y^2}{2}$$
Define
$$K^2 = 1 + 2u \left[\frac{dg(u)}{du}\right]^2$$
Then the unit vectors $\hat{u}_1$ and $\hat{u}_2$ of Figure 13-6 in the directions of the principal radii of curvature R_1 and R_2, respectively, can be determined using
$$\hat{u}_1 = \frac{\hat{a}_x y - \hat{a}_y x}{\sqrt{x^2 + y^2}}$$
$$\hat{u}_2 = \frac{\hat{a}_x x + \hat{a}_y y + \hat{a}_z \left[(x^2 + y^2)\frac{dg(u)}{du}\right]}{K\sqrt{x^2 + y^2}}$$

and the R_1 and R_2 can be found using

$$\frac{1}{R_1} = \frac{1}{K}\frac{dg(u)}{du}$$

$$\frac{1}{R_2} = \frac{1}{K^3}\left[\frac{dg(u)}{du} + 2u\frac{d^2g(u)}{du^2}\right]$$

For a paraboloidal reflector (parabola of revolution), widely used as a microwave reflector antenna, whose surface can be represented by

$$z = f - \frac{x^2 + y^2}{4f}$$

where f is the focal distance, show that the principal radii of curvature are given by

$$\frac{1}{R_1} = -\frac{1}{2f}\frac{1}{\left[1 + \frac{x^2+y^2}{4f^2}\right]^{1/2}}$$

$$\frac{1}{R_2} = -\frac{1}{2f}\frac{1}{\left[1 + \frac{x^2+y^2}{4f^2}\right]^{3/2}}$$

13.3. Derive (13-30) using the geometry of Figure 13-7.

13.4. An electric line source is placed in front of a 30° convex segment of an infinite length conducting circular arc, as shown in Figure P13-4. Assume the line source is placed symmetrically about the arc, its position coincides with the origin of the coordinate system, and it is parallel to the length of the arc. Using geometrical optics determine the following:

(a) The location of the caustic for fields reflected by the surface of the cap.
(b) The far-zone backscattered electric field at a distance of 50λ from the center of the apex of the arc.

Assume that the radius of the arc is 5λ and the incident electric field at the apex of the arc is

$$\mathbf{E}^i(\rho = 5\lambda, \phi = 0°) = \hat{a}_z 10^{-3} \text{ V/m}$$

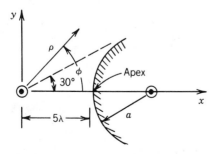

FIGURE P13-4

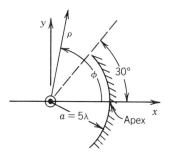

FIGURE P13-5

13.5. Repeat Problem 13.4 for a 30° concave segment of an infinite length conducting circular arc, as shown in Figure P13-5.

13.6. Repeat the calculations of Example 13-2 for $h = 0.25$ and $w = 2\lambda$.

13.7. Show that by combining (13-60a) and (13-60b) you get (13-61).

13.8. Show that evaluation of (13-48b) along C_T leads to (13-62a) and along $\text{SDP}_{\pm\pi}$ leads to (13-62b).

13.9. When the line source is in the vicinity of the edge of the wedge and the observations are made at large distances ($\rho \gg \rho'$) in Figure 13-12, show the following:

(a) Green's function of (13-38) can be approximated by
$$G \simeq \sqrt{\frac{2}{\pi\beta\rho}}\, e^{-j(\beta\rho - \pi/4)} F(\beta\rho')$$

where
$$F(\beta\rho') = \frac{1}{n}\sum_{m=0}^{\infty} \varepsilon_m J_{m/n}(\beta\rho') e^{+j(m/n)(\pi/2)}\left[\cos\frac{m}{n}(\phi - \phi') \pm \cos\frac{m}{n}(\phi + \phi')\right]$$

(b) Geometrical optics fields of (13-65) can be written as
$$F_G(\beta\rho') = \begin{cases} e^{j\beta\rho'\cos(\phi-\phi')} \pm e^{j\beta\rho'\cos(\phi+\phi')} & \text{for } 0 < \phi < \pi - \phi' \\ e^{j\beta\rho'\cos(\phi-\phi')} & \text{for } \pi - \phi' < \phi < \pi + \phi' \\ 0 & \text{for } \pi + \phi' < \phi < n\pi \end{cases}$$

(c) Diffracted fields of (13-67) can be written as
$$V_D^{i,r}(\rho', \phi \mp \phi', n) = V_D^i(\rho', \phi - \phi', n) \pm V_D^r(\rho', \phi + \phi', n)$$
$$= \frac{e^{-j\pi/4}}{\sqrt{2\pi\beta}}\,\frac{1}{n}\sin\left(\frac{\pi}{n}\right)$$
$$\times \left[\frac{1}{\cos\left(\dfrac{\pi}{n}\right) - \cos\left(\dfrac{\phi - \phi'}{n}\right)} \pm \frac{1}{\cos\left(\dfrac{\pi}{n}\right) - \cos\left(\dfrac{\phi + \phi'}{n}\right)}\right]\frac{e^{-j\beta\rho'}}{\sqrt{\rho'}}$$

13.10. A unity amplitude uniform plane wave of soft polarization is incident normally on a half-plane at an angle of 45°, as shown in Figure P13-10. At an observation point P

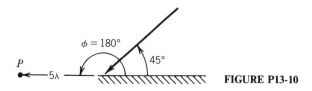

FIGURE P13-10

of $\rho = 5\lambda$, $\phi = 180°$ from the edge of the half-plane, determine the following:
(a) Incident geometrical optics field.
(b) Reflected geometrical optics field.
(c) Total geometrical optics field.
(d) Incident diffracted field.
(e) Reflected diffracted field.
(f) Total diffracted field.
(g) Total field (geometrical optics plus diffracted).

13.11. Repeat Problem 13.10 for a hard polarization uniform plane wave.

13.12. An electric line source, whose normalized electric field at the origin is unity, is placed at a distance of $\rho' = 5\lambda$, $\phi' = 180°$ from the edge of a half-plane, as shown in Figure P13-12. Using the reciprocity principle of Figure 13-16 and the results of Problem 13.9, determine at large distances the following:
(a) Incident geometrical optics electric field.
(b) Reflected geometrical optics electric field.
(c) Total geometrical optics electric field.
(d) Incident diffracted electric field.
(e) Reflected diffracted electric field.
(f) Total diffracted electric field.
(g) Total electric field (geometrical optics plus diffracted).

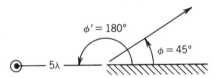

FIGURE P13-12

13.13. Repeat Problem 13.12 for a magnetic line source whose normalized magnetic field at the origin is unity. At each observation determine the magnetic field.

13.14. A unity amplitude uniform plane wave is incident normally on a 30° conducting wedge. Assume that the incident electric field is polarized in the z direction and the incident angle is 45°. Then determine at $\rho = 5.5\lambda$ the following:
(a) Incident GO electric field at $\phi = 225°$ (minus).
(b) Incident GO electric field at $\phi = 225°$ (plus).
(c) Approximate incident diffracted field at $\phi = 225°$ (minus).
(d) Approximate incident diffracted field at $\phi = 225°$ (plus).
(e) Approximate total electric field at $\phi = 225°$ (minus).
(f) Approximate total electric field at $\phi = 225°$ (plus).

Plus and minus refer to angles slightly greater or smaller than the designated values.

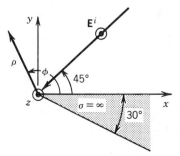

FIGURE P13-14

13.15. Repeat Problem 13.14 when the incident magnetic field of the unity amplitude uniform plane wave is polarized in the z direction. At each point determine the corresponding GO, diffracted, or total magnetic field.

13.16. A unity amplitude uniform plane wave of soft polarization is incident upon a half-plane as shown in Figure P13-16. At a plane parallel and behind the half-plane perform the following tasks.

(a) Formulate expressions for the incident and reflected geometrical optics fields, incident and reflected diffracted fields, and total field.
(b) Plot along the observation plane the total field (geometrical optics plus diffracted fields) when $y_0 = 5\lambda$, $-5\lambda \le x_0 \le 5\lambda$.
(c) Determine the total field at $y_0 = 5\lambda$ and $x_0 = 0$. Explain the result.

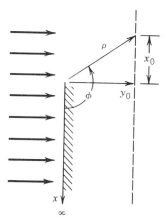

FIGURE P13-16

13.17. Repeat Problem 13.16 for a hard polarized uniform plane wave.

13.18. A unity amplitude uniform plane wave is incident normally on a two-dimensional conducting wedge, as shown in Figure P13-18.

(a) Formulate expressions that can be used to determine the incident, reflected, and total diffracted fields away from the incident and reflected shadow boundaries.
(b) Plot the soft and hard polarization normalized diffraction coefficients $(D_{s,h}/\sqrt{\lambda})$ as a function of ϕ for $n = 1.5$ and $n = 2$.
(c) Simplify the expressions of part a when $n = 2$ (half-plane).
(d) Formulate expressions for the two-dimensional scattering width of the half-plane ($n = 2$) for soft and hard polarizations.
(e) Simplify the expressions of part d for backscattering observations ($\phi = \phi'$). How can the results of this part be used to design low-observable radar targets?

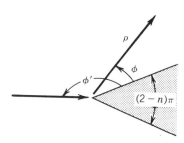

FIGURE P13-18

13.19. For the geometry of Problem 13.18 plot at $\phi = \phi'$ the normalized diffraction coefficient $(D_{s,h}/\sqrt{\lambda})$ as a function of the wedge angle $(1 \leq n \leq 2)$ when the polarization of the wave is (a) soft and (b) hard.

13.20. Repeat Problem 13.19 when the observations are made along the surface of the wedge $(\phi = 0°)$.

13.21. By approximating the integrand of the Fresnel integral of (13-63c) with a truncated Taylor series

$$e^{-j\tau^2} \simeq \sum_{n=0}^{M} \frac{(-j\tau^2)^n}{n!}$$

derive the small argument approximation of (13-74a) for the transition function.

13.22. By repeatedly integrating by parts the Fresnel integral of (13-63c), derive the large argument approximation of (13-74b) for the transition function.

13.23. Show that there is a finite discontinuity of unity amplitude with the proper polarity, similar to (13-83), along the following boundaries.
   (a) RSB of Figure 13-23a.
   (b) ISB of Figure 13-23b.
   (c) RSB of Figure 13-23b.

13.24. Compute the corresponding phases for the fields, GO and diffracted, for Figure 13-24a and for Figure 13-24b.

13.25. Repeat the calculations of Example 13-4 for a strip of width $w = 2\lambda$ and with the line source at a height of $h = 0.25\lambda$.

13.26. Repeat Example 13-5 for a hard polarized uniform plane wave.

13.27. A unity amplitude uniform plane wave is incident normally on a two-dimensional strip of width w as shown in Figure P13-27.
   (a) Formulate expressions for the fields when the observations are made below the strip along its axis of symmetry.
   (b) Compute the field when $w = 3\lambda$ and $d = 5\lambda$ for soft and hard polarizations.

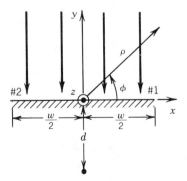

FIGURE P13-27

13.28. A transmitter and a receiver are placed on either side of a mountain that can be modeled as a perfectly conducting half-plane, as shown in Figure P13-28. Assume that the transmitting source is isotropic.
   (a) Derive an expression for the field at the receiver that is diffracted from the top of the mountain. Assume the field is soft or hard polarized.

844 GEOMETRICAL THEORY OF DIFFRACTION

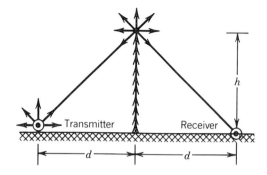

FIGURE P13-28

(b) Compute the power loss (in decibels) at the receiver that is due to the presence of the mountain when $h = 5\lambda$ and $d = 5\lambda$. Do this when the transmitter and receiver are both, for each case, either soft or hard polarized.

13.29. A rectangular waveguide of dimensions a and b operating in the dominant TE_{10} mode is mounted on a square ground plane with dimensions w on each of its sides, as shown in Figure P13-29. Using the geometry of Example 13-6 and assuming that the total geometrical optics field above the ground plane in the principal yz plane is given by

$$E_{\theta G}(\theta) = E_0 \left[\frac{\sin\left(\frac{\beta b}{2} \sin\theta\right)}{\frac{\beta b}{2} \sin\theta} \right] \frac{e^{-j\beta r}}{r} \qquad 0 \le \theta \le \frac{\pi}{2}$$

(a) Show that the fields diffracted from edges 1 and 2 in the principal yz plane are given by

$$E_{\theta 1}^d(\theta) = E_0 \frac{\sin\left(\frac{\beta b}{2}\right)}{\frac{\beta b}{2}} V_B^i\left(\frac{w}{2}, \psi_1, n_1 = 2\right) e^{+j(\beta w/2)\sin\theta} \frac{e^{-j\beta r}}{r}$$

$$\psi_1 = \frac{\pi}{2} + \theta, \quad 0 \le \theta \le \pi$$

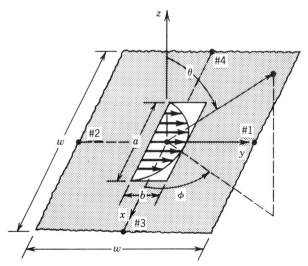

FIGURE P13-29

$$E_{\theta 2}^{d}(\theta) = E_0 \frac{\sin\left(\dfrac{\beta b}{2}\right)}{\dfrac{\beta b}{2}} V_B^i\left(\dfrac{w}{2}, \psi_2, n_2 = 2\right) e^{-j(\beta w/2)\sin\theta} \dfrac{e^{-j\beta r}}{r}$$

$$\psi_2 = \begin{cases} \dfrac{\pi}{2} - \theta & 0 \le \theta \le \dfrac{\pi}{2} \\ \dfrac{5\pi}{2} - \theta & \dfrac{\pi}{2} \le \theta \le \pi \end{cases}$$

(b) Plot the normalized amplitude pattern (in decibels) for $0° \le \theta \le 180°$ when $w/2 = 14.825\lambda$ and $b = 0.42\lambda$.

13.30. A unity amplitude uniform plane wave is incident at a grazing angle on a circular ground plane of radius a as shown in Figure P13-30.

(a) Formulate expressions for the field diffracted from the leading edge (#2) of the ground plane along the principal yz plane for both soft and hard polarizations.
(b) Locate the position of the caustic for the leading edge.
(c) Assuming the incident electric field amplitude at the leading edge of the plate is 10^{-3} V/m and its phase is zero, compute for each polarization at $r = 50\lambda$ when $a = 2\lambda$ the backscattered electric field that is due to the leading edge of the plate.

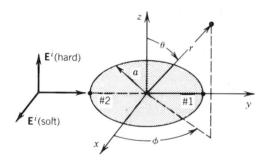

FIGURE P13-30

13.31. Show that the fields radiated by a circular loop with nonuniform equivalent electric and magnetic current I_ϕ^e and I_ϕ^m are given, respectively, by (13-105a) and (13-105b). If the equivalent currents I_ϕ^e and I_ϕ^m are uniform, show that (13-105a) and (13-105b) reduce, respectively, to (13-106a) and (13-106b).

13.32. An infinitesimal dipole is placed on the tip of a finite cone, as shown in Figure P13-32. The total geometrical optics magnetic field radiated by the source in the presence of the cone, referred to the center of the base of the cone, is given by

$$H_{\phi G} = R(\theta) e^{j\beta s \cos\theta \cos(\alpha/2)}$$

where $R(\theta)$ is the field distribution when the cone is infinite in length ($s = \infty$). For the finite length cone there is also a diffracted field forming a ring source at the base of the cone.

(a) Using (13-105b), show that the diffracted magnetic field is given by

$$H_\phi^d = b R\left(\theta = \pi - \dfrac{\alpha}{2}\right) V_B^i(s, \psi, n) \int_0^{2\pi} \cos(\phi) e^{j(\beta b/2)\sin\theta\cos\phi}\, d\phi, \qquad \psi = \dfrac{\alpha}{2} + \theta$$

where $V_B^i(s, \psi, n)$ is the incident diffraction function and ϕ is the azimuthal observation angle.

846 GEOMETRICAL THEORY OF DIFFRACTION

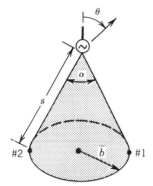

FIGURE P13-32

(b) Assuming that the base of the cone is large ($b \gg \lambda$) and the observations are made away from the symmetry axis so that $\sin\theta > 0$, show that by using the method of steepest descent of Appendix VI the integral formed by the ring source diffracted field from the rim of the cone reduces to the field diffracted from two diametrically opposite points on the rim (i.e., points #1 and #2 of Figure P13-32).

13.33. For the $\lambda/4$ monopole on the circular ground plane of Example 13-7, model the rim of the ground plane as a ring source radiator using equivalent current concepts of Section 13.3.5 to correct for the caustic formed when the observations are made near the axis ($\theta \simeq 0°$ and $180°$) of the ground plane.

(a) Derive expressions for the diffracted field from the rim using the equivalent currents.

(b) Compute and plot the pattern near the axis ($\theta \simeq 0°$ and $180°$) of the ground plane when the diameter of the ground plane is $d = 4.064\lambda$. Use this part of the pattern to complement that computed using the two-point diffraction of Example 13-7.

13.34. The normalized total geometrical optics field radiated in the principal xz plane (H plane; $\phi = 0°, 180°$) by the rectangular waveguide of Problem 13.29 (Figure P13-29) is given by

$$E_{\phi G} = E_0 \cos\theta \frac{\cos\left(\dfrac{\beta a}{2}\sin\theta\right)}{\left(\dfrac{\beta a}{2}\sin\theta\right)^2 - \left(\dfrac{\pi}{2}\right)^2} \frac{e^{-j\beta r}}{r} \qquad 0° \leq \theta \leq 90°$$

Use slope diffraction concepts of Section 13.3.6.

(a) Formulate the field diffracted along the xz plane using two-point diffraction (points #3 and #4 of Figure P13-29).

(b) Plot (in decibels) the normalized amplitude pattern when $a = \lambda/2$, $w = 4\lambda$.

13.35. An infinite magnetic line source is placed on a two-dimensional square conducting cylinder at the center of its top side. Use successive single-order diffractions on each of the edges of the cylinder.

(a) Formulate expressions for the magnetic field that would be observed at the center of the bottom side of the cylinder.

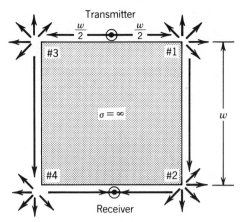

FIGURE P13-35

(b) Compute the power loss (in decibels) at the observation point that is due to the presence of the cylinder when $w = 5\lambda$. Assume far-field observation approximations.

848 GEOMETRICAL THEORY OF DIFFRACTION

COMPUTER PROGRAM: WDC

```
      SUBROUTINE WDC(CDCS,CDCH,R,PHID,PHIPD,BTD,FN)
C*************************************************************
C                                                             *
C     THIS SUBROUTINE COMPUTES THE FIRST ORDER WEDGE DIFFRACTION *
C     COEFFICIENT. IT IS BASED ON EQUATIONS (13-89A)-(13-90B)  *
C     THE ALGORITHM WAS DEVELOPED AND REPORTED IN REFERENCE (53). *
C                                                             *
C... ON INPUT                                                 *
C     R    -REAL, DISTANCE PARAMETER (IN WAVELENGTHS).        *
C     PHID -REAL, OBSERVATION ANGLE PHI (IN DEGREES).         *
C     PHIPD-REAL, INCIDENT ANGLE PHI PRIME (IN DEGREES).      *
C     BTD  -REAL, OBLIQUE INCIDENT ANGLE BETA (IN DEGREES).   *
C     FN   -REAL, WEDGE FACTOR N (DIMENSIONLESS).             *
C                                                             *
C... ON OUTPUT                                                *
C     CDCS -COMPLEX DIFFRACTION COEFFICIENT (SOFT POLARIZATION). *
C     CDCH -COMPLEX DIFFRACTION COEFFICIENT (HARD POLARIZATION). *
C                                                             *
C                                                             *
C*************************************************************
      COMPLEX D(4),CT,TERM,FT(4),FTF,CDCS,CDCH
      DATA CT,UTPI/(-0.05626977,0.05626977),0.15915494/
      DATA PI,TPI,SML/3.14159265,6.28318531,0.001/
      DTOR=PI/180.
      PHR=PHID*DTOR
      PHPR=PHIPD*DTOR
      SBO=SIN(DTOR*BTD)
      IF(ABS(FN-0.5).LT.SML) GOTO 10
      DKL=TPI*R
      UFN=1./FN
      BETAR=PHR-PHPR
      TERM=CT/(FN*SBO)
      SGN=1.
      I=0
    2 I=I+1
      DN=UFN*(0.5*SGN+UTPI*BETAR)
      N=DN+SIGN(0.5,DN)
      ANG=TPI*FN*N-BETAR
      A=2.*(COS(0.5*ANG)**2)
      X=DKL*A
      Y=PI+SGN*BETAR
      IF(ABS(X).LT.1.E-10) GOTO 3
      FT(I)=FTF(X)/TAN(.5*Y*UFN)
      GOTO 4
    3 FT(I)=(0.,0.)
      IF(ABS(COS(0.5*Y*UFN)).LT.1.E-3) GOTO 4
      FT(I)=(1.7725,1.7725)*SIGN(SQRT(DKL),Y)
      FT(I)=(FT(I)-(0.,2.)*DKL*(PI-SGN*ANG))*FN
    4 SGN=-SGN
      D(I)=TERM*FT(I)
      IF(SGN.LT.0.) GOTO 2
      BETAR=PHR+PHPR
      IF(I.LE.3) GOTO 2
      CDCS=(D(1)+D(2))-(D(3)+D(4))
      CDCH=(D(1)+D(2))+(D(3)+D(4))
      RETURN
   10 CDCS=(0.,0.)
      CDCH=(0.,0.)
      RETURN
      END
```

COMPUTER PROGRAM: SWDC

```
      SUBROUTINE  SWDC(CSDCS,CSDCH,R,PHID,PHIPD,BTD,FN)
C************************************************************************
C                                                                       *
C     THIS SUBROUTINE  COMPUTES THE SLOPE WEDGE DIFFRACTION COEFFICIENT  *
C     FOR A WEDGE. IT IS BASED ON EQUATIONS (13-111A)-(13-111C)          *
C     THIS ALGORITHM WAS DEVELOPED AND REPORTED IN REFERENCE (53).       *
C                                                                       *
C... ON INPUT                                                           *
C     R    -REAL,  DISTANCE (IN WAVELENGTHS).                           *
C     PHID -REAL,  OBSERVATION ANGLE PHI (IN DEGREES).                  *
C     PHIPD-REAL,  INCIDENT ANGLE PHI PRIME (IN DEGREES).               *
C     BTD  -REAL,  OBLIQUE INCIDENT ANGLE BETA (IN DEGREES).            *
C     FN   -REAL,  WEDGE FACTOR N (DIMENSIONLESS).                      *
C                                                                       *
C... ON OUTPUT                                                          *
C     CSDCS-COMPLEX SLOPE DIFFRACTION COEFFICIENT (SOFT POLARIZATION).   *
C     CSDCH-COMPLEX SLOPE DIFFRACTION COEFFICIENT (HARD POLARIZATION).   *
C                                                                       *
C************************************************************************
      COMPLEX D(4),CT,TERM,FT(4),FTF,FSFCT,CSDCS,CSDCH
      DATA CT,UTPI/(-0.05626977,0.05626977),0.15915494/
      DATA PI,TPI,SML/3.14159265,6.28318531,0.001/
      DTOR=PI/180.
      PHR=PHID*DTOR
      PHPR=PHIPD*DTOR
      SBO=SIN(DTOR*BTD)
      IF(ABS(FN-0.5).LT.SML) GOTO 10
      DKL=TPI*R
      UFN=1./FN
      BETAR=PHR-PHPR
      TERM=CT/(2.*FN*FN*SBO)
      SGN=1.
      I=0
   2  I=I+1
      DN=UFN*(0.5*SGN+UTPI*BETAR)
      N=DN+SIGN(0.5,DN)
      ANG=TPI*FN*N-BETAR
      A=2.*(COS(0.5*ANG)**2)
      X=DKL*A
      Y=PI+SGN*BETAR
      IF(ABS(X).LT.1.E-10) GOTO 3
      FSFCT=(0.,2.)*X*(1.-FTF(X))
      IF(ABS(X).GT.1000.) FSDCT=(1.,0.)
      FT(I)=FSFCT/(SIN(0.5*Y*UFN)**2)
      GOTO 4
   3  FT(I)=(0.,0.)
      IF(ABS(COS(0.5*Y*UFN)).LT.1.E-3) GOTO  4
      FT(I)=(0.,4.)*DKL*FN*FN
      FT(I)=FT(I)*(1.-((0.88625,0.88625)*SQRT(DKL)*(PI-SGN*ANG)))
   4  SGN=-SGN
      D(I)=TERM*FT(I)
      IF(SGN.LT.0.) GOTO 2
      BETAR=PHR+PHPR
      IF(I.LE.3) GOTO 2
      CSDCS=(D(1)-D(2))+(D(3)-D(4))
      CSDCH=(D(1)-D(2))-(D(3)-D(4))
      RETURN
  10  CSDCS=(0.,0.)
      CSDCH=(0.,0.)
      RETURN
      END
```

COMPUTER PROGRAM: FTF

```
      COMPLEX FUNCTION FTF(XF)
C******************************************************************
C   FRESNEL FUNCTION                                               *
C   THIS FUNCTION PROGRAM COMPUTES THE FRESNEL TRANSITION FUNCTION *
C   F(X) BASED ON:                                                 *
C        A. EQUATION (13-74A) FOR SMALL ARGUMENT ( X<0.3 ).        *
C        B. EQUATION (13-74B) FOR LARGE ARGUMENT ( X>5.5 ).        *
C        C. LINEAR INTERPOLATION FOR INTERMEDIATE ARGUMENTS (0.3<X<5.5). *
C   THE ALGORITHM WAS DEVELOPED AND REPORTED IN REFERENCE (53).    *
C                                                                  *
C******************************************************************
      COMPLEX FXX(8),FX(8),CJ
      DIMENSION XX(8)
      DATA XX/.3,.5,.7,1.,1.5,2.3,4.,5.5/
      DATA CJ/(0.,1.)/
      DATA FX/(0.5729,0.2677),(0.6768,0.2682),(0.7439,0.2549),
     1(0.8095,0.2322),(0.873,0.1982),(0.9240,0.1577),(0.9658,0.1073),
     2(0.9797,0.0828)/
      DATA FXX/(0.,0.),(0.5195,0.0025),(0.3355,-0.0665),
     1(0.2187,-0.0757),(0.127,-0.068),(0.0638,-0.0506),
     2(0.0246,-0.0296),(0.0093,-0.0163)/
      X=ABS(XF)
      IF(X.GT.5.5) GOTO 1
      IF(X.GT.0.3) GOTO 10
      FTF=((1.253,1.253)*SQRT(X)-(0.,2.)*X-0.6667*X*X)*CEXP(CJ*X)
      GOTO 20
   10 DO 11 N=2,7
   11 IF(X.LT.XX(N)) GOTO 12
   12 FTF=FXX(N)*(X-XX(N))+FX(N)
      GOTO 20
    1 FTF=1.+(75./(16.*X*X)-0.75)/X/X+CJ*(0.5-15./(8.*X*X))/X
   20 IF(XF.GE.0.) RETURN
      FTF=CONJG(FTF)
      RETURN
      END
```

CHAPTER 14

GREEN'S FUNCTIONS

14.1 INTRODUCTION

In the area of electromagnetics, solutions to many problems are obtained using a second-order uncoupled partial differential equation, derived from Maxwell's equations, and the appropriate boundary conditions. The form of most of these type of solutions is an infinite series, provided the partial differential equation and the boundary conditions representing the problems are separable in the coordinate system chosen. The difficulty in using these type of solutions to obtain an insight into the behavior of the function is that they are usually slowly convergent especially at regions where rapid changes occur. It would then seem appropriate, at least for some problems and associated regions, that closed form solutions would be desirable. Even solutions in the form of integrals would be acceptable. The technique known as the *Green's function* does accomplish this goal.

Before proceeding with the presentation of the Green's function solution, let us briefly describe what the Green's functions represent and how they are used to obtain the overall solution to the problem.

With the Green's function technique a solution to the partial differential equation is obtained using a unit source (impulse, Dirac delta) as the driving function. *This is known as the Green's function.* The solution to the actual driving function is written as a superposition of the impulse response solutions (Green's function) with the Dirac delta source at different locations, which in the limit reduces to an integral. The contributions to the overall solution from the general source may be greater or smaller than that of the impulse response depending on the strength of the source at that given location. In engineering terminology then, the Green's function is nothing else but the *impulse response* of a system; in system theory, this is better known as the *transfer function*.

For a given problem, the Green's function can take various forms. One form of its solution can be expressed in terms of finite explicit functions, and it is obtained based on a procedure that will be outlined later. This procedure for developing the Green's function can be used only if the solution to the homogeneous differential equation is known. Another form of the Green's function is to construct its solution by an infinite series of suitably chosen orthonormal functions. The boundary conditions determine the eigenvalues of the eigenfunctions and the strength of the sources influences the coefficients of the eigenfunctions. Integral forms can also be used to represent the Green's function, especially when the eigenvalue spectrum is continuous. All solutions, although different in form, give the same results. The form of the Green's function that is most appropriate will depend on the problem in question. *The representation of the actual source plays a significant role as to which form of the Green's function may be most convenient for a given problem.*

Usually there is as much work involved in finding the Green's function as there is in obtaining the infinite series solution. However, the major advantages of the Green's function technique become evident when the same problem is to be solved for a variety of driving sources and when the sources are in the presence of boundaries [1–8].

In this chapter we shall initially study the one-dimensional differential equation

$$[L + \lambda r(x)] y(x) = f(x) \qquad (14\text{-}1)$$

where L is the Sturm–Liouville operator and λ is a constant. It is hoped that an understanding of the Green's function method for this equation will lead to a better understanding of the equations that occur in electromagnetic field theory applied to homogeneous media such as

$$\nabla^2 \phi(\mathbf{R}) + \beta^2 \phi(\mathbf{R}) = p(\mathbf{R}) \quad \text{(scalar wave equation)} \quad (14\text{-}2a)$$

$$\nabla \times \nabla \times \psi(\mathbf{R}) + \beta^2 \psi(\mathbf{R}) = \mathbf{F}(\mathbf{R}) \quad \text{(vector wave equation)} \quad (14\text{-}2b)$$

particularly since (14-2a) and (14-2b) can often be reduced to several equations similar to (14-1) by the separation of variables technique. Before proceeding to the actual solution of (14-1) by the Green's function method, we shall first consider several examples of Green's functions in other areas of electrical and general engineering. This will be followed by some topics associated with the Sturm–Liouville operator L before we embark on the solution of (14-2a).

14.2 GREEN'S FUNCTIONS IN ENGINEERING

The Green's function approach to solution of differential equations has been used in many areas of engineering, physics, and elsewhere [9–18]. Before we embark on constructing Green's function solutions to electromagnetic boundary-value problems, let us consider two other problems, one dealing with electric circuit theory and the other with mechanics. This will give the reader a better appreciation of the Green's function concept.

14.2.1 Circuit Theory

Analysis of lumped electric element circuits is a fundamental of electrical engineering. Therefore we will relate the Green's function to the solution of a very simple lumped element circuit problem.

Let us assume that a voltage source $v(t)$ is connected to a resistor R and inductor L, as shown in Figure 14-1a. The equation that governs the solution to that

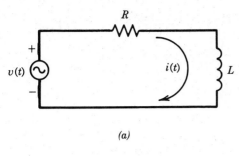

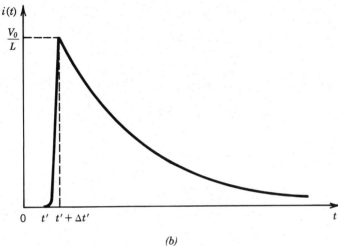

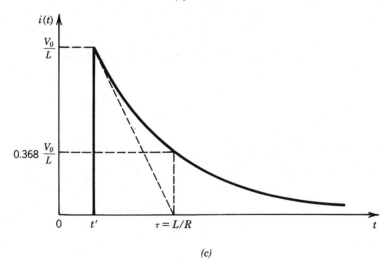

FIGURE 14-1 (*a*) *RL* series circuit, (*b*) current response, and (*c*) time constant.

circuit can be written as

$$L\frac{di}{dt} + Ri = v(t) \tag{14-3}$$

where $v(t)$ is the excitation voltage source that is turned on at $t = t'$. Initially ($t < t'$) the circuit is at rest and at $t = t'$ the voltage is suddenly turned on by an impulse V_0 of a very short duration $\Delta t'$. For $t > t' + \Delta t'$, when $v(t) = 0$, the circuit performance is governed by the homogeneous equation

$$L\frac{di}{dt} + Ri(t) = 0 \quad \text{for } t > t' + \Delta t' \tag{14-4}$$

whose solution for $i(t)$ can be written as

$$i(t) = I_0 e^{-(R/L)t} \quad \text{for } t > t' + \Delta t' \tag{14-4a}$$

where I_0 is a constant and L/R is referred to as the *time constant* τ of the circuit.
Since the voltage excitation $v(t)$ during $\Delta t'$ was an impulse V_0, then

$$\int_{t'}^{t'+\Delta t'} v(t)\, dt = V_0 \tag{14-5}$$

Therefore between $t' \leq t \leq t' + \Delta t'$ (14-3) can be written using (14-5) as

$$L\int_{t'}^{t'+\Delta t'} di + R\int_{t'}^{t'+\Delta t'} i(t)\, dt = \int_{t'}^{t'+\Delta t'} v(t)\, dt$$

$$L[i(t'+\Delta t') - i(t')] + R\int_{t'}^{t'+\Delta t'} i(t)\, dt = V_0 \tag{14-6}$$

Because $\Delta t'$ is very small we assume that $i(t)$ during the excitation $\Delta t'$ of the voltage source is not exceedingly large, and it behaves as shown in Figure 14-1b. Therefore during $\Delta t'$

$$\lim_{\Delta t' \to 0} R\int_{t'}^{t'+\Delta t'} i(t)\, dt \simeq 0 \tag{14-7}$$

so that the terms on the left side of (14-6) reduce using (14-4a) to

$$i(t') = 0 \tag{14-8a}$$

$$i(t'+\Delta t') = I_0 e^{-(R/L)(t'+\Delta t')} \stackrel{\Delta t' \to 0}{\simeq} I_0 e^{-(R/L)t'} \tag{14-8b}$$

Using (14-7) through (14-8b) we can express (14-6) as

$$LI_0 e^{-(R/L)t'} = V_0 \tag{14-9}$$

or

$$I_0 = \frac{V_0}{L} e^{+(R/L)t'} \tag{14-9a}$$

Therefore (14-4a) can be written using (14-9a) as

$$i(t) = \begin{cases} 0 & t < t' \tag{14-10a} \\ \dfrac{V_0}{L} e^{-(R/L)(t-t')} & t \geq t' \tag{14-10b} \end{cases}$$

which is shown plotted in Figure 14-1c.

If the circuit is subjected to N voltage impulses each of duration Δt and amplitude $V_i (i = 0, \ldots, N)$, occurring at $t = t_i'$, then the current response can be

written as

$$i(t) = \begin{cases} 0 & t < t'_0 \\ \dfrac{V_0}{L} e^{-(R/L)(t-t'_0)} & t'_0 < t < t'_1 \\ \dfrac{V_0}{L} e^{-(R/L)(t-t'_0)} + \dfrac{V_1}{L} e^{-(R/L)(t-t'_1)} & t'_1 < t < t'_2 \\ \vdots & \vdots \\ \displaystyle\sum_{i=0}^{N} \dfrac{V_i}{L} e^{-(R/L)(t-t'_i)} & t'_N < t < t'_{N+1} \end{cases} \quad (14\text{-}11)$$

If the circuit is subjected to a continuous voltage source $v(t)$ starting at t'_0 such that at an instant of time $t = t'$ and short interval $\Delta t'$ would produce an impulse of

$$dV = v(t')\,dt' \qquad (14\text{-}12)$$

then the response of the system for $t \geq t'$ can be expressed, provided that $i(t) = v(t) = 0$ for $t < t'$, as

$$i(t) = \int_{t'}^{t} \left[\frac{v(t')\,dt'}{L} e^{-(R/L)(t-t')} \right] = \int_{t'}^{t} v(t') \frac{e^{-(R/L)(t-t')}}{L} dt'$$

$$\boxed{i(t) = \int_{t'}^{t} v(t') G(t, t')\,dt'} \qquad (14\text{-}13)$$

where

$$\boxed{G(t, t') = \frac{e^{-(R/L)(t-t')}}{L} \quad \text{for } t > t'} \qquad (14\text{-}13\text{a})$$

In (14-13) $G(t, t')$ of (14-13a) is referred to as the *Green's function*, and it represents the response of the system for $t > t'$ when an excitation voltage $v(t)$ at $t = t'$ is an impulse (*Dirac delta*) function. Knowing the response of the system to an impulse function, represented by the Green's function of (14-13a), the response $i(t)$ to any voltage source $v(t)$ can then be obtained by convolving the voltage source excitation with the Green's function according to (14-13).

14.2.2 Mechanics

Another problem that the reader may be familiar with is that of a string of length ℓ that is connected at the two ends and is subjected to external force per unit length (load) of $F(x)$. The objective is to find the displacement $u(x)$ of the string. If the load $F(x)$ is assumed to be acting down (negative direction), the displacement $u(x)$ of the string is governed by the differential equation

$$T \frac{d^2 u}{dx^2} = F(x) \qquad (14\text{-}14)$$

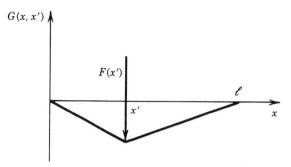

FIGURE 14-2 Attached string subjected by a load force.

or

$$\frac{d^2 u}{dx^2} = \frac{1}{T} F(x) = f(x) \qquad (14\text{-}14a)$$

where T is the uniform tensile force of the string. If the string is stationary at the two ends, then the displacement function $u(x)$ satisfies the boundary conditions

$$u(x = 0) = u(x = \ell) = 0 \qquad (14\text{-}15)$$

Initially instead of solving the displacement $u(x)$ of the string subject to the load $F(x)$, let us assume that the load subjected to the string is a concentrated load (impulse) of $F(x = x') = \delta(x - x')$ at a point $x = x'$, as shown in Figure 14-2. For the impulse load, the differential equation 14-14a can be written as

$$\frac{d^2 G(x, x')}{dx^2} = \frac{1}{T} \delta(x - x') \qquad (14\text{-}16)$$

subject to the boundary conditions

$$G(x = 0, x') = G(0, x') = 0 \qquad (14\text{-}16a)$$
$$G(x = \ell, x') = G(\ell, x') = 0 \qquad (14\text{-}16b)$$

In (14-16) $G(x, x')$ represents the displacement of the string when it is subject to an impulse load of $1/T$ at $x = x'$, and it is referred to as the *Green's function* for the string. Once this is found, the displacement $u(x)$ of the string subject to the load $F(x)$ can be determined by convolving the load $F(x)$ with the Green's function $G(x, x')$, as was done for the circuit problem by (14-13).

The solution to (14-16) is accomplished by following the procedure outlined here. Away from the load at $x = x'$, the differential equation 14-16 reduces to the homogeneous form

$$\frac{d^2 G(x, x')}{dx^2} = 0 \qquad (14\text{-}17)$$

which has solutions of

$$G(x, x') = \begin{cases} A_1 x + B_1 & 0 \le x \le x' \\ A_2 x + B_2 & x' \le x \le \ell \end{cases} \qquad \begin{matrix}(14\text{-}17a)\\(14\text{-}17b)\end{matrix}$$

Applying the boundary conditions (14-16a) and (14-16b) lead to

$$G(x = 0, x') = A_1(0) + B_1 = 0 \Rightarrow B_1 = 0 \qquad (14\text{-}18a)$$

$$G(x = \ell, x') = A_2\ell + B_2 = 0 \Rightarrow B_2 = -A_2\ell \qquad (14\text{-}18b)$$

Thus (14-17a) and (14-17b) reduce to

$$G(x, x') = \begin{cases} A_1 x & 0 \le x \le x' \\ A_2(x - \ell) & x' \le x \le \ell \end{cases} \qquad \begin{matrix}(14\text{-}19a)\\(14\text{-}19b)\end{matrix}$$

where A_1 and A_2 have not yet been determined.

At $x = x'$ the displacement $u(x)$ of the string must be continuous. Therefore the Green's function of (14-19a) and (14-19b) must also be continuous at $x = x'$. Thus

$$A_1 x' = A_2(x' - \ell) \Rightarrow A_2 = A_1 \frac{x'}{x' - \ell} \qquad (14\text{-}20)$$

According to (14-16) the second derivative of $G(x, x')$ is equal to an impulse function. Therefore the first derivative of $G(x, x')$, obtained by integrating (14-16), must be discontinuous by an amount equal to $1/T$. Thus

$$\lim_{\varepsilon \to 0} \left[\frac{dG(x' + \varepsilon, x')}{dx} - \frac{dG(x' - \varepsilon, x')}{dx} \right] = \frac{1}{T} \qquad (14\text{-}21)$$

or

$$\frac{dG(x'_+, x')}{dx} - \frac{dG(x'_-, x')}{dx} = \frac{1}{T} \qquad (14\text{-}21a)$$

Using (14-19a), (14-19b), and (14-20), we can write that

$$\frac{dG(x'_-, x')}{dx} = A_1 \qquad (14\text{-}22a)$$

$$\frac{dG(x'_+, x')}{dx} = A_2 = A_1 \frac{x'}{x' - \ell} \qquad (14\text{-}22b)$$

Thus (14-21a) leads, using (14-22a) and (14-22b), to

$$A_1 \frac{x'}{x' - \ell} - A_1 = \frac{1}{T} \Rightarrow A_1 \frac{\ell}{x' - \ell} = \frac{1}{T} \Rightarrow A_1 = \frac{1}{T} \frac{x' - \ell}{\ell} \qquad (14\text{-}23)$$

Therefore the Green's function of (14-19a) and (14-19b) can be written using (14-20) and (14-23) as

$$G(x, x') = \begin{cases} \dfrac{1}{T}\left(\dfrac{x' - \ell}{\ell}\right) x & 0 \le x \le x' \\ \dfrac{1}{T}\left(\dfrac{x - \ell}{\ell}\right) x' & x' \le x \le \ell \end{cases} \qquad \begin{matrix}(14\text{-}24a)\\(14\text{-}24b)\end{matrix}$$

The displacement $u(x)$ subject to the load $F(x)$, governed by (14-14a), can now be written as

$$\boxed{\begin{aligned} u(x) &= \int_0^\ell F(x')G(x,x')\,dx' \\ &= \frac{1}{T}\int_0^x F(x')\left(\frac{x-\ell}{\ell}\right)x'\,dx' \\ &\quad + \frac{1}{T}\int_x^\ell F(x')\left(\frac{x'-\ell}{\ell}\right)x\,dx' \end{aligned}} \qquad (14\text{-}25)$$

14.3 STURM–LIOUVILLE PROBLEMS

Now that we illustrated the Green's function development for two specific problems, one for an *RL* electrical circuit and the other for a stretched string, let us consider the construction of Green's functions for more general differential equations subject to appropriate boundary conditions. Specifically in this section we want to consider Green's functions for the one-dimensional differential equation of the *Sturm–Liouville* form [1, 7, 13].

A one-dimensional differential equation of the form

$$\boxed{\frac{d}{dx}\left[p(x)\frac{dy}{dx}\right] - q(x)y = f(x)} \qquad (14\text{-}26)$$

subject to homogeneous boundary conditions, is a *Sturm–Liouville* problem. This equation can also be written as

$$Ly = f(x) \qquad (14\text{-}27)$$

where L is the Sturm–Liouville operator

$$L \equiv \left\{\frac{d}{dx}\left[p(x)\frac{d}{dx}\right] - q(x)\right\} \qquad (14\text{-}27a)$$

Every general one-dimensional, source-excited, second-order differential equation of

$$A(x)\frac{d^2y}{dx^2} + B(x)\frac{dy}{dx} + C(x)y = S(x) \qquad (14\text{-}28)$$

or

$$Dy = S(x) \qquad (14\text{-}28a)$$

where

$$D \equiv \left[A(x)\frac{d^2}{dx^2} + B(x)\frac{d}{dx} + C(x)\right] \qquad (14\text{-}28b)$$

can be converted to a Sturm–Liouville form. This can be accomplished by following the procedure as outlined here.

First expand (14-26) and write it as

$$p(x)\frac{d^2y}{dx^2} + \frac{dp}{dx}\frac{dy}{dx} - q(x)y = f(x) \qquad (14\text{-}29)$$

Dividing (14-28) by $A(x)$ and (14-29) by $p(x)$, we have that

$$\frac{d^2y}{dx^2} + \frac{B(x)}{A(x)}\frac{dy}{dx} + \frac{C(x)}{A(x)}y = \frac{S(x)}{A(x)} \qquad (14\text{-}30a)$$

$$\frac{d^2y}{dx^2} + \frac{1}{p(x)}\frac{dp}{dx}\frac{dy}{dx} - \frac{q(x)}{p(x)}y = \frac{f(x)}{p(x)} \qquad (14\text{-}30b)$$

Comparing (14-30a) and (14-30b), we see that

$$\frac{B(x)}{A(x)} = \frac{1}{p(x)}\frac{dp(x)}{dx} \qquad (14\text{-}31a)$$

$$\frac{C(x)}{A(x)} = -\frac{q(x)}{p(x)} \qquad (14\text{-}31b)$$

$$\frac{S(x)}{A(x)} = \frac{f(x)}{p(x)} \qquad (14\text{-}31c)$$

From (14-31a) we have that

$$\frac{dp(x)}{dx} = p(x)\frac{B(x)}{A(x)} \qquad (14\text{-}32)$$

which is a linear first-order differential equation, a particular solution of which is

$$\boxed{p(x) = \exp\left[\int^x \frac{B(t)}{A(t)}\,dt\right]} \qquad (14\text{-}32a)$$

From (14-31b)

$$\boxed{q(x) = -p(x)\frac{C(x)}{A(x)} = -\frac{C(x)}{A(x)}\exp\left[\int^x \frac{B(t)}{A(t)}\,dt\right]} \qquad (14\text{-}32b)$$

and from (14-31c)

$$\boxed{f(x) = p(x)\frac{S(x)}{A(x)} = \frac{S(x)}{A(x)}\exp\left[\int^x \frac{B(t)}{A(t)}\,dt\right]} \qquad (14\text{-}32c)$$

In summary then, *a one-dimensional, source-excited, second-order differential equation of the form* (14-28) *is converted to a Sturm–Liouville form* (14-26) *by letting* $p(x)$ *be that of* (14-32a), $q(x)$ *by* (14-32b), *and* $f(x)$ *by* (14-32c).

To demonstrate, let us consider an example.

Example 14-1. Convert the Bessel differential equation

$$x^2 \frac{d^2 y}{dx^2} + x \frac{dy}{dx} + (x^2 - \lambda^2) y = 0$$

to a Sturm–Liouville form.

Solution. Since

$$A(x) = x^2$$
$$B(x) = x$$
$$C(x) = x^2 - \lambda^2$$
$$S(x) = 0$$

then according to (14-32a), (14-32b), an (14-32c)

$$p(x) = \exp\left[\int^x \frac{B(t)}{A(t)} dt\right] = \exp\left[\int^x \frac{t}{t^2} dt\right] = \exp\left[\int^x \frac{dt}{t}\right] = e^{\ln(x)} = x$$

$$q(x) = -p(x) \frac{C(x)}{A(x)} = -x \frac{(x^2 - \lambda^2)}{x^2} = -\left(\frac{x^2 - \lambda^2}{x}\right)$$

$$f(x) = p(x) \frac{S(x)}{A(x)} = 0$$

Thus using (14-26), Bessel's differential equation takes the Sturm–Liouville form

$$\frac{d}{dx}\left(x \frac{dy}{dx}\right) + \left(\frac{x^2 - \lambda^2}{x}\right) y = 0$$

As a check, when the preceding equation is expanded and is multiplied by x it reduces to the usual form of Bessel's differential equation.

14.3.1 Green's Function in Closed Form

Now that we have shown that each general second-order, source-excited differential equation can be converted to a Sturm–Liouville form, let us develop a procedure to construct the Green's function of a Sturm–Liouville differential equation represented by (14-26) or more generally by that of

$$\left\{\frac{d}{dx}\left[p(x) \frac{dy}{dx}\right] - q(x) y\right\} + \lambda r(x) y = f(x) \qquad (14\text{-}33)$$

or

$$\left[\left\{\frac{d}{dx}\left[p(x) \frac{d}{dx}\right] - q(x)\right\} + \lambda r(x)\right] y = f(x) \qquad (14\text{-}33a)$$

or

$$[L + \lambda r(x)] y = f(x) \qquad (14\text{-}33b)$$

where L is the Sturm–Liouville operator of (14-27a). In (14-33) $r(x)$ and $f(x)$ are assumed to be piecewise continuous in the region of interest ($a \leq x \leq b$) and λ is a parameter to be determined by the nature and boundary of the region of interest. *It should be noted that throughout this chapter λ is used to represent eigenvalues, and it should not be confused with the wavelength. The use of λ to represent eigenvalues is a very common practice in Green's function theory.* The differential equations 14-33 through 14-33b will possess a Green's function for all values of λ except those that are *eigenvalues* of the homogeneous equation

$$[L + \lambda r(x)] y = 0 \tag{14-34}$$

For values of λ for which (14-34) has nontrivial solutions, a Green's function will not exist. This is analogous to a system of linear equations represented by

$$Dy = f \tag{14-35}$$

which has a solution of

$$y = D^{-1} f \tag{14-35a}$$

provided D^{-1} exists (it is nonsingular). If D^{-1} is singular (does not exist), then (14-35) does not possess a solution. This occurs when $Dy = 0$, which has a nontrivial solution when the determinant of D is zero [i.e., $\det(D) = 0$].

According to (14-25) the solution of (14-33b) can be written as

$$y(x) = \int_a^b f(x) G(x, x') \, dx' \tag{14-36}$$

where $G(x, x')$ is the Green's function of (14-33) or (14-33b). Since (14-35a) is a solution to (14-35), and it exists only if D^{-1} is nonsingular, then (14-36) is a solution to (14-33b) if the inverse of the operator $[L + \lambda r(x)]$ exists. Then (14-33b) can be written as

$$y(x) = [L + \lambda r(x)]^{-1} f \tag{14-36a}$$

By comparing (14-36) to (14-36a), then $G(x, x')$ is analogous to the inverse of the operator $[L + \lambda r(x)]$.

Whenever λ is equal to an eigenvalue of the operator $[L + \lambda r(x)]$, obtained by setting the determinant of (14-34) equal to zero, or

$$\det [L + \lambda r(x)] = 0 \tag{14-37}$$

then the inverse of $[L + \lambda r(x)]$ does not exist, and (14-36) and (14-36a) cannot be valid. Thus for values of λ equal to the eigenvalues of the operator $[L + \lambda r(x)]$ the Green's function does not exist.

For a unit impulse driving function, the Sturm–Liouville equation 14-33 can be written as

$$\frac{d}{dx}\left[p(x)\frac{dG}{dx}\right] - q(x)G + \lambda r(x)G = \delta(x - x') \tag{14-38}$$

where G is the Green's function. At points removed from the impulse driving function, (14-38) reduces to

$$\left\{\frac{d}{dx}\left[p(x)\frac{dG}{dx}\right] - q(x)G\right\} + \lambda r(x) G = 0 \tag{14-38a}$$

As can be verified by the Green's function (14-24a) and (14-24b) of the mechanics problems in Section 14.2.2, the Green's functions of (14-38a) in general exhibit the following properties:

Properties of Green's Functions

1. $G(x, x')$ satisfies the *homogeneous* differential equation *except* at $x = x'$.
2. $G(x, x')$ is symmetrical with respect to x and x'.
3. $G(x, x')$ satisfies certain *homogeneous* boundary conditions.
4. $G(x, x')$ is continuous at $x = x'$.
5. $[dG(x', x')]/dx$ has a discontinuity of $1/[p(x')]$ at $x = x'$.

The discontinuity of the derivative of $G(x, x')$ at $x = x'$ ($[dG(x', x')]/dx = 1/[p(x')]$) can be derived by first integrating the differential equation 14-38 between $x = x' - \varepsilon$ to $x = x' + \varepsilon$. Doing this leads to

$$\lim_{\varepsilon \to 0} \left\{ \int_{x'-\varepsilon}^{x'+\varepsilon} \frac{d}{dx}\left[p(x) \frac{dG(x, x')}{dx} \right] dx + \int_{x'-\varepsilon}^{x'+\varepsilon} [-q(x) + \lambda r(x)] G(x, x') \, dx \right\}$$

$$= \int_{x'-\varepsilon}^{x'+\varepsilon} \delta(x - x') \, dx$$

$$\lim_{\varepsilon \to 0} \left\{ p(x) \frac{dG(x, x')}{dx} \bigg|_{x'-\varepsilon}^{x'+\varepsilon} + \int_{x'-\varepsilon}^{x'+\varepsilon} [-q(x) + \lambda r(x)] G(x, x') \, dx \right\} = 1 \quad (14\text{-}39)$$

Since $q(x)$, $r(x)$, and $G(x, x')$ are continuous at $x = x'$, then

$$\lim_{\varepsilon \to 0} \int_{x'-\varepsilon}^{x'+\varepsilon} [-q(x) + \lambda r(x)] G(x, x') \, dx = 0 \quad (14\text{-}40)$$

Using (14-40) reduces (14-39) to

$$\lim_{\varepsilon \to 0} \left\{ p(x) \left[\frac{dG(x'+\varepsilon, x')}{dx} - \frac{dG(x'-\varepsilon, x')}{dx} \right] \right\} = 1$$

$$p(x) \left[\frac{dG(x'_+, x')}{dx} - \frac{dG(x'_-, x')}{dx} \right] = 1 \quad (14\text{-}41)$$

or

$$\frac{dG(x'_+, x')}{dx} - \frac{dG(x'_-, x')}{dx} = \frac{1}{p(x)} \quad (14\text{-}41a)$$

which proves the discontinuity of the derivative of $G(x, x')$ at $x = x'$.

The Green's function must satisfy the differential equation 14-38a, the five general properties listed previously, and the appropriate boundary conditions. We propose to construct the Green's function solution into two parts: one that is valid for $a \leq x \leq x'$ and the other for $x' \leq x \leq b$ where a and b are the limits of the region of interest. For the homogeneous equation 14-33, valid at all points except $x = x'$:

1. Let $y_1(x)$ represent a nontrivial solution of the homogeneous differential equation of 14-33 in the interval $a \leq x < x'$ satisfying the boundary conditions at $x = a$. Since both $y_1(x)$ and $G(x, x')$ satisfy the same differential equation in the interval $a \leq x < x'$, they are related to each other by a constant, that is,

$$G(x, x') = A_1 y_1(x) \qquad a \leq x < x' \qquad (14\text{-}42\text{a})$$

2. Let $y_2(x)$ represent a nontrivial solution of the homogeneous differential equation of 14-33 in the interval $x' < x \leq b$ satisfying the boundary conditions at $x = b$. Since both $y_2(x)$ and $G(x, x')$ satisfy the same differential equation in the interval $x' < x \leq b$, they are related to each other by a constant, that is

$$G(x, x') = A_2 y_2(x) \qquad x' < x \leq b \qquad (14\text{-}42\text{b})$$

Since one of the general properties of the Green's function is that it must be continuous at $x = x'$, then using (14-42a) and (14-42b)

$$A_1 y_1(x') = A_2 y_2(x') \Rightarrow -A_1 y_1(x') + A_2 y_2(x') = 0 \qquad (14\text{-}43\text{a})$$

Also one of the properties of the derivative of the Green's function is that it must be discontinuous at $x = x'$ by an amount of $1/p(x')$. Thus using (14-42a) and (14-42b)

$$-A_1 y_1'(x') + A_2 y_2'(x') = \frac{1}{p(x')} \qquad (14\text{-}43\text{b})$$

Solving (14-43a) and (14-43b) simultaneously leads to

$$A_1 = \frac{y_2(x')}{p(x')W(x')} \qquad (14\text{-}44\text{a})$$

$$A_2 = \frac{y_1(x')}{p(x')W(x')} \qquad (14\text{-}44\text{b})$$

where $W(x')$ is the Wronskian of y_1 and y_2 at $x = x'$ defined as

$$\boxed{W(x') = y_1(x') y_2'(x') - y_2(x') y_1'(x')} \qquad (14\text{-}44\text{c})$$

Using (14-44a) through (14-44c) the closed form Green's function of (14-42a) and (14-42b) for the differential equation 14-33 or 14-38 can be written as

$$\boxed{G(x, x') = \begin{cases} \dfrac{y_2(x')}{p(x')W(x')} y_1(x) & a \leq x \leq x' \qquad (14\text{-}45\text{a}) \\ \dfrac{y_1(x')}{p(x')W(x')} y_2(x) & x' \leq x \leq b \qquad (14\text{-}45\text{b}) \end{cases}}$$

where $y_1(x)$ and $y_2(x)$ are two independent solutions of the homogeneous form of the differential equation 14-33 each satisfying, respectively, the boundary conditions at $x = a$ and $x = b$.

The preceding recipe can be used to construct in closed form the Green's function of a differential equation of the form of (14-33). It is convenient to use this procedure with the following provisions.

1. The solution to the homogeneous differential equation is known.
2. The Green's function is desired in closed form, instead of an infinite series of orthogonal functions as will be shown in the next section.

If this procedure is used for the mechanics problem of (14-14a), Section 14.2.2, the same answer [as given by (14-24a) and (14-24b)] will be obtained. In the next section we want to present an alternate procedure for constructing the Green's function. By this other method, the Green's function will be represented by an infinite series of orthonormal functions. Whether one form of the Green's function is more suitable than the other will depend on the problem in question. Remember, however, that the closed form procedure just derived can only be used provided the solution to the homogeneous differential equation is known.

Before we proceed, let us illustrate that the Sturm–Liouville operator L exhibits *Hermitian* (or *symmetrical*) properties [1]. These are very important, and they establish the relationships that are used in the construction of the Green's function.

Example 14-2. Show that the Sturm–Liouville operator L of (14-27a) or (14-33) through (14-33b) exhibits Hermitian (symmetrical) properties. Assume that in the interval $a \leq x \leq b$ any solution $y_i(x)$ to (14-33b) satisfies the boundary conditions

$$\alpha_1 y_i(x = a) + \alpha_2 \frac{dy_i(x = a)}{dx} = \alpha_1 y_i(a) + \alpha_2 y_i'(a) = 0$$

$$\beta_1 y_i(x = b) + \beta_2 \frac{dy_i(x = b)}{dx} = \beta_1 y_i(b) + \beta_2 y_i'(b) = 0$$

Solution. Let us assume that $y_1(x)$ and $y_2(x)$ are two solutions to (14-33) through (14-33b) each satisfying the boundary conditions. Then according to (14-27a)

$$Ly_1(x) = \frac{d}{dx}\left[p(x)\frac{dy_1(x)}{dx}\right] - q(x)y_1(x)$$

$$Ly_2(x) = \frac{d}{dx}\left[p(x)\frac{dy_2(x)}{dx}\right] - q(x)y_2(x)$$

Multiplying the first by $y_2(x)$ and the second by $y_1(x)$, we can write each using a shorthand notation as

$$y_2 Ly_1 = y_2(py_1')' - y_2 q y_1$$
$$y_1 Ly_2 = y_1(py_2')' - y_1 q y_2$$

where $'$ indicates d/dx. Subtracting the two and integrating between a and b leads to

$$\int_a^b (y_2 Ly_1 - y_1 Ly_2)\, dx = \int_a^b [y_2(py_1')' - y_1(py_2')']\, dx$$

Since
$$(y_2 py_1')' = y_2(py_1')' + y_2' py_1'$$
$$(y_1 py_2')' = y_1(py_2')' + y_1' py_2'$$

then by subtracting the two,
$$(y_2 py_1')' - (y_1 py_2')' = y_2(py_1')' - y_1(py_2')'$$

Thus the integral reduces to
$$\int_a^b (y_2 L y_1 - y_1 L y_2)\,dx = \int_a^b [(y_2 py_1')' - (y_1 py_2')']\,dx = [p(y_2 y_1' - y_1 y_2')]_a^b$$

Each of the solutions, $y_1(x)$ and $y_2(x)$, satisfy the same boundary conditions which can be written as

$$\alpha_1 y_1(a) + \alpha_2 y_1'(a) = 0 \Rightarrow \alpha_1 y_1(\alpha) = -\alpha_2 y_1'(a)$$
$$\beta_1 y_1(b) + \beta_2 y_1'(b) = 0 \Rightarrow \beta_1 y_1(b) = -\beta_2 y_1'(b)$$
$$\alpha_1 y_2(a) + \alpha_2 y_2'(a) = 0 \Rightarrow \alpha_1 y_2(a) = -\alpha_2 y_2'(a)$$
$$\beta_1 y_2(b) + \beta_2 y_2'(b) = 0 \Rightarrow \beta_1 y_2(b) = -\beta_2 y_2'(b)$$

Dividing the first by the third and the second by the fourth, we can write that

$$\frac{y_1(a)}{y_2(a)} = \frac{y_1'(a)}{y_2'(a)} \Rightarrow y_1(a) y_2'(a) = y_2(a) y_1'(a)$$

$$\frac{y_1(b)}{y_2(b)} = \frac{y_1'(b)}{y_2'(b)} \Rightarrow y_1(b) y_2'(b) = y_2(b) y_1'(b)$$

Using these relations it is apparent that the right side of the previous integral equation vanishes, and we can write it as

$$\int_a^b (y_2 L y_1 - y_1 L y_2)\,dx = [p(y_2 y_1' - y_1 y_2')]_a^b = 0$$

or

$$\boxed{\int_a^b (y_2 L y_1)\,dx = \int_a^b (y_1 L y_2)\,dx}$$

This illustrates that the operator L exhibits Hermitian (symmetrical) properties with respect to the solutions $y_1(x)$ and $y_2(x)$.

14.3.2 Green's Function in Series

The procedure outlined in the previous section can be only used to derive in closed form the Green's function for differential equations whose homogeneous form solution is known. Otherwise other techniques must be used. Even for equations whose homogeneous form solution is known, the closed form representation of the Green's function may not be the most convenient one. Therefore an alternate representation may be attractive even for those cases.

An alternate form of the Green's function is to represent it as a series of orthonormal functions. The most appropriate orthonormal functions would be those that satisfy the boundary conditions. To demonstrate the procedure let us initially rederive the Green's function of the mechanics problem of Section 14.2.2 but this time represented as a series of orthonormal functions. We will then generalize the method to (14-33) through (14-33b).

A. VIBRATING STRING

For the differential equation 14-14a, subject to the boundary conditions of (14-15), its Green's function must satisfy (14-16) subject to the boundary conditions of (14-16a) and (14-16b). Since the Green's function $G(x, x')$ must vanish at $x = 0$ and ℓ, it is most convenient to represent $G(x, x')$ as an infinite series of $\sin(n\pi x/\ell)$ orthonormal functions, that is

$$G(x, x') = \sum_{n=1}^{\infty} a_n(x') \sin\left(\frac{n\pi}{\ell}x\right) \qquad (14\text{-}46)$$

where $a_n(x')$ represent the amplitude expansion coefficients that will be a function of the position x' of the excitation source.

Substituting (14-46) into (14-16), multiplying both sides by $\sin(m\pi x/\ell)$, and then integrating in x from 0 to ℓ leads to

$$-\sum \left(\frac{n\pi}{\ell}\right)^2 a_n(x') \int_0^\ell \sin\left(\frac{n\pi}{\ell}x\right) \sin\left(\frac{m\pi}{\ell}x\right) dx = \frac{1}{T} \int_0^\ell \delta(x - x') \sin\left(\frac{m\pi}{\ell}x\right) dx \qquad (14\text{-}47)$$

Because the orthogonality conditions of sine functions state that

$$\int_0^\ell \sin\left(\frac{n\pi}{\ell}x\right) \sin\left(\frac{m\pi}{\ell}x\right) dx = \begin{cases} \dfrac{\ell}{2} & m = n \qquad (14\text{-}48\text{a}) \\ 0 & m \neq n \qquad (14\text{-}48\text{b}) \end{cases}$$

then (14-47) reduces to

$$-\left(\frac{n\pi}{\ell}\right)^2 \frac{\ell}{2} a_n(x') = \frac{1}{T} \sin\left(\frac{n\pi}{\ell}x'\right) \qquad (14\text{-}49)$$

or

$$a_n(x') = -\frac{2\ell}{\pi^2 T} \frac{1}{n^2} \sin\left(\frac{n\pi}{\ell}x'\right) \qquad (14\text{-}49\text{a})$$

Thus the Green's function of (14-46) can be expressed, using (14-49a), as

$$\boxed{G(x, x') = -\frac{2\ell}{\pi^2 T} \sum_{n=1}^{\infty} \frac{1}{n^2} \sin\left(\frac{n\pi}{\ell}x'\right) \sin\left(\frac{n\pi}{\ell}x\right)} \qquad (14\text{-}50)$$

This is an alternate form to (14-24a) and (14-24b), but one that leads to the same results, even though its form looks quite different. The displacement $u(x)$ subject to the load $F(x)$, governed by (14-14a), can now be written as

$$u(x) = \int_0^\ell F(x')G(x,x')\,dx' = -\frac{2\ell}{\pi^2 T}\sum_{n=1}^{\infty}\frac{1}{n^2}\sin\left(\frac{n\pi}{\ell}x\right)\int_0^\ell F(x')\sin\left(\frac{n\pi}{\ell}x'\right)dx'$$

(14-51)

B. STURM–LIOUVILLE OPERATOR

Let us now generalize the Green's function series expansion method of the vibrating string, as given by (14-50) and (14-51), to (14-33b) where L is a Sturm–Liouville operator and λ is an arbitrary parameter to be determined by the nature and boundary of the region of interest. We seek a solution to solve the differential equation 14-33b, or

$$[L + \lambda r(x)]y(x) = f(x) \qquad (14\text{-}52)$$

in the interval $a \leq x \leq b$ subject to the general boundary conditions of

$$\alpha_1 y(x)|_{x=a} + \alpha_2 \frac{dy(x)}{dx}\bigg|_{x=a} = \alpha_1 y(a) + \alpha_2 \frac{dy(a)}{dx} = 0 \qquad (14\text{-}52a)$$

$$\beta_1 y(x)|_{x=b} + \beta_2 \frac{dy(x)}{dx}\bigg|_{x=b} = \beta_1 y(b) + \beta_2 \frac{dy(b)}{dx} = 0 \qquad (14\text{-}52b)$$

which are usually referred to as the *mixed* boundary conditions. In (14-52a) at least one of the constants α_1 or α_2, if not both of them, are nonzero. The same is true for (14-52b). The Green's function $G(x, x')$, if it exists, will satisfy the differential equation

$$[L + \lambda r(x)]G(x, x') = \delta(x - x') \qquad (14\text{-}53)$$

subject to the boundary conditions

$$\alpha_1 G(a, x') + \alpha_2 \frac{dG(a, x')}{dx} = 0 \qquad (14\text{-}53a)$$

$$\beta_1 G(b, x') + \beta_2 \frac{dG(b, x')}{dx} = 0 \qquad (14\text{-}53b)$$

If $\{\psi_n(x)\}$ represents a complete set of orthonormal eigenfunctions for the Sturm–Liouville operator L, then it must satisfy the differential equation

$$[L + \lambda_n r(x)]\psi_n(x) = 0 \qquad (14\text{-}54)$$

subject to the same boundary conditions of (14-52a) or (14-52b) for the Green's function of (14-53), or

$$\alpha_1 \psi_n(a) + \alpha_2 \frac{d\psi_n(a)}{dx} = 0 \qquad (14\text{-}54a)$$

$$\beta_1 \psi_n(b) + \beta_2 \frac{d\psi_n(b)}{dx} = 0 \qquad (14\text{-}54b)$$

The boundary conditions of (14-54a) and (14-54b) of $\psi_n(x)$ are also used to determine the eigenvalues λ_n. In the finite interval $a \leq x \leq b$ the complete set of orthonormal eigenfunctions $\{\psi_n(x)\}$ and their amplitude coefficients must satisfy the orthogonality condition of

$$\boxed{\int_a^b \psi_m(x)\psi_n(x)r(x)\,dx = \delta_{mn} = \begin{cases} 1 & m = n \\ 0 & m \neq n \end{cases}} \qquad (14\text{-}55)$$

where δ_{mn} is the Kronecker delta function.

If the Green's function exists, it can be represented in series form in terms of the orthonormal eigenfunctions $\psi_n(x)$ as

$$G(x, x') = \sum_n a_n(x')\psi_n(x) \qquad (14\text{-}56)$$

where $a_n(x')$ are the amplitude coefficients. These can be obtained by multiplying both sides of (14-56) by $\psi_m(x)r(x)$, integrating from a to b, and then using (14-55). It can be shown that

$$a_n(x') = \int_a^b G(x, x')\psi_n(x)r(x)\,dx \qquad (14\text{-}56\mathrm{a})$$

Since $G(x, x')$ satisfies (14-53) and $\psi_n(x)$ satisfies (14-54), then we can rewrite each as

$$LG(x, x') = -\lambda r(x)G(x, x') + \delta(x - x') \qquad (14\text{-}57\mathrm{a})$$
$$L\psi_n(x) = -\lambda_n r(x)\psi_n(x, x') \qquad (14\text{-}57\mathrm{b})$$

Multiplying (14-57a) by $\psi_n(x)$, (14-57b) by $G(x, x')$, and then subtracting the two equations leads to

$$\psi_n(x)LG(x, x') - G(x, x')L\psi_n(x)$$
$$= -(\lambda - \lambda_n)G(x, x')\psi_n(x)r(x) + \delta(x - x')\psi_n(x) \qquad (14\text{-}58)$$

Integrating (14-58) between a and b we can write that

$$\int_a^b [\psi_n(x)LG(x, x') - G(x, x')L\psi_n(x)]\,dx$$
$$= -(\lambda - \lambda_n)\int_a^b G(x, x')\psi_n(x)r(x)\,dx + \int_a^b \delta(x - x')\psi_n(x)\,dx \qquad (14\text{-}59)$$

which by using (14-56a) reduces to

$$\int_a^b [\psi_n(x)LG(x, x') - G(x, x')L\psi_n(x)]\,dx = -(\lambda - \lambda_n)a_n(x') + \psi_n(x')$$
$$(14\text{-}59\mathrm{a})$$

By the symmetrical (Hermitian) property of the operator L, as derived in Example 14-2, with respect to the functions $y_1 = G(x, x')$ and $y_2 = \psi_n(x)$, the left side of (14-59) or (14-59a) vanishes. Therefore (14-59) and (14-59a) reduce to

$$-(\lambda - \lambda_n)a_n(x') + \psi_n(x') = 0 \qquad (14\text{-}60)$$

or

$$a_n(x') = \frac{\psi_n(x')}{\lambda - \lambda_n} \qquad (14\text{-}60\mathrm{a})$$

Thus the series form of the Green's function of (14-56) can ultimately be expressed as

$$\boxed{G(x, x') = \sum_n \frac{\psi_n(x')\psi_n(x)}{(\lambda - \lambda_n)}} \qquad (14\text{-}61)$$

where $\{\psi_n(z)\}$ represents a complete set of orthonormal eigenfunctions for the Sturm–Liouville operator L which satisfies the differential equation 14-54 subject to the boundary conditions of (14-54a) and (14-54b). This is also referred to as the *bilinear formula*, and it represents, aside from (14-45a) and (14-45b), the second form that can be used to derive the Green's function for the differential equation of (14-33) through (14-33b) as a series solution in the finite interval of $a \le x \le b$. It should be noted that at $\lambda = \lambda_n$ the Green's function of (14-61) possesses singularities. Usually these singularities are simple poles although in some cases the λ_n's are branch points whose branch cuts represent a continuous spectrum of eigenvalues. In those cases the Green's function may involve a summation, for the discrete spectrum of the eigenvalues, and an integral, for the continuous spectrum of the eigenvalues.

Example 14-3. A very common differential equation in solutions of transmission line and antenna problems (such as metallic waveguides, microstrip antennas, etc.) that exhibit rectangular configurations is

$$\frac{d^2\varphi(x)}{dx^2} + \beta^2\varphi(x) = f(x)$$

subject to the boundary conditions of

$$\varphi(0) = \varphi(\ell) = 0$$

where $\beta^2 = \omega^2\mu\varepsilon$. Derive in closed and series forms the Green's functions for the given equation.

Solution. For the given equation the Green's function must satisfy the differential equation of

$$\frac{d^2 G(x, x')}{dx^2} + \beta^2 G(x, x') = \delta(x - x')$$

subject to the boundary conditions of

$$G(0) = G(\ell) = 0$$

The differential equation is of the Sturm–Liouville form of (14-33) with

$$\left.\begin{array}{l} p(x) = 1 \\ q(x) = 0 \\ r(x) = 1 \\ \lambda = \beta^2 \\ y(x) = \varphi(x) \\ L = \dfrac{d^2}{dx^2} \end{array}\right\} \Rightarrow \frac{d^2\varphi}{dx^2} + \beta^2\varphi = f(x)$$

A. Closed Form Solution: This form of the solution will be obtained using the recipe of (14-45a) and (14-45b) along with (14-44c). The homogeneous differential equation for $\varphi(x)$ reduces to

$$\frac{d^2\varphi}{dx^2} + \beta^2\varphi = 0$$

Two independent solutions, one $\phi_1(x)$ valid in the interval of $0 \le x \le x'$ and that vanishes at $x = 0$, and the other $\phi_2(x)$ valid in the interval $x' \le x \le \ell$ and that vanishes at $x = \ell$, take the form of

$$\phi_1(x) = \sin(\beta x)$$
$$\phi_2(x) = \sin[\beta(\ell - x)]$$

According to (14-44c) the Wronskian can be written as

$$W(x') = -\beta\{\sin(\beta x')\cos[\beta(\ell - x')] + \sin[\beta(\ell - x')]\cos(\beta x')\}$$
$$W(x') = -\beta\sin(\beta x' + \beta\ell - \beta x') = -\beta\sin(\beta\ell)$$

Thus the Green's function in closed form can be expressed using (14-45a) and (14-45b) as

$$G(x, x') = \begin{cases} -\dfrac{\sin[\beta(\ell - x')]}{\beta\sin(\beta\ell)}\sin(\beta x) & 0 \le x \le x' \\[2mm] -\dfrac{\sin(\beta x')}{\beta\sin(\beta\ell)}\sin[\beta(\ell - x)] & x' \le x \le \ell \end{cases}$$

This form of the Green's function indicates that it possesses singularities (poles) when

$$\beta\ell = \beta_r\ell = n\pi \Rightarrow \beta_r = \omega_r\sqrt{\mu\varepsilon} = 2\pi f_r\sqrt{\mu\varepsilon} = \frac{n\pi}{\ell}$$

or

$$f_r = \frac{n}{2\ell\sqrt{\mu\varepsilon}} \qquad n = 1, 2, 3, \ldots$$

B. Series Form Solution: This form of the solution will be obtained using (14-61). The complete set of eigenfunctions $\{\psi_n(x)\}$ must satisfy the differential equation 14-54, or

$$\frac{d^2\psi_n(x)}{dx^2} + \beta_n^2\psi_n(x) = 0$$

subject to the boundary conditions of

$$\psi_n(0) = \psi_n(\ell) = 0$$

The most appropriate solution of $\psi_n(x)$ is to represent it in terms of standing wave eigenfunctions which in a rectangular coordinate system are sine and cosine functions, as discussed in Chapters 3 and 8, Sections 3.4.1 and 8.2.1.

Thus we can write, according to (3-28b) or (8-4a), that
$$\psi_n(x) = A\cos(\beta_n x) + B\sin(\beta_n x)$$

The allowable eigenvalues of β_n are found by applying the boundary conditions. Since $\psi_n(0) = 0$, then
$$\psi_n(0) = A + B(0) = 0 \Rightarrow A = 0$$

Also since $\psi_n(\ell) = 0$, then
$$\psi_n(\ell) = B\sin(\beta_n \ell) = 0 \Rightarrow \beta_n \ell = \sin^{-1}(0) = n\pi$$

or
$$\beta_n = \frac{n\pi}{\ell} \qquad n = 1, 2, 3, \ldots \text{ (for nontrivial solutions)}$$

Thus
$$\psi_n(x) = B\sin(\beta_n x) = B\sin\left(\frac{n\pi}{\ell}x\right)$$

The amplitude constant B is such that (14-55) is satisfied. Therefore
$$B^2 \int_0^\ell \sin^2\left(\frac{n\pi}{\ell}x\right) dx = 1$$
$$\frac{B^2}{2} \int_0^\ell \left[1 - \cos\left(\frac{2n\pi}{\ell}x\right)\right] dx = B^2\left(\frac{\ell}{2}\right) = 1 \Rightarrow B = \sqrt{\frac{2}{\ell}}$$

Thus the complete set of the orthonormal eigenfunctions of $\{\psi_n(x)\}$ is represented by
$$\psi_n(x) = \sqrt{\frac{2}{\ell}}\sin\left(\frac{n\pi}{\ell}x\right) \qquad n = 1, 2, 3, \ldots$$

with
$$\lambda = \beta^2 = \omega^2\mu\varepsilon$$
$$\lambda_n = \beta_n^2 = \left(\frac{n\pi}{\ell}\right)^2 \qquad n = 1, 2, 3, \ldots$$

In series form the Green's function of (14-61) can then be written as
$$G(x, x') = \frac{2}{\ell} \sum_{n=1,2,\ldots}^{\infty} \frac{\sin\left(\frac{n\pi}{\ell}x'\right)\sin\left(\frac{n\pi}{\ell}x\right)}{\beta^2 - \left(\frac{n\pi}{\ell}\right)^2}$$

which yields the same results as the closed form solution of part A even though it looks quite different analytically. It is apparent that the Green's function is symmetrical. Also it possesses a singularity, and it fails to exist when
$$\beta = \beta_r = \omega_r\sqrt{\mu\varepsilon} = 2\pi f_r\sqrt{\mu\varepsilon} = \frac{n\pi}{\ell}$$

or

$$f_r = \frac{n\pi/\ell}{2\pi\sqrt{\mu\varepsilon}} = \frac{n}{2\ell\sqrt{\mu\varepsilon}}$$

which is identical to that obtained by the closed form solution in part A.

This is in accordance with (14-34) which states that the Green's function of (14-33b) exists for all values of λ, in this case $\lambda = \beta^2 = \omega^2\mu\varepsilon$, except those that are eigenvalues of the homogeneous equation 14-34. For our case (14-34) reduces to

$$\frac{d^2\varphi(x)}{dx^2} + \beta^2\varphi(x) = 0$$

whose nontrivial solution takes the form of

$$\varphi(x) = C\sin\left(\frac{n\pi}{\ell}x\right)$$

with eigenvalues of $\beta = n\pi/\ell$, $n = 1, 2, 3, \ldots$.

It should be noted that when

$$\beta = \beta_r = \omega_r\sqrt{\mu\varepsilon} = 2\pi f_r\sqrt{\mu\varepsilon} = \frac{n\pi}{\ell} = f_r = \frac{n}{2\ell\sqrt{\mu\varepsilon}}$$

the Green's function singularity consists of simple poles. At those frequencies the external frequencies of the source match the natural (characteristic) frequencies of the system, in this instance the transmission line. This is referred to as *resonance*. When this occurs the field of the mode whose natural frequency matches the source excitation frequency (resonance condition) will continuously increase without any bounds, in the limit reaching values of infinity. For those situations no steady-state solutions can exist. One way to contain the field amplitude is to introduce damping.

14.3.3 Green's Function in Integral Form

In the previous two sections we outlined procedures that can be used to derive the Green's function in closed and series forms. The bilinear formula (14-61) of Section 14.3.2 is used to derive the Green's function of (14-33) through (14-33b) when the eigenvalue spectrum, represented by the λ_n's in (14-61), is discrete. However, often the eigenvalue spectrum is continuous, and it can be represented in (14-61) by an integral. In the limit the infinite summation of the bilinear formula reduces to an integral. This form is usually desirable when at least one of the boundary conditions is at infinity. This would be true when a source placed at the origin is radiating in an unbounded medium.

To demonstrate the derivation, let us construct the Green's function of the one-dimensional scalar Helmholtz equation

$$\frac{d^2\varphi}{dx^2} + \beta_0^2\varphi = f(x) \tag{14-62}$$

subject to the boundary (radiation) conditions of

$$\varphi(+\infty) = \varphi(-\infty) = 0 \tag{14-62a}$$

The Green's function $G(x, x')$ will satisfy the differential equation

$$\frac{d^2 G(x, x')}{dx^2} + \beta_0^2 G(x, x') = \delta(x - x') \qquad (14\text{-}63)$$

subject to the boundary conditions of

$$G(+\infty) = G(-\infty) = 0 \qquad (14\text{-}63a)$$

The complete set of orthonormal eigenfunctions, represented here by $\psi(x)$, must satisfy the differential equation

$$\frac{d^2 \psi}{dx^2} = \lambda \psi = -\beta^2 \psi \qquad (14\text{-}64)$$

where $\lambda = -\beta^2$, subject to the boundary conditions of

$$\psi(+\infty) = \psi(-\infty) = 0 \qquad (14\text{-}64a)$$

Since the source is radiating in an unbounded medium, represented here by the boundary (radiation) conditions, the most appropriate eigenfunctions are those representing traveling waves, instead of standing waves. Thus a solution for (14-64) subject to (14-64a) is

$$\psi(x, x') = C(x') e^{\mp j\beta x} \quad \begin{array}{l} - \text{ for } x > x' \\ + \text{ for } x < x' \end{array} \qquad (14\text{-}65)$$

where for an $e^{j\omega t}$ time convention the upper sign (minus) represents waves traveling in the $+x$ direction, satisfying the boundary condition at $x = +\infty$, and the lower sign (plus) represents waves traveling in the $-x$ direction, satisfying the boundary condition at $x = -\infty$. Let us assume that the waves of interest here are those traveling in the $+x$ direction, represented in (14-65) by the upper sign or

$$\psi(x, x') = C(x') e^{-j\beta x} \qquad (14\text{-}65a)$$

which represents a plane wave of amplitude $C(x')$.

The Green's function can be represented by a continuous spectrum of plane waves or by a Fourier integral of

$$G(x, x') = \frac{1}{\sqrt{2\pi}} \int_{-\infty}^{+\infty} g(\beta, x') e^{-j\beta x} \, d\beta \qquad (14\text{-}66)$$

whose Fourier transform pair is

$$g(\beta, x') = \frac{1}{\sqrt{2\pi}} \int_{-\infty}^{+\infty} G(x, x') e^{+j\beta x} \, dx \qquad (14\text{-}66a)$$

In (14-66) $G(x, x')$ is represented by a continuous spectrum of plane waves each of the form of (14-65a) and each with an amplitude coefficient of $g(\beta, x')$. Using (14-66a) we can write the transform $\tilde{\delta}(\beta - x')$ of the Dirac delta function $\delta(x - x')$ as

$$\tilde{\delta}(\beta - x') = \frac{1}{\sqrt{2\pi}} \int_{-\infty}^{+\infty} \delta(x - x') e^{+j\beta x} \, dx = \frac{1}{\sqrt{2\pi}} e^{+j\beta x'} \qquad (14\text{-}67)$$

874 GREEN'S FUNCTIONS

Thus according to (14-66) $\delta(x - x')$ can then be written using (14-67) as

$$\delta(x - x') = \frac{1}{\sqrt{2\pi}} \int_{-\infty}^{+\infty} \tilde{\delta}(\beta - x') e^{-j\beta x} \, d\beta = \frac{1}{\sqrt{2\pi}} \int_{-\infty}^{+\infty} \left(\frac{1}{\sqrt{2\pi}} e^{+j\beta x'} \right) e^{-j\beta x} \, d\beta \tag{14-67a}$$

The amplitude coefficients $g(x, x')$ in (14-66) can be determined by substituting (14-66) and (14-67a) into (14-63). Then it can be shown that

$$\frac{1}{\sqrt{2\pi}} \int_{-\infty}^{+\infty} (-\beta^2 + \beta_0^2) g(\beta, x') e^{-j\beta x} \, d\beta = \frac{1}{\sqrt{2\pi}} \int_{-\infty}^{+\infty} \left[\frac{1}{\sqrt{2\pi}} e^{+j\beta x'} \right] e^{-j\beta x} \, d\beta$$

$$\frac{1}{\sqrt{2\pi}} \int_{-\infty}^{+\infty} \left\{ (\beta_0^2 - \beta^2) g(\beta, x') - \frac{1}{\sqrt{2\pi}} e^{+j\beta x'} \right\} e^{-j\beta x} \, d\beta = 0 \tag{14-68}$$

which is satisfied provided

$$g(\beta, x') = \frac{1}{\sqrt{2\pi}} \frac{e^{j\beta x'}}{\beta_0^2 - \beta^2} \tag{14-68a}$$

Thus the Green's function of (14-66) reduces to

$$\boxed{G(x, x') = \frac{1}{2\pi} \int_{-\infty}^{+\infty} \frac{e^{-j\beta(x - x')}}{\beta_0^2 - \beta^2} \, d\beta} \tag{14-69}$$

which is a generalization of the bilinear formula of (14-61).

The integrand in (14-69) has poles at $\beta = \pm \beta_0$ and can be evaluated using residue calculus [5]. In the evaluation of (14-69) the contour along a circular arc C_R of radius $R \to \infty$ with center at the origin should close in the lower half plane for $x > x'$, as shown in Figure 14-3a, and should close in the upper half plane for $x < x'$, as shown in Figure 14-3b. This is necessary so that the contribution of the integral along the circular arc C_R of radius $R \to \infty$ is equal to zero. In general then, by residue calculus the integral of (14-69) can be evaluated using the geometry of Figure 14-3, and it can be written as

$$G(x, x') = \frac{1}{2\pi} \int_{-\infty}^{+\infty} \frac{e^{-j\beta(x - x')}}{(\beta_0^2 - \beta^2)} \, d\beta$$

$$= \mp 2\pi j \left[\text{residue} \left(\beta = \pm \beta_0 \right) \right] - \int_{C_R} \frac{e^{-j\beta(z - x')}}{\beta_0^2 - \beta^2} \, d\beta$$

$$G(x, x') = \mp 2\pi j \left[\text{residue} \left(\beta = \pm \beta_0 \right) \right] \quad \begin{array}{l} \text{upper signs for } x > x' \\ \text{lower signs for } x < x' \end{array} \tag{14-70}$$

since the contribution along C_R is zero.

It is apparent that the Green's function of (14-69) possesses pole singularities at $\beta = +\beta_0$ and $\beta = -\beta_0$. If these were allowed to contribute, then the exponentials in the integral of (14-69) for an $e^{j\omega t}$ time convention would be represented by either

$$e^{-j\beta_0(x - x')} e^{j\omega t} = e^{+j\beta_0 x'} e^{j(-\beta_0 x + \omega t)} \quad \text{for } \beta = +\beta_0 \tag{14-70a}$$

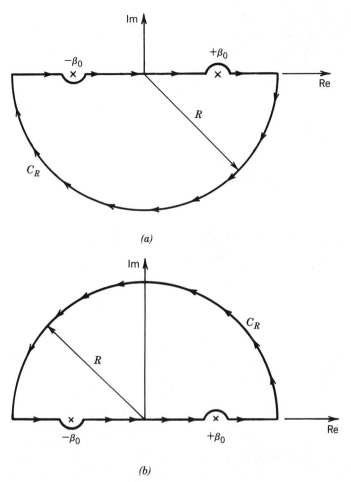

FIGURE 14-3 Residue calculus for contour integration. (*a*) $x > x'$.
(*b*) $x < x'$.

or

$$e^{+j\beta_0(x-x')}e^{j\omega t} = e^{-j\beta_0 x'}e^{j(\beta_0 x + \omega t)} \quad \text{for } \beta = -\beta_0 \qquad (14\text{-}70b)$$

Thus (14-70a) represents waves traveling in the $+x$ direction and (14-70b) represents waves traveling in the $-x$ direction. In the contour evaluation of (14-69), the contour should do the following:

1. For $x > x'$, pass the pole at $\beta = -\beta_0$ from below, the one at $\beta = +\beta_0$ from above, and then close down so that only the latter contributes, as shown in Figure 14-3*a*.
2. For $x < x'$, pass the pole $\beta = -\beta_0$ from below, the one at $\beta = +\beta_0$ from above, and then close up so that only the former contributes, as shown in Figure 14-3*b*.

Sometimes an integral Green's function can be used to represent both discrete and continuous spectra whereby part of the integral would represent the discrete spectrum and the remainder would represent the continuous spectrum. Typically the discrete spectrum would represent a finite number of propagating modes and an infinite number of evanescent modes in the closed regions and the continuous spectrum would represent radiation in open regions.

14.4 TWO-DIMENSIONAL GREEN'S FUNCTION IN RECTANGULAR COORDINATES

Until now we have considered the construction of Green's functions for problems involving a single space variable. Let us now consider problems involving two space variables, both for static and time-varying fields.

14.4.1 Static Fields

A two-dimensional partial differential equation often encountered in static electromagnetics is the Poisson equation

$$\frac{\partial^2 V}{\partial x^2} + \frac{\partial^2 V}{\partial y^2} = f(x, y) = q(x, y) \tag{14-71}$$

subject to the boundary conditions

$$V(x = 0, 0 \leq y \leq b) = V(x = a, 0 \leq y \leq b) = 0 \tag{14-71a}$$
$$V(0 \leq x \leq a, y = 0) = V(0 \leq x \leq a, y = b) = 0 \tag{14-71b}$$

In (14-71) V can represent the electric potential distribution on a rectangular structure of dimensions a along the x direction and b along the y direction and $f(x, y) = q(x, y)$ can represent the electric charge distribution along the structure. The objective here is to obtain the Green's function of the problem and ultimately the potential distribution. The Green's function $G(x, y; x', y')$ will satisfy the partial differential equation

$$\frac{\partial^2 G}{\partial x^2} + \frac{\partial^2 G}{\partial y^2} = \delta(x - x')\delta(y - y') \tag{14-72}$$

subject to the boundary conditions

$$G(x = 0, 0 \leq y \leq b) = G(x = a, 0 \leq y \leq b) = 0 \tag{14-72a}$$
$$G(0 \leq x \leq a, y = 0) = G(0 \leq x \leq a, y = b) = 0 \tag{14-72b}$$

and the potential distribution $V(x, y)$ will be represented by

$$V(x, y) = \int_0^b \int_0^a q(x', y') G(x, y; x', y') \, dx' \, dy' \tag{14-73}$$

We will derive the Green's function here in two forms.

1. *Closed form*, similar to that of Section 14.3.1 but utilizing (14-44c) and (14-45a) through (14-45b) for a two space variable problem.
2. *Series form*, similar to that of Section 14.3.2 but using basically two-dimensional expressions for (14-54), (14-55), and (14-61).

A. CLOSED FORM

The Green's function of (14-72) for the closed form solution can be formulated by initially choosing functions that satisfy the boundary conditions either along the x direction, at $x = 0$ and $x = a$, or along the y direction, at $y = 0$ and $y = b$. Let us begin here the development of the Green's function of (14-72) by choosing functions that satisfy the boundary conditions along the x direction. This is accomplished by initially representing the Green's function by a normalized single function Fourier series of sine functions that satisfy the boundary conditions at $x = 0$ and $x = a$, that is

$$G(x, y; x', y') = \sum_{m=1,2,\ldots}^{\infty} g_m(y; x', y') \sin\left(\frac{m\pi}{a}x\right) \quad (14\text{-}74)$$

The coefficients $g_m(y, x', y')$ of the Fourier series will be determined by first substituting (14-74) into (14-72). This leads to

$$\sum_{m=1,2,\ldots}^{\infty}\left[-\left(\frac{m\pi}{a}\right)^2 g_m(y; x', y') \sin\left(\frac{m\pi}{a}x\right) + \sin\left(\frac{m\pi}{a}x\right)\frac{d^2 g_m(y; x', y')}{dy^2}\right]$$
$$= \delta(x - x')\delta(y - y') \quad (14\text{-}75)$$

Multiplying both sides of (14-75) by $\sin(n\pi x/a)$, integrating with respect to x from 0 to a, and using (14-48a) and (14-48b) we can write that

$$\frac{d^2 g_m(y; x', y')}{dy^2} - \left(\frac{m\pi}{a}\right)^2 g_m(y; x', y') = \frac{2}{a}\sin\left(\frac{m\pi}{a}x'\right)\delta(y - y') \quad (14\text{-}76)$$

Equation 14-76 is recognized as a one-dimensional differential equation for $g_m(y; x', y')$ which can be solved using the recipe of Section 14.3.1 as provided by (14-44c) and (14-45a) through (14-45b).

For the homogeneous form of (14-76), or

$$\frac{d^2 g_m(y; x', y')}{dy^2} - \left(\frac{m\pi}{a}\right)^2 g_m(y; x', y') = 0 \quad (14\text{-}77)$$

two solutions that satisfy, respectively, the boundary conditions at $y = 0$ and $y = b$ are

$$g_m^{(1)}(y; x', y') = A_m(x', y') \sinh\left(\frac{m\pi}{a}y\right) \qquad \text{for } y \leq y' \quad (14\text{-}78a)$$

$$g_m^{(2)}(y; x', y') = B_m(x', y') \sinh\left[\frac{m\pi}{a}(b - y)\right] \qquad \text{for } y \geq y' \quad (14\text{-}78b)$$

The hyperbolic functions were chosen as solutions to (14-77), instead of real exponentials, so that (14-78a) satisfies the boundary condition of (14-71b) at $y = 0$ and (14-78b) satisfies the boundary condition of (14-71b) at $y = b$.

Using (14-44c) where $y_1 = g_m^{(1)}$ and $y_2 = g_m^{(2)}$, we can write the Wronskian as

$$W(y; x', y') = -\left(\frac{m\pi}{a}\right) A_m B_m \left\{\sinh\left(\frac{m\pi}{a}y'\right)\cosh\left[\frac{m\pi}{a}(b - y')\right]\right.$$
$$\left. + \cosh\left(\frac{m\pi}{a}y'\right)\sinh\left[\frac{m\pi}{a}(b - y')\right]\right\}$$

$$W(y; x', y') = -\left(\frac{m\pi}{a}\right) A_m B_m \sinh\left(\frac{m\pi b}{a}\right) \quad (14\text{-}79)$$

By comparing (14-76) with the form of (14-33), it is apparent that

$$p(y) = 1$$
$$q(y) = 0$$
$$r(y) = 1$$
$$\lambda = -\left(\frac{m\pi}{a}\right)^2 \tag{14-80}$$

Using (14-78a) through (14-80) the solution for $g_m(y; x', y')$ of (14-76) can be written by referring to (14-45a) and (14-45b) as

$$g_m(y; x', y') = \begin{cases} -\dfrac{2}{m\pi}\sin\left(\dfrac{m\pi}{a}x'\right)\dfrac{\sinh\left[\dfrac{m\pi}{a}(b-y')\right]}{\sinh\left(\dfrac{m\pi b}{a}\right)}\sinh\left(\dfrac{m\pi}{a}y\right) & (14\text{-}81a) \\[2ex] \qquad\qquad\qquad\qquad\qquad 0 \leq y \leq y' \\[2ex] -\dfrac{2}{m\pi}\sin\left(\dfrac{m\pi}{a}x'\right)\dfrac{\sinh\left(\dfrac{m\pi}{a}y'\right)}{\sinh\left(\dfrac{m\pi b}{a}\right)}\sinh\left[\dfrac{m\pi}{a}(b-y)\right] & (14\text{-}81b) \\[2ex] \qquad\qquad\qquad\qquad\qquad y' \leq y \leq b \end{cases}$$

Thus the Green's function of (14-74) can be written as

$$G(x, y; x', y') = \begin{cases} -\dfrac{2}{\pi}\displaystyle\sum_{m=1,2,\ldots}^{\infty}\dfrac{\sin\left(\dfrac{m\pi}{a}x'\right)\sinh\left[\dfrac{m\pi}{a}(b-y')\right]}{m\sinh\left(\dfrac{m\pi b}{a}\right)} \\[2ex] \qquad\qquad \times \sin\left(\dfrac{m\pi}{a}x\right)\sinh\left(\dfrac{m\pi}{a}y\right) \qquad (14\text{-}82a)\\[2ex] \qquad\qquad \text{for } 0 \leq x \leq a, \ 0 \leq y \leq y' \\[2ex] -\dfrac{2}{\pi}\displaystyle\sum_{m=1,2,\ldots}^{\infty}\dfrac{\sin\left(\dfrac{m\pi}{a}x'\right)\sinh\left(\dfrac{m\pi}{a}y'\right)}{m\sinh\left(\dfrac{m\pi b}{a}\right)} \\[2ex] \qquad\qquad \times \sin\left(\dfrac{m\pi}{a}x\right)\sinh\left[\dfrac{m\pi}{a}(b-y)\right] \qquad (14\text{-}82b)\\[2ex] \qquad\qquad \text{for } 0 \leq x \leq a, \ y' \leq y \leq b \end{cases}$$

which is a series summation of sine functions in x' and x, and hyperbolic sine functions in y' and y.

If the Green's function solution were developed by selecting and writing initially (14-74) by functions that satisfy the boundary conditions at $y = 0$ and $y = b$, then it can be shown that the Green's function can be written as a series summation of hyperbolic sine functions in x' and x, and ordinary sine functions in y' and y, or

$$G(x, y; x', y') = \begin{cases} -\dfrac{2}{\pi} \sum_{n=1,2,\ldots}^{\infty} \dfrac{\sinh\left[\dfrac{n\pi}{b}(a - x')\right] \sin\left(\dfrac{n\pi}{b}y'\right)}{n \sinh\left(\dfrac{n\pi a}{b}\right)} \\ \qquad \times \sinh\left(\dfrac{n\pi}{b}x\right) \sin\left(\dfrac{n\pi}{b}y\right) \qquad \text{(14-83a)} \\ \quad \text{for } 0 \leq x \leq x', \; 0 \leq y \leq b \\[1em] -\dfrac{2}{\pi} \sum_{n=1,2,\ldots}^{\infty} \dfrac{\sinh\left(\dfrac{n\pi}{b}x'\right) \sin\left(\dfrac{n\pi}{b}y'\right)}{n \sinh\left(\dfrac{n\pi a}{b}\right)} \\ \qquad \times \sinh\left[\dfrac{n\pi}{b}(a - x)\right] \sin\left(\dfrac{n\pi}{b}y\right) \qquad \text{(14-83b)} \\ \quad \text{for } x' \leq x \leq a, \; 0 \leq y \leq b \end{cases}$$

The derivation of this is left to the reader as an end of chapter exercise.

Example 14-4. Electric charge is uniformly distributed along an infinitely long conducting wire positioned at $\rho = \rho'$, $\phi = \phi'$ and circumscribed by a grounded ($V = 0$) electric conducting circular cylinder of radius a and infinite length, as shown in Figure 14-4. Find series form expressions for the Green's function and potential distribution. Assume free space within the cylinder.

Solution. The potential distribution $V(\rho, \phi, z)$ must satisfy Poisson's equation

$$\nabla^2 V(\rho, \phi, z) = -\frac{1}{\varepsilon_0} q(\rho, \phi, z)$$

subject to the boundary condition

$$V(\rho = a, 0 \leq \phi \leq 2\pi, z) = 0$$

Since the wire is infinitely long, the solutions for the potential will not be a function of z. Thus an expanded form of Poisson's equation reduces to

$$\frac{1}{\rho} \frac{\partial}{\partial \rho}\left(\rho \frac{\partial V}{\partial \rho}\right) + \frac{1}{\rho^2} \frac{\partial^2 V}{\partial \phi^2} = -\frac{1}{\varepsilon_0} q(\rho, \phi)$$

The Green's function $G(\rho, \phi; \rho', \phi')$ must satisfy the partial differential equation

$$\nabla^2 G(\rho, \phi; \rho', \phi') = \delta(\rho - \rho')$$

which in expanded form reduces for this problem to

$$\frac{1}{\rho} \frac{\partial}{\partial \rho}\left(\rho \frac{\partial G}{\partial \rho}\right) + \frac{1}{\rho^2} \frac{\partial^2 G}{\partial \phi^2} = \delta(\rho - \rho')$$

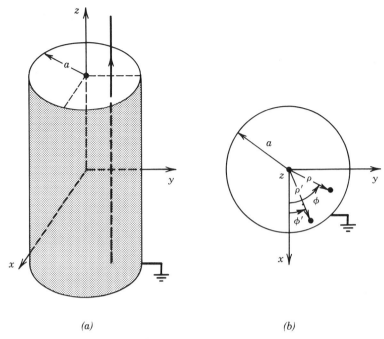

FIGURE 14-4 Long wire within a grounded circular conducting cylinder. (*a*) Wire and grounded cylinder. (*b*) Top view.

For the series solution of the Green's function, the complete set of orthonormal eigenfunctions $\{\psi_{mn}(\rho, \phi)\}$ can be obtained by considering the homogeneous form of Poisson's equation, or

$$\frac{1}{\rho}\frac{\partial}{\partial \rho}\left(\rho \frac{\partial \psi_{mn}}{\partial \rho}\right) + \frac{1}{\rho^2}\frac{\partial^2 \psi_{mn}}{\partial \phi^2} = \lambda_{mn}\psi_{mn}$$

subject to the boundary condition

$$\psi_{mn}(\rho = a, 0 \leq \phi \leq 2\pi, z) = 0$$

Using the separation of variables method of Section 3.4.2, we can express $\psi_{mn}(\rho, \phi)$ by

$$\psi_{mn}(\rho, \phi) = f(\rho)g(\phi)$$

Following the method outlined in Section 3.4.2, the functions $f(\rho)$ and $g(\phi)$ satisfy, respectively, the differential equations of (3-66a) and (3-66b), or

$$\rho^2 \frac{d^2f}{d\rho^2} + \rho \frac{df}{d\rho} + \left(\lambda_{mn}\rho^2 - m^2\right)f = 0$$

$$\frac{d^2g}{d\phi^2} = -m^2 g$$

whose appropriate solutions for this problem according to (3-67a) and (3-68b) are, respectively

$$f = AJ_m(\sqrt{\lambda_{mn}}\rho) + BY_m(\sqrt{\lambda_{mn}}\rho)$$
$$g = C\cos(m\phi) + D\sin(m\phi)$$

Since ψ_{mn} must be periodic in ϕ, then m must take integer values, $m = 0, 1, 2, \ldots$, and both the $\cos(m\phi)$ and $\sin(m\phi)$ variations (modes) exist simultaneously; see Chapter 9. Also since ψ_{mn} must be finite everywhere, including $\rho = 0$, then $B = 0$. Thus the eigenfunctions are reduced to either of two forms, that is,

$$\psi_{mn}^{(1)} = A_{mn}J_m(\sqrt{\lambda_{mn}}\rho)\cos(m\phi)$$

or

$$\psi_{mn}^{(2)} = A_{mn}J_m(\sqrt{\lambda_{mn}}\rho)\sin(m\phi)$$

The eigenvalues λ_{mn} are found by applying the boundary condition at $\rho = a$, that is

$$\psi_{mn}(\rho = a, 0 \le \phi \le 2\pi) = A_{mn}J_m(\sqrt{\lambda_{mn}}a)$$

or

$$\sqrt{\lambda_{mn}}a = \chi_{mn} \Rightarrow \lambda_{mn} = \left(\frac{\chi_{mn}}{a}\right)^2$$

where χ_{mn} represents the n zeroes of the Bessel function J_m of the first kind of order m. These are listed in Table 9-2.

The complete set of orthonormal eigenfunctions must be normalized so that

$$\int_0^{2\pi}\int_0^a \psi_{mn}^{(1)}(\rho,\phi)\psi_{mp}^{(1)}(\rho,\phi)\rho\,d\rho\,d\phi = \int_0^{2\pi}\int_0^a \psi_{mn}^{(2)}(\rho,\phi)\psi_{mp}^{(2)}(\rho,\phi)\rho\,d\rho\,d\phi = 1$$

Thus

$$A_{mn}^2 \int_0^{2\pi}\int_0^a \rho J_m(\sqrt{\lambda_{mn}}\rho)J_m(\sqrt{\lambda_{mp}}\rho)\cos^2(m\phi)\,d\rho\,d\phi = 1$$

or

$$A_{mn}^2 \int_0^{2\pi}\int_0^a \rho J_m(\sqrt{\lambda_{mn}}\rho)J_m(\sqrt{\lambda_{mp}}\rho)\sin^2(m\phi)\,d\rho\,d\phi = 1$$

Since

$$\int_0^{2\pi}\cos^2(m\phi)\,d\phi = \begin{cases} 2\pi & m = 0 \\ \pi & m \ne 0 \end{cases}$$

$$\int_0^{2\pi}\sin^2(m\phi)\,d\phi = \begin{cases} 0 & m = 0 \\ \pi & m \ne 0 \end{cases}$$

and

$$\int_0^a \rho J_m(\sqrt{\lambda_{mn}}\rho)J_m(\sqrt{\lambda_{mp}}\rho)\,d\rho = \begin{cases} \dfrac{a^2}{2}\left[J_m'(\sqrt{\lambda_{mn}}a)\right]^2 & p = n \\ 0 & p \ne n \end{cases}$$

then

$$A_{mn}^2 \varepsilon_m \frac{\pi a^2}{2}\left[J_m'\left(\sqrt{\lambda_{mp}}\,a\right)\right]^2 = 1$$

or

$$A_{mn} = \sqrt{\frac{2\varepsilon_m}{\pi}}\,\frac{1}{aJ_m'\left(\sqrt{\lambda_{mn}}\,a\right)}$$

where

$$\varepsilon_m = \begin{cases} 2 & m = 0 \\ 1 & m \neq 0 \end{cases}$$

Thus the complete set of orthonormal eigenfunctions can be written as

$$\psi_{mn}^{(1)} = \sqrt{\frac{2\varepsilon_m}{\pi}}\,\frac{1}{aJ_m'\left(\sqrt{\lambda_{mn}}\,a\right)} J_m\left(\sqrt{\lambda_{mp}}\,\rho\right)\cos(m\phi)$$

or

$$\psi_{mn}^{(2)} = \sqrt{\frac{2\varepsilon_m}{\pi}}\,\frac{1}{aJ_m'\left(\sqrt{\lambda_{mn}}\,a\right)} J_m\left(\sqrt{\lambda_{mp}}\,\rho\right)\sin(m\phi)$$

The Green's function can now be written using the bilinear formula (14-94) with $\lambda = 0$ as

$$G(\rho,\phi;\rho',\phi') = -\frac{2}{\pi}\varepsilon_m \frac{1}{a^2\left[J_m'\left(\sqrt{\lambda_{mn}}\,a\right)\right]^2} \sum_{m=0}^{\infty}\sum_{n=1}^{\infty} \frac{J_m\left(\sqrt{\lambda_{mn}}\,\rho'\right)J_m\left(\sqrt{\lambda_{mn}}\,\rho\right)}{\lambda_{mn}}$$
$$\times [\cos(m\phi)\cos(m\phi') + \sin(m\phi)\sin(m\phi')]$$

$$G(\rho,\phi;\rho',\phi') = -\frac{2}{\pi}\varepsilon_m \frac{1}{a^2\left[J_m'\left(\sqrt{\lambda_{mn}}\,a\right)\right]^2} \sum_{m=0}^{\infty}\sum_{n=1}^{\infty} \frac{J_m\left(\sqrt{\lambda_{mn}}\,\rho'\right)J_m\left(\sqrt{\lambda_{mn}}\,\rho\right)}{\lambda_{mn}}$$
$$\times \cos[m(\phi - \phi')]$$

where

$$\lambda_{mn} = \left(\frac{\chi_{mn}}{a}\right)^2$$

Both the $\cos(m\phi)$ and $\sin(m\phi)$ field variations were included in the final expression for the Green's function.

Finally the potential distribution $V(\rho, z)$ can be written as

$$V(\rho,\phi) = -\frac{1}{\varepsilon_0}\int_0^{2\pi}\int_0^a q(\rho',\phi') G(\rho,\delta;\rho',\phi')\rho'\,d\rho'\,d\phi'$$

where $G(\rho,\phi;\rho',\phi')$ is the Green's function and $q(\rho',\phi')$ is the linear charge distribution.

The Green's function of Example 14-4 can also be developed in closed form. This is done in Section 14.6.2 for a time-harmonic electric line source inside a circular cylinder. The statics solution is obtained by letting $\beta_0 = 0$.

B. SERIES FORM

For the series solution of the Green's function of (14-72), the complete set of orthonormal eigenfunctions $\{\psi_{mn}(x, y)\}$ can be obtained by considering the homogeneous form of (14-71), or

$$\frac{\partial^2 \psi_{mn}}{\partial x^2} + \frac{\partial^2 \psi_{mn}}{\partial y^2} = \lambda_{mn}\psi_{mn} \qquad (14\text{-}84)$$

subject to the boundary conditions of

$$\psi_{mn}(x = 0, 0 \le y \le b) = \psi_{mn}(x = a, 0 \le y \le b) = 0 \qquad (14\text{-}84a)$$
$$\psi_{mn}(0 \le x \le a, y = 0) = \psi_{mn}(0 \le x \le a, y = b) = 0 \qquad (14\text{-}84b)$$

Using the method of separation of variables of Section 3.4.1, we can represent $\psi_{mn}(x, y)$ by

$$\psi_{mn}(x, y) = f(x)g(y) \qquad (14\text{-}85)$$

Substituting (14-85) into (14-84) reduces to

$$\frac{1}{f}\frac{d^2f}{dx^2} = -p^2 \Rightarrow \frac{d^2f}{dx^2} = -p^2 f \Rightarrow f(x) = A\cos(px) + B\sin(px) \qquad (14\text{-}85a)$$

$$\frac{1}{g}\frac{d^2g}{dy^2} = -q^2 \Rightarrow \frac{d^2g}{dy^2} = -q^2 g \Rightarrow g(y) = C\cos(qy) + D\sin(qy) \qquad (14\text{-}85b)$$

where the system eigenvalues are those of

$$\lambda_{mn} = p^2 + q^2 \qquad (14\text{-}85c)$$

Thus (14-85) can be represented by

$$\psi_{mn}(x, y) = [A\cos(px) + B\sin(px)][C\cos(qy) + D\sin(qy)] \qquad (14\text{-}86)$$

Applying the boundary conditions of (14-84a) on (14-86) leads to

$$\psi_{mn}(x = 0, 0 \le y \le b) = [A(1) + B(0)][C\cos(qy) + D\sin(qy)] = 0 \Rightarrow A = 0 \qquad (14\text{-}87a)$$

$$\psi_{mn}(x = a, 0 \le y \le b) = B\sin(pa)[C\cos(qy) + D\sin(qy)] = 0 \Rightarrow \sin(pa) = 0$$

$$pa = \sin^{-1}(0) = m\pi$$

$$p = \frac{m\pi}{a} \qquad m = 1, 2, 3, \ldots \qquad (14\text{-}87b)$$

Similarly applying the boundary conditions of (14-84b) into (14-86) using (14-87a) and (14-87b) leads to

$$\psi_{mn}(0 \le x \le a, y = 0) = B\sin\left(\frac{m\pi}{a}x\right)[C(1) + D(0)] = 0 \Rightarrow C = 0 \qquad (14\text{-}88a)$$

$$\psi_{mn}(0 \le x \le a, y = b) = BD\sin\left(\frac{m\pi}{a}x\right)\sin(qb)] = 0 \Rightarrow \sin(qb) = 0$$

$$qb = \sin^{-1}(0) = n\pi$$

$$q = \frac{n\pi}{b} \qquad n = 1, 2, 3, \ldots \qquad (14\text{-}88b)$$

Thus the eigenfunctions of (14-85) reduce to

$$\psi_{mn}(x, y) = BD \sin\left(\frac{m\pi}{a}x\right) \sin\left(\frac{n\pi}{b}y\right) = B_{mn} \sin\left(\frac{m\pi}{a}x\right) \sin\left(\frac{n\pi}{b}y\right) \quad (14\text{-}89)$$

where the eigenvalues of (14-85c) are equal to

$$\lambda_{mn} = (p^2 + q^2) = \left(\frac{m\pi}{a}\right)^2 + \left(\frac{n\pi}{b}\right)^2 \quad \begin{matrix} m = 1, 2, 3, \ldots \\ n = 1, 2, 3, \ldots \end{matrix} \quad (14\text{-}89a)$$

To form the Green's function, the eigenfunctions of (14-89) must be normalized so that

$$\int_0^b \int_0^a \psi_{mn}(x, y) \psi_{rs}(x, y) \, dx \, dy = \begin{cases} 1 & m = r \quad n = s \\ 0 & m \neq r \quad n \neq s \end{cases} \quad (14\text{-}90)$$

Equation 14-90 is similar and it is an expanded form in two space variables of (14-55). Substituting (14-89) into (14-90) we can write that

$$B_{mn}^2 \int_0^b \int_0^a \sin\left(\frac{m\pi}{a}x\right) \sin\left(\frac{n\pi}{b}y\right) \sin\left(\frac{r\pi}{a}x\right) \sin\left(\frac{s\pi}{b}y\right) dx \, dy = 1 \quad (14\text{-}91)$$

Using (14-48a) and (14-48b) reduces (14-91) to

$$B_{mn}^2 \left(\frac{ab}{4}\right) = 1 \quad (14\text{-}92)$$

or

$$B_{mn} = \frac{2}{\sqrt{ab}} \quad (14\text{-}92a)$$

Thus (14-89) can be written as

$$\psi_{mn}(x, y) = \frac{2}{\sqrt{ab}} \sin\left(\frac{m\pi}{a}x\right) \sin\left(\frac{n\pi}{b}y\right) \quad (14\text{-}93)$$

The Green's function can be expressed as a double summation of two-dimensional eigenfunctions, or

$$G(x, y; x', y') = \sum_m \sum_n \frac{\psi_{mn}(x', y') \psi_{mn}(x, y)}{\lambda - \lambda_{mn}} \quad (14\text{-}94)$$

Equation 14-94 is an expanded version in two space variables of the bilinear equation of (14-61). Thus we can write the Green's function of (14-94) using (14-89a), (14-93), and $\lambda = 0$ as

$$G(x, y; x', y') = -\frac{4}{ab} \sum_{m=1,2,\ldots}^{\infty} \sum_{n=1,2,\ldots}^{\infty} \frac{\sin\left(\frac{m\pi}{a}x'\right) \sin\left(\frac{n\pi}{b}y'\right)}{\left(\frac{m\pi}{a}\right)^2 + \left(\frac{n\pi}{b}\right)^2} \sin\left(\frac{m\pi}{a}x\right) \sin\left(\frac{n\pi}{b}y\right)$$

$$(14\text{-}95)$$

The electric potential of (14-73), due to the electric charge density of $q(x', y')$, can be expressed as

$$V(x, y) = -\frac{4}{ab} \sum_{m=1,2,\ldots}^{\infty} \sum_{n=1,2,\ldots}^{\infty} \frac{\sin\left(\frac{m\pi}{a}x\right)\sin\left(\frac{n\pi}{b}y\right)}{\left(\frac{m\pi}{a}\right)^2 + \left(\frac{n\pi}{b}\right)^2} \\ \times \int_0^b \int_0^a q(x', y')\sin\left(\frac{m\pi}{a}x'\right)\sin\left(\frac{n\pi}{b}y'\right) dx' dy' \quad (14\text{-}96)$$

14.4.2 Time-Harmonic Fields

For time-harmonic fields a popular partial differential equation is

$$\frac{\partial^2 E_z}{\partial x^2} + \frac{\partial^2 E_z}{\partial y^2} + \beta^2 E_z = f(x, y) = j\omega\mu J_z(x, y) \quad (14\text{-}97)$$

subject to the boundary conditions of

$$E_z(x = 0, 0 \leq y \leq b) = E_z(x = a, 0 \leq y \leq b) = 0 \quad (14\text{-}97a)$$

$$E_z(0 \leq x \leq a, y = 0) = E_z(0 \leq x \leq a, y = b) = 0 \quad (14\text{-}97b)$$

In (14-97) E_z can represent the electric field component of a TM^z field configuration (mode) with no z variations inside a rectangular metallic cavity of dimensions a, b, c in the x, y, z directions, respectively, as shown in Figure 14-5. The function $f(x, y) = j\omega\mu J_z$ can represent the normalized electric current density component of the feed probe which is used to excite the fields within the metallic cavity. The objective here is to obtain the Green's function of the problem and ultimately the electric field component represented in (14-97) by $E_z(x, y)$.

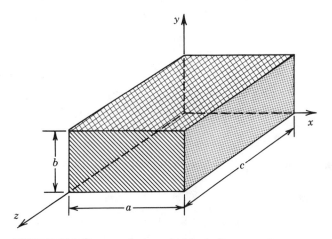

FIGURE 14-5 Rectangular waveguide cavity geometry.

The Green's function $G(x, y; x', y')$ will satisfy the partial differential equation of

$$\frac{\partial^2 G}{\partial x^2} + \frac{\partial^2 G}{\partial y^2} + \beta^2 G = \delta(x - x')\delta(y - y') \qquad (14\text{-}98)$$

subject to the boundary conditions of

$$G(x = 0, 0 \le y \le b) = G(x = a, 0 \le y \le b) = 0 \qquad (14\text{-}98a)$$
$$G(0 \le x \le a, y = 0) = G(0 \le x \le a, y = b) = 0 \qquad (14\text{-}98b)$$

and the electric field distribution $E_z(x, y)$ will be represented by

$$E_z(x, y) = j\omega\mu \int_0^b \int_0^a J_z(x', y') G(x, y; x', y')\, dx'\, dy' \qquad (14\text{-}99)$$

The Green's function of (14-98) can be derived either in closed form, as was done in Section 14.4.1A for the statics problem, or in series form, as was done in Section 14.4.1B for the statics problem. We will derive the Green's function here by a series form. The closed form is left as an end of chapter exercise for the reader.

For the solution of the Green's function, the complete set of eigenfunctions $\{\psi_{mn}(x, y)\}$ can be obtained by considering the homogeneous form of (14-97), or

$$\frac{\partial^2 \psi_{mn}}{\partial x^2} + \frac{\partial^2 \psi_{mn}}{\partial y^2} + \beta_{mn}^2 \psi_{mn} = 0 \qquad (14\text{-}100)$$

subject to the boundary conditions

$$\psi_{mn}(x = 0, 0 \le y \le b) = \psi_{mn}(x = a, 0 \le y \le b) = 0 \qquad (14\text{-}100a)$$
$$\psi_{mn}(0 \le x \le a, y = 0) = \psi_{mn}(0 \le x \le a, y = b) = 0 \qquad (14\text{-}100b)$$

Using the separation of variables of Section 3.4.1, we can represent $\psi_{mn}(x, y)$ by

$$\psi_{mn}(x, y) = f(x) g(y) \qquad (14\text{-}101)$$

Substituting (14-101) and applying the boundary conditions of (14-100a) and (14-100b), it can be shown that $\psi_{mn}(x, y)$ reduces to

$$\psi_{mn}(x, y) = B_{mn} \sin(\beta_x x) \sin(\beta_y y) = B_{mn} \sin\left(\frac{m\pi}{a} x\right) \sin\left(\frac{n\pi}{b} y\right) \qquad (14\text{-}102)$$

where

$$\beta_x = \frac{m\pi}{a} \qquad m = 1, 2, 3, \ldots \qquad (14\text{-}102a)$$

$$\beta_y = \frac{n\pi}{b} \qquad n = 1, 2, 3, \ldots \qquad (14\text{-}102b)$$

The eigenvalues of the system are equal to

$$\lambda_{mn} = \beta_{mn}^2 = (\beta_x)^2 + (\beta_y)^2 = \left(\frac{m\pi}{a}\right)^2 + \left(\frac{n\pi}{b}\right)^2 \qquad (14\text{-}103a)$$

and

$$\lambda = \beta^2 \qquad (14\text{-}103b)$$

The eigenfunctions of (14-102) must satisfy an equation similar to (14-90) but over a volume integral. That is

$$\int_0^c \int_0^b \int_0^a \psi_{mn}(x, y, z)\psi_{rs}(x, y, z)\, dx\, dy\, dz = \begin{cases} 1 & m = r \quad n = s \\ 0 & m \neq r \quad n \neq s \end{cases} \quad (14\text{-}104)$$

which leads to

$$B_{mn} = \frac{2}{\sqrt{abc}} \quad (14\text{-}104a)$$

Thus

$$\psi_{mn}(x, y) = \frac{2}{\sqrt{abc}} \sin\left(\frac{m\pi}{a}x\right) \sin\left(\frac{n\pi}{b}y\right) \quad (14\text{-}105)$$

The Green's function is obtained using (14-94), (14-103a), (14-103b), and (14-105). Doing this we can write that

$$G(x, y; x', y') = \frac{4}{abc} \sum_{m=1,2,\ldots}^{\infty} \sum_{n=1,2,\ldots}^{\infty} \frac{\sin\left(\frac{m\pi}{a}x'\right)\sin\left(\frac{n\pi}{b}y'\right)}{\beta^2 - \left[\left(\frac{m\pi}{a}\right)^2 + \left(\frac{n\pi}{b}\right)^2\right]} \times \sin\left(\frac{m\pi}{a}x\right)\sin\left(\frac{n\pi}{b}y\right) \quad (14\text{-}106)$$

and the electric field component of (14-99), due to the normalized electric current density represented by $J_n(x', y')$, can be written as

$$E_z(x, y) = \frac{4}{abc} \sum_{m=1,2\ldots}^{\infty} \sum_{n=1,2\ldots}^{\infty} \frac{\sin\left(\frac{m\pi}{a}x\right)\sin\left(\frac{n\pi}{b}y\right)}{\beta^2 - \left[\left(\frac{m\pi}{a}\right)^2 + \left(\frac{n\pi}{b}\right)^2\right]} \times \int_0^b \int_0^a J_n(x', y') \sin\left(\frac{m\pi}{a}x'\right)\sin\left(\frac{n\pi}{b}y'\right) dx'\, dy' \quad (14\text{-}107)$$

It is apparent that the Green's function possesses a singularity, and it fails to exist when

$$\beta = \beta_r = \omega_r\sqrt{\mu\varepsilon} = 2\pi f_r\sqrt{\mu\varepsilon} = \sqrt{\left(\frac{m\pi}{a}\right)^2 + \left(\frac{n\pi}{b}\right)^2} \quad (14\text{-}108)$$

or

$$f_r = \frac{1}{2\pi\sqrt{\mu\varepsilon}} \sqrt{\left(\frac{m\pi}{a}\right)^2 + \left(\frac{n\pi}{b}\right)^2} \quad (14\text{-}108a)$$

This is in accordance with (14-34) which states that the Green's function of (14-33b) exists for all values of λ (here $\lambda = \beta^2 = \omega^2\mu\varepsilon$) except those which are eigenvalues

λ_{mn} of the homogeneous equation 14-34 [here λ_{mn} given by (14-103a)]. At those eigenvalues for which (14-106) and (14-107) possess singularities (simple poles here), the frequencies of the excitation source match the natural (characteristic) frequencies of the system. This is referred to as *resonance*, and the field will continuously increase without any bounds (in the limit reaching values of infinity). For these cases no steady-state solutions exist. One way to contain the field is to introduce damping. In practice, for metallic cavities damping is introduced by the losses that are due to the nonperfectly conducting walls.

14.5 GREEN'S IDENTITIES AND METHODS

Now that we have derived Green's functions, both for single and two space variables in rectangular coordinates for the general Sturm–Liouville self-adjoint operator L, let us generalize the procedure for the development of the Green's function for the three-dimensional scalar Helmholtz partial differential equation

$$\nabla^2 \varphi(\mathbf{r}) + \beta^2 \varphi(\mathbf{r}) = f(\mathbf{r}) \qquad (14\text{-}109)$$

subject to the generalized homogeneous boundary conditions

$$\alpha_1 \varphi(\mathbf{r}_s) + \alpha_2 \frac{\partial \varphi(\mathbf{r}_s)}{\partial n} = 0 \qquad (14\text{-}109\text{a})$$

where $\mathbf{r}_s$ is on S and $\hat{n}$ is an outward directed unit vector. In electromagnetics these are referred to not only as the *mixed* boundary conditions [7] but also as the *impedance* boundary conditions. The Green's function $G(\mathbf{r}, \mathbf{r}')$ of (14-109) must satisfy the partial differential equation

$$\nabla^2 G(\mathbf{r}, \mathbf{r}') + \beta^2 G(\mathbf{r}, \mathbf{r}') = \delta(\mathbf{r} - \mathbf{r}') \qquad (14\text{-}110)$$

subject to the generalized homogeneous boundary conditions of

$$\alpha_1 G(\mathbf{r}_s, \mathbf{r}') + \alpha_2 \frac{\partial G(\mathbf{r}_s, \mathbf{r}')}{\partial n} = 0 \qquad (14\text{-}110\text{a})$$

To accomplish this we will need two identities from vector calculus that are usually referred to as *Green's first and second identities*. We will state them first before proceeding with the development of the generalized Green's function.

14.5.1 Green's First and Second Identities

Within a volume V conducting bodies with surfaces $S_1, S_2, S_3, \ldots, S_n$ are contained, as shown in Figure 14-6. By introducing appropriate cuts, the volume V is bounded by a regular surface S that consists of surfaces $S_1 - S_n$, the surfaces along the cuts,

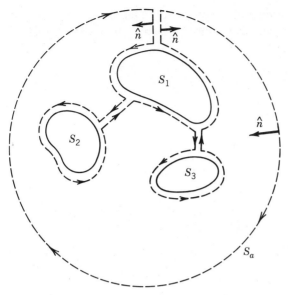

FIGURE 14-6 Conducting surfaces and appropriate cuts for application of Green's theorem.

and the surface S_a of an infinite radius sphere which encloses all the conducting bodies. A unit vector $\hat{n}$ normal to S is directed inward to the volume V, as shown in Figure 14-6.

Let us introduce within V two scalar functions ϕ and ψ which along with their first and second derivatives are continuous within V and on the surface S. To the vector $\phi\nabla\psi$ we apply the divergence theorem of (1-8), or

$$\oiint_S (\phi\nabla\psi)\cdot d\mathbf{s} = \oiint_S (\phi\nabla\psi)\cdot \hat{n}\, da = \iiint_V \nabla\cdot(\phi\nabla\psi)\, dv \quad (14\text{-}111)$$

When expanded, the integrand of the volume integral can be written as

$$\nabla\cdot(\phi\nabla\psi) = \phi\nabla\cdot(\nabla\psi) + \nabla\phi\cdot\nabla\psi = \phi\nabla^2\psi + \nabla\phi\cdot\nabla\psi \quad (14\text{-}112)$$

Thus (14-111) can be expressed as

$$\boxed{\oiint_S (\phi\nabla\psi\cdot d\mathbf{s}) = \iiint_V (\phi\nabla^2\psi)\, dv + \iiint_V (\nabla\phi\cdot\nabla\psi)\, dv} \quad (14\text{-}113)$$

which is referred to as *Green's first identity*. Since

$$(\nabla\psi)\cdot\hat{n} = \frac{\partial\psi}{\partial n} \quad (14\text{-}114)$$

where the derivative $\partial\psi/\partial n$ is taken in the direction of positive normal, (14-113) can also be written as

$$\boxed{\oiint_S \left(\phi\frac{\partial\psi}{\partial n}\right) ds = \iiint_V (\phi\nabla^2\psi)\, dv + \iiint_V (\nabla\phi\cdot\nabla\psi)\, dv} \quad (14\text{-}115)$$

which is an *alternate form of Green's first identity*.

If we repeat the procedure but apply the divergence theorem of (14-111) to the vector $\psi\nabla\phi$, then we can write, respectively, Green's first identity of (14-113) and its alternate form of (14-115) as

$$\oiint_S (\psi\nabla\phi \cdot d\mathbf{s}) = \iiint_V (\psi\nabla^2\phi)\, dv + \iiint_V (\nabla\psi \cdot \nabla\phi)\, dv \quad (14\text{-}116)$$

and

$$\oiint_S \left(\psi\frac{\partial\phi}{\partial n}\right) ds = \iiint_V (\psi\nabla^2\phi)\, dv + \iiint_V (\nabla\psi \cdot \nabla\phi)\, dv \quad (14\text{-}117)$$

Subtracting (14-116) from (14-113) we can write that

$$\boxed{\oiint_S (\phi\nabla\psi - \psi\nabla\phi) \cdot d\mathbf{s} = \iiint_V (\phi\nabla^2\psi - \psi\nabla^2\phi)\, dv} \quad (14\text{-}118)$$

which is referred to as *Green's second identity*. Its alternate form

$$\boxed{\oiint_S \left(\phi\frac{\partial\psi}{\partial n} - \psi\frac{\partial\phi}{\partial n}\right) ds = \iiint_V (\phi\nabla^2\psi - \psi\nabla^2\phi)\, dv} \quad (14\text{-}119)$$

is obtained by subtracting (14-117) from (14-115).

Green's first and second identities expressed, respectively, either as (14-113) and (14-118) or (14-115) and (14-119) will be used to develop the formulation for the more general Green's function.

14.5.2 Generalized Green's Function Method

With the introduction of Green's first and second identities in the previous section, we are now ready to develop the formulation of the generalized Green's function method of φ for the partial differential equation 14-109 whose Green's function $G(\mathbf{r}, \mathbf{r}')$ satisfies (14-110).

Let us multiply (14-109) by $G(\mathbf{r}, \mathbf{r}')$ and (14-110) by $\varphi(\mathbf{r})$. Doing this leads to

$$G\nabla^2\varphi + \beta^2\varphi G = fG \quad (14\text{-}120a)$$

$$\varphi\nabla^2 G + \beta^2\varphi G = \varphi\delta(\mathbf{r} - \mathbf{r}') \quad (14\text{-}120b)$$

Subtracting (14-120a) from (14-120b) and integrating over the volume V, we can write that

$$\iiint_V \varphi\delta(\mathbf{r} - \mathbf{r}')\, dv - \iiint_V fG\, dv = \iiint_V (\varphi\nabla^2 G - G\nabla^2\varphi)\, dv \quad (14\text{-}121)$$

or

$$\varphi(r = r') = \varphi(\mathbf{r}') = \iiint_V f(\mathbf{r}) G(\mathbf{r}, \mathbf{r}')\, dv$$

$$+ \iiint_V [\varphi(\mathbf{r})\nabla^2 G(\mathbf{r}, \mathbf{r}') - G(\mathbf{r}, \mathbf{r}')\nabla^2\varphi(\mathbf{r})]\, dv \quad (14\text{-}121a)$$

Applying Green's second identity (14-118) reduces (14-121a) to

$$\varphi(\mathbf{r}') = \iiint_V f(\mathbf{r}) G(\mathbf{r},\mathbf{r}')\, dv + \oiint_S [\varphi(\mathbf{r}) \nabla G(\mathbf{r},\mathbf{r}') - G(\mathbf{r},\mathbf{r}') \nabla \varphi(\mathbf{r})] \cdot d\mathbf{s} \quad (14\text{-}122)$$

Since $\mathbf{r}'$ is an arbitrary point within V and $\mathbf{r}$ is a dummy variable, we can also write (14-122) as

$$\boxed{\varphi(\mathbf{r}) = \iiint_V f(\mathbf{r}') G(\mathbf{r},\mathbf{r}')\, dv' + \oiint_S [\varphi(\mathbf{r}') \nabla' G(\mathbf{r},\mathbf{r}') - G(\mathbf{r},\mathbf{r}') \nabla' \varphi(\mathbf{r}')] \cdot d\mathbf{s}'}$$

(14-123)

where ∇' indicates differentiation with respect to the prime coordinates.

Equation 14-123 is a generalized formula for the development of the Green's function for a three-dimensional scalar Helmholtz equation. It can be simplified depending on the boundary conditions of φ and G, and their derivatives on S. The objective then will be to judiciously choose the boundary conditions on the development of G, once the boundary conditions on φ are stated, so as to simplify, if not completely eliminate, the surface integral contribution in (14-123). We will demonstrate here some combinations of boundary conditions on φ and G, and the simplifications of (14-123) based on those boundary conditions.

A. NONHOMOGENEOUS PARTIAL DIFFERENTIAL EQUATION WITH HOMOGENEOUS DIRICHLET BOUNDARY CONDITIONS

If the nonhomogeneous form of the partial differential equation of (14-109) satisfies the homogeneous Dirichlet boundary condition of

$$\varphi(\mathbf{r}_s) = 0 \quad \text{where } \mathbf{r}_s \text{ is on } S \quad (14\text{-}124a)$$

then it is reasonable to construct a Green's function with the same boundary condition of

$$G(\mathbf{r}_s,\mathbf{r}') = 0 \quad \text{where } \mathbf{r}_s \text{ is on } S \quad (14\text{-}124b)$$

so as to simplify the surface integral contributions in (14-123).

For these boundary conditions on φ and G both terms in the surface integral of (14-123) vanish, so that (14-123) reduces to

$$\boxed{\varphi(\mathbf{r}) = \iiint_V f(\mathbf{r}') G(\mathbf{r},\mathbf{r}')\, dv'} \quad (14\text{-}125)$$

The Green's function $G(\mathbf{r},\mathbf{r}')$ needed in (14-125) can be obtained using any of the previous methods developed in Sections 14.3.1 through 14.3.2. In many cases the bilinear form of (14-61) or (14-94) or its equivalent, in the desired coordinate system and number of space variables, is appropriate for forming the Green's function. Its existence will depend upon the eigenvalues of the homogeneous partial differential equation, as discussed in Section 14.3.1.

B. NONHOMOGENEOUS PARTIAL DIFFERENTIAL EQUATION WITH NONHOMOGENEOUS DIRICHLET BOUNDARY CONDITIONS

If the nonhomogeneous partial differential equation 14-109 satisfies the nonhomogeneous Dirichlet boundary condition

$$\varphi(\mathbf{r}_s) = g(\mathbf{r}_s) \quad \text{where } \mathbf{r}_s \text{ is on } S \tag{14-126a}$$

then we can still construct a Green's function that satisfies the boundary condition

$$G(\mathbf{r}_s, \mathbf{r}') = 0 \quad \text{where } \mathbf{r}_s \text{ is on } S \tag{14-126b}$$

For these boundary conditions on φ and G the second term in the surface integral of (14-123) vanishes, so that (14-123) reduces to

$$\boxed{\varphi(\mathbf{r}) = \iiint_V f(\mathbf{r}') G(\mathbf{r}, \mathbf{r}') \, dv' + \oiint_S \varphi(\mathbf{r}') \nabla' G(\mathbf{r}_s, \mathbf{r}') \cdot d\mathbf{s}'} \tag{14-127}$$

The Green's function $G(\mathbf{r}, \mathbf{r}')$ needed in (14-127) can be determined using any of the previous methods developed in Sections 14.3.1 through 14.3.2.

C. NONHOMOGENEOUS PARTIAL DIFFERENTIAL EQUATION WITH HOMOGENEOUS NEUMANN BOUNDARY CONDITIONS

When Neumann boundary conditions are involved, the solutions become more involved primarily because the normal gradients of $\varphi(\mathbf{r})$ are not independent of the partial differential equation. If the nonhomogeneous form of the partial differential equation 14-109 satisfies the homogeneous Neumann boundary condition of

$$[\nabla' \varphi(\mathbf{r}_s)] \cdot \hat{n} = \frac{\partial \varphi(\mathbf{r}_s)}{\partial n} = 0 \quad \text{where } \mathbf{r}_s \text{ is on } S \tag{14-128}$$

then we *cannot*, in general, construct a Green's function with a boundary condition of $[\nabla' G(\mathbf{r}_s, \mathbf{r}')] \cdot \hat{n} = [\partial G(\mathbf{r}_s, \mathbf{r}')/\partial n] = 0$. This is evident from what follows.

If we apply the divergence theorem of (1-8) to the vector $\nabla G(\mathbf{r}, \mathbf{r}')$, we can write that

$$\oiint_S \nabla G(\mathbf{r}, \mathbf{r}') \cdot d\mathbf{s} = \iiint_V \nabla \cdot \nabla G(\mathbf{r}, \mathbf{r}') \, dv = \iiint_V \nabla^2 G(\mathbf{r}, \mathbf{r}') \, dv \tag{14-129}$$

Taking the volume integral of (14-110), we can express it as

$$\iiint_V \nabla^2 G(\mathbf{r}, \mathbf{r}') \, dv + \beta^2 \iiint_V G(\mathbf{r}, \mathbf{r}') \, dv = \iiint_V \delta(\mathbf{r} - \mathbf{r}') \, dv \tag{14-130}$$

Using (14-129) reduces (14-130) to

$$\oiint_S \nabla G(\mathbf{r}, \mathbf{r}') \cdot d\mathbf{s} + \beta^2 \iiint_V G(\mathbf{r}, \mathbf{r}') \, dv = 1 \tag{14-131}$$

If we choose

$$\left[\nabla G(\mathbf{r}, \mathbf{r}')|_{\mathbf{r}=\mathbf{r}_s}\right] \cdot \hat{n} = [\nabla G(\mathbf{r}_s, \mathbf{r}')] \cdot \hat{n} = \frac{\partial G(\mathbf{r}_s, \mathbf{r}')}{\partial n} = 0 \tag{14-132}$$

as a boundary condition for $G(\mathbf{r},\mathbf{r}')$, then (14-131) reduces to

$$\beta^2 \iiint_V G(\mathbf{r},\mathbf{r}')\,dv = 1 \tag{14-133}$$

which cannot be satisfied if $\beta = 0$. When $\beta = 0$, (14-131) reduces to

$$\oiint_S \nabla G(\mathbf{r},\mathbf{r}') \cdot d\mathbf{s} = 1 \tag{14-134}$$

or

$$|\nabla G(\mathbf{r}_s,\mathbf{r}')|S_0 = 1 \tag{14-134a}$$

where S_0 is the area of the surface. This implies that a consistent boundary condition for the normal gradient of $G(\mathbf{r},\mathbf{r}')$ on S to satisfy (14-134) or (14-134a) would be

$$\nabla' G(\mathbf{r},\mathbf{r}')|_{\mathbf{r}=\mathbf{r}_s} = \frac{1}{S_0} = \nabla' G(\mathbf{r}_s,\mathbf{r}') \tag{14-135}$$

Substituting (14-128) and (14-135) into (14-123) leads to

$$\boxed{\varphi(\mathbf{r}) = \iiint_V f(\mathbf{r}') G(\mathbf{r},\mathbf{r}')\,dv' + \frac{1}{S_0}\oiint_S \varphi(\mathbf{r}')\,ds'} \tag{14-136}$$

The second term on the right side of (14-136) is a constant, and it can be dropped since $\varphi(\mathbf{r})$ is undetermined by the boundary conditions up to an additive constant.

D. NONHOMOGENEOUS PARTIAL DIFFERENTIAL EQUATION WITH MIXED BOUNDARY CONDITIONS

When the boundary conditions on $\varphi(\mathbf{r})$ are such that $\varphi(\mathbf{r}_s)$ is specified in part of the surface S and $[\nabla'\varphi(\mathbf{r}_s)] \cdot \hat{n} = \partial\phi(\mathbf{r}_s)/\partial n$ is specified over the remaining part of S, it is referred to as having *mixed* boundary conditions. Then it is desirable to construct a Green's function so that it vanishes on that part of S over which $\varphi(\mathbf{r}_s)$ is specified and its normal derivative $[\partial G(\mathbf{r}_s,\mathbf{r}')/\partial n]$ vanishes over the remaining part of S over which $\partial\psi(\mathbf{r}_s)/\partial n$ is specified. Although this is a more complex procedure, it does provide a method to derive the Green's function even under those conditions.

14.6 GREEN'S FUNCTIONS OF THE SCALAR HELMHOLTZ EQUATION

Now that we have derived the development of the generalized Green's function, let us apply the formulation to the scalar Helmholtz equation in three-dimensional problems of rectangular, cylindrical, and spherical coordinates.

14.6.1 Rectangular Coordinates

The development of Green's functions in rectangular coordinates has already been applied for one and two space variables in almost all of the previous sections. In this section we want to derive it for a three-dimensional problem. Specifically let us derive the Green's function for the electric field component E_y which satisfies the partial differential equation

$$\nabla^2 E_y + \beta_0^2 E_y = \left(\frac{\partial^2}{\partial x^2} + \frac{\partial^2}{\partial y^2} + \frac{\partial^2}{\partial z^2}\right) E_y + \beta_0^2 E_y = j\omega\mu J_y(x, y, z) \quad (14\text{-}137)$$

subject to the boundary conditions

$$E_y(x = 0, 0 \leq y \leq b, -\infty \leq z \leq +\infty)$$
$$= E_y(x = a, 0 \leq y \leq b, -\infty \leq z \leq +\infty) = 0 \quad (14\text{-}137a)$$
$$E_z(0 \leq x \leq a, y = 0, -\infty \leq z \leq +\infty)$$
$$= E_z(0 \leq x \leq a, y = b, -\infty \leq z \leq +\infty) = 0 \quad (14\text{-}137b)$$

In (14-137) E_y represents the electric field component of a TM^z field configuration (subject to $\nabla \cdot \mathbf{J} = 0$) inside a metallic waveguide of dimensions a, b in the x, y directions, respectively, as shown in Figure 14-7. Also $J_y(x, y, z)$ represents the electric current density of the feed probe that is used to excite the fields within the metallic waveguide. It is assumed that the wave is traveling in the z direction. The time-harmonic variations are of $e^{+j\omega t}$, and they are suppressed.

The Green's function must satisfy the partial differential equation

$$\nabla^2 G(x, y, z; x', y', z') + \beta_0^2 G(x, y, z; x', y', z')$$
$$= \delta(x - x')\delta(y - y')\delta(z - z') \quad (14\text{-}138)$$

subject to the boundary conditions

$$G(x = 0, 0 \leq y \leq b, -\infty \leq z \leq +\infty)$$
$$= G(x = a, 0 \leq y \leq b, -\infty \leq z \leq +\infty) = 0 \quad (14\text{-}138a)$$

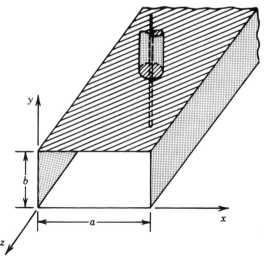

FIGURE 14-7 Rectangular waveguide excited by linear electric probe.

Since the electric field and Green's function satisfy, respectively, the Dirichlet boundary conditions (14-124a) and (14-124b), then according to (14-123) or (14-125) the electric field is obtained using

$$E_y(x, y, z) = j\omega\mu \iiint_V J_y(x', y', z') G(x, y, z; x', y', z') \, dx' \, dy' \, dz' \quad (14\text{-}139)$$

The Green's function can be derived either in closed, series, or integral form. We will choose here the series form. We begin the development of the Green's function by assuming its solution can be represented by a two-function Fourier series of sine function in x and cosine function in y which satisfy, respectively, the boundary conditions at $x = 0, a$. Thus we can express $G(x, y, z; x', y', z')$ as

$$G(x, y, z; x', y', z') = \sum_{m=1,2,\ldots}^{\infty} \sum_{n=1,2,\ldots}^{\infty} g_{mn}(z; x', y', z') \sin\left(\frac{m\pi}{a}x\right) \cos\left(\frac{n\pi}{b}y\right)$$

(14-140)

Substituting (14-140) into (14-138) leads to

$$\sum_{m=1,2,\ldots}^{\infty} \sum_{n=1,2,\ldots}^{\infty} \left[-\left(\frac{m\pi}{a}\right)^2 - \left(\frac{n\pi}{b}\right)^2 + \beta_0^2 + \frac{\partial^2}{\partial z^2}\right]$$

$$\times g_{mn}(z; x', y', z') \sin\left(\frac{m\pi}{a}x\right) \cos\left(\frac{n\pi}{b}y\right)$$

$$= \delta(x - x')\delta(y - y')\delta(z - z') \quad (14\text{-}141)$$

Multiplying both sides of (14-141) by $\sin(p\pi x/a) \cos(q\pi y/b)$, integrating between 0 to a in x and 0 to b in y, and using (8-56a), (8-56b), (14-48a) and (14-48b) we can reduce (14-141) to

$$\frac{ab}{4}\left(\frac{\partial^2}{\partial z^2} + \beta_z^2\right) g_{mn}(z; x', y', z') = \sin\left(\frac{m\pi}{a}x'\right) \cos\left(\frac{n\pi}{b}y'\right) \delta(z - z') \quad (14\text{-}142)$$

or

$$\boxed{\left(\frac{d^2}{dz^2} + \beta_z^2\right) g_{mn}(z; x', y', z') = \frac{4}{ab} \sin\left(\frac{m\pi}{a}x'\right) \cos\left(\frac{n\pi}{b}y'\right) \delta(z - z')}$$

(14-142a)

where

$$\beta_z^2 = \beta_0^2 - \left[\left(\frac{m\pi}{a}\right)^2 + \left(\frac{n\pi}{b}\right)^2\right] = \beta_0^2 - (\beta_x^2 + \beta_y^2) \quad (14\text{-}142b)$$

$$\beta_x = \frac{m\pi}{a} \quad m = 1, 2, 3, \ldots \quad (14\text{-}142c)$$

$$\beta_y = \frac{n\pi}{b} \quad n = 1, 2, 3, \ldots \quad (14\text{-}142d)$$

The function $g_{mn}(z; x', y', z')$ satisfies the single variable differential equation 14-142a, and it can be found by using the recipe of Section 14.3.1 as represented by

(14-44c), (14-45a), and (14-45b). Two solutions of the homogeneous differential equation

$$\left(\frac{d^2}{dz^2} + \beta_z^2\right) g_{mn}(z; x', y', z') = 0 \qquad (14\text{-}143)$$

of (14-142a) are

$$g_{mn}^{(1)} = A_{mn} e^{+j\beta_z z} \qquad \text{for } z < z' \qquad (14\text{-}143a)$$
$$g_{mn}^{(2)} = B_{mn} e^{-j\beta_z z} \qquad \text{for } z > z' \qquad (14\text{-}143b)$$

Using (14-44c) where $y_1 = g_{mn}^{(1)}$ and $y_2 = g_{mn}^{(2)}$, we can write the Wronskian as

$$W(z') = A_{mn} B_{mn}(-j\beta_z) e^{j\beta_z z'} e^{-j\beta_z z'} - A_{mn} B_{mn}(j\beta_z) e^{-j\beta_z z'} e^{+j\beta_z z'}$$
$$W(z') = -j2\beta_z A_{mn} B_{mn} \qquad (14\text{-}144)$$

By comparing (14-142a) with (14-33), it is apparent that

$$\begin{aligned} p(z) &= 1 \\ q(z) &= 0 \\ r(z) &= 1 \\ \lambda &= \beta_z^2 \end{aligned} \qquad (14\text{-}145)$$

Using (14-143a) through (14-145), the solution for $g_{mn}(z; x', y', z')$ of (14-142a) can be written by referring to (14-45a) and (14-45b) as

$$g_{mn}(z; x', y', z') = \begin{cases} j\dfrac{2}{ab} \dfrac{\sin\left(\dfrac{m\pi}{a} x'\right) \cos\left(\dfrac{n\pi}{b} y'\right)}{\beta_z} e^{-j\beta_z(z'-z)} & \text{for } z < z' \\ & \qquad\qquad (14\text{-}146a) \\ j\dfrac{2}{ab} \dfrac{\sin\left(\dfrac{m\pi}{a} x'\right) \cos\left(\dfrac{n\pi}{b} y'\right)}{\beta_z} e^{-j\beta_z(z-z')} & \text{for } z > z' \\ & \qquad\qquad (14\text{-}146b) \end{cases}$$

or

$$g_{mn}(z; x', y', z') = j\frac{2}{ab} \frac{\sin\left(\dfrac{m\pi}{a} x'\right) \cos\left(\dfrac{n\pi}{b} y'\right)}{\beta_z} e^{-j\beta_z|z-z'|} \quad \text{for } z < z', \ z > z'$$

$$(14\text{-}146c)$$

Thus the Green's function of (14-140) can now be expressed as

$$G(x, y, z; x', y', z') = j\frac{2}{ab} \sum_{m=1,2\ldots}^{\infty} \sum_{n=1,2\ldots}^{\infty} \frac{\sin\left(\dfrac{m\pi}{a} x'\right) \cos\left(\dfrac{n\pi}{b} y'\right)}{\beta_z}$$
$$\times \sin\left(\frac{m\pi}{a} x\right) \cos\left(\frac{n\pi}{b} y\right) e^{-j\beta_z|z-z'|} \qquad \text{for } z < z', \ z > z'$$

$$(14\text{-}147)$$

where

$$\beta_z = \begin{cases} \sqrt{\beta_0^2 - (\beta_x^2 + \beta_y^2)} & \text{for } \beta_0^2 > (\beta_x^2 + \beta_y^2) \quad (14\text{-}147\text{a}) \\ -j\sqrt{(\beta_x^2 + \beta_y^2) - \beta_0^2} & \text{for } \beta_0^2 < (\beta_x^2 + \beta_y^2) \quad (14\text{-}147\text{b}) \end{cases}$$

It is evident from (14-147) through (14-147b) that when $\beta_0^2 > (\beta_x^2 + \beta_y^2)$ the modes are propagating and when $\beta_0^2 < (\beta_x^2 + \beta_y^2)$ the modes are not propagating (evanescent). The nonpropagating modes converge very rapidly when $|z - z'|$ is very large.

Once the Green's function is formulated as in (14-147), the electric field can be found using (14-139).

14.6.2 Cylindrical Coordinates

Until now we have concentrated on developing primarily Green's functions of problems dealing with rectangular coordinates. This was done to maintain simplicity in the mathematics so that the analytical formulations would not obscure the fundamental concepts. Now we are ready to deal with problems expressed by other coordinate systems, such as cylindrical and spherical.

Let us assume that an infinite electric line source of constant current I_z is placed at $\rho = \rho'$, $\phi = \phi'$ inside a circular waveguide of radius a, as shown in Figure 14-8. The electric field component E_z satisfies the partial differential equation

$$\boxed{\nabla^2 E_z + \beta_0^2 E_z = f(\rho, \phi) = j\omega\mu I_z} \quad (14\text{-}148)$$

subject to the boundary condition

$$E_z(\rho = a, 0 \le \phi \le 2\pi, z) = 0 \quad (14\text{-}148\text{a})$$

The Green's function of this problem will satisfy the partial differential equation

$$\boxed{\nabla^2 G + \beta_0^2 G = \delta(\rho - \rho')} \quad (14\text{-}149)$$

The boundary condition for the Green's function can be chosen so that

$$G(\rho = a, 0 \le \phi \le 2\pi, z) = 0 \quad (14\text{-}149\text{a})$$

Since the boundary conditions of (14-148a) and (14-149a) on E_z and G, respectively, are of the Dirichlet type, then according to (14-123) or (14-125)

$$E_z(\rho, \phi) = \iint_S f(\rho', \phi') G(\rho, \phi; \rho', \phi') \, ds' = j\omega\mu \iint_S I_z(\rho', \phi') G(\rho, \phi; \rho', \phi') \, ds' \quad (14\text{-}150)$$

Since both the current source and the circular waveguide are of infinite length, the problem reduces to a two-dimensional one. Thus we can express initially the Green's function by an infinite Fourier series whose eigenvalues in ϕ satisfy the

898 GREEN'S FUNCTIONS

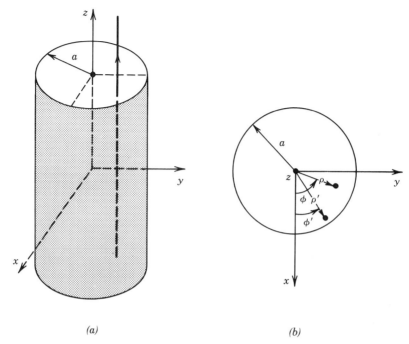

(a)   (b)

FIGURE 14-8 Electric line source within circular conducting cylinder. (*a*) Line source and conducting cylinder. (*b*) Top view.

periodicity requirements. That is

$$\boxed{G(\rho, \phi; \rho', \phi') = \sum_{m=-\infty}^{+\infty} g_m(\rho; \rho', \phi') e^{jm\phi}} \quad (14\text{-}151)$$

In cylindrical coordinates, the delta function of $\delta(\rho - \rho')$ in (14-149) can be expressed, in general, as [10, 13]

$$\delta(\rho - \rho') = \begin{cases} \dfrac{1}{\rho}\delta(\rho - \rho')\delta(\phi - \phi')\delta(z - z') & (14\text{-}152a) \\[6pt] \dfrac{1}{2\pi\rho}\delta(\rho - \rho')\delta(z - z') & \text{for no } \phi \text{ dependence} & (14\text{-}152b) \\[6pt] \dfrac{1}{2\pi\rho}\delta(\rho - \rho') & \text{for no } \phi \text{ and } z \text{ dependence} & (14\text{-}152c) \\[6pt] \dfrac{1}{\rho}\delta(\rho - \rho')\delta(\phi - \phi') & \text{for no } z \text{ dependence} & (14\text{-}152d) \end{cases}$$

In expanded form the Green's function of (14-149) can now be written, using (14-152d) and assuming no z variations, as

$$\frac{\partial^2 G}{\partial \rho^2} + \frac{1}{\rho}\frac{\partial G}{\partial \rho} + \frac{1}{\rho^2}\frac{\partial^2 G}{\partial \phi^2} + \beta_0^2 G = \frac{1}{\rho}\delta(\rho - \rho')\delta(\phi - \phi') \quad (14\text{-}153)$$

Substituting (14-151) into (14-153) leads to

$$\sum_{m=-\infty}^{+\infty}\left[\frac{\partial^2}{\partial\rho^2}+\frac{1}{\rho}\frac{\partial}{\partial\rho}-\frac{m^2}{\rho^2}+\beta_0^2\right]g_m(\rho;\rho',\phi')e^{jm\phi}=\frac{1}{\rho}\delta(\rho-\rho')\delta(\phi-\phi')$$

(14-154)

Multiplying both sides of (14-154) by $e^{-jn\phi}$, integrating both sides from 0 to 2π in ϕ, and using the orthogonality condition

$$\int_0^{2\pi} e^{j(m-n)\phi}\,d\phi = \begin{cases} 2\pi & m=n \\ 0 & m\neq n \end{cases}$$

(14-155a)
(14-155b)

reduces (14-154) to

$$2\pi\left[\frac{\partial^2 g_m}{\partial\rho^2}+\frac{1}{\rho}\frac{\partial g_m}{\partial\rho}+\left(\beta_0^2-\frac{m^2}{\rho^2}\right)g_m\right]=\frac{1}{\rho}e^{-jm\phi'}\delta(\rho-\rho')$$

or

$$\boxed{\rho\frac{d^2 g_m}{d\rho^2}+\frac{dg_m}{d\rho}+\left(\rho\beta_0^2-\frac{m^2}{\rho}\right)g_m=\frac{e^{-jm\phi'}}{2\pi}\delta(\rho-\rho')}$$ (14-156)

where the partial derivatives have been replaced by ordinary derivatives.

The function $g_m(\rho;\rho',\phi')$ satisfies the differential equation 14-156, and its solution can be obtained using the closed form recipe of Section 14.3.1 represented by (14-44c) and (14-45a) through (14-45b). The homogeneous equation 14-156 can be written by multiplying through by ρ as

$$\rho^2\frac{d^2 g_m}{d\rho^2}+\rho\frac{dg_m}{d\rho}+\left(\beta_0^2\rho^2-m^2\right)g_m=0$$ (14-157)

or

$$\rho\frac{d^2 g_m}{d\rho^2}+\frac{dg_m}{d\rho}+\left(\beta_0^2\rho-\frac{m^2}{\rho}\right)g_m=0$$ (14-157a)

which is of the one-dimensional Sturm–Liouville form of (14-26) or (14-33) (see Example 14-1) where

$$\begin{aligned} p(\rho) &= \rho \\ q(\rho) &= \frac{m^2}{\rho} \\ r(\rho) &= \rho \\ \lambda &= \beta_0^2 \end{aligned}$$

(14-158)

Equation 14-157 is recognized as being Bessel's differential equation 3-64 whose two solutions can be written according to (3-67a) as

$$g_m^{(1)} = A_m J_m(\beta_0\rho) + B_m Y_m(\beta_0\rho) \quad \text{for } \rho<\rho'$$ (14-159a)
$$g_m^{(2)} = C_m J_m(\beta_0\rho) + D_m Y_m(\beta_0\rho) \quad \text{for } \rho>\rho'$$ (14-159b)

These two solutions were chosen because the fields within the waveguide form standing waves instead of traveling waves.

Since the Green's function of (14-151) must represent according to (14-150) the field everywhere, including the origin, then $B_m = 0$ in (14-159a) since $Y_m(\beta_0\rho)$ possesses a singularity at $\rho = 0$. Also since the Green's function must satisfy the boundary condition of (14-149a), then the solution of $g_m^{(2)}$ of (14-159b) must also satisfy (14-149a). Thus

$$g_m^{(2)}(\rho = a) = C_m J_m(\beta_0 a) + D_m Y_m(\beta_0 a) = 0$$

or

$$D_m = -C_m \frac{J_m(\beta_0 a)}{Y_m(\beta_0 a)} \tag{14-160}$$

Thus (14-159a) and (14-159b) can be reduced to

$$g_m^{(1)} = A_m J_m(\beta_0 \rho) \qquad \text{for } \rho < \rho' \tag{14-161a}$$

$$g_m^{(2)} = C_m \left[J_m(\beta_0\rho) - \frac{J_m(\beta_0 a)}{Y_m(\beta_0 a)} Y_m(\beta_0\rho) \right] \qquad \text{for } \rho > \rho' \tag{14-161b}$$

Using (14-44c) where $y_1 = g_m^{(1)}$ and $y_2 = g_m^{(2)}$, we can write the Wronskian as

$$W(\rho') = \beta_0 A_m C_m \frac{J_m(\beta_0 a)}{Y_m(\beta_0 a)} [J_m'(\beta_0\rho') Y_m(\beta_0\rho') - J_m(\beta_0\rho') Y_m'(\beta_0\rho')] \tag{14-162}$$

where the prime indicates partial with respect to the entire argument $[' = \partial/\partial(\beta_0\rho')]$. Using the Wronskian for Bessel functions of (11-95), we can reduce (14-162) to

$$W(\rho') = -\frac{2}{\pi} A_m C_m \frac{J_m(\beta_0 a)}{Y_m(\beta_0 a)} \frac{1}{\rho'} \tag{14-162a}$$

Finally $g_m(\rho; \rho', \phi')$ of (14-156) can be written using (14-158), (14-161a) through (14-161b), and (14-162a) by referring to (14-45a) and (14-45b), as

$$g_m(\rho; \rho', \phi') = \begin{cases} -\dfrac{1}{4} [J_m(\beta_0\rho') Y_m(\beta_0 a) - J_m(\beta_0 a) Y_m(\beta_0\rho')] \dfrac{J_m(\beta_0\rho)}{J_m(\beta_0 a)} e^{-jm\phi'} \\ \qquad\qquad \text{for } \rho < \rho' \qquad (14\text{-}163a) \\ -\dfrac{1}{4} [J_m(\beta_0\rho) Y_m(\beta_0 a) - J_m(\beta_0 a) Y_m(\beta_0\rho)] \dfrac{J_m(\beta_0\rho')}{J_m(\beta_0 a)} e^{-jm\phi'} \\ \qquad\qquad \text{for } \rho > \rho' \qquad (14\text{-}163b) \end{cases}$$

Thus the Green's function of (14-151) can then be written as

$$G(\rho,\phi;\rho',\phi') = -\frac{1}{4} \begin{cases} \displaystyle\sum_{m=-\infty}^{+\infty} [J_m(\beta_0\rho') Y_m(\beta_0 a) - J_m(\beta_0 a) Y_m(\beta_0\rho')] \\ \qquad \times \dfrac{J_m(\beta_0\rho)}{J_m(\beta_0 a)} e^{jm(\phi-\phi')} \qquad \text{for } \rho < \rho' \qquad (14\text{-}164a) \\ \displaystyle\sum_{m=-\infty}^{+\infty} [J_m(\beta_0\rho) Y_m(\beta_0 a) - J_m(\beta_0 a) Y_m(\beta_0\rho)] \\ \qquad \times \dfrac{J_m(\beta_0\rho')}{J_m(\beta_0 a)} e^{jm(\phi-\phi')} \qquad \text{for } \rho > \rho' \qquad (14\text{-}164b) \end{cases}$$

or in terms of cosine terms as

$$G(\rho,\phi;\rho',\phi') = -\frac{1}{2}\begin{cases} \sum_{m=0}^{+\infty} \dfrac{[J_m(\beta_0\rho')Y_m(\beta_0 a) - J_m(\beta_0 a)Y_m(\beta_0\rho')]}{\varepsilon_m} \\ \qquad \times \dfrac{J_m(\beta_0\rho)}{J_m(\beta_0 a)} \cos[m(\phi-\phi')] \quad \text{for } \rho < \rho' \quad (14\text{-}165\text{a}) \\ \sum_{m=0}^{+\infty} \dfrac{[J_m(\beta_0\rho)Y_m(\beta_0 a) - J_m(\beta_0 a)Y_m(\beta_0\rho)]}{\varepsilon_m} \\ \qquad \times \dfrac{J_m(\beta_0\rho')}{J_m(\beta_0 a)} \cos[m(\phi-\phi')] \quad \text{for } \rho > \rho' \quad (14\text{-}165\text{b}) \end{cases}$$

where

$$\varepsilon_m = \begin{cases} 2 & m = 0 \qquad\qquad (14\text{-}165\text{c}) \\ 1 & m \neq 0 \qquad\qquad (14\text{-}165\text{d}) \end{cases}$$

The Green's function for this problem can also be derived using the two space variable series expansion method whereby it is represented by orthonormal expansion functions. This is left to the reader as an end of chapter exercise.

Example 14-5. An infinite electric line source of constant current I_z is located at $\rho = \rho'$, $\phi = \phi'$, as shown in Figure 14-9, and it is radiating in an unbounded free-space medium. Derive its Green's function in closed form.

Solution. Since the line source is removed from the origin, its Green's function will be a function of ϕ and ϕ'. Thus it takes the form of (14-151), and it satisfies the differential equations 14-153 through 14-157. However, since the Green's function must satisfy the radiation conditions at infinity ($G \to 0$ as $\rho \to \infty$), the two solutions to the homogeneous differential equation 14-158

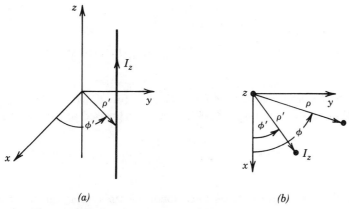

FIGURE 14-9 Electric line source displaced from the origin. (*a*) Perspective view. (*b*) Top view.

can be written as

$$g_m^{(1)} = A_m J_m(\beta_0 \rho) + B_m Y_m(\beta_0 \rho) \quad \text{for } \rho < \rho'$$
$$g_m^{(2)} = C_m H_m^{(1)}(\beta_0 \rho) + D_m H_m^{(2)}(\beta_0 \rho) \quad \text{for } \rho > \rho'$$

Because the fields must be finite everywhere, including $\rho = 0$, $g_m^{(1)}$ reduces to

$$g_m^{(1)} = A_m J_m(\beta_0 \rho) \quad \text{for } \rho < \rho'$$

In addition for $\rho > \rho'$ the wave functions must represent outwardly traveling waves. Thus for $e^{j\omega t}$ time variations $g_m^{(2)}$ reduces to

$$g_m^{(2)} = D_m H_m^{(2)}(\beta_0 \rho) \quad \text{for } \rho > \rho'$$

Using (14-44c) where $y_1 = g_m^{(1)}$ and $y_2 = g_m^{(2)}$, we can write the Wronskian as

$$W(\rho') = \beta_0 A_m D_m \left[J_m(\beta_0 \rho') H_m^{(2)'}(\beta_0 \rho') - H_m^{(2)}(\beta_0 \rho') J_m'(\beta_0 \rho') \right]$$
$$= -j\beta_0 A_m D_m \left[J_m(\beta_0 \rho') Y_m'(\beta_0 \rho') - J_m'(\beta_0 \rho') Y_m(\beta_0 \rho') \right]$$

which by using the Wronskian of (11-95) for Bessel functions can be expressed as

$$W(\rho') = -j \frac{2}{\pi \rho'} A_m D_m$$

Thus $g_m(\rho; \rho', \phi')$ of (14-156) can be written using (14-45a) through (14-45b) and (14-158) along with the preceding expressions for $g_m^{(1)}$, $g_m^{(2)}$, and $W(\rho')$ as

$$g_m(\rho; \rho', \phi') = \begin{cases} J_m(\beta_0 \rho) H_m^{(2)}(\beta_0 \rho') e^{-jm\phi'} & \text{for } \rho < \rho' \\ J_m(\beta_0 \rho') H_m^{(2)}(\beta_0 \rho) e^{-jm\phi'} & \text{for } \rho > \rho' \end{cases}$$

Thus the Green's function of (14-151) can be written as

$$G(\rho, \phi; \rho', \phi') = -\frac{1}{4j} \begin{cases} \sum_{m=-\infty}^{+\infty} J_m(\beta_0 \rho) H_m^{(2)}(\beta_0 \rho') e^{jm(\phi-\phi')} & \text{for } \rho < \rho' \\ \sum_{m=-\infty}^{+\infty} J_m(\beta_0 \rho') H_m^{(2)}(\beta_0 \rho) e^{jm(\phi-\phi')} & \text{for } \rho > \rho' \end{cases}$$

which by the addition theorem for Hankel functions of (11-69a) through (11-69b) or (11-82a) through (11-82b) can be expressed in succinct form as

$$G(\rho, \phi; \rho', \phi') = -\frac{1}{4j} H_0^{(2)}(\beta_0 |\rho - \rho'|)$$

This is the well known two-dimensional Green's function for cylindrical waves.

14.6.3 Spherical Coordinates

The development of the Green's function for problems represented by spherical coordinates is more complex, and it must be expressed in general in terms of spherical Bessel and Hankel functions, Legendre functions, and complex exponentials or cosinusoids (see Chapter 10). In order to minimize the mathematical complexities here, we will develop the Green's function of a source positioned at r', θ', ϕ', inside a sphere of radius a and with free-space, as shown in Figure 14-10.

The Green's function must satisfy the partial differential equation

$$\nabla^2 G + \beta_0 G = \delta(\mathbf{r} - \mathbf{r}') \qquad (14\text{-}166)$$

subject to the boundary conditions

$$G(r = a, 0 \leq \theta \leq \pi, 0 \leq \phi \leq 2\pi) = 0 \qquad (14\text{-}166a)$$

In spherical coordinates the delta function $\delta(\mathbf{r} - \mathbf{r}')$ in (14-166) can be expressed, in general, as [10, 13]

$$\delta(\mathbf{r} - \mathbf{r}') = \begin{cases} \dfrac{1}{r^2 \sin\theta} \delta(r - r')\delta(\theta - \theta')\delta(\phi - \phi') & (14\text{-}167a) \\[1em] \dfrac{1}{2r^2} \delta(r - r')\delta(\phi - \phi') & \text{for no } \theta \text{ dependence} \quad (14\text{-}167b) \\[1em] \dfrac{1}{2\pi r^2 \sin\theta} \delta(r - r')\delta(\theta - \theta') & \text{for no } \phi \text{ dependence} \quad (14\text{-}167c) \\[1em] \dfrac{1}{4\pi r^2} \delta(r - r') & \text{for no } \theta \text{ and } \phi \text{ dependence} \quad (14\text{-}167d) \end{cases}$$

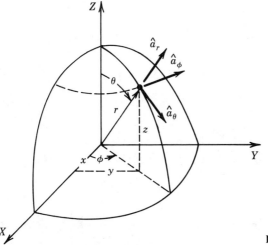

FIGURE 14-10 Spherical coordinate system.

In expanded form the Green's function of (14-166) can now be written using (14-167a) as

$$\frac{1}{r^2}\frac{\partial}{\partial r}\left(r^2\frac{\partial G}{\partial r}\right) + \frac{1}{r^2 \sin\theta}\frac{\partial}{\partial \theta}\left(\sin\theta\frac{\partial G}{\partial \theta}\right) + \frac{1}{r^2 \sin^2\theta}\frac{\partial^2 G}{\partial \phi^2} + \beta_0^2 G$$
$$= \frac{1}{r^2 \sin\theta}\delta(r-r')\delta(\theta-\theta')\delta(\phi-\phi') \qquad (14\text{-}168)$$

Since the spherical harmonics form a complete set for functions of the angles θ and ϕ, the Green's function can be represented by a double summation of an infinite series

$$G(r,\theta,\phi;r',\theta',\phi') = \sum_{n=0}^{\infty}\sum_{m=-n}^{n} g_{mn}(r;r',\theta',\phi') P_m^m(\cos\theta) e^{jm\phi}$$
$$= \sum_{n=0}^{\infty}\sum_{m=-n}^{n} g_{mn}(r;r',\theta',\phi') T_{mn}(\theta,\phi) \qquad (14\text{-}169)$$

where $T_{mn}(\theta,\phi)$ represents the tesseral harmonics of (11-214a) and (11-214b), or

$$T_{mn}(\theta,\phi) = C_{mn} P_n^m(\cos\theta) e^{jm\phi} \qquad (14\text{-}169a)$$

where

$$C_{mn} = \sqrt{\frac{(2n+1)(n-m)!}{4\pi(n+m)!}} \qquad (14\text{-}169b)$$

Multiplying (14-168) by r^2 and then substituting (14-169) into (14-168), we can write that

$$\sum_{n=0}^{\infty}\sum_{m=-n}^{n}\left[T_{mn}\frac{\partial}{\partial r}\left(r^2\frac{\partial g_{mn}}{\partial r}\right) + \frac{g_{mn}}{\sin\theta}\frac{\partial}{\partial \theta}\left(\sin\theta\frac{\partial T_{mn}}{\partial \theta}\right) - \frac{m^2}{\sin^2\theta}g_{mn}T_{mn} + (\beta_0 r)^2 g_{mn}T_{mn}\right]$$
$$= \frac{1}{\sin\theta}\delta(r-r')\delta(\theta-\theta')\delta(\phi-\phi') \qquad (14\text{-}170)$$

Dividing both sides of (14-170) by $g_{mn}T_{mn}$ we can write that

$$\sum_{n=0}^{\infty}\sum_{m=-n}^{n}\left[\frac{1}{g_{mn}}\frac{\partial}{\partial r}\left(r^2\frac{\partial g_{mn}}{\partial r}\right) + \frac{1}{T_{mn}\sin\theta}\frac{\partial}{\partial \theta}\left(\sin\theta\frac{\partial T_{mn}}{\partial \theta}\right) - \frac{m^2}{\sin^2\theta} + (\beta_0 r)^2\right]$$
$$= \frac{1}{g_{mn}T_{mn}\sin\theta}\delta(r-r')\delta(\theta-\theta')\delta(\phi-\phi') \qquad (14\text{-}170a)$$

Using (3-85b) we can write that

$$\frac{1}{T_{mn}\sin\theta}\frac{\partial}{\partial \theta}\left(\sin\theta\frac{\partial T_{mn}}{\partial \theta}\right) - \frac{m^2}{\sin^2\theta} = -n(n+1) \qquad (14\text{-}171)$$

or

$$\frac{1}{\sin\theta}\frac{\partial}{\partial \theta}\left(\sin\theta\frac{\partial T_{mn}}{\partial \theta}\right) + \left[n(n+1) - \left(\frac{m}{\sin\theta}\right)^2\right]T_{mn} = 0 \qquad (14\text{-}171a)$$

Thus (14-170a) reduces, by substituting (14-171) into it and then multiplying through by $g_{mn}T_{mn}$, to

$$\sum_{n=0}^{\infty} \sum_{m=-n}^{n} \left\{ \frac{\partial}{\partial r}\left(r^2 \frac{\partial g_{mn}}{\partial r}\right) + \left[(\beta_0 r)^2 - n(n+1)\right] g_{mn} \right\} T_{mn}$$

$$= \frac{1}{\sin\theta} \delta(r-r')\delta(\theta-\theta')\delta(\phi-\phi') \qquad (14\text{-}172)$$

From (11-214a) through (11-216d) and the definitions of the tesseral harmonics and Legendre functions, it can be shown that the orthogonality conditions of the tesseral harmonics are [19]

$$\int_0^{2\pi}\int_0^{\pi} T_{mn}(\theta,\phi) T_{pq}^*(\theta,\phi) \sin\theta \, d\theta \, d\phi = \delta_{mp}\delta_{nq} \qquad (14\text{-}173)$$

where

$$T_{pq}^*(\theta,\phi) = (-1)^p T_{(-p)q}(\theta,\phi) \qquad (14\text{-}173a)$$

$$\delta_{rs} = \begin{cases} 1 & r=s \\ 0 & r \neq s \end{cases} \qquad (14\text{-}173b)$$

Let
$$g_{mn}(r;r',\theta',\phi') = h_{mn}(r,r')\, T^*(\theta',\phi') \qquad (14\text{-}173c)$$

Multiplying both sides of (14-172) by $T_{pq}^*(\theta,\phi)\sin\theta$, integrating from 0 to π in θ and 0 to 2π in ϕ, and using the orthogonality condition of (14-173), it can be shown that (14-172) reduces to

$$\boxed{\frac{d}{dr}\left(r^2 \frac{\partial h_{mn}}{\partial r}\right) + \left[(\beta_0 r)^2 - n(n+1)\right] h_{mn} = \delta(r-r')} \qquad (14\text{-}174)$$

where the partial derivatives have been replaced by ordinary derivatives.

The function $h_{mn}(r,r')$ satisfies the differential equation 14-174, and its solution can be obtained using the closed form recipe of Section 14.3.1 represented by (14-44c) and (14-45a) through (14-45b). The homogeneous equation 14-174 can be written as

$$\frac{d}{dr}\left(r^2 \frac{dh_{mn}}{dr}\right) + \left[(\beta_0 r)^2 - n(n+1)\right] h_{mn} = 0 \qquad (14\text{-}175)$$

which is of the one-dimensional Sturm–Liouville form of (14-26) or (14-33) where

$$\begin{aligned} p(r) &= r^2 \\ q(r) &= -n(n+1) \\ r(x) &= r^2 \\ \lambda &= \beta^2 \end{aligned} \qquad (14\text{-}176)$$

Equation 14-175 is recognized as being (3-82) or (3-85a) whose solution can be represented by either (3-86a) or (3-86b). We choose here the form of (3-86a) since we need to represent the Green's function within the sphere by standing wave functions. Thus the two solutions of (14-175) can be written as

$$h_{mn}^{(1)} = A_m j_n(\beta_0 r) + B_m y_n(\beta_0 r) \qquad \text{for } r < r' \qquad (14\text{-}177a)$$

$$h_{mn}^{(2)} = C_m j_n(\beta_0 r) + D_m y_n(\beta_0 r) \qquad \text{for } r > r' \qquad (14\text{-}177b)$$

where $j_n(\beta_0 r)$ and $y_n(\beta_0 r)$ are, respectively, spherical Bessel functions of the first and second kind.

Since the Green's function of (14-169) must be finite everywhere, including the origin, then $B_m = 0$ since $y_n(\beta_0 r)$ possesses a singularity at $r = 0$. Also the Green's function must satisfy the boundary condition of (14-166a). Therefore $h_{mn}^{(2)}$ of (14-177b) at $r = a$ reduces to

$$h_{mn}^{(2)}(r = a) = C_m j_n(\beta_0 a) + D_m y_n(\beta_0 a) = 0 \qquad (14\text{-}178)$$

or

$$D_m = -C_m \frac{j_n(\beta_0 a)}{y_n(\beta_0 a)} \qquad (14\text{-}178a)$$

Thus (14-177a) and (14-177b) are reduced to

$$h_{mn}^{(1)} = A_m j_n(\beta_0 r) \qquad \text{for } r < r' \qquad (14\text{-}179a)$$

$$h_{mn}^{(2)} = C_m \left[j_n(\beta_0 r) - \frac{j_n(\beta_0 a)}{y_n(\beta_0 a)} y_n(\beta_0 r) \right] \qquad \text{for } r > r' \qquad (14\text{-}179b)$$

Using (14-44c) where $y_1 = h_{mn}^{(1)}$ and $y_2 = h_{mn}^{(2)}$, we can write the Wronskian as

$$W(r') = \beta_0 A_m C_m \frac{j_n(\beta_0 a)}{y_n(\beta_0 a)} [j_n'(\beta_0 r') y_n(\beta_0 r') - j_n(\beta_0 r') y_n'(\beta_0 r')] \qquad (14\text{-}180)$$

Using the Wronskian for spherical Bessel functions of

$$j_n(\beta_0 r') y_n'(\beta_0 r') - j_n'(\beta_0 r') y_n(\beta_0 r') = \frac{1}{(\beta_0 r')^2} \qquad (14\text{-}180a)$$

reduces (14-180) to

$$W(r') = -\frac{1}{\beta_0} A_m C_m \frac{j_n(\beta_0 a)}{y_n(\beta_0 a)} \frac{1}{(r')^2} \qquad (14\text{-}180b)$$

Finally $h_{mn}(r, r')$ of (14-174) can be written using (14-176), (14-179a) through (14-179b), and (14-180b) by referring to (14-45a) and (14-45b), as

$$h_{mn}(r, r') = \begin{cases} -\beta_0 C_{mn}^2 [j_n(\beta_0 r') y_n(\beta_0 a) - j_n(\beta_0 a) y_n(\beta_0 r')] \dfrac{j_n(\beta_0 r)}{j_n(\beta_0 a)} \\ \qquad\qquad\qquad\qquad\qquad \text{for } r < r' \qquad (14\text{-}181\text{a}) \\ \\ -\beta_0 C_{mn}^2 [j_n(\beta_0 r) y_n(\beta_0 a) - j_n(\beta_0 a) y_n(\beta_0 r)] \dfrac{j_n(\beta_0 r')}{j_n(\beta_0 a)} \\ \qquad\qquad\qquad\qquad\qquad \text{for } r > r' \qquad (14\text{-}181\text{b}) \end{cases}$$

Thus the Green's function of (14-169) can be written as

$$G(r, \theta, \phi; r', \theta', \phi') = -\beta_0 \begin{cases} \displaystyle\sum_{n=0}^{\infty} \sum_{m=-n}^{n} (-1)^m C_{mn}^2 [j_n(\beta_0 r') y_n(\beta_0 a) - j_n(\beta_0 a) y_n(\beta_0 r')] \\ \quad \times \dfrac{j_n(\beta_0 r)}{j_n(\beta_0 a)} P_n^m(\cos\theta) P_n^{-m}(\cos\theta') e^{jm(\phi-\phi')} \quad (14\text{-}182\text{a}) \\ \qquad\qquad\qquad\qquad\qquad \text{for } r < r' \\ \\ \displaystyle\sum_{n=0}^{\infty} \sum_{m=-n}^{n} (-1)^m C_{mn}^2 [j_n(\beta_0 r') y_n(\beta_0 a) - j_n(\beta_0 a) y_n(\beta_0 r')] \\ \quad \times \dfrac{j_n(\beta_0 r)}{j_n(\beta_0 a)} P_n^m(\cos\theta) P_n^{-m}(\cos\theta') e^{jm(\phi-\phi')} \quad (14\text{-}182\text{b}) \\ \qquad\qquad\qquad\qquad\qquad \text{for } r > r' \end{cases}$$

14.7 DYADIC GREEN'S FUNCTIONS

The Green's function development of the previous sections can be used for the solution of electromagnetic problems that satisfy the scalar wave equation. The most general Green's function development and electromagnetic field solution, for problems that satisfy the vector wave equation, will be to use *vectors and dyadics* [14–18]. Before we briefly discuss such a procedure, let us first introduce and define dyadics.

14.7.1 Dyadics

Vectors and dyadics are used, in general, to describe linear transformations *within a given orthogonal coordinate system*, and they simplify the manipulations of mathematical relations, compared to using tensors. For electromagnetic problems, where linear transformations between sources and fields *within a given orthogonal coordinate system* are often necessary, vectors and dyadics are very convenient to use.

A *dyad* is defined by the juxtaposition **AB** of the vectors **A** and **B**, with no dot or cross product between them. In general a dyad has nine terms and in matrix form can be represented by

$$(\mathbf{AB}) = \begin{pmatrix} A_1 B_1 & A_1 B_2 & A_1 B_3 \\ A_2 B_1 & A_2 B_2 & A_2 B_3 \\ A_3 B_1 & A_3 B_2 & A_3 B_3 \end{pmatrix} \quad (14\text{-}183)$$

A *dyadic* $\overline{\mathbf{D}}$ can be defined by the sum of N dyads. That is

$$\overline{\mathbf{D}} = \sum_{n=1}^{N} \mathbf{A}^n \mathbf{B}^n \quad (14\text{-}184)$$

In general no more than three dyads are required to represent a dyadic, that is $N_{max} = 3$.

Let us now define the vectors **A**, **C**, $\mathbf{D}_1$, $\mathbf{D}_2$, and $\mathbf{D}_3$ in a general coordinate system with unit vectors $\hat{a}_1$, $\hat{a}_2$, and $\hat{a}_3$. That is,

$$\mathbf{A} = \hat{a}_1 A_1 + \hat{a}_2 A_2 + \hat{a}_3 A_3 \quad (14\text{-}185a)$$
$$\mathbf{C} = \hat{a}_1 C_1 + \hat{a}_2 C_2 + \hat{a}_3 C_3 \quad (14\text{-}185b)$$
$$\mathbf{D}_1 = \hat{a}_1 D_{11} + \hat{a}_2 D_{12} + \hat{a}_3 D_{13} \quad (14\text{-}185c)$$
$$\mathbf{D}_2 = \hat{a}_1 D_{21} + \hat{a}_2 D_{22} + \hat{a}_3 D_{23} \quad (14\text{-}185d)$$
$$\mathbf{D}_3 = \hat{a}_1 D_{31} + \hat{a}_2 D_{32} + \hat{a}_3 D_{33} \quad (14\text{-}185e)$$

Let us now write that

$$\mathbf{C} = (\mathbf{A} \cdot \hat{a}_1)\mathbf{D}_1 + (\mathbf{A} \cdot \hat{a}_2)\mathbf{D}_2 + (\mathbf{A} \cdot \hat{a}_3)\mathbf{D}_3$$
$$= \mathbf{A} \cdot (\hat{a}_1 \mathbf{D}_1) + \mathbf{A} \cdot (\hat{a}_2 \mathbf{D}_2) + \mathbf{A} \cdot (\hat{a}_3 \mathbf{D}_3)$$
$$\mathbf{C} = \mathbf{A} \cdot (\hat{a}_1 \mathbf{D}_1 + \hat{a}_2 \mathbf{D}_2 + \hat{a}_3 \mathbf{D}_3) \quad (14\text{-}186)$$

or

$$\mathbf{C} = \mathbf{A} \cdot \overline{\mathbf{D}} \quad (14\text{-}186a)$$

where

$$\overline{\mathbf{D}} = \hat{a}_1 \mathbf{D}_1 + \hat{a}_2 \mathbf{D}_2 + \hat{a}_3 \mathbf{D}_3 \quad (14\text{-}186b)$$

In (14-186) through (14-186b) $\overline{\mathbf{D}}$ is a dyadic, and it is defined by the sum of the three dyads $\hat{a}_n \mathbf{D}_n$, $n = 1 - 3$. In matrix form (14-186) or (14-186a) can be written as

$$(C_1 \ C_2 \ C_3) = (A_1 \ A_2 \ A_3) \begin{pmatrix} D_{11} & D_{12} & D_{13} \\ D_{21} & D_{22} & D_{23} \\ D_{31} & D_{32} & D_{33} \end{pmatrix} \quad (14\text{-}187)$$

where the dyadic $\overline{\mathbf{D}}$ has nine elements.

Just like vectors, dyadics satisfy a number of identities involving dot and cross products, differentiations, and integrations. The uninformed reader should refer to the literature [10–13] for such relations.

14.7.2 Green's Functions

In electromagnetics it is often desirable to solve, using the Green's functions approach, the linear vector problem of

$$\boxed{\mathscr{L}\mathbf{h} = \mathbf{f}} \tag{14-188}$$

where $\mathscr{L}$ is a differential operator. Equation 14-188 is a more general and vector representation of (14-27). It should be noted here that the solution of (14-188) *cannot*, in general, be represented by

$$\mathbf{h}(\mathbf{r}) \neq \iiint_V \mathbf{f}(\mathbf{r}')G(\mathbf{r},\mathbf{r}')\,dv' \tag{14-189}$$

where $G(\mathbf{r},\mathbf{r}')$ is a single scalar Green's function. The relation of (14-189) would imply that a component of the source **f** parallel to a given axis produces a response (field) **h** parallel to the same axis. This, in general, is not true.

A more appropriate representation of the solution of (14-188) in a rectangular coordinate system will be to write that

$$h_x(\mathbf{r}) = \iiint_V \left[f_x(\mathbf{r}')G_{xx}(\mathbf{r},\mathbf{r}') + f_y(\mathbf{r}')G_{xy}(\mathbf{r},\mathbf{r}') + f_z(\mathbf{r}')G_{xz}(\mathbf{r},\mathbf{r}') \right] dv' \tag{14-190a}$$

$$h_y(\mathbf{r}) = \iiint_V \left[f_x(\mathbf{r}')G_{yx}(\mathbf{r},\mathbf{r}') + f_y(\mathbf{r}')G_{yy}(\mathbf{r},\mathbf{r}') + f_z(\mathbf{r}')G_{yz}(\mathbf{r},\mathbf{r}') \right] dv' \tag{14-190b}$$

$$h_z(\mathbf{r}) = \iiint_V \left[f_x(\mathbf{r}')G_{zx}(\mathbf{r},\mathbf{r}') + f_y(\mathbf{r}')G_{zy}(\mathbf{r},\mathbf{r}') + f_z(\mathbf{r}')G_{zz}(\mathbf{r},\mathbf{r}') \right] dv' \tag{14-190c}$$

which in a more compact form can be written as

$$\boxed{\mathbf{h}(\mathbf{r}) = \iiint_V \left[f_x(\mathbf{r}')\mathbf{G}_x(\mathbf{r},\mathbf{r}') + f_y(\mathbf{r}')\mathbf{G}_y(\mathbf{r},\mathbf{r}') + f_z(\mathbf{r}')\mathbf{G}_z(\mathbf{r},\mathbf{r}') \right] dv'} \tag{14-191}$$

where as in (14-185c) through (14-185e)

$$\mathbf{G}_x(\mathbf{r},\mathbf{r}') = \hat{a}_x G_{xx}(\mathbf{r},\mathbf{r}') + \hat{a}_y G_{yx}(\mathbf{r},\mathbf{r}') + \hat{a}_z G_{zx}(\mathbf{r},\mathbf{r}') \tag{14-191a}$$

$$\mathbf{G}_y(\mathbf{r},\mathbf{r}') = \hat{a}_x G_{xy}(\mathbf{r},\mathbf{r}') + \hat{a}_y G_{yy}(\mathbf{r},\mathbf{r}') + \hat{a}_z G_{zy}(\mathbf{r},\mathbf{r}') \tag{14-191b}$$

$$\mathbf{G}_z(\mathbf{r},\mathbf{r}') = \hat{a}_x G_{xz}(\mathbf{r},\mathbf{r}') + \hat{a}_y G_{yz}(\mathbf{r},\mathbf{r}') + \hat{a}_z G_{zz}(\mathbf{r},\mathbf{r}') \tag{14-191c}$$

In (14-190a) through (14-191c) the $G_{ij}(\mathbf{r},\mathbf{r}')$'s are the elements of the dyadic $\overline{\mathbf{G}}(\mathbf{r},\mathbf{r}')$ which is referred to here as the *dyadic Green's function*. In (14-191) through (14-191c) the $\mathbf{G}_i(\mathbf{r},\mathbf{r}')$'s are the column vectors of the dyadic Green's function $\overline{\mathbf{G}}(\mathbf{r},\mathbf{r}')$.

Using the notation of (14-186) through (14-186b), the solution of (14-191) can also be written as

$$\mathbf{h}(\mathbf{r}) = \iiint_V \left\{ [\mathbf{f}(\mathbf{r}') \cdot \hat{a}_x]\mathbf{G}_x(\mathbf{r},\mathbf{r}') + [\mathbf{f}(\mathbf{r}') \cdot \hat{a}_y]\mathbf{G}_y(\mathbf{r},\mathbf{r}') + [\mathbf{f}(\mathbf{r}') \cdot \hat{a}_z]\mathbf{G}_z(\mathbf{r},\mathbf{r}') \right\} dv'$$

$$= \iiint_V \mathbf{f}(\mathbf{r}') \cdot \left[\hat{a}_x \mathbf{G}_x(\mathbf{r},\mathbf{r}') + \hat{a}_y \mathbf{G}_y(\mathbf{r},\mathbf{r}') + \hat{a}_z \mathbf{G}_z(\mathbf{r},\mathbf{r}') \right] dv'$$

$$\boxed{\mathbf{h}(\mathbf{r}) = \iiint_V \mathbf{f}(\mathbf{r}) \cdot \overline{\mathbf{G}}(\mathbf{r},\mathbf{r}')\,dv'} \tag{14-192}$$

where $\overline{\mathbf{G}}(\mathbf{r}, \mathbf{r}')$ is the dyadic Green's function of

$$\overline{\mathbf{G}}(\mathbf{r}, \mathbf{r}') = \hat{a}_x \mathbf{G}_x(\mathbf{r}, \mathbf{r}') + \hat{a}_y \mathbf{G}_y(\mathbf{r}, \mathbf{r}') + \hat{a}_z \mathbf{G}_z(\mathbf{r}, \mathbf{r}') \quad (14\text{-}192a)$$

The dyadic Green's function $\overline{\mathbf{G}}(\mathbf{r}, \mathbf{r}')$ can be found by first finding the vectors $\mathbf{G}_x(\mathbf{r}, \mathbf{r}')$, $\mathbf{G}_y(\mathbf{r}, \mathbf{r}')$, and $\mathbf{G}_z(\mathbf{r}, \mathbf{r}')$ each satisfying the homogeneous form of the partial differential of (14-188), or

$$\mathscr{L} \mathbf{G}_x(\mathbf{r}, \mathbf{r}') = \hat{a}_x \delta(\mathbf{r} - \mathbf{r}') \quad (14\text{-}193a)$$

$$\mathscr{L} \mathbf{G}_y(\mathbf{r}, \mathbf{r}') = \hat{a}_y \delta(\mathbf{r} - \mathbf{r}') \quad (14\text{-}193b)$$

$$\mathscr{L} \mathbf{G}_z(\mathbf{r}, \mathbf{r}') = \hat{a}_z \delta(\mathbf{r} - \mathbf{r}') \quad (14\text{-}193c)$$

and the appropriate boundary conditions, and then using (14-192a) to form the dyadic Green's function. Any of the methods of the previous sections can be used to find the Green's function of (14-193a) through (14-193c).

An example for the potential use of the dyadic Green's function is the solution for the electric and magnetic fields that is due to a source represented by the electric current density **J**. According to (6-32a) and (6-32b) the electric and magnetic fields can be written as

$$\mathbf{H}(\mathbf{r}) = \frac{1}{\mu} \nabla \times \mathbf{A} \quad (14\text{-}194a)$$

$$\mathbf{E}(\mathbf{r}) = -j\omega \mathbf{A} - j \frac{1}{\omega \mu \varepsilon} \nabla(\nabla \cdot \mathbf{A}) \quad (14\text{-}194b)$$

where the vector potential **A** satisfies the partial differential equation 6-30 or

$$\nabla^2 \mathbf{A} + \beta^2 \mathbf{A} = -\mu \mathbf{J} \quad (14\text{-}195)$$

Using the dyadic Green's function approach, the vector potential **A** can be found using

$$\mathbf{A} = -\mu \iiint_V \mathbf{J}(\mathbf{r}') \cdot \overline{\mathbf{G}}(\mathbf{r}, \mathbf{r}') \, dv' \quad (14\text{-}196)$$

where the dyadic Green's function must satisfy the partial differential equation

$$\nabla^2 \overline{\mathbf{G}} + \beta^2 \overline{\mathbf{G}} = \overline{\delta}(\mathbf{r} - \mathbf{r}') \quad (14\text{-}197)$$

and the appropriate boundary conditions.

Because of the complexity for the development of the dyadic Green's function, it will not be pursued any further here. The interested reader is referred to the literature [10–18] for more details.

REFERENCES

1. R. Courant and D. Hilbert, *Methods of Mathematical Physics*, vol. I, Wiley, New York, 1937.
2. A. Wester, *Partial Differential Equations of Mathematical Physics*, S. Plimpton (Ed.), Second Edition, Hafner, New York, 1947.
3. P. M. Morse and H. Feshbach, *Methods of Theoretical Physics*, vols. I and II, McGraw-Hill, New York, 1953.

4. B. Friedman, *Principles and Techniques of Applied Mathematics*, Wiley, New York, 1956.
5. J. Dettman, *Mathematical Methods in Physics and Engineering*, McGraw-Hill, New York, 1962.
6. J. D. Jackson, *Classical Electrodynamics*, Wiley, New York, 1962.
7. H. W. Wyld, *Mathematical Methods for Physics*, Benjamin/Cummings, Menlo Park, CA, 1976.
8. I. Stakgold, *Green's Functions and Boundary Value Problems*, Wiley, New York, 1979.
9. R. E. Collin, *Field Theory of Guided Waves*, McGraw-Hill, New York, 1960.
10. J. Van Bladel, *Electromagnetic Fields*, McGraw-Hill, New York, 1964.
11. C.-T. Tai, *Dyadic Green's Functions in Electromagnetic Theory*, Intext Educational Publishers, Scranton, PA, 1971.
12. L. B. Felsen and N. Marcuvitz, *Radiation and Scattering of Waves*, Prentice-Hall, Englewood Cliffs, NJ, 1973.
13. D. C. Stinson, *Intermediate Mathematics of Electromagnetics*, Prentice-Hall, Englewood Cliffs, NJ, 1976.
14. C.-T. Tai, "On the eigenfunction expansion of dyadic Green's functions," *Proc. IEEE*, vol. 61, pp. 480–481, April 1973.
15. C.-T. Tai and P. Rozenfeld, "Different representations of dyadic Green's functions for a rectangular cavity," *IEEE Trans. Microwave Theory Tech.*, vol. MTT-24, pp. 597–601, September 1976.
16. A. Q. Howard, Jr., "On the longitudinal component of the Green's function dyadic," *Proc. IEEE*, vol. 62, pp. 1704–1705, December 1974.
17. A. D. Yaghjian, "Electric dyadic Green's functions in the source region," *Proc. IEEE*, vol. 68, no. 2, pp. 248–263, February 1980.
18. P. H. Pathak, "On the eigenfunction expansion of electromagnetic dyadic Green's functions," *IEEE Trans. Antennas Propagat.*, vol. AP-31, pp. 837–846, November 1983.
19. R. F. Harrington, *Time-Harmonic Electromagnetic Fields*, McGraw-Hill, New York, 1961.

PROBLEMS

14.1. Using the procedure of Section 14.3.1, as represented by (14-44c) and (14-45a) through (14-45b), derive the Green's function of the mechanics problem of Section 14.2.2 as given by (14-24a) and (14-24b).

14.2. The displacement $u(x)$ of a string of length ℓ subjected to a cosinusoidal force
$$f(x,t) = f(x)e^{+j\omega t}$$
is determined by
$$u(x,t) = u(x)e^{+j\omega t}$$
where $u(x)$ satisfies the differential equation
$$\frac{d^2 u(x)}{dx^2} + \beta^2 u(x) = f(x), \qquad \beta^2 = \omega/c$$
Assuming the ends of the string are fixed
$$u(x=0) = u(x=\ell) = 0$$
determine in closed form the Green's function of the system.

14.3. Derive in closed form the Green's function representing the electric potential distribution $V(x)$ of an infinite length, finite width conducting strip shown in Figure P14-3 subject to the boundary conditions
$$V(x=0) = V(x=w) = 0$$

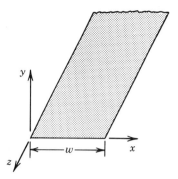

FIGURE P14-3

14.4. Derive in series form the Green's function of Problem 14.3.

14.5. Two infinite radial plates with an interior angle of α, as shown in Figure P14-5, are both maintained at a potential of $V = 0$. Determine in closed form the Green's function for the electric potential distribution between the plates.

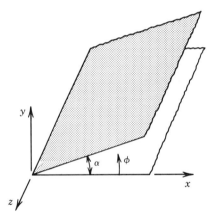

FIGURE P14-5

14.6. Repeat Problem 14.5 when the two plates are isolated from each other with the plate of $\phi = 0$ grounded while that at $\phi = \alpha$ is maintained at a constant potential V_0.

14.7. The top side of a rectangular cross section, infinite length pipe is insulated from the other three, and it is maintained at a constant potential V_0. The other three are held at a grounded potential of zero as shown in Figure P14-7. Determine in closed form the Green's function for the electric potential distribution within the pipe.

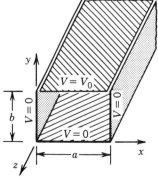

FIGURE P14-7

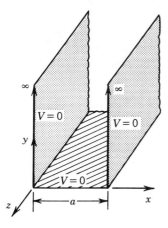

FIGURE P14-8

14.8. The three sides of an infinite length and infinite height trough are maintained at a grounded potential of zero as shown in Figure P14-8. The width of the trough is a. Determine in closed form the Green's function for the electric field distribution within the trough.

14.9. Derive the Green's function of (14-83a) through (14-83b) by initially choosing a solution for (14-74) that satisfies the boundary condition at $y = 0$ and $y = b$.

14.10. An infinitely long conducting wire positioned at $\rho = \rho'$, $\phi = \phi'$ is circumscribed by a grounded ($V = 0$) electric conducting circular cylinder of radius a and infinite length, as shown in Figure 14-4. Derive in closed-form the Green's function for the potential distribution within the cylinder. Assume free space within the cylinder.

14.11. Derive in closed form the Green's function of the time-harmonic problem represented by (14-97) subject to the boundary conditions of (14-97a) and (14-97b). This would be an alternate representation of (14-106).

14.12. An annular microstrip antenna fed by a coaxial line is comprised of an annular conducting circular strip, with inner and outer radii of a and b, placed on the top

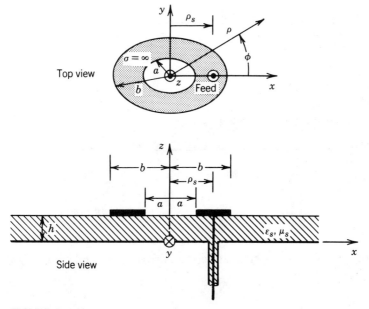

FIGURE P14-12

surface of a lossless substrate of height h and electrical parameters ε_s, μ_s as shown in Figure P14-12. The substrate is supported by a ground plane. Assuming the microstrip antenna can be modeled as a cavity with ideal open circuits of vanishing tangential magnetic fields at the inner ($\rho = a$) and outer ($\rho = b$) edges

$$H_\phi(\rho = a, 0 \leq \phi \leq 2\pi, 0 \leq z \leq h) = H_\phi(\rho = b, 0 \leq \phi \leq 2\pi, 0 \leq z \leq h) = 0$$

and vanishing tangential electric fields on its top and bottom sides, determine the Green's function for the TM^z modes (subject to $\nabla \cdot \mathbf{J} = 0$) with independent z variations within the cavity. For such modes the electric field must have only a z component of

$$\mathbf{E} = \hat{a}_z E_z(\rho, \phi)$$

which must satisfy the partial differential equation

$$\nabla^2 E_z + \beta^2 E_z = j\omega\mu J_z(\rho_f, \phi_f) \qquad \beta^2 = \omega^2 \mu \varepsilon$$

14.13. Repeat Problem 14.12 for an annular sector microstrip antenna whose geometry is shown in Figure P14-13.

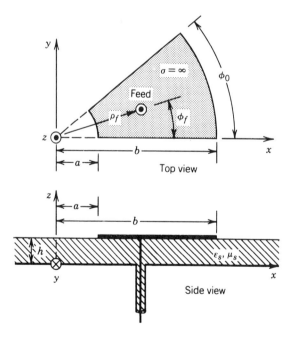

FIGURE P14-13

14.14. Repeat Problem 14.12 for a circular sector microstrip antenna whose geometry is shown in Figure P14-14.

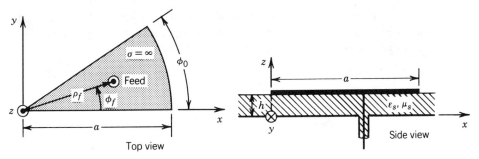

FIGURE P14-14

14.15. Derive the Green's function represented by (14-165a) and (14-165b) in terms of the two space variable series expansion method using orthonormal expansion functions.

14.16. An infinite length electric line source of constant current I_e is placed near a conducting circular cylinder of infinite length, as shown in Figure P14-16. Derive in closed form the Green's function for the fields in the space surrounding the cylinder.

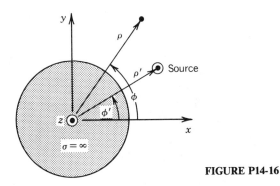

FIGURE P14-16

14.17. Repeat Problem 14.16 for a magnetic line source of constant current I_m.

14.18. An infinite length electric line source of constant electric current I_e is placed near a two-dimensional conducting wedge of interior angle 2α as shown in Figure P14-18. Derive in closed form the Green's function for the fields in the space surrounding the wedge. Compare with the expressions of (11-182a) and (11-182b).

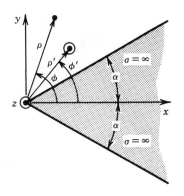

FIGURE P14-18

14.19. Repeat Problem 14.18 for an infinite length magnetic line source of constant magnetic current I_m. Compare the answers with the expressions of (11-192a) through (11-192b) or (11-193a) through (11-193b).

14.20. A point source placed at x', y', z' is radiating in free space as shown in Figure P14-20.

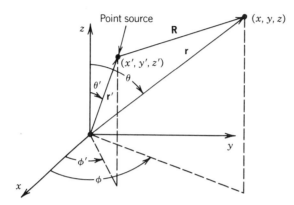

FIGURE P14-20

(a) Derive its Green's function of

$$G = -\frac{1}{4\pi}\frac{e^{-j\beta R}}{R}$$

where R is the radial distance from the point source to the observation point.

(b) By using the integral of (11-28a) or

$$\int_{-\infty}^{+\infty}\frac{e^{-j\beta R}}{R}\,dz = -j\pi H_0^{(2)}(\beta R)$$

show that the three-dimensional Green's function of the point source reduces to the two-dimensional Green's function of the line source derived in Example 14-5.

APPENDIX I

IDENTITIES

I.1 TRIGONOMETRIC

1. Sum or difference:
   a. $\sin(x+y) = \sin x \cos y + \cos x \sin y$
   b. $\sin(x-y) = \sin x \cos y - \cos x \sin y$
   c. $\cos(x+y) = \cos x \cos y - \sin x \sin y$
   d. $\cos(x-y) = \cos x \cos y + \sin x \sin y$
   e. $\tan(x+y) = \dfrac{\tan x + \tan y}{1 - \tan x \tan y}$
   f. $\tan(x-y) = \dfrac{\tan x - \tan y}{1 + \tan x \tan y}$
   g. $\sin^2 x + \cos^2 x = 1$
   h. $\tan^2 x - \sec^2 x = -1$
   i. $\cot^2 x - \csc^2 x = -1$

2. Sum or difference into products:
   a. $\sin x + \sin y = 2 \sin \tfrac{1}{2}(x+y) \cos \tfrac{1}{2}(x-y)$
   b. $\sin x - \sin y = 2 \cos \tfrac{1}{2}(x+y) \sin \tfrac{1}{2}(x-y)$
   c. $\cos x + \cos y = 2 \cos \tfrac{1}{2}(x+y) \cos \tfrac{1}{2}(x-y)$
   d. $\cos x - \cos y = -2 \sin \tfrac{1}{2}(x+y) \sin \tfrac{1}{2}(x-y)$

3. Products into sum or difference:
   a. $2 \sin x \cos y = \sin(x+y) + \sin(x-y)$
   b. $2 \cos x \sin y = \sin(x+y) - \sin(x-y)$
   c. $2 \cos x \cos y = \cos(x+y) + \cos(x-y)$
   d. $2 \sin x \sin y = -\cos(x+y) + \cos(x-y)$

4. Double and half-angles:
   a. $\sin 2x = 2 \sin x \cos x$
   b. $\cos 2x = \cos^2 x - \sin^2 x = 2\cos^2 x - 1 = 1 - 2\sin^2 x$

918 APPENDIX I IDENTITIES

c. $\tan 2x = \dfrac{2 \tan x}{1 - \tan^2 x}$

d. $\sin \dfrac{1}{2}x = \pm \sqrt{\dfrac{1 - \cos x}{2}}$ or $2 \sin^2 \theta = 1 - \cos 2\theta$

e. $\cos \dfrac{1}{2}x = \pm \sqrt{\dfrac{1 + \cos x}{2}}$ or $2 \cos^2 \theta = 1 + \cos 2\theta$

f. $\tan \dfrac{1}{2}x = \pm \sqrt{\dfrac{1 - \cos x}{1 + \cos x}} = \dfrac{\sin x}{1 + \cos x} = \dfrac{1 - \cos x}{\sin x}$

5. Series:

a. $\sin x = \dfrac{e^{jx} - e^{-jx}}{2j} = x - \dfrac{x^3}{3!} + \dfrac{x^5}{5!} - \dfrac{x^7}{7!} + \cdots$

b. $\cos x = \dfrac{e^{jx} + e^{-jx}}{2} = 1 - \dfrac{x^2}{2!} + \dfrac{x^4}{4!} - \dfrac{x^6}{6!} + \cdots$

c. $\tan x = \dfrac{e^{jx} - e^{-jx}}{j(e^{jx} + e^{-jx})} = x + \dfrac{x^3}{3} + \dfrac{2x^5}{15} + \dfrac{17x^7}{315} + \cdots$

I.2 HYPERBOLIC

1. Definitions:
   a. Hyperbolic sine: $\sinh x = \tfrac{1}{2}(e^x - e^{-x})$
   b. Hyperbolic cosine: $\cosh x = \tfrac{1}{2}(e^x + e^{-x})$
   c. Hyperbolic tangent: $\tanh x = \dfrac{\sinh x}{\cosh x}$
   d. Hyperbolic cotangent: $\coth x = \dfrac{1}{\tanh x} = \dfrac{\cosh x}{\sinh x}$
   e. Hyperbolic secant: $\text{sech } x = \dfrac{1}{\cosh x}$
   f. Hyperbolic cosecant: $\text{csch } x = \dfrac{1}{\sinh x}$

2. Sum or difference:
   a. $\cosh(x + y) = \cosh x \cosh y + \sinh x \sinh y$
   b. $\sinh(x - y) = \sinh x \cosh y - \cosh x \sinh y$
   c. $\cosh(x - y) = \cosh x \cosh y - \sinh x \sinh y$
   d. $\tanh(x + y) = \dfrac{\tanh x + \tanh y}{1 + \tanh x \tanh y}$
   e. $\tanh(x - y) = \dfrac{\tanh x - \tanh y}{1 - \tanh x \tanh y}$
   f. $\cosh^2 x - \sinh^2 x = 1$
   g. $\tanh^2 x + \text{sech}^2 x = 1$
   h. $\coth^2 x - \text{csch}^2 x = 1$
   i. $\cosh(x \pm jy) = \cosh x \cos y \pm j \sinh x \sin y$
   j. $\sinh(x \pm jy) = \sinh x \cos y \pm j \cosh x \sin y$

3. Series:

   a. $\sinh x = \dfrac{e^x - e^{-x}}{2} = x + \dfrac{x^3}{3!} + \dfrac{x^5}{5!} + \dfrac{x^7}{7!} + \cdots$

   b. $\cosh x = \dfrac{e^x + e^{-x}}{2} = 1 + \dfrac{x^2}{2!} + \dfrac{x^4}{4!} + \dfrac{x^6}{6!} + \cdots$

   c. $e^x = 1 + x + \dfrac{x^2}{2!} + \dfrac{x^3}{3!} + \dfrac{x^4}{4!} + \cdots$

I.3 LOGARITHMIC

1. $\log_b (MN) = \log_b M + \log_b N$
2. $\log_b (M/N) = \log_b M - \log_b N$
3. $\log_b (1/N) = -\log_b N$
4. $\log_b (M^n) = n \log_b M$
5. $\log_b (M^{1/n}) = \dfrac{1}{n} \log_b M$
6. $\log_a N = \log_b N \cdot \log_a b = \log_b N / \log_b a$
7. $\log_e N = \log_{10} N \cdot \log_e 10 = 2.302585 \log_{10} N$
8. $\log_{10} N = \log_e N \cdot \log_{10} e = 0.434294 \log_e N$

APPENDIX II

VECTOR ANALYSIS

II.1 VECTOR TRANSFORMATIONS

In this appendix we will indicate the vector transformations from rectangular-to-cylindrical (and vice versa), from cylindrical-to-spherical (and vice versa), and from rectangular-to-spherical (and vice versa). The three coordinate systems are shown in Figure II-1.

II.1.1 Rectangular-to-Cylindrical (and Vice Versa)

The coordinate transformation from rectangular (x, y, z) to cylindrical (ρ, ϕ, z) is given, referring to Figure II-1(b):

$$x = \rho \cos \phi$$
$$y = \rho \sin \phi$$
$$z = z \tag{II-1}$$

In the rectangular coordinate system we express a vector **A** as

$$\mathbf{A} = \hat{a}_x A_x + \hat{a}_y A_y + \hat{a}_z A_z \tag{II-2}$$

where $\hat{a}_x, \hat{a}_y, \hat{a}_z$ are the unit vectors and A_x, A_y, A_z are the components of the vector **A** in the rectangular coordinate system. We wish to write **A** as

$$\mathbf{A} = \hat{a}_\rho A_\rho + \hat{a}_\phi A_\phi + \hat{a}_z A_z \tag{II-3}$$

where $\hat{a}_\rho, \hat{a}_\phi, \hat{a}_z$ are the unit vectors and A_ρ, A_ϕ, A_z are the vector components in the cylindrical coordinate system. The z axis is common to both of them.

Referring to Figure II-2, we can write

$$\hat{a}_x = \hat{a}_\rho \cos \phi - \hat{a}_\phi \sin \phi$$
$$\hat{a}_y = \hat{a}_\rho \sin \phi + \hat{a}_\phi \cos \phi$$
$$\hat{a}_z = \hat{a}_z \tag{II-4}$$

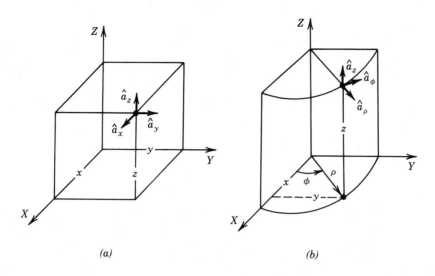

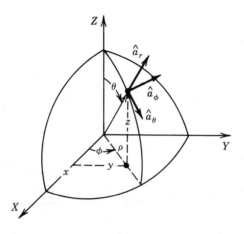

FIGURE II-1 (*a*) Rectangular, (*b*) cylindrical, and (*c*) spherical coordinate systems. (*Source:* C. A. Balanis, *Antenna Theory: Analysis and Design*; copyright © 1982, John Wiley & Sons, Inc.; reprinted by permission of John Wiley & Sons, Inc.)

Using (II-4) reduces (II-2) to

$$\mathbf{A} = (\hat{a}_\rho \cos\phi - \hat{a}_\phi \sin\phi)A_x + (\hat{a}_\rho \sin\phi + \hat{a}_\phi \cos\phi)A_y + \hat{a}_z A_z$$

$$\mathbf{A} = \hat{a}_\rho(A_x \cos\phi + A_y \sin\phi) + \hat{a}_\phi(-A_x \sin\phi + A_y \cos\phi) + \hat{a}_z A_z \quad \text{(II-5)}$$

which when compared with (II-3) leads to

$$A_\rho = A_x \cos\phi + A_y \sin\phi$$
$$A_\phi = -A_x \sin\phi + A_y \cos\phi$$
$$A_z = A_z \quad \text{(II-6)}$$

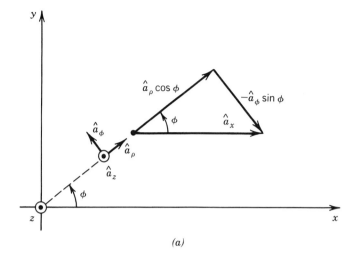

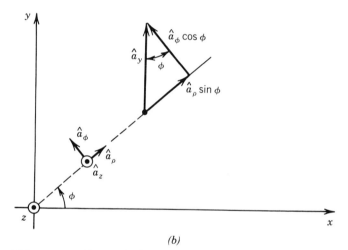

FIGURE II-2 Geometrical representation of transformation between unit vectors of rectangular and cylindrical coordinate systems. (*Source:* C. A. Balanis, *Antenna Theory: Analysis and Design*; copyright © 1982, John Wiley & Sons, Inc.; reprinted by permission of John Wiley & Sons, Inc.) (*a*) Geometry for unit vector $\hat{a}_x$. (*b*) Geometry for unit vector $\hat{a}_y$.

In matrix form, (II-6) can be written as

$$\begin{pmatrix} A_\rho \\ A_\phi \\ A_z \end{pmatrix} = \begin{pmatrix} \cos\phi & \sin\phi & 0 \\ -\sin\phi & \cos\phi & 0 \\ 0 & 0 & 1 \end{pmatrix} \begin{pmatrix} A_x \\ A_y \\ A_z \end{pmatrix} \quad \text{(II-6a)}$$

where

$$[A]_{rc} = \begin{bmatrix} \cos\phi & \sin\phi & 0 \\ -\sin\phi & \cos\phi & 0 \\ 0 & 0 & 1 \end{bmatrix} \quad \text{(II-6b)}$$

is the transformation matrix for rectangular-to-cylindrical components.

Since $[A]_{rc}$ is an orthonormal matrix (its inverse is equal to its transpose), we can write the transformation matrix for cylindrical-to-rectangular components as

$$[A]_{cr} = [A]_{rc}^{-1} = [A]_{rc}^{t} = \begin{bmatrix} \cos\phi & -\sin\phi & 0 \\ \sin\phi & \cos\phi & 0 \\ 0 & 0 & 1 \end{bmatrix} \quad \text{(II-7)}$$

or

$$\begin{pmatrix} A_x \\ A_y \\ A_z \end{pmatrix} = \begin{pmatrix} \cos\phi & -\sin\phi & 0 \\ \sin\phi & \cos\phi & 0 \\ 0 & 0 & 1 \end{pmatrix} \begin{pmatrix} A_\rho \\ A_\phi \\ A_z \end{pmatrix} \quad \text{(II-7a)}$$

or

$$A_x = A_\rho \cos\phi - A_\phi \sin\phi$$
$$A_y = A_\rho \sin\phi + A_\phi \cos\phi$$
$$A_z = A_z \quad \text{(II-7b)}$$

II.1.2 Cylindrical-to-Spherical (and Vice Versa)

Referring to Figure II-1(c), we can write that the cylindrical and spherical coordinates are related by

$$\rho = r\sin\theta$$
$$z = r\cos\theta \quad \text{(II-8)}$$

In a geometrical approach similar to the one employed in the previous section, we can show that the cylindrical-to-spherical transformation of vector components is given by

$$A_r = A_\rho \sin\theta + A_z \cos\theta$$
$$A_\theta = A_\rho \cos\theta - A_z \sin\theta$$
$$A_\phi = A_\phi \quad \text{(II-9)}$$

or in matrix form by

$$\begin{pmatrix} A_r \\ A_\theta \\ A_\phi \end{pmatrix} = \begin{pmatrix} \sin\theta & 0 & \cos\theta \\ \cos\theta & 0 & -\sin\theta \\ 0 & 1 & 0 \end{pmatrix} \begin{pmatrix} A_\rho \\ A_\phi \\ A_z \end{pmatrix} \quad \text{(II-9a)}$$

Thus the cylindrical-to-spherical transformation matrix can be written as

$$[A]_{cs} = \begin{bmatrix} \sin\theta & 0 & \cos\theta \\ \cos\theta & 0 & -\sin\theta \\ 0 & 1 & 0 \end{bmatrix} \quad \text{(II-9b)}$$

The $[A]_{cs}$ matrix is also orthonormal so that its inverse is given by

$$[A]_{sc} = [A]_{cs}^{-1} = [A]_{cs}^{t} = \begin{bmatrix} \sin\theta & \cos\theta & 0 \\ 0 & 0 & 1 \\ \cos\theta & -\sin\theta & 0 \end{bmatrix} \quad \text{(II-10)}$$

and the spherical-to-cylindrical transformation is accomplished by

$$\begin{pmatrix} A_\rho \\ A_\phi \\ A_z \end{pmatrix} = \begin{pmatrix} \sin\theta & \cos\theta & 0 \\ 0 & 0 & 1 \\ \cos\theta & -\sin\theta & 0 \end{pmatrix} \begin{pmatrix} A_r \\ A_\theta \\ A_\phi \end{pmatrix} \quad \text{(II-10a)}$$

or

$$\begin{aligned} A_\rho &= A_r \sin\theta + A_\theta \cos\theta \\ A_\phi &= A_\phi \\ A_z &= A_r \cos\theta - A_\theta \sin\theta \end{aligned} \quad \text{(II-10b)}$$

This time the component A_ϕ and coordinate ϕ are the same in both systems.

II.1.3 Rectangular-to-Spherical (and Vice Versa)

Many times it may be required that a transformation be performed directly from rectangular-to-spherical components. By referring to Figure II-1, we can write that the rectangular and spherical coordinates are related by

$$\begin{aligned} x &= r \sin\theta \cos\phi \\ y &= r \sin\theta \sin\phi \\ z &= r \cos\theta \end{aligned} \quad \text{(II-11)}$$

and the rectangular and spherical components by

$$\begin{aligned} A_r &= A_x \sin\theta \cos\phi + A_y \sin\theta \sin\phi + A_z \cos\theta \\ A_\theta &= A_x \cos\theta \cos\phi + A_y \cos\theta \sin\phi - A_z \sin\theta \\ A_\phi &= -A_x \sin\phi + A_y \cos\phi \end{aligned} \quad \text{(II-12)}$$

which can also be obtained by substituting (II-6) into (II-9). In matrix form, (II-12) can be written as

$$\begin{pmatrix} A_r \\ A_\theta \\ A_\phi \end{pmatrix} = \begin{pmatrix} \sin\theta\cos\phi & \sin\theta\sin\phi & \cos\theta \\ \cos\theta\cos\phi & \cos\theta\sin\phi & -\sin\theta \\ -\sin\phi & \cos\phi & 0 \end{pmatrix} \begin{pmatrix} A_x \\ A_y \\ A_z \end{pmatrix} \quad \text{(II-12a)}$$

with the rectangular-to-spherical transformation matrix being

$$[A]_{rs} = \begin{bmatrix} \sin\theta\cos\phi & \sin\theta\sin\phi & \cos\theta \\ \cos\theta\cos\phi & \cos\theta\sin\phi & -\sin\theta \\ -\sin\phi & \cos\phi & 0 \end{bmatrix} \quad \text{(II-12b)}$$

The transformation matrix of (II-12b) is also orthonormal so that its inverse can be written as

$$[A]_{sr} = [A]_{rs}^{-1} = [A]_{rs}^{t} = \begin{bmatrix} \sin\theta\cos\phi & \cos\theta\cos\phi & -\sin\phi \\ \sin\theta\sin\phi & \cos\theta\sin\phi & \cos\phi \\ \cos\theta & -\sin\theta & 0 \end{bmatrix} \quad \text{(II-13)}$$

and the spherical-to-rectangular components related by

$$\begin{pmatrix} A_x \\ A_y \\ A_z \end{pmatrix} = \begin{pmatrix} \sin\theta\cos\phi & \cos\theta\cos\phi & -\sin\phi \\ \sin\theta\sin\phi & \cos\theta\sin\phi & \cos\phi \\ \cos\theta & -\sin\theta & 0 \end{pmatrix} \begin{pmatrix} A_r \\ A_\theta \\ A_\phi \end{pmatrix} \quad \text{(II-13a)}$$

or

$$A_x = A_r \sin\theta \cos\phi + A_\theta \cos\theta \cos\phi - A_\phi \sin\phi$$
$$A_y = A_r \sin\theta \sin\phi + A_\theta \cos\theta \sin\phi + A_\phi \cos\phi$$
$$A_z = A_r \cos\theta - A_\theta \sin\theta \tag{II-13b}$$

II.2 VECTOR DIFFERENTIAL OPERATORS

The differential operators of gradient of a scalar ($\nabla\psi$), divergence of a vector ($\nabla \cdot \mathbf{A}$), curl of a vector ($\nabla \times \mathbf{A}$), Laplacian of a scalar ($\nabla^2\psi$), and Laplacian of a vector ($\nabla^2\mathbf{A}$) frequently encountered in electromagnetic field analysis will be listed in the rectangular, cylindrical, and spherical coordinate systems.

II.2.1 Rectangular Coordinates

$$\nabla\psi = \hat{a}_x \frac{\partial \psi}{\partial x} + \hat{a}_y \frac{\partial \psi}{\partial y} + \hat{a}_z \frac{\partial \psi}{\partial z} \tag{II-14}$$

$$\nabla \cdot \mathbf{A} = \frac{\partial A_x}{\partial x} + \frac{\partial A_y}{\partial y} + \frac{\partial A_z}{\partial z} \tag{II-15}$$

$$\nabla \times \mathbf{A} = \hat{a}_x \left(\frac{\partial A_z}{\partial y} - \frac{\partial A_y}{\partial z}\right) + \hat{a}_y \left(\frac{\partial A_x}{\partial z} - \frac{\partial A_z}{\partial x}\right) + \hat{a}_z \left(\frac{\partial A_y}{\partial x} - \frac{\partial A_x}{\partial y}\right) \tag{II-16}$$

$$\nabla \cdot \nabla\psi = \nabla^2\psi = \frac{\partial^2 \psi}{\partial x^2} + \frac{\partial^2 \psi}{\partial y^2} + \frac{\partial^2 \psi}{\partial z^2} \tag{II-17}$$

$$\nabla^2 \mathbf{A} = \hat{a}_x \nabla^2 A_x + \hat{a}_y \nabla^2 A_y + \hat{a}_z \nabla^2 A_z \tag{II-18}$$

II.2.2 Cylindrical Coordinates

$$\nabla\psi = \hat{a}_\rho \frac{\partial \psi}{\partial \rho} + \hat{a}_\phi \frac{1}{\rho}\frac{\partial \psi}{\partial \phi} + \hat{a}_z \frac{\partial \psi}{\partial z} \tag{II-19}$$

$$\nabla \cdot \mathbf{A} = \frac{1}{\rho}\frac{\partial}{\partial \rho}(\rho A_\rho) + \frac{1}{\rho}\frac{\partial A_\phi}{\partial \phi} + \frac{\partial A_z}{\partial z} \tag{II-20}$$

$$\nabla \times \mathbf{A} = \hat{a}_\rho \left(\frac{1}{\rho}\frac{\partial A_z}{\partial \phi} - \frac{\partial A_\phi}{\partial z}\right) + \hat{a}_\phi \left(\frac{\partial A_\rho}{\partial z} - \frac{\partial A_z}{\partial \rho}\right)$$
$$+ \hat{a}_z \left(\frac{1}{\rho}\frac{\partial (\rho A_\phi)}{\partial \rho} - \frac{1}{\rho}\frac{\partial A_\rho}{\partial \phi}\right) \tag{II-21}$$

$$\nabla^2\psi = \frac{1}{\rho}\frac{\partial}{\partial \rho}\left(\rho \frac{\partial \psi}{\partial \rho}\right) + \frac{1}{\rho^2}\frac{\partial^2 \psi}{\partial \phi^2} + \frac{\partial^2 \psi}{\partial z^2} \tag{II-22}$$

$$\nabla^2 \mathbf{A} = \nabla(\nabla \cdot \mathbf{A}) - \nabla \times \nabla \times \mathbf{A} \tag{II-23}$$

or in an expanded form

$$\nabla^2 \mathbf{A} = \hat{a}_\rho \left(\frac{\partial^2 A_\rho}{\partial \rho^2} + \frac{1}{\rho} \frac{\partial A_\rho}{\partial \rho} - \frac{A_\rho}{\rho^2} + \frac{1}{\rho^2} \frac{\partial^2 A_\rho}{\partial \phi^2} - \frac{2}{\rho^2} \frac{\partial A_\phi}{\partial \phi} + \frac{\partial^2 A_\rho}{\partial z^2} \right)$$

$$+ \hat{a}_\phi \left(\frac{\partial^2 A_\phi}{\partial \rho^2} + \frac{1}{\rho} \frac{\partial A_\phi}{\partial \rho} - \frac{A_\phi}{\rho^2} + \frac{1}{\rho^2} \frac{\partial^2 A_\phi}{\partial \phi^2} + \frac{2}{\rho^2} \frac{\partial A_\rho}{\partial \phi} + \frac{\partial^2 A_\phi}{\partial z^2} \right)$$

$$+ \hat{a}_z \left(\frac{\partial^2 A_z}{\partial \rho^2} + \frac{1}{\rho} \frac{\partial A_z}{\partial \rho} + \frac{1}{\rho^2} \frac{\partial^2 A_z}{\partial \phi^2} + \frac{\partial^2 A_z}{\partial z^2} \right) \qquad \text{(II-23a)}$$

In the cylindrical coordinate system $\nabla^2 \mathbf{A} \neq \hat{a}_\rho \nabla^2 A_\rho + \hat{a}_\phi \nabla^2 A_\phi + \hat{a}_z \nabla^2 A_z$ because the orientation of the unit vectors $\hat{a}_\rho$ and $\hat{a}_\phi$ varies with the ρ and ϕ coordinates.

II.2.3 Spherical Coordinates

$$\nabla \psi = \hat{a}_r \frac{\partial \psi}{\partial r} + \hat{a}_\theta \frac{1}{r} \frac{\partial \psi}{\partial \theta} + \hat{a}_\phi \frac{1}{r \sin \theta} \frac{\partial \psi}{\partial \phi} \qquad \text{(II-24)}$$

$$\nabla \cdot \mathbf{A} = \frac{1}{r^2} \frac{\partial}{\partial r} (r^2 A_r) + \frac{1}{r \sin \theta} \frac{\partial}{\partial \theta} (\sin \theta A_\theta) + \frac{1}{r \sin \theta} \frac{\partial A_\phi}{\partial \phi} \qquad \text{(II-25)}$$

$$\nabla \times \mathbf{A} = \frac{\hat{a}_r}{r \sin \theta} \left[\frac{\partial}{\partial \theta} (A_\phi \sin \theta) - \frac{\partial A_\theta}{\partial \phi} \right] + \frac{\hat{a}_\theta}{r} \left[\frac{1}{\sin \theta} \frac{\partial A_r}{\partial \phi} - \frac{\partial}{\partial r} (rA_\phi) \right]$$

$$+ \frac{\hat{a}_\phi}{r} \left[\frac{\partial}{\partial r} (rA_\theta) - \frac{\partial A_r}{\partial \theta} \right] \qquad \text{(II-26)}$$

$$\nabla^2 \psi = \frac{1}{r^2} \frac{\partial}{\partial r} \left(r^2 \frac{\partial \psi}{\partial r} \right) + \frac{1}{r^2 \sin \theta} \frac{\partial}{\partial \theta} \left(\sin \theta \frac{\partial \psi}{\partial \theta} \right) + \frac{1}{r^2 \sin^2 \theta} \frac{\partial^2 \psi}{\partial \phi^2} \qquad \text{(II-27)}$$

$$\nabla^2 \mathbf{A} = \nabla (\nabla \cdot \mathbf{A}) - \nabla \times \nabla \times \mathbf{A} \qquad \text{(II-28)}$$

or in an expanded form

$$\nabla^2 \mathbf{A} = \hat{a}_r \left(\frac{\partial^2 A_r}{\partial r^2} + \frac{2}{r} \frac{\partial A_r}{\partial r} - \frac{2}{r^2} A_r + \frac{1}{r^2} \frac{\partial^2 A_r}{\partial \theta^2} + \frac{\cot \theta}{r^2} \frac{\partial A_r}{\partial \theta} + \frac{1}{r^2 \sin^2 \theta} \frac{\partial^2 A_r}{\partial \phi^2} \right.$$

$$\left. - \frac{2}{r^2} \frac{\partial A_\theta}{\partial \theta} - \frac{2 \cot \theta}{r^2} A_\theta - \frac{2}{r^2 \sin \theta} \frac{\partial A_\phi}{\partial \phi} \right)$$

$$+ \hat{a}_\theta \left(\frac{\partial^2 A_\theta}{\partial r^2} + \frac{2}{r} \frac{\partial A_\theta}{\partial r} - \frac{A_\theta}{r^2 \sin^2 \theta} + \frac{1}{r^2} \frac{\partial^2 A_\theta}{\partial \theta^2} + \frac{\cot \theta}{r^2} \frac{\partial A_\theta}{\partial \theta} \right.$$

$$\left. + \frac{1}{r^2 \sin^2 \theta} \frac{\partial^2 A_\theta}{\partial \phi^2} + \frac{2}{r^2} \frac{\partial A_r}{\partial \theta} - \frac{2 \cot \theta}{r^2 \sin \theta} \frac{\partial A_\phi}{\partial \phi} \right)$$

$$+ \hat{a}_\phi \left(\frac{\partial^2 A_\phi}{\partial r^2} + \frac{2}{r} \frac{\partial A_\phi}{\partial r} - \frac{1}{r^2 \sin^2 \theta} A_\phi + \frac{1}{r^2} \frac{\partial^2 A_\phi}{\partial \theta^2} \right.$$

$$+ \frac{\cot \theta}{r^2} \frac{\partial A_\phi}{\partial \theta} + \frac{1}{r^2 \sin^2 \theta} \frac{\partial^2 A_\phi}{\partial \phi^2} + \frac{2}{r^2 \sin \theta} \frac{\partial A_r}{\partial \phi}$$

$$\left. + \frac{2 \cot \theta}{r^2 \sin \theta} \frac{\partial A_\theta}{\partial \phi} \right) \qquad \text{(II-28a)}$$

Again note that $\nabla^2 \mathbf{A} \neq \hat{a}_r \nabla^2 A_r + \hat{a}_\theta \nabla^2 A_\theta + \hat{a}_\phi \nabla^2 A_\phi$ since the orientation of the unit vectors $\hat{a}_r$, $\hat{a}_\theta$, and $\hat{a}_\phi$ varies with the r, θ, and ϕ coordinates.

II.3 VECTOR IDENTITIES

II.3.1 Addition and Multiplication

$$\mathbf{A} \cdot \mathbf{A} = |\mathbf{A}|^2 \tag{II-29}$$

$$\mathbf{A} \cdot \mathbf{A}^* = |\mathbf{A}|^2 \tag{II-30}$$

$$\mathbf{A} + \mathbf{B} = \mathbf{B} + \mathbf{A} \tag{II-31}$$

$$\mathbf{A} \cdot \mathbf{B} = \mathbf{B} \cdot \mathbf{A} \tag{II-32}$$

$$\mathbf{A} \times \mathbf{B} = -\mathbf{B} \times \mathbf{A} \tag{II-33}$$

$$(\mathbf{A} + \mathbf{B}) \cdot \mathbf{C} = \mathbf{A} \cdot \mathbf{C} + \mathbf{B} \cdot \mathbf{C} \tag{II-34}$$

$$(\mathbf{A} + \mathbf{B}) \times \mathbf{C} = \mathbf{A} \times \mathbf{C} + \mathbf{B} \times \mathbf{C} \tag{II-35}$$

$$\mathbf{A} \cdot \mathbf{B} \times \mathbf{C} = \mathbf{B} \cdot \mathbf{C} \times \mathbf{A} = \mathbf{C} \cdot \mathbf{A} \times \mathbf{B} \tag{II-36}$$

$$\mathbf{A} \times (\mathbf{B} \times \mathbf{C}) = (\mathbf{A} \cdot \mathbf{C})\mathbf{B} - (\mathbf{A} \cdot \mathbf{B})\mathbf{C} \tag{II-37}$$

$$\begin{aligned}(\mathbf{A} \times \mathbf{B}) \cdot (\mathbf{C} \times \mathbf{D}) &= \mathbf{A} \cdot \mathbf{B} \times (\mathbf{C} \times \mathbf{D}) \\ &= \mathbf{A} \cdot (\mathbf{B} \cdot \mathbf{D}\mathbf{C} - \mathbf{B} \cdot \mathbf{C}\mathbf{D}) \\ &= (\mathbf{A} \cdot \mathbf{C})(\mathbf{B} \cdot \mathbf{D}) - (\mathbf{A} \cdot \mathbf{D})(\mathbf{B} \cdot \mathbf{C}) \end{aligned} \tag{II-38}$$

$$(\mathbf{A} \times \mathbf{B}) \times (\mathbf{C} \times \mathbf{D}) = (\mathbf{A} \times \mathbf{B} \cdot \mathbf{D})\mathbf{C} - (\mathbf{A} \times \mathbf{B} \cdot \mathbf{C})\mathbf{D} \tag{II-39}$$

II.3.2 Differentiation

$$\nabla \cdot (\nabla \times \mathbf{A}) = 0 \tag{II-40}$$

$$\nabla \times \nabla \psi = 0 \tag{II-41}$$

$$\nabla(\phi + \psi) = \nabla\phi + \nabla\psi \tag{II-42}$$

$$\nabla(\phi\psi) = \phi\nabla\psi + \psi\nabla\phi \tag{II-43}$$

$$\nabla \cdot (\mathbf{A} + \mathbf{B}) = \nabla \cdot \mathbf{A} + \nabla \cdot \mathbf{B} \tag{II-44}$$

$$\nabla \times (\mathbf{A} + \mathbf{B}) = \nabla \times \mathbf{A} + \nabla \times \mathbf{B} \tag{II-45}$$

$$\nabla \cdot (\psi\mathbf{A}) = \mathbf{A} \cdot \nabla\psi + \psi\nabla \cdot \mathbf{A} \tag{II-46}$$

$$\nabla \times (\psi\mathbf{A}) = \nabla\psi \times \mathbf{A} + \psi\nabla \times \mathbf{A} \tag{II-47}$$

$$\nabla(\mathbf{A} \cdot \mathbf{B}) = (\mathbf{A} \cdot \nabla)\mathbf{B} + (\mathbf{B} \cdot \nabla)\mathbf{A} + \mathbf{A} \times (\nabla \times \mathbf{B}) + \mathbf{B} \times (\nabla \times \mathbf{A}) \tag{II-48}$$

$$\nabla \cdot (\mathbf{A} \times \mathbf{B}) = \mathbf{B} \cdot \nabla \times \mathbf{A} - \mathbf{A} \cdot \nabla \times \mathbf{B} \tag{II-49}$$

$$\nabla \times (\mathbf{A} \times \mathbf{B}) = \mathbf{A}\nabla \cdot \mathbf{B} - \mathbf{B}\nabla \cdot \mathbf{A} + (\mathbf{B} \cdot \nabla)\mathbf{A} - (\mathbf{A} \cdot \nabla)\mathbf{B} \tag{II-50}$$

$$\nabla \times \nabla \times \mathbf{A} = \nabla(\nabla \cdot \mathbf{A}) - \nabla^2 \mathbf{A} \tag{II-51}$$

II.3.3 Integration

$$\oint_C \mathbf{A} \cdot d\mathbf{l} = \iint_S (\nabla \times \mathbf{A}) \cdot d\mathbf{s} \quad \text{Stokes' theorem} \tag{II-52}$$

$$\oiint_S \mathbf{A} \cdot d\mathbf{s} = \iiint_V (\nabla \cdot \mathbf{A}) \, dv \quad \text{divergence theorem} \tag{II-53}$$

$$\oiint_S (\hat{n} \times \mathbf{A}) \, ds = \iiint_V (\nabla \times \mathbf{A}) \, dv \tag{II-54}$$

$$\oiint_S \psi \, d\mathbf{s} = \iiint_V \nabla \psi \, dv \tag{II-55}$$

$$\oint_C \psi \, d\mathbf{l} = \iint_S \hat{n} \times \nabla \psi \, ds \tag{II-56}$$

APPENDIX III

FRESNEL INTEGRALS

$$C_0(x) = \int_0^x \frac{\cos(\tau)}{\sqrt{2\pi\tau}} d\tau \qquad \text{(III-1)}$$

$$S_0(x) = \int_0^x \frac{\sin(\tau)}{\sqrt{2\pi\tau}} d\tau \qquad \text{(III-2)}$$

$$C(x) = \int_0^x \cos\left(\frac{\pi}{2}\tau^2\right) d\tau \qquad \text{(III-3)}$$

$$S(x) = \int_0^x \sin\left(\frac{\pi}{2}\tau^2\right) d\tau \qquad \text{(III-4)}$$

$$C_1(x) = \int_x^\infty \cos(\tau^2) d\tau \qquad \text{(III-5)}$$

$$S_1(x) = \int_x^\infty \sin(\tau^2) d\tau \qquad \text{(III-6)}$$

$$C(x) - jS(x) = \int_0^x e^{-j(\pi/2)\tau^2} d\tau = \int_0^{(\pi/2)x^2} \frac{e^{-j\tau}}{\sqrt{2\pi\tau}} d\tau$$

$$C(x) - jS(x) = C_0\left(\frac{\pi}{2}x^2\right) - jS_0\left(\frac{\pi}{2}x^2\right) \qquad \text{(III-7)}$$

$$C_1(x) - jS_1(x) = \int_x^\infty e^{-j\tau^2} d\tau = \sqrt{\frac{\pi}{2}} \int_{x^2}^\infty \frac{e^{-j\tau}}{\sqrt{2\pi\tau}} d\tau$$

$$C_1(x) - jS_1(x) = \sqrt{\frac{\pi}{2}} \left\{ \int_0^\infty \frac{e^{-j\tau}}{\sqrt{2\pi\tau}} d\tau - \int_0^{x^2} \frac{e^{-j\tau}}{\sqrt{2\pi\tau}} d\tau \right\}$$

$$C_1(x) - jS_1(x) = \sqrt{\frac{\pi}{2}} \left\{ \left[\frac{1}{2} - j\frac{1}{2}\right] - [C_0(x^2) - jS_0(x^2)] \right\}$$

$$C_1(x) - jS_1(x) = \sqrt{\frac{\pi}{2}} \left\{ \left[\frac{1}{2} - C_0(x^2)\right] - j\left[\frac{1}{2} - S_0(x^2)\right] \right\} \qquad \text{(III-8)}$$

APPENDIX III FRESNEL INTEGRALS

x	$C_1(x)$	$S_1(x)$	$C(x)$	$S(x)$
0.0	0.62666	0.62666	0.0	0.0
0.1	0.52666	0.62632	0.10000	0.00052
0.2	0.42669	0.62399	0.19992	0.00419
0.3	0.32690	0.61766	0.29940	0.01412
0.4	0.22768	0.60536	0.39748	0.03336
0.5	0.12977	0.58518	0.49234	0.06473
0.6	0.03439	0.55532	0.58110	0.11054
0.7	−0.05672	0.51427	0.65965	0.17214
0.8	−0.14119	0.46092	0.72284	0.24934
0.9	−0.21606	0.39481	0.76482	0.33978
1.0	−0.27787	0.31639	0.77989	0.43826
1.1	−0.32285	0.22728	0.76381	0.53650
1.2	−0.34729	0.13054	0.71544	0.62340
1.3	−0.34803	0.03081	0.63855	0.68633
1.4	−0.32312	−0.06573	0.54310	0.71353
1.5	−0.27253	−0.15158	0.44526	0.69751
1.6	−0.19886	−0.21861	0.36546	0.63889
1.7	−0.10790	−0.25905	0.32383	0.54920
1.8	−0.00871	−0.26682	0.33363	0.45094
1.9	0.08680	−0.23918	0.39447	0.37335
2.0	0.16520	−0.17812	0.48825	0.34342
2.1	0.21359	−0.09141	0.58156	0.37427
2.2	0.22242	0.00743	0.63629	0.45570
2.3	0.18833	0.10054	0.62656	0.55315
2.4	0.11650	0.16879	0.55496	0.61969
2.5	0.02135	0.19614	0.45742	0.61918
2.6	−0.07518	0.17454	0.38894	0.54999
2.7	−0.14816	0.10789	0.39249	0.45292
2.8	−0.17646	0.01329	0.46749	0.39153
2.9	−0.15021	−0.08181	0.56237	0.41014
3.0	−0.07621	−0.14690	0.60572	0.49631
3.1	0.02152	−0.15883	0.56160	0.58181
3.2	0.10791	−0.11181	0.46632	0.59335
3.3	0.14907	−0.02260	0.40570	0.51929
3.4	0.12691	0.07301	0.43849	0.42965
3.5	0.04965	0.13335	0.53257	0.41525
3.6	−0.04819	0.12973	0.58795	0.49231
3.7	−0.11929	0.06258	0.54195	0.57498
3.8	−0.12649	−0.03483	0.44810	0.56562
3.9	−0.06469	−0.11030	0.42233	0.47521
4.0	0.03219	−0.12048	0.49842	0.42052
4.1	0.10690	−0.05815	0.57369	0.47580
4.2	0.11228	0.03885	0.54172	0.56320
4.3	0.04374	0.10751	0.44944	0.55400
4.4	−0.05287	0.10038	0.43833	0.46227
4.5	−0.10884	0.02149	0.52602	0.43427
4.6	−0.08188	−0.07126	0.56724	0.51619
4.7	0.00810	−0.10594	0.49143	0.56715
4.8	0.08905	−0.05381	0.43380	0.49675
4.9	0.09277	0.04224	0.50016	0.43507
5.0	0.01519	0.09874	0.56363	0.49919

x	$C_1(x)$	$S_1(x)$	$C(x)$	$S(x)$
5.1	−0.07411	0.06405	0.49979	0.56239
5.2	−0.09125	−0.03004	0.43889	0.49688
5.3	−0.01892	−0.09235	0.50778	0.44047
5.4	0.07063	−0.05976	0.55723	0.51403
5.5	0.08408	0.03440	0.47843	0.55369
5.6	0.00641	0.08900	0.45171	0.47004
5.7	−0.07642	0.04296	0.53846	0.45953
5.8	−0.06919	−0.05135	0.52984	0.54604
5.9	0.01998	−0.08231	0.44859	0.51633
6.0	0.08245	−0.01181	0.49953	0.44696
6.1	0.03946	0.07180	0.54950	0.51647
6.2	−0.05363	0.06018	0.46761	0.53982
6.3	−0.07284	−0.03144	0.47600	0.45555
6.4	0.00835	−0.07765	0.54960	0.49649
6.5	0.07574	−0.01326	0.48161	0.54538
6.6	0.03183	0.06872	0.46899	0.46307
6.7	−0.05828	0.04658	0.54674	0.49150
6.8	−0.05734	−0.04600	0.48307	0.54364
6.9	0.03317	−0.06440	0.47322	0.46244
7.0	0.06832	0.02077	0.54547	0.49970
7.1	−0.00944	0.06977	0.47332	0.53602
7.2	−0.06943	0.00041	0.48874	0.45725
7.3	−0.00864	−0.06793	0.53927	0.51894
7.4	0.06582	−0.01521	0.46010	0.51607
7.5	0.02018	0.06353	0.51601	0.46070
7.6	−0.06137	0.02367	0.51564	0.53885
7.7	−0.02580	−0.05958	0.46278	0.48202
7.8	0.05828	−0.02668	0.53947	0.48964
7.9	0.02638	0.05752	0.47598	0.53235
8.0	−0.05730	0.02494	0.49980	0.46021
8.1	−0.02238	−0.05752	0.52275	0.53204
8.2	0.05803	−0.01870	0.46384	0.48589
8.3	0.01387	0.05861	0.53775	0.49323
8.4	−0.05899	0.00789	0.47092	0.52429
8.5	−0.00080	−0.05881	0.51417	0.46534
8.6	0.05767	0.00729	0.50249	0.53693
8.7	−0.01616	0.05515	0.48274	0.46774
8.8	−0.05079	−0.02545	0.52797	0.52294
8.9	0.03461	−0.04425	0.46612	0.48856
9.0	0.03526	0.04293	0.53537	0.49985
9.1	−0.04951	0.02381	0.46661	0.51042
9.2	−0.01021	−0.05338	0.52914	0.48135
9.3	0.05354	0.00485	0.47628	0.52467
9.4	−0.02020	0.04920	0.51803	0.47134
9.5	−0.03995	−0.03426	0.48729	0.53100
9.6	0.04513	−0.02599	0.50813	0.46786
9.7	0.00837	0.05086	0.49549	0.53250
9.8	−0.04983	−0.01094	0.50192	0.46758
9.9	0.02916	−0.04124	0.49961	0.53215

x	$C_1(x)$	$S_1(x)$	$C(x)$	$S(x)$
10.0	0.02554	0.04298	0.49989	0.46817
10.1	−0.04927	0.00478	0.49961	0.53151
10.2	0.01738	−0.04583	0.50186	0.46885
10.3	0.03233	0.03621	0.49575	0.53061
10.4	−0.04681	0.01094	0.50751	0.47033
10.5	0.01360	−0.04563	0.48849	0.52804
10.6	0.03187	0.03477	0.51601	0.47460
10.7	−0.04595	0.00848	0.47936	0.52143
10.8	0.01789	−0.04270	0.52484	0.48413
10.9	0.02494	0.03850	0.47211	0.50867
11.0	−0.04541	−0.00202	0.52894	0.49991
11.1	0.02845	−0.03492	0.47284	0.49079
11.2	0.01008	0.04349	0.52195	0.51805
11.3	−0.03981	−0.01930	0.48675	0.47514
11.4	0.04005	−0.01789	0.50183	0.52786
11.5	−0.01282	0.04155	0.51052	0.47440
11.6	−0.02188	−0.03714	0.47890	0.51755
11.7	0.04164	0.00962	0.52679	0.49525
11.8	−0.03580	0.02267	0.47489	0.49013
11.9	0.00977	−0.04086	0.51544	0.52184
12.0	0.02059	0.03622	0.49993	0.47347
12.1	−0.03919	−0.01309	0.48426	0.52108
12.2	0.03792	−0.01555	0.52525	0.49345
12.3	−0.01914	0.03586	0.47673	0.48867
12.4	−0.00728	−0.03966	0.50951	0.52384
12.5	0.02960	0.02691	0.50969	0.47645
12.6	−0.03946	−0.00421	0.47653	0.50936
12.7	0.03445	−0.01906	0.52253	0.51097
12.8	−0.01783	0.03475	0.49376	0.47593
12.9	−0.00377	−0.03857	0.48523	0.51977
13.0	0.02325	0.03064	0.52449	0.49994
13.1	−0.03530	−0.01452	0.48598	0.48015
13.2	0.03760	−0.00459	0.49117	0.52244
13.3	−0.03075	0.02163	0.52357	0.49583
13.4	0.01744	−0.03299	0.48482	0.48173
13.5	−0.00129	0.03701	0.49103	0.52180
13.6	−0.01421	−0.03391	0.52336	0.49848
13.7	0.02639	0.02521	0.48908	0.47949
13.8	−0.03377	−0.01313	0.48534	0.51781
13.9	0.03597	−0.00002	0.52168	0.50737
14.0	−0.03352	0.01232	0.49996	0.47726
14.1	0.02749	−0.02240	0.47844	0.50668
14.2	−0.01916	0.02954	0.51205	0.51890
14.3	0.00979	−0.03357	0.51546	0.48398
14.4	−0.00043	0.03472	0.48131	0.48819
14.5	−0.00817	−0.03350	0.49164	0.52030
14.6	0.01553	0.03052	0.52113	0.50538
14.7	−0.02145	−0.02640	0.50301	0.47856
14.8	0.02591	0.02168	0.47853	0.49869
14.9	−0.02903	−0.01683	0.49971	0.52136
15.0	0.03103	0.01217	0.52122	0.49926

APPENDIX III FRESNEL INTEGRALS   **933**

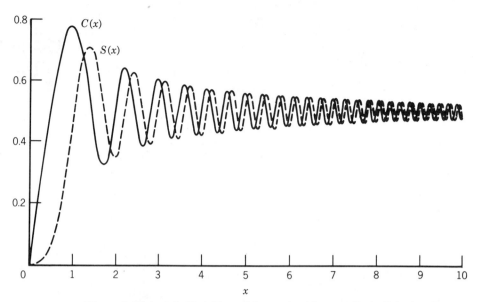

FIGURE III-1 Plots of $C(x)$ and $S(x)$ Fresnel integrals. (*Source:* C. A. Balanis, *Antenna Theory: Analysis and Design*, copyright © 1982, John Wiley & Sons, Inc. Reprinted by permission of John Wiley & Sons, Inc.)

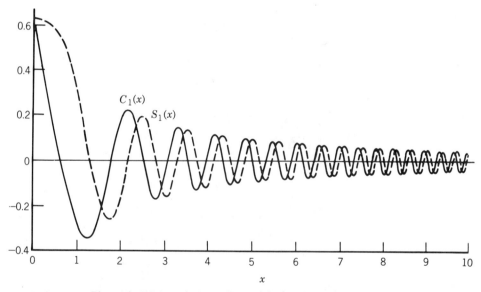

FIGURE III-2 Plots of $C_1(x)$ and $S_1(x)$ Fresnel integrals. (*Source:* C. A. Balanis, *Antenna Theory: Analysis and Design*, copyright © 1982, John Wiley & Sons, Inc. Reprinted by permission of John Wiley & Sons, Inc.)

APPENDIX IV

BESSEL FUNCTIONS

IV.1 BESSEL AND HANKEL FUNCTIONS

Bessel's equation can be written as

$$x^2 \frac{d^2y}{dx^2} + x\frac{dy}{dx} + (x^2 - p^2)y = 0 \qquad \text{(IV-1)}$$

Using the method of Frobenius, we can write its solutions as

$$y(x) = A_1 J_p(x) + B_1 J_{-p}(x) \qquad p \neq 0 \text{ or integer} \qquad \text{(IV-2)}$$

or

$$y(x) = A_2 J_n(x) + B_2 Y_n(x) \qquad p = n = 0 \text{ or integer} \qquad \text{(IV-3)}$$

where

$$J_p(x) = \sum_{m=0}^{\infty} \frac{(-1)^m (x/2)^{2m+p}}{m!(m+p)!} \qquad \text{(IV-4)}$$

$$J_{-p}(x) = \sum_{m=0}^{\infty} \frac{(-1)^m (x/2)^{2m-p}}{m!(m-p)!} \qquad \text{(IV-5)}$$

$$Y_p(x) = \frac{J_p(x)\cos(p\pi) - J_{-p}(x)}{\sin(p\pi)} \qquad \text{(IV-6)}$$

$$m! = \Gamma(m+1) \qquad \text{(IV-7)}$$

$J_p(x)$ is referred to as the Bessel function of the first kind of order p, $Y_p(x)$ as the Bessel function of the second kind of order p, and $\Gamma(x)$ as the gamma function.

When $p = n =$ integer, using (IV-5) and (IV-7) it can be shown that

$$J_{-n}(x) = (-1)^n J_n(x) \qquad \text{(IV-8)}$$

and no longer are the two Bessel functions independent of each other. Therefore a second solution is required and it is given by (IV-3). It can also be shown that

$$Y_n(x) = \lim_{p \to n} Y_p(x) = \lim_{p \to n} \frac{J_p(x)\cos(p\pi) - J_{-p}(x)}{\sin(p\pi)} \quad \text{(IV-9)}$$

When the argument of the Bessel function is negative and $p = n$, using (IV-4) leads to

$$J_n(-x) = (-1)^n J_n(x) \quad \text{(IV-10)}$$

In many applications, Bessel functions of small and large arguments are required. Using asymptotic methods, it can be shown that

$$\left. \begin{array}{l} J_0(x) \simeq 1 \\ Y_0(x) \simeq \dfrac{2}{\pi} \ln\left(\dfrac{\gamma x}{2}\right) \\ \gamma = 1.781 \end{array} \right\} \quad x \to 0 \quad \text{(IV-11)}$$

$$\left. \begin{array}{l} J_p(x) \simeq \dfrac{1}{p!}\left(\dfrac{x}{2}\right)^p \\ Y_p(x) \simeq -\dfrac{(p-1)!}{\pi}\left(\dfrac{2}{x}\right)^p \end{array} \right\} \quad \begin{array}{l} x \to 0 \\ p > 0 \end{array} \quad \text{(IV-12)}$$

and

$$\left. \begin{array}{l} J_p(x) \simeq \sqrt{\dfrac{2}{\pi x}} \cos\left(x - \dfrac{\pi}{4} - \dfrac{p\pi}{2}\right) \\ Y_p(x) \simeq \sqrt{\dfrac{2}{\pi x}} \sin\left(x - \dfrac{\pi}{4} - \dfrac{p\pi}{2}\right) \end{array} \right\} \quad x \to \infty \quad \text{(IV-13)}$$

For wave propagation it is often convenient to introduce Hankel functions defined as

$$H_p^{(1)}(x) = J_p(x) + jY_p(x) \quad \text{(IV-14)}$$

$$H_p^{(2)}(x) = J_p(x) - jY_p(x) \quad \text{(IV-15)}$$

where $H_p^{(1)}(x)$ is the Hankel function of the first kind of order p and $H_p^{(2)}(x)$ is the Hankel function of the second kind of order p. For large arguments

$$H_p^{(1)}(x) \simeq \sqrt{\frac{2}{\pi x}} e^{j[x - p(\pi/2) - \pi/4]} \quad x \to \infty \quad \text{(IV-16)}$$

$$H_p^{(2)}(x) \simeq \sqrt{\frac{2}{\pi x}} e^{-j[x - p(\pi/2) - \pi/4]} \quad x \to \infty \quad \text{(IV-17)}$$

A derivative can be taken using either

$$\frac{d}{dx}[Z_p(\alpha x)] = \alpha Z_{p-1}(\alpha x) - \frac{p}{x} Z_p(\alpha x) \tag{IV-18}$$

or

$$\frac{d}{dx}[Z_p(\alpha x)] = -\alpha Z_{p+1}(\alpha x) + \frac{p}{x} Z_p(\alpha x) \tag{IV-19}$$

where Z_p can be a Bessel function (J_p, Y_p) or a Hankel function ($H_p^{(1)}$ or $H_p^{(2)}$).

A useful identity relating Bessel functions and their derivatives is given by

$$J_p(x) Y_p'(x) - Y_p(x) J_p'(x) = \frac{2}{\pi x} \tag{IV-20}$$

and it is referred to as the Wronskian. The prime (') indicates a derivative. Also

$$J_p(x) J_{-p}'(x) - J_{-p}(x) J_p'(x) = -\frac{2}{\pi x} \sin(p\pi) \tag{IV-21}$$

Some useful integrals of Bessel functions are

$$\int x^{p+1} J_p(\alpha x) \, dx = \frac{1}{\alpha} x^{p+1} J_{p+1}(\alpha x) + C \tag{IV-22}$$

$$\int x^{1-p} J_p(\alpha x) \, dx = -\frac{1}{\alpha} x^{1-p} J_{p-1}(\alpha x) + C \tag{IV-23}$$

$$\int x^3 J_0(x) \, dx = x^3 J_1(x) - 2x^2 J_2(x) + C \tag{IV-24}$$

$$\int x^6 J_1(x) \, dx = x^6 J_2(x) - 4x^5 J_3(x) + 8x^4 J_4(x) + C \tag{IV-25}$$

$$\int J_3(x) \, dx = -J_2(x) - \frac{2}{x} J_1(x) + C \tag{IV-26}$$

$$\int x J_1(x) \, dx = -x J_0(x) + \int J_0(x) \, dx + C \tag{IV-27}$$

$$\int x^{-1} J_1(x) \, dx = -J_1(x) + \int J_0(x) \, dx + C \tag{IV-28}$$

$$\int J_2(x) \, dx = -2 J_1(x) + \int J_0(x) \, dx + C \tag{IV-29}$$

$$\int x^m J_n(x) \, dx = x^m J_{n+1}(x) - (m - n - 1) \int x^{m-1} J_{n+1}(x) \, dx \tag{IV-30}$$

$$\int x^m J_n(x) \, dx = -x^m J_{n-1}(x) + (m + n - 1) \int x^{m-1} J_{n-1}(x) \, dx \tag{IV-31}$$

$$J_1(x) = \frac{2}{\pi} \int_0^{\pi/2} \sin(x \sin\theta) \sin\theta \, d\theta \tag{IV-32}$$

$$\frac{1}{x} J_1(x) = \frac{2}{\pi} \int_0^{\pi/2} \cos(x \sin\theta) \cos^2\theta \, d\theta \tag{IV-33}$$

$$J_2(x) = \frac{2}{\pi} \int_0^{\pi/2} \cos(x \sin \theta) \cos 2\theta \, d\theta \qquad \text{(IV-34)}$$

$$J_n(x) = \frac{j^{-n}}{2\pi} \int_0^{2\pi} e^{jx \cos \phi} e^{jn\phi} \, d\phi \qquad \text{(IV-35)}$$

$$J_n(x) = \frac{j^{-n}}{\pi} \int_0^{\pi} \cos(n\phi) e^{jx \cos \phi} \, d\phi \qquad \text{(IV-36)}$$

$$J_n(x) = \frac{1}{\pi} \int_0^{\pi} \cos(x \sin \phi - n\phi) \, d\phi \qquad \text{(IV-37)}$$

$$J_{2n}(x) = \frac{2}{\pi} \int_0^{\pi/2} \cos(x \sin \phi) \cos(2n\phi) \, d\phi \qquad \text{(IV-38)}$$

$$J_{2n}(x) = (-1)^n \frac{2}{\pi} \int_0^{\pi/2} \cos(x \cos \phi) \cos(2n\phi) \, d\phi \qquad \text{(IV-39)}$$

The integrals

$$\int_0^x J_0(\tau) \, d\tau \quad \text{and} \quad \int_0^x Y_0(\tau) \, d\tau \qquad \text{(IV-40)}$$

often appear in solutions of problems but cannot be integrated in closed form. Graphs and tables for each, obtained using numerical techniques, are included.

IV.2 MODIFIED BESSEL FUNCTIONS

In addition to the regular cylindrical Bessel functions of the first and second kind, there exists another set of cylindrical Bessel functions that are referred to as the *modified* Bessel functions of the first $I_p(x)$ and second $K_p(x)$ kind. These modified cylindrical Bessel functions exhibit ascending and descending variations for increasing argument as shown, respectively, in Figures IV-5 and IV-6. For real values of the argument, the modified Bessel functions exhibit real values.

The modified Bessel functions are related to the regular Bessel and Hankel functions by

$$I_p(x) = j^{-p} J_p(jx) = j^p J_{-p}(jx) = j^p J_p(-jx) \qquad \text{(IV-41)}$$

$$K_p(x) = \frac{\pi}{2} j^{p+1} H_p^{(1)}(jx) = \frac{\pi}{2} (-j)^{p+1} H_p^{(2)}(-jx) \qquad \text{(IV-42)}$$

Some of the identities involving modified Bessel functions are

$$I_{-p}(x) = j^p J_{-p}(jx) \qquad \text{(IV-43)}$$

$$I_{-n}(x) = I_n(x) \qquad n = 0, 1, 2, 3, \ldots \qquad \text{(IV-44)}$$

$$K_{-n}(x) = K_n(x) \qquad n = 0, 1, 2, 3, \ldots \qquad \text{(IV-45)}$$

For large arguments the modified Bessel functions can be computed using the asymptotic formulas of

$$\left. \begin{array}{l} I_p(x) \simeq \dfrac{e^x}{\sqrt{2\pi x}} \\[2mm] K_p(x) \simeq \sqrt{\dfrac{\pi}{2}} \, e^{-x} \end{array} \right\} \quad x \to \infty \qquad \begin{array}{l} \text{(IV-46a)} \\[2mm] \text{(IV-46b)} \end{array}$$

Derivatives of both modified Bessel functions can be found using the same expressions, (IV-18) and (IV-19), as for the regular Bessel functions.

IV.3 SPHERICAL BESSEL AND HANKEL FUNCTIONS

There is another set of Bessel and Hankel functions which are usually referred to as the *spherical* Bessel and Hankel functions. These spherical Bessel and Hankel functions of order n are related, respectively, to the regular cylindrical Bessel and Hankel of order $n + 1/2$ by

$$j_n(x) = \sqrt{\frac{\pi}{2x}} J_{n+1/2}(x) \qquad \text{(IV-47a)}$$

$$y_n(x) = \sqrt{\frac{\pi}{2x}} Y_{n+1/2}(x) \qquad \text{(IV-47b)}$$

$$h_n^{(1)}(x) = \sqrt{\frac{\pi}{2x}} H_{n+1/2}^{(1)}(x) \qquad \text{(IV-47c)}$$

$$h_n^{(2)}(x) = \sqrt{\frac{\pi}{2x}} H_{n+1/2}^{(2)}(x) \qquad \text{(IV-47d)}$$

where j_n, y_n, $h_n^{(1)}$, and $h_n^{(2)}$ are the spherical Bessel and Hankel functions. These spherical Bessel and Hankel functions are used as solutions to electromagnetic problems solved using spherical coordinates.

For small arguments

$$\left. \begin{array}{l} j_n(x) \simeq \dfrac{x^n}{1 \cdot 3 \cdot 5 \cdots (2n+1)} \\ y_n(x) \simeq -1 \cdot 3 \cdot 5 \cdots (2n-1) x^{-(n+1)} \end{array} \right\} \quad \begin{array}{l} n = 0, 1, 2, \ldots \quad \text{(IV-48a)} \\ x \to 0 \quad \text{(IV-48b)} \end{array}$$

Another set of spherical Bessel and Hankel functions which appear in solutions of electromagnetic problems is that denoted by $\hat{B}_n(x)$ where $\hat{B}_n$ can be used to represent $\hat{J}_n$, $\hat{Y}_n$, $\hat{H}_n^{(1)}$, or $\hat{H}_n^{(2)}$. These are related to the preceding spherical Bessel and Hankel functions [denoted by b_n to represent j_n, y_n, $h_n^{(1)}$, or $h_n^{(2)}$] and to the regular cylindrical Bessel and Hankel functions [denoted by $B_{n+1/2}$ to represent $J_{n+1/2}$, $Y_{n+1/2}$, $H_{n+1/2}^{(1)}$, or $H_{n+1/2}^{(2)}$] by

$$\hat{B}_n(x) = x b_n(x) = \sqrt{\frac{\pi x}{2}} B_{n+1/2}(x) \qquad \text{(IV-49)}$$

x	$J_0(x)$	$J_1(x)$	$Y_0(x)$	$Y_1(x)$
0.0	1.00000	0.00000	$-\infty$	$-\infty$
0.1	0.99750	0.04994	-1.53424	-6.45895
0.2	0.99003	0.09950	-1.08110	-3.32382
0.3	0.97763	0.14832	-0.80727	-2.29310
0.4	0.96040	0.19603	-0.60602	-1.78087
0.5	0.93847	0.24227	-0.44452	-1.47147
0.6	0.91201	0.28670	-0.30851	-1.26039
0.7	0.88120	0.32900	-0.19066	-1.10325
0.8	0.84629	0.36884	-0.08680	-0.97814
0.9	0.80752	0.40595	0.00563	-0.87313
1.0	0.76520	0.44005	0.08826	-0.78121
1.1	0.71962	0.47090	0.16216	-0.69812
1.2	0.67113	0.49829	0.22808	-0.62114
1.3	0.62009	0.52202	0.28654	-0.54852
1.4	0.56686	0.54195	0.33789	-0.47915
1.5	0.51183	0.55794	0.38245	-0.41231
1.6	0.45540	0.56990	0.42043	-0.34758
1.7	0.39799	0.57777	0.45203	-0.28473
1.8	0.33999	0.58152	0.47743	-0.22366
1.9	0.28182	0.58116	0.49682	-0.16441
2.0	0.22389	0.57673	0.51038	-0.10703
2.1	0.16661	0.56829	0.51829	-0.05168
2.2	0.11036	0.55596	0.52078	0.00149
2.3	0.05554	0.53987	0.51807	0.05228
2.4	0.00251	0.52019	0.51041	0.10049
2.5	-0.04838	0.49710	0.49807	0.14592
2.6	-0.09681	0.47082	0.48133	0.18836
2.7	-0.14245	0.44161	0.46050	0.22763
2.8	-0.18504	0.40972	0.43592	0.26354
2.9	-0.22432	0.37544	0.40791	0.29594
3.0	-0.26005	0.33906	0.37686	0.32467
3.1	-0.29206	0.30092	0.34310	0.34963
3.2	-0.32019	0.26134	0.30705	0.37071
3.3	-0.34430	0.22066	0.26909	0.38785
3.4	-0.36430	0.17923	0.22962	0.40101
3.5	-0.38013	0.13738	0.18902	0.41019
3.6	-0.39177	0.09547	0.14771	0.41539
3.7	-0.39923	0.05383	0.10607	0.41667
3.8	-0.40256	0.01282	0.06450	0.41411
3.9	-0.40183	-0.02724	0.02338	0.40782
4.0	-0.39715	-0.06604	-0.01694	0.39793
4.1	-0.38868	-0.10328	-0.05609	0.38459
4.2	-0.37657	-0.13865	-0.09375	0.36801
4.3	-0.36102	-0.17190	-0.12960	0.34839
4.4	-0.34226	-0.20278	-0.16334	0.32597
4.5	-0.32054	-0.23106	-0.19471	0.30100
4.6	-0.29614	-0.25655	-0.22346	0.27375
4.7	-0.26933	-0.27908	-0.24939	0.24450
4.8	-0.24043	-0.29850	-0.27230	0.21356
4.9	-0.20974	-0.31470	-0.29205	0.18125

APPENDIX IV BESSEL FUNCTIONS

x	$J_0(x)$	$J_1(x)$	$Y_0(x)$	$Y_1(x)$
5.0	−0.17760	−0.32758	−0.30852	0.14786
5.1	−0.14434	−0.33710	−0.32160	0.11374
5.2	−0.11029	−0.34322	−0.33125	0.07919
5.3	−0.07580	−0.34596	−0.33744	0.04455
5.4	−0.04121	−0.34534	−0.34017	0.01013
5.5	−0.00684	−0.34144	−0.33948	−0.02376
5.6	0.02697	−0.33433	−0.33544	−0.05681
5.7	0.05992	−0.32415	−0.32816	−0.08872
5.8	0.09170	−0.31103	−0.31775	−0.11923
5.9	0.12203	−0.29514	−0.30437	−0.14808
6.0	0.15065	−0.27668	−0.28819	−0.17501
6.1	0.17729	−0.25587	−0.26943	−0.19981
6.2	0.20175	−0.23292	−0.24831	−0.22228
6.3	0.22381	−0.20809	−0.22506	−0.24225
6.4	0.24331	−0.18164	−0.19995	−0.25956
6.5	0.26009	−0.15384	−0.17324	−0.27409
6.6	0.27404	−0.12498	−0.14523	−0.28575
6.7	0.28506	−0.09534	−0.11619	−0.29446
6.8	0.29310	−0.06522	−0.08643	−0.30019
6.9	0.29810	−0.03490	−0.05625	−0.30292
7.0	0.30008	−0.00468	−0.02595	−0.30267
7.1	0.29905	0.02515	0.00418	−0.29948
7.2	0.29507	0.05433	0.03385	−0.29342
7.3	0.28822	0.08257	0.06277	−0.28459
7.4	0.27860	0.10962	0.09068	−0.27311
7.5	0.26634	0.13525	0.11731	−0.25913
7.6	0.25160	0.15921	0.14243	−0.24280
7.7	0.23456	0.18131	0.16580	−0.22432
7.8	0.21541	0.20136	0.18723	−0.20388
7.9	0.19436	0.21918	0.20652	−0.18172
8.0	0.17165	0.23464	0.22352	−0.15806
8.1	0.14752	0.24761	0.23809	−0.13315
8.2	0.12222	0.25800	0.25012	−0.10724
8.3	0.09601	0.26574	0.25951	−0.08060
8.4	0.06916	0.27079	0.26622	−0.05348
8.5	0.04194	0.27312	0.27021	−0.02617
8.6	0.01462	0.27276	0.27146	0.00108
8.7	−0.01252	0.26972	0.27000	0.02801
8.8	−0.03923	0.26407	0.26587	0.05436
8.9	−0.06525	0.25590	0.25916	0.07987
9.0	−0.09033	0.24531	0.24994	0.10431
9.1	−0.11424	0.23243	0.23834	0.12747
9.2	−0.13675	0.21741	0.22449	0.14911
9.3	−0.15765	0.20041	0.20857	0.16906
9.4	−0.17677	0.18163	0.19074	0.18714
9.5	−0.19393	0.16126	0.17121	0.20318
9.6	−0.20898	0.13952	0.15018	0.21706
9.7	−0.22180	0.11664	0.12787	0.22866
9.8	−0.23228	0.09284	0.10453	0.23789
9.9	−0.24034	0.06837	0.08038	0.24469

x	$J_0(x)$	$J_1(x)$	$Y_0(x)$	$Y_1(x)$
10.0	−0.24594	0.04347	0.05567	0.24902
10.1	−0.24903	0.01840	0.03066	0.25084
10.2	−0.24962	−0.00662	0.00558	0.25019
10.3	−0.24772	−0.03132	−0.01930	0.24707
10.4	−0.24337	−0.05547	−0.04375	0.24155
10.5	−0.23665	−0.07885	−0.06753	0.23370
10.6	−0.22764	−0.10123	−0.09042	0.22363
10.7	−0.21644	−0.12240	−0.11219	0.21144
10.8	−0.20320	−0.14217	−0.13264	0.19729
10.9	−0.18806	−0.16035	−0.15158	0.18132
11.0	−0.17119	−0.17679	−0.16885	0.16371
11.1	−0.15277	−0.19133	−0.18428	0.14464
11.2	−0.13299	−0.20385	−0.19773	0.12431
11.3	−0.11207	−0.21426	−0.20910	0.10294
11.4	−0.09021	−0.22245	−0.21829	0.08074
11.5	−0.06765	−0.22838	−0.22523	0.05794
11.6	−0.04462	−0.23200	−0.22987	0.03477
11.7	−0.02133	−0.23330	−0.23218	0.01145
11.8	0.00197	−0.23229	−0.23216	−0.01179
11.9	0.02505	−0.22898	−0.22983	−0.03471
12.0	0.04769	−0.22345	−0.22524	−0.05710
12.1	0.06967	−0.21575	−0.21844	−0.07874
12.2	0.09077	−0.20598	−0.20952	−0.09942
12.3	0.11080	−0.19426	−0.19859	−0.11895
12.4	0.12956	−0.18071	−0.18578	−0.13714
12.5	0.14689	−0.16549	−0.17121	−0.15384
12.6	0.16261	−0.14874	−0.15506	−0.16888
12.7	0.17659	−0.13066	−0.13750	−0.18213
12.8	0.18870	−0.11143	−0.11870	−0.19347
12.9	0.19885	−0.09125	−0.09887	−0.20282
13.0	0.20693	−0.07032	−0.07821	−0.21008
13.1	0.21289	−0.04885	−0.05692	−0.21521
13.2	0.21669	−0.02707	−0.03524	−0.21817
13.3	0.21830	−0.00518	−0.01336	−0.21895
13.4	0.21773	0.01660	0.00848	−0.21756
13.5	0.21499	0.03805	0.03008	−0.21402
13.6	0.21013	0.05896	0.05122	−0.20839
13.7	0.20322	0.07914	0.07169	−0.20074
13.8	0.19434	0.09839	0.09130	−0.19116
13.9	0.18358	0.11653	0.10986	−0.17975
14.0	0.17108	0.13338	0.12719	−0.16664
14.1	0.15695	0.14879	0.14314	−0.15198
14.2	0.14137	0.16261	0.15754	−0.13592
14.3	0.12449	0.17473	0.17028	−0.11862
14.4	0.10649	0.18503	0.18123	−0.10026
14.5	0.08755	0.19343	0.19030	−0.08104
14.6	0.06787	0.19986	0.19742	−0.06115
14.7	0.04764	0.20426	0.20252	−0.04079
14.8	0.02708	0.20660	0.20557	−0.02016
14.9	0.00639	0.20688	0.20655	0.00053
15.0	−0.01422	0.20511	0.20546	0.02107

APPENDIX IV BESSEL FUNCTIONS

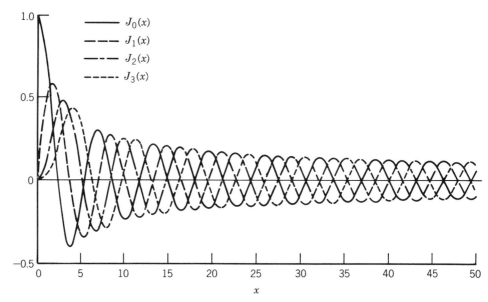

FIGURE IV-1 Bessel functions of the first kind [$J_0(x)$, $J_1(x)$, $J_2(x)$, and $J_3(x)$]. (*Source:* C. A. Balanis, *Antenna Theory: Analysis and Design*, copyright © 1982, John Wiley & Sons, Inc. Reprinted by permission of John Wiley & Sons, Inc.)

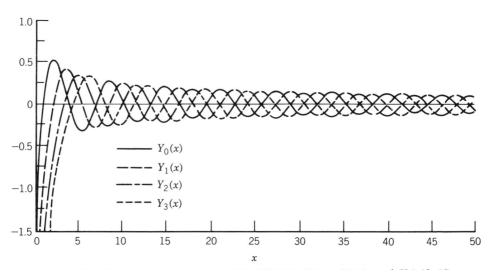

FIGURE IV-2 Bessel functions of the second kind [$Y_0(x)$, $Y_1(x)$, $Y_2(x)$, and $Y_3(x)$]. (*Source:* C. A. Balanis, *Antenna Theory: Analysis and Design*, copyright © 1982, John Wiley & Sons, Inc. Reprinted by permission of John Wiley & Sons, Inc.)

$J_1(x)/x$ function

x	$J_1(x)/x$	x	$J_1(x)/x$	x	$J_1(x)/x$
0.0	0.50000	5.0	−0.06552	10.0	0.00435
0.1	0.49938	5.1	−0.06610	10.1	0.00182
0.2	0.49750	5.2	−0.06600	10.2	−0.00065
0.3	0.49440	5.3	−0.06528	10.3	−0.00304
0.4	0.49007	5.4	−0.06395	10.4	−0.00533
0.5	0.48454	5.5	−0.06208	10.5	−0.00751
0.6	0.47783	5.6	−0.05970	10.6	−0.00955
0.7	0.46999	5.7	−0.05687	10.7	−0.01144
0.8	0.46105	5.8	−0.05363	10.8	−0.01316
0.9	0.45105	5.9	−0.05002	10.9	−0.01471
1.0	0.44005	6.0	−0.04611	11.0	−0.01607
1.1	0.42809	6.1	−0.04194	11.1	−0.01724
1.2	0.41524	6.2	−0.03757	11.2	−0.01820
1.3	0.40156	6.3	−0.03303	11.3	−0.01896
1.4	0.38710	6.4	−0.02838	11.4	−0.01951
1.5	0.37196	6.5	−0.02367	11.5	−0.01986
1.6	0.35618	6.6	−0.01894	11.6	−0.02000
1.7	0.33986	6.7	−0.01423	11.7	−0.01994
1.8	0.32306	6.8	−0.00959	11.8	−0.01969
1.9	0.30587	6.9	−0.00506	11.9	−0.01924
2.0	0.28836	7.0	−0.00067	12.0	−0.01862
2.1	0.27061	7.1	0.00354	12.1	−0.01783
2.2	0.25271	7.2	0.00755	12.2	−0.01688
2.3	0.23473	7.3	0.01131	12.3	−0.01579
2.4	0.21674	7.4	0.01481	12.4	−0.01457
2.5	0.19884	7.5	0.01803	12.5	−0.01324
2.6	0.18108	7.6	0.02095	12.6	−0.01180
2.7	0.16356	7.7	0.02355	12.7	−0.01029
2.8	0.14633	7.8	0.02582	12.8	−0.00871
2.9	0.12946	7.9	0.02774	12.9	−0.00707
3.0	0.11302	8.0	0.02933	13.0	−0.00541
3.1	0.09707	8.1	0.03057	13.1	−0.00373
3.2	0.08167	8.2	0.03146	13.2	−0.00205
3.3	0.06687	8.3	0.03202	13.3	−0.00039
3.4	0.05271	8.4	0.03224	13.4	0.00124
3.5	0.03925	8.5	0.03213	13.5	0.00282
3.6	0.02652	8.6	0.03172	13.6	0.00434
3.7	0.01455	8.7	0.03100	13.7	0.00578
3.8	0.00337	8.8	0.03001	13.8	0.00713
3.9	−0.00699	8.9	0.02875	13.9	0.00838
4.0	−0.01651	9.0	0.02726	14.0	0.00953
4.1	−0.02519	9.1	0.02554	14.1	0.01055
4.2	−0.03301	9.2	0.02363	14.2	0.01145
4.3	−0.03998	9.3	0.02155	14.3	0.01222
4.4	−0.04609	9.4	0.01932	14.4	0.01285
4.5	−0.05135	9.5	0.01697	14.5	0.01334
4.6	−0.05578	9.6	0.01453	14.6	0.01369
4.7	−0.05938	9.7	0.01202	14.7	0.01389
4.8	−0.06219	9.8	0.00947	14.8	0.01396
4.9	−0.06423	9.9	0.00691	14.9	0.01388
				15.0	0.01367

$\int_0^x J_0(\tau)\,d\tau$ and $\int_0^x Y_0(\tau)\,d\tau$ functions

x	$\int_0^x J_0(\tau)\,d\tau$	$\int_0^x Y_0(\tau)\,d\tau$	x	$\int_0^x J_0(\tau)\,d\tau$	$\int_0^x Y_0(\tau)\,d\tau$
0.0	0.00000	0.00000	5.0	0.71531	0.19971
0.1	0.09991	−0.21743	5.1	0.69920	0.16818
0.2	0.19933	−0.34570	5.2	0.68647	0.13551
0.3	0.29775	−0.43928	5.3	0.67716	0.10205
0.4	0.39469	−0.50952	5.4	0.67131	0.06814
0.5	0.48968	−0.56179	5.5	0.66891	0.03413
0.6	0.58224	−0.59927	5.6	0.66992	0.00035
0.7	0.67193	−0.62409	5.7	0.67427	−0.03284
0.8	0.75834	−0.63786	5.8	0.68187	−0.06517
0.9	0.84106	−0.64184	5.9	0.69257	−0.09630
1.0	0.91973	−0.63706	6.0	0.70622	−0.12595
1.1	0.99399	−0.62447	6.1	0.72263	−0.15385
1.2	1.06355	−0.60490	6.2	0.74160	−0.17975
1.3	1.12813	−0.57911	6.3	0.76290	−0.20344
1.4	1.18750	−0.54783	6.4	0.78628	−0.22470
1.5	1.24144	−0.51175	6.5	0.81147	−0.24338
1.6	1.28982	−0.47156	6.6	0.83820	−0.25931
1.7	1.33249	−0.42788	6.7	0.86618	−0.27239
1.8	1.36939	−0.38136	6.8	0.89512	−0.28252
1.9	1.40048	−0.33260	6.9	0.92470	−0.28966
2.0	1.42577	−0.28219	7.0	0.95464	−0.29377
2.1	1.44528	−0.23071	7.1	0.98462	−0.29486
2.2	1.45912	−0.17871	7.2	1.01435	−0.29295
2.3	1.46740	−0.12672	7.3	1.04354	−0.28811
2.4	1.47029	−0.07526	7.4	1.07190	−0.28043
2.5	1.46798	−0.02480	7.5	1.09917	−0.27002
2.6	1.46069	0.02420	7.6	1.12508	−0.25702
2.7	1.44871	0.07132	7.7	1.14941	−0.24159
2.8	1.43231	0.11617	7.8	1.17192	−0.22392
2.9	1.41181	0.15839	7.9	1.19243	−0.20421
3.0	1.38756	0.19765	8.0	1.21074	−0.18269
3.1	1.35992	0.23367	8.1	1.22671	−0.15959
3.2	1.32928	0.26620	8.2	1.24021	−0.13516
3.3	1.29602	0.29502	8.3	1.25112	−0.10966
3.4	1.26056	0.31996	8.4	1.25939	−0.08335
3.5	1.22330	0.34090	8.5	1.26494	−0.05650
3.6	1.18467	0.35775	8.6	1.26777	−0.02940
3.7	1.14509	0.37044	8.7	1.26787	−0.00230
3.8	1.10496	0.37896	8.8	1.26528	0.02451
3.9	1.06471	0.38335	8.9	1.26005	0.05078
4.0	1.02473	0.38366	9.0	1.25226	0.07625
4.1	0.98541	0.38000	9.1	1.24202	0.10069
4.2	0.94712	0.37250	9.2	1.22946	0.12385
4.3	0.91021	0.36131	9.3	1.21473	0.14552
4.4	0.87502	0.34665	9.4	1.19799	0.16550
4.5	0.84186	0.32872	9.5	1.17944	0.18361
4.6	0.81100	0.30779	9.6	1.15927	0.19969
4.7	0.78271	0.28413	9.7	1.13772	0.21360
4.8	0.75721	0.25802	9.8	1.11499	0.22523
4.9	0.73468	0.22977	9.9	1.09134	0.23448
			10.0	1.06701	0.24129

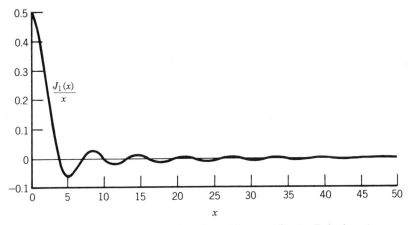

FIGURE IV-3 Plot of $J_1(x)/x$ function. (*Source:* C. A. Balanis, *Antenna Theory: Analysis and Design*, copyright © 1982, John Wiley & Sons, Inc. Reprinted by permission of John Wiley & Sons, Inc.)

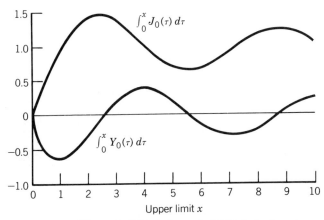

FIGURE IV-4 Plots of $\int_0^x J_0(\tau)\, d\tau$ and $\int_0^x Y_0(\tau)\, d\tau$ functions. (*Source:* C. A. Balanis, *Antenna Theory: Analysis and Design*, copyright © 1982, John Wiley & Sons, Inc. Reprinted by permission of John Wiley & Sons, Inc.)

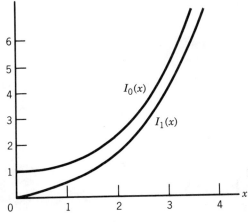

FIGURE IV-5 Modified Bessel functions of the first kind [$I_0(x)$ and $I_1(x)$].

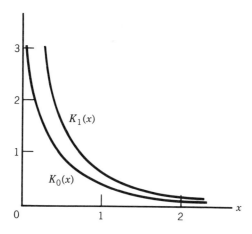

FIGURE IV-6 Modified Bessel functions of the second kind [$K_0(x)$ and $K_1(x)$].

APPENDIX V

LEGENDRE POLYNOMIALS AND FUNCTIONS

V.1 LEGENDRE POLYNOMIALS AND FUNCTIONS

The *ordinary* Legendre differential equation can be written as

$$(1 - x^2)\frac{d^2y}{dx^2} - 2x\frac{dy}{dx} + p(p+1)y = 0 \tag{V-1}$$

Its solution can be written as

$$y(x) = A_1 P_p(x) + B_1 P_p(-x) \qquad p \neq \text{integer} \tag{V-2}$$

where $P_p(x)$ is referred to as the *Legendre function of the first kind*. If p is an integer ($p = n$), then $P_n(x)$ and $P_n(-x)$ are not two independent solutions because

$$P_n(-x) = (-1)^n P_n(x) \tag{V-3}$$

Therefore two independent solutions to (V-1) for $p = n =$ integer are

$$y(x) = A_2 P_n(x) + B_2 Q_n(x) \tag{V-4}$$

where $Q_n(x)$ is referred to as the *Legendre function of the second kind*.

When $p = n$, $P_n(x)$ are also referred to as the *Legendre polynomials* of order n, and are defined by

$$P_n(x) = \sum_{m=0}^{M} \frac{(-1)^m (2n - 2m)!(x)^{n-2m}}{2^n m!(n-m)!(n-2m)!} \tag{V-5}$$

where $M = n/2$ or $(n-1)/2$, whichever is an integer. The Legendre functions

$Q_n(x)$ of the second kind are defined by

$$Q_n(x) = \lim_{p \to n} Q_p(x) = \lim_{p \to n} \frac{\pi}{2} \frac{P_p(x)\cos(p\pi) - P_p(-x)}{\sin(p\pi)} \quad \text{(V-6)}$$

The Legendre polynomials (or Legendre functions of the first kind) $P_n(x)$ can also be obtained more conveniently using *Rodrigues' formula*

$$P_n(x) = \frac{1}{2^n n!} \frac{d^n}{dx^n}(x^2 - 1)^n \quad \text{(V-7)}$$

which when expanded leads (for $n = 0, 1, 2, \ldots, 7$) to

$$\begin{aligned}
P_0(x) &= 1 \\
P_1(x) &= x \\
P_2(x) &= \tfrac{1}{2}(3x^2 - 1) \\
P_3(x) &= \tfrac{1}{2}(5x^3 - 3x) \\
P_4(x) &= \tfrac{1}{8}(35x^4 - 30x^2 + 3) \\
P_5(x) &= \tfrac{1}{8}(63x^5 - 70x^3 + 15x) \\
P_6(x) &= \tfrac{1}{16}(231x^6 - 315x^4 + 105x^2 - 5) \\
P_7(x) &= \tfrac{1}{16}(429x^7 - 693x^5 + 315x^3 - 35x)
\end{aligned} \quad \text{(V-8)}$$

If $x = \cos\theta$, the Legendre polynomials (or Legendre functions of the first kind) of (V-8) can be written as

$$\begin{aligned}
P_0(\cos\theta) &= 1 \\
P_1(\cos\theta) &= \cos\theta \\
P_2(\cos\theta) &= \tfrac{1}{4}(3\cos 2\theta + 1) \\
P_3(\cos\theta) &= \tfrac{1}{8}(5\cos 3\theta + 3\cos\theta) \\
P_4(\cos\theta) &= \tfrac{1}{64}(35\cos 4\theta + 20\cos 2\theta + 9) \\
P_5(\cos\theta) &= \tfrac{1}{128}(63\cos 5\theta + 35\cos 3\theta + 30\cos\theta) \\
P_6(\cos\theta) &= \tfrac{1}{512}(231\cos 6\theta + 126\cos 4\theta + 105\cos 2\theta + 50) \\
P_7(\cos\theta) &= \tfrac{1}{1024}(429\cos 7\theta + 231\cos 5\theta + 189\cos 3\theta + 175\cos\theta)
\end{aligned} \quad \text{(V-9)}$$

The Legendre functions $Q_n(x)$ of the second kind exhibit singularities at $x = \pm 1$ or $\theta = 0, \pi$ and can be obtained from the Legendre functions $P_n(x)$ of the first kind using the formula of

$$Q_n(x) = P_n(x)\left\{\frac{1}{2}\ln\left(\frac{1+x}{1-x}\right) - \psi(n)\right\} + \sum_{m=1}^{n} \frac{(-1)^m (n+m)!}{(m!)^2 (n-m)!}\psi(m)\left(\frac{1-x}{2}\right)^m \quad \text{(V-10)}$$

where

$$\psi(n) = 1 + \frac{1}{2} + \frac{1}{3} + \cdots + \frac{1}{n} \quad \text{(V-10a)}$$

When (V-10) is expanded it leads (for $n = 0, 1, 2, 3$) to

$$Q_0(x) = \frac{1}{2} \ln\left(\frac{1+x}{1-x}\right)$$

$$Q_1(x) = \frac{x}{2} \ln\left(\frac{1+x}{1-x}\right) - 1$$

$$Q_2(x) = \frac{3x^2 - 1}{4} \ln\left(\frac{1+x}{1-x}\right) - \frac{3x}{2} \qquad \text{(V-11)}$$

$$Q_3(x) = \frac{5x^3 - 3x}{4} \ln\left(\frac{1+x}{1-x}\right) - \frac{5x^2}{2} + \frac{2}{3}$$

or for $x = \cos\theta$ to

$$Q_0(\cos\theta) = \ln\left(\cot\frac{\theta}{2}\right)$$

$$Q_1(\cos\theta) = \cos\theta \ln\left(\cot\frac{\theta}{2}\right) - 1$$

$$Q_2(\cos\theta) = \frac{1}{4}(1 + 3\cos 2\theta) \ln\left(\cot\frac{\theta}{2}\right) - \frac{3}{2}\cos\theta \qquad \text{(V-12)}$$

$$Q_3(\cos\theta) = \frac{1}{8}(3\cos\theta + 5\cos 3\theta) \ln\left(\cot\frac{\theta}{2}\right) - \frac{5}{4}\cos 2\theta - \frac{7}{12}$$

The Legendre functions of the first $P_n(x)$ and second $Q_n(x)$ kind obey the following recurrence relations:

$$(n+1)R_{n+1}(x) - (2n+1)xR_n(x) + nR_{n-1}(x) = 0 \qquad \text{(V-13a)}$$

$$\frac{dR_{n+1}(x)}{dx} - x\frac{dR_n(x)}{dx} = (n+1)R_n(x) \qquad \text{(V-13b)}$$

$$x\frac{dR_n(x)}{dx} - \frac{dR_{n-1}(x)}{dx} = nR_n(x) \qquad \text{(V-13c)}$$

$$\frac{dR_{n+1}(x)}{dx} - \frac{dR_{n-1}(x)}{dx} = (2n+1)R_n(x) \qquad \text{(V-13d)}$$

$$(x^2 - 1)\frac{dR_n(x)}{dx} = nxR_n(x) - nR_{n-1}(x) = -(n+1)(xR_n - R_{n+1}) \qquad \text{(V-13e)}$$

where $R_n(x)$ can be either $P_n(x)$ or $Q_n(x)$.

Some other useful formulas involving Legendre polynomials $P_n(x)$ are

$$\int_{-1}^{1} P_m(x) P_n(x)\, dx = 0 \qquad m \neq n \qquad \text{(V-14a)}$$

$$\int_{-1}^{1} [P_n(x)]^2\, dx = \frac{2}{2n+1} \qquad \text{(V-14b)}$$

$$\int_{0}^{1} [Q_n(x)]^2\, dx = \frac{1}{2n+1}\left[\frac{\pi^2}{4} - \frac{1}{(n+1)^2} - \frac{1}{(n+2)^2} - \cdots\right] \qquad \text{(V-14c)}$$

which indicate that the Legendre polynomials are orthogonal in the range of $-1 \leq x \leq 1$. Also

$$P_n(-x) = (-1)^n P_n(x) \qquad \text{(V-15a)}$$

$$P_n(x) = P_{-n-1}(x) \qquad \text{(V-15b)}$$

$$Q_n(-x) = (-1)^{n+1} Q_n(x) \qquad \text{(V-15c)}$$

$$P_n(0) = \begin{cases} 0 & n = \text{odd} \\ (-1)^{n/2} \dfrac{1 \cdot 3 \cdot 5 \cdots (n-1)}{2 \cdot 4 \cdot 6 \cdots n} & n = \text{even} \end{cases} \qquad \text{(V-15d)}$$

$$P_n(1) = 1 \qquad \text{(V-15e)}$$

$$P_n(-1) = \begin{cases} 1 & n = \text{even} \\ -1 & n = \text{odd} \end{cases} \qquad \text{(V-15f)}$$

$$Q_n(1) = +\infty \qquad \text{(V-15g)}$$

$$Q_n(-1) = \begin{cases} -\infty & n = \text{even} \\ +\infty & n = \text{odd} \end{cases} \qquad \text{(V-15h)}$$

$$P_n(x) = \frac{1}{\pi} \int_0^\pi \left(x + \sqrt{x^2 - 1} \cos \psi \right)^n d\psi \qquad \text{(V-15i)}$$

$$\int P_n(x) \, dx = \frac{P_{n+1}(x) - P_{n-1}(x)}{(2n+1)} \qquad \text{(V-15j)}$$

Plots of $P_n(x)$ and $Q_n(x)$ for $n = 0, 1, 2, 3$ in the range of $-1 \leq x \leq 1$ are shown in Figures V-1 and V-2.

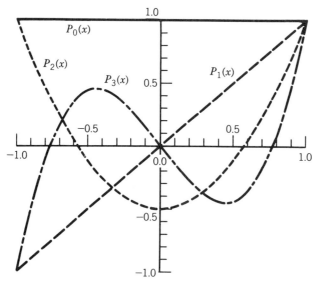

FIGURE V-1 Legendre functions of the first kind [$P_0(x)$, $P_1(x)$, $P_2(x)$, and $P_3(x)$].

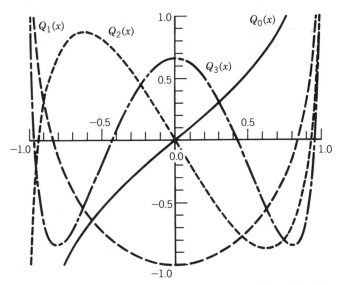

FIGURE V-2 Legendre functions of the second kind [$Q_0(x)$, $Q_1(x)$, $Q_2(x)$, and $Q_3(x)$].

V.2 ASSOCIATED LEGENDRE FUNCTIONS

In addition to the *ordinary* Legendre differential equation V-1, there also exists the *associated* Legendre differential equation

$$(1 - x^2)\frac{d^2y}{dx^2} - 2x\frac{dy}{dx} + \left[n(n + 1) - \frac{m^2}{1 - x^2}\right]y = 0 \qquad \text{(V-16)}$$

whose solution for nonnegative integer values of n and m takes the form

$$y(x) = A_1 P_n^m(x) + B_1 Q_n^m(x) \qquad \text{(V-17)}$$

where $P_n^m(x)$ and $Q_n^m(x)$ are referred to, respectively, as the *associated Legendre functions of the first and second kind*.

The associated Legendre functions $P_n^m(x)$ and $Q_n^m(x)$ of the first and second kind are related, respectively, to the Legendre functions $P_n(x)$ and $Q_n(x)$ of the first and second kind by

$$P_n^m(x) = (-1)^m (1 - x^2)^{m/2} \frac{d^m P_n(x)}{dx^m}$$

$$= (-1)^m \frac{(1 - x^2)^{m/2}}{2^n n!} \frac{d^{m+n}(x^2 - 1)^n}{dx^{m+n}} \qquad \text{(V-18a)}$$

$$Q_n^m(x) = (-1)^m (1 - x^2)^{m/2} \frac{d^m Q_n(x)}{dx^m} \qquad \text{(V-18b)}$$

The associated Legendre functions $Q_n^m(x)$ of the second kind are singular at $x = \pm 1$, as are the Legendre functions $Q_n(x)$.

When (V-18a) and (V-18b) are expanded we can write, using (V-8) and (V-11), the first few orders of $P_n^m(x)$ and $Q_n^m(x)$ as

$$P_0^0(x) = P_0(x) = 1 \qquad\qquad P_0^2(x) = 0$$
$$P_1^0(x) = P_1(x) = x \qquad\qquad P_1^2(x) = 0$$
$$P_2^0(x) = P_2(x) = \tfrac{1}{2}(3x^2 - 1) \qquad\qquad P_2^2(x) = 3(1 - x^2)$$
$$P_3^0(x) = P_3(x) = \tfrac{1}{2}(5x^3 - 3x) \qquad\qquad P_3^2(x) = 15x(1 - x^2)$$
$$\vdots \qquad\qquad\qquad\qquad \vdots$$

$$P_0^1(x) = 0 \qquad\qquad P_0^3(x) = 0$$
$$P_1^1(x) = -(1 - x^2)^{1/2} \qquad\qquad P_1^3(x) = 0 \qquad\qquad \text{(V-19a)}$$
$$P_2^1(x) = -3x(1 - x^2)^{1/2} \qquad\qquad P_2^3(x) = 0$$
$$P_3^1(x) = -\tfrac{3}{2}(5x^2 - 1)(1 - x^2)^{1/2} \qquad P_3^3(x) = -15(1 - x^2)^{3/2}$$
$$\vdots \qquad\qquad\qquad\qquad \vdots$$

$$Q_0^0(x) = Q_0(x) = \frac{1}{2}\ln\left(\frac{1 + x}{1 - x}\right)$$

$$Q_1^0(x) = Q_1(x) = \frac{x}{2}\ln\left(\frac{1 + x}{1 - x}\right) - 1$$

$$Q_2^0(x) = Q_2(x) = \frac{3x^2 - 1}{4}\ln\left(\frac{1 + x}{1 - x}\right) - \frac{3x}{2}$$

$$\vdots$$

$$Q_0^1(x) = 0$$

$$Q_1^1(x) = -(1 - x^2)^{1/2}\left[\frac{1}{2}\ln\left(\frac{1 + x}{1 - x}\right) + \frac{x}{1 - x^2}\right] \qquad \text{(V-19b)}$$

$$Q_2^1(x) = -(1 - x^2)^{1/2}\left[\frac{3x}{2}\ln\left(\frac{1 + x}{1 - x}\right) + \frac{3x^2 - 2}{1 - x^2}\right]$$

$$\vdots$$

$$Q_0^2(x) = 0$$
$$Q_1^2(x) = 0$$

$$Q_2^2(x) = (1 - x^2)^{1/2}\left[\frac{3}{2}\ln\left(\frac{1 + x}{1 - x}\right) + \frac{5x - 3x^2}{(1 - x^2)^2}\right]$$

$$\vdots$$

It should be noted that

$$P_n^0(x) = P_n(x) \qquad\qquad \text{(V-20a)}$$
$$Q_n^0(x) = Q_n(x) \qquad\qquad \text{(V-20b)}$$

$$P_n^m(x) = 0 \qquad m > n \qquad \text{(V-20c)}$$

$$Q_n^m(x) = 0 \qquad m > n \qquad \text{(V-20d)}$$

$$P_n^m(-x) = (-1)^{n-m} P_n^m(x) \qquad \text{(V-20e)}$$

$$P_n^m(x) = P_{-n-1}^m(x) \qquad \text{(V-20f)}$$

$$Q_n^m(-x) = (-1)^{n+m+1} Q_n(x) \qquad \text{(V-20g)}$$

$$P_n^m(1) = \begin{cases} 1 & m = 0 \\ 0 & m > 0 \end{cases} \qquad \text{(V-20h)}$$

$$P_n^m(0) = \begin{cases} (-1)^{(n+m)/2} \dfrac{1 \cdot 3 \cdot 5 \cdots (n+m-1)}{2 \cdot 4 \cdot 6 \cdots (n-m)} & n+m = \text{even} \\ 0 & n+m = \text{odd} \end{cases}$$
$$\text{(V-20i)}$$

$$Q_n^m(0) = \begin{cases} 0 & n+m = \text{even} \\ (-1)^{(n+m+1)/2} \dfrac{2 \cdot 4 \cdot 6 \cdots (n+m-1)}{1 \cdot 3 \cdot 5 \cdots (n-m)} & n+m = \text{odd} \end{cases}$$
$$\text{(V-20j)}$$

$$\left. \frac{d^q P_n^m(x)}{dx^q} \right|_{x=0} = (-1)^q P_n^{m+q}(0) \qquad \text{(V-20k)}$$

$$\left. \frac{d^q Q_n^m(x)}{dx^q} \right|_{x=0} = (-1)^q Q_n^{m+q}(0) \qquad \text{(V-20}\ell\text{)}$$

Orthogonality relations of $P_n^m(x)$ in the range of $-1 \leq x \leq 1$ are

$$\int_{-1}^{1} P_n^m(x) P_l^m(x) \, dx = 0 \qquad n \neq l \qquad \text{(V-21a)}$$

$$\int_{-1}^{1} [P_n^m(x)]^2 \, dx = \frac{2}{2n+1} \frac{(n+m)!}{(n-m)!} \qquad \text{(V-21b)}$$

$$\int_{-1}^{1} \left[\frac{dP_n(x)}{dx} \right]^2 dx = n(n+1) \qquad \text{(V-21c)}$$

and useful recurrence formulas are

$$(n+1-m) R_{n+1}^m(x) - (2n+1) x R_n^m(x) + (n+m) R_{n-1}^m(x) = 0 \quad \text{(V-22a)}$$

$$R_n^{m+2}(x) + \frac{2(m+1)x}{(1-x^2)^{1/2}} R_n^{m+1}(x) + (n-m)(n+m+1) R_n^m(x) = 0 \quad \text{(V-22b)}$$

where $R_n^m(x)$ can be either $P_n^m(x)$ or $Q_n^m(x)$.

When m is not an integer in the associated Legendre differential equation V-16, the solutions become more complex and can be expressed in terms of *hypergeometric functions*. These solutions are beyond this book, and the reader is referred to the literature.

APPENDIX V LEGENDRE POLYNOMIALS AND FUNCTIONS

x	$P_0(x)$	$P_1(x)$	$P_2(x)$	$P_3(x)$
-1.00	1.00000	-1.00000	1.00000	-1.00000
-0.99	1.00000	-0.99000	0.97015	-0.94075
-0.98	1.00000	-0.98000	0.94060	-0.88298
-0.97	1.00000	-0.97000	0.91135	-0.82668
-0.96	1.00000	-0.96000	0.88240	-0.77184
-0.95	1.00000	-0.95000	0.85375	-0.71844
-0.94	1.00000	-0.94000	0.82540	-0.66646
-0.93	1.00000	-0.93000	0.79735	-0.61589
-0.92	1.00000	-0.92000	0.76960	-0.56672
-0.91	1.00000	-0.91000	0.74215	-0.51893
-0.90	1.00000	-0.90000	0.71500	-0.47250
-0.89	1.00000	-0.89000	0.68815	-0.42742
-0.88	1.00000	-0.88000	0.66160	-0.38368
-0.87	1.00000	-0.87000	0.63535	-0.34126
-0.86	1.00000	-0.86000	0.60940	-0.30014
-0.85	1.00000	-0.85000	0.58375	-0.26031
-0.84	1.00000	-0.84000	0.55840	-0.22176
-0.83	1.00000	-0.83000	0.53335	-0.18447
-0.82	1.00000	-0.82000	0.50860	-0.14842
-0.81	1.00000	-0.81000	0.48415	-0.11360
-0.80	1.00000	-0.80000	0.46000	-0.08000
-0.79	1.00000	-0.79000	0.43615	-0.04760
-0.78	1.00000	-0.78000	0.41260	-0.01638
-0.77	1.00000	-0.77000	0.38935	0.01367
-0.76	1.00000	-0.76000	0.36640	0.04256
-0.75	1.00000	-0.75000	0.34375	0.07031
-0.74	1.00000	-0.74000	0.32140	0.09694
-0.73	1.00000	-0.73000	0.29935	0.12246
-0.72	1.00000	-0.72000	0.27760	0.14688
-0.71	1.00000	-0.71000	0.25615	0.17022
-0.70	1.00000	-0.70000	0.23500	0.19250
-0.69	1.00000	-0.69000	0.21415	0.21373
-0.68	1.00000	-0.68000	0.19360	0.23392
-0.67	1.00000	-0.67000	0.17335	0.25309
-0.66	1.00000	-0.66000	0.15340	0.27126
-0.65	1.00000	-0.65000	0.13375	0.28844
-0.64	1.00000	-0.64000	0.11440	0.30464
-0.63	1.00000	-0.63000	0.09535	0.31988
-0.62	1.00000	-0.62000	0.07660	0.33418
-0.61	1.00000	-0.61000	0.05815	0.34755
-0.60	1.00000	-0.60000	0.04000	0.36000
-0.59	1.00000	-0.59000	0.02215	0.37155
-0.58	1.00000	-0.58000	0.00460	0.38222
-0.57	1.00000	-0.57000	-0.01265	0.39202
-0.56	1.00000	-0.56000	-0.02960	0.40096
-0.55	1.00000	-0.55000	-0.04625	0.40906
-0.54	1.00000	-0.54000	-0.06260	0.41634
-0.53	1.00000	-0.53000	-0.07865	0.42281

ASSOCIATED LEGENDRE FUNCTIONS

x	$P_0(x)$	$P_1(x)$	$P_2(x)$	$P_3(x)$
−0.52	1.00000	−0.52000	−0.09440	0.42848
−0.51	1.00000	−0.51000	−0.10985	0.43337
−0.50	1.00000	−0.50000	−0.12500	0.43750
−0.49	1.00000	−0.49000	−0.13985	0.44088
−0.48	1.00000	−0.48000	−0.15440	0.44352
−0.47	1.00000	−0.47000	−0.16865	0.44544
−0.46	1.00000	−0.46000	−0.18260	0.44666
−0.45	1.00000	−0.45000	−0.19625	0.44719
−0.44	1.00000	−0.44000	−0.20960	0.44704
−0.43	1.00000	−0.43000	−0.22265	0.44623
−0.42	1.00000	−0.42000	−0.23540	0.44478
−0.41	1.00000	−0.41000	−0.24785	0.44270
−0.40	1.00000	−0.40000	−0.26000	0.44000
−0.39	1.00000	−0.39000	−0.27185	0.43670
−0.38	1.00000	−0.38000	−0.28340	0.43282
−0.37	1.00000	−0.37000	−0.29465	0.42837
−0.36	1.00000	−0.36000	−0.30560	0.42336
−0.35	1.00000	−0.35000	−0.31625	0.41781
−0.34	1.00000	−0.34000	−0.32660	0.41174
−0.33	1.00000	−0.33000	−0.33665	0.40516
−0.32	1.00000	−0.32000	−0.34640	0.39808
−0.31	1.00000	−0.31000	−0.35585	0.39052
−0.30	1.00000	−0.30000	−0.36500	0.38250
−0.29	1.00000	−0.29000	−0.37385	0.37403
−0.28	1.00000	−0.28000	−0.38240	0.36512
−0.27	1.00000	−0.27000	−0.39065	0.35579
−0.26	1.00000	−0.26000	−0.39860	0.34606
−0.25	1.00000	−0.25000	−0.40625	0.33594
−0.24	1.00000	−0.24000	−0.41360	0.32544
−0.23	1.00000	−0.23000	−0.42065	0.31458
−0.22	1.00000	−0.22000	−0.42740	0.30338
−0.21	1.00000	−0.21000	−0.43385	0.29185
−0.20	1.00000	−0.20000	−0.44000	0.28000
−0.19	1.00000	−0.19000	−0.44585	0.26785
−0.18	1.00000	−0.18000	−0.45140	0.25542
−0.17	1.00000	−0.17000	−0.45665	0.24272
−0.16	1.00000	−0.16000	−0.46160	0.22976
−0.15	1.00000	−0.15000	−0.46625	0.21656
−0.14	1.00000	−0.14000	−0.47060	0.20314
−0.13	1.00000	−0.13000	−0.47465	0.18951
−0.12	1.00000	−0.12000	−0.47840	0.17568
−0.11	1.00000	−0.11000	−0.48185	0.16167
−0.10	1.00000	−0.10000	−0.48500	0.14750
−0.09	1.00000	−0.09000	−0.48785	0.13318
−0.08	1.00000	−0.08000	−0.49040	0.11872
−0.07	1.00000	−0.07000	−0.49265	0.10414
−0.06	1.00000	−0.06000	−0.49460	0.08946
−0.05	1.00000	−0.05000	−0.49625	0.07469

APPENDIX V LEGENDRE POLYNOMIALS AND FUNCTIONS

x	$P_0(x)$	$P_1(x)$	$P_2(x)$	$P_3(x)$
−0.04	1.00000	−0.04000	−0.49760	0.05984
−0.03	1.00000	−0.03000	−0.49865	0.04493
−0.02	1.00000	−0.02000	−0.49940	0.02998
−0.01	1.00000	−0.01000	−0.49985	0.01500
0.00	1.00000	0.00000	−0.50000	0.00000
0.01	1.00000	0.01000	−0.49985	−0.01500
0.02	1.00000	0.02000	−0.49940	−0.02998
0.03	1.00000	0.03000	−0.49865	−0.04493
0.04	1.00000	0.04000	−0.49760	−0.05984
0.05	1.00000	0.05000	−0.49625	−0.07469
0.06	1.00000	0.06000	−0.49460	−0.08946
0.07	1.00000	0.07000	−0.49265	−0.10414
0.08	1.00000	0.08000	−0.49040	−0.11872
0.09	1.00000	0.09000	−0.48785	−0.13318
0.10	1.00000	0.10000	−0.48500	−0.14750
0.11	1.00000	0.11000	−0.48185	−0.16167
0.12	1.00000	0.12000	−0.47840	−0.17568
0.13	1.00000	0.13000	−0.47465	−0.18951
0.14	1.00000	0.14000	−0.47060	−0.20314
0.15	1.00000	0.15000	−0.46625	−0.21656
0.16	1.00000	0.16000	−0.46160	−0.22976
0.17	1.00000	0.17000	−0.45665	−0.24272
0.18	1.00000	0.18000	−0.45140	−0.25542
0.19	1.00000	0.19000	−0.44585	−0.26785
0.20	1.00000	0.20000	−0.44000	−0.28000
0.21	1.00000	0.21000	−0.43385	−0.29185
0.22	1.00000	0.22000	−0.42740	−0.30338
0.23	1.00000	0.23000	−0.42065	−0.31458
0.24	1.00000	0.24000	−0.41360	−0.32544
0.25	1.00000	0.25000	−0.40625	−0.33594
0.26	1.00000	0.26000	−0.39860	−0.34606
0.27	1.00000	0.27000	−0.39065	−0.35579
0.28	1.00000	0.28000	−0.38240	−0.36512
0.29	1.00000	0.29000	−0.37385	−0.37403
0.30	1.00000	0.30000	−0.36500	−0.38250
0.31	1.00000	0.31000	−0.35585	−0.39052
0.32	1.00000	0.32000	−0.34640	−0.39808
0.33	1.00000	0.33000	−0.33665	−0.40516
0.34	1.00000	0.34000	−0.32660	−0.41174
0.35	1.00000	0.35000	−0.31625	−0.41781
0.36	1.00000	0.36000	−0.30560	−0.42336
0.37	1.00000	0.37000	−0.29465	−0.42837
0.38	1.00000	0.38000	−0.28340	−0.43282
0.39	1.00000	0.39000	−0.27185	−0.43670
0.40	1.00000	0.40000	−0.26000	−0.44000
0.41	1.00000	0.41000	−0.24785	−0.44270
0.42	1.00000	0.42000	−0.23540	−0.44478
0.43	1.00000	0.43000	−0.22265	−0.44623

ASSOCIATED LEGENDRE FUNCTIONS

x	$P_0(x)$	$P_1(x)$	$P_2(x)$	$P_3(x)$
0.44	1.00000	0.44000	-0.20960	-0.44704
0.45	1.00000	0.45000	-0.19625	-0.44719
0.46	1.00000	0.46000	-0.18260	-0.44666
0.47	1.00000	0.47000	-0.16865	-0.44544
0.48	1.00000	0.48000	-0.15440	-0.44352
0.49	1.00000	0.49000	-0.13985	-0.44088
0.50	1.00000	0.50000	-0.12500	-0.43750
0.51	1.00000	0.51000	-0.10985	-0.43337
0.52	1.00000	0.52000	-0.09440	-0.42848
0.53	1.00000	0.53000	-0.07865	-0.42281
0.54	1.00000	0.54000	-0.06260	-0.41634
0.55	1.00000	0.55000	-0.04625	-0.40906
0.56	1.00000	0.56000	-0.02960	-0.40096
0.57	1.00000	0.57000	-0.01265	-0.39202
0.58	1.00000	0.58000	0.00460	-0.38222
0.59	1.00000	0.59000	0.02215	-0.37155
0.60	1.00000	0.60000	0.04000	-0.36000
0.61	1.00000	0.61000	0.05815	-0.34755
0.62	1.00000	0.62000	0.07660	-0.33418
0.63	1.00000	0.63000	0.09535	-0.31988
0.64	1.00000	0.64000	0.11440	-0.30464
0.65	1.00000	0.65000	0.13375	-0.28844
0.66	1.00000	0.66000	0.15340	-0.27126
0.67	1.00000	0.67000	0.17335	-0.25309
0.68	1.00000	0.68000	0.19360	-0.23392
0.69	1.00000	0.69000	0.21415	-0.21373
0.70	1.00000	0.70000	0.23500	-0.19250
0.71	1.00000	0.71000	0.25615	-0.17022
0.72	1.00000	0.72000	0.27760	-0.14688
0.73	1.00000	0.73000	0.29935	-0.12246
0.74	1.00000	0.74000	0.32140	-0.09694
0.75	1.00000	0.75000	0.34375	-0.07031
0.76	1.00000	0.76000	0.36640	-0.04256
0.77	1.00000	0.77000	0.38935	-0.01367
0.78	1.00000	0.78000	0.41260	0.01638
0.79	1.00000	0.79000	0.43615	0.04760
0.80	1.00000	0.80000	0.46000	0.08000
0.81	1.00000	0.81000	0.48415	0.11360
0.82	1.00000	0.82000	0.50860	0.14842
0.83	1.00000	0.83000	0.53335	0.18447
0.84	1.00000	0.84000	0.55840	0.22176
0.85	1.00000	0.85000	0.58375	0.26031
0.86	1.00000	0.86000	0.60940	0.30014
0.87	1.00000	0.87000	0.63535	0.34126
0.88	1.00000	0.88000	0.66160	0.38368
0.89	1.00000	0.89000	0.68815	0.42742
0.90	1.00000	0.90000	0.71500	0.47250
0.91	1.00000	0.91000	0.74215	0.51893

x	$P_0(x)$	$P_1(x)$	$P_2(x)$	$P_3(x)$
0.92	1.00000	0.92000	0.76960	0.56672
0.93	1.00000	0.93000	0.79735	0.61589
0.94	1.00000	0.94000	0.82540	0.66646
0.95	1.00000	0.95000	0.85375	0.71844
0.96	1.00000	0.96000	0.88240	0.77184
0.97	1.00000	0.97000	0.91135	0.82668
0.98	1.00000	0.98000	0.94060	0.88298
0.99	1.00000	0.99000	0.97015	0.94075
1.00	1.00000	1.00000	1.00000	1.00000

x	$Q_0(x)$	$Q_1(x)$	$Q_2(x)$	$Q_3(x)$
−1.00	−∞	+∞	−∞	+∞
−0.99	−2.64665	1.62019	−1.08265	0.70625
−0.98	−2.29756	1.25161	−0.69109	0.29437
−0.97	−2.09230	1.02953	−0.45181	0.04408
−0.96	−1.94591	0.86807	−0.27707	−0.13540
−0.95	−1.83178	0.74019	−0.13888	−0.27356
−0.94	−1.73805	0.63377	−0.02459	−0.38399
−0.93	−1.65839	0.54230	0.07268	−0.47419
−0.92	−1.58903	0.46190	0.15708	−0.54880
−0.91	−1.52752	0.39005	0.23135	−0.61091
−0.90	−1.47222	0.32500	0.29736	−0.66271
−0.89	−1.42193	0.26551	0.35650	−0.70582
−0.88	−1.37577	0.21068	0.40979	−0.74148
−0.87	−1.33308	0.15978	0.45803	−0.77066
−0.86	−1.29334	0.11228	0.50184	−0.79415
−0.85	−1.25615	0.06773	0.54172	−0.81259
−0.84	−1.22117	0.02579	0.57810	−0.82653
−0.83	−1.18814	−0.01385	0.61131	−0.83641
−0.82	−1.15682	−0.05141	0.64164	−0.84264
−0.81	−1.12703	−0.08711	0.66935	−0.84555
−0.80	−1.09861	−0.12111	0.69464	−0.84544
−0.79	−1.07143	−0.15357	0.71770	−0.84259
−0.78	−1.04537	−0.18461	0.73868	−0.83721
−0.77	−1.02033	−0.21435	0.75774	−0.82953
−0.76	−0.99622	−0.24288	0.77499	−0.81973
−0.75	−0.97296	−0.27028	0.79055	−0.80799
−0.74	−0.95048	−0.29665	0.80452	−0.79447
−0.73	−0.92873	−0.32203	0.81699	−0.77931
−0.72	−0.90764	−0.34650	0.82804	−0.76265
−0.71	−0.88718	−0.37010	0.83775	−0.74460
−0.70	−0.86730	−0.39289	0.84618	−0.72529
−0.69	−0.84796	−0.41491	0.85341	−0.70481
−0.68	−0.82911	−0.43620	0.85948	−0.68328
−0.67	−0.81074	−0.45680	0.86446	−0.66078
−0.66	−0.79281	−0.47674	0.86838	−0.63739
−0.65	−0.77530	−0.49606	0.87130	−0.61321

ASSOCIATED LEGENDRE FUNCTIONS

x	$Q_0(x)$	$Q_1(x)$	$Q_2(x)$	$Q_3(x)$
−0.64	−0.75817	−0.51477	0.87326	−0.58830
−0.63	−0.74142	−0.53291	0.87431	−0.56275
−0.62	−0.72500	−0.55050	0.87446	−0.53662
−0.61	−0.70892	−0.56756	0.87378	−0.50997
−0.60	−0.69315	−0.58411	0.87227	−0.48287
−0.59	−0.67767	−0.60018	0.86999	−0.45537
−0.58	−0.66246	−0.61577	0.86695	−0.42754
−0.57	−0.64752	−0.63091	0.86319	−0.39942
−0.56	−0.63283	−0.64561	0.85873	−0.37107
−0.55	−0.61838	−0.65989	0.85360	−0.34254
−0.54	−0.60416	−0.67376	0.84782	−0.31387
−0.53	−0.59015	−0.68722	0.84141	−0.28510
−0.52	−0.57634	−0.70030	0.83441	−0.25628
−0.51	−0.56273	−0.71301	0.82682	−0.22745
−0.50	−0.54931	−0.72535	0.81866	−0.19865
−0.49	−0.53606	−0.73733	0.80997	−0.16992
−0.48	−0.52298	−0.74897	0.80075	−0.14129
−0.47	−0.51007	−0.76027	0.79102	−0.11279
−0.46	−0.49731	−0.77124	0.78081	−0.08446
−0.45	−0.48470	−0.78188	0.77012	−0.05634
−0.44	−0.47223	−0.79222	0.75898	−0.02844
−0.43	−0.45990	−0.80224	0.74740	−0.00080
−0.42	−0.44769	−0.81197	0.73539	0.02654
−0.41	−0.43561	−0.82140	0.72297	0.05357
−0.40	−0.42365	−0.83054	0.71015	0.08026
−0.39	−0.41180	−0.83940	0.69695	0.10658
−0.38	−0.40006	−0.84798	0.68338	0.13251
−0.37	−0.38842	−0.85628	0.66945	0.15803
−0.36	−0.37689	−0.86432	0.65518	0.18311
−0.35	−0.36544	−0.87209	0.64057	0.20773
−0.34	−0.35409	−0.87961	0.62565	0.23187
−0.33	−0.34283	−0.88687	0.61041	0.25552
−0.32	−0.33165	−0.89387	0.59488	0.27864
−0.31	−0.32055	−0.90063	0.57907	0.30124
−0.30	−0.30952	−0.90714	0.56297	0.32328
−0.29	−0.29857	−0.91342	0.54662	0.34474
−0.28	−0.28768	−0.91945	0.53001	0.36563
−0.27	−0.27686	−0.92525	0.51316	0.38591
−0.26	−0.26611	−0.93081	0.49607	0.40558
−0.25	−0.25541	−0.93615	0.47876	0.42461
−0.24	−0.24477	−0.94125	0.46124	0.44301
−0.23	−0.23419	−0.94614	0.44351	0.46074
−0.22	−0.22366	−0.95080	0.42559	0.47781
−0.21	−0.21317	−0.95523	0.40748	0.49420
−0.20	−0.20273	−0.95945	0.38920	0.50990
−0.19	−0.19234	−0.96346	0.37075	0.52490
−0.18	−0.18198	−0.96724	0.35215	0.53918
−0.17	−0.17167	−0.97082	0.33339	0.55275

APPENDIX V LEGENDRE POLYNOMIALS AND FUNCTIONS

x	$Q_0(x)$	$Q_1(x)$	$Q_2(x)$	$Q_3(x)$
−0.16	−0.16139	−0.97418	0.31450	0.56559
−0.15	−0.15114	−0.97733	0.29547	0.57769
−0.14	−0.14093	−0.98027	0.27632	0.58904
−0.13	−0.13074	−0.98300	0.25706	0.59964
−0.12	−0.12058	−0.98553	0.23769	0.60948
−0.11	−0.11045	−0.98785	0.21822	0.61856
−0.10	−0.10034	−0.98997	0.19866	0.62687
−0.09	−0.09024	−0.99188	0.17903	0.63440
−0.08	−0.08017	−0.99359	0.15932	0.64115
−0.07	−0.07011	−0.99509	0.13954	0.64711
−0.06	−0.06007	−0.99640	0.11971	0.65229
−0.05	−0.05004	−0.99750	0.09983	0.65668
−0.04	−0.04002	−0.99840	0.07991	0.66027
−0.03	−0.03001	−0.99910	0.05996	0.66307
−0.02	−0.02000	−0.99960	0.03999	0.66507
−0.01	−0.01000	−0.99990	0.02000	0.66627
0.00	0.00000	−1.00000	0.00000	0.66667
0.01	0.01000	−0.99990	−0.02000	0.66627
0.02	0.02000	−0.99960	−0.03999	0.66507
0.03	0.03001	−0.99910	−0.05996	0.66307
0.04	0.04002	−0.99840	−0.07991	0.66027
0.05	0.05004	−0.99750	−0.09983	0.65668
0.06	0.06007	−0.99640	−0.11971	0.65229
0.07	0.07011	−0.99509	−0.13954	0.64711
0.08	0.08017	−0.99359	−0.15932	0.64115
0.09	0.09024	−0.99188	−0.17903	0.63440
0.10	0.10033	−0.98997	−0.19866	0.62687
0.11	0.11045	−0.98785	−0.21822	0.61856
0.12	0.12058	−0.98553	−0.23769	0.60948
0.13	0.13074	−0.98300	−0.25706	0.59964
0.14	0.14092	−0.98027	−0.27632	0.58904
0.15	0.15114	−0.97733	−0.29547	0.57769
0.16	0.16139	−0.97418	−0.31450	0.56559
0.17	0.17167	−0.97082	−0.33339	0.55275
0.18	0.18198	−0.96724	−0.35215	0.53919
0.19	0.19234	−0.96346	−0.37075	0.52490
0.20	0.20273	−0.95945	−0.38920	0.50990
0.21	0.21317	−0.95523	−0.40748	0.49420
0.22	0.22366	−0.95080	−0.42559	0.47782
0.23	0.23419	−0.94614	−0.44351	0.46075
0.24	0.24477	−0.94125	−0.46124	0.44301
0.25	0.25541	−0.93615	−0.47876	0.42461
0.26	0.26611	−0.93081	−0.49607	0.40558
0.27	0.27686	−0.92525	−0.51316	0.38591
0.28	0.28768	−0.91945	−0.53001	0.36563
0.29	0.29857	−0.91342	−0.54662	0.34474
0.30	0.30952	−0.90714	−0.56297	0.32328
0.31	0.32054	−0.90063	−0.57907	0.30124

ASSOCIATED LEGENDRE FUNCTIONS

x	$Q_0(x)$	$Q_1(x)$	$Q_2(x)$	$Q_3(x)$
0.32	0.33165	−0.89387	−0.59488	0.27865
0.33	0.34283	−0.88687	−0.61041	0.25552
0.34	0.35409	−0.87961	−0.62565	0.23187
0.35	0.36544	−0.87210	−0.64057	0.20773
0.36	0.37689	−0.86432	−0.65518	0.18311
0.37	0.38842	−0.85628	−0.66945	0.15803
0.38	0.40006	−0.84798	−0.68338	0.13251
0.39	0.41180	−0.83940	−0.69695	0.10658
0.40	0.42365	−0.83054	−0.71015	0.08026
0.41	0.43561	−0.82140	−0.72297	0.05357
0.42	0.44769	−0.81197	−0.73539	0.02654
0.43	0.45990	−0.80225	−0.74740	−0.00080
0.44	0.47223	−0.79222	−0.75898	−0.02844
0.45	0.48470	−0.78189	−0.77012	−0.05633
0.46	0.49731	−0.77124	−0.78081	−0.08446
0.47	0.51007	−0.76027	−0.79102	−0.11279
0.48	0.52298	−0.74897	−0.80075	−0.14129
0.49	0.53606	−0.73733	−0.80997	−0.16992
0.50	0.54931	−0.72535	−0.81866	−0.19865
0.51	0.56273	−0.71301	−0.82682	−0.22746
0.52	0.57634	−0.70030	−0.83441	−0.25628
0.53	0.59014	−0.68722	−0.84141	−0.28510
0.54	0.60416	−0.67376	−0.84782	−0.31387
0.55	0.61838	−0.65989	−0.85360	−0.34254
0.56	0.63283	−0.64561	−0.85873	−0.37107
0.57	0.64752	−0.63091	−0.86319	−0.39942
0.58	0.66246	−0.61577	−0.86695	−0.42754
0.59	0.67767	−0.60018	−0.86999	−0.45537
0.60	0.69315	−0.58411	−0.87227	−0.48286
0.61	0.70892	−0.56756	−0.87378	−0.50997
0.62	0.72500	−0.55050	−0.87446	−0.53662
0.63	0.74142	−0.53291	−0.87431	−0.56275
0.64	0.75817	−0.51477	−0.87327	−0.58830
0.65	0.77530	−0.49606	−0.87130	−0.61321
0.66	0.79281	−0.47674	−0.86838	−0.63739
0.67	0.81074	−0.45680	−0.86446	−0.66078
0.68	0.82911	−0.43620	−0.85948	−0.68328
0.69	0.84795	−0.41491	−0.85341	−0.70481
0.70	0.86730	−0.39289	−0.84618	−0.72529
0.71	0.88718	−0.37010	−0.83775	−0.74460
0.72	0.90765	−0.34649	−0.82804	−0.76265
0.73	0.92873	−0.32203	−0.81699	−0.77931
0.74	0.95048	−0.29665	−0.80452	−0.79447
0.75	0.97296	−0.27028	−0.79055	−0.80799
0.76	0.99622	−0.24288	−0.77499	−0.81973
0.77	1.02033	−0.21435	−0.75774	−0.82953
0.78	1.04537	−0.18461	−0.73868	−0.83721
0.79	1.07143	−0.15357	−0.71770	−0.84259

x	$Q_0(x)$	$Q_1(x)$	$Q_2(x)$	$Q_3(x)$
0.80	1.09861	−0.12111	−0.69464	−0.84544
0.81	1.12703	−0.08711	−0.66935	−0.84555
0.82	1.15682	−0.05141	−0.64164	−0.84264
0.83	1.18814	−0.01385	−0.61131	−0.83641
0.84	1.22117	0.02579	−0.57810	−0.82653
0.85	1.25615	0.06773	−0.54172	−0.81259
0.86	1.29334	0.11227	−0.50184	−0.79415
0.87	1.33308	0.15978	−0.45803	−0.77066
0.88	1.37577	0.21068	−0.40979	−0.74148
0.89	1.42192	0.26551	−0.35651	−0.70582
0.90	1.47222	0.32500	−0.29737	−0.66271
0.91	1.52752	0.39005	−0.23135	−0.61091
0.92	1.58903	0.46191	−0.15708	−0.54880
0.93	1.65839	0.54231	−0.07268	−0.47419
0.94	1.73805	0.63376	0.02458	−0.38400
0.95	1.83178	0.74019	0.13888	−0.27357
0.96	1.94591	0.86807	0.27707	−0.13540
0.97	2.09230	1.02953	0.45182	0.04409
0.98	2.29755	1.25160	0.69107	0.29435
0.99	2.64664	1.62017	1.08264	0.70624
1.00	$+\infty$	$+\infty$	$+\infty$	$+\infty$

APPENDIX VI

THE METHOD OF STEEPEST DESCENT (SADDLE-POINT METHOD)

The method of steepest descent (saddle-point method) is used to evaluate for large values of β, in an approximate sense, integrals of the form

$$I(\beta) = \int_C F(z) e^{\beta f(z)} \, dz \qquad \text{(VI-1)}$$

where $f(z)$ is an analytic function and C is the path of integration in the complex z plane, as shown in Figure VI-1. The philosophy of the method is that, within certain limits, the path of integration can be altered continuously without affecting the value of the integral provided that, during the deformation, the path does not pass through singularities of the integrand. The new path can also be chosen in such a way that most of the contributions to the integral are attributed only to small segments of the new path. The integrand can then be approximated by simpler functions over the important parts of the path and its behavior can be neglected over all other segments. If during the deformation from the old to the new paths singularities for the function $F(z)$ are encountered, we must add (a) the residue when crossing a pole and (b) the integral, when encountering a branch point, over the edges of an appropriate cut where the function is single-valued.

In general, we can write (VI-1) as

$$I(\beta) = \int_C F(z) e^{\beta f(z)} \, dz = I_{\text{SI}} + I_{\text{SDP}} \qquad \text{(VI-2)}$$

where I_{SI} takes into account the contributions from the singularities and I_{SDP} from the steepest-descent path. In this Appendix our concern will be the I_{SDP} contribution of (VI-2) or

$$I_{\text{SDP}} = \int_{\text{SDP}} F(z) e^{\beta f(z)} \, dz \qquad \text{(VI-3)}$$

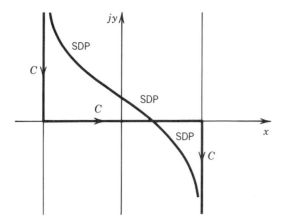

FIGURE VI-1 C and SDP paths.

where now $F(z)$ is assumed to be a well behaved function and $f(z)$ to be analytic in the complex z plane ($z = x + jy$).

Assuming that β is real and positive, we can write

$$f(z) = U(z) + jV(z) = U(x, y) + jV(x, y) \tag{VI-4}$$

where U and V are real functions, so that the integrand of (VI-3) can be written as

$$F(z)e^{\beta f(z)} = F(x, y)e^{\beta U(x, y)}e^{j\beta V(x, y)} \tag{VI-5}$$

If $f(z)$ is an analytic function, the Cauchy–Riemann conditions state that

$$\frac{df}{dz} = \frac{\partial U}{\partial x} + j\frac{\partial V}{\partial x} = -j\frac{\partial U}{\partial y} + \frac{\partial V}{\partial y} \tag{VI-6}$$

or

$$\frac{\partial U}{\partial x} = \frac{\partial V}{\partial y} \tag{VI-6a}$$

$$\frac{\partial U}{\partial y} = -\frac{\partial V}{\partial x} \tag{VI-6b}$$

If there exists a point $z_s = x_s + jy_s$ where

$$\left.\frac{df}{dz}\right|_{z=z_s} \equiv f'(z = z_s) = f'(z_s) = 0 \tag{VI-7}$$

then

$$\frac{\partial U}{\partial x} = \frac{\partial V}{\partial y} = \frac{\partial U}{\partial y} = \frac{\partial V}{\partial x} = 0 \quad \text{at } x = x_s, y = y_s \tag{VI-8}$$

The surfaces $U(x, y) = $ constant and $V(x, y) = $ constant satisfy (VI-8) but do not have an absolute maximum or minimum at (x_s, y_s). The Cauchy–Riemann conditions (VI-6a) and (VI-6b) also tell us that, for a first-order saddle point

$[f''(z_s) \neq 0]$

$$\frac{\partial^2 U}{\partial x^2} = \frac{\partial^2 V}{\partial x \, \partial y} = \frac{\partial}{\partial y}\left(\frac{\partial V}{\partial x}\right) = \frac{\partial}{\partial y}\left(-\frac{\partial U}{\partial y}\right) = -\frac{\partial^2 U}{\partial y^2}$$

$$\frac{\partial^2 U}{\partial x^2} = -\frac{\partial^2 U}{\partial y^2} \qquad \text{(VI-9a)}$$

$$\frac{\partial^2 V}{\partial y^2} = \frac{\partial^2 U}{\partial y \, \partial x} = \frac{\partial}{\partial x}\left(\frac{\partial U}{\partial y}\right) = \frac{\partial}{\partial x}\left(-\frac{\partial V}{\partial x}\right) = -\frac{\partial^2 V}{\partial x^2}$$

$$\frac{\partial^2 V}{\partial y^2} = -\frac{\partial^2 V}{\partial x^2} \qquad \text{(VI-9b)}$$

Because of (VI-8), (VI-9a), and (VI-9b) neither $U(x, y)$ nor $V(x, y)$ has a maximum or a minimum at such a point z_s, but a *minimax* or saddle point. If $U(x, y)$ has an extremum at z_s, then ΔU is positive for some changes in x and y and negative for others (a positive slope in one direction and negative at right angles to it) whereas ΔV remains constant. The same holds if $V(x, y)$ has an extremum. Thus the lines of most rapid increase or decrease of one part of the complex function $f(z) = U(x, y) + jV(x, y)$ are constant lines of the other.

The magnitude of the exponential factor $e^{\beta U(x, y)}$ of (VI-5) may increase, decrease, or remain constant depending on the choice of the path through the saddle point z_s. To avoid $U(x, y)$ contributing in the exponential of (VI-5) over a large part of the path, we must pass the saddle point in the fastest possible manner. This is accomplished by taking the path of integration through the saddle point and leaving it along the line of the most rapid decrease (steepest descent) of the function $U(x, y)$.

Referring to Figure VI-2, let us choose a path P through the saddle point z_s with differential length ds. Then

$$\frac{dU}{ds} = \frac{\partial U}{\partial x}\frac{\partial x}{\partial s} + \frac{\partial U}{\partial y}\frac{\partial y}{\partial s} = \frac{\partial U}{\partial x}\cos\gamma + \frac{\partial U}{\partial y}\sin\gamma \qquad \text{(VI-10)}$$

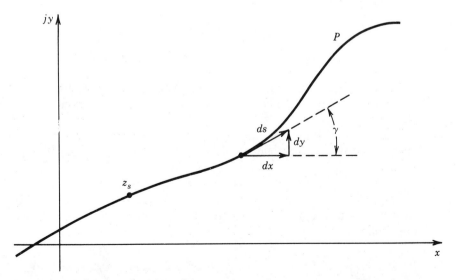

FIGURE VI-2 Steepest descent path in complex z plane.

where γ is the angle between ds and the x axis. The function dU/ds is a maximum for values of γ defined by

$$\frac{\partial}{\partial \gamma}\left(\frac{\partial U}{\partial s}\right) = \frac{\partial^2 U}{\partial s^2} = \frac{\partial}{\partial \gamma}\left[\frac{\partial U}{\partial x}\cos\gamma + \frac{\partial U}{\partial y}\sin\gamma\right] = 0 \qquad \text{(VI-11)}$$

or

$$\frac{\partial}{\partial \gamma}\left(\frac{\partial U}{\partial s}\right) = -\sin\gamma\left(\frac{\partial U}{\partial x}\right) + \cos\gamma\left(\frac{\partial U}{\partial y}\right) = 0 \qquad \text{(VI-11a)}$$

Using the Cauchy–Riemann conditions (VI-6a) and (VI-6b), we can write (VI-11a) as

$$\frac{\partial}{\partial \gamma}\left(\frac{\partial U}{\partial s}\right) = -\sin\gamma\left(\frac{\partial V}{\partial y}\right) + \cos\gamma\left(-\frac{\partial V}{\partial x}\right) = -\left[\frac{\partial V}{\partial y}\sin\gamma + \frac{\partial V}{\partial x}\cos\gamma\right] = 0$$

$$\frac{\partial}{\partial \gamma}\left(\frac{\partial U}{\partial s}\right) = -\left(\frac{\partial V}{\partial y}\frac{\partial y}{\partial s} + \frac{\partial V}{\partial x}\frac{\partial x}{\partial s}\right) = -\left(\frac{dV}{ds}\right) = 0 \qquad \text{(VI-12)}$$

Thus V = constant for paths along which $U(x, y)$ changes most rapidly (and vice versa), so the *steepest amplitude path* is a *constant phase path*. These are known as *steepest ascent* or *descent* paths. We choose the steepest descent path, thus the name *method of steepest descent*. Since β is real and positive, the exponential $\exp[\beta U(x, y)]$ of (VI-5) will decrease rapidly with distance from the saddle point and only a small portion of the integration path, including the saddle point, will make any significant contributions to the value of the entire integral.

To find the path of steepest descent, we form a function

$$f(z) = f(z_s) - s^2 \qquad \text{(VI-13)}$$

where z_s is the saddle point and s is real ($-\infty \leq s \leq +\infty$). The saddle point corresponds to $s = 0$. Using (VI-4) we can write (VI-13) as

$$U(z) = U(z_s) - s^2 \qquad \text{(VI-13a)}$$
$$V(z) = V(z_s) \quad \text{steepest descent path} \qquad \text{(VI-13b)}$$

Since the imaginary part remains constant, while the real part attains maximum at $s = 0$ and decreases for other values, the path of steepest descent is described by (VI-13b).

To evaluate the integral of (VI-3), we first find the saddle point z_s by (VI-7). Next we express $f(z)$, around the saddle point z_s, by a truncated Taylor series

$$f(z) \simeq f(z_s) + \tfrac{1}{2}(z - z_s)^2 f''(z_s) \qquad \text{(VI-14)}$$

since $f'(z_s) = 0$. The double prime indicates a second derivative with respect to z. Substitution of (VI-14) into (VI-3) leads to

$$I(\beta) = \int_{\text{SDP}} F(z)e^{\beta f(z)}\,dz \simeq e^{\beta f(z_s)}\int_{\text{SDP}} F(z)e^{(\beta/2)(z-z_s)^2 f''(z_s)}\,dz \qquad \text{(VI-15)}$$

Letting

$$-\beta(z - z_s)^2 f''(z_s) = \xi^2 \tag{VI-15a}$$

$$dz = \frac{d\xi}{\sqrt{-\beta f''(z_s)}} \tag{VI-15b}$$

we can write (VI-15), by extending the limits to infinity, as

$$I(\beta) \simeq \frac{e^{\beta f(z_s)}}{\sqrt{-\beta f''(z_s)}} \int_{-\infty}^{+\infty} F(z) e^{-\xi^2/2} d\xi \tag{VI-16}$$

Assuming that $F(z)$ is a slow varying function in the neighborhood of the saddle point, we can write (VI-16), by replacing $F(z)$ by $F(z_s)$, as

$$I(\beta) \simeq \frac{e^{\beta f(z_s)}}{\sqrt{-\beta f''(z_s)}} 2F(z_s) \int_0^\infty e^{-\xi^2/2} d\xi$$

$$I(\beta) \simeq \frac{e^{\beta f(z_s)}}{\sqrt{-\beta f''(z_s)}} F(z_s) 2\sqrt{\frac{\pi}{2}} = \sqrt{\frac{2\pi}{-\beta f''(z_s)}} F(z_s) e^{\beta f(z_s)} \tag{VI-17}$$

If more than one saddle point exists, then (VI-16) can be written as

$$I(\beta) \simeq \sqrt{\frac{2\pi}{\beta}} \sum_{s=1}^{N} \frac{F(z_s)}{\sqrt{-f''(z_s)}} e^{\beta f(z_s)} \tag{VI-18}$$

where N is equal to the number of saddle points. The summation assumes, through the principle of superposition, that the contribution of each saddle point is not affected by the presence of the others.

Equation VI-18 accounts for the contribution to the integral (VI-3) from first-order saddle points $[f'(z_s) = 0$ but $f''(z_s) \neq 0]$. For second-order saddle points $[f'(z_s) = 0$ and $f''(z_s) = 0]$, the expression is different. For general forms of $f(z)$, the determination of all the steepest descent paths may be too complicated.

If a constant level path is chosen such that $|\exp[\beta U(x, y)]|$ remains constant everywhere and $\exp[j\beta V(x, y)]$ varies most rapidly away from the saddle points, the evaluation of the integral can be carried out from contributions near the saddle points. Since the phase factor $\exp(j\beta V)$ is stationary at and near the saddle points, and oscillates very rapidly in the remaining parts of the path, it makes the net contributions from the other parts, excluding the saddle points, very negligible. This is known as the method of *stationary phase*, and it may not yield the same result as the method of steepest descent because their corresponding paths are different. The two will lead to identical results if the constant level path can be continuously deformed to the steepest descent path. This is accomplished if the two paths have identical terminations and there are no singularities of $f(z)$ in the region between the two paths.

INDEX

A

Acceptors, 64
Aluminum:
    atom, 43
    charge density, 66
    conductivity, 66
    mobility, 66
Ampere's law, 6
Angular frequency:
    resonant, 76
    natural, 76
Array factor, 320
Astigmatic tube of rays, 749, 750, 754, 766
Atom, 42–44
    aluminum, 43
    germanium, 43
    hydrogen, 43
    silicon, 43
Atomic number, 42
Attenuation constant, 107, 113–115, 146
    effective, 214–220
Axial ratio, 164, 168, 237–243

B

Backscattered, 578
Basis functions, 673, 682, 683–687
    entire domain, 685–686
        Hermite, 686
        Legendre, 686
        Maclaurin, 686
        Tschebyscheff, 686
    subdomain, 683–687
        piecewise constant, 683–684
        piecewise linear, 684–685
        piecewise sinusoid, 685–686
        truncated cosine, 685, 687
Bednorz, J. G., 69
Bessel functions, 119, 121, 471–474, 477–478, 934–946
    asymptotic forms, 935
    derivative, 936
    differential equation, 118, 934
    first kind, 119, 121, 934–935
    integrals, 936–937, 945
        graphs, 945
        tables, 944
    graphs, 942, 945
    modified, 937–938
        first kind, 937–938
        graphs, 945–946
        second kind, 937–938
    second kind, 119, 121, 934–935
    spherical, 125–126, 543, 551–552, 938
        tables, 939–941, 943
        zeroes, 559, 561
    zeroes:
        derivative, 472
        function, 478
Biconical transmission line, 552–557
    characteristic impedance, 556–557
    modes:
        TE^r, 552–554
        TEM^r, 554–557
        TM^r, 554
Bilinear formula, 869
Binomial impedance transformer, 231–233
Bistatic, 578
    circular cylinder:
        TE^z, 612–614, 625
        TM^z, 607–608, 619–620
    rectangular plate:
        TE^x, 591–592
        TM^x, 592–594
    sphere, 655–656
    strip:
        TE^z, 585–586
        TM^z, 582–583
Boundary conditions, 13–20, 24, 26–28
    finite conductivity, 13–16, 20, 26–27
    infinite conductivity, 16–19, 20, 26
    sources along boundaries, 19–20, 26
Brewster angle, 180, 193–196
    polarization:
        horizontal, 193–194
        parallel, 194–196
        perpendicular, 193–194
        vertical, 194–196

C

Canonical problem, 744
Capacitor, 10, 11, 12
Cauchy–Riemann conditions, 964, 966

Caustic, 315, 316, 749, 753, 760, 765–769, 817–820
Cavities:
  circular, 492–499
  circular dielectric, 517–524
  quality factor Q, 388
  rectangular, 388–394
  spherical, 557–562
Charge density:
  bound, 45–49
  electric, 3, 12, 31
  magnetic, 3, 12, 31
Charge distribution, 671–677
  straight wire, 671–676
  bend wire, 675, 677
Chebyshev (see Tschebyscheff)
Chu, P. C., 70
Circuit relations, 8–12, 31
Circular cavity, 492–499
  coupling, 492
  dissipated power, 497–498
  dominant mode, 495
  modes:
    TE^z, 492–494
    TM^z, 494–495
  quality factor, 495, 498
  resonant frequency, 494–496
  stored energy, 497
  wave numbers, 493, 494
Circular cylinder scattering, 602–634
  line-source:
    current density, 628, 633
    electric (TM^z polarization), 626–630
    far-field pattern, 629–630, 633–634
    magnetic (TE^z polarization), 630–634
  normal plane wave incidence:
    current density, 605–606, 610–611
    far-zone field, 606, 611–612
    scattering width, 607–608, 612–613
    small radius approximation, 606, 608, 610–611, 613
    TE^z, 608–614
    three-dimensional approximation, 608, 614
    TM^z, 603–608
  oblique plane wave incidence, 614–626
    far-zone field, 618–619, 624–625
    scattered field, 618, 623–624
    scattering width, 619, 625
    small radius approximation, 619–620, 625
    TE^z, 620–625
    three-dimensional approximation, 620, 625
    TM^z, 614–620

Circular waveguide, 470–491, 499–535
  attenuation, 485–489
  attenuation constant, 485, 486–489, 491
  bandwidth, 474, 479
  coupling, 484
  current density, 488
  cutoff frequency, 473, 479
  cutoff wave number, 473
  degenerate modes, 479
  dominant mode, 479
  eigenfunction, 472, 478
  eigenvalue, 472, 478
  guide wavelength, 473, 479
  mode patterns, 380–381
  modes:
    TE^z, 470–477
    TM^z, 477–484
  phase constant, 472, 473, 478, 479
  reference table, 490–491
  wave impedance, 475, 483
Clausius–Mosetti equation, 84
Complex angles, 200–201, 210–213
Computer codes:
  electromagnetic surface patch (ESP), 730–731
  Fresnel transition function (FTF), 789, 791–792, 835, 850
  mini-numerical electromagnetics code (MININEC), 730
  numerical electromagnetics code (NEC), 729–730
  Pocklington's wire radiation and scattering (PWRS), 729, 737–742
    radiation, 729
    scattering, 729
  slope wedge diffraction coefficient (SWDC), 830, 835, 849
  two-dimensional radiation and scattering (TDRS), 727–728
    circular, elliptical or rectangular cylinder, 728
    strip, 727
  wedge diffraction coefficient (WDC), 789, 810, 834, 848
Conduction band, 65
Conductivity, 7, 11, 59–63, 77–78
  alternating, 77–78
  equivalent, 77–78
  static, 77–78
  table, 62
Conductors, 7, 59–63
  good, 80
Continuity equation:
  differential form, 4–6, 25

integral form, 6, 12, 25, 31
Constitutive:
  parameters, 7, 31
  relations, 7, 12, 31
Copper:
  charge density, 66
  conductivity, 66
  mobility, 66
Critical angle, 180, 196–206
  polarization:
    horizontal, 196–206
    parallel, 206
    perpendicular, 196–206
    vertical, 206
Critical temperature, 68–70
Curie temperature, 47
Current, 9–12, 59–63
Current density, 2–4, 12, 31, 60, 78–80
  conduction electric, 3–4, 12, 31, 78
  conduction magnetic, 3–4, 84
  convection, 60
  displacement electric, 3–4, 12, 31, 78
  displacement magnetic, 3–4, 12, 31, 84
  impressed electric, 3–4, 78
  impressed magnetic, 3–4, 84
  source electric, 3
  source magnetic, 3
  total:
    electric, 78
    magnetic, 84
Cylindrical wave:
  Bessel functions addition, 600–602
  Hankel functions addition, 597–600, 602
  theorems, 595, 597–602
  transformations, 595–596, 602

D

Damped:
  critically, 75
  over, 75, 81
  under, 75, 81
Debye equation, 83
Delta-gap source, 717, 722
Dielectric:
  constant, 51
  good, 80
  hysteresis, 78
Dielectric circular waveguide, 506–519, 527–535
  eigenvalue equation, 512
  loosely bound surface waves, 509
  mode patterns, 515–519
  modes:
    dipole, 507
    dominant, 507, 513, 516–517
    evanescent, 512
    EH, 507, 512–517
    HE, 507, 512–517
    HEM, 507–517
    hybrid, 507–517
    TE^z, 507
    TM^z, 507
  normalized diameter, 517–518
  tightly bound surface waves, 509
  wave numbers, 510–517
Dielectric covered ground plane, 441–444
  cutoff frequency, 443
  modes:
    even, 441–444
    odd, 441–444
    surface wave, 443
    TE^z, 441–444
    TM^z, 441–444
  wave numbers, 443–444
Dielectric covered rod, 527–535
Dielectric resonator (circular), 517–524
  modes:
    TE^z, 520–521
    $TE_{01\delta}$, 522–524
    TM^z, 521–522
  PEC, 522–524
  PMC, 518–524
  resonant frequency, 521
  wave numbers, 520–524
Dielectrics, 7–8, 44–51
  anisotropic, 7–8, 71
  dispersive, 7–8, 71
  homogeneous, 7–8, 71
  inhomogeneous, 7–8, 71
  isotropic, 7–8, 71
  linear, 7–8, 71
  nondispersive, 7–8, 71
  nonhomogeneous, 7–8, 71
  nonisotropic, 7–8, 71
  nonlinear, 7–8, 71
Dielectric slab waveguide, 414–441
  critical angle, 431–433
  cutoff frequency, 422, 428
  graphical solution, 423–427
  modes:
    air-substrate, 431–433
    capacitive surface wave, 428
    even, 416–431
    inductive surface wave, 420
    odd, 416–431
    substrate, 431–433

Dielectric slab waveguide (*Continued*)
    TE^z, 427–431, 439–441
    TM^z, 416–431, 436–439
    waveguide, 431–433
    polarization:
        parallel, 436–439
        perpendicular, 439–441
    wave:
        numbers, 420–431
        impedance, 420–428
Differential equation:
    coupled, 117, 122–123
    uncoupled, 117
Diffracted field:
    hard, 788
    incident, 780, 782–783, 785, 790
    reflected, 781–783, 787, 790
    soft, 788
Diffraction coefficients, 785–807
    hard polarization, 788–790, 809
    incident, 785–787, 790, 810, 817
    Keller's, 790–791
    reflection, 787–788, 790, 810, 818
    soft, 788–790, 809
Diffraction plane, 809
Dipole:
    electric, 318–321, 645–646
    horizontal, 321–323
    magnetic, 321–323
    moment:
        electric, 45
        magnetic, 52–53
    torque, 53
    vertical, 318–321
Dirac delta function, 855
Dispersion:
    equation, 77
    normal, 81
Distance parameter, 810, 818
    incident, 818
    reflected, 818
Divergence theorem, 5
Donors, 64
Doping, 64
Duality theorem, 310–312
Dyadic reflection coefficient, 753

E

Effective radius, 534
Eikonal surface, 745–748
    cylindrical, 747
    plane, 747
    spherical, 748
Electrets, 46

Electric field integral equation (EFIE), 680, 695–707
    Green's function, 697
    two-dimensional EFIE, 698–707
        TE^z, 702–707
        TM^z, 698–702
Electric potential, 671–675
Electrons:
    bound, 63
    free, 63
Energy, 20–23, 28–31
    capacitor, 31
    conservation of, 21, 29
    electric, 22, 30–31
    density:
        electric, 22, 31
        magnetic, 22, 31
    inductor, 31
    magnetic, 22, 30–31
Energy velocity, 135–136
Equiphase, 113
Equivalent currents, 824–828
    electric, 825–826
    magnetic, 825–826
Equivalents, 310, 329–346
Expansion functions, 673, 682

F

Faraday's law, 5, 12, 31
Far-field radiation, 280–282
Ferrites, 85–94
Fermat's principle, 744
Ferroelectrics, 46–47
Fiber optics cable, 205, 415, 524–527
    attenuation, 524
    discrete modes, 524
    graded-index, 524–526
    multimode, 524–526
    normalized diameter, 518
    single-mode, 524–526
    step-index, 524–526
Field intensity:
    electric, 3, 12, 31
    magnetic, 3, 12, 31
Field relations, 8–12, 31
Flux:
    density:
        electric, 3, 12, 31, 49
        magnetic, 3, 12, 31
    magnetic, 8–9, 12
Forbidden band, 65
Fresnel coefficients:
    reflection:
        horizontal, 187

parallel, 191
perpendicular, 187
vertical, 191
transmission:
horizontal, 187
parallel, 191
perpendicular, 187
vertical, 191
Fresnel integrals, 929–933
graphs, 933
tables, 930–932
Fresnel transition function (FTF), 789, 791–792, 850
large argument, 792
small argument, 791
Fringe wave, 670

G

Galerkin's method, 690, 692
Gallium arsenide:
charge density, 66
conductivity, 66
mobility, 66–67
Gauss's law, 6
Geometrical optics, 570, 670, 694, 743–764
amplitude, 746–750
cylindrical, 749
plane, 749
spherical, 749
astigmatic rays, 749, 750, 754
caustic, 749, 753, 760
conservation of energy flux, 746–749
divergence factor, 752, 755
dyadic reflection coefficient, 753
eikonal surface, 745–748
cylindrical, 747, 749
plane, 747, 749
spherical, 748–749
Luneberg–Kline series, 751–753, 763–764
normal section, 756–757
phase, 751–753
primary wave front, 745, 746
principal radii of curvature, 755–759
ray optics, 744
region, 657, 759
secondary wave front, 745, 746
spatial attenuation, 752, 755
spreading factor, 752, 755
variational differential, 745
Geometrical optics field:
incident, 779, 782
reflected, 780, 782
Geometrical theory of diffraction, 570, 634, 670, 694–696, 701–702, 743–850

amplitude, 765–769, 811, 819
astigmatic tube of rays, 766
caustic, 765–769
caustic distance, 767, 819, 820
computer codes, 834–835
diffraction:
coefficients, 785–810
curved edge, 814–823
normal incidence, 769–807
oblique incidence, 807–814
plane, 809
straight edge, 769–814
Dirichlet boundary condition, 770
distance parameter, 810
divergence factor, 767, 768, 807, 808
equivalent currents, 824–827
hard polarization, 772
higher order diffractions, 830–834
incidence plane, 809
law of diffraction, 769–770
method of steepest descent, 771, 773, 782
conventional, 773–781
Pauli–Clemmow, 781–782
multiple diffractions, 830–834
Neumann boundary condition, 770
phase, 765–769
polarization, 765–769
reciprocity, 773–774
self-consistent method, 830
shadow boundary:
incident, 769, 770
reflection, 769, 770
slope diffraction, 828–830
soft polarization, 772
spatial attenuation factor, 767, 768
spreading factor, 767, 768
successive scattering procedure, 830
Germanium, 63
atom, 43
charge density, 66–68
conductivity, 66–68
mobility, 66–68
Good conductors, 150–151
Good dielectrics, 149–151
Graded-index, 415, 524–526
Green's function, 851–910
bilinear formula, 869
circuit theory, 852–855
closed form, 860–865, 870
continuity, 862
derivative, 862
discontinuity, 862
dyadic, 907–910
dyadics, 907–908
Green's functions, 909–910

Green's function (*Continued*)
  generalized method, 890–893
    homogeneous Dirichlet boundary conditions, 891
    homogeneous Neumann boundary conditions, 892
    mixed boundary conditions, 893
    nonhomogeneous boundary conditions, 892
  Helmholtz equation, 893–907
    cylindrical coordinates, 897–902
    rectangular coordinates, 893–897
    series form, 895–897
    spherical coordinates, 903–907
  identities, 888–890
    first, 889
    second, 890
  impulse response, 851
  integral form, 872–876
  mechanics, 855
  orthonormal eigenfunctions, 868, 869, 871
  properties, 862
  two-dimensional rectangular coordinates, 876–888
    closed form, 877–882
    series form, 883–885
    static, 876–885
    time-harmonic, 885–888
  vibrating string, 855–858, 866–867
  Wronskian, 863
Group velocity, 135–136, 143–144
Gyromagnetic ratio, 87

H

Hallén's integral equation, 717, 720–722
Hankel functions, 119, 121, 935–936
  derivative, 936
  spherical, 938
Hard polarization, 645, 772
Helmholtz wave equation, 107
  cylindrical, 116
  rectangular, 108
  spherical, 121
Holes, 63
Huygen's principle, 329–334
Hybrid modes, 394–410, 507–519
  dielectric circular waveguide, 507–519
  filled rectangular waveguide, 394–397
  longitudinal section electric (LSE), 394–397
  longitudinal section magnetic (LSM), 397
  partially filled rectangular waveguide, 398–410
Hydrogen atom, 43

I

Identities:
  hyperbolic, 918–919
  logarithmic, 919
  trigonometric, 917–918
Image theory, 310, 314–323
Impedance:
  directional, 142, 152, 154
  intrinsic, 133, 147, 150
  wave, 133–135, 142, 147, 150, 151
Index of refraction, 51
Induction:
  equivalent, 310, 334–338
  theorem, 310, 334–338
Inductor, 8–9, 11–12
Integral equations, 570, 670–731, 743
  basis functions (*see also* basis functions), 673, 682, 683–687
    entire domain, 685–686
    subdomain, 683–686
  delta-gap, 717, 722
  diagonal terms, 688
  expansion functions, 673, 682
  Galerkin's method, 690, 692
  Hallén's, 717, 720–722
  linear integral operator, 682
  magnetic frill generator, 717, 722–726
  Pocklington's, 717–720
  point-matching, 681–683
  self terms, 688
  testing functions, 689
  weighting functions, 689
  weighted residuals, 690, 693
Isolated poles, 773, 778–779

K

Kamerlingh Onnes, 68
Keller's diffraction:
  coefficients, 790–791
  functions, 783
Kirchhoff's:
  current law, 9–10, 12, 31
  voltage law, 8–9, 12, 31
Kramers–Kronig relations, 83
Kronecker delta function, 714, 868

L

Larmor precession frequency, 86
Law of diffraction, 769
Legendre, 123, 125, 543, 553–554, 558–561, 947–962
  associated functions, 951–953

differential equation, 123
functions, 125, 543, 553–554
polynomials, 125, 558–561, 947–951
  graphs, 950–951
  tables, 954–962
Line source:
  electric, 571–573
    far-field, 573
  magnetic, 573–574
    far-field, 574
  strip, 574–576
    far-field, 576
Longitudinal section electric (LSE) modes, 394–404
  filled rectangular waveguide, 394–397
    cutoff frequency, 397
    wave numbers, 396, 397
  partially filled waveguide, 398–404
    cutoff frequency, 401–404
    wave numbers, 401–404
Longitudinal section magnetic (LSM) modes, 394, 397, 404–410
  filled rectangular waveguide, 397
    cutoff frequency, 397
    wave numbers, 397
  partially filled waveguide, 404–410
    cutoff frequency, 406–409
    wave numbers, 405–408
Lorentz reciprocity theorem, 325
Lossless media, 130–145
Lossy media, 145–154
Loss tangent:
  electric, 78–80
    alternating, 78
    effective, 78
    static, 78
  magnetic, 84
    alternating, 84
  table, 79
Love's equivalence principle, 330
Luneberg–Kline series expansion, 751–753, 763–764
  equation:
    conditional, 751
    eikonal, 751
    transport, 751

M

Magnetic field integral equation (MFIE), 695–696, 707–717
  Green's function, 708
  two-dimensional MFIE, 707–717
    TE^z, 710–716
    TM^z, 709–710

Magnetic frill generator, 717, 722–726
Magnetic material, 7–8, 51–59
  model:
    atomic, 85
    phenomenological, 85
Magnetics, 7–8, 51–59
  anisotropic, 7–8, 71
  antiferromagnetic, 57–59
  diamagnetic, 57–59
  dispersive, 7–8, 71
  ferrimagnetic, 57–59
  ferromagnetic, 57–59
  homogeneous, 7–8, 71
  inhomogeneous, 7–8, 71
  isotropic, 7–8, 71
  linear, 7–8, 71
  nondispersive, 7–8, 71
  nonhomogeneous, 7–8, 71
  nonisotropic, 7–8, 71
  nonlinear, 7–8, 71
  paramagnetic, 57–59
  torque, 53, 86–87
Magnetization, 51–59
  current, 55
  current density:
    surface, 54–55
    volume, 55–56
Maxwell's equations, 2–6, 25
  differential form, 2–5, 25
  integral form, 5–6, 25
Method of steepest descent, 634, 771, 963–967
  Cauchy–Riemann conditions, 964–966
  constant phase path, 966
  conventional, 773–781, 963–967
  Pauli–Clemmow, 781–782
  steepest amplitude path, 966
  steepest descent path, 776–777, 780–782, 964–966
Microstrip, 444–447, 449–457
  boundary-value problem, 455–457
  characteristic impedance, 450–455
  dispersion, 454–455
  effective dielectric constant, 450–455
  shielded configuration, 456
  spectral domain, 457
Microwave cooking, 78
Mie region, 657, 759
Mobility:
  electron, 63, 65–67
  hole, 65–67
Modal techniques, 570
Modes, 129
  plane waves, 130
  transverse electric (TE), 273–276

Modes (*Continued*)
  transverse electromagnetic (TEM), 129
  transverse magnetic (TM), 269–273
  uniform plane wave, 130
Molecule, 42
Moment method, 670–731, 743
  basis functions (*see also* basis functions), 673, 682–687
    entire domain, 685–686
    subdomain, 683–686
  collocation, 681–683
  delta-gap, 717, 722
  expansion functions, 673, 682
  diagonal terms, 688
  Galerkin's method, 690, 692
  linear integral operator, 682
  magnetic frill generator, 717, 722–726
  nondiagonal terms, 689
  point-matching, 681–683
  self terms, 688
  testing functions, 689
  weighting functions, 689
  weighted residual, 690, 693
Monostatic, 578
  circular cylinder:
    TE^z, 608
    TM^z, 608
  rectangular plate:
    TE^x, 591–592
    TM^x, 592, 594
  sphere, 656–658
  strip:
    TE^z, 583, 585–586
    TM^z, 582–583, 587
Mueller, K. A., 69
Multiple diffractions, 830–834

N

Neutrons, 42, 43
Nonpolar, 46–47
Normal section, 756–757

O

Ohm's law, 12, 31
Optical fiber cable (*see also* fiber optics cable), 205
Orthonormal eigenfunctions, 868, 869, 871, 880–884

P

Pattern multiplication, 320
Pauli–Clemmow modified method of steepest descent, 781–782
Perfect electric conductor (PEC), 331, 337, 342
Perfect magnetic conductor (PMC), 331
Permeability, 7, 11, 51–55
  complex, 84–85
  effective, 91
  relative, 55, 85
  static, 55
  table, 55
  tensor, 93–94
Permittivity, 7, 11, 44–51
  complex, 73, 76–77, 83
  principal, 71
  relative, 50, 77
  static, 50
  table, 50
  tensor, 71
Phase constant, 107, 113–115, 134, 140, 146
  effective, 214–220
Phase velocity, 112–113, 135–136, 143–144
Physical:
  equivalent, 310, 338–346
  optics, 570, 694–696, 701–702, 743
  optics equivalent, 310, 338–346
  theory of diffraction, 570, 694–696, 701–702, 743
Plane waves, 130
Pocklington's integrodifferential equation, 717–720
Poincaré sphere, 168–173, 237–243
Point-matching, 681–683
Polar, 46–47
Polarization (wave), 87–88, 154–173, 236–243
  axial ratio, 164, 168
  circular, 155, 158–163, 239–243
  clockwise, 87, 155, 158–161, 241–242
  counterclockwise, 87, 155, 161–163, 239–243
  elliptical, 155, 163–168, 242–243
  left-hand, 87, 155, 161–163, 166, 169, 239–243
  linear, 135, 156–158
  Poincaré sphere, 168–173
  right-hand, 87, 155, 158–161, 165, 169, 241–242
  state, 172
Polarization (electric), 44–51
  complex, 76, 80, 83
  dipole, 46
  electronic, 46
  ionic, 46
  molecular, 46
  orientational, 44
  remnant, 47

Polarization (electric) (*Continued*)
  steady-state, 76
  vector:
    electric, 45–51
    magnetic, 52–56
Power, 20–23, 28–31
  conservation of, 22
  dissipated, 22, 30, 31
  exited, 21, 22, 30
  supplied, 22, 30
Power density, 21, 28–29, 135–136, 144–145, 183
Poynting vector, 21, 28–29
Precession frequency, 86
Principal radii of curvature, 755–759
  incident wave front, 755, 756, 818
  reflected wave front, 755, 756, 818
Principles, 320, 329
Propagation, 129–154
  lossless media, 130–145, 180–206
  lossy media, 145–154, 206–220
  oblique axis, 138–145, 151–154
  principal axis, 131–138, 145–151
Propagation constant, 107, 113–115, 146
Protons, 42–43

Q

Quality factor Q:
  circular cavity, 495, 498
  rectangular cavity, 388, 390–392
  spherical cavity, 562–565
Quanta, 43

R

Radar cross section (RCS), 578
  conversion to two-dimensional:
    normal incidence, 578, 591, 594, 608, 614
    oblique incidence, 620, 625
Radial waveguides, 499–504
Radiation equations, 282–305
  cylindrical coordinates, 300–305
  far-field, 285-288
  near-field, 282–284
  rectangular coordinates, 288–300
Rayleigh region, 657, 759
Reaction theorem, 310, 326–327
Reciprocity theorem, 310, 323–325, 773–774
Rectangular cavity, 388–394
  coupling, 388
  current density, 391
  dissipated power, 391
  modes:
    dominant, 390

TE^z, 388–392
TM^z, 392–394
quality factor, 388, 390–392
resonant frequency, 390, 393, 394
stored energy, 390
wave numbers, 389, 390, 393
Rectangular plate scattering, 586–594
  backscattered, 591
  bistatic, 591, 594
  E-plane, 590
  H-plane, 590
  monostatic, 594
  TE^x, 586, 588–592
  TM^x, 588, 592–594
Rectangular waveguide, 352, 387
  attenuation, 376–384
  attenuation constant, 379, 381, 383
  bandwidth, 360, 361, 364, 365
  coupling:
    electric field, 383
    magnetic field, 384
  cutoff frequency, 357, 361, 363, 364
  cutoff wave number, 357, 363
  degenerate modes, 357
  dominant mode, 360, 366–372
  eigenfunction, 355, 362
  eigenvalue, 355, 362
  evanescent waves, 358
  guide wavelength, 359, 363
  hybrid modes, 394–410
  losses:
    conduction, 376–381
    dielectric, 381–384
    ohmic, 376–381
  mode patterns, 365
  modes:
    TE_{10}, 366–373
    TE^x, 366, 394
    TE^y, 366, 394–397
    TE^z, 353–362
    TM^x, 366, 394
    TM^y, 366, 394, 397
    TM^z, 353–366
  phase constant, 357, 358, 363
  power, 374–376
  power density, 374–376
  reference table, 380–381, 386–387
  wave impedance, 358, 363
Reflection, 180–243
  coefficient, 181, 182, 187, 191
  horizontal polarization, 185–189
  lossless media, 180–206
  lossy media, 206–220
  multiple interfaces, 220–236
  normal incidence, 180–185

Reflection (*Continued*)
  oblique incidence, 185–206
  parallel polarization, 189–193
  perpendicular polarization, 185–189
  vertical polarization, 189–193
Relaxation time constant, 61, 84
Residue calculus, 777–779
Resistivity, 63
Resistor, 11, 12, 31
Resonance region, 657, 759
Resonators (*see* cavities)
Ridged waveguide, 457–461
  dual, 457–461
  quadruple, 457
  single, 457–461

S

Saddle point method, 634, 963–967
Saddle points, 773, 779–782, 964, 966, 967
Scattered fields, 328, 335
Scattering, 570–669
  backscattered, 578–591
  bistatic, 578, 591, 594
  circular cylinder (*see also* circular cylinder scattering), 602–634
  field:
    direct, 677
    incident, 570
    reflected, 678
    scattered, 570, 678
  monostatic, 578
  radar cross section (RCS), 577–578
    conversion to two-dimensional, 578
    pattern, 578
  rectangular plate (*see also* rectangular plate scattering), 586–594
  specular, 578
  scattering width (SW), 577–578
    conversion to three-dimensional, 578
  sphere (*see also* sphere scattering), 650–658
  strip (*see also* strip scattering), 578–586
  wedge (*see also* wedge scattering), 634–645
Scattering equations, 282–305
  cylindrical coordinates, 300–305
  far-field, 285–288
  near-field, 282–284
  rectangular coordinates, 288–300
Scattering width (SW), 577
  conversion to three-dimensional:
    normal incidence, 577, 591, 594, 608, 614
    oblique incidence, 620, 625
Schelkunoff, S. A., 329

Self-consistent method, 830
Semiconductors, 7, 63–68
  n-type, 64
  p-type, 64
Separation of variables:
  cylindrical coordinate system, 118
  rectangular coordinate system, 109, 114
  spherical coordinate system, 123
Shadow boundary:
  incident, 769, 770
  reflected, 769, 770
Silicon, 63
  atom, 43
  charge density, 66–68
  conductivity, 66–68
  mobility, 66–68
Silver:
  charge density, 66
  conductivity, 66
  mobility, 66
Skin depth, 149–151, 209
Slope diffraction, 828–830
  hard, 828–829
  soft, 828–829
Slope wedge diffraction coefficient (SWDC), 830, 849
Snell's law:
  reflection, 187, 191
  refraction, 187, 191
Soft polarization, 645, 772
Specular, 578
Sphere scattering, 650–658
  bistatic, 655–656
  far zone field, 655
  geometrical optics region, 657, 759
  Mie region, 657, 759
  monostatic, 656–658
  plane wave incidence, 650–658
  radar cross section, 657–658
  Rayleigh region, 657, 759
  resonance region, 657, 759
  TE^r, 652–653
  TM^r, 652–653
Spherical cavity, 557–562
  current density, 563
  degenerate modes, 559, 561
  dissipated power, 563
  dominant mode, 559, 561, 562
  modes:
    TE^r, 557–560
    TM^r, 560–562
  quality factor, 562–565
  resonant frequency, 559, 560
  stored energy, 563
  wave numbers, 559, 560

Spherical wave:
  orthogonalities, 647–648
  theorems, 648, 650
    Hankel functions addition, 650
  transformations, 648–650
Spherical waveguide, 543–557
  biconical transmission line, 552–557
  construction of solutions, 543–552
  Helmholtz wave equation, 550–552
  modes:
    TE^r, 547–549
    TM^r, 549–550
Standing wave:
  pattern, 137, 138, 147, 149, 184
  ratio (SWR), 137
Standing waves, 136–138
  attenuating, 110, 121
  cylindrical coordinate system, 120, 121
  rectangular coordinate system, 110, 111
  spherical coordinate system, 125
Stationary phase, 967
Steepest descent path, 775–777
Step-index, 415, 524–526
Stokes' theorem, 5
Stray:
  capacitance, 10, 12
  inductance, 9, 12
Stripline, 444–449
Strip scattering, 578–586
  bistatic, 582–583, 585–586
  electric line source, 574–576
  monostatic, 582–583, 586
  plane wave, 578–587
    TE^z, 583–587
    TM^z, 578–583
Sturm–Liouville:
  equation, 860–861
  Hermitian properties, 864
  operator, 851, 867–869
  problem, 858–860
  symmetrical properties, 864, 868
  symmetry, 862
Successive scattering procedure, 830
Superconductors, 68–70
Surface equivalence theorem, 310, 329–334
  electric current density, 329–331, 336–337
  magnetic current density, 329–331, 336–337
Surface wave, 198–206, 533, 534
  loosely bound, 518
  slow, 200
  tightly bound, 200
Susceptibility:
  complex, 8
  dipole, 80
  electric, 49, 80
  electronic, 80
  ionic, 80
  magnetic, 54
  tensor, 93

T

TE$_{10}$, 366–373
  bouncing plane waves, 370
  current density, 368
  cutoff frequency, 368
  group velocity, 372
  guide wavelength, 368
  phase constant, 367, 370–372
  wave impedance, 368
TEM, 129, 261–269
  cylindrical coordinates, 266–269
  rectangular coordinates, 261–265
Testing functions, 689
Theorems, 310–328
Time-harmonic fields, 23–31
  boundary conditions, 24, 26–28
  energy, 28–31
  Maxwell's equations:
    differential form, 24–25
    integral form, 24–25
  power, 28–31
Transition regions, 783, 792–796
Transmission, 180–243
  binomial transformer, 231–233
  coefficient, 181–182, 187
  horizontal polarization, 185–189
  lossless media, 180–206
  lossy media, 206–220
  multiple interfaces, 220–236
  normal incidence, 180–185
  oblique incidence, 185–206, 235–236
  parallel polarization, 189–193
  perpendicular polarization, 185–189
  quarter-wavelength transformer, 230–231
  Tschebyscheff impedance transformer, 189–235
  vertical polarization, 189–193
Transverse electric (TE) modes, 273–276
  cylindrical coordinates, 275–276
    TE^z, 275
  rectangular coordinates, 274–275
    TE^x, 274
    TE^y, 275
    TE^z, 274
Transverse magnetic (TM) modes, 269–273
  cylindrical coordinates, 272–273
    TM^z, 272–273

Transverse magnetic (TM) modes (*Continued*)
  rectangular coordinates, 269–272
    TM^x, 271
    TM^y, 271
    TM^z, 269–270
Transverse resonance method (TRM), 410–414
  equation, 412
  modes:
    LSE, 413
    LSM, 413–414
    TE, 413
    TM, 413–414
  wave number, 412
Traveling waves, 111, 120, 121, 184
  attenuating, 110, 121
  cylindrical coordinate system, 120, 121
  rectangular coordinate system, 110, 111
  spherical coordinate system, 125
Tschebyscheff impedance transformer, 231–233

U

Uniform plane wave, 130–154
  energy density, 135, 136, 144, 145
  energy velocity, 135, 136, 143, 144
  group velocity, 135, 136, 143, 144
  lossless media, 131–145
  lossy media, 145–154
  phase velocity, 135, 136, 143, 144
  power density, 135, 136, 144, 145, 183
  standing waves, 136–138
  wave impedance, 133–135, 142, 147, 151
Uniform theory of diffraction (UTD) (*see also* geometrical theory of diffraction), 789
Uniqueness theorem, 310, 312–314

V

Valence:
  band, 65
  electrons, 42, 43
  shell, 42, 43
Vector analysis, 920–928
  differential operators, 925–927
    cylindrical coordinates 925–926
    rectangular coordinates, 925
    spherical coordinates, 926–927
  identities, 927–928
    addition, 927
    differentiations, 927
    integration, 928
    multiplication, 927
  transformations, 920–925
    cylindrical-to-spherical (and vice versa), 923–924
    rectangular-to-cylindrical (and vice versa), 920–923
    rectangular-to-spherical (and vice versa), 924–925
Vector potentials:
  electric, 254, 257–259, 260–261, 278–280
  Hertz, 254
  magnetic, 254, 256–257, 260–261, 276–280
Virtual source, 315, 316
Voltage, 8–9, 12
Voltage standing wave ratio (VSWR), 137
Volume equivalence theorem, 310–328
  current density:
    electric, 328
    magnetic, 328

W

Watson transformation, 634
Wave equation:
  lossless media, 107
  lossy media, 107
  solution:
    cylindrical coordinate system, 116–121
    rectangular coordinate system, 108–116
    source-free and lossless media, 108
    source-free and lossy media, 113
  time-harmonic fields, 106–107
  time-varying fields, 104–106
Waveguides, 352–388, 394–461
  circular (*see also* circular waveguide), 470–519, 524–537
  dielectric (*see also* dielectric circular waveguide; dielectric covered ground plane; dielectric covered rod; dielectric slab waveguide), 414–444, 506–517, 524–537
  microstrip (*see also* microstrip), 444–447, 449–457
  radial, 499–504
  rectangular (*see also* rectangular waveguide), 352–387
  ridged, 457–461
  spherical (*see also* spherical waveguide), 543–557
  stripline, 444–449
  wedged plates, 504–506
Wedge:
  exterior, 783, 807
  interior, 807
Wedge diffraction (*see also* geometrical theory of diffraction), 743–850

Wedge diffraction coefficient (WDC), 789, 810, 848
Wedged plates waveguide, 504–506
Wedge scattering, 634–645
   electric line source, 635–639, 642–645
      current density, 637
      far-zone field, 637–638
      Green's function, 644–645
   hard polarization, 645
   magnetic line source, 639–645
      far-zone field, 641–642
      Green's function, 644–645
   soft polarization, 645
   TE^z polarization, 639–642
      Green's function, 644–645
   TM^z polarization, 635–639
      Green's function, 644–645
Weighting functions, 689